ModTech International Conference: Modern Technologies in Industrial Engineering IX (ModTech 2021)

IOP Conference Series: Materials Science and Engineering
Volume 1182

Eforie, Romania
23 - 26 June 2021

ISBN: 978-1-7138-4642-0
ISSN: 1757-8981

Printed from e-media with permission by:

Curran Associates, Inc.
57 Morehouse Lane
Red Hook, NY 12571

CURRAN ASSOCIATES INC. proceedings.com

Some format issues inherent in the e-media version may also appear in this print version.

Institute of Physics
Dirac House, Temple Back
Bristol BS1 6BE UK

Phone: 44 1 17 929 7481
Fax: 44 1 17 920 0979

techtracking@iop.org

Additional copies of this publication are available from:

Curran Associates, Inc.
57 Morehouse Lane
Red Hook, NY 12571 USA
Phone: 845-758-0400
Fax: 845-758-2633
Email: curran@proceedings.com
Web: www.proceedings.com

TABLE OF CONTENTS

IOP Conf. Series: Materials Science and Engineering 1182 (2021) 011001 doi:10.1088/1757-899X/1182/1/011001

Preface

The ModTech Professional Association, as main organizer together with "Gheorghe Asachi" Technical University of Iasi and Constanta Maritime University, organized the 9th online edition of the ModTech2021 International Conference, between 23 and 26 June 2021. Other prestigious institutions, from Romania and abroad, such as the Society of Powder Technology of Japan, Silesian University of Technology of Gliwice, Poland, "Alecu Russo" University of Balti, Republic of Moldova, Technical Sciences Academy of Romania, "Lucian Blaga" University of Sibiu, "Dunarea de Jos" University of Galati, "Stefan cel Mare" University of Suceava and The General Association of the Engineers in Romania (AGIR) were also co-organizers. The ModTech Professional Association currently has more than 400 members from Romania and abroad and 13 branches in Australia, Greece, Spain, Malaysia, India, Japan, Mexico, South Korea, Iraq, Poland, Turkey, Serbia, and Republic of Moldova.

The ModTech2021 International Conference was also online this year because of Covid-19: meeting restrictions. Continuity in the annual development of the ModTech International Conference, the wish expressed by many participants, following the consultation carried out approximately 3 months before the conference. The organizing committee respected their desire to hold the conference online, the desire justified by the following reasons:
- the need to publish (disseminate) the research results within some research projects, an important stage of the annual reporting;
- the annual evaluation of the teachers also follows the activity of participation in conferences, the publication in the ISI Proceedings are well rated;
- career promotion also depends on the participation in international conferences (especially ISI indexed) in a given period of time; the existence of a period of interruption of participation in such conferences could diminishes the level of appreciation of the activity from this point of view;
-at the start of the consultation action of the participants, were already a large number of abstracts received and accepted.

The online platform for the activities of the conference had all the stages. The ModTech Professional Association, the main organizer of the ModTech2021 International Conference, has signed a contract with the Techsoup Association, which collaborates with Google. Among other contractual conditions there was also the one regarding the storage space. The next step was to create the virtual page (figure 1) on Google Meet.

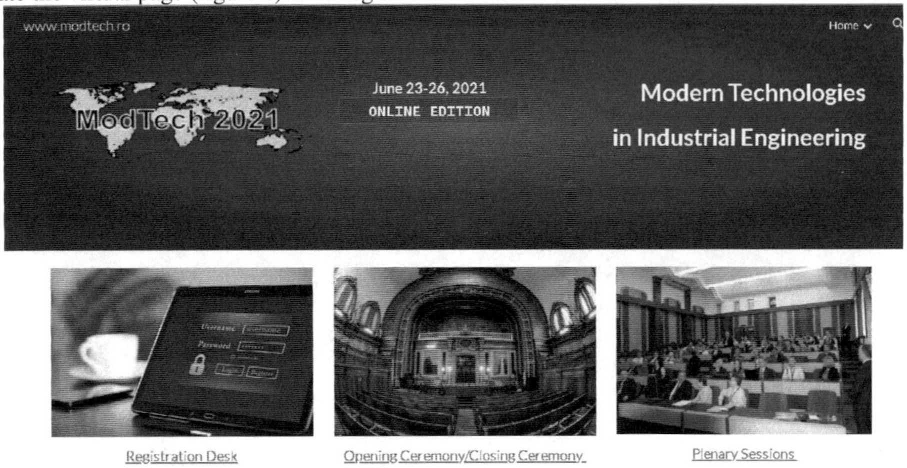

Fig.1. The virtual page of ModTech2021 International Conference

To familiarize participants with how to operate on this platform was developed and submitted in pdf format - ModTech2021_Online_User_Guide.

Also, for this purpose, in the week before the conference, one simulation of all activities were organized in which participants could enter the platform and practice how to operate for the online presentation of papers under the assistance of technical secretaries on each section of the conference. The program of the conference with the schedule presentations was posted on the virtual page for each section (figure 2), sent by email to all participants and entered in the Book of abstracts, available from the account of each participant.

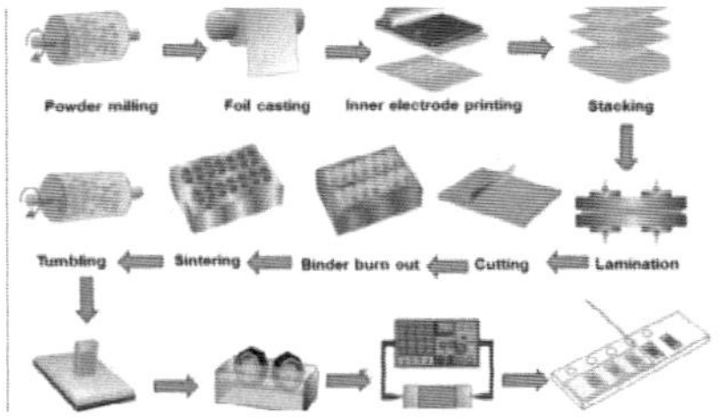

Section A

Engineering of Manufacturing Processes

Novel Manufacturing Methods; Advances in Nontraditional Manufacturing Processes; Virtual Manufacturing; Advanced Methods and Tools for Computer Integrated Manufacturing; New Industrial Applications; Smart Manufacturing; Energy Efficiency in Manufacturing

Presidents: **Prof. Viorel Paunoiu**

"Dunarea de Jos" University of Galati, Romania

&

Lecturer Eugen Herghelegiu

"Vasile Alecsandri" University of Bacau, Romania

Section F

Micro- and Nano- Technologies

New Developments in Micro/Nano Scale Processes; Fabrication of Nanostructure and Materials; Nanofilms, Nano-bubbles, Nano-droplets, Nanofluids

President: **Assoc. Prof. Alexandr Ojegov**

"Alecu Russo" Balti State University, Republic of Moldova

Technical Secretary: PhD.Eng. Simona Mazurchevici

Email: simona.mazurchevici@modtech.ro

a)

< Back

Section A & F

ONLINE MEETING

All papers will be supported online according to the schedule in the tables below.

Please respect the schedule of the support being present at least 30 minutes before the estimated time.

28 : 21 : 38 : 32
Days Hrs Mins Secs

Sections A&F

Thursday, June 24th, 2021

Time	Authors and coauthors names	Abstract ID & Paper title
11:30-11:50	Yogesh Rajpal, Sinead Marie Ui Mhurchadha, Paul Quinn & Ramesh Raghaven	A-1: A comparative study on the effect of multi-machine manufacturing on a dditively manufactured part qualiti
11:50-12:10	Jae-Ho Jeon	A-2: Behavior of templates during texturing of piezoceramics
12:10-12:25	Reggie Davidrajuh	A-8: GPenSIM: A new tool for modeling and performance analysis of large discrete manufacturing systems
12:25-12:35	Damit-Sorin Ionel & Constantin Gheorghe Opran	A-11: Dynamic Lean 4.0 cyber-physical system for polymeric products manufacturing in Industry 4.0
12:35-12:45	Pavel Topala, Serghei Talpa & Natalia Pinzaru	A-12: Dryer-separator with vibration-fluidized bed and process for drying and separation
12:45-12:55	Pavel Topala, Alexandr Ojegov, Vitalie Besliu, Petru Stoicev & Valeriu Crecium	A-13: Study of anti-wear properties of graphite films layed on metal surfaces by applying pulsed electric discharge machining
12:55-13:05	Mihail Melenciuc	A-14: Practical and theoretical contributions on CO2 modified environment pears drying
13:05-13:15	Pavel Topala, Dorin Guzgan & Alexandr Ojegov	A-15: Current-voltage characteristics of thermo-electronic emission process
13:15-13:25	Marcin Zuk & Sebastian Stano	A-16: Laser welding of 4330V steel with a micro-additive of vanadium
13:25-13:35	Sorin R. Sintea & Catalin C. Pomazan	A-17: Cloud connected power inverter
13:30-15:00	**Own World Lunch Break**	
15:00-15:20	Gianni Nicoletto	A-3: Integrated workflow for metal additive manufacturing: from part design to part qualification
15:20-15:40	Natalia Kamanina	F-1: Materials surfaces and their modifications via nanotechnology approaches
15:40-15:55	Prachand Man Pradhan	A-9: Partially infilled frame behavior under lateral loading
15:55-16:10	Jaime Taha-Tijerina, Ruben Calderon & Barbara Rodriguez	F-2: Optimization and nanoreinforcements of lubricant concentration for steel sheet forming process
16:10-16:20	Daniela Ionescu & Gabriela Apreotesei	F-3: Metamaterial optical filter with maximal absorption coefficient

24.06.2021 25.06.2021 26.06.2021

Fig.2. Example for A&F sections: a)-details of A&F sections; b)-the schedule papers presentation

Each section of the conference was chaired by a chair assisted by a technical secretary (who technically ensured the smooth running of the business: an acceptable number of open microphones - to avoid polyphony and accept the connection of participants without a registered account on the platform, start / stopping conference recordings, etc.)
To identify the papers, each paper was found based on an ID in Book of Abstracts (ID consisting of a section and an order number) but also on the online platform.
For each paper presented during the plenary session, between 30 minutes and 40 minutes were allocated (including Q&A). For each paper during the sections, 10 minutes were allocated.
The announcement of each paper was made by the president of the section, who invited the author to present his paper, in PowerPoint.
Discussions/comments respectively Q/A existed for all papers presented and were led by the president of each section. From a technical point of view, there was no problem for the Q/A side.

The participants of the 9th edition were from 25 countries. The ModTech2021 International Conference got together 155 specialists in the field of modern material manufacture and processing technologies both from Romania and from other world countries with interesting presentations about the latest information on the research topics tackled by the conference.

The Octavian Pruteanu excellence award was awarded during the opening ceremony to Prof. Maria Eugenia Rabanal Jimenez, Carlos III University of Madrid, Spain, Prof. Makio Naito, Osaka University, Japan and Prof. Mohd Khairol Anuar Mohd Ariffin, Putra University, Malaysia.

The papers presented by the participants both during the three plenary sessions and in the different sections tackled the following topics: manufacturing process engineering; breakthroughs concerning composite materials and technologies; mechanical process characterization, modeling and simulation; integrated manufacture and robotics; technology transfer; micro and nanotechnologies; maritime engineering.

The papers presented during the ModTech2021 conference will the published in the International Journal of Modern Manufacturing Technology and IOP Conference Series - Materials Science and Engineering, Great Britain.

IOP Editors ModTech2021:
Maria Eugenia Rabanal Jimenez, Anna Timofiejczuk, Kyratsis Panagiotis, Mohd Khairol Anuar Mohd Ariffin, Simona-Nicoleta Mazurchevici, Harshit Dave, Pavel Topala, Dumitru Nedelcu.

IJMMT Special issue 3-2021 Guest Editors:
Esteban Broitman, Rainer Gadow, Constantin Carausu

Conference Committees

Honorary Presidents

Professor Dan Cascaval, Ph.D., Rector of Gheorghe Asachi Technical University of Iasi, Romania
Professor Octavian Pruteanu, Ph.D., DHC, Vicepresident of ModTech Professional Association Iasi, Romania.
Professor Mihai Mihaita, Ph.D., DHC, President of Technical Sciences Academy of Romania and The General Association of the Engineers in Romania.
Professor Makio Naito, Ph.D., President of The Society of Powder Technology, Japan.
Professor Jerzy Swider, Ph.D., Director of the Institute of Engineering Processes Automation and Integrated Manufacturing Systems, Silesian University of Technology, Gliwice, Poland.
Professor Valentina Pritcan, Ph.D., Vice Rector of Alecu Russo State University of Balti, Republic of Moldova.

IOP Conf. Series: Materials Science and Engineering 1182 (2021) 011001 doi:10.1088/1757-899X/1182/1/011001

IOP Conf. Series: Materials Science and Engineering 1182 (2021) 011002 doi:10.1088/1757-899X/1182/1/011002

Peer review declaration

All papers published in this volume of *IOP Conference Series: Materials Science and Engineering* have been peer reviewed through processes administered by the Editors. Reviews were conducted by expert referees to the professional and scientific standards expected of a proceedings journal published by IOP Publishing.

- **Type of peer review: Single-blind**
- **Describe criteria used by Reviewers when accepting/declining papers. Was there the opportunity to resubmit articles after revisions?**

The criteria used by the reviewers were as follows (with Excellent, Good, Fair, Poor and Very Poor highlight for each criteria):
-Significance of actual state of the problem;
-Originality of the work;
-Quality of research methodology;
-Validity of the results and interpretation;
-Clarity of the conclusions;
-Validity of the future results;
-Adequacy of recognition of the quotations of previous work;
-Quality of the language.
The reviewer's recommendations were:
-Accepted as it is;
-Accepted with minor revisions;
-Resubmitted with major revisions;
-Rejected.
Yes, of course was the possibility for all authors to resubmit articles after reviewer's comments.

- **Conference submission management system:**

  The Conference management system consist in our platform www.modtech.ro with the database included such as: REGISTER, LOGIN, USERS INFORMATION (email, city/country, uploading paper time, options (automatically email to the each author, view & edit details about the participant), delete the user account), NEWSLETTER (inform in a same time all the participants), FILES (uploading abstracts and full papers, message for accepted/denied).

  Also, the modtech.ro website has the information about ORGANIZERS, TOPICS, CONFERENCE COMMITIES, PUBLICATION ETHICS AND MALPRACTICE STATEMENT, ModTech ATTENDANCE POLICY, CALL FOR PAPERS (Abstract template, Paper template, Poster template, Copyright, Attendance Application Form), REGISTRATION, IMPORTANT DATES, PUBLICATIONS, BOOK OF ABSTRACTS, PLENARY/KEYNOTE/INVITED SPEAKERS, VENUE, NEWS, CONTACT.

- **Number of submissions received:** 131

IOP Conf. Series: Materials Science and Engineering 1182 (2021) 011002 doi:10.1088/1757-899X/1182/1/011002

- **Number of submissions sent for review:** 128 (3 paper rejected after the plagiarism report)
- **Number of submissions accepted:** 125 (IOP-91; IJMMT-20; 14-withdrawn due to the loss of SCOPUS indexation by the IOP)
- **Acceptance Rate (Number of Submissions Accepted / Number of Submissions Received X 100):** 95.4%
- **Average number of reviews per paper:** 2
- **Total number of reviewers involved:** 39
- **Any additional info on review process (ie plagiarism check system):** Was used the PLAGIARISM DETECTOR PRO v. 1861
- **Contact person for queries:** Dumitru Nedelcu, Phone: +40771346968, Email: office.modtech@gmail.com

To the theory and practice of creating modern roller machines

A Abdukarimov and S Madaminov
Institute of Mechanics and Seismic Stability of Structures named after M.T. Urazbaev, Academy of Sciences of the Republic of Uzbekistan, 31 Durmon Street, 100125, Tashkent, Uzbekistan

E-mail: aabdusalam54@gmail.com

Abstract. The article describes the areas of application of roller machines, trends in the development of their designs, research work on the modernization and creation of new roller machines. Special attention given to the roller machines with a variable center distance of the working shafts. The descriptions of the gear-lever differential transmission mechanisms created by authors are given. In particular, the correct applications of the created mechanisms in roller machines of six types are shown; they are 1. One of the working shafts has the ability to rotate around its own axis, and the second working shaft, in addition to rotation around its own axis, moves relative to the first working shaft along the line, passing through the center of the axes of rotation of both working shafts; 2. Both working shafts have the ability to rotate around their own axes and move symmetrically along a line passing through the same axes of rotation relative to each other; 3. One of the working shafts has the ability to rotate around its own axis, and the second working shaft, in addition to rotation around its own axis, moves relative to the first working shaft along an arc with a certain radius of curvature; 4. Both working shafts can rotate around their own axes and move symmetrically relative to each other along an arc with some symmetrical centers of rotation; 5. One of the working shafts has the ability to rotate around its own axis, and the second working shaft, in addition to rotation around its own axis, moves relative to the first working shaft along a complex trajectory; 6. Both working shafts can rotate around their own axes and move symmetrically relative to each other along a complex trajectory.

1. Introduction

Currently, in the industry of our Republic and abroad, roller machines are widely used for various purposes. In metallurgy, roller machines are used for the metal treatment under pressure. In mechanical engineering, roller machines are used in expansion and rolling operations. Roller machines are also widely used in agriculture, in light industry (textile, leather, and paper production). In particular, in the tannery, fleshing and planing machines to remove hair and clean the grain leather surface; double belt machines, machines for squeezing and spreading leather and fur, staking and softening machines, machines for processing leather by pressure, grinding and dedusting machines and units, machines for leather dyeing and many others are the roller machines [1].

Despite the wide applicability and great variety, roller technological machines are poorly researched [2-4].

The main advantage of roller machines is their fit into production lines and the possibility of continuous or piece-by-piece continuous execution of the technological process.

All roller machines can be divided into two groups according to the degree of freedom of roll modules:
1. roller machines with a constant roll-to-roll distance between the working shafts;
2. roller machines with varying roll-to-roll distances between the working shafts.

The first group of roller machines is mainly used in the metallurgical industry, and the second group is used in light industry, agriculture, etc.

In tanneries, a roller pressing machine VOPM-1800-K is used for pressing chrome leather semi-finished product [1]. An AORU-1800-K, developed by VNIILTekMash is also a well-known unit [1], designed for squeezing and distributing pigskin chrome semi-finished leather and bovine semi-finished leather. To squeeze out bovine leather semi-finished products, a squeezing machine produced by "SWIT" company is used [1]. To soften the leather semi-finished product, the staking-softening machine TMPH-1800-K is used [1]. For rolling hard skins, the PK-P machine is used [1]. To squeeze the moisture out of chrome-tanned leather, squeezing rollers machines MOV-K and MOP-1800-K [1] are used, in which the squeezing of semi-finished products is performed by smooth steel shafts covered by a rubber coating. To wring out semi-finished leather from a bovine hide, a feed-through squeezing machine RS-40 by Menechetti (Italy) [1] and 594 machine by "Turner" are used.

In agriculture, for the primary processing of bast fibers, a roller machine is used with corrugated working shafts [1]. The harvesting apparatus of a cotton picker for picking raw cotton also presents a roller technological machine [5].

In all of the above listed and many other roller machines, due to the features of the technological process, the center distance of the working shafts changes.

Roller technological machines used in various industries do not fully meet the agrotechnical and technological requirements imposed on them, such as work without geometric sliding between the processed material and the working shafts, ensuring the required pressure on the processed material, ensuring the required velocity parameters, constancy in dimensions and positions of working bodies, etc. This results in poor product quality, low productivity and durability of the machine. This happens not only because of the wrong choice of the scheme and design errors of the working body of these machines, but also because of the flaws and wrong choice of their actuators, especially inter-roll transmission mechanisms [2 - 4]. The research carried out by the authors makes it possible to conclude that in some roller machines the transmission mechanisms are selected and designed without considering the structural, kinematic, dynamic properties of these mechanisms. Such design flaws are due to insufficient research of roller technological machines and their actuating mechanisms, in general, and inter-roller transmission mechanisms, in particular, as well as due to the lack of methods for structural, kinematic, dynamic analysis and synthesis of these transmission mechanisms.

2. Materials

Currently, the laboratory "Theory of Mechanisms and Machines"of the Institute of Mechanics and Seismic Stability of Structures named after M.T. Urazbaev of the Academy of Sciences of the Republic of Uzbekistan conducts research and design study aimed at improving the design and methods of analysis and synthesis of roller machines, their mechanisms and working bodies [6-10]. Articles [11-14] are devoted to solutions of contact interaction in two-roll modules. Mathematical models of roll contact curves, friction stresses and contact stress distribution patterns were obtained. The variety of operations performed by roller machines has not allowed creating a unified system for their design, which is associated with the difference in technological tasks and processes occurring in the contact zone of the rollers with the material being processed [15]. In the contact zone, each point of the surface at a given moment of time corresponds to the velocity of the material and the contact stress determined by the magnitude and direction, which together forms a vector field. A necessary condition for the machine to perform a certain function is the creation of stress and velocity fields with a given value and permissible deviation on the contact surface of the rollers. Thus, the main task of designing roller machines is to ensure the above conditions, for which it is necessary to know the fields and their influence on the performance of the given functions by the mechanism, and the characteristics of the fields that most affect the function to be performed. To date, issues related to the stress field in the contact zone are partially solved, although without considering the velocity characteristics, while the problem of velocity characteristics and their effect on the stress of the contact zone, depending on the functional purpose of machine, awaits its solution. Consequently, the main task of designing a roller machine is to create a machine design that provides for the required value and permissible uniformity of velocities and stresses in the contact zone of the working rollers to perform technological tasks. Hence, the development of calculation methods, new designs of actuators, especially inter-roller treansmission mechanisms, which increase labor productivity, improve the quality of the product, save energy and resources in machining processes, is an urgent problem.

In the context of solving these problems, a number of studies were completed and published [1-15]. However, the insufficient choice of schemes of inter-roll transmission mechanisms, the lack of an effective classification of these mechanisms, the working bodies of roller technological machines and roller machines in general, considering the agrotechnical and technological requirements for these machines, do not give designers the opportunity to correctly choose the schemes of the working bodies of these machines and their transmission mechanisms.

In order for the designers to facilitate the correct choice of the scheme of the working bodies of the roll machines according to the technological requirements and the correct choice of the transmission mechanism for the selected scheme of the working bodies, we have compiled a classification of the roll modules (working bodies of the roll machines).

The structures of roller modulus with variable inter-axial distance of working shafts depending on the types of paths with displaced centers of rotation of movable working shafts relative to base box, may be of the following types (figure 1):

I. One of working shafts has an ability of axial rotation, and the second working shaft besides axial rotation has an ability to move relative to the first working shaft along the line crossing the center of axes of rotation of both working shafts;

II. Both working shafts have an ability for axial rotation and symmetric movement relative to each other along the line crossing the same axes of rotation;

III. One of working shafts has an ability for axial rotation and the second one, besides axial rotation, has an ability to move relative to the first working shaft along the arc with a certain radius of curvature;

IV. Both working shafts have an ability for axial rotation and symmetric movement relative each other along the arc of circle with certain symmetric centers;

V. One of working shafts has an ability for axial rotation, and the second working shaft, besides axial rotation has an ability to move relative to the first working shaft along a complex path;

VI. Both working shafts have an ability for axial rotation and symmetric movement relative to each other along complex path.

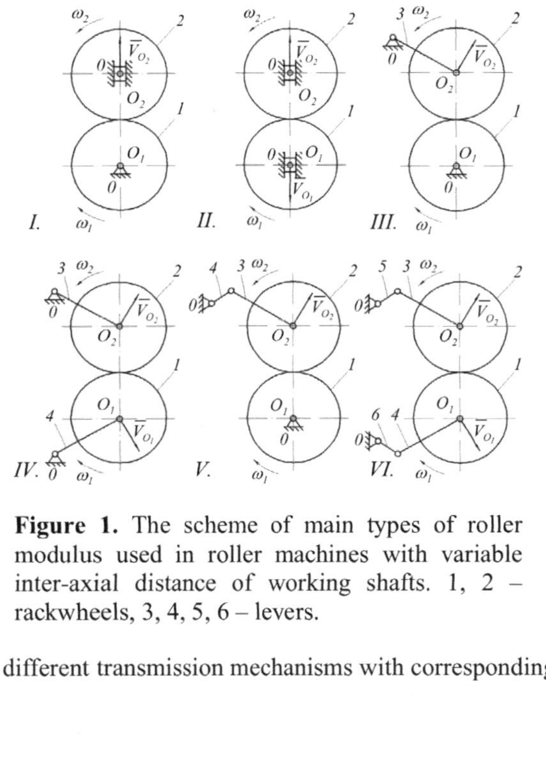

Figure 1. The scheme of main types of roller modulus used in roller machines with variable inter-axial distance of working shafts. 1, 2 – rackwheels, 3, 4, 5, 6 – levers.

To satisfy technological requirements to such roller modulus it is necessary to construct different transmission mechanisms with corresponding characteristics.

The stages of creating these machines are varied. They depend on such factors as urgency, preparedness, availability of similar research and design work, inventions, etc. One of the most important stages in the creation of machines and mechanisms is the synthesis of mechanisms, on the basis of which the structural - kinematic diagram of the machine is developed. N.I. Levitsky wrote that synthesis is the main stage in the creation of machines, after the creation of the structural diagram of the machine; further stages do not significantly affect the structure and it does not change substantially [16]. After the synthesis of the basic scheme, the next important working body is the frame, then the other schemes of actuators and devices are developed. Then comes the stage of design, working and technological documentation, and finally a mechanism or machine is manufactured. Further modernization mainly takes place on the basis of primary structural diagrams, and sometimes the primary diagram, due to the lack of research and development, inventions, rationalization proposals and other ideas in this area, may not fully meet the technological requirements.

Many years of experience in creating the scientific foundations of roller technological machines for the light industry, in particular for the leather industry (in this area, roller machines are used everywhere) have shown that:
- some roller machines do not fully meet the technological requirements for these machines, due to errors inherent in the structural-kinematic diagram showing the location of the working bodies on the frame with their possible degrees of freedom;
-in some roller machines, although the structural-kinematic diagram showing the location of the working bodies on the frame with their possible degrees of freedom meets the presented technological requirements, however, the actuators, especially the inter-roll transmission mechanisms, do not meet the requirements for inter-roll transmission mechanisms, dictated by the technological process. This discrepancy between the actuators and the technological requirements leads to the fact that the whole machine does not meet the requirements [1 - 4]. This is especially common in roller machines with a variable center distance of the working shafts, that is, in roller machines in which at the time of the technological process the center distance of the working shafts changes.

One gets the impression that the designers at the time of designing have ignored the phenomenon of changing the center distance of the working shafts. There are many examples such as: a gear-lever transmission mechanism for the machine for processing *basts of kenaf stalks* [1 - 4]; a gear-lever transmission mechanism used in the harvesting machine of a cotton picker; a gear transmission mechanism used in the VOPM-1800-K squeezing machine; a gear-lever transmission mechanism used in staking-softening machine TMPH-1800-K; chain transmission mechanism of the squeezing machine of the “SWIT” company [1 - 4]. The above mentioned mechanisms are designed to transfer torque from one working shaft to another with a constant gear ratio as well as ensure the constancy of the gear ratio only with a constant center distance of the working shafts, and at the moment of changing the center distance of the working shafts, the gear ratio changes, which leads to a violation of the implementation of agrotechnical and technological requirements for the machines.

Such violations, ultimately, lead to a deterioration in the quality of the processed material, sometimes to their damage, and to a decrease in productivity and a decrease in the durability of machines [1 - 4]. Such shortcomings in the design of roller machines are apparently due to the absence of such materials as reference books on roller machines, atlases on the design and effective classification of roller machines and their inter-roller transmission mechanisms. These materials cannot be developed without a thorough study of roller technological machines and their actuating mechanisms in general, inter-roller transmission mechanisms in particular, without the development of methods for structural, kinematic, dynamic analysis and synthesis of inter-roller transmission mechanisms.

In addition, the aforementioned shortcomings in the design of roller machines may be due to the lack of a systematic approach to the design of these machines, which is as follows: 1) when choosing a structural diagram of a machine, the potential functionality of mechanisms of a particular structure is taken into account in advance, which is then implemented at the stage of kinematic synthesis [17]; 2) the kinematic synthesis of mechanisms is performed considering the whole complex of provisions of the technical specifications for its design, including all kinds of constructive and kinematic constraints, and the quality criteria of motion transmission, not only based on the requirement to reproduce a given motion of the executive body (working rollers); 3) kinematic analysis is considered not only as a stage in the study of synthesized mechanisms, but as a procedure repeatedly used in the process of optimization synthesis of mechanisms. At this stage, in the case of a comprehensive analysis of the "roll module-actuators" system, some discrepancy between the kinematics of the machine and the technical requirement may be revealed, which implies the choice or development of a new structural diagram of the machine. The systematic approach to design also lies in the fact that for all three stages, appropriate systems of algorithms must be developed, with which the automation of the design of roller machines is performed, which, in turn, is impossible without appropriate classifications of roller machines and their inter-roller transmission mechanisms.

We have developed the classifications of roller machines and their inter-roll transmission mechanisms [18]. We have also developed and investigated a gear-lever differential transmission mechanism in relation to a leather roller squeezing machine, in which one of the working shafts has one degree of freedom (rotation around its own axis), and the second shaft has two degrees of freedom (rotation around its own axis and reciprocation along the axis passing perpendicular to the processed material) [19]. The structure, kinematics and dynamics of the developed transmission mechanism were theoretically investigated. The synthesis of this mechanism was conducted for the above-mentioned squeezing machine. It was proven that a squeezing machine with a newly developed toothed-lever differential transmission does not have the above disadvantages. However, as shown above, existing roller machines are diverse. They differ both in purpose and structure, and their transmission mechanisms are also diverse. All the disadvantages of roller machines and their transmission mechanisms, described above, apply to other roller machines. Due to the fact that the purpose and structure of various roller machines are different, despite some identical disadvantages of different roller machines, the ways to eliminate these disadvantages are also different. Therefore, it is necessary to develop a generalized design theory applicable to various roller machines and their mechanisms.

Modern roller machines are a system of mechanisms, consisting of a set of complex working bodies (rollers), executive, auxiliary mechanisms and devices interconnected in a certain way and designed to perform mutually coordinated movements dictated by various technological, agrotechnical requirements. In this system of mechanisms, as in any other, the parameters of functioning, i.e. position, velocity and acceleration of the executive links of individual mechanisms, control and monitoring devices must be coordinated, which is achieved by a purposeful control impact on the object. With an Automatic Control System (ACS), when the impact is realized without direct human involvement [20], the conditions in roller machines have radically changed towards improving the quality of the processed material and saving resources and energy. The MMM laboratory conducted research on the development and implementation of the ACS in production. One of these developments is a system for continuous monitoring and control of uniform feeding of sheet material into the processing zone of the working rollers, which includes tracking and control facilities, a guidance means and a tool for adjusting the tension of the sheet material. Currently, research is underway to determine the geometric, kinematic, and dynamic parameters of the system, which will provide the mechanisms with interconnected operations in a given mode.

In the above five transmission examples, the three transmission mechanisms and the one transmission newly developed by authors, are gear-link transmissions. In recent years, foreign specialists in the field of mechanical engineering and instrument making have shown increasing interest in the theory and practice of using gear-lever mechanisms. This is due to the fact that with gear-lever mechanisms, it is possible to implement diverse and complex patterns of links motion. Numerous combinations of the lever kinematic chain and the kinematic chain composed of gear wheels are diverse and allow us to obtain a wide range of movements of the output link.

Some remarkable properties of gear-lever mechanisms recently revealed, make it possible to consider them as one of the most promising units for the creation of modern machines and devices. [21, 22] Therefore, today the development of roller machines with ACS, inter-roll transmission mechanisms, in particular inter-roll transmission mechanisms of gear-lever type and the development of methods for their structural, kinematic and dynamic analysis and synthesis, in order to provide designers and engineers and calculators of machine-building factories, scientific workers of research and educational institutes with recommendations for the calculation and design of these machines and their mechanisms, is an urgent problem in the theory and practice of creating roller machines.

3. Conclusions

From the above arguments, the following needs arise:

- to develop new, modern and efficient roller machines.
- to develop automated control systems (ACS) for roller technological machines, including programmed control systems (PCS).
- to develop existing and new methods for calculating the geometric, kinematic, dynamic parameters of the ACS actuating mechanisms and their synthesis;
- to develop new transmission mechanisms of roller machines with characteristics that meet certain technological or agrotechnical requirements for roller machines;
- to develop existing and new methods of structural, kinematic, dynamic analysis and synthesis of newly developed and existing transmission mechanisms;
- to research and develop the existing classifications of roll modules and their actuators;
- to develop a classification of roll modules, taking into account the technological and agrotechnical requirements for these modules;
- to develop a classification of inter-roll transmission mechanisms taking into account the design of roller modules and technological, agrotechnical requirements for them.

4. References

[1] Burmistrov A.G. 2006 *Machines and devices for the production of leather and fur*. (Moscow: KolosS).

[2] Bahadirov G.A. 2010 *Mechanics of a squeezing roller pair* (Tashkent: Fan).

[3] Abdukarimov A. 1995 *Analysis and synthesis of transmission mechanisms of roller machines with variable center distance of the working shafts* (Tashkent: IMSS).

[4] Kuznetsov G.K., Smirnov B.I. 1967 *Design of machines for primary processing of bast fibers* (Moscow: Mechanical engineering).

[5] Bahadirov G.A., Ravutov Sh.T., Abdukarimov A. and Toshmatov E. 2021 Development of the methods of kinematic analysis of elliptic drum of vertical-spindle cotton harvester *IOP Conf. Series: Earth and Environmental Science* **1030** (2021) 012160 doi:10.1088/1755-899X/1030/1/012160.

[6] Bahadirov G.A., Sultanov T.Z. and Abdukarimov A. 2020 Kinematic analysis of tooth-lever differential transmission mechanisms *IOP Conf. Series: Earth and Environmental Science* **614** (2020) 012102 doi:10.1088/1755-1315/614/1/012102.

[7] Bahadirov G.A., Sultanov T.Z. and Abdukarimov A. 2020 Comparative analysis of two gear-lever differential inter-roller transmission mechanisms *IOP Conf. Series: Earth and Environmental Science* **614** (2020) 012101 doi:10.1088/1755-1315/614/1/012101.

[8] Abdukarimov A., Saydokulov I.Kh., Abdullazhanov A.A. 2020 Experimental study of the transmission mechanisms of roller machines *J. Problems of mechanics* (Tashkent) **1-2** 150-154.

[9] Abdukarimov A., Saydokulov I.Kh., Abdullajonov A.A. 2020 Experimental analysis of roller machine transmission mechanisms *ACADEMICIA: An International Multidisciplinary Research Journal* (India) **10** 206-213.

[10] Khusanov K. 2020 Equations of motion of mechanical systems with nonlinear nonholonomic servoconstraints *IOP Conference Series Materials Science and Engineering* **869** (2020) 072021 doi: 10.1088/1757-899X/869/7/072021.

[11] Khurramov, Sh.R. 2020 Simulation of the form of contact curves rolls in two-roll modules

IOP Conf. Series: Earth and Environmental Science **614** (2020) 012096 doi:10.1088/1755-1315/614/1/ 012096.

[12] Khurramov, Sh.R. 2020 Some questions of the contact interaction theory in two-roll modules *Journal of Physics: Conf. Series* **1546** (2020) 012132 doi:10.1088/1742-6596/1546/ 012132.

[13] Khurramov, Sh.R., Kurbanova F.Z. 2020 Distribution of contact voltages in asymmetric two-roll module *IOP Conf. Series: Earth and Environmental Science* **614** (2020)012098 doi:10.1088/1755-1315/614/1/ 012098.

[14] Khurramov, Sh.R., Khalturayev F.S. 2020 Simulation of contact voltages in two-roll modules *IOP Conf. Series: Earth and Environmental Science* **614** (2020)012097 doi:10.1088/1755-1315/614/1/ 012097.

[15] Petrovsky V.S., Kuznetsov G.K. 1982 Analysis of the nature of the interaction of roll mechanisms with the processed material *of Universities Textile industry technology* **8**, 93-96.

[16] Levitskaya O.N., Levitsky N.I. 1985 *The course of the theory of mechanisms and machines* (Moscow: High School).

[17] Peytsakh E.E., Nesterov V.A. 1988 *Planar Linkage Design System* (Moscow Mechanical Engineering)-232.

[18] Abdukarimov A. 2014 Classification of two-roll modules used in roller machines with variable center distance of the working shafts *Repots of AS RUz.* (Tashkent) **5** 34-36.

[19] Patent for invention RUz No. IAP 04934. *Gear-lever transmission mechanism of roller machines* (options) Option 3 Abdukarimov A., Bahadirov G.A., Nabiev A.M., Saydakhmetova N.B., Saidokulov I.Kh., Abdukarimov A.A. Official Bulletin 2014 **8**.

[20] Frolov K.V., Popov S.A., Musatov A.K. et al. 2001 *Theory of mechanisms and mechanics of machines*: (Moscow: High School) 496.

[21] Fateev N.A. 2008 *Determination of angular velocities and angular accelerations of links of flat gear-link mechanisms* SevKavGTU 218 www.ncstu.ru.

[22] Shashkin A.S. 1987 *Adjustable toothed linkage and roller mechanisms* (Voronezh, Voronezh University Press) 248.

IOP Conf. Series: Materials Science and Engineering 1182 (2021) 012002 doi:10.1088/1757-899X/1182/1/012002

Kinematics of the transmission gear mechanism for roller machines with different diameters of drive shaft

A Abdukarimov, A Abdullajonov and A Abdukarimov
Institute of Mechanics and Seismic Stability of Structures named after M.T. Urazbaev, Academy of Sciences of the Republic of Uzbekistan, 31 Durmon Street, 100125, Tashkent, Uzbekistan

E-mail: alishanster@gmail.com

Abstract: The article describes the arrangement, principle of operation and kinematic analysis of a ten-link gear-lever differential transmission mechanism with a parallelogram lever contour for roller machines with different diameters of the working shafts. The purpose of the kinematic analysis of the transmission mechanism is to determine the velocities and accelerations of the mechanism links, the gear ratio of the mechanism, and to show the constancy of the gear ratio of the transmission mechanism when the center distance of the driving and driven tooth gears changes. An important condition for the normal operation of roller machines with different diameters of the working shafts is the coincidence of the gear ratio of the working shafts and the gear ratio of the transmission mechanism between these shafts. Initially, the kinematics of the mechanism was performed using the analytical method. In the analytical study of the mechanism, the centroid method was used. Dependencies for determining the kinematic parameters of the ten-link gear-lever transmission mechanism are derived. The advantages and disadvantages of this method are shown. Further, to compare the results of the study, the kinematics of the mechanism is performed by the graphic-analytical method. The method developed by the authors was used in the graphic-analytical study.

1. Introduction

In roller machines (squeezing-washing, staking, corrugated crimping machines - for primary processing of bast crops, and squeezing, staking-softening machines - for processing leather and leather semi-finished products) the center distance of the working shafts changes during operation [1-3]. Normal operation of corrugated shafts in pairs takes place when the groove of one shaft is located in the center of the cavity of the other, which is possible only with the synchronous rotation of the corrugated shafts regardless of the change in their center distance [1]. Drives and transmission mechanisms between the working shafts, currently used in some roller machines, while changing their center-to-center distances, do not ensure the equality of the linear velocities of the surfaces of the working shafts [4]. In order to eliminate the above disadvantage in the transmission mechanisms of existing two-roll machines with unequal diameters of the working shafts, we have developed a four-wheel gear-lever differential transmission mechanism (GLDTM), which provides an equal linear velocity of the surfaces of these working shafts, regardless of the changes in their center distance, i.e. e., providing the synchronous rotation of the working shafts regardless of the changes in their center distance and the difference in the diameters of the working shafts [5]. One of the most promising mechanisms for creating modern machines and devices is a gear-lever mechanism [6-8]. Therefore, we have invented and investigated one of the types of differential gear-lever transmission mechanisms.

2. The arrangement and principle of operation of the roller machine with GLDTM

Figure 1 shows a structural-kinematic diagram of a roller machine with such transmission mechanism. The machine consists of two working shafts *1* and *2*.

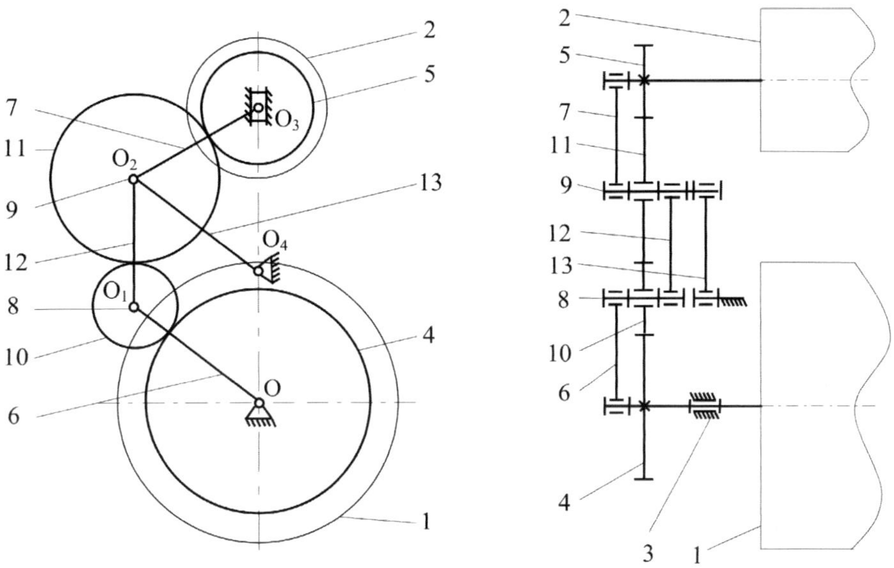

Figure 1. Structural and kinematic diagram of a roller machine with a gear-lever differential transmission mechanism. *1, 2*-working shafts; *3*-rack; *4*-drive gear; *5*-driven gear; *6, 7, 12, 13*-levers; *8, 9*-axles of intermediary gears; *10, 11*-intermediary gears.

The upper working shaft has the ability to rotate around its own axis and move vertically along a line passing along the axis of two working shafts, and the lower working shaft is mounted on rack *3* and has the ability to rotate around its own axis. At the output ends of working shafts *1* and *2*, gears *4* and *5* are rigidly fixed and levers *6* and *7* are hinged, which are the supports for axles *8* and *9*. Intermediary gears *10* and *11* are freely installed on axles *8* and *9*. Axles *8* and *9* are pivotally connected to lever *12*. Axle (*9*) of the intermediary wheel (*11*), the nearest to the movable working shaft (*2*), is kinematically connected to the rack (*3*) by means of lever *13*. Levers *6, 12* and *13* form a lever parallelogram $OO_1O_2O_4$.

3. Analytical kinematics of the GLDTM

The kinematic analysis of the mechanism was performed to determine the angular and linear velocities and accelerations of the mechanism links depending on the angular velocity of the driving link (gear *1*) and the linear velocity and acceleration of the center of rotation of the free working shaft (*2*), and to prove the equality of linear velocities of the surfaces of the working shafts at the point of contact of these shafts with the processed material. Kinematic analysis was performed according to the centroid theory by the method of the instantaneous center of rotation of the links. Let the diameters of the working shafts be D_1 and D_4 (figure 2), the pitch diameters of the gears - d_1, d_2, d_3, d_4 and the working shaft *1* rotates at angular velocity ω_1. The free working shaft moves at velocity $\overline{V}_{O_3}$. Considering that in most roller machines the diameters of the working shafts are unequal to each other, as a special case, the transmission mechanism for such a machine. Consequently,

$$D_1 \neq D_4 \tag{1}$$

The following conditions must be met for the linear velocities of the surface of the roller $\overline{V}_{b_1}$ and $\overline{V}_{b_4}$ to be equal to each other at $D_1 \neq D_4$:

$$d_1 \neq d_2 \neq d_3 \neq d_4 \tag{2}$$

or

$$\frac{D_1}{D_4} = \frac{d_1}{d_4} = \frac{d_3}{d_2} \tag{3}$$

where - d_1, d_2, d_3, d_4 are the pitch diameters of gear wheels.
At a constant center distance of the working shafts $(O_1O_3 = const)$, i.e. in the absence of a linear velocity of the axles of the free working shaft $(\overline{V}_{O_3} = 0)$, the velocity of the centers of rotation of the intermediary gears is also zero $(\overline{V}_{O_1} = 0,\ \overline{V}_{O_2} = 0)$; the compliance with conditions *2, 3* or *4* ensures the equality of the linear velocities of the surfaces of the working shafts ($\overline{V}_{b_1} = \overline{V}_{b_4}$).
It is necessary to prove that $\overline{V}_{b_1} = \overline{V}_{b_4}$ when the center distance of the working shafts changes, that is, in the presence of velocity $\overline{V}_{O_3}$.
So,

$$\overline{V}_{b_1} = \omega_1 \cdot R_1, \tag{4}$$

$$\overline{V}_{b_4} = \omega_4 \cdot R_4, \tag{5}$$

then, taking into account equality (1), and $D = 2R$, it suffices to prove that $\omega_1 = \omega_4$.

Let is consider the lever contours of the gear-lever mechanism. From the rocker-slider contour, the velocity of point O_2 $(\overline{V}_{O_2})$ can be determined depending on the velocity of point O_3 $(\overline{V}_{O_3})$

$$\overline{V}_{O_2} = \overline{V}_{O_3} + \overline{V}_{O_2O_3}, \tag{6}$$

$$\overline{V}_{O_2} = \frac{\overline{V}_{O_3}}{2\sin\alpha} \tag{7}$$

Given that the lever contour $OO_1O_2O_4$ is a parallelogram, we can write

$$\overline{V}_{O_2} = \overline{V}_{O_1}. \tag{8}$$

Velocity (V_a) of the contact point (a) of gears 1 and 2 can be determined by the following equation

$$\overline{V}_a = \omega_1 r_1 \tag{9}$$

IOP Conf. Series: Materials Science and Engineering 1182 (2021) 012002 doi:10.1088/1757-899X/1182/1/012002

$$\overline{V}_{b_1} = \omega_1 \cdot R_1 \qquad (10)$$

where r_1 –is the pitch radius of gear wheel *1*.

Knowing the velocities of two points of a rigid body, we can determine its instantaneous center of rotation [6]. Knowing the velocities of two points $(O_1 \textit{ and } a)$ of gear wheel *2*, we determine the instantaneous center of rotation (P_1) of this gear wheel; and the angular velocity of this wheel is determined by the following equation:

$$\omega_2 = \frac{\overline{V}_{O_1}}{P_1 O_1} \qquad (11)$$

Connecting the point of contact (b) of gears *2* and *3* with point P_1, we can determine the velocity of point *b* by the following equation

$$\overline{V}_b = \omega_2 \cdot bP_1 \qquad (12)$$

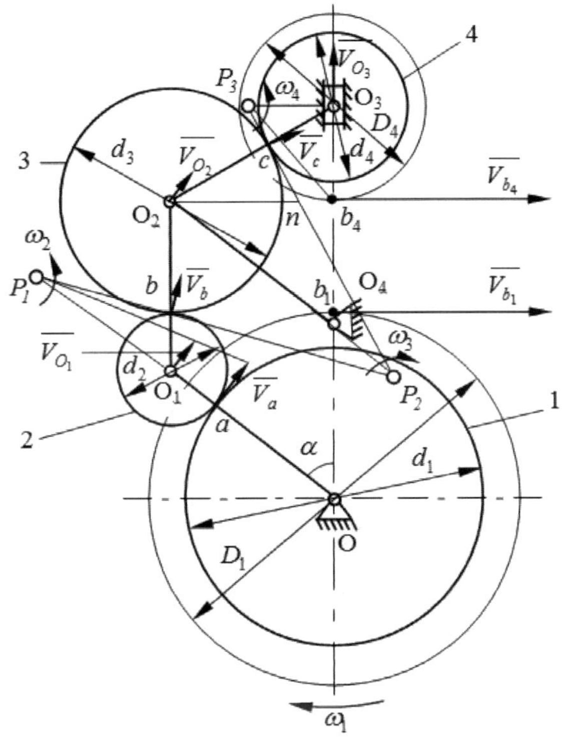

Figure 2. Design diagram of the kinematics of a roller machine with a gear-lever differential transmission mechanism. 1-drive gear; 2, 3-parasitic gears; 4-driven gear.

Knowing the velocities of two points of wheel *3* $(\overline{V}_b \textit{ and } \overline{V}_{O_2})$, we determine the instantaneous center of rotation (P_2) of this wheel; its angular velocity (ω_3) is determined by the equation

$$\omega_3 = \frac{\overline{V}_b}{bP_2}. \qquad (13)$$

Connecting the point of contact (c) of gears 3 and 4 with point P_2, we determine the velocity of point *(c)* by

$$\overline{V}_c = \omega_3 \cdot cP_2. \qquad (14)$$

Knowing the velocities of two points of wheel *4* $(\overline{V}_c \textit{ and } \overline{V}_{O_3})$, we determine the instantaneous center of rotation of this wheel (P_3); and its angular velocity (ω_4) is determined by the following equation

$$\omega_4 = \frac{\overline{V}_c}{cP_3}. \tag{15}$$

Taking into account equations (11), (12), (13) and (14), equation (15) takes the form

$$\omega_4 = \overline{V}_{O_1} \frac{bP_1 \cdot cP_2}{P_1O_1 \cdot bP_2 \cdot cP_3}. \tag{16}$$

The following relation can be written for wheel *2*

$$\omega_2 = \frac{\overline{V}_a}{P_1O_1 + r_2} = \frac{\overline{V}_{O_1}}{P_1O_1}. \tag{17}$$

where r_2 – is the pitch radius of the gear wheel *2*.
From relation (17) we determine V_{O_1}

$$\overline{V}_{O_1} = \overline{V}_a \frac{P_1O_1}{P_1O_1 + r_2}. \tag{18}$$

Using (10), equation (18) can be written in the following form

$$\overline{V}_{O_1} = \omega_1 \frac{r_1 \cdot P_1O_1}{P_1O_1 + r_2}. \tag{19}$$

Using (19), equation (16) takes the following form

$$\omega_4 = \omega_1 \frac{r_1 \cdot bP_1 \cdot cP_2}{(P_1O_1 + r_2) \cdot bP_2 \cdot cP_3}. \tag{20}$$

In equation (20), the values of $P_1O_1, bP_1, bP_2, cP_2, cP_3$ are determined by solving the triangles $cO_2P_2, O_2P_2b, P_1O_1b$. Using the derived equations ((8) - (20)) to determine the angular and linear velocities of the characteristic points of all links, and taking into account the relation of $\overline{V}_{O_3}$ with angle α, we determine the angular and linear accelerations of the characteristic points of all links.
Let us proceed to the proof of $\overline{V}_{b_1} = \overline{V}_{b_4}$, from the similarity of triangles O_2P_2b and O_1P_1b we can write

$$\frac{r_2}{r_3} = \frac{bP_1}{bP_2}. \tag{21}$$

With equations (12) and (21), equation (13) can be written in the following form

$$bP_2 = \frac{bP_1 r_3}{r_2}. \tag{22}$$

P_1O_1 is determined from equation (18)

$$P_1O_1 = \frac{\overline{V}_{O_1} \cdot r_2}{\overline{V}_a - V_{O_1}}. \tag{23}$$

From triangle cO_2P_2, by the cosine theorem, we write

$$cP_2 = \sqrt{r_3^2 + (O_1P_1)^2 + 2 \cdot r_3 (O_1P_1) \cos 2\alpha} \tag{24}$$

Consider similar triangles P_3O_3c and $O_2c\,n$.
From the similarity of triangles, we can write the following relation

$$\frac{r_3}{r_4} = \frac{cn}{cP_3}, \tag{25}$$

Hence

$$cP_3 = \frac{r_4(cn)}{r_3}. \tag{26}$$

From triangles O_2cP_2, nO_2P_2, nO_2c we determine nO_2

$$nO_2 = \frac{\sqrt{r_3 \cdot O_1P_1 \cdot \left[(r_3 + O_1P_1)^2 - (cP_2)^2\right]}}{O_1P_1 + r_3}. \tag{27}$$

Applying the cosine theorem to triangle $O_2\,c\,n$ we determine $c\,n$

$$cn = \sqrt{r_3^2 + (nO_2)^2 - 2r_3(nO_2)\sin\alpha}\quad, \tag{28}$$

With (27), equation (28) takes the following form

$$cn = \frac{r_3}{r_3 + O_1P_1} cP_2. \tag{29}$$

Substituting equation (29) into equation (26), we obtain

$$cP_3 = \frac{r_4 . cP_2}{r_3 + O_1P_1}. \tag{30}$$

With (1), (2), (21), (24) and (30), equation (20) takes the following form

$$\overline{V}_{b_4} = \omega_1 \frac{r_1 \cdot bP_1 \cdot cP_2}{(P_1O_1 + r_2) \cdot bP_2 \cdot cP_3} \tag{31}$$

Consequently,

$$\overline{V}_{b_1} = \overline{V}_{b_4} \tag{32}$$

The gear-lever differential transmission mechanism considered makes it possible to provide similar linear velocities of the contact surfaces of the working shafts, regardless of the change in their center-to-center distance. The results of the analysis and the derived equations make it possible to investigate the dynamics and to perform the geometric and dynamic synthesis of this transmission mechanism.

4. Conclusions

The gear-lever differential gear mechanism considered in the article, with the correct choice of the scheme of the two-roll module and the geometric parameters of the intermediary driving and driven gears, allows us to provide the same linear velocities of the contact surfaces of the working shafts and the processed material, regardless of the change in their center distance and the difference in the diameters of the working shafts. The results of the analysis and the equations derived make it possible to study the dynamics and to perform the geometric and dynamic synthesis of this transmission mechanism.

5. References

[1] Smirnov B.I., Kuznetsov G.K. 1967 Design of machines for primary processing of bast fibers. *M.: Mechanical Engineering* 270.
[2] Burmistrov A.G. 2006 Machine and apparatus for the production of leather and fur. *M.: Kolos.* 384.
[3] Bahadirov G.A. 2010 Mechanics of a squeezing roller pair. *Tashkent: "Fan". 156 p.*
[4] Amanov T.Yu., Abdukarimov A., Nabiev A.M. 2007 Kinematic study of the process of dehydration of wet materials *J. Problems of mechanics* **6** 43-45.
[5] Invention certificate 1632047 A 1 USSR, MKI S14 B 1/00, 1/02. Transfer mechanism of roller machines / A. Abdukarimov, T.Yu. Amanov, G.A. Bahadirov, No. 4621476/12; Appl. 19.12.88. Reg. 01.02.90.
[6] Frolov K.V., Popov S.A., Musatov A.K. et al. 2001 Theory of mechanisms and mechanics of machines *M.: Higher school 496.*
[7] Volmer, J., 1969 Getriebetechnik. Lehrbuch. *VEB-Verlag "Technic". Berlin.*
[8] Levitsky, N.I., et al. 1974 Theory and application of gear-lever mechanisms. *M.: Nauka* 135-136.

IOP Conf. Series: Materials Science and Engineering 1182 (2021) 012003 doi:10.1088/1757-899X/1182/1/012003

The control of Industrial emission via the subsoil green filters

S J B Al-Maliki[1], K Jaafar[2] and B I Wahab[3]
[1]College of Engineering, Al-Iraqia University, Iraq
[2]Al-Karkh University, Iraq
[3]College of Science, Atmospheric Science Department, Mustansiriyah University, Iraq

E-mail: salambash2000@yahoo.com

Abstract: The vast consumption of fossil fuel as an energy source for the various industrial, commercial and municipal activities, has led to the production of huge air pollutants, in a manner that needs more modern and effective efforts and technologies to prevent their hazardous consequences. This project aims to the use of the subsoil of nearby grounds as a natural filter to distribute these pollutants based on the soil`s high porosity and their ability to handle huge amounts of the gaseous effluents of the artificial projects. Soil properties such as the mechanical properties, porosity and ground water content were tested for different depths for two types of soils, in order to assess their readiness for the tentative use as a gaseous filter and to avoid the possible negative effects for the injection of gaseous pollutants on their biologic systems. An experimental system was made using a 5 KVA power generator, the gaseous emissions of which would be distributed through the subsoil via a buried pipe. The project also, included the precautions for the possible machine backpressure that may damage it. The wet clay type of soil has shown remarkable better performance than the dry sand type.

1. Introduction

The resulting wastes was always an issue for almost all types and all capacities of human being activities, that is due to their potential negative impacts on the receiving bodies for these wastes, whether these would be the atmosphere, surface waters, or lands. For centuries, it was a classical practice to use landfills as, to some extent, an easy and cheap management procedure for the industrial and municipal solid wastes. However, that was found the cause of additional pollution problems for the underground environment such as the farming areas, the ground-water wells and the possible seepage towards the rivers and lakes [1,2]. Many progresses and improvements were considered and applied to avoid these complications by numerous researchers who tested the practice of subsoil as filters to remedy the pollution loads on the environment due to the various types of domestic, commercial and industrial solid wastes, residues of radioactivity linked activities and the hazardous wastes. The land burial disposal of low/intermediate radioactivity level wastes, was the classical practice for many countries since the flourish on atomic/nuclear related industries, such as USA, UK, India, China, Canada, France, and the former Soviet Union as the primary choice for the protection their communities from the prospective risks of these wastes through the application of firm strategic, regulative and organizational actions that assure the safety of the recipient environment [3-5]. However, the airborne contaminants of the industrial or commercial emissions including the various transportation means, have been a big concern due to their vast ability to cover wide range of lands, as they were renowned of the content of large amounts of criteria pollutants, such as the hydrocarbons HC, and carbon monoxide CO, that may cause several health and economic adverse effects [6]. Researchers from various institutes studied the potentials of green means as safe handling mechanism against these gaseous releases to reach the allowable limits of exposure as were agreed by several relevant entities such as listed in table 1. Non edible tree leaves such as the Common myrtle tree were studied by AlMaliky, 2010 [7], and Kapoor, Bamniay and Kapoor, 2012 [8], and verified substantial control efficiency regarding the reduction of the concentration of the gaseous emissions from commercial and industrial sources to reach 73% and 82% of the emitted Nitrogen dioxide NO_2 and Carbon monoxide CO gases, respectively via the absorption process.

The main problem here is to test and prove the subsoil mediums capacity to filter the contaminates from the gaseous effluent streams of industrial sources, depending on the porous characteristic of these soils.

Table 1. Exposure Limits for Gaseous Pollutants according to the National Ambient Air Quality Standards NAAQS [9].

Pollutant	Primary/ Secondary	Averaging Time	Level	Form
CO	primary	8 hours	9 ppm	Not to be exceeded more than once per year
		1 hour	35 ppm	
NO_2	primary	1 hour	100 ppb	98th percentile of 1-hour daily maximum concentrations, averaged over 3 years
	primary and secondary	1 year	53 ppb (*)	Annual Mean

(*) The level of the annual NO_2 standard is 0.053 ppm. It is shown here in terms of ppb for the purposes of clearer comparison to the 1-hour standard level.

2. Laboratory arrangements

Two types of subsoils are selected to test their filtration potential of the gaseous contaminants released from a 5KW, Gasoline electrical power generator. These are dry sand soil and wet clay soil that are the most abundance in Baghdad, capital of Iraq. The external hosepipe of the power generator is elongated to allow the burial at various depths (20 cm-80 cm) underground each of these two soil types as demonstrated by figure 1, to test the effect of each burial depth in reducing the ground and aerial pollution index from each of the gaseous pollutants under study; CO, and NO_2, via a TG-501 gas analyzer above ground surface and at flat distances; 30cm, 60cm 90 cm and 120cm, from the end of the subsurface pipe end, after the laps of four hours of operation per day in an open area. As a protection measure for the hosepipe end, a wooden box is fixed above it to avoid and accidental soil avalanche that may clog it. ERGIL Storagetech™'s Model 193 Vacuum Relief, is utilized and connected to the pressure vessel and set to actuate at 1.3 psi that was measured as the backpressure limit, so as to elude the expected adverse effects on the power generator during the study period.

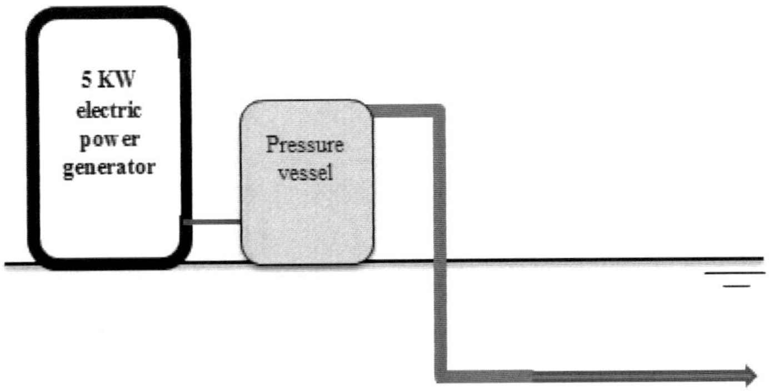

Figure 1. Laboratory setup for subsurface disposal of industrial gaseous releases.

3. Results and Analysis

The first important issue that need to be assured is the porosity of the subsoil layers that would decide the ability of these regions to disseminate the gas streams of the hosepipe end, that is according to ASTM D7263-21[10]. Figure 2, demonstrate these values for both subsoil types and it tells that the wet clay type has higher porosities at all tested depths as compared to these of the dry sand soil, in an indication to higher distribution performance for the former soil type regarding the contaminated gas release. Also, it is determined that the porosity is relatively less at higher depths to these at shallow depths for both types of soils, as it goes to half its value at 80 cm depth as compared to that at 20 cm depth, and this might be attributed to the relative soil condensation at depths far from the surface. This leads to the conclusion that no significant enhancement would be expected with further burial depths for the hosepipe.

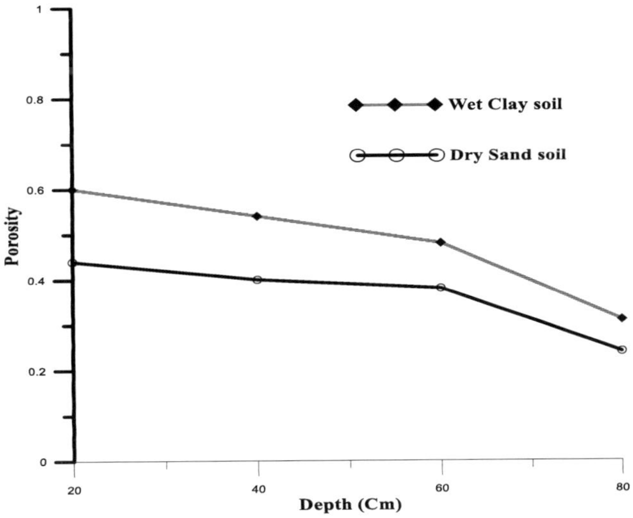

Figure 2. Variation of subsoil porosity with underground depth for the tested soils.

The statistical analysis of variance ANOVA-two way test for the results of the gas injections at variable underground depths confirmed very clear worth of depth and horizontal flat dimensions for the subsurface gas injection from the pollution release source as influences impacting the surface concentrations of the understudy gases with 95% confidence interval. This is demonstrated for the dry sand soil and the wet clay soil in table 2. That is also, clarified in the weak probabilities compared with the 0.05 significance zone, for all factors in the two types of subsoils, which might be attributed to the consequent gas concentrations above the surface as a result for the injection through the subsurface layers. These thorough statistical analyses are tabulated in table 2 for both subsoil types understudy.

Table 2. The ANOVA-two way statistical results for the two understudy subsoil types.

Factors	CO				NO_2			
	F-test		Probability		F-test		Probability	
	dry sand	Wet clay	dry sand	Wet clay	Dry sand	Wet clay	Dry sand	Wet clay
Subsurface depth	13.68	8.89	0.016	0.034	10.32	18.93	0.026	0.009

The performance of gas injection is best demonstrated by the gas concentrations; CO and NO_2 at the flat surface distances from the pollution source as a result of each injection depth. These records are demonstrated by figures 3 and 4 regarding the wet clay soil and figures 5 and 6 regarding the dry sand soil. In general, all these figures show significant reduction in the tested gas concentrations as the injection depth increases from zero (surface) to 60 cm. where more than 73% of the initial CO gas concentration is dissipated through the surrounding wet clay subsoil, while it is about 52% in the dry sand soil. From the other side, it is reduced to about 75% and 62% respectively for the NO_2 gas. beyond that, the reductions tend to be moderate to insignificant as the depths approach 80 cm, in an important sign that no further profitable gain might be expected with more injection depth, in addition to the risk of activating the backpressure valve or jeopardizing the machine (generator) as a result of possible clog. This might be attributed to the above mentioned lower porosities at these depths for both types of subsoils as a result of layers compactness. In an additional effect of the porosity on this injection remedy approach, it is clear from the mentioned results, that the wet clay of higher porosity all through the tested depths, proved better reduction performance for both tested gases. The records at flat distances for each tested depth, determine that the surface gas concentrations are highly affected by distancing away from the pollution source, as these records almost approach zero values at distances of 120 cm for both tested gases in both tested subsoils. However, the wet clay subsoil is determined to cause 8 times reduction as compared to the dry sand soil in handling the CO gas, and about 2.4 times regarding the NO_2.

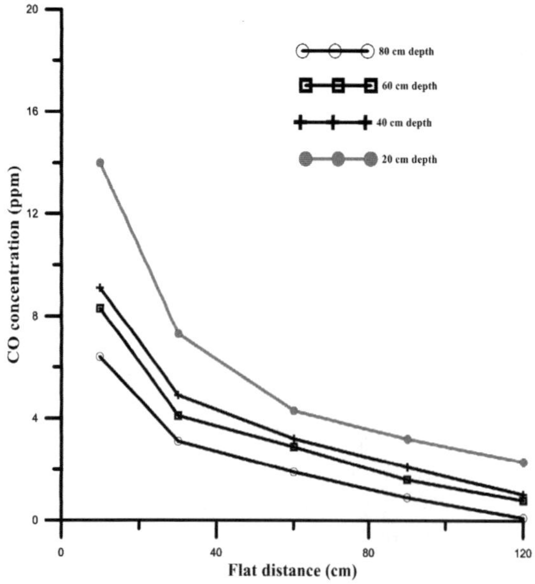

Figure 3. The concentration of CO at flat distances vs. the injection depths-Wet clay.

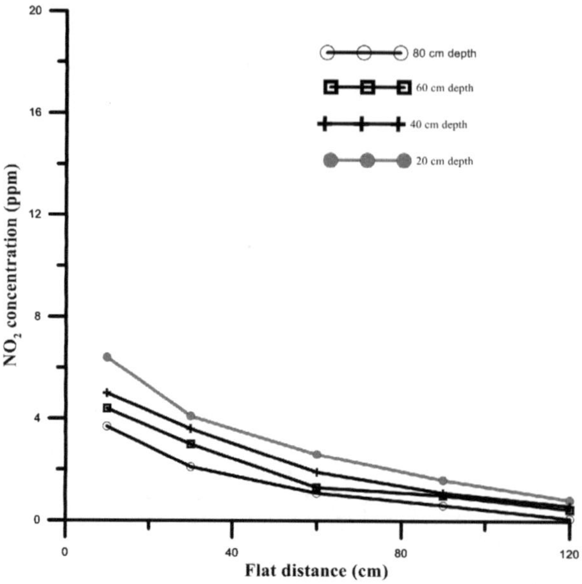

Figure 4. The concentration of NO_2 at flat distances vs. the injection depths-Wet clay,

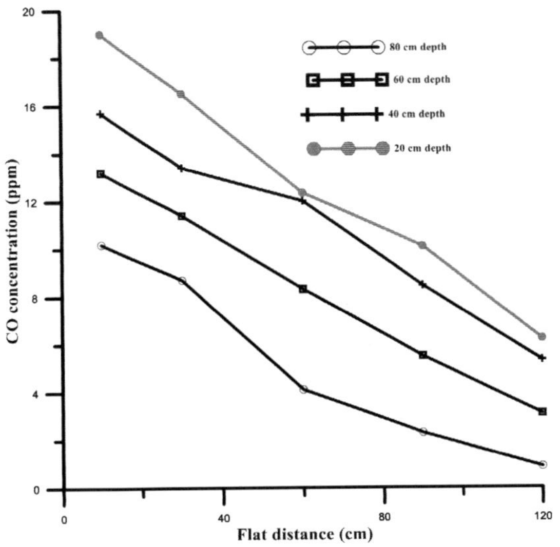

Figure 5. The concentration of CO at flat distances vs. the injection depths-Dry sand.

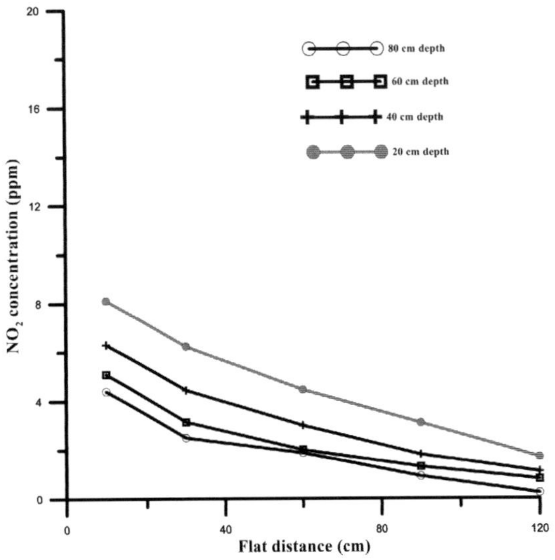

Figure 6. The concentration of NO_2 at flat distances vs. the injection depths-Dry sand.

4. Conclusions

The dissipation of potential hazardous gas releases from the various industrial and commercial sources through the underground soils is studies and tested as one of the promising effective, green, and low cost solutions to produce more livable environment for the coming generations. This task is achieved in a laboratory sized approach to inject the gaseous emissions of a middle sized home power generator into various depths at two types of soils; wet clay and dry sand, that represent the most prevailing soil types in Baghdad, the capitol of Iraq. The injection depth and flat distance away from the pollution source is tested for both subsoil to determine their mutual impacts on the surface concentrations of the gases CO and NO_2. The gas stream subsoil injection proved a significant reduction performance for both soils against both gases understudy, as the final surface concentrations about 120 cm away from the pollution source approached the value of zero, although the wet clay subsoil is determined to be much significantly better filter media for CO and remarkable for the filtration of NO_2 as compared to the NO_2 gas. These remarks are highly reinforced by the statistical analysis ANOVA that showed high correlation between the vertical and horizontal injection dimensions.

5. References

[1] Al-Maliky Salam J. Bash, ElKhayat Zainab Qahtan 2012 Kitchen Food Waste Inventory for Residential Areas in Baghdad City *Modern Applied Science* **6**(8) pp.45-51.

[2] M. D. Vaverkova et al 2013 Environmental impact of landfill on soils- The example of the Czech republic *Polish journal of soil science* V. L/1 DOI: 10.17951/pjss/2017.50.1.93.

[3] AlKhayat, Z.Q., Bash AlMaliki, S.J. 2020 Effects of the deterioration of construction materials industry on the radioactivity background in Baghdad *Iraq Kuwait Journal of Science* **47**(1) pp. 97–102.

[4] Klaus Brun and Rainer Kurz 2019 *Compression Machinery for Oil and Gas* Gulf Professional Publishing DOI 10.1016/C2017-0-02064-5.

[5] IAEA 2010 Radiation protection and the management of radioactive waste in the oil and gas industry *Training course series* No. 40 Vienna, Austria.

[6] Lee Jechan 2010 *A Study on Performance and Emissions of a 4-stroke IC Engine Operating on Landfill Gas with the Addition of H_2, CO and Syngas* MSc thesis Department of Earth and

Environmental Engineering HKSM Fu Foundation School of Eng. and App. Science, Columbia Univ. New York.
[7] Al-Maliky, S.J.B 2015 Control of gaseous pollution via the leaves of non-edible trees *IOP Conference Series: Materials Science and Engineering* **95**(1), 012081.
[8] Kapoor C., Bamniya B., and Kapoor K. 2012 Natural and effective control of air pollution through plants- studies on a tree species: Holoptelea integrifolia L. *Journal of Mitigation and Adapt. Strategies for Global Change*. **17**(7) 793-803 Springer doi: 10.1007/s11027-011-9344-4.
[9] NAAQS 2021 https://www.epa.gov/criteria-air-pollutants/naaqs-table accessed on 15.02.2021.
[10] ASTM D7263-21 2021 Standard Test Methods for Laboratory Determination of Density and Unit Weight of Soil Specimens, *ASTM International*, West Conshohocken, PA.

Local flow assessment of the Japan Bulk Carrier using different turbulence models

A Bekhit[1] and F Popescu[2]
[1]"Dunarea de Jos" University of Galati, Faculty of Naval Architecture, 47 Domneasca Street, 800008, Galati, Romania
[2]"Dunarea de Jos" University of Galati, Faculty of Engineering, 47 Domneasca Street, 800008, Galati, Romania

E-mail: adham.bekhit@ugal.ro

Abstract. Ship resistance and powering represent the most important aspects in the initial design stage of the ship. Based on their estimation the basic milestone for selecting the main engine and the propulsion system is established. The majority of ships in the international fleet nowadays rely on the screw propeller working in the wake zone behind the ship. The wake flow of the ship has a direct impact on the propeller performance and the propulsion efficiency. Accurate prediction of the nominal and effective wake is crucially important to provide a proper understanding of the flow where the propeller will perform. From this point of view, the wake flow of the Capesize Japan Bulk Carrier (JBC) is assessed using a viscous flow Computational Fluid Dynamics (CFD) method. Numerical simulations are performed to predict the nominal and effective wake of the ship by making use of the viscous flow solver ISIS_CFD of the FINE^{TM}/Marine software provided by NUMECA. The solver is based on the finite volume method to build the spatial discretization of the transport equation to resolve the Reynolds-Averaged Navier-Stokes (RANS) equations. Closure to turbulence is achieved using different turbulence models in order to investigate their accuracy in predicting the complex wake flow of the ship. Two-phase flow approach is used to model the air-water interface where the Volume of Fluid method is implemented to capture the free-surface. The results for both nominal and effective wake are assessed against the experimental data provided by the National Maritime Research Institute (NMRI) and Yokohama National University in Japan that were presented in the seventh Workshop on CFD in ship hydrodynamics (Tokyo2015). The results validation showed a reasonable agreement compared to the experimental data for both nominal and effective wake. As it was expected, some turbulence models showed to be more accurate in predicting ship wake, especially the Shear Stress Transport (K-ω SST) and Explicit Algebraic Reynolds Stress (EASM) Models. A special investigation of the flow vortices is also taken into consideration.

1. Introduction

Ship resistance and powering are the two most important aspects in the initial design stage of a ship. Their accurate prediction is crucially important, since the following aspects regarding ship operability, speed and performance may be directly or indirectly influenced by their estimation. Once the ship powering is predicted, the following stage is to select the main engine and propulsion system. The vast majority of commercial ships in the international fleet are relying on the rotating propeller that is working in the wake zone behind the ship. This imposes another challenge for the naval architect to predict accurately the flow configurations in the propeller working zone in order to avoid any undesired problems such as noise, vibration, cavitation, and/or loss of propulsion. Heading from this point of view, the importance of the wake flow analysis is augmented and rather becomes extremely crucial for the high-block coefficient ships, as in case of tankers and bulk carriers where a significant flow separation and vortex formations are expected to occur. For this reason, this study investigates the local flow of the Japan Bulk Carrier (JBC hereafter) ship on three different levels: the first is for the nominal wake, i.e. the ship is not equipped with the propeller; the second is for the propeller open water (POW) condition and finally the effective wake were the propeller is connected and working behind the ship. The study is based on the Computational Fluid Dynamic (CFD) method that becomes more popular in the ship hydrodynamic field as the quality of the predicted results in the recent two decades are competing head to head with the classic experimental method from the accuracy point of view. Besides, one may consider an advantage of the CFD compared to the experimental method regarding its flexibility and recently the associated low cost, which resulted after the significant development in the numerical modelling and the associated revolutionary progress in the computational power. These improvements resulted in less physical modelling assumptions, the possibility to use finer grids and less turnaround simulation time. Turbulence modelling has played a significant role in predicting the flow configurations in CFD for ship hydrodynamics. Implementation of turbulence models such as Reynold's-Averaged Navier-Stokes (RANS), Large Eddy Simulation (LES) and Hybrid RANS-LES models increased the reliability of CFD and helped understanding more sophisticated phenomena such as flow separation and hull-fluid interactions. In this study, the RANS method are applied to solve and predict the wake flow of the JBC ship model in a resistance, open water propeller, and self-propulsion simulations. Closure to turbulence is achieved by applying different turbulence models that are available in the ISIS-CFD solver used in this study, such as for resistance simulation the Spalart-Allmaras, K-ω, the baseline K-ω BL, the shear stress transport K-ω SST, the enhanced K-ω SST 2003 and finally, the Explicit Algebraic Stress Model (EASM) were used. While for the POW and self-propulsion simulations, the EASM and the DES models were applied. More details about the turbulence models can be found in [1-3].

2. Geometry and analysis conditions

The JBC (Japan Bulk Carrier) is a capesize bulk carrier equipped with a stern duct that plays the role of a wake equalizing duct or Energy Saving Device (ESD). National Maritime Research Institute (NMRI), Yokohama National University and Ship Building Research Centre of Japan (SRC) are jointly involved in the design of the ship hull, the duct, propeller and rudder. The full scale ship does not exist; however, a geometrically similar model was built and tested in the towing tank experiments planned and performed at NMRI, SRC and Osaka University. The test included the resistance tests, self-propulsion tests and PIV measurements of stern flow fields and the test results were presented for CFD validation and verification in the Tokyo 2015 Workshop on CFD in ship Hydrodynamics [4, 5]. The geometry of the 7.0 m ship model and the main dimensions for ship, duct and propeller are given in figure 1 and table 1, respectively.

The analysis conditions include three simulations: the first is dedicated for studying the local flow of the ship with and without the wake equalizing duct in resistance estimation, the second is for the propeller in the open water condition, while the final is for the self-propulsion condition for the ship with and without the duct using two different approaches; the infinite blade actuator disk and the 3D rotating propeller based on the sliding grid approach. More details regarding the two models were previously presented by the author in [6]. The difference between the previous study and the current is regarding the turbulence models used for the analysis and the use of finer grid to capture more details in the wake zone.

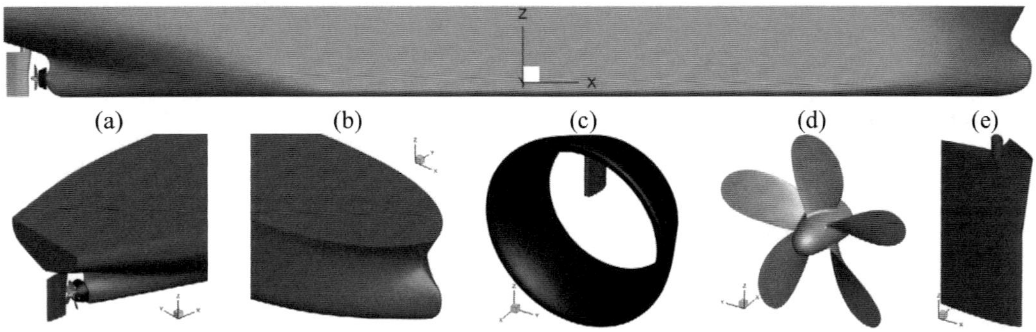

Figure 1. JBC model geometry highlighting: (a) stern, (b) fore, (c) duct, (d) propeller and (e) rudder.

Table 1. Principal particulars for the JBC ship model, duct and propeller.

	Particulars	Unit	Value
Ship	Length between Perpendiculars (L_{PP})	[m]	7.0
	Beam (B)	[m]	1.125
	Depth (D)	[m]	0.625
	Draft (T)	[m]	0.4125
	Volumetric Displacement (∇)	[m^3]	2.787
	Wetted Surface Area (S_0) (without ESD)	[m^2]	12.225
	Wetted Surface Area (S_0) (with ESD)	[m^2]	12.272
	Block Coefficient (C_B)	[-]	0.858
	LCB (%L_{PP}), fwd+	[-]	-2.548
Duct	Duct Outlet Diameter ($0.55D_p$)	[m]	0.11165
	Duct Cord Length ($0.3D_p$)	[m]	0.0609
	Duct Angle of Attack	20	
	Duct Foil Section	NACA4420	
Propeller	Propeller Diameter, D_p	[m]	0.203
	Propeller expanded area ratio, A_E/A_0	0.5	
	Number of blades, Z	5	
	Propeller rotation direction	Clockwise	

3. Numerical Simulation Framework

3.1. Numerical solver

The numerical simulations performed in this study are carried out using the ISIS-CFD solver of the software FINE^{TM}/Marine provided under the NUMECA suite. The solver is based on the finite volume method to build the spatial discretization of the transport equation to solve the Reynolds-Averaged Navier Stokes Equations (RANSE) [7]. The solver implies a face-based spatial discretization that constructs the fluxes face by face, which makes it optimum for using arbitrary shape control volumes with arbitrary number of faces. Hence, it relies on the use of unstructured grids to provide more flexibility for discretising complex hull geometries, ship appendages, etc. The temporal discretization is cell-centred based on a second order three-level scheme. The velocity field is obtained from the momentum equation; while the pressure is extracted from the mass constraint or continuity equation transformed into a pressure equation conform a velocity pressure coupling based on a Rhie and Chaw

SIMPLE type algorithm [8]. Closure to turbulence is achieved through various types of turbulence models that are provided in the solver, as previously described. Multi-phase flow is used to model the free-surface interface based on the Volume of Fluid (VOF) interface capturing technique, where incompressible and non-miscible flow phases are modelled by introducing conservation equations of the volume fraction for each volume phase/fluid [7]. Convection and diffusion terms in the governing equations are discretized using second-order upwind and central differencing scheme, respectively.

3.2. Governing equations
The time averaged continuity and momentum equations for the incompressible flow with external forces, can be written in tensor form, in the Cartesian coordinate system as:

$$\frac{\partial(\rho\bar{u}_i)}{\partial x_i} = 0 \tag{1}$$

$$\frac{\partial(\rho\bar{u}_i)}{\partial t} + \frac{\partial}{\partial x_j}\left(\rho\bar{u}_i\bar{u}_j + \rho\overline{u_i'u_j'}\right) = -\frac{\partial\bar{p}}{\partial x_i} + \frac{\partial\bar{\tau}_{ij}}{\partial x_j} \tag{2}$$

where $\bar{u}_i$ is the relative averaged velocity vector of flow between the fluid and the control volume, $\overline{u_i'u_j'}$ is the Reynolds stresses, $\bar{p}$ is the mean pressure and $\bar{\tau}_{ij}$ is the mean viscous stress tensor components for Newtonian fluid under the incompressible flow assumption, and it can be expressed as

$$\bar{\tau}_{ij} = \mu\left(\frac{\partial\bar{u}_i}{\partial x_j} + \frac{\partial\bar{u}_j}{\partial x_i}\right) \tag{3}$$

3.3. Computational domain and boundary conditions
For resistance, the computational domain is having a rectangular prism configuration that represents only the half body of the ship in order to reduce the computational cost and effort, taking into consideration the symmetry of the hull and the fact that the ship is moving only in the forward direction. The dimensions of the computational domain in the Cartesian coordinate system in (*x*–*y*–*z*) directions are set as (5.0L_{PP}–2.0L_{PP}–2.0L_{PP}) as illustrated in figure 2(a). As for the self-propulsion simulation, the same dimensions in resistance are typically implied except for the fact that the symmetry condition is not applicable here and the full ship is introduced in the domain. For this reason, the lateral dimension of the self-propulsion domain in *y*- direction is set as 4.0L_{PP}, as figure 2(b) bears out. The undisturbed free-surface level is set at the design draft of the ship located at 0.4125m from the ship baseline. A global earth-fixed reference frame is set for the resistance and self-propulsion simulations at the point (0, 0, and 0) which results from the intersection of the ship baseline with the aft perpendicular (A.P). The boundary conditions are set for the boundaries such that: the flow inlet is set at 2.0L_{PP} from the reference point, while the outlet is set at 3.0L_{PP} from the reference. The far field side boundaries are set at 2.0L_{PP} from the ship centreline and finally, the prescribed pressure top and bottom boundaries are located at 0.5L_{PP} and 1.5L_{PP} from the undisturbed free-surface level, respectively. Symmetry boundary is set at the ship centreline only in the resistance simulation. No slip condition is chosen for the entire ship hull, except the deck where it is assumed to remain in air during the simulation; hence it was set to a slip condition. Besides, for the self-propulsion simulation, also the propeller shaft is set to a slip condition to avoid any discontinuity at the interface zone between the hull and propeller for the sliding grid.

On the other hand, for the POW simulation, the domain is having a cylindrical configuration whose dimensions are 6.0 times the propeller diameter (D_P) in the circumferential direction and 8.0 times the propeller diameter in the longitudinal direction as illustrated in figure 2(c). A global reference frame is set at the point (0, 0 and 0) which results from the intersection of the propeller generator line and the shaft axis. The velocity inlet condition is set at the inflow boundary located at 2.0D_P upstream. The frozen pressure condition is applied on the exit boundary located at 6.0D_P downstream. The velocity inlet condition is also applied on the domain cylindrical wall. For the sake of capturing more details in the wake zone, a cylindrical refinement zone is added in the vicinity of the propeller extended from $0.25D_P$ upstream and $4.0D_P$ downstream, while its diameter is set to $1.5D_P$ distributed evenly above and underneath the propeller shaft axis. No slip condition is applied on the propeller blades, boss and cap, while the slip condition is used for the propeller shaft.

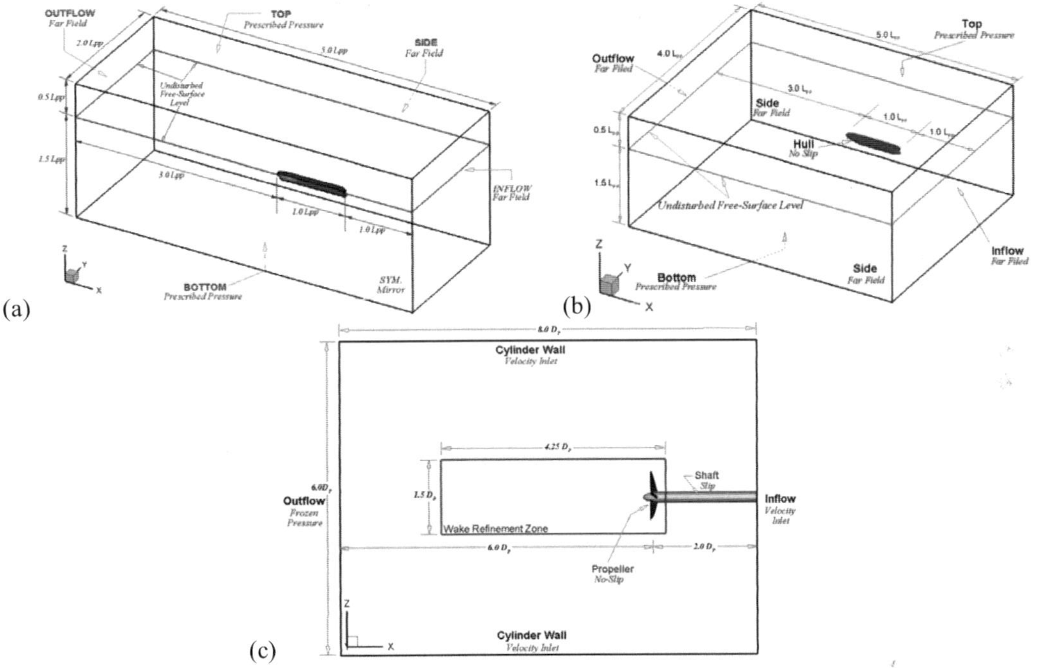

Figure 2. Computational domains and boundary conditions for the three simulation cases (a) resistance, (b) self-propulsion, (c) POW.

3.4. Discretization Grid

The grids are generated for the three simulation conditions using the automatic unstructured hexahedral grid generator HEXPRESS^{TM} provided in the Fine^{TM}/MARINE package. In order to capture more details of the flow in the zones of concern, special set of local refinement zones were created in the free-surface, forward and more intensely in the wake zone of the ship and propeller. The number of grid cells used for discretization is summarized in table 2, while the grid configurations are presented in figure 3 showing the forward, aft and free-surface refinement zones for the resistance simulation in figure 3(a), the aft zone, propeller and duct in the self-propulsion simulation in figure 3(b) and finally the propeller and wake refinement zone in figure 3(c).

Table 2. Number of grid cells for simulation conditions and approaches.

Simulation	Ship Resistance		Self-Propulsion				POW
	JBC w/o ESD	JBC w. ESD	A.D.[a] w/o. ESD	A.D. w. ESD	S.G.[b] w/o. ESD	S.G. w. ESD	
Number of grid cells (M)	23.563	25.077	19.883	21.750	25.846	27.353	34.633

[a]A.D.= The Actuator Disk approach, [b]S.G.=Sliding Grid approach.

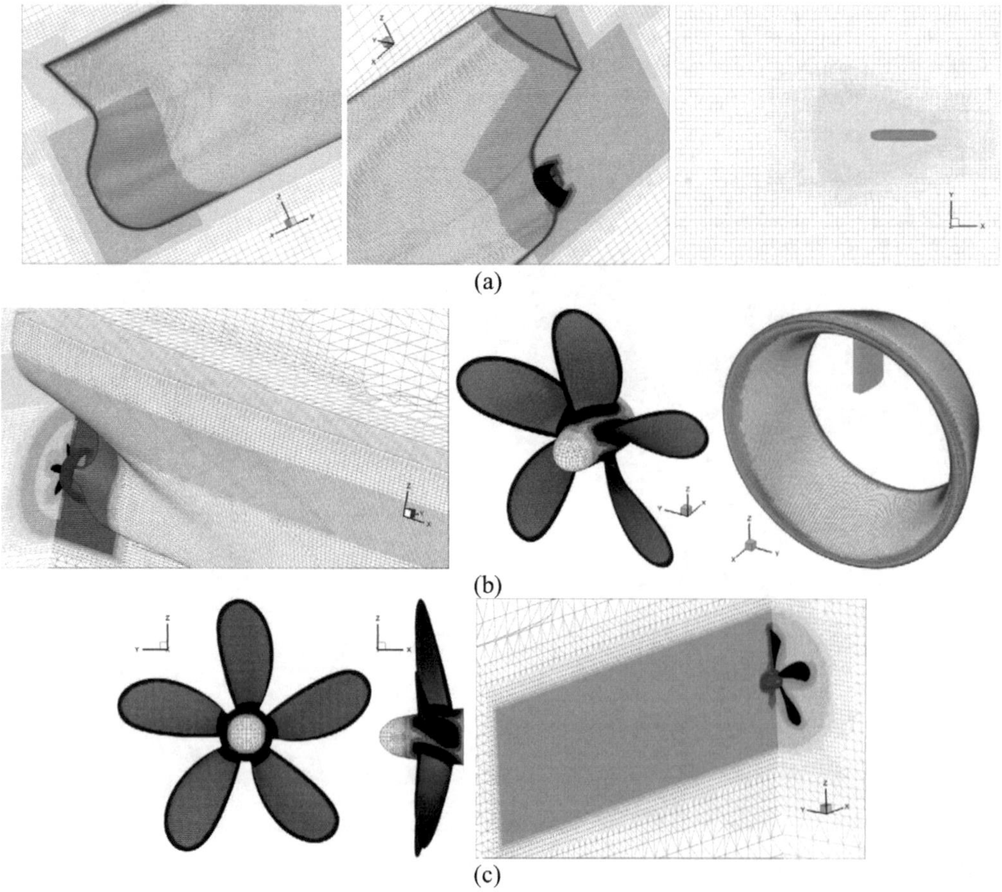

Figure 3. Discretization grids for simulation cases: (a) resistance (b) self-propulsion, and (c) POW.

3.5. Solution strategy and available resources

All the computations are performed on a High Performance Computing (HPC) machine with available 120 cores at 2.5 up to 3.3 GHz. For resistance prediction, the simulation is performed for 30 seconds to ensure adequate numerical convergence for resistance and vertical motions. The flow is accelerated based on a steady quasi-static approach for a selected period based on the ship speed to satisfy the condition $T_{acc} = 2L_{PP}/U$ in order to avoid any numerical instabilities in the beginning of the simulation. 10 non-linear iterations are used with a time step Δt is chosen based on the ITTC formula [9] for RANSE simulations with *two-equations* turbulence models as $\Delta t = 0.005U/\ L_{PP}$ also kept the same for the one-*equation* Spalart-Allmaras model, while for the EASM model the time step is set as $\Delta t = 0.001L_{PP}/U$. In the self-propulsion the simulation is performed in two levels; in the first, the same strategy used in the resistance is applied for the first run to stabilize the ship resistance and vertical motion with only 8 non-linear iterations and also for 30 seconds, while in the second run, unsteady simulation with 12 non-linear iterations is used with a smaller time step to stabilize the propeller thrust that should provide 200 time-steps/propeller rotation.

As for the POW simulation, unsteady approach with a rotating frame technique is applied to reduce the simulation time, with a time step set to provide 100 time steps/ propeller rotation and only 5 non-linear iteration and total simulation time is 6 seconds that was increased to 10 for the low advance velocity J=0.1~0.3, where $J=U/nD_P$ (to stabilize the flow). For the DES simulation in self-propulsion and POW cases, the time step was chosen with an order of magnitude with 10^{-1} less than that used in the EASM model. The total simulation time for each case is within 16–118 physical hours, which are definitely influenced by the grid resolution and the corresponding time-step.

4. Results and Discussions

4.1. Resistance simulation

For the resistance simulation, corresponding to the tank test where the local flow velocity contours were presented for three measuring sections as reported in the Tokyo 2015 workshop named as sections S2, S4 and S7 which are measured at distances 0.2625m, 0.11m and 0.0m, respectively from the A.P. as illustrated in figure 4. The numerical results for the axial velocity contours in x- direction are also presented for the same section in figures 5, 6 and 7 showing the computed results after 30 seconds of simulation using different turbulence models and arranged based on the accuracy of the obtained results compared to the experimental data (EFD).

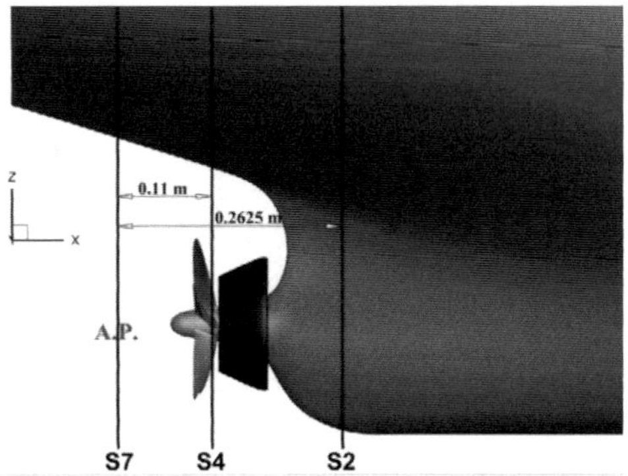

Figure 4. PIV measuring sections.

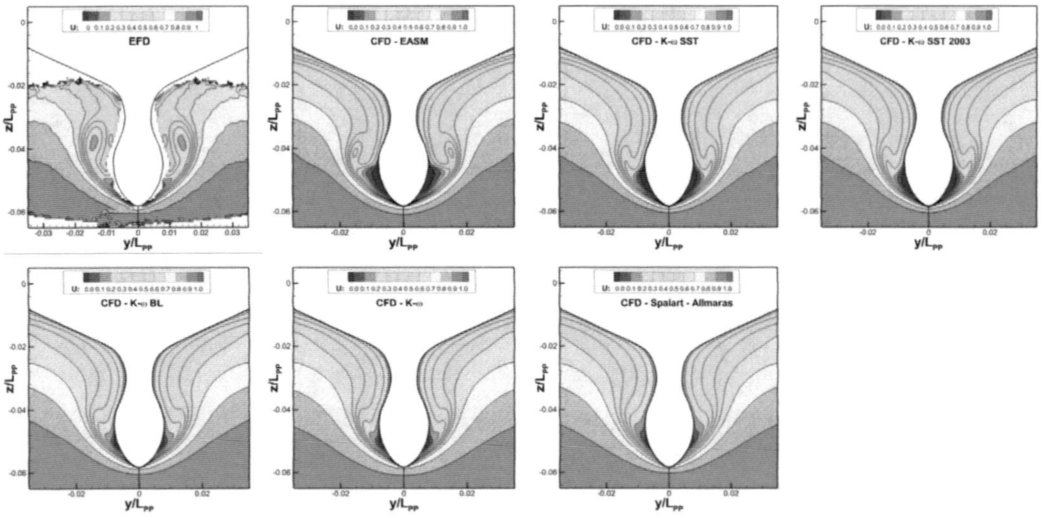

Figure 5. Section S2 EFD and CFD results for the axial velocity contours computed at T=30 s for ship without ESD using different turbulence models.

Following the flow configuration of the obtained and measured results, one can observe that the flow is characterized by the formation of an intense stern bilge vortex which leads to a noticeable distortion of the axial velocity contours at the measuring sections. This distortion results from the transport of low momentum fluid from the vicinity of the hull to the centre of the flow field under the effect of an intense longitudinal vortex. Besides, it can also be observed the existence of a secondary counter-rotating vortex close to the vertical plane of symmetry. Which leads to the formation of the so-called *hook-shape* of the velocity contours which is clearly visible in the towing tank data and the CFD results.

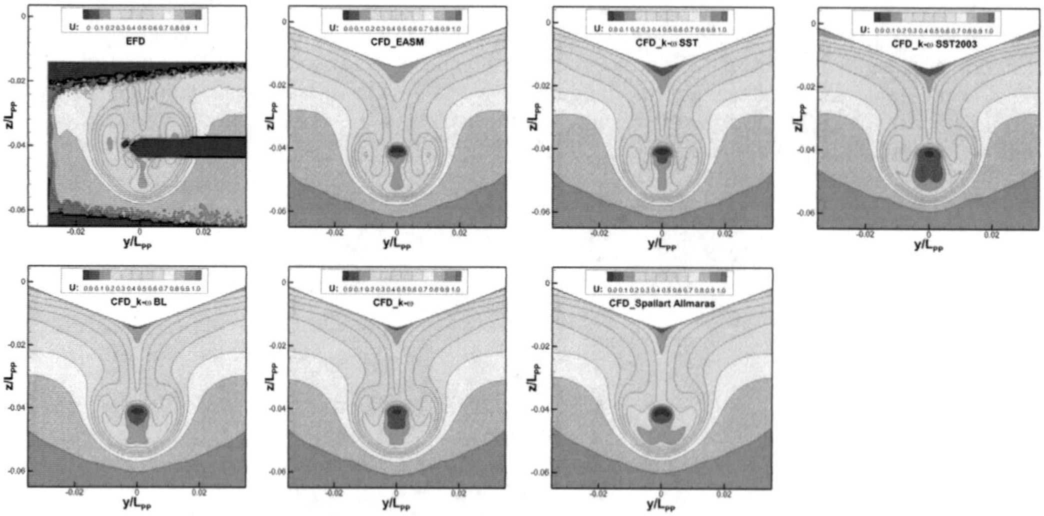

Figure 6. Section S4 EFD and CFD results for the axial velocity contours computed at T=30 s for ship without ESD using different turbulence models.

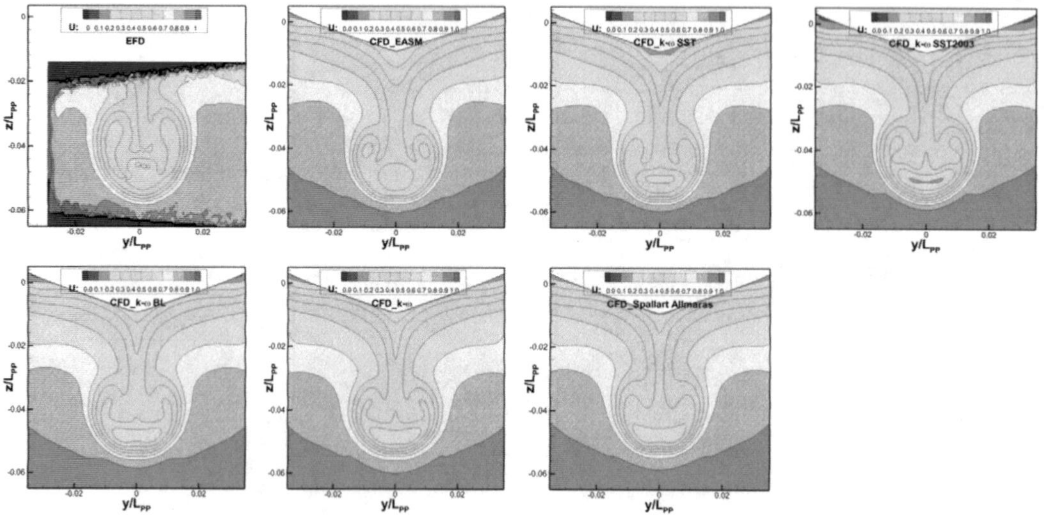

Figure 7. Section S7 EFD and CFD results for the axial velocity contours computed at T=30 s for ship without ESD using different turbulence models.

The significant challenge in the prediction of the local flow of this type of ships is to capture accurately this deformation of the velocity contours as it will be important since this phenomena happens in the vicinity of the propeller plane and it will significantly affect the propeller performance prediction in the further study in the self-propulsion condition.

The obtained results show the accuracy of the turbulence models to predict the hook-shape velocity contours with different levels. For example, the EASM, K-ω SST models showed to be the most accurate in predicting the velocity contours compared to the classic K-ω and Spalart-Allmaras models where they came in the last places. Nevertheless, it can be said that even with the under predicted velocity contours, the hook-shape was still captured with a sufficient level of accuracy for design purposes, especially compared to the previously used methods such as the methodical series and empirical based approaches where the local flow was totally not considered.

Having the fact that the EASM model gave the best obtained results, the local velocity contours configuration of the case when the ship is equipped with the ESD is represented in figure 8, showing a promising agreement between the computed results and the provided EFD data.

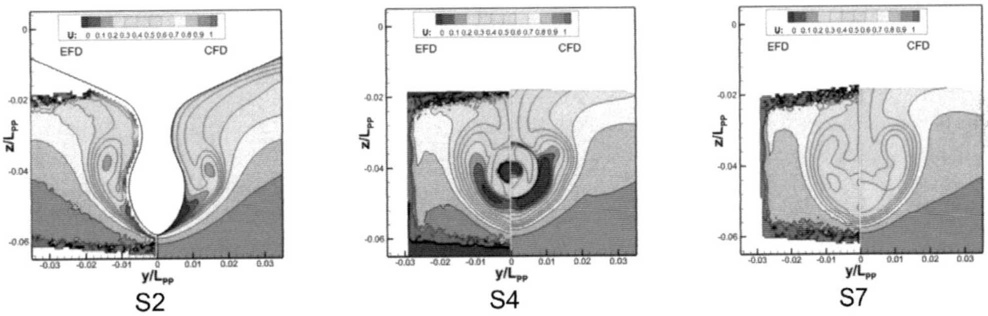

Figure 8. Comparison between the streamwise velocity contours measured and computed at T=30 s using EASM turbulence model for ship with ESD for sections S2, S4 and S7.

In order to take a closer look on the flow configuration and the vortices development in the wake zone of the hull, another proposal by the Workshop was given as a comparative data for the vortex formation based on the Q^*-second invariant of the axial velocity that may help visualizing the vortex formation in the stern zone, as it is illustrated in figure 9.

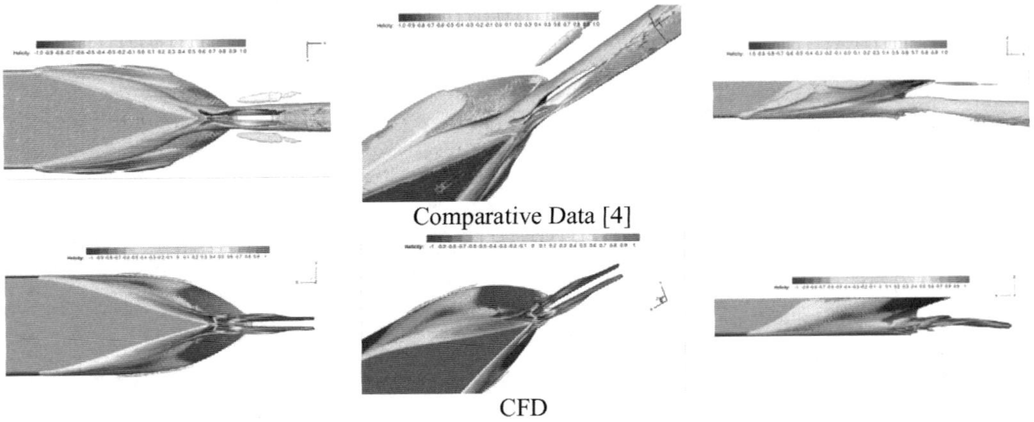

Figure 9. Workshop data for the second invariant iso-surface Q^*=25 colored by helicity and the corresponding CFD results computed at T=30 s using EASM model.

Though the comparison between the CFD results and the provided data shows an under predicted vortices structure; yet, the main vortices can be observed clearly in the vicinity of the ship boss and the propeller zone. The flow separation starts early at the bilge and transferred to the centreline as described before. Two counter-rotating vortices from the port and starboard sides can be clearly visualized and also underneath that, contra-rotating small size vortex tubes can be noticed. One more drawback in the obtained data is regarded to the choice of the length of the stern refinement zone by the author, which seems to be insufficient to capture the continuous vortices as in the provided data. However, this is with less important as the main configuration of the vortices were well predicted.

4.2. Propeller Open Water

The propeller open water test was repeated numerically for 8 advance ratio J to investigate the capability of the CFD method in predicting the local flow of the propeller in the POW condition. The complete set of results including grid convergence study and more details was reported by the author in [10] based on the EASM turbulence model. This study includes the DES simulation for J=0.2 in order to capture more details in the flow.

The axial velocity contours computed at 10 seconds in the propeller centreline plan in the longitudinal direction can be visualized in figure 10(a) where the velocity contours are showing a maximum value at the trace of the tip vortices with a significant fluctuation downstream around the core of the vortices. The corresponding pressure is also presented in figure 10(b) showing the pressure peaks in the vicinity of the propeller and downstream again with respect to the formed vortices. The turbulent viscosity and turbulent kinetic energy are presented in figure 10(c) and (d), respectively. All the four components can represent the formation of the wake flow vortices as three significant vortices can be observed; the first is generated at the propeller blade tip, the second is at the propeller blade root and finally the third is at the propeller hub, which can also be observed completely in figure 11.

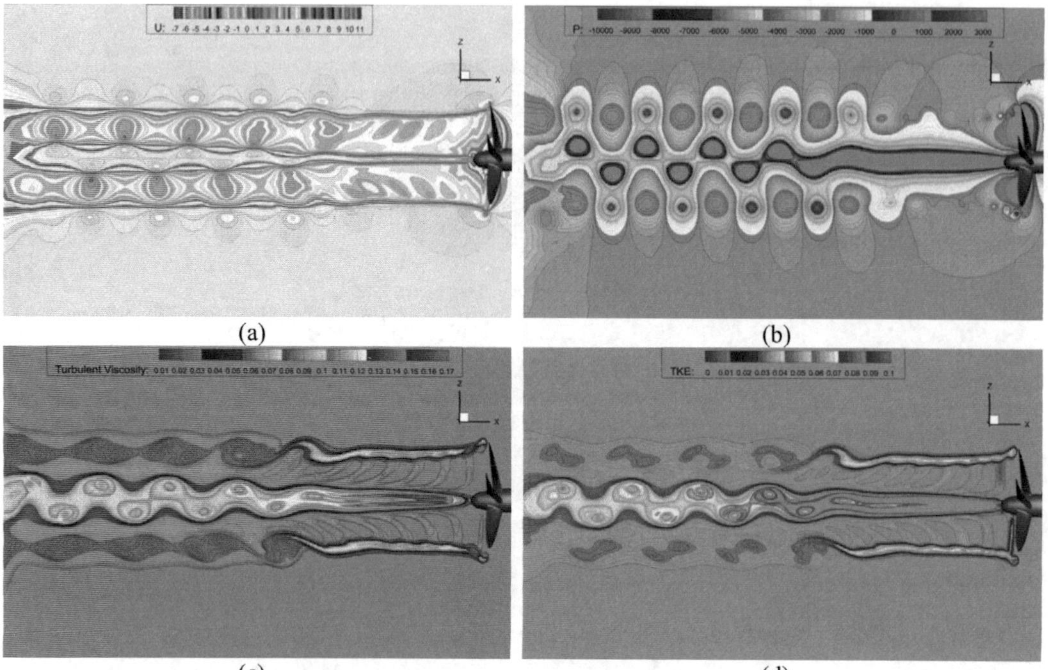

Figure 10. Wake flow configuration for the POW at J=0.2 showing: (a) axial velocity contours, (b) pressure, (c) turbulent viscosity and (d) TKE.

It can also be observed a significant fluctuation of the hub vortices that is apparently resulting from the interaction between the high momentum flow transferred downstream from the propeller vicinity to the low momentum flow in the far filed. This can be visualized clearly at the hub vortices in the downstream zone before the exit. It can also be noticed that the propeller tip and root vortices are combined together due to the viscous effect generating a thick vortex tube before it is transferred to the flow exit. It is worth mentioning that the grid quality should be enhanced for capturing more details in the wake zone for this analysis.

Figure 11. Second invariant $Q^*=50$ computed at 10s, $J=0.2$ and colored by nondimentional helicity.

4.3. Self-propulsion simulation

The self-propulsion simulation was performed based on two different approaches as previously described; the first is based on the actuator disk (A.D.) method which represents a quick and accurate solution for predicting the self-propulsion performance of the ship and propeller without a need of propeller modelling, besides the fact that the required simulation time and effort is significantly reduced. Unfortunately, it does not provide full details about propeller blade loading and cavitation inception, etc. The method was qualitatively and quantitatively assessed and presented by the author in [6] showing a reliable and accurate results.

In order to perform the self-propulsion using the actuator disk method, two levels of simulation must be performed, the first is to obtain the nominal wake by a standard resistance simulation, and then the actuator disk is introduced in the simulation to provide the prediction of the effective wake. Having the data from the resistance test and introducing the open water data in the solver, the self-propulsion coefficients and characteristics can be obtained based on the thrust or torque identity method.

The EFD data for the axial velocity contours measured at the section located in the propeller reference plan is plotted in figure 12 for the ship with and without the ESD at the nominal and effective wake condition, along with the computed results for the three relative velocity components V_a, V_r, V_{theta} for the actuator disk also located at the propeller reference plan at a distance 0.10148 m from the A.P. The comparison shows that the obtained results have a well congruence with the EFD data as figure 12 bears out. The measured and the computed results show the effect of the ESD on the flow configuration at the propeller vicinity, which indicate that the axial velocity contours are having more circular configuration compared to the ship without ESD. This means that the flow entering the propeller is more uniform, which can positively influence and enhance the propulsion efficiency. At the core of the velocity contours, a slight discrepancy can be observed which is related to the fact that in this study, the propeller hub was not connected to the ship boss during the self-propulsion simulation with the A.D. approach. Nevertheless, the effect is insignificant and the computed velocity contours resembles well with EFD data.

Similarly, the results for the axial velocity contours for the ship without ESD computed based on the 3D modelled propeller with the help of the sliding grid technique can be observed in figure 13 for the measuring sections S4 and S7, which means before and after the propeller, respectively. The comparison also reveals the fact that the computed results are within a good agreement with the measured data.

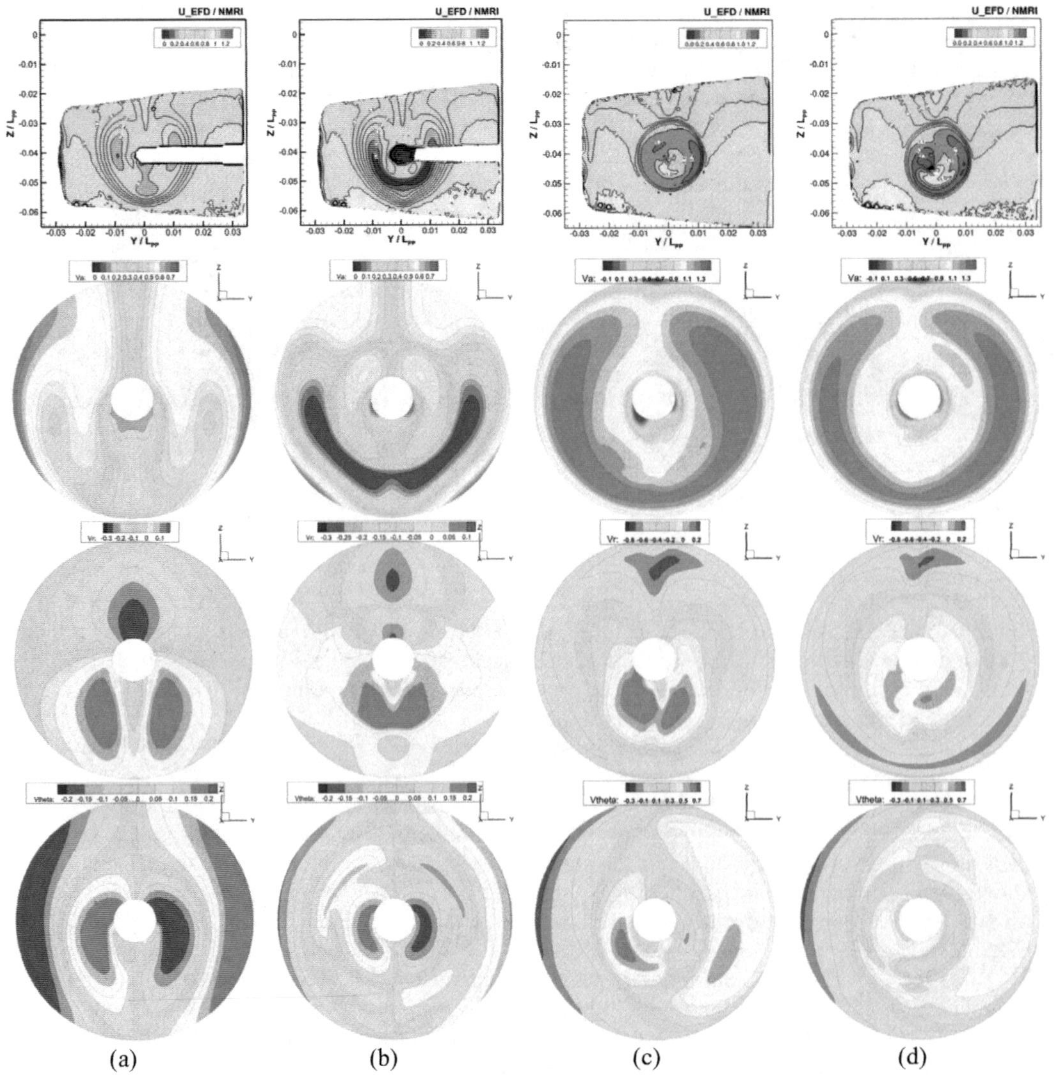

Figure 12. CFD results for nominal (a and b) and effective (c and d) axial velocity contours computed at T=30 s for ship with and without ESD using actuator disk method based on K-ω SST model.

Few years after the workshop was held, the rudder configuration was uploaded on the official Workshop website. This was a good opportunity to study the self-propulsion simulation for the ship with rudder as a primary step for a further research interest regarding the propeller rudder interaction. The local flow stream-wise velocity contours were brought to attention in figure 14 as a comparison between the ship with and without rudder. It was observed in the study that the rudder influences the propeller performance by enhancing the thrust forces due to the interference with the propeller as a part from the tangential velocity is transformed into an axial velocity due to the existence of rudder, which resulted in an enhanced thrust outcome. Also, the Q^* second invariant is also plotted in figure 14 showing the previously mentioned propeller tip, root and hub vortices for the ship without rudder and one more vortex formation is resulting from the rudder lower trailing edge separation. More details about this study can be found in [11].

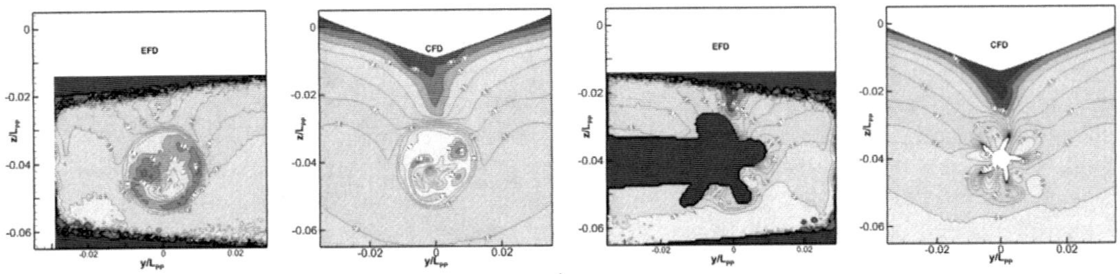

S4 | S7

Figure 13. Comparison between the streamwise velocity contours measured and computed at T=33 s using EASM turbulence model for ship with ESD for sections S4 and S7.

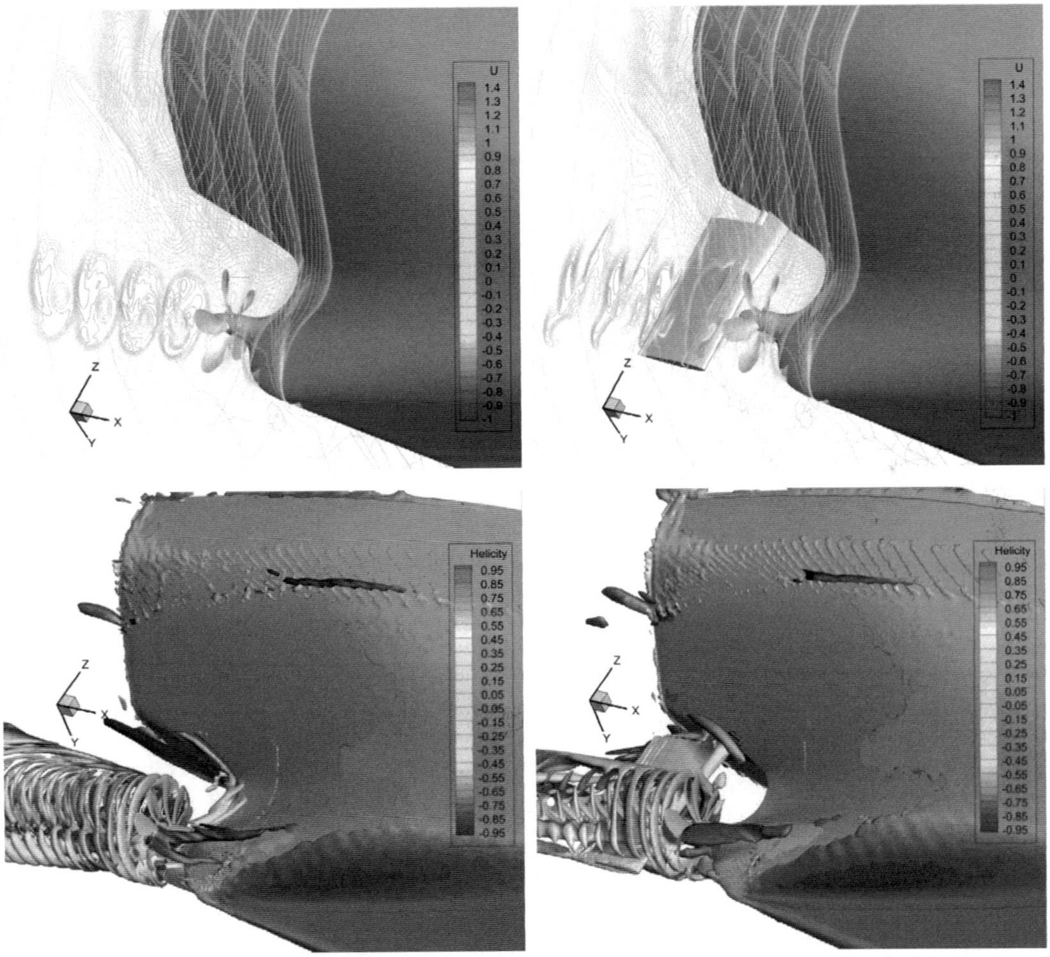

Figure 14. Comparison between the stream-wise velocity contours computed at T=30 s using DES turbulence model for ship without ESD and with/without rudder and the second invariant Q^*=50 colored by nondimentional helicity.

5. Conclusions

The local flow assessment for the Japan Bulk Carrier JBC ship model of a 7.0m length was computed based on a viscous CFD method, presented and compared to the EFD data provided in the Workshop on CFD in ship hydrodynamics held in Tokyo 2015. The comparison was made for three different conditions including the resistance, POW and the self-propulsion conditions using different turbulence models. The assessment of the CFD results showed a general good agreement with the experimental data and showed the capability of the CFD method to capture accurately the challenging local flow of a high-block coefficient ship with significant flow separation. The ship was tested on two levels with respect to the existence or the absence of the ESD which is aimed at working as a wake equalizing duct in order to enhance the flow heading the propeller. The effect of the duct was properly calculated and presented in the study. Having the main objective of the current study and its outcome, we may conclude the following main points:

- In the resistance simulation, the turbulence models used for computing and predicting the flow separation showed to be sufficiently accurate with different level of reliability. EASM model showed to have the highest agreement with the EFD data, while Spallart-Allmaras showed to have the lowest. It is still worth mentioning that the simulation time for EASM is within 3 times longer than the other turbulence models. K-ω SST models showed a good balance between accuracy and simulation time.
- The stern vortices were computed and presented compared to the data provided in the Workshop showing a good agreement in predicting the vortical structures formations.
- In the POW simulation, the flow configuration was computed and presented showing the flow development in the propeller vicinity and in the downstream. The results captured accurately the formation of tip, root, and hub vortices and their development and interaction during the simulation time.
- The Self-propulsion test was presented based on both A.D. method and fully discretized propeller. The advantage of the A.D. over the fully discretized propeller is regarding the low simulation time and the simple application. The disadvantage is regarding the fact that the propeller blade loading and cavitation occurrence cannot be predicted based on this model. The results obtained using the A.D. were within a satisfying agreement compared to EFD data. As for the fully discretized propeller, the results were in a good match with the EFD data.

In general, the study was successful in predicting the local flow and the flow separation for the JBC ship model based on various turbulence models. Further investigation is planned for ship-propeller-rudder interaction in manoeuvring simulations continuing the previous research effort represented in [12-15].

6. References

[1] Wilcox D C 2006 *Turbulence Modeling for CFD (3rd Ed.)*(USA: D C W Industries).
[2] Ferziger J H, Peric M 2002 *Computational Methods for Fluid Dynamics (3rd Ed.)* (NY: Springer).
[3] Menter, F R, Kuntz M, and Langtry R, 2003 *Ten Years of Industrial Experience with the SST Turbulence Model*, Turbulence, Heat and Mass Transfer 4, ed: Hanjalic K, Nagano K, and Tummers M, Begell House, Inc., pp. 625 - 632.
[4] Tokyo 2015 *A Workshop on CFD in Ship Hydrodynamics* https://t2015.nmri.go.jp/ accessed on 18.03.2021.
[5] Hirata N 2015 http://www.t2015.nmri.go.jp/Presentations/Day1-AM2-JBC-TestData1-Hirata.pdf accessed on 18.03.2021.
[6] Bekhit A S 2018 *IOP Conf. Ser.: Mater. Sci. Eng.* **400** 042004 1–11.
[7] Deng G B, Queutey M and Visonneau M 2010 *J. Hydrodynamics*. **22**(5) 476–481.
[8] Queutey P, Deng G B, Wackers J, Guilmineau E, Leroyer A and Visonneau M 2012 *J. Ship Technology Research* **59**(2) 44-58.

[9] International Towing Tank Conference (ITTC) 2011, *Practical guidelines for ship CFD applications.* **Rev.1** 7.5-03-02-03 1–18
[10] Bekhit A S, Lungu A 2019, AIP Conf. Proc. **2116**, 450007 (2019)
[11] Bekhit A S, Pacuraru F, Pacuraru S 2020, AIP Conf. Proc. **2293**, 420091 (2020)
[12] Pacuraru F, Lungu A 2009, AIP Conf. Proc. **1281** (1), 111-114
[13] Pacuraru F, Lungu A, Ungureanu C, Marcu O 2010, IOP Conf. Ser.: Earth and Env. Sc. **12**(1), 012032
[14] Pacuraru F, Lungu A, Marcu O 2011, AIP Conf. Proc. **1389**(1), 191-194

Acknowledgement
This work was carried out in the framework of the research project DREAM (Dynamics of the REsources and technological Advance in harvesting Marine renewable energy), supported by the Romanian Executive Agency for Higher Education, Research, Development and Innovation Funding – UEFISCDI, grant number PN-III-P4-ID-PCE-2020-0008.

IOP Conf. Series: Materials Science and Engineering 1182 (2021) 012005 doi:10.1088/1757-899X/1182/1/012005

Numerical Investigation of the Shallow Water Effect on the Total Resistance, Vertical Motion and Wave Profile of a Container Ship Model

A Bekhit[1] and F Popescu[2]
[1]"Dunarea de Jos" University of Galati, Faculty of Naval Architecture, 47 Domneasca Street, 800008, Galati, Romania
[2]"Dunarea de Jos" University of Galati, Faculty of Engineering, 47 Domneasca Street, 800008, Galati, Romania

E-mail: adham.bekhit@ugal.ro

Abstract. A ship sailing in shallow water is affected by the interaction between the moving hull and the seabed in different forms such as: significant increase in total resistance, increase in the sinkage and trim that may result in squatting effect, a change in wave pattern and increased wave amplitude, a change in the propeller wake field and an altered propeller and manoeuvring performance. In order to investigate the influence of shallow water on a container ship that is assumed to be subjected to work in different water depths, the KRISO container ship model is analysed in both deep and shallow water, with a special focus on the change in resistance, vertical motion and wave profile. A viscous flow simulation is performed first to predict the ship performance in deep water and the results are compared with the experimental data that are available in the public domain. Then, the critical shallow water condition is investigated and the obtained results are compared against those formerly obtained in deep water. The numerical simulations are performed using the viscous flow solver ISIS-CFD of the FINE^{TM}/Marine software provided by NUMECA. The solver is based on the finite volume method to build the spatial discretization of the transport equation in order to solve the Reynolds-Averaged Navier-Stokes (RANS) equations. Closure to turbulence is achieved using the Menter Shear Stress Transport (K-ω SST) model, while the free-surface is captured through an air-water interface based on the Volume of Fluid (VOF) method. The comparison between the numerically obtained results and the available EFD data showed a satisfactory congruence.

1. Introduction

The maritime transport represents one of the most important factors in the world trade activities. Over 80 per cent of global trade by volume and more than 70 per cent of its value being carried on board ships, as reported in the Review of Maritime Transport 2017 [1]. The inland water transport can enhance significantly the quality and the amount of transported goods with a cheaper, more safe and effective transportation mode. The only drawbacks may be regarding the fear of pollution hazards and shallow water problems, that may tend to cause damage to the ship and its propulsion system. For this reason, and in order to reduce the grounding incidents of the ship it is very important to study the ship performance in shallow water. A ship sailing in shallow water is affected by the interaction between the moving hull and the seabed in different forms such as: significant increase in total resistance, increase in the sinkage and trim that may result in squatting effect, a change in wave pattern and

IOP Conf. Series: Materials Science and Engineering 1182 (2021) 012005 doi:10.1088/1757-899X/1182/1/012005

increased wave amplitude, a change in the propeller wake field and an altered propeller and manoeuvring performance. For this purpose, studying the shallow water effect on a ship is crucially important to provide sufficient powering for those ships that were especially designed for working in shallow waters and coastal zones where the water depth is limited.

In such complicated circumstances, the experimental approach is inconvenient and not practicable from cost and risk points of view. The best alternative is to use the Computational Fluid Dynamics (CFD) to investigate the ship performance in shallow water, as it provides a significant flexibility for these conditions and the obtained results are highly reliable, especially with the significant development recorded in the past two decades in physical modelling and computational power.

Heading form this perspective, this study investigates the shallow water effect on a container ship that is assumed to be sailing in shallow water condition with different water depth based on a CFD viscous flow method. Six different cases with respect to the water depth including six different ship velocities in each case were investigated with special focus on the ship resistance, vertical motions and free-surface flow. In order to insure the accuracy of the computed results, a validation test against the available Experimental Fluid Dynamics (EFD) data was performed especially for the computed total resistance and showed a proper agreement with the available data.

2. Geometry and analysis conditions

The current study is applied on the ship model of the Korea Research Institute of Ships and Ocean Engineering (KRISO) Container Ship (KCS). The ship was conceived to provide benchmark data for both explication of flow physics and CFD validation for a modern container hull with bulb bow and stern. The full scale ship does not exist; however, several models were built and tested by various internationally well recognized towing tank organizations with different scopes to predict ship resistance, free-surface, local flow configurations, seakeeping and maneuverability. The results for these tests can be found in the public domain for CFD verification and validation purposes.

In this case study, the ship model of scale 1/31.6 is used for validating this simulation, which has a length between perpendiculars of L_{pp}=7.2786m and a draft of 0.3418m. Full details of the ship characteristics are tabulated in table 1 for both model and full scale, while the geometry of the ship is depicted in figure 1.

Table 1. Particulars of the KCS ship model.

Parameter	Unit	Full Scale	Model Scale
Scale, λ	-	1	31.6
Length, L_{PP}	[m]	230.0	7.2786
Beam, B	[m]	32.2	1.0190
Draft, T	[m]	10.8	0.3418
Depth, D	[m]	19.0	0.6013
Block coefficient, C_B	-	0.6505	0.6505
Mid-ship section coefficient, C_M	-	0.9849	0.9849
Displacement of volume, ∇	[m^3]	52030	1.6490
Wetted surface area w/o rudder, S_0	[m^2]	9424	9.4379

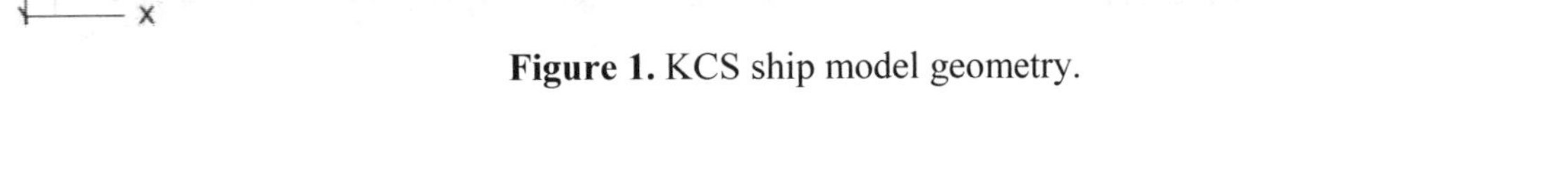

Figure 1. KCS ship model geometry.

The analysis conditions include Six different cases; every case consists of six velocities corresponding to Froude number ($Fr = V/\sqrt{gL_{PP}}$) range from Fr=0.108 to 0.282, where V represents the ship velocity and g is the gravitational acceleration and L_{PP} stands for the ship length. In table 2, the simulation cases are categorized with respect to the ship velocity, the corresponding Froude number, water depth H, and water depth to draft ratio H/T.

Table 2. Simulation cases and corresponding ship velocity.

Ship Velocity		Water Depth (H)					
		1.5L_{PP}	1.0L_{PP}	0.5L_{PP}	0.25L_{PP}	0.125L_{PP}	0.0625 L_{PP}
V[m/s]	Fr [-]	H=10.92 m	H=7.28 m	H=3.64 m	H=1.82 m	H=0.91 m	H=0.455 m
		H/T=35	H/T=15.95	H/T=10.65	H/T=5.32	H/T=2.66	H/T=1.33
0.92	0.108	*C1.1	C2.1	C3.1	C4.1	C5.1	C6.1
1.28	0.152	C1.2	C2.2	C3.2	C4.2	C5.2	C6.2
1.65	0.195	C1.3	C2.3	C3.3	C4.3	C5.3	C6.3
1.92	0.227	C1.4	C2.4	C3.4	C4.4	C5.4	C6.4
2.20	0.26	C1.5	C2.5	C3.5	C4.5	C5.5	C6.5
2.38	0.282	C1.6	C2.6	C3.6	C4.6	C5.6	C6.6

*C=Simulation Case Number

3. Numerical Simulation Framework

3.1. Numerical solver

The numerical simulations performed in this study are carried out using the ISIS-CFD solver of the software FINE™/Marine provided by the NUMECA. The solver is based on the finite volume method to build the spatial discretization of the transport equation to solve the Reynolds-Averaged Navier Stokes Equations (RANSE) [2]. The spatial discretization is face-based in which the fluxes are constructed face by face, which means that any arbitrary shape control volumes with arbitrary number of faces can be used. Hence, it is suitable for the use of unstructured grids to provide more flexibility for discretising complex hull geometries, ship appendages, etc. The temporal discretization is cell-centred based on a second order three-level scheme. The velocity field is obtained from the momentum equation; while the pressure is extracted from the mass constraint or continuity equation transformed into a pressure equation conform a velocity pressure coupling based on a Rhie and Chaw SIMPLE type algorithm [3]. Closure to turbulence is achieved through the blended Menter's Shear Stress Transport K-ω SST turbulence model. Multi-phase flow is used to model the free-surface interface based on the Volume of Fluid (VOF) interface capturing technique, where incompressible and non-miscible flow phases are modelled by introducing conservation equations of the volume fraction for each volume phase/fluid [2]. Convection and diffusion terms in the governing equations are discretized using second-order upwind and central differencing scheme, respectively.

3.2. Governing equations

The time averaged continuity and momentum equations for the incompressible flow with external forces, can be written in tensor form, in the Cartesian coordinate system as:

$$\frac{\partial(\rho\bar{u}_i)}{\partial x_i} = 0 \tag{1}$$

$$\frac{\partial(\rho\bar{u}_i)}{\partial t} + \frac{\partial}{\partial x_j}\left(\rho\bar{u}_i\bar{u}_j + \rho\overline{u_i'u_j'}\right) = -\frac{\partial\bar{p}}{\partial x_i} + \frac{\partial\bar{\tau}_{ij}}{\partial x_j} \tag{2}$$

where $\bar{u}_i$ is the relative averaged velocity vector of flow between the fluid and the control volume, $\overline{u_i'u_j'}$ is the Reynolds stresses, $\bar{p}$ is the mean pressure and $\bar{\tau}_{ij}$ is the mean viscous stress tensor components for Newtonian fluid under the incompressible flow assumption, and it can be expressed as

$$\bar{\tau}_{ij} = \mu\left(\frac{\partial \bar{u}_i}{\partial x_j} + \frac{\partial \bar{u}_j}{\partial x_i}\right) \tag{3}$$

3.3. Computational domain and boundary conditions
The computational domain is having a rectangular prism configuration as schematically presented in figure 2. The main dimensions of the computational domain in (*x*–*y*–*z*) direction are (5.0L_{PP}–2.0 L_{PP}–(0.5+*h*) L_{PP}), where *H* is the depth factor which varies based on the assumed water depth such that (*H*=1.5, 1.0, 0.5, 0.25, 0.125Lpp and 0.0625L_{pp}, as illustrated in figure2. The inflow boundary is located at 2.0 L_{pp} upstream, the exit boundary is located at 3.0 L_{pp} downstream, 2.0L_{pp} between the ship symmetry plan and the side boundary, the top boundary is located at 0.5L_{pp} above the undisturbed free-surface level, which is located at z=0.3148m from baseline, as it is highlighted in blue in figure 2. Taking into account the symmetry of the hull and considering the fact that only vertical motion is taken into consideration, only half of the ship is presented in the numerical simulations to simplify the problem and reduce the number of grid cells required, and hence the computational effort.

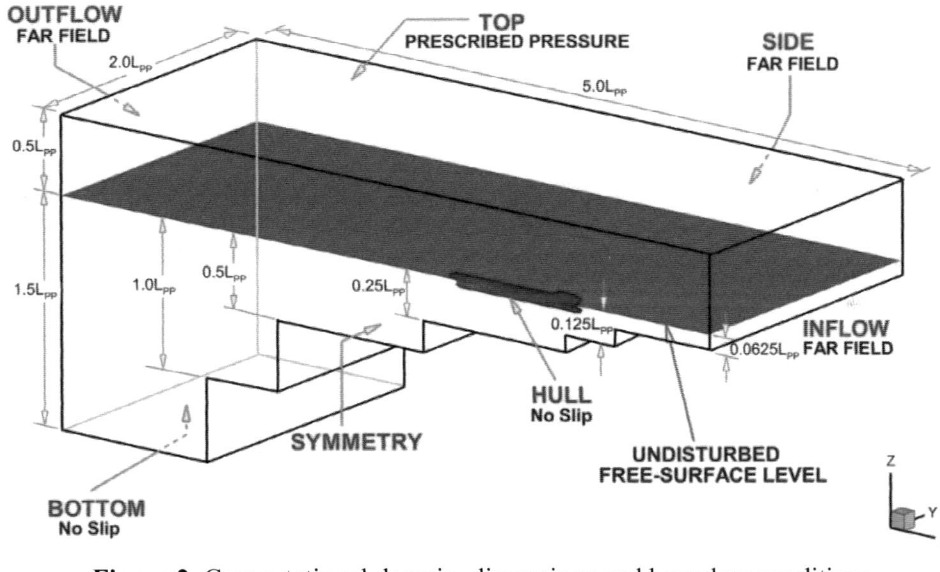

Figure 2. Computational domain, dimensions and boundary conditions.

The boundary conditions are shown in figure 2 for the computational domain boundaries, which implies a velocity inlet upstream, while the far-field is applied on the downstream. The mirror boundary condition is applied on symmetry sides of the domain at y=0, the prescribed pressure condition is applied on the top boundary, no slip condition is imposed on both, ship hull and bottom boundary and finally, the slip condition is selected for the deck, assuming that it remains in air during the simulation, where the viscous effect of air compared to that in water can be disregarded. The rigid body motion is measured with respect to an earth fixed reference frame located at the point (0, 0,0) located at the intersection between the ship aft perpendicular and base line.

3.4. Discretization Grid
The grids are generated using the automatic unstructured hexahedral grid generator HEXPRESS™ provided in the Fine™/MARINE package. The detailed number of grid cells for every simulation case

is tabulated in table 3, while figure 3 shows the computational grid for the case number C3, where H is $0.5L_{PP}$. Two levels local refinements are applied on the free-surface by imposing a global surface refinement at the undisturbed water level for the entire domain with relatively coarser grid, while a dense grid is applied on the Kelvin pattern zone with a wider opening angle of 80° for the shallow water cases to capture more details of the generated wave by the ship, as shown in figure 3.c. Nevertheless, for case 1, the Kelvin angle is retained at 20°. The refinement for the Kelvin pattern is chosen to provide 60 grid cells per wave length in x- and y-directions such that $\Delta x = \Delta y = 2\pi Fr^2 L_{WL}/60$; while in z-direction, the thickness of the refinement zone is chosen to cover the stagnation wave amplitude defined by $\zeta_{max} = 0.5U^2/g$. The cell size in z-direction for the free-surface refinement sector is chosen as $\Delta z = L_{WL}/1000$. The bottom zone is divided in two sections, both sections are treated with wall modelled approach having the first distance to the wall corresponding to $y^+=60$. The first section which is located underneath the hull is treated with a thicker viscous layer with 8levels, while only 4 viscous layers were imposed for the downstream section. The hull is also treated with a wall-modelled approach having $y^+=60$.

Table 3. Number of grid cells for simulation cases.

	Water Depth (H)					
Number of grid cells (M)	$1.5L_{PP}$	$1.0L_{PP}$	$0.5L_{PP}$	$0.25L_{PP}$	$0.125L_{PP}$	$0.0625L_{PP}$
	4.188	5.847	10.487	17.355	18.196	18.686

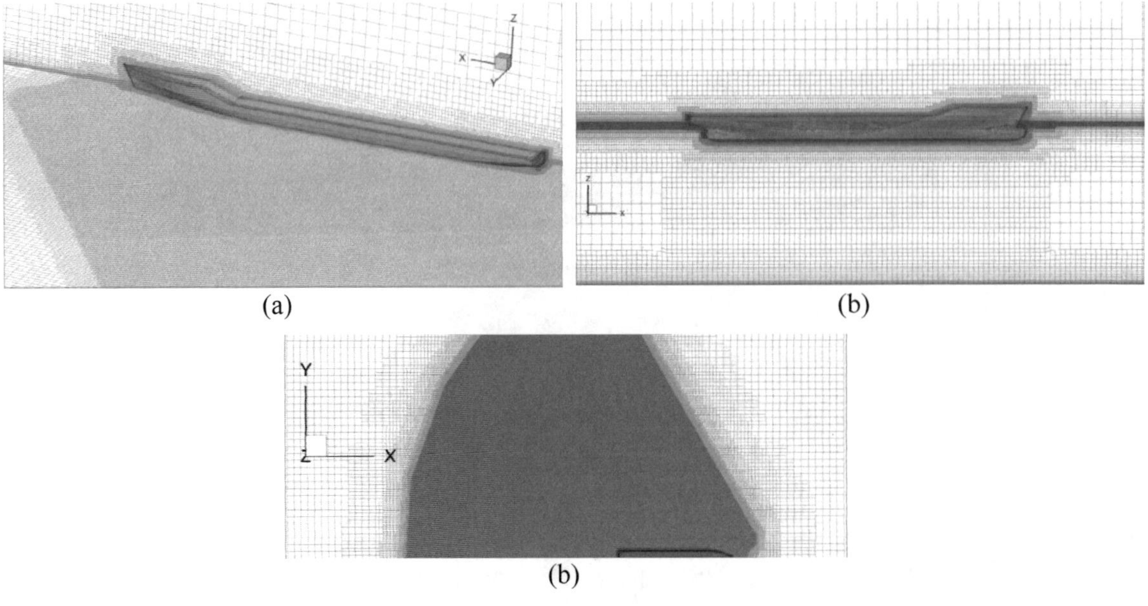

Figure 3. Discretization grids showing: (a) 3D view, (b) side view and (c) Top view.

3.5. Solution strategy and available resources

All the simulations are performed with the help of a High Performance Computing (HPC) machine with available 120 cores at 2.5 up to 3.3 GHz. For resistance prediction, the simulation is performed for 30 seconds to ensure adequate numerical convergence for resistance and vertical motions. The flow is accelerated based on a steady quasi-static approach for a selected period based on the ship speed to satisfy the condition $T_{acc} = 2L_{PP}/U$ in order to avoid any numerical instabilities in the beginning of the simulation. 10 non-linear iterations are used with a time step Δt is chosen based on the ITTC formula for RANSE simulations with *two-equations* turbulence models as $\Delta t = 0.005\ L_{PP}/U$ [3,4].

4. Results and Discussions

4.1. Resistance Results

In the scope of validating the computed results against the available EFD data retrieved from [5,6], the computed total resistance coefficient is represented against the EFD in table 4 for case number C1 where the water depth is set at H=1.5L_{PP}. The obtained results showed a good agreement with the EFD data with an absolute error range between 0.35 and 2.0%. This can highlight the accuracy of the computed results.

Table 4. Computed total ship resistance coefficient compared to EFD.

Ship Velocity		C_Tx10^3 [-]		Error% C_T [-]
V [m/s]	Fr [-]	EFD	CFD	$\varepsilon\% = 100(CFD - EFD)/EFD$
0.92	0.108	3.796	3.723	-1.925
1.28	0.152	3.641	3.604	-1.006
1.65	0.195	3.475	3.463	-0.357
1.92	0.227	3.467	3.496	0.829
2.20	0.26	3.711	3.675	-0.960
2.38	0.282	4.501	4.463	-0.850

Based on the obtained results the comparison between the different cases with respect to the water depth is listed in table 5 showing the effect of the shallow water on the increment in ship resistance, while figure 4 shows the bar chart diagram for the ship resistance based on the water depth. The theoretical principle remains valid for the computed results, since the ship resistance increases as the water depth decreases. The only exception is between the three cases where H=1.5L_{PP}, H=1.0L_{PP} and H=0.5L_{PP} where the values remain within a certain level even for the high speed. This might be related to the fact that H/T ratio is reasonable and the change is insignificant in comparison with the water depth; yet, it might be an influence of the numerical errors that result in this fluctuation. This might be verified in the future with a grid convergence study.

Table 5. Computed total ship resistance for different water depth.

Ship Velocity	Total Ship Resistance R_T [N]					
V [m/s]	H=1.5L_{PP}	H=1.0L_{PP}	H=0.5L_{PP}	H=0.25L_{PP}	H=0.125L_{PP}	H=0.0625L_{PP}
0.92	15.042	15.106	15.084	16.124	16.762	21.404
1.28	28.19	28.356	28.52	31.944	32.206	35.438
1.65	45	44.912	44.822	47.694	48.102	Ship Grounded
1.92	61.516	61.398	60.958	70.114	73.862	
2.20	84.916	84.936	84.154	106.224	119.926	
2.38	120.67	120.626	120.602	138.620	144.962	

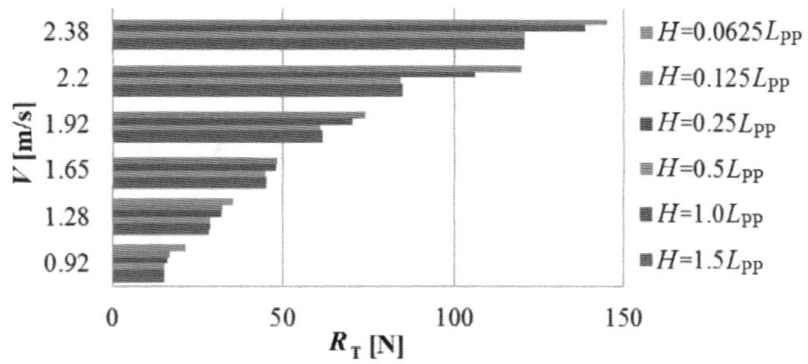

Figure 4. Resistance comparison based on the water depth

4.2. Vertical motions

The sinkage and trim for the six cases were computed and represented in table 6 and 7 showing the fact that the ship sinkage and trim increases with the ship speed in shallower water due to the speed effect which tends to create a low pressure zone that consequently results in a suction force that pulls the ship bottom towards the seabed. It is worth mentioning that there some values in the shallower water that do not comply with this criteria, this might be related to iterative errors as the fluctuation of the computed values increases significantly as the ship speed increases. The values recorded represent an average for the last 10% of the simulation time that lasted for 60 seconds. The authors think that this time should be increased, especially for the shallowest cases to enhance the convergence criteria.

Table 6. Computed sinkage for different water depth.

Ship Velocity	Sinkage [m]					
V [m/s]	H=1.5L_{PP}	H=1.0L_{PP}	H=0.5L_{PP}	H=0.25L_{PP}	H=0.125L_{PP}	H=0.0625L_{PP}
0.92	0.351	0.350	0.352	0.297	0.295	0.325
1.28	0.349	0.348	0.353	0.313	0.313	0.341
1.65	0.346	0.347	0.346	0.332	0.328	Ship Grounded
1.92	0.343	0.342	0.342	0.335	0.337	
2.20	0.339	0.338	0.338	0.341	0.344	
2.38	0.335	0.335	0.335	0.346	0.349	

Table 7. Computed trim for different water depth.

Ship Velocity	Trim [°]					
V [m/s]	H=1.5L_{PP}	H=1.0L_{PP}	H=0.5L_{PP}	H=0.25L_{PP}	H=0.125L_{PP}	H=0.0625L_{PP}
0.92	0.022	0.021	0.022	0.023	0.024	0.050
1.28	0.048	0.049	0.049	0.055	0.060	0.151
1.65	0.090	0.087	0.088	0.111	0.124	Ship Grounded
1.92	0.128	0.128	0.128	0.172	0.196	
2.20	0.156	0.157	0.157	0.232	0.285	
2.38	0.172	0.174	0.175	0.261	0.383	

4.3. Free-surface

It is well known that the ship sailing in shallow water encounters a significant effect on the free-surface topology that results in changing the Kelvin wave pattern. For this purpose, the wave pattern in shallow water is represented in this section only for water depths H=0.5L_{PP}, H=0.125L_{PP} and H=0.0625L_{PP}. A special form of the computed dated based on the water depth of H=0.25L_{PP} was also investigated for higher ship speeds up to the squatting condition to investigate the change in wave profile correspondingly.

4.3.1. Case C3 with H=0.5L_{PP}. The free-surface topologies for the cases C3.3, C3.4, C3.5 and C3.6 which are corresponding to the ship velocity of V=1.65, 1.92, 2.20 and 2.38, respectively are depicted in figure 5. The Kelvin pattern seems to not be affected even for the high speed ship, which actually supports the previously presented results where it was showing that for the water depth of H=0.5L_{PP} the effect of depth is insignificant and gives similar values for resistance and vertical motions like the H=1.5L_{PP} case.

4.3.2. Case C5 with H=0.125L_{PP}. The corresponding free-surface topology for this case can be visualized in figure 6, where the effect of the increased ship speed results in a noticeable deformation in the Kelvin wave pattern envelope angle, which tends to increase as the ship speed increases. This complies positively with the theoretical principles regarding the shallow water effect on free-surface.

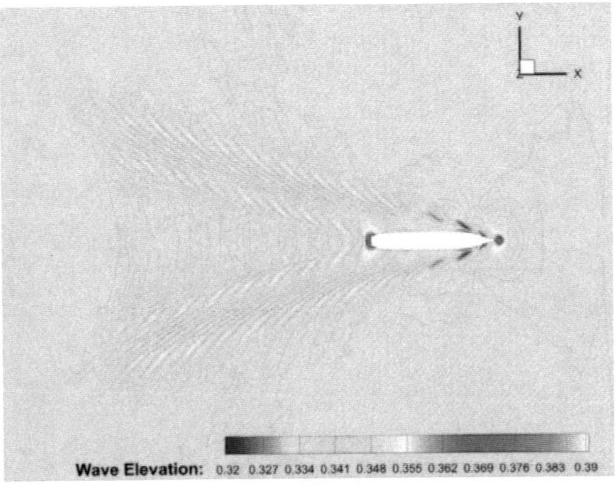

V=1.65 m/s

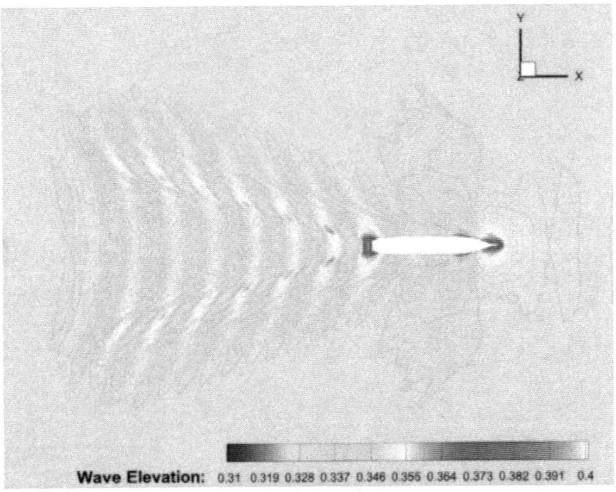

V=1.92 m/s

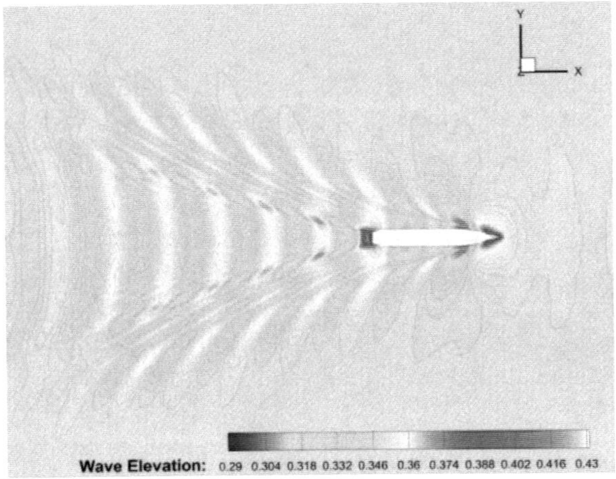

V=2.20 m/s

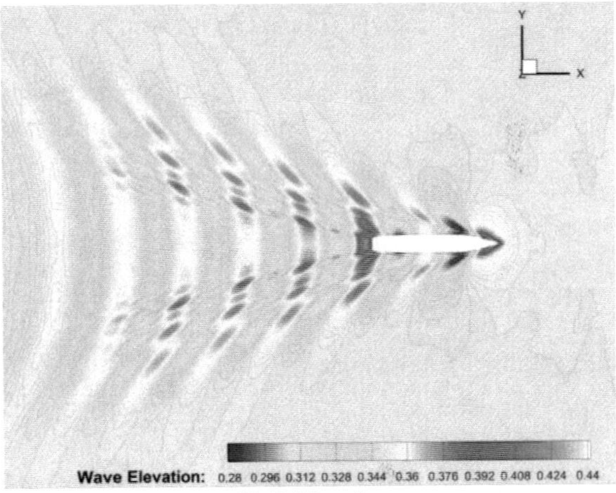

V=2.38 m/s

Figure 5. free-surface profile for case C3 with H=0.5L_{PP}.

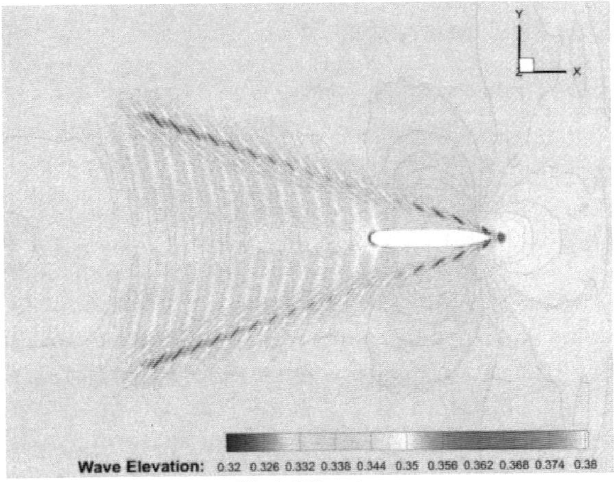

V=1.28 m/s

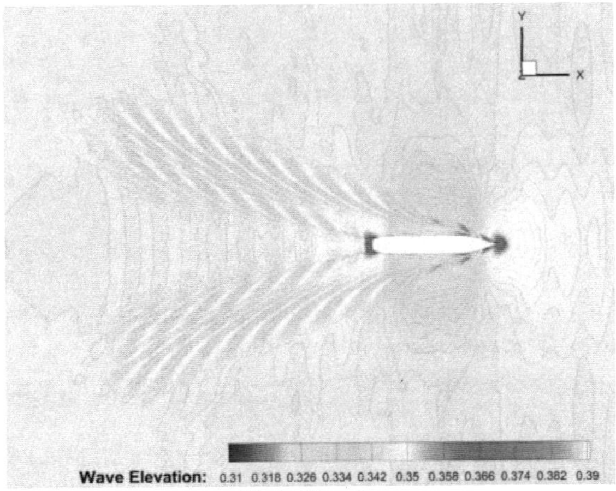

V=1.65 m/s

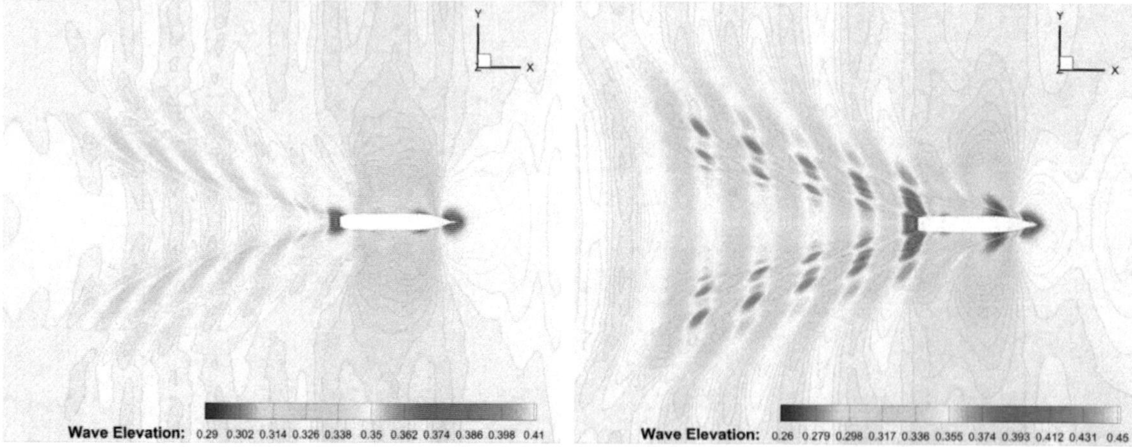

V=1.92 m/s *V*=2.20 m/s

Figure 6. free-surface profile for case C3 with *H*=0.5L_{PP}.

4.3.3. Case C6 with H=0.0625L_{PP}. The free-surface topology for case C6 is depicted for two ship speeds in figure 7. The same comments that were introduced for case C5 remains also valid for this case.

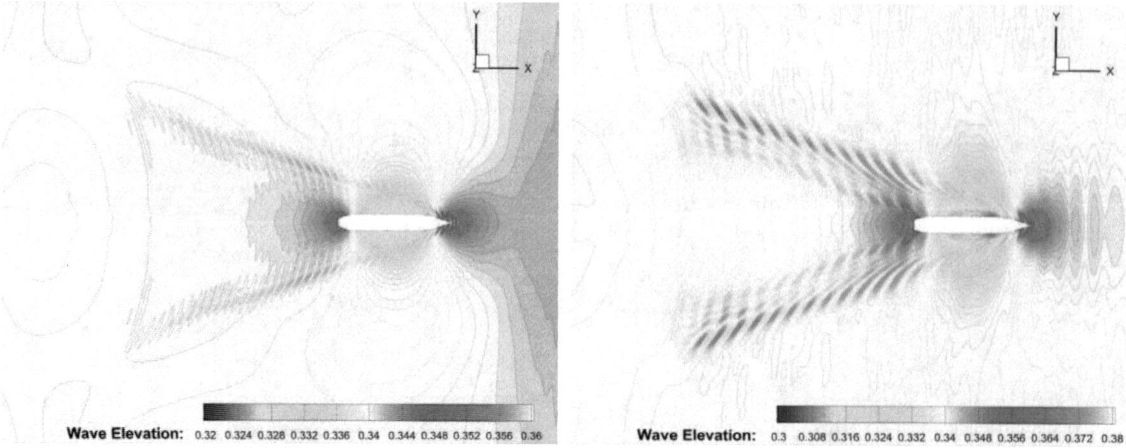

V=0.92 m/s *V*=1.28 m/s

Figure 7. free-surface profile for case C6 with *H*=0.0625L_{PP}.

4.3.4. Case C7 with H=0.1L_{PP}.* In order to highlight the significance of ship velocity on the wave pattern for even higher speeds, an extra simulation case on a coarser grid with 3.415M cells was introduced especially after the grounding of the case C6 for the higher set of speeds. Another water depth was selected between cases C5 and C6 having the water depth within H=0.1L_{PP}. for this purpose two extra ship velocities were calculated which are V=3.17 and 4.12 m/s which are close to the critical squatting condition corresponding to Froude number based on depth close to unity. The deformation in the wave pattern is significant and the opening angle of Kelvin pattern is increased dramatically. Also wave profile and wave crests and troughs distribution seem to be highly affected.

5. Conclusions

The shallow water effect on the KCS benchmark ship model based on a CFD viscous flow solver was presented with respect to the influence of the water depth on the total ship resistance, the vertical motions and finally the wave profile. Six different simulation cases assigned based on the water depth with six different ship speeds were analysed and investigated individually for every condition, in order to highlight their influence on the obtained results. In the scope of validating the numerically obtained results, a direct comparison validation test was performed against the EFD data showing a good agreement between CFD and EFD results with an error less than 2%.

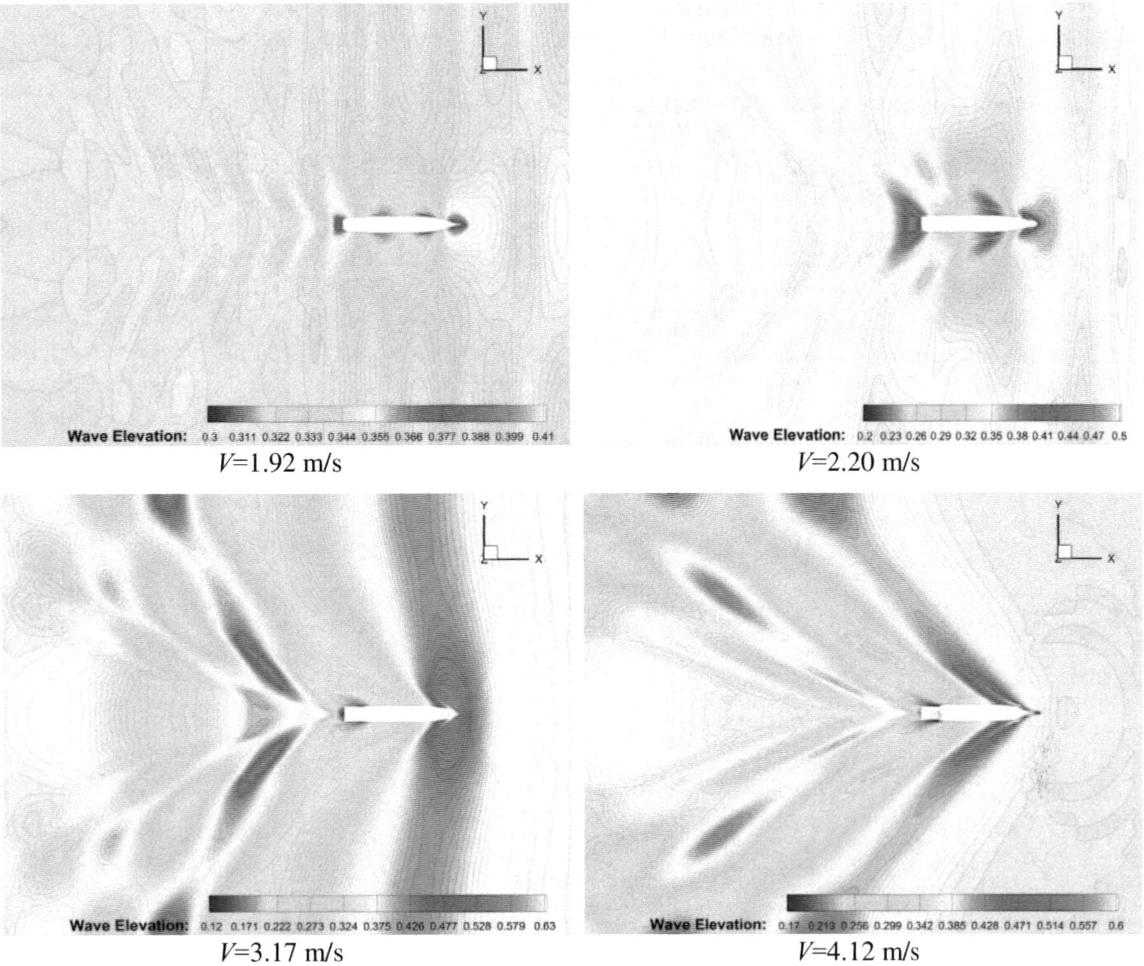

V=1.92 m/s *V*=2.20 m/s

V=3.17 m/s *V*=4.12 m/s

Figure 8. Free-surface profile for case C7* with $H=0.1L_{PP}$.

The total ship resistance results for every case were compared to the deep water outcomes showing the increment in total resistance with respect to the reduction of water depth. It was concluded that until the water depth of $H=0.5L_{PP}$ the effect of depth was still insignificant resulting in almost similar resistance values.

For the vertical motions, the sinkage and trim were computed and showed to comply with the theoretical approach for the shallow water effect except for some discrepancies that resulted from the iterative error. This might be important to increase the total simulation time, especially for the shallowest water cases to improve the convergence criteria.

Free-surface results showed to have a proper agreement with the theoretical shallow water effect for all the simulated and presented results. It may be concluded that the grid density for the coarsest grids could be enhance to capture more accurately the wave profile for the smallest ship speeds.

The future plan is to investigate the influence of the shallow water on the wake flow characteristics, propulsion efficiency and the manoeuvrability of the ship.

6. References

[1] United Nations Conference on Trade and Development (UNCTAD) 2017 *Review of Maritime Transport 2017*, United Nation, New York and Geneva.
[2] Deng G B, Queutey M and Visonneau M 2010 *J. Hydrodynamics*. **22**(5) 476–481.
[3] Queutey P, Deng G B, Wackers J, Guilmineau E, Leroyer A and Visonneau M 2012 *J. Ship Technology Research* **59**(2) 44-58.
[4] International Towing Tank Conference (ITTC) 2011, *Practical guidelines for ship CFD applications*. **Rev.1** 7.5-03-02-03 1–18.
[5] Tokyo 2015 A Workshop on CFD in Ship Hydrodynamics https://t2015.nmri.go.jp/ accessed on 15.02.2021.
[6] Tokyo 2015 A Workshop on CFD in Ship Hydrodynamics https://t2015.nmri.go.jp/Presentations/Day2-AM2-KCS-Resistance_SP-Kim.pdf accessed on 15.02.2021.

Acknowledgement

This work was carried out in the framework of the research project DREAM (Dynamics of the REsources and technological Advance in harvesting Marine renewable energy), supported by the Romanian Executive Agency for Higher Education, Research, Development and Innovation Funding – UEFISCDI, grant number PN-III-P4-ID-PCE-2020-0008.

IOP Conf. Series: Materials Science and Engineering 1182 (2021) 012006 doi:10.1088/1757-899X/1182/1/012006

Monitoring console using Virtual Reality for automotive industry

N Belu[1], L M Ionescu[2], E L Nitu[1] and A G Mazare[2]
[1] University of Pitesti, Manufacturing and Industrial Management Department, Targul din Vale Street No.1, Romania
[2] University of Pitesti, Electronics, Computer and Electrical Engineering Department, Targul din Vale Street No.1, Romania

E-mail: nadia_belu2001@yahoo.com

Abstract. One of the technologies used in Industry 4.0 is virtual reality and augmented reality in the industrial applications. These technologies allow: the introduction of the operator in the production process with minimal risk, in the training process (learn manufacturing), the realization of maintenance works with a much more detailed vision of the equipment and the process, the realization of consoles to monitor the manufacturing process and equipment with extended facilities compared to any type of console that has been used before. In this paper we will present an example of a monitoring console using virtual reality. As we will show, the virtual environment accessible through the virtual reality headset can be used as a console with an extended viewing angle in which indicators on the status of the monitored process can be displayed. The indicated values come from IoT equipment with current sensors and temperature sensors. Also, by navigating, using teleport tool, in the virtual environment you can get an overview from several angles of the monitored equipment (in our case a chassis welding workstation) with minimal risk. The environment contains both a modeled image (a 3D sketch of the equipment) and real images of the equipment captured through industrial cameras placed next to the real equipment. In the virtual environment the two "live" video streams are displayed as 2D screens. With this modern monitoring tool, the work standards will be developed that the operator can use in training and in the work process. The system presented in the paper is implemented with an Ocululs Rift kit with headset and touch consoles for capturing the movements of operators and their contacts with virtual objects. The virtual environment is fully implemented using Unity. The system will model workstation for chassis welding. Two IoT devices will be used: one with temperature sensor and the other with current sensor. Two industrial video cameras whit high resolution will also be used to capture images from the workstation. As a result, along with the presentation of the functional system will be presented a model of operational standard made by captures from Unity when exploring the virtual environment.

1. Introduction

One of the most modern tools and environments for displaying information is virtual reality and extended reality (augmented reality, virtuality). It has multiple fields of applicability: entertainment, education, sports, medicine but also in industry. One of the components of Industry 4.0 is the use of virtual reality, extended reality and virtuality [1] in production processes.

On the one hand, virtual reality can be used in learn manufacturing. Virtual courses have been built through which production lines can be explored and virtual manufacturing processes can be visualized [2]. Through Touch devices the learner can interact with virtual objects. The use of virtual reality in learn manufacturing is a field still in research but there are also commercial solutions that can be applied for specific manufacturing processes. The difference between a training on a "classic" 2D environment and one done with virtual reality is related to the great interactivity we find in the virtual reality environment, the increased amount of information we get by navigating virtual reality and the proximity of real operating scenarios. This is also the reason why this type of training will certainly be the main educational method of the future.

On the other hand, virtual reality and augmented reality are also used in maintenance work on industrial equipment [3] or in training for maintenance work [4]. Augmented reality allows the operator to be guided by the functional diagram of the machine superimposed over the real image that the operator sees. Also, through virtual reality the operator can see the consequences of his actions on the machine before they are applied. Such maintenance tools are studied in very recent research and attempts are being made to implement them in commercial solutions.

Last but not least, virtual reality can also be used as a method of monitoring and exploration in the production process in design [5] or remote control [6]. The operator can monitor the production process with detailed images of the schematics of the components involved that can be present, in a virtual way, inside the process. All this would be impossible in real environment, so virtual reality extends the monitoring capacity to the operator.

The paper features a virtual reality-based monitoring interface that can be used for both training and equipment monitoring. The interface will have the following features:

- It will implement a virtual environment in which the equipment is modeled (chassis welding installation);
- It will allow interaction with the equipment via Touch input devices. It is allowed to hold the tools or parts "by hand" and place them in another location;
- It will allow the monitoring of real data coming from two IoT devices that measure temperature and current consumption. The connection to the equipment is made by implementing some "driver" functions.
- Allows to monitor two images from two industrial cameras. Thus in the virtual environment we can have "real" images of the working station.
- It will allow the operator to move in the virtual environment through the options of physical movement (by movement) and by teleportation.

Some of the features listed above are specific to virtual reality systems: modeling equipment, interaction and manipulation of virtual objects, moving in the virtual environment. Another part are specific, innovative elements that have been brought and that will be presented in this paper: taking real data from IoT devices and "bringing" them into the virtual environment where they can be viewed, viewing virtual images in the virtual environment, modeling of specific chassis welding equipment in the virtual environment. The structure of this article is as follows: the next section is dedicated to the presentation of the system, section 3 will present the virtual environment as it was built and section 4 will present how to test this interface. The paper ends with conclusions and future directions of development.

2. Presentation of the interface

Along with the virtual reality kit - the Oculus headset and Touch controllers - which is the main component of the monitoring interface, there are the following components - as shown in figure 1.

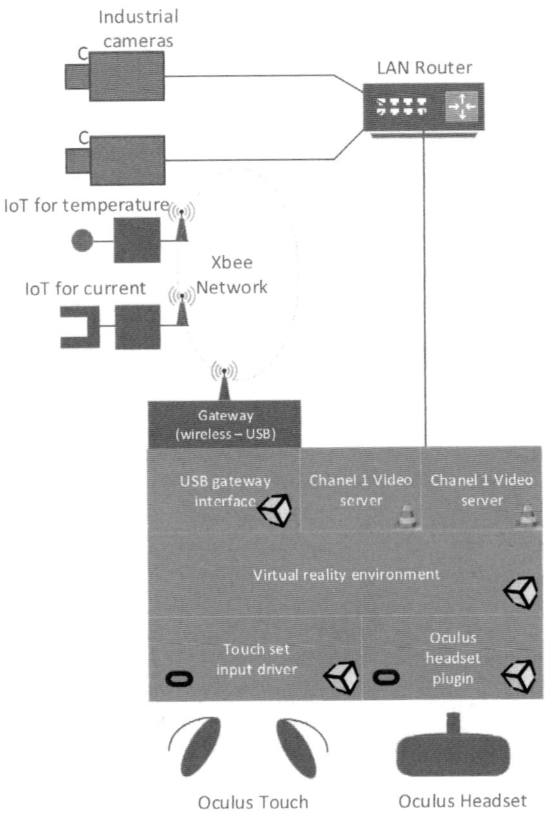

Figure 1. Monitoring interface – block diagram.

The two IoT devices for the acquisition of "real" data on the spot via temperature sensors and power consumption are based on the Waspmote 1.2 board from Libelium. The board contains an ATMEGA 320 microcontroller, an interface to which the sensor (temperature or current pliers' sensor) is connected and a shield together with the XBee transceiver - which communicates in the 2.4GHz free band at distances of about 1 km.

Both devices acquire data from sensors, package it, and transmit it over the XBee wireless LAN to the server. The network is a scalable one, it allows the connection of several IoT devices.

Another component consists of the two industrial chambers. The virtual reality kit also has a headset camera that allows the overlapping of real images with virtual ones (augmented reality). In this case, it is desired to acquire images from the industrial process from two different positions. This is why two industrial chambers were used. These are IP cameras wired to the network where the PC server station is located. The cameras are 720p HD and allow monitoring in daylight conditions but also in low IR lighting conditions. This allows them to transmit images from the welding process when the machine is operated or when it is switched off.

The PC server station is the one that runs both the virtual environment displayed on the virtual reality kit and the interface for acquiring data from the sensors and the part for acquiring video streams from the cameras and providing it to the real environment.

For the virtual environment was used Unity 2018.4.20 as design environment. It will also be used for running application. The gateway is the device connected to the PC (via USB) through which data is taken from the XBee wireless network. A NodeJS server is used to retrieve data from USB from the

IOP Conf. Series: Materials Science and Engineering 1182 (2021) 012006 doi:10.1088/1757-899X/1182/1/012006

gateway and transmit it as a web service to Unity. Finally, the video streams, in RTSP format, reach the PC via the intranet and are then taken over by two VLC instances and transformed into http streams. The latter can be integrated into Unity - just like a movie that is provided on YouToube and has its own address.

3. Presentation of the virtual environment

The virtual environment was made entirely in Unity. Several stages were completed of its realization.

3.1. Equipment modeling

First of all, the equipment used was modeled. Its modeling was done by using 3D primitives: cubes, parallelepipeds, cylinders and spheres as well as by applying textures - free textbooks from the internet were used for textures. Also, in the model of the electric motor that drives the arm of the welding station and for the robot arm itself, a free model from the internet was used. The models used in the virtual environment are illustrated in figure 2.

Figure 2. Models for the virtual environment.

3.2. Implementation of actions to catch virtual objects and manipulate them

To take inputs from Touch controllers, "driver" interfaces have been built in Unity. Such an interface handles events from a Touch controller. Specifically, when a controller is moved or pressed it converts the value returned from the controller sensors to real numbers. These real numbers are called FloatActions and are associated with "hand" objects. In this way, by moving a controller or by operating it, the effects will be seen on the hand model.

The construction of the drivers' scripts is done in code (C #) and the association with hand is done in the Unity graphical interface.

The contact between the hand object and a virtual object and the implementation of more complex actions of grabbing the virtual objects and even throwing them is done by introducing objects already made for this purpose (prefabs) and associating them with virtual objects. The properties of these prefabricated objects allow actions of grabbing and moving virtual objects by associating the vector transform.rotation and tranform.position from the hand object and the grabbed object. To generate a speed when throwing objects, a script (built) with hand objects will be assigned. The steps taken to implement the action of catching and manipulating virtual objects are shown in figure 3.

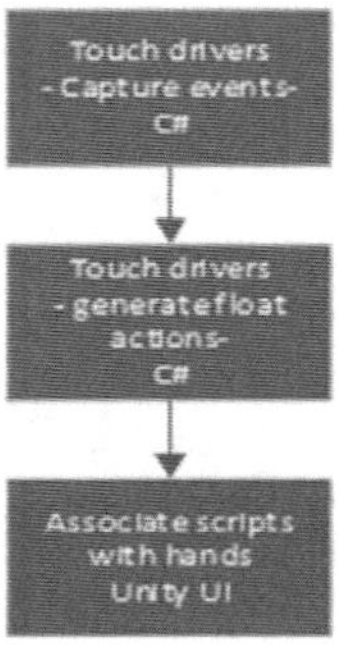

Figure 3. Implementing catch action for virtual objects.

Figure 4 shows how the capture and manipulation action is perceived in the virtual environment.

Figure 4. Objects handling in virtual environment.

3.3. The movement in virtual environment

There are two forms in which the movement has been implemented in the virtual environment: physical movement - as in the real environment and movement by teleportation.

Physical movement occurs by default when the virtual camera is associated with the head set. The movements of the head are taken over and converted into movements of the main camera in the virtual environment because the headset has accelerometers on all three axes. In the same way, the movements of the person holding the headset are taken over and converted into movements in a virtual environment on the same scale. So, moving in the real environment, we also move in the virtual environment. However, physical movement has some drawbacks. First of all, it is limited due to the cables through which the headset is connected to the PC. Secondly, it is not safe: the person can hit objects that exist in the perimeter where the movement takes place. This is also the reason why a second form of travel was chosen, namely by teleportation.

Unity has the tools to easily implement teleportation. An object of the Teleporter Instant type can be associated with one of the Touch controllers (in our case the one on the left hand). A cursor is also associated with the same controller. By operating the joystick, a cursor is displayed that indicates the place where the movement will be made and by releasing the joystick, the "teleportation" takes place in that place. This method of movement allows you to travel in the virtual environment over long distances.

Figure 5 shows how the teleporter is associated in Unity.

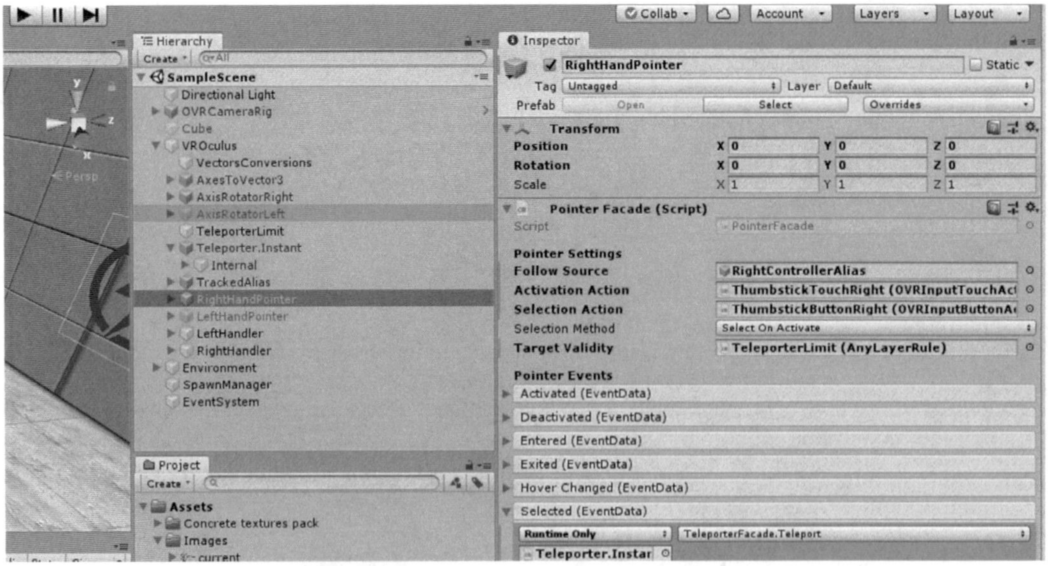

Figure 5. Teleporter object and teleporter limits definition in Unity.

The aspects presented so far are typical for a VR application. The following elements are specific to the type of application we have developed and presented in this paper.

3.4. Display data from IoT devices

In Unity, it is allowed to connect to the USB ports of the PC in order to retrieve the data that a device transmits there. Specifically, communication is implemented at the level of a C # script that can generate actions which are then interpreted at the Unity level. A general organization chart (without going into the details of the implementation) is presented in figure 6.

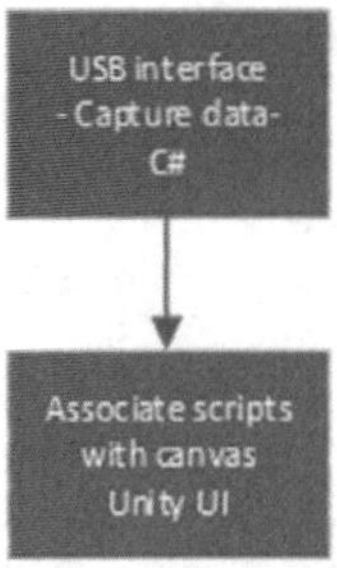

Figure 6. General flowchart for connecting to the IoT gateway.

To display the values from the sensors, two canvas objects were built with the world-space rendering mode (so related to the virtual environment) and which associated the CenterEyeAnchor event camera. Images containing text and icons have been attached to these objects. They have been dimensioned at the level of some displays. Their perception in the virtual environment is in the form of transparent displays on which the numbers and letters of the measured values appear - as if on some glass surfaces. The effect is very interesting in the sense that the image is 2D. Image editing was done in Paint.Net using transparency.

3.5. Display video streams

Images taken from on-site industrial cameras were captured by video mini-servers using VLC. Its role is to turn the video stream from RTSP format - incompatible with the Unity environment into HTTP stream - the same type of stream that is used for Youtube movies. So, there will be two VLC instances that will run for the two video streams. For using this solution with a larger number of cameras, it can be purchased a professional Video Management System (VMS) application.

The integration of images is done in the same way as data from IoT devices: canvases of the same type have been created but here links to video streams have been associated. In this way, you can see real images on the spot in the virtual environment.

Canvases are objects that will not be affected by the physical laws of the virtual environment: they can be placed anywhere, including can float in the air - so did the two indicators from IoT devices. Thus, the monitoring panel made in the virtual environment can be placed in positions impossible to reach in reality with maximum ergonomic comfort for the operator.

4. Testing the monitoring interface

Figure 7 shows the final shape that was given to the interface - of course this shape can be adjusted at any time to be able to adapt to another type of process.

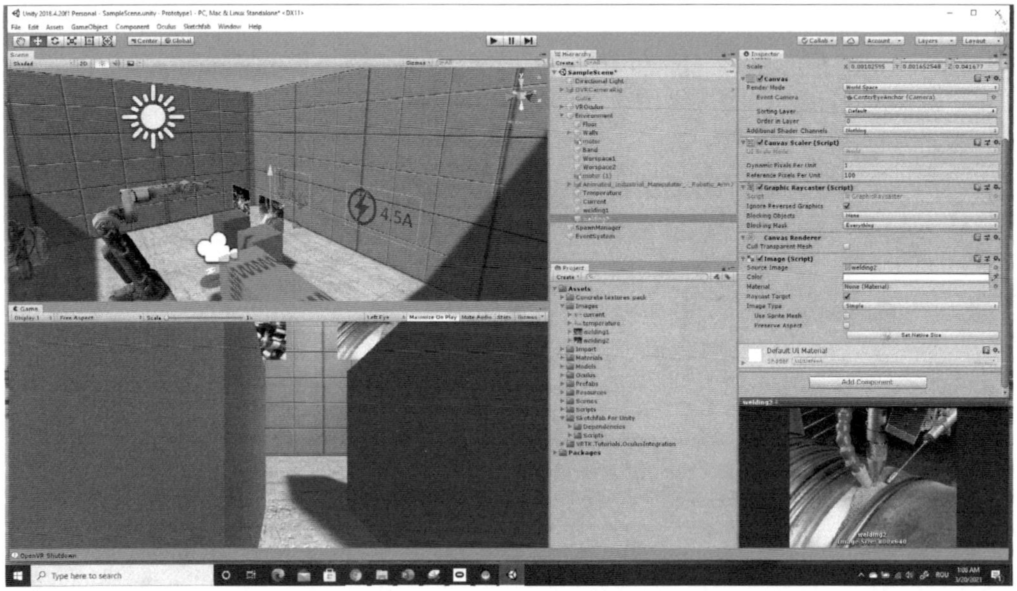

Figure 7. The final interface before running.

The scenario here is as follows:

The operator is placed next to the conveyor belt on which are provided the parts to be welded by the robot. It must ensure the correct placement of the parts to be welded. In this phase, the pieces are red and blue cubes. The blue cubes must be placed on the blue support and the red cubes on the red support. The movement of the operator to place the cubes on the two supports is a physical one. The operational scenario is a simple one, the pieces appear randomly - simulating the production flow.

To the left of the operator is the robot - inoperative until the part is placed correctly (on the correct support). To the right of the operator is the belt drive motor - its movement is limited by these machines - as is the real case.

At the level of the two supports on which the parts are placed, there are real video streams. By placing the part, the operator can observe the real support and can monitor the real welding process. In front of the conveyor belt there are indicators on the temperature at the working point and the current consumption of the equipment.

In figure 8 we attached several images during the operation.

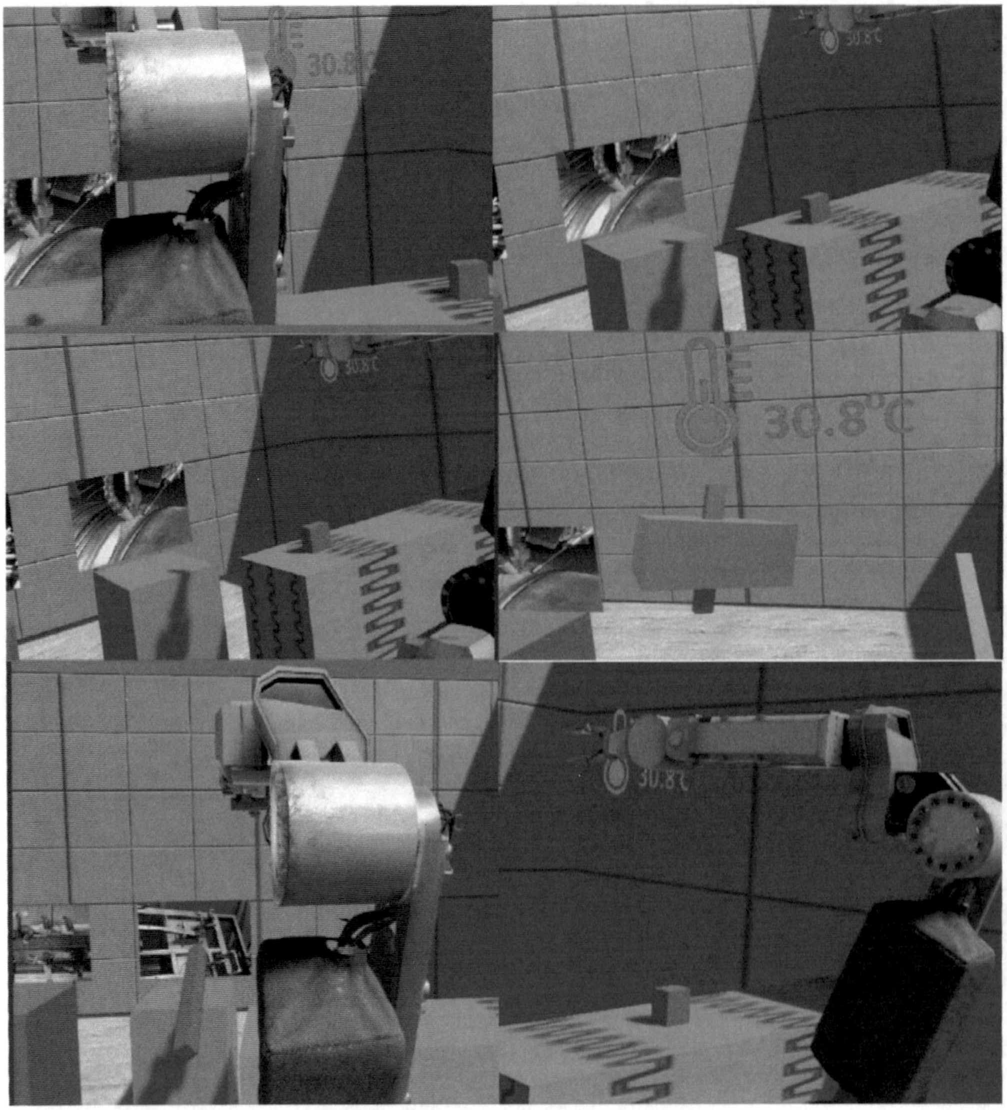

Figure 8. Operating in virtual environment.

As you can see, the operator handles virtual objects safely - so the platform can be used for a training scenario. Placing the equipment, using as realistic models as possible makes the operator get ready for the activity he will perform, but it can also be a very useful support for analyzing the ergonomics of the work point.

On the other hand, as you can see, we are talking about a very interactive monitoring platform. As you can see, we have real data on the spot: measured values or even images. In this way, the platform can be used for maintenance operations: monitoring the status, performing virtual actions that can then be actually reproduced with the visualization of the results. Also, the actions that take place in the real environment (e.g, the automated welding process) can be monitored in detail and transposed into the virtual environment.

5. Conclusions

As we have seen, the use of virtual reality greatly expands the possibilities of training, analysis of workplace ergonomics, monitoring for maintenance or production flow operation. The paper presents typical features of VR applications: integration of 3D models for equipment (as accurate as possible), capturing operator movements, actions on virtual objects, moving in the virtual environment, combined with own monitoring solutions: acquisition of real data from IoT devices and displaying them in the virtual environment, purchasing video streams and displaying them in the virtual environment. All these features "collected" in a VR monitoring platform are used in the industrial environment. The scenario they applied is related to a chassis welding process in which the operator has to supply the components (parts) which are then welded by an industrial robot. The prototype presented here was used for operator training but can also be a tool for analyzing workplace ergonomics or an interactive monitoring platform for maintenance work or production flow operation. In the future, the applicability of the platform will be extended to various other manufacturing areas and new forms of virtual environment - real environment interaction will be implemented.

6. References

[1] Farshid M, Paschen J, Eriksson T, Kietzmannc J 2018 Go boldly!: Explore augmented reality (AR), virtual reality (VR), and mixed reality (MR) for business, *Business Horizons* **61** Issue 5 p. 657-663.

[2] Salah B, Abidi MH, Mian SH, Krid M, Alkhalefah H, Abdo A. 2019 Virtual Reality-Based Engineering Education to Enhance Manufacturing Sustainability in Industry 4.0 *Sustainability* **11**(5):1477.

[3] Eschen H, Kötter T, Rodeck R, Harnisch M, Schüppstuhl T 2018 Augmented and Virtual Reality for Inspection and Maintenance Processes in the Aviation Industry *Procedia Manufacturing* **19** p. 156-163.

[4] Numfu M, Riel A, Noel F 2019 Virtual Reality Based Digital Chain for Maintenance Training *Procedia CIRP* **84** p 1069-1074.

[5] Berg, L.P., Vance, J.M 2017 Industry use of virtual reality in product design and manufacturing: a survey *Virtual Reality* **21** p. 1–17.

[6] Lipton J I, Fay A J, Rus D 2018 Baxter's Homunculus: Virtual Reality Spaces for Teleoperation in Manufacturing *IEEE Robotics and Automation Letters* **3**(1) p. 179-186.

IOP Conf. Series: Materials Science and Engineering 1182 (2021) 012007 doi:10.1088/1757-899X/1182/1/012007

The possibility of physical implementation of active vibration reduction

K Białas
Silesian University of Technology University, Faculty of Mechanical Engineering, Institute of Engineering Processes Automation and Integrated Manufacturing Systems, Konarskiego 18A, 44-100 Gliwice, Poland

E-mail: katarzyna.bialas@polsl.pl

Abstract. This work presents non-classical method of design of mechanic systems with subsystem reducing vibrations. The purpose of this paper is also introduces synthesis of mechanic system with reducing vibrations understand as design of this type of systems. This work presents method of reduction unwanted vibration by using mechanical elements, electrical elements, magnetorheological damper ect. The use of such a method enables the analysis and synthesis of mechanical systems irrespective of the type and number of the elements of such a system. The purpose of this paper is also introduces synthesis of mechanical system which is the reverse task of dynamics. The result of synthesis is obtaining system meeting the defined requirements. The system was consisted from mechanical elements and sub-system realized by many types of elements. This subsystem is an active or semiactive subsystem which enables overcoming limitations which occur if passive elements are used. The majority of vibration occurring in devices and machines is harmful and has a disadvantageous effect on their condition. Harmful impact of vibration is caused by the occurrence of increased stresses and the loss of energy, which results in faster wear machinery. Introduced in this work approach adopted makes it possible to undertake actions aiming at the elimination of phenomena resulting in the unwanted operation of machinery or generation of hazardous situations in the machinery environment. Thank to the approach, can be conducted as early as during the designing of future functions of the system as well as during the construction of the system.

1. Introduction

Vibrations are among the most frequent physical phenomena observed in everyday life. Typically, vibrations have an adverse effect on their environment. The negative effects include the impact on plant, machinery and structure operation and life. Humans also experience the effect of vibrations. Exposure to a vibration environment can be a cause of multiple diseases. For this reason, many research institutes are involved in studies on vibration damping and isolation. Designers take the vibration potential and vibration damping options into account in building and designing new buildings or structures. Another task for designers and structural engineers is to adapt existing plant and machinery to the conditions in which they operate. Vibration reduction can be used in the case of existing systems, or by predicting the behaviour of the system, it is possible to design systems with vibration reduction. The paper presents two non-classical methods of designing systems with regard to the requirements imposed on them.

The non-classical approach to circuit design is also called structural-parametric synthesis. After obtaining systems that meet the desired requirements, active vibration reduction can be applied.

Four theoretical examples of the implementation of active subsystems made of various types of elements are presented [1-9].

2. Non-classical method of design of mechanical system

Synthesis, conceived as the design of vibrating mechanical systems, consists of a development of system structure and parameters depending on the desired frequency spectrum. System properties can be recorded with the use of dynamic characteristics transformed into dynamic response or rigidity and, following transformation, into mobility (1,2) or response (2,3). Two main methods of vibrating mechanical system synthesis are known: distribution of characteristic function by continued fraction and distribution of characteristic function by partial fractions.

Mobility of semi-specified systems:

$$V(s)=H\frac{c_k s^k + c_{k-1}s^{k-2}+\ldots+c_0}{d_l s^l + d_{l-1}s^{l-2}+\ldots+d_1 s} \tag{1}$$

where:

k – even order of numerator, l – order of denominator, $k\text{-}l = 1$, H – any positive real number.

Mobility of restrained systems:

$$V(s)=H\frac{c_k s^k + c_{k-1}s^{k-2}+\ldots+c_1 s}{d_l s^l + d_{l-1}s^{l-2}+\ldots+d_0} \tag{2}$$

where:

k – odd order of numerator, l – order of denominator, $k\text{-}l = 1$, H – any positive real number.

Slowness of semi-specified systems:

$$U(s)=H\frac{d_l s^l + d_{l-1}s^{l-2}+\ldots+d_1 s}{c_k s^k + c_{k-1}s^{k-2}+\ldots+c_0} \tag{3}$$

where:

l – odd order of numerator, k – order of denominator, $l\text{-}k = 1$, H – any positive real number.

Slowness of restrained systems:

$$U(s)=H\frac{d_l s^l + d_{l-1}s^{l-2}+\ldots+d_0}{c_k s^k + c_{k-1}s^{k-2}+\ldots+c_1 s} \tag{4}$$

where:

l – even order of numerator, k – order of denominator, $l\text{-}k = 1$, H – any positive real number.

In order to obtain the structure and parameters of inert and elastic elements of a dynamic system one applies two basic methods [1]:

- decomposition of a characteristic function into a continued fraction (5),
- decomposition of a characteristic function into simple fractions (6).

$$V(s)=\frac{c_1}{s}+m_1 s+\cfrac{1}{\cfrac{s}{c_2}+\cfrac{1}{m_2 s+\ldots+\cfrac{1}{\cfrac{s}{c_n}+\cfrac{1}{m_n s}}}} \tag{5}$$

$$U(s)\frac{1}{H}=\frac{c_1}{s}+m_1 s+\cfrac{1}{\cfrac{s}{c_2}+\cfrac{1}{m_2 s}}+\ldots+\cfrac{1}{\cfrac{s}{c_n}+\cfrac{1}{m_n s}} \tag{6}$$

where:
c – elastic elements, m – inertial elements.

3. Active reduction of vibration – physical realization

Physical realizations of active systems can consist of various elements. These may include mechanical, electric, hydraulic, pneumatic components, etc. Example arrangements involving the particular types of components are shown on the figures 1 to 4. An active system made of mechanical components can have the form of kinematic excitation (figure 1). In the case of kinematic excitation, the elements capable of changing their values are displacements y_i, while elastic elements c_i remain unchanged [2].

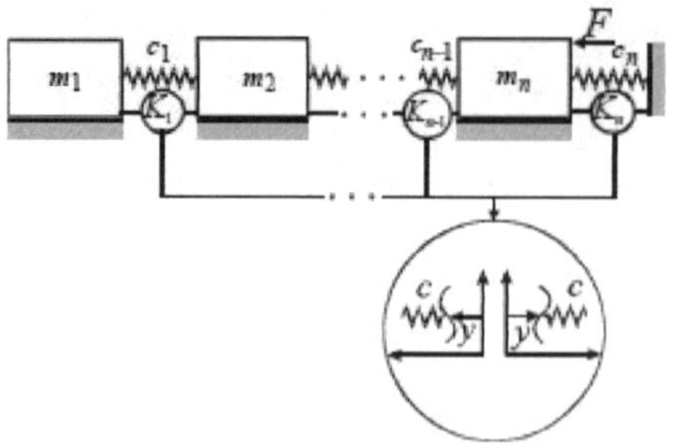

Figure 1. A model of the system with active elements – kinematic excitation.

Another possible arrangement for a physical realization of active components is the use of electric items such as a solenoid. When sinusoidal voltage is applied to the solenoid, the polarity and strength of the magnetic field produced by the solenoid will change, and axial movement of the core follows (figure 2) [2-4].

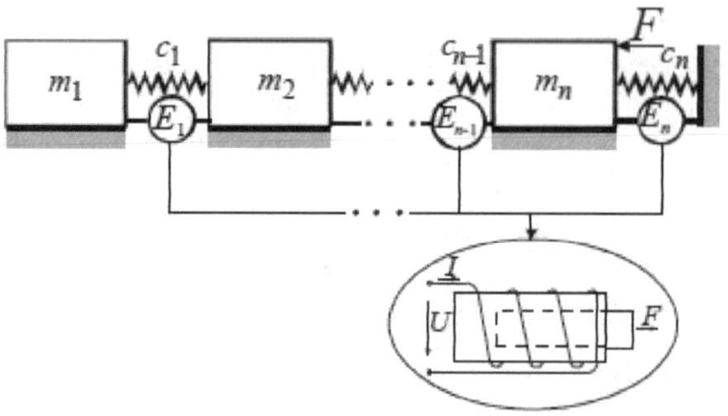

Figure 2. A model of the system with electric elements – moving-coil transducer.

The implementation of active vibration damping through piezoelectric components is presented on the drawing. A piezoelectric oscillator is a solid body, typically in the proper shape, cut out of piezoelectric material in the appropriate manner.

The most commonly used piezoelectric materials include crystalline quartz, shaped as a circular or square plate. The plate is positioned between electrodes. Due to the properties of the piezoelectric materials, they are capable of producing vibrations (figure 3).

Two piezoelectric effects are known. The first one of these is related to the presence of mechanical tension occurring in response to electric charges. An inverse piezoelectric effect consists of a change of crystal dimensions in response to the applied electric field.

The mechanical energy into electrical energy conversion and the electrical energy into mechanical energy conversion that occurs in piezoelectric materials has opened an extensive range of options for using such materials in multiple fields of engineering [9-10].

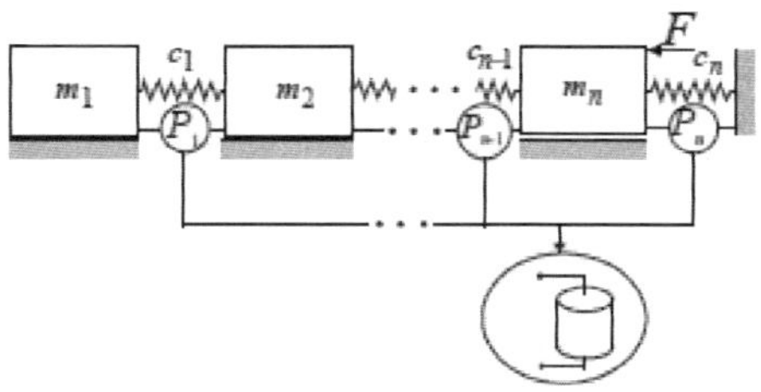

Figure 3. A model of the system with active elements – piezoelectric.

Magnetorheological dampers are used as the solution in realization of active components.

A system incorporating magnetorheological dampers as active components is shown on the figure 4 [11].

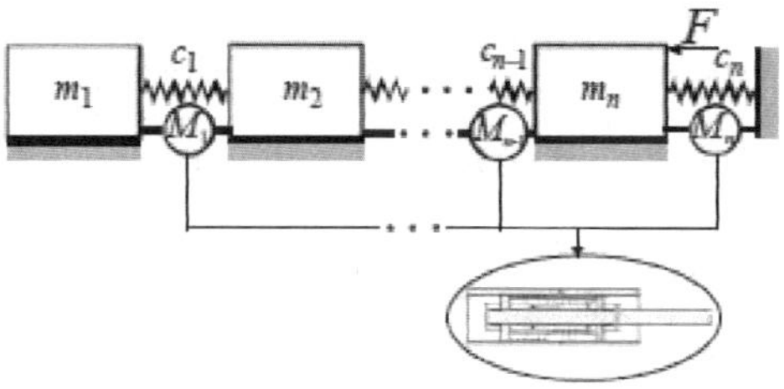

Figure 4. A model of the system with active elements – magnetorheological dampers.

Magnetorheological dampers are frequently used in such important applications as the construction industry and the automotive industry.

The first devices working with the use of magnetorheological fluid dampers were the dampers used in large drum-type washing machines. The function of magnetorheological dampers was the damping of undesired vibration occurring during the work of these washing machines. Construction is another industry area where magnetorheological dampers are used. They are commonly applied here to protect structures against seismic action. Within the construction industry, magnetorheological dampers are also used in bridge structures to prevent displacement. Dampers of this type are used in suspension bridges to prevent rope displacement. The automotive industry is among those sectors in which magnetorheological dampers play an important role. Magnetorheological fluid was first used in the driver seat construction in 1996. Specifically, it was the arrangement implemented in a truck driver's seat. Drivers of these vehicles are exposed to physiological effects of vibrations, related to the prevalence of many diseases. As a result of further development and research of magnetorheological fluid, it was used in passenger car suspension structure springs. The first solutions of this type appeared on the market under the MagneRide brand, developed by Delphi and Lord.

Before the magnetorheological fluid solutions, solenoid valves were used in shock absorbers. Work was also started on modifying car suspensions through the use of solutions based on magnetorheological fluid properties and control options. In 2002, mass production of a passenger car with a suspension fitted with magnetorheological dampers was launched in the Cadillac Seville. This opened the process of replacing the old structural suspension solutions with new ones, particularly popular in high profile models. This is due to the relatively high prices of magnetorheological liquid as the primary medium in these solutions. After 2002, other car maker brands started to implement these solutions, including Audi, Chevrolet Corvet, Cadillac Roadster.

Where magnetorheological dampers are used, it is important to choose and adjust them precisely to specific suspensions mounted in specific car makes and models. There are many factors affecting the final versions of these dampers, such as the car size and weight, technology, and the production process. Control of such dampers is based on solenoids, electric power or permanent magnets. Magnetorheological fluids are recognized among smart materials, and their rheological properties are sensitive to the present magnetic field.

Multiple technology solutions can be distinguished for magnetorheological damper fluids. A shock absorber very often consists of a piston, cylinder, and magnetorheological fluid. A permanent magnet is positioned in a wall seat, consisting of two similar rings in a symmetrical arrangement. Through the permanent magnet, a magnetic field is generated, affecting the magnetorheological fluid and changing the rheological properties, which leads to shock absorption (figure 5) [11-14].

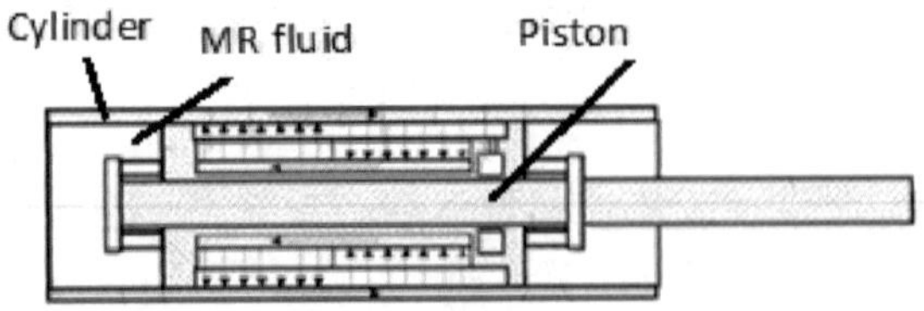

Figure 5. Magnetorheological damper.

Magnetorheological fluid is composed of two basic ingredients, i.e. the medium and ferromagnetic particles, with such additional items as surfactants. Table 1 presents the example parameters of magnetorheological fluid.

Table 1. Parameters of magnetorheological fluid.

Parameter	Magnetorheological fluid
Viscosity without field	0.2÷0.3 Pa·s
Yield point	50÷100 kPa
Maximum current	50÷100 kPa
Density	3÷4 g/cm3
Response time	<10 ms
Typical power ratings	2÷25 V, 1÷2 A
Field limitations	Saturation
Operating temperature	50 do 150 °C

Based on a Bounc Wen model of a magnetorheological damper, the force generated by such a damper can be determined (7).

$$F_t = f_0 + c\dot{x} + k_0 x + \alpha z \quad (7)$$

where:
f_0 – battery impact
$c\dot{x}$ – viscous resistance of fluid flow through an opening

k_0x – elastic properties of the fluid
αz – magnetic field effect on the fluid

4. Conclusions

The paper presents a non-traditional design engineering method for vibrating mechanical systems. The author describes certain active vibration reduction methods. The paper presents the options for active components realization. Examples of such realizations are the applications of mechanical, electrical, piezoelectric and magnetorheological elements. In order to verify the operation of an active system, it is important to analyse the basic system, the active subsystems, the sensors and the mutual effects among all these components.

5. References

[1] Białas K., Buchacz A., Dzitkowski T. 2009 Synthesis of vibrating active mechanical systems with dumping in view of polar graphs and structural numbers. *Gliwice Monograph* 230, Silesian University of Technology Press, (in Polish).
[2] Białas K. 2012 Mechanical and Electrical Subsystems as Implementation of Active Reduction of Vibration *International Journal of Modern Manufacturing Technologies*, **IV**(1) 29-34.
[3] Bialas K. 2013 Electrical Elements in Reduction of Mechanical Vibrations *Applied Mechanics and Materials* **371** 657-661.
[4] Białas K., Buchacz A. 2014 The Influence of Changing the Parameters of Electrical Components Implementing the Active Reduction of Vibration *Applied Mechanics and Materials* **657** 614-618.
[5] Białas K., Buchacz A. 2015 Reduction of vibration by using mechatronical subsystem. *Book Series: IOP Conference Series-Materials Science and Engineering* **95** 012051.
[6] Buchacz A., Płaczek M. 2010 The Analysis of Vibrating Systems Based on the Exact and Approximate Method *International Journal of Modern Manufacturing Technologies* **II**(1) 19-24.
[7] Engel Z., Kowal J. 1995 *Vibro-accoustic processes control*. Kraków AGH Press, (in Polish).
[8] Michałowski S. 1994 *Active systems in machines construction*. Cracow Publication by Cracow University of Technology, Monograph 171, (in Polish).
[9] Onwubolu G. C. 2005 *Mechatronics Principles and Applications*. Elsevier Butterworth-Heinemann.
[10] Serridge M., Licht T. R. 1987 *Piezoelectric accelerometer and vibration preamplifier handbook*, Brüel & Kjaer, Glostrup.
[11] Skalski P. 2017 *Magnetoreological dampers*. Aviation Institute Warsaw, Energy Transport and Conversion Centre, Department of Transport (in Polish).
[12] Ławniczak A., Milecki A. 1999 *Electro- and magnetoreological liquids and their use in technology*. Poznań University of Technology Publishing House (in Polish).
[13] Sapiński B. 2005 Real-time control for a magnetorheological shock absorber in a driver seat *Journal of Theoretical and Applied Mechanics* **43**, 631-653.
[14] Yao G.Z., Yap F.F., Chen G., Li W.H., Yeo S.H. 2002 MR damper and its application for semi-active control of vehicle suspension system *Mechatronics* 963-973.

Research regarding strip cladding of heat-resistant SA 387 Gr.11Cl.2 steel type

A M Bogatu, C Rontescu, R C Diacu and D-T Cicic
University Politehnica of Bucharest, Faculty of Industrial Engineering and Robotics, Quality Engineering and Industrial Technologies Department, Independenţei Street No. 313, Romania

E-mail: dumitru.cicic@gmail.com

Abstract. The continuous development of industrial processes in the petrochemical field requires the use of materials with anticorrosive properties, able to withstand working conditions at high temperatures but also in highly corrosive environments. The basic materials with anticorrosive properties have a high cost price which leads to the need to find alternative manufacturing solutions. In such situations, the use of bimetal plates or the option of welding cladding with an anticorrosive layer of the base material is used. The paper analyses the possibility of strip cladding using the submerged arc welding process of a reactor made of SA 387Gr11Cl.2 basic material, a heat-resistant alloy steel. To achieve the deposition of the appropriate anticorrosive layer (stainless steel type AISI 347) it was necessary to use a buffer layer made of austenitic stainless-steel type 309L. In order to eliminate the internal tensions appeared after the cladding process, the sample was subjected to a stress relieving heat treatment. After the heat treatment, the sample was subjected to non-destructive examinations (visual and penetrant testing) but also to destructive examinations specific to the approval tests of cladding technology (Metallographic tests, Vickers hardness test, bend test, ferrite number, chemical composition). Metallographic tests of cross-sectional phase structures were carried out in macroscopic and microscopic terms. In the area of the deposited material can be observed a typical austenitic casting structure resulting from the use of a high linear energy, specific to the automatic submerged arc welding. Following the performance of hardness measurements made in the characteristic areas of the deposition, it can be seen that there was a slight increase in the values obtained in the heat affected zone, without exceeding the prescribed values for the materials used. Following the analysis of the results obtained from the specific examinations and tests to the plated samples, it can be seen that they are within the limits prescribed by the standards in force regarding the approval of the cladding technology used.

1. Introduction

Modern industrial processes require the use of high values of technological parameters (temperature, pressure, etc.) as well as different working environments, some of which are highly corrosive. Erosion, high temperature, high pressure, vibrations during operation can cause corrosion of materials, leaks and working failure of products [1-4]. In order to meet the technological requirements, materials with high mechanical properties and high thicknesses are needed (in order to meet the requirements of high pressure) but also materials with anticorrosive properties, in contact with the working environment.

Taking into account the high cost price in the case of using materials with anticorrosive properties, the solution chosen in such cases is to use bimetal plates obtained by roll bonding, in the case of shell components manufacturing. For elliptic head manufacturing, taking into account the drawing process used, the bimetal plates is not accepted. In such cases, after drawing the base material, the anticorrosive layer is deposited with the help of cladding processes. Thus, cladding can be done using welding cladding procedures (laser welding, submerged arc welding, explosive welding, gas tungsten arc welding, plasma arc welding, etc.). The welding cladding process is a demanding process that requires a series of tests to obtain the desired quality of deposition [5-8].

The corrosion properties of the layers deposited by welding are given by the alloying elements contained by the filler materials. Taking into account the fact that the deposition is made on base material type C-Mn steel or low alloy steel base material, the resulting dilution must be as small as possible [9-13]. Table 1 presents the values of the dilution coefficient obtained in the case of using different welding processes [2]. In the case of submerged arc welding (SAW) the mechanical properties of the deposited layer depend on the properties of the chosen wire/flux combination but also on the dilution of the welding process [14,15].

Table 1. Dilution percentage for different welding processes [2].

Welding process	Typical dilution (%)
Shielded metal arc welding (SMAW)	25-30
Flux cored arc welding (FCAW)	20-25
Submerged arc welding (SAW), wire	25-35
Submerged arc welding (SAW), strip: 60 × 0.5 mm	18-20
Electroslag welding (ESW), strip: 60 × 0.5 mm	7-12

The main objective of the paper was to present the necessary steps to qualify the welding technology of the base material SA 387 Grade 11 Class 2 (a heat-resistant alloy steel) with a layer of steel with anticorrosive properties type AISI 347 (stainless steel) using the SAW welding method with strip electrode. The cladding is necessary for the deposition of the anticorrosive layer of an elliptical head from the component of a pressure vessel reactor.

2. Experimental data

The base material used for the reactor manufacturing was SA 387Gr11Cl.2 steel. Corrosion protection of shell plates was achieved by using bimetal plates (SA 387Gr11Cl.2 and AISI 347) manufactured by roll bonding. For the ellipsoidal heads, the welding option with SOUDOTAPE 347 strip electrode was chosen (60x0.5mm). To achieve the deposition it was necessary to use a buffer layer made of austenitic stainless steel type SOUDOTAPE 309L (60x0.5mm). The flux used for both types of band electrode was RECORD INT 109. The chemical composition as well as the mechanical properties of the materials used for cladding, according to the standards in force, are presented in table 2 and table 3 [14-16].

Table 2. Chemical composition of basic materials and filler material [14-16].

	Weight Percent							
Materials	**C**	**Si**	**Mn**	**P**	**S**	**Cr**	**Mo**	**Ni**
SA387Gr11CL2	0.04-0.17	0.5-0.8	0.4-0.65	0.025	0.025	1.0-1.5	0.4-0.7	-
SUDOTAPE 309L	0.013	0.4	1.69	0.016	0.001	23.7	0.1	13.6
SUDOTAPE 347	0.010	0.41	1.70	0.017	0.001	19.7	0.07	10.5

Table 3. Mechanical properties of base materials and filler material [14-16].

Materials	Yield Strength (MPa)	Tensile Strength (MPa)	Elongation (%)
SA387Gr11CL2	310	515-690	18
SOUDOTAPE 309L	365	580	38
SOUDOTAPE 347	470	640	35

For the qualification of the welding cladding procedure, the sample made of the basic material SA387Gr11CL2, with dimensions 30x300x400 mm, was mechanically cleaned in order to remove the dust, oil, rust and other foreign particles and preheated to a preheating temperature of 150°C, according to the preliminary specification of the welding procedure (figure 1).

a) b)

Figure 1. Metal plate subjected to the: a) mechanical cleaning process; b) preheating process.

After cleaning the sample and performing the preheating, the layers were deposited with the two types of tape electrode, using the technological parameters indicated in table 4. The thickness of the deposited layers must be at least 6 mm.

Table 4. The parameters used in the experiments.

No.crt.	Parameter	Value	
		SOUDOTAPE 309L	SOUDOTAPE 347
1	Welding current [A]	680-730	650-700
2	Arc voltage [V]	29-30	29-30
3	Filler material [mm]	60x0.5	60x0.5
4	Travel speed [cm/min]	15	16
5	Flux type	RECORD INT 109	RECORD INT 109

3. Results

After finishing the cladding process (figure 2), the samples were subjected to non-destructive testing (visual, penetrant and ultrasound testing – figure 3) [17-19]. After performing the examinations, no imperfections were detected, the piece being subjected to a heat treatment of stress relieving in order to minimize residual stresses in the structure. The heat treatment was performed with the following parameters: heating temperature 650-680°C, heating rate maximum 60°C/h, cooling rate maximum 60°C/h, holding time 240 minutes. After the completion of the heat treatment, the piece was subjected to non-destructive and destructive examinations specific to cladding samples.

Figure 2. Cladding plate.

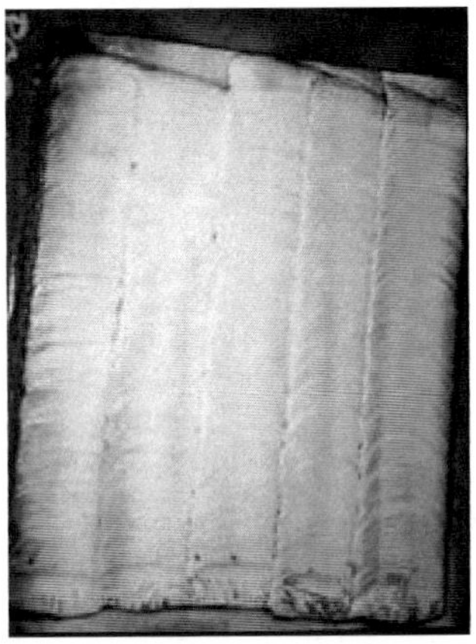

Figure 3. Penetration testing image.

In order to visualize the geometric characteristics and the structure of the welding deposition areas, the sample was cut, cleaned, degreased and subjected to the metallographic preparation process (grinding, polishing, washing, drying and etching. For a detailed analysis of the plated sample, two specimens were taken, one cut perpendicular to the cladding direction and the second parallel to the cladding direction (figure 4) [20]. Metallographic images were taken using the Olympus GX51 optical microscope. The magnification used for image capture was 10x for macrographs and x 100 for micrographic images. Following the analysis of the metallographic samples, it was found that the layer deposited in two passes has a thickness between 6.5 and 7.5 mm.

Chemical analysis of the deposited material was done at a distance of 3 mm from the top of the sample. The values of the resulting chemical composition are presented in table 5.

Table 5. The chemical composition of the deposited layer

	Weight Percent							
Materials	C	Si	Mn	P	S	Cr	Mo	Ni
Deposit material	0.023	0.682	1.45	<0.005	0.0015	19.0	0.364	9.77

Following the analysis of the values presented in table 5, it can be seen that they differ from the values of the filler material. The values obtained depend on the dilution of the welding process but also on the number of layers deposited.

(a) x10 (b) x10

Figure 4. Macrographic samples: (a) – perpendicular to the cladding direction; (b) - parallel to the cladding direction.

From the analysis of the micrographic images it can be observed that the structure of the basic material is a ferrito-perlitic one, with plastic deformation strings and small metallic inclusions (figure 5(a)). The heated affected zone (HAZ) has a ferritic-pearlitic with bainite structure grain size 2-3 and normalizing grain size 10. In the area of the deposited material a typical austenitic casting structure can be observed.

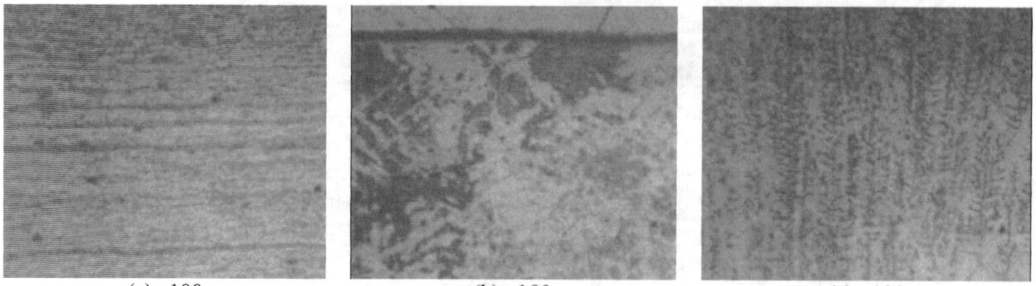

(a) x100 (b) x100 (c) x100

Figure 5. Resulted welded joints: (a) – base material; (b) – HAZ; (c) – cladding zone.

Following the analysis of the results of the additional destructive tests (side bend, ferrite content, Vickers hardness test HV10) necessary for the qualification of the welding cladding technology, the sample was declared compliant in relation to the specific test standards and a procedure qualification record and a welding procedure specification were issued, necessary for use in technology developed in production. Figure 6 presents images during welding cladding an elliptic head of the reactor made from the SA 387 Grade 11 Class 2 base material. The cladding was made using the parameters of certified technology.

Figure 6. Welding cladding of the elliptic head made using the certified technology.

4. Conclusions

The use of welding cladding technologies combines the advantages offered by different classes of materials to meet the ever-increasing demands of industrial technological processes. In this case, the high mechanical properties of the non-alloy material are doubled by the anticorrosive properties of the stainless material.

The dilution of the welding process significantly influences the mechanical properties of the deposited layer. In order to obtain properties as close as possible to those of the filler material, it is recommended to use a dilution as small as possible or to deposit several layers of material.

Considering the nature of the cladding process used, it can be seen that the structure of the material deposited by cladding is a typical austenitic casting. It is recommended to use a low linear energy to avoid overheating of the material and to reduce the risk of excessive growth of grains in the HAZ.

5. References

[1] Bin L et al. 2018 *IOP Conf. Ser.: Mater. Sci. Eng.* **307** 012035.
[2] Chattopadhyay P, van der Mee V and Zhang Z 2019 Hybrid electroslag cladding (H-ESC): an innovation in high speed electroslag strip cladding, *Welding in the World* **63** 663–672.
[3] W Fan et al. 2020 *Mater. Res. Express* **7** 046513.
[4] Mandina M, Magnasco M, and Zappavigna G 2014 The electro-slag cladding process: Application, main metallurgical and testing issues in the fabrication of reactors in large thickness 2Cr–1Mo–V. *Weld. Int.*, **28**, 380–396.
[5] Kolařík L, Kolaříková M and Janovec J 2016 Electroslag cladding of stainless austenitic layer by strip electrode, *Metal 2016*, May 25th - 27th, Brno, Czech Republic, EU.
[6] Pak S, Rigdal S, Karlsson L and Gustavsson, A C 2013Electroslag and submerged arc stainless steel strip cladding *Anti-Corros. Methods Mater.* **45** pp 41–47.
[7] Vitaliy Ivanov and Elena Lavrova 2018 *J. Phys.: Conf. Ser.* **1059** 012020.
[8] Prasad V V S, Rao A S Prakash U, Baligidad R G 2002 *Sci. Technol. Weld. Join.* **7** 102–106.
[9] Zhang Fulong et al. 2020 *Mater. Res. Express* **7** 076501.
[10] Ustundag O et al. 2019 Improvement of filler wire dilution using external oscillating magnetic field at full penetration hybrid laser-arc welding of thick materials *Metals* **9**.
[11] Kuo I C, Chou C P, Tseng C F and Lee I K 2009 Submerged Arc Stainless Steel Strip Cladding - Effect of Post-Weld Heat Treatment on Thermal Fatigue Resistance, *JMEPEG* **18** 154–161.
[12] Qian S, Dai Y, Guo Y, Zhang Y 2021 Microstructure and wear resistance of multi-layer Ni-based alloy cladding coating on 316L SS under different laser power *Materials* **14**.
[13] Chivu O et al. 2016 The effects of reconditioning by welding of crankshafts in automotive industry *Metalurgija*, **55** 55-58.
[13] Cicic DT, Rontescu C, Amza C G and Chivu OR 2015 The combined effect of chemical elements on the properties of a layer deposited by welding, *Revista de chimie*, **66** 1299-1301.
[14] ASME II Part A:2019 – *Ferrous material specifications*.
[15] EN ISO 14343:2017 *Welding consumables - Wire electrodes, strip electrodes, wires and rods for arc welding of stainless and heat resisting steels - Classification*.
[16] EN ISO 14174:2019 - *Welding consumables - Fluxes for submerged arc welding and electroslag welding – Classification*.
[17] EN ISO 17637:2016 - *Non-destructive testing of welds - Visual testing of fusion-welded joints*.
[18] EN ISO 3452-1:2013 - *Non-destructive testing - Penetrant testing - Part 1: General principles*.
[19] EN ISO 17640:2017 - *Non-destructive testing of welds - Ultrasonic testing - Techniques, testing levels, and assessment*.
[20] EN ISO 17639:2003 *Destructive tests on welds in metallic materials - Macroscopic and microscopic examination of welds*.

ModTech 2021 IOP Publishing
IOP Conf. Series: Materials Science and Engineering 1182 (2021) 012009 doi:10.1088/1757-899X/1182/1/012009

Drag Based Wind Turbine Lenz Type Manufacturing and Assembly

I O Bucur[1], I Malael[1], B Duran[2] and D Preda[2]
[1]Romanian Research and Development Institute for Gas Turbine - COMOTI, CFD Department, Iuliu Maniu Boulevard, No. 220D, 061126, Bucharest, Romania
[2]S.C. Rolix Series Impex SRL, Basarabiei Blvd, nr. 256, District 3, Bucharest, Romania

E-mail: ioana.bucur@comoti.ro

Abstract. Due to the current technological evolution trend, the use of renewable energies seems to be the best suited solution for the environment protection against pollution. This paper aims to present the methodology used for the manufacturing and assembling processes of a drag based vertical axis wind turbine, the steps for both processes being addressed broadly. The manufactured vertical axis wind turbine can successfully function in wind conditions with speeds of up to 16m/s and manages to provide up to 5kW of power. The manufacturing materials used for the execution of the blades include metal sheets, mounted on a structural frame for better stiffness. The wind turbine rotor is comprised of three blades, that are installed on a shaft connected to a gear box which will transfer the torque to the permanent magnet generator. The whole process is described in the paper and the result consists of the installed turbine, that is ready to function in real environment operating conditions. The work carried out within this paper is relevant to the general know-how regarding wind turbine manufacturing and installation, as the authors highlight the main impediments they had to overcome when developing, executing and installing the described model.

1. Introduction

Considering the worldwide energy scenario, it is safe to assume that in the next years most countries will face massive attempts to progress from common energy sources to renewables. This statement is reinforced by international major steps in the direction of green energy, supported by programs such as the European Green Deal [1] or by documents as the Paris Agreement [2]. Wind energy represents the second most efficient renewable source of electrical power, according to the data provided in the "Renewables 2019 Global Status Report" [3]. The extraction of kinetic energy from the wind and its transition into mechanical or electrical energy is realized using specifically designed systems for this task. Such systems are commonly called wind turbines and are divided into two large categories according to the direction of the rotating shaft: horizontal axis wind turbines (HAWTs) and vertical axis wind turbines (VAWTs). The two are compared in [4], the main highlights being that HAWTs can produce larger amounts of energy compared to VAWTs, but they need a specific work environment (isolated areas with high wind speeds), whereas VAWTs are more efficient in low wind velocities and in turbulent flows, being suited for instalment in urban areas and adjacent zones.

In order to improve the efficiency of VAWTs and to help them surpass the performances of HAWTs, new configurations have been developed and tested. Most common configurations include

Savonius, Darrieus and H-Darrieus wind turbines and derived from these, new geometries have been developed, such as Crossflex wind turbine, combined Savonius-Darrieus rotor, Zephyr turbine or Lenz [5]. This paper describes the manufacturing process of the latter mentioned configuration, emphasizing on the used methodology, materials, and equipment.

The Lenz wind turbine geometry is derived from a combination between the Savonius and Darrieus type and is better suited for functioning in reduced wind speeds. The advantages of a Lenz type VAWT include low cost fabrication, as its design is not very complicated, reliability, improved starting behaviour and good performances in low winds [6]. A mathematical model for calculation activities that present as an outcome the geometric parameters for a Lenz turbine is described in paper [7]. The design procedure for this type of turbine is discussed and future work includes the manufacturing of such a model, with applications in urban areas. In paper [8] a Lenz VAWT with three blades was designed and manufactured using aluminium sheets. The system was validated through an experimental study. Furthermore, the model was tested with and without deflector vanes, used as a method to control air flow direction and optimize the baseline geometry. The study concluded that a deflector system can improve the overall performances of the turbine, when installed properly, the angle of instalment having a great influence on the turbine's efficiency. Some experimental campaigns, such as the one presented in [9], study the blade number influence for the discussed configurations. The authors design and manufacture a five-bladed Lenz wind turbine, using aluminium sheets and fiberglass. Sivamani et al. experimentally investigate a two-stage Lenz wind turbine with three blades, for a velocity range from 5 to 7 m/s. As results, they discuss power and moment coefficient variations [10]. The model investigated was manufactured using a solid mild steel material for the shaft and disks, whereas for the blades plywood and aluminium sheets were used.

The most common materials for the manufacturing of the blades for the discussed turbine geometry include thin metallic sheets (often aluminium alloys sheets) and steel for the other components, as can be concluded from the discussed papers. The fabrication methods frequently include cutting processes and rolling.

In this paper, for the 3D modelling of the Lenz turbine and final assembly, the SolidWorks software is employed, as this CAD tool is largely used for the modelling of various wind turbine configurations, such as HAWT [11], Savonius [12], H-Darrieus [13] or even combined geometries, such as Darrieus-Savonius, as discussed in paper [14]. In the following sections the 3D model for the Lenz type VAWT is presented, as well as the design of the full assembly. Then the methodology for its manufacturing is detailed, including the technological methods used, materials needed and necessary equipment. In the end the full assembly is presented.

2. 3D Model of the Lenz turbine and the assembly

This section will briefly present the Lenz wind turbine integrated in a fully operable assembly, discussing the main parts of the construction. Firstly, the 3D model of the blades was realized using the SolidWorks CAD software, after evaluating the geometrical parameters of the turbine according to the mathematical model mentioned in the introduction. The blade consisted of two lids (figure 1(a)), one disposed at each end of the blade; two ribs used to connect the turbine's arms to the blades (figure 1(b)); four rods designed to stiffen the blade (figure 1(c)) and, finally, the blade structure with a thickness of 0.5mm (figure 1(d)). The components are illustrated in figure 1.

a) Lid b) Rib

c) Rod

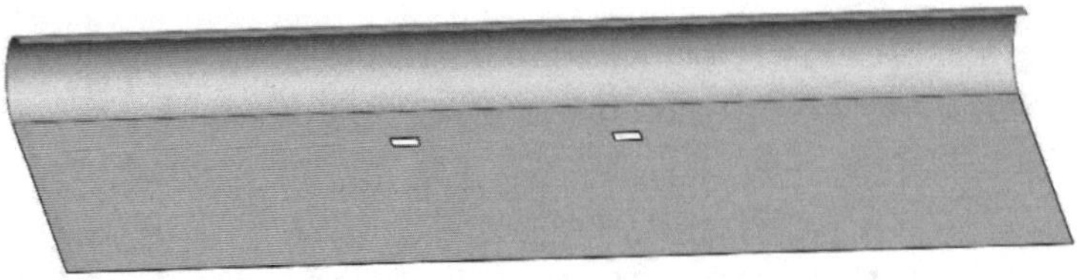

d) Blade structure

Figure 1. Lenz blade components.

The final assembly for the blade is represented in figure 2, where the elements mentioned above can be easily identified. The blade's height was 3m, whereas the turbine's diameter was 2.8m. For the purpose of connecting the blades to the shaft, an arm system was designed. The arms were effectuated using a rectangular pipe of 80x40 mm. As the blade is fixated in two different points, using the ribs described previously, the arms were disposed in the manner depicted in figure 3.

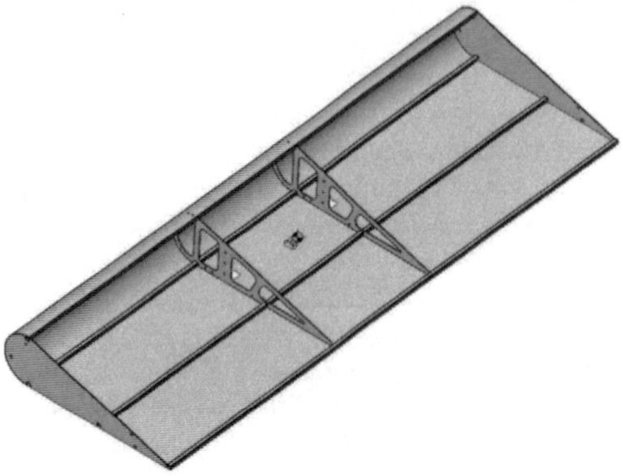

Figure 2. Lenz blade assembly.

Figure 3. The arm system used for supporting the blades.

The blades mounted on the on the arm system are illustrated in figure 4.

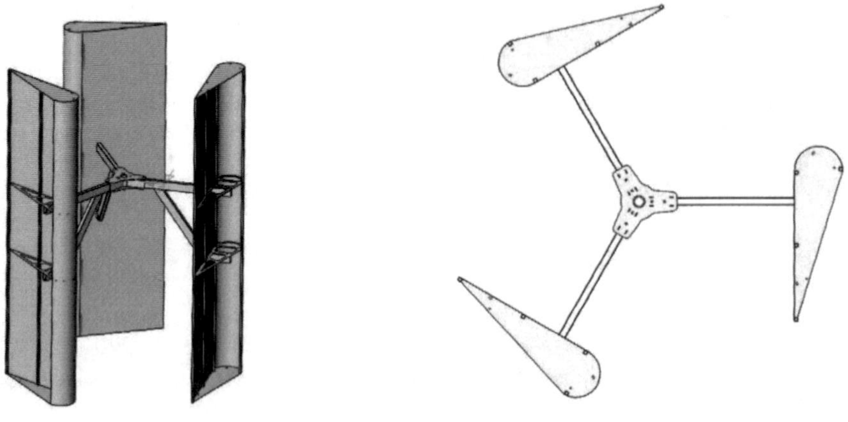

Figure 4. Blades installed on the arm system.

IOP Conf. Series: Materials Science and Engineering 1182 (2021) 012009 doi:10.1088/1757-899X/1182/1/012009

In figure 5, the complete assembly for the Lenz wind turbine is illustrated. The full assembly mainly consists of: a support tower (s5), which for better backing is reinforced with four metal tie beams (s6) connected to the main pillar by four metallic poles (s4); the structure of the turbine – the turbine's upper tower (s3), the generator's box (s8), the arms of the blades (s7) mounted into a stand (s9) and the blades (s1). In order to ensure the blades' stability while functioning, they were secured with metallic beams (s10). As it can be observed in figure 5(c), the main pillar has three components: the base supporting tower (s5), the big shaft (s2) and the small one (s3). The integrity of the whole structure is, as prior mentioned, secured using four additional tie beams.

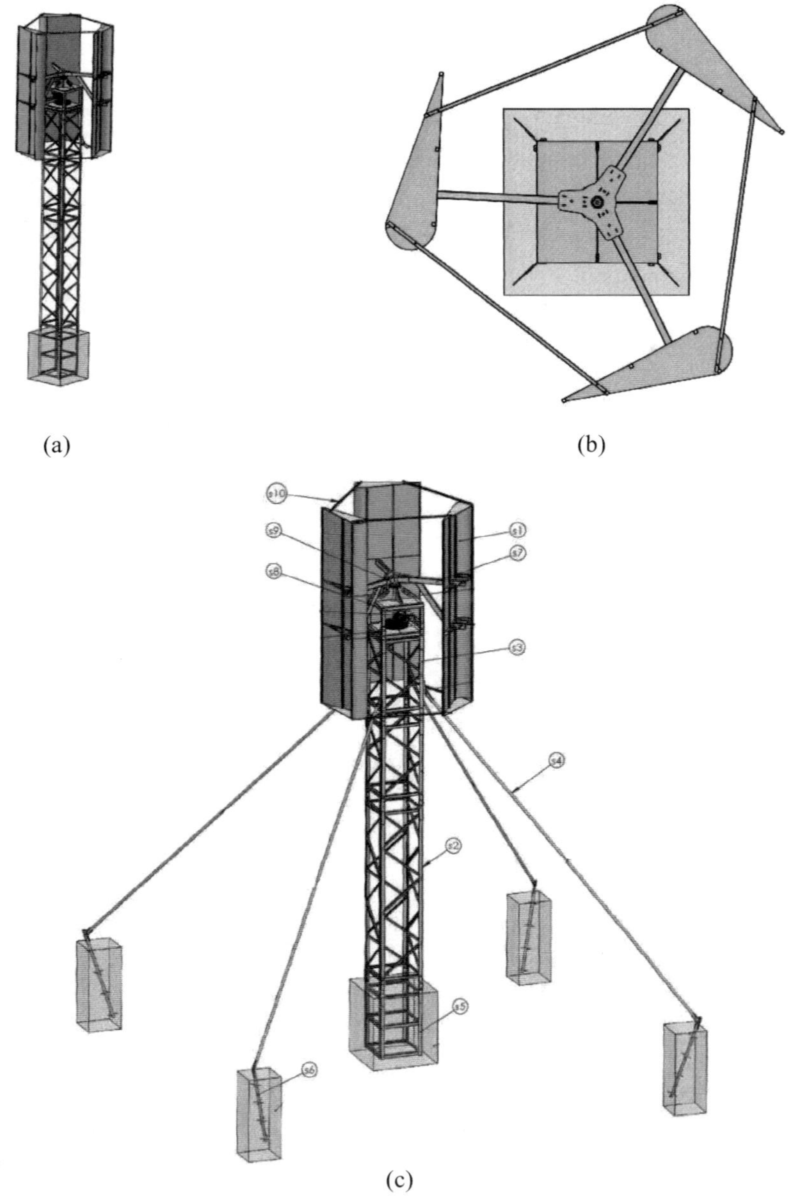

Figure 5. Lenz wind turbine model and assembly.

3. Manufacturing process

The main used materials for the manufacturing of the discussed metallic construction include rectangular pipes – S275JR steel, square pipes – S275JR steel, laminated U profiles – S275JR steel, bended U profiles – S275JR steel and laminated L profiles – steel S275JR.

Some fundamental methods used while manufacturing the wind turbine and the other components of the full assembly incorporate cutting processes, metal machining and welding.

The manufacturing process began with the cutting of the semi-fabricates, using a plasma cutting plant, as illustrated in figure 6(a) and other metal machining processes such as turning (figure 6(b)), milling (figure 6(c)) and drilling (figure 6(d)).

c) Milling process d) Drilling process

Figure 6. Metal machining processes employed in the manufacturing of the semi-fabricates for the Lenz wind turbine.

The specimens obtained as a result of the specified machining processes might present some excess material remains that should be removed. For the deburr process the selected finishing manufacturing process was vibratory finishing, as it was a rapid and low cost solution. In order to permanently join necessary parts electric resistance welding and pressure welding were used, as illustrated in figure 7.

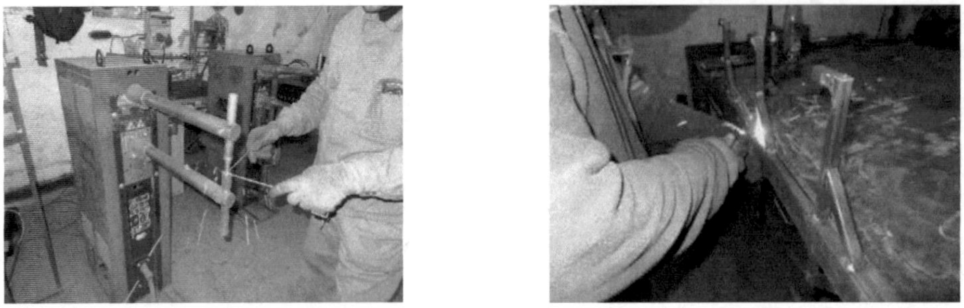

Figure 7. Pressure welding (left) and electric resistance welding (right).

A protective zinc coating was applied to the manufactured parts in attempt to prevent rusting. Some parts after the galvanization process are figured in the figure 8.

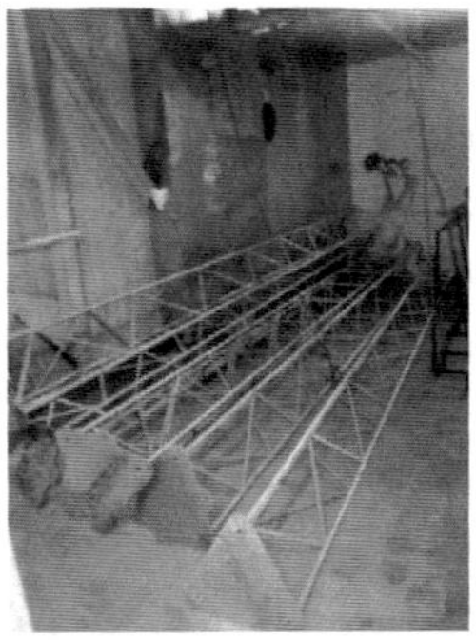

Figure 8. Turbine towers (left) and blade (right) after applying the protective zinc coat.

The manufactured blades and parts for the upper part of the main pillar are presented in figure 9 before and after painting.

Figure 9. Blades and upper parts from the main pillar before and after the painting procedure.

A generator was designed and manufactured, the result being depicted in figure 10. It was placed on the superior part of the designed supporting pillar, as discussed in section 2 and illustrated in figure 5, in the mentioned chapter.

Figure 10. Assembled generator.

Apart from the main components listed and discussed above, a wind turbine system also contains some fundamental electronical devices. The essential electronics used for the developed Lenz wind turbine include an electronic tool for data control (figure 11(a)), a power source (figure 11(b)), an electrodynamic brake provided with a voltage switch module from AC to DC (figure 11(c)), a GPRS (General Packet Radio Services) module (figure 11(d)) which allows the monitorization of the mechanical parameters of the wind turbine (example: tower vibration), a main control data module (figure 11(e)), a temperature sensor for the generator (figure 11(f)), and a weather module (figure 11(g)). All the electronics are illustrated in figure 11.

Figure 11. Lenz wind turbine electronics.

The weather module incorporates sensors for the supervision of the following parameters: wind speed, wind direction, precipitations, humidity, and air temperature.

4. Results

After manufacturing all the parts and applying the necessary mentioned treatments, testing the generator, and revising all the electronics in order to ensure they operate properly, the final assembly was done. The result is presented in figure 12.The foundation for the instalment of the main supporting tower, backed with the four tie beams was represented by a 25m^2 area. For the placement of the Lenz wind turbine on the pillar a crane is used, as pictured in figure 13.

Figure 12. Lenz turbine assembly.

Figure 13. Crane lifting the top of the wind turbine.

The final result can be observed in figures 14 and 15, respectively. The latter figure provides a closer look at the upper part of the wind turbine, and the generator together with the weather station can be noticed.

Figure 14. Lenz wind turbine installed.

Figure 15. Lenz wind turbine – closer look.

5. Conclusions

This paper displays in detail the main stages needed for the development, manufacturing and installing of a Lenz vertical axis wind turbine. Firstly, the 3D modelling of the wind turbine is detailed, followed by the manufacturing methods and in the end the assembly process is described. The used materials and equipment were provided in the manufacturing section. The result consisted in the assembled and installed Lenz wind turbine, capable of generating up to 5kW of power. The metallic structure was provided with a lighting protecting system and the tower was connected to a certified electrical grounding. This paper contributes to the common expertise regarding the manufacturing and instalment of wind turbines, as it concisely depicts the phases included in such a broad process.

6. References

[1] *EU Climate action and the European Green Deal.* Available online: https://ec.europa.eu/clima/policies/eu-climate-action_en (accessed on 2/02/2021).

[2] *Paris Agreement* 2015, Paris, France. Available online: https://unfccc.int/documents/9097 (accessed on 2/02/2021).

[3] *Renewables 2019 Global Status Report*, Paris: REN21 Secretariat, 2019. ISBN 978-3-9818911-7-1.

[4] Johari M K, Jalil M A A and Shariff M F M 2018, Comparison of horizontal axis wind turbine (HAWT) and vertical axis wind turbine (VAWT), *International Journal of Engineering & Technology* **7** 74-80.

[5] Bhutta M M A, Hayat N, Farooq A U, Ali Z, Jamil S R and Hussain Z 2012, Vertical axis wind turbine – A review of various configurations and design techniques, *Renewable and Sustainable Energy Reviews* **16** 1926-1939.

[6] Nishioka A H and de Almeida O 2018, Study, Design and test of a LENZ-type wind turbine, *International Journal of Advanced Engineering Research and Science* **5**.

[7] Gohil H P and Patel S T 2014, Design procedure for Lenz type vertical axis wind turbine for urban domestic application, *International Journal for Scientific Research & Development* **2.**

[8] Deori B, Barman S, Das S, Hussain M, Basumatary S M and Sharma K K 2015, Experimental Study on the Performance of Lenz Vertical Axis Wind Turbine, *Journal of Material Science and Mechanical Engineering* **2** 62-64.

[9] Mongkunkeaw T, Kotpai S, Rachsale N, Nakarungsu S, Thungsuk N, Yuji T and Chansri P 2015, A Lenz Wind Turbines Five Blades for Produced Electricity, *2nd Asian Conference on Electrical Installation & Applied Technlology*.

[10] Sivamani S, Premkumar M T, Sohail M, Mohan T and Hariram V 2017, Experimental data on load test and performance parameters of a Lenz type vertical axis wind turbine in open environment condition, *Data in Brief* **15** 1035-1042.

[11] Hosseini S F and Moetakef-Imani B 2017, Innovative approach to computer-aided design of horizontal axis wind turbine blades, *Journal of Computational Design and Engineering* **4** 98-105.

[12] Belmili H, Cheikh R, Smail T, Seddaoui N and Biara R W 2017, Study, design and manufacturing of hybrid vertical axis Savonius wind turbine for urban architecture, *Energy Procedia* **136** 330-335.

[13] Ferroudji F, Khelifi C and Meguellati F 2016, Modal Analysis of a Small H-Darrieus Wind Turbine Based on 3D CAD, FEA, *International Journal of Renewable Energy Research* **6.**

[14] Ferroudji F, Khelifi C, Meguellati F and Koussa K 2017, Design and Static Structural Analysis of a 2.5kW Combined Darrieus-Savonius Wind Turbine, *International Journal of Engineering Research in Africa* **30** 94-99.

Acknowledgments

This work was carried out within POC – Competitiveness Operational Program, supported by the EU and Romanian Minister of Research and Innovation funds, project number POC 9/01.09.2016, MySmis 105890, ID P_40_309.

IOP Conf. Series: Materials Science and Engineering 1182 (2021) 012010 doi:10.1088/1757-899X/1182/1/012010

Analysis of Carbon Fibers Treatment Technology to Obtain Composites Materials with Metal and Non-Metal Matrix

R Caliman
"Vasile Alecsandri" University of Bacau, Faculty of Enginering, Department of Engineering and Management, Mechatronics, Calea Marasesti 157, Romania

E-mail: rcaliman@ub.ro

Abstract. When designing any composite material, the compatibility between the component elements must be considered, a compatibility that can be seen from a physical and chemical point of view. Chemical compatibility refers to the existence or development to a small extent of reactions between components. Thus, at high temperatures, the diffusion processes intensify, and fragile compounds can form, which cancel the direct connection between the components, resulting in a significant decrease in the mechanical strength of the composite material. A successful process of manufacturing carbon fiber-reinforced composites requires that the fiber to be protected, usually with a coating, during their manufacture and use. The paper aims to analyze the process of Chemical Vapor Deposition (CVD) which is a successful process of manufacturing carbon fiber reinforced composites. They require that the fiber be protected with a coating during their manufacture and use.

1. Introduction

Complementary materials, used to reinforce the energy of the matrix or to induce the self-lubricating property of the composite material, differ from each other in chemical nature and configuration [1-4].

According to the configuration, the complementary materials are divided into two main categories: fibers and particles, each category including numerous other types, differentiated by size, by length / diameter ratio and by chemical composition in cross section [5-8].

Compared to fibers, particles are easier to make and incorporate into the matrix material. Instead, the fibers are irreplaceable if the aim is to obtain a composite with high toughness [9].

The fibers are used as reinforcing elements, having the role of taking over a large part of the stresses to which the matrix material is subjected. Depending on the nature of the matrix and the purpose pursued, the fibers are made of organic substances, metals, ceramics or pairs of such materials, of different shapes and sizes [10-15].

An important criterion for the classification of fibers (figure 1) is the ratio between length and diameter. Depending on the structure, the fibers can be polycrystalline, monocrystalline or amorphous.

2. Fiber production techniques

Fiber production methods have developed in parallel with the growing interest in these materials. Although there is a wide variety of manufacturing processes, current technologies can fall into the following three main distinct categories [16,17]:

- *melt extraction,* used to produce 50-60% of glass and alumina-silica fibers, or melt extraction, used to obtain fibers from amorphous alloys.

- *solid state transformation,* a method used in the production of carbon fibers, silicon carbide, alumina-silica, or zirconium oxide, both for the continuous ones (multifilaments) and for the usual discontinuous ones. The process is based on the use of a precursor in the form of fibers, transformed after a heat treatment into the material to be obtained in the end.

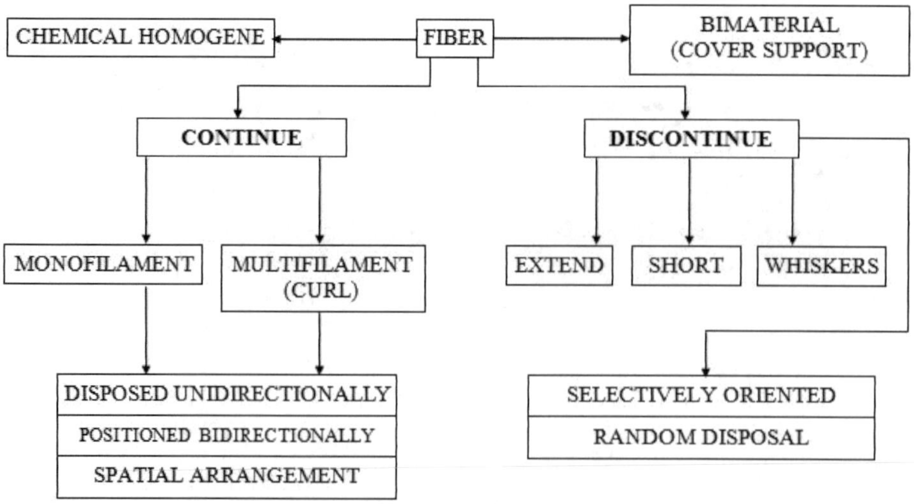

Figure 1. Classification of fibers used in the production of composite materials [14].

- *vapor growth,* a method used to produce very short (monocrystalline) fibers or continuous monofilament fibers. Monofilament fibers are obtained by depositing a material on backing fibers (such as depositing silicon carbide or boron on tungsten or carbon backing fibers) [17].

Carbon fibers have become popular lately due to their many advantages: outstanding mechanical properties, low cost, high temperature stability, good chemical compatibility with organic matrices and the possibility of using a wide range of raw materials. The term carbon fibers mean fibers that contain more than 80% C in the form of amorphous carbon and graphite and have a density between 1.50 and 1.96 g / cm^3 (the density of graphite being 2.265 g/cm^3), which means the existence of a porosity of 16.5 ... 18%. The porosity is created by microcavities elongated much in the direction of the fiber [18-20].

3. Surface treatments specific to the preparation of carbon fiber, to obtain polymeric composites.

The basic physical parameters that influence the formation of the matrix-fiber bond are the specific surface and the degree of surface roughness, both parameters characterizing the surface topometry and effectively determining the contact surface between the matrix and the fiber.

The chemical composition of the surface, in the case of carbon fiber, is different from that of the interior, so that at the surface are present oxygen atoms (in variable numbers), up to a depth of 3 μm and hydrogen atoms. Thus, in order to obtain an adhesion between fiber and resin, it is necessary to obtain a physico-chemical reactivity at the interface, simultaneously with the achievement of optimal compositions of matrix materials and their modifications, at sintering [21].

Theoretically, a tight adhesion of the fiber-matrix is achieved by advanced wetting of the fiber surface with the liquid resin.

Carbon fibers, even in the case of high-performance ones, cannot be used as such to reinforce polymeric (or metallic) matrices. For carbon fibers, the means by which good adhesion can be achieved is their surface oxidation treatment [22].

The most important methods for such treatments are [23]:

- electrochemical oxidation;
- dry oxidation;
- wet oxidation.

Anodic oxidation allows the strict maintenance of the necessary oxidation conditions, as well as a rigorous control, without difficulties, on the speed of the process.

Most experiments on the electrochemical oxidation of carbon fibers aim to produce physico-chemical changes at the surface, which ensure the removal of some of the surface defects and, at the same time, to generate functional groups on the surface of the fibers [24].

Anodic oxidation of carbon fibers is performed under "mild" conditions, which do not produce essential morphological changes on their surface. The improvement in adhesion between oxidized carbon fibers and the polymeric matrix can be explained by the formation of chemical bonds, by physical adhesion or mechanical interactions. The reaction media in which the anodic oxidation is carried out are determined by the subsequent applications of the composites made with these fibers. Thus, solutions of various concentrations of sulfuric acid, sodium hydroxide or potassium sulphate may be used. Treatment time is also variable, from 1 minute to 10 minutes.

The dry oxidation of carbon fibers is performed in air or oxygen, at temperatures between 500 ÷ 800° C. The dry oxidative treatment time varies in a wide range of values (from 30 seconds to 2 hours), depending on the nature of the fibers and their subsequent applications.

The third process, that of wet oxidation, is the most used, because it does not require the existence of complex treatment facilities. The solutions used in this process are generally inorganic acids having oxygen atoms in their molecular structure. Thus, we currently work with nitric acid of various concentrations. The duration of wet oxidative treatments is in the range of 10 minutes ÷ 150 hours.

The improvement of the mechanical properties of oxidized carbon fiber composites is due to the chemical activation of the fiber surface.

The mechanical properties decrease in size due to damage to the carbon fiber during thermal oxidation. The maximum mechanical strength of the composite is obtained after short oxidations, while for long oxidation times, the effect of fiber damage is predominant, leading to the impossibility of embedding in the matrix.

The analyzes performed on the treated carbon fibers, in order to determine the structural changes occurred, showed that, in the case of wet oxidation, acid and non-acid groups are formed on the surface of the component filaments of the polyfilament carbon cable. On dry oxidation the formation of non-acidic groups was confirmed.

In the case of anodic oxidation, it is found that appropriate results are obtained, if working under the conditions of the transition range between non-corrosive and corrosive treatment.

All the surface treatments applied to the carbon fibers presented above, lead to the formation of reactive oxide groups on the surface of the fibers. The improvement of the fiber-matrix adhesion can be directly correlated with the increase of their concentration at the surface of the fibers.

4. CVD coatings applied to carbon fibers intended for metallic composite materials.

Among the new composite materials, metal matrix composites reinforced with carbon fibers have experienced a special development, given the possibilities of use they have in various fields. A successful process of manufacturing carbon fiber-reinforced composites requires that the fiber be protected, usually with a coating, during their manufacture and use.

Chemical Vapor Deposition (CVD) is a modern technology for fiber coating. The experience gained so far shows that there can be large differences between the theoretical and real values that have the properties of a composite material. The results of the experiments show that the mechanical strength is much lower than expected.

The reason for such a difference is, in part, the result of the fiber-matrix interaction. At high working temperatures, the fiber can react with the matrix, losing some of its properties. Residual stresses as well as thermal instability of fibers can also contribute to this phenomenon.

In order to keep the fibers intact during processing and to produce the desired surface properties, the process by which the composite material is manufactured almost always comprises a process by which the fiber is coated. The coating protects the fiber during the operation in which the fiber is combined with the matrix. Simultaneously, the wetting and binding of the fiber with the matrix is favored; In some cases, the coating is a sacrificial layer, which is destroyed during processing. In general, the structure, composition and morphology of the coating affect the strength of the composite after manufacture.

Chemical vapor deposition (CVD) and its variants are used for fiber coating. A typical coating has a thickness of about 2 μm. Figure 2 shows a hot-walled vertical reactor for fiber coating. The installation contains the reactive gas supply system and the deposition cell, into which the carbon fiber coating substrate is introduced.

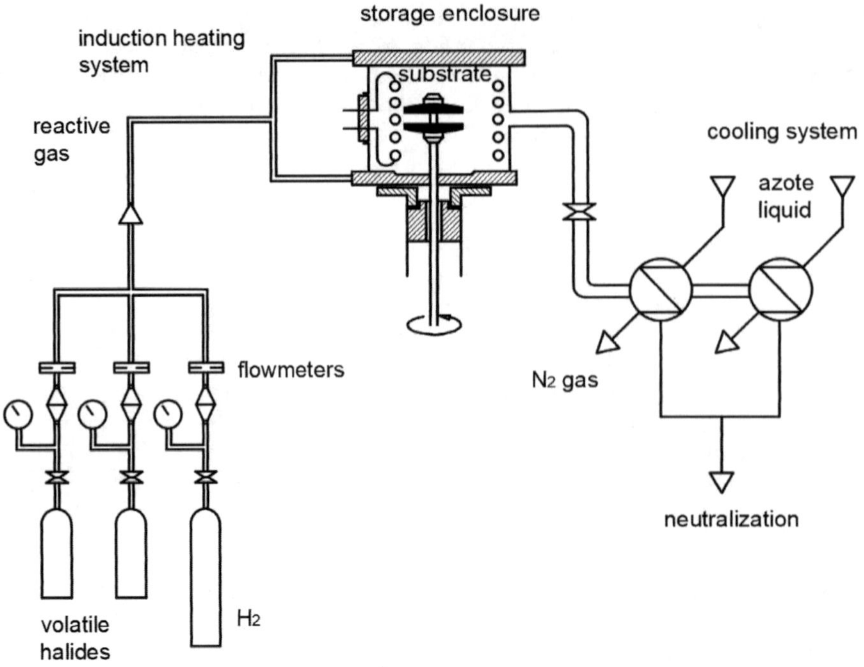

Figure 2. Scheme of the chemical vapor deposition plant [4].

An obvious problem is the oxidation of carbon fibers during and after processing; the phenomenon can be ameliorated by coating the fibers with carbides or metal oxides. The most successful carbon fiber reinforced composites were produced from pyrocarbon coated fibers, SiC or BN, by a CVD process.

The CVD process can be defined as the transport of solid particles by vapors, the chemical reactions being inherent in this process.

An example of a fiber coating CVD reactor is the cylindrical reactor through which the fiber is moved as gaseous reactants circulate around it. The gases and the surface of the fiber are heated to the reaction temperature. A hot-walled reactor can be used, as in figure 2, or the fiber can be electrically heated. The reactions occur in the gas phase and several chemical species can form. Thermodynamic equilibrium may or may not be reached in CVD reactions. If homogeneous nucleation occurs in the gas phase, solid nuclei will result. Such nucleation products (sometimes called "snow") are usually unwanted in the deposit on the layer.

The reaction products diffuse to the surface, producing surface reactions, which lead to the formation of a deposit on the surface of the fiber. The limiting factor in the deposition process may be the reaction kinetics or the diffusion of vapor species to the surface. At higher temperatures (above 1300° C for xylans), the process tends to be controlled by diffusion. Low temperature deposition rates are limited by reaction kinetics. Some reactors may operate in a transient mode, in which the deposition rate is also determined by the reaction kinetics and vapor diffusion.

5. Conclusions

Due to the current demand for advanced materials, the use of materials with a monolithic structure and, implicitly, of metal matrix composites has been widely used, especially for those applications that require a high specific resistance to high temperatures or chemical attack.

The Chemical Vapor Deposition process is the preferred method for coating fibers, due to its advantages like low deposition temperatures and high purity coatings. Various microstructures can occur, deposits can be made on felt or bundles thus the induced thermal stresses are low.

Among the critical parameters of the coating process we mention: temperature distribution in the reactor, reactant concentration and pressure, gas velocity distribution, system geometry.

In addition to these critical parameters, they can also be mentioned: fiber feed speed, reaction kinetics in gases and at the fiber surface, thermodynamics of gas phase reactants and solid products and gas transport coefficient such as diffusivity and viscosity.

6. References

[1] Shao G, a.o. 2019 Mechanical properties of graphene nanoplates reinforced copper matrix composites prepared by electrostatic self-assembly and spark plasma sintering *Materials Science and Engineering* **739** 329-334.

[2] German R M 2004 *Powder Metallurgy* no. 8.

[3] Zang G-H a.o. 2019 Microstructures and mechanical properties of alumina whisker reinforced copper matrix composites prepared by hot-pressing and hot isostatic pressing *Materials Research Express* **6**(11) 116513.

[4] Ispas S 1987 *Composite materials* (in romanian), Ed. Tehnica.

[5] Liu Q, a.o. 2019 Effects of morphological characteristics of graphite fillers on the thermal conductivity of the graphite/copper composites fabricated by vacuum hot pressing sintering *Vacuum* **167** 199-206.

[6] Piatti G 1982 Benefits in the field of Composite Materials (translation from English into Russian) *Metallurgia* Moscova 42 – 58.

[7] Berghezan A 1984 *Syllabus of constructions and personalities in the preparation of compositional materials with metallicevskoi matrici*, Laboratoire de metallurgie physique, Belgique (transtation in Russsian) 9 – 25.

[8] Ayyappadas C a.o. 2017 An investigation on the effect of sintering mode on various properties of copper-graphene metal matrix composite *Advanced Powder Technology* **28**(7) 1760-1768.

[9] Hutchings M I 1994 *Materials Science and Technology* **10** 513 – 517.

[10] Nautiyal H a.o. 2019 Copper matrix composites reinforced by rGO-MoS2 hybrid: Strengthening effect to enhancement of tribological properties *Composites Part B: Engineering* **173** 106931

[11] Berdal A and Hoier R 1991 *Proceedings of the 12th International Symposion on Material Science in Metal Matrix Composites: Processiong, Microstructures & Properties* Ed. Hasen N. Riso Nat. Lab. Hoskilde Denmark.

[12] Nayan N a.o. 2017 Processing and characterization of spark plasma sintered copper/carbon nanotube composites *Materials Science and Engineering: A* **682** 229-237

[13] Gadea S, Petrescu M 1981 *Metallurgy and metal studies* (in romanian) Ed. Didactica şi

Pedagogica Bucharest **2** 261 – 281.
[14] Stefanik P, Sebo P. 1993 *Jornal of Materials Science Letters* **12** 1083 – 1085.
[15] Chawla K K 2005 High-Performance Fiber Reinforcements in Composites *Journal of Minerals Metals & Materials* 47.
[16] Davim JP, Marques N, Baptista AM 2001 Effect of carbon fibre reinforcement in the frictional behaviour in a water lubricated environment *Wear* Elsevier Science **251** 1100-1104.
[17] Caliman R, Olaru I 2006 An Abrasive Wear Model for Composite Materials *MOCM* 12 **1** Romanian Academy Branch office Iaşi 29-34.
[18] * * * 2002 *Carbone Lorraine Materiaux carbone-carbone*, Prospect freins.
[19] * * * 2002 *Carbone Lorraine Freinage a haute energie-Les composites carbon-carbon*. Prospect.
[20] * * * 2002 *NASCO Aircraft Brake Inc. - Original Equipment Brake and Wheel Systems*, INFO.
[21] * * * 2002 *Carbone Lorraine Airospatiale. AEROLOR Carbon - carbon composites* developed by the Gie, INFO Product Line.
[22] * * * 2002 *Carbone Lorraine Airospatiale. Carbon-carbon composites and CERAMETAL sintered metal*, Prospect.
[23] * * * 2002 *Carbone Lorraine - Materiaux frittes - materiaux de friction*. Prospect.
[24] * * * 2004 *The Automotive Application of Discontinuously Reinforced TiB-Ti Composites (Overview)*, T. Saito 33-36.

IOP Conf. Series: Materials Science and Engineering 1182 (2021) 012011 doi:10.1088/1757-899X/1182/1/012011

Study on the Kinetics of the Chemical Vapor Deposition Process for the Coating of Carbon Fibers from the Structure of Composites Materials with Metal and Non-Metal Matrix

R Caliman
"Vasile Alecsandri" University of Bacau, Faculty of Enginering, Department of Engineering and Management, Mechatronics, Calea Marasesti 157, Romania

E-mail: rcaliman@ub.ro

Abstract. In order to keep the fibers intact during processing and to produce the desired surface properties, the process by which the composite material is manufactured almost always comprises a process by which the fiber is coated. The coating protects the fiber during the operation in which the fiber is combined with the matrix. Simultaneously, the wetting and binding of the fiber with the matrix is favored; In some cases, the coating is a sacrificial layer, which is destroyed during processing. In general, the structure, composition and morphology of the coating affect the strength of the composite after manufacture. Thus, the Chemical Vapor Deposition (CVD) process and its variants are used for fiber coating. The reaction products diffuse to the surface, producing surface reactions, which lead to the formation of a deposit on the surface of the fiber. The limiting factor in the deposition process may be the reaction kinetics or the diffusion of vapor species to the surface.

1. Introduction

The matrix-reinforcing agent interface plays a key role in the manufacture of composite materials. The quality of the interface determines the structural integrity, the response to the action of the environment and the physical and mechanical properties of the composite material [1-3].

The interface includes the contact area between the matrix and the fibers and the immediately adjacent domain. The nature and strength of the interfacial bond depend on the structure, the surface characteristics of the fibers (roughness, specific surface area, porosity, crystal size, the presence of chemically active functional groups) and the structural characteristics of the matrix (chemical composition, macromolecule conformation) [4-7].

The literature presents different models of adhesion which, alone or combined, provide a satisfactory interpretation of the phenomena at the interface. The adhesion between the matrix and the fibers was explained in terms of mechanical and thermodynamic adhesion, chemical compatibility, by electrostatic attraction forces and based on the phenomenon of macromolecular interdiffusion [8].

The interdependence between maintenance technology and surface phenomena.

From a practical point of view, complex, interdependent correlations between the interactional aspects, properties and compositional microstructure must be analyzed starting from the technology of maintenance and processing of the composite material [9].

In composites obtained by technologies based on the coupling of solid state components (by powder metallurgy, by diffusion), the interfacial effects have a cumulative influence on the properties and do not contribute directly in the different stages of the technological process. The initial stages of preparation, mixing and preconsolidation of the components in powder form influence the interface through the surface quality of the powders, the adsorbed or deposited layers [10].

In composites obtained by technologies derived from liquid matrix casting processes, the interfacial effects are influenced, to a greater extent and directly, by the following factors: the chemical composition of the matrix, the wetting conditions of the complementary material by the matrix expressed by contact angle, interphase voltage balance, component reactivity, temperature, processing atmosphere, solidification conditions [11].

2. Chemical Vapor Deposition-CVD

Chemical vapor deposition is a well-known advanced technology that consists of the deposition of a solid on a heated surface, as a result of a chemical reaction in the gas phase, which occurs near the surface of the substrate. Figure 1 shows a vertical walled reactor for fiber coating [12].

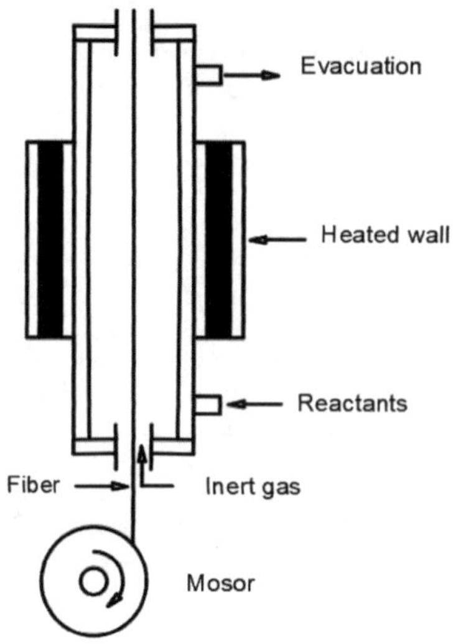

Figure 1. Warm wall reactor for carbon fiber coating.

2.1 Kinetics of the CVD process

As a result of the chemical interactions between the matrix and the complementary material, a reaction zone with non-uniform thickness is formed at the interface, with composition and structure determined by the diffusion processes of the elements [13].

Quantitative estimation of the thickness of the reaction layer is difficult, as it depends on specific microstructural parameters (eg structural defects interrupt the diffusion circuit, leading to an uneven thickness of the reaction zone), as well as the processing mode and parameters of the composite material [14].

The thickness of the reaction layer - x, has a parabolic increase as a function of time - t, dependence expressed by the equation (1):

$$x = k \cdot t^{1/2} \tag{1}$$

where k is the parabolic growth rate constant of the reaction layer.

When achieving optimal coatings, it is important to predict (or control) the coating speed and coating distribution, as well as the composition and microstructure of the coating.

Coverage speed is a macroscopic parameter. The composition and microstructure of the coating will govern the interaction with the matrix material, during manufacture and after its realization. The presence of grains of irregular shape and size, cracks or inclusions in the coating, will tend to reduce the effectiveness of the coating in the final product [15].

In order to achieve the analytical modeling of the carbon fiber coating phenomena, for their coating a complex analysis of the hot wall reactor was performed, intended for fiber coating. In this analysis, the diffusion equation was numerically integrated along a 2 cm long reactor. Figure 2 shows the profile of the coating arrangement, calculated for the typical conditions in this reactor. According to theoretical estimates, the deposition rate decreases sharply near the reactor outlet. But the actual deposition rates in fiber reactors are variable; the range 1 ÷ 20 μm is considered a common domain.

2.2 Thermal stability of copper coatings on carbon fibers

The performance of metal matrix and fiber reinforced composites is largely controlled by the adhesion between the fibers and the matrix. Interface phenomena are also important in material processing. For example, carbon fiber-reinforced copper matrix composites can be applied as contact materials and substrate for semiconductor devices [16, 17].

Copper poorly moistens the carbon fiber, so the interface of the carbon-copper composite is extremely weak. In order to create the possibility for the carbon fiber to take over the maximum load, it is essential to ensure an appropriate interaction between the matrix and the carbon fiber. Increasing the resistance of the interface is generally achieved by two means:

- favoring the reaction of the fiber with the matrix;
- ensuring the dissolution of the fiber in the matrix.

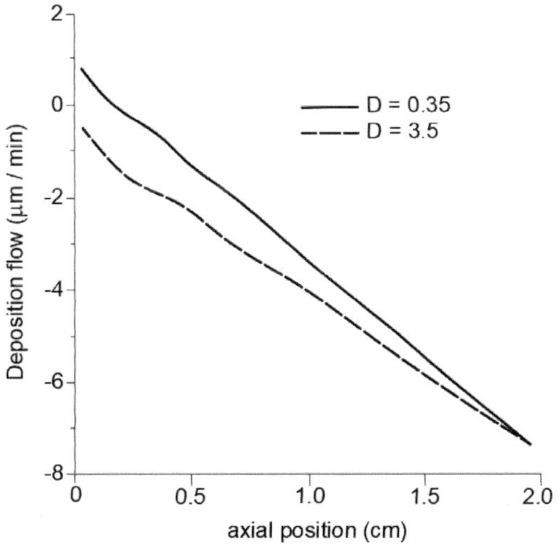

Figure 2. Deposition profiles, calculated for the hot wall reactor, intended for fiber coating; D is the fiber diffusivity (cm^2/s).

There are studies on increasing the resistance at the interface, on Fe and Ni layers deposited on copper-coated carbon fibers. Iron diffuses through the Cu layer to the carbon fibers, reacting with them and forming Fe_3C (bonds by chemical reaction). In contrast, Ni coating is converted to solid Cu-Ni solution, which is able to dissolve small volumes of carbon fiber (dissolved bonds) [18-21].

By methods that do not involve the use of electric current, it is possible to deposit on the carbon fiber a copper layer with a thickness of several tens of microns. A thicker layer can be deposited by galvanic coating. The compactness of the Cu coating, after heating to high temperatures, becomes unstable, starting to exfoliate. This exfoliation of copper occurs due to the surface diffusion process, accompanied by a reduction in surface energy. Therefore, prior oxidation of carbon fiber is recommended.

Oxidation can be performed in air at temperatures of 400 ÷ 700°C for 240 seconds. In order to establish the influence of the prior oxidation process of the fiber, both oxidized fibers and non-oxidized copper fibers were covered (the layer thickness being 0.2 μm). Also, a second set of experiments was performed with oxidized fibers in air, at 600°C and subsequently covered with Cu, by the galvanic method, the thickness of the deposit being 1 μm. Both sets of fibers were then treated up to 1000°C in a hydrogen atmosphere for 600 s.

Microscopic investigations have shown that the limit temperature for exfoliating Cu coating is higher for oxidized carbon fiber at a higher temperature. Oxidation temperatures of 700°C and above lead to a smaller fiber diameter. The coating of Cu on oxidized carbon fibers at this temperature becomes inhomogeneous, generating the appearance of discontinuous bands, after a treatment made at a temperature of 900°C.

In the case of galvanic deposition with a thickness of 1 μm, made on a non-oxidized fiber, it is observed that the Cu coating began to detach from the fiber after an effective treatment at 900°C. After an effective treatment at 1000°C, the coating was partially detached from the fiber. In the case of galvanically coated oxidized fibers, even at 1100°C, the coating is cracked but not spheroidally exfoliated (figure 3).

The limit temperature for a Cu coating on carbon fibers depends on the thickness of the coating (figure 4). The thicker the thickness, the higher the limit temperature.

Figure 3. Graphite fiber coated with chemically deposited copper from the vapor phase.

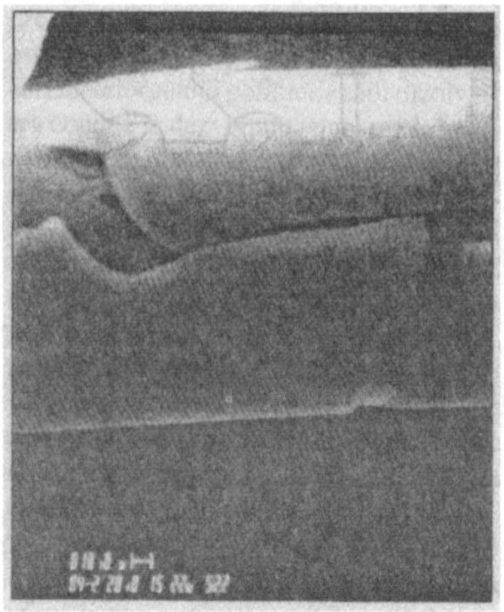

Figure 4. SiC coating chemically deposited from vapor phase on C fiber (coating cracks during property testing).

2.3. The efficiency of Ni coatings on carbon fibers

Coatings with Ni by the CVD method, on carbon fibers, lead to obtaining fibers that provide protection against interaction with electromagnetic fields and fields with radio frequencies, when used as polymeric composite materials.

In the case of a polyfilament carbon cable with 12000 filaments, with a diameter of 7 μm/filament, a coating of 20% Ni (mass percentages) can be achieved, resulting in a thickness of less than 1μm. The use, in the form of short fiber of carbon fiber coated with Ni, provides a protection of 75 dB, in the case of polycarbonate composites.

For unidirectionally layered or woven composite structures, Ni-coated conductive fiber embedded in heat-cured resins also provides light protection. Figure 5 shows the efficiency of Ni-coated carbon fiber protection.

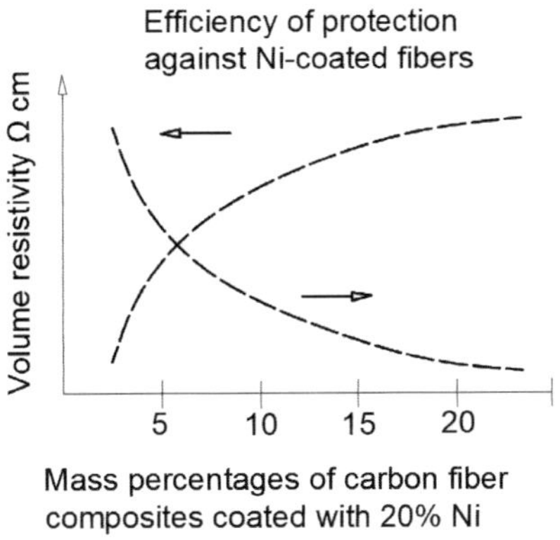

Figure 5. Efficiency of protection against nickel-coated carbon fibers.

3. Conclusions

Experience gained so far shows that there can be large differences between theoretical and real values that have the properties of a composite material. This is especially noticeable when used as polymeric composites.

No change in coating morphology was observed at temperatures up to 400°C. The exfoliation limit value of the Cu coating is determined by the oxidation temperature applied to the carbon fiber: the higher the oxidation temperature, the higher the limit temperature.

An oxidation temperature of 700°C and above accelerated the reaction between atmospheric oxygen and carbon, which led to a change in the appearance of the fiber surface and decreased the fiber diameter.

Another conclusion reached would be that Ni coatings by the CVD method on carbon fibers lead to fibers that provide protection against interaction with electromagnetic fields and radio frequency fields.

4. References

[1] Zang G-H a.o. 2019 Microstructures and mechanical properties of alumina whisker reinforced copper matrix composites prepared by hot-pressing and hot isostatic pressing, *Materials Research Express* **6**(11) 116513.

[2] Ashbee K H 1993 *Fundamental Principles of Fiber Reinforced Composites* 2nd edition (Technomic Publishing Company, Lancaster, PA) p 440.

[3] ASM Handbook 2001 *Composites* Vol. 21 (ASM International, Materials Park, OH).

[4] Chawla, K K 1998 *Composite Materials Science and Engineering* 2nd edition (Springer-Verlag, NewYork).

[5] Ayyappadas C a.o. 2017 An investigation on the effect of sintering mode on various properties of copper-graphene metal matrix composite *Advanced Powder Technology* **28**(7) 1760-1768.

[6] Hollaway L 1994 *Handbook of Polymer Composites for Engineers* Technomic Publishing Company (Lancaster, PA).

[7] Nayan N a.o. 2017 Processing and characterization of spark plasma sintered copper/carbon nanotube composites *Materials Science and Engineering: A* **682** 229-237

[8] Mallick P K 1993 *Fiber- Reinforced Composites*, Materials, Manufacturing and Design 2nd edition (Marcel Dekker, New York).

[9] Peters S T 1998 *Handbook of Composites*, 2nd edition (Springer-Verlag, New York).

[10] Strong A B 2008 *Fundamentals of Composites: Materials, Methods and Applications*, Society of Manufacturing Engineers (Dearborn, Michigan) p 610.

[11] Woishnis W A. 1993 *Engineering Plastics and Composites*, 2nd edition (ASM International, Materials Park, OH).

[12] Caliman R, Olaru I 2006 *An Aluminum Matrix Ni-Ti Fiber-Reinforced Smart Composite*, MOCM 12, vol.1, Romanian Academy, Branch office Iaşi 35-38.

[13] Samal P K and Newkirk J W 2015 *Metals Handbook*, vol. 7 Powder Metallurgy ASM International p 907.

[14] German R M 2018 *Powder Metallurgy Science* no. 8.

[15] Chawla K K 2005 High-Performance Fiber Reinforcements in Composites *Journal of Minerals, Metals & Materials* 47.

[16] Barrera E V, Lozano K, 2000 New Technologies, New Composites *Journal of Minerals, Metals & Materials* 32.

[17] Suraj R 2001 Metal-Matrix Composites for Space Applications *Journal of Minerals, Metals & Materials*.

[18] Kunze J M, Bampton C C 2001 Challenges to Developing and Producing MMCs for Space Applications *Journal of Minerals, Metals & Materials* 22-25.

[19] Grácio J, Davim JP, Hua Fan Q, Ali N. 2002 New Developments on Tribology: Theoretical Analysis and Application to Industrial Process *Proceedings 8th Portuguese Conference of Tribology* (University of Aveiro).

[20] Marques N, Davim JP 2002 Tribological comparative study of a conventional and composite material to biomedical applications *Key Engineering Materials, Trans Tech Publication Lda* **230/232** 487-490.

[21] Marques N, Davim JP. 2001 Tribological comparative study of a conventional and composite material to biomedical applications, *Proc. 10th Meeting of the Portuguese Materials Society and First International Materials Symposium* **A** 175 (Coimbra, Portugal).

Aspects regarding 3d modelling and finite element analysis of a crusher bucket

G F Căpățână and M A Potîrniche
"Dunărea de Jos" University of Galați, Engineering and Agronomy Faculty of Brăila, Research Center for Mechanics of the Machines and Technological Equipments, Calea Călărașilor 29, 810017 Brăila, România

E-mail: gcapatana@ugal.ro

Abstract. The highest consumption of solid materials, especially rocks extracted from quarries, has the construction materials industry. These materials can be used either in natural form (gravel, sand), or after a reduced processing (marble, large stone, for foundations), or after a complex processing (lime, plaster, cement, sandstone, clay). The crusher buckets are used to grind materials in order to reach optimal or convenient sizes of processed materials in order to carry out a good technological process. The study of the shredding processes and of the equipment related to these processes is done in order to reduce the energy consumption necessary to reduce the dimensions of the raw materials to dimensions necessary for further processing. This study is of particular importance due to the fact that the shredding operation is energy intensive. This paper presents aspects regarding the geometric modelling of a crusher bucket and the finite element analysis of the bucket housing. All the components of the bucket were modelled in the Inventor software and then assembled to obtain the assembly. The modelling was based on an example of a crusher bucket taken from the literature. Finite element analysis was performed to highlight the eigenmodes of the crusher bucket in order to estimate the eigenfrequencies range of this equipment.

1. Introduction

The construction materials industry consumes a large amount of rocks extracted from quarries. They can be used in natural form or in processed form. The rock-type materials are crushed to reach optimal dimensions in order to carry out a good technological process. The optimal dimensions of the materials allow the development of the technological process for each product at a maximum efficiency [1].

Solid materials subjected to crushing initially have very varied geometric shapes and dimensions and physical-mechanical properties specific to their nature. The purpose of crushing raw materials is to reduce their size.

The purpose of designing a crushing machine is to determine the conditions necessary to increase the probability of crushing particles of certain dimensions and to obtain a distribution of the desired dimensions to the final product. The crushing process must be carried out in such a way that the processed material does not suffer undesirable changes, such as impurification or excessive heating [2].

The operations necessary for crushing solid materials are crushing, grinding and granulation.

The crusher is the machine used for crushing and it breaks the hard materials into smaller pieces. If it is desired that the crushing result has rounded shapes, the granulation operation is resorted to. When the final result is desired to be of a fine granularity, the materials are ground with the help of the crushing mills [2, 3].

2. Theoretical considerations on crushers

According to the main way of the piece of material solicitation, the crushing machines can be divided into two main classes [2]:

- machines that perform crushing by compression and friction (jaw crushers, rotary crushers, roller crushers, rolling mills);
- machines that perform crushing by shock (impact crushers with hammers, rotary drum mills with free bodies, jet mills).

Quite indicatively, a distinction can be made between crushers and crushing mills according to the input size of the lumps or particles. If the input size exceeds 0.02-0.025 m, it is called a crusher, and if the size is below this value, the machine is called a crushing mill.

The object of this paper is a jaw crusher.

Jaw crushers are characterized by gripping pieces of material between two robust pieces of large mass, whose surfaces, through approach, exert a compressive force on the material.

The jaw crusher is used for crushing various high hardness materials (limestone, ferroalloys, various ores).

The crushers are used if, following a single crushing step, a fairly well-dispersed mixture of crushed particles must be obtained.

The crushers grind brittle materials very well and are less effective for wet products or those that contain a large amount of fat. The jaw crushers are mainly used for coarse crushing but in some cases also for medium crushing [2, 3].

They are most often used in the mining industry, the industry of materials for construction but also in the food industry, especially the sugar industry, for crushing lime stone that enters into the technological process of sugar manufacturing.

The figure 1 shows a model of a jaw crusher made by FLSmidth, year of manufacture 2017 [4].

Figure 1. FLSmidth jaw crusher [4]
1 - abrasion resistant plates; 2 - the assembly of frame with screws; 3 - mobile jaw; 4 - mobile jaw switching system; 5 - hydraulic cylinder of the locking system; 6 - safety bolt; 7 - spherical roller bearings with multi-pass labyrinth system.

The advantages of jaw crushers are [1,2]:

- simple construction and maintenance;
- high operational safety;
- low mass and low cost.

The disadvantages of jaw crushers are [1,2]:

- cyclic operation, with large masses in oscillation, which cannot be fully balanced and determines the vibrant operation with high noise;
- the need for a heavy fly-wheel and an expensive foundation.

3. Case study

For the virtual modeling of the crushing equipment, an existing model in the specialized literature was taken as an idea. The model is produced by Meccanica Breganzese (MB) and the type of crushing bucket is BF 135.8 (figure 2) [5].

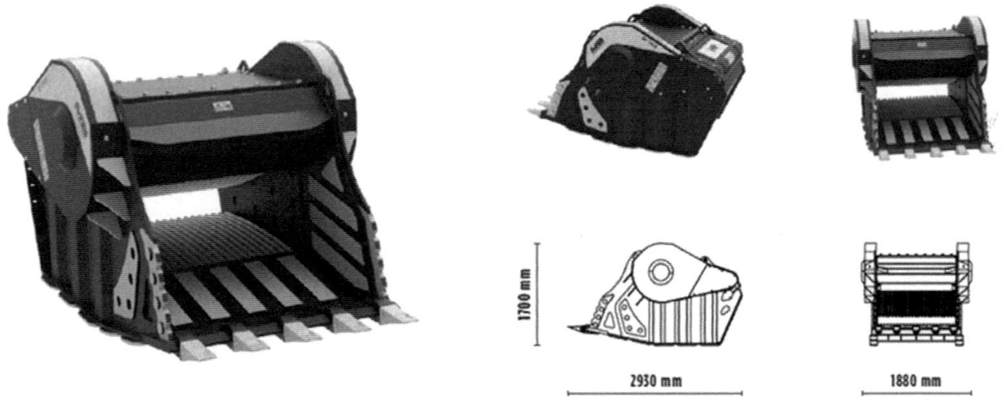

Figure 2. MB BF 135.8 crusher bucket [5].

It was chosen that the 3D modeling of the crushing bucket be made in the Inventor software.

Inventor has a multitude of advanced 3D modeling tools. A 3D model is based on a sketch of the main profile of the piece.

The realization of the sketch is extraordinarily convenient even compared to AutoCAD because it involves a series of simple steps: free drawing of a contour close to the final one, geometric restriction of the lines that compose the contour and dimensional restriction by linear and angular dimensions of the contour [6].

As the essence of the Inventor is parametric design, it can always go back to any contour to edit and redefine the constraints.

In order to generate the bucket model proposed in this paper, it started with the execution of the main sketch, necessary to model the crusher bucket housing.

Depending on the geometry and shape of the component parts, primary tools were used such as: Extrude, Revolve, Sweep, Loft, Shell, Fillet and the 3D modeling of the parts was completed (figure 3).

It was continued with the modification of the bucket in terms of shape, by cutting or adding material in the areas where the components are to be assembled and stiffening in areas where the bucket presents a tendency to deform. All these operations were performed using well-defined sketches. In addition to the commands already specified, the following were also used: Mirror, Hole, Circular Pattern (figure 4).

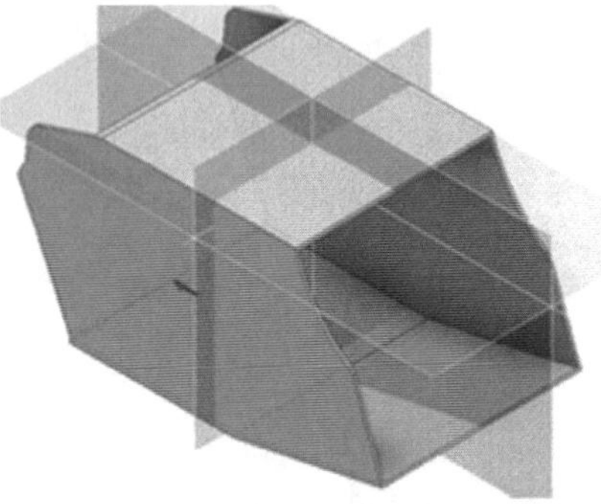

Figure 3. 3D modeling of the crusher bucket housing.

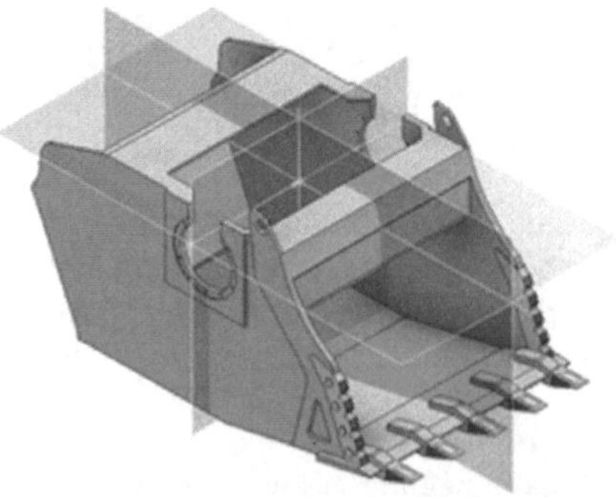

Figure 4. Modification by processing and stiffening of the crusher bucket housing.

The component parts of the crushing bucket were generated in turn: the mobile jaw and its shield, the fixed jaw and its shield, the eccentric shaft, the flywheel, the bearings of the mobile jaw support bearings, seals, the elastic system, the adjustment system, the coupling system.

By assembling all these component elements resulted the final assembly of the crushing bucket, which is shown in figure 5.

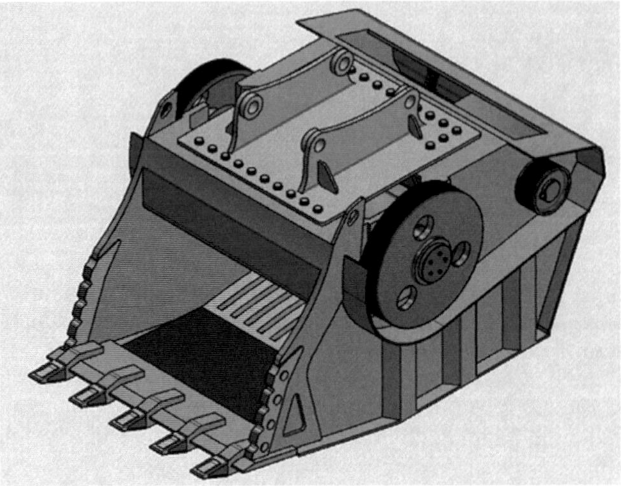

Figure 5. The final assembly of the crushing bucket.

The housing of the crushing bucket, modified by processing and stiffening shown in figure 4 is the object of the finite elements analysis that will be performed in the next step.

The finite element method is a convenient way to obtain approximate solutions to engineering problems. The method is certainly a convenient and necessary tool in design and research calculations.

In this paper, a modal analysis will be performed, which results in the behavior of the crusher bucket housing in the first 10 eigenmodes.

Finite element analysis was performed in the Inventor software, which provides users with analysis tools for 3D models. By performing a frequency analysis, it was desired to highlight the behavior of the crusher bucket housing in terms of the displacements of the nodes of the structure in the first 10 eigenmodes. The 3D model of the bucket was automatically discretized by the Inventor software.

After completing the discretization operation, the boundary conditions were specified (in this case, performing a frequency analysis, it is only necessary to define a support set). Thus, the crusher bucket was fixed on the mounting holes of the coupling system at the housing (figure 6).

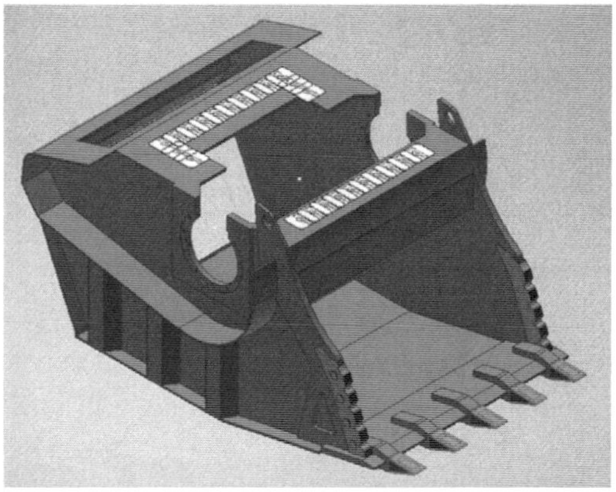

Figure 6. Defining of the support set in Inventor.

From the set of obtained results, it is interesting to visualize the behavior of the crusher bucket housing in the first 10 eigenmodes. From the existing types of frequency response, it was chosen to visualize the behavior of the crusher bucket housing when the response is a displacement. The displacement of the structure nodes in the first 10 eigenmodes is shown in figure 7.

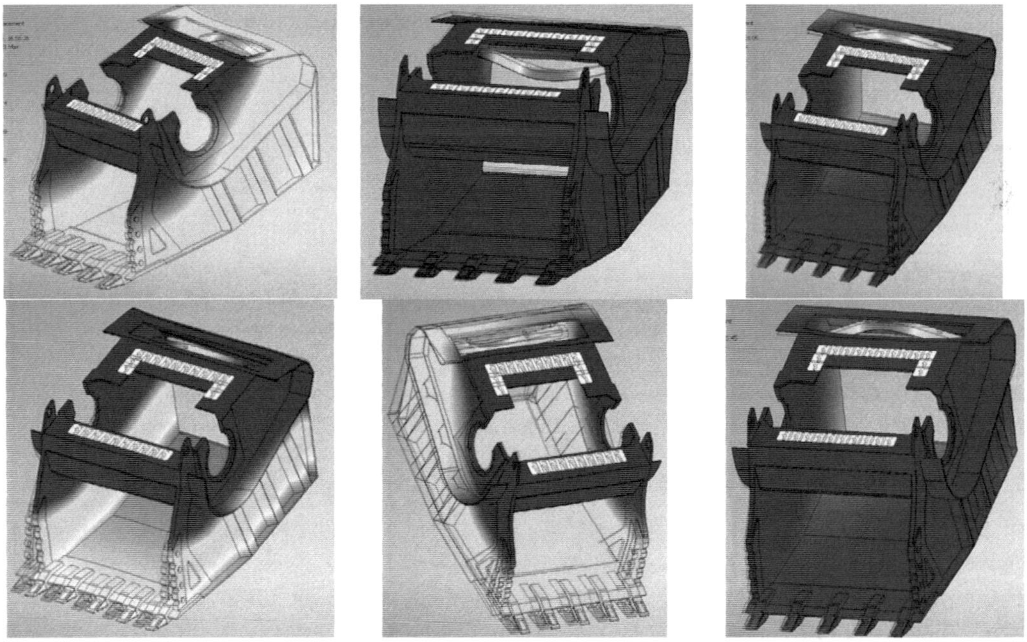

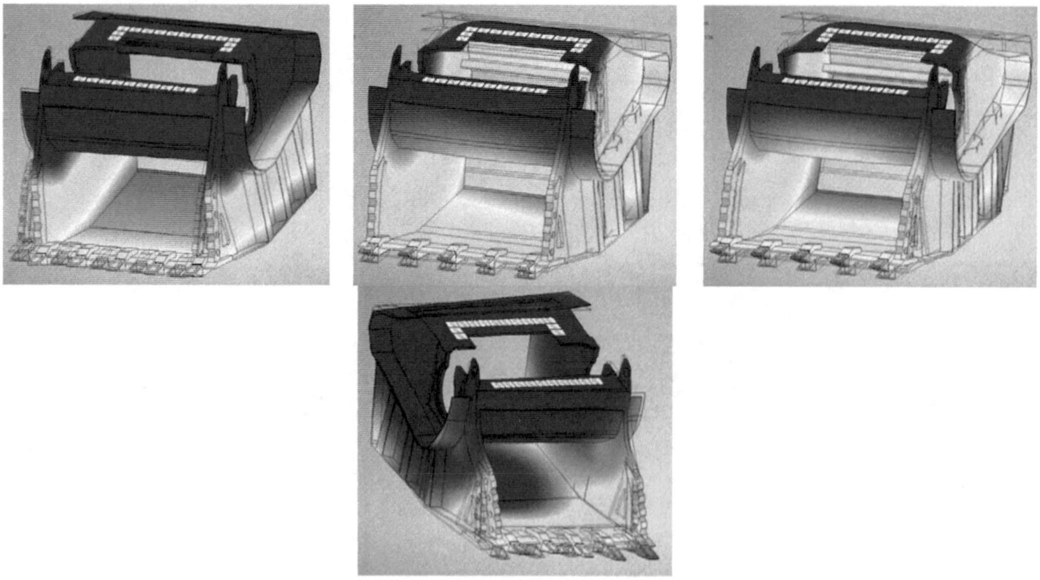

Figure 7. Visualization of the behavior of the crusher bucket housing in the first 10 eigenmodes with highlighting the displacement of the structure nodes.

The following is a detailed drawing containing the values of the displacements of the structure nodes (figure 8) and the values of the eigenfrequencies for the 10 eigenmodes for which the analysis was performed (figure 9).

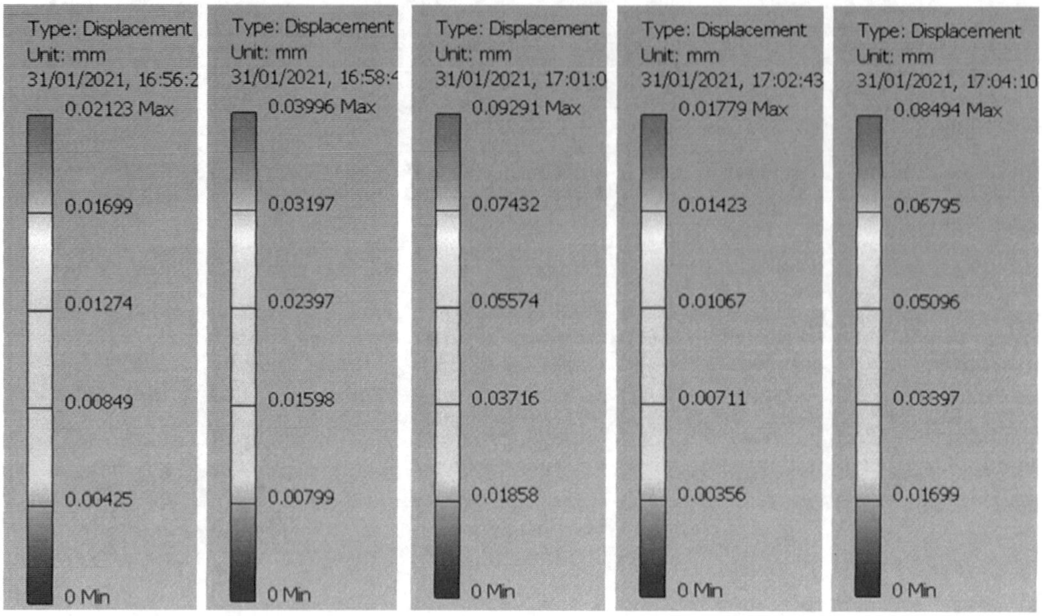

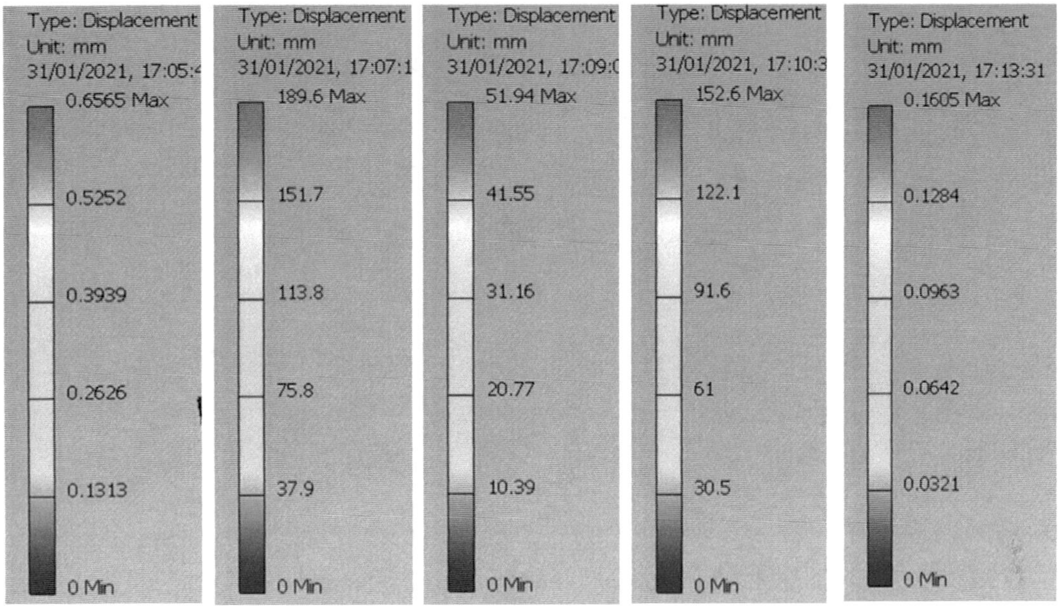

Figure 8. The values of the displacements of the nodes of the crusher bucket housing in the first 10 eigenmodes.

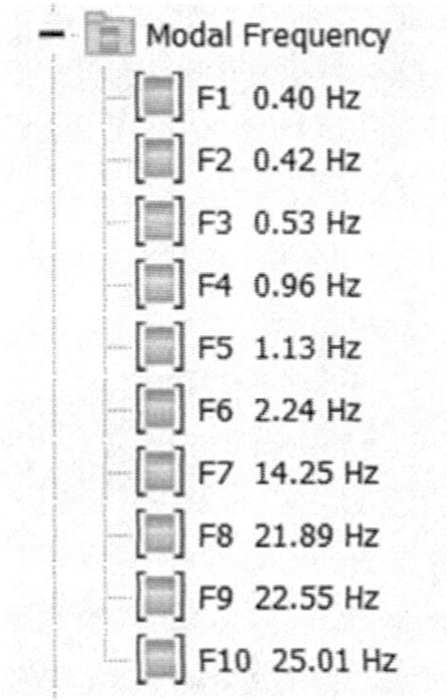

Figure 9. Eigenfrequencies values for the first 10 eigenmodes.

With the values of the maximum displacements and of the eigenfrequencies presented in the figures 8 and 9 a graph has been drawn up that illustrates the dependence of the two variables. The chart was made using the Excel application from the Office software and is shown in figure 10.

IOP Conf. Series: Materials Science and Engineering 1182 (2021) 012012 doi:10.1088/1757-899X/1182/1/012012

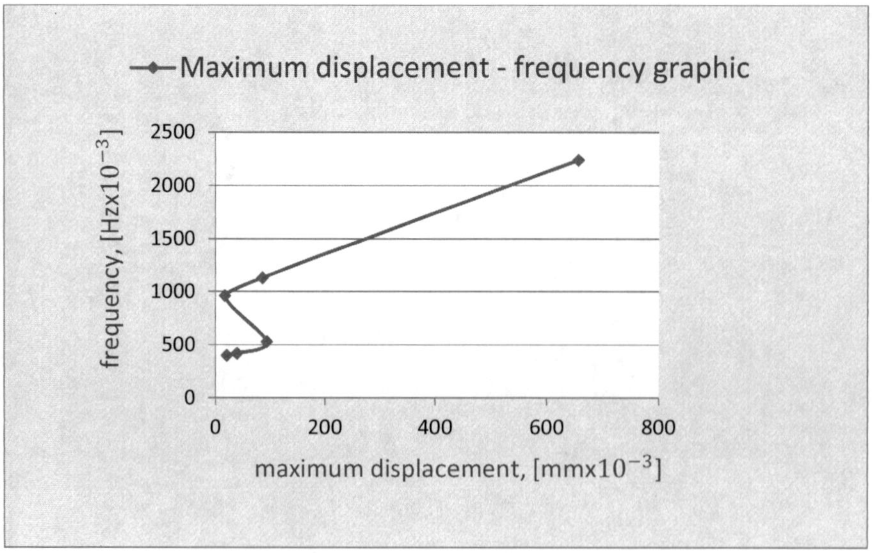

Figure 10. Maximum displacement – frequency graphic.

4. Conclusions

The analysis of the eigenmodes of the crusher bucket housing was imposed by the need to estimate the eigenfrequencies range of this type of equipment. This information is useful to highlight the possible occurrence of resonance phenomena due to the overlap of frequencies generated by the dynamics of the system during work with those in the eigenfrequency range of the structure.

It is well known that mechanical resonance can cause violent vibrations or even complete destruction in the case of incorrectly constructed structures. When it is designed a structure of any type, engineers must ensure that the mechanical resonant frequencies of the component parts are not equal to the oscillating frequencies of the motors or other oscillating parts, a phenomenon known as destructive resonance [7].

Avoiding destructive resonance is a major goal for building any type of structure. As a countermeasure, dampers can be placed to absorb the resonant frequencies and thus dissipate the accumulated energy. Also, the structures are made so that resonances occur at frequencies that are difficult to reach.

5. References

[1] Gheorghe E 1982 *Utilaje pentru mărunțirea materialelor solide* Editura Matrixrom (București).

[2] https://www.wikiwand.com/en/Crusher accessed in 2.2021.

[3] Muscă (Anghelache) G D 2020 *Mașini de ridicat și transportat - Note de curs, partea I* Editura Galați University Press - GUP ISBN 978-606-696-191-2, ISBN 978-606-696-192-9.

[4] https://flsmidth-prod-cdn.azureedge.net/-/media/brochures/brochures-products/crushing-and-sizing/2017/tst_jawcrusher.pdf accessed in 2.2021.

[5] www.mbcrusher.com/it/it/prodotti/benne-frantoio/bf-135.8 accessed in 2.2021.

[6] Goanță A M and Anghelache D G 2016 Aspects on Transfer of Aided - Design Files *IOP Conference Series:Materials Science and Engineering* **145**(4) 042001 doi: 10.1088/1757-899X/145/4/042001 WOS:000396437600053.

[7] https://ro.wikipedia.org/wiki/Rezonan%C8%9B%C4%83 accessed on 2.2021.

Study on the behavior of benchmark container ships in regular waves

A-M Chirosca[1], L Rusu [1] and F Pacuraru[2]
[1]"Dunarea de Jos" University of Galati, Faculty of Engineering, Department Mechanical Engineering 111 Domnească Street, RO-800201, Galati, Romania
[2]"Dunarea de Jos" University of Galati, Faculty of Naval Architecture, Department Naval Architecture 111 Domnească Street, RO-800201, Galati, Romania

E-mail: ana.chirosca@ugal.ro

Abstract. The study of the behaviour of ships in the real sea has become a real challenge in recent years due to climate change, especially in the European sea basin. One of the main categories of types of ships that cross these waters are transport ships. The development of the economy and trade have a major influence on this type of ships, which must be studied and improved in order to be fully exploited. Investigating the ship's behavior is not a new topic, but the changing environment of the transport ship's specifications requires continuous study. An alternative to model testing is the use of numerical simulation through which performance can be fully evaluated using only computers, without the need to produce a scale model for testing. Due to improved computer performance, but also the disadvantages of towing testing, numerical simulations have become increasingly popular for determining ship characteristics. The shorter time in which results can be obtained, as well as the ease of controlling and changing the input parameters, contributes to the preference to use numerical simulations, to the detriment of basin tests. In this paper, the behavior of a Post-Panamax benchmark container ship in regular waves was studied. Numerical simulations were performed using SHIPFLOW MOTIONS, a solver that treats potential nonlinear flows for regular waves and provides accurate results by the free surface potential flow panel method. For our case, the simulations were performed for a range of six speeds corresponding to Froude numbers ranged between 0.174 and 0.218, for five nondimensional wavelengths (λ/Lpp) and for four wave heights. Because body discretization is one of the main factors for obtaining accurate results, an additional study was conducted in calm waters on body refinement and the influence that the mesh has on results, for a range of ten grids with different densities.

1. Introduction

Climate change has become more pronounced in recent years, which is a real challenge to counteract its effects. Evaluation of the climate change impact on the sea state conditions and on the safe design of ship structures were also performed [1, 2]. Navigation in Europe's maritime areas is affected by extreme weather conditions [3]. Maritime traffic is dominated by transport vessels (figure 1), especially general cargo ships [4], which provide maritime transport in Europe but also outside Europe. Since this type of ship is of great interest, but also due to the need to adapt the ship's performances to the ever-changing navigation conditions, these types of ships have been studied in order to improve their particulars, characteristics, and their hulls shape.

Over the years several transport ships have been studied using numerical simulations, such as numerical and experimental analysis of added resistance of ships in waves [5] with Reynolds-Averaged Navier – Stokes methods, the examination of speed performance for various types of commercial ships [6] also based on RANS methods, comparative seakeeping analysis [7] using different methods, etc.

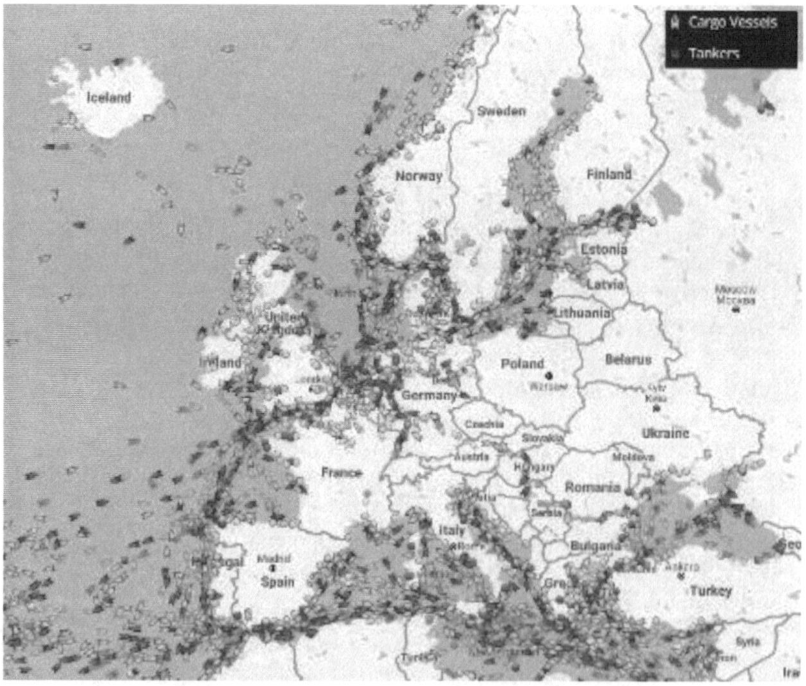

Figure 1. Maritime traffic in Europe [8].

One of the biggest challenges is predicting the behavior of ships in the real sea. Prediction of seakeeping performances is essential from an early design stage, to ensure that the ship will be able to meet the operation requirements and fulfil its mission in real sea conditions. But studying the behavior of a ship involves several problems to solve, such as: nonlinearities induced by fluid viscosity and hydrodynamic pressure, interference between ship's body and waves, and time-domain procedures for analysing the ship's response to oscillations [7]. Mathematical models can be solved numerically, offering approximate solutions to very complex problems that are most of the time impossible to treat analytically. Because of this numerical approach are becoming increasingly popular and this is also supported by the enhanced accuracy due to the increase of computing capabilities. The use of CFD (Computational Fluid Dynamics) methods is constantly increasing due to its advantages compared to towing testing. The main advantages of numerical simulations are lower costs, there is no need for a ship model, the configuration parameters can be changed very easily, and the results can be obtained in a much shorter time. But to use these methods we need a high-performance computer with multiple processors with which we can significantly reduce the computation time.

In this paper, a grid convergence test for three grids with panels of different densities was performed to demonstrate that the results did not depend on the fineness of the discretization grid. A series of simulations in calm water were performed to validate the results obtained with those of the experimental method. Also, a numerical approach of the seakeeping performance of a container ship in regular waves, for a range of six speeds, five wavelengths and four wave heights using a 3D fully nonlinear time-domain Boundary Element Method (BEM) implemented in the commercial software SHIPFLOW Motions is presented.

IOP Conf. Series: Materials Science and Engineering 1182 (2021) 012013 doi:10.1088/1757-899X/1182/1/012013

Regarding this hull, results for model test basin are available only for resistance, propulsion performances and roll decay. In a previous study, the experimental results were compared with different numerical approach results [9]. The mean difference over the speed domain, obtained with SHIPFLOW was 2.011%.

2. Computing strategy

For this study the hull of a benchmark container ship was studied. The hull is a typical container ship developed at Institute of Ship Technology, Ocean Engineering and Transport Systems (ISMT) [10] and presented in figure 2.

Figure 2. Benchmark container ship hull.

The container ship, also known as Duisburg Test Case, is a modern 14000 TEU Post-Panamax vessel with a bulbous bow, large stern, and a single screw. Bare hull, without rudder and propeller has been considered for the numerical simulations. In table 1 are presented the main characteristics of the vessel in design loading condition, both for model and full scale [9, 10].

Table 1. DTC ship main characteristics.

Main particulars		Full scale	Model
Length between perpendiculars	L_{pp} [m]	355	5.976
Waterline breadth	B_{WL} [m]	51	0.859
Draught midship	T_m [m]	14.5	0.244
Trim angle	Ʋ [°]	0	0
Volume displacement	V [m^3]	173467	0.827
Block coefficient	c_B [-]	0.661	0.661
Wetted surface	S_w [m^2]	22032	6.243
Design speed	v_d [m/s]	12.86	1.668

The prediction for ship motions and additional resistance in waves is presented in this paper. In waves, ships develop a much higher resistance than resistance in calm water, which is called additional resistance and is generated by energy dissipation due to ship motions and due to the reflection of incident waves [11].

The purpose of this paper is to investigate the behaviour of the Duisburg Test Case container ship (DTC) in regular waves at six speeds, from 1.335 m/s to 1.668 m/s for several nondimensional wave lengths and nondimensional wave heights presented in table 2. The simulations were performed for regular head waves based on the 5th order Stokes wave theory and ran until the convergence was reached.

Simulations were performed on a desktop computer with four cores of 3.1 GHz. Additionally, simulations were performed to study the discretization grid, for a panel density range from 0.2 to 1.2, with a step of 0.1, on a desktop computer with 10 cores. The time of the simulations for the discretization grids was from 42 minutes to 8 hours, and the time of the simulations on regular waves lasted around 4-6 hours per case. The total number of simulations was 129.

Table 2. Test cases.

v [m/s]	Method	λ/L	H/L
1.335		0.50	0.0056
1.401		0.75	0.0133
1.469	BEM	1.00	0.0169
1.535		1.25	0.0225
1.602		1.50	
1.668			

The 3D fully nonlinear time-domain boundary element method was used in this study. This method was used through commercial software SHIPFLOW Motions, a product of FLOTECH International AB., developed to compute ship motions in regular and irregular waves. The BEM, or so-called panel method, deals with potential boundary value problems. This method has the advantage of considering any nonlinearity of body geometry and making it possible to perform viscous and extraneous simulations. Another advantage is the short time through which results can be obtained, compared to other methods.

3. Grid convergence test

When dealing with number simulations, the discretization of the ship's body is an essential factor in obtaining accurate results. Therefore, before performing the simulations for seakeeping analysis, a study of the convergence of the discretization grid was performed, based on Richardson extrapolation [12] in order to investigate the numerical simulation error.

Numerical simulations on calm water at maximum speed were performed for three grids: coarse (corresponds to a panel density factor with the value 0.9), medium (corresponds to a panel density factor with the value 1.0) and fine (corresponds to a panel density factor with the value 1.1).

In table 3 are presented the number of panels and the number of nodes distributed on hull and free surface for the grids used for grid convergence test. The simulations were performed on desktop machines with ten cores.

Table 3. Computational grids.

	Coarse	Medium	Fine
Density factor	0.9	1.0	1.1
c_w *1000	0.676	0.721	0.792
R_w [N]	5.677	6.060	6.651
No. of panels	32446	40882	49301
No. of nodes	33266	41800	50311

The refining grid parameter (r_G) was considered 1.26. S1, S2 and S3 correspond to the solution returned by the simulation for the grids used in this calculation. With the help of these parameters we calculated R_G which represents the convergence ratio and has the following expression:

$$R_G = \varepsilon_{21} / \varepsilon_{32} \tag{1}$$

where ε_{12}=S_2-S_1 and ε_{12}=S_3-S_2.

Depending on the value R_G, three convergence conditions are possible [12]: monotonic convergence when R_G is between 0 and 1, oscillatory convergence when R_G is less than 0 and divergence when R_G is greater than 1. The value of R_G is 0.647 for c_w and 0.646 for R_w,. The value is between 0 and 1, which means that we are in the case of monotonic convergence.

The next step is to apply the generalized Richard extrapolation [12] which estimates the error δ_G and the order of precision p_G.

$$p_G = \frac{\ln\left(\frac{\varepsilon_{32}}{\varepsilon_{21}}\right)}{\ln(r_G)} \tag{2}$$

$$\delta_G = \frac{\varepsilon_{21}}{r_G^{\ p_G}} \tag{3}$$

The last step is to calculate the correction factor and with its help the uncertainty. For the calculation of the correction factor, the theoretical order of the network convergence (p_{th}) was considered equal to two.

$$c_G = \frac{r_G^{\ p_G} - 1}{r_G^{\ p_{th}} - 1} \tag{4}$$

$$U_G = \left(|c_G| + |1 - c_G|\right)|\delta_G| \tag{5}$$

According to the results presented, we can conclude that the solutions obtained after running in SHIPFLOW are grid independent. For the regular wave simulations performed in this paper, the fine grid was chosen.

4. Results and Discussion

This section presents the results of regular wave simulations over the full range of speeds, heights, and wavelengths, as well as the influence of wave height, wavelength, but also speeds. To validate the results obtained by numerical simulations, a comparison of the model resistance with the CFD resistance was made. The results of the towing tests used in this paper come from the model tests performed in the SVA Potsdam model test tanks and are available to the public [10].

In table 4 are presented the results obtained from the calm water simulations for the entire speed range, as well as the results of the towing tests. Figure 3 shows the comparison between the model test resistance and the total resistance obtained in SHIPFLOW. The total resistance in the simulations consists of the wave resistance R_w added with the friction resistance R_F. We can see that the values of the results are almost equal, the mean error difference being 3.36%. In the last column of table 4, the percentage difference between the resistance obtained on the model and the one in SHIPFLOW was calculated, for each speed.

Table 4. Total resistance results.

v	R_w – calm water (CFD)	R_F – calm water (CFD)	R_T – calm water (CFD)	Percentage Difference
[m/s]	**[N]**	**[N]**	**[N]**	**[%]**
1.335	1.33	17.56	18.89	7.39
1.401	1.86	19.19	21.06	4.65
1.469	2.56	20.96	23.51	2.63
1.535	3.71	22.74	26.45	0.03
1.602	5.30	24.61	29.91	3.13
1.668	6.06	26.53	32.59	2.35

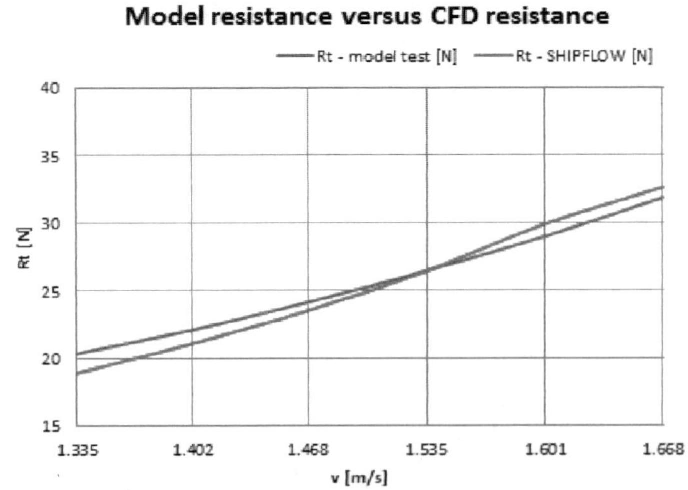

Figure 3. Comparison between model resistance and numerical simulation resistance.

Numerical simulations were performed for a range of speeds from 1.335 m /s to regime speed, 1.668 m /s, for a range of nondimensional wavelengths from 0.5 to 1.5 with a step of 0.25, and for a range of nondimensional wave heights from 0.0056 to 0.0225. Figure 4 and figure 5 illustrates the effect of wavelength at regime speed (1.668 m/s) on the coefficient of additional resistance in waves and on the wave resistance on regular waves when the wave height is constant. The coefficient of additional resistance in waves was calculated as follows:

$$c_{aw} = \frac{R_w}{\rho \cdot g \cdot \zeta_A^2 \cdot B^2 / L_{pp}} \tag{6}$$

where:

- $R_w = R_{aw} - R_{w0}$; R_w [N] represents the wave resistance in regular waves, R_{aw} [N] represents the total wave resistance and R_{w0} [N] represents the wave resistance in calm water;
- ρ [kg/m^2] is the water density;
- g [m/s^2] is the gravitational acceleration;
- ζ_A [m] is the wave amplitude; ζ_A =H/2.

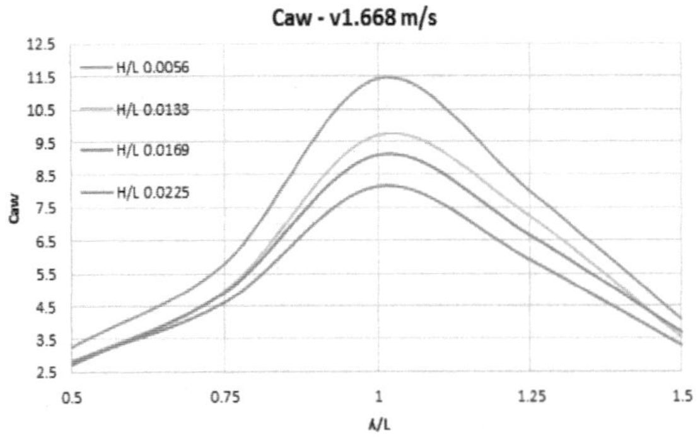

Figure 4. Influence of wavelength at regime speed on the coefficient of additional resistance.

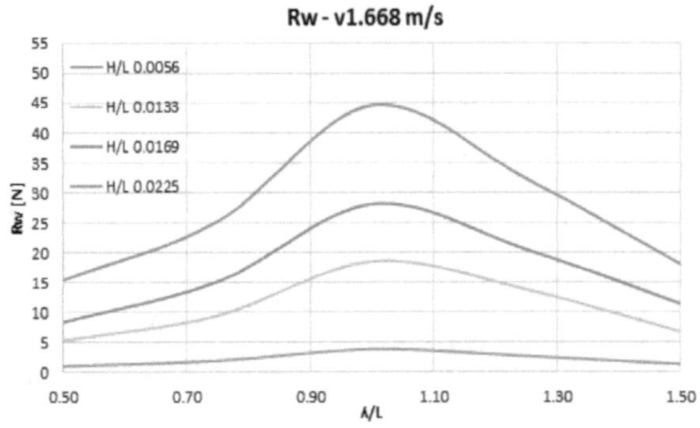

Figure 5. Influence of wavelength at regime speed on the wave resistance.

The influence of the wave resistance on the total resistance is presented as a percentage in table 5 (with green are represented the lowest values and with red the highest values). As the wave height increases, the contribution of the wave resistance increases, from 11.64% at the lowest wave height and lowest wave length to 67.41% at the highest wave height and wave length 5.976 m. Another observation would be that the percentage increase is slightly lower at higher speeds.

Table 5. Influence of wave resistance.

v[m/s]	λ/L	H/L			
		0.0056	**0.0133**	**0.0169**	**0.0225**
	0.5	11.64%	28.12%	35.93%	47.72%
	0.75	15.40%	40.34%	49.19%	60.38%
1.335	**1**	21.35%	47.79%	57.16%	67.41%
	1.25	16.87%	38.44%	47.49%	57.34%
	1.5	13.06%	29.57%	36.28%	46.43%
	0.5	13.28%	27.48%	35.19%	46.59%
	0.75	16.02%	38.77%	47.53%	58.40%
1.401	**1**	21.02%	47.55%	57.12%	67.34%
	1.25	17.21%	39.03%	47.53%	57.40%
	1.5	13.30%	29.85%	36.69%	46.17%
	0.5	14.97%	27.26%	34.14%	44.75%
	0.75	17.68%	37.49%	46.12%	56.83%
1.469	**1**	22.34%	47.84%	56.84%	67.02%
	1.25	18.10%	39.51%	47.91%	57.68%
	1.5	14.79%	29.95%	37.01%	46.56%
	0.5	16.55%	28.37%	35.04%	45.29%
	0.75	19.71%	36.98%	45.30%	55.54%
1.535	**1**	23.64%	47.32%	56.06%	65.63%
	1.25	20.49%	40.67%	48.84%	58.25%
	1.5	17.25%	30.59%	37.86%	47.16%
	0.5	19.78%	29.95%	35.87%	45.25%
	0.75	22.14%	37.49%	45.51%	55.19%
1.602	**1**	26.35%	47.92%	56.25%	65.86%
	1.25	23.47%	42.18%	49.23%	58.95%
	1.5	20.57%	31.83%	39.01%	47.38%
	0.5	21.27%	30.09%	35.30%	44.85%
	0.75	23.23%	36.98%	44.49%	54.16%
1.668	**1**	27.25%	48.13%	56.28%	65.62%
	1.25	24.84%	42.81%	50.04%	59.10%
	1.5	21.92%	32.62%	39.74%	47.64%

The influence of the wave generated by the ship is shown in figure 6(a) where the pitch RAO for variable wave lengths is illustrated. The distance between pitch curves calculated for the H/L=0.0056 and the other curves plotted for higher wave hight, emphasise the non-linear effect of the wave generated by the ship on the pitch especially for higher speed. One can see that with the increase of the length, the influence of the wave also increases. Figure 6(b) shows the wave resistance in the case of variable wave heights, which increases with the increase of the ship-wave meeting frequency.

Figure 7 shows the free surface for the lowest studied wave height (H/L = 0.0056) and the highest studied wave height (H/L=0.0225), for the entire speed range, at the wavelength λ/L =1.

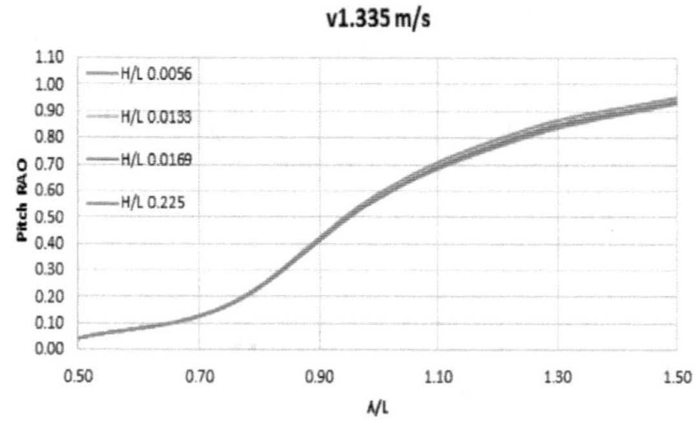
v1.335 m/s
H/L 0.0056
H/L 0.0133
H/L 0.0169
H/L 0.225
Pitch RAO
λ/L

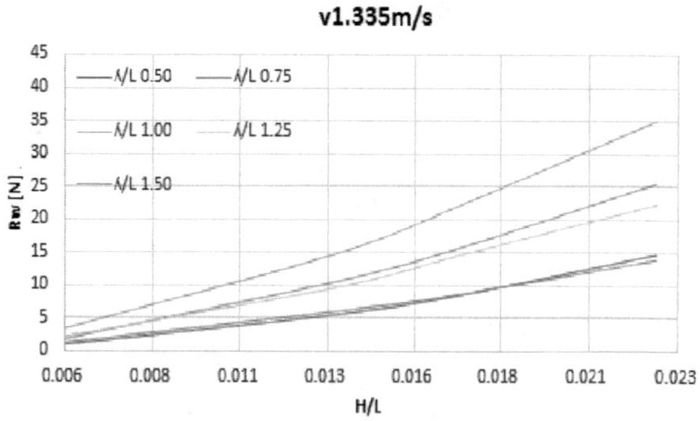
v1.335m/s
λ/L 0.50
λ/L 0.75
λ/L 1.00
λ/L 1.25
λ/L 1.50
Rw [N]
H/L

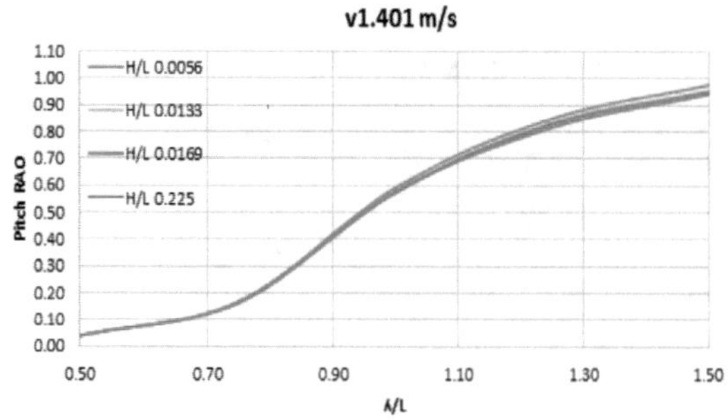
v1.401 m/s
H/L 0.0056
H/L 0.0133
H/L 0.0169
H/L 0.225
Pitch RAO
λ/L

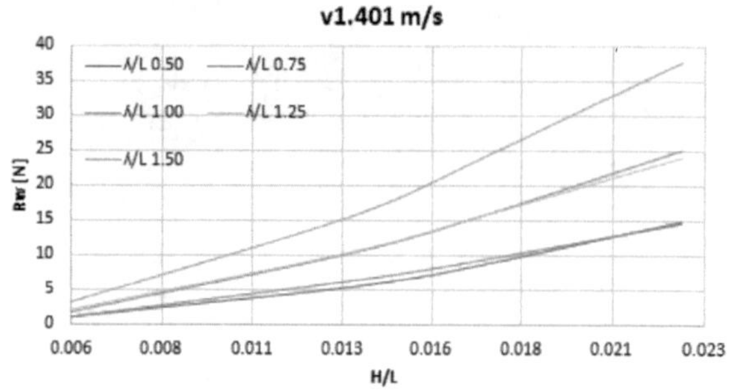
v1.401 m/s
λ/L 0.50
λ/L 0.75
λ/L 1.00
λ/L 1.25
λ/L 1.50
Rw [N]
H/L

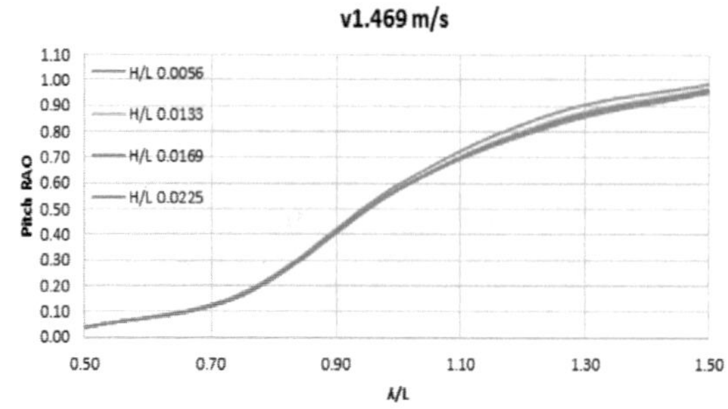
v1.469 m/s
H/L 0.0056
H/L 0.0133
H/L 0.0169
H/L 0.0225
Pitch RAO
λ/L

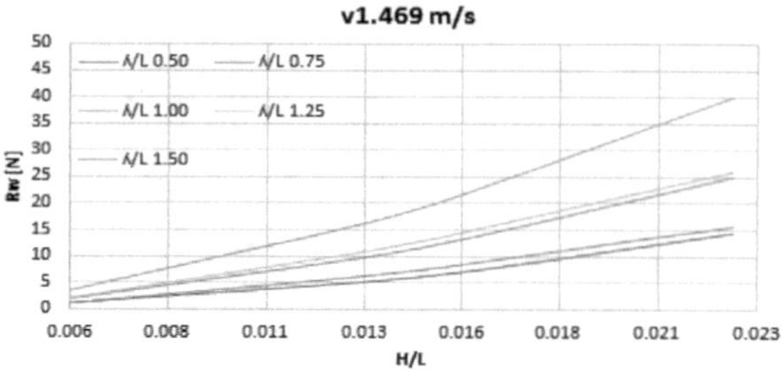
v1.469 m/s
λ/L 0.50
λ/L 0.75
λ/L 1.00
λ/L 1.25
λ/L 1.50
Rw [N]
H/L

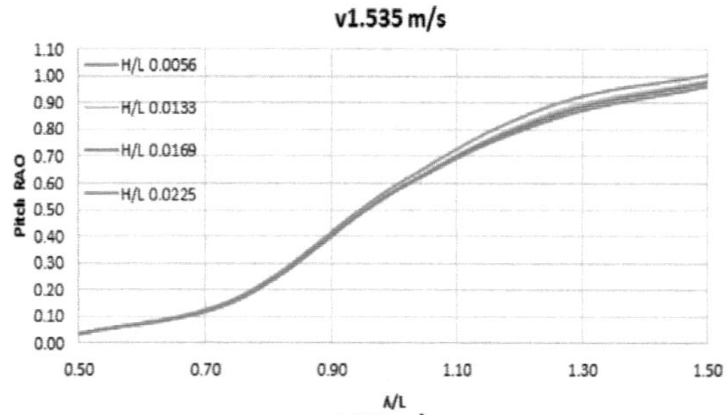
v1.535 m/s
H/L 0.0056
H/L 0.0133
H/L 0.0169
H/L 0.0225
Pitch RAO
λ/L

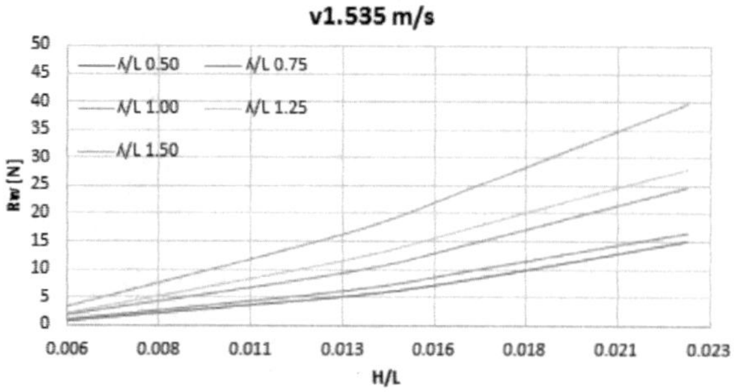
v1.535 m/s
λ/L 0.50
λ/L 0.75
λ/L 1.00
λ/L 1.25
λ/L 1.50
Rw [N]
H/L

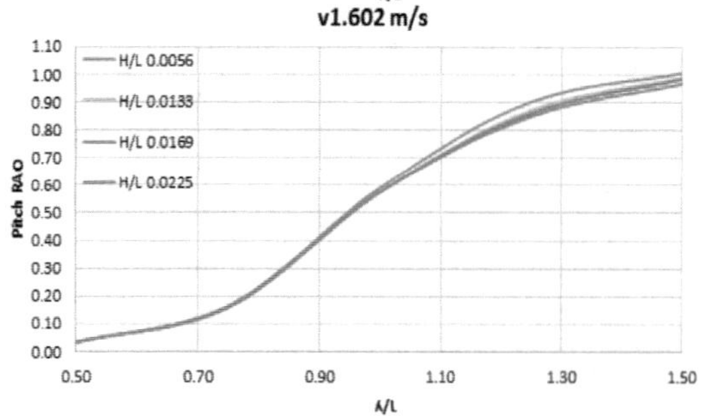
v1.602 m/s
H/L 0.0056
H/L 0.0133
H/L 0.0169
H/L 0.0225
Pitch RAO
λ/L

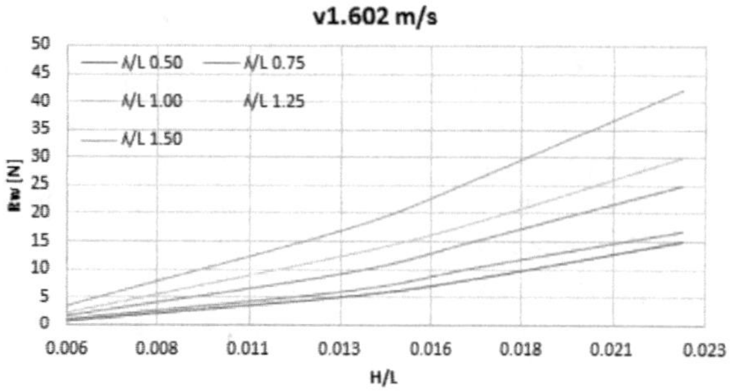
v1.602 m/s
λ/L 0.50
λ/L 0.75
λ/L 1.00
λ/L 1.25
λ/L 1.50
Rw [N]
H/L

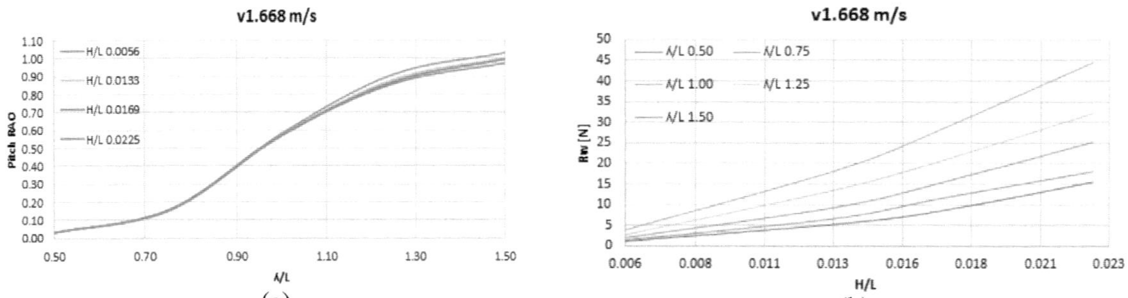

(a) (b)

Figure 6. Influence of wavelength on pitch RAO (a) and influence of wave height on wave resistance (b).

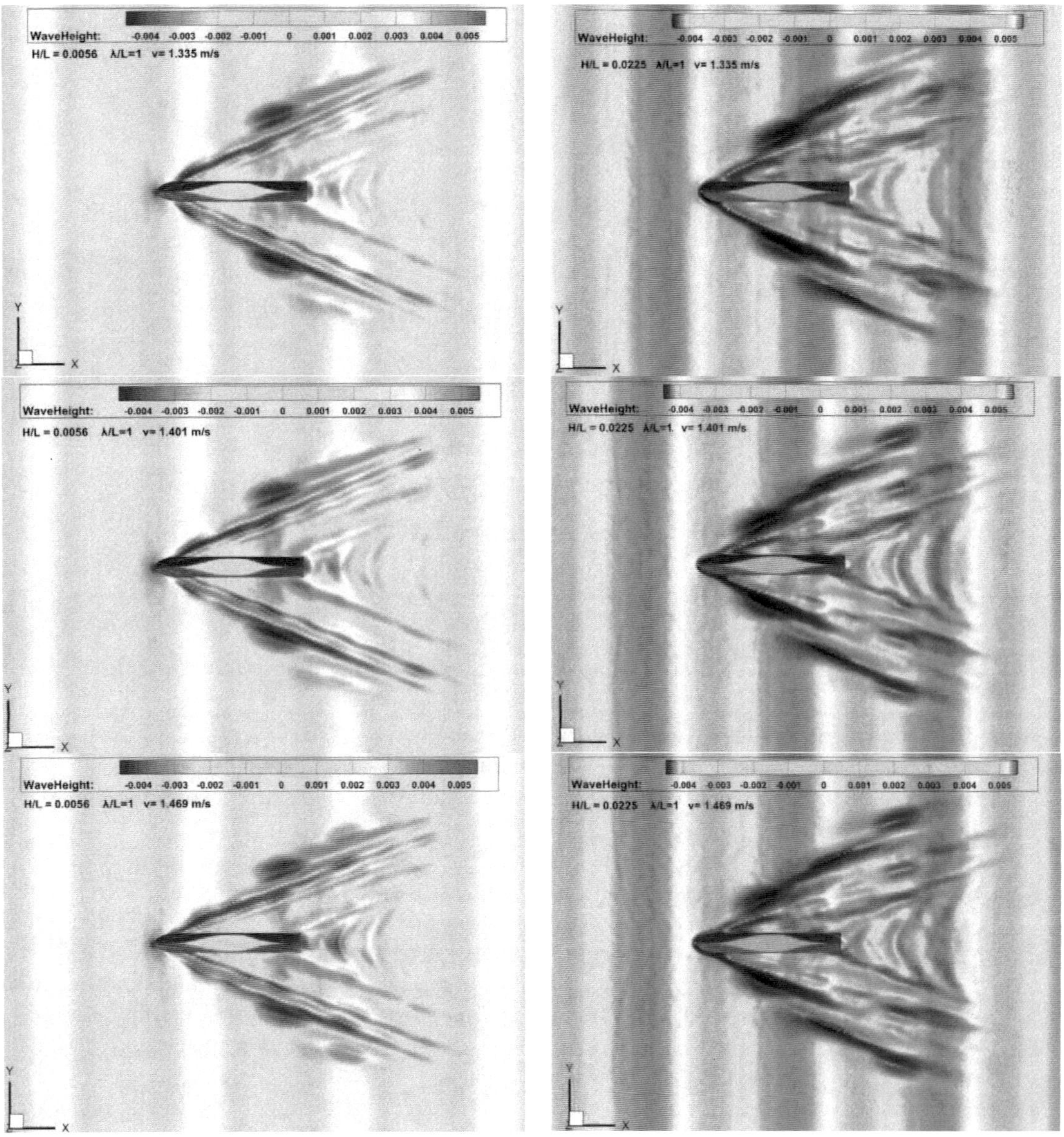

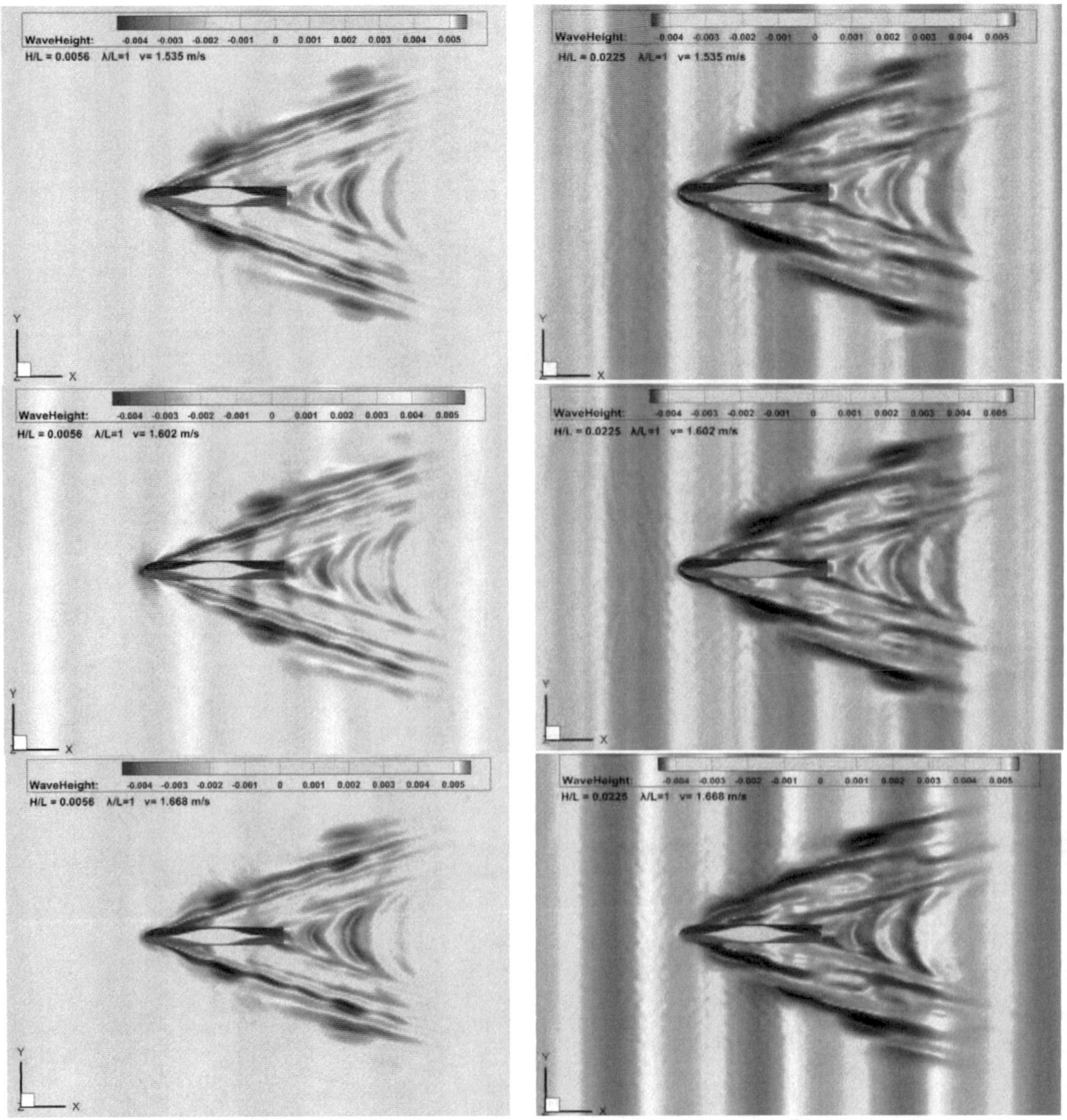

(a) (b)

Figure 7. Distribution of the wave height for the lowest wave height (a) and the highest wave height (b).

5. Conclusions

The behavior of ships on waves is a topical issue in continuous research and development, so that ships will be able to sail at full capacity, regardless of changing weather conditions. CFD methods have advanced with the increased performance of computers, and in the future, it is desired to replace expensive towing tests with numerical simulations, which are easier to perform.

In this paper, a series of numerical simulations were made to see how a benchmark container ship behaves on regular waves, at different wave heights and lengths. A total of 129 numerical simulations were performed, of which 3 for grid testing, 6 for validation of results in calm water and 120 for the study of behavior on regular waves.

A convergence study was performed for three grids: coarse, medium and fine. According to the grid convergence test, the convergence ratio value is in the case of monotonic convergence, and the solutions obtained are grid independent.

The results in calm water were validated by comparing them with the results of the towing tests. The average error across the entire speed range was 3.36%.

The study of regular waves was performed to identify the influence of wave height, wavelength, but also the speed of the ship. As expected, with the increase of the wavelength, it was possible to observe the increase of the wave contribution, which has an influence on the total resistance from 11% to 67%. The lowest influence was found for the case $\lambda / L = 0.5$ and $H/L=0.0056$, and the highest influence was found for the case $\lambda / L = 1$ and $H / L=0.0225$. Influences of 65-67% were also found for the entire speed range in the case of $H / L=0.0225$ for λ /L from 0.75 to 1.25. Also, as the height of the waves increases, another factor that affects the wave resistance is the increased frequency of the ship-wave encounter.

Studying the behavior of ships on waves is necessary to be able to study the effect that waves have on the movements of the ship, as well as the effect of the ship's wave. At the same time, sudden climate change significantly affects the navigation capabilities of ships and, therefore, in the future, we want to study the behavior of the container ship on irregular waves and for different heading angles.

6. References

[1] Bitner-Gregersen E M *et al.*, 2018 Climate change and safe design of ship structures, *Ocean Engineering* **149** pp. 226-237.

[2] Aarnes O J *et al.* 2017 Projected changes in significant wave height toward the end of the 21st century: Northeast Atlantic, *Journal of Geophysical Research: Oceans* **122**(4) 3394-3403.

[3] Bernardino M, Rusu L and Guedes Soares C 2020 Evaluation of extreme storm waves in the Black Sea, *Journal of Operational Oceanography*, 1-15. doi.org/10.1080/1755876X. 2020.1736748.

[4] Chirosca A., Rusu L. 2020 Statistical analysis of the types of ships that have crossed the European ports in the last decade *20th International Multidisciplinary Scientific GeoConference SGEM 2020* **20** pp 249-256 ISBN: 978-619-7603-17-0, ISSN: 1314-2704.

[5] Choi, JE; Min, K.S.; Kim, J.H.; Lee, S.B.; Seo, H.W. 2010 Resistance and propulsion characteristics of various commercial ships based on CFD results *Ocean engineering* **37**(7) 549-566.

[6] O. el Moctar, S. Sigmund, Th. E. Schellin 2015Numerical and experimental analysis of added resistance of ships in waves *Proceedings of the ASME 2014 34th International Conference on Ocean, Offshore and Arctic Engineering, OMAE2015-42403*St. John's, NL, Canada.

[7] Pacuraru, F.; Domnisoru, L.; Pacuraru, S. 2020 On the Comparative Seakeeping Analysis of the Full Scale KCS by Several Hydrodynamic Approaches. *J. Mar. Sci. Eng.* **8** 962 https://doi.org/10.3390/jmse8120962.

[8] Map data@2020 GeoBasis-DE/BKG (@2019), Google, Inst. Geogr. Nacional, https://www.marinetraffic.com, accessed on January 2021.

[9] Chiroşcă, A. M.; Rusu, L. 2021 Comparison between Model Test and Three CFD Studies for a Benchmark Container Ship *J. Mar. Sci. Eng.*, **9**, 62. https://doi.org/10.3390/jmse9010062.

[10] El. Moctar, O. Shigunov, V. and Zorn, T. 2012 Duisburg Test Case: Post-panamax container ship for benchmarking, *Ship Technology Research Schiffstechnik*, **59**(3), pp. 50-64.

[11] Ageno, E., Begovic, E, Bruzzone, D, Galli, A, Gualeni, P 2015 A boundary element method for motions and added resistance of ships in waves *Transactions of Famena* **39** 1-12.

[12] International Towing Tank Conference (ITTC), *Uncertainty Analysis in CFD Verification and Validation Methodology and Procedures* 7.5-03-01-01. 2017, pp. 1–13. Available online: https://www.ittc.info/media/8153/75-03-01-01.pdf (accessed on 20 February 2021).

Acknowledgment: The work was carried out in the framework of the research project DREAM (Dynamics of the REsources and technological Advance in harvesting Marine renewable energy), supported by the Romanian Executive Agency for Higher Education, Research, Development and Innovation Funding – UEFISCDI, grant number PN-III-P4-ID-PCE-2020-0008.

Obtaining new products in the PVC/AL joinery industry applying the TIPS method - case study

V A Ciubotariu[1], C M Radu[1], C C Grigoraş[2] and V Zichil[2]
[1] Vasile Alecsandri University of Bacau - Romania, Department of Engineering & Management of Industrial Systems, Calea Marasesti, No. 157, 600115, Bacau, Romania
[2] Vasile Alecsandri University of Bacau - Romania, Department of Engineering and Management, Mechatronics, Calea Marasesti, No. 157, 600115, Bacau, Romania

E-mail: vlad.ciubotariu@ub.ro

Abstract. Obtaining a new product, in the current stage, can no longer be achieved by waiting for it by chance, as it happened in the past, but by systematically, perseveringly, and scientifically looking for it. This paper is a case study on the application of a systematic methodology for generating solution concepts in the case of the PVC and AL joinery accessories industry. Considering that one of the requirements was that the new connector part should mount on several standardized profiles, by approaching the TIPS tools kit, a modular connector with superior strength characteristics (3%) and low complexity was obtained. Also, by rethinking the product, a reduction of 15% of the required material (PA type 6) was obtained, as well as a decrease by approximately 11% of the tools cost necessary for the manufacturing process. For the preliminary evaluation of the new product, finite element analysis was necessary and the SolidWorks Simulation 2019 software was used. The numerical data was compared with the experimental data provided by the beneficiary, the differences being within the limit of 3.5%. These data encourage the beneficiary to study in detail the new product from several points of view, considering the option of introduction into manufacturing.

1. Introduction

Product design / development engineers are accustomed to paradox, ambiguity and contradiction. Any technical problem has at least one contradiction and for many people it remains unnecessarily associated with uncertainty, difficulty of choice and compromise. This is unnecessary because there are a finite number of practical solutions to any contradiction, namely 40 Inventive Principles for solving problems [1]. A contradiction is a conflict between solutions, where solutions with opposite effect are desired, or if one solution improves one feature of the system, another will undergo negative changes to it. When a function of a system is presented as a good solution and the adversative conjunction "but" is used, it means that something is wrong in that system.

The Theory of Inventive Problem Solving (TIPS), which is applied in this study, is a method specific to engineering fields where a systematic methodology for generating solution concepts is proposed. Therefore, two types of contradictions are considered: (a) technical contradictions: improving a characteristic, but another is worsening; (b) physical contradictions: opposite solutions are desired. Physical contradictions are solved by separating features in different ways that ensure the existence of both effects. Solutions can be separated according to time, space, size and condition [2].

To solve contradictions, the TIPS method, holds two extremely powerful and intercorrelated tools that cannot be used without one another: *Contradiction Matrix* which is based on 39 technical parameters, used to solve technical contradictions, as in the initial version created by G Altshuller [1] or 48 technical parameters, as in the later versions of the method [3]; *40 Inventive Principles*, used to solve physical contradictions and should be separated according to the four characteristics mentioned above.

Joinery involves the action of combining pieces of material to produce more complex products. Some joints use fasteners, binders or adhesives, while others use only elements of the material. The characteristics of these joints are strength, flexibility, appearance, etc. and derives from the properties of the materials involved and the purpose of the joint. Therefore, different joinery methods are used to meet different requirements [4, 6]. For example, the joinery used to build an enclosure may be different from that used to make windows or doors, although some concepts overlap.

PVC and AL are perfect materials suitable for joineries. As a matter of fact, internal and external joineries are under the control of various types of aggression, such as bad weather, repeated handling and other types of stresses [5]. Careful choice of a good material is essential. Indeed, PVC has a high thermal, phonic and safety performances. Moreover, PVC withstands the most difficult climatic conditions and adapts perfectly to external joineries [6, 7].

Each product has an impact on the environment, and not only, throughout its life cycle. One of the main challenges for product designers is the lack of tools to incorporate environmental decisions in their designs [3]. This study aims to develop a new product, with superior strength characteristics, with a reduced geometric complexity using the TIPS principles. Throughout the product design activity, a lower geometric complexity leads to much simpler tools and devices. This has a considerably reduced impact on several components of the environment or the economy.

2. Problem context and discussion

A local manufacturer from the PVC and AL joinery industry requested that on the basis of an existing product to obtain a component that meets all the requirements imposed by this industry, but be very versatile in terms of mounting on at least three profiles made by established manufacturers in the field.

Thus, for a product in the group of accessories used in PVC and AL joinery – connector – the possibility of mounting it on several standardized profiles is achievable by fastening it using screws or clips. These components must also ensure the necessary strength in operation without excessive consumption of materials or cost of tools and devices required for their manufacture.

The original part offered by the local manufacturer is a connector like the one shown in figure 1. This connector is generally used in the case of mounting PVC poles on assemblies that include AL sills. The requirement is to obtain a product which matches at least three different arrangements.

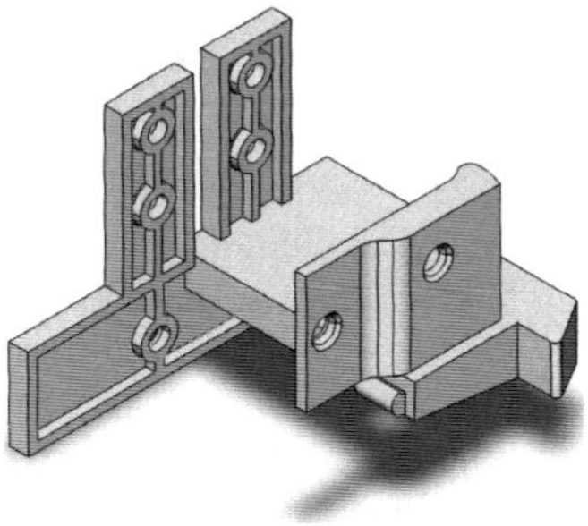

Figure 1. Type of connector made of PVC used in the case of screw mounting of PVC poles on assemblies that include AL sills.

Before exposing the problem, the manufacturer had two approaches: a new product could be made by additive manufacturing, but the production cost was much too high or the adaptation of existing products by further processing, but the appearance of the finished product was far from being satisfactory for the required level of quality.

There is a desirable needed characteristic – versatility – that appears to be associated with something costly or unwanted – weight or shape complexity – which in the field of TIPS that is a contradiction. It would appear that we are constrained to remain on a compromise area. Ideally it is wanted a component that has both good versatility and low weight or shape complexity.

3. TRIZ approach for solving the problem

In this case-study, it was decided to use one of the classic tools from the TIPS toolkit for solving contradictions: the Contradiction Matrix (figure 2). The method is a three-step systematic process [1]:

- identification of the contradictions in the problem;
- using a statistically derived look-up table – the Contradiction Matrix – to determine which generic Inventive Principle have been successfully used in the past to solve contradictions of the generic type of those in the present problem;
- taking the generic Inventive Principles suggested and apply them to the present problem.

3.1. Identifying the contradictions

In the present case-study, two initial technical contradictions ware identified: a versatile product but not heavy and a versatile product which has no complex shape.

For the second contradiction it could be asked why the product has a complex shape. The answer may be related to the arrangement of the assembly where the part has its functionality or to the tools and devices used in the production process. This in turn suggest the possibility of yet another type of contradiction added to the problem: a product that is versatile but not expensive to manufacture.

These contradictions can be expressed as a graph, which typically takes the form presented in figure 3. As the versatility of the possible product is improved (marked by the "+" sign) the weight or complexity of it becomes worse (marked by the "–" sign), and the other way around. The general idea of the method is to obtain a solution which finds itself under the curve and ideally closer to the area where the two "+" signs are shown.

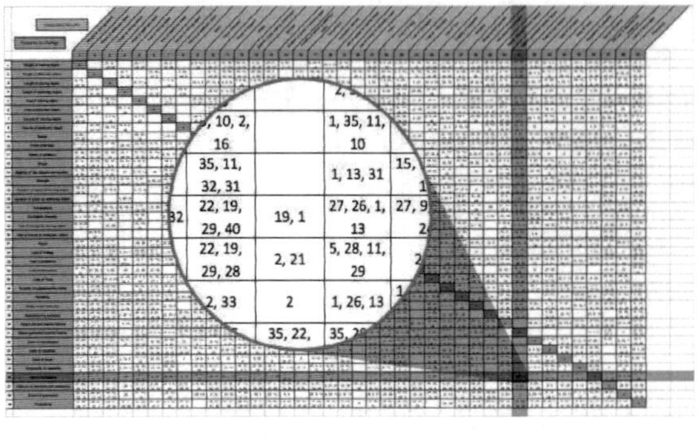

Figure 2. TIPS - Contradiction Matrix [9]

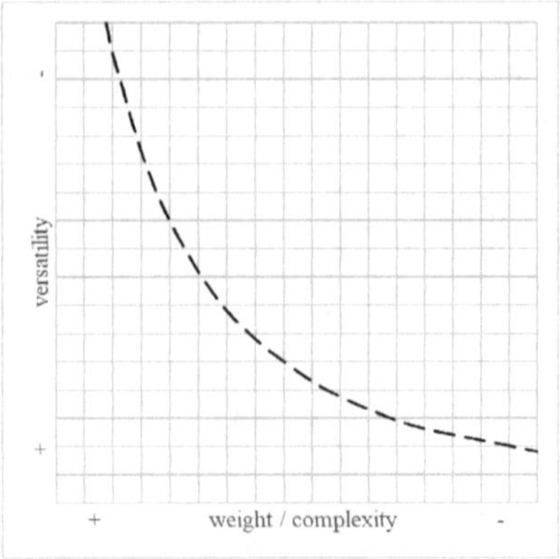

Figure 3. Possible compromise solutions [1]

3.2. Using the Contradiction Matrix

Before the contradiction matrix can be used it is necessary to map the specific parameters – versatility, weight, complexity – onto the generic parameters used by this statistically derived look-up table. In the present study, the contradiction matrix is not explicitly shown because of its large display. For this particular case it was not very difficult to arrive at an appropriate mapping as:

- Versatility matches element number 35 from the Contradiction Matrix as *adaptability or versatility*;
- Weight matches element number 2 as *weight of stationary object*;

- Shape complexity matches both element number 12 as *shape* and element number 32 as *ease of manufacture*.

Thus, the three real-world contradictions, when mapped onto matrix parameters become: adaptability or versatility vs. weight of stationary object, adaptability or versatility vs. shape and adaptability or versatility vs. ease of manufacture. Each of these contradictions are used with the Contradiction Matrix to obtain suggested Inventive Principles.

3.2.1. First contradiction. In this particular case the improving parameter or what is wanted to get better is *adaptability or versatility* and the worsening parameter or that is wanted not to get worse is *weight of stationary object.* These two parameters are used to cross-index the Contradiction Matrix to obtain the following four Inventive Principles that statistically have been found to be the most successful ways of obtaining good results [1]:

- *Periodic action* where the separation principle for solving the physical contradiction is *time*;
- *Dynamics* where the separation principle for solving the physical contradiction is *time*;
- *Pneumatics and hydraulics* where the separation principles for solving the physical contradictions are *time* and *condition*;
- *Partial or excessive action* where the separation principle for solving the physical contradiction is *time*.

3.2.2. Second contradiction. What is wanted to get better is *adaptability or versatility* and what is wanted not to get worse is *shape.* These two parameters are used to cross-index the contradiction matrix to obtain the following four Inventive Principles that statistically have been found to be the most successful ways of obtaining good results [1]:

- *Dynamics* where the separation principle for solving the physical contradiction is *time*;
- *Thermal expansion* where the separation principle for solving the physical contradiction is *time*;
- *Segmentation* where the separation principles for solving the physical contradictions are *space*, *time* and *scale*;
- *Anti-weight* where the separation principle for solving the physical contradiction is *scale*.

3.2.3. Third contradiction. In this particular case the improving parameter is *adaptability or versatility* and the worsening parameter is *ease of manufacture.* These two parameters are used to cross-index the contradiction matrix to obtain the following three Inventive Principles that statistically have been found to be the most successful ways of obtaining good results [1]:

- *Segmentation* where the separation principles for solving the physical contradictions are *space*, *time* and *scale*;
- *The other way around* where the separation principles for solving the physical contradictions are *space* and *scale*;
- *Porous materials* where the separation principle for solving the physical contradiction is *condition*.

3.3. Appling the suggested Inventive Principles

Having derived some suggested generic Inventive Principles, the next task is to apply these to the particular problem. This is part of a recurring theme in TRIZ where the specific problem is extracted, it is generalized in order to access known generic solutions and finally the general solution is applied to the particular problem.

As Inventive Principles *Segmentation* and *Dynamics* ware suggested by the Contradiction Matrix twice, particular attention was given to this as a solution trigger for this problem. The following step in the problem-solving process is to fully understand these suggestions. The names of the 40 Inventive Principles are just convenient labels and it is important to appreciate the full definition of each principle before attempting to apply it.

Analyzing the first inventive principle – *Segmentation* – it must be taken into account that it refers mainly to the following:

- division of an object into independent parts;

- making an object sectional – easy to assemble or disassemble;
- increase the degree of fragmentation, etc.

In the same way, the second inventive principle – *Dynamics* – refers mainly to the following:

- change the object or outside environment for optimal performance at every stage of operation;
- divide an object into parts capable of movement relative to each other;
- change from immobile to mobile;
- increase the degree of free motion, etc.

Following the analysis and revision of all the suggestions extracted from the Inventive Principles, a potential solution materialized in the construction of a connector consisting of two distinct components, each having its role in the final assembly. The two components are assembled to each other by using clips, and through a smaller number of screws the final assembly that includes both the PVC pole and the AL sills is made.

Because the local manufacturer indicated the fact that the interior vertical restraint of the initial part is actually not necessary and makes the assembly sequences mandatory in a specific manner, the new product has no such restraint, making the tools necessary in the manufacturing process much simpler, therefore more efficient cost wise.

The modular connector proposed to solve the local manufacturer's problem is shown in figure 4, and an overall example in which the proposed solution could be integrated in is shown in figure 5.

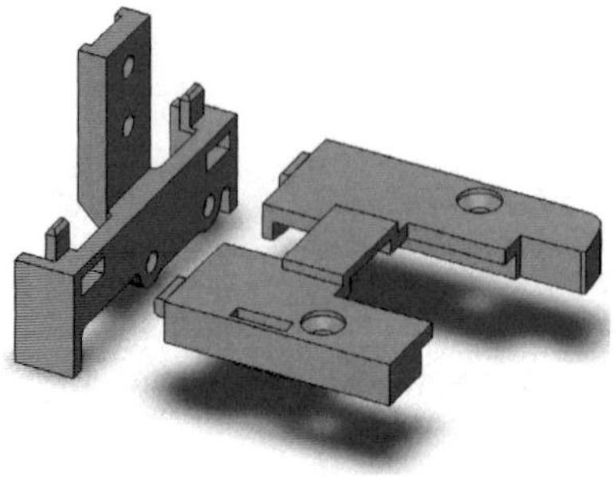

Figure 4. Modular connector proposed as solution for the problem in discussion.

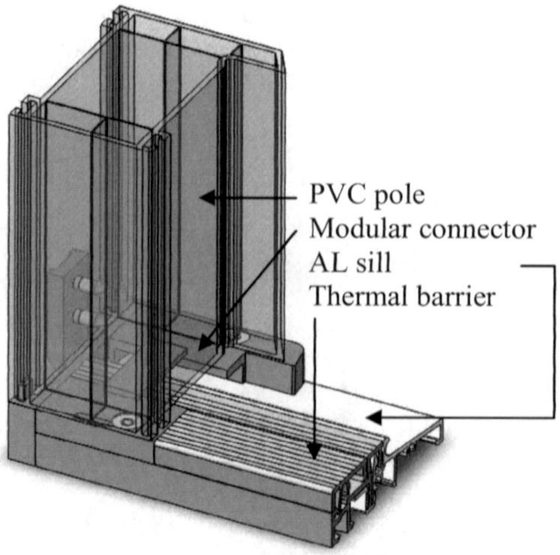

Figure 5. Example of assembly in which the modular connector could be integrated in.

4. Numerical evaluation of the new product

Regarding the in-service behaviour of the new product, the aim was to carry out a comparative study between the data provided by the local manufacturer and those obtained by using finite element analysis. The numerical evaluation was performed through the SolidWorks Simulation 2019 module, which is an extension to SolidWorks package.

Given the fact that the local manufacturer has requested certain tests to verify the behaviour of the proposed product, the following are the general data included in the finite element analysis. Therefore, at the suggestion of the local manufacturer, the new product was subjected to just two analysis as follows: a nonlinear static analysis and a drop test study.

The comparison between the two products takes into account the mass of the components, the complexity of the geometric elements and the results obtained from the numerical analysis.

4.1. Simulation setup

The material used by the local manufacturer for such components is generally PA Type 6. The mechanical properties of this material are shown in table 1. The material in question is a polymer assumed to be linear elastic isotropic.

In the case of the nonlinear static analysis the boundary conditions are shown in figure 6 for the initial part and figure 7 for the proposed part respectively. The curve which best describes the time varying load applied on the analysed parts is shown in figure 8. The constraints are applied to all screw holes intended for fastening onto the support assembly.

Table 1. Mechanical properties of PA Type 6.

Property	Value	Units
Elastic Modulus	2620	MPa
Poisson's Ratio	0.34	-
Shear Modulus	970.4	MPa
Mass Density	1120	kg/m^3
Tensile Strength	103.65	MPa
Yield Strength	90	MPa
Thermal Conductivity	0.233	W/(m·k)
Specific Heat	1601	J/(kg·J)

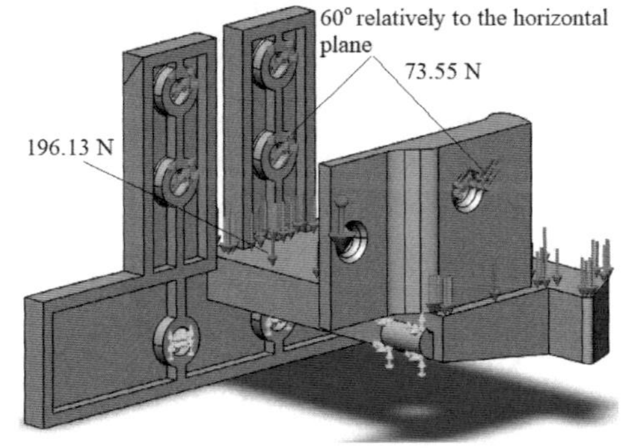

Figure 6. Boundary conditions for the analysis of the initial part

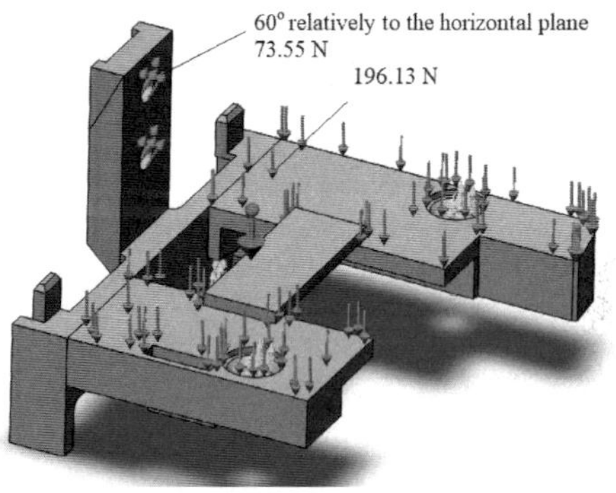

Figure 7. Boundary conditions for the analysis of the new proposed part

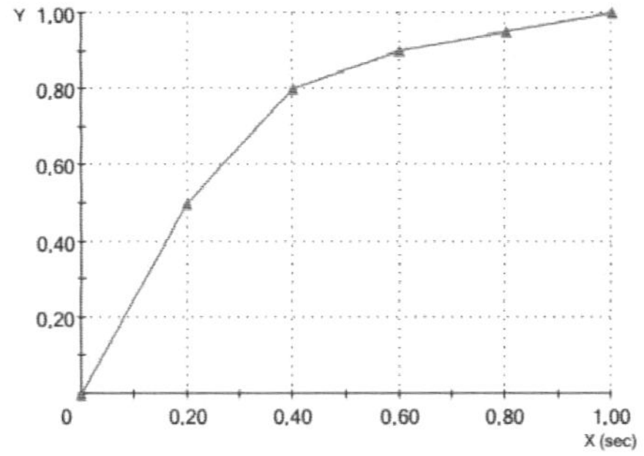

Figure 8. Time curve for nonlinear variation of the load applied on the vertical restraints of both types of analysed parts.

IOP Conf. Series: Materials Science and Engineering 1182 (2021) 012014 doi:10.1088/1757-899X/1182/1/012014

A curvature-based mesh was used and the specific parameters ware as it follows: maximum tetrahedral element size was 4 mm, minimum element size was 0.8 mm, minimum number of elements in a circle was 8, the element size growth ratio was 1.6 and the number of Jacobian points was 16.

Regarding the modelling of the new proposed part, two components were used. The contact conditions between these two components are of the “no penetration” type, taking into account a coefficient of friction of 0.2. The mesh interaction between the components is of the "surface to surface" type.

Regarding the drop test analysis, the time varying stresses and deformations due to an initial impact of the part with a rigid planar surface are calculated. As the part deforms, secondary internal and external impacts are also calculated, locating critical weaknesses or failure points, as well as stresses and displacements. The maximum gravitational force equivalent, or g-force, experienced by individual part is one of the primary unknowns before a drop test. This is a critical parameter since many mechanical parts are not rated for further use above a specific maximum g-force. Therefore, the time varying accelerations at any location within the part can be measured [8].

In the drop test study, the hight from which the parts are dropped on a rigid floor is considered to be 1500 mm from the centroid of the part. The only loading applied on the parts is the gravity. Both products ware considered to be dropped on the side towards the aluminium sill.

4.2. Results and discussions

Regarding the quantity of material used for designing the new part, the software estimate is 16.10 grams in comparison with the initial part which has 19.02 grams. Therefore, the new part uses with 15.35% less material.

Taking into account the input size of tetrahedral elements, the initial part was discretized with a total number of 19956 elements, while the new part was discretized with a total number of 17373 elements.

In the case of nonlinear static analysis, both connectors were subjected to the same loading cases. Figures 9, 10 graphically describe final results of the analysis regarding von Mises stress. Even though the new modular connector has only 2.72 MPa decrease in stress values (96.68 MPa) as opposed to the initial part (99.40 MPa), the decrease in weight also suggests better results in operation. An explanation for this can be given based on the asymmetry of material distribution, especially in the case of the original part.

Numerical analysis does not detect any stresses affecting the sitting surface of any of the products.

The final results of the analysis regarding displacements indicate that the new part has a maximum displacement of just 4.01 mm, while the initial part has a maximum displacement of 7.32 mm. Therefore, a decrease of about 44.8% of the total displacement was obtained with the new part.

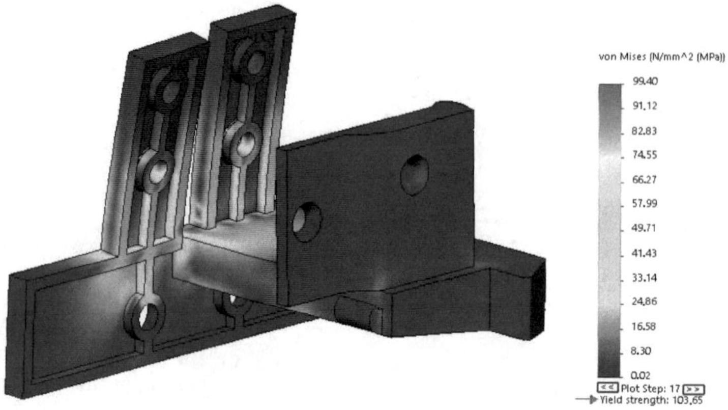

Figure 9. von Mises stress result in nonlinear static analysis of the initial part.

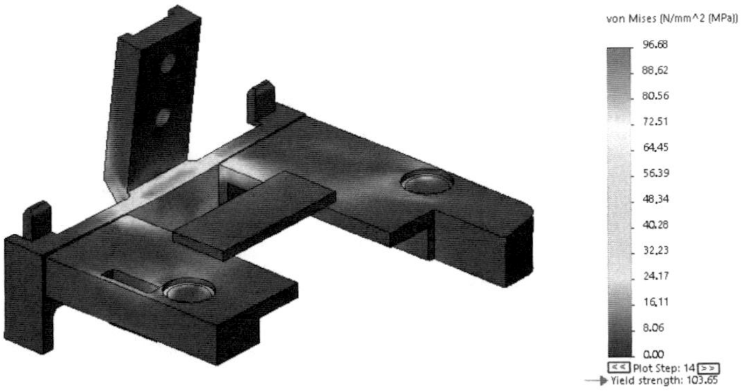

Figure 10. von Mises stress result in nonlinear static analysis of the new proposed part.

Regarding the drop test analysis, figures 11, 12 graphically describe final results of ESTRN: Equivalent Strain. Lower maximum strain values found (0.035) suggest that the new part seems to be stiffer. This effect may occur due to the lower overall mass of the new part as well as the somewhat symmetrical distribution of mass over the entire width of it.

The maximum values of displacement caused by the drop, show that the new part presents 5.26% higher values (0.40 mm) than the original part (0.38 mm).

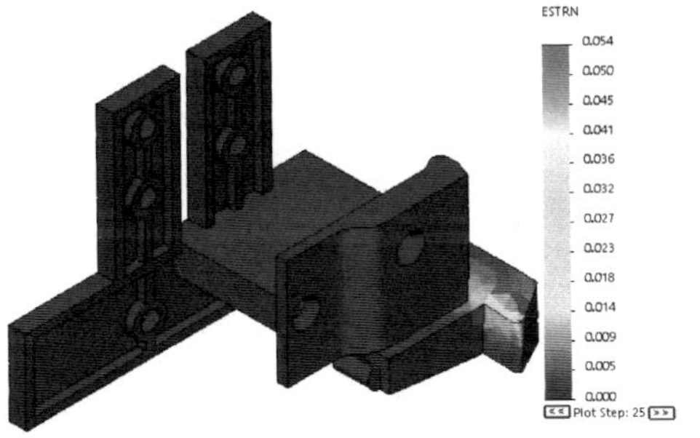

Figure 11. ESTRN: Equivalent Strain in drop test analysis of the initial part

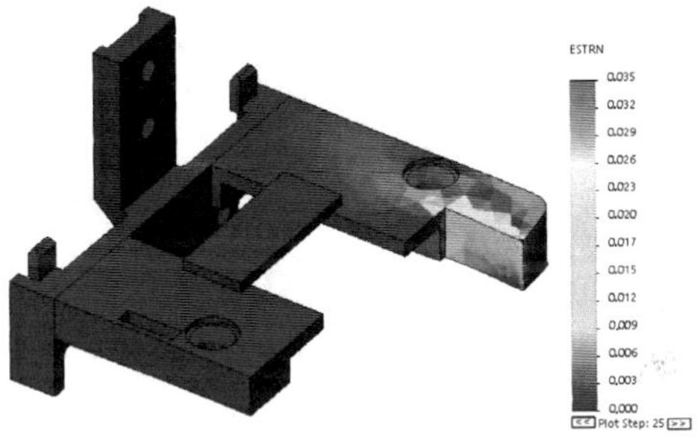

Figure 12. ESTRN: Equivalent Strain in drop test analysis of the new proposed part

5. Conclusions

The decrease with 15.35% in weight of the new modular connector and the decrease in stress values with 2.73% as opposed to the initial part, suggests better results in operation of the new product in terms of the loading generated by the PVC column. Also, from the drop test analysis, the decrease with 35.16% of the maximum strain values found suggests that the new part seems to be stiffer.

The local manufacturer analyzed the proposal for the new product and estimated a cost of production tools with approximately 10.8% lower in comparison to those used for the initial product. Also, from the experimental tests, the company concluded that the difference between experimental and numerical analyzes is between -1.28% and 2.22%.

Therefore, by using certain elements of the TIPS method, such as the Contradiction Matrix and the 40 Inventive Principles and also taking into account the requirements of the beneficiary, a new modular product with superior technical characteristics and much better economic efficiency was possible to obtain.

6. References

[1] Gadd K 2011 *TRIZ for Engineers: enabling inventive problem solving* (Chichester: Wiley) 3-21 ISBN 9780470741887.

[2] Hua Z, Yang J, Coulibaly S and Zhang B 2006 Integration TRIZ with problem-solving tools: a literature review from 1995 to 2006 *International Journal of Business Innovation and Research* **1**(1–2) 111–128.

[3] Sheng I L S and Kok-Soo T 2010 Eco-efficient product design using theory of inventive problem solving (TRIZ) principles *American Journal of Applied Sciences* **7**(6) 852–858.

[4] Baker P, Curtis R, Kennedy C and Wood C 2010 Thermal performance of traditional windows and low-cost energy-saving retrofits *APT Bulletin: The Journal of Preservation Technology* **41**(1) 29-34.

[5] Morin P and Morrison J 2004 The geometry of carpentry and joinery *Discrete Applied Mathematics* **144**(3) 374-380 ISSN 0166-218X.

[6] Knop K and Ulewicz R 2020 Analysis of the impact of technological innovation on improving the quality of door joinery, *31st International Scientific Conference ICWST, Sustainability of Forest-based Industries in the Global Economy* 327-331 ISBN 9789535782285.

[7] Matotek R 2019 Exterior joinery - production, installation and quality control, *Gradevinar*, **71**(12) 1219 ISSN: 0350-2465.

[8] Weber M and Verma G 2016 *SolidWorks Simulation Black Book* (Eastman: CadCamCaeWorks) ISBN 9781523393749.

[9] https://www.designorate.com/practice-guide-to-solve-problems-with-triz/ accessed on: 12.04.2021.

Acknowledgments

SC ECO-Traiding SRL is acknowledged for its undisputed involvement in obtaining these results. Also, based on the research contract no. 2/2019 between the above-mentioned company and Vasile Alecsandri University of Bacau, the company allowed the use of these data while retaining the rights of possession of the geometries related to the products presented in this case-study.

Analysis of robotic welding possibilities of a car chassis assembly

M N Coman[1], C Rontescu[2], A M Bogatu[2] and D T Cicic[2]
[1]Piroux Romania, Calea Câmpulungului 55A, Mioveni, Romania
[2]University Politehnica of Bucharest, Faculty of Industrial Engineering and Robotics, Quality Engineering and Industrial Technologies Department, Independenţei Street No. 313, Romania

E-mail: dumitru.cicic@gmail.com

Abstract. The continuous increase of the requirements regarding the quality of the welded joints (elimination of the influence of the human factor) but also the increase of the productivity of the technological processes imposed the analysis of the possibilities of implementing robotic welding processes. The introduction of robotic production lines is an increasingly common thing in the automotive industry. This paper presents the stages necessary for the realization in welded construction of a crossbeam, part of the chassis of a car. The elaborated welding technology aimed to achieve the transition from semi-mechanized welding of the assembly to robotic welding using the MAG (Metal Activ Gas) welding process. The basic material from which the crossbar is made is S355 MC steel, steel for constructions with high mechanical properties. For the robotic welding of the assembly, the design of the device necessary for the orientation and fixing of the components was made. The welding technology developed was validated by performing standard samples that were subjected to visual examination and penetrant testing but also to destructive examinations which consisted in measuring the hardness in the characteristic points of the welding bead and also by measuring its geometric characteristics. Following the analysis of the initial dimensional values obtained, it resulted that the incorrect positioning of the welding head in relation to the welded components led to the appearance of geometric defects of the welding seams, although the technological parameters of the welding regime were appropriate. After making adjustments on the position of the welding head, accepted characteristics were obtained which led to the decision to validate the welding technology used.

1. Introduction

The continuous and rapid evolution of technological processes has required the development of new technologies and equipment capable of satisfying the increasingly complex requirements of the industrial environment. Welding processes are an important component in the machine building industry, with the advantages of low cost, good performance of connection and easy operation, etc [1-3]. Fusion welding is a process used in the automotive industry to achieve highly required non-removable assemblies, as is the case with chassis components. During welding, the basic materials are subjected to a thermal heating-cooling cycle, which influences their structure and properties. The Metal Activ Gas (MAG) process is one of the most used welding processes with applicability in many industrial fields having the advantages of a high productivity while offering a very good quality of the resulting welded joints. The main technological parameters of the welding regime (welding current, arc voltage, welding speed) directly influence the geometric characteristics of the resulting welding seams. The position of the welding head in relation to the parts to be assembled influences the shape of the welding bead but also the penetration depth for each component [4-6]. The geometric characteristics of the resulting weld bead can be highlighted with the help of macrographic examination. Following the analysis of the obtained results, recommendations can be made regarding the modification of the values of the welding regime parameters so that the resulting joints comply with the imposed quality requirements.

Increasing demands on the quality of welding seams have required the elimination, as far as possible, of the human factor in welding processes and the use of mechanized or robotic systems. The introduction of industrial robots eliminates the disadvantages of manual or semi-machined welding and meets the new quality and productivity requirements of technological processes [7-9]. Robot programming can be done online or offline and ensures process repeatability under the specified initial conditions [10-13].

The main objective of the paper was to obtain information on the geometric characteristics of welded seams obtained in the case of robotic welding, using the MAG process, of a chassis type car component (fusion line position, weld metal and heat-affected zone) as well as to analyse the interdependence between these characteristics and the technological parameters of the welding regime. Following the analysis of the initial results, recommendations were made regarding the modification of the values of the welding parameters so as to obtain the desired quality level of the welded joints, in order to validate the chosen welding technologies. In the case analysed in the paper, the car chassis consists of several cross members and assembly elements. Each cross member is welded on a robotic workstation following that in the end the final assembly of the car chassis to be performed.

2. Experimental data

A robotic cell was used for the realization of the welded joints which has in its component a two-axis positioner, welding fixture, Air Luquide welding power sources DIGIWAVE 500 and the welding robot Fanuc Arc Mate 120iC (six-axis servo motor), whose working area, indicated in figure 1, ensures the accessibility to all the necessary joining areas.

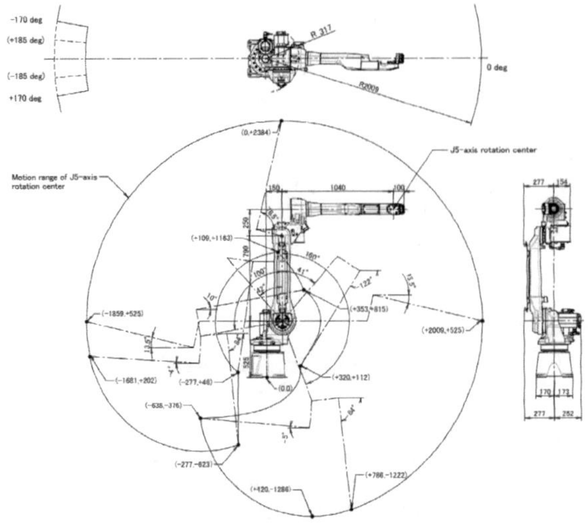

Figure 1. Fanuc ARC Mate 120iC working range.

The positioner together with the afferent devices ensures the fastening and fixing of the components of the chassis cross member in order to be joined by welding and with the help of the external axis of rotation the relative position of the welding head with respect to the components to be joined is ensured. The robotic welding cell used to weld the cross member is presented in figure 2.

Figure 2. Auto chassis welding workstation.

The main basic materials used to make the components of the car chassis cross assembly (figure 3 and figure 4) were steels: S355MC (fine grain structural steel thermomechanically rolled and special cold forming) for the main structure of the chassis and S355J2H (structural steel for hollow section) for additional components. The filler material used to make the welded joints was solid wire type G3Si1 according to EN ISO 14341: 2020 and the shielding gas used was the mixture type M2.0 according to EN ISO 14175: 2008 [14-17].

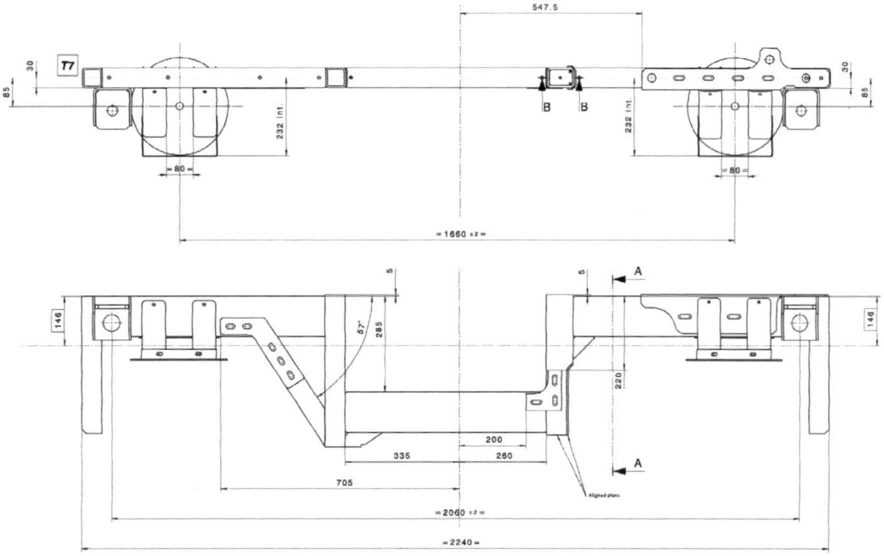

Figure 3. Nominal dimensions of the car chassis cross member.

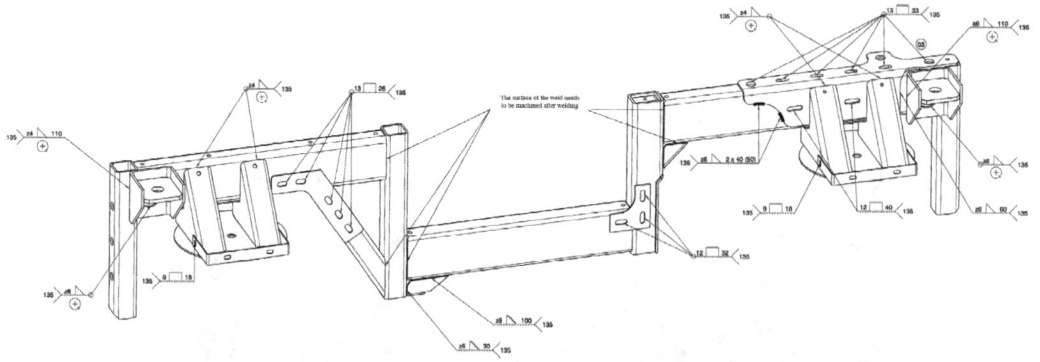

Figure 4. The welded joints necessary for the realization in welded construction of the car chassis cross member.

The chemical composition as well as the mechanical properties of the materials used to make the welded joints shown in figure 4, according to the standards in force, are presented in table 1 and table 2 [14-17].

Table 1. Chemical composition of basic materials and filler material [14-17].

	Weight Percent								
Materials	**C**	**Si**	**Mn**	**P**	**S**	**Al**	**V**	**Nb**	**Ti**
S355MC	0.12	0.50	1.50	0.025	0.020	0.015	0.20	0.09	0.15
S355J2H	0.22	0.55	1.60	0.030	0.030	-	-	-	-
G3Si1	0.06-0.14	0.80-1.0	1.4-1.6	0.025	0.025	-	-	-	-

Table 2. Mechanical properties of base materials and filler material [14-17].

Materials	Tensile Strength (MPa)	Yield Strength (MPa)	Elongation (%)
S355MC	950	880	14
S355J2H	895	828	10
G3Si1	470-600	380	20

For welding the chassis using the MAG process, depending on the thickness of the components, 5 types of joints were used, according to EN ISO 9692/1 [18]. The components were laser cut and mechanically machined to ensure the geometric configuration required for assembly. Two types of joints used to make the welded joints are presented in figure 5, where MB represents the base material.

After the mechanical processing of the joints, the components were fixed in devices and welded in the robotic cell using the working parameters presented in table 3. Some of the resulting welded joints are presented in figure 6.

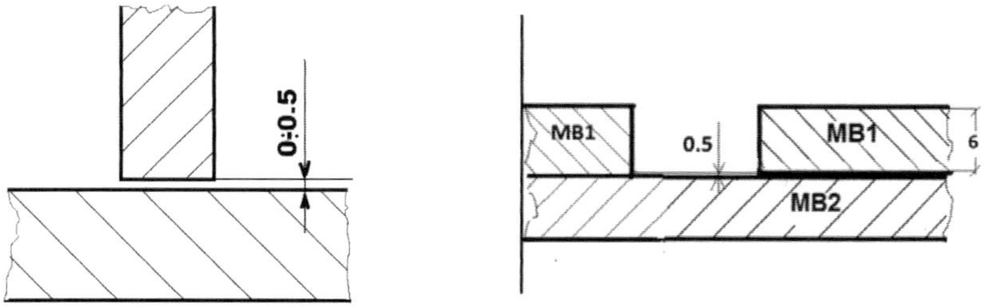

Figure 5. Types of joints used to weld the car chassis.

Table 3. The parameters used in the experiments.

No.crt.	Parameter	Value
1	Welding current [A]	210-220
2	Arc voltage [V]	22-24
3	Filler material diameter [mm]	1.2
4	Gas flow [l/min]	18-20
5	Travel speed [cm/min]	60
6	Type of gas protection	Ar+10%CO_2

3. Results

After finishing the welding process, the samples were subjected to visual and dimensional examination [19] and penetrant testing [20]. Following the analysis of the obtained results, it was observed that the samples do not show surface imperfections or that are related to the surface.

Figure 6. Resulted welded joints.

In order to avoid overheating of the samples and a possible structural change during cutting in order to perform the metallographic analysis [21], the samples were cut with low cutting speed on a device with continuous cooling.

Following the cutting process, the samples were first cleaned from impurities and then subjected to polishing using different granulations of metallographic paper; the final polishing was made with abrasive diamond paste. In order to reveal structural details the samples were etched with Nital 4%. [22].

After the metallographic preparation of the obtained samples, they were examined macroscopically with the help of an optical microscope, photographed and introduced in a computer program for measuring the geometric characteristics of the welded seams. The quality evaluation of the welded joints was made in accordance with the internal norms of the beneficiary. From the analysis of the welded joints obtained (figure 7) it can be seen that some of the samples showed a series of imperfections such as: surface pore, lack of penetration, intermittent undercut, incorrect weld toe, excessive asymmetry of fillet weld. The acceptance criteria for imperfections are presented in table 4. Possible causes of these types of non-conformities were: incorrect positioning of the welding head relative to the welding axis, free length of improper electrode wire or too low value of welding current intensity.

Table 4. Acceptance criteria for the welding imperfections (extras).

No.	Imperfection	Value
100	Crack	Not permitted
2017	Surface pore	Not permitted
402	Lack of penetration	Not permitted
505	Incorrect weld toe	$\alpha \geq 100^0$
512	Excessive asymmetry of fillet weld	$h \leq 2mm+0.5a$
5012	Intermittent undercut	$h \leq 0.1t$, but max. 0.5mm
617	Incorrect root gap for fillet welds	$h \leq 0.5mm+0.2a$, but max. 3mm

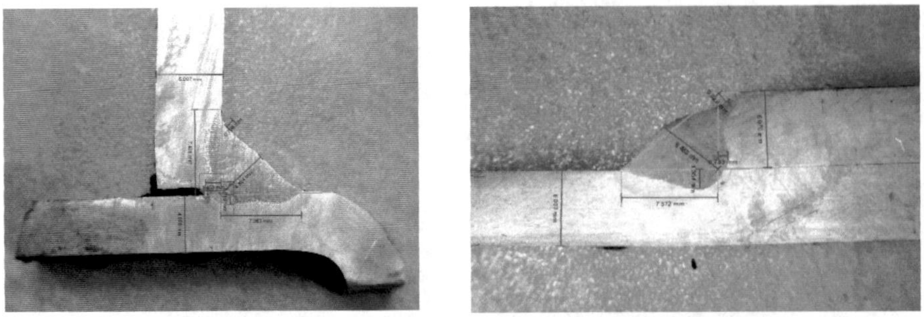

Figure 7. Metallographic evidence of non-compliant welded joints.

Taking into account the imperfections, a series of changes were proposed to the parameters of the welding regimes: increasing the value of the welding current in the case of thicker components, changing the position of the welding head depending on the geometric configuration of each joint and synchronizing the positioner movement with the welding head in order to ensure better accessibility at the welded joint. After making the proposed changes regarding the values of the technological parameters, the welded samples were remade on a test cross member. After the examinations of all the welded joints made with the improved technology (figure 8) no imperfections were detected, which led to the validation of the welding technologies used in this case. The robotic welding programs made were saved for use in series production.

Figure 8. Metallographic samples of compliant welded joints obtained after adjusting the welding process parameters.

The analysis of the geometric characteristics of the welded seams, with the help of computer programs, as can be seen from figure 7 and figure 8, provides the necessary information in order to decide their acceptance and the welding technology depending on the required level of acceptance of imperfections, in accordance with an international, national or internal norm, depending on the requirements of the beneficiary.

4. Conclusions

By using qualified robotic welding technologies and by adapting and validating them to the working conditions in the welding workshops, the human factor in the welding process has been eliminated. The introduction of robotic welding has led to high quality welded joints with a much lower chance of imperfections occurrence.

Changes to the welding regime parameters can be easily made and the validated welding technology can be repeated by resuming the working programme. Welding paths can be easily changed by programming the robot online or offline.

The costs of the initial investment necessary to purchase the robotic cell can be amortized quickly by reducing the costs necessary to fix any imperfections that may occur in welded joints due to the human operator, but also by reducing manufacturing times.

With the help of computer programs designed to analyse metallographic images it was possible to make measurements of the geometric characteristics of the welded seams and, by comparison with the values included in the acceptability criteria, the decision was made to validate the welding technologies use in the paper.

5. References

[1] Wan Z, Sperling D, Wang Y S. 2015 China's electric car frustrations. *J. Transport Research Part D:Transport and Environment.* **34** 116-121.

[2] Yuan X L, Liu X, Zuo J 2015 The development of new energy vehicles for a sustainable future: A review. *J. Renewable and sustainable Energy Reviews.* **42** 298-305.

[3] Delucchi M A, Yang C, Burke A F, Ogden J M, Kurani K, Kessler J, Sperling D 2014 An assessment of electric vehicles: technology, infrastructure requirements, greenhouse-gas emissions, petroleum use, material use, lifetime cost, consumer acceptance and policy initiatives. *J. Philos. Trans. R. Soc. A*. 20120325.

[4] I. Voiculescu, et al. 2004 Dimensional analysis of robotized welds, *U.P.B. Sci. Bull. Series B*, **66**, **2-4**, pp105 -112.

[5] Luciane B et al. 2017 Seam Tracking and Welding Bead Geometry Analysis for Autonomous Welding Robot, *IEEE Latin American Robotics Symposium*, Curitiba, Brazil, pp 1-6.

[6] A. S. Shleenkov et al. 2020 Metallographic examination of defects in welds performed by butt welding, *AIP Conference Proceedings* **2313**(1), 060036.

[7] Zhang Yan 2019 Discussion on the welding performance of automatic welding of industrial robot and manual welding, *Materials Science and Engineering* **637** 012004.
[8] Jianlong Huang et al. 2018 *IOP Conf. Ser.: Mater. Sci. Eng.* **394** 042070.
[9] Li Yang et al. 2019 *J. Phys.: Conf. Ser.* **1311** 012048.
[10] G Navaneeswar Reddy and M VenkataRamana 2018 *IOP Conf. Ser.: Mater. Sci. Eng.* **330** 012096.
[11] Bin Zhang et al. 2017 *IOP Conf. Ser.: Mater. Sci. Eng.* **207** 012036.
[12] Min Hu 2018 *IOP Conf. Ser.: Mater. Sci. Eng.* **307** 012073.
[13] Popa R and Ghionea A 2013 Robotic welding cells structure optimization, *U.P.B. Sci. Bull., Series D* **75**(1) pp 107-120.
[14] EN 10149-2-2013 - *Hot rolled flat products made of high yield strength steels for cold forming - Part 2: Technical delivery conditions for thermomechanically rolled steels.*
[15] EN 10210-1:2006 *Hot finished structural hollow sections of non-alloy and fine grain steels. Technical delivery requirements.*
[16] EN ISO 14341:2020 - *Welding consumables - Wire electrodes and weld deposits for gas shielded metal arc welding of non-alloy and fine grain steels – Classification.*
[17] EN ISO 14175:2008 - *Welding consumables - Gases and gas mixtures for fusion welding and allied processes.*
[18] EN ISO 9692-1:2014 - *Welding and allied processes - Types of joint preparation — Part 1: Manual metal arc welding, gas-shielded metal arc welding, gas welding, TIG welding and beam welding of steels.*
[19] EN ISO 17637:2016 - *Non-destructive testing of welds - Visual testing of fusion-welded joints.*
[20] EN ISO 3452-1:2013 - *Non-destructive testing - Penetrant testing - Part 1: General principles.*
[21] EN ISO 17639-2014 - *Destructive tests on welds in metallic materials - Macroscopic and microscopic examination of welds.*
[22] Chivu, O et al. 2016 The effects of reconditioning by welding of crankshafts in automotive industry, *Metalurgija*, **55**(1) pp 55-58.

The nonlinear ship rolling and safe basin erosion in stochastic beam seas

D Deleanu and C L Dumitrache
Constanta Maritime University, Department of General Engineering Sciences, Mircea cel Batran street, No. 104, 900663, Romania.

E-mail: dumitrudeleanu@yahoo.com

Abstract. A short number of steep breaking waves hitting a ship from the side may affect its dynamic behaviour and determine extreme roll angles and, eventually, capsizing. Even the capsize is a rare event, its consequences for the ship and crew are often fatal. Generally, the ship rolling in regular or stochastic beam seas could be described by a second-order non-linear differential equation with the roll angle as dependent variable. Non-linearity comes from the restoring and damping moments, which are usually represented by polynomials of roll angle or of its time derivative. In the paper, we used such a model equation to estimate the ship rolling and capsizing in a stochastic beam sea. The sea action was simulated by a harmonic function with random frequency and phase. The roll equation was solved using a simple, fast and accurate iterative scheme based on Taylor expansion which proved to be very competitive in terms of accuracy with more elaborate methods and which allowed a substantial reduction of the CPU time. Due to these properties of the scheme, we were able to conduct an extensive investigation on fractal erosion of safe basins and to represent the boundaries between capsizing and non-capsizing regions in wave frequency – wave amplitude plane and normalized integrity curves for different combinations of random wave parameters. The restoring and damping coefficients corresponded to a vehicle ferry model, considered to be either with or without bilge keels.

1. Introduction

In ship motion analysis, the study of large amplitude nonlinear rolling is crucial because it is closely linked with capsizes dynamics. For such roll analysis, linear approximation is no longer valid and, as a consequence, obtaining closed – form solutions becomes difficult or even impossible.

The nonlinear effects occur mainly due to nature of restoring, damping and external loadings' moments. Various nonlinear models for ship rolling have been proposed in the last decades. In most of them, the restoring moment is approximated by quadratic, cubic or even higher – order polynomials of roll angle. The damping is usually inserted in the model equation by means of linear – quadratic or linear – cubic terms in the angular roll velocity. Finally, the sea loads could be set for regular or stochastic waves [1 – 3].

Capsizing or *keeling over* is that catastrophic situation in which the ship is turned on its side or is upside down. Both theoretical and experimental studies have identified several mechanisms for ship capsize, including pure resonant rolling, parametric excitation, broaching and loss of stability at a wave crest. It was emphasized the fact that more attention should be paid to capsize under transient, rather than steady – state conditions. This situation corresponds to a short train of regular/random waves hitting the ship from the side in otherwise calm whether conditions. Dangerous large – amplitude motions and, finally, capsizing can appear when the ship is affected by no more than 8 – 10 sufficiently steep waves. Tools of nonlinear dynamics such as safe basins, integrity curves or Lyapunov exponents are involved in an attempt to understand the mechanisms behind ship capsizing [4 – 8].

If the rolling in beam waves is considered, it is possible to ignore the coupling of rolling and the other five degrees of freedom of ship motion. It results that for solving the rolling problem and thus the capsizing scenario, we can numerically integrate a second – order nonlinear differential equation with the roll angle as dependent variable. Numerical integrators incorporated in modern software packages, as *ode45* in Matlab, encounter efficiency problems for the unbounded solutions associated with capsizing. This is why, in this paper, we have developed our own software that uses a simple and accurate iterative scheme based on Taylor expansion to sole a typical nonlinear roll equation [9]. The coefficients for damping and restoring moments correspond to an intact vehicle ferry model equipped or not with bilge keels.

2. The nonlinear roll equation

The differential equation of ship roll motion under the random beam seas is given by

$$(I_{44} + A_{44})\ddot{\theta} + B_{44}\dot{\theta} + B_{44q}\dot{\theta}^3 + \Delta \cdot GZ(\theta) = M_{sea}(t) \tag{1}$$

where $I_{44} + A_{44}$ represents the sum between ship's moment of inertia in roll and added mass inertia, B_{44} and B_{44q} denote the linear and nonlinear damping coefficients, Δ the ship displacement, $GZ(\theta)$ the arm of static stability and $M_{sea}(t)$ the torque caused by random beam waves. Additionally, θ represents the roll angle and a dot over denotes the time differentiation.

Considering a cubic polynomial approximation for $GZ(\theta)$, namely $C_1\theta + C_3\theta^3$, and dividing by the total moment of inertia, one obtains the following equation:

$$\ddot{\theta} + d_1\dot{\theta} + d_3\dot{\theta}^3 + k_1\theta + k_3\theta^3 = \xi(t) \tag{2}$$

where

$$d_1 = \frac{B_{44}}{I_{44} + A_{44}}, d_3 = \frac{B_{44q}}{I_{44} + A_{44}}, k_1 = \frac{\Delta C_1}{I_{44} + A_{44}}, d_3 = \frac{\Delta C_3}{I_{44} + A_{44}}, \xi(t) = \frac{M_{sea}(t)}{I_{44} + A_{44}}$$

The excitation of the stochastic beam waves is shaped by a harmonic function with random frequency and phase:

$$\xi(t) = \mu\, cos(\omega t + \psi), \psi = \sigma B(t) + \Gamma \tag{3}$$

where ω is the center frequency, μ and σ the amplitude and the intensity of the excitation, $B(t)$ a Wiener process and Γ a random variable with the uniform distribution in $[0, 2\pi]$ (see [10]).

For numerical simulation, the damping and restoring coefficients we chosen for a real vehicle ferry ship [4]. If the ship was equipped with bilge keels then $d_1 = 0.0476542\, s^{-1}, d_3 = 3.765\, s$, otherwise $d_1 = 0.01265913\, s^{-1}, d_3 = 0.4954\, s$. For the restoring coefficients, a cubic spline interpolation of the static stability curve results in $k_1 = 0.69199703\, s^{-2}, k_3 = -0.5392039\, s^{-2}$. The initial conditions are of the form $\theta(0) = \theta_0$ and $\dot{\theta}(0) = \dot{\theta}_0$.

3. The iterative scheme

In order to benefit from a better flexibility in the integration of equation (2) and to obtain a reasonably short CPU time, especially in the transition area from bounded to unbounded solutions associated with capsizing, we developed a software based on a simple and accurate scheme proposed in [11].

Adapted to our problem, the scheme reads as

$$\begin{aligned} \theta_n &= \theta_{n-1} + \dot{\theta}_{n-1}\Delta t + 0.5\, \ddot{\theta}_{n-1}(\Delta t)^2 \\ \theta_{n+1} &= \theta_{n-1} + 2\, \dot{\theta}_{n-1}\Delta t + 2\, \ddot{\theta}_{n-1}(\Delta t)^2 \\ \dot{\theta}_n &= (\theta_{n+1} + \theta_{n-1})/(2\, \Delta t) \quad, n \geq 2 \\ \ddot{\theta}_n &= -d_1\dot{\theta}_n - d_3\dot{\theta}_n^3 - k_1\theta_n - k_3\theta_n^3 + \xi\big((n-1)\Delta t\big) \end{aligned} \tag{4}$$

The starting values are

$$\theta_1 = \theta(0), \dot{\theta}_1 = \theta(0), \ddot{\theta}_1 = \mu - d_1\dot{\theta}(0) - d_3\dot{\theta}^3(0) - k_1\theta(0) - k_3\theta^3(0) \tag{5}$$

The scheme (4) is $O(\Delta t^2)$ of accuracy.

4. Numerical simulations

In this section, with the help of scheme (4), we carried out a thorough numerical investigation on the solutions of roll equation (2) for different combinations (ω, μ, σ) of wave parameters. The computations were performed with the Matlab software package.

To begin with, let us form an image of the influence exerted by the intensity σ of the random phase ψ on the shape of the wave moment. Figure 1 presents the graph of $\xi(t)$ for $\omega = 0.9, \mu = 0.3, \sigma \in \{0.2, 0.6\}$, $\Delta t = 0.0014, t \in [0, 4 \cdot 2\pi/\omega]$. The continuous dark line is associated with the periodic wave $f = \mu\, cos(\omega t)$, that is the function ξ with $\sigma = 0$. It is obvious that for small σ function ξ is a narrow – band process, the difference between the regular and random waves being relatively insignificant. The shape of the waves' moments separates more and more with the increase of σ.

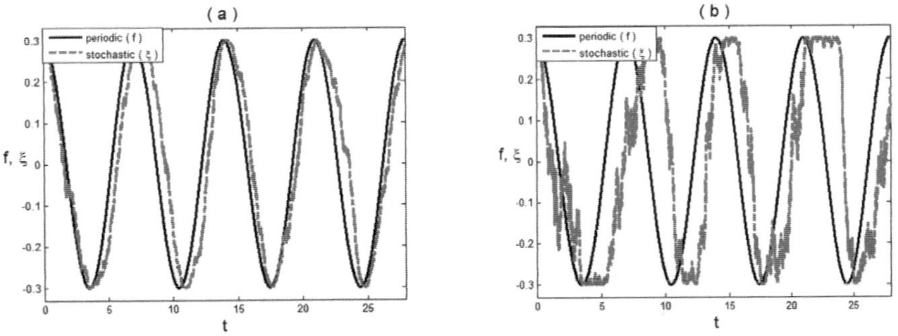

Figure 1. The time evolution of function $\xi(t)$ for $\omega = 0.9, \mu = 0.3$, $\Delta t = 0.0014$ and $(a)\ \sigma = 0.2$; $(b) \sigma = 0.6$. f represents $\xi(t)$ with $\sigma = 0$.

The size of the step Δt influences to some extend the stochastic wave moment, as illustrated in figure 2. In order to keep the solution obtained with (4) as close as possible to the one provide by Matlab's solver ode45, in the following computations the value $\Delta t = 0.001$ was used.

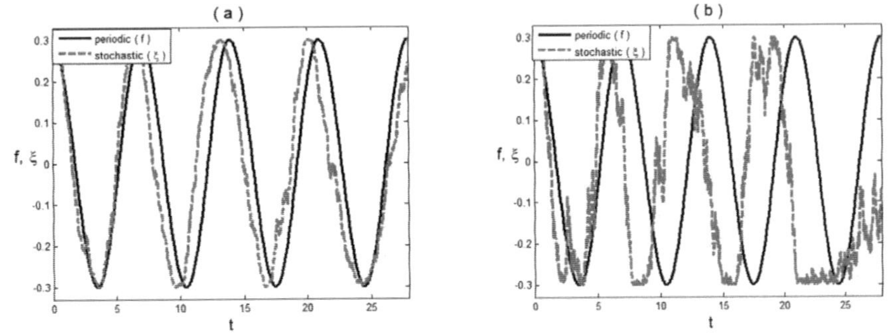

Figure 2. The same as in figure 1 but for $\Delta t = 0.0028$.

As we mentioned in the introductory section, the probability of capsizing for given wave parameters is higher in the transient regime. If capsize does not occur within 8 – 10 cycles of forcing than it is unlikely to appear later. The cumulative conditions $\theta > 1.5\, rad$ and $t < 10 \cdot 2\pi/\omega$ were the basis of the representations in figure 3, where the dark (green) colour corresponds to the pairs (ω, μ) for which the solutions of equation (2) are bounded and the light one (yellow) to the pairs in which the ship is in a capsized state. The four panels, which describe the behaviour of the ship without bilge keels, indicate that, as the degree of stochasticity increases, lower and lower amplitudes μ are sufficient for the ship to capsize (for a given ω). A similar situation is encountered for the ship with bilge keels in operation.

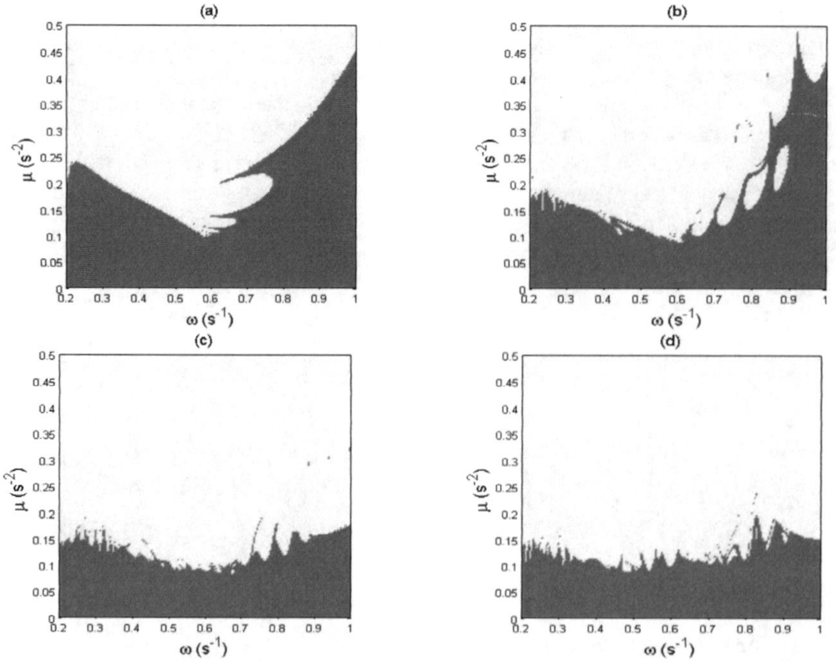

Figure 3. Separation of the plane (ω, μ) in regions with non-capsizing effect (dark area) or capsizing effect (light area) for the ship not equipped with bilge keels:
(a) $\sigma = 0$; *(b)* $\sigma = 0.2$; *(c)* $\sigma = 0.4$; *(d)* $\sigma = 0.4$.
The initial conditions were $\theta(0) = \dot{\theta}(0) = 0$.

Safe basin concept has been introduced in 1990s for the study of nonlinear ship rolling and capsize. The safe basin denotes the set of initial conditions in the phase plane $(\theta, \dot{\theta})$ which define the separatrix between capsize and non-capsize areas and illustrate the high sensitivity of capsize to initial conditions. In fact, for a given set of parameters (ω, μ), the safe basin of attraction is formed by all initial conditions $(\theta(0), \dot{\theta}(0))$ that do not lead to capsize. The size, shape and locations of safe basins provide valuable information about the engineering integrity of the ship. The numerical simulation showed that an increasing of forcing amplitude μ leads to a process of fractal erosion of the safe basin, finished with a sharp decrease of safe area. We will prove this in the following by examples for three ω values.

The initial conditions were selected from a vast set having 40,401= 201 x 201 elements, obtained by dividing the rectangle $[-1.2,1.2]\times[-1.0,1.0]$ in equally spaced segments. The running time was chosen equal to $10T$, with $T = 2\pi/\omega$. Each point is tested against the escape criterion $\theta > 1.5\, rad$, and a colored small rectangle around the initial condition is used to indicate the time after which this criterion was satisfied. It turns out that the white region will correspond to the safe basin.

<u>**Case 1**</u>: $\omega = 0.9\, rad/s$ $(10T = 69.77\, s)$

Figure 3 indicates that this is a relatively safe frequency, because the ship capsizing occurs for large forcing μ. Keeping fixed initial conditions, one observes a rapid decrease in the amplitudes μ required for ship to capsize with the wave randomness increase, as illustrated in figure 4.

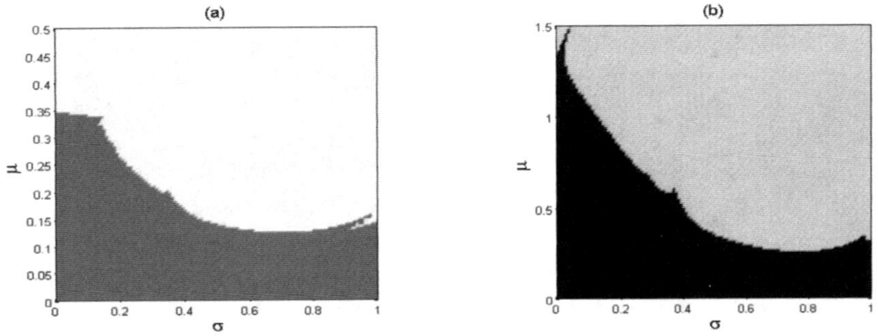

Figure 4. Separation of the plane (σ, μ) in regions with non-capsizing effect (dark area) or capsizing effect (light area) for the ship (*a*) without bilge keels; (*b*) with bilge keels. The central frequency was $\omega = 0.9\,rad/s$ and $\theta(0) = \dot{\theta}(0) = 0$.

a) *Ship without bilge keels*

For a perfect sinusoidal excitation ($\sigma = 0$) and $\mu = 0$, the safe basin contains 66% from the tested initial conditions (see figure 5(a)). By increasing μ the basin erodes slightly from outside but the important loss occurs from the inside. Here, the start is given by a few thin "fingers" that begin to invade into the safe basin (see figure 5(b)). If μ continues to grow, these objects thicken and finally cover the entire safe area (figures 5(c) and (d)).

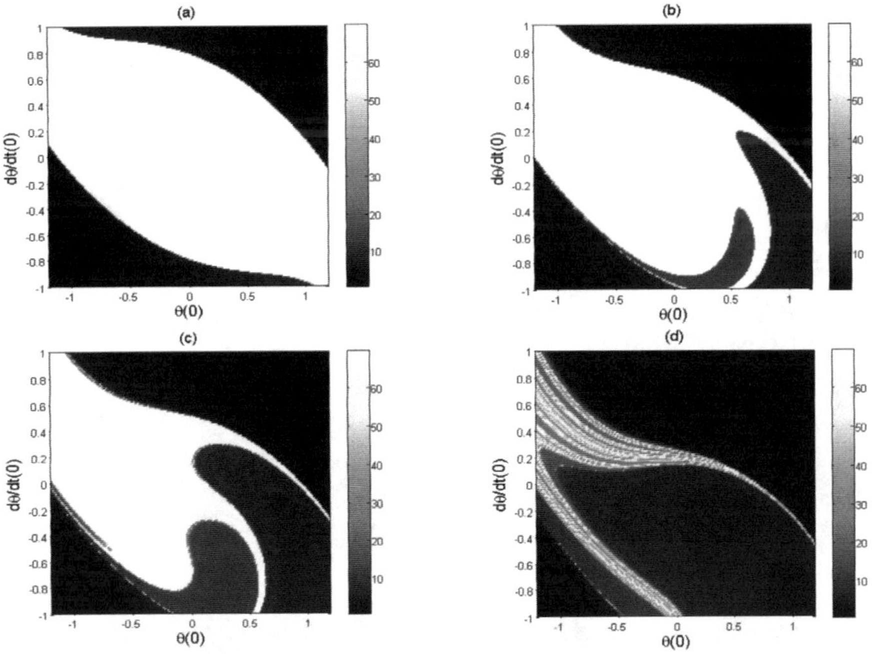

Figure 5. Safe basin's fractal erosion for the ship without bilge keels if $\omega = 0.9$ and $\sigma = 0$. (*a*) $\mu = 0$; (*b*) $\mu = 0.2$; (*c*) $\mu = 0.32$; (*d*) $\mu = 0.62$. The colour bar shows the time required for the criterion $\theta < 1.5\,rad$ to be violated.

For $\sigma \neq 0$, this internal erosion occurs and develops for lower amplitudes μ. The appearance of invasive formations in the safe basin may differ with the increase of σ, as shown in figure 6.

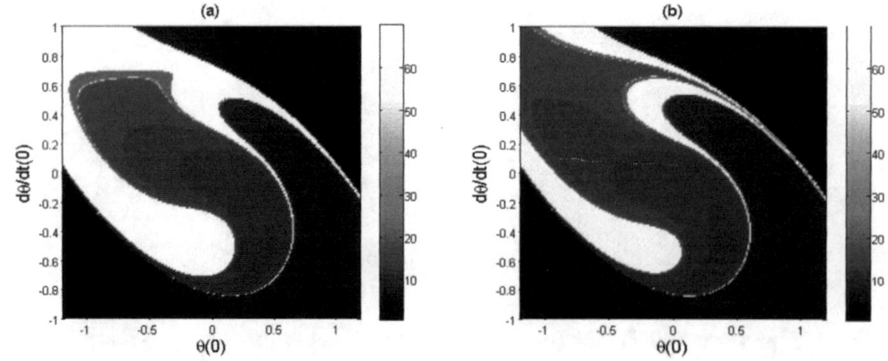

Figure 6. Safe basin's fractal erosion for the ship without bilge keels if $\omega = 0.9$ and $\sigma = 0.4$. (*a*) $\mu = 0.2$; (*b*) $\mu = 0.22$.

b) *Ship with bilge keels*

The damping supplement introduced by bile keels has two immediate effects: the widening of the safe basin (which reaches 88% of the tested initial conditions for $\sigma = \mu = 0$) and doubling or even tripling the value of the amplitude μ required for overturning (depending on the initial conditions). Thus, figure 7 shows a basin in the shape of a curvilinear parallelogram that erodes only from outside, through stripes of initial conditions that pass into the capsizing area with the increase of μ.

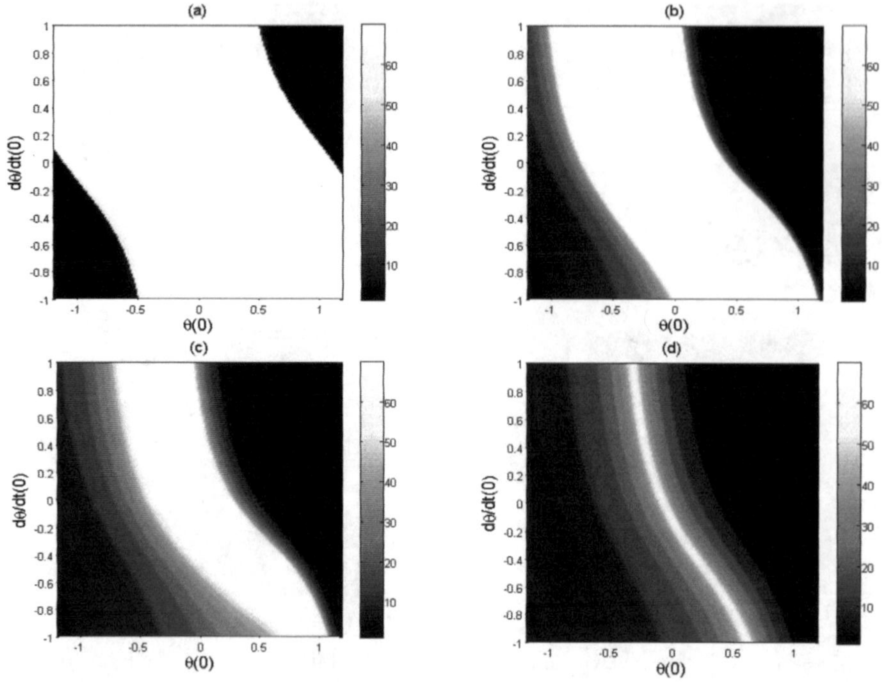

Figure 7. Safe basin's fractal erosion for the ship with bilge keels if $\omega = 0.9$ and $\sigma = 0.0$. (*a*) $\mu = 0.0$; (*b*) $\mu = 1.1$; (*c*) $\mu = 1.2$; (*d*) $\mu = 1.6$.

The increase of the degree of stochasticity of the waves causes the disappearance of the safe basin to occur at much lower values of amplitude μ (see figure 4(b)). The erosion mechanism may follow other pathways, as shown in figure 8. It is also worth noting that very small increases in amplitude μ have a tremendous effect on the safe basin's size.

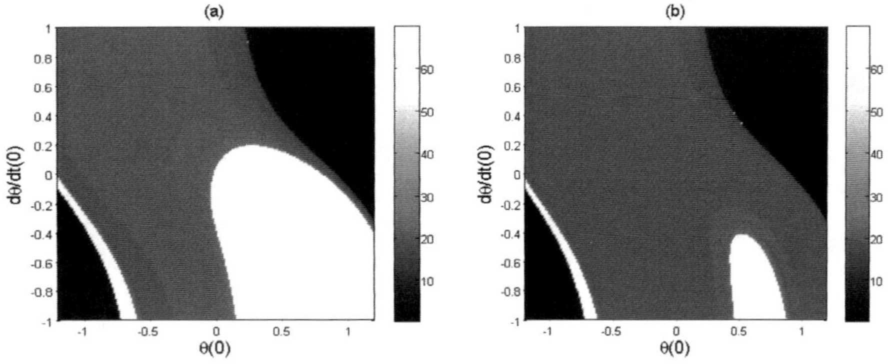

Figure 8. Safe basin's fractal erosion for the ship with bilge keels if $\omega = 0.9$ and $\sigma = 0.6$. (a) $\mu = 0.3135$; (b) $\mu = 0.315$.

Case 2: $\omega = 0.6\, rad/s$ $(10T = 104.7\, s)$

Figure 3 tells us that this is one of the most dangerous central wave frequencies for the ship, capsizing occurring for a minimal forcing amplitude μ. On the other hand, the degree of stochasticity influences only to a small extend the transition to boundless solutions of the roll equation, as shown in figures 3 and 9.

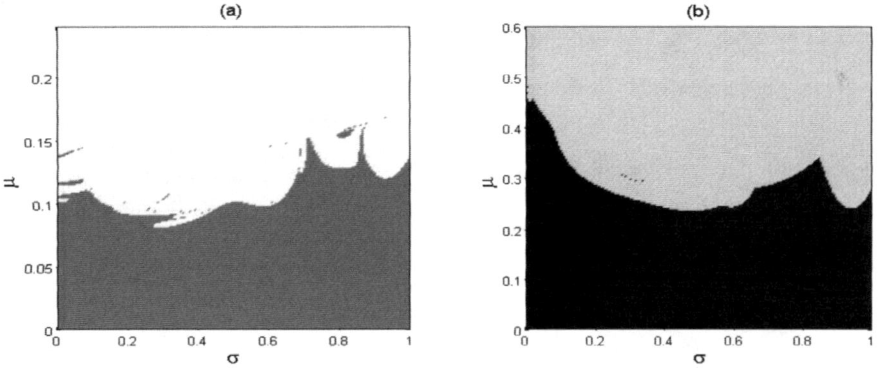

Figure 9. Separation of the plane (σ, μ) in regions with non-capsizing effect (dark area) or capsizing effect (light area) for the ship (a) without bilge keels; (b) with bilge keels. The central frequency was $\omega = 0.6\, rad/s$ and $\theta(0) = \dot{\theta}(0) = 0$.

Which is really remarkable for $\sigma \neq 0$, no matter if the ship is equipped with bilge keels or not, is the unfailing and then the collapse of the safe basin when exceeding a threshold of the amplitude μ. Figures 10 and 11 represent two examples in this sense, the erosion mechanism undergoing only small changes compared to the case $\omega = 0.9$.

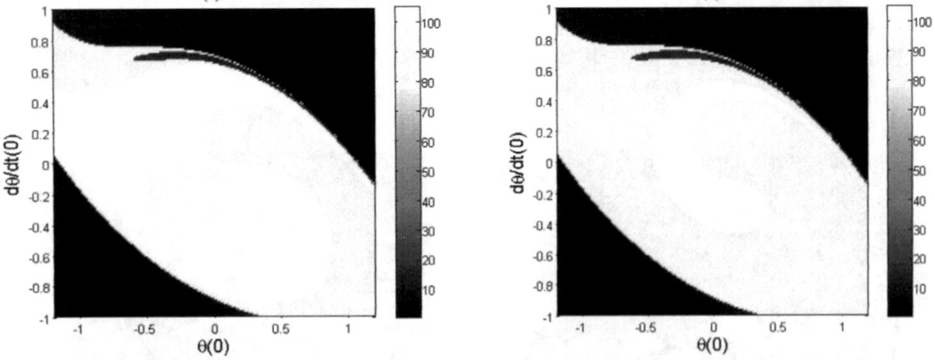

Figure 10. Safe basin's fractal erosion for the ship without bilge keels if $\omega = 0.6$ and $\sigma = 0.4$. (*a*) $\mu = 0.088$; (*b*) $\mu = 0.0885$.

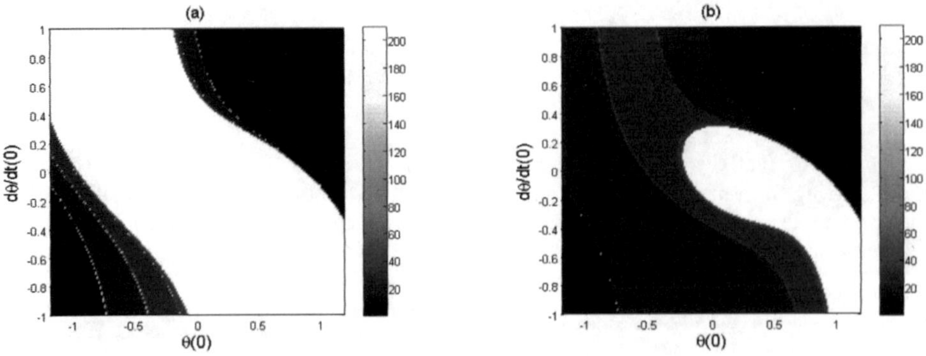

Figure 11. Safe basin's fractal erosion for the ship with bilge keels if $\omega = 0.6$ and $\sigma = 0.4$. (*a*) $\mu = 0.272$; (*b*) $\mu = 0.273$.

The *integrity curves* show the relative influence of wave excitation's amplitude μ on capsize relative to vessel safety in the absence of incident waves. They could be generated by plotting the safe area, normalized to unity at $\mu = 0$. In order to have a common reporting base, this normalization was done relative to the case of the ship equipped with bilge keels. This situation corresponds to a maximum safe basin. Figures 12 and 13 present the integrity curves associated to the ship without/with bilge keels for $\omega \in \{0.3, 0.6, 0.9\}$ and $\sigma \in \{0.0, 0.2, 0.4, 0.6\}$. They were obtained by interpolating 10 to 20 safe basins' sizes for different forcing amplitudes μ.

Analysing the two figures we find that, as σ grows, the values of amplitude μ for which the safe basin is completely empty are closer and closer for the different values of the frequency ω. The sea stochasticity has a levelling effect on the influence of parameter ω. At the same time, the additional damping due to the bilge keels considerably increases the area below the normalized integrity curves. As an example, for $\omega = 0.6$ and $\sigma = 0.4$, the ratio of these areas is four in favour of the ship equipped with bilge keels.

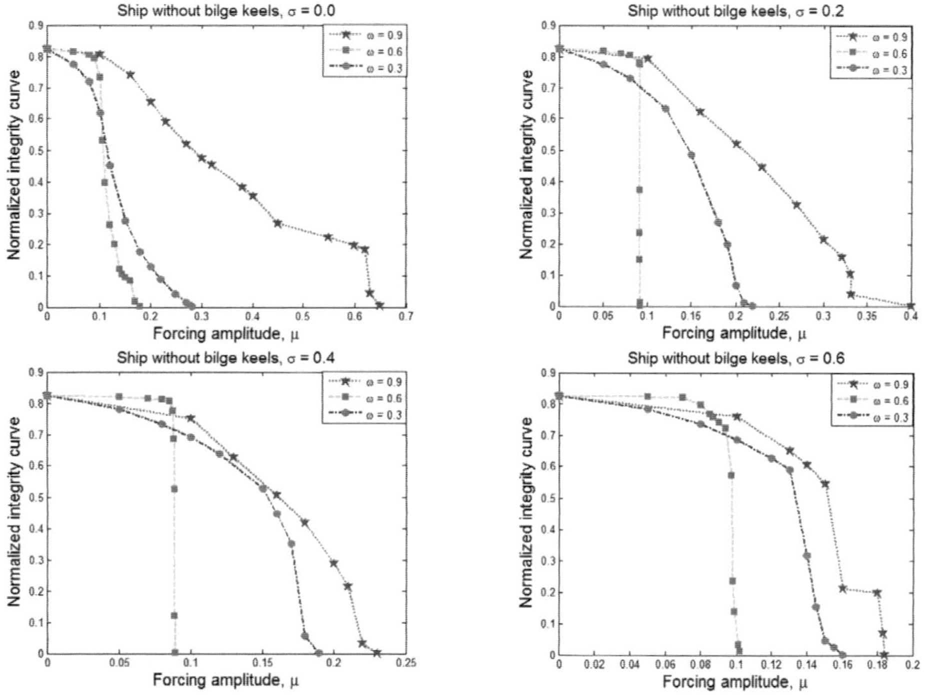

Figure 12. Normalized integrity curves for the ship without bilge keels.

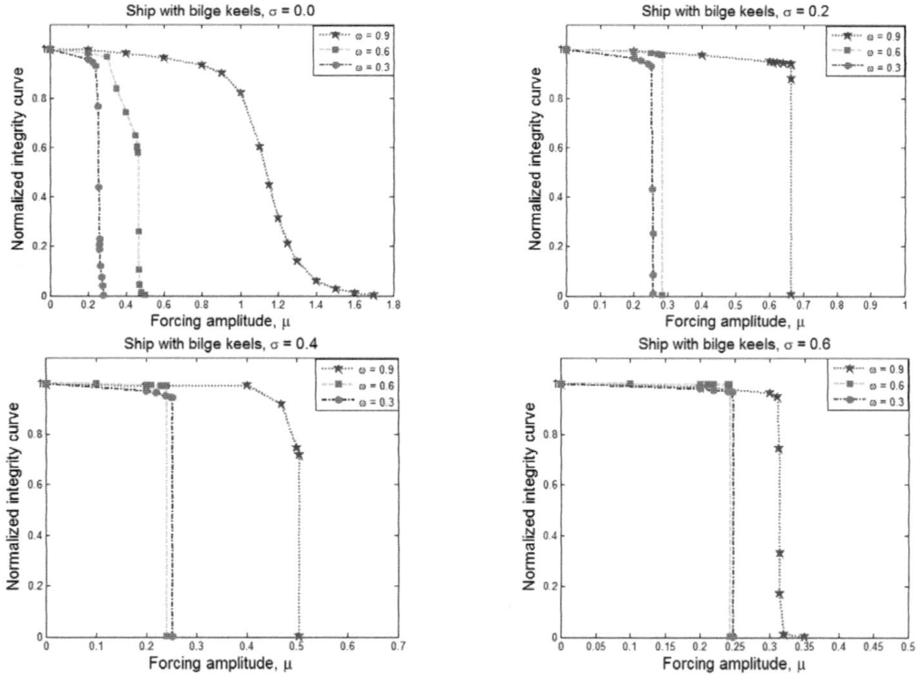

Figure 13. Normalized integrity curves for the ship with bilge keels.

In order to highlight even more clearly the influence of sea stochasticity on the size of the safe basin as well as its sudden erosion when reaching a limit value of amplitude μ, figure 14 shows the normalized integrity curves for $\omega = 0.6$ and $\sigma \in \{0.0, 0.2, 0.4, 0.6\}$. Once the profile of the waves' disturbing moment has moved away from the purely sinusoidal one (regular sea), the integrity curves almost overlap and the surface below them is only slight different from that of a rectangle.

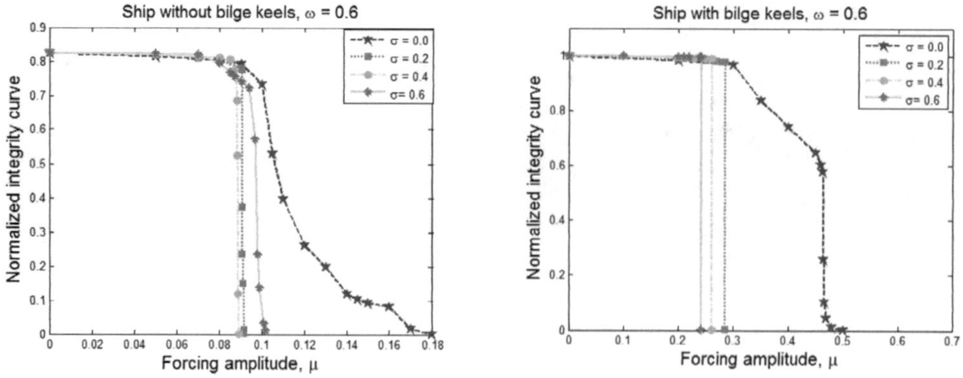

Figure 14. Influence of sea stochasticity on the safe basin for a fixed central frequency ω.

5. Conclusions

In the present study, a simple, fat and flexible iterative scheme based on Taylor expansion was applied to a nonlinear equation for estimating the roll motion of a ship in a stochastic beam sea. The speediness of the scheme was a determining factor in revealing several mechanisms responsible for eroding the safe basin of the ship and in construction of the so-called normalized integrity curves.

The main conclusions of the study are as follows:

a) As the amplitude of the sea excitation increases, the safe basins erode both from the outside and from the inside. How this erosion occurs depends on the excitation central frequency and on the damping moment;

b) For a regular beam sea, there are frequencies for which the ship is more prone for capsizing. In a stochastic beam sea, as the degree of randomness increases, a leveling effect of the excitation frequency's influence on the ship's response occurs;

c) In a stochastic sea there is, for each central frequency, a threshold for the excitation amplitude at which the safe basin suddenly erodes. These limits tend to approach with the removal of the purely sinusoidal perturbation associated with the regular sea;

d) The usefulness of the bilge keels in preventing the ship capsizing was well evidenced by the obtained numerical results.

6. References

[1] Taylan M 2000 The effect of nonlinear damping and restoring in ship rolling *Ocean Engineering* **27**(9) 921 – 932.

[2] El Bassouny A F 2007 Nonlinear analysis for a ship with a general roll damping model *Physica scripta* **75**(5) 691 – 701.

[3] Li S and Wang K 2017 Chaos analysis of ship rolling motion in stochastic beam seas *Journal of Vibroengineering* **19**(8) 6403 – 6411.

[4] Pedisic Buca M and Senjanovic I 2006 Nonlinear ship rolling and capsizing *Brodogradnja* **57**(4) 321 – 331.

[5] Hinz T 2007 Mathematical models in description of capsizing scenarios *Archives of Civil and Mechanical Engineering* **7**(3) 125 – 134.
[6] Deleanu D 2014 On a geometric approach of safe basin's fractal erosion. Application to the symmetric capsize equation *Constanta Maritime University Annals* **22** pp 125 – 130.
[7] Deleanu D 2015 On approximating the periodic solutions of capsize equation *International Journal for science, technics and innovations for the industry* Year **IX** Issue 9 pp 9 – 12.
[8] Scurtu I C, Clinci C and Popa A 2018 Water interference effect on ships due to square shaped object shielding *IOP Conference Series: Earth and Environmental Sciences* **172**(1) 012030.
[9] Deleanu D, Panaitescu M and Panaitescu FV 2020 An accurate and fast iterative scheme for estimating the ship rolling and capsizing in regular waves *IOP Conference Series: Mater. Sci. Eng.* **916** 012025.
[10] Liu L, Tang Y and Li H 2009 Random jumping of ships with water on deck in beam seas *Ship Building in China* **150** pp 1 – 11.
[11] Okasha El Nady A and Lashin M A 2016 Approximate solution of nonlinear Duffing oscillator using Taylor expansion *Journal of Mechanical Engineering and Automation* **6**(5) 110 – 116.

Computational fluid dynamics simulations for composite rocket propellant optimization

F-M Dîrloman[1], L-C Matache[2], T Rotariu[2], T-V Ţigănescu[3], D Zvîncu[2], M-I Ungureanu[1] and O Iorga[4]
[1]Military Technical Academy „Ferdinand I", Doctoral School „Defense and Security Systems Engineering", Blvd. George Coşbuc, No. 39-49, 050141, Bucharest, Romania
[2]Military Technical Academy „Ferdinand I", Faculty of Integrated Weapons Systems, Engineering and Mechatronics, Blvd. George Coşbuc, No. 39-49, 050141, Bucharest, Romania
[3]Military Equipment and Technologies Research Agency, Aeroportului Street, No. 16, 077025, Clinceni, Ilfov, Romania
[4] Scientific Research Center for CBRN Defence and Ecology, Olteniţei Street, No225, 041309, Bucharest, Romania

E-mail: florin.dirloman@mta.ro

Abstract. When designing a rocket engine configuration, both in terms of propellant grain, combustion chamber and nozzle geometry, one of the most convenient approach is using Computational Fluid Dynamics (CFD) Simulation. Numerical simulation is an alternative method of scientific investigation, which substitutes large number of experiments that often imply high financial burden and are also dangerous for the personnel involved. The numerical approach is often more useful than consecrated experimental method because it provides complete data that cannot be directly observed or measured, or it is difficult to highlight by other means. In this study we focused on applying CFD simulation to composite rocket propellants in a rocket engine with convergent-divergent nozzle configuration using Ansys Fluent Software. An ammonium nitrate (AN) based composite rocket propellant having four components was analyzed: oxidizer, metallic fuel, binder and catalyst agent. Explo5®, a thermochemical software, was also used to calculate the equilibrium compositions of the combustion products in the combustion chamber. It turned out that the results obtained on the basis of the simulation are consistent with those of the experimental testing. The data collected so far will be used to optimize the grain configuration of the composite rocket propellant.

1. Introduction

Solid rocket composite propellants are solid heterogeneous mixtures, consisting of three basic components: oxidizer, metallic fuel and binder. The usual oxidizer for solid rocket propellants is ammonium perchlorate (NH_4ClO_4), but in the last decades, ammonium nitrate (NH_4NO_3) and ammonium dinitramide, are under intense study due to their remarkable performance and green character [1]. In the case of metallic fuel, taking into account the compatibility with the oxidizer and the performance character, the most suitable solutions seems to be aluminum (Al) and magnalium (Al-Mg), an aluminum-magnesium alloy (50 : 50 wt.%) [1]. The last element of the propellants is the one creating the matrix that incorporates the oxidizer-metallic fuel mixture, providing homogeneity, mechanical strength and protection towards external agents. Currently the most widely used binders, are hydroxyl-terminated polybutadiene (HTPB), an inert polymer and glycidyl azide polymer (GAP), an energetic polymer [1]. In addition to the aforementioned purpose, the binder also serves as combustible element.

In order to determine the most suitable heterogeneous mixture, before performing the experimental part, rocketry specialists use various optimization software programs. Computational Fluid Dynamics (CFD) Simulation represents one of the most convenient approach prior to propellant and motor rocket design. Numerical simulation is an alternative method of scientific investigation, which substitutes large number of experiments that often imply high financial burden and are also dangerous for the personnel involved.

The main purpose of this study was to determinate the variation of some important parameters, pressure, temperature and gas flow velocity, for a proposed AN-based composite rocket propellant, in a rocket motor chamber with convergent-divergent nozzle configuration using Ansys Fluent Software [2, 3].

2. Computational procedure

Details on theoretical equations governing the rocket motor, internal geometry, mesh configuration, set-up conditions and combustion modeling of the employed composite propellant are given hereafter.

2.1. Mathematical model

Fluent software can solve the problem in two ways, based on pressure or density parameters. Given that the problem analyzed is of great complexity, it seems that the pressure-based solution seems to be the most reliable choice. Due to its design, it offers a faster solution to the problem. The mathematical model behind the program running in the Fluent Software is indicated below.

The mass conservation equation in the case of a two-dimensional, axial-symmetric configuration has the following form:

$$\frac{\partial \rho}{\partial t}+\frac{\partial}{\partial x}(\rho v_x)+\frac{\partial}{\partial r}(\rho v_r)+\rho v_r r^{-1}=S_m \tag{1}$$

where: t_x- axial coordinate; r -radial coordinate; v_x - axial velocity; v_r- radial velocity; ρ - propellant density; S_m -mass source of material.

We can consider the impulse conservation equations, for:

- axial direction

$$\begin{aligned}&\frac{\partial}{\partial t}(\rho v_x)+r^{-1}\frac{\partial}{\partial x}(r\rho v_x^{\,2})+r^{-1}\frac{\partial}{\partial r}(r\rho v_r v_x)=\\&-\frac{\partial p}{\partial x}+r^{-1}\frac{\partial}{\partial x}\left[r\mu\left(2\frac{\partial v_x}{\partial x}-\frac{2}{3}(\nabla\vec{v})\right)\right]+r^{-1}\frac{\partial}{\partial r}\left[r\mu\left(\frac{\partial v_x}{dr}+\frac{\partial v_r}{dx}\right)\right]+F_x\end{aligned} \tag{2}$$

- radial direction

$$\begin{aligned}&\frac{\partial}{\partial t}(\rho v_r)+r^{-1}\frac{\partial}{\partial x}(r\rho v_r v_x)+r^{-1}\frac{\partial}{\partial r}(r\rho\, v_r^{\,2})=\\&-\frac{\partial p}{\partial r}+r^{-1}\frac{\partial}{\partial x}\left[r\mu\left(\frac{\partial v_r}{dx}+\frac{\partial v_x}{dr}\right)\right]+r^{-1}\frac{\partial}{\partial r}\left[r\mu\left(2\frac{\partial v_r}{\partial r}-\frac{2}{3}(\nabla\vec{v})\right)\right]\\&-2\mu v_r r^{-2}+\frac{2}{3}\mu r^{-1}(\nabla\vec{v})+\rho v_z^{\,2}r^{-1}+F_r\end{aligned} \tag{3}$$

where: F_x - axial force [N]; F_r - radial force [N]; μ - dynamic viscosity of combustion products [Pa·s].

The *energy conservation equation* is represented as:

$$\frac{\partial}{\partial t}(\rho E)+\frac{\partial}{\partial x}[v_x(\rho E+p)]+r^{-1}\frac{\partial}{\partial r}[rv_r(\rho E+p)]=$$
$$\frac{\partial}{\partial x}\left[\left(k+c_p\mu(Pr_t)^{-1}\right)\cdot\frac{\partial T}{\partial x}+u\cdot(\tau_{ij})_{eff}\right]+r^{-1}\frac{\partial}{\partial r}\left[r\left(k+c_p\mu(Pr_t)^{-1}\right)\cdot\frac{\partial T}{\partial x}+ru\cdot(\tau_{ij})_{eff}\right] \tag{4}$$

where: E - total energy [J]; $(\tau_{ij})_{eff}$ -viscous heating [J]; Pr_t -Prandtl number, c_p-specific heat capacity at constant pressure [J·kg^{-1}·K^{-1}], k-thermal conductivity [W·m^{-1}·K^{-1}], u-local flow velocity [m·s^{-1}].

Due to the high pressures and temperatures of the gases, the law of the ideal gas does not describe in a satisfactory manner the real phenomenon. As a consequence, the real gas state law will be used.

$$p\left(\rho^{-1}-\alpha\right)=RT \tag{5}$$

where: p - gas pressure [Pa]; ρ - gas density [kg·m^{-3}]; α - gas covolume [m^3·kg^{-1}]; R - gas constant [J·K^{-1}·mol^{-1}]; T - gas absolute temperature [K].

The burning law velocity has the form:

$$u_a=u_1p^{v} \tag{6}$$

where: v - burning exponent; u_1 - burning coefficient.

The mathematical model presented has a number of 5 equations with 5 indeterminate variables: v_x, v_r, p, T, ρ.

Compressible flows can be characterized by the value of the Mach number:

$$M=uc^{-1} \tag{7}$$

$$c=(\gamma RT)^{-\frac{1}{2}} \tag{8}$$

$$\gamma=c_pc_v^{-1} \tag{9}$$

where: c - sound velocity in gas [m·s^{-1}]; γ - specific heat ratio, c_v-specific heat capacity at constant volume [J·kg^{-1}·K^{-1}].

Typically, compressible flows are characterized by the maximum gas pressure and temperature, as follows:

$$p_0p^{-1}=\exp\left(\int_T^{T_0}c_pT^{-1}dTR^{-1}\right) \tag{10}$$

For c_p= constant, the equation is reduced to:

$$p_0p^{-1}=\left[1+(\gamma-1)2^{-1}M^2\right]^{\frac{\gamma}{\gamma-1}} \tag{11}$$

$$T_0T^{-1}=1+(\gamma-1)2^{-1}M^2 \tag{12}$$

For compressible gases, the law of the ideal gas is represented in the form:

$$\rho=\left(p_{op}+p\right)\left(\Re M_W^{-1}T\right)^{-1} \tag{13}$$

where: p_{op} - pressure under operating conditions; p - relative static local pressure; $\Re$ - universal gas constant; M_w - molecular mass.

Due to the existence of turbulent flow in the combustion chamber, the kinetic energy dissipates into thermal energy, in other words the gas velocity is influenced by turbulence, in the sense of decreasing it compared to a laminar flow. The influence of turbulent flow is modeled in FLUENT with the transport equation for viscosity as:

$$\frac{\partial}{\partial t}(\rho\widetilde{v})+\frac{\partial}{\partial x_i}(\rho\widetilde{v}u_i)=G_v+\sigma_{\widetilde{v}}^{-1}\left\{\frac{\partial}{\partial r}\left[(\mu+\rho\widetilde{v})\frac{\partial\widetilde{v}}{\partial x_j}\right]+C_{b2}\rho\left(\frac{\partial\widetilde{v}}{\partial x_j}\right)^2\right\}-Y_v+S_{\widetilde{v}} \quad (14)$$

where: $\widetilde{v}$ -turbulent kinematic viscosity; u_i -velocity of species i; G_v -variable that represents the turbulence production due to kinematic viscosity; Y_v -variable that represents the turbulence destruction due to kinematic viscosity; $\sigma_{\widetilde{v}}$, C_{b2} -constants; $S_{\widetilde{v}}$ -heat source.

2.2. *Internal motor and propellant geometry modeling*
The 2-D sketch of the rocket motor internal geometry and propellant configuration was generated using Gambit Software. Depending on the desired propellant combustion mechanism: neutral, progressive or regressive, there are different types of grain configurations, such as circular, tubular, star shaped [1]. In this study circular grain geometry was used. The dimensions and the geometry of the sketch are depicted in table 1.

Table 1. Sketch dimensions.

Internal rocket chamber		Propellant configuration	
Characteristics	Value	Characteristics	Value
Length	80 mm	Length	40 mm
Diameter	50 mm		
Nozzle length	6 mm	Outer diameter	45 mm
Throat diameter	4 mm		
Divergence angle	150 degree	Inner diameter	10 mm
Convergent angle	135 degree		

2.3. *Mesh generation*
The rocket motor structure and the propellant configuration have been modeled with SHELL triangular elements, as shown in figure 2. The mesh characteristics are presented in table 2.

Table 2. Mesh characteristics.

Time	**Characteristics**				
(s)	Level	Cells	Faces	Nodes	Partitions
0	0	5380	8910	3843	1

2.4. *Set up*
In setup, single precision with serial processing option was selected. The general conditions, the material characteristics, the boundary conditions and the combustion models were solved as presented below.

2.4.1 General. In order to calculate the compressible effects, due to the C-D nozzle, Density-based solver type was selected. The flow is symmetric about X axis so 2-D space was set to Axisymmetric.

The effects of the gravity were neglected. The energy equation was turned ON for the all analysis; viscous model is S-A (Spalart–Allmaras).

2.4.2 Material characteristics. The analysis was carried out for both Air and Gas as working fluids. The characteristics of the working fluids are depicted in table 3.

Table 3. Materials characteristics.

Characteristics	Gas
Density (ρ)	ideal gas
Thermal conductivity (k)	user defined (Appendix 1)
Molecular viscosity (μ)	user defined (Appendix 1)
Ratio of specific heats (ϒ)	piecewise-linear

2.4.3 Boundary conditions. The following boundary conditions were used for the analysis (table 4):

Table 4. Boundary conditions.

Characteristics	Values
Inlet (Mass flow-Inlet)	user defined (Appendix 2)
Inlet Temperature	user defined (Appendix 3)
Operating conditions pressure	0bar
Outlet Pressure, temperature	1bar, 300K

2.5. Combustion modeling

The combustion process of a rocket composite propellant is a complex process. In order to evaluate the combustion products a thermochemical software, Explo5®, was used [4]. The analyzed propellant formulation is: NH_4NO_3 (49%), KNO_3 (10%, potassium nitrate), GAP (20%), Al (20%) and Fe_2O_3 (iron oxide-1%). NH_4NO_3 is the oxidizer which evolves in gases species (NO_2, O_2, N_2O) necessary for combustion of Al, metallic fuel, and GAP, binder and organic fuel. The iron oxide is used as catalytic agent. The results are presented in figure 1, for the chamber, inlet, throat section and outlet.

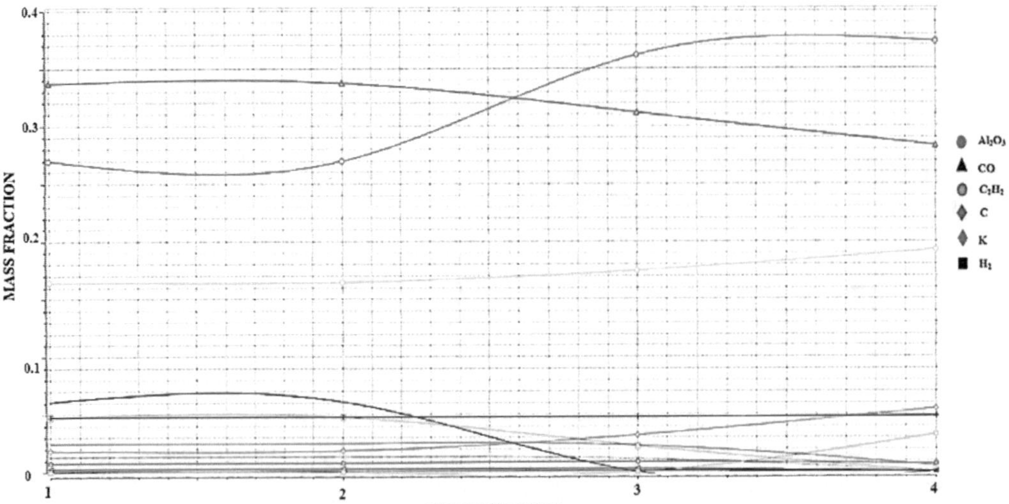

Figure 1. Mass fraction of the combustion products.

IOP Conf. Series: Materials Science and Engineering 1182 (2021) 012017 doi:10.1088/1757-899X/1182/1/012017

3. Results and discussions

The flow simulation was carried out under Steady state condition. Contours of pressure, temperature and velocity were obtained in different points of interest by using four gauges, depicted in figures 3 to 5. The positions of the gauges, ignition charge and propellant are given in figure 2.

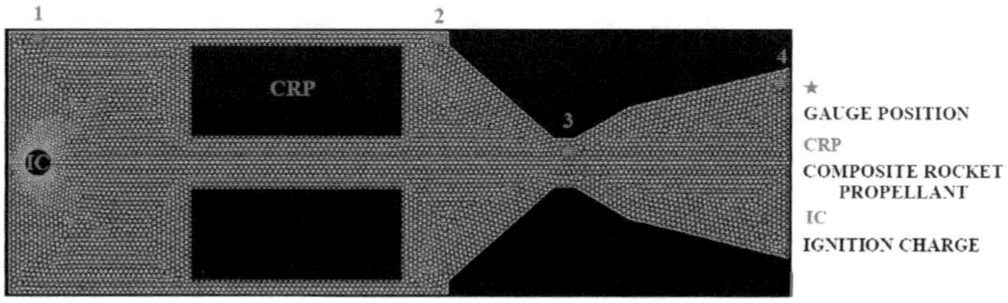

Figure 2. Mesh elements position.

The pressure and temperature are maximum at the inlet and begin to decrease until the outlet. The maximum values are 46.4 bar and 2700 K, while the minimums are 0.465 bar and 865 K.

Figure 3. Pressure contour.

Figure 4. Temperature contour.

When it comes to velocity, it can be concluded that this parameter is inversely proportional to the two discussed above, because it is minimal near the convergent zone and starts to increase to the end of the divergent zone, while in the throat section it is nearly Mach 1. The maximum values obtained for magnitude velocity is 2640 m/s.

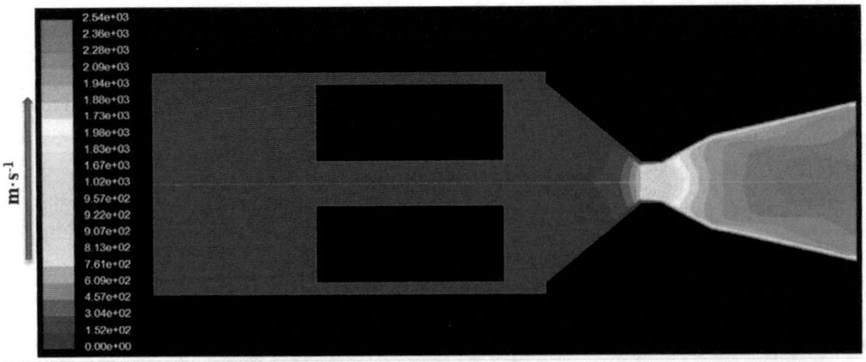

Figure 5. Velocity contour.

Based on the simulation, the variations of pressure and temperature versus time were recorded, as presented in figure 6.

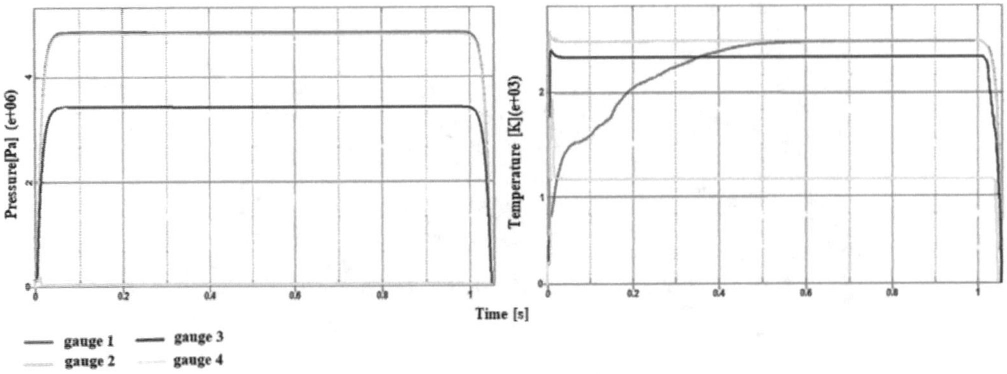

Figure 6. Pressure and temperature variation versus time in the simulation.

In order to validate the pressure values obtained based on the simulation, some experimental tests were carried by real firing of a propellant grain on a stand burner. The experiments were performed on a sub-scale Rocket Motor Tests TRM 35 and the composite propellant samples were ignited with 1gram of black powder. The testing installation is depicted in figure 7.

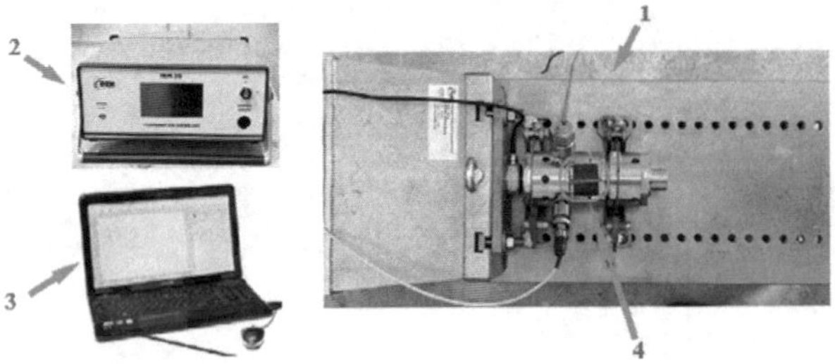

Figure 7. Testing installation.
(1-rocket stand; 2-ignition, pressure and thrust system acquisition; 3-remote control data and acquisition; 4-sample)

The time dependence of pressure determined in the experiments is presented in figure 8, where can be observed that the maximum experimental pressure recorded was 48.92 bar. But this value sums up the contribution corresponding to the ignition charge. Considering that, to determine the average pressure value of the propellant, this value was not taken into account. The average pressure value in the combustion chamber was 44.56 bar. Thus the pressure obtained by Computational Fluid Dynamics (CFD) is a good approximation of the experimental one, the difference being of about 4%.

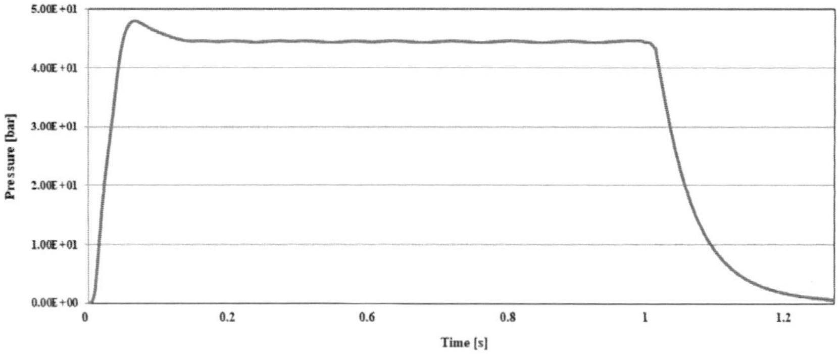

Figure 8. Pressure-time dependence.

4. Conclusions and perspectives

A numerical simulation model was developed for the calculation of the gas flow dynamic in the rocket motor, which was applied to a propellant formulation consisting of PSAN, GAP, Al and Fe_2O_3. It turned out that the results obtained are in accordance with those resulted from experimental tests.

The data obtained so far create a perfect background for a more extensive computational simulation, to determine the combustion behavior more accurately. In the development phase of composite mixtures, the model can be applied to a variety of configurations and the results can be compared for optimization. Numerical modeling of green solid composite propellants for civilian and military applications based on various components can be further performed.

5. References

[1] Jai Prakash Agrawal 2010 *High Energy Materials, Propellants, Explosives and Pyrotechnics* (Weinheim: WILEY-VCH Verlag GmbH & Co.KGaA) chapter 6 pp 209-316.

[2] *** 2013 *Ansys Fluent User's Guide*, Available online:http://www.pmt.usp.br/ACADEMIC/martoran/NotasModelosGrad/ANSYS%20Fluent%20Users%20Guide.pdf (accessed 19.01.2020).

[3] *** 2013 *Ansys Fluent UDF Manual*, Available online: http://www.pmt.usp.br/academic/martoran/notasmodelosgrad/ANSYS%20Fluent%20Theory%20Guide%2015.pdf (accessed 19.01.2020)

[4] SUCESKA, M., *Explo 5 Users Guide, Version 6.05 /Edition 1/2018.*

IOP Conf. Series: Materials Science and Engineering 1182 (2021) 012017 doi:10.1088/1757-899X/1182/1/012017

Appendix 1

```
DEFINE_PROPERTY(cell_thermal_conductivity, c, t)
{
        real  T, lam;
            real p1 = -9.004e-17;
            real p2 = 2.077e-12;
            real p3 = -1.748e-8;
            real p4 = 1.256e-4;
            real p5 = 3.243e-3;


            T = C_T(c,t);

            lam = p1*pow(T,4) + p2*pow(T,3) + p3*pow(T,2) + p4*pow(T,1) + p5;

            return lam;
}

DEFINE_PROPERTY(cell_viscosity, c, t)
{
        real mu, T;
            real p1 = -1.158e-19;
            real p2 = 1.984e-15;
            real p3 = -1.299e-11;
            real p4 = 6.5e-8;
            real p5 = 7.019e-6;


            T = C_T(c,t);
            mu = p1*pow(T,4) + p2*pow(T,3) + p3*pow(T,2) + p4*pow(T,1) + p5;

            return mu;
 }
```

Appendix 2

```
DEFINE_PROFILE(Ma_flu_elp, t, nv)
{
                float time, m_flux, x0, y0, x1, y1, d0, temp, Sx, Sy, Xcc, Ycc, Dar,
Psi, R, pres, ropb, vardo, niu, Alelp_real, Alelp_sim, rap_Alelp;
                int Kid;
                int Ntrs = 0;
                int Nelp = 1;
                int Nel_red;
                int in_int3 = 13;
                real Ar;
                real Aria = 0;
                real xbaza = 0;
                real ybaza = 0;


                Domain *d;
                Node *v;
                face_t f;
                cell_t c0;
                Thread *tb;
                real NV_VEC (area);
     vardo = 0.084e-06;
     niu = 0.726;
     ropb = 1.60*1e+3;



         time = RP_Get_Real("flow-time");

                 Sx = 0;
                 Sy = 0;
                 Kid = 0;

         begin_f_loop(f,t)
     {
            Kid = Kid + 1;
       v = F_NODE(f,t,0);
            x0 = NODE_X(v);
            y0 = NODE_Y(v);
            Sx = Sx + x0;
            Sy = Sy + y0;
         }
         end_f_loop(f,t)

       Xcc = Sx/Kid;
       Ycc = Sy/Kid;

       Sx = 0;
       Sy = 0;


         begin_f_loop(f,t)
         {
       v = F_NODE(f,t,0);
            x0 = NODE_X(v);
```

IOP Conf. Series: Materials Science and Engineering 1182 (2021) 012017 doi:10.1088/1757-899X/1182/1/012017

```
            y0 = NODE_Y(v);
        v = F_NODE(f,t,1);
            x1 = NODE_X(v);
            y1 = NODE_Y(v);
        Dar = ((x1-Xcc)*(y0-Ycc)-(x0-Xcc)*(y1-Ycc))/2;
            Aria = Aria + Dar;
         }
         end_f_loop(f,t)

                 Aria = Aria*1e6;

         Nel++;

         if(Nel==1)
                 Aria0 = Aria;

  begin_f_loop(f,t)
         {
        pres = F_P (f,t);
            temp = F_T(f,t);

            if(Aria>=1)
            {
                    if(temp>=500&&pres>0)
                    {
                            m_flux = ropb*vardo*pow((pres),niu);
                    }

                         else
                         {
                            m_flux = 1e-6;
                         }
            }

           if(Aria<1)
           {
                    m_flux = 1e-6;
           }

                 F_PROFILE(f,t,nv) = m_flux;
         }
         end_f_loop(f,t)

 }
```

Appendix 3

```
DEFINE_PROFILE(Temp_elp, t, nv)
{

        cell_t c0;
        Thread *t0;

 face_t f;
 real NV_VEC (area);

  float time, temp0, temp1;
  real Temp_c, Pres_c;


  time = RP_Get_Real("flow-time");


        begin_f_loop(f,t)
        {

       c0 = F_C0(f,t);
           t0 = F_C0_THREAD(f,t);
           Temp_c = C_T(c0,t0);
       Pres_c = C_P(c0,t0);
           temp0 = F_T(f,t);


                if(temp0>500)
                        {
               temp1 = temp0+100;
                   if(temp1>2300)
                          temp1 = 2300;
                        }
                else
               temp1 = 0.2*Temp_c+0.8*temp0;


                F_PROFILE(f,t,nv) = temp1;
        }
        end_f_loop(f,t)


  }
```

IOP Conf. Series: Materials Science and Engineering 1182 (2021) 012018 doi:10.1088/1757-899X/1182/1/012018

Aspects regarding the behaviour and comfort of reusable wool protective masks in the context of the COVID 19 pandemic

M Dochia[1], R Kaminszky[2], M S Fogorasi[3] , A Bucevschi[3], I Barbu[3], A Popa[3] and C Nicolaescu[4]

[1]"Aurel Vlaicu" University of Arad, The Institute for Research- Development-Innovation in Technical and Natural Sciences, Revoluţiei Avenue, 77, Arad, Romania
[2]"SWS" Simleu Silvaniei, Cehei Street, 100/A, Simleu Silvaniei, Salaj, Romania
[3]"Aurel Vlaicu" University of Arad, Faculty of Engineering, Department of Automation, Industrial Engineering, Textile and Transportation , Revoluţiei Avenue, 77, Arad, Romania
[4]"Aurel Vlaicu" University of Arad, Faculty of Economics, Department of Economic Disciplines, Revoluţiei Avenue, 77, Arad, Romania

E-mail: mfogorasi@yahoo.com

Abstract. In the current context of involvement in the fight against COVID 19 wearing protective masks are strictly necessary to limit the spread of the virus SARS-CoV2. Since the appearance of this virus, solutions have been sought to cover the needs of masks for the population and also to improve their wearing comfort compared to nonwoven fabric (TNT) masks. These masks are made of chemical fibres and it is recommended to be worn for a maximum of 4-8 hours and then they must be discarded. In this study, two types of masks from wool and silk blended knitted fabrics were manufactured and analysed. The masks were initially subjected to manual ironing sterilization and disinfection. Subsequently, several analyses were performed: specific mass, density, abrasion and pilling resistance, colourfastness to crocking, colour fastness to accelerated laundering, air permeability, bacterial filtration efficiency and breathability. The analysed masks can be washed and disinfected at home and are reusable. This aspect offers a strong sense of safety for the user, but also has beneficial effects on environmental protection. Even so, they do not correspond to bacterial filtration efficiency. Eventually, the comfort perception of about 300 volunteers was also taken into consideration.

1. Introduction

In the current context of involvement in the fight against COVID 19 wearing protective masks are strictly necessary, even imposed by the authorities, both inside and outside, to limit the spread of the virus SARS-CoV2 [1-3]. The wearing of facial masks is likely to become a habit in the future as a measure of protection against other viruses even if they are not as aggressive [4, 5]. Since the appearance of this virus, solutions have been sought to cover the needs of masks for the population and also to improve their wearing comfort compared to TNT masks (nonwoven textiles) sold through the pharmaceutical network [6-9]. The TNT masks are made of chemical fibres and it is recommended to be worn for a maximum of 4-8 hours and then they must be discarded. Also, the increase in masks utilization has led to waste accumulation and environmental issues.

For a while, another problem was the lack of medical masks and their auction at a very high price. Thus, the masks made of textile materials as personal protection have become a habit among the population. For non-medical masks realization, alternatives were searched in using various materials with filtration efficiency, breathability and low environmental impact. A series of studies were performed in which alternative materials that can capture particles and droplet were analysed [10]. Kin-Fai Ho et al. have reported that 100 % cotton towel or 70 % cotton with 30 % polyester material masks could have 40–60 % filtration capability for polydisperse sodium chloride (NaCl) aerosol particles [11].

Even if a lot of research has been done on facemasks, the number of studies on the choice of materials, especially those that have taken into account the impact on the environment due to waste and their recycling possibility is still limited. Decreasing the use of PPE and replacing materials with reusable alternatives will lead to less waste and less pollution. Moreover, the design, selection of material, number o layers and shape of the non-medical masks should be more investigated. These masks are usually done by small companies or homemade, without seeing aspects like quality, design, materials used, comfort and others.

Reusable masks or anti-virus masks could solve mask shortages or environmental issues in the future. The life of the reusable masks will be extended and anti-virus masks will annihilate the virus on the mask and will prevent a secondary infection generate by virus retention [12].

Therefore, in this study, an interdisciplinary group of researchers in collaboration with the economic environment analysed two types of self-designed facial masks made from wool and silk blended plain knitted fabric and evaluate the perception of about 300 volunteers in terms of comfort they provide.

2. Materials and methods

Two types of wool and silk blended plain knitted fabric supplied by Silvania Worsted Spinning SRL with different thicknesses and colours (grey and brown) were used to confectioned facial masks. After the industrial manufacture, the self-designed facial masks were initially subjected to manual ironing sterilization and disinfection treatment at 125 °C with steam at 110 °C using industrial ironing boards equipped with professional irons in laboratories of the Engineering Faculty at the Aurel Vlaicu University of Arad. Laboratory analyses of the masks were performed at the Research- Development-Innovation in Technical and Natural Sciences Institute from "Aurel Vlaicu" University of Arad and to Techtex SRL laboratory. Subsequently, several analyses were performed: the specific mass (grams per square meter), the density of the knitted fabrics, the abrasion and pilling resistance, colourfastness to crocking by Crockmeter Method, colour fastness to accelerated laundering, air permeability, bacterial filtration efficiency (BFE) and breathability. The BFE and breathability results were compared with those of disposable TNT masks.

2.1. GSM (grams per square meter) of knitted fabrics

Fabrics weight was measured according to ASTM D 3776 / D3776M - 09a "Standard Test Methods for Mass Per Unit Area (Weight) of Fabric" on ten circular specimens taken from different parts of the knits using the GSM cutter device. The circular samples were weighed using an analytical balance and subsequent calculation of the weight in grams per square meter was done.

2.2. The density of knitted fabrics

The number of wales and courses per unit distance for both types of knitted fabrics were determined on five samples by using a pick glass and a counting device following ASTM D 3887-96(2008) „Standard Specification for Tolerances for Knitted Fabrics".

2.3. Abrasion and pilling resistance

The evaluation of the abrasion resistance was performed with the Martindale Abrasion and Pilling Tester - model YG401B. Two samples from each type of knitted fabrics were measured for weight loss and colour change. The analyses were made under ASTM D4966-12- "Standard Test Method for Abrasion Resistance of Textile Fabrics (Martindale Abrasion Tester Method)".

After some preliminary analysis, the number of rubbing cycles for the two types of knitted fabrics was established. An important factor for establishing this number, along with the behaviour of the knit in the rubbing process, was the mass/m^2. Thus, for the grey knit with a mass of 155.50 g/m^2, the number of rubbing cycles was established between 1000 - 5000 rubbing cycles with the step of 1000 cycles and for the brown knit with a mass of 220.43 g/m^2 in the range of 10000 - 20000 rubbing cycles with the step of 5000 cycles.

The colour change evaluation was performed with Datacolor 500 colourimeter. On each sample, the measurement was done at three different points. The index used to assess colour change was the colour difference between the abraded samples and the untested control sample. Based on the colour difference, the equivalent grade of the greyscale for colour change was calculated. This evaluation was made following AATCC Evaluation Procedure 7-2009- "Instrumental Assessment of the Change in Colour of a Test Specimen".

To assess the pilling resistance of the knitted fabrics, the same standard ASTM D4966-12- "Standard Test Method for Abrasion Resistance of Textile Fabrics (Martindale Abrasion Tester Method)" was used. To give a rate, the tested samples on the Martindale device were compared with a set of five photographs 105 mm numbered 1 to 5, illustrating different degrees of pilling from "very severe pilling" to "no pilling".

2.4. Colourfastness to Crocking: Crockmeter Method

The evaluation of colourfastness to crocking of the dyed knitted fabrics was done according to the standard AATCC Test Method 8-2007- "Colourfastness to Crocking: Crockmeter Method" on the Crock Meter Y571D. This test method is designed to determine the amount of colour transferred from the surface of coloured textiles to other surfaces as a result of rubbing with an arm that mimics the human finger. The tests were performed on both dry and wet crocking methods. The colour change of the tested samples and colour transferred to a standard white test cloth was evaluated with Datacolor 500 colourimeter according to AATCC Evaluation Procedure 7-2009- "Instrumental Assessment of the Change in Colour of a Test Specimen". Colourfastness grade was assigned based on the colour difference between specimens that had been submitted to the colourfastness tests and the colour of an identical untreated specimen.

2.5. Colourfastness to accelerated laundering

The accelerated laundering test is to evaluate the colourfastness to laundering of textiles. The evaluation of knitted fabrics samples was performed according to the AATCC Test Method 61-2009 - "Colourfastness to Laundering: Accelerated" with a Launder - Ömeter, model M228AA. The use of the accelerated washing method is equivalent to five cycles of home laundering. The Datacolor500 colourimeter and the standard - AATCC Evaluation Procedure 7-2009 - "Instrumental Assessment of the Change in Colour of a Test Specimen" were used to evaluate the colour change of the samples subjected to washing.

This evaluation covers two aspects: the colour change of the material subjected to washing and the amount of colour transferred from the samples surface to other surfaces of a fabric consisting of 6 types of fibres (acetate, cotton, nylon, silk, viscose, wool), called "multifiber".

2.6. Air permeability

The air permeability assessment was performed according to ASTM D 737 - 04 - "Standard Test Method for Air Permeability of Textile Fabrics" with the Digital Air Permeability Tester, model YG461E. The measurements conditions were: pressure (P) = 100 Pa, nozzle diameter (Φ) = 14.5 mm, test area of the sample 20 cm^2. For an accurate evaluation of the material permeability, 10 measurements were made in different areas on each type of knitted fabrics.

2.7. Bacterial filtration efficiency (BFE) and breathability
The test method for bacterial filtration efficiency (BFE) was performed according to the European standard SR EN 14683 + AC: 2019, section 5.2.2. and Annex B. The method consisted of determining the aerosol penetration coefficient with particle diameter between 0.3 – 10 μm on a Mask Bacterial Filtration Efficiency Tester, model GK-1000 from Ningbo Dahe Instrument Manufacturer.

Section 5.2.2. and Annex C from the same standard was used to evaluate the face masks breathability (differential pressure). The GT-RA04 Face Mask Differential Pressure Tester was used to determine the air exchange pressure of the face masks material by measuring the differential pressure required to draw air through a measured surface area of 4.9 cm^2 from the mask at a constant airflow rate of 8 l/min.

3. Results and Discussion

3.1. GSM (grams per square meter) and density of the knitted fabrics
Table 1 and table 2 present the GSM values and density for both types of analysed knitted fabric and represent the average of 10 measurements. The mass/m^2 was an important factor for establishing the number of rubbing cycles for abrasion and pilling resistance testing.

Table 1. GSM of the knitted fabrics.

Grey knitted fabric [g/m^2]	Brown knitted fabric [g/m^2]
155.50	220.43

Table 2. The density of the knitted fabrics.

Grey knitted fabric		Brown knitted fabric	
Wales/centimeter (wpc)	Courses/centimeter (cpc)	Wales/centimeter (wpc)	Courses/centimeter (cpc)
12	20	12	18

3.2. Abrasion and pilling resistance
To evaluate the abrasion resistance, the samples were weighed before and after each step of rubbing cycles. The values for the initial mass, the final mass and calculated weigh loss are presented in table 3 and table 4.

Table 3. The weight loss after abrasion resistance tests for grey knitted fabric.

Samples	Number of rubbing cycles	Initial mass - m_i [g]	Final mass – m_f [g]	Weight loss [%]
1	1000	0.220	0.217	1.36
	2000	0.214	0.210	1.87
	3000	0.220	0.212	3.64
	4000	0.213	0.202	5.16
	5000	0.225	0.209	7.11
2	1000	0.211	0.209	0.95
	2000	0.212	0.205	3.30
	3000	0.228	0.219	3.95
	4000	0.227	0.215	5.29
	5000	0.215	0.198	7.91

Table 4. The weight loss after abrasion resistance tests for brown knitted fabric.

Samples	Number of rubbing cycles	Initial mass - m_i [g]	Final mass – m_f [g]	Weight loss [%]
1	10000	0.254	0.241	5.39
	15000	0.254	0.240	5.51
	20000	0.254	0.236	7.08
2	10000	0.241	0.229	4.98
	15000	0.241	0.227	5.80
	20000	0.241	0.225	6.64

From the values obtained for both types of knitted fabric, it can be seen that the weight loss was directly proportional to the number of rubbing cycles, being between 7 - 8 % after 5000 cycles for the grey knit and 6 - 7 % for the brown one after 20000 cycles. Also, a decrease in material thickness was found, but without any perforations.

Because after the abrasion resistance tests, the pilling phenomenon appeared on the tested side of the fabrics, it was decided that for the colour change evaluation, the samples should be measured both on the front side of the material and on the backside of the material. In figures 1 to 4 are presented the colour difference – ΔE values and the attributed colourfastness grade for the front and backside of both grey and brown knitted fabric samples subjected to abrasion resistance testing.

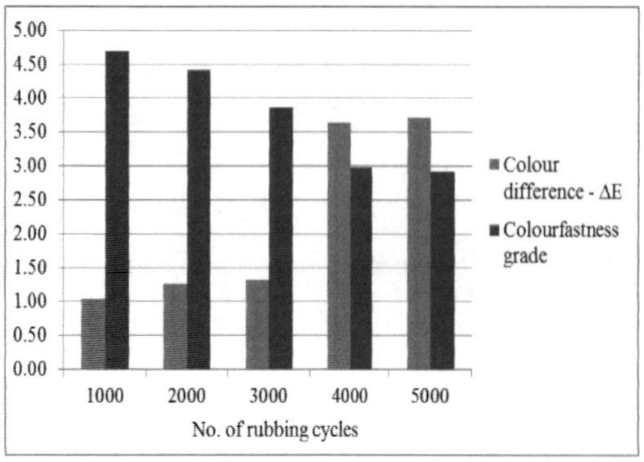

Figure 1. The colour difference – ΔE and colourfastness grade for the front side of grey knitted fabric.

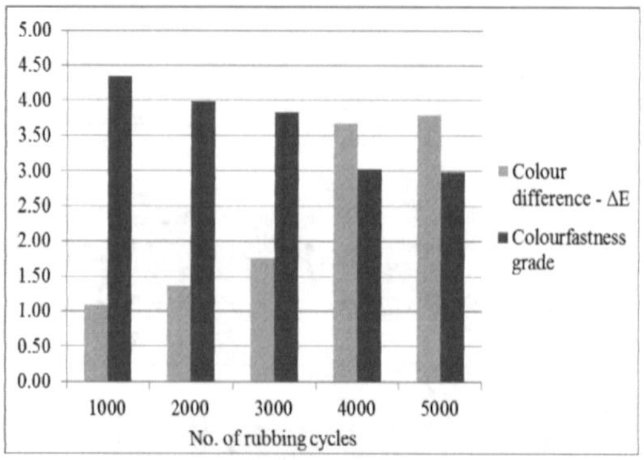

Figure 2. The colour difference – ΔE and colourfastness grade for the backside of grey knitted fabric.

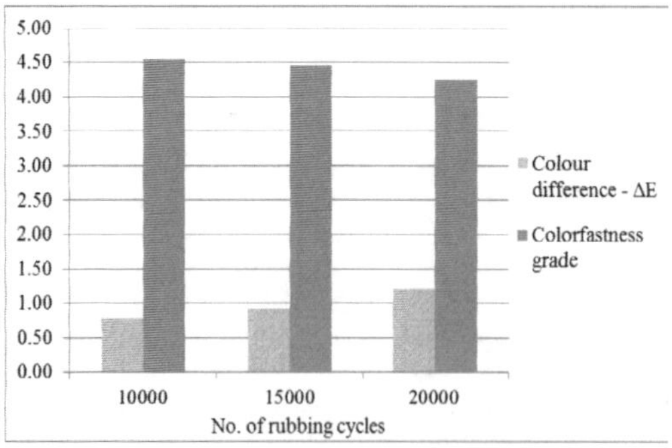

Figure 3. The colour difference – ΔE and colourfastness grade for the front side of brown knitted fabric.

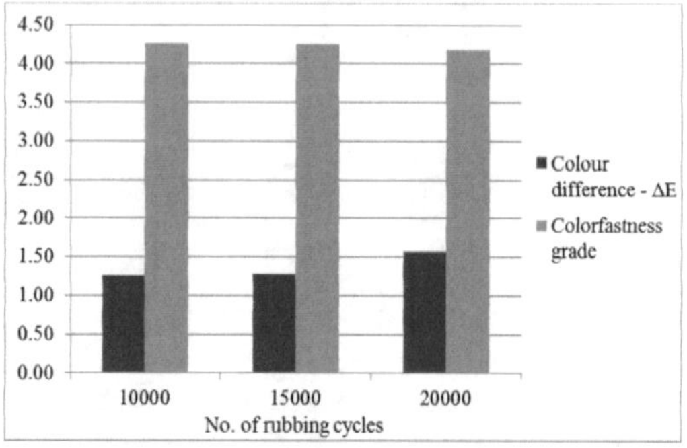

Figure 4. The colour difference – ΔE and colourfastness grade for the backside of brown knitted fabric.

From the obtained values, it can be observed that for both the front and backside of the grey knitted fabric, the “endpoint” was reached at 4000 rubbing cycles when the colour change of the material was evaluated with grade ~ 3 according to greyscale. The "endpoint" for the brown knitted fabric was not reached even at 20000 rubbing cycles. According to the greyscale, the assigned grade was ~ 4.

After pilling resistance testing, the knitted fabrics were compared with a set of five photographs numbered 1 to 5, illustrating different degrees of pilling as presented in figure 5.

Grade 5: no pilling Grade 4: slight pilling Grade 3: moderate pilling

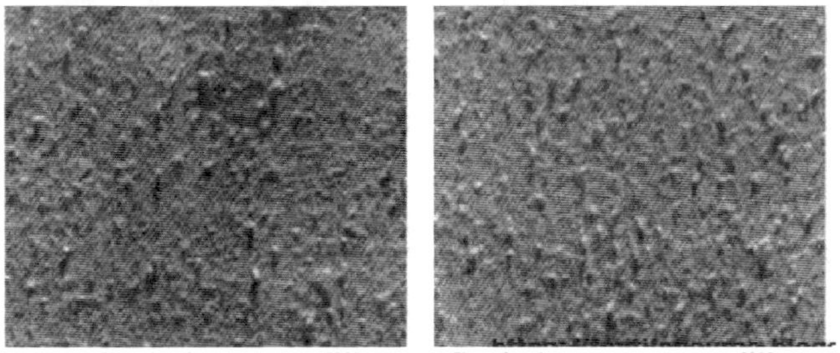

Grade 2: severe pilling Grade 1: very severe pilling

Figure 5. Photographic standards for pilling evaluation.

The appearance of both types of knitted fabric subjected to pilling resistance testing is shown in figure 6. After comparison with the photographic standards, the grey knit was marked with grade 4 and the brown knit with grade 3. As a result, the brown knitted fabric has a more pronounced tendency to form pilling compared to the grey one.

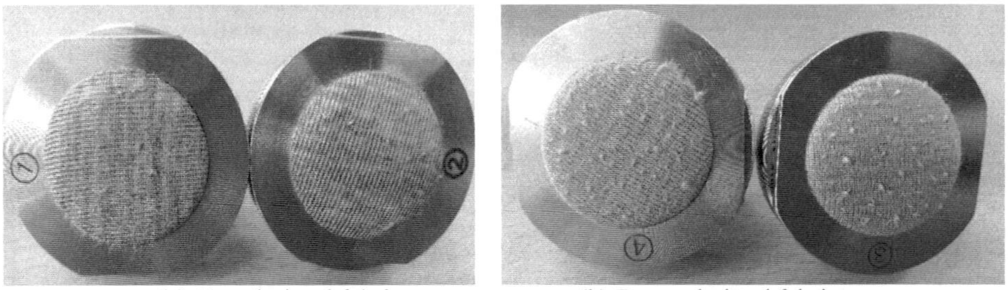

(a) Grey knitted fabric (b) Brown knitted fabric

Figure 6. The appearance of knitted fabric samples after pilling resistance testing.

3.3. Colourfastness to Crocking: Crockmeter Method

The colourfastness to dry and wet crocking of knitted fabrics samples was evaluated by measuring the colour transferred to a white standard test cloth and by measuring the colour change of the tested samples. Colourfastness grade was assigned based on the colour difference between specimens that had been submitted to the colourfastness tests and the colour of an identical untreated specimen.

Tables 5 and 6 present the data obtained after dry and wet crocking of the grey knitted fabric.

Table 5. The colourfastness to dry crocking of the grey knitted fabric.

Samples	The colour change of the knitted fabric samples		The colour transferred to white test cloth	
	Colour difference ΔE	Colourfastness grade	Colour difference ΔE	Colourfastness grade
1	1.22	4.28	1.47	4.14
2	1.21	4.29	0.81	4.52

Table 6. The colourfastness to wet crocking of the grey knitted fabric.

Samples	The colour change of the knitted fabric samples		The colour transferred to white test cloth	
	Colour difference ΔE	Colourfastness grade	Colour difference ΔE	Colourfastness grade
1	1.62	4.05	5.34	2.32
2	1.75	4.02	6.68	2.03

For the brown knitted fabric, the colour change and colourfastness values after dry and wet crocking are presented in table 7 and table 8.

Table 7. The colourfastness to dry crocking of the brown knitted fabric.

Samples	The colour change of the knitted fabric samples		The colour transferred to white test cloth	
	Colour difference ΔE	Colourfastness grade	Colour difference ΔE	Colourfastness grade
1	0.61	4.64	0.72	4.58
2	0.78	4.54	0.79	4.54

Table 8. The colourfastness to wet crocking of the brown knitted fabric.

Samples	The colour change of the knitted fabric samples		The colour transferred to white test cloth	
	Colour difference ΔE	Colourfastness grade	Colour difference ΔE	Colourfastness grade
1	0.46	4.73	2.79	3.36
2	0.74	4.56	6.82	2.00

The data presented, show an insignificant colour change of the grey and brown knitted fabrics samples after both dry and wet crocking tests. Thus, the grades assigned according to the greyscale were between 4.02 - 4.73.

Similar values were obtained for the amount of colour transferred to white test cloth in case of dry crocking, the assigned grades being between 4.14 - 4.58.

A different behaviour was noticed for wet crocking, where the colour transfer was more accentuated and the colourfastness grades were 2.00 - 3.36.

3.4. Colourfastness to accelerated laundering

The colour change (colour difference – ΔE) and assigned colourfastness grade of grey and brown knitted fabrics after accelerated laundering are shown in figure 7.

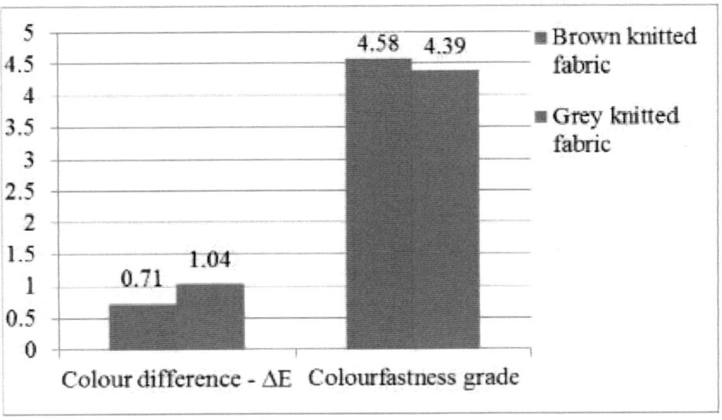

Figure 7. The colour difference and colourfastness grade to accelerated laundering of the grey and brown knitted fabrics.

As the data show, for both types of knitwear, there is a good colour resistance to accelerated laundering, the assigned colourfastness grades being 4.39 for grey knitted fabric and 4.58 for brown ones.

The amount of transferred colour on the "multifiber" support of knitted fabrics samples subjected to accelerated laundering is presented in figures 8 and 9. For the grey knit there was a low transfer on the acetate, nylon and viscose fibres followed by the wool and cotton. An accentuated transfer of the dye occurred on the silk fibre, the attributed colourfastness grade being 3.09.

For the brown samples, the values related to colour transfer shows better resistance to accelerated laundering, the attributed grades being over 4 for most fibres, except for the cotton fibre were the colourfastness grade was 3.75.

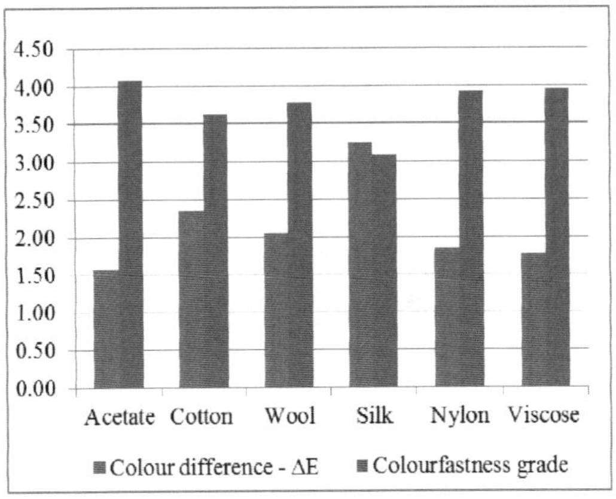

Figure 8. The colour difference and colourfastness grade to accelerated laundering of the grey knitted fabrics.

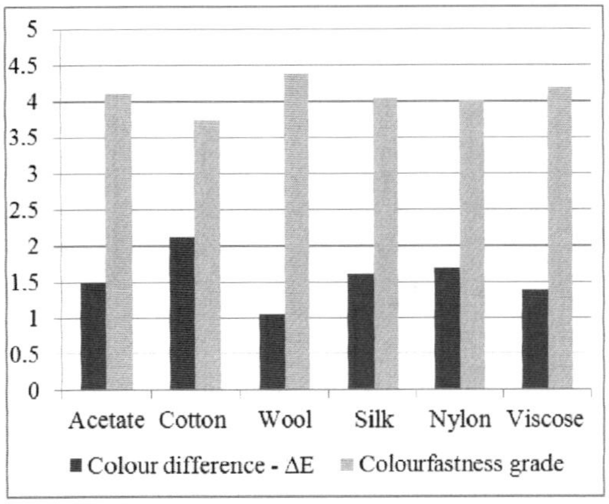

Figure 9. The colour difference and colourfastness grade to accelerated laundering of the brown knitted fabrics.

3.5. Air permeability
The air permeability values obtained for the ten measured areas of grey and brown knitted fabrics are presented in table 9.

Table 9. The air permeability of grey and brown knitted fabrics.

Samples	Air permeability (mm/s)	
	Grey knitted fabric	Brown knitted fabric
1	10900	14300
2	12700	12700
3	12100	12900
4	13900	12700
5	11200	10400
6	10400	10300
7	11000	10200
8	10800	12100
9	11600	10400
10	10500	10600
Average values	**11510**	**11660**

From the air permeability values presented in table 10, a certain structural non-uniformity of the tested knitted fabrics can be observed. Even if the two types of analyzed knitted fabrics have different thicknesses, the air permeability is similar. This may be due to the similar density of the knits as well as the porosity.

3.6 Bacterial filtration efficiency (BFE) and breathability
The evaluation of the results regarding BFE and breathability of the analyzed knitted fabric masks was made in comparison with the results obtained for the medical face masks. Figure 10 shows the bacterial filtration efficiency (BFE) and breathability of grey and brown knitted fabric masks compared to TNT medical face masks.

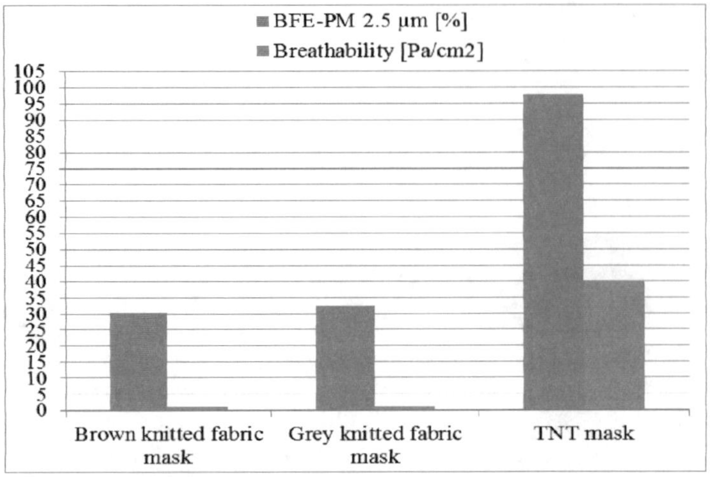

Figure 10. The bacterial filtration efficiency (BFE) and breathability of the grey and brown knitted fabrics masks compared to TNT medical face masks.

The results obtained for the bacterial filtration efficiency at the particle diameter of 2.5 μm show a low efficiency of both grey and brown knitted fabrics masks (~ 30 – 32 %) compared to TNT medical face masks (98 %). Instead, good breathability can be observed with low values of 1.41 -1.46 Pa/cm^2 for the pressure drop compared to 40 Pa/cm^2 for TNT masks.

3.7. The comfort perception

Conclusions on the wearing comfort of the masks were formulated on the basis of a questionnaire. The questionnaire consists of 16 questions. The questionnaire was completed by 258 respondents (148 women and 110 men) with an average age of 36,4 years. The analysis of the questionnaire shows that there have been no dimensional changes or comfort changes after wearing and washing the masks at home, for none of the two variants (90% of respondents). 75% of respondents did not feel any skin irritation when wearing masks and no special odor. Related to the comfort offered by the knitted masks in comparison with TNT masks more than 50% of respondents consider them more comfortable.

4. Conclusions

The analysed masks made of natural fibres compared to disposable ones can be washed and disinfected at home and are reusable. This aspect offers a strong sense of safety for the user, but also has beneficial effects on environmental protection. TNT masks are made of chemical fibres and it is recommended to be worn for a maximum of 4-8 hours and then they must be discarded. Of course, these natural fibre masks can be manufactured at higher costs, both in terms of raw materials and labour, than those of TNT, but through reuse, the user's financial effort is lower overall.

Despite all these benefits the knitted fabric masks do not fulfil de requirements regarding bacterial filtration efficiency according to European standard SR EN 14683 + AC: 2019 for medical masks. Eventually, the comfort perception of about 300 volunteers could also be taken into consideration.

5. References

[1] Cortes M F, Sanchez Espinoza E P, V´asconez Noguera S L, Alves Silva A, Sielfeld Araya de Medeiros M E, Villas Boas L S, Ferreira N E, Tozetto-Mendoza T R, Alves Morais F G, Santiago de Queiroz R, Coracini Tonacio de Proenca A, Guimaraes T, Guedes A R, Letaif H. L S, Cardoso Montal A., Mendes-Correa M C, John V M, Levin A S Figueiredo Costa S 2020 Decontamination and reuse of surgical masks and respirators during COVID-19 pandemic *Int. J. Infect. Dis.* Reference: IJID 4974.

[2] Zorko D J, Gertsman S, O'Hearn K, Timmerman N, Ambu-Ali N, Dinh T, Sampson M, Sikora L, McNally J D and Choong K 2020 Decontamination interventions for the reuse of surgical mask personal protective equipment: a systematic review *J. Hosp. Infect.* **106** 283-294.

[3] Juan Carlos Rubio-Romero, María del Carmen Pardo-Ferreira, Juan Antonio Torrecilla-García, Santiago Calero-Castro, 2020 Disposable masks: Disinfection and sterilization for reuse, and non-certified manufacturing, in the face of shortages during the COVID-19 pandemic *Saf. Sci.* **129** 104830.

[4] Correspondence 2020 Reusable masks for COVID-19: A missing piece of the microplastic problem during the global health crisis *Mar. Pollut. Bull.* **161** 111777.

[5] Sangkham S 2020 Face mask and medical waste disposal during the novel COVID-19 pandemic in Asia *Case Studies Chemical and Environmental Engineering* **2** 100052.

[6] Zhang Z, Ji D, He H and Ramakrishna S 2021 Electrospun ultrafine fibers for advanced face masks *Mater. Sci. Eng. R* **143** 100594.

[7] Kumara A, Kamal Sharmab K and Dixita A R 2020 Role of graphene in biosensor and protective textile against viruses *Med. Hypotheses* **144** 110253.

[8] Swennena G R J, Pottelb L and Haers P E 2020 Custom-made 3D-printed face masks in case of pandemic crisis situations with a lack of commercially available FFP2/3 masks *Int. J. Oral Maxillofac. Surg.* **49**(5) 673-677.
[9] Palmieria V, De Maiod F, De Spirito M and Papi M 2021 Face masks and nanotechnology: Keep the blue side up *Nano Today* **37** 101077.
[10] Hartanto B W and Mayasari D S 2021 Environmentally friendly non-medical mask: An attempt to reduce the environmental impact from used masks during COVID 19 pandemic *Sci. Total Environ.* **760** 144143.
[11] Ho K-F, Linb L-Y, Weng S-P, Chuang K-J 2020 Medical mask versus cotton mask for preventing respiratory droplet transmission in micro environments *Sci. Total Environ.* **735** 139510.
[12] D. Ji, L. Fan, X. Li, S. Ramakrishna 2020 Addressing the worldwide shortages of face masks *BMC Mater.* **2** 1–11.

Acknowledgements
The study was carried out under a research contract funded by Silvania Worsted Spinning S.R.L from Simleu Silvaniei, Salaj, Romania.

Study on the behaviour and comfort of reusable knitting suits of wool blends in the context of the Covid 19 pandemic

M Dochia[1], R Kaminszky[2], M S Fogorasi[3], D Farima[4], I Barbu[3], C Nicolaescu[5], M Pustianu[3], A Popa[3] and A Bucevschi[3]

[1]"Aurel Vlaicu" University of Arad, The Institute for Research- Development-Innovation in Technical and Natural Sciences, Revoluţiei Avenue, 77, Arad, Romania

[2] SWS Simleu Silvaniei, 100/A Cehei Street, Simleu Silvaniei Salaj, România

[3]Aurel Vlaicu" University of Arad, Faculty of Engineering, Department of Automation, Industrial Engineering, Textile and Transportation , Revoluţiei Avenue, 77, Arad, Romania

[4]"Gheorghe Asachi Technical University", Faculty of Industrial Design and Business Management, Department of Knitting and Garment, 55 Profesor Dimitrie Mangeron Avenue, Iasi, Romania

[5]"Aurel Vlaicu" University of Arad, Faculty of Economics, Department of Economic Disciplines, Revoluţiei Avenue, 77, Arad, Romania

E-mail: mfogorasi@yahoo.com

Abstract. Nowadays, humanity is facing serious challenges due to the disruption of the COVID-19 pandemic, affecting dramatically our healthcare system along with the economic structure, social and cultural life. Healthcare personnel treating patients with infections like COVID-19 are exposed themselves to infection. Therefore, it is compulsory to be protected from contaminated body fluids using personal protective equipment (PPE). This equipment made of synthetic fibers proved to be uncomfortable. Therefore, we developed a reusable suit composed of a blend of wool and silk to be wear under full-body protective clothing aiming to improve their comfort and safety. Firstly, we focused on the analysis of some dimensional characteristics: specific mass, porosity, thickness followed by comfort characteristics of knitted fabrics. Accordingly, the test methods comprise measurements for hydrophilic properties, air and water vapour permeability, thermal conductivity and dimensional stability to repetitive washing. After material analysis, a significant number of two-piece suits composed of blouse and trousers were industrially manufactured and delivered to healthcare personnel to be tested. The results are completed and supported with the perception of the healthcare staff involved in the study via a survey. Finally, the results and implications for future research are presented.

1. Introduction

Nowadays, humanity is facing serious challenges due to the disruption of the Covid-19 pandemic, affecting dramatically our healthcare system along with the economic structure, social and cultural life [1]. Healthcare personnel treating patients with infections like Covid-19 are exposed themselves to infection. Thus, it is compulsory to be protected from contaminated body fluids using personal protective equipment (PPE) [2].

But this equipment proved to be uncomfortable to the wearer in extended use since it consists of synthetic fibers. The current application of textiles in the medical field and past pandemics are well known. Presently, there is a major concern in evolving in the forthcoming future protective clothing based on nonwoven micro-and nanoporous materials. The main benefits of these new materials relay in their, breathability, lightweight, lower price and comfort [1].

Under these circumstances, we take into account the possibility to develop a reusable suit to be wear by healthcare staff in the Covid sections under the full-body protective clothing aiming to improve their comfort and safety. Why is comfort so important? Because better comfort impedes stress and tiredness.

The main objective of the study envisages whether a simple protective cloth composed of a blend of natural fibres wool and silk (95% wool and 5% silk) is appropriate to enhance both ergonomic and physiological comfort. The study was developed by an interdisciplinary team, formed by researchers and engineers from the Textile Industry field, accounting field and healthcare staff from the Covid hospitals.

As a complex issue, *ergonomic comfort is* considered how well the clothing suits, if it allows the wearer to perform easily all the necessary movements. It is quite simple to estimate by trying the garment on. Physiological comfort involves skin sensory comfort and thermo-physiological comfort [3].

Thermo-physiological comfort refers to a pleasant thermal and humid condition involving both thermo-regulating and moisture-regulating capacity in addition to the thermal insulation capacity of the clothing. All these abilities are related to the environmental and working conditions. Furthermore, thermo-physiological comfort is influenced by the fabric composition and structure and garment construction [4-6].

Since the style and cut of the two-piece ensembles conceived by our team do not restrict movements and do not exert any constraints upon the body, the study investigated the parameters representative for their thermo-physiological comfort. Because people using similar clothes, in identical physical ambiance may perceive various comfort degrees, our study was completed with the results of a survey regarding the subjective perception of the comfort level of about 51 volunteers from medical staff.

2. Material and Methods

Within this experimental part, we focused firstly on the analysis of some dimensional characteristics like mass per square meter, porosity, thickness followed by comfort characteristics of knitted fabrics designed to be used for the suits.

Accordingly, the test methods accomplished comprise measurements for hydrophilic properties, air and water vapour permeability [7], thermal conductivity and dimensional stability to repetitive washing. Since we refer to a reusable ensemble that will be subjected to several washing processes, dimensional stability is also an important aspect, because wool fibre is a precious natural material very sensitive to shrinking.

2.1. Materials

Fabric subjected to testing is a blend of wool/silk 95/5% knitted fabric, supplied by SILVANIA WORSTED SPINNING SRL. Why a blend of natural fibers?

Because wool continues to be a unique apparel fibre with a remarkable array of technical features. These properties refer to softness, warmth and coolness, breathability, moisture absorption and buffering resilience, odour absorption, softness, stain and flame resistance, elasticity, biodegradability and recyclability. Due to these excellent attributes, wool can be employed either for a worsted suit or knitted outerwear and/or for technical applications [8,9].

Besides, the adaptability of wool has been revealed by several leading-edge clothing products designed for the sportswear area, including outerwear, mid-layer, and underwear garments, that require superior moisture management [10,11].

Further, the presence of silk in the blend contributes to enhanced softness.

IOP Conf. Series: Materials Science and Engineering 1182 (2021) 012019 doi:10.1088/1757-899X/1182/1/012019

2.2. Methods

2.2.1. Specific mass. The specific mass is defined as the mass, in grams, of a unit of area (m^2) of the material. The determinations were performed according to D 3776/D3776M – 09a "Standard test methods for mass per unit area (weight) of fabric" on circular test specimens taken from different parts of the knitted fabric using the GSM cutter. The values given in the table are the average of 10 determinations.

2.2.2. Fabric density. Assisted by a counting glass, the stitch density of the knitted fabric, expressed in wales and course per centimeter, was measured, according to ASTM D 3887-96(2008)"Standard Specification for Tolerances for Knitted Fabrics"

2.2.3. Porosity. Porosity is a parameter with a significant influence on both the absorption and desorption of textile materials. An increase in porosity leads to enhanced vapour permeability and permeability to air [7]. The porosity of the analysed sample Pz[%] was determined by the pycnometric method.

2.2.4. Fabric thickness. The thickness of the knitted fabric was assessed using a textile micrometer model DM 100.

2.2.5. Hydrophilic properties. The evaluation of the hydrophilic properties of the analysed knitted fabric was carried out according to the AATTCC Test method standard 79-2007. By this method, a drop of water is delivered from a burette to the surface of the material subjected to analyses. The time required for the water absorption represents the wetting time.

2.2.6. Air permeability. The air permeability measurement was accomplished according to the ASTM D 737-04), Standard Test Method for Air Permeability of Textile Fabrics. For this purpose, the ATL2 Metrimpex apparatus was used. The working parameter was set for a difference in air pressure Δp =5mCOl water and a temperature of T=20 °C. Δp – represents the pressure difference between the two sides of the knitted fabric and T the average air temperature adopted for each season and used for calculations. The recorded values of the air permeability (P_a) were expressed in [m^3/min m^2].

Additionally, air permeability index (i) and airflow passing resistance Rpa [mm m^2 h/kg] have been considered for the assessment. The air permeability index (i) [kg/m^2·h] has been determined with the equation (1) :

$$i = 60 \cdot Pa \cdot \gamma \left(\frac{kg}{m^2 \cdot h} \right) \quad (\text{kg/m}^2\text{h}) \tag{1}$$

where: γ — represent the air density [kg/m^3];
Pa - air permeability corresponding to the knitted fabric [m^3/min·m^2].

The air density shall be calculated with the equation (2):

$$\gamma = \frac{\gamma_0}{1 + \frac{t}{273}} \tag{2}$$

where: γ_0 — air density at temperature 0 °C [γ0=1.293 kg/m^3];
t - the temperature of the air passing through the textile material (temperature of 20 °C).

IOP Conf. Series: Materials Science and Engineering 1182 (2021) 012019 doi:10.1088/1757-899X/1182/1/012019

2.2.7. Water vapour permeability. Measurements of water vapour permeability were performed using the upright cup method (Water Vapour Permeability - ASTM E 96- Method B). The samples were positioned and sealed above a cup filled with distilled water at a depth of ¾ from the height of the cup. After that, the samples were placed in a standard atmosphere (23 ± 1 ^{0}C and RH 50 ± 2 %) with ventilation in a Pol-Eko-Aparatura oven. The test was carried out for 6 hours. The gravimetric method was used to evaluate the change in mass every two hours. The specimens were weighed on a sensitive Kern balance, model ABT220-4M and the difference between the mass at 2, 4 and 6 hours of vapour exposure and initial mass represents the ability of the knitted fabric to transfer perspiration. Vapour permeability has been also evaluated through water vapour transmission coefficient (WVT) μ[g/m^2·h].

2.2.8. Thermal conductivity. The thermal insulation properties of the knitted fabric were assessed via the coefficient of thermal conductivity λ[kcal/m·h·C] using the λ Tex Tester, with a sintered copper plate, on the principle of the skin model. Using the analytical method, the indirect indicator heat transfer resistance Rt [m^2·h·^{o}C/kcal] has been calculated for a complete assessment of the comfort properties.

2.2.9. Dimensional stability after home laundering. The dimensional stability of the suits manufactured from knitted fabric to home laundering was assessed taking into consideration the method describes in AATCC Test Method 150-2003 Dimensional Changes of Garments after Home Laundering. The washing was carried out on the automatic washing machine Zanussi HydroLine, model ZWF 1438. Laundering of textiles is a process designed to remove dirt and/or stain by washing with an aqueous detergent solution. The treatment operation normally includes washing, rinsing, and drying of the material. Dimensional stability refers to the ability of fabrics or knitted fabrics, garments (specimens) to keep the shape and dimensions following domestic maintenance procedures (washing, cleaning, ironing). The change is usually indicated as a percentage of the initial dimension of the textile specimen. From the point of view of the dimensional stability of the knitted fabric, the first washing is of particular importance. Thus, factors like washing temperature, washing agents, treatment duration, and mechanical action exert an important influence upon the washing process and dimensional stability. Accordingly, for washing the knitted fabric suit of 95% wool/ 5% silk, an appropriate program designed for wool clothes was chosen. The washing conditions were as follows: 50 ml of commercial detergent, washing temperature 30 °C, 500 rpm centrifugation, and treatment time 51 minutes. After the washing procedure, the suits have been dried free at ambient temperature.

2.2.10. Softening. Before the testing, the suits were subjected to treatment with a commercial softener for a soft handle.

Each person from the healthcare system involved in the study received a suite and a set of information regarding the chemical composition of the textile material, washing guidance and its use.

3. Results and discussion

3.1. Specific mass (grams per square meter), density and thickness of the knitted fabrics

The characteristics of the knitted fabric assessed by standard methods are comprised in table 1.

Table 1. The specific mass, fabric density, and thickness of wool-knitted fabric blend with silk.

		Knitted fabric 95% wool + 5% silk
Mass/m^2 [g/m^2]		155.50
Density	Courses/centimetre (cpc)	12
	Wales/centimetre (wpc)	20
Thickness [mm]		0.76

3.2. Evaluation of dimensional stability after home laundering

To better highlight dimensional changes after household washing, 10 washing/drying cycles have been accomplished with measurements taken after each washing cycle. Since the suit should be worn as a mid-layer garment being in contact with the skin, washing after each wear is mandatory. We selected a higher number of washing/drying cycles than the one specified in the standard for the following reasons: the suit is made of knitted fabric that is susceptible to deformation and on the other hand the wool fiber has an accentuated tendency to shrink by repeated washing.

The following references were used to accomplish the measurements:

- Length of trousers, front rise and back rise;
- Sleeve lengths of the blouse (on the back from the shoulder to the bottom hem)
- The body length of the blouse;
- Width at the chest of the blouse, both front, and backside.

Figure 1 and figure 2 illustrate the dimensional changes in the length of the pant after each washing cycle (Wc). The measurements were taken both in the wet state of the material immediately after washing and after drying in a free state at ambient temperature.

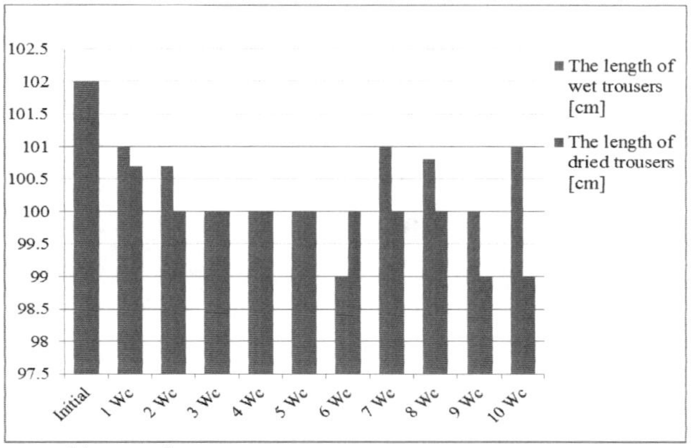

Wc- washing cycles

Figure 1. The length of the wet and dried trousers after each cycle of washing/drying.

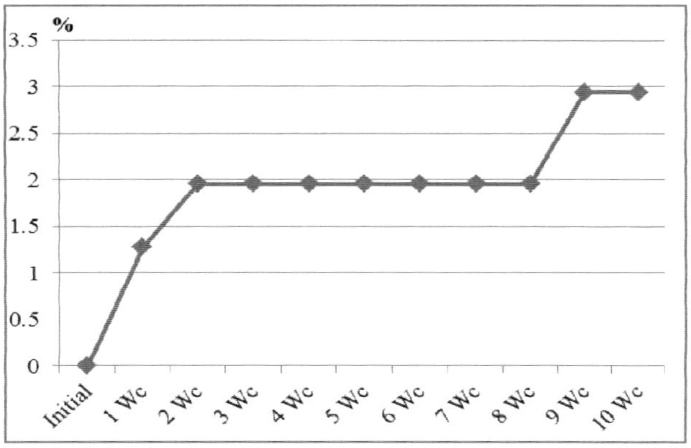

Wc- washing cycles

Figure 2. Shrinkage of trousers length after each wash/dry cycle.

IOP Conf. Series: Materials Science and Engineering 1182 (2021) 012019 doi:10.1088/1757-899X/1182/1/012019

The measurement reveals a change in the length of the trousers after the first wash, this being reduced from 102 cm by 1.3 cm, meaning a shrinkage of 1.27% from the initial length. An increase of the shrinkage was also noticed after the second wash cycle to 1.96%. After this, the length is maintained constant until the washing cycle 9 characterized by shrinkage of 2.94 %.

The same value of 2.94 % was obtained after the 10th wash cycle. About the dimensional change of the pant in a wet state, an unstable behavior is observed with shrinkage of the material up to the washing cycle 6, after which the material elongation occurs at cycle 7 and 8, followed by shrinkage after cycle 9 and again an elongation after washing cycle 10.

The dimensional changes recorded for the blouse following each washing cycle are presented in figures 3, 4, 5, 6 and 7. Figures 3 and 4 show changes in the sleeve of the blouse after 10 washing cycles.

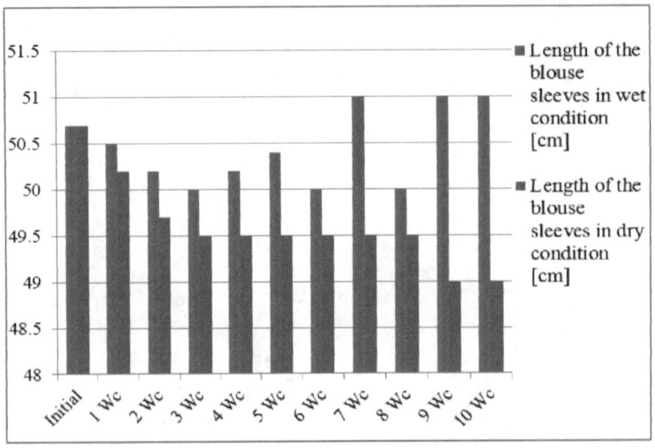

Wc- washing cycles

Figure 3. Length of the blouse sleeves in wet and dry condition after each washing/drying cycle.

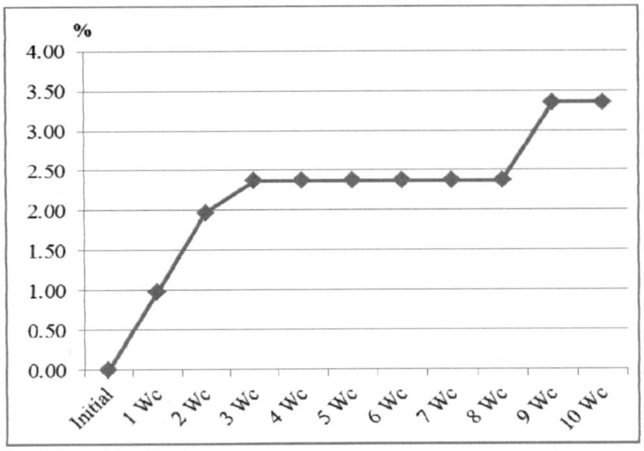

Wc- washing cycles

Figure 4. The blouse sleeve shrinkage in length after each washing/drying cycle.

Concerning the changes in the sleeve of the blouse with an initial length of 50.7 cm, a shortening of 0.5 cm was observed after the first wash, corresponding to shrinkage of 0.99%. The shrinkage increased to 1.97% (1 cm) after the second wash and to 2.37% (1.2 cm). after the third wash. This value went on constant for the 4-8 washing cycles after which on the 9 and 10 cycles the shrinkage increased to 3.35% (1.7 cm). Dimensional changes in the wet sleeve show the same variable behavior as the pants, observing several contractions alternated with elongations.

Figures 5 and 6 illustrate the changes in the length of the blouse after each washing cycle.

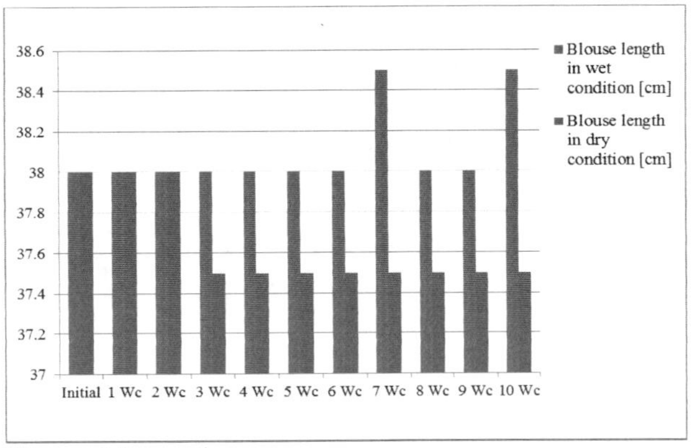

Wc- washing cycles

Figure 5. Blouse length in wet and dry conditions after each washing/drying cycle.

From the presented figures, it can be observed that the length of the blue has been kept at the same value as the initial length until after the second washing cycle when it has suddenly decreased by 0.5 cm (1.32% shrinkage).

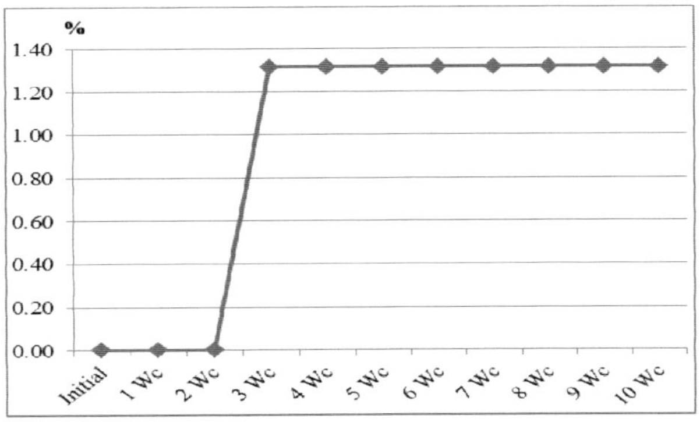

Wc- washing cycles

Figure 6. Shrinkage of the blouse length after each washing/drying cycle.

From washing cycle 3 to 10 no further change has occurred. As regards the wet length of the blouse, a more stable behavior is observed compared to that of the sleeve or pants, with only 2 slight elongations after washing cycle 7 and washing cycle 10. Finally, analysis of the dimensional stability of the blouse after repeated washings considered the assessment of the width. The values obtained after each washing cycle are given in figure 7.

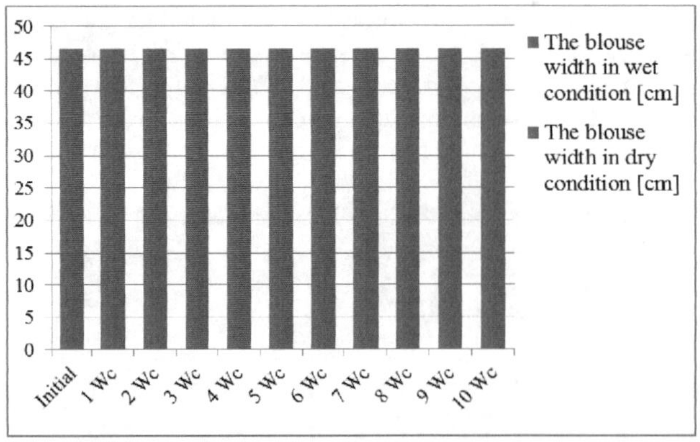

Wc- washing cycles

Figure 7. The width of the wet and dry blouse after each washing/drying cycle.

Considering figures 7 it can be observed that the width of the blue has not changed from the initial width (46.5 cm), with the contraction being 0 %. Repeated washings have led to an increase in density in wales direction and compactness of the fabric, thus increasing the number of yarn-to-yarn contact points and the number of contact surfaces between the yarns, all leading to dimensional changes.

The length of the pants was within ± 2% of the acceptable limits only up to the 9 washing cycle. As regards the blouse, the length of the sleeve exceeds the allowed limit after washing cycle 3, due to the shrinkage of 2.37 %. For the length of the blouse, the shrinkage does not exceed 1.32% even after the 10th washing cycle. The width of the blouse was constant after all 10 washing cycles.

After each washing cycle, the mass change of the suit was also followed. From figure 8, a mass decrease of 0.83% can be observed after the first wash, reaching a maximum of 1.54% after the 4th wash and then begins to rise above the initial mass after the seventh wash. This can be ascribed to the fact that after the first washing a removal of impurities and other substances present on the material occurred.

Besides, the increase in mass from washing 8 may be attributed to several successive washing steps that have led to fiber swelling and improvement of the material's hydrophilic features, which makes it easier to regain the humidity from the environment.

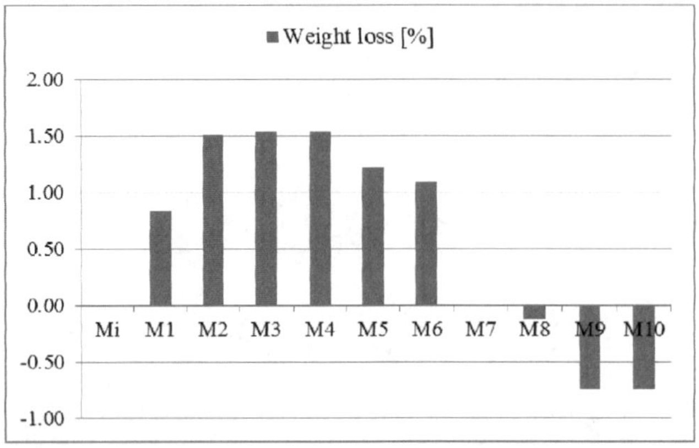

Figure 8. Weight of the suit after each washing/drying cycle.

3.3. Hydrophilic properties

Clothing in general, but particularly the first layer that comes into contact with the skin, should properly dampness the moisture in excess from the undergarment area. This requires that materials designed for the first layer of clothing are hydrophilic, being capable of absorbing or adsorbing moisture and ensuring its migration into or onto the surface of the material.

Average values of five determinations accomplished on different areas of the material have been considered. The results reveal that the knitted fabric in its original state did not absorb any water at all, but its wettability increased with each wash, attaining a wetting time about 30 seconds after the washing cycle 5. This wetting time could be noticed also after the 10th wash cycle.

For a material to be considered hydrophilic, the wetting time should be approximately 10 seconds. In the present case, no proper hydrophilic feature has been obtained for the analyzed knitted sample. Since the wool and silk fibers are very hygroscopic, the explanation for this result may be ascribed to former treatments applied to the material which generate its hydrophobic behaviour.

3.4. Air permeability

Air permeability is an influential determinant in the performance of textile material designed for clothing and technical application. Factors affecting the air permeability include the yarn properties, fabric structure (knitted, woven, nonwoven) and properties such as thickness, density, porosity, together with finishing treatments.

To evaluate the air permeability of the knitted fabric we choose as reference a 100% cotton knitted fabric with similar characteristics designed for an undergarment. According to the values from table2, our suit presents comparable air permeability with reference (60.29 m^3/min m^2) [12].

Table 2. Parameters for porosity and air permeability parameters of the knitted fabric.

Porosity [%]	Air permeability [m^3/min m^2]	Air permeability index [kg/m^2h]	Airflow passing resistance [mm m^2 h°/kg]
89	60.68	4386.46	0.00016
83.33	58.71	4227.12	0.00012

For a comprehensive evaluation of the air permeability features, another relevant parameter was airflow passing resistance, calculated for undergarments (consisting of 1 layer) by using the values for fabric thickness and the calculated air permeability index. The lower the airflow passing resistance is the better the comfort properties of the analysed suit are.

3.5. Water vapour permeability

Water vapour permeability characteristic is one of the crucial factors in determining clothing comfort as it represents the ability to transfer perspiration. To assess the transfer of water vapour by diffusion, the permeability of vapours (Pv) has been used as a direct indicator. The recorded values are depicted in table 3, representing the mass change at every 2 hours of the tested samples. The increasing of the exposure time leads to a more significant mass difference compared to the initial one.

The results suggest the blend's ability to transfer humidity from the body to the outer layer of the suit.

Table 3. Parameters for assessment of water vapour permeability.

Samples of single-layer knitted fabric	Pv - 2h [g]	Pv - 4h [g]	Pv - 6h [g]
P1	0.286	0.593	0.286
P2	0.067	0.796	0.067

Figure 9 presents the water vapour transmission (WVT) coefficient (μ) at 2, 4 and 6 hours for the single-layer knitted fabric samples.

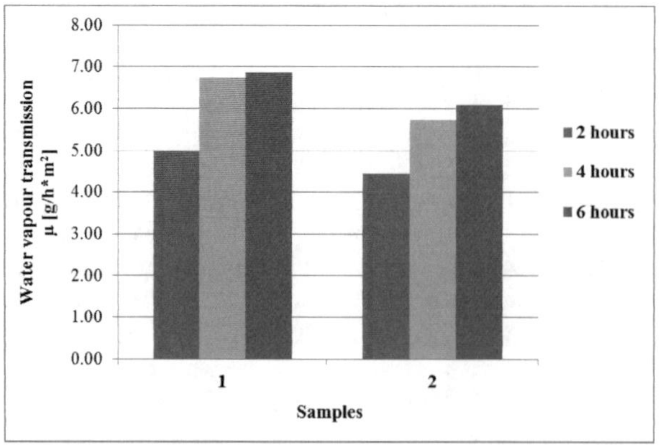

Figure 9. Evolution of water vapour transmission coefficient.

A significant increase in the water vapour transmission (WVT) coefficient (~ 4.7) was noticed after 2 hours for the knitted fabric. After four hours, a further increase was observed, but less pronounced, the WTF coefficient being 6.2. From 4 to 6 hours according to the WTF coefficient (6.4) the water vapour transmission was slower. Even so, after 6 hours the material still can transfer moisture.

3.6. Thermal conductivity

The thermal isolation of the human body by clothing intends to establish a thermal regime that allows the development of various processes of the body, inside the comfort limits. Besides other functions, the garment competes in establishing a thermal balance among the produced and released quantity of heat. The heat transfer by conduction takes place only in case of a difference between the body temperature and the outside temperature [13].

The defining parameters for heat transfer of the knitted fabric are depicted in table 4. These values reveal a good thermal conductivity with low heat transfer resistance according to the expected values of the end-use of the conceived ensemble. Besides, the heat transfer resistance is even lower than that of the reference (0.0537 m^{2}h^{o}C/kcal) [12].

Table 4. Thermal conductivity of the knitted fabric.

Samples	Thermal conductivity- λ [kcal/mh^{o}C]	Heat transfer resistance - Rt [m^{2}h^{o}C/kcal]
1	0.0414	0.01787
2	0.0378	0.02063

3.7. The comfort perception via survey

Conclusions on the wearing comfort of the knitting suits were formulated based on a questionnaire. This questionnaire was answered by 51 medical staff, with an average age between 47 and 48 years (13 men and 38 women). Over 90% of respondents appreciated the suit, consisting of a blouse and pants, as a comfortable one.

At the request of explaining the choice made regarding the comfort conferred by the suit, the vast majority were satisfied with the comfort conferred. Only a few responded that it irritates the skin and gives a warm feeling. Almost 90% of the respondents appreciated that this suit ensured good thermophysiological comfort during all the wearing. More than 90% of the respondents appreciated that from a sensory point of view, the suit gave the feeling of soft/velvety/silky.

Regarding the dimensional changes following washing treatments, at home, almost 40% of respondents reported dimensional changes.

4. Conclusions

Within the study, we developed a reusable suit composed of a blend of natural fibres (wool/silk 95/5%) addressed to the healthcare staff in the Covid sections. This suit should be worn under the full-body protective clothing that proved to be uncomfortable, intending to improve their termo-physiological comfort. The accomplished investigation methods envisaged whether our suit fulfills or not the termo-physiological comfort properties. The results of the study confirm the fact that the conceived suit can ensure appropriate termo-physiological comfort to the healthcare personnel.

Besides, the results of the research are consistent with the perception of the medical personnel involved in the study via the survey.

5. References

[1] Ivanoska-Dacikj A and Stachewicz U 2020 Smart textiles and wearable technologies – opportunities offered in the fight against pandemics in relation to current COVID-19 state*Rev. Adv. Mater. Sci.* **59** 487–505.

[2] Verbeek J H, Rajamaki B, Ijaz S, Sauni R, Toomey E, Blackwood B, Tikka C, Ruotsalainen J Hand Kilinc Balci F S 2020 Personal protective equipment for preventing highly infectious diseases due to exposure to contaminated body fluids in healthcare staff *Cochrane Database of Systematic Reviews* **5** Art. No.: CD011621.

[3] 2011 *Handbook of Medical Textiles* (Philadelphia:Woodhead Publishing Series in Textiles) pp 443-460, Editors. Bartels.

[4] Raeve A D, Vasile S and Cools J 2018 Selected factors influencing wear comfort of clothing: case studies. *J Textile Eng Fashion Technol*. **4**(1) 66-71.

[5] Özkan E T and Kaplangiray B M 2019 Investigating thermophysiological comfort properties of polyester knitted fabrics *J Textile Eng Fashion Technol.* **5**(1) 50–56.

[6] Marolleau A, Salaun F, Dupont D, Gidik H and Ducept S 2017 Influence of textile properties on thermal comfort *IOP Conf. Ser.: Mater. Sci. Eng.* **254** 182007.

[7] Negru D, Buhu L, Loghin E, Dulgheriu I and Buhu A 2017 Absorption and moisture transfer through knitted fabrics made of natural and man-made fibers *Industria Textila* **68**(4) 269-74.

[8] https://www.woolmark.com/fibre/ (accessed 06.03.2021).

[9] Rippon J A, Christoe J R, Denning R J, Evans D J, Huson M G, Lamb P R, Millington K R and Pierlot A P 2016 Wool: Structure, Properties, and Processing *Encyclopedia of Polymer Science and Technology* John Wiley & Sons.

[10] Fangueiro R, Gonçalves P, Soutinho F and Freitas C 2009 Moisture management performance of functional yarns based on wool fibres, *Indian Journal of Fibre & Textile Research* **34** 315-20.

[11] Kumar P, Sinha S K and Ghosh S 2015 Moisture management behaviour of modified polyester wool fabrics *Fashion and Textiles* **2**

[12] Cîrja J, Ivaşco V, Scripcenco A, Ceban I and Plugaru V Experimental research on thermophysiological comfort characteristics of knitwear for underwear products http://repository.utm.md/handle/5014/2221

[13] Dulgheriu I, Ionescu I, Ionesi D and Dragomir A 2015 Evaluation and calculation model for heat transfer equilibrium through clothing articles *Industria Textila* **66**(2) 59-66

Acknowledgements

The study was carried out under a research contract funded by Silvania Worsted Spinning S.R.L from Simleu Silvaniei, Salaj, Romania.

IOP Conf. Series: Materials Science and Engineering 1182 (2021) 012020 doi:10.1088/1757-899X/1182/1/012020

On the long-term hydroelastic global structural response evaluation of a 100000 tdw oil-tanker

L Domnisoru
"Dunarea de Jos" University of Galati, Department of Naval Architecture,
Domneasca Street 111, 800201, Galati, Romania

E-mail: leonard.domnisoru@ugal.ro

Abstract. In this paper, the hydroelastic analysis for a 100000 tdw oil-tanker, with a length of 240 m, operation reference speed of 15 knots, on two loading cases full cargo and ballast, is developed. For a realistic approach of the dynamic response, the oil-tanker is considered under random waves' loads, requiring several steps in the process of the short-term and long-term global structural analysis. The random waves have Longuet-Higgins formulation, with first-order harmonics by ITTC power density spectrum, with maximum wave significant height 12 m, and second-order interference waves' harmonics. The short-term hydroelastic analysis includes the linear response by frequency-domain solutions, the non-linear response by time-domain solutions, the global strength significant response by FFT Fast Fourier Transformation spectral approach. The long-term analysis includes the hydroelastic global strength assessment for Word Wide Trade averaged wave's height histogram, the fatigue hull structure evaluation, over a 25 years exploitation reference time, applying the cumulative damage ratio approach. The stress hot-spot factors, up to 1.5, influence is considered. The oil-tanker model by own program DYN is developed, modules for hydroelastic response and statistical processing, applicable in any ship's design stage, stressing out to the 100000 tdw oil-tanker long-term operation limits on the two relevant loading cases.

1. Introduction

The ships with a length over 200 m have an elastic hull structure, with the fundamental natural global frequencies lower than 1 Hz [1, 2]. The global structural evaluation for large ships includes the vibrations response by advanced hydroelasticity approach [3-5]. The hydroelastic global response includes linear and non-linear oscillation and vibration components. Hydrodynamic phenomena slamming, whipping, springing by hydroelasticity are modeled, providing an extended evaluation of the long-term ships' operation capabilities in extreme sea state conditions [6-8].

The hydroelastic numerical deterministic and statistic simulations focus on a 100000 tdw oil-tanker ship [9]. The tanker has significant geometric non-linearities at the fore part, V shape with bulb, and at the aft part, U shape. Amidships an extended prismatic shape is modeled (section 3).

The hydroelastic theoretical method (section 2) implemented in own program DYN [10], for a head Longuet-Higgins wave excitation with multi-harmonic interference components, includes several computational modules previously validated by experimental tests [11].

The 100000 tdw oil-tanker hydroelastic response evaluation includes three main criteria: bottom slamming and deck flooding occurrence prediction by short-term most probable vertical maximum displacements, long-term stresses check out by yielding stress limit, fatigue strength evaluation by cumulative damage approach, thus providing the ship's operation limits assessment (sections 4, 5).

2. Fundaments of long-term ship's hydroelastic response in irregular waves

The ship's hydroelastic response combines oscillations and vibrations hull global dynamic analyses, involving short-term linear frequency domain solutions and non-linear time-domain solutions, followed by long-term prediction response and structural fatigue analysis. The theoretical method implemented in own code DYN [10] has four main analysis steps with interlinked numerical modules.

- First analysis step is the short-term linear hydroelastic ship's response by module STABY (DYN) [10].

For the hydroelastic analysis as excitation head wave model Longuet-Higgins (L&H) [10] is considered. The wave elongation $\zeta_v(x,t)$ (1) includes first-order components $\zeta_v^1(x,t)$ (2) with the ITTC spectrum [12] and second-order positive and negative interference components $\zeta_v^{2(\pm)}(x,t)$ (3).

$$\zeta_v(x,t)=\zeta_v^1(x,t)+\zeta_v^{2(+)}(x,t)+\zeta_v^{2(-)}(x,t)=\sum_{(\omega_e)}\left[\zeta_v^C(x,\omega_e)\cos\omega_e t-\zeta_v^S(x,\omega_e)\sin\omega_e t\right]\ ;\ x\in[0,L] \quad (1)$$

$$\zeta_v^1(x,t)=\sum_{i=1}^{m}\left[\zeta_{iC}^1(x,\omega_e^i)\cos(\omega_e^i t)-\zeta_{iS}^1(x,\omega_e^i)\sin(\omega_e^i t)\right]\ ;\ \omega_e^i=\omega_i+\frac{\omega_i^2}{g}v\ ;\ m=250 \quad (2)$$

$$\zeta_v^{2(\pm)}(x,t)=\sum_{i=1}^{m}\sum_{j=1}^{m}\left[\zeta_{i,jC}^{2(\pm)}(x,\omega_e^{i,j(\pm)})\cos(\omega_e^{i,j(\pm)}t)-\zeta_{i,jS}^{2(\pm)}(x,\omega_e^{i,j(\pm)})\sin(\omega_e^{i,j(\pm)}t)\right]\ ;\ \omega_e^{i,j(\pm)}=\omega_e^i\pm\omega_e^j\neq 0 \quad (3)$$

where: g gravity acceleration; v , L ship speed and length; $\omega_i, i=1,m$ wave circular-frequencies by ITTC spectrum [12]; t time; x longitudinal coordinate; $\omega_e\in\{\omega_e^i, \omega_e^{i,j(+)}, \omega_e^{i,j(-)}\}, i,j=1,m$.

The global vibration response analysis requires idealizing the hull by an elastic beam finite element model [10]. The ship's stations idealization uses multi-parametric conformal transformation [10]. The strip theory hydrodynamic formulation includes the ship's hull vertical deflection.

The linear hydroelastic response $w_{lin}(x,t)$ (4), according to the modal analysis approach, includes the heave (r=0) and pitch (r=1) oscillation modes and the first three vibration modes (r=2, 3, 4) [10].

$$w_{lin}(x,t)=\sum_{r=0}^{4}w_r(x)\cdot p_r^{(1)}(t)\ ;\ \{p^{(1)}(t)\}=\{p_r^{(1)}(t), r=0,4\}^T\ ;\ x\in[0,L] \quad (4)$$

where: $w_r(x), r=0,4$ displacement modal forms; $\{p^{(1)}(t)\}$ first-order (linear) modal coordinates.

The linear hydroelastic equations system (5), under irregular Longuet-Higgins wave, has a frequency domain solution for each wave circular-frequency ω_e component [10]. The linear solution corresponds to the short-term linear oscillation and springing response components (8).

$$[A(\omega_e)]\{\ddot{p}^{(1)}(t)\}+[B(\omega_e)]\{\dot{p}^{(1)}\{t\}\}+[C(\omega_e)]\{p^{(1)}(t)\}=\sum_{(\omega_e)}\left[\{F_v^C(\omega_e)\}\cos\omega_e t-\{F_v^S(\omega_e)\}\sin\omega_e t\right]\rightarrow\{p^{(1)}(t)\} \quad (5)$$

where: $[A(\omega_e)], [B(\omega_e)], [C(\omega_e)]$ the structural plus hydrodynamic inertial, damping, restoring matrices for wave ω_e ; $\{F_v^C(\omega_e)\}, \{F_v^S(\omega_e)\}$ the Longuet-Higgins generalized wave excitation force terms.

- Second analysis step is the short-term non-linear hydroelastic response by module TRANZY (DYN) [10].

The non-linear hydroelastic response $w_{nl}(x,t)$ (6) by the modal analysis approach formulation, corresponds to the short-time non-linear oscillation and springing, transitory whipping, due to bottom and side slamming, response components (8).

$$w_{nl}(x,t)=\sum_{r=0}^{4}w_r(x)\cdot\left[p_r^{(1)}(t)+p_r^{(2)}(t)\right]\ ;\ x\in[0,L] \quad (6)$$

where: $\{p^{(2)}(t)\}=\{p_r^{(2)}(t), r=0,4\}^T$ second-order modal coordinates non-linear hydroelastic response.

The non-linear hydroelastic system includes terms for the ship's hull geometrical non-linearities and the bottom slamming impact pressure. Therefore, a non-linear hydrodynamic strip-theory formulation is required. Using the first-order modal coordinates (4) and the Longuet-Higgins wave elongation (1), the second-order modal coordinates (6) result from a non-linear iterative time-domain solution approach by the equations system (7). For each iteration, the simulation time is 80 s, with a time step of 0.01s, using the β-Newmark method for the direct time-domain integration approach [10].

$$\left\lfloor A\right|_{\omega_{osc}}^{\omega_2}\left\rfloor\{\ddot{p}^{(2)}(t)\}+\left\lfloor B\right|_{\omega_{osc}}^{\omega_2}\left\rfloor\{\dot{p}^{(2)}\{t\}\}+\left\lfloor C\right|_{\omega_{osc}}^{\omega_2}\left\rfloor\{p^{(2)}(t)\}=\{F_{nl}(t,\{\ddot{p}^{(2)}(t)\},\{\dot{p}^{(2)}(t)\},\{p^{(2)}(t)\})\}\rightarrow\{p^{(2)}(t)\} \quad (7)$$

where: $\left\lfloor A\right|_{\omega_{osc}}^{\omega_2}\left\rfloor, \left\lfloor B\right|_{\omega_{osc}}^{\omega_2}\left\rfloor, \left\lfloor C\right|_{\omega_{osc}}^{\omega_2}\right\rfloor$ are the structural plus hydrodynamic inertial, damping, restoring matrices, for $\omega_{osc}=(\omega_0+\omega_1)/2$ the average oscillation natural circular-frequency and ω_2 the first-order vibration natural circular frequency; $\{F_{nl}(t)\}$ the non-linearities generalized force.

Based on linear (5) and non-linear (7) solutions, the power density spectra and the short-term statistical linear and non-linear hydroelastic response components (8, 9, 10, 11) by FFT Fast Fourier Transformation spectral method result [10].

$$RMS_Y^\chi\big|_\alpha(x,h_{1/3})=\sqrt{\int_{\omega_{e1\alpha}}^{\omega_{e2\alpha}} S_Y^\chi(\omega_e,x)\big|_{h_{1/3}}\,d\omega_e} \; ; \; RMS_Y^\chi\big|_{hel}(x,h_{1/3})=\sqrt{\sum_{(\alpha)}\left[RMS_Y^\chi\big|_\alpha(x,h_{1/3})\right]^2} \; ; \; x\in[0,L] \quad (8)$$

where: $\alpha\in\{osc,vib1,vib2,vib3\}$ oscillation and vibration response; osc includes both heave and pitch components that have close natural frequencies; hel includes all the hydroelastic components; $\omega_{e1\alpha},\omega_{e2\alpha}$ the circular-frequencies reference limits for hydroelastic components domain; $S_Y(\omega_e)$ power density spectrum for any hydroelastic response type; $h_{1/3}$ wave significant height; $\chi\in\{lin,nl\}$ linear and non-linear hydroelastic response.

$$Z_{sw}(x)=d_{aft}+(d_{fore}-d_{aft})\frac{x}{L}; \; w_{RMS}^\chi\big|_\alpha(x,h_{1/3})=RMS_w^\chi\big|_\alpha(x,h_{1/3}); \; x\in[0,L] \quad (9)$$

$$Z_{deck}^\chi(x,h_{1/3})=Z_{sw}(x)+w_{RMS}^\chi\big|_{hel}(x,h_{1/3})\le Z_D(x); \; Z_{keel}^\chi(x,h_{1/3})=Z_{sw}(x)-w_{RMS}^\chi\big|_{hel}(x,h_{1/3})\ge Z_K(x) \quad (10)$$

where: w_{RMS} vertical displacement most probable amplitude (9); d_{aft},d_{fore} aft and fore draught in still water; Z_{deck},Z_{keel} total short-term displacement at deck and keel into base plane reference; Z_D,Z_K vertical limit references at deck and keel line for the oil-tanker lines (figure 1) for bottom slamming and deck flooding occurrence evaluation (figures 10, 16).

$$Q_{1/3}^\chi\big|_{\alpha,hel}(x,h_{1/3})=2\cdot RMS_Q^\chi\big|_{\alpha,hel}(x,h_{1/3}); \; Q_{1/3+sw}^\chi(x,h_{1/3})=Q_{sw}(x)+Q_{1/3}^\chi\big|_{hel}(x,h_{1/3}); \; x\in[0,L] \quad (11)$$

where: $Q\in\{M,T\}$ bending moment or shear force in ship's vertical plane; $Q_{1/3}$ the bending moment or shear force significant amplitudes; $Q_{sw},Q_{1/3+sw}$ still water and total short-term bending moment or shear force in ship's vertical plane.

The short-term significant deck normal stress $\sigma_{xD1/3}^\chi\big|_{\alpha,hel}^x$, bottom normal stress $\sigma_{xB1/3}^\chi\big|_{\alpha,hel}^x$ and neutral axis tangential stress $\tau_{xzNN1/3}^\chi\big|_{\alpha,hel}^x$ result from expression (12).

$$\sigma_{xD,B1/3}^\chi\big|_{\alpha,hel}(x,h_{1/3})=K_s\frac{M_{1/3}^\chi\big|_{\alpha,hel}(x,h_{1/3})}{W_{D,B}(x)}; \; \tau_{xzNN1/3}^\chi\big|_{\alpha,hel}(x,h_{1/3})=K_sT_{1/3}^\chi\big|_{\alpha,hel}(x,h_{1/3})\cdot K_{tnn}(x) \quad (12)$$

where: $W_{D,B}(x)=W_{D0,B0}\cdot f_{W_{D,B}}(x)$ deck and bottom bending modules with distribution functions $f_{W_{D,B}}(x)$ and amidships W_{D0},W_{B0} values; $K_{tnn}(x)=K_{tnn0}\cdot f_{K_{tnn}}(x)$ the maximum shearing stress coefficient at neutral axis with distribution function $f_{K_{tnn}}(x)$ and amidships K_{tnn0} value (figure 2, table 2); $K_s=1.0\div 1.5$ the average stress hot-spot coefficient.

- Third analysis step is the long-term hydroelastic response by module LTP (DYN) [10].

The long-term hydroelastic analysis uses the averaged Word Wide Trade (WWT) histogram of the wave's significant height $p(h_{1/3i}), i=1,12$ (figure 5) [12]. The sea state long-term occurrence probability $P_{LTP}(R)$ is function to the reference exploitation period R (year), with the values presented in table 1 for the WWT long-term wave's significant height $h_{1/3}$ histogram.

Table 1. Sea state long-term occurrence probability $P_{LTP}(R)$ for WWT wave's height histogram.

R(years)	1 hour	1 day	1 week	1 month	1 year	5 years	10 years	15 years	20 years	25 years
P_{LTP}(R)	$2.2\cdot 10^{-3}$	$9.1\cdot 10^{-5}$	$1.3\cdot 10^{-5}$	$3.0\cdot 10^{-6}$	$2.5\cdot 10^{-7}$	$5.0\cdot 10^{-8}$	$2.5\cdot 10^{-8}$	$1.7\cdot 10^{-8}$	$1.2\cdot 10^{-8}$	$1.0\cdot 10^{-8}$

The long-term hydroelastic linear and non-linear stresses $Y_R^\chi\big|_{hel}^x \in \left\{\sigma_{xD-R}^\chi\big|_{hel}^x, \sigma_{xB-R}^\chi\big|_{hel}^x, \tau_{xyNN-R}^\chi\big|_{hel}^x\right\}$ result by solving the non-linear equation (13) for each reference period R (table 1) and WWT wave's significant height histogram, based on the short-term stresses (12). The total stress values (14) result by adding the still water stresses ($\sigma_{xDsw}, \sigma_{xBsw}, \tau_{xzNNsw}$) to the long-term hydroelastic stress (13), with assessment by yielding stress limit criteria (σ_{adm}, τ_{adm}, table 2).

$$P_{LTP}(R)=\sum_{i=1}^{12} p(h_{1/3i})\cdot Exp\left[-\frac{2\cdot\left(Y_R^\chi\big|_{hel}^x\right)^2}{\left(Y_{1/3}^\chi\big|_{hel}^x(h_{1/3i})\right)^2}\right]\to Y_R^\chi\big|_{hel}^x\ ;\ Y_{1/3}^\chi\big|_{hel}^x\in\left\{\sigma_{xD1/3}^\chi\big|_{hel}^x,\sigma_{xB1/3}^\chi\big|_{hel}^x,\tau_{xyNN1/3}^\chi\big|_{hel}^x\right\} \quad (13)$$

$$\sigma_{xD,B-R}^\chi(x)=\sigma_{xD,Bsw}(x)+\sigma_{xD,B-R}^\chi\big|_{hel}^x\le\sigma_{adm}\ ;\ \tau_{xzNN-R}^\chi(x)=\tau_{xzNNsw}(x)+\tau_{xzNN-R}^\chi\big|_{hel}^x\le\tau_{adm}\ ;\ x\in[0,L] \quad (14)$$

- Fourth analysis step is the long-time hydroelastic fatigue analysis by module FAT (DYN) [10].

For the hydroelastic fatigue analysis, the damage cumulative ratio D approach (15, 16) is applied [12], and the predicted safety exploitation life R_E (17) results. For reference, we consider R=25 years for long-term WWT wave's height histogram and deck shell but-welded structural details [12].

$$D_\alpha^\chi(x)=\sum_{i=1}^{12}\frac{n_\alpha(h_{1/3i})}{N_\alpha^\chi(h_{1/3i},x)}\ ;n_\alpha(h_{1/3i})=p(h_{1/3i})\cdot 3.1536\cdot 10^7 f_\alpha R\,;N_\alpha^\chi(h_{1/3i},x)=F_{SN}\left(2f_M f_W\sigma_{xD1/3}^\chi\big|_\alpha^x(h_{1/3})\right) \quad (15)$$

$$D^\chi(x)\big|_{full,ball}=D_{osc}^\chi(x)\big|_{full,ball}+D_{vib1}^\chi(x)\big|_{full,ball}+D_{vib2}^\chi(x)\big|_{full,ball}+D_{vib3}^\chi(x)\big|_{full,ball}\ ;\ \chi\in\{lin,nl\} \quad (16)$$

$$D^\chi(x)=0.5\cdot D_{full}^\chi(x)+0.5\cdot D_{ball}^\chi(x);\ \ R_E^\chi(x)=R/D^\chi(x);\ \ x\in[0,L] \quad (17)$$

where: f_α oscillations and vibrations natural frequencies (table 2); f_M equivalent symmetric cycle factor [12]; f_W welding quality factor [12]; F_{SN} stress-strain fatigue curves for a structural detail [12]; *full, ball* loading cases full-cargo and ballast considered with the same occurrence probability 50%.

The fatigue analysis is done for each stress hot-spot coefficient K_s, for the linear and non-linear hydroelastic response, for several relevant stations (x/L=0.25, 0.50, 0.75). A safe structural detail, by fatigue criteria, should ensure $D\le 1$, $R_E\ge 25$ years, but at least an exploitation life of $R_{E\min}\ge 5$ years that covers one time between periodic ship's hull class inspections [12].

3. The 100000 tdw oil-tanker numerical model

The hydroelastic analysis focuses on a 100000 tdw oil-tanker (table 2) with the 3D hull-lines from figure 1, modeling with accuracy the aft and fore hull geometric non-linearities. The prismatic shape with vertical sides extends over 50% of the ship hull, corresponding to the cargo-hold compartments. The hull-lines model is conformal for the linear and non-linear strip theory hydrodynamic formulation (section 2). Table 2 presents the 100000 tdw oil-tanker characteristics [9].

The hull structure preliminary design is according to the DNVGL rules [12]. Figure 2 presents the strength data simplified distribution functions and the amidships structure at the cargo-hold compartments. The 100000 tdw oil-tanker has a double hull into a longitudinal structural system, with a longitudinal central bulkhead. At the center structural part, the strength characteristics have a simplified uniform distribution, with the preliminary design values in table 2.

The hull structure is made of shipbuilding steel class A and AH32 (deck), isotropic material, according to the DNV-GL rules [12], with the yielding stress limit admissible values in table 2. For the vibration modes, the average structural damping is β=0.001 [10].

Table 2. Oil-tanker 100000 tdw characteristics.

Main data		Loading cases			Natural frequencies f (Hz)					Steel material	
L (m)	240.000	Case	Full cargo	Ballast	Mode	Full cargo		Ballast		E (N/m^2)	$2.1\cdot10^{11}$
B (m)	42.000	Δ (t)	126457.16	81763.34	*add.mass*	*dry*	*wet*	*dry*	*wet*	G (N/m^2)	$8.1\cdot10^{10}$
H (m)	21.300	x_G (m)	126.457	126.732	(0) osc_1	-	0.097	-	0.107	ρ_{mat}(t/m^3)	7.85
Stations	165(41)	y_G (m)	0	0	(1) osc_2	-	0.101	-	0.111	β	0.001
g (m/s^2)	9.81	d_{aft} (m)	15.000	10.500	(2) vib_1	0.811	0.664	1.027	0.810	R_{eH} (MPa)	315_{AH32}
ρ_{water} (t/m^3)	1.025	d_m (m)	15.500	10.000	(3) vib_2	1.878	1.517	2.359	1.828	σ_{adm} (MPa)	263_{AH32}
v (kn)	15	d_{fore} (m)	15.000	9.500	(4) vib_3	3.111	2.482	3.917	2.961	R_{eH} (MPa)	235_A
I_{yy0} (m^4)	417.481	e_D (m)	11.895	e_B (m)	9.405	A_{fz0} (m^2)	2.564	$h_{DNV}(m)$	10.285	σ_{adm} (MPa)	205_A
A_0 (m^2)	5.635	W_{D0} (m^3)	35.096	W_{B0} (m^3)	44.391	$1/K_{tmn0}$(m^2)	1.522	$h_{1/3}$ (m)	0-12	τ_{adm} (MPa)	120_A

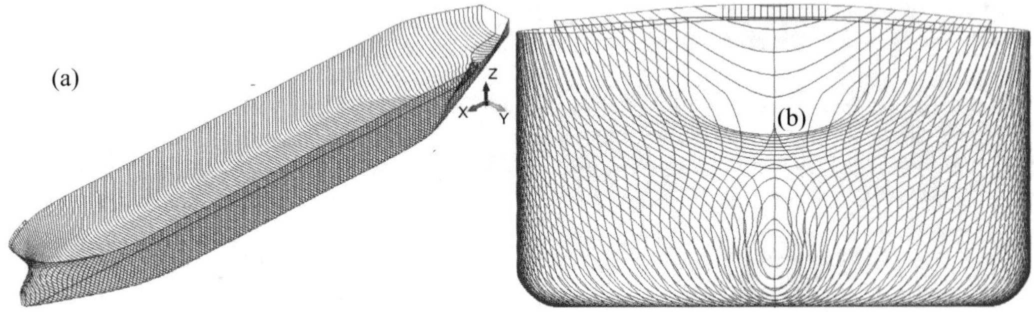

Figure 1. Oil-tanker 100000 tdw hull-lines model, (a) 3D view, (b) fore to aft view [9].

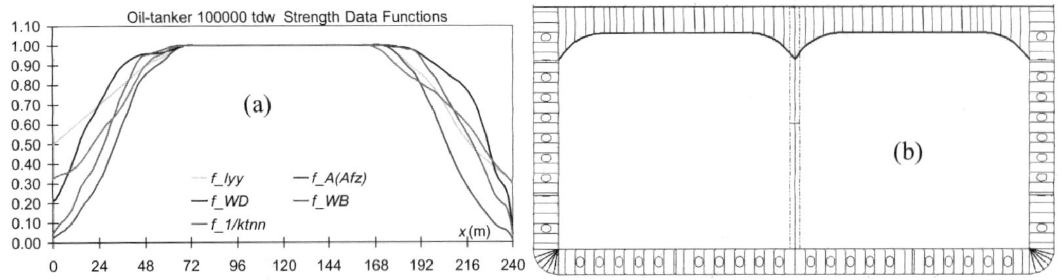

Figure 2. Oil-tanker 100000 tdw, (a) strength data distribution functions, (b) hull structure amidships.

For the 100000 tdw oil-tanker global hydroelastic analysis, we select two relevant loading cases, full cargo and standard ballast, with the mass distributions m_x (figure 4), displacement, draughts in table 2. The oil-tanker hull structure is with a 1D equivalent beam numerical model [10] (section 2).

Figure 3 presents the hydrodynamic added mass m_{33} and damping N_{33} per unit length function to the motion frequency and instantaneous draught ($z \leq 21.3$ m) at station x/L=0.5. The natural heave and pitch oscillation frequencies, by a linear seakeeping method [10], have similar values with an average of 0.104 Hz (table 2). The natural vibration frequencies, by the Jacobi method [10], have first-order frequencies higher than 0.660 Hz (table 2). The hydrodynamic added masse and damping have changes in the oscillation domain. In the vibration domain, the added masses are no longer function to the frequency, the hydrodynamic damping becomes negligible, sustaining the formulation in (7).

Figure 5 presents the averaged WWT Word Wide Trade [12] wave significant height histogram, describing the long-term occurrence probability of the irregular waves. Figure 6 shows the short-term Longuet-Higgins wave time record (1) and FFT amplitude spectrum, reference $h_{1/3}$=10.285 m.

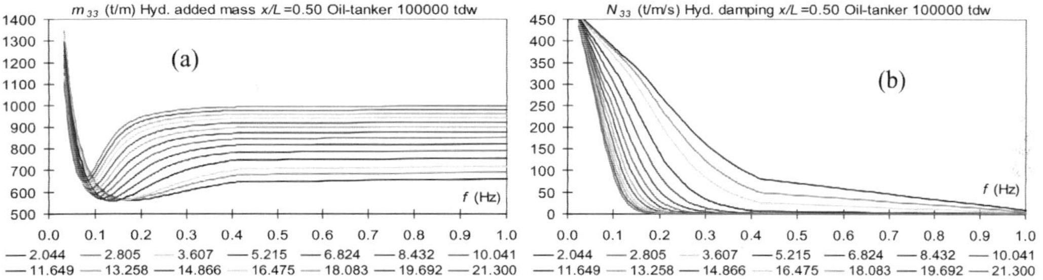

Figure 3. Oil-tanker 100000 tdw, reference section x/L=0.5, (a) m_{33} (t/m) hydrodynamic added mass per unit length, (b) N_{33} (t/m/s) hydrodynamic damping per unit length.

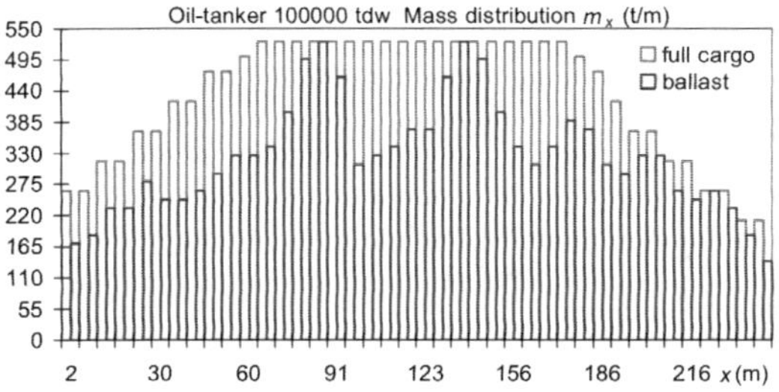

Figure 4. Oil-tanker mass distribution m_x (t/m)

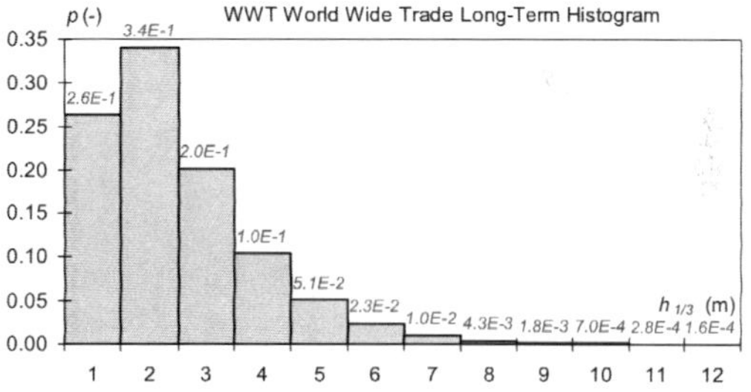

Figure 5. WWT wave $h_{1/3}$ height histogram [12].

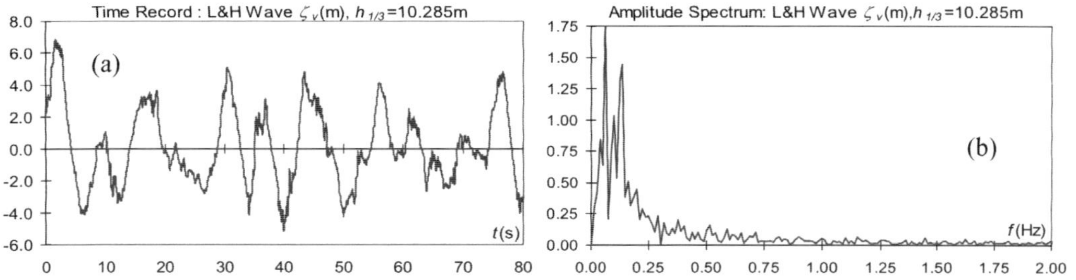

Figure 6. Longuet-Higgins (L&H) wave ζ_v (m), reference $h_{1/3}$=10.285 m, (a) time record, (b) FFT amplitude spectrum.

4. The short-term hydroelastic analysis of the 100000 tdw oil-tanker

4.1. The oil-tanker short-term hydroelastic response in full-cargo case

This section presents a selection of the relevant results for the short-term hydroelastic analysis of the oil-tanker, in full-cargo case (table 2), as following:

- Figure 7 presents the vertical displacement w (m) time record and FFT amplitude spectrum, at x/L=0.25 section, non-linear solution (6), $h_{1/3}$=10.285 m, full-cargo case.
- Figures 8 and 9 present the vertical bending moment M (kNm) time record and FFT amplitude spectrum, at x/L=0.50 section, linear and non-linear solutions, $h_{1/3}$=10.285 m, full-cargo case.
- Figure 10 presents the total short-term vertical displacement at deck Z_{deck} (m) and keel Z_{keel} (m) (10), into base plane reference, non-linear solution, x=0-240m, $h_{1/3}$=0-12 m, full-cargo case.
- Figure 11 presents the total short-term vertical bending moment $M_{1/3+SW}$ (kNm) and vertical shear force $T_{1/3+SW}$ (kNm) (11), non-linear solution, x=0-240m, $h_{1/3}$=0-12 m, full-cargo case.
- Figure 12 presents the significant vertical bending moment $M_{1/3}$ (kNm), oscillation, vibrations modes (11), linear and non-linear solution, x=0-240m, $h_{1/3}$=0-12 m, full-cargo case.
- Table 3 presents the most probable vertical displacement w_{RMS} (m) (9), at x/L=0.25, 0.50, 0.75, the ratios between first, second, third vibrations and oscillation modes, $h_{1/3}$=10.285 m, full-cargo.
- Table 4 presents the significant vertical bending moments $M_{1/3}$ (kNm) and the significant shear forces $T_{1/3}$ (kNm) (11), at x/L=0.25, 0.50, 0.75, including the ratios between first, second, third vibrations and oscillation modes, for reference $h_{1/3}$=10.285 m, full-cargo case.

For the hydroelastic vertical displacement response, the FFT amplitude spectrum (figure 7) points out as the main component oscillations (f_{osc}≈0.099 Hz) rigid body modes. The most probable vertical deflection on vibration modes is in the range of 21.4÷81.1 mm (x/L=0.25-0.75), which represent 2.98÷3.03% (f_{vib1}=0.664 Hz), 2.32÷2.36% (f_{vib2}=1.517 Hz), 1.30÷1.33% (f_{vib3}=2.482 Hz) from the most probable oscillation displacement (table 3). The differences between the non-linear and linear most probable vertical displacements are 3.68% ÷ 6.71% (table 3) on average. The total most-probable vertical displacement points out that bottom slamming occurs at both end parts, aft $h_{1/3}$>4m, fore $h_{1/3}$> 5m (figure 10). Side slamming occurs at both ends due to the ship's flare forms (figure 1). The green sea (deck flooding) occurs only at the forepeak, for $h_{1/3}$>10m (figure 10).

For the hydroelastic vertical bending moment response, the FFT amplitude spectra (figures 8, 9) reveal that besides the oscillations, heave, pitch (f_{osc}=0.097÷0.101 Hz) rigid body modes, the first-order vibration mode (f_{vib1}=0.664 Hz) is also noticeable (figure 12). The total short-term vertical bending moment (figure 11) records maximum values around the amidships sections (x/L=0.50) at the cargo-holds compartments. Comparing the significant vertical bending moment values (figure 12), at selected sections x/L=0.25, 0.50, 0.75, for oscillation mode as reference (table 4), results that the vibration modes represent 20.76%÷38.00% first-order (vib_1), 6.50%÷7.72% second-order (vib_2), 2.00%÷2.67% third-order (vib_3), supporting the linear and non-linear results in figures 8, 9, 12.

The total short-term vertical shear force (figure 11) records maximum values around the aft-quarter sections (x/L=0.25) at the ship's structural transition region from cargo holds compartments to the engine compartment. For the significant vertical shear forces, at selected sections x/L=0.25, 0.50, 0.75, the vibration modes represent 16.50%÷30.34% first-order (vib_1), 9.08%÷11.23% second-order (vib_2), 2.00%÷3.07% third-order (vib_3) from the significant oscillation modes shear force.

The differences between the non-linear and linear significant vertical bending moments (table 4) are 2.96% (osc), 88.40% (vib_1), 22.07% (vib_2), 34.80% (vib_3) on average. The differences between the non-linear and linear significant vertical shear forces (table 4) are 1.42% (osc), 83.98% (vib_1), 24.38% (vib_2), 49.05% (vib_3) on average. Thus, the non-linear solutions provide a noticeable increase of the bending moments (figure 12) and shear forces on vibration modes, especially on the first-order.

Comparing the ship's responses in full-cargo case (tables 3, 4), for the hydroelastic phenomena result that the steady linear springing is small and the non-linear springing is medium. The whipping is high due to the occurrence of the bottom and side slamming at both ship's extremities.

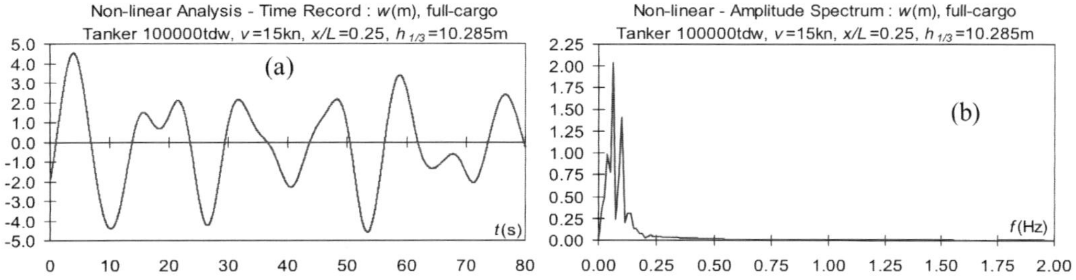

Figure 7. Oil-tanker 100000 tdw, non-linear analysis, *w* (m) vertical displacement, full-cargo, v=15 kn, x/L=0.25, L&H wave $h_{1/3}$=10.285 m, (a) time record, (b) FFT amplitude spectrum.

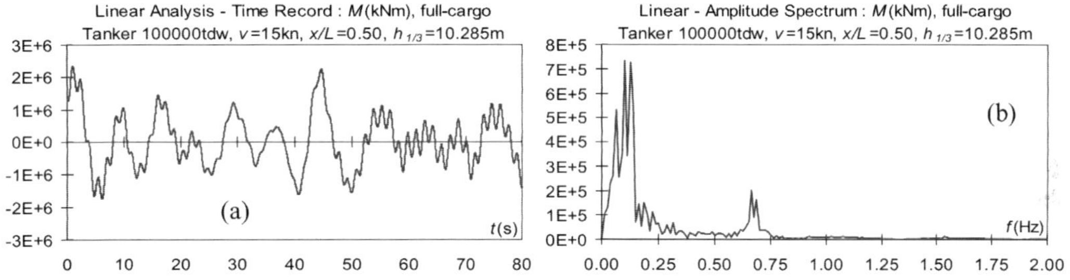

Figure 8. Oil-tanker 100000 tdw, linear analysis, *M* (kNm) vertical bending moment, full-cargo, v=15 kn, x/L=0.50, L&H wave $h_{1/3}$=10.285 m, (a) time record, (b) FFT amplitude spectrum.

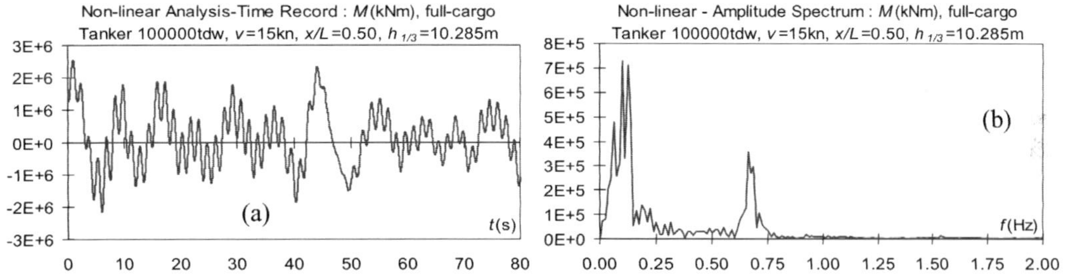

Figure 9. Oil-tanker 100000 tdw, non-linear analysis, *M* (kNm) vertical bending moment, full-cargo, v=15 kn, x/L=0.50, L&H wave $h_{1/3}$=10.285 m, (a) time record, (b) FFT amplitude spectrum.

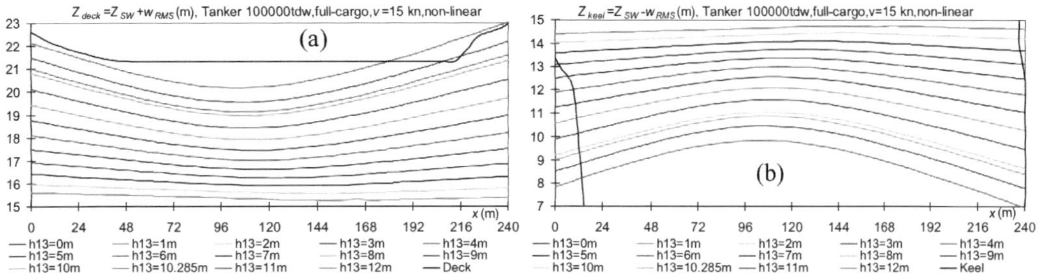

Figure 10. Oil-tanker 100000 tdw, non-linear analysis, full-cargo, v=15 kn, L&H wave $h_{1/3}$=0-12 m, total short-term vertical displacement, (a) Z_{deck} (m) deck reference, (b) Z_{keel} (m) keel line reference.

Table 3. Oil-tanker 100000 tdw, full-cargo, short-term most-probable vertical displacements w_{RMS}(m), ratios between vibrations and oscillation modes, for reference wave $h_{1/3}$=10.285 m.

Section	x(m)	Analysis	w_{oscRMS}(m)	$w_{vibRMS(1)}$(m)	%$vib1/osc$	$w_{vibRMS(2)}$(m)	%$vib2/osc$	$w_{vibRMS(3)}$(m)	%$vib3/osc$
0.25L	60	Linear	1.872	$5.29\cdot10^{-2}$	2.83%	$4.15\cdot10^{-2}$	2.22%	$2.31\cdot10^{-2}$	1.24%
0.50L	120		1.595	$4.91\cdot10^{-2}$	3.07%	$3.83\cdot10^{-2}$	2.39%	$2.14\cdot10^{-2}$	1.34%
0.75L	180		2.577	$7.82\cdot10^{-2}$	3.03%	$6.08\cdot10^{-2}$	2.36%	$3.38\cdot10^{-2}$	1.31%
Average % :			-	-	**2.98%**	-	**2.32%**	-	**1.30%**
0.25L	60	Non-linear	2.057	$5.72\cdot10^{-2}$	2.78%	$4.51\cdot10^{-2}$	2.19%	$2.54\cdot10^{-2}$	1.23%
0.50L	120		1.599	$5.13\cdot10^{-2}$	3.19%	$3.98\cdot10^{-2}$	2.47%	$2.25\cdot10^{-2}$	1.40%
0.75L	180		2.605	$8.11\cdot10^{-2}$	3.11%	$6.30\cdot10^{-2}$	2.41%	$3.55\cdot10^{-2}$	1.36%
Average % :			*(3.68%)*	*(5.44%)*	**3.03%**	*(5.40%)*	**2.36%**	*(6.71%)*	**1.33%**
Phenomena		Non-linear	Bottom slamming		Side slamming		Green sea		*ref.* $h_{1/3}$= *10.285m*
aft	0		$h_{1/3}$ > 4 m		yes		no		
fore	240		$h_{1/3}$ > 5 m		yes		$h_{1/3}$ > 10 m		

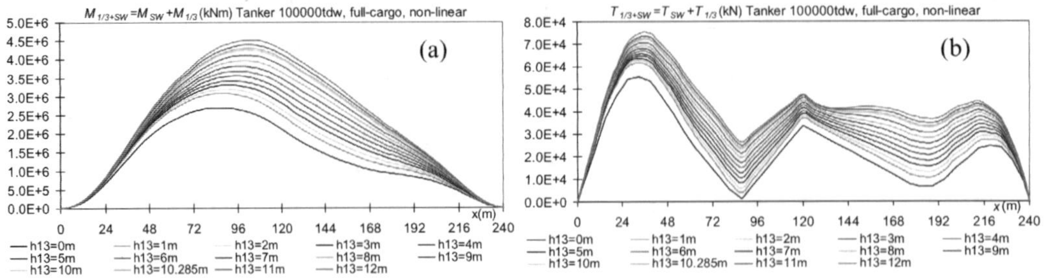

Figure 11. Oil-tanker 100000 tdw, non-linear analysis, full-cargo, v=15 kn, L&H wave $h_{1/3}$=0-12 m, (a) $M_{1/3+SW}$(kNm) total short-term bending moment, (b) $T_{1/3+SW}$(kN) total short-term shear force.

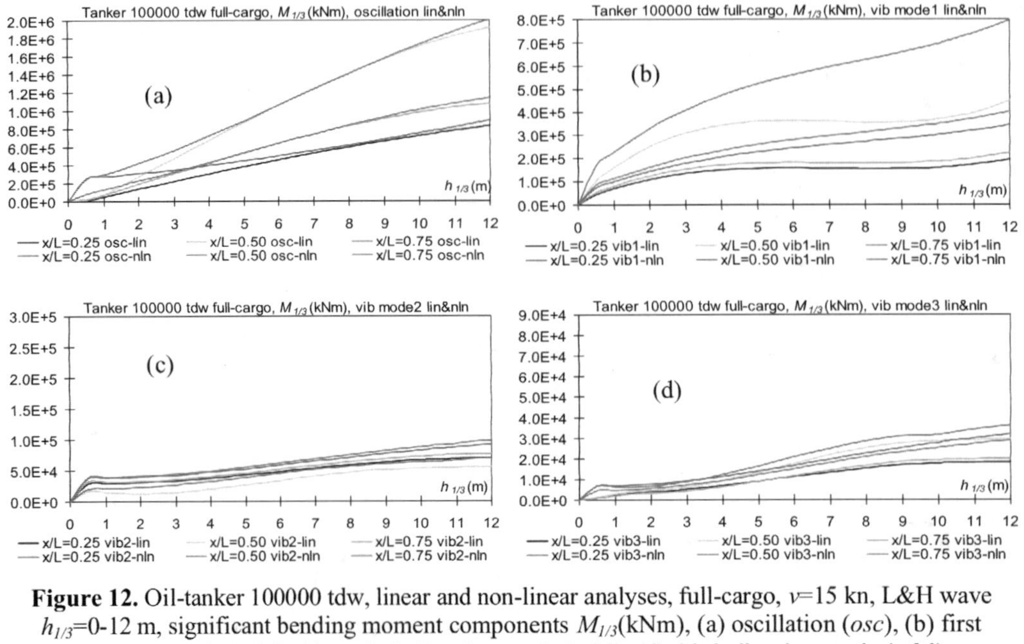

Figure 12. Oil-tanker 100000 tdw, linear and non-linear analyses, full-cargo, v=15 kn, L&H wave $h_{1/3}$=0-12 m, significant bending moment components $M_{1/3}$(kNm), (a) oscillation (*osc*), (b) first vibration mode (*vib1*), (c) second vibration mode (*vib2*), (d) third vibration mode (*vib3*).

Table 4. Oil-tanker 100000 tdw, full-cargo, short-term significant bending moments $M_{1/3}$(kNm) and shear forces $T_{1/3}$(kN), ratios between vibrations and oscillation modes, wave $h_{1/3}$=10.285 m.

Section	x(m)	Analysis	$M_{osc1/3}$(kNm)	$M_{vib1/3(1)}$(kNm)	*%vib1/osc*	$M_{vib1/3(2)}$(kNm)	*%vib2/osc*	$M_{vib1/3(3)}$(kNm)	*%vib3/osc*
0.25L	60	Linear	$7.41 \cdot 10^{+5}$	$1.63 \cdot 10^{+5}$	22.00%	$6.75 \cdot 10^{+4}$	9.11%	$1.78 \cdot 10^{+4}$	2.40%
0.50L	120	Linear	$1.75 \cdot 10^{+6}$	$3.74 \cdot 10^{+5}$	21.37%	$5.33 \cdot 10^{+4}$	3.05%	$2.87 \cdot 10^{+4}$	1.64%
0.75L	180	Linear	$9.99 \cdot 10^{+5}$	$1.89 \cdot 10^{+5}$	18.92%	$7.34 \cdot 10^{+4}$	7.35%	$1.95 \cdot 10^{+4}$	1.95%
Average% :			-	-	**20.76%**	-	**6.50%**	-	**2.00%**
0.25L	60	Non-linear	$7.71 \cdot 10^{+5}$	$3.07 \cdot 10^{+5}$	39.82%	$8.39 \cdot 10^{+4}$	10.88%	$2.84 \cdot 10^{+4}$	3.68%
0.50L	120	Non-linear	$1.78 \cdot 10^{+6}$	$7.07 \cdot 10^{+5}$	39.72%	$6.38 \cdot 10^{+4}$	3.58%	$3.23 \cdot 10^{+4}$	1.81%
0.75L	180	Non-linear	$1.03 \cdot 10^{+6}$	$3.55 \cdot 10^{+5}$	34.47%	$8.97 \cdot 10^{+4}$	8.71%	$2.58 \cdot 10^{+4}$	2.51%
Average% :			*(2.96%)*	*(88.40%)*	**38.00%**	*(22.07%)*	**7.72%**	*(34.80%)*	**2.67%**
Section	x(m)	Analysis	$T_{osc1/3}$(kN)	$T_{vib1/3(1)}$(kN)	*%vib1/osc*	$T_{vib1/3(2)}$(kN)	*%vib2/osc*	$T_{vib1/3(3)}$(kN)	*%vib3/osc*
0.25L	60	Linear	$2.14 \cdot 10^{+4}$	$4.67 \cdot 10^{+3}$	21.82%	$1.10 \cdot 10^{+3}$	5.14%	$3.56 \cdot 10^{+2}$	1.66%
0.50L	120	Linear	$1.27 \cdot 10^{+4}$	$9.26 \cdot 10^{+2}$	7.29%	$2.29 \cdot 10^{+3}$	18.03%	$3.14 \cdot 10^{+2}$	2.47%
0.75L	180	Linear	$2.50 \cdot 10^{+4}$	$5.10 \cdot 10^{+3}$	20.40%	$1.02 \cdot 10^{+3}$	4.08%	$4.64 \cdot 10^{+2}$	1.86%
Average% :			-	-	**16.50%**	-	**9.08%**	-	**2.00%**
0.25L	60	Non-linear	$2.18 \cdot 10^{+4}$	$8.81 \cdot 10^{+3}$	40.41%	$1.33 \cdot 10^{+3}$	6.10%	$4.22 \cdot 10^{+2}$	1.94%
0.50L	120	Non-linear	$1.28 \cdot 10^{+4}$	$1.61 \cdot 10^{+3}$	12.58%	$2.88 \cdot 10^{+3}$	22.50%	$6.36 \cdot 10^{+2}$	4.97%
0.75L	180	Non-linear	$2.54 \cdot 10^{+4}$	$9.66 \cdot 10^{+3}$	38.03%	$1.29 \cdot 10^{+3}$	5.08%	$5.85 \cdot 10^{+2}$	2.30%
Average% :			*(1.42%)*	*(83.98%)*	**30.34%**	*(24.38%)*	**11.23%**	*(49.05%)*	**3.07%**
Phenomena			Linear springing small		Non-linear springing medium		Whipping high		*ref.* $h_{1/3}$= *10.285m*

4.2. The oil-tanker short-term hydroelastic response in ballast case

Similar to the full-cargo case, this section presents a selection of the relevant results for the short-term hydroelastic analysis of the oil-tanker for the ballast case (table 2).

- Figure 13 shows the vertical displacement w (m) time record and FFT amplitude spectrum, at x/L=0.25 section, non-linear results (6), reference L&H wave $h_{1/3}$=10.285 m, in ballast case.
- Figures 14 and 15 present the vertical bending moment M (kNm) time record and FFT amplitude spectrum, at x/L=0.50 section, linear and non-linear results, L&H wave $h_{1/3}$=10.285 m, ballast case.
- Figure 16 presents the total short-term vertical displacement at deck Z_{deck} (m) and keel line Z_{keel} (m) (10), non-linear results, x=0-240m, L&H wave $h_{1/3}$=0-12 m, ballast case.
- Figure 17 presents the vertical total short-term bending moment $M_{1/3+SW}$ (kNm) and shear force $T_{1/3+SW}$ (kNm) (11), non-linear results, x=0-240m, L&H wave, $h_{1/3}$=0-12 m, ballast case.
- Figure 18 presents the significant vertical bending moment $M_{1/3}$ (kNm), oscillation, vibrations modes (11), linear and non-linear results, x=0-240m, L&H wave $h_{1/3}$=0-12 m, ballast case.
- Table 5 presents the most-probable vertical displacement w_{RMS} (m) (9), at x/L=0.25, 0.50, 0.75 reference sections, the ratios between first, second, third vibrations and oscillation modes, for reference Longuet-Higgins (1) wave $h_{1/3}$=10.285 m, ballast case.
- Table 6 presents the significant vertical bending moments $M_{1/3}$ (kNm) and the shear forces $T_{1/3}$ (kNm) (11), at x/L=0.25, 0.50, 0.75 reference sections, the ratios between first, second, third vibrations and oscillation modes, for reference L&H wave $h_{1/3}$=10.285 m, ballast case.

Analog to the full-load case, the hydroelastic vertical displacement response (figure 13) has the oscillation modes (f_{osc}≈0.109 Hz) dominant. Taking as reference the most-probable oscillation displacement, for vibration modes the most-probable vertical deflections are in the range of 16.8÷69.8 mm (x/L=0.25-0.75), representing 2.91÷2.94% (f_{vib1}=0.810 Hz), 2.33÷2.37% (f_{vib2}=1.828 Hz), 1.28÷1.32% (f_{vib3}=2.961 Hz) (table 5). The non-linear and linear most probable vertical displacements have 1.47% ÷ 4.28% (table 5) on average differences. The bottom slamming occurs at both end parts for any wave significant height, $h_{1/3}$>0m (figure 16), due to smaller draught as for full-cargo case (table 2). Due to the ship's flare forms at extremities (figure 1), the side slamming occurs at both ends. In the ballast case, the deck flooding (green sea) does not occur (figure 16).

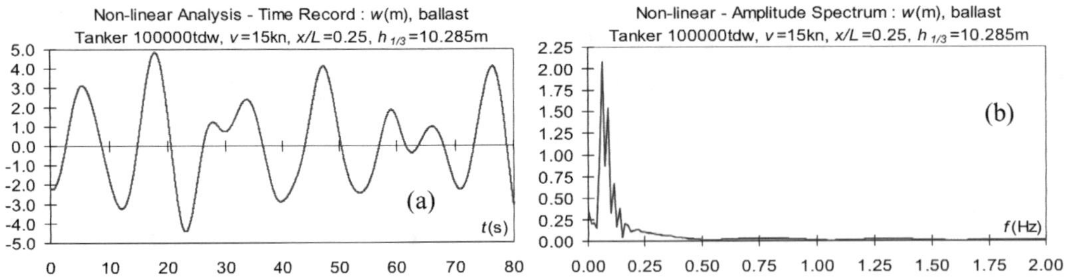

Figure 13. Oil-tanker 100000 tdw, non-linear analysis, w (m) vertical displacement, ballast, v=15 kn, x/L=0.25, L&H wave $h_{1/3}$=10.285 m, (a) time record, (b) FFT amplitude spectrum.

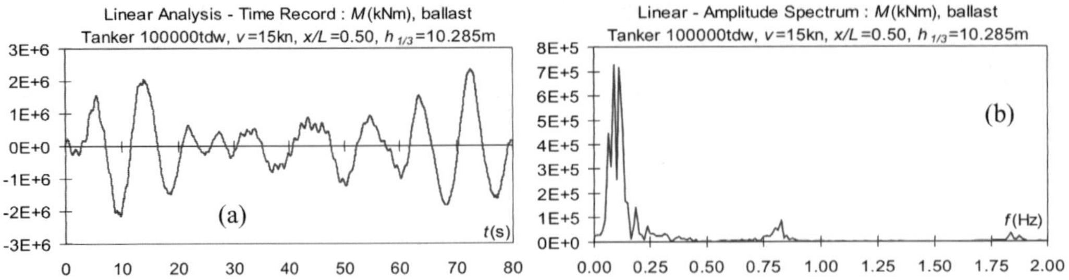

Figure 14. Oil-tanker 100000 tdw, linear analysis, M (kNm) vertical bending moment, ballast, v=15 kn, x/L=0.50, L&H wave $h_{1/3}$=10.285 m, (a) time record, (b) FFT amplitude spectrum.

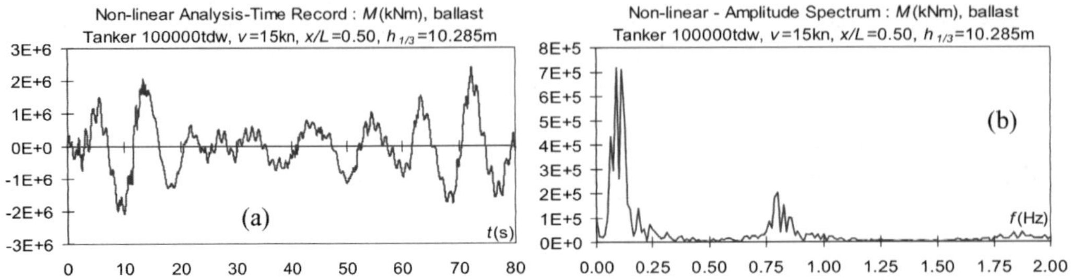

Figure 15. Oil-tanker 100000 tdw, non-linear analysis, M (kNm) vertical bending moment, ballast, v=15 kn, x/L=0.50, L&H wave $h_{1/3}$=10.285 m, (a) time record, (b) FFT amplitude spectrum.

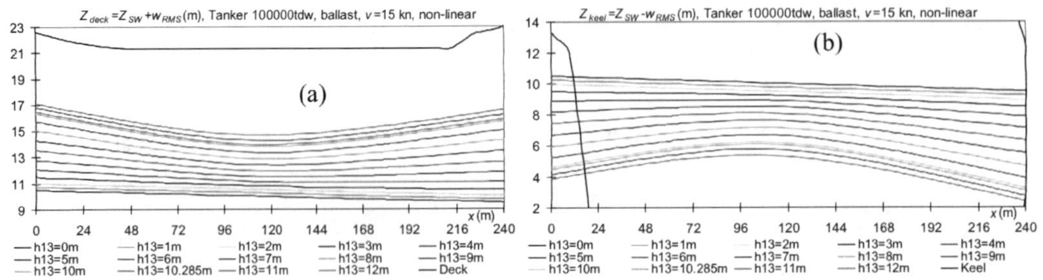

Figure 16. Oil-tanker 100000 tdw, non-linear analysis, ballast, v=15 kn, L&H wave $h_{1/3}$=0-12 m, total short-term vertical displacement (a) Z_{deck} (m) deck reference, (b) Z_{keel} (m) keel line reference.

Table 5. Oil-tanker 100000 tdw, ballast, short-term most-probable vertical displacements w_{RMS}(m), ratios between vibrations and oscillation modes, for reference wave $h_{1/3}$=10.285 m.

Section x(m)		Analysis	w_{oscRMS}(m)	$w_{vibRMS(1)}$(m)	%vib1/osc	$w_{vibRMS(2)}$(m)	%vib2/osc	$w_{vibRMS(3)}$(m)	%vib3/osc
0.25L	60	Linear	1.989	$6.87\cdot10^{-2}$	3.45%	$5.55\cdot10^{-2}$	2.79%	$3.05\cdot10^{-2}$	1.53%
0.50L	120		1.373	$3.81\cdot10^{-2}$	2.78%	$3.06\cdot10^{-2}$	2.23%	$1.68\cdot10^{-2}$	1.23%
0.75L	180		2.360	$5.87\cdot10^{-2}$	2.49%	$4.68\cdot10^{-2}$	1.98%	$2.57\cdot10^{-2}$	1.09%
Average % :			-	-	**2.91%**	-	**2.33%**	-	**1.28%**
0.25L	60	Non-linear	1.998	$6.98\cdot10^{-2}$	3.49%	$5.66\cdot10^{-2}$	2.83%	$3.16\cdot10^{-2}$	1.58%
0.50L	120		1.387	$3.96\cdot10^{-2}$	2.85%	$3.19\cdot10^{-2}$	2.29%	$1.77\cdot10^{-2}$	1.27%
0.75L	180		2.429	$6.04\cdot10^{-2}$	2.49%	$4.80\cdot10^{-2}$	1.98%	$2.67\cdot10^{-2}$	1.10%
Average % :			*(1.47%)*	*(2.81%)*	**2.94%**	*(2.93%)*	**2.37%**	*(4.28%)*	**1.32%**
Phenomena		Non-linear	Bottom slamming		Side slamming		Green sea		*ref.* $h_{1/3}$= *10.285m*
aft	0		$h_{1/3}>0$ m		yes		no		
fore	240		$h_{1/3}>0$ m		yes		no		

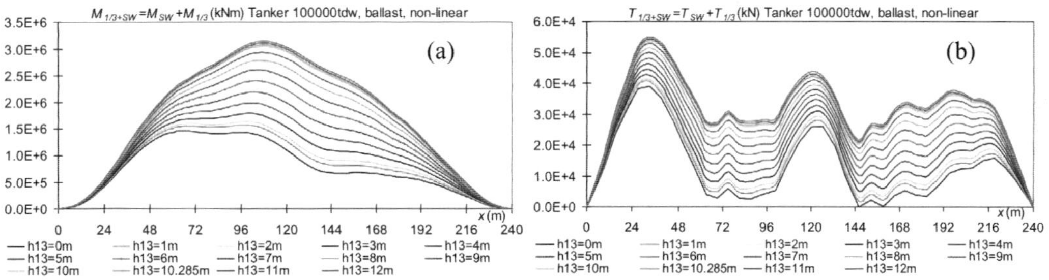

Figure 17. Oil-tanker 100000 tdw, non-linear analysis, ballast, v=15 kn, L&H wave $h_{1/3}$=0-12 m, (a) $M_{1/3+SW}$(kNm) total short-term bending moment, (b) $T_{1/3+SW}$(kN) total short-term shear force.

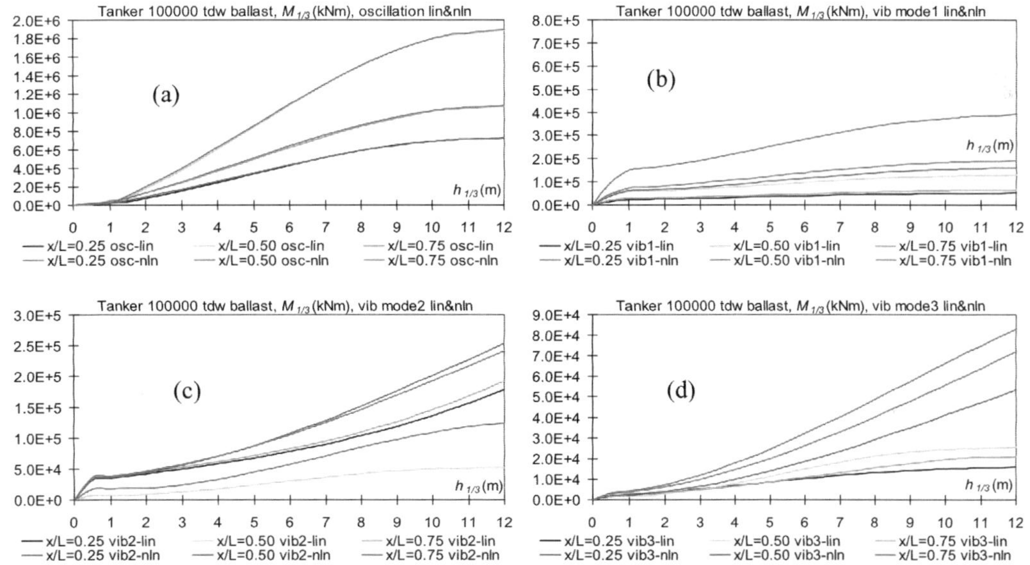

Figure 18. Oil-tanker 100000 tdw, linear and non-linear analyses, ballast, v=15 kn, L&H wave $h_{1/3}$=0-12 m, significant bending moment components $M_{1/3}$(kNm) (a) oscillation (*osc*), (b) first vibration mode (vib1), (c) second vibration mode (vib2), (d) third vibration mode (vib3).

Table 6. Oil-tanker 100000 tdw, ballast, short-term significant bending moments $M_{1/3}$(kNm) and shear forces $T_{1/3}$(kN), ratios between vibrations and oscillation modes, reference wave $h_{1/3}$=10.285 m.

Section	x(m)	Analysis	$M_{osc1/3}$(kNm)	$M_{vib1/3(1)}$(kNm)	*%vib1/osc*	$M_{vib1/3(2)}$(kNm)	*%vib2/osc*	$M_{vib1/3(3)}$(kNm)	*%vib3/osc*
0.25L	60	Linear	$6.95 \cdot 10^{+5}$	$4.98 \cdot 10^{+4}$	7.17%	$1.42 \cdot 10^{+5}$	20.43%	$1.52 \cdot 10^{+4}$	2.19%
0.50L	120		$1.81 \cdot 10^{+6}$	$1.25 \cdot 10^{+5}$	6.91%	$5.04 \cdot 10^{+4}$	2.78%	$2.45 \cdot 10^{+4}$	1.35%
0.75L	180		$1.02 \cdot 10^{+6}$	$6.35 \cdot 10^{+4}$	6.23%	$1.52 \cdot 10^{+5}$	14.90%	$2.00 \cdot 10^{+4}$	1.96%
Average % :			-	-	**6.77%**	-	**12.70%**	-	**1.83%**
0.25L	60	Non-linear	$6.96 \cdot 10^{+5}$	$1.54 \cdot 10^{+5}$	22.13%	$2.09 \cdot 10^{+5}$	30.03%	$4.28 \cdot 10^{+4}$	6.15%
0.50L	120		$1.83 \cdot 10^{+6}$	$3.76 \cdot 10^{+5}$	20.55%	$1.12 \cdot 10^{+5}$	6.12%	$6.88 \cdot 10^{+4}$	3.76%
0.75L	180		$1.03 \cdot 10^{+6}$	$1.85 \cdot 10^{+5}$	17.96%	$2.00 \cdot 10^{+5}$	19.42%	$5.81 \cdot 10^{+4}$	5.64%
Average % :			*(0.74%)*	*(113.06%)*	**20.21%**	*(67.00%)*	**18.61%**	*(184.30%)*	**5.18%**
Section	x(m)	Analysis	$T_{osc1/3}$(kNm)	$T_{ib1/3(1)}$(kNm)	*%vib1/osc*	$T_{ib1/3(2)}$(kNm)	*%vib2/osc*	$T_{ib1/3(3)}$(kNm)	*%vib3/osc*
0.25L	60	Linear	$2.13 \cdot 10^{+4}$	$1.50 \cdot 10^{+3}$	7.04%	$2.48 \cdot 10^{+3}$	11.64%	$2.80 \cdot 10^{+2}$	1.31%
0.50L	120		$1.37 \cdot 10^{+4}$	$5.50 \cdot 10^{+2}$	4.01%	$5.25 \cdot 10^{+3}$	38.32%	$3.39 \cdot 10^{+2}$	2.47%
0.75L	180		$2.66 \cdot 10^{+4}$	$1.75 \cdot 10^{+3}$	6.58%	$1.55 \cdot 10^{+3}$	5.83%	$4.21 \cdot 10^{+2}$	1.58%
Average % :			-	-	**5.88%**	-	**18.60%**	-	**1.79%**
0.25L	60	Non-linear	$2.14 \cdot 10^{+4}$	$4.58 \cdot 10^{+3}$	21.40%	$3.50 \cdot 10^{+3}$	16.36%	$9.85 \cdot 10^{+2}$	4.60%
0.50L	120		$1.45 \cdot 10^{+4}$	$1.42 \cdot 10^{+3}$	9.80%	$7.46 \cdot 10^{+3}$	51.45%	$1.26 \cdot 10^{+3}$	8.69%
0.75L	180		$2.67 \cdot 10^{+4}$	$5.24 \cdot 10^{+3}$	19.63%	$2.87 \cdot 10^{+3}$	10.75%	$1.45 \cdot 10^{+3}$	5.43%
Average % :			*(2.23%)*	*(187.64%)*	**16.94%**	*(56.13%)*	**26.19%**	*(255.96%)*	**6.24%**
Phenomena			Linear springing small		Non-linear springing medium		Whipping higher		*ref.* $h_{1/3}$=10.285m

The FFT vertical bending moment spectra (figures 14, 15) make visible the first-order (f_{vib1}=0.810 Hz) and second-order (f_{vib2}=1.828 Hz) modes by linear and non-linear hydroelastic solutions (figure 18). The natural frequencies in the ballast case are higher in comparison to the full-cargo case (table 2). Thus, the hydrodynamic damping becomes effective zero (figure 3) for second-order vibration mode, the bending response becoming visible for that mode (figures 14, 15).

Analog to the full-cargo case, the oscillation vertical bending moment is the dominant (f_{osc}=0.107÷0.111 Hz) modal component (figures 14, 15). The maximum total short-term vertical bending moments, for any wave condition, are obtained at the amidships sections x/L=0.50 (figure 17). For the selected sections (x/L=0.25, 0.50, 0.75), the vibration significant vertical moment modes represent 6.77%÷20.21% first-order (vib_1), 12.70%÷18.61% second-order (vib_2), 1.83%÷5.18% third-order (vib_3), in compare to the reference oscillation component (table 6, figure 18).

For this loading case, the total short-term vertical shear force (figure 17) has maximum values at aft-quarter, amidships, fore-quarter sections (x/L=0.25, 0.50, 0.75). The vibration significant vertical shear forces at the selected sections with maximum values represent 5.88%÷16.94% first-order (vib_1), 18.60%÷26.19% second-order (vib_2), 1.79%÷6.25% third-order (vib_3), from the oscillation mode reference values (table 6).

The non-linear and linear significant vertical bending moments have 0.74% (osc), 113.06% (vib_1), 67.00% (vib_2), 184.30% (vib_3) on average differences (table 6). The non-linear and linear significant vertical shear forces have 2.23% (osc), 187.64% (vib_1), 56.13% (vib_2), 255.96% (vib_3) on average differences (table 6). Thus, the non-linear hydrodynamic solutions lead to a significant increase for the vibration modes on the bending moments (figure 18) and shear forces on vibration modes.

Analog to the full-cargo case, the linear springing is small, the non-linear springing is medium (tables 5, 6). Because the impact bottom and side slamming can occur in any wave condition (figure 16), with hydrodynamic damping considered zero for frequencies over 1 Hz (figure 3), the transitory whipping is higher (figure 18).

The total short-term bending moments and shear forces in the full-cargo case are higher than in the ballast case (figures 11, 17) so that the first loading case (table 2) is the extreme case in this study for the assessment of the 100000 oil-tanker tdw strength capabilities.

5. The long-term hydroelastic and fatigue analysis of the 100000 tdw oil-tanker

5.1. The oil-tanker long-term hydroelastic and fatigue analysis in full-cargo case

This section presents a selection of the relevant results for the long-term hydroelastic (13) and fatigue (15) analysis of the oil-tanker, WWT waves significant height histogram, in the full-cargo case.

- Figure 19 presents the total long-term hydroelastic deck normal-stress $\sigma_{xD\text{-}R}$ (MPa) and neutral axis tangential-stress $\tau_{xzNN\text{-}R}$ (MPa) (14), non-linear solution, x=0-240m, R≤25years, K_s=1.2, full-cargo case.
- Figure 20 presents the trend-line for the maximum total long-term hydroelastic deck normal-stress $\sigma_{xD\text{-}R}$ (MPa) and neutral axis tangential-stress $\tau_{xzNN\text{-}R}$ (MPa) (14), linear and non-linear solutions, R=0÷25 years, reference hot-spot stress K_s=1.2, full-cargo case.
- Table 7 presents the evaluation of the yielding stress limit criteria (14) with admissible values in table 2. The assessment includes the total long-term deck, bottom, neutral axis stresses (14), linear and non-linear solutions, hot-spot stress factor K_s=1.0÷1.5, for R=25 years reference, full-cargo case.
- Table 8 presents the evaluation of the fatigue criteria (15, 16) and safety exploitation life R_E (17) prediction. The assessment has as reference the butt-welded deck shell structural detail at section x/L=0.50, linear and non-linear solutions, hydroelastic and only oscillation responses, hot-spot stress factor K_s=1.0÷1.5, welding quality standard and high quality (h.q.), full-cargo case.

The total long-term stresses (13, 14) have significant increase over the whole reference period R=25 years, being maximum amidships at deck shell and aft-quarter at the neutral axis of side, double side, longitudinal bulkhead shells (figures 19, 20).

The yielding stress criteria evaluation (table 7) points out that critical strength level results. Thus, K_s≥1.3 for the non-linear hydroelastic solutions, K_s≥1.4 for the linear hydroelastic solutions are the hot-spot stress factor limits. For the K_s=1.5 factor, the yielding stress criteria (14) imposes to reduce the long-term reference period significantly under 25 years for the full-cargo case.

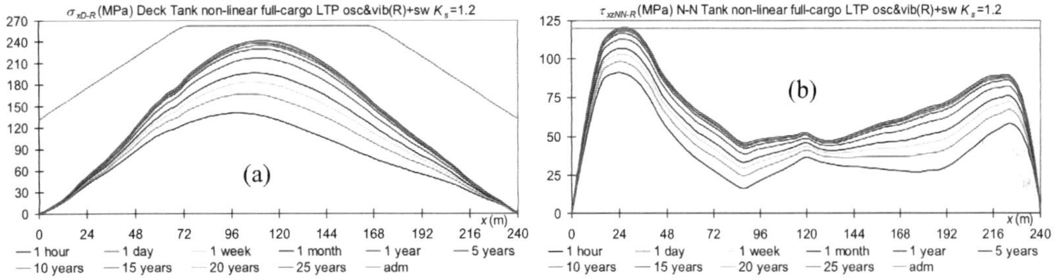

Figure 19. Oil-tanker 100000 tdw, non-linear analysis, full cargo, v=15 kn, WWT wave histogram, K_s=1.2, long-term (a) $\sigma_{xD\text{-}R}$ (MPa) deck normal-stress, (b) $\tau_{xzNN\text{-}R}$ (MPa) neutral axis tangential-stress.

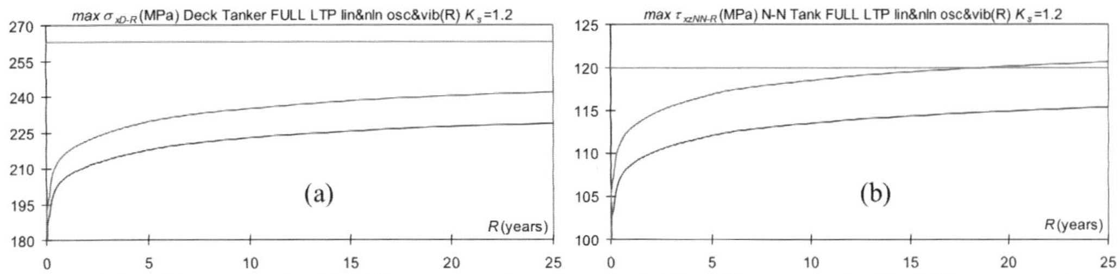

Figure 20. Oil-tanker 100000 tdw, linear and non-linear analyses, full cargo, v=15 kn, WWT wave histogram, K_s=1.2, long-term maximum stress (a) $\sigma_{xD\text{-}R}$ (MPa) deck, (b) $\tau_{xzNN\text{-}R}$ (MPa) neutral axis.

Table 7. Oil-tanker 100000 tdw, full-cargo case, total long-term maximum deck, bottom, neutral axis stresses (MPa), WWT wave histogram, linear and non-linear analysis, for R=25 years reference.

Panel selection Stress (MPa) (14)	DYN	adm (MPa)	K_s=1.0 S(MPa)	K_s=1.0 S/adm	K_s=1.0 R(years)	K_s=1.1 S(MPa)	K_s=1.1 S/adm	K_s=1.1 R(years)	K_s=1.2 S(MPa)	K_s=1.2 S/adm	K_s=1.2 R(years)
$\sigma_{xD\text{-}R}$ deck	Linear	263	190.70	0.726	>25	209.77	0.798	>25	228.84	0.871	>25
$\sigma_{xB\text{-}R}$ bottom	Linear	205	150.77	0.735	>25	165.85	0.809	>25	180.93	0.883	>25
$\tau_{xzNN\text{-}R}$ n-n axis	Linear	120	96.12	0.801	>25	105.73	0.881	>25	115.34	0.961	>25
$\sigma_{xD\text{-}R}$ deck	Non-linear	263	201.59	0.767	>25	221.74	0.844	>25	241.90	0.920	>25
$\sigma_{xB\text{-}R}$ bottom	Non-linear	205	159.38	0.777	>25	175.32	0.855	>25	191.25	0.933	>25
$\tau_{xzNN\text{-}R}$ n-n axis	Non-linear	120	100.51	0.838	>25	110.57	0.921	>25	120.62	1.005	**≈25**
Panel selection Stress (MPa) (14)	DYN	adm (MPa)	K_s=1.3 S(MPa)	K_s=1.3 S/adm	K_s=1.3 R(years)	K_s=1.4 S(MPa)	K_s=1.4 S/adm	K_s=1.4 R(years)	K_s=1.5 S(MPa)	K_s=1.5 S/adm	K_s=1.5 R(years)
$\sigma_{xD\text{-}R}$ deck	Linear	263	247.91	0.943	>25	266.99	1.016	**<25**	286.06	1.088	**<25**
$\sigma_{xB\text{-}R}$ bottom	Linear	205	196.01	0.956	>25	211.08	1.030	**<25**	226.16	1.103	**<25**
$\tau_{xzNN\text{-}R}$ n-n axis	Linear	120	124.95	0.997	>25	134.56	1.121	**<25**	144.17	1.201	**<25**
$\sigma_{xD\text{-}R}$ deck	Non-linear	263	262.06	1.011	**<25**	282.22	1.074	**<25**	302.38	1.151	**<25**
$\sigma_{xB\text{-}R}$ bottom	Non-linear	205	207.19	1.041	**<25**	223.13	1.088	**<25**	239.07	1.166	**<25**
$\tau_{xzNN\text{-}R}$ n-n axis	Non-linear	120	130.67	1.089	**<25**	140.72	1.173	**<25**	150.77	1.256	**<25**

Table 8. Oil-tanker 100000 tdw, full-cargo case, long-term fatigue analysis, WWT wave histogram, linear and non-linear analysis, oscillations and vibrations, K_s=1.0-1.5, reference section x/L=0.50, deck shell.

Model type	Welding quality	D_{osc}	D_{vib1}	D_{vib2}	D_{vib3}	D_{hel}	R_{Ehel} (year)	R_{Eosc} (year)	D_{osc}	D_{vib1}	D_{vib2}	D_{vib3}	D_{hel}	R_{Ehel} (year)	R_{Eosc} (year)
Hot-spot stress factor				K_s =1.0							K_s =1.1				
Linear	standard	$4.5\cdot10^{-1}$	$3.9\cdot10^{-2}$	$2.5\cdot10^{-7}$	$1.6\cdot10^{-8}$	0.494	>25	>25	$7.4\cdot10^{-1}$	$6.8\cdot10^{-2}$	$4.4\cdot10^{-7}$	$2.8\cdot10^{-8}$	0.812	>25	>25
Non-linear	standard	$5.1\cdot10^{-1}$	$2.2\cdot10^{-1}$	$1.3\cdot10^{-6}$	$3.1\cdot10^{-8}$	0.733	>25	>25	$8.4\cdot10^{-1}$	$3.8\cdot10^{-1}$	$2.2\cdot10^{-6}$	$5.5\cdot10^{-8}$	1.229	**20.3**	>25
	h.q.	$1.8\cdot10^{-1}$	$7.2\cdot10^{-2}$	$4.2\cdot10^{-7}$	$1.0\cdot10^{-8}$	0.252	>25	>25	$3.1\cdot10^{-1}$	$1.3\cdot10^{-1}$	$7.3\cdot10^{-7}$	$1.8\cdot10^{-8}$	0.432	>25	>25
Hot-spot stress factor				K_s =1.2							K_s =1.3				
Linear	standard	1.1	$1.1\cdot10^{-1}$	$7.4\cdot10^{-7}$	$4.7\cdot10^{-8}$	1.251	**20.0**	**22.0**	1.7	$1.9\cdot10^{-1}$	$1.2\cdot10^{-6}$	$7.6\cdot10^{-8}$	1.857	**13.5**	**15.0**
Non-linear	standard	1.3	$6.4\cdot10^{-1}$	$3.7\cdot10^{-6}$	$9.2\cdot10^{-8}$	1.953	**12.8**	**19.1**	2.0	1.0	$6.0\cdot10^{-6}$	$1.5\cdot10^{-7}$	2.995	**8.3**	**12.8**
	h.q.	$4.9\cdot10^{-1}$	$2.1\cdot10^{-1}$	$1.2\cdot10^{-6}$	$3.0\cdot10^{-8}$	0.706	>25	>25	$7.6\cdot10^{-1}$	$3.4\cdot10^{-1}$	$2.0\cdot10^{-6}$	$4.9\cdot10^{-8}$	1.104	**22.7**	>25
Hot-spot stress factor				K_s =1.4							K_s =1.5				
Linear	standard	2.4	$2.9\cdot10^{-1}$	$1.9\cdot10^{-6}$	$1.2\cdot10^{-7}$	2.653	**9.4**	**10.6**	3.3	$4.5\cdot10^{-1}$	$2.9\cdot10^{-6}$	$1.9\cdot10^{-7}$	3.716	**6.7**	**7.7**
Non-linear	standard	2.8	1.6	$9.5\cdot10^{-6}$	$2.3\cdot10^{-7}$	4.446	**5.6**	**8.9**	3.9	2.5	$1.5\cdot10^{-5}$	$3.6\cdot10^{-7}$	6.434	**3.9**	**6.4**
	h.q.	1.1	$5.4\cdot10^{-1}$	$3.1\cdot10^{-6}$	$7.7\cdot10^{-8}$	1.666	**15.0**	**22.2**	1.6	$8.3\cdot10^{-1}$	$4.8\cdot10^{-6}$	$1.2\cdot10^{-7}$	2.457	**10.2**	**15.4**

The fatigue analysis takes as reference a butt-welded structural detail from the deck shell, amidships at x/L=0.5, where the long-term stresses (figure 19) are maximum.

For the non-linear hydroelastic response, by the fatigue strength criteria (15, 16, 17), the exploitation life has restrictions for $K_s \geq 1.1$, R_E=3.9 years (K_s=1.5), standard welding quality, and improving for high-quality welding $K_s \geq 1.3$, R_E=10.2 years (K_s=1.5) (table 8). Also, for the linear hydroelastic response for $K_s \geq 1.2$, R_E=6.7 years (K_s=1.5) restrictions result.

For the non-linear oscillation response, by the fatigue strength criteria (15, 16, 17), the exploitation life has restrictions for $K_s \geq 1.2$, R_E=6.4 years (K_s=1.5), standard welding quality, and improving for high-quality welding $K_s \geq 1.4$, R_E=15.4 years (K_s=1.5) (table 8). Also, for the linear oscillation response for $K_s \geq 1.2$, R_E=7.7 years (K_s=1.5) restrictions result. Thus, the fatigue analysis based on hydroelastic response leads to higher restraints in comparison to the fatigue analysis based on oscillation response.

For standard welding quality, taking as reference the non-linear hydroelastic response, the hot-spot stress must be lower than $K_s \leq 1.4$ so that the exploitation life R_E=5.6 >5 years is over the regular inspection period [12]. A selection of high-quality welding for the full-cargo case is increasing the exploitation life up to twice of the inspection period R_E=10.2>10 years (K_s=1.5).

5.2. The oil-tanker long-term hydroelastic and fatigue analysis in ballast case

This section presents a selection of the relevant results for the long-term hydroelastic (13) and fatigue (15) analysis of the oil-tanker, WWT waves significant height histogram, in the ballast case.

- Figure 21 presents the total long-term hydroelastic deck normal-stress $\sigma_{xD\text{-}R}$ (MPa) and neutral axis tangential-stress $\tau_{xzNN\text{-}R}$ (MPa) (14), non-linear solution, stress hot-spot factor K_s=1.2, ballast case.
- Figure 22 presents the trend-line for the maximum total long-term hydroelastic deck normal-stress $\sigma_{xD\text{-}R}$ (MPa) and neutral axis tangential-stress $\tau_{xzNN\text{-}R}$ (MPa) (14), K_s=1.2, ballast case.
- Table 9 presents the evaluation of the yielding stress limit criteria (14) with admissible values in table 2 and total long-term deck, bottom, neutral axis stresses (14), ballast case.
- Table 10 presents the evaluation of the fatigue criteria (15, 16) and safety exploitation life R_E (17) prediction for butt-welded deck shell structural detail at section x/L=0.50 reference, ballast case.

Analog to the full-cargo case, the total long-term stress is maximum at deck amidships (figure 21). The total long-term transversal stress is maximum at the aft and fore-quarter neutral axis (figure 21). The influence of the reference period for $R \leq 10$ years is significant on the total long-term stresses (13, 14), continuing to increase the stress values up to R=25 years (figure 22).

In the ballast case, the yielding stress criteria (table 9) leads to no restrictions of R=25 years for any stress hot-spot factor K_s=1.0-1.5, due to lower total long-term stresses (figures 21, 22) in comparison to the full-cargo case (figures 19, 20).

Analog to the full-cargo case, for the ballast case a deck shell butt-welded detail, at x/L=0.5 with maximum long-term stress (figure 21), is the reference structure for the fatigue analysis.

In the ballast case, for the non-linear hydroelastic response, by the fatigue strength criteria (15, 16, 17), the exploitation life has restrictions for $K_s \geq 1.3$, R_E=12.5 years (K_s=1.5), standard welding quality. For high-quality (h.q.) welding no restrictions R_E>25 years (K_s=1.5) (table 10) result. Also, it results in restrictions for the linear hydroelastic response for $K_s \geq 1.3$, R_E=13.5 years (K_s=1.5).

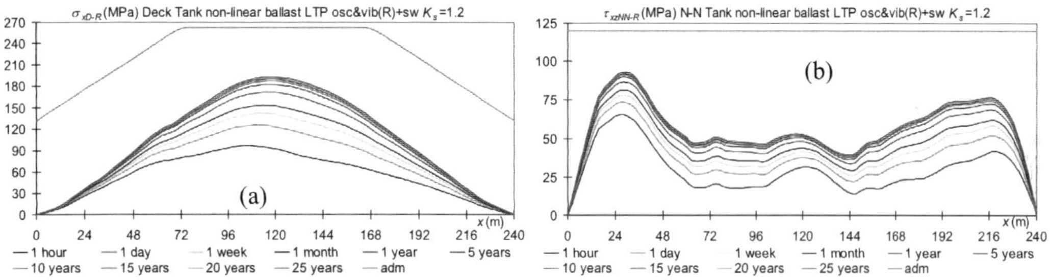

Figure 21. Oil-tanker 100000 tdw, non-linear analysis, ballast, v=15 kn, WWT wave histogram, K_s=1.2, long-term (a) $\sigma_{xD\text{-}R}$ (MPa) deck normal-stress, (b) $\tau_{xzNN\text{-}R}$ (MPa) neutral axis tangential-stress.

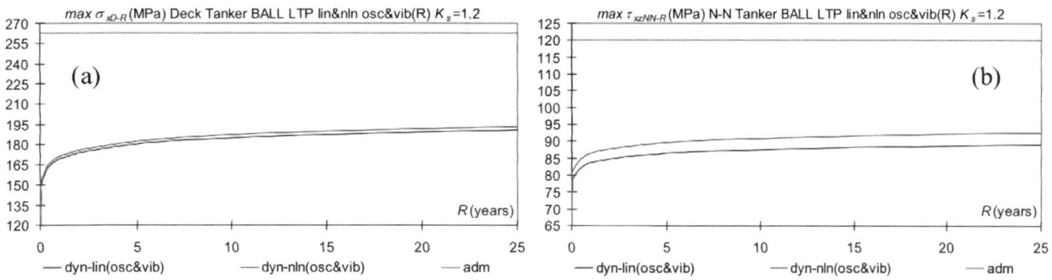

Figure 22. Oil-tanker 100000 tdw, linear and non-linear analyses, ballast, v=15 kn, WWT wave histogram, K_s=1.2, long-term maximum stress (a) $\sigma_{xD\text{-}R}$ (MPa) deck, (b) $\tau_{xzNN\text{-}R}$ (MPa) neutral axis.

In ballast cases, for the non-linear oscillation response, by the fatigue strength criteria (15, 16, 17), the exploitation life has restrictions for K_s≥1.3, R_E=12.8 years (K_s=1.5), standard welding quality, for high-quality (h.q.) welding no restrictions R_E≥25 years (K_s=1.5) (table 10) occur. For the linear oscillation response restrictions for K_s≥1.3, R_E=13.6 years (K_s=1.5) occur.

The exploitation life is more than twice of the inspection period R_E=12.5>10 years (K_s=1.5) for standard welding quality and without restrictions R_E≥25 years for high-quality welding condition, taking as reference the non-linear hydroelastic response.

Table 9. Oil-tanker 100000 tdw, ballast, total long-term maximum deck, bottom, neutral axis stresses (MPa), WWT wave histogram, linear and non-linear solutions, for *R*=25 years reference.

Panel selection Stress (MPa) (14)	DYN	adm (MPa)	$K_s=1.0$ S(MPa)	S/adm	R(years)	$K_s=1.1$ S(MPa)	S/adm	R(years)	$K_s=1.2$ S(MPa)	S/adm	R(years)
σ_{xD-R} deck	Linear	263	158.88	0.605	>25	174.77	0.665	>25	190.66	0.725	>25
σ_{xB-R} bottom		205	125.62	0.613	>25	138.18	0.674	>25	150.74	0.735	>25
τ_{xzNN-R} n-n axis		120	74.22	0.619	>25	81.65	0.680	>25	89.07	0.742	>25
σ_{xD-R} deck	Non-linear	263	160.99	0.613	>25	177.09	0.674	>25	193.19	0.735	>25
σ_{xB-R} bottom		205	127.28	0.621	>25	140.01	0.683	>25	152.74	0.745	>25
τ_{xzNN-R} n-n axis		120	77.12	0.643	>25	84.83	0.707	>25	92.55	0.771	>25
Panel selection Stress (MPa) (14)	DYN	adm (MPa)	$K_s=1.3$ S(MPa)	S/adm	R(years)	$K_s=1.4$ S(MPa)	S/adm	R(years)	$K_s=1.5$ S(MPa)	S/adm	R(years)
σ_{xD-R} deck	Linear	263	206.55	0.786	>25	222.44	0.846	>25	238.33	0.907	>25
σ_{xB-R} bottom		205	163.30	0.797	>25	175.86	0.858	>25	188.43	0.919	>25
τ_{xzNN-R} n-n axis		120	96.49	0.796	>25	103.91	0.866	>25	111.34	0.928	>25
σ_{xD-R} deck	Non-linear	263	209.28	0.807	>25	225.38	0.858	>25	241.48	0.919	>25
σ_{xB-R} bottom		205	165.46	0.804	>25	178.19	0.869	>25	190.92	0.931	>25
τ_{xzNN-R} n-n axis		120	100.26	0.835	>25	107.97	0.900	>25	115.68	0.964	>25

Table 10. Oil-tanker 100000 tdw, ballast, long-term fatigue analysis, WWT wave histogram, linear and non-linear analysis, oscillations and vibrations, K_s=1.0-1.5, reference section *x/L*=0.50, deck shell.

Model type	Welding quality	D_{osc}	D_{vib1}	D_{vib2}	D_{vib3}	D_{hel}	R_{Ehel} (year)	R_{Eosc} (year)	D_{osc}	D_{vib1}	D_{vib2}	D_{vib3}	D_{hel}	R_{Ehel} (year)	R_{Eosc} (year)
Hot-spot stress factor		K_s =1.0							K_s =1.1						
Linear	standard	$3.0\cdot10^{-1}$	$3.0\cdot10^{-5}$	$1.2\cdot10^{-7}$	$5.4\cdot10^{-9}$	0.301	>25	>25	$4.8\cdot10^{-1}$	$5.1\cdot10^{-5}$	$2.0\cdot10^{-7}$	$9.1\cdot10^{-9}$	0.476	>25	>25
Non-	standard	$3.2\cdot10^{-1}$	$5.4\cdot10^{-3}$	$3.8\cdot10^{-6}$	$5.9\cdot10^{-7}$	0.321	>25	>25	$5.0\cdot10^{-1}$	$9.0\cdot10^{-3}$	$6.4\cdot10^{-6}$	$9.8\cdot10^{-7}$	0.509	>25	>25
linear	h.q.	$1.1\cdot10^{-1}$	$1.8\cdot10^{-3}$	$1.3\cdot10^{-6}$	$1.9\cdot10^{-7}$	0.108	>25	>25	$1.8\cdot10^{-1}$	$3.0\cdot10^{-3}$	$2.1\cdot10^{-6}$	$3.2\cdot10^{-7}$	0.179	>25	>25
Hot-spot stress factor		K_s =1.2							K_s =1.3						
Linear	standard	$7.1\cdot10^{-1}$	$8.1\cdot10^{-5}$	$3.2\cdot10^{-7}$	$1.4\cdot10^{-8}$	0.715	>25	>25	1.0	$1.2\cdot10^{-4}$	$4.9\cdot10^{-7}$	$2.2\cdot10^{-8}$	1.011	**24.7**	**25.0**
Non-	standard	$7.5\cdot10^{-1}$	$1.4\cdot10^{-2}$	$1.0\cdot10^{-5}$	$1.6\cdot10^{-6}$	0.766	>25	>25	1.1	$2.2\cdot10^{-2}$	$1.6\cdot10^{-5}$	$2.4\cdot10^{-6}$	1.086	**23.0**	**23.5**
linear	h.q.	$2.8\cdot10^{-1}$	$4.7\cdot10^{-3}$	$3.3\cdot10^{-6}$	$5.1\cdot10^{-7}$	0.281	>25	>25	$4.1\cdot10^{-1}$	$7.3\cdot10^{-3}$	$5.1\cdot10^{-6}$	$7.9\cdot10^{-7}$	0.419	>25	>25
Hot-spot stress factor		K_s =1.4							K_s =1.5						
Linear	standard	1.4	$1.9\cdot10^{-4}$	$7.4\cdot10^{-7}$	$3.3\cdot10^{-8}$	1.400	**17.9**	**17.9**	1.8	$2.7\cdot10^{-4}$	$1.1\cdot10^{-6}$	$4.9\cdot10^{-8}$	1.847	**13.5**	**13.6**
Non-	standard	1.5	$3.3\cdot10^{-2}$	$2.3\cdot10^{-5}$	$3.6\cdot10^{-6}$	1.513	**16.5**	**16.9**	1.9	$4.9\cdot10^{-2}$	$3.4\cdot10^{-5}$	$5.3\cdot10^{-6}$	1.997	**12.5**	**12.8**
linear	h.q.	$5.9\cdot10^{-1}$	$1.1\cdot10^{-2}$	$7.7\cdot10^{-6}$	$1.2\cdot10^{-6}$	0.596	>25	>25	$8.1\cdot10^{-1}$	$1.6\cdot10^{-2}$	$1.1\cdot10^{-5}$	$1.7\cdot10^{-6}$	0.826	>25	>25

5.3. The oil-tanker long-term fatigue analysis in combined full-cargo and ballast cases

For a realistic evaluation of the 100000 tdw oil-tanker safety exploitation life (17), the results from fatigue analysis in full-cargo (table 8) and ballast (table 10) must be combined, considering equal occurrence probability of both loading cases.

Table 11 presents the evaluation of the fatigue criteria (15, 16) and safety exploitation life R_E (17) prediction for butt-welded deck shell structural detail, at section *x/L*=0.50 reference, for the combined full-cargo and ballast cases (17).

Table 11. Oil-tanker 100000 tdw, combined full-cargo and ballast cases, long-term fatigue analysis, WWT wave histogram, linear and non-linear analysis, K_s=1.0-1.5, reference section x/L=0.50, deck shell.

Model type	Welding quality	D_{osc}	D_{hel}	R_{Ehel} (year)	R_{Eosc} (year)	D_{osc}	D_{hel}	R_{Ehel} (year)	R_{Eosc} (year)	D_{osc}	D_{hel}	R_{Ehel} (year)	R_{Eosc} (year)
Hot-spot stress factor		K_s =1.0				K_s =1.1				K_s =1.2			
Linear	standard	0.378	0.397	>25	>25	0.610	0.644	>25	>25	0.925	0.983	>25	>25
Non-	standard	0.414	0.527	>25	>25	0.672	0.869	>25	>25	1.030	1.360	**18.4**	**24.3**
linear	h.q.	0.143	0.180	>25	>25	0.241	0.305	>25	>25	0.385	0.493	>25	>25
Hot-spot stress factor		K_s =1.3				K_s =1.4				K_s =1.5			
Linear	standard	1.341	1.434	**17.4**	**18.6**	1.232	2.027	**12.3**	**20.3**	2.555	2.781	**9.0**	**9.8**
Non-	standard	1.507	2.041	**12.3**	**16.6**	2.139	2.979	**8.4**	**11.7**	2.924	4.215	**5.9**	**8.6**
linear	h.q.	0.587	0.761	>25	>25	0.856	1.131	**22.1**	>25	1.218	1.641	**15.2**	**20.5**

The long-term fatigue analysis for combined full-cargo and ballast cases (table 11) leads to:

- for linear hydroelastic response, standard welding quality, restrictions for K_s≥1.3, R_E=9.0 years = 1.8· T_R (K_s=1.5), over 80% to the technical inspection period T_R=5 years [12];
- for linear oscillation response, standard welding quality, restrictions for K_s≥1.3, R_E=9.8 years = 1.96· T_R (K_s=1.5), almost twice the technical inspection period;
- for non-linear hydroelastic response, standard welding quality, restrictions for K_s≥1.2, R_E=5.9 years = 1.18· T_R (K_s=1.5), close to the technical inspection period;
- for non-linear oscillation response, standard welding quality, restrictions for K_s≥1.2, R_E=8.6 years = 1.72· T_R (K_s=1.5), over 72% to the technical inspection period;
- for non-linear hydroelastic response, high-quality welding, restrictions for K_s≥1.4, R_E=15.2 years = 3.04· T_R (K_s=1.5), three times the technical inspection period;
- for non-linear oscillation response, high-quality welding, restrictions for K_s≥1.5, R_E=20.5 years = 4.10· T_R (K_s=1.5), four times the technical inspection period.

The fatigue results are pointing out that the hydroelastic response is delivering more restrictions in comparison to the oscillations response. For the long-term ship's behavior improvement, the best solution is the consideration of the high-quality welding process solution, so that the oil-tanker safety exploitation life is around R_E=15.2 years ≈ 3· T_R in the extreme case of stress hot-spot factor K_s=1.5.

6. Conclusions

For ships with elastic hull girder, the long-term safe exploitation life evaluation requires an advanced hydroelastic approach with coupled oscillations and vibrations formulation. This study involves the own hydroelastic code, validated by towering tank experimental tests [11]. For best accuracy, the hydroelastic approach includes the Longuet-Higgins model wave excitation, short-term and long-term hydroelastic response, with linear and non-linear solutions, enhanced by fatigue analysis (section 2).

The hydroelastic analysis includes hydrodynamic phenomena as bottom slamming, side slamming, steady springing and transitory whipping [10]. Thus, it can generate non-linear vibration components up to 5 Hz that requires a significant amount of computational time.

This study focuses on a 100000 tdw oil-tanker with 240 m length and first-order natural frequency lower than 1 Hz (section 3). Thus, the hull girder has much sensitivity on global vibration modes, which qualifies this ship for a long-term hydroelastic assessment.

The short-term hydroelastic analysis (section 4) points out that the oscillation component represents the main response component, with small changes from the linear to the non-linear solutions. The global vibration components record a significant increase in the non-linear solution in comparison to the linear solutions. For the sectional efforts in the full-cargo case, the non-linear first-order vibration component becomes significantly higher in comparison to the other two vibration modes. In the ballast case, the non-linear sectional efforts of second and first-order vibration components become significant due to higher natural vibration frequencies in comparison to the full-cargo loading case that practically leads to zero hydrodynamic damping in the vibration domain. Including the still water sectional efforts, the total short-term hydroelastic bending moments and shear forces result higher for the full-cargo case in comparison to the ballast case. The bottom and side slamming occur for mid-sea state conditions in the full-cargo case and any sea state in the ballast case. The deck flooding (green sea) occurs only in the full-cargo case, at the forepeak, for extreme sea state conditions. For both cases, the linear springing is small; the non-linear springing becomes medium due to the geometric non-linearities. The transitory whipping is high for both loading cases due to much reduced hydrodynamic and structural damping on vibration modes.

The long-term hydroelastic stresses (section 5) have a significant increase function to the reference period. In the full-cargo case, the stresses hot-spot larger than 1.3 lead to restrictions by the yielding stress criteria, without restrictions in the ballast case. For both cases, the total long-term deck stress is maximum at amidships over the cargo-hold compartments. Thus, the selected reference structure for fatigue analysis is a butt-welded deck shell structural detail.

The long-term fatigue analysis (section 5) points out that the vibration components have a negative effect on the fatigue strength, reducing the expected exploitation life. More restrictions occur in the full-cargo case in comparison to the ballast case, due to higher stress values. More differences result between the linear and non-linear solutions, the stresses hot-spot values having a negative influence. For combined loading cases, stress hot-spot 1.5, non-linear solutions, the safe exploitation life is 5.9 years (standard welding) and 15.2 years (high-quality welding) (table 11), covering at least one inspection time reference (5 years).

In conclusion, the long-term hydroelastic analysis of the 100000 tdw oil-tanker points out that the best accuracy results by the non-linear hydroelastic solutions. For the increase of the predicted long-term safe exploitation life, the high-quality welding process is the best solution. Also, by design, the structural details stress hot-spot should reduce as much as possible.

Further studies shall continue to study the influence of the oil-tanker speed and other loading cases on the long-term hydroelastic response. An extended analysis of the stress hot-spot factors for the oil-tanker shall continue by 3D structural models. The sensitivity assessment of the DYN code for hydroelastic analysis shall continue by other experimental tests on new ship types.

7. References

[1] ABS 2018 *Guidance Notes on Ship Vibration* (Houston: American Bureau of Shipping).
[2] ISSC 2018 *Proceeding of the 20th ISSC Congress* (Amsterdam: IOS Press).
[3] Hirdaris S and Chunhua G 2005 *Review and Introduction to Hydroelasticity of Ships. Report No. 8* (London: Lloyd's Register).
[4] ICHMT 2012 *Proceedings of the 6th International Conference on Hydroelasticity in Marine Technology* (Tokyo: The University of Tokyo).
[5] Tilander S, Patey M and Hirdaris S 2020 Springing Analysis of a Passenger Ship in Waves *MDPI: Journal of Marine Science and Technology* **8**(492) 1-18.
[6] ABS 2010 *Guidance Notes on Whipping Assessment for Container Carriers* (Houston: ABS).
[7] DNVGL 2015 *Fatigue and Ultimate Strength Assessment of Container Ships Including Whipping and Springing* (Novik: Det Norske Veritas).
[8] LR 2018 *Global Design Loads of Container Ships and Other Ships Prone to Whipping and Springing* (London: Lloyd's Register).
[9] Dumitru D 2014 *Compendium of Ships Forms* (Galati:"Dunarea de Jos" University Foundation).
[10] Domnisoru L 2001 *Ship Dynamics. Oscillations and Vibrations* (Bucharest: Technical House).
[11] Domnisoru L and Domnisoru D 2000 Experimental Analysis of Springing and Whipping Phenomena *IOSPRESS: International Shipbuilding Progress* **47**(450) 129-140.
[12] DNVGL 2020 *Rules for Classification. Ships. Wave Loads* (Novik: Det Norske Veritas).

Acknowledgements

This research and *DYN* code in the frame of Naval Architecture Research Centre, from the "Dunarea de Jos" University of Galati, are developed.

Computational NX Fluid Structure Interaction (FSI) analysis on naval three way ball valve

C L Dumitrache and D Deleanu
Maritime University of Constanta, Department of General Engineering Sciences, Mircea cel Batran street, No. 104, Constanta, 900663, Romania

E-mail: ldumitr@yahoo.com

Abstract. This article represents the continuation of a study that was presented in another paper using the NX Siemens CAD program and FSI (fluid structure interaction analysis) for a complex valve encountered in the naval domain, a ball valve known as (three way ball valve - TWB). In designing of this valve presented in the previous work, some drawings (sketches) of the component parts presented in engineering guidelines were used, all dimensions being determined by the author. In the FSI analysis the same NX Siemens program was used to discretize the solid bodies of the valve coming into contact with the fluid, materials were chosen for these bodies, boundary conditions were set and the pressure field exerted by the fluid on the solid bodies was imported. Finally a series of conclusions were drawn about the speeds, pressures of the fluid and how they are distributed on the walls of the valve, determining the so-called structural analysis "fluid structure interaction".

1. Introduction

These TWB ball valves (figure 1) are among the least expensive, having a wide range of applications for all sizes, including very large sizes for oil pipelines and in the maritime industries. They are used for liquids in general (water, oil, liquid chemicals). These TWBs are used for a wide range of diameters, at full opening and their operation is simple and easy.

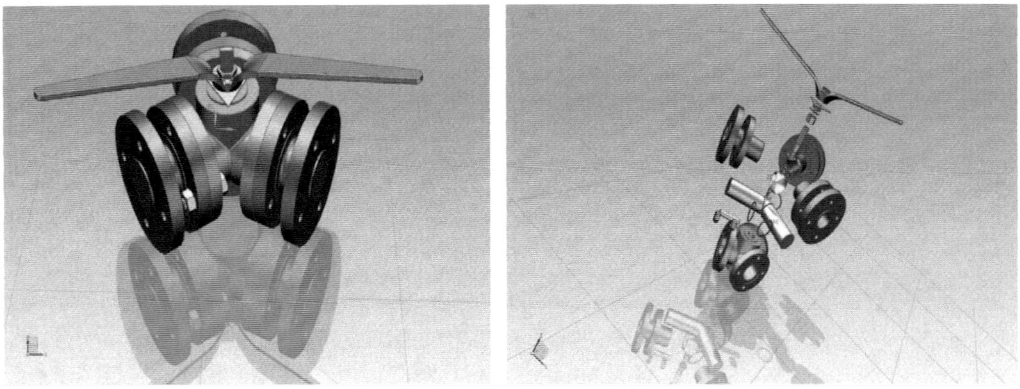

Figure 1. The three way ball valve (TWB), 3D NX assembly and exploded viewing [1].

IOP Conf. Series: Materials Science and Engineering 1182 (2021) 012021 doi:10.1088/1757-899X/1182/1/012021

The basic geometric element is a ball located in the body of the valve that has three flanges (three ways). The ball has a hole made along two axes arranged at an angle of 120° degrees. The complete operating motion is rotating the ball at 120° degrees. The ball can also be rotated at various angles so we can adjust the fluid flow.

The main parts of the TWB are [1]:

- the valve body;
- the body connectors;
- the ball;
- the threaded stem;
- the seats;

Full fluid flow is obtained when the hole inside the ball is aligned with the inlet and outlet flanges in the direction of flow by 120°. When the stem is rotated by 120° with the help of a lever or wrench, the flow is blocked in the third flange, and the locking washer may be incorporated to limit the movement of the lever, or continuous rotation may be possible. Larger ball valves can be actuated by means of reduction gearing or electric motor drives (actuators) [3].

In all cases, the opening/closing torque is low, because the frictional forces involved are only between the ball and the seats and the friction offered by the stem o-rings and lower- upper stem seals.

2. Computational Fluid Dynamic (CFD) analysis

In a previous paper [1] were presented CAD models made with NX Siemens of the component parts of TWB valve consulting engineering guidelines of some manufacturing companies [2], all dimensions of these models being original. All these models were assembled using geometric and dimensional constraints giving birth to the final assembly named "ans_valvulabila.prt" file.

Further, the volume of fluid was created which will interact with certain parts of the valve component. For this, enter the file named "ans_valvulabila.prt", deselect the component parts that are not in direct contact with the fluid and then enter the "Part Navigator"; the geometry of the pieces that come into contact with the fluid is copied using the "Wave geometry linker" button and then they are merged into a final volume with "Unite". By pressing the "Delete Face" button we will delete all the surfaces of the previously created parts except the surfaces that are wet with fluid if we activate the "Boss and Pocket Face" option. In this way the volume of fluid is created using specific tools listed in the NX program.

Two situations were considered for this flow study:

- First situation - when the valve is fully opened and the inlet flange is aligned with the outlet flange in the 120° flow direction and the third flange is blocked;
- Second situation - when the valve is partially opened and the liquid passes through two outlet flanges.

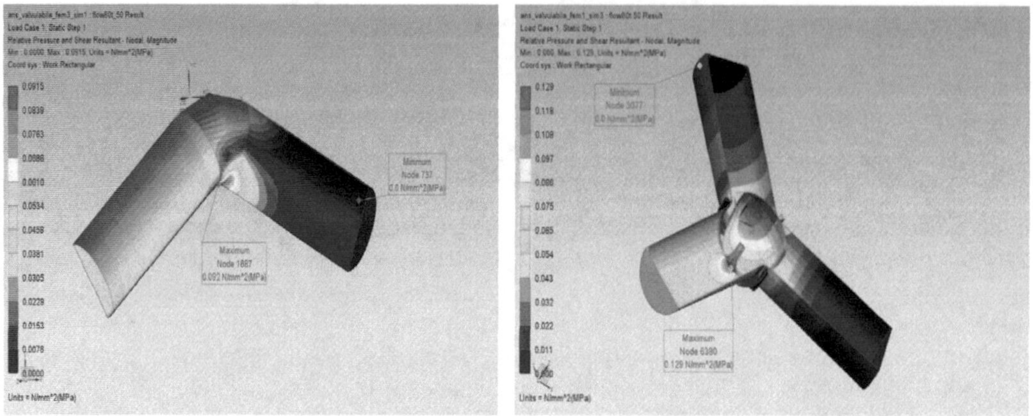

Figure 2. Relative Press and Shear Resultants a.-first situation; b.-second situation.

2.1. Discretization of volume fluid and boundary condition

With the volume of fluid obtained previously opens the "Advanced Simulation" module of the NX Siemens program. For the discretization of the fluid volume, tetrahedral elements of the TET(4) type were used. It was considered water with a temperature of 80˚C with an inlet flow rate of 50 m^3/h. At the exhaust flanges it is considered to be air at ambient temperature and pressure.

Then I clicked Start→All Aplications→Design Simulation which will open a toolbar with the "New Fem Simulation" window that created two renamed files: ans_valvulabila_fem3; ans_valvulabila_fem3_sim1. Added to these is the "Idealized Part" window that generated another ans_valvulabila_fem3_i.prt file. We have redefined the "Solution" window with the name "flow80t_50", the solver chosen is "NX Thermal/Flow", the type of analysis "Coupled Thermal Flow", and the solution will be of the type "Thermal-Flow". After pressing the "Solve" button and after completing the calculation operations in the directory structure, a series of data files about speeds, fluid pressures presented in figures 2 and 3 have appeared. The maximum speed values, pressures for the two study situations read from figures 2 and 3 are presented in table 1.

Table 1. Velocities, Relative pressure and Shear Resultants, Total pressure.

	Velocities [m/s]	Relative Pressure and Shear Resultants [MPa]	Total Pressure [MPa]
First situation	14.95	0.0915	0.108
Second situation	15.38	0.129	0.143

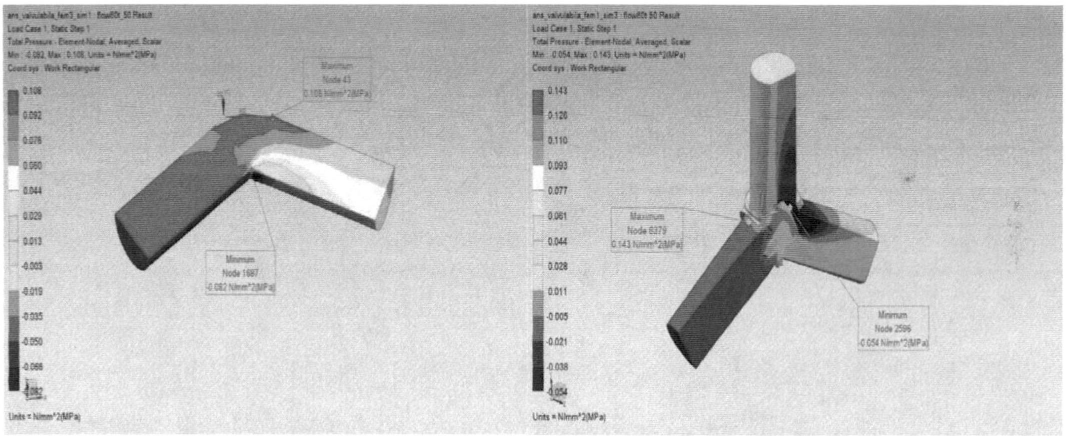

Figure 3. Total Pressure a.-first situation; b.-second situation.

3. Fluid Structure Interaction (FSI)Analysis

FSI analysis considers how the fluid interacts with the material of the valve walls. There are two types of FSI analysis [4]:

- one way FSI type when the information obtained from the fluid flow simulation (CFD) is transferred to the valve wall material structure;
- two way FSI type when the data from CFD flow simulation are transmitted to the structure of the wall material but also the information from the structure level influences the fluid flow.

The analysis presented in this article is the one way FSI type. We will transfer the data obtained from the CFD simulation to the specific structural models made for this TWB valve. The structural models consist of the parts of the valve assembly that come into contact with the fluid. These are the three body connectors with flanges, the seats fixed on them, the ball that is operated by the lever and the body valve. The ball is the structural element that is in contact with the three seats and when operated by the lever it is rubbed by the three seats. The selected materials are presented in table 2.

Table 2. Materials and Mechanical Properties.

Materials / Mechanical Properties	Yield Strength [MPa]	Ultimate Tensile Strength [MPa]
Nodular Cast Iron (the seats and the ball)	250	400
Steel Rolled (the body connectors and the body valve)	235	340

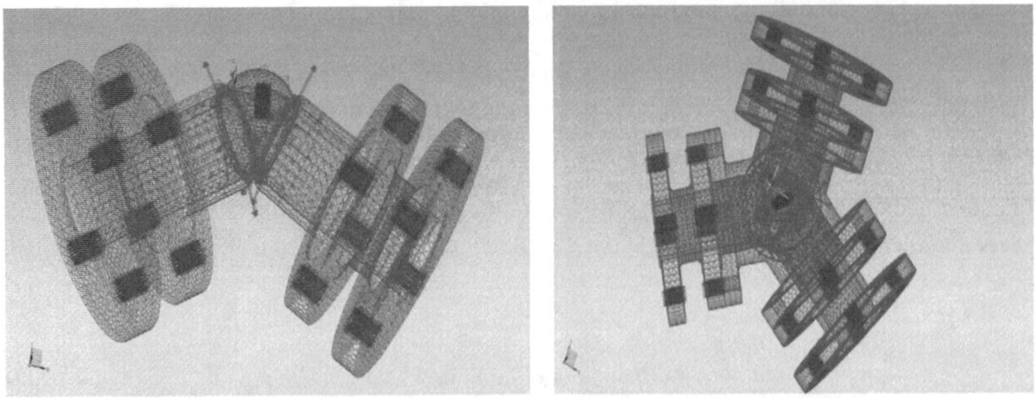

Figure 4. The mesh discretization for a.-first situation; b.-second situation.

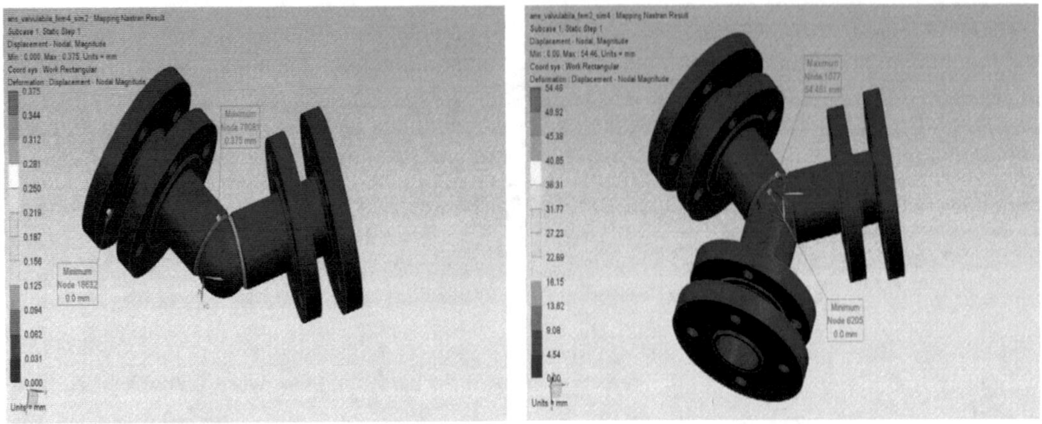

Figure 5. The displacements for a.-first situation; b.-second situation.

The choice of cast iron for rubbing parts was dictated by the carbon in the cast iron structure nodularly arranged in a structure that plays the role of lubricant and thus exhibits a better wear resistance than a carbon steel. It is a cheaper material than bronzes and has a good corrosion resistance.

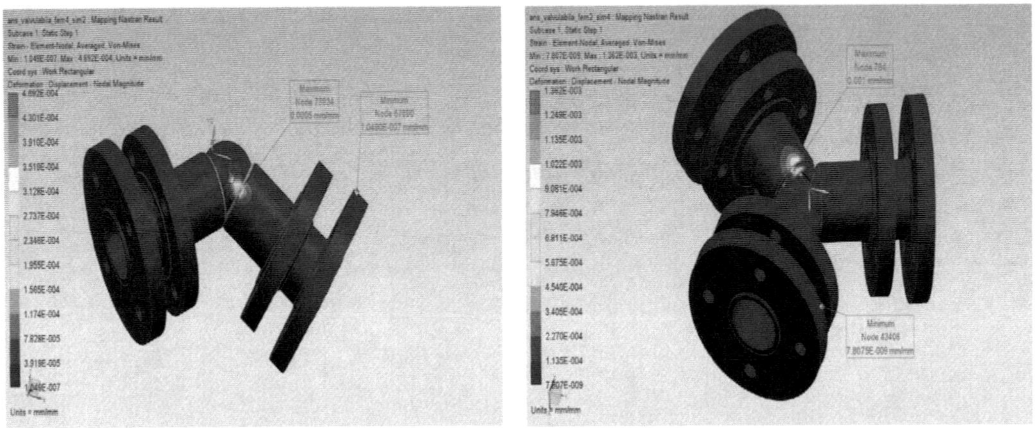

Figure 6. The deformations (strains) for a.-first situation; b.-second situation.

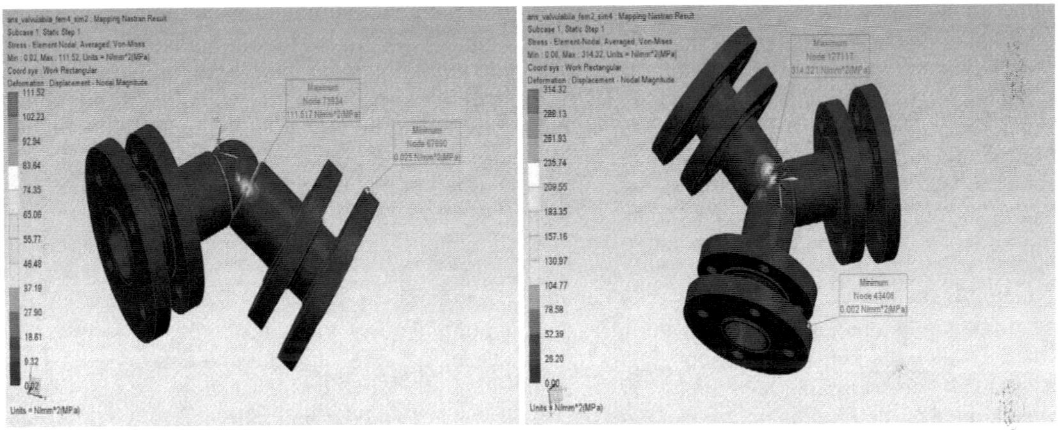

Figure 7. The Misses Stresses for a.-first situation; b.-second situation.

In mesh discretization it should be taken into account that the structural models are made up of several parts of the assembly. The first structural model consists of two body connectors with flanges, two seats and a ball (figure 4, first situation), to the second, a seat, a body connector with flange and the valve body (figure 4, second situation) are added. For discretization, tetrahedral elements CTETRA (10) with element size of 5 and 3 mm were used. The solver chosen is "NX Nastran" and the type of analysis is "Structural".

The "NX Thermal / Flow" solver was chosen at the "New Solution" and the "Mapping" option for the type of analysis to be able to transfer from the CFD analysis the fluid pressure and force field (Relative Pressure and Shear Resultants - figure 2) oriented normally to the surface of the walls through which the fluid flows. Choose the file with the ".bun" extension that contains the result of the CFD analysis and refer to the "Optional Output" to create a "Mapping Nastran" solution.

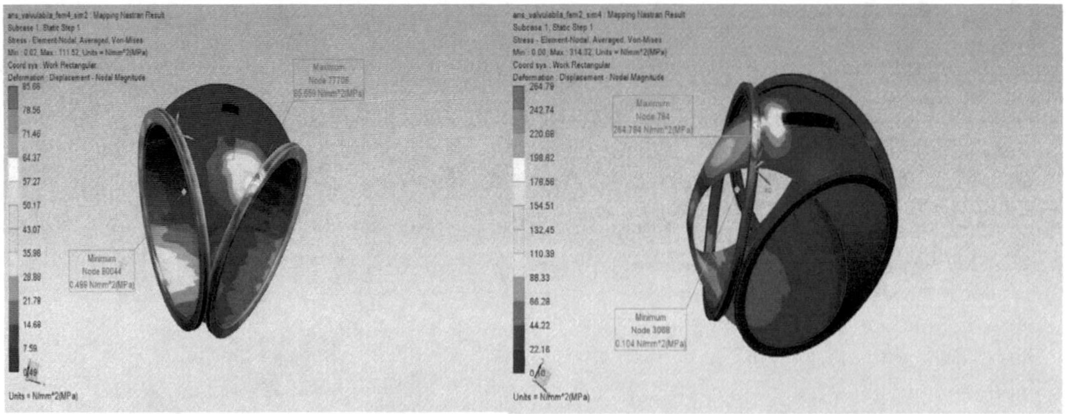

Figure 8. The ball and seats Misses Stresses a.-first situation; b.-second situation.

At "Simulation Object" the surfaces of the parts (the ball, seats, flanged body connectors) in contact were highlighted by "Face Contact" and this fact is shown in figures 4 through those brown arrows.

Fixed constraints were also applied on the flange holes where the screws come with nuts and on the rectangular hole executed on the surface of the ball where the threaded stem that is rotated through the lever comes fixed. Fixed constraints are coloured in blue and there are presented in the figure 4.

The "Solve" button was pressed and the following results were obtained for the structural analysis in the figure 5 – the displacements, figure 6 – the deformations (strains) and figure 7 - the Misses stresses.

4. Conclusions

From the static structural analysis it is evident that the most significant distribution of stresses is at the level of the body flanged connectors (figure 7) and at the level of the ball and seats (figure 8).

Usually, these component parts are the most exposed to both friction by actuating the valve lever but also by the direct action of the fluid. All the results were obtained taking into account the influence of the temperature. Generally the biggest problems in operation are given by seats that have to be changed quite often due to wear and the stress cracking corrosion phenomenon. Problems also occur on the ball as well as on the body flanged connectors.

The aim is to use materials (metal alloys) with much better mechanical properties such as bronze and brass alloys. The periodic change of fixing and sealing gaskets is also considered. An important conclusion can be drawn from the structural analysis of stresses; it is indicated that this type of valve should work as far as possible in the fully open position, so that the inlet and outlet flanges are aligned in the flow direction by 120°. In this situation the loads of seats and the ball are much smaller.

5. References

[1] Dumitrache C., et. al, 2015 Naval Plug Valve Design and Computer Fluid Dynamic Analysis, *Constanta Maritime University Annals* Year XVI **23**, ISSN 1582 3601.

[2] *Ball Valves Types, Construction, Applications and Advantages*, http://www.pipingguide.net/2013/10/ball-valves-types-construction.html accessed date: March 10 2021.

[3] T. Cristopher Dickenson, *Valves, Piping and Pipelines*, Third Edition, 1999, Published by Elsevier Advanced Technology, ISBN 185617252X

[4] J. A. Valverde, S. H. Frankel, G. P. Salvador 2004 Three-dimensional control valve with complex geometry: CFD modelling and experimental validation, *34th AIAA Fluid Dynamics Conference and Exhibit* Portland, Oregon, https://doi.org/10.2514/6.2004-2422.

Exhaust Gas Recirculation (EGR) valve, design and computational fluid dynamic analysis

C L Dumitrache[1], B Hnatiuc[2] and D Deleanu[1]
[1] Maritime University of Constanta, Faculty of Electromechanics, Department of General Engineering Sciences, Mircea cel Batran street, No. 104, Constanta, 900663, Romania
[2] Maritime University of Constanta, Faculty of Electromechanics, Department of Electrical Engineering Sciences, Mircea cel Batran street, No. 104, Constanta, 900663, Romania

E-mail: ldumitr@yahoo.com

Abstract. All EGR systems used on contemporary cars operate on the principle of "external" exhaust gas recirculation. A certain proportion of the engine exhaust gases is extracted and transmitted back into the fresh air-gasoline mixture by means of a control valve. Exhaust gas recirculation valve is generally controlled by a pneumatic or mechanical system designed to dose the exhaust gases in accordance with certain factors such as: engine speed; intake manifold pressure; engine temperature. This paper presents the design steps of the EGR valve using NX SIEMENS. After making the 3D model, it was imported into ANSYS where a study of the flow of fluids (flue gases CO_2 and liquid coolant) was made. In this study it was considered that the valve was actuated and the exhaust gases enter the engine while they are also cooled by the liquid coolant. From the fluid flow modeling process it can be seen that the coolant does not take much of the flue gas temperature, its role being more to maintain a proper temperature of the body valve.

1. Introduction

There are active measures aimed at combating the formation of pollutants (NOx) by optimizing combustion, their effect being focused on the processes in the combustion chamber. Contradictory influences of some factors on the types of legislated pollutants were observed, but also contradictory effects of the same factor on the same pollutant, in the case of different engines. In general, measures to reduce pollutants have led to relatively small increases in fuel consumption, which means finding a compromise [1].

Within the group of listed pollutants (HC, CO, NOx, PT-particles) there is a need for a compromise between NOx values, on the one hand, and PT, CO, HC values, on the other hand. The inverse dependence of these quantities is due to the temperatures developed in the combustion chamber. Measures to reduce NOx and other pollutants simultaneously are relatively few and are usually applied outside the combustion chamber.

The increased amount of waste gas in the cylinder leads to a decrease in the amount of fresh air sucked in, with a decrease in NOx and an increase in smoke; the flue gases remaining in the cylinder reduce the amount of O_2 available, slowing down NOx formation reactions. The favorable influence of waste gases on reducing NOx formation has led to the Exhaust Gas Recirculation method, abbreviated

IOP Conf. Series: Materials Science and Engineering 1182 (2021) 012022 doi:10.1088/1757-899X/1182/1/012022

EGR, which consists in reintroducing a fraction of the flue gases into the cylinder. Increasing the degree of recirculation is very efficient in terms of reducing NOx.

Using EGR reduces the amount of air in the fresh fluid in the cylinder and there is a tendency to increase the smoke and the duration of combustion, which leads to increased fuel consumption; these trends are stronger with increasing load, so that EGR proves to be a very good measure of NOx reduction, but only under the conditions of proper correlation with engine load and within acceptable limits of increasing fuel consumption [2, 3].

2. Experimental research using EGR systems

A research team [1] designed and executed an EGR installation that tested it on two Romanian engines with different litres 392-L4-DT and 1035-L6-DTI:

2.1. Experimental determinations on the 392-L4-DT engine

The main scheme of the installation is represented in figure 1:

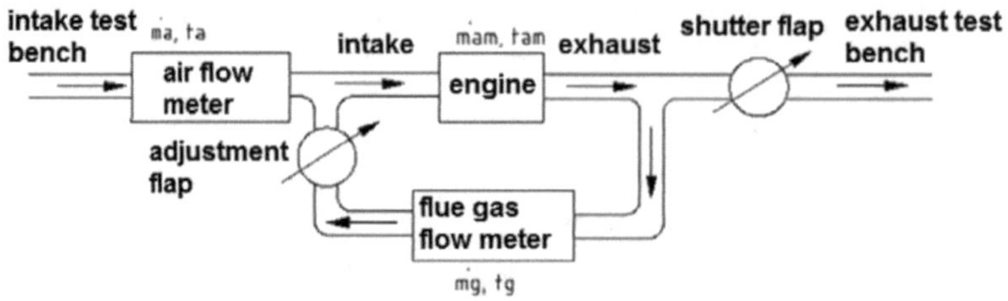

Figure 1. Engine 392-L4-DT test bench [1].

The exhaust system of the test bench had the possibility to be partially blocked in order to allow a certain part of the flue gases to be recirculated under the influence of pressure differences. The originality of the installation consists in fact that it is built in such a way that it can test engines with different litre' values, without being dependent on the constructive or functional type of the recirculation valve.

For the measurement of specific quantities, the measuring installations of the test bench were used such as:

- speed transducer;
- gravimetric vessel for determining fuel consumption;
- thermocouples, manometers;
- differential flow meter with laminar element MERIAM

The flue gas analysis was performed with the BECKMAN 1945-1954 type analysis installation composed of:

- flame ionization detector for unburned hydrocarbons (model 402) HFID;
- infrared absorption analyzer for carbon monoxide (model 854) NDIR;
- chemiluminescence analyzer for nitrogen oxides (model 951) HCLA.

The smoke index was determined with an AVL Dismoke 435 opacimeter. The research aimed to compare some sets of engine parameters:

- pollutant emissions - NOx, CO, HC;
- smoke index;
- fuel consumption;
- performance - engine torque, power

determined in operation, with and without flue gas recirculation.

The degree of recirculation was used as a recirculation parameter. The case without recirculation or with zero degree of recirculation was considered as a reference.
In order to be able to assess the degree of flue gas recirculation, a calibrated diaphragm was used to determine the flow, a calculation that takes into account the procedure described in standard 7347/3 [4]. The mass flow rate was calculated considering the variation of the gas density with temperature.

The flue gas composition, according to [5] was considered as follows: CO_2 = 13%, H_2O = 11%, N_2 = 76%. The recirculated gas flow, D_g, measured in [kg/s] was determined with the relation:

$$D_g = \alpha \cdot \varepsilon \cdot S \cdot \sqrt{2 \cdot \Delta p \cdot \rho} \tag{1}$$

where: α - flow coefficient ($\alpha = 0.6$), ε - expansion coefficient ($\varepsilon = 0.999$), S – diaphragm section ($S = \pi d^2 / 4 = \pi\ 28.3772^2 / 4$), Δp - pressure drop per diaphragm [mm col H_2O], ρ - flue gas density [kg/m^3].

Depending on the desired degree of recirculation, the recirculation pipe was partially blocked from the control valve, and the recirculated flow rate was calculated with the above formula, based on the pressure drop reading. The samples aimed to identify the recirculation effect according to a recirculated gas flow dictated by a recirculation degree (GR) of 10%, 20% and 30%, having as reference the case of zero recirculation. The definition of the degree of recirculation is as follows:

$$GR = \frac{D_g}{D_g + D_{air}} [\%] \tag{2}$$

where Dg - recirculated gas flow rate determined by relation (1), and D_{air} - fresh air flow rate allowed in the engine [kg/s].

The values of the calculation quantities and the gaseous pollutant emissions were calculated for a degree of recirculation of 0%, 10%, 20% and 30% respectively, and the reductions of the pollutants are summarized in table 1.

Table 1. Pollutant emissions depending on the degree of recirculation [1].

Pollutant [g/kwh]	**GR=0%**	**GR=10%**	**GR=20%**	**GR=30%**
NOx	13.4	11.0	6.2	3.2
NOx reduction*[%]	-	17.8	54	77.6
CO	5.8	7.1	9.7	10.4
CO increase*[%]	-	23.5	67.5	79.6
HC	0.58	0.57	0.54	0.63
HC reduction*[%]	-	-1.7	6.8	-8.6
Particle PT	0.7	0.9	1.7	2.4
Particle PT increase*[%]	-	17.6	135	229

* in percentage compared to GR = 0%

The NOx reduction potential was confirmed by recirculating the flue gas using the described experimental installation. The NOx reduction is considerable, up to 77%, which is a lot for this pollutant.

Flue gas recirculation leads to a significant increase in particles and CO emissions. Overall, reducing NOx is more important than increasing other pollutants because they can be reduced by a passive method outside the engine, for example by using the oxidation catalyst.

2.2. Experimental determinations on the 1035-L6-DTI engine

The research was similar to the previous one with the observation that the engine had cylinders of 10.35 litres, being supercharged and having intermediate cooling, and the recirculation installation included a recirculated gas cooler, positioned as in figure 2:

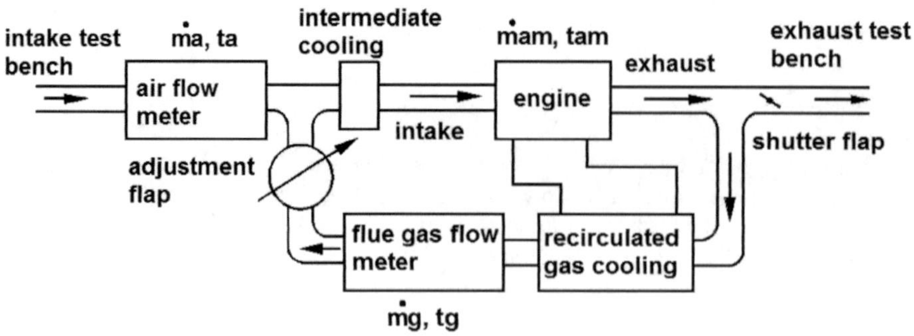

Figure 2. Engine 1035-L6-DTI test bench [1].

The behaviour of the system at different speeds was investigated, choosing for example the results from the speed of 1600 rpm processed in figure 3; it can be seen how NOx emissions decrease with increasing recirculation and how smoke ("fum" in the figure 3) emissions increase; the cooling of the recycled gases maintained the performance gain of the engine resulting from the application of the intermediate cooling of the intake air.

The most important conclusions of the research are [1]:

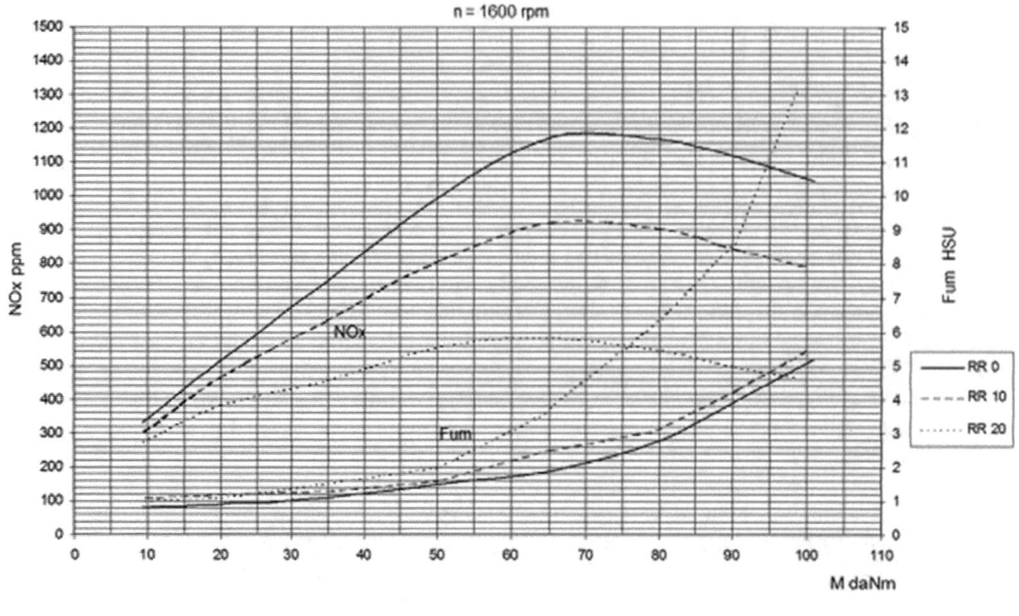

Figure 3. The correlation between NOx and smoke (Fum) versus engine torque M [1].

- Gas-cooled recirculation is an effective NOx-lowering technique if a few precautions are taken to avoid increasing smoke. The largest decrease in NOx occurs at high recirculation levels, which increases the emission of smoke and CO. For a degree of recirculation of 20% with the cooling of the flue gases that varies depending on the load from 220-490˚C to 50-60˚C an average decrease of NOx of 46% is obtained.
- Even if the CO increases for a recirculation degree GR = 20%, its level does not exceed the EURO III pollution limit. At low loads below 50%, recirculations of 20% can be practiced, at higher loads, no.
- HC emission is less sensitive to varying degrees of recirculation.

3. The CAD models of EGR parts

The idea of writing this article started from the change of the EGR valve from the personal car (figure 4). Having the piece in my hands, I measured the overall dimensions and the other dimensions inside the piece, then, I translated them into a 3D model made with NX SIEMENS. This valve is found on almost all current cars that have gasoline and diesel engines.

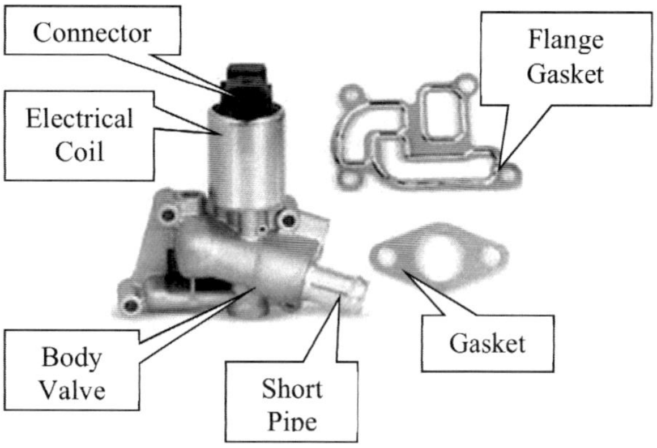

Figure 4. EGR valve model of Opel Astra 1.4 [6].

In the Service and Repair Manual [6] is written that the system is designated to recirculate small quantities of exhaust gas into the inlet tract, and therefore into the combustions process. This process reduce the level of unburnt hydrocarbons present in the exhaust gas before it reaches the catalytic converter. The system is controlled by the fuel-injection/emission control unit (ECU) using the information from its various sensors, via the EGR valve. On 1.4 litre model, EGR valve is an electrically-operated valve mounted on the inlet manifold or the left-hand end of the cylinder head.

When designing this valve using NX Siemens, all the dimensions measured directly on the real part were taken into account. The valve has been designed in parts such as:

- body valve;
- electrical coil through which the valve is actuated;
- the connector through which the valve is electrically operated;
- short pipe to which the hose through which the coolant passes is connected

which were later constrained into an assembly.

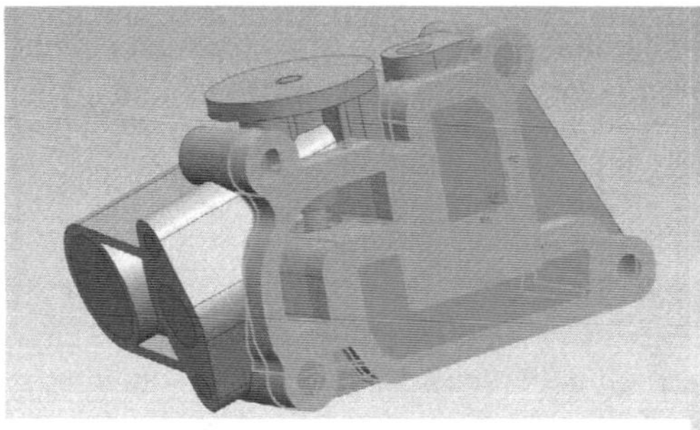

Figure 5. EGR valve body flange.

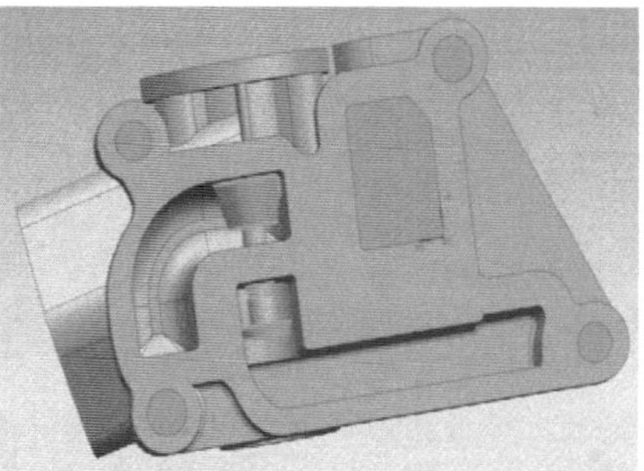

Figure 6. EGR valve shaft.

The real design challenge was the valve body. It was divided into simple geometric bodies which were later assembled using the "Unite" command on the valve body flange (figure 5). All geometric bodies were created using the "Extrude" command. In the same way, the valve shaft was created, having the open position, thus allowing the passage of the flue gases inside the engine, where the combustion reactions took place (figure 6).

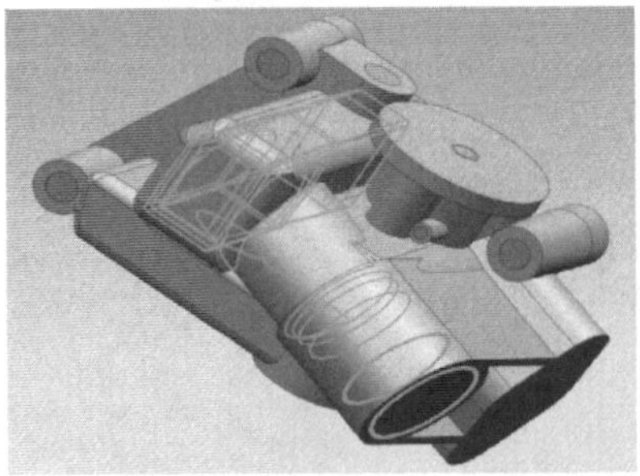

Figure 7. Circuits of coolant.

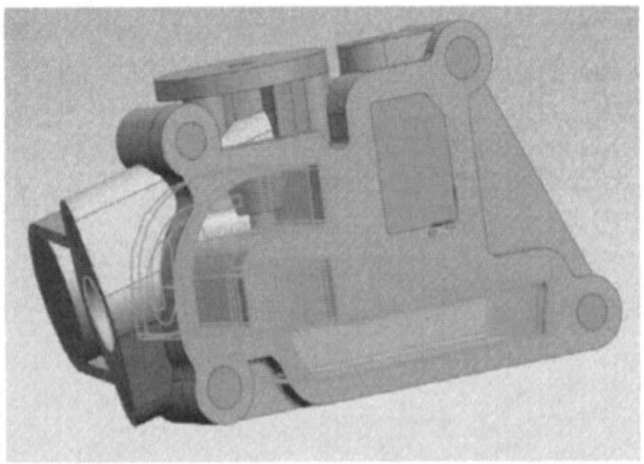

Figure 8. Circuits of flue gases.

After the creation of the geometric bodies, the circuits of the two fluids, the coolant (figure 7) and flue gases (figure 8) were created. The same "Extrude" command was used only in the areas through which the fluids pass, material was extracted using the "Subtract" option at the "Extrude" command. The material extractions left sharp edges and surfaces that were rounded using the "Edge Blend" and "Face Blend" commands.

The other parts of the valve, the electrical coil and the short pipe are two separate ".prt" files and were made with the same "Extrude" command. The assembly was completed. The body valve is the basic component on which the other parts are assembled using position constraints such as "Concentric", "Allign" and 'Touch".

Finally, using a "True Shading Editor" view (figure 9, 10), specific colours were defined for each part:

- the body valve the colour of an aluminium alloy;
- ebonite (black) or hard plastic connector;
- electrical coil covered with a thin steel sheet;
- short pipe aluminium alloy

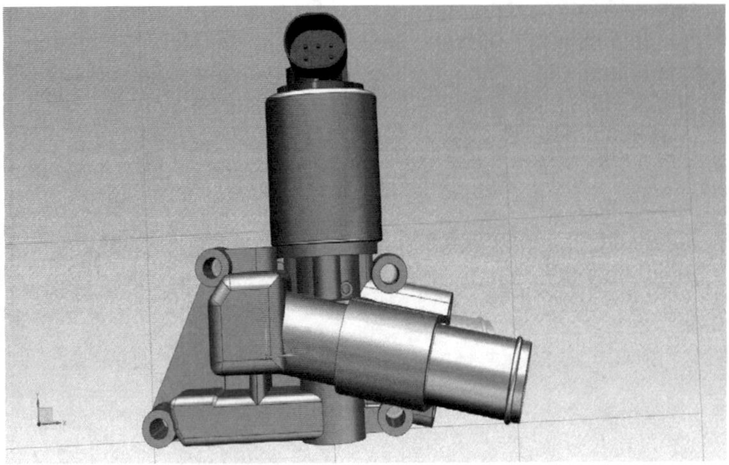

Figure 9. EGR valve, CAD assembly.

Figure 10. EGR valve, CAD assembly - flange viewing.

4. The Fluent flow analysis

After making the 3D assembly, it was imported into Ansys to model the flow of fluids with different states of aggregation which are:

- the exhaust gases that will enter in small quantity in the engine' combustion space;
- the rest of the flue gas is removed from the engine through the exhaust;
- the coolant in order to reduce the overheating of the body valve and to protect the electrical coil that acts to open or close the valve;

Prior to the mesh discretization operation, it is important to rename the fluid-solid separation surfaces for both flue gases and coolant. For the modelling operation, carbon dioxide (CO_2) was used as exhaust gas and water was used for the coolant. The solid-liquid separation surfaces are also renamed, mentioning that for the body valve and for the short pipe, an aluminium-based alloy was used as material.

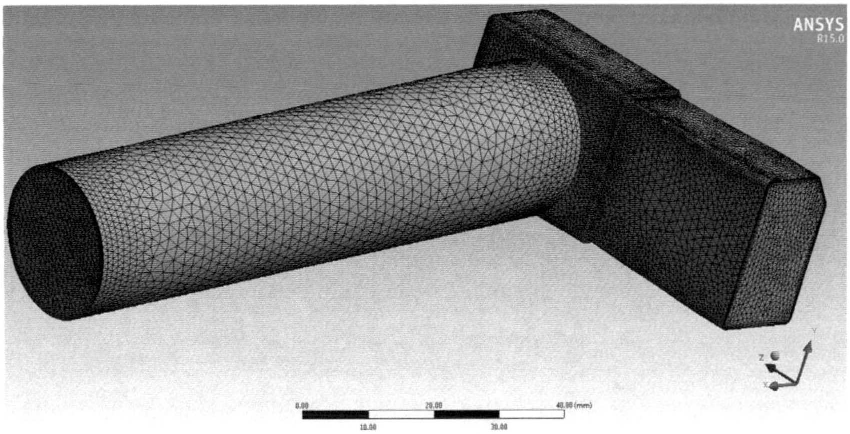

Figure 11. The coolant meshed domain.

After renaming the separation surfaces, it is good for each solid component in the 3D assembly to be defined as domains using the "Body Element" option in the toolbar.

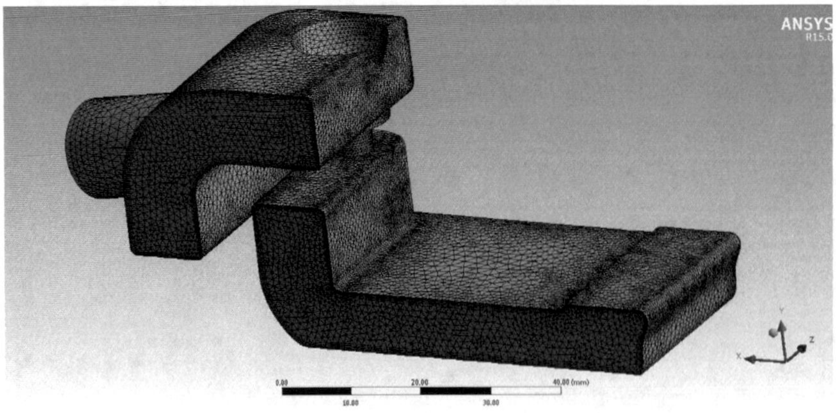

Figure 12. The exhaust gas meshed domain.

With the "Fill" option, the spaces occupied by the coolant and the exhaust gas were well defined (figures 11, 12). As with solid bodies, the volumes occupied by fluids were named after which, the discretization operations were performed. It is recommended that when applying discretization operations on fluid volumes to apply the "Inflation" operation to the fluid-solid separation surfaces, where five successive layers of fluid with a thickness of 1 mm between them have been created. This can be seen in the figures 11 and 12. After the discretization operation with finite volumes, 176614 nodes and 603956 microvolumes were obtained for the field occupied by the flue gases, and for the field occupied by the cooling liquid 130241 nodes and 482182 microvolumes.

Another operation that must be taken into account in the operation of discretization with finite volumes of the fluid domains is that on the "inlet" and "outlet" surfaces to apply a dimensioning of the "Face Sizing" type surfaces to which "Element Size" has the value 1 mm.

The finite element discretization operations of parts of the valve were performed differently from the volumes occupied by the fluids. The assembly consisting of the body valve and the short pipe will contain after meshed operation 30069 nodes and 94155 elements for the short pipe and 411111 nodes and 2158618 elements (figure 13).

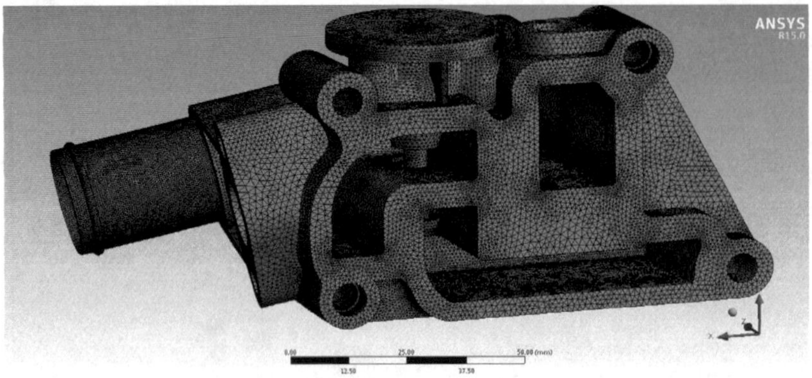

Figure 13. The body valve meshed domain.

Then I entered the Setup menu from Fluent, a "Check Mesh" is made from where a "Quality Report" is obtained and here a parameter called "Orthogonal Quality" (range from 0 to 1) is deduced. If the value of this parameter is closer to zero then the quality of the meshed zone is poor. If its value is less than 0.01 then the mesh operation must be resumed. In our article, for this parameter the value 0.02 was obtained. In our case, the explanation why this parameter has such a small value is primarily due to the complexity of the 3D model.

Ansys Fluent was used to model the flow of fluids in different states of aggregation (flue gases and coolant). From various sources [7] it is specified that the valve will not open at idle or when the engine is cold. The valve will open periodically and not when you press the accelerator pedal to the floor but when driving the car at low and medium loads. When you do not press the accelerator pedal or when you press fully, the EGR valve will not open, but it is possible to open between these marks if the engine is warm enough. The EGR valve has a sensor that will decide when the valve opens and the driver will not have to do anything.

Unfortunately, EGR breaks down and this is a problem related to the operation of cars. These valves work directly with the engine pollutants, the gases are cooled by a special system before passing through the EGR valve. An expected flue gas temperature before entering the EGR is 130°C, the temperature that was taken as a boundary condition for modelling the flue gas flow from this article. It is specified that one of the causes of failure of this valve is that of driving in the city for a long time. The common fault of these valves is clogging the valve and locking it in a certain position and this will affect the proper operation of the engine in the sense that the engine will emit a lot of smoke and will not respond to acceleration as in its good times.

As it is specified in the technical book of the car [5] the flue gases will be recirculated in small quantities. We considered as boundary conditions for the flue gases the following:

- for the flue gases entering the valve we set for pressure inlet the value 49000 Pa = 0.5 bars;
- as flue gases we set the carbon dioxide CO_2 which has the inlet temperature in the valve T=130°C;
- as can be seen in the design of the valve body (figure 13), the flue gas circuit forks into two circuits, one of them passes through the valve on a short path, and the other path is longer and passes through the hole left free by the valve opening;
- at the outlet of the flue gas towards the exhaust, a pressure outlet type condition with a value of 29400 Pa = 0.3 bars was set, registering a small pressure loss because the circuit of these gases to the outlet of the valve is short without losing much of the heat that was set at T=120°C;

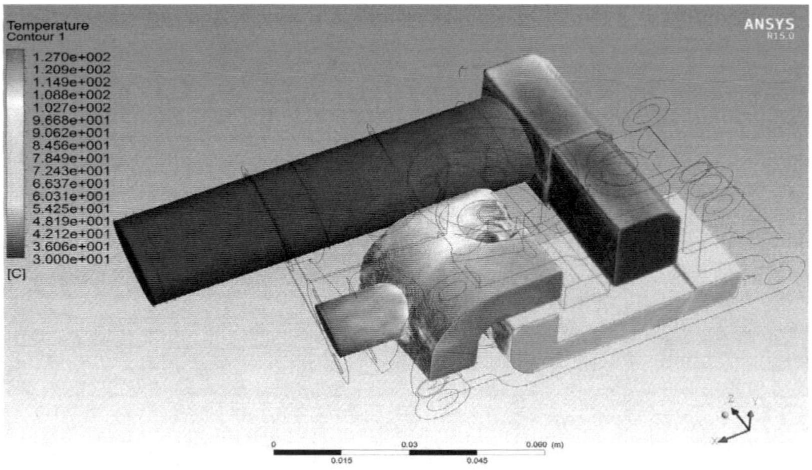

Figure 14. The temperatures volumes of coolant and flue gases

- the flue gases used for recirculation will enter the combustion space of the engine once the valve is opened. For these, a pressure outlet type boundary condition with the value 19600 Pa = 0.2 bar was used;
- the modelling of the flow of these gases presented in this article was made in the conditions in which it is considered that the valve is already open and the gases enter in small quantities back into the engine. The circuit of these gases is longer than that of the gases coming out towards the exhaust (figure 14). The orifice through which the gases come out (diameter = 6 mm) after opening the valve makes the gas blanket to be thinned so that its pressure increases easily and the gas cooling becomes easier (figure 16). The temperature at the exit of these gases from the body valve and at the return to the engine was set at T=90°C.
- for the coolant entering the valve, the value 39200 Pa = 0.4 bar was set for the pressure inlet;
- the cooling water circuit is long enough and covers almost all areas of the valve (figures 14);
- I set as coolant the water that at the entrance to the valve has a temperature of T=30°C because it does not come directly from the radiator, it goes through a longer circuit and it is possible to register a slight increase in it.
- when the water comes out of the body valve I set the pressure outlet with the value 37240 Pa = 0.38 bar, a slight pressure drop because the water circuit inside the valve is without obstacles and the temperature is T=96°C.

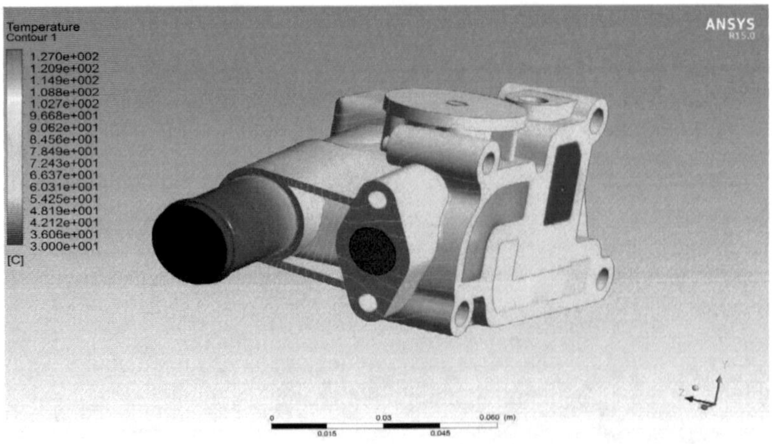

Figure 15. The temperature contours of body valve and fluids

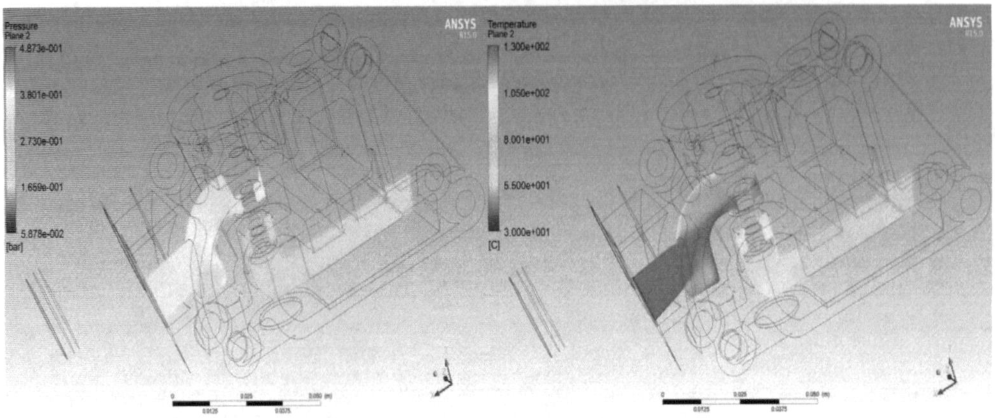

a b

Figure 16. The section planes: a – flue gases pressure; b – flue gases temperature.

5. Conclusions

The 3D assembly made with NX Siemens was made with all dimensions measured on the real EGR valve from an Opel Astra 1.4. In the process of designing the EGR valve, the real challenge was to make the valve body due to its complexity.

For the flow of fluids, information presented in various sources such as Service and Repair Manual as well as other sources that were mentioned in the article were used.

In the first part of the paper were presented some researches that have shown that at low degrees of flue gas recirculation it is possible to achieve a significant reduction of NOx-containing pollutants.

At the discretization processes with finite volumes of fluids a series of difficulties were encountered because the mesh operation had to be performed with great caution because in the "Run Calculation" operation we often had warnings such as "Reverse flow in faces" which generated fall. The following conclusions can be drawn in modelling fluid flow:

- the coolant does not take much of the flue gas temperature, its role being more to maintain a proper temperature of the body valve (figure 15) that does not adversely affect the operation of the coil and the sensor that opens the valve;
- the flue gases entering the EGR valves passing through the short circuit to the exhaust will not be consistently cooled;
- the flue gases entering through the open hole of the valve will form some thin gas beds which will cool faster (figure 16, b) after which they will enter the engine.

At the initial "Run Calculation" operation, a "Number of Iteration" equal to 20 was chosen and it was followed if there is a convergence of solutions, after which we moved to a "Number of Iteration" of 5000.

6. References

[1] Sandu V 2015, Habilitation Thesis, *Contributions to the Optimization of Heavy Diesel Engines*, Transylvania University of Brasov, 181 pages.

[2] Aistleitner K., et all, 1999, *The Turbocharging and Exhaust Gas Recirculation Control for the New BMW V8 Direct-injection Diesel Engine*, MTZ 9/1999.

[3] Schmitt F et all, 2001, *The Potential of Different Exhaust Recirculation Systems*, Autotechnology apr. 2001.

[4] STAS 7347/1, 1983, *Metoda micşorării locale a secţiunii de curgere. Măsurarea cu diafragme şi ajutaje.*

[5] Leonăchescu N, et all, 1977, *Probleme de termotehnica*, Editura Didactică şi Pedagogică, Bucharest.

[6] Legg A K, Mimi L, Randall M 2003 *Vauxhall Astra and Zafira, Service and Repair Manual*, Hayness Publishing, 301 pages, Sparklord, Yeovil, Somerset BA22 7JJ, England ISBN 1 85960 758 6.

[7] *Ce este EGR si la ce este folosita aceasta piesa* – Cavaleria.ro https://youtu.be/ISaNnONpY9o, accessed date: March 10 2021.

IOP Conf. Series: Materials Science and Engineering 1182 (2021) 012023 doi:10.1088/1757-899X/1182/1/012023

Study of the functional parameters of the freshwater generation system, as an integral part of the waste heat recovery system for a tanker ship

C Faităr, L-C Stan, N Buzbuchi and M Baciu
Constanta Maritime University, Faculty of Naval Electromechanics, 104 Mircea cel Bătrân Street, 900663 Constanţa, Romania

E-mail: liviustan14@yahoo.com

Abstract. A ship needs fuel for both travel and on-board operations. In the most general case, the fuel is converted on board the ship into energy in the form required for its end use: mechanical power for propulsion, electrical power for auxiliary board systems and thermal power for thermal needs. In this paper, I propose to deal with both elements of theoretical calculation and real elements of the operation of the freshwater generation system from the main engine of the ship. I chose to treat this system because it is the main subassembly of the residual heat recovery system from the main engine through cooling water. In the theoretical calculation, The results of the paper will be focus on calculation of the geometric and functional parameters of the fresh water generator. In the final part of the chapter, I will refer to the real parameters of the freshwater generation system where I will draw operation diagrams to highlight as concretely as possible the operation of this system on board the ship.

1. Introduction

Nowadays, the main methods of producing fresh water from sea water are:

- MED -Multiple- effect distillation;
- MSF -Multi stage flash distillation;
- VC-Vapour compression;
- BWRO-Brackish water reverse osmosis;
- SWRO - Sea Water Reverse Osmosis [1].

This paper deals with aspects of theoretical calculation regarding the energy flows of the naval desalination plant. At the same time, emphasis was placed on the real parameters of the desalination plant and real operating graphs were drawn.

Energy consumption for different methods are presented inside the table 1:

Table 1. Energy consumption for different producing fresh water systems.

Energy type	MED	MSF	VC	BWRO	SWRO
Steam pressure [Mpa]	0.02-0.04	25-35			
Thermal energy [kWh / m^3]	4.5-6	9.5-11			
Electrical energy [kWh / m^3]	1.2-1.8	3.2-4	8-12	0.3-2.8	2.5-4
Total energy [kWh / m^3]	5.7-7.8	12.7-15	8-12	0.3-2.8	2.5-4

2. General description of the freshwater generator

The production of fresh water on board is necessary activity for ships with unlimited navigation area and involves its quality in terms of use by various consumers. An important parameter is salt concentration. As a process, distillation takes place with the change of phase, which involves a high energy consumption and is already considered classic. For the reference system, seawater vaporization is performed under vacuum at a temperature t <100 ° C. Main characteristics are shown in table 2 [2].

Table 2. KM 50 characteristics.

Current no.	Characteristics	Value
1	Type	KM 50
2	Capacity	40 tones / day
3	Heat consumption	1140.5 kW
	Cooling sea water	
4	Sea water cooling inlet temperature	32 °C
5	Sea water cooling outlet temperature	40.3 °C
6	Sea water flow	115 m^3 / h
7	Maximum pressure losing	0.0343 MPa
	Sea water cooling from main engine	
8	Cooling water from the main engine to fresh water generator inlet	75 °C
9	Cooling water from the main engine to fresh water generator outlet	62.1 °C
10	Cooling water flow	76 m^3 / h
11	Maximum pressure losing	0.0294 MPa
	Other components	
12	Condenser pressure	0.38 MPa
13	Heat exchanger - heater	0.44 MPa
14	Heat exchanger - vaporizer	0.10 MPa
	Distillate pump	
15	Flow	2.5 m^3 / h
16	Head	0.36 MPa
17	Power	1.5 kW
	Ejector pump	
18	Flow	36 m^3 / h
19	Head	0.38 MPa
20	Power	7.5 kW

Figure 1. Fresh water generator SASAKURA KM50 [Horaisan tanker ship].

The reference desalination plant is shown in figure 1. Once it has reached boiling temperature, is at a lower pressure than atmospheric pressure, salt water goes through a partial stage of evaporation and generates a mixture of steam and salt. The brine is separated from the steam and extracted by means of the combined air / brine ejector [3].

The vessels pass through the separation zone and enter the second plate channel (each channel being composed of plates) from the condensation section. Cooling water is distributed through the other channel and absorbs heat which is transferred by the steam from the condensing zone. Finally, the fresh water generated is extracted by means of the water pump and pumped to the fresh water tank [4].

3. Calculation of theoretical energy flows on the freshwater generator

The saturation temperature can be determined by the Antoine empirical equation for the known pressure in the vessel and the difference between condenser and evaporator. The curves are present depending on the boiling point of the salt water. Because the temperature differences between the evaporator and the condenser (adică δT_{evap} și δT_{cond}) are in addition to heating and coolant flows, the main generators of heat exchange, which explains how the vaporisation intensity will increase and condensation will be reduced with the fall pressure in the vessel and vice versa.

It is obvious that an optimal pressure in the vessel can be determined, taking into account, in particular, the external conditions and the conditions of the distiller, but due to the optimal pressure regulation system, a certain interval will appear, usually within which the changes will vary [5].

The energy balance of the distiller (as shown in figure 2), can be presented in terms as follows:

$$\Delta E = U \tag{1}$$

where ΔE [kW] represents the difference between the inlet and outlet heat process, and
U [kW] represents the internal energy of the whole distiller which can be defined more precisely as:

$$\Delta Q_{heat/HT} + U_{SW} - \left(\Delta Q_{cond} + U_{brine} + U_{dist} + U_{gas} + Q_{rad}\right) = m_{cas} c_{cas} \Delta t_{cas} \; [kW] \tag{2}$$

If we compare ΔQ_{cond}, U_{brine} and U_{dist}, the last two terms to the left of the equation are negligible. The equation will become:

$$\Delta Q_{heat/exch} + U_{i/o} = m_{cas} c_{cas} \Delta t_{cas} \; [kW] \tag{3}$$

where the first term on the left represents the difference between the heat introduced into the evaporator exchange process and the heat removed from the condenser exchange process; the second term represents the difference between the internal energy of the inlet seawater and the outlet energy of the brine and distilled water. From the beginning of the exchange process, the differences in heat were positive [6].

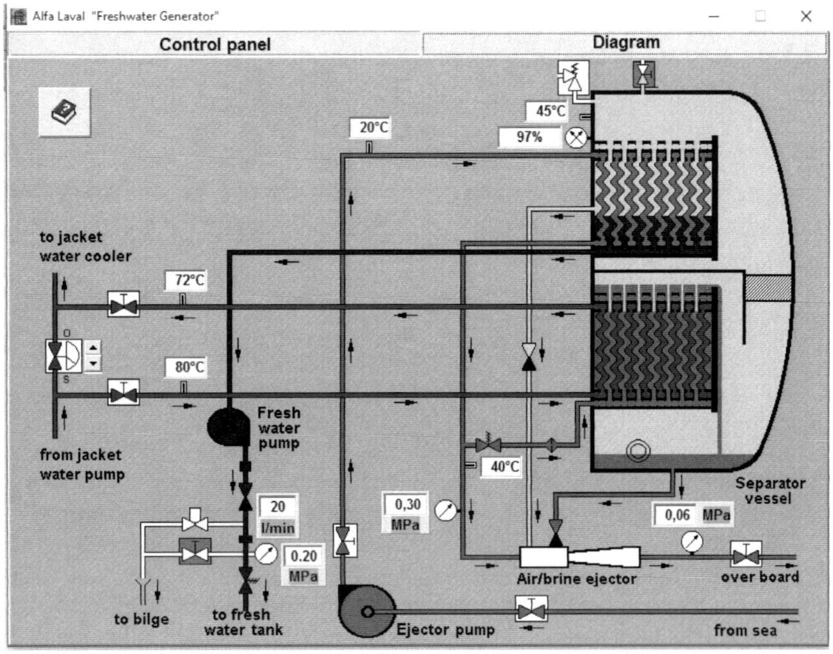

Figure 2. Fresh water generator – working principle.

Figure 2 shows the operation of the naval distiller through the electronic command and control program. The light blue part represents the energy flow from the main engine and has an inlet temperature of 80 °C and has the role of heating (vaporisation). The green part represents the energy flow from seawater and has the role of cooling (condensation). The condensation temperature is 20 °C. The dark blue colour is the distillate from vaporization and condensation. The higher the quantity of distillate, the lower the quality (higher salinity) and, conversely, the lower the amount of distillate, the higher the quality (lower salinity). At the same time, we have a pink brine that is evacuated with the help of an ejector that works with sea water. Also, the energy flows for auxiliary systems are shown in figure 3 [7].

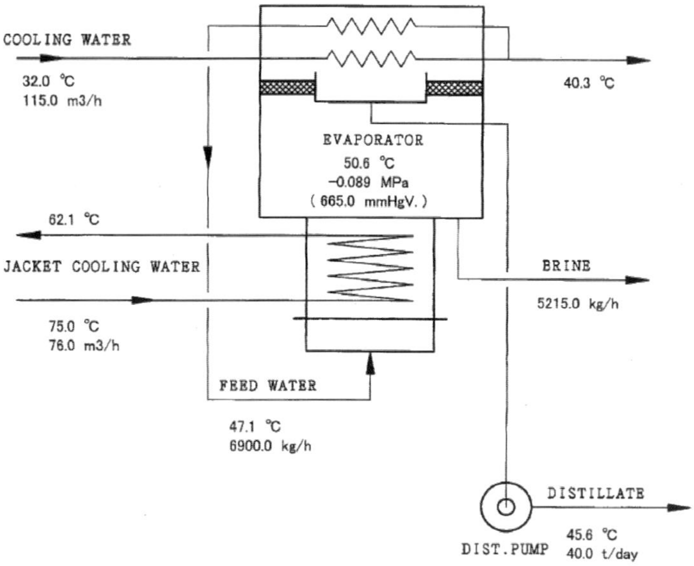

Figure 3. Energy flow [8].

The thermal flow of sea water cooling

Seawater has been used for long time as a cooling fluid in heat exchangers to reduce fresh water usage in industry and power plants.

The sea water pump flow is:

$$\dot{V}_{cool} = \frac{115}{3600} = 0.0319 \frac{m^3}{s}$$

The specific heat is:

$$c_{sw} = 4.19 \frac{KJ}{KgK}$$

The inlet temperature of sea water is:

$$t_{in} = 32^0 C$$

The outlet temperature of sea water is:

$$t_{out} = 40.3^0 C$$

The sea water density is:

$$\rho_{SW} = 1025 \frac{Kg}{m^3}$$

The energy flow of cooling sea water is:

$$\dot{Q}_{cool} = \dot{V}_{cool} \cdot \rho_{sw} \cdot c_{sw} \cdot (t_{out} - t_{in}) = 1137.12 \left[kW \right] \tag{4}$$

Enegy flow of cooling water from main engine

The fresh water pump flow is:

$$\dot{V}_{fw} = \frac{76}{3600} = 0.0211 \frac{m^3}{s}$$

The Specific heat is:

$$c_{fw} = 4.19 \frac{KJ}{KgK}$$

The inlet temperature of fresh water is:

$$t_{in} = 75^0 C$$

The outlet temperature of fresh water is:

$$t_{out} = 62.1^0 C$$

The water density is:

$$\rho_{fw} = 1000 \frac{Kg}{m^3}$$

The energy flow of cooling fresh water is:

$$\dot{Q}_{fw} = \dot{V}_{fw} \cdot \rho_{fw} \cdot c_{fw} \cdot (t_{out} - t_{in}) = 1140.47 \; [kW] \tag{5}$$

4. Calculation of real energy flows on the freshwater generator

For calculation will use 4 and 5 formulas.

Energy flow of sea water

$$\dot{Q}_{cool} = \dot{V}_{cool} \cdot \rho_{sw} \cdot c_{sw} \cdot (t_{out} - t_{in}) \; [kW] \tag{6}$$

Energy flow of fresh water

$$\dot{Q}_{fw} = \dot{V}_{fw} \cdot \rho_{fw} \cdot c_{fw} \cdot (t_{out} - t_{in}) \; [kW] \tag{7}$$

In table 3 the actual energy flows were calculated according to the flow rates and temperatures of the working fluids using equations (6) and (7).

Table 3. Real energy flows in fresh water generator.

Crt. no.	Fresh water flow [m^3/day]	Cooling water flow [m^3/day]	Cooling water flow [m^3/h]	Energy flow cooling water [kW]	Maximum energy flow sea water [kW]	Water flow main engine [m^3/day]	Water flow main engine [m^3/h]	Water energy flow main engine [kW]	Maximum energy theoretical flow sea water main engine [kW]	Sea water temp [°C]		Water temp main engine [°C]	
										in	out	in	out
1	33	94.9	0.026	1358.2	1137.1	62.7	0.017	1313.6	1140.5	32	44	78	60
2	32	92.0	0.026	1207.3	11371	60.8	0.017	1273.8	1140.5	31	42	78	60
3	21	60.4	0.017	720.3	1137.1	39.9	0.011	928.8	1140.5	28	38	78	58
4	29	83.4	0.023	1094.1	1137.1	55.1	0.015	1154.3	1140.5	26	37	75	57
5	31	89.1	0.025	1063.2	1137.1	58.9	0.016	1234.0	1140.5	30	40	75	57
6	33	94.9	0.026	1131.8	1137.1	62.7	0.017	1167.6	1140.5	31	41	73	57
7	31	89.1	0.025	1063.2	1137.1	58.9	0.016	1165.4	1140.5	32	42	76	59
8	34	97.8	0.027	1282.8	1137.1	64.6	0.018	1353.4	1140.5	32	43	78	60
9	36	103.5	0.029	1358.2	1137.1	68.4	0.019	1273.8	1140.5	32	43	76	60
10	36	103.5	0.029	1358.2	1137.1	68.4	0.019	1273.8	1140.5	32	43	76	60
11	19	54.6	0.015	521.3	1137.1	36.1	0.010	798.3	1140.5	32	40	76	57
12	25	71.9	0.020	1286.2	1137.1	47.5	0.013	718.7	1140.5	31	46	76	63
13	36	103.5	0.029	1728.6	1137.1	68.4	0.019	1034.9	1140.5	31	45	76	63
14	38	109.3	0.030	1955.0	1137.1	72.2	0.020	1092.4	1140.5	30	45	76	63
15	36	103.5	0.029	1358.2	1137.1	68.4	0.019	955.3	1140.5	29	40	76	64
16	28	80.5	0.022	1056.4	1137.1	53.2	0.015	1176.5	1140.5	29	40	76	57
17	27	77.6	0.022	1018.7	1137.1	51.3	0.014	1134.4	1140.5	30	41	76	57
18	28	80.5	0.022	1056.4	1137.1	53.2	0.015	1176.5	1140.5	30	41	74	55
19	27	77.6	0.022	1018.7	1137.1	51.3	0.014	1134.4	1140.5	31	42	76	57

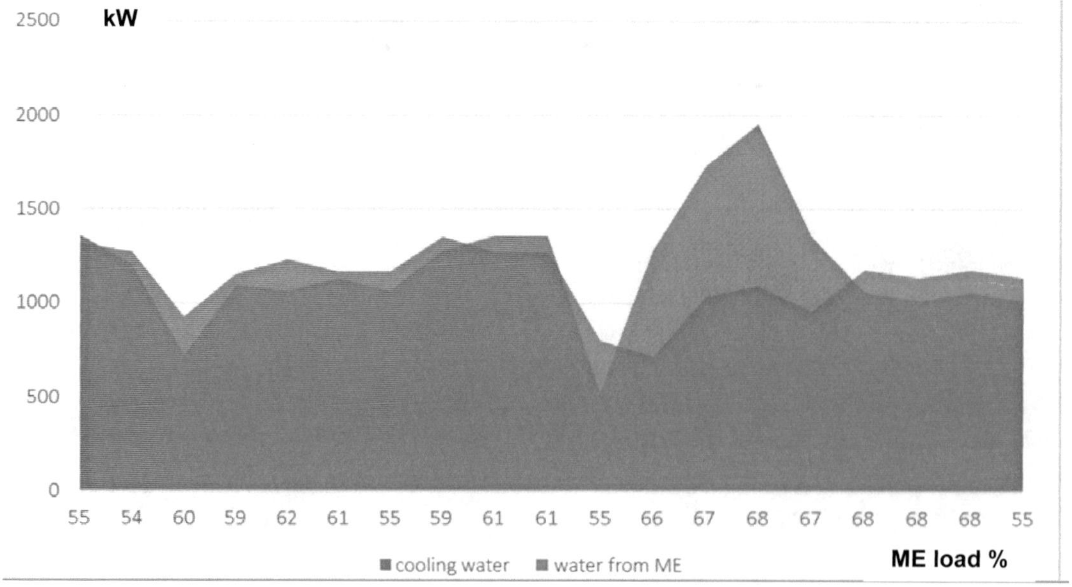

Figure 4. Sea water energy (blue). Fresh water energy (red) – kW.

In figure 4 the energy flows for technical water as well as for salt water. At the same time, the graph provides a comparative image between the second flow with values in kW [9].

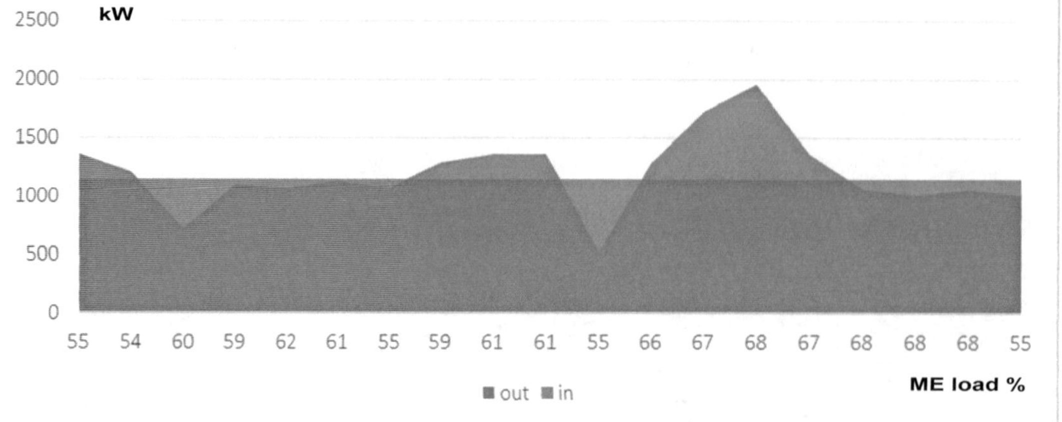

Figure 5. Sea water energy (blue). Water cooling theoretical energy flow (red) in kW.

In figure 5 the energy flow from sea water is compared to the maximum theoretical flow calculated. Several peaks can be observed that exceed the maximum theoretical flow, this does not significantly influence the condensation [10].

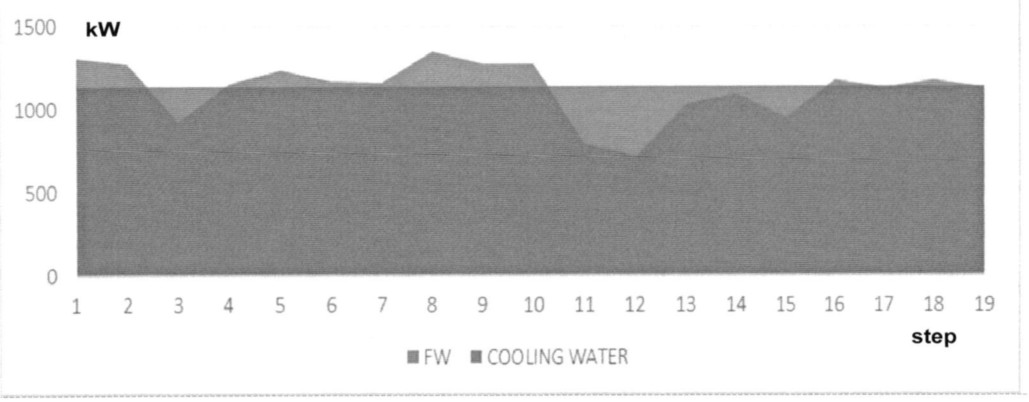

Figure 6. Fresh water energy (red). Water cooling theoretical energy flow in the main engine (blue) in kW.

In figure 6 the energy flow from the main engine is compared to the calculated maximum theoretical flow. Several peaks can be observed that exceed the maximum theoretical flow. Most of the time the engine is running below the maximum value of the flow from its cooling water.

5. Conclusions

The terms in the energy flow equations change depending on the external and internal variable conditions:

- the average rate of heating flow is constant at a certain position of the by-pass control valve, but the temperature changes according to the load changes of the propulsion engine, with the quality of the system controlling this temperature and with the heat exchange coefficient.
- the amount of condensing cooling seawater is constant for the position of the control valve, while the heat exchange coefficient changes with the seawater temperature at the inlet.
- the rate of energy lost with brine depends on the intensity of evaporation.
- the rate of energy released suitable with the distillate also depends on the intensity of evaporation.

Of all the values of the variable energy flows that determine the operation of the distiller, referring primarily to the quantity and quality of the distillate generated, the most important are the rate of heat supplied by the heating fluid and the working pressure (vacuum) [11].

A considerable increase in the amount of distillate can be achieved by increasing the rate of heat supplied with the heating fluid. An unpleasant consequence of such an operation is a lower quality of the distillate, ie its increased salinity.

The heat from the main engine through its cooling fluid is the deciding factor in the quality and quantity of distillate produced.

6. References

[1] 15748-1:2002, ISO 15748-1:2002 - *Ships and marine Technology (Nave şi tehnologii marine) - Alimentarea cu apă potabilă a navei şi structurilor navale; Planificare şi proiectare.*

[2] I. 19458:2006, ISO 19458:2006 - *Water quality (Calitatea apei) - Mostre pentru analiza microbiologică.*

[3] I. 17025:2005, ISOIIEC 17025:2005 - *General requirements for the competence of testing and calibration laboratories (Cerințe generale pentru calibrarea laboratoarelor şi testarea competenței).*

[4] Cipollina, A., Micale, G., Rizzuti, L. 2009*Seawater desalination: conventional and renewable energy process,.* Springer Science & Business Media.

[5] M. H. I. 2012 *Ltd, Mitsubishi Heavy Industry Ltd - MJ12-FW Generator* (Including Instruction Book).

[6] Alfa Laval, https://www.alfalaval.com/products/process-solutions/fresh-water-generation/multi-effect-desalination/mep-multi-effect-plate-evaporator/. Accessed on 24 of april 2021.
[7] Kakac, S., Liu, S. 2002 *Heat Exchangers Selection, Rating and Thermal Design*, CRC:USA..
[8] http://mfame.guru/heat-recovery-system-operation-and-experience Accessed on 24 of april 2021.
[9] Muley, A., Manglik, R. M. 1999 Experimental study of turbulent flow heat transfer and pressure drop in a plate heat exchanger with chevron plates, *ASME Journal of Heat Transfer*, **121**(1), 110-117.
[10] Cieslinski, J., Fiuk, A., Typinski, K. 2016 *Heat transfer in plate heat exchanger channels: Experimental validation of selected correlation equations.Archives of Thermodynamics,* **37**(3), 19-29.
[11] Soto, J. A., Martinez, M. T.,Magana, J. L. 2015 *Theoretical-experimental study of the effect of the external plates on the thermal performance of a plate heat exchanger*, Chemical Engineering and Science, **3**(1), 12-18.

Acknowledgements
„This work is supported by the project ANTREPRENORDOC, in the framework of Human Resources Development Operational Programme 2014-2020, financed from the European Social Fund under the contract number 36355/23.05.2019 HRD OP /380/6/13 – SMIS Code: 123847.

IOP Conf. Series: Materials Science and Engineering 1182 (2021) 012024 doi:10.1088/1757-899X/1182/1/012024

Study of the functional parameters of main engine turbocharger for a tanker ship

C Faităr, D Jugănaru, N Buzbuchi, L-C Stan and V Poenaru
Constanta Maritime University, Faculty of Naval Electromechanics, 104 Mircea cel Batran Street, 900663 Constanta, Romania

E-mail address: liviustan14@yahoo.com

Abstract. To correctly understand the requirements of a system, we must first study its schematics. In the case of energy systems, they require a detailed diagram of energy performance. Heat losses can be classified into high-temperature losses, medium-temperature and low-temperature losses. On board ships with low-speed propulsion engines, heat recovery systems operate in the range of 100–400 °C. Many residual energy recovery systems from internal combustion engines are under development: for example the MAN WHR development program for Tier III technologies. In this paper, I will treat both theoretical calculation elements and real elements of the operation of the turbocharging system, for the main reference engine. I chose to treat this system because it is a main subassembly for the engine, in terms of operation and thus its efficiency. In the theoretical calculation, I will start from the initial calculation data and I will expose the geometric and functional parameters of the turbocharger. The exhaust gases up to a temperature of 650 °C passes continuously through the turbine and heat its components, without its system of contraction cooling. In particular, the shaft bearing must withstand high operating temperatures without ever breaking the lubrication device.

On the compressor side, the air is heated to over 200 °C. High temperatures lead to extreme thermal loads of the material in many locations. Speeds are extremely high: METxxSE turbochargers, as in figure 1, run at speeds between 10000 and 35000 rpm, depending on size. In this respect, the tangential speeds of 560 m / s and even more are reached at the compressor turbine, which rises to 1.7 times the speed of sound or 2000 km / h. The efficiency of the operation of the turbocharger system related to the main propulsion engine depends on the internal processes that take place in its operation, and also on the environmental conditions. In this way, the direction of processing on the installation is dictated by the engine load, taking into account the extreme situations, corresponding to the operation of the ship in special conditions (overload, tropical temperature, arctic temperature etc.).

1. Introduction

For turbocharging the engine MITSUBISHI-UE MDE 7UEC85LSII was chosen a turbocharger consisting of a centrifugal compressor with radial blades with a stage, which is driven by an axial turbine driven by exhaust gases from the engine, the engine having a total of 2 turbochargers. The gases discharged from the engine are led through the exhaust manifolds to the gas inlet housing (water-cooled) and from there, through the nozzle ring, to the turbine, after which they are discharged to the chimney. The air required to supercharge the engine is sucked through the suction housing and

IOP Conf. Series: Materials Science and Engineering 1182 (2021) 012024 doi:10.1088/1757-899X/1182/1/012024

expelled at high pressure through the spiral housing into the engine air cooler, from where, through the air ducts, it passes into the cylinders [1].

2. General description of the engine turbocharger

The main features of the main engine turbocharger are (figure 1):

- maximum allowed speed: 11960 rpm;
- maximum continuous permissible temperature: 580 °C;
- permissible overload temperature for a short time: 610 °C;
- lubrication method: external lubrication under pressure;
- lubricant type: SAE 20, 30, 40;
- lubrication pressure: 0.06 ... 0.15 MPa;
- standard temperature of the lubricating oil at the inlet: 35 .... 60 °C;
- standard temperature of the lubricating oil at the inlet: 85 .... 90 °C;
- total weight: 11250 Kg;
- paint: heat-resistant silver-based [2].

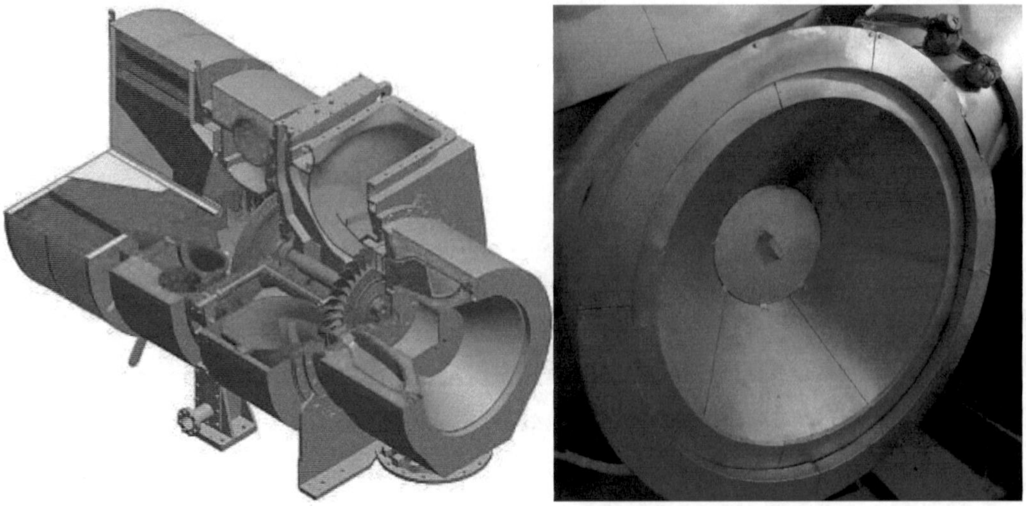

Figure 1. Turbocharger type MET83SE.

3.Turbocharger calculation

3.1.Specific mass air flow calculation for gas exchange

$$d_{asg} = m_{tsg} \cdot \alpha_{sg} \cdot c_e \left[\frac{Kgair}{kWh}\right] \tag{1}$$

where:

- $m_{tsg} = 14.2 \frac{kgair}{kg\,fuel}$ - theoretical air mass for 1 Kg fuel burn;
- $\alpha_{sg} = 2$ - excess air coefficient for gas exchange;
- $c_e = 0.1636 \frac{kg\,fuel}{kwh}$ - specific fuel consumption.

To ensure good quality combustion of the fuel, this combustion is performed with a certain amount of excess air. The ratio between the actual amount of air L, which returns to 1 Kg of fuel, and the amount of air theoretically required is called the excess air coefficient. Decreasing the value is one of the effective measures to force the engine cycle. However, the possibilities of reducing the value of this parameter depend on the degree of perfection of the fuel mixture formation and combustion, determined by the architecture of the combustion chamber and the direction of the fresh fluid and fuel in this enclosure. When reducing the value, the consequences of increasing the thermal demands of the cycle must also be taken into account. Exergy flows in turbocharger are shown in figure 2 [3].

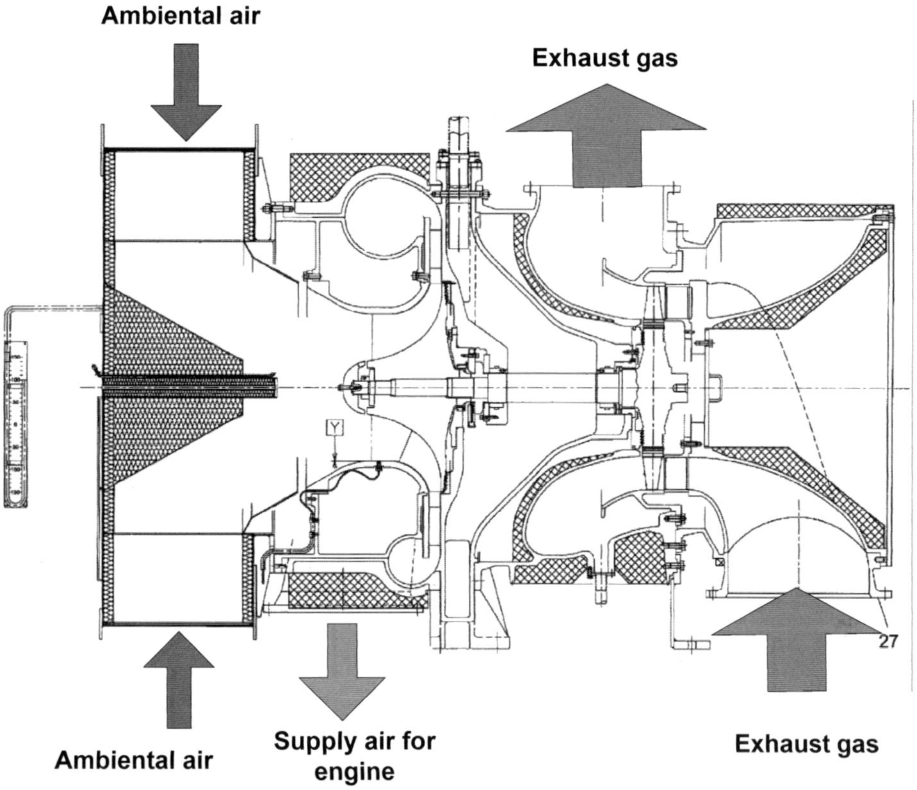

Figure 2. Energy flows in turbocharger [4].

According to the experimental data and theoretical parameters as in figure 3, the excess air coefficient at the first load varies within the limits:

- Low speed engine with natural suction: $\alpha = 1.8 \ldots 2$;
- fast speed engine with natural suction: $\alpha = 1.3 \ldots 1.7$;
- supercharged engine: $\alpha = 1.7 \ldots 2.2$.

I adopted: $\alpha_{sg} = 2$ [4]

$$d_{asg} = 14.2 \cdot 2 \cdot 0.1636 = 4.64 \frac{Kgair}{kWh} \quad (2)$$

3.2. Calculation the mass flow of air required for gas exchange at maximum capacity

$$d_{hsg} = d_{asg} \cdot P_e \left[\frac{Kgair}{h} \right] \quad (3)$$

$$d_{hsg} = 4.64 \cdot 27020 = 125541.4 \frac{Kgaer}{h} = 34.87 \frac{Kgair}{s} \tag{4}$$

3.3. Calculation of the volume air flow required for gas exchange at maximum capacity

$$d_{hvsg} = \frac{d_{asg} \cdot P_e}{\rho_a} \left[\frac{m^3 aer}{h} \right] \tag{5}$$

$$d_{hsg} = 104617.8 \frac{m^3 air}{h} = 29.6 \frac{m^3 air}{s} \tag{6}$$

$\rho_a = 1.2 \frac{Kgair}{m^3}$ - air density at + 20 °C, 70% humidity and 101.3 kPa pressure.

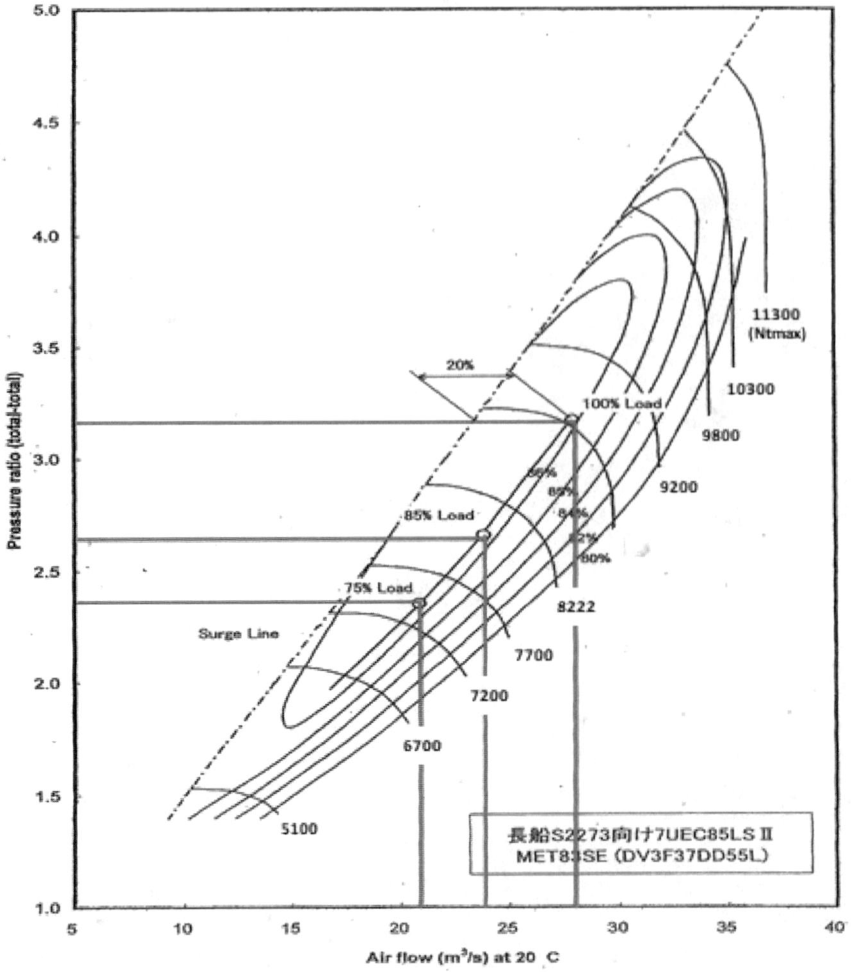

Figure 3. Calculation diagram for air flow versus engine load (at 100% load is a flow of 28 m³ air / s) [5].

4. Drawing real operating graphs of the reference engine turbochargers

The turbocharger system parameters are displayed on the local control panels and are centralized in the command, control and monitoring system of the main engine. The control panels for the auxiliary units of the main propulsion engine are located as close as possible to the equipment in service. The control and indication of the turbocharger installation parameters are also performed by specific control and protection panels. The panels include the indication, alarm and automatic shutdown of the installation where appropriate. Parameters are centralised in table 1 and table 2 [6].

Table 1. Turbochargers speed

Crt.no	Turbocharger 1 [rpm]	Turbocharger 2 [rpm]	Main engine [rpm]
1	4130	4110	55
2	4100	4080	54
3	5810	5800	60
4	5730	5720	59
5	5980	5970	62
6	5900	5980	61
7	4380	4350	55
8	5735	5720	59
9	6020	6020	61
10	6010	6000	61
11	4350	4330	55
12	7450	7440	66
13	7490	7490	67
14	7520	7520	68
15	7420	7410	67
16	7490	7500	68
17	7510	7510	68
18	7600	7610	68
19	4350	4330	55
20	7350	7340	68
21	7425	7415	67

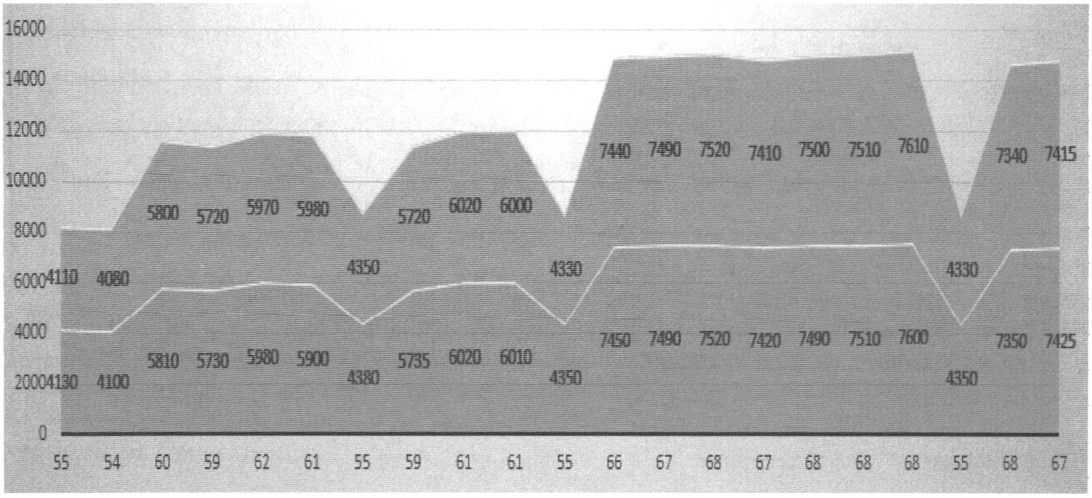

Figure 4. Turbochargers speed versus engine speed

Figure 4 shows the speed of the 2 turbochargers related to the main engine depending on its speed. You can easily see the very close speeds of the turbochargers, the deviation between the 2 speeds of the compressors being between 0.99 ... 1% [7].

Table 2. Oil temperatures versus engine speed.

Crt.no.	Inlet oil temperature turbocharger 1 [°C]	Outlet oil temperature turbocharger 1 [°C]	Inlet oil temperature turbocharger 2 [°C]	Outlet oil temperature turbocharger 2 [°C]	Main engine speed [rpm]
1	44	50	44	50	55
2	44	50	44	50	54
3	44	54	44	53	60
4	44	54	44	53	59
5	43	55	43	54	62
6	43	54	43	54	61
7	43	51	43	50	55
8	44	54	44	53	59
9	46	57	46	56	61
10	45	56	45	56	61
11	43	50	43	50	55
12	44	59	44	59	66
13	45	60	45	60	67
14	45	61	45	60	68
15	46	61	46	60	67
16	45	61	45	60	68
17	45	61	45	60	68
18	44	60	44	59	68
19	43	50	43	50	55
20	45	60	45	59	68
21	46	61	46	60	67

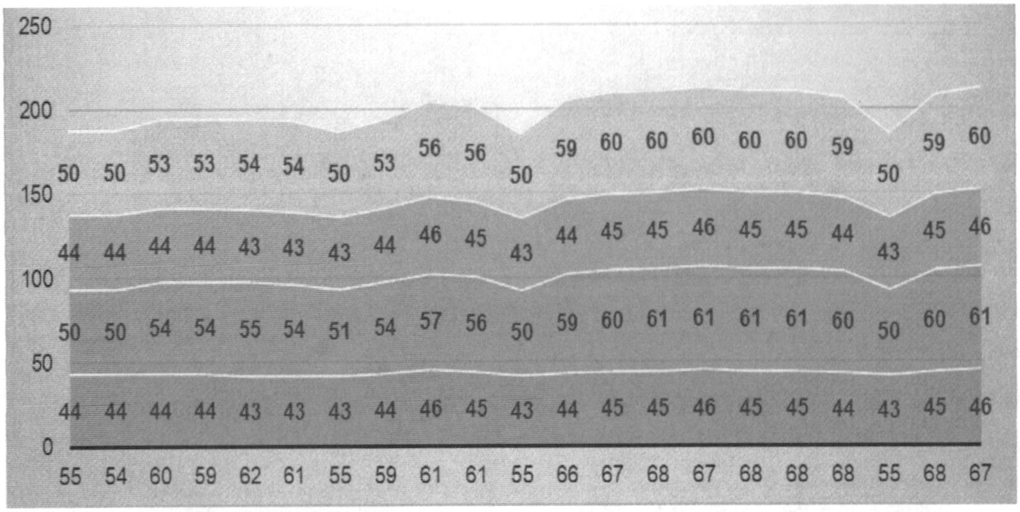

Figure 5. Oil temperatures (inlet–outlet) of turbochargers versus engine speed.

Figure 5 represents the lubrication temperatures of the turbochargers at the inlet and outlet of the turbocharger. The average temperature difference for turbocharger number 1 is 11.8 °C, and for turbocharger number 2 is 11.19 °C. Having very close temperatures, the 2 turbochargers act almost identically on the engine, increasing its efficiency [8].

5. Conclusions

Starting from the fact that, out of the total amount of energy contained in the burned fuel in a diesel engine, only 20% is used for the effective propulsion of the ship, shipbuilders act in all ways to reduce energy losses and increase propulsion efficiency. Their attention is focused on both the engine and the engine itself and the ship. The component that aims to enrich the mixture that will be consumed by the engine, is the centrifugal compressor driven by the gas turbine. The turbocharging process ensures the recovery of a fraction of the energy contained in the flue gases emitted from the engine, which represents approximately 30-40% of the energy released in the combustion process. The internal combustion engine is a machine characterized by the presence of a stationary flow of the working fluid, as a result of which the turbine of the turbocharger group must be able to operate in such conditions. The efficiency of the operation of the turbocharger system related to the main propulsion engine depends on the internal processes that take place in its operation and also on the environmental conditions. In this way, the direction of processing on the installation is dictated by the engine load, taking into account the extreme situations, corresponding to the operation of the ship in special conditions (overload, tropical temperature, arctic temperature, etc.). The choice of one or more turbocharger groups on the ship is determined according to the operating conditions of the ship and implicitly of the engine, the volume of labour embedded in the installations, the location, closely correlated with the problems of size, weight, noise. Taking into account the thermodynamically and gas-dynamic approaches of the centrifugal compressor evolutions presented above, we can say that, at any operating regime, its performance is highlighted by the degree of compression π_C and the efficiency η_C, dependent in turn on the torque pressure - temperature values. The characteristic curves of the compressor, turbine and engine for the case of joint operation are determined experimentally, individually for each machine, based on the similarity ratios of the curved curves for such machines. In the latter case, the method gives exact results, and in the second case, it will give approximate results. If the characteristics of the turbine and the compressor are given in generalized parameters, it will be possible to calculate by the relations of their definition, the values that are of interest. The concrete way of establishing the common points of operation of the units that make up the supercharged diesel engine depends on the scheme of grouping the respective units. Determining the operation of the compressors, the heat engine and the turbine (which together form what is called a supercharged engine), is done individually for each unit, determining the working regime in common with the other component units of the assembly.

6. References

[1] https://eur-lex.europa.eu/legal-content/RO/TXT/?uri=CELEX%3A32008R0440 Accessed on 28 of february 2021.

[2] L. MITSUBISHI HE.AVV INDUSTRIES, 2012, *MB06-Main Engine Final Drawing - Horaisan Ship.*

[3] KEIICHI, S. 2006 Development of High-efficiency Turbocharger MET-MA Series for Large-Bore Diesel Engine, *Nagasaki Shipyard & Machinery Works: Technical Review* **43**(1).

[4] Cong, G., Gerasimos, T., Peilin, Z., Hui, G. 2014 Computational investigation of a large containership propulsion, *Appl. Energy 2014*, **130**, 370–383.

[5] https://moam.info/organizational-morphing-technology_5b8dc802097c472a0f8b46aa.html Accessed on 28 of february 2021.

[6] Wahlstrom, J., Eriksson, L. 2011 Modelling diesel engines with a variable-geometry turbocharger and exhaust gas, *Proc. Inst.Mech. Eng. Part D J. Automob. Eng.* **225**, 960–986.

[7] Yin, J., Su, T., Guan, Z., Chu, Q., Meng, C., Jia, L, Wang J. and Zhang Y 2017 Modeling and Validation of a Diesel Engine with urbocharger for Hardware-in-the-Loop Applications, *Energies,* **10**(5), 685; doi:10.3390/en10050685.

[8] https://inmateh.eu/INMATEH_1_2013/8_Marinuc%20M.pdf Accessed on 28 of february 2021.

Acknowledgements

„This work is supported by the project ANTREPRENORDOC, in the framework of Human Resources Development Operational Programme 2014-2020, financed from the European Social Fund under the contract number 36355/23.05.2019 HRD OP /380/6/13 – SMIS Code: 123847.

IOP Conf. Series: Materials Science and Engineering 1182 (2021) 012025 doi:10.1088/1757-899X/1182/1/012025

XPS, SEM, AFM, and Nano-Indentation characterization for powder recycling within additive manufacturing process

N E Gorji, R O'Connor and D Brabazon
I-Form, the SFI Research Centre for Advanced Manufacturing, Dublin City University, Dublin 9, Ireland

E-mail: nima.gorji@i-form.ie

Abstract. Powder recycling and reducing the waste metallic powder is EU's key provision in waste framework directive (2008/98/EC). The aim of this investigation is to analyse the correlation between the surface and morphology properties of (virgin and recycled) powders and the microstructure and mechanical properties of the 3D printed parts (made of three powders). Two biomedical Tibia implants have been 3D printed from virgin and 3-5 times recycled powders of stainless steel 316L. For this, the surface composition and microstructure of the powders has been characterized and correlated to the nanoindentation measurements carrier out on these implants. X-ray surface spectroscopy (XPS) has been used to analyse the oxidation level on the powder's surface revealing less than 10% more oxygen on the surface of recycled powders. SEM analysis shows less than 5 μm difference in powder size distribution even though the shape and circularity of the recycled powders seem to be affected under several reusing cycles. The size of the powder particles does not show much difference but satellites and binding between the powders increased in recycled powder. The hardness and effective modulus of the parts from recycled powders are significantly smaller than the virgin-made implants, which could be due to higher porosity present in the recycled powder or due to oxygen increment on recycled powder. The surface roughness (AFM analysis) has slightly increased on part made of recycled powders. However, the overall morphology shows little difference between the two parts.

1. Introduction

The recycling of metal powder is possible and valuable in terms of saving costs or generating revenue for companies and contributing to a sustainable environment and society through conserving energy and resources and decreasing pollutants. Additive Manufacturing (AM) is finding a massive capacity in industry, which obliges to reduce the cost associated with the process for financial viability. Powder Bed Fusion (PDF) or 3D printing of metallic parts often produces powder leftover. Reusing these in the manufacturing process will make the process more sustainable by cutting the cost related to powder feedstock which is just as important as the machine when producing high-performance parts [1,2]. Powder recycling has been the focus of many research groups as an alternative to reuse, and repurpose the leftover powder in the 3D printing machines. Characterization of metallic powders for manufacturing new parts has been suggested by designing new powder mixing strategies or post-treatment of recycled powders (e.g. plasma spheroidization) to prepare them for subsequent printing cycles [3]. It is conventionally believed that reusing the powders will deteriorate the mechanical properties of the part by affecting the quality of the meltpool, creating porosity in the layers and the microstructure morphology [4]. However, the recent developments have indicated negligible impact from reusing the leftover powder in powder bed fusion process by following the proper recycling strategies [5]. The amount of powder required to print a part is not equal to the part's dimension. The whole powder bed (10 cm radius and 15 cm height) must be filled with the powder in order to print a sample cube of 5cm^3, for example. Clearly, a huge amount of powder is required for a large dimension build chamber regardless of the size of the parts but only a few percentage of the powder is used and a majority of the powder is leftover in the build and must be recycled for sieving and subsequent use [6]. Powell et al. reported that powder degradation after a number of recycling cycles is the major problem in powder recycling process [7]. Therefore, the powder manufacturers may also study the powder degradation mechanisms or utilizing the end-of-life powder for PBF applications. Powder degradation will reflect in the quality of the manufactured parts where the low quality powders (with different size, surface properties and shape) [8]. Currently, an optimal number of recycling times is evaluated for every powder, printing machine and part application [9]. We have previously reported on the importance of surface composition of fresh and recycled powders for AM applications. Here, we present our latest results on surface characterization, surface roughness, microstructure and mechanical stress-strain analysis on both powders (fresh & recycled) and parts manufactured from fresh and 4-5 times recycled stainless Steel 316L powders. Characterization of powders and parts for surface quality, microstructure and mechanical properties and identifying a sensible relationship with the AM parameters is a key in developing the PBF to more sustainable greener technology.

2. Materials and methods

The gas atomised 316L stainless steel powder used in this study was purchased from Carpenter Additive with average particle size of 30µm. The surface composition of the powder was scanned by the lab-based VG Microtech electron energy analyser (XPS system), at base pressure of 1×10^{-9} mbar, and a conventional X-ray source Mg K_α ($h\nu$= 1486.6 eV). The XPS curves were analysed using CASAxps software 2.3.22. No surface treatment or chemical cleaning were performed on the fresh and recycled powders prior to XPS analysis to prevent surface contamination. The XPS acquisition parameters were set as: beam size of 1 cm^2, pass energy 20 eV, step of 0.1 eV and 30° take-off angle with a resolution of 1.2Ev. Mmicrostructure morphology, shape, circularity and size distribution of the powders were conducted using EVO-LS15 Scanning Electron Microscope (SEM) and Hitachi S5500 Field Emission SEM was used to image the microstructure of the parts. ImageJ software was used to extract the size distribution and circularity of the particles. Both the parts and powders were examined. Parts were mounted on resin and polished before the SEM analysis. Instead, the powders were sparkled on a carbon pad. The recycled powders were sieved for particles bigger than 50 μm before characterization even though some tabular shaped particles or spatters succeed the sieve and were present in the sieved. Nanoindentation was performed on the parts in different locations with a force of 1000 µN kept for up to 10 seconds using a Bruker HYSITRON TI Premier. Parts were mounted and polished with Metkon Forcipol polisher. The pore areas on the powders were identified by the optical microscope of the nanoindentation machine. The surface roughness of the parts was measured using Atomic Force Microscopy (AFM)- Bruker Dimension ICON. AFM images were processed by Gwyddion software. Our open access powder bed fusion printer is Aconity Mini machine manufactured by Aconity3D Aachen, Germany, has been used to print the biomedical Tibia biomedical implants. The machine build chamber's size is 145 mm diameter by 150 mm in height and has a 200 W fibre laser (λ=1068 nm). Continues monitoring was used to ensure the oxygen level remained below 20 ppm. Argon gas with 99.99% purity was passed though the build chamber during part production using a circulation fan. Both implants were located at the centre of the powder bed. The part made of virgin powder was printed alone and the part made of recycled powder was printed together with small artefacts parts (max. 8 mm height) at a large distance from the implant. Printing parameters were set at 0.12 mm hatch spacing, 0.3 m/s laser velocity, 170 W laser beam power, and 80 µm hatch spacing. The implants were 3D printed with 2000 layers of 50 µm thickness in vertical direction. The implant was 100 mm in height, 10 mm long, and 2 mm thick.

3. Results and discussions

Figure 1(a) shows the **surface composition XPS spectra** of the virgin and recycled powders. Both spectra show a dominant concentration of oxygen and carbon on the surface of particles as usually is observed in XPS performed on metal powders which have been exposed to air ambient where the oxygen and carbon are absorbed by the medium, and the carbon pad substrate may also contribute to the carbon peak. Si is the third element present on the surface after oxygen and carbon. The presence of Si is confirmed by Si 2p and Si 2s peaks at around 101 eV and 150 eV, respectively, in both powders as we also reported in our previous investigations [10]. We calculated the concentration of elements using XPS spectra in figure 1 as presented on table 1. The oxygen concentration increases from over 28% on virgin powder to 35% on recycled powder. The other three main elements (Mn around 639 eV, Cr, Fe around 710 eV) are present on surface with very low concentration. Ni is not present on the surface. More details about the concertation of different elements on the surface (e.g. O, Mn, Cr, Fe) scanned in a specific ranges will be presented elsewhere. We have measured the Ni concentration using HAXPES measurements carrier out at a synchrotron facility in a previous study [11]. Figure 1(b), presents the oxygen level on both virgin and recycled powders where the difference is less than 10%. The Si concentration on the surface of the powder is usually not given much importance. However, this can provide a good information about the meltpool and the quality of heat distribution imposed to the parts during manufacturing. Si concentration increases on the surface by out-diffusion of the element to the surface followed by formation of SiO layer which impedes the light absorption or oxygen penetration into bulk of the part or out-diffusion of other elements (Cr, Fe) towards the surface [12].

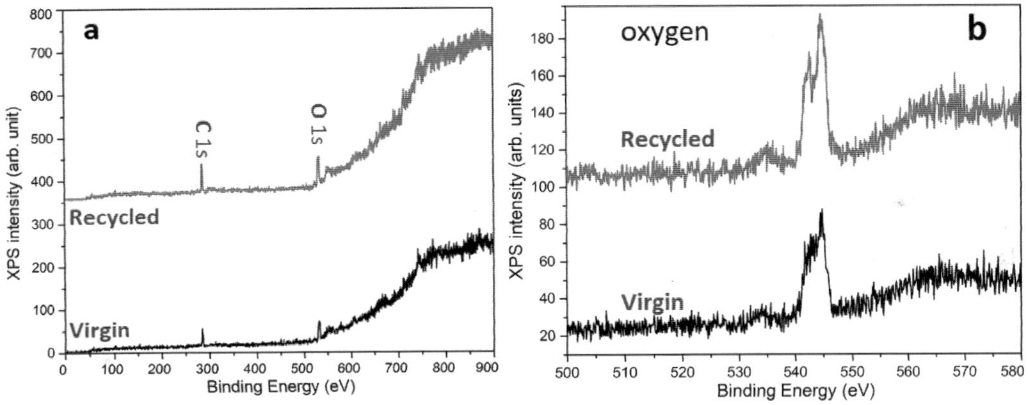

Figure 1. (a) the XPS measurements on both virgin and 4-5 times recycled powders, (b) XPS range scan of oxygen concentration on virgin and recycled powders.

Table 1. The concentration of surface elements extracted from XPS spectra.

	O	C	Si	Mn	Cr	Fe
virgin	28.6	40	18	2.5	4.3	7
recycled	35	44.7	5.8	4.2	2	8.5

The **SEM microstructure** and morphology of the virgin and recycled powders has been illustrated in figure 2(a) and 2(b). The recycled powders were collected after 4-5 times reusing the printer. A good spherecity and circularity of the particles is observed for the virgin powder, which is the typical morphology of gas atomized powders even though some bonded particles and satellite surfaces are also observed among the virgin powder particles. The recycled powder has a slightly different morphology from the virgin powder and more of an aggregated morphology, lower circularity and more non-uniformly shaped particles (elongated) with more satellites and rough surfaces. This may influence the morphology of the printed parts during heat absorption, melting, and cooling process [13]. The SEM images were also taken from the parts manufactured from the virgin and recycled powders as shown in figure 2(c) and figure 2(d). The morphology is roughly similar. However, the part made of recycled powder shows some porous areas or areas which were not even melted. This could be an indication of the oxidation impact on laser absorption on the surface of the powder particles and the formation of Si layer on the surface of recycled powders. The average size of virgin and recycled powders was measured around 29.8 μm and 32μm, which is a negligible difference.

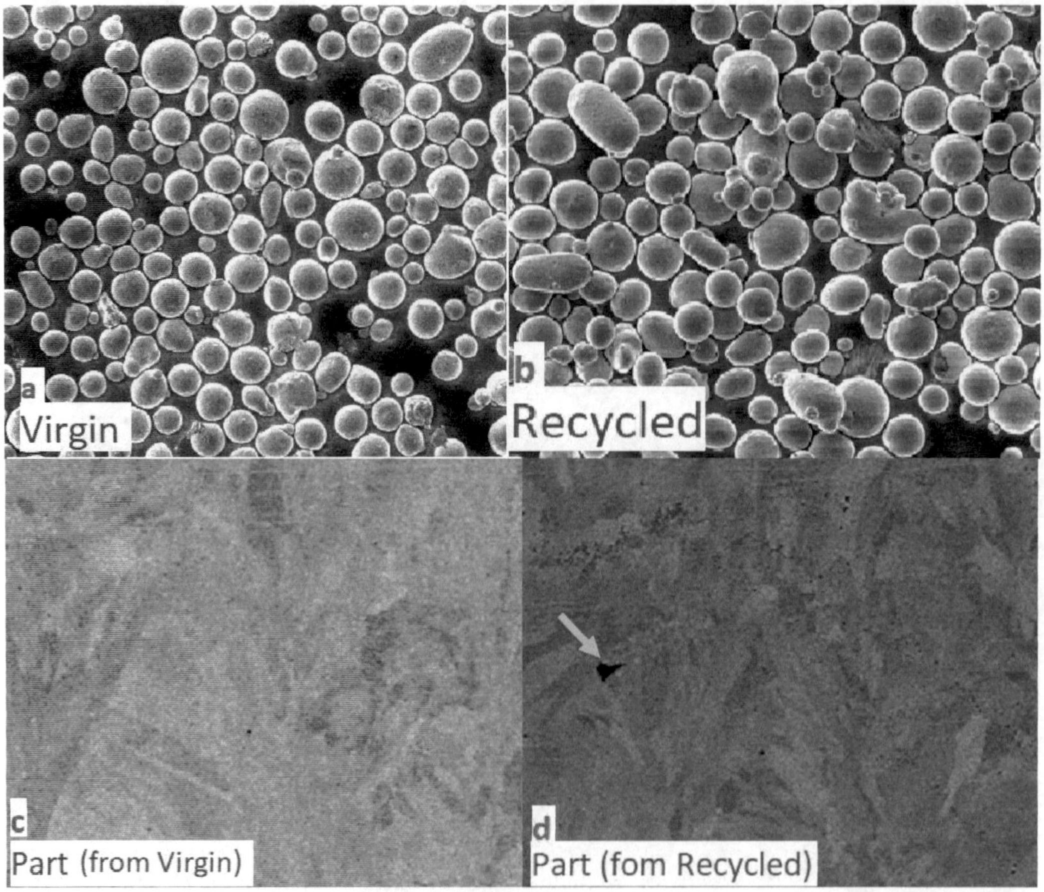

Figure 2. SEM images from (a) virgin and (b) recycled powders and parts manufactured by virgin (c) and (4-5 times) recycled (d) powders. The arrow in (d) indicates a small porosity area or non-melted area on the part manufactured from recycled powder. All images taken at 20μm resolution.

Nanoindentation measurements has been carried out in order to assess the mechanical properties of both parts. Number of nano-indents has been applied on the surface of the parts on different areas in order to obtain an average value of the hardness (H) and effective modulus (E_r) of every part. The results are shown in figure 3 and the average value of H and E_r are presented in table 2 where the parts manufactured from recycled powders show significantly lower H and E_r values (H= 5.34 GPa vs. 3.78 GPa and E_r=189.3 vs. 149). The more fragile mechanical properties of the part from recycled powders is due to higher porosity or weaker microstructure which has a root in the recycled powders with higher oxygen on surface or broader size distribution as discussed in previous figures [14]. This might be better understood by topography measurements as presented in following.

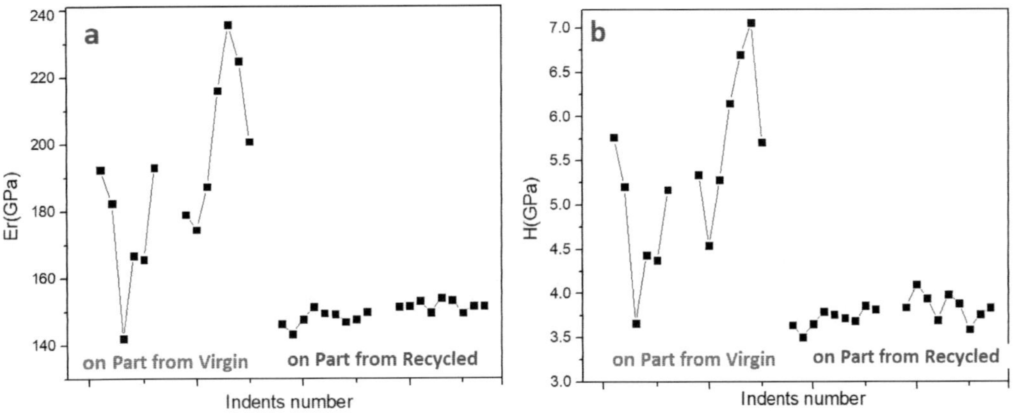

Figure 3. The Effective Modulus (a) and Hardness (b) values obtained from the nanoindentation measurements on both parts manufactured using virgin and recycled powders.

Table 2. The nanoindentation results on both parts printed from virgin and recycled powders.

Parameters	Hardness, H	Eff. Modulus, E_r
Part from virgin	5.34	189.3 GPa
Part from Recycled	3.78	149.8 GPa

AFM surface topography analysis has been performed on both parts to calculate the surface roughness and for topography analysis of the polished surfaces. This will further help to understand the nanoindentation results. Figure 4 presents the surface roughness calculated along the red line drawn on the surface of Part printed from recycled powders. The surface roughness has been calculated more for the recycled powder (0.58 μm vs. 0.34 μm). The quality of the surface is rougher for the part from recycled powder and with more contamination or deformation as shown in the 3D map. The AFM map are in agreement with the topography results presented in [15]. There are more areas on Part from recycled powder, which has not been melted and is not removed by polishing or is opened out after polishing due to the porosity beneath the layers.

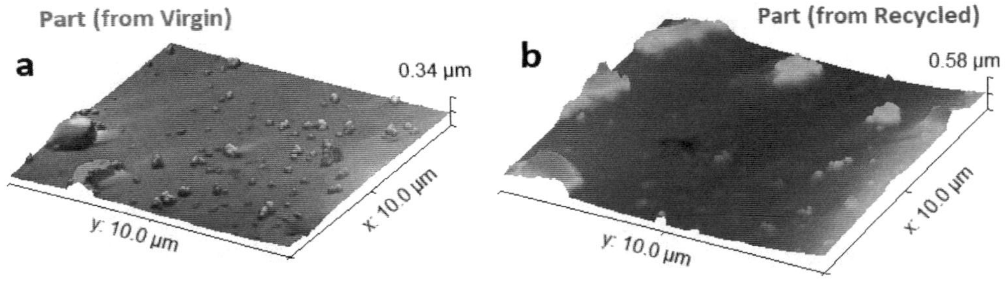

Figure 4. AFM Roughness analysis on the parts printed from (a) virgin and (b) recycled powders. The topography of the recycled powder shows more disturbance of the microstructure.

4. Conclusions

The surface composition analysis (XPS) and microstructure characterization (SEM), topography (AFM) and nanoindentation measurements of the virgin and 4.5 times recycled stainless steel 316L powders, as well as the tibia implants 3D printed from these powders, using the Aconity Mini 3D printer were presented. The recycled powder shows less than 10% more oxygen on the surface and particles with less circularity but less than 5 μm broader size distribution. Nevertheless, the mechanical properties of hardness and effective modulus of the parts from recycled powder were significantly lower than those manufactured from virgin powder (H= 5.34 GPa vs. 3.78 GPa and E_r=189.3 vs. 149). AFM measurements support the nanoindentation results and more rough and deformed topography is observed from the surface of the part from recycled powder (0.58 μm vs. 0.34 μm). A more detailed analysis is ongoing to disclose further the correlation between the properties of the recycled powders and the weaker mechanical properties of the 3D printed parts. This requires identifying the porosity population inside the powders and the printed parts using CT scan analysis as well as EDX measurements from the bulk of the parts to reveal the presence of Ni or Cr. Synchrotron analysis would also be very useful to understand the recrystallization process especially for the part manufactured from the recycled powder.

5. References

[1] Acquesta A, Monetta, T. 2021 *Materials Science Forum.* ***1016*** *MSF* 225-230
[2] Gatto M.L, Groppo R, Bloise N, Fassina L, Visai L, Galati M, Iuliano L, Mengucci P. 2021 *Materials,* **14**(1) 224 1-20 https://doi.org/10.3390/ma14010224
[3] Harkin R. Wu H, Nikam S, Quinn J, McFadden S. 2020 *Metals,* **10**(12) 1700 1-14.
[4] Nezhadfar P.D, Soltani-Tehrani A, Sterling A, Tsolas N, Shamsaei N. *2018 Solid Freeform Fab. Proceed.* 29*th* *Ann. International,* 1292.
[5] Soltani-Tehrani A, Pegues J, Shamsaei N. 2020 *Additive Manufacturing,* **36** 101398.
[6] Heiden M.J, Deibler M.A, Rodelas J.M, Joepke J.R, Tung D.J, Saiz D.J, Jared B.H, 2019 *Additive Manufacturing* **25** 84.
[7] Jacob G, Brown C, Donmez A, Watson S, Slotwinski J 2016 *NIST Adv. Manufac. Series* **100**-6
[8] Powell D, Rennie A.E.W, Geekie L, Burns N. 2020 *Journal of Cleaner Production* **268** 122077.
[9] Lutter-Günther M, Gebbe C, Kamps T, Seide C, Reinhart G 2018 *Production Engineering* **12** 377.
[10] Gorji N E, O'Connor R, Mussatto A, Snelgrove M, González P, Brabazon D. 2019 *Materialia* **8**100489.
[11] Gorji N E, O'Connor R, Brabazon D. 2018 *IOP Conference Series: Materials Science & Engineering* 591 **1** 012016.
[12] Yakout M, Elbestawi M, Veldhuis S 2018 *Int. J. Adv. Manuf. Technol.* **95** 1953.
[13] Kim H, Lee S, 2021 *Journal of Iron and Steel Research International* 28 **1** 111-121.
[14] Chandrasekar S, Coble J.B, Yoder S, Nandwana P, Dehoff R.R, Paquit V.C, Babu S.S. 2020 *Additive Manufacturing* **32** 100994.
[15] Tupac Y. H, Armani A. 2021 *Journal of Materials Engineering and Performance,* in press.

Acknowledgment

This research was supported by Science Foundation Ireland (SFI) under Grant Number 16/RC/3872 and is co-funded under the European Regional Development Fund. N.E. Gorji acknowledges the New Foundation Award (No. P60687) received from Irish Research Council (2019-2020).

Analysis of precipitates in heat treated thermo-mechanically processed steel with a high yield strength

J Górka[1], T Poloczek[1], A Kotarska[1] and W Jamrozik[2]
[1]Silesian University of Technology, Mechanical Engineering Faculty, Department of Welding Engineering, Konarskiego 18a, 44-100 Gliwice, Poland
[2]Silesian University of Technology, Mechanical Engineering Faculty, Department of Fundamentals of Machinery Design, Konarskiego 18a, 44-100 Gliwice, Poland

E-mail: aleksandra.kotarska@polsl.pl

Abstract: The aim of this study was to analyse the precipitates and determine their impact on the structure of thermomechanically processed high strength S700MC steel. The tested samples were subjected to a heating process at temperatures from 100°C up to 1300°C with a 100°C step. The chemical composition analysis in the micro-areas of the samples, thin-film tests and quantitative measurements of carbonitride precipitates were performed. In order to describe the grain size, a method of calculating the number of grains cut through the measuring line was used. In order to determine the effect of heat treatment on strength and plastic properties, Charpy V-notch, Vickers hardness and static tensile tests were carried out. The study showed that S700MC steel is characterized by a non-equivalent, fine-grained bainitic-ferritic structure. The test showed a significant impact of heat treatment on the strength and plastic properties of S700MC steel.

1. Introduction

In recent years, there has been growing interest in the complexity and operational expectations for steel structures. These conditions can be met through the use of high strength steel with good plastic properties, corrosion resistance and also due to the way they are joined with the appropriate weldability. In the early 1960s, publications by Hall and Petch showed that in ferritic-pearlitic steels the fine ferrite grain causes a simultaneous increase in strength and a decrease of the brittle transition temperature of steel products. Since then, it is the most important metallurgical parameter in the production of structural steels. The production of high-strength steel, in addition to strong grain refinement, required the use of other steel strengthening mechanisms, i.e. precipitation strengthening. Therefore, in order to obtain the required material properties, the production of C-Mn structural steels with Nb, V, Ti alloy additions began which allowed to increase the yield strength up to 500MPa. These elements are micro-additives in steel, i.e. their concentration does not exceed 0.22%. Steels containing micro-additives are called microalloyed steels or high-strength low-alloy steels. In order to obtain the fine-grained structure, these steels are most often produced in a controlled rolling process followed by controlled cooling. The combination of these processes is called thermomechanical processing. The application of the described process and new metallurgical technologies allowed to obtain high strength steels while maintaining good plastic properties of the material. The use of thermomechanical treatment instead of conventional microstructure forming methods like normalizing heat treatment saves energy and allows to achieve greater grain refinement. Application of steel with high yield strength allow to produce of slimmer and lighter structure elements while maintaining the same load capacity and reducing welding time and costs. Technical and economic aspects have made these steels one of the basic structural materials. Therefore, improving the methods of their manufacture, analyzing precipitation processes during the production of these steels and during welding processes is extremely important from the point of view of the manufacture and construction of steel structures in the 21st century. Thermo-mechanical control process (TMCP) consists on continuous control of the temperature and deformation of manufactured steel products during rolling in order to obtain a material with high strength properties. During this treatment, two main effects are used: the impact of fine-grained structure on the increase of strength and plastic properties and limitation or delay of recrystallization due to the small amounts of alloying elements added to the material. Avoiding or limiting the recrystallization process is caused by micro-alloying elements, which also form secondary phases that inhibit the recovery process. Thanks to this, the austenite structure is characterized by strong grain refinement, high dislocation density and a significant number of crystals of new structural components directly from rolling temperature during a controlled cooling process. While the steel is cooled in the air, a fine grained ferrite with a small amount of perlite is obtained. When accelerated control cooling process is applied a fine grained polygonal or coniferous ferrite is obtained, while the pearlitic transformation is replaced by bainitic and partly martensitic transformation (depending on the temperature of the rolling end and cooling speed). Accelerated cooling after a controlled rolling process causes fragmentation of the ferrite grain and contributes to solid solution strengthening and precipitation hardening by dispersed particles. Steels produced in this way can reach a yield point of up to 700 MPa [1-30].

2. Materials and methodology

The tests were carried out on thermomechanically processed high-strength low-alloy S700MC steel of 10 mm thick. Chemical composition and properties of investigated material are shown in table 1.

Table 1. Actual chemical composition of 10 mm thick S700MC steel.

Contents of chemical elements, % by weight					
C	Mn	Si	S	P	Al
0.056	1.68	0.16	0.005	0.01	0.027
Nb	Ti	V	N*		C_e**
0.044	0.12	0.006	72		0.33
* - N: content in ppm; nitrogen was identified using the high-temperature extraction method					
** - C_e – carbon equivalent					
Total content of Nb, V and Ti should amount to a maximum of 0.22%					
Mechanical properties					
Tensile strength R_m, MPa	Yield point R_e, MPa	Elongation A_5, %	Toughness, J/cm^2 (-20 °C)		
822	768	19	135		

S700MC steel is characterized by non-equivalent, fine-grained bainitic-ferritic structure with a relatively low carbon equivalent (at the level of 0.33%), which should indicate its good weldability. The high strength properties of S700MC steel are obtained by a thermomechanical treatment process combined with the strengthening by micro-additives, such as Ti, Nb and a small amount of V. The total content of alloying micro-additives is 0.17 wt%, which does not exceed the permissible value of 0.22 wt% in steel with micro additives. S700MC steel has a very low carbon concentration (0.056 wt%), which may limit the impact of austenite transformation under the influence of welding thermal cycle which affects on the strength and plastic properties of welded joints.

2.1 Heat treatment

In order to determine the effect of temperature on the mechanical and plastic properties of S700MC steel, heat treatment was performed on investigated material. The samples of tested steel were heated at temperature ranges from 100°C to 1300°C with 100°C gradation. The heat treatment was carried out on a 24-channel resistance annealing machine WO6524 with a rated power of 65 kVA designed by LMS S.C., equipped with temperature controllers P62. Each sample was placed on a separate heating mat, then the samples were insulated and heat treated. The time for soaking the samples was 1 hour and cooling process took place in the air. The method of individual identification of samples is shown in table 2.

Table 2. Designation of heat treated samples.

Soaking temperature in accordance to sample Identification number [°C]						
A	B	C	D	E	F	G
100	200	300	400	500	600	700
H	I	J	K	L	M	
800	900	1000	1100	1200	1300	

2.2 Precipitates analysis

After the heat treatment process, each sample on cross-section was subjected to metallographic examination (etching – Nital). Additionally, the examinations were provided on the sample which after thermomechanical rolling was not heat treated and was marked with the symbol MR. Five microphotographs were taken on each specimen using a digital camera coupled with a light microscope at 1000x magnification. Microphotographs were taken near the corners of the sample and in its central part, as shown in figure 1.

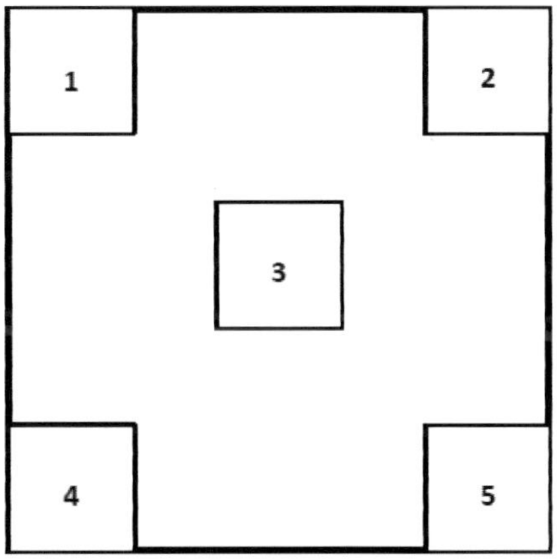

Figure 1. Scheme of areas where individual micrographs of sampes were taken.

Chemical composition researches in the micro-areas of the samples were performed on a Zeiss SUPRA 35 scanning electron microscope equipped with an EDS made by EDAX company which allows quantitative analysis of chemical elements present in the sample. The backscattered electrons (BSE) techniques of observation was used. Tests of thin films were performed on a High-resolution scanning transmission electron microscope (HRSTEM) Titan 80-300kV made by FEI, equipped with XFEG electron gun with Schottky field emission. Quantitative measurements of carbonitride precipitates were carried out on automatic image analyser – NiS – Elements BR 3.1. Using this analyser the number of carbonitride precipitates was calculated on all 5 images and other parameters such as area, circumference, length and width of obtained precipitates were also measured. Additionally, the percentage share of precipitates to the surface area on each microphotography was calculated. The percentage share of carbonitrides was calculated according to the following formula:

$$A_A=(\sum^{n}_{i=1}\cdot a_i/A)\cdot 100\% \quad (1)$$

IOP Conf. Series: Materials Science and Engineering 1182 (2021) 012026 doi:10.1088/1757-899X/1182/1/012026

where:
A_A – percentage share of carbonitrides on investigated area,
$\sum^{n}_{i=1} \cdot a_i$ – sum of carbonitrides area on observation area,
A – observation area (microphotography area).

The grain size was measured on one selected microphotography of each sample in accordance with the international standard PN-EN ISO 643. In order to evaluate the grain size, the method of counting number of grains cut by the measuring line was used. The measuring line was a section of a straight line of equal length to the image placed on each sample at the same microregion. Counted number of grains "N" formed the basis for determining "NL" parameter, i.e. the average number of grains per measuring line length, which allowed to determine the parameter "l" (average chord lengths cut off by grains on measuring line). The calculation procedure is presented below.

Parameters definition:
N – number of grains cut with measuring line,
L - length of measurement line after division by magnification,
$\bar{N}$ - average number of grains cut with a measuring line,
L_T – actual length of measuring line.

Based on the above-mentioned parameters:

$$N_L = N/L_T \tag{2}$$

N_L – average number of cut grains per unit of measuring line length.

Then the parameter N_L was used to calculate the value "l":

$$l = 1/N_L \tag{3}$$

l – average chord lengths cut off by grains on a measuring line.

3. Results and discussion

3.1 Analysis of basic material

Chemical composition analysis of S700MC steel on Optical Emission Spectroscopy analyser (OES) confirmed compliance of chemical composition with the PN EN 10149-2 standard. In tested steel, the carbon concentration is 0.056% wt, titanium – 0.12% wt, niobium – 0.044% wt, vanadium – 0.006% wt. The sum of micro-additives responsible for strengthening mechanism does not exceed the maximum permissible concentration of 0.22% wt. Content of nitrogen (70 ppm) based on the high-temperature extraction process confirms compliance of investigated material with delivered material certificate. Studies of thin films on Transmission Electron Microscope have shown that the strengthening mechanism of S700MC steel is determined by precipitates (Ti,Nb)(C,N) which size is of few to several nanometers. The precipitation hardening process is mainly caused by dispersion of a few nanometre precipitates, which are produced in ferrite during cooling of steel (figure 2). Particles larger than dozen nanometers do not make a significant contribution to the precipitation hardening however, they limit growth of recrystallized austenite grain, favouring the formation of fine-grained structure.

3.2 Microstructure analysis

Based on the heat-treatment test carried out on S700MC steel, it was found that in heating temperature range from 100 to 600°C the bainitic structure undergoes tempering and from 500°C the process of single grains recrystallization begins and strain hardening is reduced. After exceeding 800°C process of complete recrystallization is observed and the proportion of ferrite in the structure increases. Further increase in temperature above 1000°C causes strong grain growth (figure 3). The microscopic tests confirmed high temperature stability of Ti(C,N) and (Ti,Nb)(C,N) precipitates.

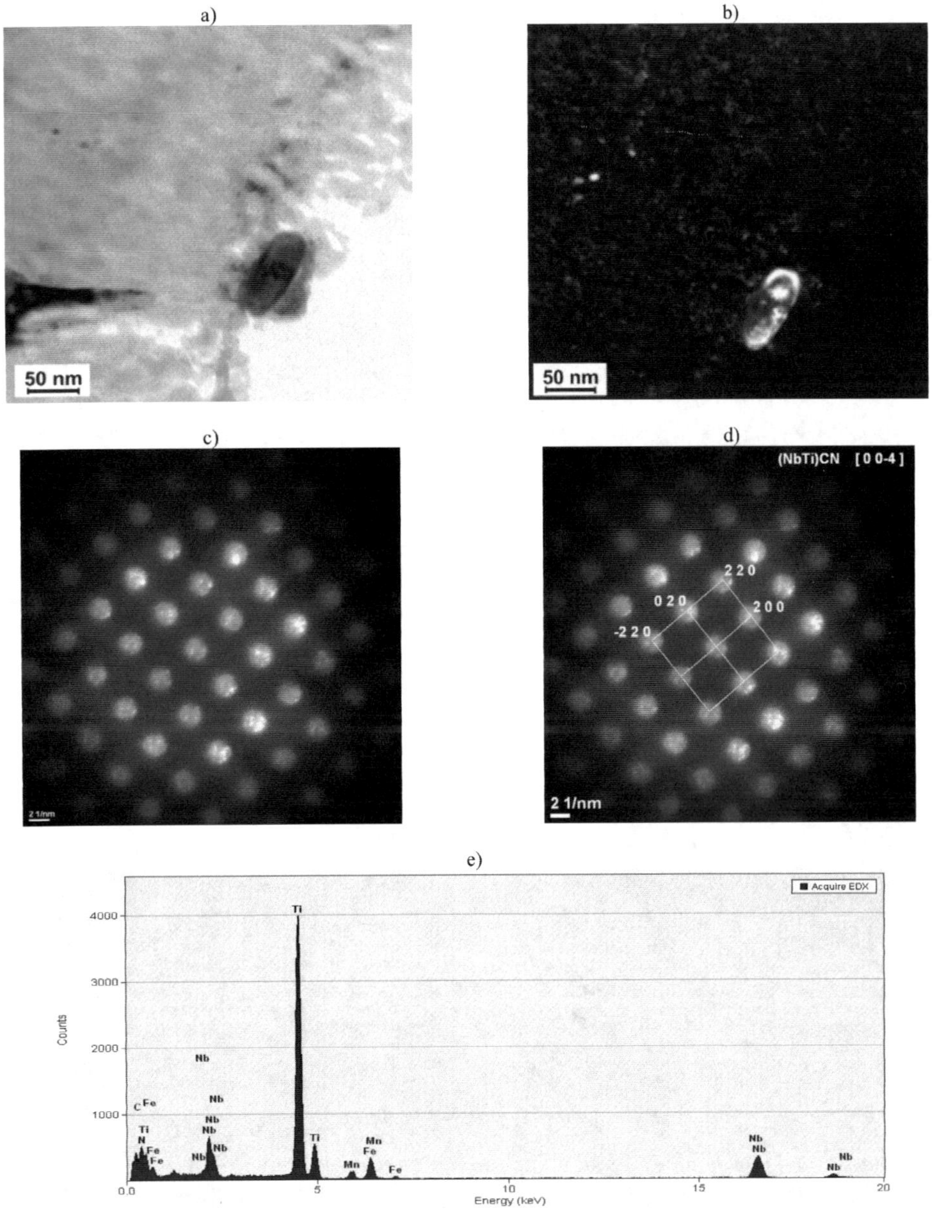

Figure 2. TEM micrograph with the corresponding selected area diffraction pattern illustrating carbonitride (Ti,Nb)(C,N) with small spherical precipitates responsible for precipitation strengthening of S700MC steel: (a) Bright-field; (b) Dark field; (c, d) selected area diffraction pattern; (e) EDX spectrum

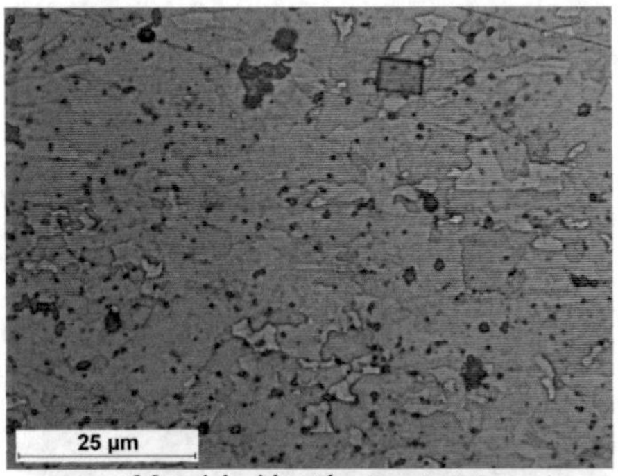

Material without heat treatment

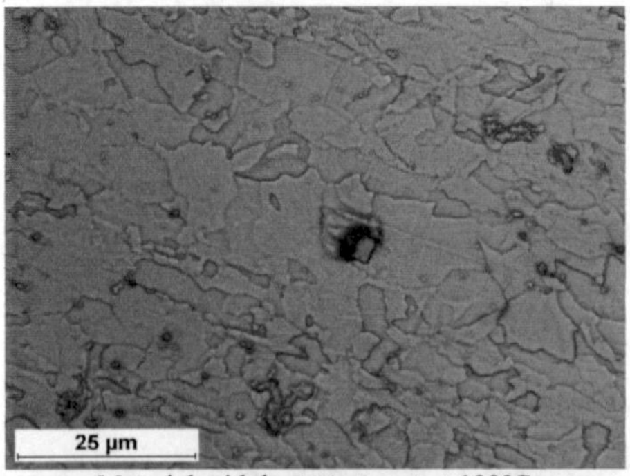

Material with heat treatment at 100°C

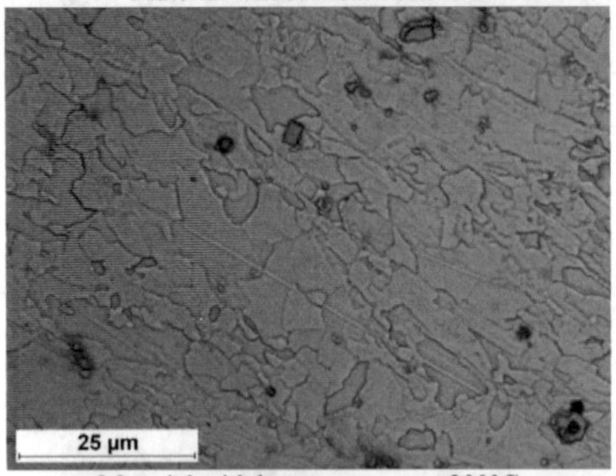

Material with heat treatment at 200°C

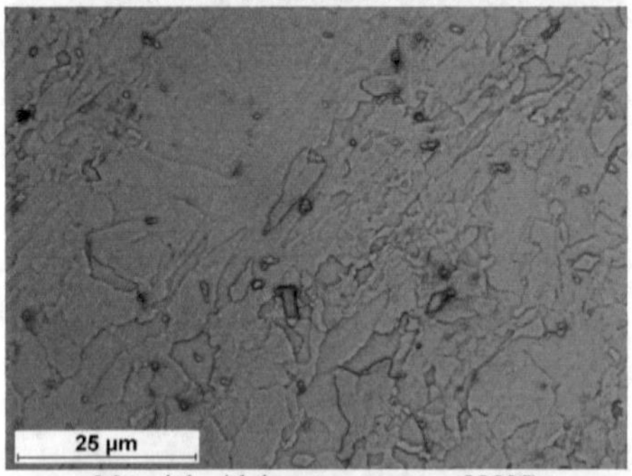

Material with heat treatment at 300°C

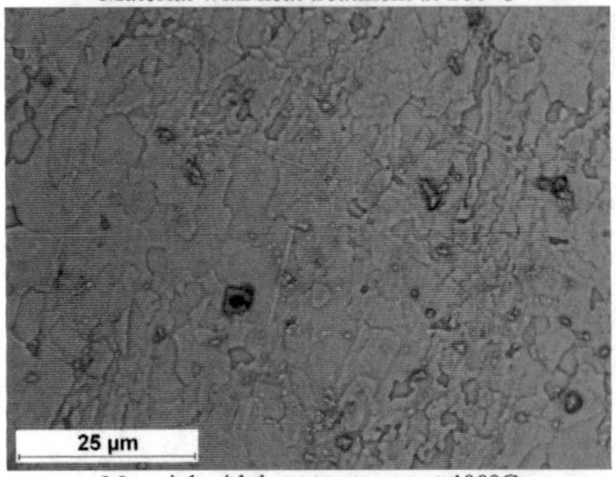

Material with heat treatment at 400°C

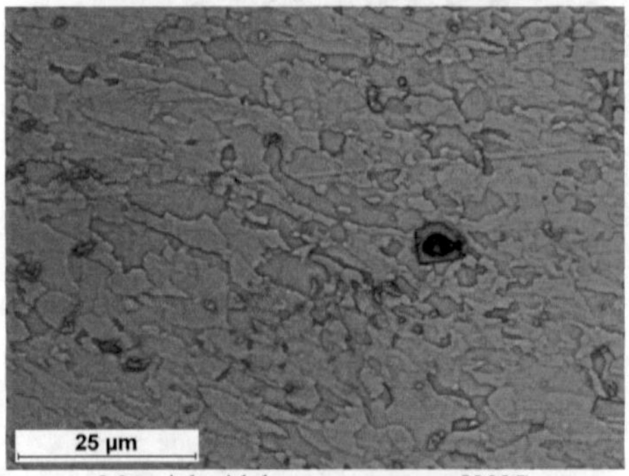

Material with heat treatment at 500°C

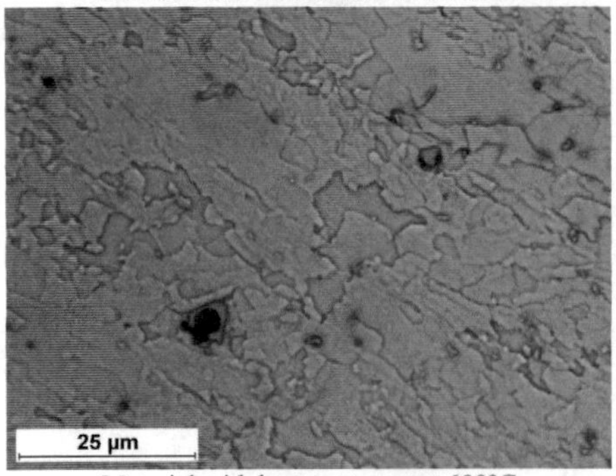

Material with heat treatment at 600°C

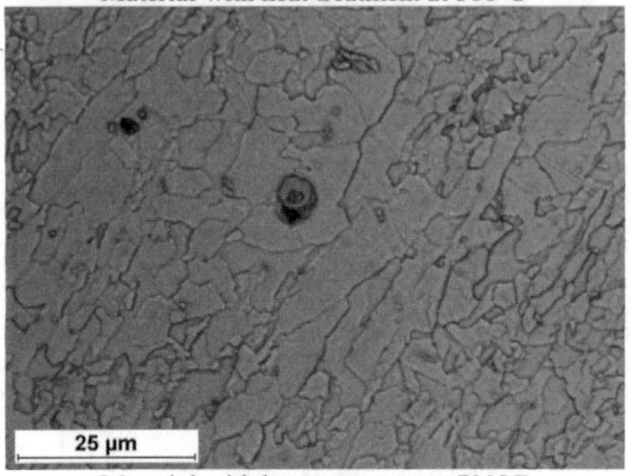

Material with heat treatment at 700°C

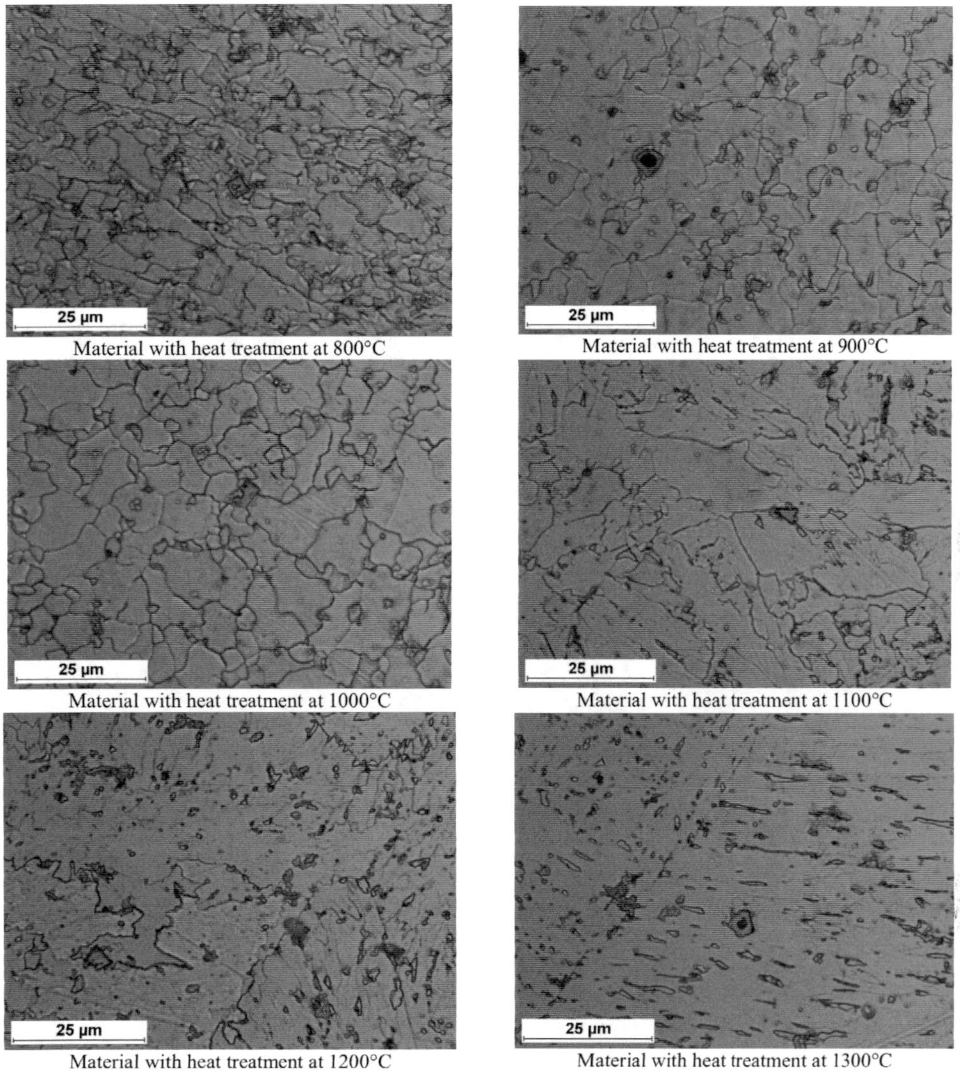

Figure 3. Microstructure of heat-treated samples.

3.3 Quantitative share of precipitates

Investigated precipitates had relatively large sizes, the average length of their cross sections $\bar{d}$ ranged from 4.29 to 9.64 μm. The large size of carbonitride phases indicates that they had to be formed during steel production process at high temperatures. This is evidenced by their considerable size already in the MR sample that was not heat treated. The number of particles on the plane ranged from 22.1 up to 40 mm^{-2}. table 3. The percentage share of carbonitrides was in the range of 0.327÷0.821%. Its highest value – 0.821% was registered for sample D heat treated at 400°C, while the lowest value was observed in sample G (0.340%) and H (0.327%) heat treated at 700 and 800°C. Distribution of this parameter is variable, i.e. it increases and decreases without a clear trend line. In samples from J to K, the percentage share of carbonitrides is at a similar level, table 3. The distribution of this parameter is shown in figure 4.

3.4 Grain size

The grain size was measured in accordance with international standard PN-EN ISO 643. For each sample the grain size was characterized by average chord lengths "l", obtained results were in the range from 1.77 up to 15.4 μm, figure 5. Parameter of average chord lengths of samples MR to H remains at a relatively low level. The smallest grain size value of 1.77 μm was observed for sample H, heat-treated at 800°C. This is probably the result of the appearance of fine grains due to the austenite phase transformation. After reaching the smallest value of grain size in sample H, in subsequent samples heat treated at higher temperatures a sudden increase of this parameter was observed. This is due to the recrystallization phenomena and growth of austenite grains. The largest grain size was observed for samples L (8.06 μm) and M (15.40 μm) heat treated at 1200°C and 1300°C, respectively.

Table 3. Average precipitation lengths and number of particles on microphotography on each sample.

Sample designation	Heating temperature to [°C]	Average length of precipitates d [μm]	Number of precipitates on investigated area Na [mm^{-2}]
MR	n/a	6.85	316.5
A	100	6.91	271.3
B	200	4.29	406.9
C	300	7.08	361.7
D	400	9.64	271.3
E	500	6.00	316.5
F	600	7.97	316.5
G	700	4.81	316.5
H	800	5.65	316.5
I	900	4.58	406.9
J	1000	9.29	226.1
K	1100	5.50	271.3
L	1200	7.30	226.1
M	1300	6.52	316.5

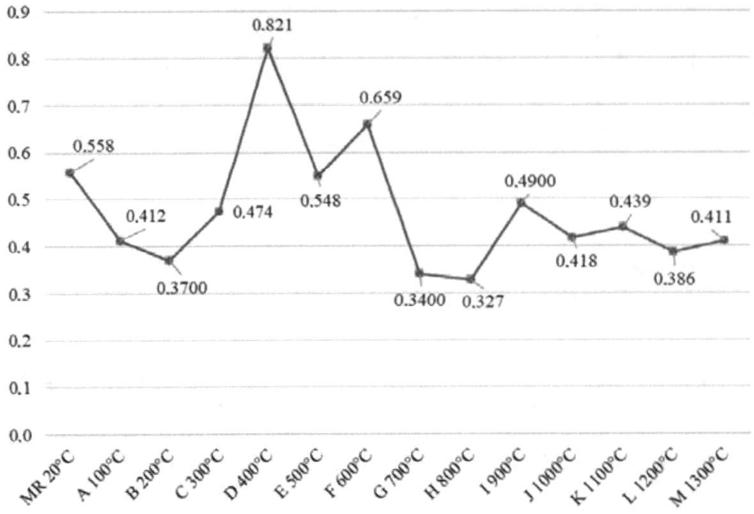

Figure 4. Distribution of percentage share of carbonitrides.

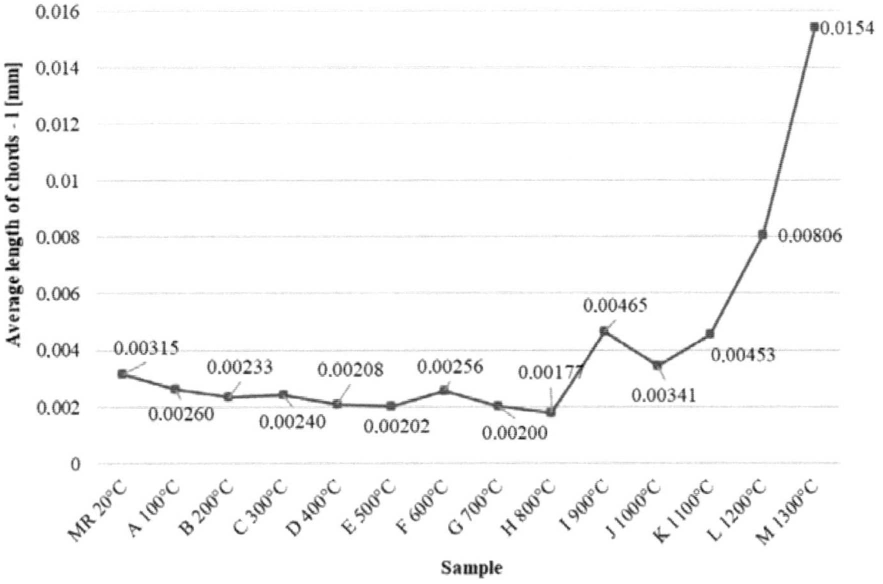

Figure 5. Graph showing average chord lengths as a parameter to determine grain size.

4. Conclusions

S700MC steel is characterized by an unbalanced, fine-grained bainitic-ferritic structure. High strength properties of tested material are ensured by thermomechanical processing combined with strengthening process caused by micro-additives such as Ti, Nb. Precipitation hardening is mainly caused by the dispersion of few nanometers precipitates, which are formed in ferrite during cooling process. Microscopic studies showed the presence of large carbonitride precipitates, their average size ranged from 4.29 to 9.4 μm. The significant size of the carbonitride phases indicates that they had been formed during the production process of steel at high temperatures, due to the high diffusion coefficient of the elements that form the precipitates. The large size and small amount of carbonitride phases in the range from 226.1 to 406 mm^{-2} indicates that grain boundaries rarely encounter them during growth. Therefore, the micro-additives in S700MC steel are not fully used to control precipitation hardening process and grain growth. The increase of average chord lengths parameter in the temperature range from 900 to 1300°C is associated with recrystallization phenomena and grain growth. Recrystallization and subsequent grain growth could start at higher temperatures if there were more niobium dissolved in a solution or a higher amount of precipitates.

5. References

[1] Bhadeshia, H. K 2001 *Bainite in Steels,* pp. 265-268, IOM Communications Ltd, Cambridge.
[2] Xue, E., Shan, Y., Zheng, S. and Lou, S 2006 Microstructural characteristic of low carbon microalloyed steels produced by termomechanical controlled process *Material Science and Engineering A* **438-440** pp. 285-287.
[3] Opiela, M. 2015 Thermodynamic analysis of the precipitation of carbonitrides in microalloyed steels *Materiali in Tehnologije* **49**(3) 395-401.
[4] Medina, S. F., Gomez, M. and Gomez, P. 2010 Effects of V and Nb on static recrystallization of austenite and precipitate size in microalloyed steels *Journal of Materials Science* **45** 5553-5557.
[5] Grajcar, A. 2014 Thermodynamic analysis of precipitation processes in Nb-Ti-microalloyed Si-Al TRIP steel *Journal of Thermal Analysis and Calorimetry* **118** 1011-1020.

[6] Efron, L. I., Morozov, D. and Goli-Oglu, E. A. 2012. Influence of rolling temperature on the austenite structure and properties of low-carbon microalloyed steel *Steel in Translation* **42** 456-461.

[7] Lee, C., Shin, H. and Park, K. 2012 Evaluation of high strength TMCP steel weld for use in cold regions *Journal of Constructional Steel Research* **74** 134-139.

[8] Wang, S. H., Chiang, C. C. and Chan, L. I. 2003 Effect of initial microstructure on the creep behaviour of TMCP EH36 and S490MC steels *Materials Science and Engineering: A* **344** 288-295.

[9] Shin, Y. T., Kang, S. W. and Lee, H. W.2006 Fracture characteristics of TMCP and QT steel weldments with respect to crack length *Materials Science and Engineering: A* **434** 365-371.

[10] Górka, J. 2016 Microstructure and properties of the high-temperature (HAZ) of thermo-mechanically treated S700MC high-yield-strength steel *Materiali in Tehnologije* **50**(4) 617-621.

[11] Nishioka, K. and Ichikawa, K. 2012 Progress in termomechanical control of steel plates and their commercialization *Science and Technology of Advanced Materials* **13** 023001.

[12] Winczek, J., Gawronska, E., Gucwa, M. and Sczygiol, N. 2019 Theoretical and experimental investigation of temperature and phase transformation during SAW overlaying *Applied Sciences* **9**(7) 1472.

[13] Jahazia, M. and Egbali, B., 2000 The influence of hot rolling parameters on the microstructure and mechanical properties of an ultra-high strength steel *Journal of Materials Processing Technology* **103**(2) 276-279.

[14] Porter, D., Laukkanenb, A., Nevasmaa, P., Rahka, K. and Wallin, K. 2004 Performance of TMCP steel with respect to mechanical properties after cold forming and post-forming heat treatment *International Journal of Pressure Vessels and Piping* **81** 867-877.

[15] Górka, J., Janicki, D., Fidali, M. and Jamrozik, W. 2017 Thermographic Assessment of the HAZ Properties and Structure of Thermomechanically Treated Steel *International Journal of Thermophysics* **38** 183.

[16] Węgrzyn, T., Piwnik, J., Hadryś, D. and Wszołek, Ł. 2017 Low alloy steel structures after welding with micro-jet cooling *Archives of Metallurgy and Materials* **62**(1) 115-118.

[17] Górka, J. 2017 Welding Thermal Cycle-Triggered Precipitation Processes in Steel S700MC Subjected to the Thermo-Mechanical Control Processing *Archives of Metallurgy and Materials* **62** 331–336.

[18] Kik, T., Moravec, J. and Novakova, I. 2017 New method of processing heat treatment experiments with numerical simulation support *IOP Conference Series-Materials Science and Engineering* **227** 012069.

[19] Kik, T. and Górka, J., 2019 Numerical Simulations of Laser and Hybrid S700MC T-Joint Welding *Materials* **12**(3) 516.

[20] Hashimoto, S., Ikeda, S., Sugimoto, K. and Miyake, S. 2004 Effects of Nb and Mo addition to 0.2%C-1.5%Mn steel on mechanical properties of hot rolled TRIP-aided steel sheets *ISIJ International* **44**(9) 1590–1598.

[21] Wang, X. D., Huang, B. X., Wang, L. and Rong, Y. H. 2008 Microstructure and mechanical properties of microalloyed high-strength transformation-induced plasticity steels *Metallurgical and Materials Transactions A* **39** 1-7.

[22] Skowrońska, B., Szulc, J., Chmielewski, T., Sałaciński, T. and Swiercz, R. 2018 Properties and microstructure of hybride Plasma+MAG welded joints of thermomechanically treated S700MC steel *27th Anniversary International Conference on Metallurgy and Materials (METAL)* Czech Republic.

[23] Timokhaina, I. B., Hodgson, P. D. and Pereloma, E. V. 2000 Effect of microstructure on the stability of retained austenite in transformation-induced-plasticity steels *Metallurgical and Materials Transactions A* **35** 2331-2341.

[24] Opiela, M. 2014 Elaboration of thermomechanical treatment conditions of Ti-V and Ti-Nb-V

microalloyed forging steels *Archives of Metallurgy and Materials* **59** 1181-1188.

[25] Grajcar, A., Różański, M., Kamińska, M. and Grzegorczyk, B. 2016 Effect of gas atmosphere on the non-metallic inclusions in laser-welded TRIP steel with Al and Si additions *Materiali in Tehnologije* **50**(6) 945–950.

[26] Żuk, M., Górka, J., Czupryński, A. and Adamiak, M. 2016 Properties and structure of the weld joints of quench and tempered 4330V steel *Metalurgija* **55**(4) 613–616.

[27] Yasuharu, S., Osamu, M. and Hiroshi, T. 1991 Mechanical properties and retained austenite in intercritically heat-treated bainite-transformed steel and their variation with Si and Mn additions *Metallurgical Transactions A* **22** 489–498.

[28] Suder, W., Ganguly, S., Williams, S. and Yudodibroto, B. 2017 Root stability in hybrid laser welding *Journal of Laser Applications* **29**(2) 022410.

[29] Fydrych, D., Labanowski, J. and Rogalski, G. 2013 Weldability of high strength steels in wet welding conditions *Polish Maritime Research* **20**(2) 67-73.

[30] Rodriguesa, P., Perelomab, E. and Santos, C. 2000 Mechanical properties of an HSLA bainitic steel subjected to controlled rolling with accelerated cooling *Materials Science and Engineering A* **283** 136–143.

IOP Conf. Series: Materials Science and Engineering 1182 (2021) 012027 doi:10.1088/1757-899X/1182/1/012027

High-pressure water jet cutting of S235JR steel alloy. Influence of process parameters on dimensional accuracy

CC Grigoraş, B Chiriţă, G Brabie, V Zichil, E Herghelegiu, C Tâmpu and VA Ciubotariu
"Vasile Alecsandri" University of Bacau, Faculty of Engineering, Calea Mărăşeşti 157, 600115, Bacau, Romania

E-mail: cosmin.grigoras@ub.ro

Abstract. This paper aims to determine the influence that water jet cutting parameters have on dimensional accuracy. The input parameters of this study were the workpiece material, that was made from 19 mm thick S235JR steel alloy plate, using a medium to high cutting pressure (2000, 2500, and 3000 bar), with a variable standoff distance (1, 2, and 3 mm) and a programmed quality of the cut from Q1 to Q5. A statistical interpretation of the results was conducted, and the experimental plan was obtained using the Response Surface Method. The samples were analysed for entrance and exit width of cut and kerf angle. A total number of 45 samples were measured, and the results were interpreted using an ANOVA analysis for statistical significance (p-value) and fit statistics (R2). The results have shown what values are optimum for the input parameters to obtain precision cutting quickly.

1. Introduction

The water jet cutting process was used around 1850 in the mining industry. It took around 80 years for the first industrial manufacturing machines to be developed and low pressure on soft material [1]. After the Second World War, rapid technological development can be observed when Luxembourg's first high-pressure water jet cutting machine was manufactured for plastics [2]. In 1958, the first high-pressure water jet equipment was developed only two years later to cut at 690 MPa hard material such as PH15 [3]. In present days water cutting using and abrasive mix is used when dealing with high strength material, complex shapes, superior surface finish and even waste reduction [4].

Water jet cutting (WJC) is considered a non-conventional but efficient process that allows cutting soft or hard, thin or thick materials with no thermal deformation than laser or plasma cutting [5]. It is also known as a "cold" process for its ability to cut without generating heat; thus, no heat-affected zones emerge, avoiding localized damage, preserving structural and chemical integrity [6]. Another advantage of the WJC is that it is an environmentally friendly process, has a low cutting price, reduces workpiece waste, and has a low setup time. Nevertheless, its key feature is the ability to cut at low or high pressure while adjusting the traverse speed, standoff distance, and angle of attack as needed, depending on the workpiece material [7-9]. The manufacturing industry benefits from this cutting process at its total capacity when using full 5-axis capabilities. This adaptation can assure that holes can be cut with chamfers included, and also complex parts such as elicoidale gear or turbine blades can be executed [10]. Like any other industrial process, the WJC presents disadvantages; one such downside is the inability to cut with high accuracy thick, rigid materials. A direct cause is generated by not accurately controlling the abrasive water jet [11]; thus, the cut surface present deviations from perpendicularity that emerge from differences resulted from how much material the water jet removes entering the material compared to them when it exits. Complex kinematic systems for compensation have been developed to overcome these deviations, using comprehensive geometrical cutting models implemented in the machine's software [12], but these solutions are often expensive and cannot be adapted on existing machines.

2. Research methods

While some cutting parameters, such as quality of the cut and traverse speed, are implemented in the equipment's software, others have to be programmed manually, such as water pressure, stand-off distance, material type, and thickness—this fast method for programming industrial equipment, especially if the parts do not have tight dimensional tolerances. If accuracy is needed, the process parameters can be adjusted as necessary. This viable solution is highly time-dependent, as an improved quality can be obtained when cutting with high pressure and slower traverse speeds, depending on the material [13,14]. It is essential to implement fast calculation methods that can offer optimum process parameters, starting from the part's dimensional accuracy. A statistical approach was used to obtain the necessary mathematical dependencies between the process parameters (cutting pressure, stand-off distance, and quality of the cut) and the accuracy of the cut (responses), analysing the entrance and exit width of cut and the kerf angle, as shown in figure 1.

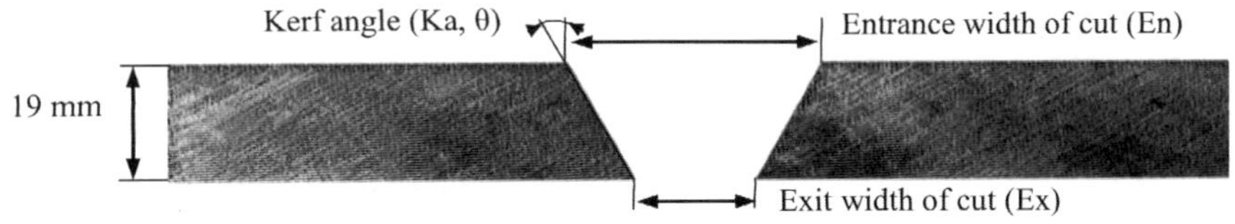

Figure 1. WJC process analysed parameters and material thickness.

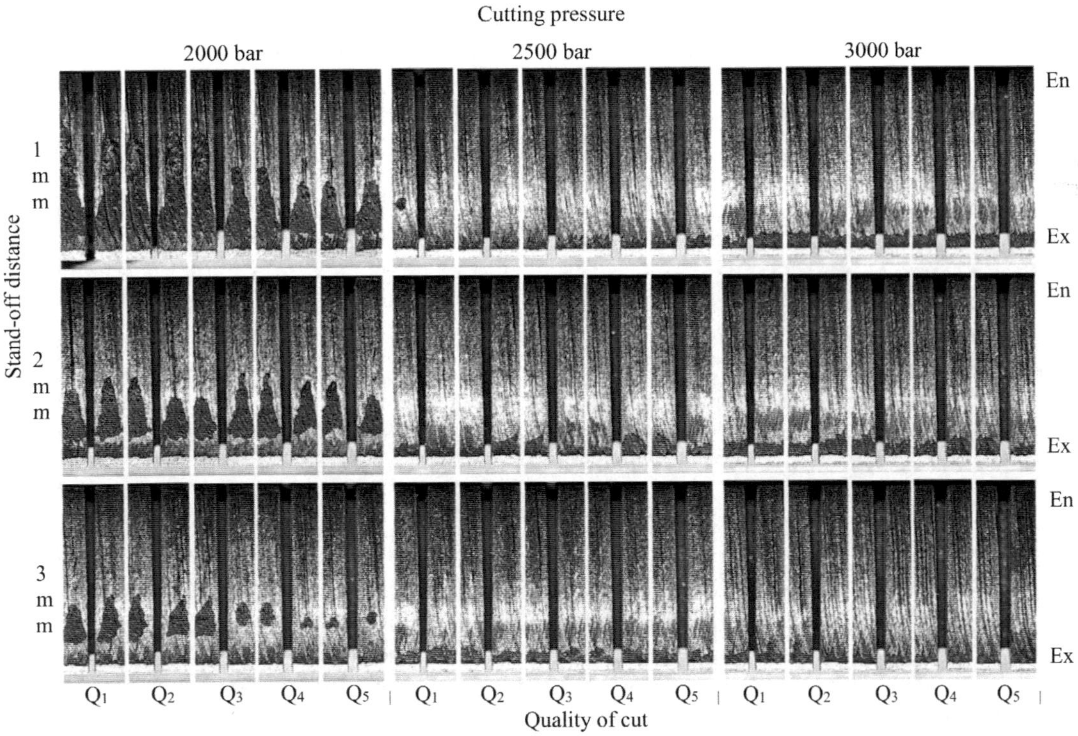

Figure 2. WJC input parameters as a process matrix and S235JR cut samples, highlighting the entrance (E_n) and exit areas (E_x).

A 19 mm thick S235JR steel alloy plate was used for the study with a total of 45 cuts executed, consistent with the Response Surface Method design of experiments (DOE) plan given by the Design-Expert software.

The design type used in this case was a central composite with a reduced cubic model. The Analysis of variance (ANOVA) for each design factor response was conducted using the automatic model selection, using a forward p-value criterion; thus, only the significant input factors and their interaction were taken into account; the Box-Cox power transformation indicated no recommended transformations. The factors taken into consideration were: the pressure (P) from 2000 up to 3000 [bar], the stand-off distance (S) of 1, 2, and 3 [mm], and all the five levels of cutting quality available presets (Q1-5). The samples' complete arrangement can be noted in figure 2, sorted as a process matrix, from where the comparison of the entrance and exit width of cut can be noted. A Mitutoyo PH-A14 profile projector was used to measure these distances with high precision. The data was inserted into Microsoft Excel, where using equation (1), the kerf angle was calculated.

$$\Theta = \tan^{-1}\left(\frac{E_{n\,width} - E_{x\,width}}{2t}\right) \tag{1}$$

3. Statistical interpretation results

The interdependency of the process parameters gives the complexity of the WJC. These parameters tend to deviate during the manufacturing process; therefore, the final part may result out of tolerances. Thus we consider that the significance of the data represents an important step in any analysis. A procedure that can be easily implemented using an ANOVA analysis determines if the process parameters impact the result and to what extent. This method has as main outputs the mean square (the sum of squares divided by the degrees of freedom), F-value (test for comparing the source's mean square to the residual mean square), p-value (the probability of getting a result at least as extreme as the one that was observed), coefficient of determination (R2, measures the amount of variation around the mean explained by the model) with its particular forms (adjusted-R2, predicted- R2, adequate precision) and the mathematical models. Among these parameters, the most important for this experimental study is the p-value, which indicates if we can reject a null hypothesis, if its value is smaller than 0.05, and the R2 value that has to be greater than 0.85 so that one can be confident that a prediction of this model offers accurate results.

3.1. ANOVA results and mathematical models

The ANOVA analysis results for the p-value are highlighted in table 1; the data is divided into groups based on the responses that indicate that each model (entrance, exit width of cut, and kerf angle) is significant if the p-value is lower than 0.0001. The analysis considered the process parameters and their interaction concerning the responses; therefore, the p-value was calculated for each one. It can be noted that the cutting pressure (P) has less influence on the entrance width of the cut compared to the exit width and kerf angle. This result is predictable, as the minimum pressure of 2000 bar is more than sufficient for cutting steel; the high pressure being instead of a way of cutting thick plates, as it can be observed the exit width of cut, where this parameter is highly significant, therefore for the kerf angle. Taking into consideration that the conical shape of the water jet changes with the distance, the analysis finds that the stand-off distance (S) model predicts that the entrance width of cut and kerf angle is affected, while the exit width is significant when in interaction with the pressure, as the p-value indicated being 0.0063. The quality of the cut (Q) for each response is very significant, with values below 0.0001, data that indicates a strong correlation between the process parameters and the dimensional accuracy of the final part, in terms of the entrance, exit width of cut, and kerf angle.

Table 1. WJC ANOVA results in terms of the p-value for each response.

	Source	Model	P	S	Q	PS	PQ	SQ	P^2	S^2	PSQ	PS^2	P^2S
p-value	E_n	***<0.0001***	0.3789	*<0.0001*	*<0.0001*	0.2832	0.5929	0.0536	-	0.0105	0.0147	0.0062	-
	E_x	***<0.0001***	*<0.0001*	0.7668	*<0.0001*	0.0063	0.0916	-	0.16	-	-	0.0312	-
	K_a	***<0.0001***	*<0.0001*	*<0.0001*	*<0.0001*	0.0872	-	-	-	-	-	-	-

As with any statistical data interpretation, the variance analysis offers predictions, as the effects of the process parameters on the responses have been determined. The confidence in the predictions is given by the coefficient of determination and its particular form. In this experimental study, the result, highlighted in table 2, indicate, with a high degree of trust, that the mathematical connections are accurate and can consequently describe the process with precision, as it can be noted from the R2 values of each response, that is between 0.9535 and 0.9868. For every response, the Adjuster R2 is in reasonable agreement with the Predicted R2, with a difference smaller than 0.2, while the adequate precision that measures the signal to noise ratio indicates that, if a ratio greater than four is obtained, the results offer an adequate signal; the overall status given by the Design-Expert software is that the model can be used to navigate the design space.

Table 2. ANOVA results in terms of R^2, Adj. R^2, Pred. R^2 and Adeq. Precision of each response.

Source	$\mathbf{R^2}$	**Adjusted** $\mathbf{R^2}$	**Predicted** $\mathbf{R^2}$	**Adeq. Precision**
E_n	0.9816	0.9648	0.8985	27.8712
E_x	0.9868	0.9813	0.9710	45.0139
K_a	0.9535	0.9447	0.9329	35.2979

Takin into consideration that the measured values of the entrance width of cut are between 0.92-1.25 [mm], the ones for the exit width are from 0.62 to 1.08 [mm], and the kerf angle is in the range of 0.106 to 0.588 [°], the mathematical models for the responses are highlighted from equation (2) to equation (8). The models are specific for each quality level as this factor was declared as a categoric discrete type. The mathematical models describe the relationship between the entrance width of the cut concerning the process parameters equations (2), (3), (4), (5), and (6). It can be noted that the stand-off distance (S) has a major impact on the result, as a single parameter and in interaction with the cutting pressure, at any quality level. The exit width of the cut mathematical model, presented in equation (5), highlights that changes in this response for each quality level, are influenced by coefficients $c_{1\text{-}5}$ and $k_{1\text{-}5}$ ($c_1 = 0.042$, $c_2 = -0.119$, $c_3 = 0.045$, $c_4 = 0.112$, $c_5 = 0.254$, $k_1 = 4.91$, $k_2 = 5.48$, $k_3 = 5.31$, $k_4 = 5.35$, $k_5 = 5.11$). This response can be obtained at a required value as minor changes have to be applied. This is available also for the kerf angle as the mathematical model from equation (8) indicates that it's values has minor variations with coefficient $e_{1\text{-}5}$ ($e_1 = 0.538$, $e_2 = 0.498$, $e_3 = 0.383$, $e_4 = 0.311$, $e_5 = 0.227$).

$$En_{width(Q1)} = 1.042 - 5.3*10^{-5}P - 0.267S + 1.12*10^{-4}PS + 0.109S^2 - 3.8*10^{-5}PS^2 \quad (2)$$

$$En_{width(Q2)} = 1.008 - 3.0*10^{-5}P - 0.248S + 1.07*10^{-4}PS + 0.109S^2 - 3.8*10^{-5}PS^2 \quad (3)$$

$$En_{width(Q3)} = 1.483 - 2.0*10^{-4}P - 0.248S + 1.77*10^{-4}PS + 0.109S^2 - 3.8*10^{-5}PS^2 \quad (4)$$

$$En_{width(Q4)} = 1.343 - 1.47*10^{-4}P - 0.370S + 1.62*10^{-4}PS + 0.109S^2 - 3.8*10^{-5}PS^2 \quad (5)$$

$$En_{width(Q5)} = 1.373 - 1.5*10^{-4}P - 0.363S + 1.62*10^{-4}PS + 0.109S^2 - 3.8*10^{-5}PS^2 \quad (6)$$

$$E_{x(Q_{1-5})} = c_{1-5} + k_{1-5}*10^{-4}P + 0.491S - 3.44*10^{-4}PS - 9.47*10^{-8}P^2 + 6.4*10^{-8}P^2S \quad (7)$$

$$K_{a(Q_{1-5})} = c_{1-5} - 5.9*10^{-5}P + 9.1*10^{-5}S + 2.4*10^{-5}PS \tag{8}$$

3.2. Models graph interpretation

The dependency of the dimensional accuracy of the responses concerning the process parameters is presented in figure 3. The values are displayed as average over the quality levels, using a blue to a red gradient, specific for small to high numerical values. As mathematical models resulting from the ANOVA analysis, the graphs can predict with high confidence, the responses presented are for pressures in the range of 1500 - 3500 [bar] and stand-off distances up to 4 [mm]. As the stand-off distance increases, the entrance width increases more than the exit width, leading to kerf angles greater than 0.5 [°]. The ideal results are obtained when using very high pressure with a stand-off distance smaller than 1 [mm]; this can be problematic to implement in practice as it is complicated to be obtained; thus, the indicated value for this parameter is around 2 [mm], to avoid any nozzle damage. Therefore, an entrance width of cut of 1.1 [mm] is in reasonable limits with an exit width of 0.85 [mm] when using cutting pressure of 2500 [bar] with a stand-off distance of 2 [mm]; this lead to an approximate kerf angle of 0.35 [°]. Estimations show that when cutting with a pressure of 3500 [bar], the improvement in dimensional accuracy is minimal.

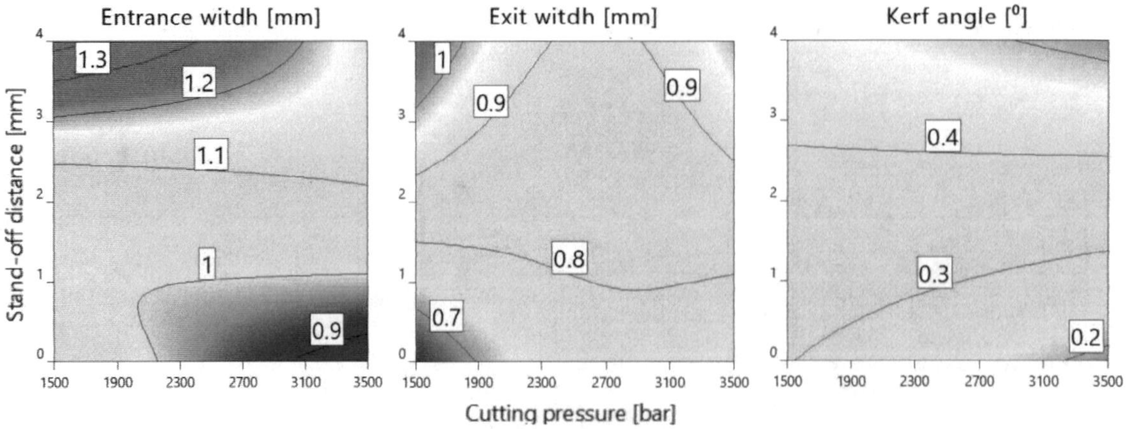

Figure 3. Entrance, exit width of cut and kerf angle values distribution function of the stand-off distance and cutting pressure, displayed as average over the quality levels.

4. Conclusions

Abrasive particles from the water jet gain kinetic energy at the exit from the nozzle but are dissipated as the traveling distance increases, leading to a drop in the quality cut, when comparing the entrance and exit width of cut, therefore a higher kerf angle that results in a higher deviation from perpendicularity. It was noted that with the increase of this distance, the kerf angle increase above 0.5 [°]. A smaller kerf angle is ideal but hard to implement; this can be accomplished using a height sensor to avoid any nozzle damage. Analysis of the experimental data indicated an optimum stand-off distance of 2 [mm].

The optimum cutting pressure is in the range of 2300 and 2700 [bar], considering that this parameter is best interpreted in interaction with the stand-off distance. Minor deviations are obtained when cutting with a 1500 [bar] pressure at 1 [mm] stand-off distance, thus improving dimensional accuracy.

The quality level has a significant impact on the results, as superior edge finish was observed at the exit cut when using Q5, compared to when using the Q1 quality that leaves indentations in the cut surface. The result's analysis indicated that a clean-cut could be obtained when using the Q3 level, as it also assures an optimal cutting time.

The stability of the WJC process in terms of part accuracy can be reduced to the kerf angle's value, as it shows the deviation of the entrance width from the exit width. Therefore, a minor kerf angle leads to minor deviations from perpendicularity, reducing the need for further machining. As indicated in this paper, the optimum process parameters are indicated in terms of part dimensional accuracy. The mathematical models ensure that the parameters can be calculated so that a minor kerf angle can be obtained, leading to the desired entrance and exit widths. Thus we consider that is an efficient way of deducing the process parameters for high part accuracy.

5. References

[1] Fourness, C., 1933 *Paper Metering, Cutting, and Reelin*, W. Machine, Editor, USA.
[2] Johnson, C.O., 1956 *Method for Cutting Up Plastic and Semi-Plastic Masses*. Luxembourg.
[3] Schwacha, B.G., 1958 *Liquid Cutting of Hard Metals*.
[4] Y. Natarajan, P. K. Murugesan, M. Mohan, S. A. Liyakath and A. Khan 2020 Abrasive Water Jet Machining process: A state of art of review *Journal of Manufacturing Processes* **49** 271-322 ISSN 1526-6125 https://doi.org/10.1016/j.jmapro.2019.11.030.
[5] Krajcarz, D. 2014 Comparison Metal Water Jet Cutting with Laser and Plasma Cutting *Procedia Engineering* **69** 838-843.
[6] H.T. Liu 2017 Precision machining of advanced materials with waterjets *IOP Conf. Series: Materials Science and Engineering* **164** 012008.
[7] Cha, Y., T.-M. Oh, and G.-C. Cho 2019 Waterjet Erosion Model for Rock-Like Material Considering Properties of Abrasive and Target Materials *Applied Sciences* **9**(20): p. 4234.
[8] Gauert, C.D.K., et al., 2013 A progress report on ultra-high-pressure waterjet cutting underground: the future of narrow reef gold and PGE mining *Journal of the Southern African Institute of Mining and Metallurgy* **113** p. 441–448.
[9] Ozcelik, Y., et al. 2011 Optimization of working parameters of water jet cutting in terms of depth and width of cut *Proceedings of the Institution of Mechanical Engineers, Part E: Journal of Process Mechanical Engineering* **226**(1) p. 64-78.
[10] Löschner, P., K. Jarosz, and P. Niesłony 2016 Investigation of the Effect of Cutting Speed on Surface Quality in Abrasive Water Jet Cutting of 316L Stainless Steel. *Procedia Engineering* **149** p. 276-282.
[11] Zhang, S., Wu, Y. & Wang, S. 2015 An exploration of an abrasive water jet cutting front profile. *Int J Adv Manuf Technol* **80** 1685–1688 https://doi.org/10.1007/s00170-015-7154-7.
[12] J. Zeng, A. Henning, 2009 *Kerf characterization in abrasive waterjet cutting,* American WJTA Conference and Expo.
[13] Miao, X., et al. 2018 The optimal cutting times of multipass abrasive water jet cutting *The International Journal of Advanced Manufacturing Technology* **97**(5): p. 1779-1786.
[14] Yuvaraj, N. and M.P. Kumar, 2017 Surface integrity studies on abrasive water jet cutting of AISI D2 steel *Materials and Manufacturing Processes* **32**(2) p. 162-170.

Acknowledgements

This work was supported by a grant from the Romanian Ministry of Research and Innovation, CCCDI – UEFISCDI, project number PN-III-P1-1.2-PCCDI-20170446 / 82PCCDI / 2018, within PNCDI III.

Analysis of defects arising during exploitating of elements of rolling bearings

A Gwiazda[1], S Topolska[2] and W Rakowska[3]
[1] Silesian University of Technology, Faculty of Mechanical Engineering, Institute of Engineering Processes Automation and Integrated Manufacturing Systems
Konarskiego 18A, 44-100 Gliwice, Poland
[2] Silesian University of Technology, Faculty of Mechanical Engineering, Institute of Engineering Materials and Biomaterials
Konarskiego 18A, 44-100 Gliwice, Poland
[3] Łukasiewicz – Welding Institute, Department of Non-Destructive Testing
Bł. Czeslawa 16-18, 44-100 Gliwice, Poland

E-mail: aleksander.gwiazda@polsl.pl

Abstract. Bearings are damaged most often as a result of operation under improper lubrication conditions (no correlation between lubricant parameters and operating conditions). this often leads to the operating temperature of the grease being exceeded. Such operating conditions are either the result of the bearing operating under overload conditions (too high load causing the dropping point to be exceeded, leading to lubricant leakage and deterioration of the bearing operating conditions), or the result of improper lubrication of the bearings (application of too little grease and lack of re-lubrication of the bearing or the application of improper grease). The article presents an exemplar analysis of the influence of the exploitation process on their slow degradation.

1. Introduction

Designing heavy-duty engineering systems has always been problematic. One of the elements that were a variable of such projects is the time of failure-free operation and the frequency of servicing. In the case of engines exposed to severe operating conditions, one of the most important structural nodes, in terms of reliability, are the bearings assemblies. Rolling element bearings are the most important part of the bearings assemblies' components to be found in industrial rotating equipment. They are found in machinery used in different industries. Particularly important is the selection the rolling element bearings in the case of heavy duty machinery systems [1-3].

A bearing itself is a mechanical device designed to reduce friction. It is a part of a machine allowing cooperating elements turns or slides. There are generally two types of bearings: plain (journal) ones and rolling ones. They may come in many sub-varieties. Because of friction reduction they are the most important components of many technical systems. Hence it is very important to have the factors like: carrying capacity and reliability, at high level of sureness. Bearings used in heavy duty machinery must be able to support: the outer loads imposed on the shaft, the momentum generated by the engine, and the loads of the driven machinery. The other task of bearings is to keep the axial movement as well as the lateral deflection of the shaft within acceptable ranges. It helps to maximize the durability of the whole technical system. One must say that bearings operation acts, in a significant way, on the operation and efficiency of the whole system. On the other hand, the signals in the form of bearing faults main indicator of decreasing performance of the system being monitored within the life-cycle management system. It can be detected on the basis of increasing the bearings vibration [4-6].

Rolling bearing operation is affected by a number of different factors. To them one can include: friction, wear and lubrication mechanisms, fluid dynamics and lubricant rheology, material properties, and contact mechanics. Hence the changes in rolling bearings states occur because of three factors: bearing elements plastic deformation (static), bearing elements contact wear (dynamic), and bearing elements interaction (fatigue). The most common is the contact wear. In this process the particles are torn from the base material and mixed into the lubricant. This converts the lubricant into an abrasive and accelerates the wear of the bearing. The level of wear could be monitored by analyzing the levels of vibrations being the result of surface degradation. Rolling contact wear and occurring at a later stage of bearing operation the rolling contact fatigue can be diagnosed by combining measured monitoring data. It also allows drawing conclusions that could support the day-to-day operation of the bearings to extend their life. One of the basic problems related to the construction of bearing nodes is just their durability. If bearings nodes are designed and used correctly, rolling bearings will last for their intended life. Bearings, however, often fail prematurely due to avoidable errors. Reasons for such premature failure include poor bearing assembly design, poor assembly, or poor bearing performance. This results in insufficient lubrication, the ingress of foreign bodies or excessive heating. The characteristics of a defective wear process in bearings can be defined by these wear symptoms [7-8].

2. Analysis of motor bearings operation conditions. Case study

Induction asynchronous motors (IAM) have a basic role in many industrial branches. They create a very sophisticated system that contains of three main elements: stator, rotor and bearings. Though any working IAM is subjected to faults related with these components. Between the possible faults of bearings one should say that nearly 40% of faults is detected in systems with IAM [9]. In the systems with IAM, bearings are subjected to large stress caused by switching the converter. This is why bearings are more frequents subjected to degradation and to lowering their efficiency. Wear, inn such a class of systems, is one of the common causes of faults in bearings. It could be improved by proper control system [10-11].

A case study was carried out because there was a failure of the technical system (heavy duty presses), individually designed press with very high pressure. It turned out that the high-powered motor bearings had failed. In order to determine the cause of the failure, these bearings were tested. Two bearing sets were provided for the tests, as each bearing node was designed as a two-bearing one (figure 1 and 2). Each set of the bearing arrangement included a ball bearing with dimensions of Ø 80 mm / Ø 140 mm SLF 6216.C3.J20A1 142N and a cylindrical roller bearing with dimensions of Ø 80 mm / Ø 140 mm NSK NU2216E JAPAN Z as well as a pressure disk.

Figure 1. First set of bearings and a pressure disk.

Figure 2. Second set of bearings and a pressure disk.

The individuality of the project consisted in replacing the hydraulic drive with an electromechanical drive in which a high-power motor was applied. However, after one year of operation, in accordance with the documentation, the system failed. One must say the bearings nodes, in this system was designed as maintenance-free. Therefore, the bearing failure was surprising. Hence the research devoted to explaining this situation.

Figure 3 shows traces of charred grease observed after disassembling the bearings. As the permissible operating temperature of the grease was 180°C, it means that this temperature was exceeded in the bearing nodes.

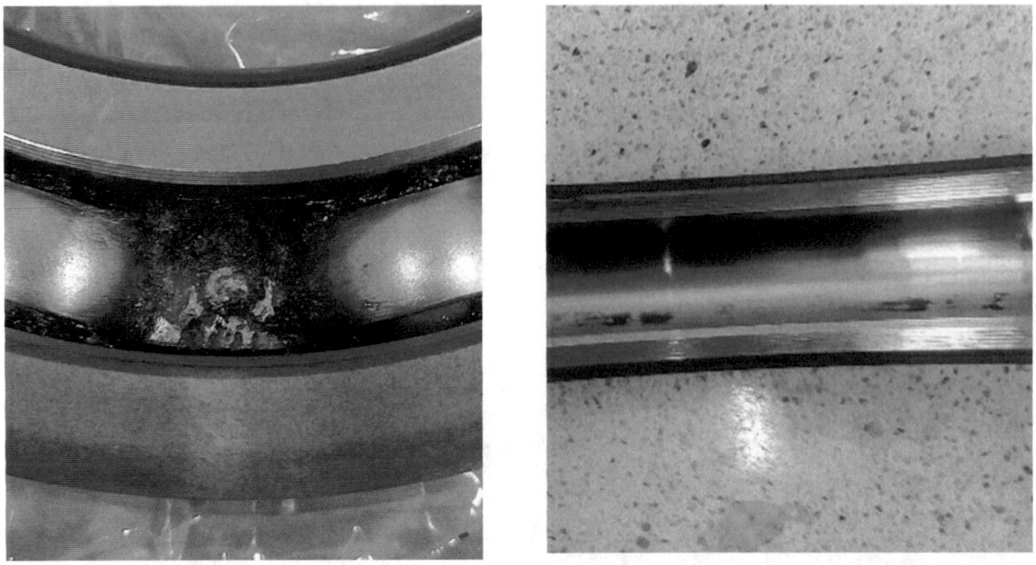

Figure 3. Carbonized grease on the bearing cage and outer race (ball bearing).

In order to determine other forms of destruction of bearing elements, they were inspected. As a result, discoloration of the working surfaces of the bearings was found under normal conditions (figure 4, two top photos.). It should be noted that discoloration of the cage, rolling elements and raceways of the rings occurs due to the reaction with the lubricant and high temperature. Its causes are: insufficient lubrication, oil stains due to reactions with the lubricant, or high temperature. In addition, on both races of the roller bearings, transverse stripes are visible, corresponding to damage such as false Brinell imprinting's of low intensity (figure 4, two bottom photos). These imprintings appear as recessed areas due to vibration and sway wear at the points of contact between the raceway and the rolling elements. This is one type of fretting corrosion [12-14]. The causes are: oscillation and vibration of a non-operating bearing at times such as transport, low-amplitude oscillating motion, or insufficient lubrication.

Figure 4. Discoloration of the working surfaces of bearing elements and false Brinell impressions.

Microscopic examination of the working surfaces of the bearings was also carried out (figure 5). The green color was chosen as it best reflects the details of the surface.

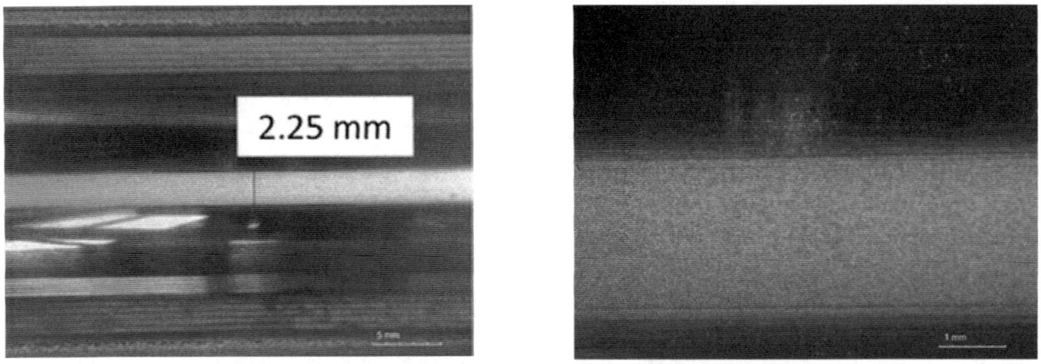

Figure 5. Microscopic photo of the bearing streak abrasion.

Analyzing the photos taken at various magnification levels, the following defects of the tested bearings are listed; smear rubbing, scratches and the material applied as a result of charring of the grease (figure 6). All of the detected defects in the working surfaces of the bearings are the result of poor lubrication of these surfaces. Additionally, the deposited carbonization products of the lubricant indicate that the faulty lubrication was the result of exceeding its operating temperature. It is worth noting that in one photo you can see the bright blue color of iron oxide bloom [15]. This means that in this place the temperature reached values of 310-315 °C. As the limit temperature of the grease was 180 °C, it should be assumed that the local exceedances of this temperature were even 135 °C.

Taking into account the identified damage to the working surfaces of the bearings, it turned out to be essential to find the cause of this state of affairs, as the metallographic tests themselves did not indicate the primary cause, only confirming that the limit temperature of the grease was exceeded.

bearing streak abrasion and scratch (mag. 28x)

bearing streak abrasion and blue deposition (mag. 28x)

scratches and deposited material (mag. 28x)

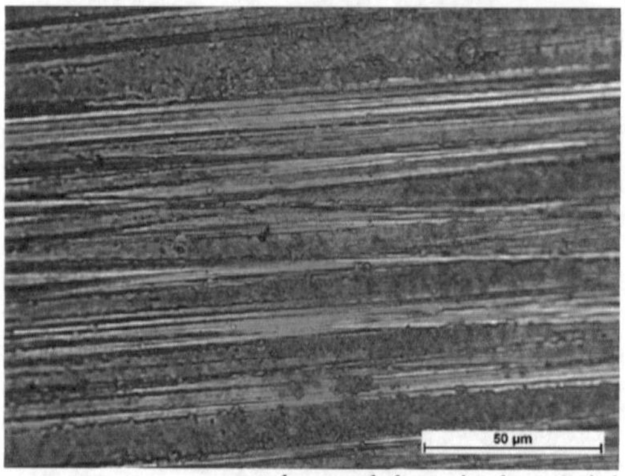

scratches and deposited material (mag 500x)

Figure 6. Microscopic photos of the bearing damages.

A qualitative analysis in the form of a weighted Ishikawa diagram (figure 7) was used to analyze the cause of failure of the tested technical system operating in difficult conditions [16-17]. Classical 6M + E graph was drawn up. During the analysis, the influence of six internal factors and the environment were taken into account. The first internal factor was men. In this regard, it was found that the operators had no knowledge of the system in use. They have been advised that it is maintenance free. In terms of material, it was found that the design of the bearing nodes was based on standard bearings closed in non-opening bushings. In terms of the factor machine it was found to be an innovative solution with elements that had never been verified in operation. Additionally, the factor determining the system operation was its performance. In terms of the method, it was found that the system was not monitored at all, due to the information about its unattended operation. In addition, the motor shaft bearings were subjected to a periodic load (continuous cycles of starting and stopping the motor with a power of 200 kW lasting for a few minutes alternately). In terms of measurement, it turned out that the safety system temperature sensors were mounted on the stator of the motor and not near the bearing nodes. The number of sensors was 2, which was too few to monitor the operation of such a large engine. Finally, in terms of management (of system maintenance), it was found that the maintenance department also relied on a record of maintenance-free system operation. The last element was the environment. It turned out that the ambient temperature (in the production hall) was about 50 °C or even a little more, which already caused the system to operate at an elevated temperature. Additionally, there was a lot of dust.

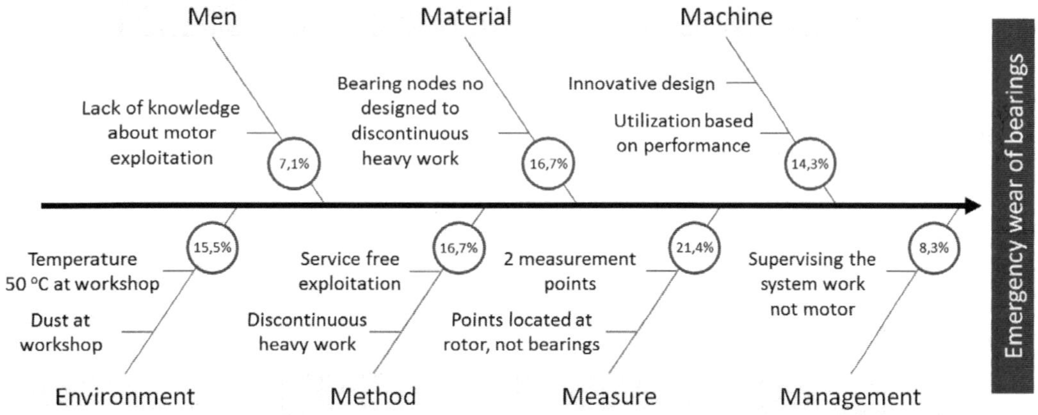

Figure 7. Weighted Ishikawa diagram of the analyzed problem.

The identified causes were subject to quantitative analysis in order to determine the structure of their impact. As shown in figure 7, the main cause of the failure should be seen in an improperly designed motor safety subsystem, which only monitored the problem of stator overheating. However, the monitoring of bearing overheating was not included. In addition, the selection of bearings itself and the failure to take into account the operating conditions in the design and the impact of the ambient temperature on the reduction of the safety margin in the scope of the bearing operating temperature were found to be incorrect.

3. Conclusions

The presented results of research on the cause of bearing failure in a 200 kW electric motor have shown that even experienced design teams may not analyze all possible aspects of the operation of complex technical systems, especially in the case of unit designs. In the described case, high-power electromechanical drives replaced hydraulic cylinders for driving heavy industrial presses. However, the specificity of the operation of these devices contributed to a significant deterioration of the operating conditions of electric motors, including their bearing nodes. Additionally, the project did not take into account the influence of the temperature in the production hall in the summer, which exceeded 50 °C (sometimes reaching almost 60 °C - own measurements). Dust also deteriorated the operating conditions of the system. However, it did not directly affect the operating conditions of the bearings, as they were in non-dismountable bushings. Generally, therefore, the general causes of failure should be looked for in poorly prepared design assumptions of the analyzed system. This resulted in the adoption of design assumptions with a much lower load on the system with the forces generated by the system operation process. This resulted in an excessive minimization of the size of individual structural nodes and an erroneous assumption about the possible, maintenance-free operation of the bearing nodes.

However, wrong design of the work monitoring system should be considered a special cause of the system failure. It was developed on the basis of two temperature sensors which, however, measured the operating temperature of the stator and rotor of the motor, and not the operating temperature of the bearings. This is strange because in the standard designs of such electric motors, the number of temperature sensors is at least 4, and two of them are located near the bearing nodes [18-19].

The presented analysis shows how important it is to study the functionality of various components in different environments, to create a methodology for designing such systems and build a knowledge base in this area [20-22].

4. References

[1] Saruhan H, Sandemir S, Çiçek A and Uygur I 2014 *J. App. Res. Tech.* **12(3)** 384-395 (doi: 10.1016/S1665-6423(14)71620-7).

[2] Kumaran S S, Velmurugan P and Tilahun S 2020 *J. Crit. Rev.* **7(9)** 492-501 (doi: 10.31838/jcr.07.09.99).

[3] Muzakkir S M and Hirani H 2015 *Int. J. Eng. Res.* **4(3)** 133-136 (doi: 10.17950/ijer/v4s3/311)

[4] Reddya M R and Srinivas J 2016 *Procedia Eng.* **144** 825-832 (doi: 10.1016/j.proeng.2016.05.093).

[5] Abu-Zeid M A. and Abdel-Rahman S. M. 2013 *Alex. Eng. J.* **52** 241–248 (doi: 10.1016/j.aej.2013.02.002).

[6] Patel V N, Tandon N and Pandey R K 2014 Procedia Tech. 14 312-319 (doi: 10.1016/j.protcy.2014.08.041).

[7] Shrivastava A and Wadhwani S 2013 *Int. J. Sci. Res.* **2(5)** 256-259 (doi: 10,36106/ijsr).

[8] Kim Y-H, Tan A C C, Mathew J and Yang B-S 182-191 in: Mathew J, Kennedy J, Ma L, Tan A and Anderson D. (eds) 2006 *Engineering Asset Management*. (London: Springer) (doi: 10.1007/978-1-84628-814-2_21).

[9] Önel I and Benbouzid M 2008 *IEEE ASME Trans. Mechatron.* **13(2)** 257-262 (doi: 10.1109/TMECH.2008.918535).

[10] Lysek K, Gwiazda A and Herbuś K 2019 *IOP Conf. Ser.: Mat. Sci. Eng.* **591** 1-6 (doi: 10.1088/1757-899X/591/1/012055).

[11] Nalepa B and Gwiazda A 2020 *IOP Conf. Ser.: Mat. Sci. Eng.* **916** 1-6 (doi: 10.1088/1757-899X/916/1/012072).

[12] Topolska S and Łabanowski J 2017 *Arch. Metall. Mater.* **62(4)** 2107 (doi: 10.1515/amm-2017-0312).

[13] Topolska S and Łabanowski J 2017 *IOP Conf. Ser.: Mat. Sci. Eng.* **227** 1-6 (doi: 10.1088/1757-899X/227/1/ 012131).

[14] Adamiak M, Czupryński A, Kopyść A, Monica Z, Olender M and Gwiazda A 2018 *Metals* **8(2)** 1-7 (doi: 10.3390/met8020142).

[15] Djilali K-A and Hebbar A 2020 *Sciendo* **22(4)** 1247-1260 (doi: 0.2478/mme-2018-0096).

[16] Malinowska E 2010 Equilibrium. Q. J. Econ. Econ. Pol. 5(2) 165–177 (doi: 10.12775/EQUIL.2010.033).

[17] Fabiś-Domagała J 2017 *Sciendo* **114(8)** 193 – 198 (doi: 10.4467/2353737XCT.17.141.6892.

[18] Hryniewicz P, Banaś W, Gwiazda A, Foit K, Sękala A and Kost G 2015 *IOP Conf. Ser.: Mat. Sci. Eng.* **95** 1-8 (doi: 10.1088/1757-899X/95/1/012086).

[19] Hryniewicz P, Banaś W, Foit K, Gwiazda A and Sękala A 2017 *IOP Conf. Ser.: Mat. Sci. Eng.* **227** 1-7 (doi: 10.1088/1757-899X/227/1/012061).

[20] Gwiazda A, Sękala A, Banaś W, Topolska S, Foit K and Monica Z 2017 *IOP Conf. Ser.: Mat. Sci. Eng.* **227** 1-6 (doi: 10.1088/1757-899X/227/1/012054).
[21] Michalec M, Svoboda P, Křupka I and Hartl M 2021 Eng. Sci. Technol., Int. J. 2 1-23 (doi: 10.1016/j.jestch.2021.01.010).
[22] Wasilczuk M and Wasilczuk F 2020 Eng. Fail. Anal. 115 1-10 (doi: 10.1016/j.engfailanal.2020.104651).

IOP Conf. Series: Materials Science and Engineering 1182 (2021) 012029 doi:10.1088/1757-899X/1182/1/012029

Finite element modelling of a plan polymeric composite product under impact and thermal conditions

S M Ion, D Nitoi, O Chivu and C G Opran
University Politehnica of Bucharest, Splaiul independentei no. 313, sect.6, Bucharest, Romania

E-mail: dan.nitoi@upb.ro

Abstract. Polymeric composites products are undeniable trends today, especially in the most innovative fields and with a permanent degree of novelty. The specialized literature presents a very wide range of their applications in the avant-garde fields such as the aerospace field and that of the automotive industry. Paper presents the results of research on modelling and impact simulation of these types of products besides a transient temperature analyse using the finite element method. Starting from the experimental part through which a flat composite product is made, the authors research its behaviour on impact and at the same time try to make a mathematical model by the finite element method. This comes to complete the experimental research by providing information on the mechanical stress that occur during experimental testing. The properties of the materials both those used in the realization of the experimental product and in the simulation are those offered by the manufacturers. The input data used in the finite element simulation tries to provide an image as close as possible to the probable behaviour of the product in case of car crash. Considering obtained results, large-scale industrial application is envisaged for automotive polymeric composite parts leading to the increase of the safety of the traffic participants in optimal vehicle manufacturing conditions.

1. Introduction

The manufacturing methods for fibre reinforced materials are largely dependent on the matrix. Such materials are generally heated, melted and injected or pressed into a mould. The final geometry of the component is therefore determined by the mould. Fibre reinforced materials are used in a variety of applications that require very low-weight materials with particularly good mechanical properties. This includes the energy sector, the aerospace industry, military and automotive industry. Reducing the weight of airplanes and cars/trucks offers the highest potential to lower fuel consumption and CO_2 emissions, thus forcing the manufacturers to explore materials alternatives with better specific mechanical properties [1].

This paper refers to the production, testing and modeling and simulation of the characteristics of a flat type composite product. The first part of the article presents the execution of the mould in which the production of the composite material will take place, presenting its main elements. In the following the composite product that will be used in the automotive industry is presented after which it will be tested similar to the various conditions that may be encountered automotive industry exploitation. The last part of the research presents the finite element analysis of the way in which the studied product responds to the applied conditions with answers close to the experimental results.

Considering polymeric composite product manufacturing, the Resin Transfer Molding process (RTM), is the most used technology to create large parts made by polymer matrix composites because:

the production time is relative small; the consumption of materials is reduced to minimum; the obtained parts are in a constant tolerance of ±0.5 mm after de-molding; both sides of the part are shiny and smooth; determinate resistance of the wall; it can be used with gelcoat on both sides or only on one; the reinforcement materials can be fiber glass, carbon fiber, sandwich materials, or metallic inserts; it does not need pressing during the curing of the resin; reduced cycle time, and low styrol emissions [3,4]. Polymeric composite constituents used to manufacture the plane composite structure used in experiments has the following properties: gelcoat is Enguard SD, grey 25620, produced by Ashland; finish fiber is Finishmat D77 0.3 mm thick, produced by Lantor; reinforcement fiber is Combiflow of 1150 g/m2 specific density and 2.6 mm thick, produced by Ahlstrom; polyester resin is Enydyne I 69277 A, HDT ISO 75-2 A, 92, produced by Ashland 0.3 mm thick. A plane-shaped research demonstration mould is designed, built and used to produce 4-mm thick fiber glass – polyester resin composite samples manufactured by vacuum-resin transfer molding (V-RTM) process (figure 1). A single layer of glass fiber fabric is placed between the outer and inner molds and then resin is transferred into the mold through nozzles shown in the picture. Working parameters are: temperature +25°C; vacuum pressure for mold closing -0.8 bar; vacuum pressure for resin flow -0.5 bar [2].

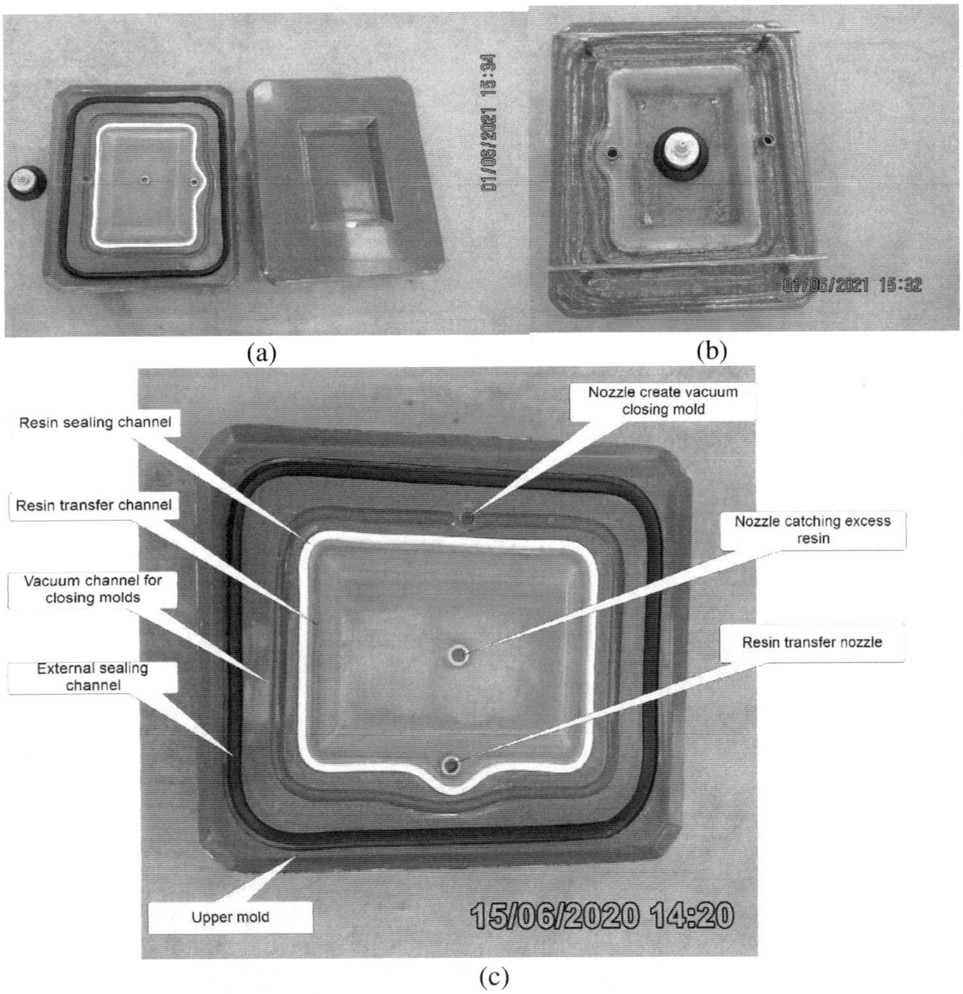

Figure 1. Composite product manufacturing: (a) mould elements; (b) closed mould with composite product; (c) mould geometry and structure.

A plane-shaped research demonstration mold is designed, built and used to produce 3.2 mm thick fiber glass – polyester resin composite samples manufactured by vacuum-resin transfer molding (V-RTM) process. A single layer of glass fiber fabric is placed between the outer and inner molds and then resin is transferred into the mold through nozzles shown in the picture. Working parameters are: temperature +25°C; vacuum pressure for mold closing -0.8 bar; vacuum pressure for resin flow -0.5 bar.

2. Impact testing of the plane composite product

The experimental tests performed on the flat composite product focused on the impact tests. These tests were performed on an Instron type machine. The impact tests were performed from several heights, namely H = 300, 400, 500, 600, 750, 800 mm. Figure 2 shows the force-energy-time graphs made by the software of the equipment used [3, 5-7].

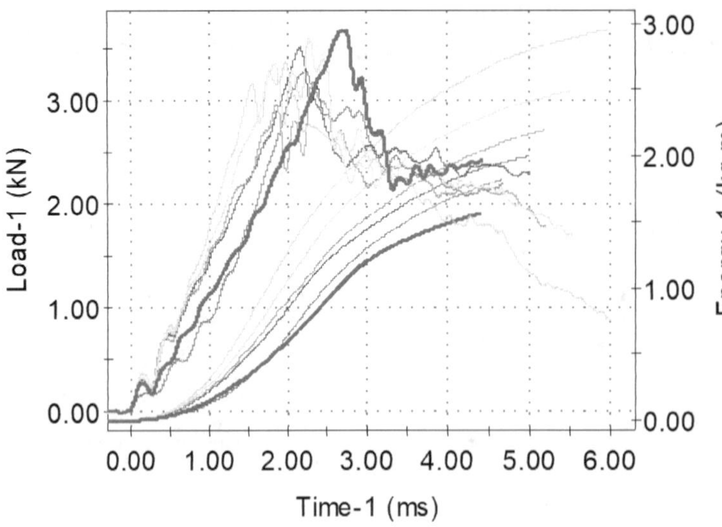

Figure 2. Testing graphs for the part in different testing conditions.

Figure 3(a) shows the fixing of the composite specimen on the table of the impact test machine. Following the shock test, the specimens were destroyed as can be seen in figure 4(b). For all working heights, figure 4(c) shows the specimens resulting from the tests.

(a)

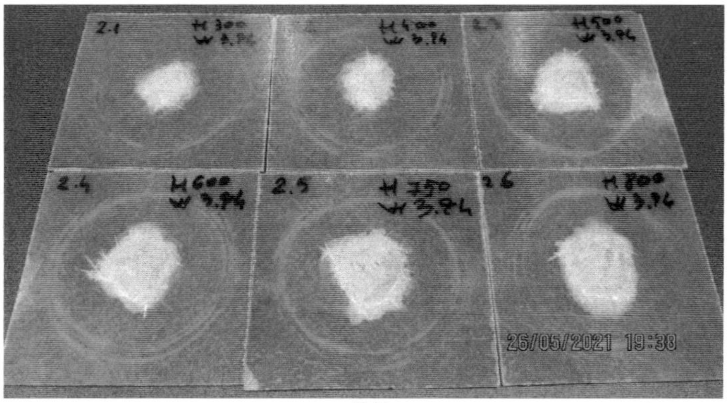

(b)

Figure 3. Testing the product: (a) composite part fixing; (b) testing of composite parts in different conditions H = 300 mm, H = 400 mm, H = 500 mm, H = 600 mm, H = 750 mm H = 800 mm.

According to the impact tests performed, in table 1 are presented the main results for six types of tests performed, namely for the heights h = 400, 500, 600, 700, 750, 800 mm. The mass of the punch is m = 3.94 kg.

Table 1. Impact test results.

Test no.	Specimen ID	Weight [Kg]	Drop Height [mm]	Peak Load [KN]	Deflection at Peak Load-1 [mm]	Energy to max load-1 [kg m]
1	1.1	3.9400	500.0000	3.2951	6.4428	0.8266
2	1.2	3.9400	600.0000	3.5116	6.1800	0.9590
3	1.3	3.9400	700.0000	3.1915	7.0515	1.1347
4	1.4	3.9400	800.0000	3.3776	7.2806	1.2493
5	1.5	3.9400	750.0000	3.6079	7.2573	1.2932
6	1.6	3.9400	400.0000	3.6864	6.7343	1.0541
Average		3.9400	625.0000	3.4450	6.8244	1.0862
Median		3.9400	650.0000	3.4446	6.8929	1.0944
Minimum		3.9400	400.0000	3.1915	6.1800	0.8266
Maximum		3.9400	800.0000	3.6864	7.2806	1.2932

3. Finite element modeling of the impact test and thermal behavior

The prediction of the behaviour in certain conditions of mechanical stress represents a very important stage in the design of any product used in the automotive industry, fact for which, in addition, the input data for a structural analysis are presented. Establishing the material properties to characterize as well as possible the behaviour of the composite product is the most important element in the proposed finite element analysis [4, 5, 8-10]. Depending on the properties of the materials provided by the manufacturers and presented previously, in table 2 are presented the material characteristics used as input data.

Table 2. Mechanical properties of the composite structure layers.

Material	Ex [GPa]	Ey [GPa]	Ez [GPa]	υ	Gx [GPa]	Gy [GPa]	Gz [GPa]	ρ [Kg/m³]
Finishmat 30%	62	62	15	0.22	2.1	3.2	2.1	913
Resin	4.5	4.5	4.5	0.35	3.2	3.2	3.3	1120
Combyflow 30%	45	45	11	0.22	4.5	2.2	4.5	1194

Using this material structure, a composite product with Lxl = 100 x100 mm was manufactured and modelled by FEM. This simulation includes the analysis of impact test behaviour in two situations. In the first the free fall height is h = 400 mm and in the second h = 800 mm. Figure 4(a) presents the test result in the first case (h = 400 mm) on both sides of the composite plate and in figure 4(b) the FEM analysis. From the FEM analysis of deformations on the OZ axis, with blue colour a maximum deformation UZ_t = 6.8 mm is observed. The experimental tests presented in table 1 gives for the same drop height the deflection on the peak load (line 6, table 1) the value UZ_{exp} = 6.73 mm, very close to theoretical value. Table 3 presents the output data for the impact test from drop height = 400 mm.

Table 3. Testing values for impact test from drop heigh = 400 mm.

Specimen ID	Weight [Kg]	Drop Height [mm]	Maximum load-1 [kN]	Time to max load-1 [ms]	Impact velocity-1 [m/s]	Impact energy-1 [kg m]	Total energy-1 [kg m]	Total time-1 [ms]
1.6	3.9400	400.00	3.6864	2.7112	2.7793	1.5517	1.5696	4.4263

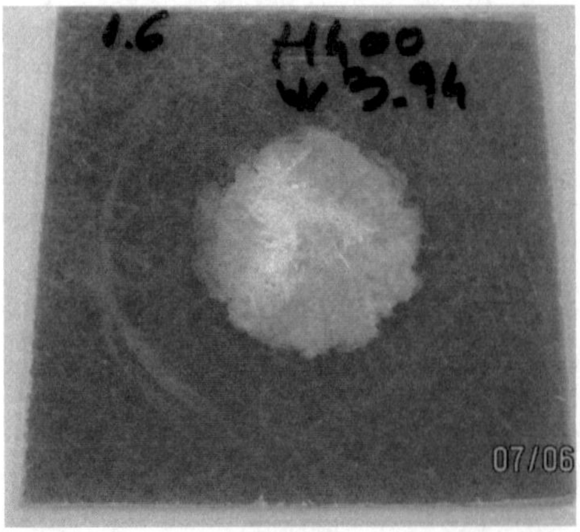

(a)

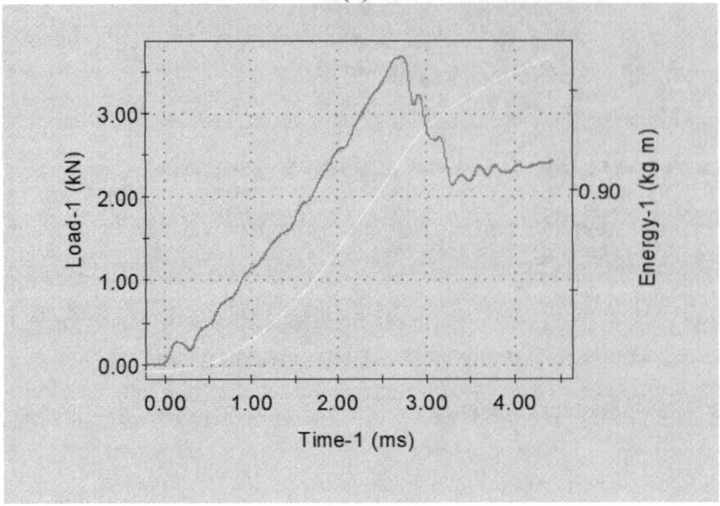

(b)

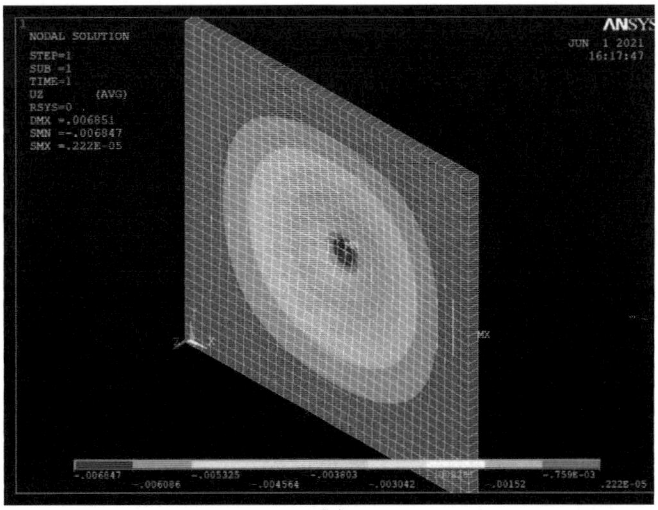

(c)

Figure 4. (a) impact test result for h = 400 mm; (b) Load – time – energy diagram for the 400 drop heigh; (c) FEM analysis Uz = 6.84 mm displacement.

The second analysis refers to the case of the test from the height h = 800 mm. Figure 6(a) shows the two sides of the composite plate in the case of testing and in figure 6b the resulting deformation in the case of FEM. The FEM calculates the displacement UZ_t = 7.83 mm. Testing machine measured (line 4, table 1) UZ_{exp} = 7.28 mm, very close to theoretical value.

Table 4. Testing values for impact test from drop heigh = 800 mm

Specimen ID	Weight [Kg]	Drop Height [mm]	Maximum load-1 [kN]	Time to max load-1 [ms]	Impact velocity-1 [m/s]	Impact energy-1 [kg m]	Total energy-1 [kg m]	Total time-1 [ms]
1.4	3.9400	800.0000	3.3776	1.9666	3.9371	3.1138	2.9503	6.0120

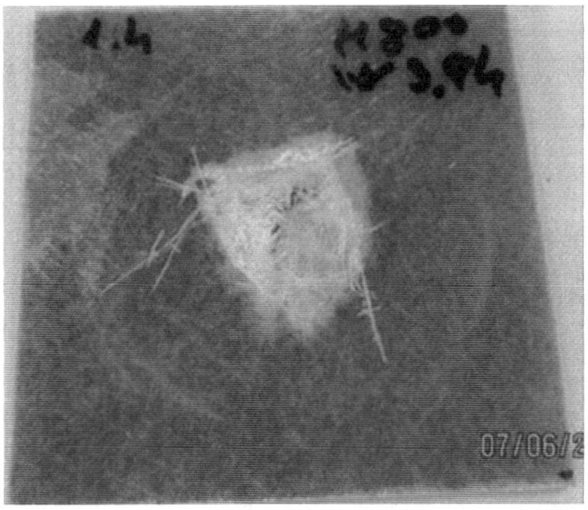

(a)

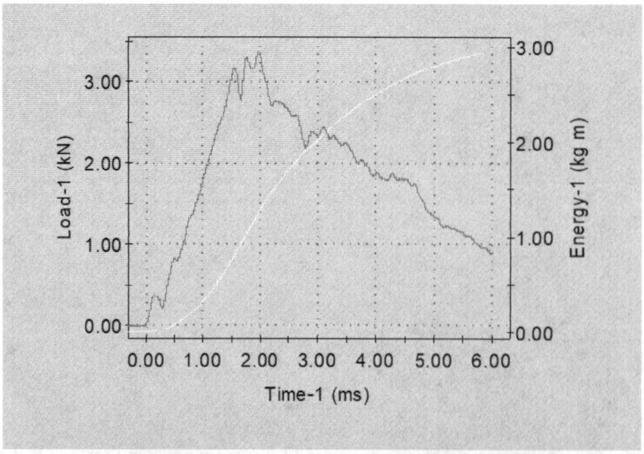

(b)

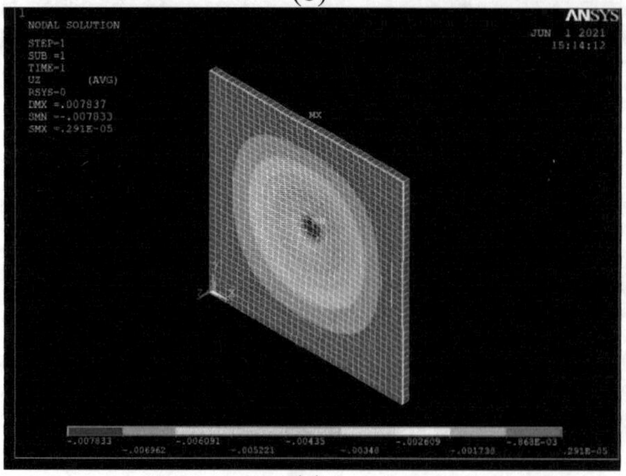

(c)

Figure 5. (a) impact test result for h = 800 mm; (b) Load – time – energy diagram for the 800 mm drop height; (c) FEM analysis Uz = 7.8 mm displacement.

The study of the stresses that appear in the composite structure offers a very useful image in the analysis of the appearance of cracks or destruction of the studied piece. Without the use of tensometric gauges the FEM offers results very close to reality and an extremely useful overview of the map of the different types of stresses used in the strength of materials. Figure 6 shows the analysis of the state of stresses of type S2 (the one that represents in a balanced way the state of tensile and compression stress in the material). From the analysis of the two images can be seen how in the first test case (h=400 mm) the maximum value of the stresses is S2 = 0.76 E9 N/m^2 (figure 6(a)). For the second analysis (h = 800 mm) the value maximum stresses is S2 = 0.87 E9 N/m (figure 6(b)). Both values of the stresses are located right in the zone of the values of the elasticity modules of the materials that constitute the composite structure. The fact that both parts were destroyed can be explained by these high values of stresses and by the possible occurrence of defects or non-uniformities in the material [11].

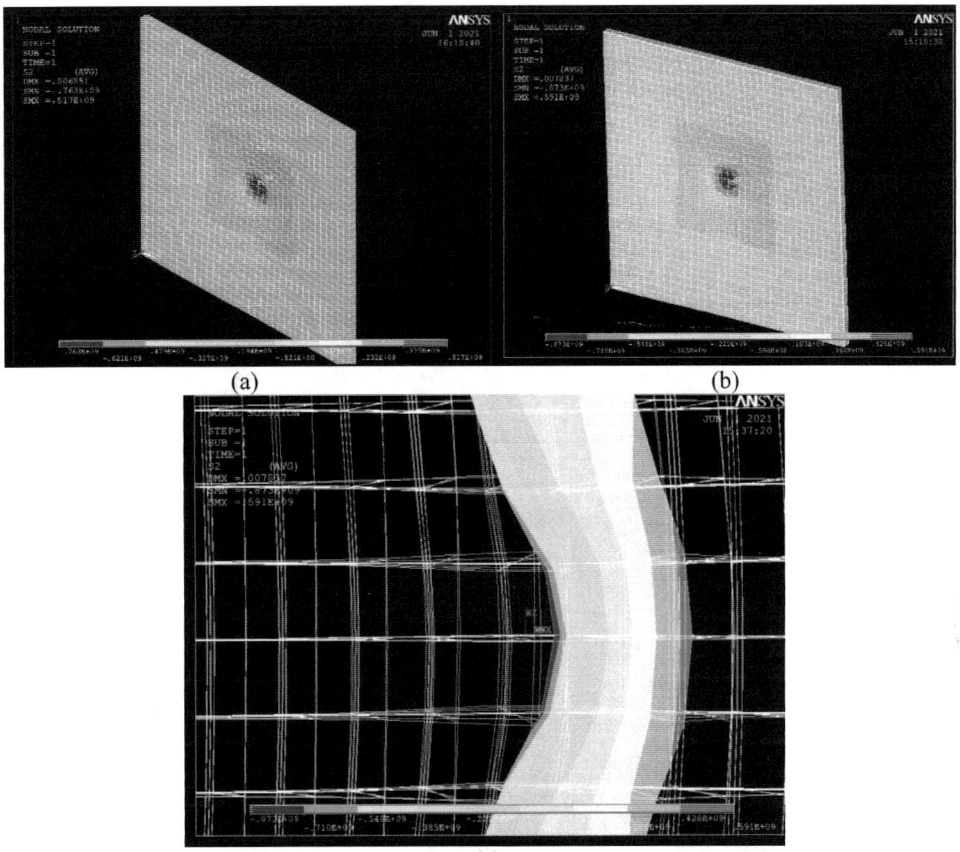

(c)

Figure 6. S2 stress analysis: a – $S2_{max}$ = 0.76 E9 N/m^2 for h = 400 mm testing case; b – $S2_{max}$ = 0.87 E9 N/m^2 for h = 800 mm testing case; c – S2 stress section.

Figure 8 shows the calculation of the shear stresses in the XOY plane. As can be seen, in this plan, there is an alternation of tension and compression stresses inside the structure which may explain its destruction during the tests. In the first case of test the maximum value of the stresses is Sxy = 0.8 E6 N/m^2 and for the second case Sxy = 0.1 E7 N/m^2 (figure 7). Even if the stress values are not high, their arrangement and alternation can cause the material to crack and break because the position and adhesion of glass fiber. Stress distribution is the same on both situations, but the values are different.

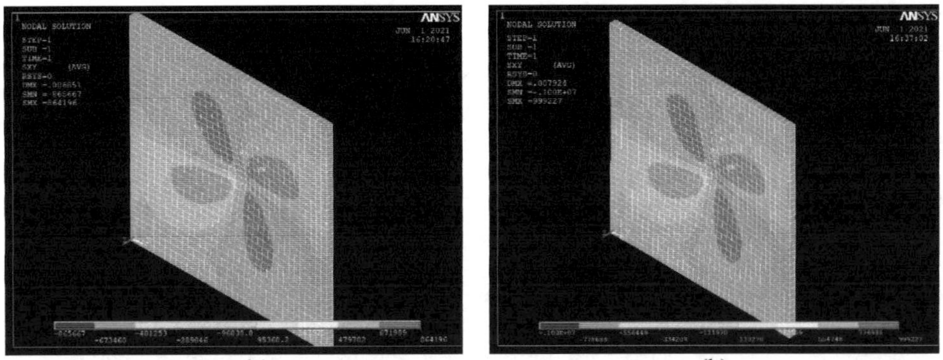

(a) (b)

Figure 7. SXY shear stress analysis. a – $S2_{max}$ = 0.76 E9 N/m^2 for h = 400 mm testing case; b – $S2_{max}$ = 0.87 E9 N/m^2 for h = 800 mm testing case.

From the point of view of the exploitation of such a composite product, for example in the automotive industry, where the temperatures do not generally exceed 90^0 C, a thermal analysis was performed. Material properties used as input data are presented in table 5. Figure 8 presents the equilibrium thermal analysis from where it can be seen how on the opposite side of the surface heated to $T = 90^0$ C the temperature remains at that of the environment, about $T = 20^0$ C. It is thus found that the composite part can function as a good thermal insulator for automotive industry.

Table 5. Thermal properties of the composite structure layers.

Material	Thermal conductivity [W/mK]	Specific heat [J/Kg K]
Glass fibre	1.2	7.35
Epoxy resin	0.35	1000

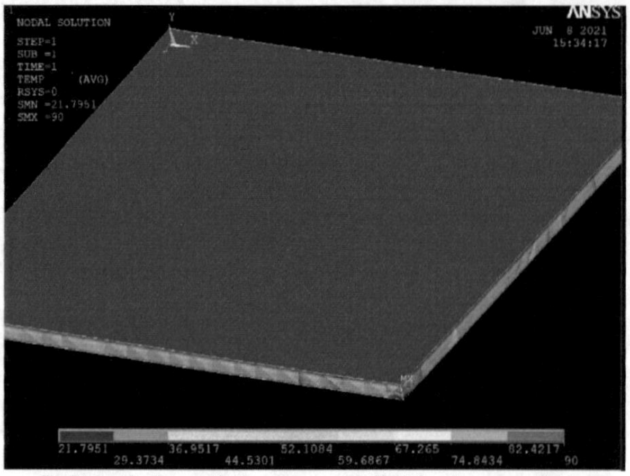

Figure 8. Simulation of the thermal field when applying a temperature of 90^0 C on a surface of the composite product.

4. Conclusions

The paper seeks to make a new contribution in the field of products obtained from composite materials that are used in the automotive industry. It is presented here the way in which the product is made by using an injection mold whose components are showed. Further, research presents several impact tests to which the composite product has been subjected, tests that are similar to what can happen in the event of a car accident. In the last part of the article are presented several results obtained by modeling and simulation starting from the input data such as geometric shape, properties of materials that are part of the product and the impact force acting on it are analyzed displacements and stress appears in the structure of the material. The results of the calculations offered by the program regarding the displacements are very close to those obtained experimentally in the two analyzed situations, namely the impact from the height of h = 400 mm respectively h = 800 mm. This proves that the working assumptions initially considered are valid. Furthermore, from the point of view of the analysis of the stress state that appears as a result of the application of the impact force, it can be seen that their values are at the limit of the elastic modulus for both fiberglass and resin used to produce composite product. The values of the calculated stresses do not exceed the modulus of elasticity being even at their lower limit. In the case of a homogeneous and isotropic structural material, the appearance of stress that do not exceed its modulus of elasticity would not have led to its rupture or destruction. The physical destruction of composite products, however, taking place in conditions where the stresses do not exceed the modulus of elasticity shows that there may still be defects inside the structures that can lead to their cracking. This, in practical applications regarding composite materials, a safety factor must be taken regarding the occurrence of defects that can lead to their possible cracking.

In the second part of the simulation was considered an equilibrium thermal analysis from which it can be seen that this type of material can be considered a good thermal insulator because applying a constant temperature T = 90° C on one of its surfaces leads only to an insignificant increase in temperature on the opposite surface at ambient temperature.

5. References

[1] Lengsfeld H., Wolf-Fabris F., Kramer J., Lacalle J., Altstadt V., 2016 *Composite technology-Prepregs and Monolithic Part Fabrication Technologies* (Hanser Publisher).

[2] Ion S. M., Opran C., Lamanna G. 2020 Mathematical Model of Vacuum–Resin Transfer Molding for Polymeric Composites Products *Macromolecular symposia*, **389**(1) doi: 10.1002/masy.201900108.

[3] Melro A. R., Camanho. P. P., Andrade Pires F. M., Pinho S. T. 2013 *International Journal of Solids and Structures* **50** 1906-1915.

[4] Doitrand A., Fagiano C., Irisarri F., Hirsekorn M. 2015 *Composites Part A: Applied Science and Manufacturing*, **73**, 143-154.

[5] Chawla N., Chawla K. 2006 *Journal of Material Science* **41** 395-40.

[6] Tsai S. W., Melo J. D. 2015 *Composite Materials Design and Testing: Unlocking mystery with invariants* (Stanford Aeronautics & Astronautics JEC Group).

[7] Tsai S. W. 2018 *Strength & Life of Composites* 2008 (JEC Group).

[8] Lamanna G., Opran C., Perfetto D. 2018 Development of an Established FE Model for the Simulation of LVI Tests on GFRP Plates, *9th International conference on "Times of polymers and composites* doi.org/10.1063/1.5045998.

[9] Herath M. T., Bandyopadhyay K., Logan J. D. 2010 *Modelling of delamination damage in composite beams;* 6 th Australasian Congress on Applied Mechanics, ACAM 6, 12-15.

[10] Michel J. C., Suquet P. M. 1993 *On the Strength of Composite Materials: Variational Bounds and Computational Aspects;* Topology Design of Structures 355-374.

[11] Feng W. 1991 *A Failure Criterion for Composite Materials;* J. Composite Materials, **25** 88-100.

IOP Conf. Series: Materials Science and Engineering 1182 (2021) 012030 doi:10.1088/1757-899X/1182/1/012030

Dynamic Lean 4.0 Cyber-Physical System for Polymeric Products Manufacturing in Industry 4.0

D-S Ionel and C Gh Opran

University Politehnica of Bucharest, 313 Splaiul Independentei, 060042, Bucharest, Romania

E-mail: sorin.ionel@gmail.com

Abstract. In the polymer products industry, the adoption of the intelligent production system offers solutions to increase automation and reduce the complexity of manufacturing processes. Simultaneously with the emergence and development of the Industry 4.0 concept and techniques, numerous approaches have appeared in order to implement Lean principles and tools at the intelligent production system level, with the main purpose of increasing efficiency and eliminating waste. Lean 4.0 architectures rely on the physical and digital resources of Industry 4.0, which can be active (directly involved in the production processes of the product) and passive (which have a supporting role). The implementation of Lean 4.0 principles is based on Lean 4.0 tools, defined at cybernetic level, and on the corresponding physical assets, which intervene in the management of the production flow. The objective of this paperwork is to present the realization and implementation framework of the cyber-physical system Lean 4.0 at the manufacturing cell level into an Industry 4.0 production system.

1. Industry 4.0

Advanced manufacturing technologies and latest developement in communication and information technology have led to increased performance of manufacturing systems and production environment. These developments have been incorporated into the Industry 4.0 concept, which is considered the new industrial revolution. According to this concept, by integrating advanced technologies at the value chain level, a higher level of productivity and economic performance is achieved [1].

The Industry 4.0 concept is applicable by defining the cyber-physical system (CPS), which encompasses the physical production system resources, their digital representation and the cybernetic system that allows interconnection and integration of intelligent and cognitive capabilities. The CPS design involves advanced technologies embedded in sensors, actuators, the Industrial Internet of Things (IIoT) platform, newest communication systems capabilities and the latest developments in collecting and processing big data [2]. The CPS description highlights the physical space, represented by the physical objects, the processes and the environment to be supervised and controlled, the cyberspace, which refers to information, services and decision-making units, as well as the IT infrastructure, which connects these two spaces and which includes IIoT, Internet of Services (IoS) and network communication technologies (equipment, devices, computer applications, protocols, programming or communication languages) (figure 1) [3].

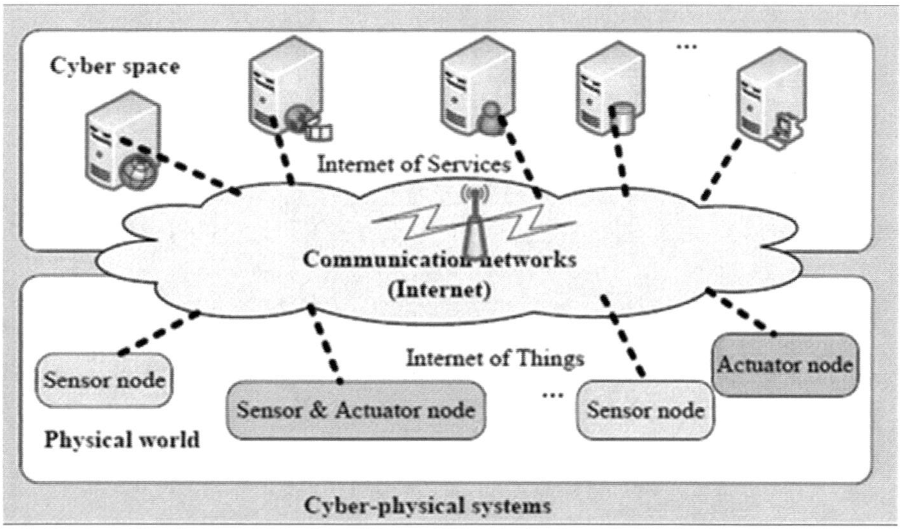

Figure 1. Holistic view of cyber-physical systems [3].

The CPS implementation allows transparency of information and decisions decentralization, wich are the main features of structured and intelligent manufacturing systems, who allow processes real-time monitoring at the cyber system level, formulates and apply decisions at the physical system level, but also cooperation between value chain participants. Also CPS technologies allow real-time production flow changes, according to customer needs and preferences [4].

For each digitally defined physical element or assembly at the CPS level, a holon is created (a structure that behaves simultaneously as a system and subsystem). To be implemented, holons needs to be complemented by agents and functional blocks (FB). The agent is a standalone software application which represents a physical or logical object of the defined system, capable of acting independently or interacting with other agents. FBs are also software applications that incorporate commands and algorithms needed to perform a task through the logical controller. At the Industry 4.0 level, agents are the Asset Administration Shells (AAS). By adopting AAS and FB technologies, CPS becomes suitable for collecting, storing, processing data, as well as for applying emerges decisions at the physical level of manufacturing system (figure 2). In this way, CPS reacts to environment changes (becomes adaptive and flexible), the system becomes proactive, autonomous and interact or cooperate with other CPD, and learning algorithms can be applied. Thus, CPS becomes a holoarchy (a hierarchical group of holons), this feature extending to the entire production system, which allows it to be fully or partially integrated into other holoarchies [5]. The CPS main component is the Asset Administration Shell (AAS), which is a standardized digital representation of an physical manufacturing resource and has the role to achieving interoperability between different applications that manage the manufacturing system. It allows the identification of the resource, integrates the operating models and technical descriptions of its functionality [5].

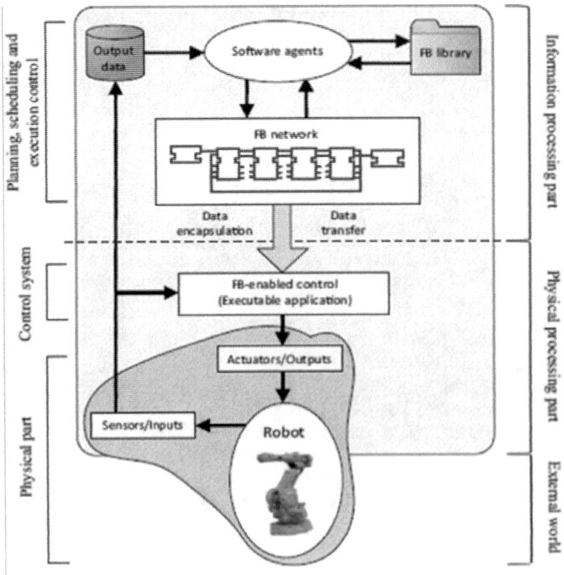

Figure 2. Robot Holon, example [5].

The AAS application can represent products, machines, equipment, manufacturing systems or production subsystems at the cybernetic system level [6]. AAS consists of a header and a body. The header includes information for communication with external systems, also for identification and functionality of subcomponents. The body integrates the operating models of the asset, which can be descriptions, applications or references to other components. Each model has its own header that is part of the body manifest. While the header is standardized, which allows the integration of AAS at CPS level, the operation modules are organized as digital object memories (DOMe) that refer to an aspect, function or digital component defined at CPS level (figure 3).

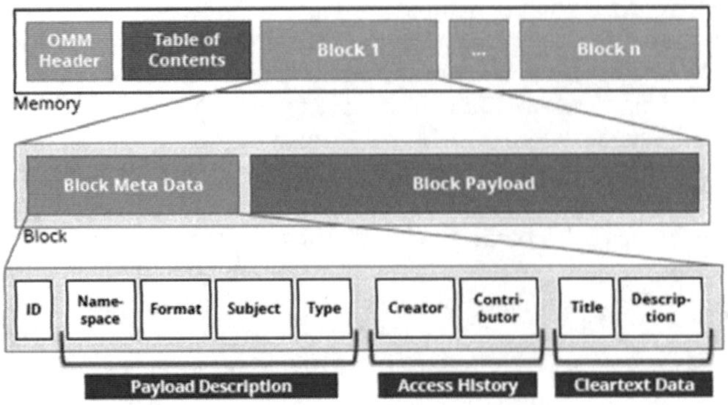

Figure 3. DOMe Model [8].

These DOMs are managed at the component manager level, which allows the loading of useful modules at the level of the manufacturing CPS [7]. This AAS modules organizing method provides a data model for sharing and a framework for storing CPS data. DOMe includes a unique identification code in the form of a URL (Uniform Resource Locator), which is a reference for both instances of the asset (physical and digital representation), and allows the asset data to be stored on a dedicated server [8]. In this way, the information included in the memory model of the represented object (OMM) is not limited by the storage space and is organized in blocks that are accessed according to the workload plan. These blocks are defined by a header containing the unique ID, and the description, which is included in a table of contents (ToC), and includes the description of the relationships between objects. OMMs also allow the ontological relations definition at the DOMe level through Resource Description Framework (RDF) and Web Ontology Language (OWL) standards, organized in a triple (subject, predicate and object), represented by URI and a declaration of validity [8]. The importance of organizing the operating models of an AAS in this way lies in the fact that they can be semantically processed and digital libraries can be set up, organized in databases, from which the OMM model can be extracted where the predicate is specific to the use case or for a particular relationship between objects, without extracting the entire content. Thus, the OMM library will be included in a modularized server, who will run a combination of components, depending on the desired manufacturing application (figure 4) and which can be accessed locally or remotely, if the data storage is distributed [8].

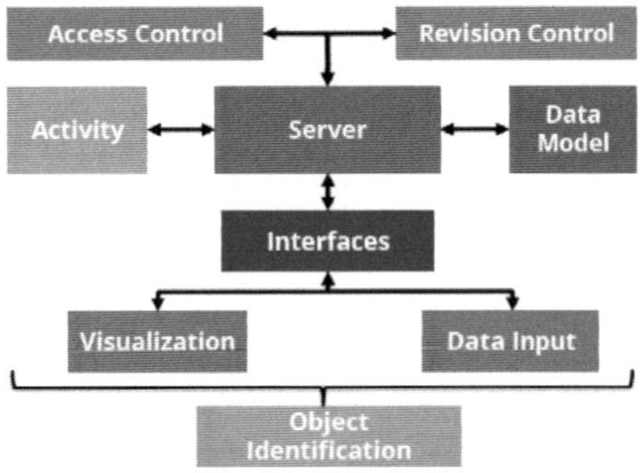

Figure 4. OMM Server components [8].

This type of AAS architecture definition provides the necessary level of abstraction to define a digital manufacturing system configuration in which its elements are interconnected, flexible and self-adaptable. In this scenario, the role of AAS is to control the manufacturing physical components, as a semantically defined software application, who provides a solution in relation to the variation, specificity and specialization of machines, acording to different standards under their operating functions are applied. The AAS ability to nest software components or other AASs, allows CPS functionality in different layers, the distribution of capabilities for querying, collecting and processing data and information, as well as organizing the information flow in a two-layer architecture, control plan (by AAS) and the data plan, which can be connected directly to a network of servers [7].

2. Lean Manufacturing 4.0

The dynamics of the polymer products industry have guided companies' efforts to identify the most efficient manufacturing management methods to achieve the desired competitiveness. The production objectives have been oriented the manufacturing systems towards the principles described in the Lean Management concept in order to respond to the market complex requirements, to performance continuous improvement and to increase flexibility. The adoption of Lean Management principles and Lean Manufacturing (LM) techniques have improved performance by increasing productivity, increasing quality, eliminating losses and creating value for customers [9].

The Industry 4.0 concept development underline similarities with LM in areas such as decentralization, transparency or flexible manufacturing structures, common tools and manufacturing management techniques [9]. One of the recent theories related to adaptive production systems is Lean Automation (LA), whose compatibility with the Industry 4.0 concept is ensured by a common set of fundamental technologies that can be implemented at the level of LM tools [9]. These technologies provide the data needed to measure Industry 4.0 performance indicators and matrices for evaluating

LM indicators, such as takt time, cycle time of production, lead time, work in progress, waiting, motion time, the one-piece flow rate, the workload, quality index. The implementation of CPS at LA level results in a Lean 4.0 production system, and leads to a flexible, transparent, self-managing system that eliminates non-value-added activities, with stable production cycles and a high level of standardization [10]. In order to achieve Lean Manufacturing 4.0 production system, it is necessary to separate the manufacturing resources actively involved in the production process from the passive ones, which have a supporting role within the manufacturing subsystems. In this way, the technologies and services necessary for Lean 4.0 intelligent manufacturing are selected, and the manufacturing system are being updated at CPS level with the physical and digital elements that need to be changed according to the production scenario, including with the specific LP tools [11].

This forms a technological layer that acts as an interface dedicated to the management of active resources of the production process, which is defined at the cyber level by the sum of AASs that form CPS. These AASs include applications, codes and routines that allow to emebed of additional instructions necessary both to operate from an Industry 4.0 perspective and to define and implement LM4.0 principles and tools. These tools are defined as Lean AAS (LAAS), which are applied to CPS processes through the instructions contained therein. LAAS tools will be grouped and managed through a meta-LAAS, defined by the requested production situation and which will assess the evolution, determine and transmit in real-time to the cyber system the adjustment commands, for interpretation and to be applied to the physical system through the function blocks [11]. An Lean 4.0 tool is conditioned by a complex set of factors distributed across product manufacturing process, depending on relevant key manufacturing performance. Defining Lean 4.0 tools is done thgrough a specific LAAS who contains DOMe related to internal references of active or passive AAS and the instructions stored in ToC. Each AAS involved in production process incorporates DOMe Lean blocks who manage or measure performance of supervised phisical asset (figure 5).

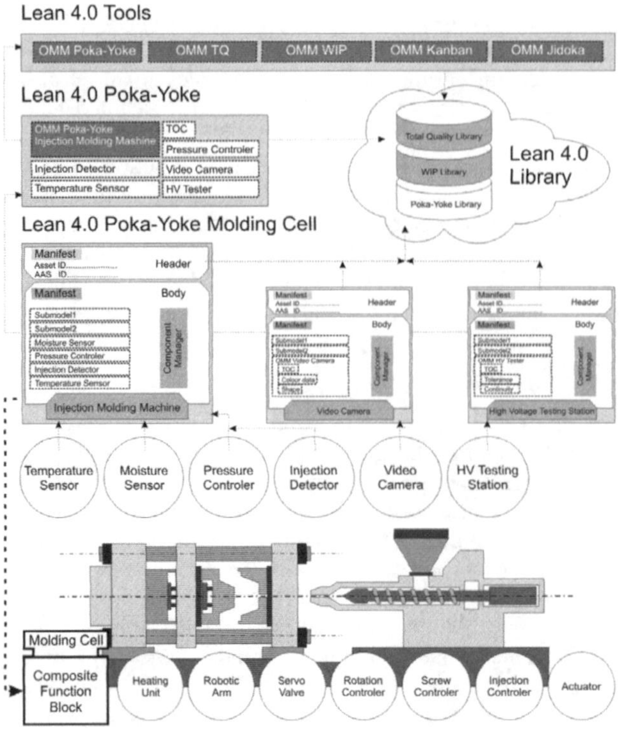

Figure 5. Lean Poka-Yoke tool [12].

In this way, Lean tools are applied at the manufacturing cell level, parameters are checked, deviations are corrected and will be actualized according to manufacturing scenario, with new instructions, incorporate applications, and applied different algorithms [12].

3. Dynamic Lean 4.0 CPS for Polymeric Products Manufacturing in Industry 4.0

The architecture of a polymer product manufacturing system, according to the principles of Lean Manufacturing 4.0 and Lean Automation is based on defining the physical and digital resources actively involved in the production process, at CPS level, according to the Industry 4.0 concept. The basic structure of CPS is AAS, which ensures interoperability and horizontal and vertical integration by standardizing and communicating the parameters, functions and characteristics of physical assets at the level of the cyber system. This is achieved by the intrinsic characteristics included in the header, as well as by defining the operating modules contained in the AAS body. Defining the functional modules as DOMe and organizing the information about the asset represented in the form of WMO, allows to structure them in digital libraries, which can be distributed across several databases and storage media. Through semantic processing, CPS becomes a dynamic and flexible structure, which uses only the operating modules relevant to the manufacturing situation and the production process, extracted from the manufacturing system databases, which can also be a distributed structure, locally, in the cloud or mixed. The databases characteristics and properties allow, by query and export, to generate of simplified database tables, that can include only the data necessary to define a specific CPS for each product that will be produced. Also, defining of Lean 4.0 instruments is made at the AAS level selected for specific CPS by embeded modules, specific to the resources involved to the manufacturing process (figure 6). These modules are accessed from sub-databases, generated and integrated acording to the specific manufacturing scenario CPS. Each Lean 4.0 tool will nest distributed modules, call functions, collect information and formulate decisions necessary for self-optimization and self-management of the manufacturing sistem.

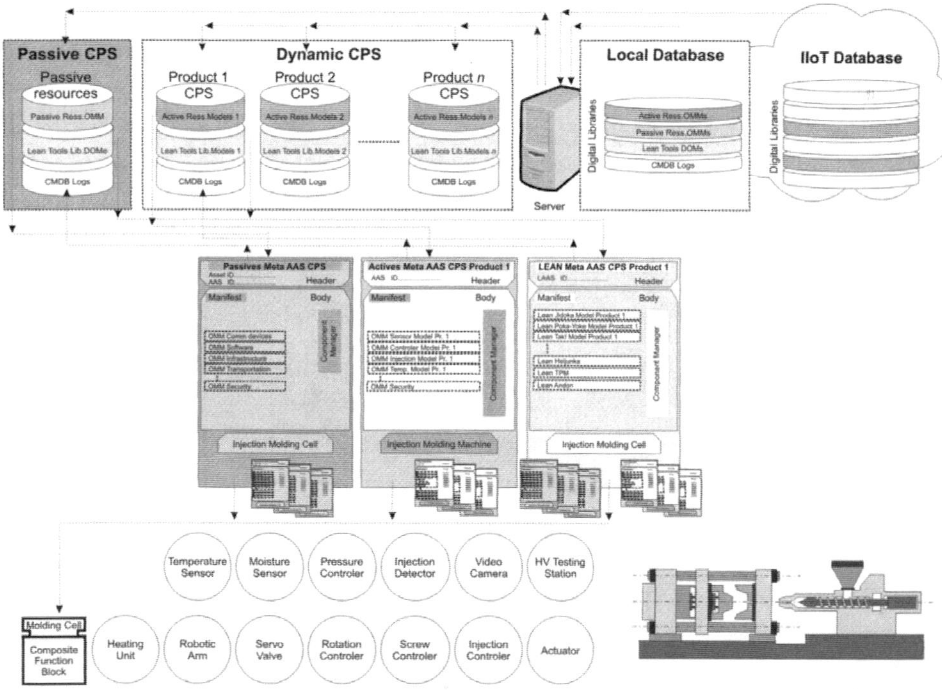

Figure 6. Dynamic Cyber Phisical Framewok for Manufacturing Lean 4.0 System.

4. Conclusions

The new elements of this paperwork are offered by the proposed solution for creating a dynamic CPS, witch is flexible and adaptable to each manufacturing situation of polymeric products, and who incorporates solutions to define a Lean Cyber Phisical System that preserves the necesary dynamic characterisctics and allows the application of Lean Manufacturing principles, individualised on planned manufacturing processes.

5. References

[1] Salkin C., Oner M., Ustundag A., Cevikcan E. 2018 *Industry 4.0: Managing The Digital Transformation.* Springer International Publishing Switzerland.

[2] Mohanta B., Pragyan Nanda P.,Patnaik S. 2020 *Management of V.U.C.A* Patnaik S., *New Paradigm of Industry 4.0. Internet of Things, Big Data & Cyber Physical Systems* Springer Nature Switzerland AG.

[3] Yao X., Zhou J., Lin Y., Li Y., Yu H., Liu Y. 2019 Smart Manufacturing Based on Cyber-Physical Systems and Beyond, *Journal of Intelligent Manufacturing* **30**(8), 2805-2817.

[4] Dasgupta J. 2020 Imparting Hands-on Industry 4.0 Education at Low Cost Using Open Source Tools and Python Eco-System, in Patnaik S. *New Paradigm of Industry 4.0. Internet of Things, Big Data & Cyber Physical Systems.* s.l. : Springer Nature Switzerland AG.

[5] Wang L., Wang X.V. 2018 *Cloud-Based Cyber-Physical Systems in Manufacturing.* Cham, Switzerland : Springer International Publishing AG.

[6] Boss B., Lin S. 2018 *Digital Twin and Asset Administration Shell Concepts and Application in the Industrial Internet and Industrie 4.0,* https://www.iiconsortium.org, accessed 01 05 2021.

[7] Tantik E., Anderl R. 2017 Potentials of the Asset Administration Shell of Industrie 4.0 for Service-Oriented Business Models., *Procedia CIRP* **64**, 363 – 368.

[8] Haupert J. 2014 *OMM: A Structure Model for Digital Object Memories (DOMe),* https://www.w3.org/2014/02/wot/papers/haupert.pdf, accesed 01.05.2021

[9] Satoglu S., Ustundag A., Emre Cevikcan E., Durmusoglu M.B. 2018 Lean Production Systems for Industry 4.0, in Ustundag A. Cevikcan E. *Industry 4.0: Managing The Digital Transformation. s.l. : Springer International Publishing Switzerland*, 43-59.

[10] Ghouat M., Haddout A, Benhadou M. 2021 Impact of Industry 4.0 Concept on the Levers of Lean Manufacturing Approach in Manufacturing Industries, *IJAME* 8523 – 8530.

[11] Ionel D.S., Opran C.G., Vălimareanu B.C. 2020 Lean Manufacturing 4.0 - dynamic physical and cybernetic system for Industry 4.0 *IOP Conf. Series: Materials Science and Engineering* **916**, 10.1088/1757-899X/916/1/012048.

[12] Ionel D.S., Lamanna G., Opran C.G. 2021 Lean 4.0 Dynamic Tools for Polymeric Products Manufacturing in Industry 4.0, *Macromolecular Symposia* **396**, https://doi.org/10.1002/masy.202000316.

Metamaterial optical filter with maximal absorption coefficient

D Ionescu and G Apreotesei
Gh. Asachi Technical University of Iasi, Romania, Department of Telecommunications and Informational Technologies, Carol I Blvd. 11, 700506 Iasi, Romania, [2] Department of Physics, Dimitrie Mangeron Blvd. 67, 700050, Iasi, Romania

E-mail: danaity@yahoo.com

Abstract. Tunable transmission and absorption coefficients in visible and infrared range have been obtained for metamaterial filters with a structure of metal on dielectric nanocomponents. Materials were stacks of nanoconstituents placed in alternate layers, which present a maximum or a minimum of the transmission coefficient and convenient values of the absorption coefficient when the periodicity is interrupted in a controlled manner. A simulational set-up with the metamaterial sample in a channel waveguide was conceived for obtaining the *S* parameters at the field propagation, using the HFSS program. The plasmon dispersion at the metal / dielectric interface was taken into account for calculating the frequency dependent surface plasmon wave vector, which can be adjusted by modifying the refractive index and the electric permittivities of the constituents. Transmission coefficients have been calculated for different structure geometries. The optimal metamaterial configurations have been chosen, in function of the desired filtering effect, for a range of incident field wavelength of 600 – 900 nm. A transmission variation of about 50 … 80 % at the central frequency, in comparison with the transmission effects in the side bands has been demonstrated for the proposed metamaterial filters.

1. Introduction

For obtaining filtering effects in visible and infrared range are necessary structures with nanometric components. Metamaterials are perfect candidates, offering tunable properties by modifying the structure geometry and constituents nature [1, 2].

Our task is to synthesize a metamaterial filter structure by modifying the design for obtaining tunable absorption and transmission coefficients in a desired frequency range. We have proposed a geometrical structure of metal on dielectric nanocomponents, analyzed by simulation techniques using the HFSS program.

2. Metamaterial structure for filtering

2.1. Samples design

The proposed metameterial filter structure consists of a stack of 2 – 4 layers including repetitive unit cells which implement a specific transfer function of the filter type. We have considered here a range of incident field wavelength of 600 – 900 nm (about 300 – 500 THz).

In this case, the unit cell dimensions are of tens to hundreds of nanometers and include metal on dielectric nanoconstituents.

The role of multiple layers is to offer the possibility of interrupting the periodicity of the neighbour unit cells by subtracting some unit cells in a controlled manner. This structure generates an anomaly of some physical parameters. This anomaly can be a maximum or a minimum and we have speculated the extreme of the transmission coefficient of the unit cells and convenient values of the absorption coefficients of the stack of layers.

We have considered unit cell dimensions of about ¼ of minimum wavelength: 120 - 180 nm. The periodicity interruptions were at dimensions of 10 to 20 % higher than maximum wavelength of the applied external field: at 1000 nm (1 μm) in our case, on both axes in plane. In the corresponding positions in the layer, the unit cells have been extracted. The result is to increase the absorption effect in the metamaterial. Global dimensions of the metamaterial filter attached to the visible range waveguides and transmission lines were of a few micrometers order.

The 3D structure of the metamaterials was reproduced with the HFSS program. Metallic segments were included in a dielectric substrate, in order to obtain distributed capacitances, inductances, respectively resistances, necessary for a specified filter effect. The unit cell structure depends on the filter type. In the figures 1 and 2 we have illustrated the unit cell geometry and the equivalent scheme for a narrow band filter, respectively a stop band filter.

The metallic segments dimensions were: length, l, of about 20 nm; width, b, of about l / 10 (2 nm) and width of the resistive paths, t, of about l / 20 (1 nm), adjustable by simulations until the desired transfer function has been obtained. Spacing between the unit cells in plane was comparable with l dimension.

Different central frequencies of the filters have been tested, corresponding to different values of the circuit elements, respectively to different dimensions of the geometrical elements inside the unit cell. We have illustrated here the design strategy for the metamaterial filter having a central frequency f_0 = 410.9 THz. (L = 0.05 pH, C = 0.003 fF)

The narrow band filter has the unit cell given in figure 1(a). Metallic segments (represented in the picture) were inserted in the dielectric substrate. Its equivalent circuit in presented in figure 1(b) and the simulation results in Orcad in figure 1(c).

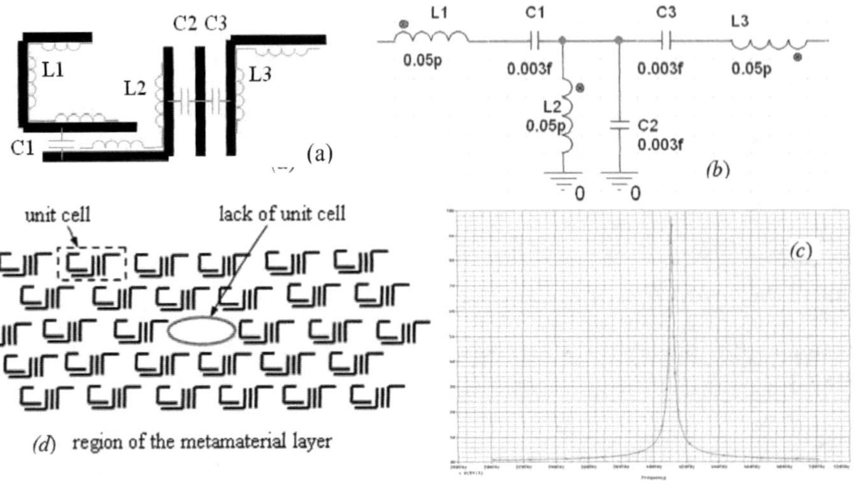

Figure 1. Metamaterial model for a narrow band filter with central frequency of 410.9 THz. a) geometry of the unit cell; b) equivalent circuit of the unit cell; c) transfer function; d) layer region.

The unit cell proposed by us replaces the following classic structure given in literature [2, 3], having the same frequency behaviour in the considered domain of about 300 – 500 THz:

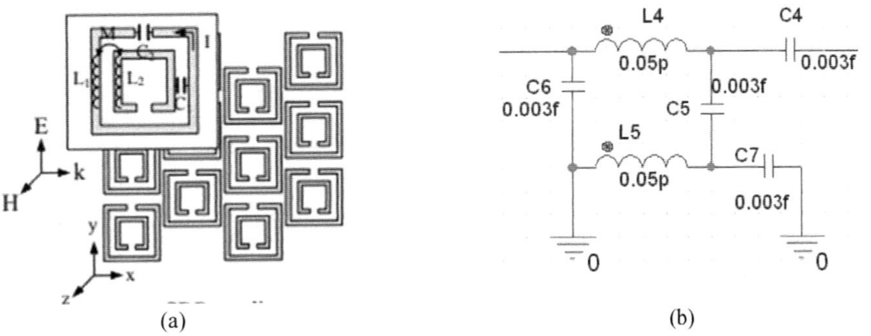

Figure 2. Classic metamaterial layer for filtering. (a) unit cell succession; (b) equivalent circuit.

Our structure removes the disadvantage of impedance loops which can be perturbed by the external fields.

The stop band filter based on metamaterial structure has the unit cell given in figure 3. The filtering effect is very selective and the valley magnitude is amplified by multiplication of the unit cell in the metamaterial layers. The same strategy of controlled removing of the unit cells in the layers has been applied. Notations for the dimensions of the metallic segments have been illustrated on the picture.

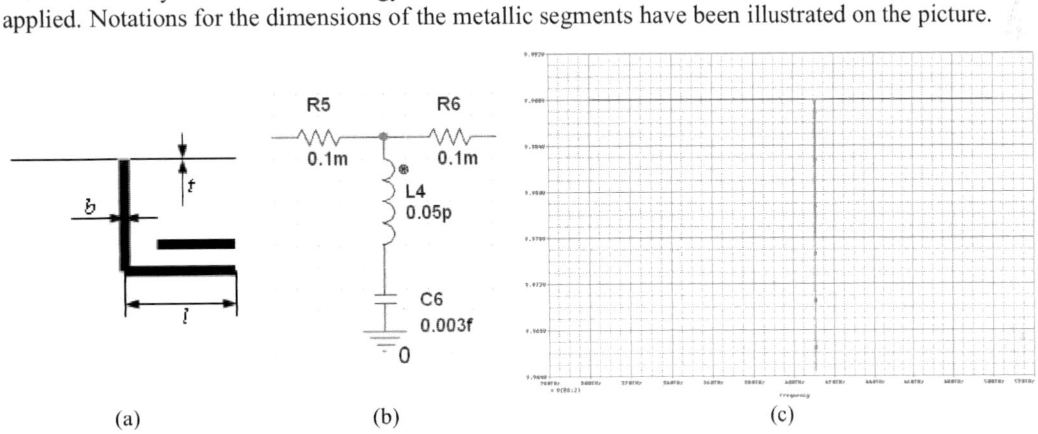

Figure 3. The unit cell of the metamaterial stop band filter. a) unit cell geometry; b) equivalent scheme of the unit cell; c) transfer function.

2.2. Testing configuration

The metamaterial structure of the filters imposes a specific analysis of the field propagating through the sample. We have considered a metal / dielectric combination of materials, with metallic segments included in a dielectric substrate. Configuration of the segments generates capacitances, inductances and electrical resistances distributed on the length, which can be adjusted by modifying the dimensions. The proper combination of the circuit elements has to be set in each case, in order to obtain a specific equivalent scheme, which will have the desired transfer function, with a peak, respectively valley as sharp as possible.

The plasmon dispersion at the metal / dielectric interface was taken into account for calculating the frequency dependent surface plasmon wave vector, which can be adjusted by modifying the refractive index and the electric permittivities of the constituents. For a silver on Si_3N_4 combination (Au and Ag have the real part of the permittivity negative within a certain frequency range of UV to NIR) we have:

$$k_{sp} = \frac{2\pi}{\lambda} n_{diel} \sqrt{\frac{\varepsilon_{metal}}{\varepsilon_{metal} + \varepsilon_{diel}}} \quad (1)$$

where k_{sp} is the surface plasmon wave vector, n_{diel} is the refractive index of the dielectric and ε_{metal} , respectively ε_{diel} are the electric permittivity of the metal, respectively dielectric [4, 5].

We have considered for the metamaterial analysis a simulational set-up with the metamaterial sample in a multi-mode ridge channel waveguide (d = 4-6 μm, g = 2.5 μm) (figure 4). The S parameters have been calculated at the field propagation through the sample. The propagation mode was TE_{10}. With help of the HFSS program, the samples analysis was performed in the domain of 600 ... 900 nm for the incident field wavelength.

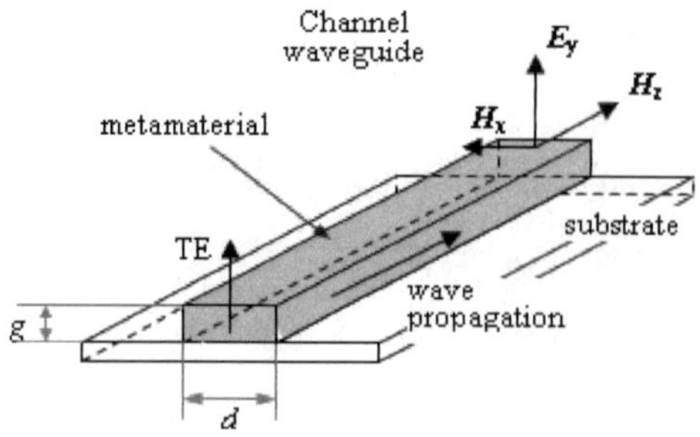

Figure 4. Simulational set-up for determining the metamaterial physical parameters.

The normalized frequency (V number, which determines the number of modes) of this rectangular waveguide can be written as [2, 6]:

$$V = \frac{2\pi}{\lambda} d \sqrt{n_{meta}^{\ 2} - n_{cladding}^{\ 2}} \tag{2}$$

where n_{meta} and $n_{cladding}$ are the refraction index of the metamaterial slab, respectively substrate and d is the metamaterial slab width. For the TE mode, $E_z = 0$, $Ex >> Ey$, and $H_z \neq 0$. The considered channel waveguide is a nonplanar one and has strong optical confinement because it is surrounded on three sides by air, which is a low-index material ($n_{clading} = n_{air} \approx 1$). The waveguide substrate is of SiO_2.

Transmission and absorption coefficients have been calculated with the following relations [2, 7]:

$$T = \left|S_{21}\right|^2 \ ; \ A = 1 - \left|S_{11}\right|^2 - \left|S_{21}\right|^2 \tag{3}$$

using the S parameters determined by simulation.

3. Results for filter parameters

For the two types of synthesized filters, narrow band, respectively stop band filter, the transfer function corresponding to the 3D metamaterials structure has to be determined. The transfer function evolution is described in our case by the transmission and absorption coefficients of the structures.

These coefficients have been calculated and represented on parametrical graphs, in function of the structure geometry. The optimal geometrical parameters and metamaterial sample configuration have been chosen, in function of the desired results in a frequency range corresponding to incident field wavelength of 600 – 900 nm.

The central frequency of the filter can be tuned by modifying the geometrical dimensions of the metal segments in the metamaterial structure and their spacing. The considered metal / dielectric materials combination was Ag on Si_3N_4. Other combinations reported in literature can be used [2, 4], like Ag or Au on SiO_2, Si_3N_4; Ag, Al on quartz, Al on polymer, etc. The same analysis method for structure optimization can be applied.

The transmission coefficient for the two types of considered filters, marrow band, respectively stop band filter, calculated on the basis of the simulation data are indicated in figure 5. Simulations have been developed for a unit cell.

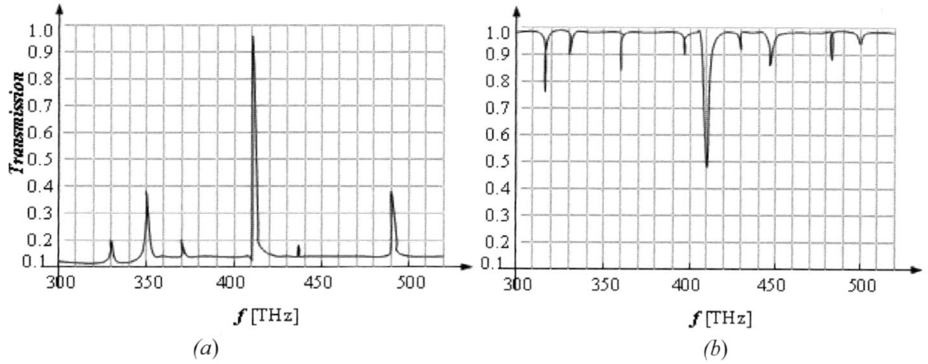

Figure 5. Transmission coefficient for the narrow band filter (a), respectively stop band filter (b), with central frequency of 410.9 THz. Results calculated using the simulation data.

Results obtained using the 3D structural simulation are in good agreement with the results obtained for the transfer function in Orcad (see figures 1 and 3).

We are dealing with a tunable structure, for which the central frequency of the filter can be changed by modifying the dimensions of the metallic segments in the unit cell and their spacing. The same set of results, by 3D simulation and simulation of the equivalent circuit can be performed in the same manner for each central frequency is set.

For optimizing the metamaterial structure dimensions for a given central frequency of the filters, some parametric curves have been obtained by structural simulation. In figure 6 is indicated the evolution of the transmission coefficient in function of the electric permittivity of the dielectric substrate of the metamaterial, for filters having different values of the central frequency, in the considered range.

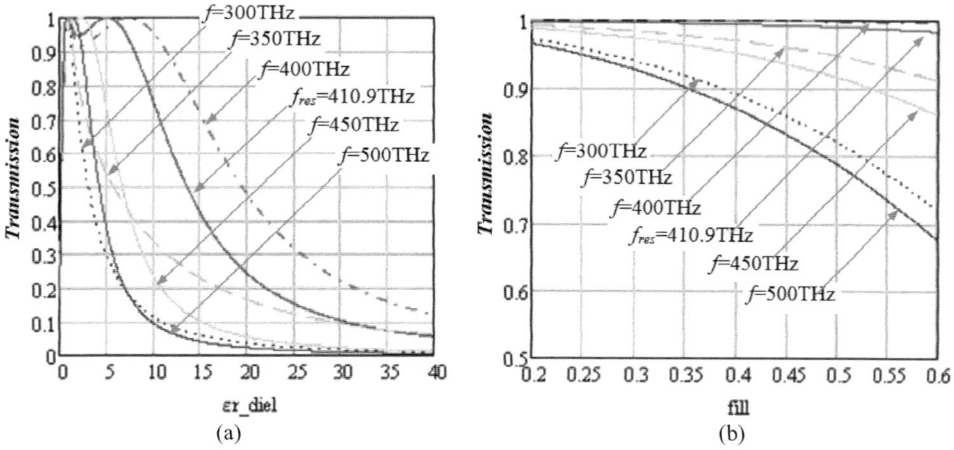

Figure 6. Transmission of the metamaterial sample in function of: (a) the electric permittivity of the dielectric inside, $\varepsilon_{r\ diel}$ and (b) the filling factor with metal in the structure.

A maximum can be observed on each curve, indicating the optimal value for the dielectric substrate relative permittivity. In our case, we have analyzed a filter with the central frequency of 410.9 THz using a Si_3N_4 dielectric substrate having $\varepsilon_{r\_diel} = 10$. According to the graphs (see the red curve), the optimal value for ε_{r_diel} is about 5, but for an ε_{r_diel} of 10, high enough transmission coefficient can be obtained (between 0.7 and 0.8).

One observes that low permittivities substrates are recommended for synthesizing the metal /dielectric filters ($\varepsilon_{r\_diel} < 5 \ldots 10$, depending on f_0).

The graphs in figure 6(b) indicate that the transmission decreases when the metal content in the metamaterial layers increases. A filling content of about 0.2 was considered for exemplification in our study, ensuring a transmission coefficient high enough. Consequently the recommendation for metamaterial synthesizing is to dispose the metallic segments not very dense in the layer. In the same time, a dense disposal of metallic segments can generate parasitic unwanted coupling between elements, which has to be avoided.

4. Conclusions

Transmission coefficients and transfer function have been determined for metamaterial filters having metal segments included in a dielectric substrate. Segments disposal was set in order to obtain convenient impedances (capacitive, inductive, respective resistive) in the unit cell, corresponding to a certain type of filter.

Very selective narrow band, respectively stop band filters were set, in the frequency domain of about 300 – 500 THz (600 – 900 nm). Filter performances depends on structure geometry. The optimal geometrical parameters and metamaterial sample configuration have been chosen, in function of the desired results in a frequency range. A transmission variation of about 50 … 80 % at the central frequency, in comparison with the transmission effects in the side bands has been demonstrated for the structure. The method offers the possibility of a new structure design, having the physical parameters in a convenient range.

5. References

[1] Sun D, Qi L, Liu Z 2020 *Results in Physics* **16** 102887.
[2] Yun S 2011 *Novel Optical Metamaterials, Absorbers, and Filters Based on Periodic Nanostructures*, Dissertation, College of Engineering, The Pennsylvania State University.
[3] Yue L, Ji S, Yan B, Tung N T, Lam V D, Wang Z 2016 Progress in Electromagnetic Research *Symposium (PIERS)*, Shanghai 2303-2303.
[4] Toudert J, Camelio S, Babonneau D 2005 *Reviews on Advanced Materials Science* **10**(2) 123-127.
[5] Pacheco-Peña V, Engheta N 2020 *Nanophotonics* **9**(2) 379–391.
[6] Jia X Q, Chen Q, An Q, Zheng Y J, Fu Y Q 2020 AIP Advances 10, 055018.
[7] Zhu Z, Zhang X, Gu J, Singh R, Tian Z, Han J, Zhang W 2013 *IEEE Transactions on Terahertz Science and Technology* **3**(6) 832-837.

IOP Conf. Series: Materials Science and Engineering 1182 (2021) 012032 doi:10.1088/1757-899X/1182/1/012032

Determination of Johnson-Cook material constants for Copper using traction tests and inverse identification

D M Iordache[1], E R Oprescu[1], C I Malea[1], E L Nițu[1], M O Crăcănel[1] and C Bădulescu[2]
[1] University of Pitesti, Faculty of Mechanics and Technology, Manufacturing and Industrial Management Department, Târgul din Vale Street No.1, Romania
[2] Dupuy de Lôme Research Institute, UBS-UBO-ENSTA Bretagne-ENIB, UMR CNRS 6027, 2, François Verny Street, 29806 Brest Cedex 9, France

E-mail : monica.iordache@upit.ro

Abstract. Copper and its alloys have specific properties - electrical and thermal conductivity, ductility, mechanical strength, corrosion resistance - which make it a material used in various industrial applications. However, these materials are difficult to weld by conventional processes due to their high thermal conductivity and high rate of oxidation at temperatures close to melting. To be able to join copper-based materials, friction stir welding seems to be a promising possibility. The quality of the assembly obtained strongly depends on the parameters of the process chosen to generate a temperature close to the optimum value. So, to find these parameters, it is necessary to carry out experimental test campaigns which are generally very expensive in terms of time and equipment. An alternative will be to use the numerical simulations of the FSW process to find the optimal parameters. Numerical simulation contributes also to a better understanding of the influence of input parameters on the phenomena in the process and the connections made. To develop a numerical model, it is necessary to use a constitutive equation that defines the behavior of the material throughout the process. The most commonly constitutive equation used for FSW modeling is Johnson-Cook. The paper describes the strategy of identifying the constants of the material, respectively the method of inverse identification of the parameters. The constants of the thermo-mechanical behavior (both elastic and plastic) are identified by coupling the finite element method with Abaqus and Matlab software. This strategy tries to minimize the quadratic deviation between the response of the model and the experimental tests. Tensile tests were carried out on quasi-pure copper CU-DHP samples, at temperatures of 22°C, and 500°C, for speeds of 3 mm/min and 30 mm/min, respectively. The validation of the model thus identified was carried out on tensile tests at a temperature of 300°C, by comparing the curves obtained with the Johnson-Cook model and those obtained experimentally.

1. Introduction

Copper welding is difficult using conventional welding processes due to the high thermal diffusivity, which is at least ten times higher than that of most steels. In order to reduce temperature losses, it is advantageous to use a welding process that is carried out at lower temperatures. The FSW process is characterized as a feasible welding process for joining copper [1-5].

In general, experimental investigations have a high cost, with a period of execution and implementation, often much too long. For the prediction of the mechanisms that take place during the welding processes, solutions are established based on the finite element method, specific to the mechanics of continuous media, but also to describe the behaviors of the investigated materials, through certain constitutive equations.

Tensile tests for different speeds and temperatures are required to identify the parameters of a material. The most commonly used model of plastic behavior of the material for FSW welding processing is the Johnson-Cook model [6-8].

The investigated material is a deoxidized copper with phosphorus content, between 0.015-0.04 % P (Cu-DHP), which has high thermal conductivity, with good behavior in FSW friction welding processes [1-5].

This paper describes the strategy by which the material constants of Cu-DHP for the elastic and plastic domains are determined, using the tensile test and the inverse identification method. For this, tensile stresses were achieved for speeds of 3 mm/min and 30 mm/min at temperatures of 22°C and 500°C. The validation of the model being thus achieved by comparing the curves obtained experimentally with those obtained following the use of the Johnson-Cook model.

2. The experimental and numerical procedure

2.1 Material and specimen geometry

Copper and copper-based alloys have electrical and thermal conductivity, mechanical strength, and formability and corrosion resistance being used in a wide range of engineering applications. Friction stir welding (FSW) is a solid-state welding process performed at temperatures lower than the melting temperature of the base material. This feature makes it suitable for welding pure copper and its alloys, whose thermal diffusivity is much higher than that of most steel alloys. The Johnson-Cook law constants determined in this paper will be used in the simulation of the copper joint by FSW. To obtain simulation results close to the experimental ones, the material must be precisely defined.

Table 1. Properties of DHP Copper.

Material	Density [kg/m^3]	Young's modulus [GPa]	Poisson's Ratio, ν	Thermal conductivity, k [W/m°C]	Specific heat, c_e [J/Kg °C]	Thermal Expansion, [10^{-6}/ °C]
Cu-DHP	8913	117.2	0.33	388	385	16.8

The material analyzed is Cu-DHP, which is a deoxidized phosphorus copper with a purity of 99.9%, which has only 0.015 - 0.04% phosphorus. The values of its properties that were used in the development of numerical simulations were taken from the property tables of the material [9], table 1. In the numerical model, the plastic behavior of pure copper is established by the Johnson-Cook empirical law (can see in equation 1) which is integrated into Abaqus / Explicit, which characterizes the relationship between flow stress σ, plastic deformation rate ε, and temperature T, according to the following relationship [10]:

$$\bar{\sigma} = \left[A + B \cdot \left(\bar{\varepsilon}^{pl}\right)^n\right]\left[1 + C \cdot \ln\left(\frac{\dot{\bar{\varepsilon}}_{pl}}{\dot{\varepsilon}_0}\right)\right]\left[1 - \left(\frac{T - T_{ref}}{T_{melt} - T_{ref}}\right)^m\right] \quad (1)$$

with: $\bar{\sigma}$ is normal strain rate; $\bar{\varepsilon}^{pl}$ the effective plastic strain; $\dot{\bar{\varepsilon}}_{pl}$ is the effective plastic strain rate; $\dot{\varepsilon}_0$ is the normalizing strain rate; A, B, C, n, T_{melt} and m are material constants; the n exponent takes into consideration the hardening of the material, while m depends on its melting; C is influenced by the speed of deformation; T_{ref} is the ambient temperature, which is 22°C in this case and at which we determine the parameters A, B, n; T_{melt} is the material's melting temperature.

The Johnson-Cook constants initially used in the simulation are shown in table 2 [3]:

Table 2. Johnson-Cook constants.

Material	T_{melt} [°C]	A [MPa]	B [MPa]	C	n	m
Cu-DHP	1083	90	292	0.025	0.31	1.09

The identification of the elastic parameters of the material and a part of the constants of the Johnson-Cook law (A, B, n) is done by the inverse method, using the simulation of the tensile test. The identification of the other constants of the Johnson-Cook law of behavior will be made starting from the results of the traction test, performed at different speeds and temperatures. The geometry of the specimen used in the simulation corresponds to the experimental one and is presented in figure 1.

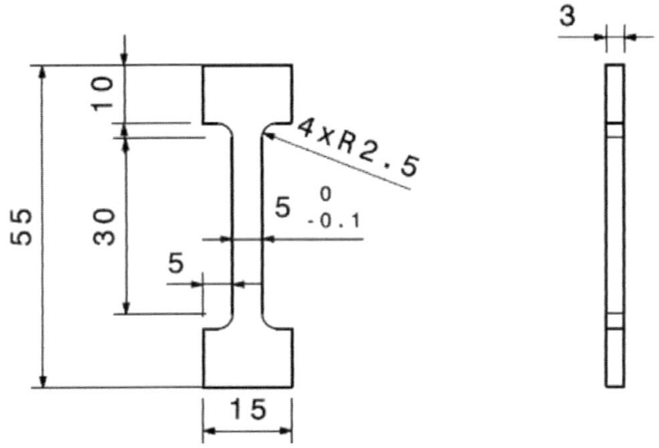

Figure 1. Standard specimen.

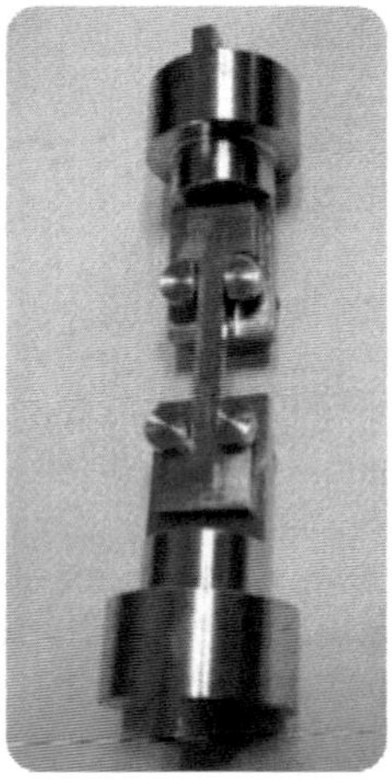

Figure 2. Experimental assembly.

The specimen is mounted with bolts, figure 2. The bolts were preferred to the detriment of the screws, because the material of the test piece, namely copper, at different temperatures and over time, can oxidize so that the thread of the screws can deform. Because the assembly is made for tensile stress, the lower part of the device will be mounted in the traction machine and will remain fixed, and the upper part of the device will move in the axial direction.

2.2 Experimental procedure

To identify all the parameters of the Johnson-Cook model, tests are required at least two load speeds and at two different temperatures. Thus, tensile tests were performed at temperatures of 22°C, 300° C and 500°C, respectively, and at speeds of 3 mm/min and 30 mm/min, respectively.

Figure 3 shows the experimental device. It consists of the hydraulic traction machine INSTRON 1342 (100KN), on which an oven is mounted. The specimen is placed inside the oven, which is fixed to the bottom and top of the machine using two rods and the cooling system. The specimen is positioned between the two contact rods utilizing the fastening system. The cooling system aims to ensure the ambient temperature of the extremities (fasteners), to highlight temperature changes.

To know the values of the temperatures that take place during the process, three thermocouples were inserted: one in the upper part of the machine, one in the lower part, and another inserted in the oven, thermocouple that comes into contact with the specimen, to be able to check and measure the actual temperature of the specimen during the test.

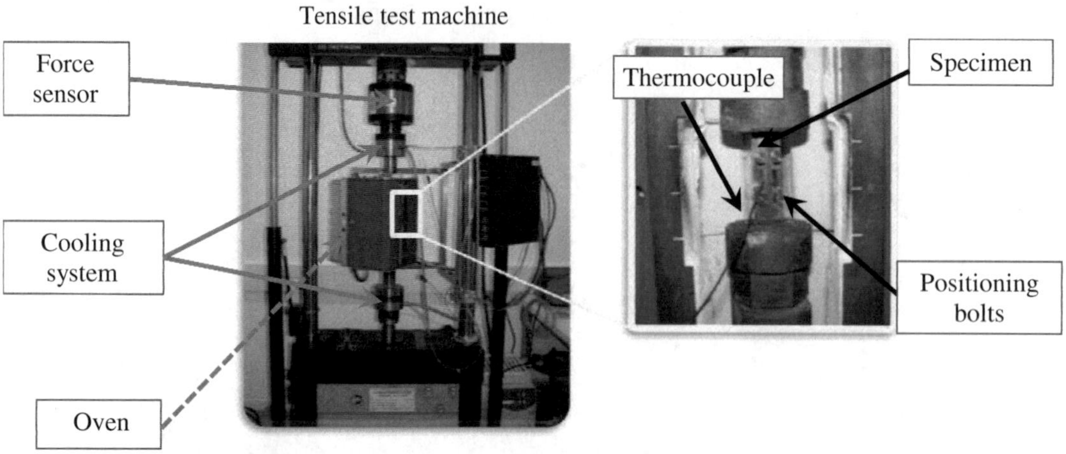

Figure 3. Experimental device.

2.3 Numerical procedure

The tensile test is the test most often used to characterize the mechanical behaviour of a progressively stressed material with a low or medium loading speed. This test allows, among other things, the study and identification of the physical mechanisms of plastic deformation, which govern the mechanisms of plasticity. In order to analyze the behavior of the material, both elastic and plastic properties, several simulations of tensile tests, at different speeds and temperatures were performed using Abaqus software. Simulations were performed at temperatures of 25 °C, 300 °C and 600 °C, respectively, and at speeds of 0.1 mm/s, 1 mm/s, and 10 mm/s, respectively.

The geometry used in the simulation corresponds to the geometry used experimentally. Thus, the cylinders at the bottom are embedded, and those at the top were applied a displacement of 10 mm, figure 4.

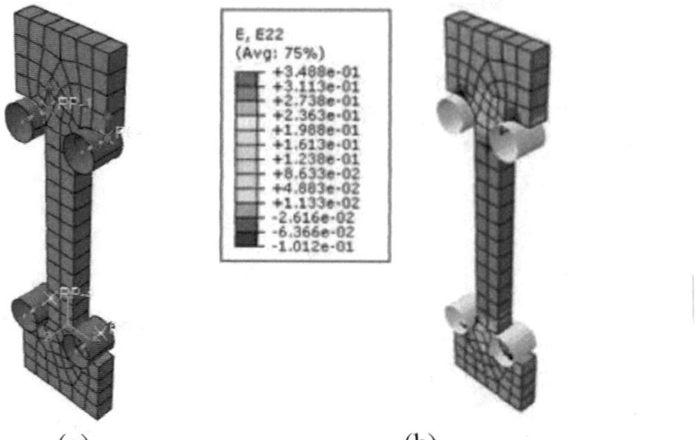

(a) (b)

Figure 4. Simulation of the tensile test
a) boundary conditions; b) simulation results.

3. Inverse identification of material constants in Johnson Cook's law of behavior

The simulation results were processed using Matlab software, and the stress-strain curves were obtained in order to analyze the elastic behavior of the material, respectively the way in which the values of Young's modulus of temperature and deformation velocity were influenced, figure 5 and figure 6.

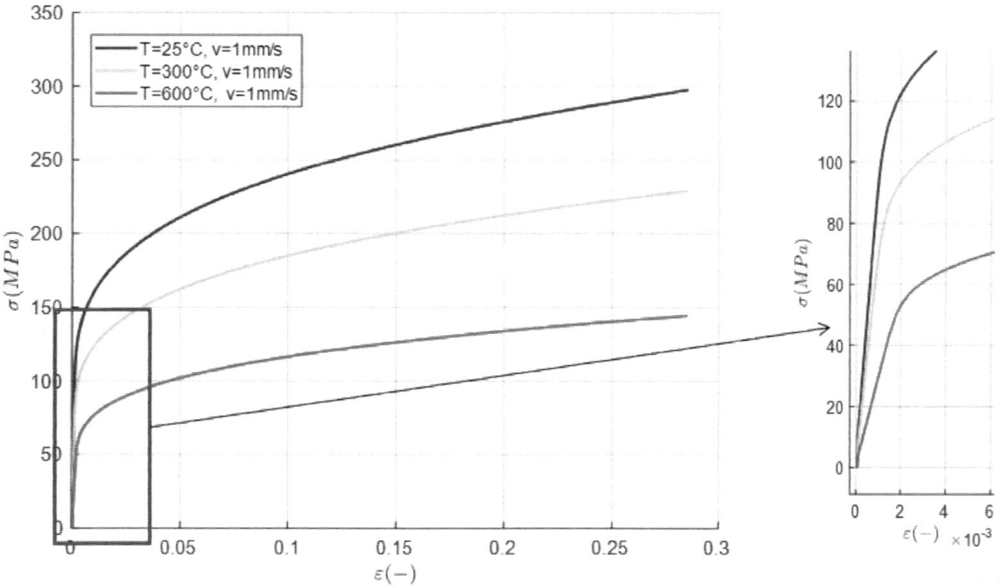

Figure 5. The influence of temperature on the modulus of elasticity E.

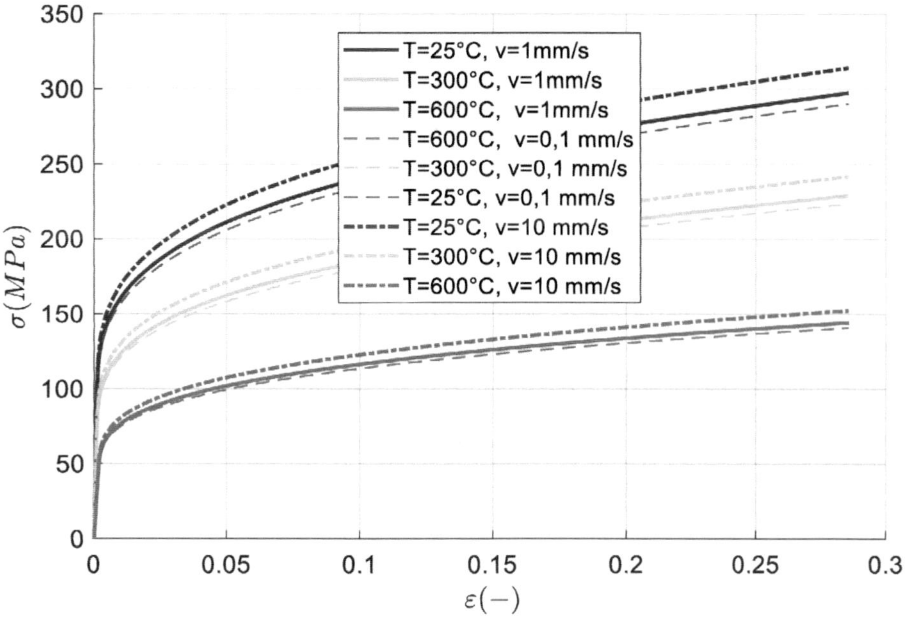

Figure 6. Stress-strain curves obtained from the simulation at different speeds and temperatures.

It turned out that the value of the modulus of elasticity decreases with temperature.

The stress-strain curves obtained from the experimental data from the tensile tests of three specimens at ambient temperature (22 °C) are shown in figure 7.

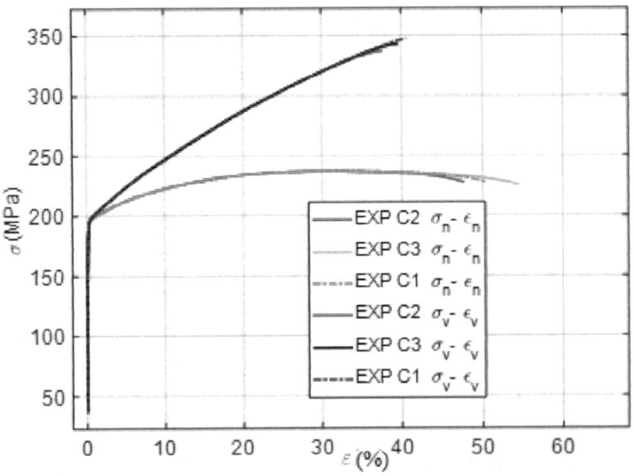

Figure 7. Experimental stress-strain curves (nominal values σn - εn; real values σv – εv).

Real experimental curves were used for the inverse identification of the parameters. The initial values of the Young E_0 module and the Poisson ratio v_0 were entered to initiate the procedure. To identify the elastic parameters, the inverse identification strategy was used, which can be summarized in the following steps:
- an optimization algorithm generated a pair of parameters (E_i-υ_i). This pair of parameters was integrated into a calculation code using the finite element method, to simulate the tensile test.
- the result of the calculation with finite elements, allowed to trace the elastic macroscopic behavior of the sample associated with the pair of parameters and was compared with the experimental one.
- this comparison was made by defining an error function: $E_r = f(E_i,\upsilon_i)$. The optimization algorithm identified a pair of values (E_f, υ_f) that minimize the function E_r built in the previous step.

The fminsearch function will generate another set of values of Ei and vi, which will be transferred to finite element calculus in Abaqus. Abaqus will identify the existence of a file with parameter values (the one generated from Scilab) and will launch the elastic calculation. After performing the calculation, the stress and strain values will be exported to build the simulated stress-strain curve. The data obtained from the finite element calculation will be compared with the experimental data and the error will be calculated. If the error is less than the tolerance defined in the fminsearch function, we will get the parameters you are looking for. Otherwise, the Scilab will generate another data set until the Er <Tol condition is met. When the condition is met, we will obtain the final values, respectively parameters Ef and vf, figure 8.

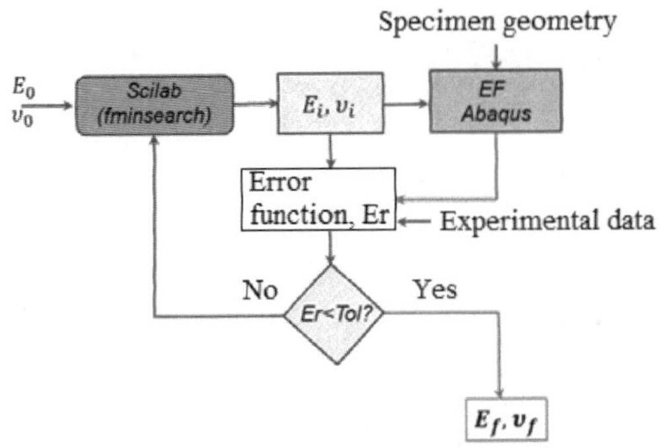

Figure 8. Logic scheme of optimization.

For example, for the experimental curve C2, from Scilab was obtained the graph with the optimization path, figure 9 and the graph of the curves σ-ε_{xx} and σ-ε_{yy} (actual and experimental values), figure 10.

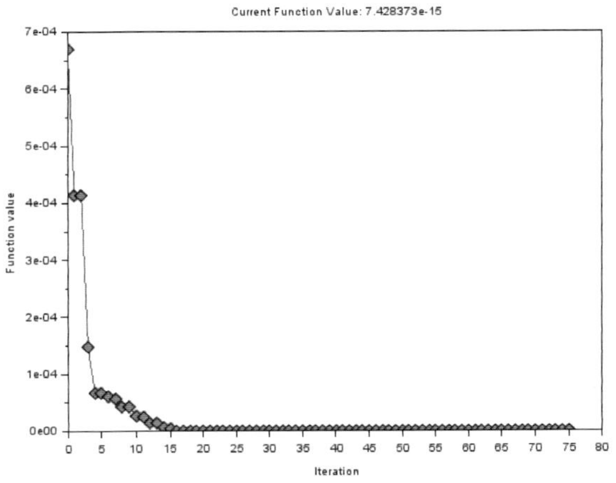

Figure 9. The optimization path.

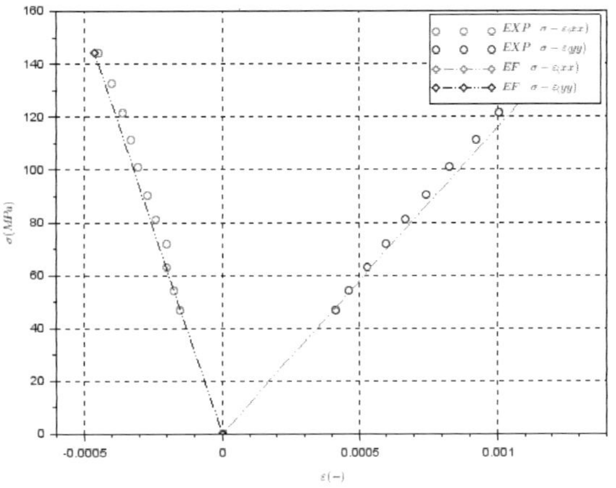

Figure 10. Curves σ-ε_{xx} and σ-ε_{yy}.

The same was done for curves C1 and C3. Thus the following mean values of E and ν were obtained:

$$E = 109364.387 \text{ [MPa]} \tag{2}$$

$$\bar{\nu} = \frac{\nu c_1 + \nu c_2 + \nu c_3}{3} = \frac{0.401 + 0.372 + 0.292}{3} = 0.355 \tag{3}$$

Figure 12 shows the real experimental curves obtained at different deformation rates and temperatures. From the evolution of the curves, it can be seen that the effect of deformation speeds is important at high temperatures, at ambient temperature can be considered negligible.
To identify the constants of Johnson Cook's law of behavior, the Curve Fitting (cftool) application from Matlab was used. Thus, the stress-strain curves in the plastic field were plotted using experimental data and the constants of the Johnson-Cook model were identified in table 3.

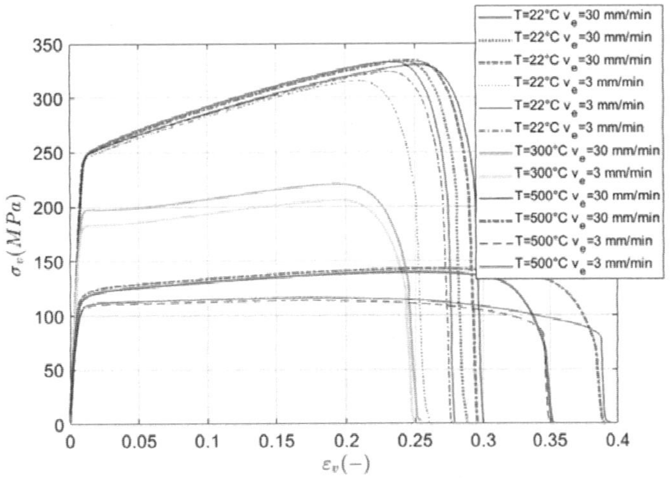

Figure 11. Characteristic curves in real deformation coordinate ε_v - real stress σ_v.

Table 3. Constants of Johnson-Cook's model for DHP copper.

Material	Tmelt(^{O}C)	Tref (^{O}C)	A (MPa)	B (MPa)	C	n	m
DHP-Cu	1083	22	250	250.4	0.0137	0.81	0.73

Johnson-Cook curves were plotted compared to the experimental ones. The solid lines, red and blue, represent the experimental data, and the dotted lines represent the curve drawn with the determined Johnson-Cook model. Figure 12 shows that there is a good correlation at 22 °C. Also, the effect of deformation speed is modeled relatively well.

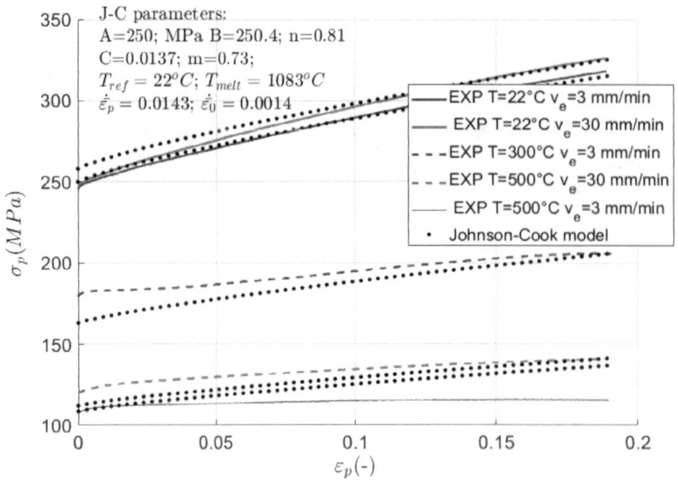

Figure 12. The results obtained by identifying the Johnson-Cook parameters.

4. Use of Johnson-Cook law constants establish in FSW simulation

4.1 Geometrical model and mesh

Friction stir welding (FSW) is a joining process, considered in the solid-state because the temperature during the process is lower than the melting temperature of the material. Although there is important recent progress in numerical simulation of the Friction Stir Welding process, most models have more limitations mainly due to the long computational time, but also due to the definition of the material behavior and the friction conditions. In this study, an efficient numerical model is developed to simulate the butt joint of copper sheets, starting from the FEM model presented in [11]. The simulation was performed with ABAQUS/CAE code using the Coupled Eulerian-Lagrangian (CEL) formulation, the workpieces to be welded are included in the Eulerian domain, figure 13.

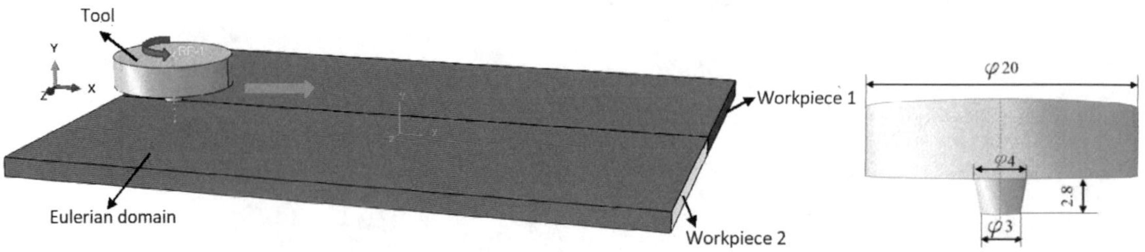

Figure 13. Model geometry.

The tool is considered a rigid body, it has a monobloc structure, with a shoulder diameter Φ20 mm, and the tool pin is conical and smooth, based on the size of 4 mm, and at the top 3 mm, with a length of 2.8 mm. Welding plates are three-dimensional, solid, deformable bodies with dimensions of 100 mm x 40 mm x 3 mm. The Eulerian domain includes the workpiece.

The FSW simulation was performed using variable friction coefficient, starting from 0.2071. The heat exchange between the environment and the Eulerian domain is modeled by a heat transfer coefficient $h=30\ W/m^2\cdot{}^{\circ}C$, which is assigned to all surfaces of the Eulerian domain. Boundary conditions imposed in the plunged phase: speed of 60 mm/min on the Y axis, with a rotation speed of 1000 rpm. The degrees of freedom on the X-axis and the Z-axis is 0.

In the welding phase, a speed of 90 mm/min is applied on the X-axis. The rotation speed remains the same throughout the process.

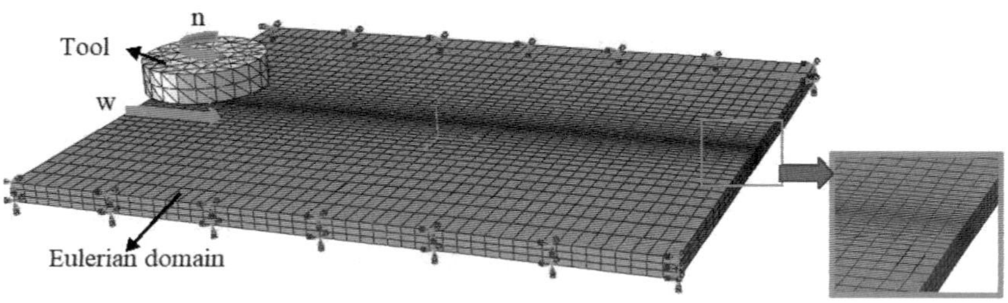

Figure 14. Mesh model.

Eulerian domain was meshed using 6000 thermally coupled 8-node Eulerian elements with 8364 nodes, figure 14. Mesh was performed progressively, increasing the width of the element from the joint line to the edge. The tool is a Lagrangian rigid body and was meshed using 785 quadratic tetrahedral elements of type C3D10M.

4.2 Simulation results

The evolution of temperature is analyzed taking into account the results in the literature. Suitable temperatures for a successful butt joining process (FSW) of copper workpieces were found between 460°C and 530°C, figure 15 [11, 12].

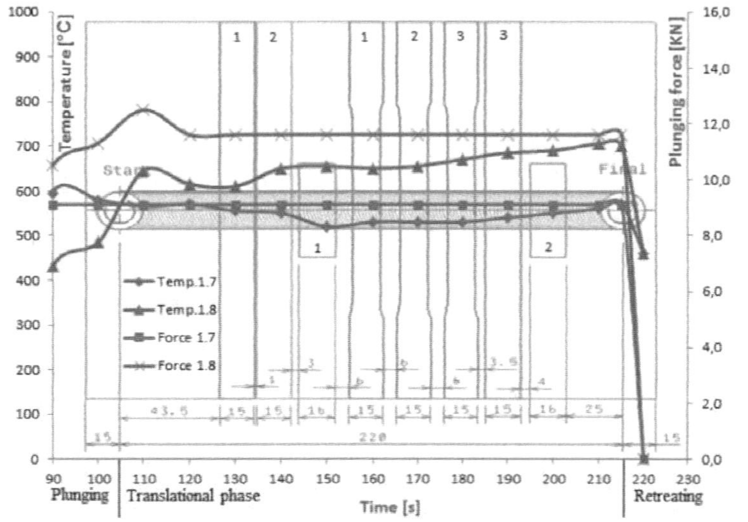

Figure 15. Analysis of temperature and axial forces [11].

Good mechanical properties of the joint are obtained if temperatures of approx. 550°C [13]. The results obtained, figure 16, were compared with the recommended temperature.

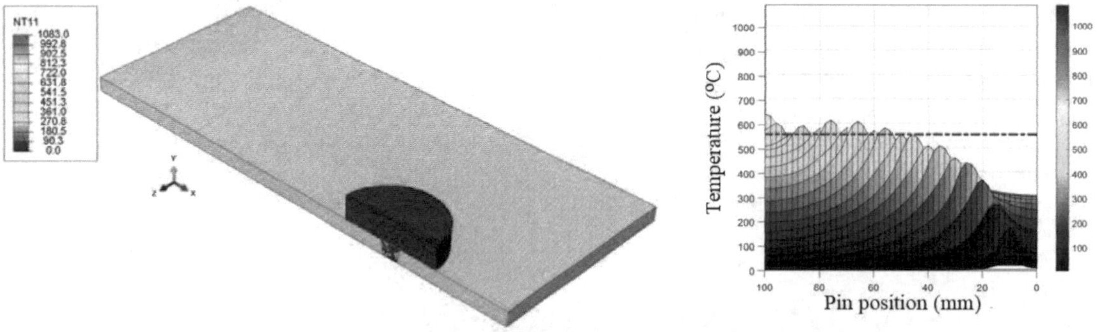

(a) (b)

Figure 16. Temperature profile.
a) Temperature distribution; b) Temperature evolution versus pin profiles.

As can be seen from the previous figure, the temperature increases progressively, but there are not many values above the optimum temperature, and the maximum temperature reaches about 700 °C (does not reach the melting temperature of Copper 1083 °C).

It is observed that the temperature reached when using the variable friction coefficient is close to the recommended one. For this reason, the simulations were performed using a coefficient of friction that varies with temperature and Johnson-Cook law constants establish.

5. Conclusions

Numerical simulation of processes that are still in development can reduce research costs. The accuracy of the results is influenced by several factors and the accuracy with which the material is defined. The paper presents the identification of Johnson Cook's law constants for Copper DHP using the inverse identification method and the tensile test at different speeds and temperatures.

The inverse method is used to identify the parameters of the elastic mechanical behavior starting from the tensile test. The data obtained from the finite element calculation were compared with the experimental data and the error was calculated. When an error less than the defined tolerance resulted, the identified parameters were retained. Thus, 3 parameters were identified for 3 experimental curves and the average value was retained ($E = 109364.4$ MPa, $\upsilon = 0.355$).

Using the real stress-strain curves at different temperatures and strain rates and the Matlab program, Johnson-Cook law constants were identified. Comparing the experimentally determined curves with those calculated using the Johnson-Cook law, it was observed that there is a good correlation at 22 °C.

Using the elastic parameters and the established Johnson Cook law and variable coefficient of friction, the joint of some copper plates was simulated with the help of FSW. Temperature values close to the optimum temperature indicated in the literature were obtained.

6. References

[1] Nagabharam P, Srikanth R D, Manoj Kumar J, Gopikrishna N 2018 *Mater. Today* **5**(1) 1264.
[2] Milicic M, Gladović P, Bojanić R, Savković T., Stojic N. 2016 *Metalurgija* **55** 107.
[3] Karrar G., Shuaib A.N., Al-Badour F.A., Merah N., Mahgoub A.K. 2014 *Int'l Mechanical Engineering Congress & Exposition Montreal.*
[4] Iordache M, Ducu C., Niţu, E L, Iacomi D, Plaiasu A G 2017 *MATEC Web of Conferences* **112** 04005.
[5] Iordache M, Ducu C, Niţu E L, Iacomi D, Plaiasu A G, Pasare M 2017 *Rev de Chim* **68** 459.

[6] Sahlot P, Singh A K, Badheka V J et al. 2019 *Trans Indian Inst Met* **72** 1339.
[7] Pashazadeh H, Teimournezhad J, Masoumi A 2014 *Mater Des* **55** 619.
[8] Pashazadeh H, Masoumi A, Teimournezhad J 2013 *Mater Des* **49** 913.
[9] Nicholson D W 2008 *Finite element analysis: Thermomechanics of solids.* CRC Press.
[10] Johnson G R, Cook W H 1983 *7th International Symposium on Ballistics*, Netherlands, **541**.
[11] Constantin M A, Iordache M D, Nitu E L, et al. 2020 *IOP Conf Ser: Mater. Sci. Eng.* **916**.
[12] Hwang Y M, Fan P L, Lin C. H. 2010 *J Mater Process Technol* **210**(12), 1667.
[13] Constantin M A, Boşneag A, Niţu E L 2018 *MATEC Web Conf* **178**.

Acknowledgements
This work was supported by a grant of the Romanian Ministry of Research and Innovation, CCCDI-UEFISCDI, project number PN-III-P3-3.1-PM-RO-FR-2019-0048/01.07.2019

IOP Conf. Series: Materials Science and Engineering 1182 (2021) 012033 doi:10.1088/1757-899X/1182/1/012033

The influence of Stress Concentration Factors and SN curves in assessing the reliability of jacket marine platforms

R Joavina[1], **M Popa**[2] **and A Jenaru**[1]

[1] Constanta Maritime University, Faculty of Navigation and Naval Transport, 104 Mircea cel Batran Street, 900663 Constanta, Romania

[2] "Ovidius" University of Constanta, Faculty of Civil Engineering, 22B Unirii Street, Constanța 900527, Romania

E-mail: raduj68@yahoo.com

Abstract: The design of offshore structures, generally located in a hostile environment, with complex and repetitive loads, with low and difficult intervention possibilities in case of accidents, is a challenge for engineer teams, starting from the design phase and ending with that of decommissioning. In addition to the approximations generated by the structural calculation, the specification of the level of effort concentration and the relationship between effort and fatigue life are also sources of uncertainties in estimating fatigue degradation. Analysis of the sensitivity of fatigue degradation to stress concentration factors (SCF) and S-N curves is an important factor in obtaining realistic results. The paper seeks to establish the optimal yield theory that can be used to evaluate the structural response to cyclic loads associated with computational assumptions by comparing fatigue conditions due to fatigue at the nodes of a complex lattice-type marine drilling structure, to minimize assessment errors of structural reliability. A structural model of jacket marine platform with dimensional and material characteristics corresponding to Romanian marine platforms of "Gloria" type it is analysed, for which is evaluate and compared the fatigue life generated by hydrodynamic loads, using stress concentration factors given by Kellog and Kuang, SN curves: AWS-X, modified AWS-x and respectively BS-F. The results show that for the self-erecting platforms with lattice beam type, with small dimension elements, the stress concentration factors determined by Kellog are closer to reality while Kuang factors gives more covering results.

1. Introduction

Metal fatigue is a complex mechanism, characterized by a general reduction in the resistance capacity of structural elements to cyclic loads due, in fact, to the propagation of pre-existing cracks. A metallic element can yield at a relatively low stress if this stress is applied a large number of times, the most important parameters in the fatigue failures of welded structures being the number of loading cycles and the size of the extreme stresses σ_{min} and σ_{max}.

Concentrations of stresses produced at certain points of the joints (hot spots) can be up to 20 times higher than the nominal stresses. The position of these points depends both on the geometric characteristics of the node and on the type of loads, respectively on the shape, location and dimensions of a local defect (it is of interest for analyses performed by the methods of Rupture Mechanics).

As a detailed assessment of local stresses in joints, either by approximating by FEM or experimental measurements, is expensive and time consuming, these stresses are estimated by multiplying the nominal stresses of the element by stress concentration factors (SCF) as a ratio of hot spot stress σ_{max} and nominal stress σ_N in joint [1].

The relationship between stress fluctuation and damage can be found using Miner's rule:

$$D_i = \frac{n_i}{N_i} \tag{1}$$

where: N_i - the number of failure cycles, for the same stress cycle repeated successively

n_i - the number of stress cycles of value corresponding to the block i performed

D_i – grade of strength reduction

The most common model used to determine N_i is the S-N approximation.

$$N_i = \left(\frac{2\sigma_i B}{k} \right)^m \tag{2}$$

where: σ_a - stress amplitude (N/mm^2)

k and m - determined constants (modelled as random variables)

B - random variable introduced in order to be able to estimate the uncertainty in the calculation of the stress amplitude.

SN curves (British Welding Institute curves - BS-F, American Welding Society curves - AWS-X and modified curves - AWS-x) are experimentally determined curves and serve as a measure of metal fatigue resistance to reversible loads, and gives the stress range applied to the number of cycles until the metal breaks.

In offshore jacket structures case, the total degradations in the yielded element D_{tot}, are obtained by summing the fatigue damages of a yielded element for a given state of the sea D_i, during the service life of the structure, speaking about a long term distribution of the sea state.

$$D_{tot} = \int_T \int_{H_{1/3}} \int_\theta \frac{T_L}{T_{mp}(t,h,\theta)} D_i(t,h,\theta) P_{T/H_{1/3}}\left(t/h\right) P_{H_{1/3}}(h) P_\theta(\theta) dt dh d\theta \tag{3}$$

where: - $P_{H_{1/3}}(h)$ - marginal probability density as a function of significant wave height H_s

- $P_{T/H_{1/3}}\left(t/h\right)$ - conditional probability density depending on the period of the spectral peak of the wave T_p and H_s
- $P_\theta(\theta)$ - probabilistic density depending on the direction of wave propagation θ
- T_L - total service life
- T_{mp} - the stress cycles period in the considered sea state

The reverse of the total damage gives the fatigue life provided for welded joints. In order to compensate the errors introduced by the simplifying calculation hypotheses, subunit values are adopted for the limit damages, which values are chosen depending on the contribution of the respective node to the safety of the structure.

2. Structural modelling. Assessment of structural response and fatigue degradation

Self-erecting steel platforms are constituted as a triaxle system of beams, consisting of vertical, horizontal and diagonal tubular elements connected by K and T type connections (figure 1).

The offshore metal structure is therefore a continuous system with an infinite number of dynamic degrees of freedom and is idealized in the present study as a spatial lattice beam, using the dimensions and characteristics of the structure elements corresponding to Gloria platform.

Figure 1. Offshore jacket in Constanta shipyard (made by autors on November 2020).

The response of the structure is determined by spectral probabilistic analysis. Local stresses are calculated by time integration (using a step of 0.02s) and using Kuang and Kellog factors for stress concentrators [2].

In order to customize the results, the study uses hydrographic and meteorological data from the Romanian exclusive economic sea zone, taking into account the specific characteristics of the Black Sea - rapid development of the water surface, with waves sometimes exceeding 10m and with a higher frequency of storms on autumn-winter period, mainly from the N and NE direction.

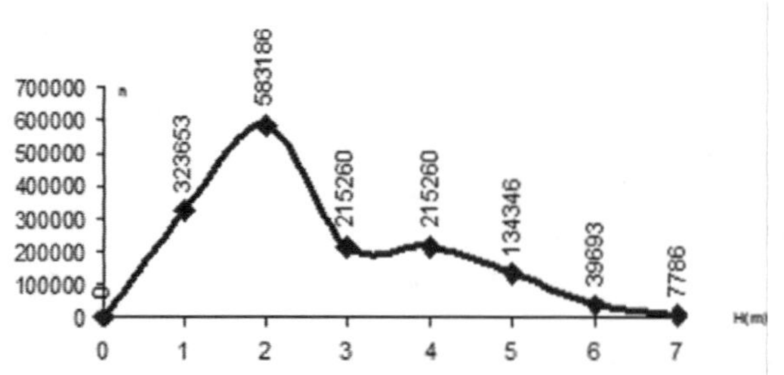

Figure 2. Example of distribution of wave high on N direction on 10m izobath (processing of the data provided by Romanian Institute of Marine Research).

It is to be noted that in the area of the continental shelf, where the influence of small bottoms disappears and the fetch conditions are much better, the wave heights increase (for example, for a maximum height of 5m on the 10m isobaths, offshore waves reached 9.5m) (figure 2).

The calculation of hydrodynamic loads is performed using the Morrison relation, considering the influence of the spectral characteristics of the sea surface [3, 4].

To allow the evaluation of differentiated loads depending on the wavelength of the wave, the calculation is performed separately on each leg, the effects overlapping for each time step.

To increase the calculation accuracy, directional spectra where used, defined on cardinal and intercardinal points, each state of the sea being considered corresponding to the respective direction.

The fatigue analysis is focused on estimating the structural damages of the joints subjected to the maximum efforts in the area of breaking the waves N1 and respectively the area of the mud line N2 on the vertical elements of the legs (figure 3).

It is mainly interesting the behavior of the joints of the vertical elements with the horizontal ones, here the maximum degradation being ascertained on the occasion of the periodic inspections. The description of the modeling and structural calculation was detailed in a previous presentation [5].

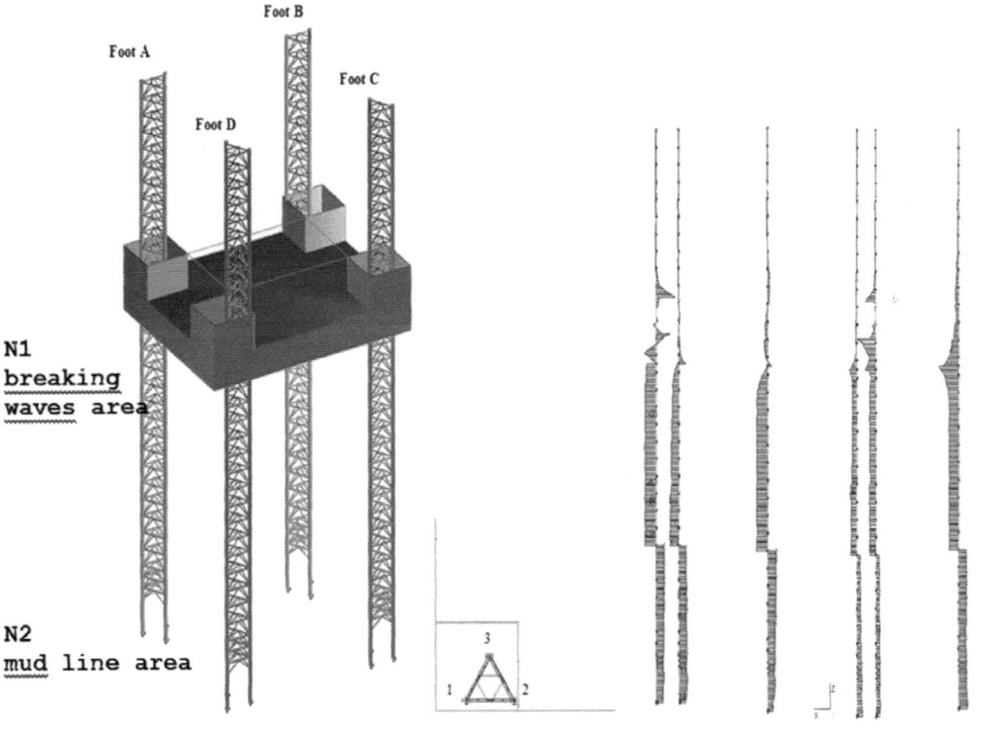

Figure 3. The structural model of the jacket, the positioning of the bars in the foot structure and the shape of σ_{min} and σ_{max} diagrams for the vertical bars 1, 2 and 3 for a wave train of average height 7m.

2.1. The effects of SCF choice in assessing fatigue degradation

The determination of the number of cycles of failure and degradation from fatigue on sea states and cardinal directions was performed using the MIRA-Fatig program [6]. It is to be noted that the structural degradations due to fatigue for the Black Sea, are generated in their great majority, by medium sea conditions, with wave heights, H, between 1-5m (figure 2).

To evaluate the effect of SCF variation on fatigue degradation on welded joints, stress concentration factors as given by Kellog and Kuang were used. Degradation is assessed using the modified AWS-X curve.

The differences between the results obtained with the help of the stress concentration factors given by Kellog and Kuang differ by about 6% for the N1 level and 17% for the N2 level, the factors proposed by Kellog being more covering.

In the following, the influence of SCF uses proposed by Kuang compared to those proposed by Kellog in assessing annual percentage fatigue damage for joints between horizontal and vertical 1, 2 and 3 bars of foot A in areas N1 and N2 is presented. The graphs show the annual percentage degradations on cardinal and intercardinal directions of wave action, as well as the cumulative degradations. The table inside the graph shows the cumulative annual percentage degradation by wave heights.

The use of SCFs proposed by Kuang for A1N1 joints determines a difference in the assessment of cumulative fatigue degradations from 0.5056094 to 0.47699, representing 5.66% (figure 4).

The annual degradation on cardinal directions is 0.386742e-2 for N-S, 0.0781513 for E-W and 0.390631 NE-SW representing 0.82% N-S, 16.53% E-W and 82.65% NE-SW of the total annual degradation value (using a average wave histogram of years 2008 - 2018).

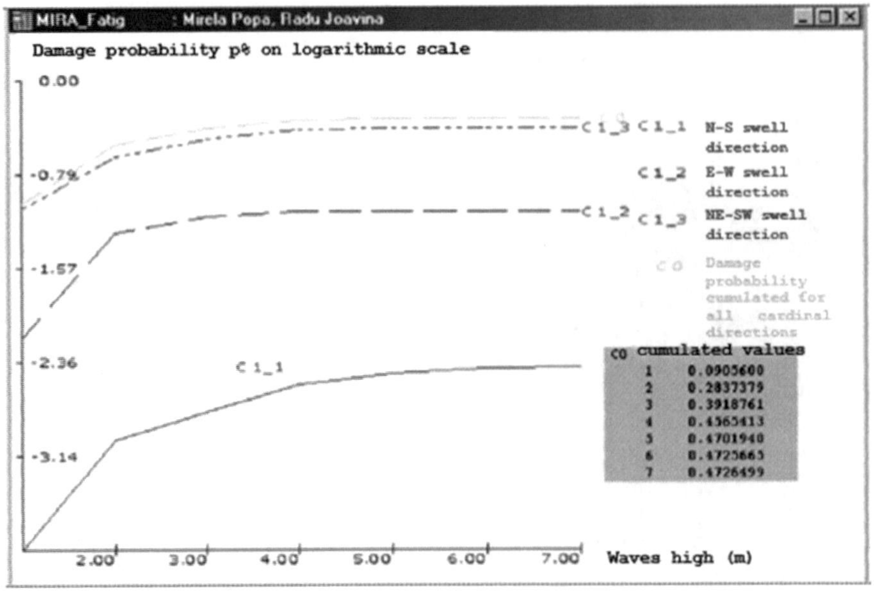

Figure 4. A1N1 joints fatigue degradation.

For the A1N2 joints (figure 5), a difference in the assessment of the cumulative fatigue degradation for Kellog stress concentrators from 0.297732842 to 0.2460602 representing 17.35%, is determined for those proposed by Kuang.

The annual degradation by cardinal directions is 0.0125814 for N-S, 0.214663 for E-W and 0.0188137 NE-SW (representing 5.1% N-S, 87.2% E-W and 7.7% NE-SW of the total annual degradation value).

For the A3N2 joint the values are slightly similar.

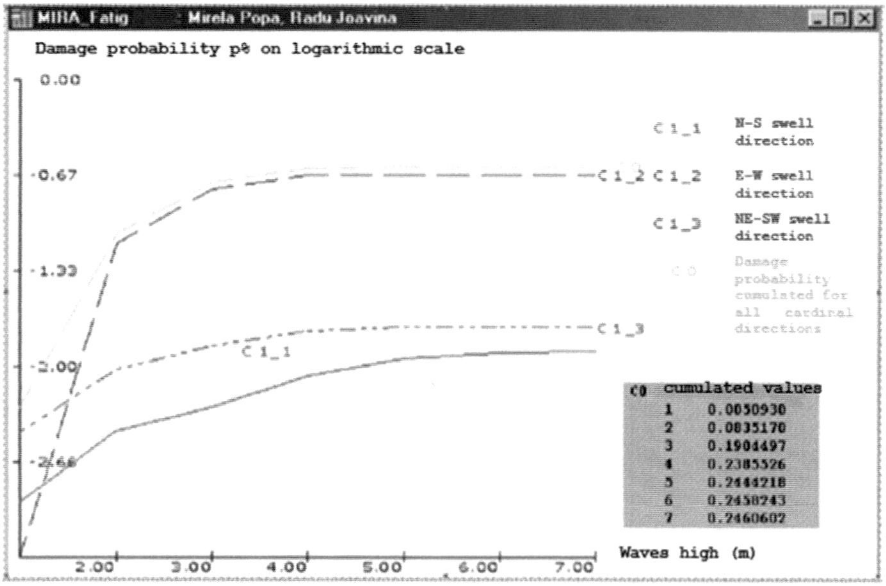

Figure 5. A1N2 joints fatigue degradation.

For the A3N1 joints the difference between the cumulative fatigue degradations compared to the SCF given by Kellog is from 1.9307082 to 1.8162824 or 5.92% (figure 6).

The annual degradation by cardinal directions is 0.167866e-2 for N-S, 0.998039 for E-W and 0.816564 NE-SW (representing 0.4% N-S, 55% E-W and 44.6% NE-SW of the total annual degradation value). The estimated lifespan for the A1N1 joint is 55 years.

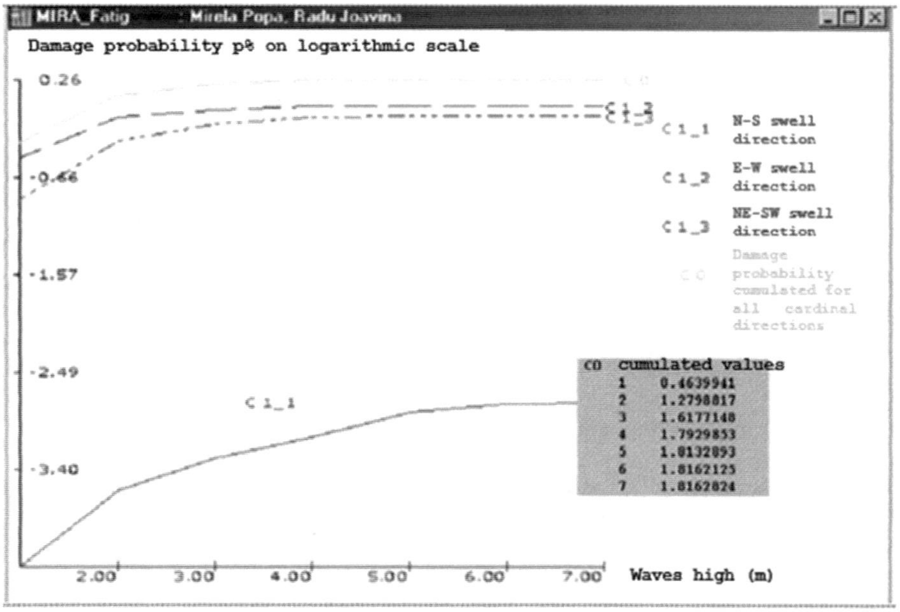

Figure 6. A3N1 joints fatigue degradation.

IOP Conf. Series: Materials Science and Engineering 1182 (2021) 012033 doi:10.1088/1757-899X/1182/1/012033

2.2. The effects of S-N curve types in assessing fatigue degradation

Fatigue degradation analysis is performed using 3 types of S-N curves (figure 7): AWS-X, modified AWS-x and BS-F. Local efforts are assessed using SCF given by Kellog. The degradation in the A1N1 joints obtained by using the modified AWS-x curve is 1.17 times the degradation associated with the AWS-X curve; the corresponding values for A1N2, A3N1 and A3N2 joints are approximately 1.43; 1.08 and 1.25 higher. The magnitude of the stresses caused by different sea conditions are generally of greater value for the A3N1 and A3N2 joints (internal vertical element) compared to those on the A1 and A2 external vertical elements. As the AWS-X and modified AWS-x curves are identical for high magnitude stresses, the differences appearing and accentuating as the stress amplitudes decrease; the variation in the degradation percentage using these curves for A1N1 and A3N1 joints is smaller. In the other cases the differences in fatigue degradation are brought about by efforts with low amplitude values, with fatigue cycles greater than 2 x 10^6 where the difference between the two S-N curves is noticeble.

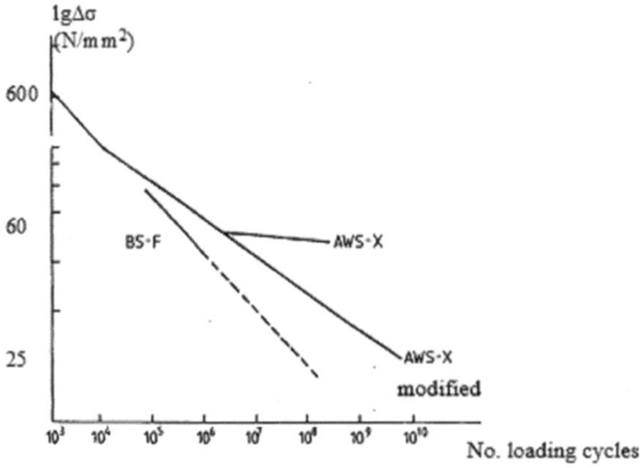

Figure 7. S-N curves as for DNV regulations [7].

The difference in fatigue degradation using the modified AWS-x and BS-F curves is significant. The BS-F curve is more conservative, giving lower fatigue cycles and therefore greater degradation. Fatigue degradation at the A1N1 joints using the BS-F curve is 5.17 times the degradation associated with the modified AWS-x curve. The corresponding values at A1N2, A3N1 and A3N2 joints are around 6.36; 1.98 and 3.73. The results differ because for a high amplitude of efforts the difference in the number of cycles at failure using these two curves is smaller than that for the value of low amplitude efforts.

3. Conclusions

The results show that for Gloria type self-erecting platforms of lattice beam, with small dimensions of the elements, the stress concentration factors determined by Kuang is recomanded to be used when the knowledge of the hydrometeorological conditions at the site are detailed and an approximation of the loads can be made as close as possible to their real evolution over time. It also involves the designer's experience regarding the respective jacket model. On the other hand, given that the long-term assessment of loads is less reliable, the SCF proposed by Kellog offers better coverage.

As the life of the structure is governed by the maximum accepted degradation of the joint, both the AWS-X curves and the modified version can be used in estimating fatigue degradations but the choice between modified AWS-X or BS-F curves can lead to significantly different results.

IOP Conf. Series: Materials Science and Engineering 1182 (2021) 012033 doi:10.1088/1757-899X/1182/1/012033

The use of Kellog's SCF factors together with BS-F curves, significativ increase the reliability coeficient of the stucture, but in the same time with the costs related to the construction. So the owner of the jacket should balance between the working life time of the structure and the economic challange.

4. References

[1] Gupta A., Singh R.P. 1986 *Fatigue Behavior of Offshore Structures* (New York, Springer-Verlag).

[2] Teng H. Hsu 1984 *Applied Offshore Structural Engineering* (Houston, Texas, Gulf Publishing Company, Book Division).

[3] Chakrabarti S.K 1994 *Hydrodynamics of Offshore Structures* (Southampton, Boston, Computational Mechanics Publication).

[4] Sarpkaya T., Isaacson M. 1981 Mechanics *of Wave Forces on Offshore Structures (*New York, Van Nostrand Reinhold Co.).

[5] Joavină R, Zăgan S, Zăgan R, Popa M 2017 Stress state reassessment of Romanian offshore structures taking into account corrosion influence *IOP Conference Series: Materials Science and Engineering* **227** 012066.

[6] Joavina R, Mirela P, Ţepeş-Onea F 2017 *Fatigue failure probability evaluation for truss structures – MIRA-Fatig computing program* (Oradea, „Modern Technologies for the 3rd Millennium" , TMPMIII).

[7] DNVGL-RP-0005:2014-06 RP-C203: *Fatigue design of offshore steel structures.*

Design of production systems using computer modelling of human-robot interaction

A Kampa, G Gołda and D Konysz
Silesian University of Technology, Faculty of Mechanical Engineering, Institute of Engineering Processes Automation and Integrated Manufacturing Systems, Konarskiego 18A, 44-100 Gliwice, Poland

E-mail: adrian.kampa@polsl.pl

Abstract. Designing a production system is a complex process that requires many factors to be taken into account, including the interaction between a human and a machine. The industrial robot is a specific type of a handling machine, which can work similarly to a human, and can be used in different areas in a factory. Therefore, there is a need to consider all important human and robotic factors early in the manufacturing process and industrial system design. Especially the computer simulation of the production processes with the use of Discrete Event Simulation method, carried out for the needs of a new project, allows us to already include the process flow at the industry system planning stage, before real implementation. The purpose of this work is developing a framework of a digital twin of a manufacturing system. An example of a newly designed manufacturing system for the production of elements from the automotive industry is presented, as well as simulation models developed in the FlexSim program, which enables predefined parametrized objects for humans and robots. An area requiring improvement was defined, and the possible effects of these actions on the overall process were reengineered, considering the parameters of availability, performance, quality, and failure rate that make up the OFE (Overall Factory Effectiveness) metric. The information obtained during the simulation allowed us to define the location of the bottleneck in the process, and thus determine what to focus on during the implementation of the manufacturing line. In the future, the models will be expanded in perspective of using the digital twin in Industry 4.0.

1. Introduction

The current concepts of project management in the field of production processes organization are striving to minimize design changes in phases occurring later than the planning phase, thanks to which savings are achieved (because the costs of introducing modifications to the design increase with the advancement of work on it). Traditional design gets conducted in a series. Each next task starts when the predecessor ends. Carrying out a project this way significantly extends the time of its implementation, as well as increases the risk of mistakes being made at particular stages of design due to limited communication. The concurrent design increases the likelihood of detecting critical points early in the process [1]. That is also related to the concept of front-loading [2], which consists of introducing changes at the beginning of the project and verifying the project at an early stage. That way, the largest number of necessary modifications is introduced at the planning stage, when the cost is the lowest. During the feasibility phase of a project, savings of about 20-40% of the total costs of this phase are possible [3]. Verification of the project assumptions at an early stage is possible with the use of modelling and computer simulation methods. The simulation gets performed on a model that is physically or mathematically equivalent to an object or event in a real system [3]. The model is a simplified version of reality, usually limited to the elements necessary for the experiment. Simulations get widely used in the field of production, which is a complex system of people and machines interacting, because researching real systems is often impossible or unjustified economically. Simulations are widely used to test new concepts, verify designs, and monitor progress. Simulations allow for presenting the project assumptions transparently to a wider audience, thus allowing more people to be involved in the early phase of the project. A diverse audience increases the probability of detecting and modifying weak points in the early phase of the project when their modification is the cheapest [3]. A simulation created to solve a problem is a project with its own life cycle. There are five stages in the simulation life cycle: goal definition, modelling, analysis, implementation, and evaluation of results. The process begins with understanding the model, followed by the model building phase, followed by the checking part. The results obtained by the simulation are entered into the action plan [3].

One of the problems is simulating the work of operators, in which the specificity of human work should be taken into account. The human factor is characterized by considerable variability and human errors, which, in principle, are impossible to predict, and may contribute to the disorganization of the production process. Due to the individual characteristics of each person, it is troublesome to introduce strict parameters defining the operator's work into the model. The most common is the averaging of operating parameters in the form of a time standard or a set time [4]. Because of instability of human operated processes, the industrial robots are often used especially for repeatable and monotonous activities. Therefore, the collaboration of humans and robots should also be taken into consideration [5]. Typically, simplified models of manufacturing systems, which give an approximate representation of reality, are used. More detailed models are difficult to build and analyse, but they can give better results [6]. Especially the category of advanced models known as "digital twin" [7], offers many possibilities of use including reducing costs, risk and design time, complexity and reconfiguration time; improving after-sales service, efficiency, maintenance decision making, security, safety and reliability, manufacturing management, processes and tools; enhancing flexibility and competitiveness of the manufacturing system [8]. For the effective operation of the digital twin, all elements of the production system structure must be mapped, including the human operator, which can also be digitally modelled [9]. Therefore, the purpose of this work is developing a framework for a digital twin of a manufacturing system. In the future, models will be expanded depending on the life cycle of the production system and the industry 4.0 perspective.

2. The description of the problem

The management board of the company dealing in the production of automotive components has decided to launch a new production line in one of the plants in Poland [10]. The production line was dedicated to the production of one product reference. Initial findings included a production line consisting of one automatic welding machine, two calibrators, three welding robots, as well as leakage testing and geometry control. The manufacturing process consisted of seven operations presented in table 1.

Preliminary analysis of the time required to complete each operation showed that the stations would have to be partially automated, and three operators should be enough to run the line without loss of efficiency. The assignment of operators to workstations is as follows:

- Operator 1: OP10, OP20, OP30,
- Operator 2: OP40, OP50,
- Operator 3: OP60, OP70.

Figure 1 shows the planned location of the stations together with the assigned operators. Machines OP10, OP40, OP50 are equipped with two windows in which the components are placed and connected with the welding robots. While the operator fixes the components on one window, the welding process takes place on the other. As a result, the welding process of the robot is not interrupted by the time the operator manipulates the workpiece.

Table 1. The description of the production process [10].

Operation	Description
OP10	Welding the inlet cone with the flange and the outlet cone
OP20	Welding of sleeves for the inlet cone and body
OP30	Calibrating the inlet diameter of the inlet cone and the inlet and outlet diameters of the outlet cone
OP40	Circular welding of the inlet and outlet cone to the body
OP50	Welding of pressure pipes and brackets
OP60	System tightness check
OP70	System geometry control

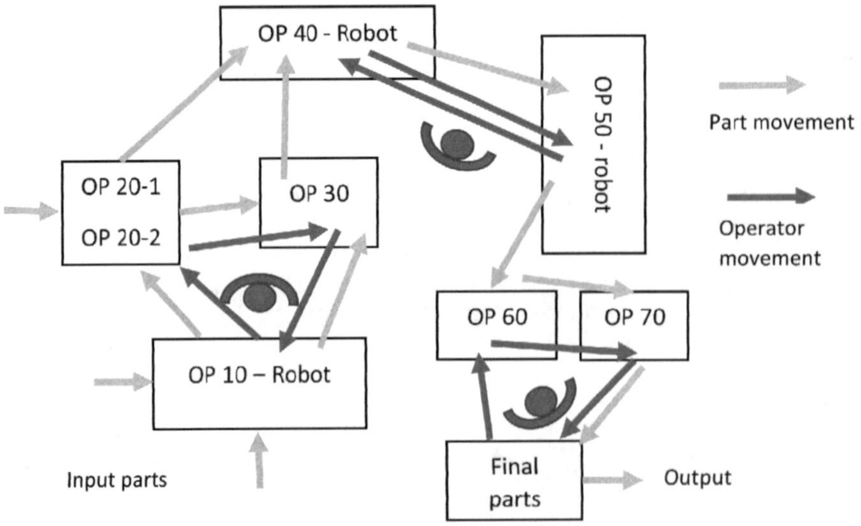

Figure 1. Initial layout of material flow and operator movements [10].

The first operator places the inlet and outlet cones on the OP10. Then they place the inlet cone sleeves and the body on the OP20, from where they take the welded body. Afterward, the operator feeds it to the OP40 and goes with the cone to the OP30. On the OP30, it calibrates the diameter of both cones, which also passes on to the OP40. Operator 2 mounts a cone and body system on the OP40, then adds tubes to the system and supports on the OP50 and places it on the OP60. There, the operator 3 performs a leak test (OP60), a geometry check (OP70) and then packs the product ready for shipment.

3. Modelling and simulation in FlexSim

The FlexSim Simulation Software package enables modelling of real-life systems and discrete-time simulations with visualisation in 3D technology, including virtual reality experience.

When creating the simulation, the time of operations performed by employees was based on MTM-UAS (Methods-Time Measurement – Universal Analysing System) analyses [4], and machine cycle times based on technological data have been taken into account. Additional information on the causes

IOP Conf. Series: Materials Science and Engineering 1182 (2021) 012034 doi:10.1088/1757-899X/1182/1/012034

and duration of machine downtime was determined based on existing, similar production processes that served as a reference for the simulation. To avoid excessive complexity of the simulation and thus to make the simulation results more readable, the following simplifications were introduced. All components and blanks in the process have been replaced with a single element that determines the sequence of material flow in the process. The robot stations were modelled in the form of two machine-type objects. One of them is responsible for the simulation of loading by the operator, the other for the simulation of the automated process.

The time needed to load the components in the machine and start the machine, as well as for all operator movements, was calculated based on the MTM-UAS analysis [11]. This statistic was introduced into the model as the setup time per one workpiece. To simulate the variability of the operators' work, a normal distribution was used with the mean value and standard deviation. In turn, the machine cycle times (processing time) were determined based on technologic data. The assumed production capacity is at least 180 pieces per shift, with 7.25 hours of effective working time. The constructed model is shown in figure 2.

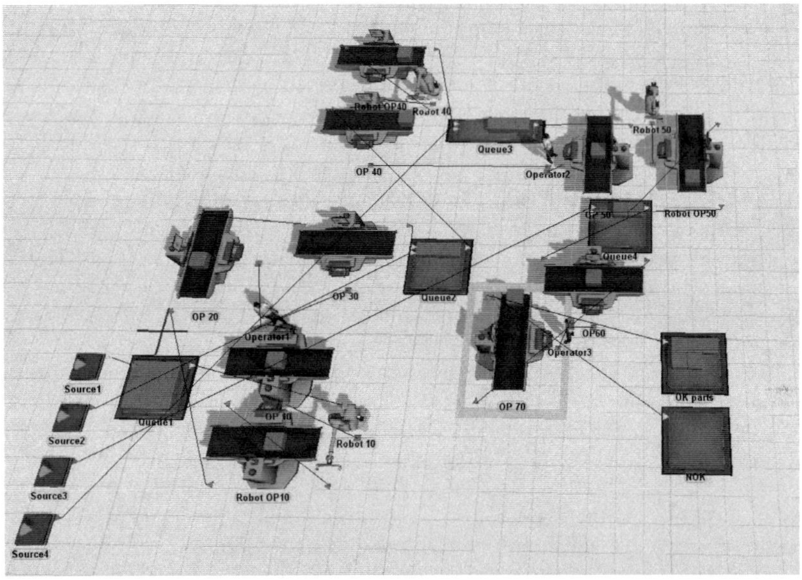

Figure 2. Model of the production system in FlexSim software [10].

The process time values for subsequent operations are presented in table 2,

Table 2. The values of process time parameters per workpiece [10].

Operation	Setup time [s]	Stand. Dev. [s]	Processing time [s]
10	41	2	108
20	31	2	20
30	0	0	29
40	44	2	120
50	56	2	100
60	44	2	60
70	0	0	60

Table 2 includes the mean value and standard deviation of a normal distribution, for setup time. The standard deviation was obtained from the time study of similar processes [10]. The processing time of the machine was based on technological data.

The next step was to introduce planned downtime on the line. The downtime consisted of:

- 5 minutes at the beginning of the shift, for the work organization,
- 30 minutes break in the fourth working hour,
- 10 minutes at the end of the shift dedicated to cleaning the workstation.

4. Simulation experiment

The duration of the simulation scenario was set to 8 hours of work, which is one shift. After entering the data, simulation no. 1 was conducted to determine the maximum production, which is not affected by any disruptions and no downtime other than planned. The average efficiency of the line was 219 pcs/shift with the calculated maximum efficiency of 238 pcs/shift, the obtained OFE (Overall Factory Efficiency) index [12] was 92%. The high result was caused by no quality losses or failures included in the first simulation. The collective simulation results are presented in table 3.

The model was supplemented with a simulation of machine downtime by adding failure rates. For the purposes of the simulation, the definition of failure was extended, including machine downtime caused by adjustments to welding parameters and replacement of the welding machine consumables. Two scenarios of machine downtime and a case of unplanned downtime (failure) of operators were created. For OP10, OP20, OP40, the mean time to failure was calculated as $MTBF_1$ (Mean Time Between Failures) = 3600s exponentially. The time range from the start of failure to machine restart was set to $MTTR_1$ (Mean Time To Repair) = 300-900s in uniform distribution. The second scenario concerned the OP50 as a machine more prone to downtime. The reason for this decision was the high complexity of the machine, a greater number of welded parts of the product, and high loads on the machine mechanisms caused by the weight of its elements. Therefore, the $MTTR_2$ was determined in the range of 300-1800s. The failure rate was defined for employees based on photos of the working day MTBFp = 3 hours and MTTRp = 600-1020 seconds.

To supplement the simulation with the possibility of a formation of scrap pieces at the OP70 output, a rule to send products to warehouses was introduced. As a principle, products leaving the operation had a 95% chance of being placed in the finished product warehouses and a 5% chance of being placed in the defect zone. The proportion of scrap was determined based on a process similar to the one described.

During simulation no. 2, which covered 8 hours of work, on average 178 good-quality finished products were produced, which constitutes 75% of the maximum efficiency of OFE and nine defective ones. The analysis of the machines' operation showed that the bottleneck in the process is the OP50.

As the obtained line efficiency was lower than the assumed capacity by two pcs/shift, the technology department analysed the FMEA (Failure Mode Effect Analysis) in terms of elimination of the causes of failure. Based on this analysis, the failure rates for OP50 were changed. After making changes, the parameters influencing machine downtime are as follows: $MTBF_3$ = 3600s, $MTTR_3$ = 300-1300s. The analysis also showed the possibility of reducing the maximum duration of the failure of the OP40 robot from 900s to 600s. Therefore, simulation number 3 was performed.

Table 3. Summary of simulation results: Pavg- average production in pieces [pcs] per shift 8h, Pmin - minimum production, Pmax maximum production, Plimit - production limit under ideal conditions, LL-lower limit, UL- upper limit, 95% confidence level, 20 simulation runs [10].

Simul. No	Pmin [pcs]	LL [pcs]	Pavg [pcs]	UL [pcs]	Pmax [pcs]	Std. dev. [pcs]	Plimit [pcs]	OFE
1	210	217.8	219.3	220.7	224	3.8	238	0.921429
2	167	176.0	178.4	180.9	191	5.3	238	0.749580
3	178	186.3	188.2	190.1	197	4.9	238	0.790756

During simulation no. 3, an average of 188 pieces of the finished product was obtained within 8 hours, thus achieving 79% of the maximum efficiency. During the simulation, ten units more than during the previous simulation have been produced, and eight units more than the required 180, obtaining additionally nine scrap items.

5. Conclusions

We can observe increasing interaction between virtual and physical spaces, through emerging IoT solutions and Big Data analysis with computer simulation. The use of advanced simulations allowed us to verify the designed production line in terms of possible production efficiency problems. The possibility of emulation of a shop-floor control system and connectivity of simulation models combined with optimisation features and SQL databases access makes FlexSim a promising choice for prototyping and experimenting with the digital twin concept. The next stage may be to supplement the simulation with the data base including expected volumes of orders on a weekly or monthly basis, which would allow us to determine the required level of production, both short and long term. Therefore, the direction of further research includes modelling and simulation of digital twins in the perspective of Industry 4.0.

6. References

[1] Ma Y, Chen G, Thimm G 2008 Paradigm Shift: Unified and Associative Feature-based Concurrent Engineering and Collaborative Engineering, *Journal of Intelligent Manufacturing*, **19**, 625–641.

[2] Thomke S, Fujimoto T 2000 The Effect of Front-Loading Problem-Solving on Product Development Performance. *Journal of Product Innovation Management*, **17**, 128-142.

[3] Beaverstock M, Greenwood A, Lavery E, Nordgren W 2017 *Applied Simulation: Modeling and Analysis Using FlexSim* (FlexSim Software Products Inc., Orem).

[4] Caragano G, Fisher H 2010 *MTM – First Time Right* (International MTM Directorate).

[5] Gołda G, Kampa A, Paprocka I 2018 Analysis of human operators and industrial robots performance and reliability *Management and Prod. Eng. Rev.* **9**(1), 24-33.

[6] Krenczyk D, Kalinowski K, Ćwikła G, Kempa W, Grabowik C, Paprocka I 2020 The design and analysis of material handling systems using simulation *International Journal of Modern Manufacturing Technologies* Special Issue, **XII**(3), 65-71.

[7] Luściński S 2018 Digital twinning for smart industry *inProc. EAI*, pp 1–9, doi: 10.4108/eai.6-11-2018.2279986.

[8] Jones D, Snider C, Nassehi A, Yon J, Hicks B 2020 Characterising the Digital Twin: A systematic literature review *CIRP Journal of Manuf. Scie. and Tech.* **29**, 36-52.

[9] Buldakova T, Suyatinov S 2019 Hierarchy of Human Operator Models for Digital Twin, *2019 International Russian Automation Conference (RusAutoCon)*, Sochi, Russia, 2019, pp 1-5, doi: 10.1109/RUSAUTOCON.2019.8867602.

[10] Konysz D 2019 Modeling and simulation of the production process on a selected example of a production system. *Master Thesis. Silesian University of Technology*, Gliwice (in Polish).

[11] Kampa A, Konysz D 2021 Metody projektowania i normowania procesów pracy z uwzględnieniem czynnika ludzkiego na przykładzie MTM-UAS. *Inżynieria zarządzania. Cyfryzacja Produkcji. Aktualności badawcze 3* pod red. R. Knosali. PWE, Warszawa, 101-110.

[12] Muthain K, Huang S 2006 A review of literature on manufacturing systems productivity measurement and improvement, *Int. J. Industrial and Systems Engineering* **1**(4), 461-484.

IOP Conf. Series: Materials Science and Engineering 1182 (2021) 012035 doi:10.1088/1757-899X/1182/1/012035

Selected methods for material identification on the border of geological layers for the automation of industrial processes of obtaining raw materials

P Kiljan and K Kalinowski
Silesian University of Technology, Faculty of Mechanical Engineering, Stanisława Konarskiego 18A, 44-100 Gliwice, Poland

E-mail: kilpio86@gmail.com

Abstract. Production plants wishing to become competitive on the local and global market should offer their products at the lowest possible price and, at the same time, in the highest possible quality. One of the key tasks for the engineering and technical staff working on them is improving the organization of work and the degree of use of means of production, allowing to increase work efficiency. The latest technological achievements focused on the concept of the *Industry 4.0* create new opportunities for further improvement of production processes, their optimization, reduction of costs, or reduction of the nuisance of physical work and mental effort of employees. Many industries use both renewable and nonrenewable natural resources available on earth. Acquiring some of them requires considerable resources and is very burdensome for people. Especially regarding resources deep underground. In the near future, we will also use material resources obtained from outside the Earth, which is an additional challenge. Therefore, the integration of the latest technologies, especially in the field of automation and autonomous robots, IoT and Big data is becoming an essential research area.
The paper presents an overview of technological methods and solutions of material identification that can be used in the processes of automated sourcing of materials on an industrial scale. Particular attention was paid to techniques that can be used by specialized, autonomous robots working in conditions that are hard and unsafe for humans. Discussed recognition systems can use advanced methods and the latest technologies. They are most often implemented using special sensors installed on or close to the executive element of a machine. The differences in rocks hardness and compressive strength allowed for the development of methods using acoustic waves and also methods using the measurement of vibration of the cutting tools and machine parts on which they are installed. Another method uses the natural features of rocks adjacent to the seam being extracted - it measures differences in the level of radioactivity. On this basis, a method was developed based on the recognition of low gamma radiation activity. Methods based on the use of radars or thermal imaging cameras testing the temperature of cutting knives installed on the cutting body were also proposed. Another interesting solution is the optical touch sensor that recognizes different types of rock layers. Difficult working conditions, also for machines, in which there is a high concentration of stone dust limiting visibility, the presence of water, and high noise generated by working devices, reveals a number of advantages and disadvantages in the above methods to use them in autonomous, robotic solutions.

1. Introduction

During several years, Polish and global companies are constantly changing, the reason for these changes is the wider implementation of information technologies. The idea of Industry 4.0 is

IOP Conf. Series: Materials Science and Engineering 1182 (2021) 012035 doi:10.1088/1757-899X/1182/1/012035

becoming more and more popular and it already has, or will have a significant impact on most companies and their employees in the near future. The influence of similar technological revolutions was visible in many areas, not only in industry, but also in science, medicine, agriculture and trade, among others. We can distinguish several most important elements without which the idea of Industry 4.0 could not function, they are: computing clouds, large data sets, Internet of Things, augmented reality, simulation, digitization, all kinds of autonomous solutions, human-robot cooperation, and integration of sensors with artificial intelligence. The Industry 4.0 concept can broadly be defined as the transformation of separate automated production plants into fully automated and optimized production environments. Production processes are combined within the existing systems in enterprises. It integrates sensors, devices and IT systems into the value chain within the company boundaries. Such factories have the ability to autonomously exchange information through the use of Internet communication protocols, thus reacting in real time to potential errors and adapting to changes in consumer demand for products. Smart factories improved in this way ensure the production of high-quality products at competitive prices, and the investment in them quickly pays off [1].

Highly automated production processes should be properly monitored and supervised to avoid and detect potential failures as well as to obtain the desired final product. For this purpose, sensors are used as "sense receptors" in the current technology, but also directly by humans in the assessment or measurement of physical quantities. In fact, a measuring sensor is the first element of a measuring chain, most often consisting of several transducers. Sensors used in industrial automation systems can form complex subsystems, for example, nodes or networks. The sensor node collects signals from the sensors, processes them, stores them, and forwards them to other sensor nodes and to the control station using an agreed protocol. A sensor network is a complex system that integrates multiple independent sensors with individual or common signal processing electronics [3].

There are six basic types of signals measured by sensors [3,4]: mechanical, thermal, electric, magnetic, radiation (including electromagnetic radio waves, microwaves), and chemical (figure 1). The range of measured quantities is closely related to the operating principle of the sensor used. Defining the measured values and their measurement enables obtaining important information about the implemented process, the condition of machines, tools and equipment.

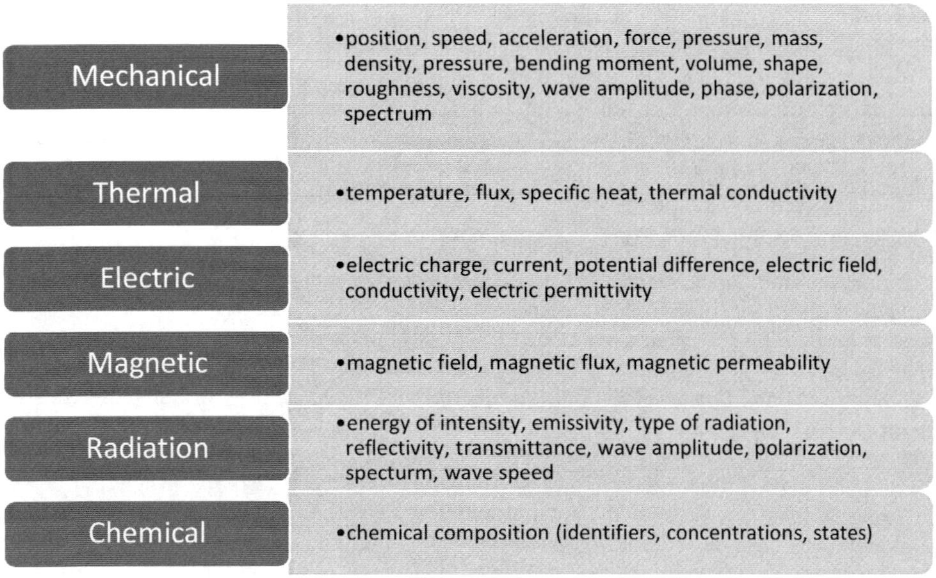

Figure 1. Measured values in the production process [2, 3].

IOP Conf. Series: Materials Science and Engineering 1182 (2021) 012035 doi:10.1088/1757-899X/1182/1/012035

One of the consequences of technological development is the increased demand for raw materials. The amount of raw materials used, especially metals, is constantly increasing, so it is necessary to search for new deposits of raw materials. Raw materials are sought at ever greater depths, such as copper deposits, in new areas where no mining has been conducted so far. Space is also very attractive and more and more realistic in the near future. Advanced research is already being carried out to identify and recognize the possibilities of acquiring its resources. Due to the huge distances to travel in space, the resources available on the Moon, Mars and the asteroid belt are considered in the first place. These raw materials do not necessarily have to be transported to the Earth - their processing and use can also be carried out in space, as is planned, for example, during the expansion of bases on Mars, where sending all necessary materials from Earth is very expensive. Lunar regolith is rich in helium isotopes, which is an ideal fuel for a thermonuclear reaction. Asteroids are rich in iron, nickel, cobalt and platinum group metals. As automation develops, the tasks of sourcing raw materials in this way are becoming more and more real. [4]

One of the fundamental problems of the above-described issues in the area of the autonomous and automatic process of obtaining natural resources is the ability to correctly identify the boundary between geological layers during extraction, to obtain the desired raw material in the least polluted form. Various methods can be used to identify the boundary between the useful layer and gangue, based on the assessment of rock properties such as: hardness, natural radiation, electrical conductivity, reflectance. The article presents an overview of the most popular (on Earth) methods of recognizing the boundary between geological layers of rocks. The selection of an appropriate method of recognizing rocks and the boundary of individual layers and the corresponding hardware implementation in the form of sensors is crucial in developing effective automation tools for the operation process under given environmental conditions (e.g. temperature, pressure, dust, and chemical composition of the atmosphere).

2. Methods of recognizing the boundary between geological layers

2.1. Natural gamma radiation

Radioactivity is the ability of the nuclei of some elements to spontaneously convert into others, through the gradual decay and reduction of atomic nuclei. Radioactivity is most often associated with the emission of alpha and beta particles and gamma radiation. The α, β and γ rays differ from each other in the degree of penetration:

- Alpha rays are blocked by an aluminum plate with a thickness of 0.04 mm or an 8 cm air ring
- Beta rays pass through a 5 cm thick aluminum plate
- Gamma rays have the greatest penetration, penetrating through an iron layer 30 cm thick and able to ionize the air

Radioactive elements with their isotopes are arranged into five families: uranium-radium, actinium, thorium, neptunium, and potassium. In terms of radioactivity, there are three types of rocks that are characterized by:

- low radioactivity - sands, sandstones, limestone, dolomite, salts, anhydrites
- medium radioactivity - clay, clays shales
- high radioactivity - clay shales, bentonite

The method of recognizing the boundary of rock layers based on the natural occurrence of γ radiation can be most used in the detection of coal and rock. Based on the research [5], it was found that the rocks surrounding the coal seams are characterized by an increased value of radiation. The concentration of radioactive isotopes observed in coal rocks are several times higher than the values recorded in coal. Table 1 shows the range of variability of the specific activity of natural radioactive isotopes occurring in Poland. Coal radiation values are several times lower than the rock layers surrounding the seam. The difference in radioactivity can therefore be used to develop the concept of recognizing the boundary between the layers. Recognition can be carried out using a detector installed on the combine arm. For the correct determination of the limit, it is necessary to take samples and determine the radioactivity value for coal and rock. Based on the determined radioactive background, the sensor recording radiation from rocks with the use of appropriate software controlling the combine's arm will enable its automatic operation [5-7].

Table 1. The range of variability of the specific activity of natural radioactive isotopes in coals and rocks [8].

Layers	^{226}Ra	^{228}Ra	^{40}K
		Bq/kg	
Coal	7-48	5-19	24-93
Mudstone i shales	51-120	59-117	393-1070

2.2. Vibration measurement method

Each process of obtaining natural resources is carried out by installed machines, their work generates vibrations that can be measured. Vibrations are usually low amplitude and low frequency vibrations up to several dozen Hz. Vibration measurement is more and more often used in measurement diagnostics of machines and devices, allowing for early detection of failures. The most popular sensor for measuring vibration is the accelerometer. There are three types of accelerometers, the names of which are derived from the measurement method used in them: piezoelectric PE, variable capacitance VR, piezoresistative PR [8].

Piezoelectric accelerometers are the most popular type of sensor used in the industry to measure vibration. They use the piezoelectric phenomenon in crystals, when a mechanical load occurs, an electric charge is created in the crystal. Systems of this type are characterized by high sensitivity and are used in seismic measurements, crash tests, and destructive tests carried out in extremely difficult conditions [8,9].

Capacitive transducers made in MEMS (Mikroelectromechanical System) technology use changing capacitance depending on the position of the moving plates. These are the smallest and cheapest sensors available on the market, used primarily in mobile devices and consumer electronics. Their disadvantage is the low accuracy of measurement for signals with higher frequencies and amplitudes, therefore they are not used in industry. Due to their small size, they are used mainly in mobile devices and in consumer electronics [8, 10-12].

Piezo-resistance accelerometers - they measure the value of electrical resistance, which varies depending on the mechanical load. They are characterized by a wide measurement band allowing the registration of high-frequency vibrations, and due to the ability to measure slow-varying signals, they are used in simple inertial navigation systems [8, 12].

Laser vibration sensors are used when the sensor cannot be installed on a running device. The principle of operation is based on comparing the laser beam reflected from the tested object and reaching the photodetector. The laser vibrometer provides non-contact measurement, eliminating interference caused by the weight of the sensors and the measurement temperature. The measurement can be made with high accuracy from a long distance [8, 12].

The rock layer obtained by the mining machine causes it to vibrate. Each geological layer that is exploited has different vibration characteristics depending on the hardness of the rock. The rock identification system consists in installing accelerometers on the mining device. Vibrations to which the machine is made during operation are recorded by sensors and monitored to detect when the desired layer is left [13, 14].

2.3. Acoustic method

The acoustic wave method is similar to the vibration method. Sound recording is used instead of measuring vibration by an accelerometer. Rock layers have different values of compressive strength. During mining, a rock with a higher compressive strength generates a higher amplitude of the measured signal. It is therefore possible to distinguish rock layers with higher and lower strength. To make it possible, we need to complete samples of signals for given rock layers, the differentiation of which we are interested in. Such samples serve as training samples for the neural network. The neural network is designed to compare training samples with samples measured during mining so as to control the mining process in the intended layer [15, 16].

2.4. Optical method

Widely used vision and thermal imaging cameras can be used to define the rock layer. A camera installed on the cutting device records the image or the temperature of the cutting knives. The recorded image is analyzed to detect the difference in the textures of the rock layers. Texture analysis is about finding and extracting from an image the texture characteristics of various objects. The next step is to calculate the value of the texture features based on the gray space coexistence matrix method. There are vectors of texture features such as energy, contrast, correlation and entropy. The energy mainly reflects the gray distribution of the image and the roughness of the texture. If the energy is high, the texture is rough. For rocks with a smoother texture, the energy is lower. Figure 2 shows the differences in rock texture using the example of mudstone and coal layers. Recognition can be by means of a neural network trained with a texture function. The difficulty for this method is keeping the lens clean, which is not an easy work when mining rocks [17, 13].

Figure 2. Picture of different textures of rocks with the example of mudstone and coal.

2.5. Ground penetrating radar

The GPR method belongs to the geophysical radiofrequency methods. Currently, the most widely used are pulse georadars. The measuring equipment consists of a central unit and two transmitting and receiving antennas. The radar emits an electromagnetic pulse of relatively high peak power and short duration. The depth range increases with the amount of energy of the generated pulse. The GPRs use antennas with frequencies from 10 MHz to 2 GHz. Antennas with a frequency of 400 MHz enable the registration of useful information up to a depth of approx. 8 m. Antennas with higher frequencies of about 1 GHz can only record up to about 1 m, however, at the expense of a smaller depth, a higher resolution of the order of centimeters is obtained. Waves propagating in the geological center are reflected, refracted and attenuated. The effectiveness of GPR operation depends on two parameters of electrical conductivity and the dielectric constant. High values of electrical conductivity reduce the radar's ability to penetrate. Electrical conductivity mainly depends on the water content, the amount of dissolved salt, the density and temperature of the material. The dielectric constant measures the ability of a material to store an electric charge when an electric field is applied to it. The condition for effective registration of useful information is the contrast against the dielectric constant between the examined rock layers. The greater the contrast, the greater the amplitude of the reflected wave. For air, the dielectric constant is about 1, for rock layers it ranges from a few to a dozen. The distance of the GPR from the measured layers should be reduced to a minimum and its value should not exceed 10 cm. By measuring the time needed to reflect the radar wave from the transmitting antenna to the receiving antenna, the thickness of the examined rock layer can be determined. Table 2 presents the values of electric permittivity and electric resistance for rocks [13, 18-20].

Table 2. Dielectric properties of selected rocks [21, 22].

Group of rock	Rock	Resistivity [Ωm]	Relative dielectric permittivity
Pyrogenic	granite	10^2-10^7	5-9
	syenite	10^2-10^5	7-14
	diorite	10^2-10^8	8-9
	gabbro	10^2-10^5	18
	basalt	10^2-10^6	12
	peridotite	10^2-10^3	7
Metamorphic	quartzite	10-10^5	7
	gneiss	10^2-10^7	8-15
	marble	10^2-10^5	8
Sedimentary	limestone	10^2-10^3	8-15
	dolomite	10^2-10^6	-
	sandstone	10-10^8	9-11
	sand	10-10^{10}	4-6
	clay	10-10^6	7-12
	slate	10-10^2	6-8
	coal	10^2-10^3	3-15
	lignite	10-10^3	4-5

3. Development of measurement results

Data from sensors installed on rock mining devices must be subjected to appropriate analysis so that they can be used to identify what rock is being cut. Proper selection of the sampling frequency is the basis of digital signal processing. According to the sampling theorem, the maximum signal frequency must not exceed half the sampling frequency. The signals received from the sensors are highly non-linear and non-stationary. The most frequently used methods of digital signal processing for non-stationary signals are: Short Time Fourier Transform, wavelet transform, Wigner-Ville transform [23]. For each signal, features representing relevant information about the signal are extracted. The final step is to train the neural network through the collected test samples. At the stage of training the neural network, the test samples of specific rocks are compared with the samples supplied by the sensor. A neural network can be trained to solve tasks such as recognition, sound, texture and vibration.

4. Conclusions

Identifying the boundaries of geological strata is an important task in the development of an automatic and autonomous system for obtaining raw materials. The article discusses the most popular groups of methods and the tools used in them. The selection of the appropriate method very strongly depends on the type of material obtained and the environmental conditions in which the extraction is to be carried out. Sensors detecting natural gamma radiation can be used mainly for detecting the boundary of carbon and rock. It has a good detectability, but requires the collection of many measurement samples for background determination. With the global trend to move away from coal-based energy, this method may no longer be useful. Recognition by means of acoustic waves and vibrations is highly effective, they are relatively cheap methods that do not require the use of complicated technical solutions. Recognition based on the analysis of camera images has good effectiveness, but has little chance of being used in dusty conditions. The use of radar is the most expensive solution with high detectability. However, the closest possible distance between the sensor and the measured rock layers should be ensured, which is not always possible. Another problem when developing an autonomous system for obtaining raw materials is the need to process a very large amount of data obtained from various sensors in a short time. Artificial intelligence tools are suitable for such analyzes. In further research, it is expected to develop conceptual versions of the recognition system using various AI tools, especially artificial neural networks.

5. References

[1] M. Sąsiadek and J. Basil 2018 Świadomość i poziom wdrożenia koncepcji Przmysł 4.0 w wybranych polskich i czeskich przedsiębiorstwach w *Innowacje w Zarządzaniu i Inżynierii Produkcji*, Opole,.
[2] W. Skoczyński 2019 Rola sensorów w systemach wytwórczych *Napędy i sterowanie,* Czerwiec.
[3] J. Fraden 2010 Handbook of Modern Sensors, London: Springer-Verlag New York.
[4] K. Łuszczek 2011 Poszukiwania nowych zasobów surowców w układzie słonecznym *Prace Naukowe Instytutu Górnictwa Politechniki Wrocławskiej* pp. 85-94.
[5] T. Okulski 2008 Zawartość uranu i toru w węglach polskich i amerykańskich. *Polityka energetyczna* **11**(1) pp. 323-334.
[6] S. Plewa 1953 Zastosowanie badań promieniotwórczości skał w otworach wiertniczych *Przegląd Geologiczny* 1(8), pp. 22-25.
[7] Z. Ningbo and L. Hangyou 2018 Radiation characteristic of natural gamma-ray from coal and gangue for recognition in top coal caving *Scientific Reports*.
[8] D. Tomaszewski *ElektronikaB2B* [Online]. Available: https://elektronikab2b.pl/technika/50568-przeglad-roznych-rodzajow-czujnikow-do-pomiaru-wibracji (access date: 10 03 2021).
[9] *WAT –Wydział Elektroniki Instytut Systemów Elektronicznych* [Online]. Available: https://zoise.wel.wat.edu.pl/dydaktyka/Czujniki%20i%20przetworniki/CW5_ Przetworniki_piezoelektryczne_teoria_2015.pdf. (access date: 10 03 2021).
[10] J. Balcerzak 2015*Rozpoznawanie wzorców ruchowych przy użyciu żyroskopu i akcelerometru* Warszawa: Politechnika Warszawska.
[11] M. Andrejašic. [Online]. Available: http://docplayer.net/20641887-University-of-ljubljana-faculty-for-mathematics-and-physics-department-of-physics-seminar-mems-accelerometers-author-matej-andrejasic.html. (access date: 10 03 2021).
[12] K. Jamielniak. [Online]. Available: https://docplayer.pl/8798050-Sensoryka-3-czujniki-drgan.html . (access date: 10 03 2021).
[13] K. Basu 1997 *Fasibility of an integrated thin seam coal mining and waste disposal system,* Blacksburg: Master of sience.
[14] Z. Guoxin, W. Zengcai and Z. Lei 2016 Recognition of rock–coal interface in top coal caving through tail beam vibrations by using stacked sparse autoencoders *Jurnal of vibroengineering* **18**(7).
[15] X. Jing, W. Zhongbin, T. Chao, L. Daohua, W. Baigong, S. Zhen and T. Yanbing 2018 Cutting Pattern Identification for Coal Mining Shearer through Sound Signals Based on a Convolutional Neural Network *Symmetry* **10**(12).

[16] L. Yong, C. Gang, C. Xihui and L. Chang 2019 Coal–rock interface recognition based on permutation entropy of LMD and supervised Kohonen neural network *Current sience* pp. 96-103.
[17] M. Hui-ling and M. Li 2016 Characteristic Analysis and Recognition of Coal-Rock Interface Based on Visual Technology *International Journal of Signal Processing, Image Processing and Pattern Recognition* pp. 61-68.
[18] J. Ralston and A. Strange 2015 An Industrial Application of Ground Penetrating Radar for Coal mining Horizon Sensing w *ISAP*, Hobart.
[19] S. Miao, X. Liu, Z. Liu and L. Zhang 2018 Ground Penetrating Radar Based Experimental Simulation and Signal Interpretation on Coal-Rock Interface Detection.
[20] J. Karczewski. [Online]. Available: http://home.agh.edu.pl/~karcz/zasadaDzialania.html. (access date: 01 03 2021).
[21] S. Gierlotka 2019 Własności elektryczne i magnetyczne skał *Energetyka* pp. 325-328.
[22] R. Drożdżak and K. Twardowski 2010 Przenikalność dielektryczna ośrodków porowatych-czynniki wpływające na jej zmienność *Wiertnictwo, Nafta, Gaz* **27** pp. 111-120.
[23] T. P. Zieliński 2014 *Cyfrowe przetwarzanie sygnałów. Od teorii do zastosowań*. Warszawa: Wydawnictwa Komunikacji i Łączności.
[24] W. Żyłka, M. Żyłka and Z. Szczerba 2017 Czujniki w praktyce kształcenia inżynierskiego. *Edukacja-Technika-Informatyka*.

IOP Conf. Series: Materials Science and Engineering 1182 (2021) 012036 doi:10.1088/1757-899X/1182/1/012036

Dielectric barrier discharge analysis from the point of view of supply voltage and reactor topology

C V Lungu[1] and B Hnatiuc[2]
[1]"Dunărea de jos" University from Galați - România, Department of Automation and Electrical Engineering, Str. Domnească, No. 47, 800008, Galați, România.
[2]Constanta Maritime University, Faculty of Marine Engineering, 104 Mircea cel Batran Street, 900663 Constanta, Romania.

E-mail: lungu.cristian.victor@gmail.com

Abstract. This paper aims to provide a comprehensive analysis of the output of DBD reactors considering 2 different power supplies that provide rectangular command voltage with variable duty cycle and frequency. The experiments presented were performed with these power supplies considering four geometries of DBD reactor. At this stage the input parameters of the power supplies are: frequency of control pulses, duty cycle and high voltage level applied on the DBD reactors by choosing a certain number of turns for the primary windings of the high voltage coil. The parameters considered as outputs from the reactor-power supply assembly were: the high voltage output and the electric current from the primary induction coil.

1. Introduction

Dielectric-barrier discharges (DBDs), also referred to as barrier discharges or silent discharges have found a number of interesting industrial applications in addition to the historical ozone generation. The generation of powerful coherent infrared radiation in CO2 lasers and of incoherent ultraviolet (UV) or vacuum ultraviolet (VUV) excimer radiation in excimer lamps are examples of more recent developments. [1]

Typical electrode configurations of planar and cylindrical dielectric-barrier discharges are given in figure 1. DBDs are characterized by the presence of one or more insulating layers in the current path between metal electrodes in addition to the discharge gap(s).[1]

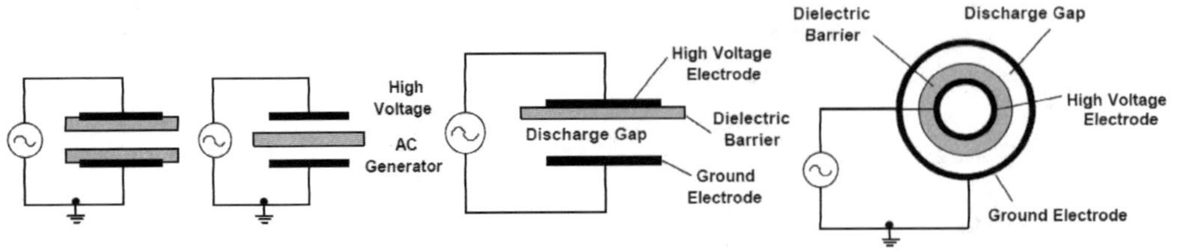

Figure 1. Common dielectric-barrier discharge electrode configurations.[1]

This paper focuses on the DBD discharge by analysing the electrical parameters using two different power supply topologies. The first power supply topology is half bridge topology and the command circuit is built around TL494 integrated circuit. The second power supply is a fly-back topology with the command circuit built with the help of UC/SG 2525 integrated circuit.

2. Experimental setup

The experiments performed with the power supplies considered all four geometries of DBD type reactors. The high voltage transformer secondary consists of 2500 windings placed on a ferrite core. At this stage the input parameters of the sources that were adjusted are:
-Frequency of control pulses - by means of potentiometers connected to the specific inputs of the command integrated circuits;
-Duty cycle - by means of potentiometers connected to the specific inputs of the command integrated circuits;
-The high voltage level applied on the DBD reactors was done by choosing a certain number of turns in the primary winding of the high voltage coil (18 and 22 windings), which allowed the modification of its transformation ratio.

The experimental setup block schematic is presented in figure 2. The supply voltage for the power section is 96V for the TL494 power supply and 48V for the SG2525. The command supply voltage used for TL494 was 12V because of the integrated circuit limitations. The SG2525 is more permissive from this point of view, thus the supply voltage for the command part used was 20V.

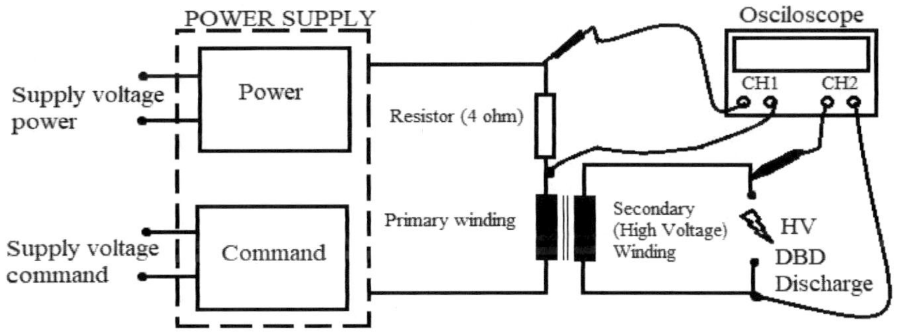

Figure 2. Block schematic of the experimental setup.

2.1. Experimental adaptation for all 4 DBD discharge electrode configurations

The 4 DBD discharge topologies are presented in figure 3. The dimensions for reactor 1 to reactor 3 are: the inside diameter of the ground electrode of 10 mm, the outside diameter of the ground electrode of 16.6 mm, the diameter of the HV electrode of 7.8 mm. The dielectric material used was glass.

The first reactor presented is the one consisting of air-dielectric configuration. The dimensions for this reactor are: the discharge space (D1.1) = 2 mm, the thickness of the dielectric (D1.2) = 1.5 mm.

The second reactor presented is the one consisting of dielectric-air-dielectric configuration. The dimensions for this reactor are: the discharge space (D2.2) = 1 mm, the thickness of the dielectric (D2.1 and D2.3) = 1.5 mm.

The third reactor presented is the one consisting of air-dielectric-air configuration. The dimensions for this reactor are: the discharge space (D3.1 and D3.3) = 1 mm, the thickness of the dielectric (D2.2) = 1.5 mm.

The fourth and the last reactor presented has circular configuration. The dimensions for this reactor are: the inside diameter of the ground electrode of 7 mm, the outside diameter of the ground electrode of 9 mm, the discharge space (D4.2) of 1.5 mm, the thickness of the dielectric (D4.3) of 1.5 mm, the diameter of the HV electrode (D4.1) of 1.7 mm and D4.4 = 17 mm.

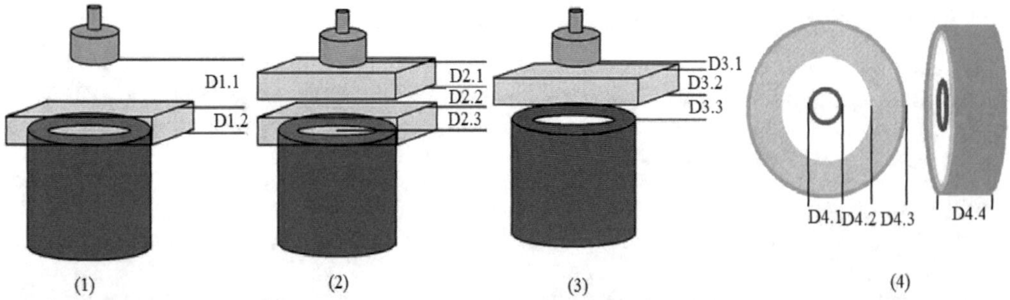

Figure 3. Experimental setup representation of the common dielectric-barrier discharge topology configurations.

2.2. Schematic of the TL494 power supply

The circuit built around the integrated TL494 is versatile and can control a half bridge or full bridge of MOSFETs or IGBTs through a gate control transformer (GDT), see figure 4. This should make a driver who is able to control the flyback transformers found in CRT TVs and computer monitors.

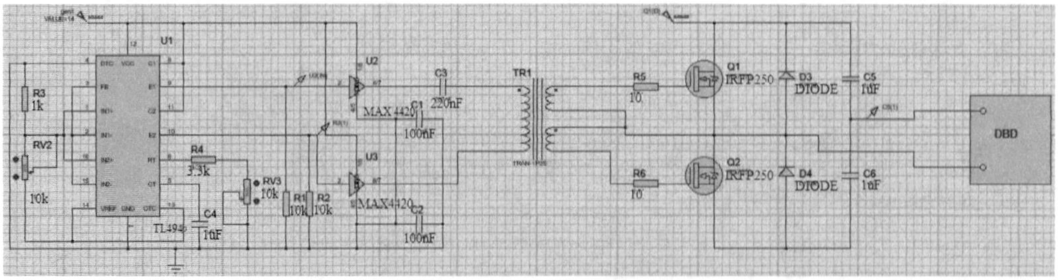

Figure 4. Complete schematic of the power supply built with TL494.[2, 3]

TL494 uses a 5% dead time to ensure correct switching and at frequencies above 150 kHz, this minimum downtime is higher. The output control on pin 13 is set to a reference voltage of 5 Volts on pin 14, which makes the two outputs of transistors operate in push-pull mode, which will be used to drive each of their ICs inverted MOSFET driver.

2.3. Schematic of the SG2525 power supply

The SG2525 integrated circuit is used in the standard PWM generator / oscillator connection where the oscillation frequency is determined by C2, R7 and RV2. The RV2 potentiometer can be adjusted to obtain exact frequencies according to the required application specifications, see figure 5.

The frequency range for this source is from 6 kHz to 60 kHz, between the two outputs at pin 11 and pin 14. The two outputs oscillate alternately in push (totem) mode, driving the connected Mosfet transistor in the flyback configuration.

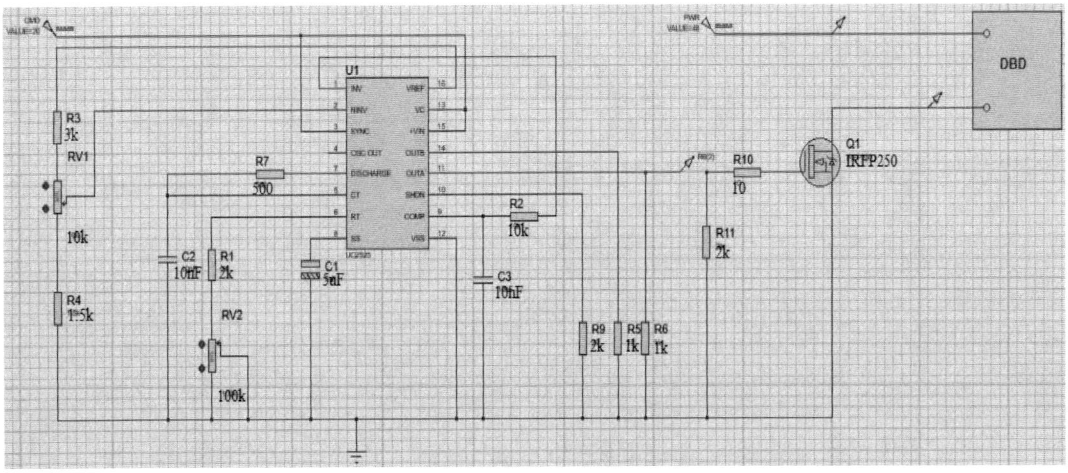

Figure 5. Complete schematic of the power supply built with SG2525.[2, 3]

3. Experimental results

The parameters measured taken under the previously described conditions are presented in table 1. Some examples of waveforms extracted from the oscilloscope are presented below.

Table 1. Experimental values (V_{rms} and I_{rms}) extracted from the data collected.

Reactor	No. of	Measuread parameters	TL 494 Power Supply				SG 2525 Power Supply			
			Frequency 30kHz/ PWM 20%	Frequency 30kHz/ PWM 50%	Frequency 60kHz/ PWM 20%	Frequency 60kHz/ PWM 50%	Frequency 30kHz/ PWM 20%	Frequency 30kHz/ PWM 50%	Frequency 60kHz/ PWM 20%	Frequency 60kHz/ PWM 50%
Reactor 1	**18**	V_{rms}	2717.569	2727.58	1162.284	1159.388	4553.1059	4489.2843	3207.94	7536.677
		I_{rms}	4.071	4.084	2.38597	2.4018	2.9111	4.45934	3.78236	4.20724
	22	V_{rms}	2717.569	2553.1022	1252.2647	1249.340786	558.5887	4481.6842	5752.63	6765.545
		I_{rms}	3.296	3.296144	2.020988	1.988655	0.981465	2.012173	3.355672	3.793
Reactor 2	**18**	V_{rms}	2960.316	3001.958	1424.688	1424.688	4416.59	3500.722	2074.972	6745.254
		I_{rms}	4.24546	4.174517	7.004	7.004	4.289	5.6561	1.6747	4.0823
	22	V_{rms}	2789.77	2681.95	1272.366	1263.24917	4855.165	4000.4833	2113.895	4550.017
		I_{rms}	3.43	3.39011	2.02099	2.011816	3.88648	2.85483	3.308715	4.846112
Reactor 3	**18**	V_{rms}	3858.57	4062.65	2070.7	2073.4893	4429.2972	4178.0155	3407.00455	7511.226
		I_{rms}	4.12024	4.129	2.7036265	2.7943	4.043663	5.07913	2.38528	4.7055
	22	V_{rms}	3619.062	3619.0621	1859.666	1859.666	4260.93104	5089.36211	2642.208	7955.3852
		I_{rms}	3.35512	3.355	2.23	2.23	2.956	2.26996	2.174424	4.224112
4	**18**	V_{rms}	2712.47	2720.6	1255.39	1308.8	3489.6	4942.123	4942.123	5466.649187
		I_{rms}	3.8164	3.84	2.446	2.419	1.50014	2.3467	2.3467	5.3075305
	22	V_{rms}	1308.8	2795.67	1402.727	1672.35	3764.2423	4626.25	5091.5564	981.353
		I_{rms}	2.42	3.1954	2.148	2.3294	1.428426	1.88241	3.9093	3.796

3.1. Experimental results for TL494 power supply

The measured discharge parameters from the TL494 power supply are more stable compared to the SG2525 power supply. The power section uses a DC voltage of 96V and a half bridge topology which provides a +/- output, see figures 6-11.

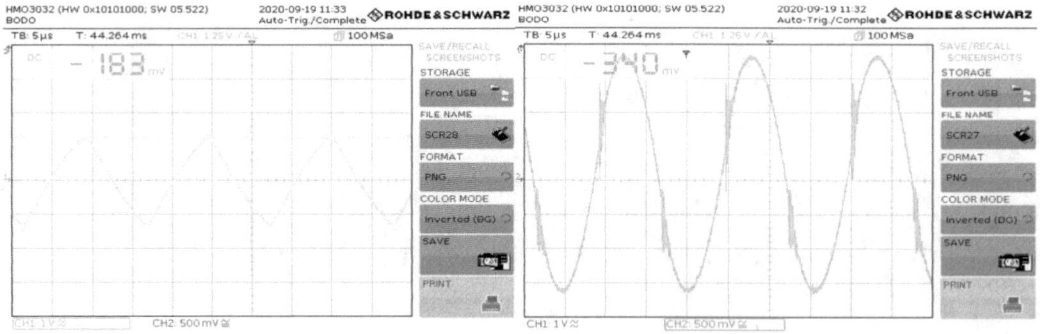

Figure 6. Measured electrical parameters using the TL494 power supply, input voltage from a 4 ohm resistor (a), output voltage (b), DBD reactor 1, 22 primary windings, Frequency 60kHz, PWM 20%.

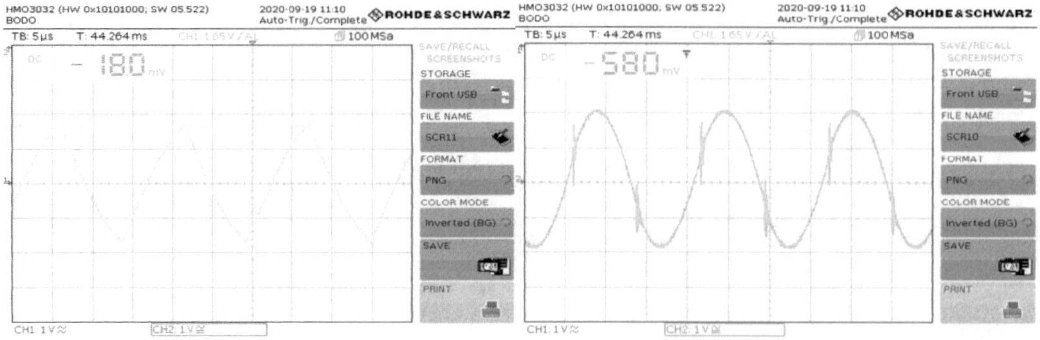

Figure 7. Measured electrical parameters using the TL494 power supply, input voltage from a 4 ohm resistor (a), output voltage (b), DBD reactor 2, 18 primary windings, Frequency 60kHz, PWM 20%.

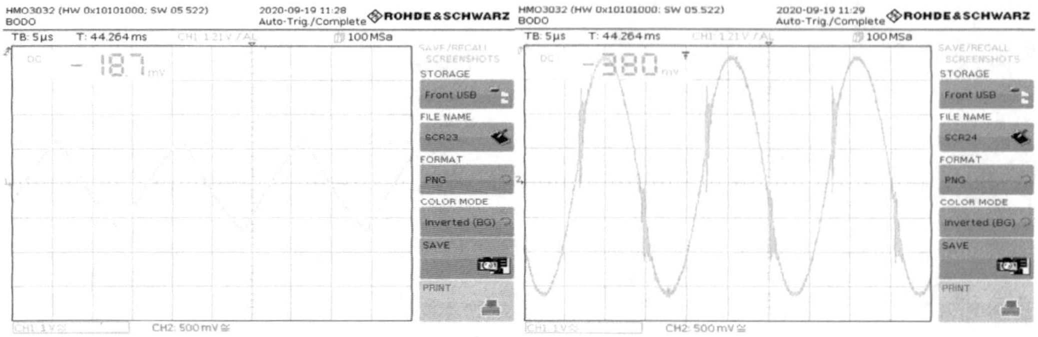

Figure 8. Measured electrical parameters using the TL494 power supply, input voltage from a 4 ohm resistor (a), output voltage (b), DBD reactor 2, 22 primary windings, Frequency 60kHz, PWM 50%.

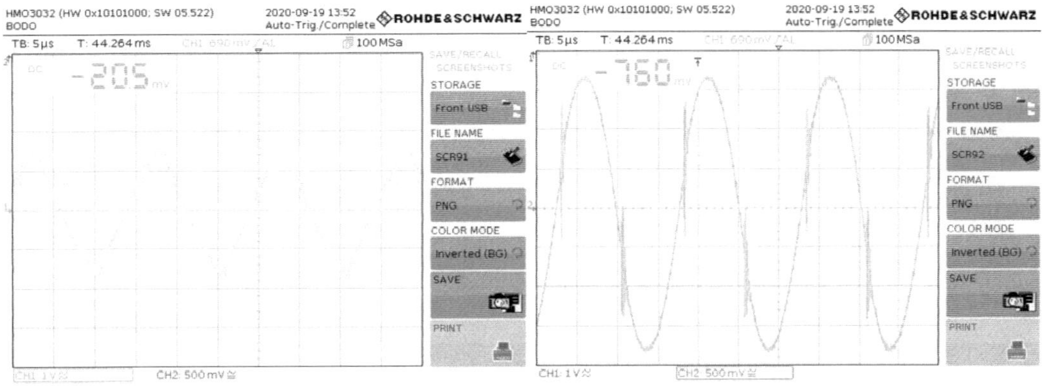

Figure 9. Measured electrical parameters using the TL494 power supply, input voltage from a 4 ohm resistor (a), output voltage (b), DBD reactor 4, 18 primary windings, Frequency 60kHz, PWM 20%.

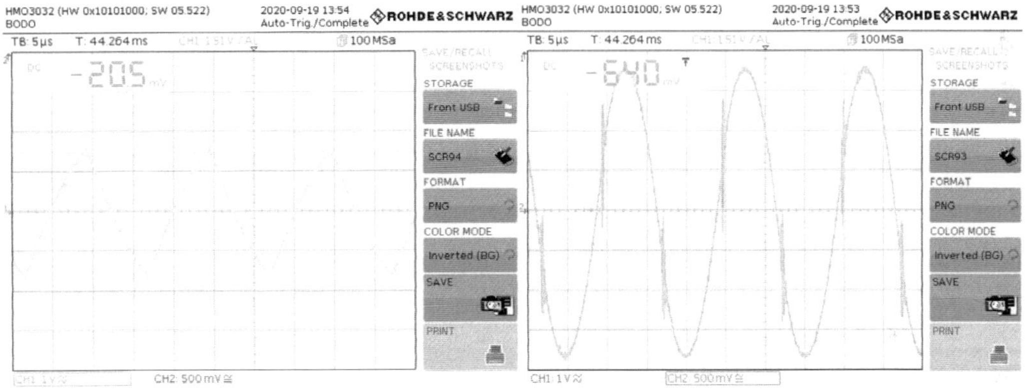

Figure 10. Measured electrical parameters using the TL494 power supply, input voltage from a 4 ohm resistor (a), output voltage (b), DBD reactor 4, 18 primary windings, Frequency 60kHz, PWM 50%.

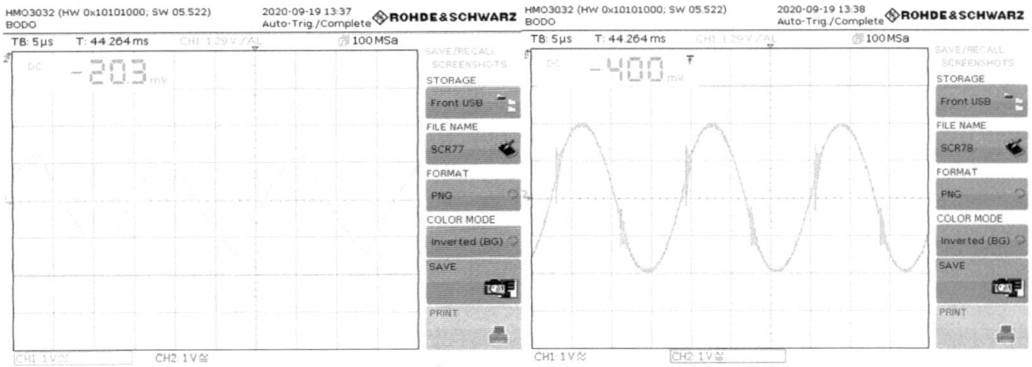

Figure 11. Measured electrical parameters using the TL494 power supply, input voltage from a 4 ohm resistor (a), output voltage (b), DBD reactor 4, 22 primary windings, Frequency 60kHz, PWM 20%.

3.2. Experimental results for SG2525 power supply

The SG2525 power supply uses a DC voltage of 48V and a flyback topology. The output is not as regular as the TL494 power supply also the discharge is not regular either, see figures 12 - 14.

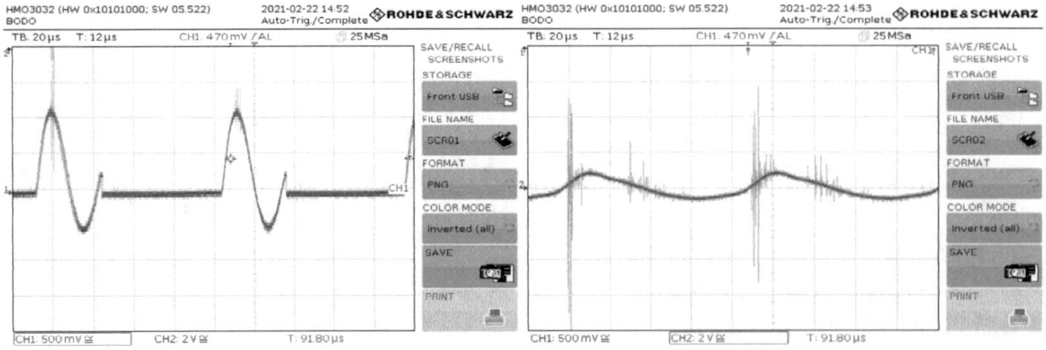

Figure 12. Measured electrical parameters using the SG2525 power supply, input voltage from a 4 ohm resistor (a), output voltage (b), DBD reactor 1, 22 primary windings, Frequency 30kHz, PWM 20%.

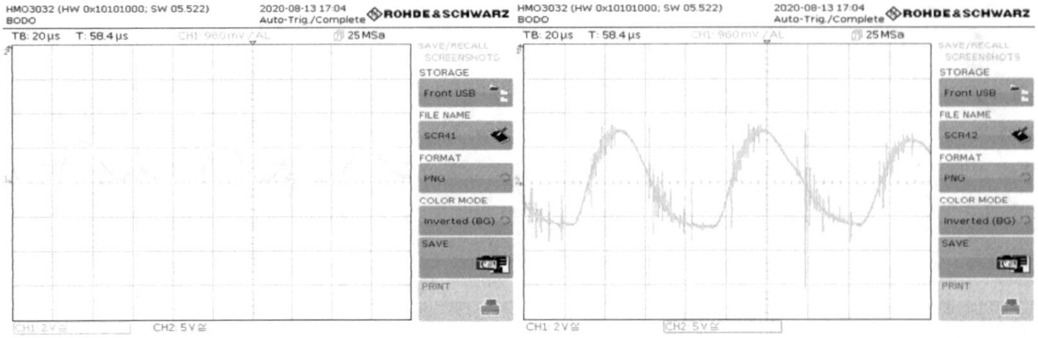

Figure 13. Measured electrical parameters using the SG2525 power supply, input voltage from a 4 ohm resistor (a), output voltage (b), DBD reactor 4, 18 primary windings, Frequency 30kHz, PWM 50%.

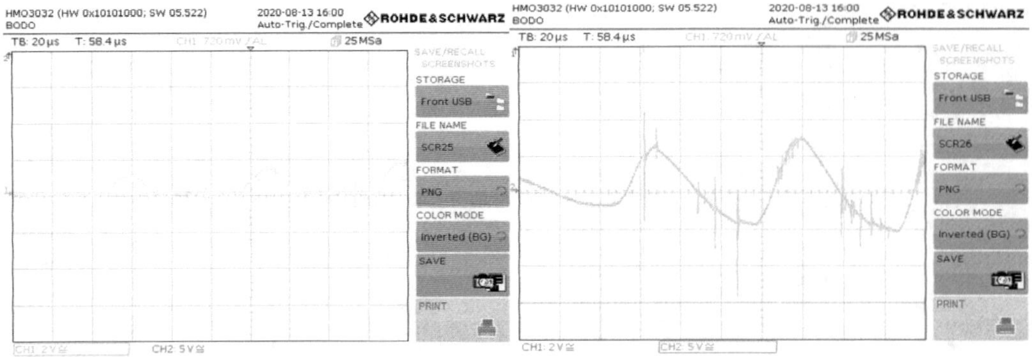

Figure 14. Measured electrical parameters using the SG2525 power supply, input voltage from a 4 ohm resistor (a), output voltage (b), DBD reactor 4, 22 primary windings, Frequency 30kHz, PWM 20%.

It was observed that the current values in the primary of the high voltage coil are in the range of 1 A - 8 A, and the voltage values measured in the secondary of the high voltage coil are in the range 1.1 kV - 8 kV.

When using the source with TL494 the differences of the electric current values between a PWM of 20% and a PWM of 50% are insignificant. When using the SG2525 power supply there are important differences between the current values for the two PWM factors.

4. Conclusions

Power supplies provide an understanding of the types of output circuits that operate with different types of electric discharges reactors. As can be seen in the previous figures, the presence of the dielectric limits the electric current of a DBD discharge. Thus, the reactor configuration dielectric - air - dielectric had the weakest DBD discharge.

The best display of the DBD discharge (the more intense electric discharge) took place in the tubular reactor, mainly due to the use of the internal copper electrode and the external aluminium electrode which ensured a good emissivity of the electrons.

DBD electrical discharges, due to many simultaneously plasma channels and the fact that the discharge goes out quickly, require a high frequency for their maintenance. On the other hand during the experimental part, the frequency had to be limited due to ferrite saturation and MOSFET limitation.

Since both power supplies operate at close frequencies and the same width factor, also considering that each operates at different supply voltages for the power section, the difference in amplitude is due to the difference between the half bridge topology and the flyback topology.

5. References

[1] Kogelschatz U., Eliasson B., Egil W. 2020 Fundamentals and applications of dielectric-barrier discharges, ABB Corporate Research Ltd, 5405 Baden, *Switzerland proceedings of the 23rd ICPIG*, pp 1–20.

[2] Lungu C.V., Hnatiuc B. 2020 Simulation of power supplies used for nonlinear electrical discharges, *Technium: Romanian Journal of Applied Sciences and Technology* **2**(1) 79–84 DOI: https://doi.org/10.47577/technium.v2i1.45.

[3] Lungu C. V. and Hnatiuc B. 2020 Power supplies for DBD used for surface treatments, *Proc. SPIE 11718, Advanced Topics in Optoelectronics, Microelectronics and Nanotechnologies* X, 1171832, https://doi.org/10.1117/12.2572232.

Acknowledgement

This work was supported by a grant of the Romanian National Authority for Scientific Research and Innovation, CCCDI - UEFISCDI, project number COFUND-ICT-AGRI-FOOD-MERIAVINO-1, within PNCDI III.

Development of machine steering with use of PLC as introduction to 4.0 Industry

K Lysek, A Gwiazda and K Herbuś
Silesian University of Technology, Faculty of Mechanical Engineering, Institute of Engineering Processes Automation and Integrated Manufacturing Systems, Konarskiego 18A, 44-100 Gliwice, Poland

E-mail: kamil.lysek@polsl.pl

Abstract. Article was written in order to make an attempt at developing a simple steering system for numerically controlled machines, with use of virtual modelling environment and an actual driver together with HMI panel. It was made as a response to demand of centralized system for steering a smart factory containing many machines. In the article, creation of a simple single machine steering system was described, which can be upscaled to the entire production hall. Used steering in this case is setting of position in an absolute system, in which only 3 variables X, Y and Z are known, and the operator can perform a correction of location and speed. Obviously, in real conditions these values will be set by a different program - like a post processor generating G-code. Usage of one data communication server can allow integration of devices within one production, and this allow creation of a smart factory suitable for 4.0 industry. In order to obtain such communication, an OPC server was used. Furthermore, the process of creating and connecting machines or programs to it was briefly presented.

1. Introduction

1.1. Hardware in the loop

Undoubtedly one of most important issues regarding modern factories is the optimization of machine movement system and production. Nowadays factories are built in the shortest possible time in order to minimize losses and to make profits as fast as possible. For this purpose, simulation of an entire production process of a given line in order to optimize it is helpful. This simulation can be done without using real steering devices. In this case, the software-in-the-loop simulation is used, although in the later step, it is recommended to perform a hardware-in-the-loop simulation. Work scheme of such loop based can be presented using schematic way [1-4], what is presented in figure 1:

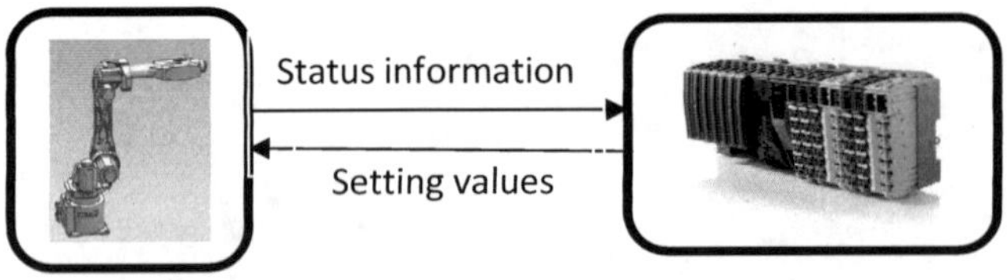

Figure 1. Scheme of Hardware in the loop.

Entire process can be described this way: controller (right figure) receives information about current state of simulated object (left figure), and then does a correction using positioning movements. Extremely important advantage of this way of simulation is the possibility of the most precise mapping of a real signal flow in real-time without risk of damaging the executive elements [4-6].

1.2. Industry 4.0

In a nutshell, Industry 4.0 is the next revolution of industrial process, which was firstly mentioned back in 2011, and thanks to which it is assumed that factories can run 30% faster with 25% efficiency growth [7, 8].

More specifically, it will consist of integrating many devices currently working in an automated way, but without mutual communication between themselves. Undoubtedly in order to allow such communication, some data management center should be started, like an OPC server, which will be used in development of the current article [7, 9, 10].

Mutual communication between devices will also allow to better manage the factory as a whole, and increase the automatization, but also customization of the production. Great example of the last one can be a factory in which the model in a computer will generate a G-code, then communicating, for example, through an OPC server with a robot, fix the machined object on a machine, and then the machine also being an integral part of the base in factory would machine the detail basing on the provided code. Even though this vision still is a long way off, there already are companies that in an automated way implemented production that is to certain extent customizable. Example of such company can be Misumi, which using this solution is able to deliver its products to clients all around the world in a few days, allowing the clients to choose from 80 sextillions (10^{21}) configurations. Example view from a factory of this manufacturer can be seen in figure 2 [11, 12].

Figure 2. Misumi factory [11].

2. Preparation of the machine tool - declaration of objects types

Entirety of work will be done in the following software: NX by Siemens and Automation Studio by B&R. First one will be responsible for the simulated milling machine, and the second one for the controller. The machine will be controlled using 3 coordinates using the absolute method, that is, the specific coordinates of a specific machine axis will be given. The milling machine that will be used as an example of a controlled machine is a simple machine made of aluminum. Whole steering can be described using scheme in figure 3.

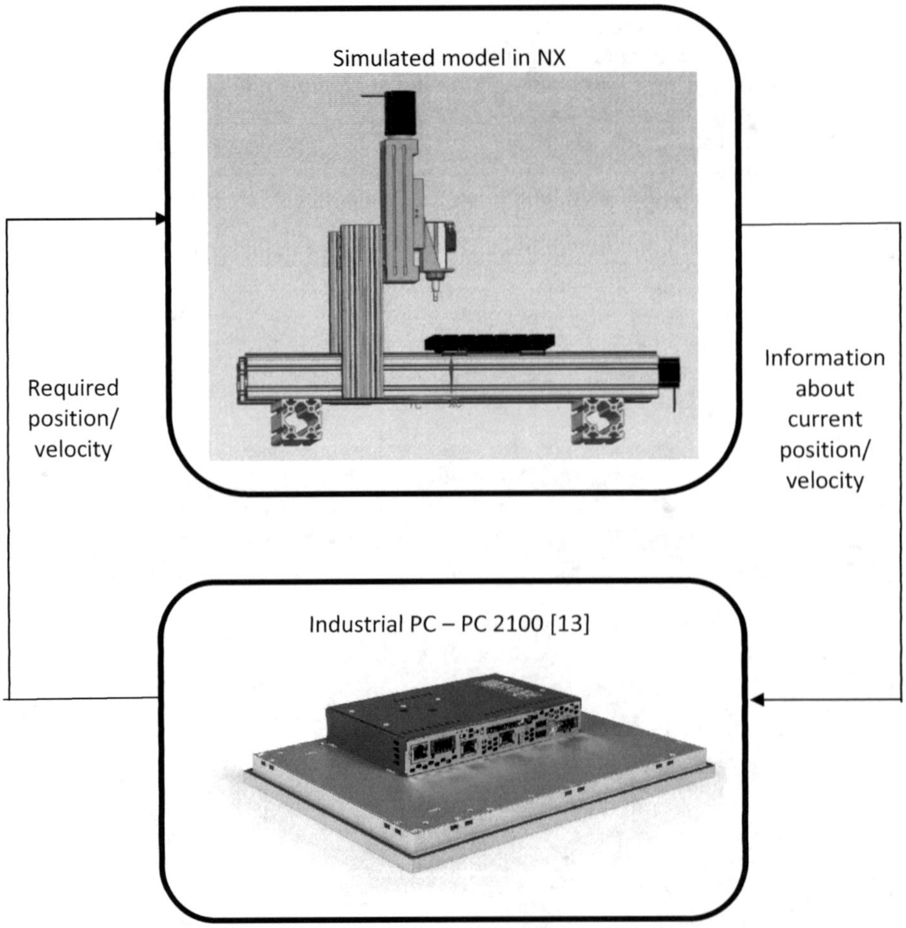

Figure 3. Scheme of communication.

Information shared as required position and speed are transferred via an Ethernet cable from a PLC/industrial PC to a PC, specifically to the model in NX program. The other way transferred are the information about current position and speed of an object on a specific axis, what allows to control the executed move and monitoring the derogations. When it comes to model preparation, individual elements were declared as rigid body, what can be seen in figure 4.

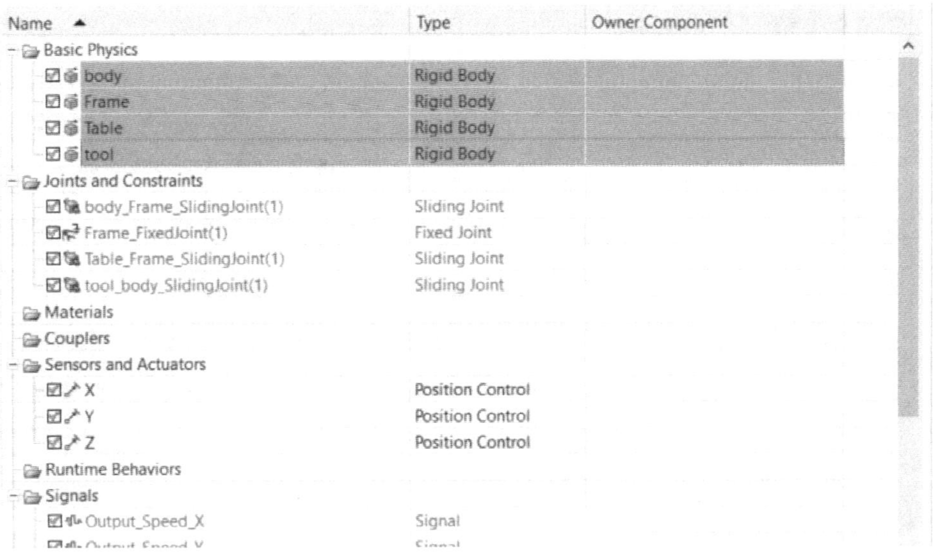

Figure 4. Declared bodies in machine.

Milling machine frame was declared as a fixed joint, while the moving elements such as the table was declared as sliding joint, as seen in figure 5.

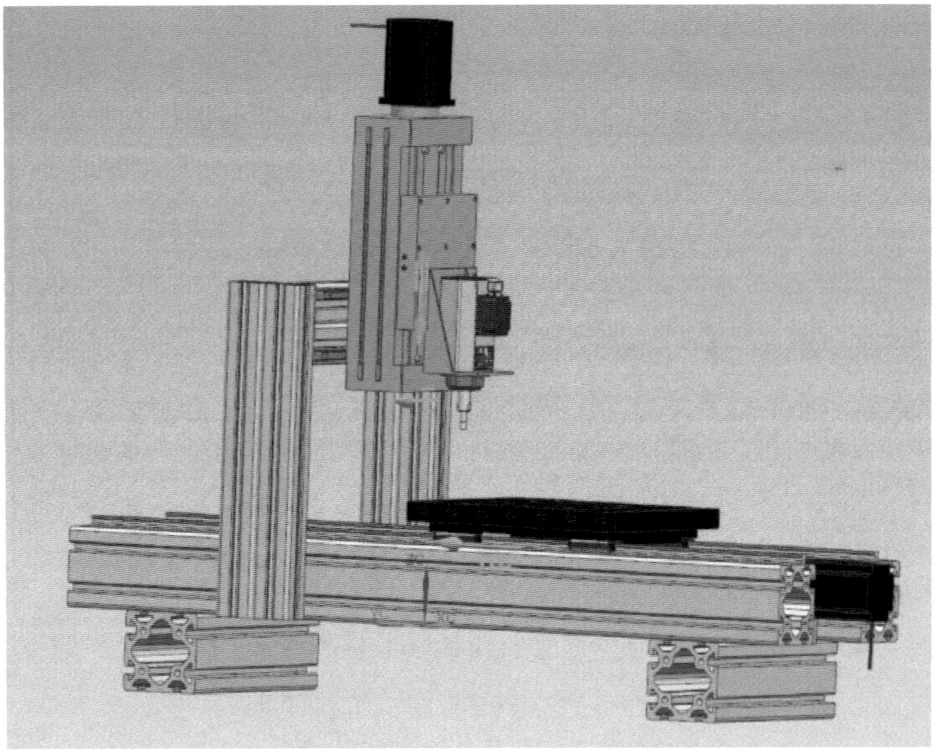

Figure 5. Joints of milling machine.

IOP Conf. Series: Materials Science and Engineering 1182 (2021) 012037 doi:10.1088/1757-899X/1182/1/012037

3. Obtaining communication between model and PLC

In order to allow information transfer between PLC and the model, an auxiliary system had to be chosen. Due to many advantages such as scalability and independence from Windows, an OPC UA server was chosen. It allows to communicate with many different objects using different system architecture. In order to activate the OPC server option, it must be enabled in the controller - example using the B&R controller can be seen in figure 6. It is also important to pay attention to the port number, as setting it will be required in order to correctly connect the synchronized simulation. Furthermore, communication on the first security level, that is without authentication should be enabled. Using this security level means that all certificates will be recognized as valid [14-16].

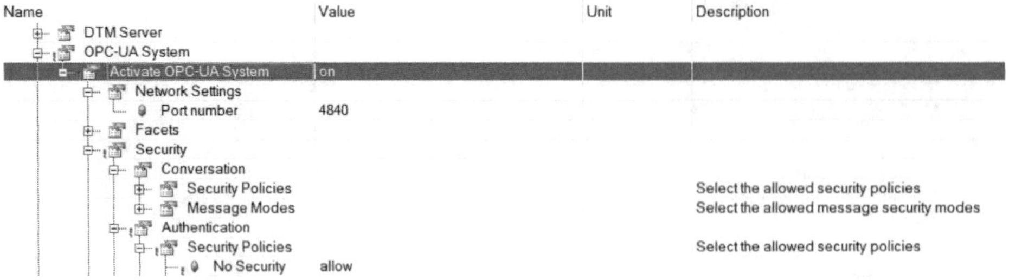

Figure 6. Activating the OPC UA server in the controller.

In the entire process a real controller is used, as seen in figure 7,

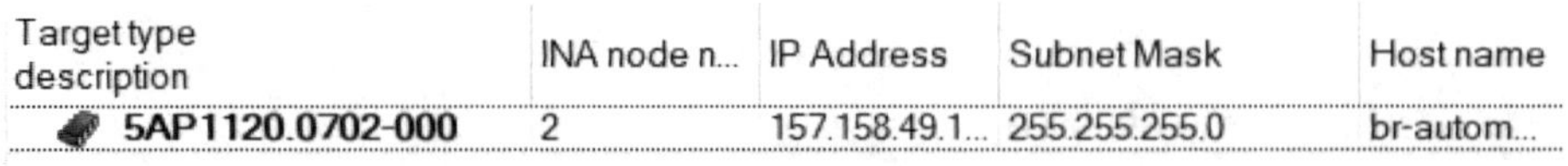

Figure 7. Connected PLC.

In order to connect both modules, an OPC server with a previously set port and selected security should be added, as in figure 8. In the given example the port is 4840.

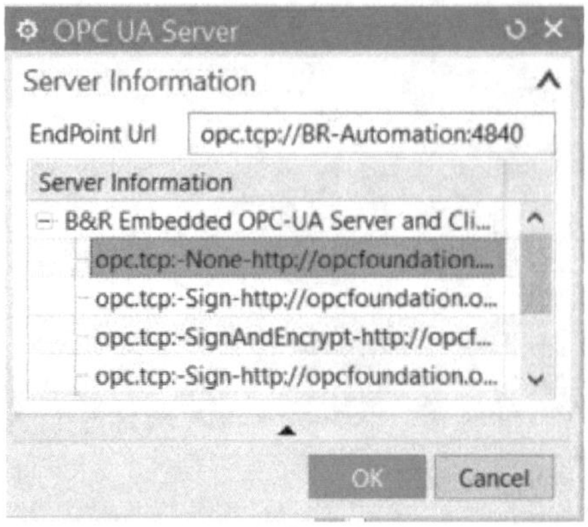

Figure 8. Adding an OPC UA server.

Then, the signals from the chosen OPC server should be mapped with those appearing in the model. Individual connections were shown in figure 9. As seen, controlled will be both the speed and the position on specific axes.

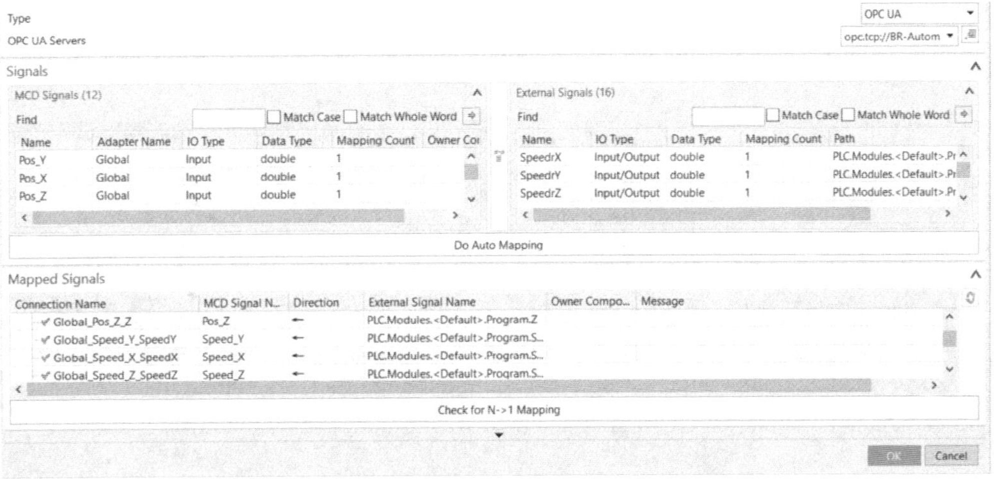

Figure 9. Signal mapping between OPC and simulation.

4. Results

Control happens using the HMI panel visible in figure 10. As seen, various speed and exact position on a specific axis of a machine can be set in it.

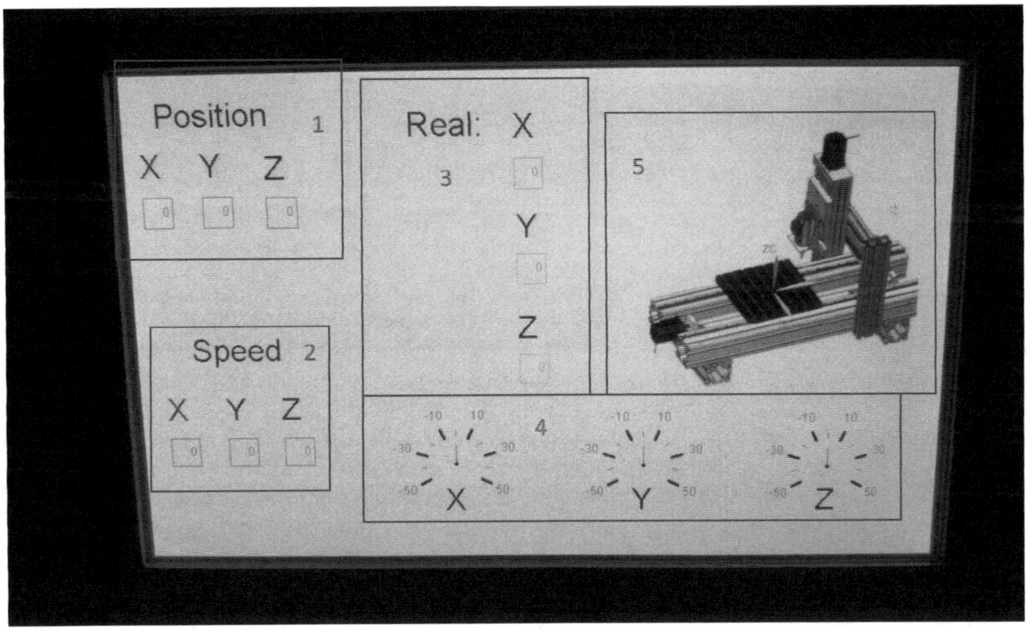

Figure 10. HMI Panel.

It can divided into 5 areas (like in figure 10):

1. Position given by the operator
2. Speed given by the operator
3. Real positions
4. Real speed

5. Picture with marked axes of the machine

Juxtaposition of the panel and obtained shift is shown in figures 11 and 12. Values are displayed in mm (position) or mm/s (velocity).

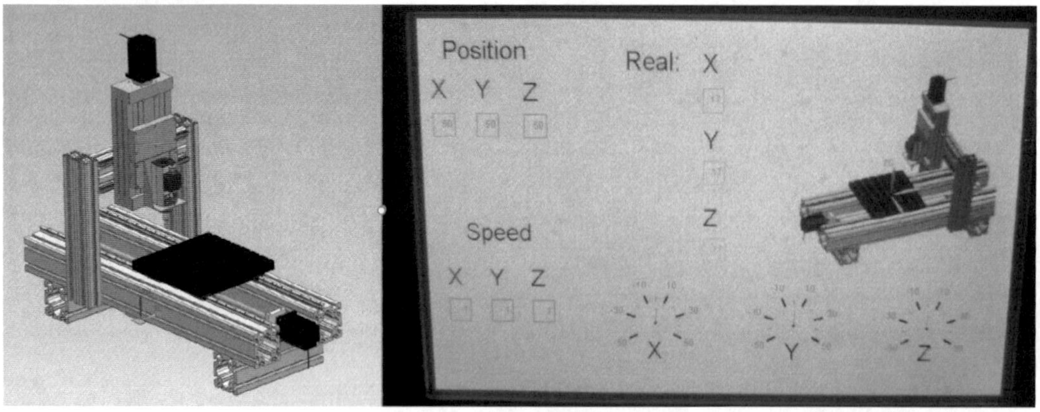

Figure 11. HMI panel and simulation of machine work – current data from robots moves.

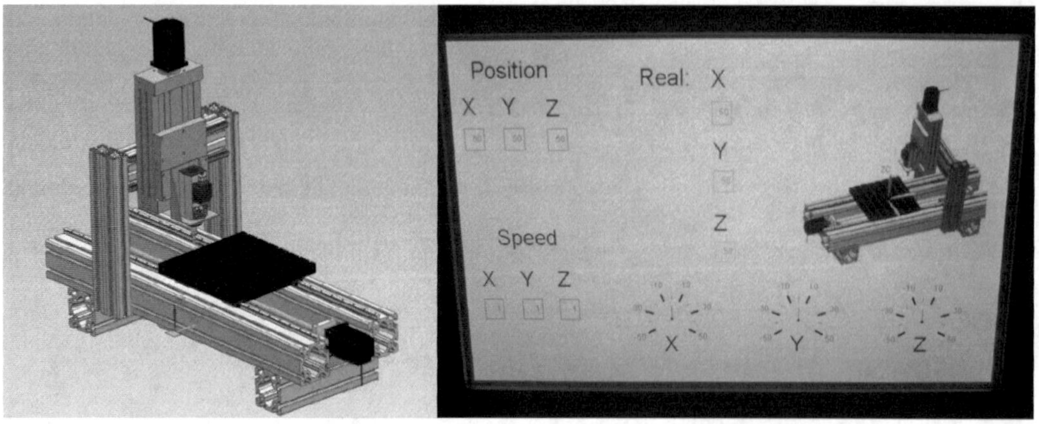

Figure 12. HMID and simulation of machine work – the end of movement.

The comparison of figures 11 and 12 shows that the milling machine was moving in a way that was planned. As can be seen, the speed values displayed on the gauges are equal to the set ones (except the end position where they indicate 0). Moreover, the end position is equal to the specified one.

5. Conclusions

As part of the work, the connection was made between the simulated model and the controller, using the PLC. This connection was made successfully, and the machine is controlled to full extent in a global coordinate system. Basing on the obtained connection, person performing the simulation is able to read the given and the current location of an element and give their own in the operator panel.

Scope for development of current work is seen in the integration of a bigger number of devices in order to make a virtual factory, and in extending the possibilities of controlling a machine. As it was mentioned, it would be ideal to read the feed setting values using the program that will read them from the G-code, previously generated by the post-processor.

6. References

[1] Lysek K, Gwiazda A, Herbuś K 2019 communication between cad systems and the PLC controller *International Journal of Modern Manufacturing Technologies* **XI**(3) 122-127 ISSN 2067–3604, Special Issue.
[2] https://www.br-automation.com/pl/produkty/plc-systems/, accessed date: 19.03.2021.
[3] Pietrusiewicz K 2015 Projektowanie mechatroniczne. Technika Hardware-in-the-loop a założenia Industry 4.0, *Napędy i sterowanie* No. 4 pp. 92-98.
[4] Davies R 2015 *Industry 4.0. Digitalisation for productivity and growth.*
[5] Chi G.Lee1Sang C.Park 2014 Survey on the virtual commissioning of manufacturing systems - *Journal of Computational Design and Engineering* **1**(3) 213–222.
[6] Nalepa B, et al 2018 *IOP Conf. Ser.: Mater. Sci. Eng.* **400** 052007.
[7] Dmowski J et al. 2016 *Przemysł 4.0 PL Szansa czy zagrożenie dla rozwoju innowacyjnej gospodarki?* BCG, Warszawa.
[8] Ślusarzyk B 2018 Industry 4.0: are we ready? *Polish Journal Of Management Studies* **17**(1).
[9] Kipper L M et al 2019 Scopus scientific mapping production in industry 4.0 (2011–2018): a bibliometric analysis, *International Journal of Production Research.*
[10] https://portalprzemyslowy.pl/utrzymanie-ruchu-produkcja/it-dla-przemyslu-erp-cmms-cad-druk-3d/opc-ua-w-inteligentnej-fabryce-uniwersalny-jezyk-urzadzen-automatyki/, accessed date: 20.03.2021.
[11] https://us.misumi-ec.com/configurable-components//, accessed date: 20.03.2021.
[12] https://blog.misumiusa.com/the-internet-of-things-industry-4-0/, accessed date: 20.03.2021.
[13] https://www.br-automation.com/en-gb/products/industrial-pcs/panel-pc-2100-single-touch/, accessed date: 19.03.2021.
[14] https://www.br-automation.com/en-gb/technologies/opc-ua/, accessed date: 19.03.2021.
[15] https://www.promotic.eu/pl/pmdoc/Subsystems/Comm/OPC/OPCUA.htm, access date: 19.03.2021.
[16] https://opcfoundation.org/about/opc-technologies/opc-ua/, accessed date: 19.03.2021.

IOP Conf. Series: Materials Science and Engineering 1182 (2021) 012038 doi:10.1088/1757-899X/1182/1/012038

Analysis of innovative decarbonation technologies by methanation

A C Macarie[1], I Balcu[1], P Sfirloaga[1], F V Panaitescu[2], M Panaitescu[2] and G Lazaroiu[3]

[1]National Research - Development Institute for Electrochemistry and Condensed Matter, 144 Dr. A. Paunescu Podeanu Street, 300569, Timisoara, Romania

[2]Constanta Maritime University, Faculty of Electromechanics, 104 Mircea cel Batran Street, 900663 Constanta, Romania

[3]Polytechnic University of Bucharest, 313 Splaiul Independenţei Street, 060042, Sector 6, Bucharest, Romania

E-mail address: ionel_balcu@yahoo.com

Abstract: In the development process of Romania, the energy sector for 2030 contributes essentially through its influence on the quality of life, the environment, the competitiveness of the economy, as well as the climate [1]. Thus, the energy sector from Romania must become more economically efficient, less polluting, and more technologically advanced.This paper presents the analysis of innovative methanation decarbonation technologies, the following aspects were pursued: methanisation, the reactions involved and the importance of the methanisation process, methanisation technologies and methane use, decarbonisation through methanisation, the importance of catalysts in methanation processes, decarbonisation with Carbon Capture Storage (CCS), catalyst enabled processes. Behind modern methanation decarbonization technologies is a history of over 100 years based on research and development [2]. At present, methanization is in the attention of scientists, industry, and energy companies, because: fossil fuels are a finite resource that will soon be exhausted; the energy system is constantly changing, and renewable energies are playing an increasingly important role in it; increasing levels of CO2 in the atmosphere contributes to global climate change.

1. Introduction

With the publication of the Energy Union Strategy (under the Paris Agreement on the supply of clean energy throughout the European Union), the EU has taken on an important role in combating climate change. in terms of energy and climate in 2030: reduction of domestic greenhouse gas emissions by at least 40% by 2030, compared to 1990; renewable energy consumption of 32% in 2030; energy efficiency improvement by 32.5% in 2030; interconnection of the electricity market at a level of 15% by 2030 [1].

In the process of Romania's development, the energy sector for 2030 contributes essentially through its influence on the quality of life, the environment, the competitiveness of the economy, as well as the climate (figure 1). Thus, the energy sector in Romania must become more economically efficient, less polluting and more technologically advanced.

Romania will largely replace electricity production capacities by 2030 with new ones, due to the fact that they are economically inefficient and ecologically inadequate based on advanced technologies [2].

Solar energy has seen phenomenal growth in recent years due to both technological improvements that have led to cost reductions and government policies that support the development and use of renewable energy [3].

Solar energy has undergone an impressive technological change. While early solar technologies consist of small-scale photovoltaic (PV) cells, recent technologies are concentrated solar energy (CSP) and also large-scale photovoltaic systems that feed into electricity grids.

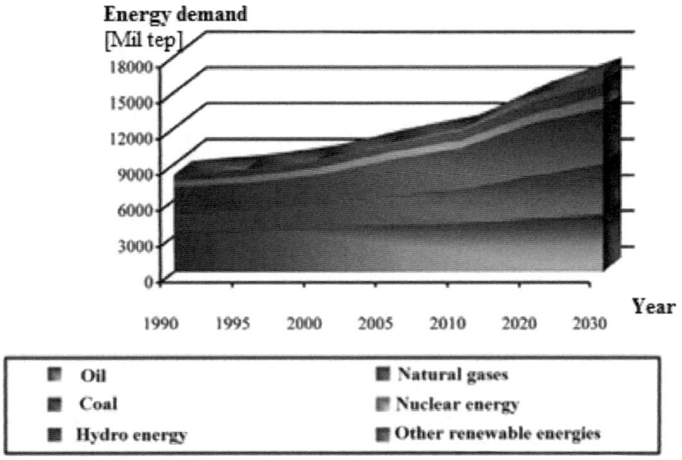

Figure 1. Global energy demand for the period 1990-2030.

2. Methods

In this paper, an experimental study of the methanation process was performed with an installation that aims to determine the optimal operation and find suitable conditions for carbon dioxide methanation catalysts. The methanation test facility was designed and commissioned to investigate the catalytic methanation of carbon dioxide and hydrogen. The process flow chart is shown in the following figure 2.

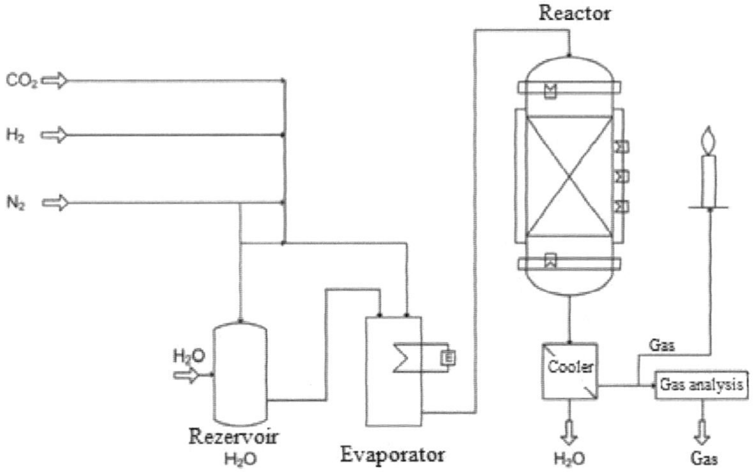

Figure 2. Process flow chart.

Due to the extremely exothermic nature of the methanation reaction, the design of the reactor is crucial to have thermal control over the system. There are basically two types of state-of-the-art reactors that can be used for this process: two-phase fixed bed reactors and reactors, with fluidized bed. Regardless of the choice of reactor, the heat generated by the methanation reaction must be continuously removed from the reactor.

For carbon dioxide methanation an increase in the molar gas fraction of methane gas of about 1% in the product gas induces a temperature rise of about 60 K.

The fluidized bed reactor is presented by a large transfer of mass and heat due to the good mixing of gases and catalyst particles in the fluidized bed which results in an almost isothermal process [4]. To remove the reaction heat, a heat exchanger is installed inside the reactor. However, fluidization leads to a shredding of the catalyst due to high mechanical stress. Moreover, fluidized bed reactors are limited by the fluidization rate which restricts their flexibility.

Fixed bed reactor- Water is introduced into an evaporator to form water vapor. The reaction gases are also introduced into this evaporator. The mixture of CO_2 and H_2 and water vapor is introduced into the heated fixed bed reactor. Heating as well as proper insulation of the reactor was necessary to minimize heat loss and to achieve quasi-adiabatic process conditions.

Upon leaving the reactor, the gas was cooled to condense the water contained in the reaction mixture. Pressure valves were installed on the reactor to control the pressure in the reactor. The composition of the gas produced was analyzed after leaving the reactor.

High reaction temperature and sensitive catalysts are the main reasons why methanization is considered a difficult process to control, with a history based on research and technological development [5].

A good catalyst, in addition to the property of dissociation of water with the formation of oxygen and hydroxide - reactive intermediates adsorbed on the surface of the catalyst - should be able and break the C-C bond, especially in the case of opening aromatic rings. Adsorbed reactive intermediates can combine with fragments of organic compounds and form CO and CO_2, while the release of adsorbed hydrogen atoms leads to the formation of hydrogen molecules. In addition, a suitable gasification catalyst must favor the steady-state development of the water-gas change reaction and the hydrogenation of CO and CO_2 with CH_4 formation [6, 7].

The temperature released in the reaction can lead to the sintering of the metal as a catalyst, which in turn leads to a decrease in its total surface area and to a decrease in its activity. Catalyst deactivation can also occur due to coke deposits, a situation in which the metal surface is blocked by carbon accumulation [8, 9]. In order to avoid these problems, the aim is to identify an efficient catalyst for the methanation reaction, which has high thermal stability and resistance to coke deposits.

According to Bartholomew, deactivating Ni-based methanization catalysts can be mechanical, chemical and thermal.

We recall three mechanical methods (fouling, wear and crushing), two chemical methods (poisoning and vapor-solid reactions) and a thermal method (thermal degradation), used to deactivate catalysts.

3. Experimental researches

3.1. The reagent mechanism

- Equilibrium concentrations of carbon oxides 10 ppm
- CO inhibits CO_2 methane.

This reaction takes place according to a two-stage mechanism. In a first step, CO_2 and H_2 are converted to CO and steam by the reaction (equation (1)):

$$CO_2 + H_2 \rightarrow CO + H_2\ O \tag{1}$$

In the second stage, CH_4 is synthesized from CO and H_2 (equation (2)):

$$CO + 3H_2 \rightarrow CH_4 + H_2O \tag{2}$$

- Very high intrinsic reaction rates.

3.2 Methanation reactions and operating conditions

- CO2 methanation reaction is an exothermic catalytic reaction that usually takes place at temperatures between 150° C and 550° C (equation (3)).

$$CO_2 + 4H_2 \rightarrow CH_4 + 2H_2O + 165\ kJ/mol \quad (3)$$

- The reactions are extremely exothermic:

76°C for every 1% CO converted;
66°C for every 1% CO2 converted.

- Typical inlet temperatures:15-40°C;
- Carbon dioxide concentration at the inlet: 0.1-1.0% vol.

3.3. Pre-reduced Methanation Catalyst

- Catalyst first reduced, then partially stabilized:
  –Faster plant start-up;
  –Higher catalyst activity.
- Reduction taken place under controlled conditions
  –Higher Ni surface area;
- Lower inlet temperatures possible.
  – Minimum inlet temperature 170°C (338°F) ;
- Catalyst is ready to go.

4. Results and interpretations

The Ni-based catalyst was physico-chemically analyzed before and after use by the following analysis techniques: X-ray diffractometer (using a PANalytical diffractometer, with Cu-Kradiations (λ = 0.15406 nm) in a 2θ range from 20 ° to 80 ° , and a scan rate of 2 ° / min), scanning electron microscopy and Edx (*Scanning Electron Microscope Inspect S*) and atomic force microscopy (*Nanosurf® EasyScan 2 Advanced Research Model*).

4.1. Analyzes for Ni-based catalyst before use

4.1.1. X-ray diffraction - The structural analysis of the Ni-based catalyst, before use, is shown in figure 3. The diffraction spectrum was refined with the X Pert HighScore Plus program, using JCPDS card no.: 00-001-1260. The diffraction peaks for were indexed to the cubic phase, Fm-3mspace group, with lattice parameters a = b = c = 3. 5175 Å, and a unit cell volume of V = 43.52 Å3.

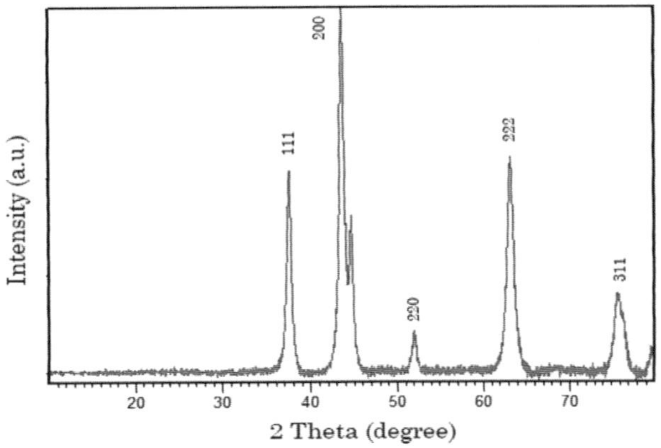

Figure 3. The structural analysis of the Ni-based catalyst.

4.1.2. Scanning electron microscopy/Edx - Electron microscopy images and semi-quantitative analysis by Edx are shown in figure 4.

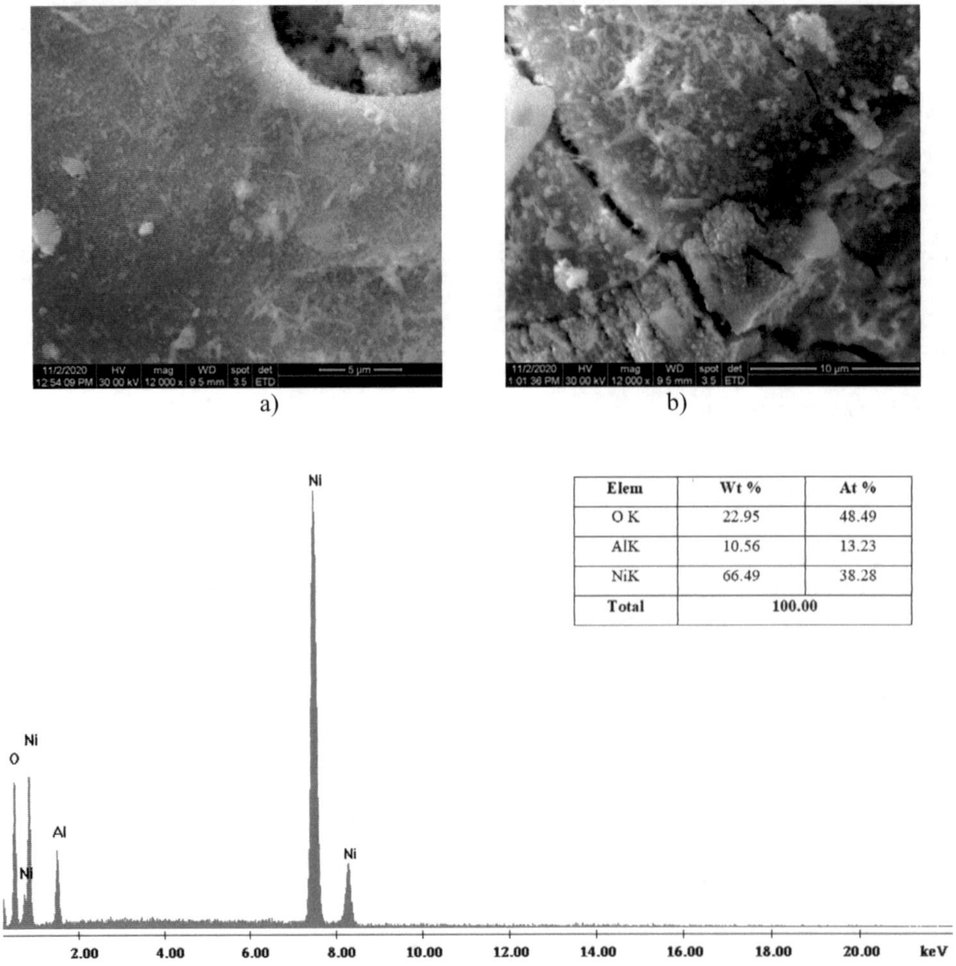

Elem	Wt %	At %
O K	22.95	48.49
AlK	10.56	13.23
NiK	66.49	38.28
Total	**100.00**	

c)

Figure 4. SEM images (a, b) and EDX spectru (Insert quantification) (c) of the Ni-based catalyst.

From these images it can be seen that the particles are strongly agglomerated in asymmetric formations, and the semiquantitative analysis showed that the predominant element is Ni (66.49 Wt%).

4.1.3. Atomic force microscopy - Figure 5 present the 2D and 3D topography for the Ni-based catalyst. The dimensions of the particles that are present on the surface were determined by using the *Nanosurf EasyScan 2* software (figure 6).

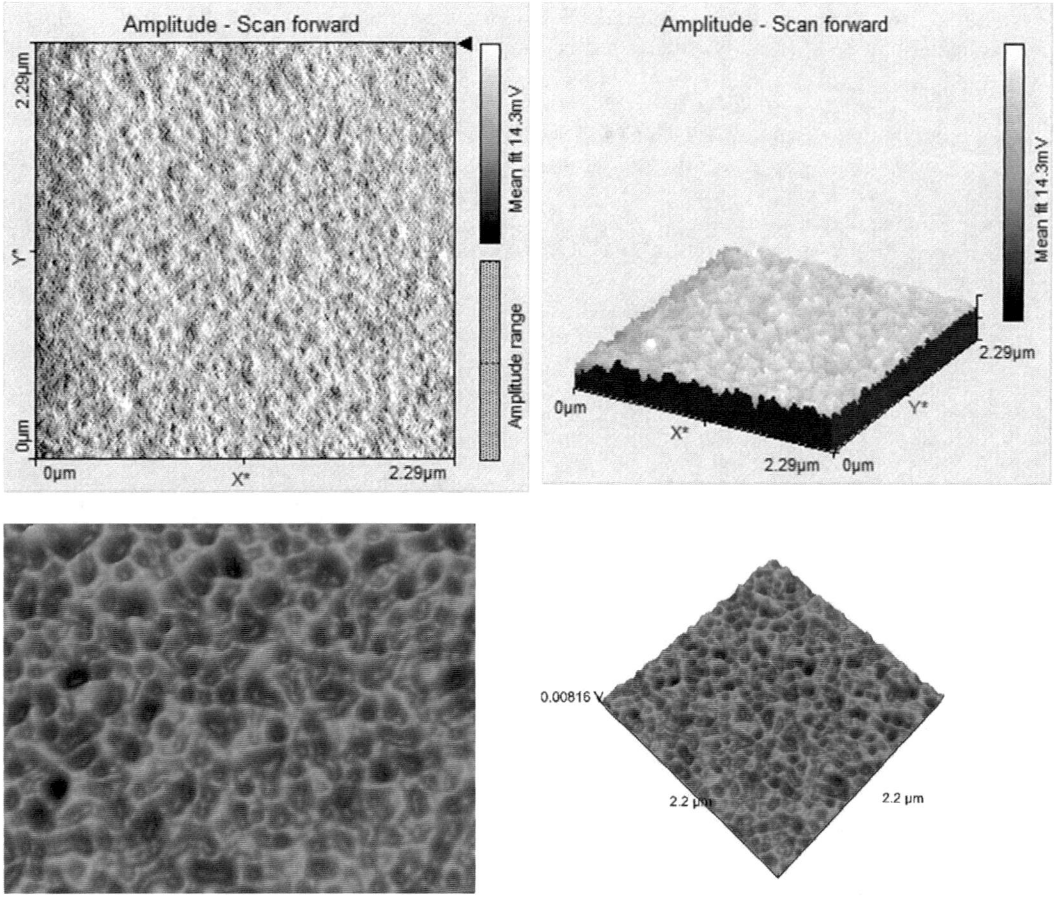

Figure 5. 2D and 3D AFM surface images of the Ni-based catalyst.

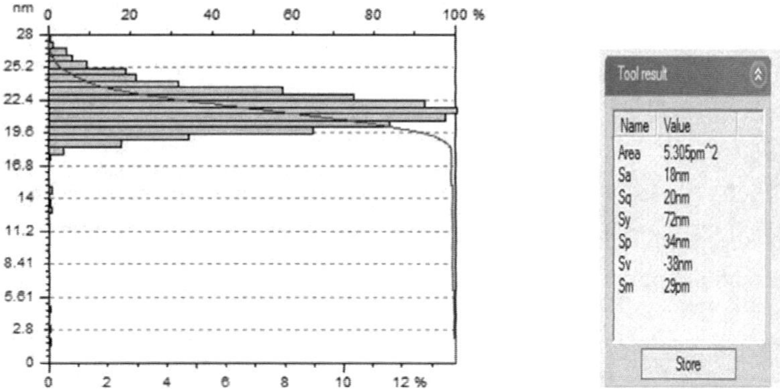

Figure 6. The particle size distribution for the Ni-based catalyst..

The surface roughness - the average roughness (S_a) and the mean square root roughness (S_q) were calculated also from AFM analysis data. The investigated/measured surface had an area of 5.305 pm^2 (figure 6).

4.2. Analyzes for Ni-based catalyst after use X-ray diffraction
Figure 7 shows the X-ray diffraction spectrum for the Ni-based catalyst after use. The diffraction spectrum was refined with the *X Pert High Score Plus* program, using JCPDS card no.: 01-087-0712. The difraction peaks for were indexed to the cubic phase, Fm-3m space group, with lattice parameters a = b = c = 3.5238Å, and a unit cell volume of V = 43,76Å3.

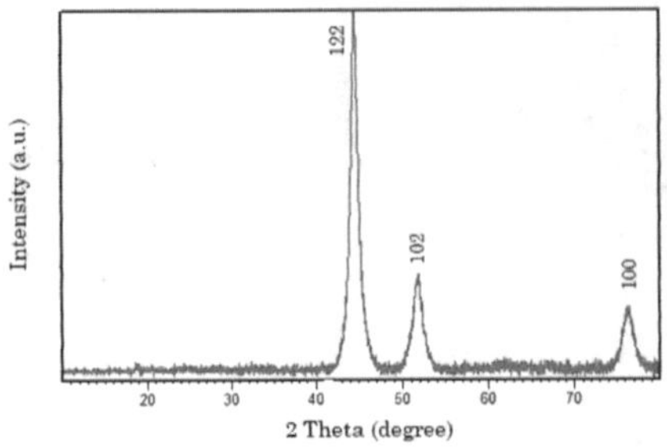

Figure 7. X-ray diffraction spectrum for the Ni-based catalyst after use.

4.2.1. Scanning electron microscopy/Edx - Figure 8 shows the surface morphology for the Ni-based catalyst after use. Thus, at the higher magnification (6000X) it can be seen that the particles are very agglomerated in asymmetric formations.

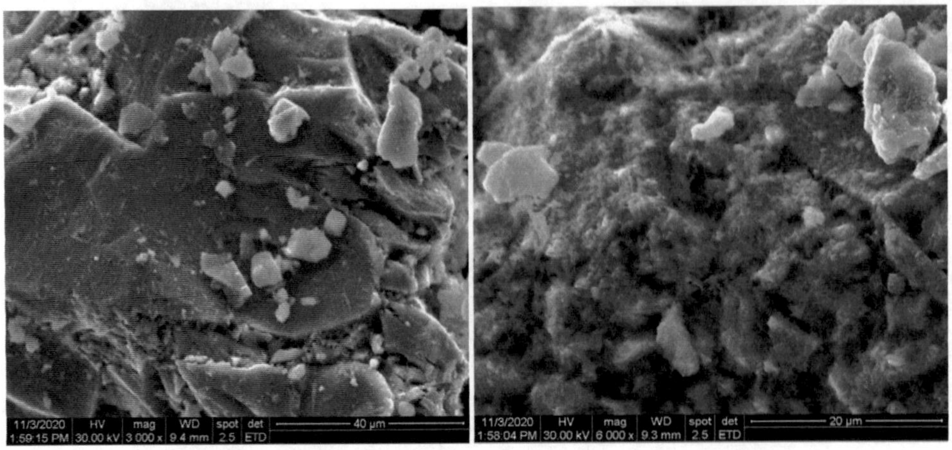

Figure 8. The surface morphology for the Ni-based catalyst after use.

4.2.2. Semi-quantitative analysis by Edx - The elemental analysis and quantification for the studied material is presented in figure 9. As can be seen from the EDAX spectrum, the predominant element is Ni, in mass percentages of 84.01%.

IOP Conf. Series: Materials Science and Engineering 1182 (2021) 012038 doi:10.1088/1757-899X/1182/1/012038

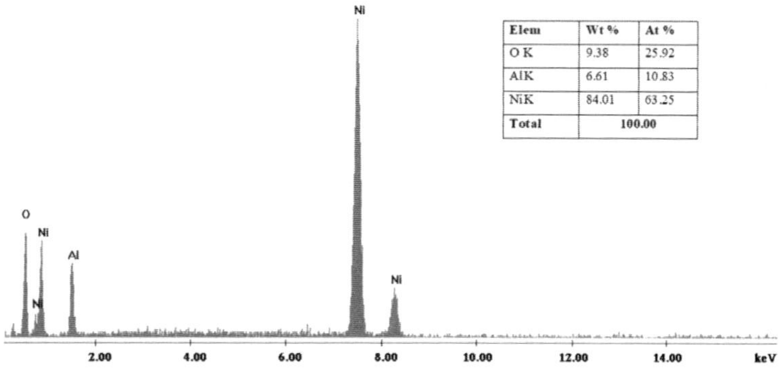

Elem	Wt %	At %
O K	9.38	25.92
AlK	6.61	10.83
NiK	84.01	63.25
Total	100.00	

Figure 9. The elemental analysis and quantification for the Ni-based catalyst after use.

4.2.3. Atomic force microscopy - 2D and 3D topography for the Ni-based catalyst after use are presented in Figure 10. Although atomic force microscopy analysis was performed at the same scale of 2.29 µm x 2.29 µm for the Ni-based catalyst before and after use, it can be seen that in the case of the Ni-based catalyst used the particles are much smaller.

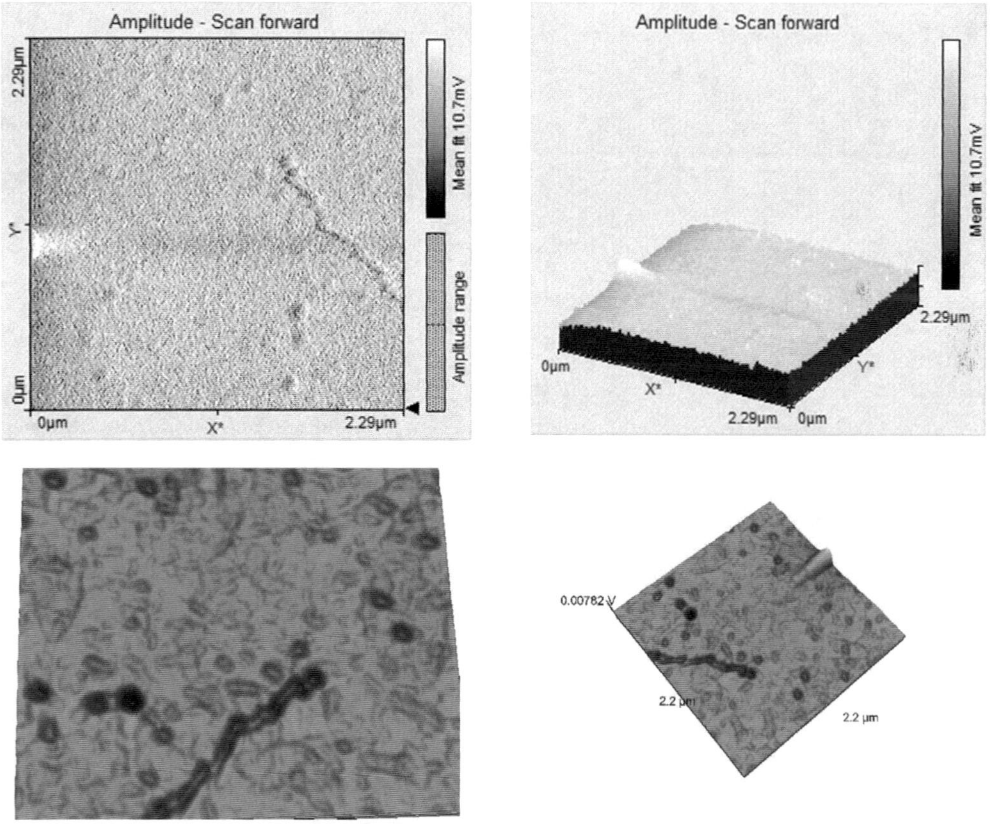

Figure 10. 2 D and 3D AFM surface imagesfor the Ni-based catalyst after use.

The surface roughness - the average roughness (S_a) and the mean square root roughness (S_q) were calculated also (figure 11), and the investigated/measured surface had an area of 5.305 pm^2 (figure 11).

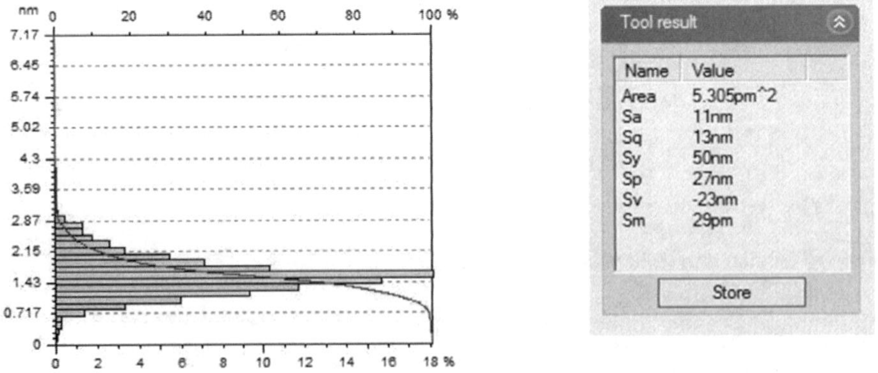

Figure 11. The particle size distribution for the Ni-based catalyst after use.

The figure 12 below shows the amount of H_2 during the methanation reaction.

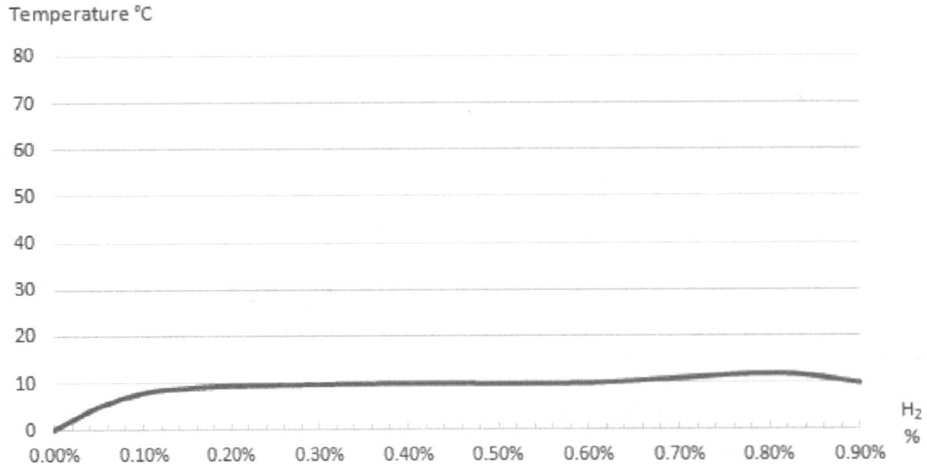

Figure 12. The amount of H_2 during the methanation reaction.

During the methanation reaction it can be observed that for a temperature variation of up to 10 °C the percentage of H_2 increases up to 0.20%, after which it has a stationary level of up to 0.70%. For slightly sensitive temperatures above 10 ^{0} C, the hydrogen variation is up to 0.85%, after which it decreases.

The figure 13 below shows the resulting amount of methane.

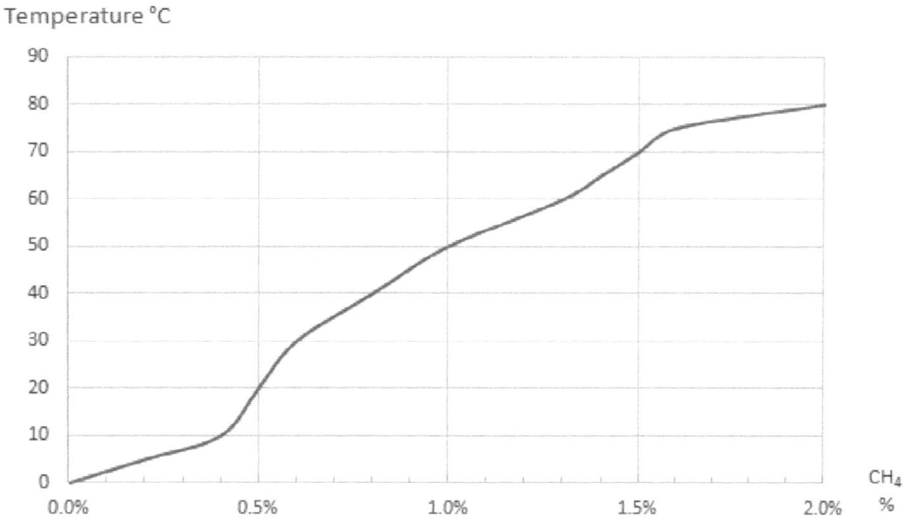

Figure 13. The resulting amount of methane.

As can be seen from the graph, the percentage of CH_4 increases in proportion to temperature. The maximum value for CH_4 is 2% for a temperature of 80° C.

5. Conclusions

From a global viewpoint, research in this field encompasses the following:

- To counter global warming, the reaction between hydrogen and CO_2 has attracted attention.
- This method is effective for CO_2 reduction measures.
- Carbon dioxide methanation requires a catalyst to be active at relatively low temperatures and selective towards methane.
- Nickel based catalysts are most widely investigated, and commercial catalysts are typically nickel on alumina support .
- Superficial gas velocity - The increase of the inlet gas flow rate enhances the productivity of the reactor for fast reactions but restricts conversion for slower reactions that requires more contact time, which suggests that the flow rate can be optimized for specific reactions.

6. References

[1] https://ec.europa.eu/energy/sites/ener/files/documents/ro_final_necp_main_ro.pdf, accessed on March 10, 2021.

[2] https://energie.gov.ro/wp-content/uploads/2016/12/Strategia-Energetica-a-Romaniei-2016-2030_FINAL_19-decembrie-2.pdf, accessed on March 16, 2021.

[3] Jachin G, Ruoss F, Karjunen h, Schaffert J, Tynjälä T 2020 Cost benefits of optimizing hydrogen storage and methanation capacities for Power-to-Gas plants in dynamic operation, *Applied Energy* **257**, 113967.

[4] Schaaf T, Grunig J, Schuster M, Orth A 2014 *Chem. Ing. Tech.* **86**(4), 476–485.

[5] Ronsch S, Schneider J, Matthischke S, Schluter M, Gotz M, Lefebvre J, Prabhakaran P, Bajohr S 2016 *Review on methanation,* From fundamentals to current projects, Fuel, **166**, 276-296.

[6] Peterson A A, Vogel F, Lachance R.P., Froling M., Antal Jr. M.J., Tester J.W. 2008 Thermochemical biofuel production in hydrothermal media: A review of sub- and supercritical water technologies *Energy and Environmental Science* **1** 32-65.
[7] Frusteri F, Frusteri L, Costa F, Mezzapica A, Cannilla C, Bonura G 2017 Methane production by sequential supercritical gasification of aqueous organic compounds and selective CO_2 methanation *Applied Catalysis A: General* **545** 24-32.
[8] Gao J, Wang Y, Ping Y, Hu D, Xu G, Gu F, Su F 2012 A thermodynamic analysis of methanation reactions of carbon oxides for the production of synthetic natural gas *RSC Advances* **2** 2358-2368.
[9] Lu J, Fu B, Kung M C, Xiao G, Elam J W, Kung H H, Stair P C 2012 Coking- and sintering-resistant palladium catalysts achieved through atomic layer deposition *Science* **335** 1205-1208.

Acknowledgements
This work was supported by a grand of the Romanian Ministry of Research and Innovation CCCDI-UEFISCDI, project PN-III-P1-1.2-PCCDI-2017-0404/ 31PCCDI / 2018, *Holistic on the Impact of Renewable Energy Sources on Environment and Climate*-HORESEC, within PNCDI III.

Energy saving for wireless sensor network in manufacturing and administrative projects via Smart Nodes

A A Mahawish
Computer Engineering Department, College of Engineering, Al-Iraqia University, Baghdad, Iraq

E-mail: Salambash2000@yahoo.com

Abstract. The cost-effectiveness of the operation of Wireless Sensor Network (WSN) in manufacturing and administrative projects can highly boost the production of lower goods and services that in turn enhance the marketing ability. The lifetime increase of the batteries by reducing energy wastage is the main challenge and hence, in the current paper, there is an attempt to save energy via transmitting packets as quickly as can be expected while avoiding congestion and drop packets. The time synchronization allows nodes to sleep and wake at the same time to avoid consuming energy. The suggested algorithm, would, successfully deliver a packet with a fairly high rate and reduce dropping packets. Besides, it manages to avoid congestion and saves energy when reduce the retransmission the dropped packets.

1. Introduction

The Wireless Sensor Network (WSN) made from many nodes can be deployed in an environment to make sensing phenomena. This network has widely been used because of its low cost, low power, and ease of deployment. These nodes generate packets that may be continuous, based on an event or based on query then send these packets to the sink node. In WSN, the performance depends on many factors, one of them is the loss packet due to congestion. The congestion phenomena are the main reason for the loss or drop packets.

Each node has a buffer which works as store and forward transmission, temporary store packet when the output link capacity less than the number of incoming packets to avoid loss packet due to congestion. The buffer size must be managed probably because a large buffer size makes a large delay and a small buffer size will not be able to store enough packets.

Figure 1 shows the general architecture of the node in WSN. The sensing unit is small in size and can sense the phenomena and send this data to a processing unit. Then after processing, the signal needs to be transmitted or received using a communication unit to allow each node to connect with another, usually using an RF (radio frequency) channel. The node parts consume the power from the battery "limit in energy". The sending packet needs energy as signal power to be able to send this packet [1].

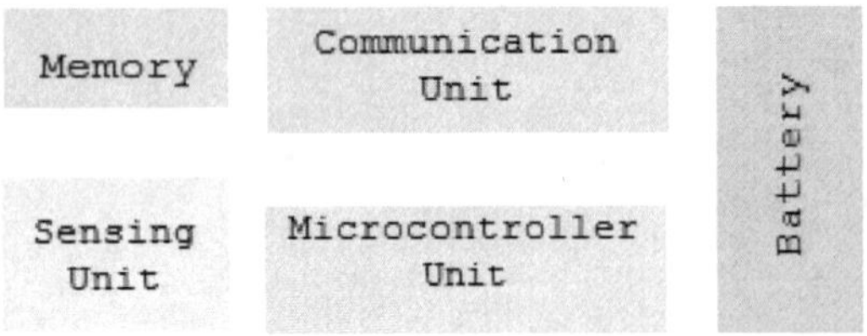

Figure 1. Sensor Node [2].

There is much effort that has been done to develop the wireless communications and electronics parts over many years, which made the networks becoming cheaper, low-power consumption, and high resolution of sensors. A WSN can be constructed from many nodes that are deployed in a specific area for a specific task. The nodes work as a collaborative network to gather information from the surrounding area and then performing simple processing on the collected data. The data travel from node to node to sending it to the destination node. The main properties for a sensor network are as follows:

- Auto-organized
- Short-ranged signal communicating.
- Deployment plus cooperative effort with each other.
- Changing network topology because of node fails.
- Power provides by Battery for transmit signal, memory, sensor, and processing

Within WSN, the node broadcasts the signal to all neighbour nodes to communicate with each other. The connection topology of these nodes can be changed because of the node's fail or even shut down due to battery lack. Subsequently, the node ought to be auto organized and able to work autonomously. This type of node has some drawbacks such as energy feeding, low processing abilities, and small memory available. A primary issue that needs to be concentrated on in WSN is their ability to be scalable [2], their association technique for connection between nodes [3], and the restricted power supply to the node. Based on the nature of WSN, applications make the battery attached with the node difficult to be replaced or recharged. Many applications domain used WSN as shown in figure 2. For example, the WSN can be used in medical fields [4] "e.g., patient observation" or even used in disaster response [5]. Also, it can be used in wildfire alarm. So, there is a requisite of managing the energy-consuming to increase the lifetime of the node. So, the dropped packet is one of the important ways to save energy for the node.

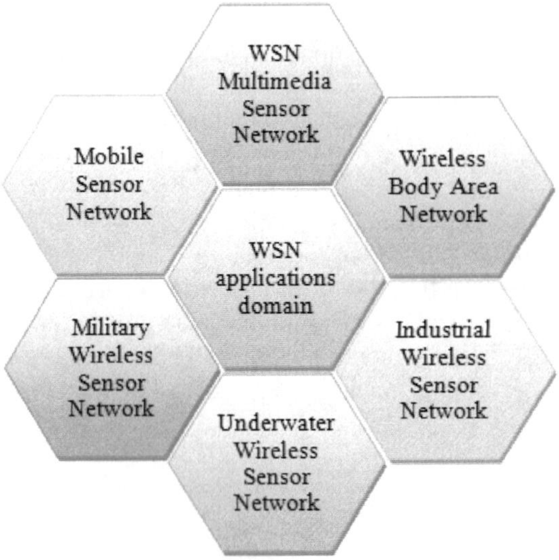

Figure 2. Applications of WSN.

The node buffer is a queue preceding a link that has a limited capacity. The buffer is capable of holding an infinite number of packets in a manner of the store and forward transmission. The buffer has a limited capacity, so when the sender rate is higher than the transmission rate of the outgoing link, the buffer will overflow which causes a packet drop. The queue management is used to avoid buffer overflow and to reduce packet dropping. The performance of WSN depends on several metrics [6] such as latency, energy saving, reliability, throughput, network lifetime, adaptively, and, robustness. In this study, we reduce the latency and increase the throughput to achieve energy saving as well as increase the lifetime of the network. While the issues affecting network performance are sensor node constraints, Dynamic network topology, Scalability, Heterogeneous traffic, Heterogeneous node's processor, redundant information, Unreliable wireless channel and, congestion in the network. In this study, we focus on network congestion and the node's processor. The Congesting occurs in nodes when the buffer overflows, which means the packet drop will happen. The Congestion is costly because the sending packet needs to be retransmitted and energy-consuming will increase. To avoid congestion, we need to implement an algorithm in WSN nodes to control the average queue size that avoids possible queue overflow.

To build and achieve our WSN network, the Zigbee [7] was used as a wireless device attached with two microcontrollers processing. The first one is PIC microcontrollers (Programmable Interface Controllers), the PIC is an electronic circuits that can be programmed to carry out a vast range of tasks [8]. The second one is ARM microcontroller, which is a series of cores optimized for power efficiency and deterministic operation. It is widely used in microcontrollers (MCUs) as a processing units and can also be found embedded into multi-core microprocessors [9]. The Zigbee uses a 2.4 GHz band that has many features such as 11 channels available, baud rate, transmission speed, and transmission power signal. Besides this section, the organization for this work can be given as follows: Section 2: presents a few previous works on congestion control and saving energy. Section 3 describes the algorithm designed for congestion control by using heterogeneous wireless nodes and how time synchronization is achieved between these nodes. Whereas, Section 4 labels the major implementing for Time synchronizing. Section 5, explains the experimentation outcomes of the Congesting Controlling Algorithm and Time synchronizing. Lastly, the findings are concluded in Section 7.

2. Relative works

Many works have been done on the main problem in WSN, which is how to improve Quality of service (QoS) and optimize power consumption. The QoS is required to meet the less packet delay, less packet drop, or increase traffic throughput, while the power consumption must be reduced to increase the lift time of the WSN node. In [10], the paper studied and reviewed the latest state-of-the-art congestion control protocols. These protocols are divided into three categories, i.e., traffic-based, resource-based, and hybrid. Traffic-based protocols are further subdivided, based on their hop-by-hop or end-to-end delivery modes. In [11], it was managed scheduling within the Medium Access Control (MAC) layer. Also, Radio Duty Cycle as an energy conservation technique was used. The design of an energy-efficient routing protocol entitled "Position Responsive Routing Protocol" (PRRP). PRRP is designed to minimize energy consumed in each node by (1) reducing the amount of time in which a sensor node is in an idle listening state and (2) reducing the average communication distance over the network [12]. The states of energy-consuming are as follows: Idle listening state: in this state the nodes use energy through many activities like sending, gathering, and waiting for a connection. They likewise consume substantially more energy through inactive and rest modes.

Collisions state: these events happen when multiple nodes send packets simultaneously. The result of the collision is dropping packets and retransmitting the required. The transmission packets will consume power, so the retransmission. Such loss in energy will occur in each node and the total energy loss will depend on the number of nodes in the path [13]. Regarding repeated transmission, strategies like arbitrary delays are important to avoid synchronize retransmit and fast return to collision event. Reducing the protocol overhead state: it indicates transmitting the packet headers plus controlling message consumes power, such information, in the end, isn't exploitable, and it ought to be left as a minor effect. Strategies, which decrease the convention overhead by making a small transmitting interval, the prompt lesser amount of energy utilization, in turn, this also assists in subsequently sparing assets, while this increases latency [14]. In outcome, a positive incentive for the transmission time frame relies upon the recurrence of progress.

3. Methodology

The suggested network of this study is created as heterogeneous networking via incorporating various Wireless types of Node such as PIC Wireless, ARM Wireless as well as PC Wireless Nodes, figure 3 below shows that.

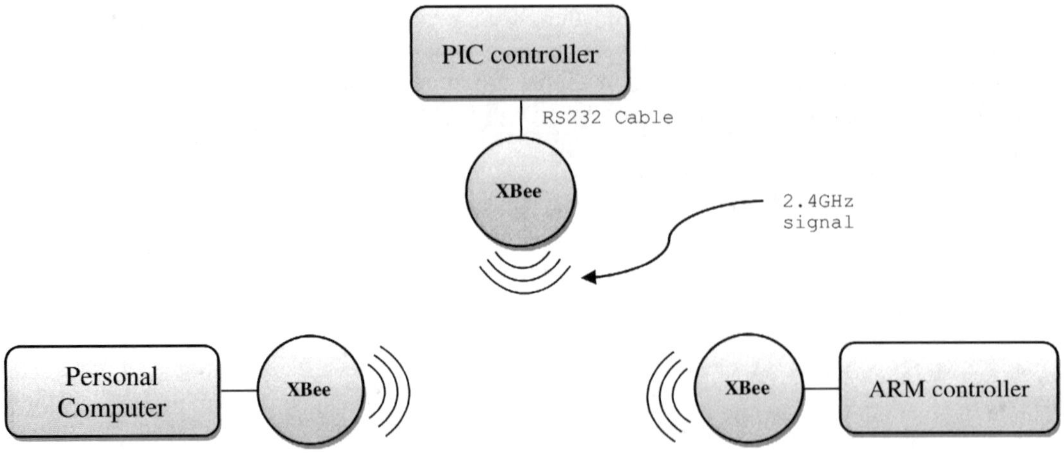

Figure 3. Heterogeneous networking.

Within heterogeneous networks there are three types of nodes, the PIC node includes a PIC microcontroller chip attached with XBee as wireless. The instruction set of PIC chip that belongs to Harvard RISC (Reduce Instructing Set Computers) and the PIC dual bus architecture give a quick and adaptable design. The PIC is attached with Wireless Device by serial cable "RS-232". While the ARM Wireless node contains an ARM microprocessor with Wireless Device. The ARM microprocessor instruction set is also based on RISC architecture. The PC Wireless Node includes the PC connect with Wireless Device through the serial cable. These three types of nodes connect using a 2.4 GHz wireless signal. The designed network needs to apply time synchronizing between sensing nodes especially when these nodes being built up from different architectures. The time synchronization allows nodes to be active and sleep within synchronize time and this way the node be active only when its time turns on to receive the data from other nodes. The receiving node can go to sleep mode after its data sent to other nodes. So the time synchronization will participate in reducing energy-consuming in heterogeneous wireless nodes network.

There are different protocols present to apply to synchronize time, one of them was suggested by Ganeriwal et. al. which is called networking-wide time synchronizing protocols for a sensing network, Ganeriwal et. al. named that as Time Synchronizing Protocols to a Sensing Network (TPSN) [15]. This protocol has two stages of working: "level discovery stage" plus "synchronizing stage". The former stage aims at creating hierarchic topologies within networks, as any node can be allocated a level. Level 0 is called root and only one node can be allocated, this node usually is the sink node. Regarding the later stage, the level k node synchronizes to the level k-1 node. The result at the end of the synchronizing stage, the synchronization for the whole nodes can be done to the root node, and it can achieve the networking-wide synchronizing [15].

Implement the TPSN Algorithm:
First level *discovery stage*: such a stage can be operated only one time on deploying the networks. Firstly, a node ought to be resolved to be a top node "root node". This node can represent a sink within the sensing networking. The algorithms would make synchronization for each node to an exterior period, when this sink isn't accessible, sensing nodes could occasionally assume control over the usefulness of the root node. A current head electing algorithm may be utilized for this intermittent root node electing pace. The top-level "level 0" is allotted for the root node and then can start the level revelation stage by communicating a level disclosure packet. The node sends packets which include the identification as well as the level of this node known as the root node. When getting such packets, any receiver node assigns itself to beneath the sender node. At that point, these nodes which are beneath the sender node update the receiving packet with its identification as well as its level. When nodes are allotted levels, they dispose of additional approaching level revelation packets. Such a transmission chain continues via the networks, so the stage can be finished if the entre nodes get allocated a level.

Second level *synchronizing stage:* The main constructing chunk for the synchronizing stage is represented by two-way messaging exchanging amid nodes' pairs. When considering the clock drifting amid nodes' pairs as constant within the little timing duration through one message exchanging, the propagating delaying can be likewise presumed to be constant for the two ways.
To start to synchronize protocol, implement the two-way messaging exchanging amid node A and node B. Node A starts the synchronizing procedure via transmitting the synchronizing pulse packets within period T1 (consistent with local timing of the node). Such packet contains the level number of node A, in addition to the T1 value. Node B measures such packet (consistent with the local timing of the node) within period T2 = T1+d+Δ, as Δ represents the timing drifting among nodes, while d represents the pulse propagating delay. Node B sends an acknowledging packet within period T3, this incorporates the level of node B as well as T1, T2, along with T3 values. At that time, Node A could measure the timing drifting along with propagating delay by using the equations (1) and (2), then making the synchronization between the Nodes, from A to B [15].

$$\Delta = \frac{(T_2 - T_1) - (T_4 - T_3)}{2} \quad (1)$$

$$d = \frac{(T_2 - T_1) + (T_4 - T_3)}{2} \quad (2)$$

Since the Wireless Node of this work using RS-232, so the node receives serial data only, which it waits a specific time to packetize and then to transmit the data. The packetizing timeout value is used to make the incoming packet waits for a specific time to ensure the group of packets received and then transmit again, this value is completely related to the baud rate of sending data. When packetizing value can be so small, this might lead to difficulties in which the packet needed to be transmitted together will disjoint. To test and configure the nodes, there is a need to send information as single transmitting data, yet during perform the calculating, the transmission of the initial information group is done and disjoining will occur. To solve this issue the incrementing the value of packetizing delay was used, our information could be ensured to stay jointly also to be sent within a single transmitting.
The RO (Packetization Timeout) command available in XBee allows the value of packetization to change, the default value is 3, and this means that you would have to wait for 3 x (the time it takes to send all the data through serial data) before you send another packet or before you put your XBee in sleep mode. For example, I was telling you before it would be 3 x (10 ms) = 30 ms. The value of RO is set to zero which means any data "character" received by the node will send it immediately without waiting. For example, if you are transmitting a group of 10 characters and you configure the serial data to have a baud rate of 9,600 bps, that means that for one character you send through serial data would last approximate 1.04 ms per character so every data is a group of 10 bits (with start and stop bit). So

sending 10 characters in one packet needs to have 10.4 ms per packet. Now the XBee needs some time to packetizing this data and sends it all as a packet instead of sending each character every time you send it. XBee can increase such value to FF (255 in decimal).

Congestion control algorithm:
To set a value to the parameter of each node in WSN, these parameters are (Buffer size, Packetizing delay, threshold, Baud rate, selecting pin, register selecting, port initializing, I/O direction). The sending data starts in Wireless Node 1 to the Wireless Node 2
Let
N: Number of packets.
th1, th2: the threshold to avoid packet drop.

Initialize variables

At beginning the sending data without packetizing delay, so the Packetizing delay value gets set to 0. Which means that the data sending speed is high.

```
Loop 1: Set Packetization delay to zero
      If (N>= th1)
      {
        Set Packetization delay by 3
        {
          If (N<= th2)
          {
            Go to Loop 1
          }
        }
      }
```

The value of RO will affect power-consumption because the data will wait for a period of time before send it again and go to sleep mode. So, any increase in RO means increasing in power-consuming. While a decrease within the values for such parameter causes extra packet losing. The threshold (*th*) value is used to count the average number of packets the group sends at the same time without the drop, its works like congestion controlling. The *th* value is determined based on link capacity and buffer size. There are two values of *th* (*th1* & *th2*), the th1 is used to specify the minimum number of the packet to send in one time (as window size in transport protocol), to best utilize the available resources. While the *th2* that specifies the maximum number of the packet that causes sending in same time beyond that value, the drop packets will occur. When performing practice experimentations on Wireless Networking the value of the threshold can be assigned.

In experimentation practice, suppose that we have 50 packets that need to be sent. If these packets are sent as one group "by increasing RO value" within the networking, the drop packets will occur, it was obtained merely 13 packets received when sending such packets group from Node 1 to Node 2. Thus it was set threshold *th1* as 10 packets. After successful receiving 10 packets at Wireless Node 2, the value of RO gets incremented to 3ms and carry on sending the next group of packets, yet there is a want for saving energy by avoiding drop packets, the RO value decreasing to initial value 0 after transmitting packet group reach to 15 packets, so the value of threshold *th2* becomes 15. If the packets within-group are less than 15 they will be sent at high speed at one time, while if they are greater than 10 the packets will be divided from the large group into a smaller group "15 packets" in each group and reduce the sending speed to avoid the congestion. In this mechanism, we can use heterogeneous nodes and allow them to send the packet at a fair speed with avoiding drop packets.

4. Implementing

This part presents the outcomes of experimental work using the designed network as illustrated in figure 2. Such outcomes present the Time Synchronizing reading as well as the influence of the Congestion Controlling Algorithm on a network along with the way of avoiding congestion throughout communicating. The Time Synchronization algorithm was applied on the Nodes of PC, PIC, and ARM for synchronizing the time among these nodes. The outcomes can be given as follows:

Table 1. Sending and receiving time effect based on different no. of characters between two nodes.

No. of send character	Node 1		Node 2	
	send	receive	send	receive
	T1	T4	T2	T3
1	1	551	2	2
10	1	601	2	2
50	1	721	2	50
200	1	931	2	330
300	1	1101	2	490
400	1	1261	2	720
500	1	1421	2	930
1000	1	2911	2	1920

Regarding table 1, the time in millisecond (*ms*) is required to send a different number of packet groups between two nodes. The T_1 is the time of sending a packet from node 1, while T_4 the receiving back the same packets to Node 1. Same for T_2 *and* T_3 at Node 2. The table 2 shows the time in *ms* for drifting and propagation delay group of character between two nodes.

Table 2. Relative Drift and Propagation delay effect based on different no. of characters between two nodes.

No. of send character	Relative Drift	Propagation delay
1	274	275
10	299	300
50	335	336
200	300	301
300	305	306
400	270	271
500	245	246
1000	495	496

5. Result analysis

Now after performing the synchronization protocol and calculating the drifting between nodes, the proposed algorithm can be applied to the network. Three Wireless Nodes exist through which the data get sent (PC, ARM, and PIC Nodes). It was received 100% of packets as before the applied algorithm and there is no need to apply the Congestion Controlling Algorithm at Node 1. By using Node 2 and before using the algorithm there are only 49% packets received while after using the algorithm there are 98% received packets as illustrated within figure 4, out of 50 packets only packet number 32 is dropped after applying the suggested algorithm, which means improving the throughput when using the algorithm on Node 2.

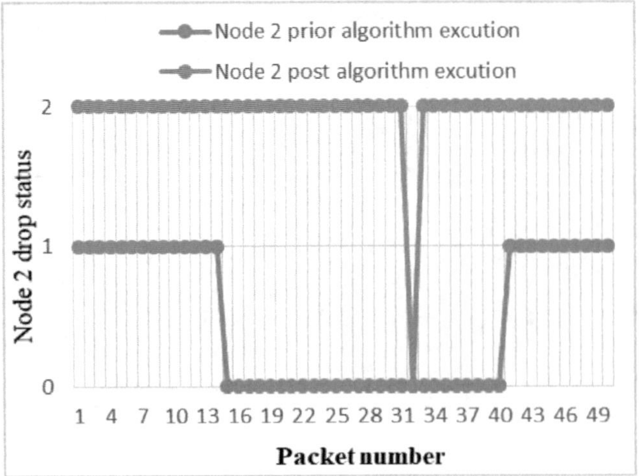

Figure 4. Results for Node number 2.

Regarding Node 3 there are only 25% of packets were received before the algorithm's application, while there are only two packets were dropped which means 96% of packets got received after the algorithm submit, see figure 5.

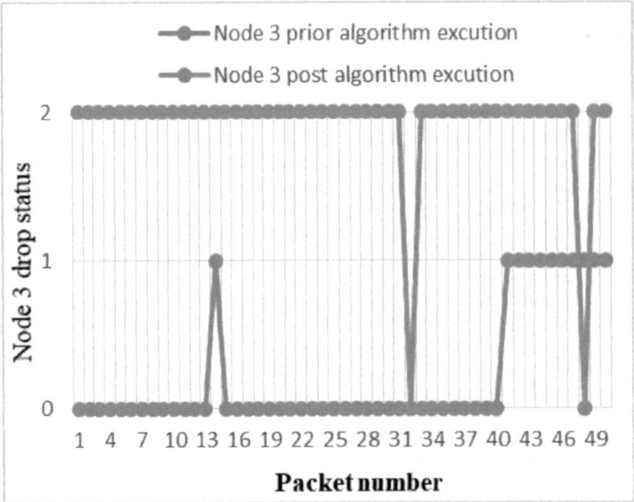

Figure 5. Result for Node 3.

Regarding figure 6, the chart shows the packet throughput for different types of WN "Wireless Node". It was sent 50 packets. In Node 1 the all packets are received, while the Node 2 there are 49% dropped and the Node 3 there are 75% dropped due to congestion, which means the resend packet required and loss of energy will occur.

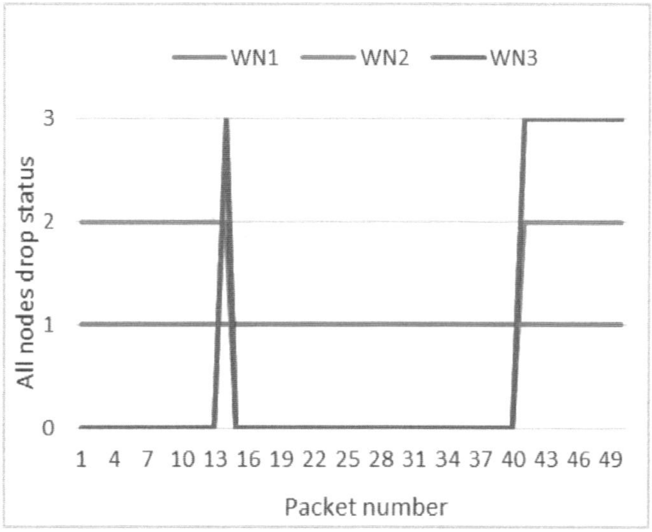

Figure 6. Result before applying the algorithm.

To save energy needs to reduce the dropped packet, so the suggested algorithm after applying the throughput of packets is improved and thus reduces the wastage of energy for retransmitted.

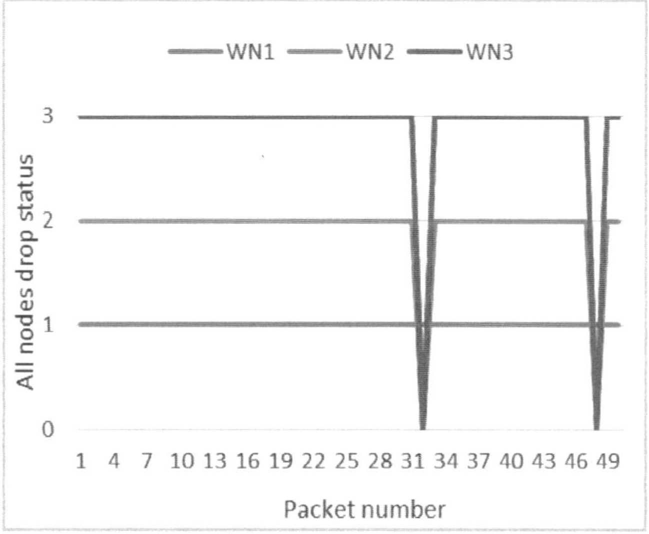

Figure 7. Result after applying the algorithm.

In figure 7, the outcome after using the congestion controlling algorithm as shown only node 2 and node 3 drops two packets out of 50 packets sent. That means the algorithm reduces the number of dropping packets and improves the performance of the network. For Node 1 it was obtained all 50 packets whereas for Node 2 it was obtained 98% and Node 3 was obtained 96%.

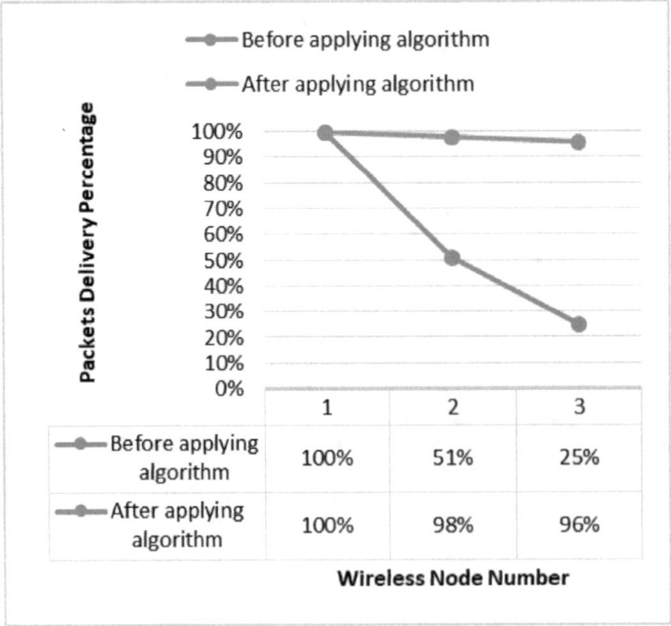

Figure 8. Result in percentage.

Figure 8 shows the percentage without using and using the suggested algorithm. Node 1 achieved 100% of the packets that were received in both cases. However, for Wireless Node 2 it was received just 51% of packets in the case without using the algorithm, while it was received 98% of packets in the case of using the algorithm. Similarly, it was received 25% of packets for Wireless Node 3 in the case without using the algorithm, yet it was obtained 96% of packets in the case of using the algorithm.

In this section the effect of different product are not mentioned specifically, the reason is to create network with heterogeneous node to simulate the real networks operations in the internet. The XBee as end device in each node will deal as overall with different packets processing of these products. As observation from result the XBee can control on the data rate as well as the delay to avoid congestion in the network and eliminate the drop packets.

6. Conclusions

In the current investigation, there is an attempt to save energy via transmitting packets as quickly as can be expected while avoiding congestion and drop packets. By using different type of microcontroller the time synchronization is required for allowing nodes to sleep and wake at in same time to avoid consuming energy. The suggested algorithm, as illustrated previously, delivers successfully packet with a fairly high rate and reduces dropping packets. This study show the avoid congestion and saves energy by using packetization to control on number of sending packets and the delay in queue which led reduce the retransmission packets as a result of reduce drop its.

7. References

[1] G. Anastasi, M. Conti, M. D. Francesco and A. Passarella 2009 *Energy conservation in wireless sensor networks: A survey* (ELSEVIER) **7**(3), pp. 537–568.

[2] B. Omodunbi, O. Arulogun and J. Emuoyibofarhe 2013 *A Review of Energy Conservation in Wireless Sensor Networks* (IISTE) **3**(5), p. 8.

[3] M. Khedher, J. Lloret and A. Douik 2016 *Enhanced group-based wireless ad-hoc sensor network protocol* (IJDSN) **12**(7), pp. 1-18.

[4] U. G. a. J. Bakal 2018 Healthcare Monitoring System Based on Wireless Sensor Network for Cardiac Patients *Biomed Pharmacol J,* **11**(3).
[5] L. Parra, L. García, S. Sendra and J. Lloret 2018 *The Use of Sensors for Monitoring the Feeding Process and Adjusting the Feed Supply Velocity in Fish Farms* (Hindawi) **2018**(1), p. 14.
[6] T. K. a. D. Kumar 2020 A survey on QoS mechanisms in WSN for computational intelligence based routing protocols, *Wireless Networks*, **26**, p. 2465–2486.
[7] XBee S1 802.15.4, Online Available: https://www.digi.com/resources/library/data-sheets/ds_xbeemultipointmodules, accessed on 18 February 2021.
[8] PIC18F46K22 datasheet Microchip Online Available: http://ww1.microchip.com/downloads/en/DeviceDoc/41412F.pdf, accessed on 18 February 2021.
[9] "nxp.com," LPC2131/2/4/6 2012. Online Available: https://www.nxp.com/docs/en/user-guide/UM10120.pdf, accessed on 18 February 2021.
[10] M. A. Jan, S. R. U. Jan, M. Alam, A. Akhunzada and I. U. Rahman 2018 A Comprehensive Analysis of Congestion Control Protocols in Wireless Sensor Networks (*Springer*) **2018**(1), pp. 456–468.
[11] M. Ojo, D. Adami and S. Giordano 2016 Performance evaluation of energy saving MAC protocols in WSN operating systems (*IEEE*) p. 11.
[12] N. Zaman, L. T. Jung and M. Yasin 2016 Enhancing Energy Efficiency of Wireless Sensor Network through the Design of Energy Efficient Routing Protocol (*Hindawi*) p. 16.
[13] B.V.Shruti, M.N.Thippeswamy and V. K 2019 Energy Efficient Medium Access Control Protocols for Wireless Sensor Networks (*IJATCSE*) 8(1), p. 13.
[14] A. Chugh and S. Panda 2019 Energy Efficient Techniques in Wireless Sensor Networks (*Recent Patents on Engineering*) 13(1), p. 9.
[15] S. M. Lasassmeh and J. M. Conrad 2010 Time synchronization in wireless sensor networks: A survey (*IEEE*) p. 11.

Partially autonomous two wheel vehicle for transport of persons at industrial plants – concept of solution

J Malaka, A Gorol and J Nowicki
Silesian University of Technology, Faculty of Mechanical Engineering, ul. Konarskiego 18A, 44-100 Gliwice, Poland

E-mail: julian.malaka@polsl.pl

Abstract: Within the research, the needs and problems occurring in the movement of the workers at industrial plants were specified and characterised. There were developed concepts of systems necessary for the automatic operation of the vehicle in the manner under consideration. The authors carried out the review and preliminary selection of technical means, used in both manually controlled vehicles and the autonomous ones, possible to adapt in order to implement the proposed solution. The paper considers the principles of the development in transport of persons at industrial plants. The results of the analysis performed are observations and assumptions about the possibility of integrating well-known mobile platform, navigation and power supply in an innovative way. It states the fundamentals for the plan of the future work which leads to actualising the device, including designing, constructing and programming of it.

1. Introduction

The movement of workers in industrial plants sometimes involves travelling long distances. In many places, different means of transport are used, either fully mechanical (muscle-based) or with independent propulsion. Due to the specifics of the environment in question, battery electric systems are usually chosen for the latter. Currently, two-wheeled vehicles are among the most popular ones.

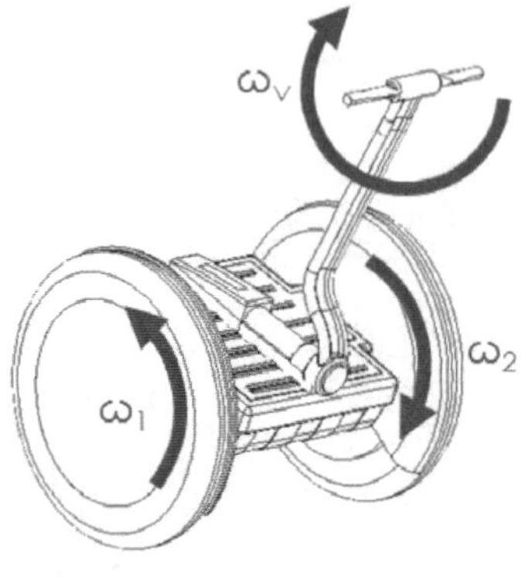

Figure 1. High manoeuvring capabilities of two-wheeled vehicle.

They are characterised by high manoeuvrability, mobility, compact design and ease of use. As shown in figure 1, the vehicle rotation ω_v is achieved by the wheels opposite rotation ω_1 and ω_2. However, the device may not be available whenever it is needed due to the necessity of periodical energy replenishment (e.g., by connecting to the grid or changing batteries) or leaving it in a remote location by the previous user. As a solution to these problems, the research team proposes to develop an electric vehicle with enhanced functionality compared to the currently known. The basic mode of operation is to be the classic manual control by the driver, with all the features of standard operation. However, when the vehicle is not being used to transport a worker, its autonomy is activated. It can then move to a loading zone or go to wherever it is needed at the time. The efficient and safe implementation of such a mode of operation can be achieved using techniques known from AGVs or mobile robots. Implementation of modern means of energy transfer to the vehicle would eliminate the need for human intervention in the whole process.

1.1 Vehicle operating conditions

Autonomous driving modes require complex and sophisticated technical solutions. Depending on the requirements of the customer and the target area of the vehicle, there are different solutions to the problem. By making assumptions about the destination:

- the vehicle will not leave the confined spaces of the workplace,
- the total area of the plant will oscillate around 10 000 m^2,
- the vehicle will navigate in a dynamic environment – moving plant workers,
- special transport routes will be set up for vehicles,

the following concept of construction and operation of the device can be proposed.

2. Construction of vehicle

Two-wheeled vehicle of "segway" type keeps balance during the ride thanks to the adjustable toque of the wheels. Keeping constant lean angle φ is achieved by manipulating torques M_1 and M_2 which generate balancing force F acting on mass centre, which is presented in figure 2.

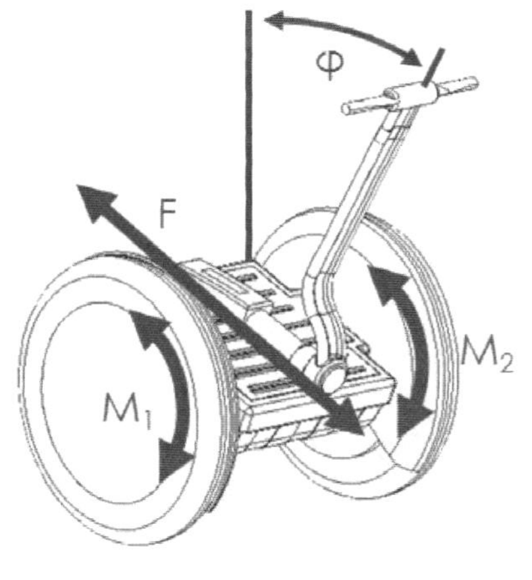

Figure 2. Keeping constant lean angle of vehicle.

The system is a hybrid combination of technical solutions used in industry and other areas of life.

2.1 Sensor equipment

The two-wheelers will be equipped with a variety of sensors (shown in figure 3) to help determine current position and distance from obstacles.

When operating in autonomous mode, it is necessary for the vehicle to know its exact location in relation to individual workstations on the shop floor, base stations and other two-wheelers. Operation in closed facilities does not allow the use of GPS, for which reason localization will take place via Wi-Fi networks and tags in the form of QR codes, infrared emitters or RFID receivers [1].

It is necessary to specify video cameras placed one on each side of the vehicle, especially the front one, of high resolution, which will read the unique QR codes distributed in the workplace, allowing the position of the vehicle to be identified unambiguously. The QR signs will be placed on poles or walls at vehicle steering wheel height. They will not significantly interfere with the appearance of the hall but will ensure the proper level of localization of two-wheelers. In addition, this type of solution does not generate high operating costs, as the signs are printed and do not require illumination for correct reading.

2.2 Energy sources

A set of supercapacitors and batteries will constitute the power supply system, which elements are presented in figure 4.

An unquestionable advantage of using supercapacitors is the possibility to recharge them instantly. They have high efficiency (about 90%), wide operating temperature range (-40°C ÷ 60°C), are maintenance free, resistant to deep discharge and their operating cost is very low, because repeated charging and discharging does not significantly affect their capacity. The disadvantage of this solution is low current capacity, which makes it necessary to use a separate energy storage, in this case a battery pack [2].

The biggest advantage of using batteries is their high capacity and current efficiency. They are able to power a vehicle much longer than supercapacitors, so they will serve as the main energy storage. However, they have a number of disadvantages such as long charging times, reduced efficiency during repeated charging and discharging, the need for periodic replacement and therefore higher operating costs. The combination of these two types of power supply and their synergic action, results in the mutual elimination of the disadvantages of each solution.

The power supply system will be equipped with an inverter maintaining voltage at a constant level, a charging module protecting batteries and supercapacitors against excessive discharge and a module allowing inductive charging.

Together with the two-wheeler, the set will form a docking station or a network of stations equipped with a wireless inductive charging system.

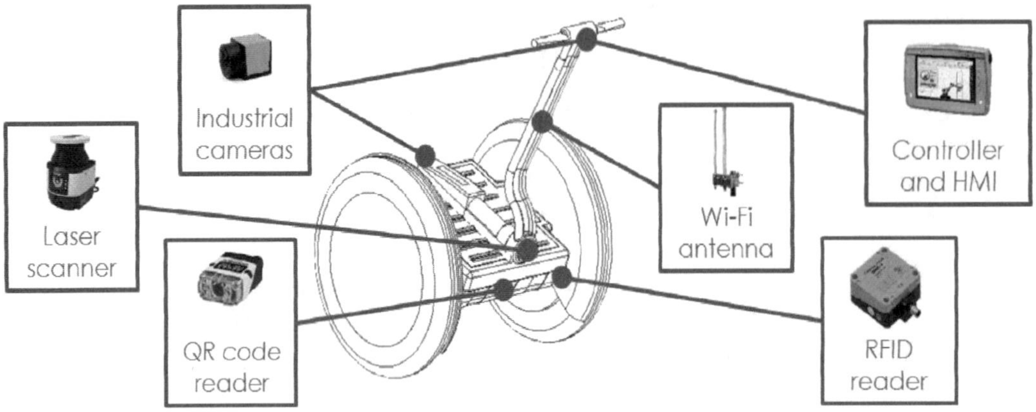

Figure 3. Location of sensor, communication and control equipment in vehicle.

3. Vehicle operation

The vehicle will be suitable for driving indoors, on level pavement and outdoors, provided that the terrain does not have irregularities in excess of 3 cm.

The minimum range of the vehicle on one full charge is 10km. The vehicle will be travelling in the vicinity of other people, so the driving speed with the driver will be limited to 20km/h, and in autonomous mode to 8km/h. This will allow for safety and at the same time efficient movement around large-scale sites.

3.1 Autonomy in driving and vehicle navigation

The minimum range of the vehicle on one full charge is 10 km. The vehicle will be travelling in the vicinity of other people, so the driving speed with the driver will be limited to 20 km/h, and in autonomous mode to 8 km/h. This will allow for safety and at the same time efficient movement around large-scale sites [3].

In addition, by connecting the vehicles to a server that monitors their operation and manages their movement, it is possible to generally locate the vehicle based on the strength of the signal to individual routers located on the premises.

In order to develop autonomous systems, a suitable solution is to implement artificial intelligence, which, on the basis of data on the movement of vehicles and the environment, will make it possible, for example, to avoid collisions with pedestrians or to plan the optimal route to the charging station.

The vehicle will recognise its surroundings using a laser scanner. In a short period of time it will send out laser beams which will determine the location, shape, size and distance of possible obstacles. An additional environmental recognition and safety system will be a thermal imaging camera. If a person appears on the route, the system will distinguish him/her from the rest of the environment by detecting body temperature. The vehicle will stop and then slowly avoid the obstacle. The process is of gathering the surroundings data is presented in figure 5.

The application of the above proposed solutions for vehicle navigation and obstacle detection and appropriate response, meets the requirements set for the design of an autonomous vehicle to navigate within the work premises. If it is necessary to travel outside the hall, the autonomous mode will be unavailable until the vehicle is located again at a point that is the limit of the autonomous operation area.

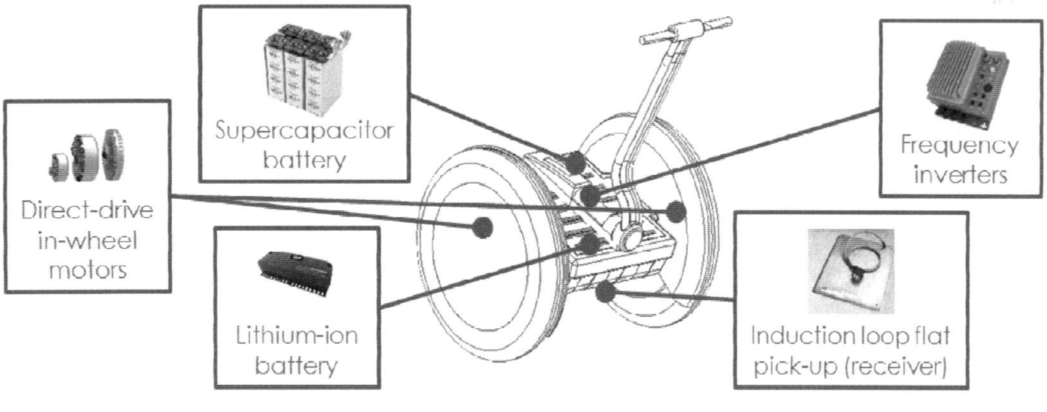

Figure 4. Location of power equipment in vehicle.

Table 1. Summary of major problems and their solutions related to autonomous operation of unoccupied vehicle [1, 4-6].

No.	Problem	Identified classes of useable equipment to be adapted from mobile robotics	Description of solution
1.	Finding unoccupied vehicles and checking their battery levels by a person	Human-machine mobile interfaces, local positioning systems, wireless communication	The person receives the localisation of vehicles and their battery levels on a mobile device.
2.	Getting a mean of transport by a person who is far from the vehicle	Human-machine mobile interfaces, wireless communication	The person calls the vehicle using a mobile device which transmits their localisation to the vehicle controller via a wireless network. The unoccupied vehicle reaches the person autonomously.
3.	Automated verifying if the unoccupied vehicle range meets the requirements for executing a task	Area mapping, energy consumption modelling, local positioning systems	The person specifies the destination while their localisation is read automatically. The driving task is calculated in respect of energy consumption.
4.	Automated positioning, way finding and self-navigating of an unoccupied vehicle	Area mapping, area scanners, distance sensors, local positioning systems	The vehicle scans the surrounding, creates a map of it and locates itself as well as routes and other objects on the map.
5.	Automated position verifying – assuring that the unoccupied vehicle is next to the particular object	Distance sensors, tag readers	The vehicle scans the RFID or QR tag placed on the refence object and measures the distance from obstacles.
6.	Automated reacting to people around by an unoccupied vehicle which moves	Artificial intelligence, machine vision, thermography	The vehicle recognises specific motions as well as the temperature distribution of people and modify the velocity or change the position in the case of collision anticipation.
7.	Automated reacting to fixed obstacles as well as objects which move by an unoccupied vehicle	Area scanners, distance sensors, tag readers, wireless communication	The vehicle identifies the object by shape scanning, tag scanning or wireless communication and adapt the motion parameters to moving around it.

3.2 Recharging

The power system monitors the battery charge level in real time. In the event of the charge level dropping below the specified threshold, the vehicle will signal and suggest that the user replace the vehicle with another - charged - vehicle or park the vehicle. When the charge level drops below the critical threshold, further driving will not be possible in order to protect the cells from accelerated wear (one can see the process in figure 6).

The supercapacitors will be responsible for quickly recharging the vehicle when there is a need to quickly recharge the batteries. This will take about two minutes, after which it will be possible to drive about a kilometre to get to work and to the nearest charging station. The main energy storage facility will be a set of batteries that will take around 15 minutes to fully charge [2].

The charging system will work in such a way that the supercapacitors will be charged first, from which the batteries will then be charged. This system will allow the vehicle to be recharged unattended. Thanks to the autonomous mode, when the two-wheeler is close to being discharged, it will drive itself from its parking place to the charging station.

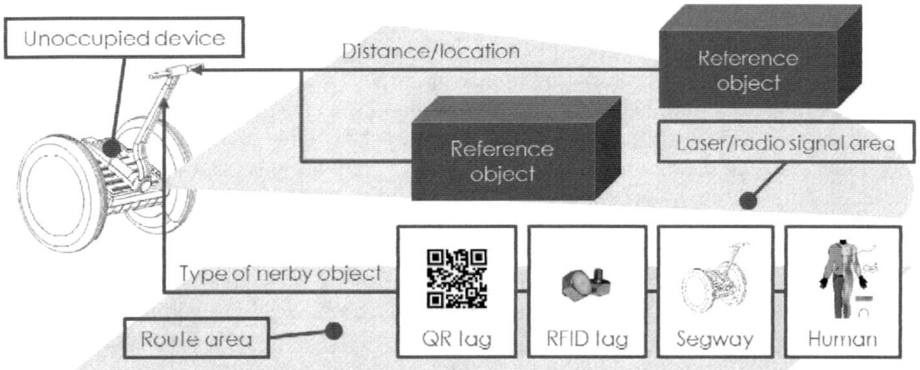

Figure 5. Sources of data in automated vehicle positioning system.

3.3 Proposed modes of operation

The vehicle is capable of operating in different states of activity as required. The standard state of the power system is moving with a person on board. During this operating mode, the batteries are gradually discharged due to the current drawn by the motors and the lack of charging. This is due to the assumption that the vehicle can not only move along strictly defined routes (where induction loops for charging the batteries are installed).

The next operating condition is the return of the vehicle to the parking point. The vehicle moves completely autonomously, taking the most optimal route to its destination. The vehicle moves along predefined paths equipped with an integrated inductive charging system under the ground surface. As it moves, the two-wheeler is charged via the aforementioned induction loop using supercapacitor fast-charging technology, which enables it to return to the base even when the batteries are deeply discharged.

The next mode of operation of the power system is charging in the docking station. The vehicle is located in the docking station, which is primarily responsible for charging the two-wheeler to the appropriate level and maintaining it. The vehicle does not consume any significant amount of energy, due to the advanced software of the power supply system enabling the power consumption to be reduced to a minimum.

The last mode of operation of the power supply system is the so-called "standby mode". The vehicle enters this mode while in the docking station or "waiting" for the driver to return. In this state, the system does not consume much power due to the intelligent power saving mode. When in the docking station, the battery charge is maintained at 80-100%.

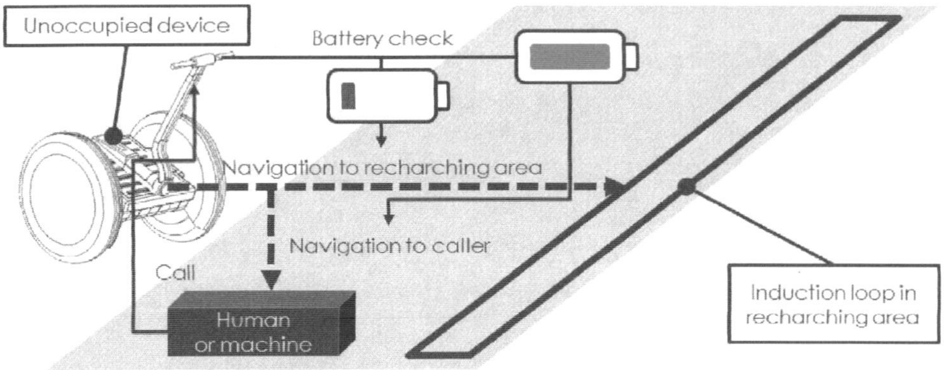

Figure 6. Autonomous power management in vehicle.

4. Conclusions

The work led to the formulation and consideration of innovative concepts for the technical means of transport under consideration in the area of in-plant transport, but which can also be transferred to non-industrial applications. The analyses carried out concerned an innovative solution, namely a partially autonomised means of transport. A two-wheeled vehicle without a user is supposed to arrive safely at various places in the industrial hall independently on demand. This leads to easier and faster passenger transport. The respective mobile unit is supposed to be connected to other machines in the environment as well, which determines the safety and efficiency of its operation. Autonomy can also manifest itself in manual operation, assisting the user in preventing collisions and maintaining safe movement parameters, or in selecting and optimising routes. Similar solutions are used in highly automated industrial plants with a large number of mobile robots. The research mainly included the identification of needs, accompanying problems and possibilities of their solution. The activities performed form the basis for future work involving the design and construction of an innovative device that meets the requirements identified so far.

5. References

[1] https://en.wikipedia.org/wiki/Wi-Fi_positioning_system, last accessed: 2021-04-19.

[2] https://www.seweurodrive.pl/rozwiazania/aplikacje/mobilny_transport_i_ruch/samojezdne_wozki_agv/agv_system/wozki_agv.html, last accessed: 2021-04-19.

[3] http://www.ahg.com/business-mobile-apps-blog/mobile-asset-tracking-technologies.html, last accessed: 2021-04-19.

[4] Malaka J. 2019 *Autonomous robot control system for automation of manipulations, Mechatronics 2017 - Ideas for industrial applications*, Springer International Publishing, Berlin, ISBN 978-3-319-10990-9, red.: Jerzy Świder, Sławomir Kciuk, Maciej Trojnacki, 298-307.

[5] Malaka J., Hetmańczyk M. 2020 *Numerical co-simulation in Industry 4.0 - creating digital twin of belt transmission with NX and Matlab*, International Journal of Modern Manufacturing Technologies **12**(3) 12/2020.

[6] Malaka J.: *Autonomiczny transport w przemyśle 4.0*, Utrzymanie Ruchu, 01/2020.

IOP Conf. Series: Materials Science and Engineering 1182 (2021) 012041 doi:10.1088/1757-899X/1182/1/012041

The effect of appendages on ship resistance

A Mandru and F Pacuraru
"Dunarea de Jos" University of Galati-Romania, Faculty of Naval Architecture, Departament of Naval Architecture, 47 Domneasca Street, 800008, Galati, Romania

E-mail address: florin.pacuraru@ugal.ro

Abstract: The availability of the robust commercial computational fluid dynamics (CFD) software and the increasing power of high-performance computers (HPC) boosts the use of CFD techniques for numerical investigations of ship hydrodynamimcs performaces. Nowadays, CFD tools provide a rather accurate solution for complex physical phenomena, such as wave braking, turbulence, flow detachment and overturning. Consequently, the ability to capture this fenomena allows a detailed insight into the flow mechanism that trigger and sustain them.
The paper focuses on a numerical investigation of flow around hull a fully appended ONR Tumblehome model 5613. Due to special operation conditions, naval ships are equipted with a large number of appendages such as sonar domes, brackets, twin propeller shafts and twin rudders. Position and alignment of these appendages are essential in order to avoid the increased resistance, decreased propulsive efficiency and cavitation. NUMECA/FineMarine commercial code has been used to evaluate the flow field around ONRT (Office of Naval Research Tumblehome) hull and to estimate the effect of the appendages on ship hydrodynamics performances. Comparison with experimental towing tank results showed good agreement with a difference of up to 2%.

1. Introduction

Due to the widespread availability of robust commercial Computational Fluid Dynamics (CFD) software and high-performance computers, CFD is increasingly being used to solve fluid engineering problems and ship hydrodynymics is no exception. Assuming that the flow solvers are capable of dealing with practical complex geometries as well as taking into account complex physical phenomena such as turbulence and free surface, CFD-based design tools can provide a fairly accurate solution to different ship hydrodynamic problems. The naval architect uses viscous free-surface flow calculations to improve the hull forms for achieving a homogeneous velocity distribution, determining the optimum inclination angle for shaft brackets, studying the hydrodynamic interactions between various appendages and arrangements, and to investigate flow separation regions. The key issues in appendages drag investigation is to provide details about the resistance, interaction and scaling of appendages.

The paper is focused on a numerical investigation using RANS computation to solve the viscous flow around a fully appended combatant ship hull. To better understand the effects of different configurations of the appendages on the wake structure in the propeller disk, several sets of calculations was performed. The NUMECA/FineMarine commercial code has been employed to compute the flow field structure surrounding the ship hull, as well as the forces acting on the bare hull and appendages. The effect of appendages on ship drag of displacement vessels has been studied by

Iwan et al. [1] which considered the scale effects of appendages drag, [2] introduced an Appendage effective Index to assess the effectiveness of an appendage, Lasky [3] developed a procedure for scaling of appendages drag. Numerically investigation of the flow around the ship with appendages has been reported in [4, 5]. Detailed analysis of the flow around the juncture between a plate and a foil considering different inclination angle of the foil have been investigated by [6, 7].

2. Ship hull and computation conditions

The paper considers the investigation of the ONR Tumblehome (ONRT) which is a preliminary design for a new surface combatant that is available to the public for fundamental research. The 1/49 scaled ship model is used as one of the benchmark cases in Tokyo 2015 CFD workshop in ship hydrodynamics. Extensive experiments were performed at IIHR Hydraulics Wave Basin Facility for this ship model. The ONRT's bow shapes are dominated by the presence of a prominent sonar at the bottom, and at the ship's stern there are two shaft lines, supported by brackets, with two propellers and two rudders. The model of the ONRT equipped with appendages is shown in figure 1, and the principle geometric characteristics are listed in table 1.

Figure 1. ONR Tumblehome.

Table 1. ONR Tumblehome principle geometric characteristics.

Dimensions	Symbol	Units	Model scale	Full scale
Length of waterline	L_{WL}	m	3.147	154
Maximum beam of waterline	B_{WL}	m	0.384	18.78
Depth	D	m	0.266	14.5
Draft	T	m	0.112	5.494
Displacement	Δ	kg/t	72.6	8507
Wetted surface area (fully appended)	S_0	m^2	1.5	NA
Bloc coefficient	C_B	–	0.535	0.535
LCB (aft of FP)	LCB	m	1.625	NA
Vertical center of gravity (from keel)	KG	m	0.156	NA
Metacentric height	GM	m	0.0422	NA
Speed	v	m/s	1.11	7.77

In table 2 are presented the test cases for calm water resistance analysis. Six different sets of simulations were carried out for 5 speeds corresponding to Froude numbers between 0.20 and 0.40. The first series of computations were performed in the case of the ONR Tumblehome without appendages in order to investigate the flow characteristics around the bare hull. The purpose of the next fore configurations is to determine the influence of each appendage on the ONRT total resistance.

The last set of simulations considers the investigation of the flow characteristics around the ONR Tumblehome fully appended.

Table 2. Calm water resistance computed test cases.

Configurations	Description	Wetted surface area [m²]	List of speeds [m/s]	Coresponding Froude number
ONRT	ONR Tumblehome bare hull	1.310		
ONRT_bilge_keel	ONR Tumblehome with bilge keel	1.402	1.11	0.20
ONRT_skeg	ONR Tumblehome with skeg	1.343	1.39	0.25
ONRT_rudder	ONR Tumblehome with rudder	1.363	1.67	0.30
ONRT_shaft	ONR Tumblehome with shaft	1.357	1.94	0.35
ONRT_appended	ONR Tumblehome fully appended	1.537	2.22	0.40

3. Numerical Approach

3.1. Governing equations

The averaged continuity and momentum equations for incompressible flows involving external forces (1 and 2) can be written in tensor form in the Cartesian coordinate system as:

$$\frac{\partial(\rho\overline{u_i})}{\partial x_i}=0 \tag{1}$$

$$\frac{\partial(\rho\overline{u}_i)}{\partial t}+\frac{\partial}{\partial x_j}\left(\rho\overline{u}_i\overline{u}_j+\rho\overline{u_i^{'}u_j^{'}}\right)=-\frac{\partial\overline{p}}{\partial x_i}+\frac{\partial\overline{\tau}_{ij}}{\partial x_j} \tag{2}$$

where $\overline{u}_i$ is the relative averaged velocity vector of flow between the fluid and the control volume, $\overline{u_i^{'}u_j^{'}}$ is the Reynolds stresses, $\overline{p}$ is the mean pressure and $\overline{\tau}_{ij}$ is the mean viscous stress tensor components for Newtonian fluid under the incompressible flow assumption, and it can be written as in equation (3):

$$\overline{\tau}_{ij}=\mu\left(\frac{\partial\overline{u}_i}{\partial x_j}+\frac{\partial\overline{u}_j}{\partial x_i}\right) \tag{3}$$

The NUMECA/FineMarine commercial code has been employed to compute the flow solution around the ONR Tumblehome. The implicit solver uses the finite volume method to build the spatial discretization for the governing eqution in order to solve the incompressible steady RANSE in a global approach Guilmineau et al [8]. The k-ω SST turbulence model with wall function formulation is used for turbulence closure in this paper. Velocity-pressure coupling is handled with pressure equation formulation (SIMPLE) using a face-based approach. The velocity field is obtained from the momentum conservation equations and the pressure field is extracted from the mass conservation constraint, or continuity equation, transformed into a pressure-equation. In the case of turbulent flows, additional transport equations for modelled variables are discretized and solved using the same principles Duvigneau et al [9]. A second-order upwind scheme and a central difference scheme, respectively, are used to discretize the RANSE's convection and diffusion terms.

For resistance computations, the flow is accelerated using a steady quasi-static approach over a time-span that corresponds to the relationship between the ship's length and the inflow velocity such that $T_{acc} = 2.0L_{pp}/U_{\infty}$. The omputations time step is calculated with respect to the equation $\Delta t = 0.005L_{pp}/U_{\infty}$ for k-ω SST turbulence model as it is suggested by the ITTC recommended procedures for CFD [10]. Free-surface capturing strategy is based on multi-phase flow approach using Volume of Fluid method with high-resolution interface schemes. Incompressible and nonmiscible flow phases are modelled using conservation equations for each volume fraction of phase/fluid, Queutey and Visonneau [11].

3.2. Computational domain

Dimension of the computational domain (figure 2) have been generated as $-3 \le x/L_{WL} \le 1$, $0 \le y/L_{WL} \le 1.5$, $-1.5 \le z/L_{WL} \le 0.5$, where the L_{WL} is the length of waterline. The origin of the Cartesian coordinate axis has been selected as the point where fore perpendicular of model intersects the waterline. The negative x-axes have been defined in the direction of flow, positive y-axes, starboard beam and z-axes upward directions. Only half of hall has been used in resistance simulation in order to reduce the computational costs. The centreline plane has been defined as the boundary condition of symmetry. The inlet, outlet and side boundaries have been defined as far field. The top and bottom of computational domain has been defined as prescribed pressure. Wall-function condition has been adopted on the hull surface and slip condition for the deck and superstructure.

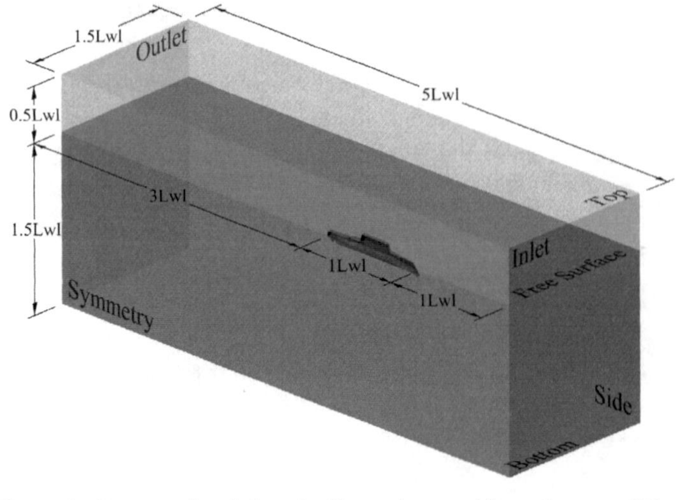

Figure 2. Computational domain dimensions and boundary condition.

3.3. Grid generation

An unstructured hexahedral mesh has been generated to cover the entire computational domain along the ONR Tumblehome model. The grid topology is a H-H type. Surface grid refinement at the bow and stern area has been considered. A box refinement has been used to define a finer mesh in the area where ship waves system develops. An analytical weighting mesh deformation approach is employed as long as both trim and sinkage were solved during the computations.

4. Results and dicussions

In experimental ship hydrodynamics, there is a common practice to measure the drag of the appendages as the difference in drag conducting resistance tests with a ship-model in the bare hull and the appended condition. In the present study, several pilot calculations were performed for the flow around the bare hull to investigate the flow characteristics, but also determine the bare hull resistance which is considered the reference value for the comparison of different appendages configuration. For the bare hull, the most important task is to capture the vortex structures that develop in the boundary layer of the ship. Unlike full-shaped ships, such as tanks and bulk carriers, where the flow is strongly influenced by bilge vortice, fast military ships do not experience flows with such vortex structures. In the case of the ONRT combatant ship model, the main flow characteristic is the vortex generated by the sonar dome located at the bow of the ship, as it is depicted in figure 3.

Analysing the variation of the total resistance coefficient against the Froude number, one can see from the table 3 and figure 4, that large increase in the slope of both curves appear at Froude 0.35, which is caused mainly by the significant increase of the trim as presented in figure 5. The trim is related to the position of wave trough/crest at the ship extremities, which is demonstrated by representation of wave pattern generated by the ship for Froude numbers between 0.2 and 0.4 plotted in figure 6.

Figure 3. Vortex structure generated by the sonar dome.

Table 3. ONRT bare hull computed results.

Fn	v [m/s]	R_T [N]	C_T	z [m]	θ [deg]
0.20	1.11	3.48	4.30E-03	0.083	0.067
0.25	1.39	5.49	4.35E-03	0.082	0.096
0.30	1.67	8.57	4.72E-03	0.079	0.088
0.35	1.94	11.76	4.75E-03	0.077	0.162
0.40	2.22	19.04	5.89E-03	0.073	-0.105

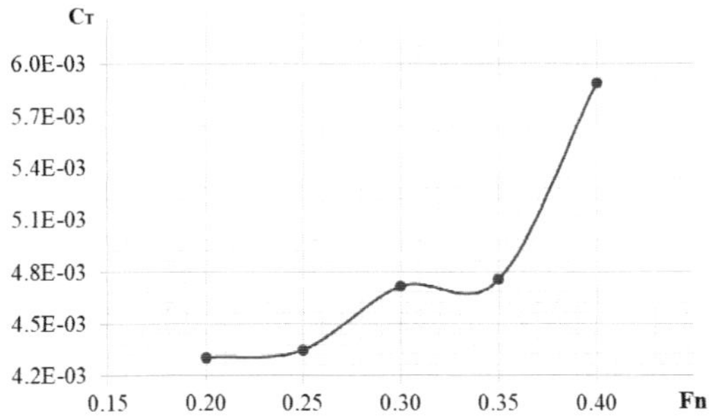

Figure 4. ONRT bare hull total resistance coefficient.

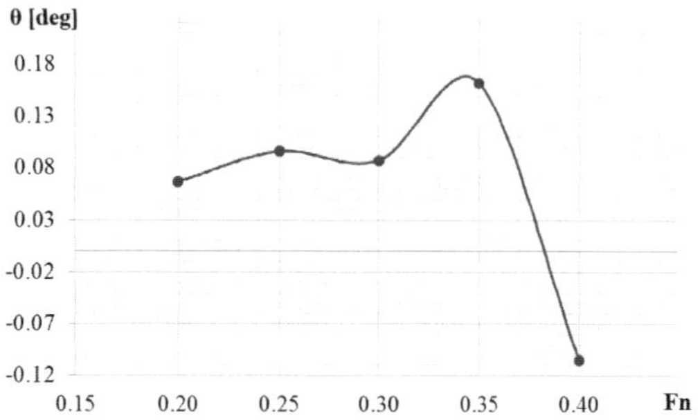

Figure 5. ONRT bare hull trim.

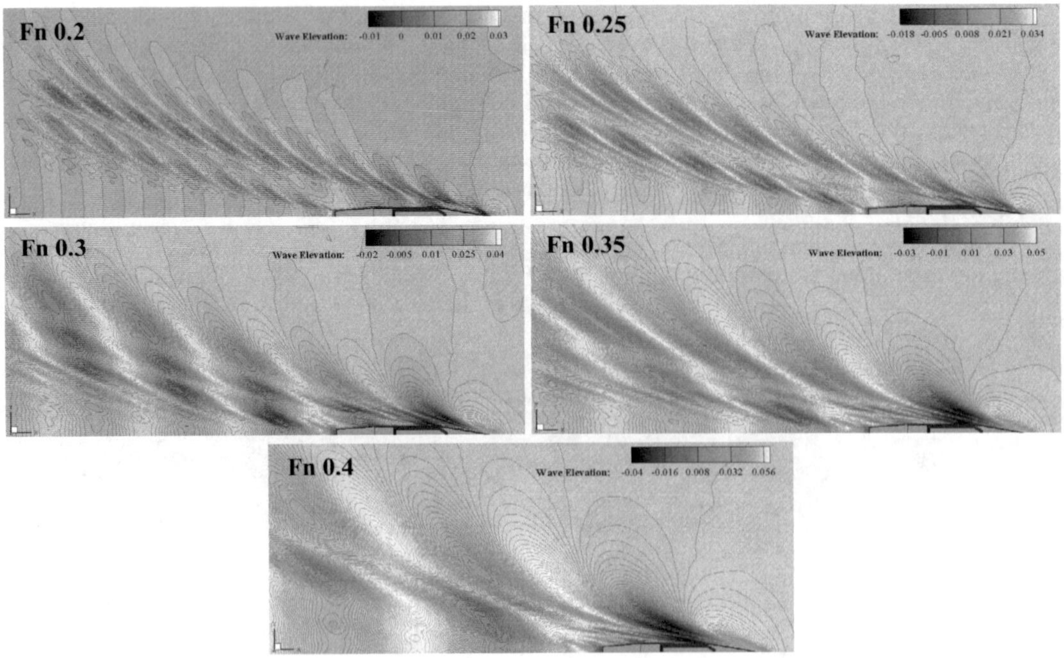

Figure 6. ONRT bare hull wave Elevation.

The results provided by Delen and Bal [11] were used as for the validation of the calculations performed. In the following, it is proposed the quantitative validation of the numerical solution obtained for the hull whith skeg configuration. Table 4 and figure 7 presents a comparison between the calculated forward resistance coefficients and those obtained experimentally. The percentage difference between the coefficients of the total resistance to advance (C_T), determined numerically and experimentally, is less than 2%.

Table 4. Comparison of ONRT skeg configuration EFD vs CFD.

Fn	R_T [N]		C_T		ε [%]
	EFD	CFD	EFD	CFD	
0.20	3.64	3.57	4.39E-03	4.30E-03	1.96
0.25	5.82	5.71	4.49E-03	4.41E-03	1.82
0.30	9.04	8.86	4.84E-03	4.75E-03	1.85

Five different sets simulations were carried out for five speeds corresponding to Froude numbers between 0.2 and 0.40 to compute the flow around five different configurations of the hull and appendages: hull with skeg (ONRT_skeg), hull with bilge keel (ONRT_bilge_keel), hull with shaft (ONRT_shaft), hull with rudder (ONRT_rudder) and full appended hull (ONRT_appended), in order to account the effect of each appendage on ship resistance. Bare hull results have been considered as a reference case for all the comparisons. Ship resistance curves for each configuration investigated are plotted in figure 8. Table 5 includes the relative variation of ship resistance for each appendage configuration. Analysis of computational results revealed that the main contribution in drag, of about 8 procent, is added by the rudder and the shaft. Another conclusion that can be drawn is that the influence of the interference between the appengages is not significant. Moreover, for each configuration the relative difference in resistance drops at Forude number 0.4, fact which is related to the significant increase in ship resistance at that speed, which is mainly caused by the wave resistance. A comparison between bare hull and full appended in terms of variation of total ship resistance coefficient (figure 9) and trim angle (figure 10) showed the same behaviour. The maximum effect of the appandages on the ship drag is 28.7 % and it appear for the Froude number 0.35.

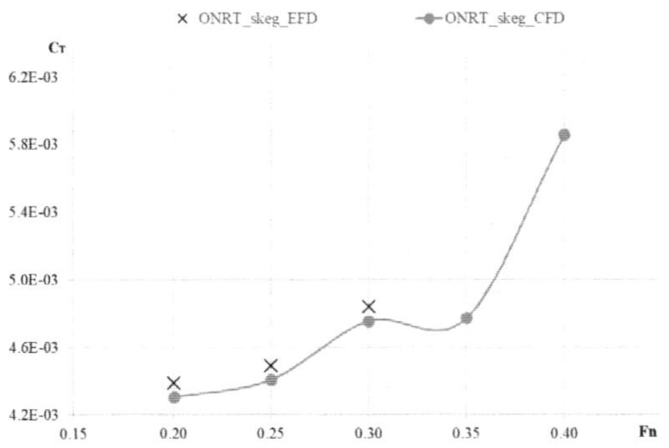

Figure 7. Comparison of total resistance coefficient EFD vs CFD.

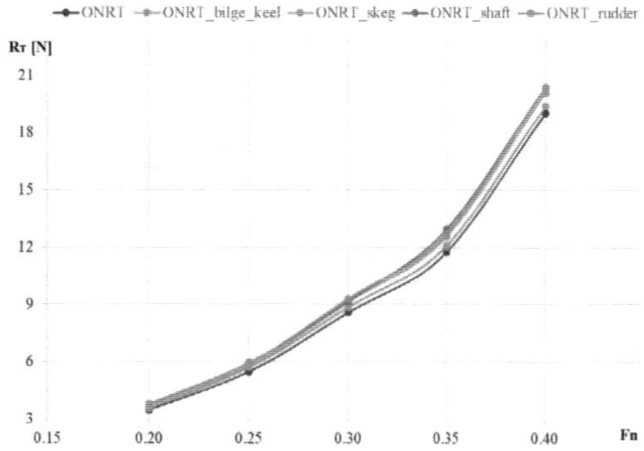

Figure 8. Resistance curves computed for each configuration.

Table 5. Relative resistance difference.

Fn	R_T [%]				
	ONRT_bilge_keel	ONRT_skeg	ONRT_shaft	ONRT_rudder	ONRT_appended
0.20	6.70	2.50	8.89	8.88	26.98
0.25	5.53	3.94	8.33	8.18	22.97
0.30	6.48	3.30	7.02	8.50	25.29
0.35	7.12	2.88	10.28	8.44	28.71
0.40	5.70	1.91	7.06	7.06	21.73

Another area of interest in terms of flow mechanism is the stern of the ship, because in this area are located the propulsion and steering systems. To gain more insight on appendages hydrodynamics, the iso-surface second invariant Q*=25 colored is depicted in figure 11 in order to capture the interferes of the vortex structures generated appendages and their junction with the hull.

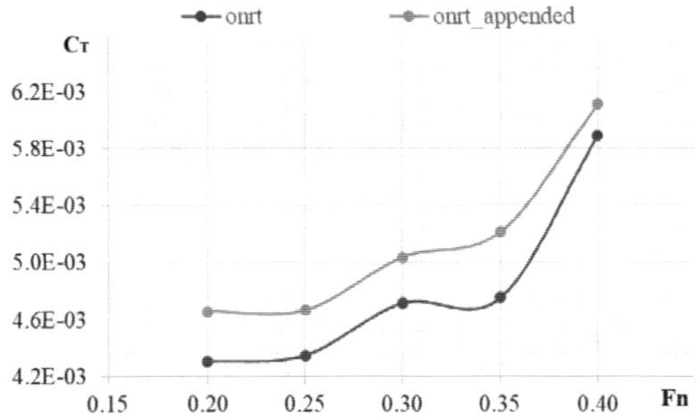

Figure 9. Comparison of total resistance coefficient bare hull vs fully appended

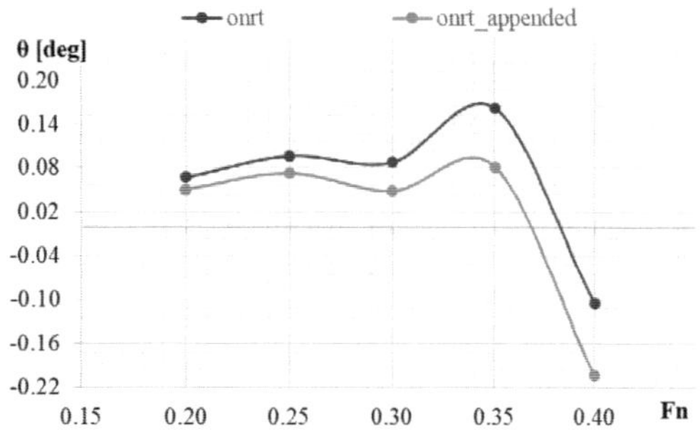

Figure 10. Comparison of trim ONRT bare hull vs ONRT fully appended

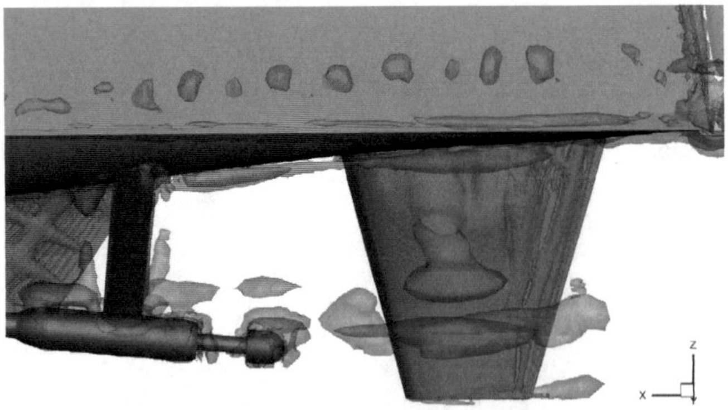

Figure 11. Vortex structure at stern area.

5. Conclusions

A sistemetic study of appendages influence on ship resistance at model scale has been performed based on RANS-VOF simulation. Six appendages configurations at five different Forude number have been investigated.

The predicted total resistance coefficient computed for hull with skeg were compared with the experiment of Delen and Bal [12], showed a good agreement with a difference of less than 2%.

Analysis of computational results revealed that the main contribution in drag, of about 8 procent, is added by the rudder and the shaft. Another conclusion that can be drawn is that the influence of the interference between the appengages is not significant. The maximum effect of the appandages on the ship drag is 28.7 % and it appears for the Froude number 0.35.

To gain more insight on appendages hydrodynamics the vortical structuctures generated by the appendages have been investigated.

Further studies will focus on prolusion and seakeeping simulation, but also on full scale investigation.

6. References

[1] Iwan MK, Muhammad SR, Yaseen AA 2020 Scaling of RV Athena's Appendage Drag Using CFD, *Advancement in Emerging Technologies and Engineering Applications, Springer*, Editor Chun Lin Saw.

[2] Shiju J, Kareem K, Praveen PC, Korulla M, Panigrahi PK Ship Hull Appendages: A Case Study *International Journal of Inovative Research and Development* **1**(10) 74-89.

[3] Lasky MP 1980 *An investigation of appendage drag*, David Taylor Naval Ship Research and Development Center Bethesda MD, report.

[4] Oliva RA, Pérez RL, Pérez FL 2013 A contribution to appendage drag extrapolation using computational tools *Proc. Developments in Maritime Transportation and Exploitation of Sea Resources*.

[5] Ozdemir CO and Barlas B 2018 Free surface flow simulation around an appended ship hull *Brodogradnja* **69**(3) pp 25-41.

[6] Lungu A and Ungureanu C 2008 Numerical Study of a 3-D Juncture Flow *AIP Conference Proceedings* **1048**(1) pp 839-842.

[7] Lungu A and Ungureanu C 2009 Numerical simulation of the turbulent flow around a strut mounted on a plate *AIP Conference Proceedings* **1168**(1) 647-650.

[8] Guilmineau E, Deng GB, Leroyer A, Queutey P, Visonneau M, Wackers J 2015 Influence of the turbulence closures for the wake prediction of a marine propeller. *Proc. of the 4th International Symposium on Marine Propulsors* Austin, Texas, USA pp.177-183.

[9] Duvigneau R, Visonneau M, Deng G.B 2003 On the role played by turbulence closures in hull shape optimization at model and full-scale *J. Marine Science and Technology* **8**(1), 1–25.
[10] International Towing Tank Conference (ITTC) 2014 ITTC – Recommended Procedures and Guidelines *Practical guidelines for ship CFD applications* 7.5-03-02-03, Revision 01.
[11] Queutey P, Visonneau M 2007 An interface capturing method for free-surface hydrodynamic flows *Computers & Fluids* **36**(9) pp 1481–1510.
[12] Delen C and Bal S 2019 Uncertainty analysis of numerical and experimental resistance test for ONR Tumblehome *Sustainable Development and Inovations in Marine Technologies* CRC Press.

Acknowledgement
The research study presented in this article was developed in the frame of "Dunarea de Jos" University of Galati, Naval Architecture Research Centre.

IOP Conf. Series: Materials Science and Engineering 1182 (2021) 012042 doi:10.1088/1757-899X/1182/1/012042

The importance of the technical specification of works for the ships repair processes in the maritime shipyards

M G Manea[1], R Zăgan[2] and E Manea[3]
[1] "Mircea cel Bătrân" Naval Academy, Marine Engineering Faculty, Department of Tactics and Naval Weapons, Fulgerului Str., No.1, Constanta, RO
[2]Maritime University of Constanta, Electromechanics Faculty, Department of Engineering Science, Mircea cel Bătrân Str., No.104, Constanta, RO
[3]Constanta Shipyard SA, Greece Branch Office, Port Str., No.1, Constanta, RO

E-mail: greti.manea@anmb.ro

Abstract. The technical specification purpose is to include the ship maintenance and improvement works needed to be carried out onboard in a repair shipyard, so that after these works performing to allow for the classification society certification, thus confirming the ship satisfactory technical condition and the relevant requirements of the classification rules were applied, thus confirming the applied management maintenance also. The technical specification has a very important role for the evaluations, done by the shipyard in preparing price and time quotation and the owner / manager, done for the comparison analyzing based on the evidence that the maintenance and improvement workflows performing on board the ship in a shipyard for works depend on the technical specification quality of preparation, which means that had to provide the necessary information in relation to the volume and type of works needed to be carried to the ship hull and structures, machinery, equipment, plants and systems supported by the necessary technical information plans approved by the classification society and the works sheets documentation referring to the followings: work location / equipment thereof number of pieces; types of the works required to be performed on location / equipment; works estimated required to be undertaken on location / equipment. The advantage of a quality technical specification is the limitation of the possibility of ordering additional works during the execution of works for maintenance, upkeeping and improvements on board the shipyard, the costs of the shipyard and the ship being thus maintained at the estimated budgets. The paper purpose is to highlight the advantages of a good quality technical specifications which consists of limiting the possible means of developing additional orders during the execution of maintenance and improvement works on board, thus the shipyard and ship costs being possible to be kept within the original estimated budget margins.

1. Introduction

The notion of performance in the shipbuilding and ship repair industry must be adapted to the complex and difficult to manage processes within the shipyards.

Performance is not an end in itself but the vector that leads to results. The results depend on the level of performance at which employees evolve, performance management being the process that facilitates the creation of an environment in which people feel motivated to make the most of their skills and qualities [1-5].

Performance indicators must be used to measure the effectiveness of the management system processes. Performance indicators measure three elements – activity, profitability and productivity – and by monitoring the results obtained, the activities performed and the costs generated, is the standard of measurement that allows examining and highlighting the viability of the strategies chosen to achieve the proposed objectives [1-5].

Where the results of process monitoring and measurement are not in accordance with the requirements, appropriate corrective action shall be taken.

The use of performance indicators in the activity of an integrated management system is generally characterized by: the obligation to proceed with the activities; repeatability of execution; monitoring activities; correct performance measurement; the possibility of identifying non-compliant processes, with an impact on costs; the application of corrective measures achieving important savings; conducting the entire activity in a systematic and planned manner; predictable and better development; employee accountability [6,7].

Currently, the performance of organizations has a fundamental role in the context in which market competition is becoming tighter, thus, the chance of survival in this competition increases considerably for those organizations that discover and minimize their vulnerabilities very quickly and, moreover, implements high-performance management tools that make it easier for them to detect, explain and solve various shortcomings of management activities, the main objective being to increase competitiveness.

Based on the requirements of the shipping companies, resulting from the updating of the objectives regarding the maintenance costs, shipyard specialised intot maintenance works, in order to maintain competitiveness in conditions of economic efficiency, will analyze their performance indicators with reference to [8-11]:

- the percentage of contracting to the number of offers sent;
- percentage of reduction of manufacturing costs / ship;
- duration of the dock cycle by types of ships;
- number of delay days in delivery of ships;
- the percentage of repetition of technical inspections for defects / ship;

so that, through the corrective measures adopted, the basic objectives are not affected:

- constant loading of the equipped docks with ships;
- constant loading of all labor specializations represented by the permanent employees of the shipyard;
- long-term collaboration with shipping companies that are in the portfolio of traditional customers;
- the possibility of starting a collaboration with new shipping companies so that the customer portfolio can be expanded.

The performance indicators necessary to be analyzed as a matter of priority are based on the information of the bidding and programming-planning activities carried out in the shipyards for the execution of the maintenance works, as they are considered to be the leaders for approving and signing a contract.

For the bidding activity, the performance indicators emerge from [8-11]:

- the number of offers sent to customers and the number of offers accepted for contracting the execution of maintenance works depending on: geographical area; the types of ships (oil tanks, bulk carriers, container ships) and their dimensions;
- the number of offers accepted for contracting the execution of maintenance works for which the final invoice was below the estimated budget in the offer of maintenance and repair works related to the maintenance of the ship elaborated on the basis of the technical specification of works drawn up by the owner / technical manager according to: customer profile; geographical area; the volume and type of canceled works.

For the planning-programming activity, the performance indicators emerge from:

- number of requests for unfulfilled offers due to lack of validity for docking in the period requested by customers;
- the number of contracted vessels that have exceeded the agreed completion deadline;
- the number of contracted ships that were delayed on arrival at the shipyard for the timely execution of maintenance works.

Bidding and scheduling activities carried out in the shipyards for the execution of marine ship maintenance works are the result of estimates that operate with fixed data conditioned by variables belonging to a wide range of conditions and limitations such as: the level of the maritime transport market; weather conditions at certain times in the geographical area where the shipyard is located; the type, capacity and age of the vessel; differences in the volume of the final work performed compared to the initial estimate due to the existing technical conditions found on the ship's systems, installations and equipment after the start of the maintenance works [8-11].

The performance indicators characteristic of a shipyard for the maintenance of ships could be those recommended in table 1 [17].

Table 1. Performance indicators characteristic of a shipyard for the maintenance of ships.

Objective	Performance indicators	Content		
		Financial		Non - Financial
Management	Availability of resources			Dock utilization rate Duration of the dock cycle by ship types and dimensions Number of days delayed in delivery of ships because of dock and quay works
	Costs	Average production costs per ship	Percentage reduction in manufacturing costs per ship	Labor consumption for a type of work performed on different types of ships Consumption of materials for a type of work performed on different types of ships Number of works not accepted for delivery to the client Number of works submitted more than 2 times until the client's acceptance
		Directly costs related to sales / allocated costs (indirect)	Commissions paid to agents / brokers	
		Penalties for ship delivery delays	Value of penalties	Number of ships by type / size / customer for which late payment penalties have been paid
		Additional works	Price negotiations	Number of offers for the price of additional works accepted at discount values different from the contractual one

Table 1. continuation

Objective	Performance indicators	Content		
		Financial		Non - Financial
Strategic	Risk - opportunity	Financial situation of customers	The amount of invoices not collected according to the contractual payment obligations	
			Late invoices for payment by types of customers	
	Market share			The total number of direct relations customers of the site
				The total number of customer relations through brokers
				Number of new customers
				Number of new customers returning by ship to the yard
Operational	Activity			Number of customer visits
				Number of customers lost
				Order productivity / day
				Number of suppliers / number of alternatives
				Number of pollution incidents / ship
				Number of work accidents / ship
	Products			Percentage of works claimed under warranty
				Percentage of defective materials
				Percentage of materials that arrived late
	Procedures			Percentage of repeat technical inspections for defects / ship
	Eficiency			Efficiency of customer visits = number of orders obtained / number of visits made
				Contracting efficiency in the number of offers sent to customers
				Energy, water, gas consumption / ship
				Effectiveness of maintenance work performed on workspaces and equipment
				Percentage of defects investigated

IOP Conf. Series: Materials Science and Engineering 1182 (2021) 012042 doi:10.1088/1757-899X/1182/1/012042

Management of maintenance / repair operations on board ships within a shipyard represents theprocess of managing the resources necessary to carry out the operations included in the technical specification. This process (presented in table 2 [17]) comprises two fundamental approaches [3-6]: strategic management: refers to the design of the complex system of maintenance / repair operations of a ship in a shipyard (establishing performance requirements; establishing the volume of works; establishing the technologies for the execution of the works; establishing the necessary number of machinery and equipment; establishing locations for machinery and equipment; establishing the necessary workforce; establishing management and control methods); tactical management: ensures the operation of the system at the designed parameters (work planning; management of works; stock management; quality control).

Table 2. Correlation of information in the management of repair operations.

STRATEGIC MANAGEMENT	INFORMATION TO BE CORRELATED
The ship works volume and the performance requirements establishing	• information provided by the marketing department based on the ship maintenance and repairs works quoatation prepared based on the prices and estimated time evaluation required for perfoming works mentioned by the technical specification prepared by the ship owner/technical manager; • factors that limit the capacity of performing work at a ship according to information provided by the Planning / Scheduling Department analyze referring to the maintenance work volume by the works types, the shipyard own work force assessment and evaluation and possibly suplimentary workforce requirements necessary to be subcontracted to the shipyard specialized collaborators
the technology for works execution establishing	• information provided by the Tehnical Department referring to the maintenance work volume by the categories of professions, estimates for the manpower by categories of professions, estimate materials needed for maintenance work on ship; checking the materials existence of the shipyard stocks for the considered leading works
determining the plant and equipment locations and the necessary manpower	• the ship maintenance works execution time analyze according to information provided by the Planning / Scheduling Department estimates for the required time estimates for the drydocked and berthed ship maintenace works performing, the number of locations for execution of works on ship, machinery and equipment location on the ship set locations; the number of work teams and each team sizing; performance criteria for execution of the ship works referring to the the allocation of maintenance works to the workshops and the work teams and timetable for the maintenance works completion based on technological sequences need to be conducted to fit the assumed by the shipyard quoted deadline for completion as approved by the ship Owner / Technical Manager
the works planning;	• All information provided by Planning / Scheduling Department analysis with the Production Department for: the ship maintenace works allocation by the categories of professions to the workshops and the work teams based on technological sequences need to be developed; allocation of tasks and equipment

Table 2. continuation

STRATEGIC MANAGEMENT	INFORMATION TO BE CORRELATED
management and control methods	• planning procedures based on the ship maintenance works graph detailed execution as the distribution to the workshops and the working teams within the Production Department where it records any possible constraints and limitations that may arise in implementation of the supply scheme of material shortages in the shipyard stocks and to hire by subcontracting of additional manpower necessary to comply with the deadline for completion assumed the shipyard being analyzed and possible alternatives; • control procedures provided as per the shipyard quality standards
the works planning;	• All information provided by Planning / Scheduling Department analysis with the Production Department for: the ship maintenace works allocation by the categories of professions to the workshops and the work teams based on technological sequences need to be developed; allocation of tasks and equipment
management of the works	• Department Planning / Programming together with Production Department, Technical Department, Purchasing Department, Subcontractors Department in predetermined intervals perform analyzes, regarding the development of the ship maintenance works detailed execution chart in connection with the timing of the works execution, positioning within manpower and material consumption estimated budget, occurred constraints and limitations in carrying out ship maintenance works recorded and propose corrective action, being analyzed possible alternatives
stock management	• during ships maintenance works perfoming Production Department inform Technical Department on any additional materials needed toward the originally allocated by bills of materials, it frees bills of additional materials and, where appropriate, inform the Supply Department for running actions to supply material deprivation stocks shipyard, with time supply that not to endanger the delivery of the ship
quality control	• Production Department, based on the ship's maintenance works detailed execution chart drafts and submits to the Quality Control Department the QSO inspection chart by technological phases inspection and completion of the ship for a final delivery

There are many unforeseen situations that affect the industrial environment – affecting lead times and / or resources (material, financial, human) – and thus complicates the project planning process.

2. The role of the technical specification of works in improving the performance of shipyards

The role of the technical specification is to include in its content the maintenance, upkeep and improvement works necessary to be performed on the ship in a repair shipyard, so that following the performance of these works to obtain the certification of the classification society, being thus confirmed that the ship is in a proper technical condition and the maintenance management is properly applied. Therefore, the period of carrying out the maintenance works of the ship at the dock and quay in a shipyard is a key indicator of performance in achieving the objectives of the two management systems involved in the process, as presented in table 3 [17].

Table 3. The objectives of the management systems regarding the period of carrying out the maintenance works of the ship at the dock and at the quay.

Objective	Ship Owner / Technical Manager	Shipyard for the execution of ship maintenance works
Strategic	Confirmation of the application of the maintenance management system on board the ship	Constant loading of equipped docks with ships Constant loading of the existing workforce
Market share	Number of ships that have carried out maintenance work on the same site	Number of traditional customers performing ship maintenance work only on site Number of new customers who returned to the shipyard after the first collaboration
Availability of resources	The period of time during which the ship will be taken out of operation for performance	Dock utilization rate
	Supply of specific spare parts and materials to the ship	Duration of the doc cycle
Costs	Number and volume of additional work ordered above the initial technical specification	Consumption of labor by types of works
	Number and volume of work performed on board during the ship's stay at the shipyard	Consumption of materials by types of works
	Number of papers not accepted for submission to the Classification Society	Number of works not accepted for delivery to the client
	Number of days delay in delivery of the ship when performing dock and quay work	Penalties for delays in the delivery of the ship
Operational	Productivity / day Number of pollution incidents / ship Number of work accidents / ship	
Eficiency	Consumption of energy, water, fuel, oil, per ship	Consumption of energy, water, gas for ship works

When drawing up the technical specification, the conditions will be mentioned, according to the standards of the integrated management systems, and as far as possible, details regarding certain special access measures necessary to be taken for carrying out works and the type of cleaning work required to be performed on the fuel, oil and ballast tanks, and others shall be indicated from the point of view of environmental management, as well as the estimation of the quantities of waste necessary to be discharged from these locations in the shipyard. Carrying out maintenance and repair works on board the ship in the shipyard according to the technical specification thus implies the application of the standards of integrated management systems with reference to quality, health, occupational safety and environment, according to the presentation in figure 1 [17].

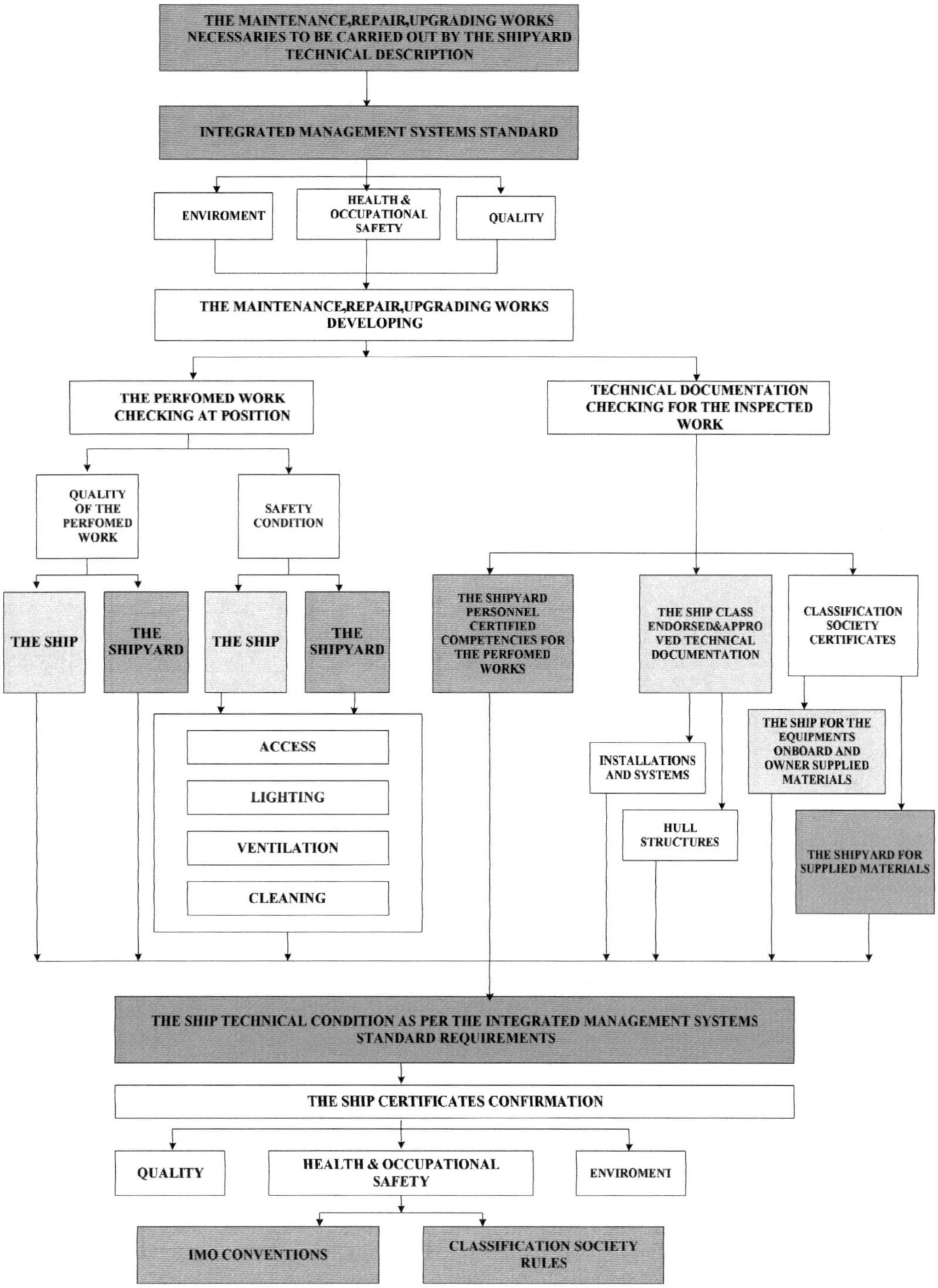

Figure 1. The role of the technical specification in the management of maintenance works in the shipyard.

From the point of view of quality management, for each work that requires the final confirmation of the classification society, it is verified:

- the documentation necessary to carry out the work with reference to:
    - class certificates for materials used by the shipyard and for those made available to it by the ship;
    - class certificates of the ship's equipment which are the subject of maintenance work on board the ship at the shipyard, this being the responsibility of the ship;
    - the plans and documentation approved and approved by the classification society for the body structures, systems, mechanisms and equipment involved in carrying out maintenance works on board the ship in the shipyard, this being the responsibility of the ship;
    - the certificates of competence of the shipyard personnel performing maintenance works that are subject to these requirements by the classification society;
- the quality of the work performed at the position.

From the point of view of occupational health and safety management, it is verified that the conditions for preparing the work in terms of access, lighting, ventilation are ensured. From the point of view of the management of the environmental conditions, the ensuring of the conditions of cleaning and observance of the requirements for storage and recycling of material waste is verified and of the waste resulting from the maintenance works on board the ship in the shipyard, being registered the waste taken from the ship, the emptying of the sewage and residue tanks.

The importance of the technical specification is given by the fact that it is the basis for drawing up estimates regarding the cost and duration of the execution of maintenance, upkeep and improvement works on board the ship in a shipyard. The shipyard draws up the price and time offer following the evaluation of the technical specification drawn up by the owner / technical manager of the ship from the point of view of the following considerations:

- Availability:
    - docks, quays, workspaces, equipment necessary for body work and structures, machines, installations and systems;
    - specialized workforce on the types of works included in the technical specification;
    - materials necessary for the execution of the works included in the technical specification.
- Based on the analysis of availabilities, they can be evaluated:
    - labor consumption by types of works;
    - required subcontracting of labor for certain types of work;
    - necessary supply of materials for the execution of certain works;
    - internal costs for the execution of the types of works included in the technical specification based on the consumption of labor and materials;
    - execution time of the types of works included in the technical specification.

The shipowner / technical manager of the ship following the analysis of the price and time offer prepared by the shipyard can evaluate the following:

- the necessary and time for the supply of spare parts and specific materials to the ship;
- the period of time in which the ship will be taken out of operation between the performance of the works from the technical specification which also includes the time required for the ship to arrive from the last port of unloading to the shipyard and the time required for the ship to arrive from the shipyard at the next port of loading;
- estimated expenditure budget comprising:
    - expenses related to the execution of the works mentioned in the technical specification for maintenance, upkeep and improvements on board the ship in the shipyard;
    - expenses related to the need, in own charge, to supply spare parts and specific materials to the ship for the execution of the works mentioned in the technical specification;

- fixed daily expenses on the ship during the execution of the works mentioned in the technical specification for maintenance, upkeep and improvements on board the ship in the shipyard;
- expenses related to the supervision of the execution in the shipyard of the works mentioned in the technical specification for maintenance, upkeep and improvements on board the ship by the Classification Society in order to extend the class certificates.

In conclusion, the technical specification has a very important role in the evaluations made both by the shipyard at the preparation of the price and time offer and by the owner / technical manager of the ship.

3. Conclusions

Over time, the authors of this paper have been concerned with the subject, looking for solutions to improve the processes of the maintenance and repair of ships in the shipyards [12-16].

It becomes obvious that the execution of the execution processes for the maintenance, upkeep and improvement works on board the ship in the shipyard depends on the quality of the technical specification that must ensure the necessary information about the volume and types of work required to be performed on the body and structures, machines, equipment, installations and systems supported by the necessary technical information from the plans approved by the classification society and from the technical documentation regarding to: technical description location / equipment, their number; the types of works necessary to be performed on location / equipment; the estimated quantitative volume of the works necessary to be performed on the location / equipment. The advantage of a technical specification of quality, consists in limiting the possibilities of appearing the orders of additional works during the execution of the works for maintenance, upkeep and improvements on board the ship in the shipyard, therefore, there is a good chance that the period for carrying out the maintenance work of the ship at the dock and quay will be completed according to the initial estimate of the shipyard, the costs of the shipyard and the ship being also possible to be maintained at the level of the initially estimated budgets.

Additional works during the execution of works for maintenance, upkeep and improvements on board the ship at the shipyard, can not be excluded, considering that when checking the condition of the hull and the structure of the ship in the dock, following the measurements of sheet metal thicknesses, the appearance or evolution of the corrosion of the ship's bottom sheets can be identified and any deformations, ruptures or other damage to the hull structure may also be identified. Also, regarding the machines, installations and equipment, following the performance of verifications and measurements, non-conformities can be identified and need to be corrected by performing parts replacement or repair work on certain subassemblies, which are additional to the original technical specification.

In the case of a technical specification with an unsatisfactory volume of information, a significant volume of orders may appear for additional work, with major implications in increasing the period of maintenance of the ship at the dock and quay and the costs of both parties involved in the process. For the shipyard, the additional works have a negative influence, due to the need for additional labor to cover the additional volume of works commissioned for the ongoing project and due to the immobilization over the estimated term of the technical capacities, thus there is a major risk of being negatively affected and the development of other projects undertaken during that period. For the ship's owner / technical manager, the additional work has a negative influence, due to the additional costs required to cover the additional volume of works ordered and due to exceeding the period of maintenance of the ship at the dock and quay initially estimated, which may have a negative influence on the further development of the commercial obligations assumed by the ship.

4. References

[1] Neely, A, Adams, C 2002 *The Performance Prism: The Scorecard for Measuring and Managing Business Success* (Financial Times, Prentice Hall).

[2] Neely A 2007 *Business performance measurement: unifying theories and integrating practice*, 2-nd edition (Cambridge University Press).
[3] Gulati, R, Smith, R 2009 *Maintenance and reliability best practices* (New York, USA: Industrial Press, Inc.).
[4] Anghelache C, Mitruţ C 2009 *Econometry. Theory, Sintesys and Case Studies* (Bucharest: Artifex Pub. House.
[5] Coetzee, J L 2004 *Maintenance* (Victoria, USA: Trafford Publishing).
.[6] Burns, E, Tom, Stalker, G M 1961 The Management of Innovation, *University of Illinois at Urbana-Champaign's Academy for Entrepreneurial Leadership Historical Research Reference in Entrepreneurship.*
[7] Mărăscu-Klein, V, Toma, V 2007 *Maintenance management* (Braşov, Romania: Ed. Transilvania University Pub. House).
[8] Arun, KD, Makaraksha, S 2015 Modeling and analysis of ship repairing time *Journal of Ship production and Design, SNAME Journal* **31**(1), p 1.
[9] Castrillon Dussan, R 2007 Shiprepair competition: drivers and opportunities *World Maritime University, Dissertations.*
[10] Lyngstol, T 2002 Improvement of Ship Drydock Specifications: A Case Study, *Master Thesis, University of New Orleans and Norwegian University of Science & Technology.*
[11] Pinjala, SK, Pintelon, L, Vereecke, A 2006 An empirical investigation on the relationship between business and maintenance strategies *Int. Jour. of Production Economics* **104**(1) p 214 Great Britain.
[12] Manea E, Militaru C, Zăgan R, Chiţu M G 2016 Improving Organizational Performance through the Application of Integrated Management Systems in Maintenance Activities in the Shipyards *Annals of Maritime University of Constanta, Romania, year XVI* **24**, p 57 p 221.
[13] Manea E, Chiţu M G 2013 Ships New Buildings, Repairs & Conversions Outlook 2009-2012. Trends *Annals of Maritime University of Constanta, Romania year XIV* 19, p 57.
[14] Manea E, Manea M G 2018 The Risk Concept and the Impact on the Organizational Performance of Maritime Shiprepairs Shipyards. *Advanced Engineering Forum 2019* **34** 300–308 https://doi.org/10.4028/www.scientific.net/aef.34.300.
[15] Manea E, Manea M G 2019 The Influence of the Deadweight in the Projection of the Duration of the Maritime Ships Mentenancy Works. *Advanced Engineering Forum 2019* **34** 292–299 https://doi.org/10.4028/www.scientific.net/aef.34.292.
[16] Manea E, Zăgan R, Manea M G, Militaru C (coordinator) 2018 Improving Shipyard Maintenance and Performance, *Vol. 1: Quality Management Applied to Improving Shipyard Maintenance and Performance*, (Constanţa: Dobrogea Pub. House).
[17] Manea E, Militaru C (scientific coordinator) 2016 *The organizational performances improvement by the time period estimation for the ships maintenance works carried out in the shiprepairs shipyards*, PhD Thesis, "Politehnica" University of Bucharest.

Comments regarding the use of Monte Carlo method in the analysis of maritime ship maintenance works

M G Manea[1], R Zăgan[2] and E Manea[3]
[1] "Mircea cel Bătrân" Naval Academy, Marine Engineering Faculty, Department of Tactics and Naval Weapons, Fulgerului Str., No.1, Constanta, RO
[2]Maritime University of Constanta, Electromechanics Faculty, Department of Engineering Science, Mircea cel Bătrân Str., No.104, Constanta, RO
[3]Constanta Shipyard SA, Greece Branch Office, Port Str., No.1, Constanta, RO

E-mail: greti.manea@anmb.ro

Abstract. Elaboration of a mathematical structure, together with a list of correspondences between the mathematical symbols and the objects of the concrete situation considered, led to what is called the mathematical model. The Monte Carlo simulation method is increasingly applied to the analysis of stochastic problems or in conditions of risk, when the same direction of action can have several consequences, the probabilities of which can be estimated. The Monte Carlo simulation can be used to analyze the flow of processes involved in the maintenance works of a ship in a shipyard, because it represents a system characterized by both input and output variables that can interact and therefore the same course of action may have several consequences. Thus, the execution of a certain type of maintenance work on the ship may lead to an increase or decrease in the volume of work initially estimated depending on: class society inspections, limiting the budget for works due to the crisis on the maritime transport market, labor force resources of the shipyard (limited in certain periods with peak load). Consequently, a stochastic model is outlined in which randomness is present, and the states of the variables are not described by unique values, but rather by probability distributions. The paper highlights, through concrete examples, some applications of the Monte Carlo method for the analysis of maintenance works performed for seagoing vessels.

1. Introduction

Models are much simpler representations of reality from which they retain the essential elements. Modeling is a method of studying processes and phenomena that are achieved by substituting the real object of research with a model.

Mathematical models arose from the need to describe and formally study the behavior of a category of real systems, in order to control and direct their future activity. The elaboration of a mathematical structure, together with a list of correspondences between the mathematical symbols and the objects of the concrete situation considered, led to what is called a mathematical model.

Random numbers are used in applications where it replaces the values of the random variable with a set of values that have its statistical properties, numerical simulation realistically reproducing certain elements of the simulated system. Within very long random number strings, some numbers may be repeated, but they meet certain requirements that bring them closer to the random ones, in which case they are called pseudo-random numbers.

IOP Conf. Series: Materials Science and Engineering 1182 (2021) 012043 doi:10.1088/1757-899X/1182/1/012043

There are several processes for producing random numbers, namely: tables with random numbers, physical processes, arithmetic processes [1].

1.1. Aspects regarding mathematical modeling

In general, a model M of an S system is another S'system which, in some respects, is equivalent to S, but which is easier to study. By an S system is meant the following set structure [2]:

$$S = \{T, X, U, V, Y, \varphi, \eta\} \tag{1}$$

In equation (1) the meaning of the terms is the following:

T is the base time used for timing and ordering events and it is a real number if the system is continuous time or integer if the system is discrete time;

X represents the set of entries in the system;

U is the set of system input segments (associated with the function $u: T \rightarrow X$, graph of the function u over an interval $[t_0, t]$, which means $u(t_0, t) = \{t, u(t) \,|t_0 \leq t \leq t_1\}$);

V is the set of system states (the state is a concept of modeling the internal structure of the system that contains its history and which affects its present and future and, together with the shape of the inputs, uniquely determines the outputs of the system);

Y is the set of system outputs;

φ is the response function of the system $\varphi: X \times V \rightarrow Y$; if at an input $u([t_0, t])$ the system is in the state $\sigma_{t_0} \in V$, then the system output is $Y = \varphi(u([t_0, t_1]), \sigma_{t_0})$;

η it is the transition function of states, so if the input $u([t_0, t_1])$ finds the system in the state $\sigma_{t_0} \in V$, then it transforms him into the state $\sigma_{t_1} = \eta(u([t_0, t_1]), \sigma_{t_0})$.

Knowledge of inputs $u \in U$ and their corresponding answers $y \in Y$, represents the behavior of the system.

A model of a system must meet the following three conditions [3,4]: the model must reflect as accurately as possible the represented reality; the model must be a simplification of the represented reality; the model is, by its essence, an idealization of the represented reality.

In the process of mathematical modeling, the components of the system are associated with certain variables / parameters, some known (controllable), called input, others unknown (uncontrollable), called output. The connections and interactions between the components of the system or the connections of the system with the outside are transposed in the mathematical model through functional relations (equations and / or identities).

The purpose of the model is to express uncontrollable variables according to controllable variables, so that performance criteria are met.

Sometimes it is not possible to express in the form of equations all the necessary connections, conditions and interdependencies, which is why some of them are described by logical conditions or procedures that can only be handled by computer.

The mathematical model completed with such procedures is a simulation model that, starting from values of controllable variables (generated with special algorithms), will produce values of uncontrollable variables, offering variants from which to choose the best.

Therefore, the simulation model produces experiments on the system it simulates, which allows the choice of those values of the variables and input parameters that lead to the desired performance.

Three types of simulation models are used in many scientific fields [2]:

- imitative models - transpose reality on another scale, larger or smaller, in order to observe the behavior of that reality;
- analogical models - are specific to a process or phenomenon whose behavior is not known and, in order to be studied, a realistic model of a phenomenon or process that presents analogies is used;

- symbolic models - use various symbols to represent the characteristics of a reality, the correlations leading to the writing of appropriate mathematical relationships and thus to the creation of an abstract (mathematical) model.

Simulation models can also be classified as such [5]:

- static models (do not explicitly take into account the time variable; reflects invariant and timeless situations and states; solutions can also be obtained analytically) or dynamic models (take into account the variation and interaction over time of the variables considered; incorporates time as a fundamental quantity, being a state variable; it is solved using the simulation technique);
- deterministic models (all variables are non-random; operative characteristics are equations of a certain form; the solutions of these models are obtained analytically) or stochastic models (contain one or more random input variables and therefore one of the operative characteristics is given by a density function; random inputs lead to random outputs; events do not occur with certainty, but with a certain probability);
- discrete models (changes in the states of variables are made at discrete moments of time) or continuous models (changes in the states of variables occur continuously).

In the process of mathematical modeling [4], the model is representative of the physical system if the causal condition is met, which leads to the classification of the elements into: input elements (cause) that form the input vector $x = (x_1, x_2, \dots, x_m)$ and output elements (effect) that form the output vector $y = (y_1, y_2, \dots, y_n)$, both vectors being generally random.

The simulation step is, by definition, a stage in which all input variables take constant values during program execution. Of great importance in building the simulation model is the process of "moving" the system over time – for this it is necessary to introduce a special variable called "simulation clock", which measures the flow of real time in which the system is simulated, in order to maintain the correct order in time of events.

The simulation models also contain: functional relations (identities and / or equations) and operative characteristics (used to express through mathematical relations the interactions of the variables and the behavior of the system).

After constructing the simulation model, the simulation itself, as an experiment, consists in varying the values of the variables and input parameters of the system and deduce based on the model, as a result of the calculations, their effects on the output variables.

There are two types of simulation: discrete, if the model variables can have only certain discrete values and continuous, if the model variables can have any value on certain real intervals.

The construction of simulation models is a broad process that, in general, involves the following steps [3]:

- defining the problem, the stage in which the objectives of the simulation are established;
- primary data collection, analysis, interpretation and processing;
- formulation of the simulation model; estimating the input parameters of the model;
- model performance evaluation and parameter testing;
- description of the simulation algorithm and writing of the calculation program;
- model validation;
- planning simulation experiments;
- analysis of simulated data.

1.2. Aspects regarding the generation of random number

It is called a generator, an arithmetic method of obtaining random numbers based on a recurrence relation of the form $X_{n+1} = f(X_n, X_{n-1}, \dots, X_{n-m})$, where $n > m, m \geq 0, (X_n)_{n \in N}$ are natural numbers, assuming that the vector of the initial values $X_0, X_1, \dots, X_m$ it is fixed in advance.

In order for an arithmetic process to be called a generator, it must meet the following conditions: be simple and fast; have as long a period as possible (produce strings of numbers of any length, without repeating the numbers of the string); to produce stochastically independent numbers from each other; to produce numbers whose distribution is uniform.

Building a random number generator [1] is quite difficult, on the one hand due to the high periodicity and on the other hand due to the statistical requirements of stochastic independence and uniformity. The idea of using arithmetic procedures for the algorithmic generation of numbers that have qualities close to random ones belongs to John von Neumann [3] who proposed a particular method known as the "middle part method". Knuth [4], concluded that random number generators should not be constructed by intuitive methods, but should be based on rigorous mathematical theories, generally based on recurrent methods., and Lehmer [5] initiated congruent methods that produced good results.

Let $\{\Omega, K, P\}$ be a field of probability and $X: \Omega \longrightarrow \mathrm{R}$ a random variable.

Either way, $F: \mathrm{R} \longrightarrow [0,1], \mathrm{F(x)} = \mathrm{P(X} < x) = \{\mathrm{P}\{\omega; \mathrm{X}(\omega) < x\}\}$.

Generating a random variable with the computer means choosing a number of elementary events $\omega_1, \omega_2, \dots, \omega_n$ and determine values $X_i = X(\omega_i)$ of the random variable.

The values $X_1, X_2, \dots, X_u$ represents a selection on the random variable X.

A generation is an algorithm that is able to produce an X_i and, iterating that algorithm, it should be able to produce $X_0, X_1, \dots, X_n$ so that they are stochastically independent and identically distributed.

The variables $X_0, X_1, \dots, X_n$are stochastically independent if

$$F(x_1, x_2, \dots, x_{1n}) = \prod_{i=1}^{n} F_i(x_i), \tag{2}$$

In equation (2) $F(x_1, x_2, \dots, x_{1n}) = P(X_1 < x_1, X_1 < x_1)$.

The variables $X_0, X_1, \dots, X_n$ are identically distributed only if they have the same distribution $F_i(x) = F_j(x), i \neq j$.

Usually, random numbers are the values of a selection related to a random variable U that has a uniform distribution. If U is a uniform discrete random variable, then all its values are equally likely. If U is a uniform continuous random variable, then it admits a shape distribution density

$$f(x) = \begin{cases} k, x \in (a,b) \\ 0, x \notin (a,b) \end{cases}, \tag{3}$$

In equation (3) k is determined by the condition (4)

$$1 = \int_{-\infty}^{\infty} f(x)dx = \int_{a}^{b} kdx = k(b-a) \Rightarrow k = \frac{1}{b-a} \tag{4}$$

and from a uniform distribution function (5)

$$F(x) = \begin{cases} 0, x \leq a \\ \frac{x-a}{b-a}, a < x < b \\ 1, x \geq b \end{cases} \tag{5}$$

In this case it is said that the random variable U has a uniform distribution on (a,b). Uniform random numbers on $(0,1)$ are most important for numerical simulation.

Considering C a family of random variables for which generation algorithms are known, the problem arises of determining an algorithm T that transforms a set of random variables into a set of random variables $X = T(C_1, C_2, \dots, C_n)$.

In other words, starting from a family of random variables for which efficient computational generation algorithms are known, it is required to determine the algorithm that generates random variables that are more difficult to generate.

For the generation of random variables, in the literature are used multiple methods: the inverse transform method; rejection method; methods specific to the given distribution; approximate methods; method of composition (mixing); composition-rejection method; particular methods for generating random variables; other methods of generating random variables.

2. Presentation of the Monte Carlo method

If a deterministic problem is associated with a random (probabilistic) model and by generating random variables functionally related to the solution, model experiments are performed and information about the solution of the deterministic problem is provided, the Monte Carlo simulation is used.

The expression "Monte Carlo method" [6] it is synonymous with the method of statistical experiments. The method is based on some conclusions resulting from the limit theorems of probability theory. To calculate a random variable we start from a random variable that has a uniform distribution over the range [0,1] from the constructive probability field, field with a shape application $\beta = f(x), x \in [0,1]$, with the property that $\int_0^{-\infty} f(x)dx = 1$.

The method involves estimating the parameters of the distribution of a random variable based on its achievements. Thus, the main problem is solved by the Monte Carlo method [6] consists in estimating the average value $M(\xi)$ of a random variable ξ depending on a permissible error and a given probability. If θ is the parameter associated with ξ what is to be represented by the value $\widehat{\omega}$ estimable on the basis of a sample $\omega = (\omega_1, \omega_2, \dots, \omega_n)$, then the estimator $f(\omega)$ will have to satisfy the condition $M(f(\omega) - \theta) \leq \varepsilon$, where ε is an acceptable deviation of the estimator from the theoretical random variable θ.

In the Monte Carlo method, the simple or weighted arithmetic mean is often used as an estimator.

For a random variable ξ with normal distribution, with average m and dispersion σ^2, likelihood function in relation to m, is

$$F(x_1, x_2, \dots, x_n; m; \sigma^2) = \frac{1}{(2\pi)^{n/2}\sigma^n} \prod_{i=1}^{n} exp\left[-\frac{1}{2}\left(\frac{x_i - m}{\sigma}\right)^2\right]; x_i, i = \overline{1,n} \tag{6}$$

In equation (6) $F(x_1, x_2, \dots, x_n; m; \sigma^2)$ is maximum in relation to m and σ^2 for achievements $x_1, x_2, \dots, x_n$ of ξ if condition (7)

$$\begin{cases} \frac{\partial lnF}{\partial m} = 0 \\ \frac{\partial lnF}{\partial \sigma^2} = 0 \end{cases} \tag{7}$$

$$\widehat{m} = \frac{1}{n}\sum_{i=1}^{n} x_i$$

whence it results (8)

$$\hat{\sigma}^2 = \frac{1}{n}\sum_{i=1}^{n}(x_i + \widehat{m}) \tag{8}$$

For the random variable with normal distribution, the mean m is the maximum likelihood estimator. Sample quality can be assessed by concordance tests that measure the proximity of the empirical distribution to the theoretical distribution..

In order for the results obtained with the Monte Carlo method to be conclusive, the following aspects must be taken into account [7]:

- if after n simulations the frequency of occurrence of a certain event is f^*, then, in reality, it will be within limits $f = f^* \pm \sqrt[2]{\frac{f^*(1-f^*)}{n}}$;

- so that the maximum permissible error in determining the probability of occurrence of the considered event does not exceed a given value Δ, it is necessary that the expression be made a number of times given by the relation $n = \frac{4p(1+p)}{\Delta^2}$, p fbeing the searched value of the probability of occurrence of an event;
- if after n experiments a statistical mean value m was determined for a random variable, then the average value will be within the limits$\overline{m} = m \pm \Delta$;
- if v_i is a particular value of a variable V and it is desired to obtain the average value m with an error of at most Δ, then the experiment must be repeated with number of times $n = \frac{4D_v}{\Delta^2}$, where $D_v = \frac{1}{n_1}\sum_{i=1}^{n_1} V_1^2 - (m)^2$ represents the dispersion of the size v, determined from the results of the first series of experiments n_1;
- the accuracy of the Monte Carlo method can be estimated statistically with a finite degree of certainty (is considered sufficient 0.990 – 0.997);
- an increase in the accuracy of the Monte Carlo method by one order of magnitude, increases the calculation time by two orders of magnitude.

The Monte Carlo method is approximate, it can be successfully adapted to economic models, the accuracy can be estimated correctly only as the calculations are performed, being determined by the number of independent tests (inversely proportional to the probability) and their variation.

3. Applications of the Monte Carlo method in the analysis of ship maintenance processes

Based on the results obtained by the Monte Carlo method, in the simulation processes are obtained various evaluations, hierarchies, which allow the substantiation of decisions in the management and control of economic processes. Various authors have been interested in the subject and the literature [1,6-16] highlights the following areas of applicability: complex storage processes (storage area is limited, there are penalties for lack of stock); waiting processes - events are interconditioned; repair processes; analysis of the critical path for complex projects - estimating the parameters of the distribution of the total duration, determining the frequency of the critical character of the activities; work processes - scheduling activities, loading equipment, launching activities; macro-economic processes - the study of flows between branches, problems of economic growth.

In examples 1-4 are presented some examples, contribution made by the authors of the presented paper.

3.1. Example for determined static model

Request: a shipyard that performs repair / maintenance works of seagoing ships can carry out maintenance works in one month on three types of electric motors in the installations and systems of ships contracted for maintenance works: 4 pcs / hour from the first type of electric motors, 2 pcs / hour from the second type and 3 pcs / hour from the third type. The benefit obtained from the execution of these maintenance works on electric motors is presented as follows: 12 USD for the first type of electric motors, 25 USD for the second type and 18 USD for the third type. Shipyard capacity is limited to 250 electric motors of the first type, 120 of the second type and 80 of the third type. Knowing that they work 160 hours a month, it is required to draw up the monthly work schedule, which should contain the number of electric motors of each type to be repaired, so that the benefit obtained is maximum.

Solving: the construction of the model can be done this way: is denoted by x_1, x_2, x_3 the number of devices of each type and is obtained the system (9) for a static and deterministic model.

$$\begin{cases} 4x_1 + 2x_2 + 3x_3 \leq 160 \\ x_1 \leq 250 \\ x_2 \leq 120 \\ x_3 \leq 80 \\ x_1, x_2, x_3 \geq 0 \\ max(12x_1 + 25x_2 + 18x_3) \end{cases} \tag{9}$$

3.2. Example for stochastic model

Request: a shipyard for maintenance work on ships must order daily a number of oxygen cylinders necessary to carry out sheet metal replacement work on ship structures so that the gain from their use is maximum.

Every day, the shipyard orders a certain number of oxygen cylinders, using some or all of them. Each one generates a certain gain. Unused oxygen cylinders can be returned, but they generate a certain loss.

The number of oxygen cylinders used varies from day to day and the probability of using a certain number of oxygen cylinders in a day can be estimated based on previous use.

Solving: the construction of the model can be done as follows: n = the number of oxygen cylinders ordered each day; c = the gain from the use of an oxygen cylinder; l = loss due to an unused oxygen cylinder; r = requested number of oxygen cylinders used in a day; $p(r)$ = the probability that r oxygen cylinders will be used on any given day; P = the gain obtained in one day.

If in a day the number of oxygen cylinders required (demand) is greater than or equal to the number of oxygen cylinders ordered ($r \geq n$), the gain will be $P(r \geq n) = nc$. If in a day the number of oxygen cylinders used (demand) is less than the number of oxygen cylinders ordered ($r < n$), the gain will be $P(r < n) = rc - (n - r)l$. The average daily gain will be as in equation (10)

$$\bar{P} = \sum_{r=0}^{n} p(r)[rc - (n - r)l] + \sum_{r=n+1}^{\infty} p(r)nc \tag{10}$$

This is a stochastic model because: n = controllable variable; r = uncontrollable variable; c, l = constants. By solving this model we must find n for which $\bar{P}$ is maximum.

3.3. Example 1 for the Monte Carlo method

Request: in the electrical works workshop of a shipyard, the main electrical distribution panels are checked for the equipment of ships contracting docking works. The number of electrical contacts checked and cleaned on board of the ships under repair works, for 100 days, is shown in the table 1.

Table 1. Data recording.

Number of electrical contacts checked (pcs / day)	Number of cases
120	2
200	5
220	11
280	15
320	20
410	15
450	12
490	13
520	5
590	2

The management of the shipyard wants to know the performances obtained in time at the electrical ship repair workshop to establish a competitive price compared to other shipyards in the area.
Solving: performance of the electrical works workshop as regards the verification of the connections to the contact terminals of a main electrical distribution panel on board a ship can be determined with the Monte Carlo simulation method, in which the performance evaluation operation is done starting from the observations made during the 100 days (as in table 2).

Table 2. Calculus for the probability and the cumulative probability.

No.	Number of electrical contacts checked (pcs / day)	Probability of achievement	Cumulative probability
1	120	0.02	0.02
2	200	0.05	0.07
3	220	0.11	0.18
4	280	0.15	0.33
5	320	0.20	0.53
6	410	0.15	0.68
7	450	0.12	0.80
8	490	0.13	0.93
9	520	0.05	0.98
10	590	0.02	1.00

The cumulative probability is plotted in figure 1.

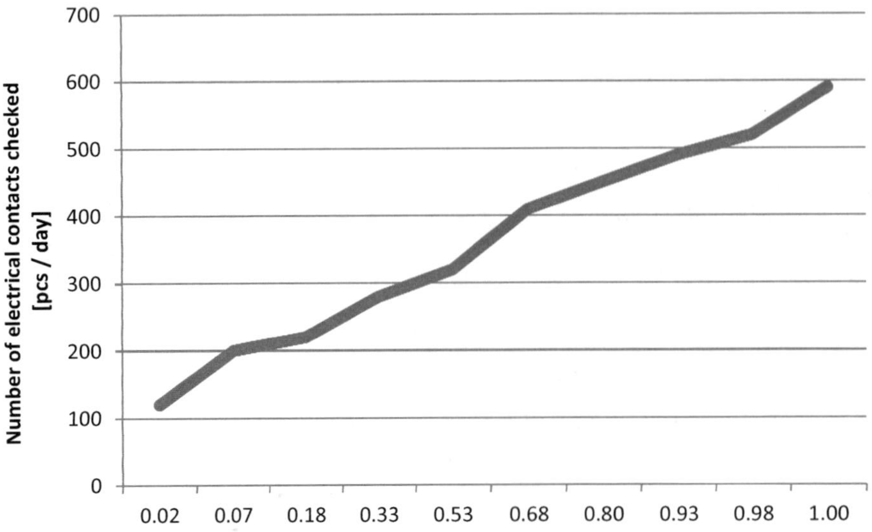

Figure 1. Cumulative probability.

It extract 20 numbers from a table 3 with random numbers in the range [0,1] and calculate the average $\bar{X}$, arithmetic mean deviation σ^2, coefficient of variation C_V and the confidence interval.

Table 3. Data recording refer to the number of electrical contacts checked (pcs / day).

No.	No. randomly	Number of electrical contacts checked (pcs / day)	$(x_i - \bar{X})$	$(x_i - \bar{X})^2$
1	0.0317	200	-174.5	30450.25
2	0.9369	520	145.5	21170.25
3	0.3406	320	-54.5	2970.25
4	0.0200	120	-254.5	64770.25
5	0.9650	520	145.5	21170.25
6	0.6568	410	35.5	1160.25
7	0.7571	450	75.5	5700.25
8	0.6174	410	35.5	1260.25
9	0.1511	220	-154.5	23870.25
10	0.0306	200	-174.5	30450.25
11	0.2333	280	-94.5	8930.25
12	0.6603	410	35.5	1260.25
13	0.5447	410	35.5	1260.25
14	0.9134	490	115.5	13340.25
15	0.7861	450	75.5	5700.25
16	0.6969	450	75.5	5700.25
17	0.9277	490	115.5	13340.25
18	0.2388	280	-94.5	8930.25
19	0.5773	410	35.5	1260.25
20	0.6971	450	75.5	5700.25
	Total	7490		268495.00

$$\bar{X} = \frac{7490}{20} = 374.5;$$

$$\sigma^2 = \frac{268495}{20} = 13424.75;$$

$$\sigma = 115.86;$$

$$C_V = \frac{115.86}{374.5} = 0.309$$

The t distribution (Romanovski test) is used to verify the hypothesis regarding the average number of electrical contacts verified at a main electrical distribution panel. The interval indicating the number of electrical contacts checked daily is $\left(\bar{x} \mp t_{\frac{\sigma}{2},N-1} \frac{\sigma}{\sqrt{N}}\right)$.

Performing the calculations, the number of electrical contacts verified as between 320 contacts / day and 428 contacts / day is obtained.

The number of experiments considered (20) makes this interval too large.

Repeating the calculations (EXCEL software was used) for other numbers of experiments, stabilizes around 360 contacts / day, which represents the performance over time of the electrical workshop (as in table 4).

Table 4. Medium value / number of experiments.

No.experiments	400	2500	4500	6000	7500	9000
Medium value	359.425	359.332	361.693	359.297	360.817	360.960

IOP Conf. Series: Materials Science and Engineering 1182 (2021) 012043 doi:10.1088/1757-899X/1182/1/012043

3.4. Example 2 for the Monte Carlo method

Request: the valve repair shop within the ship repair mechanical section uses 500x500x0.5 mm klinger sheet metal to replace the gaskets on the valves dismantled from the ships. For a period of 100 days, the consumption for this assortment was followed, respectively pcs / day. Following the observations, the data on registered consumptions were centralized, according to the data presented in the table 5.

Table 5. Registered data.

Number of klinger sheet metal 500x500x0.5 mm (pcs / day)	No. cases
5	2
10	5
15	3
21	7
25	12
30	10
32	11
35	15
37	15
40	20

The supply department wants to know the optimal daily volume of klinger sheet metal consumption, to cover the demand and avoid the situation of remaining with the product in stock.

Solving:

It is possible to determine the volume of consumption required using the Monte Carlo simulation method, in which the estimation operation is carried out based on observations on the registered orders obtained during the 100 days.

In table 6 was calculate the probability and the cumulative probability.

Table 6. Calculus of the probability and the cumulative probability.

No.	Number of klinger sheet metal 500x500x0.5 mm (pcs / day)	Probability of achievement	Cumulative probability
1	5	0.02	0.02
2	10	0.05	0.07
3	15	0.03	0.10
4	21	0.07	0.17
5	25	0.12	0.29
6	30	0.10	0.39
7	32	0.11	0.50
8	35	0.15	0.65
9	37	0.15	0.80
10	40	0.2	1.00

The cumulative probability is plotted in figure 2.

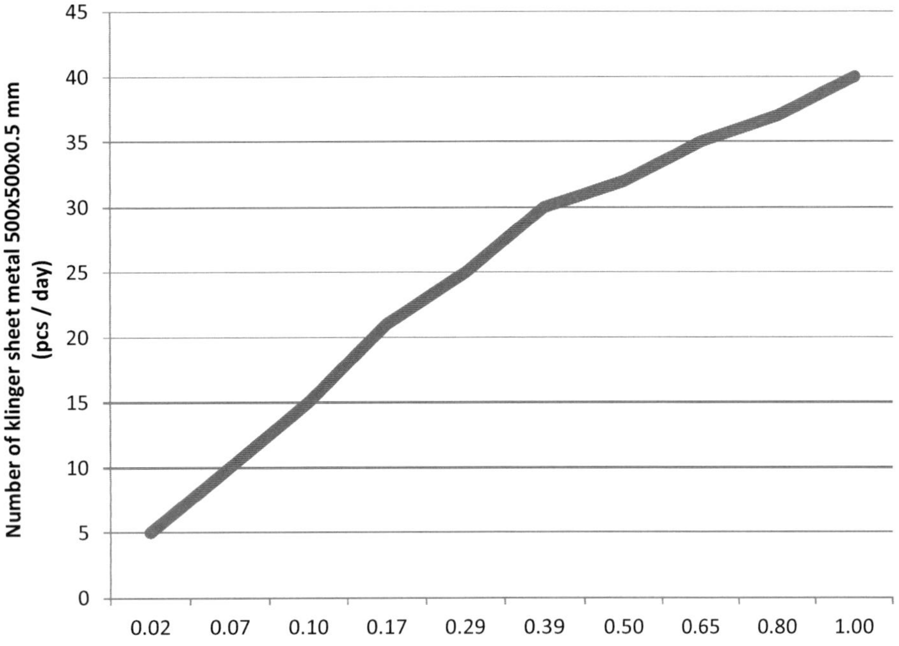

Figure 2. Cumulative probability.

It extract 20 numbers from a table 7 with random numbers in the range [0,1] and calculate the average $\bar{X}$, arithmetic mean deviation σ^2, coefficient of variation C_V and the confidence interval.

Table 7. Data recording refer to the number of klinger sheet metal.

No.	No. randomly	Number of klinger sheet metal 500x500x0.5 mm (pcs / day)	$(x_i - \bar{X})$	$(x_i - \bar{X})^2$
1	0.3217	30	-0.2	0.04
2	0.6274	35	4.8	23.04
3	0.9183	40	9.8	96.04
4	0.2741	25	-5.2	27.04
5	0.4320	32	1.8	3.24
6	0.6272	35	4.8	23.04
7	0.7100	37	6.8	46.24
8	0.2304	25	-5.2	27.04
9	0.8901	40	9.8	96.04
10	0.5327	35	4.8	23.04
11	0.3421	30	-0.2	0.04
12	0.7082	37	6.8	46.24
13	0.1973	25	-5.2	27.04
14	0.6327	35	4.8	23.04
15	0.4268	32	1.8	3.24

16	0.8391	40	9.8	96.04
17	0.1010	21	-9.2	84.64
18	0.0321	15	-15.2	231.04
19	0.3771	30	-0.2	0.04
20	0.0200	5	-25.2	635.04
	Total	604		1511.2

$$\bar{X} = \frac{604}{20} = 30.2$$

$$\sigma^2 = \frac{1511.2}{20} = 79.53;$$

$$\sigma = 8.91;$$

$$C_V = \frac{\sigma}{\bar{X}} = \frac{8.91}{30.2} = 0.295$$

The t distribution (Romanovski test) is used to verify the hypothesis regarding the klinger sheet metal consumption. The interval indicating the number of klinger sheet metal consumption per day is $\left(\bar{x} \mp t_{\frac{\sigma}{2},N-1} \frac{\sigma}{\sqrt{N}}\right)$.

Number of experiments, N = 20; degrees of freedom, g = N-1 = 19; permissible error, α=0.05; confidence interval, (1- α)=0.95.

It is obtained results from equations (11) and (12):

$$\bar{X} - t_{\sigma/2,N-1} \frac{\sigma}{\sqrt{N}} = 30.2 - 2.093 \frac{8.917}{4.47} = 26.3 \quad (11)$$

$$\bar{X} + t_{\sigma/2,N-1} \frac{\sigma}{\sqrt{N}} = 30.2 + 2.093 \frac{8.917}{4.47} = 34.37 \quad (12)$$

The daily consumption of klinger sheet metal is between 26 pcs / day and 34 pcs / day.

The number of experiments considered (20) makes this interval too large.

Repeating the calculations (EXCEL software was used) for other numbers of experiments, stabilizes around e 30 sheets / day, which represents the optimum klinger sheet metal consumption per day (as in table 8).

Table 8. Medium value / number of experiments.

No.experiments	3500	1750	3500	5000	6500	8500
Medium value	29.875	29.642	31.482	29.467	30.737	30.972

4. Conclusions

Monte Carlo simulation can be used to analyze the flow of processes involved in the maintenance of a ship in a shipyard, whereas it is a system characterized by input variables and output variables that can interact and therefore the same direction of action can have several consequences.

Thus, the execution of a certain type of maintenance work on the ship (replacements of body structures, treatment of interior surfaces) may lead to an increase or decrease in the volume of work initially estimated depending on: class inspections, limitation of the budget for works due to the crisis on the maritime transport market, labor resources of the shipyard (limited in certain periods with peak load).

Consequently, a stochastic model is outlined in which the random character is present, and the states of the variables are not described by unique values, but rather by probability distributions.

5. References

[1] Pletea A, Popa L 1999 *Probability theory* (Iaşi: "Gh. Asachi" Technical University Pub. House).
[2] Anghelache C, Mitruţ C 2009 *Econometrics. Theory, Synthesis and Case Studies* (Bucharest: Artifex Pub. House).
[3] Hîncu D, Florescu M 2006 *Modeling and simulation of economic processes*, (Bucharest: „România de Mâine" Pub. House).
[4] Klein L, Welfe A, Welfe W 2003 *Principles of macroeconomic modeling*, (Bucharest: Echonomical Pub. House).
[5] Mohora C, Coteţ C E, Pătraşcu 2001 *Simulation of production systems - Simulation of processes, material and information flows,* (Bucharest: Agir Pub. House).
[6] Petrişor E, *Monte Carlo Simulation* 2006 (Timişoara: Politehnica Pub. House).
[7] Manaţe D, Fărcaş 2010 Model for use Monte Carlo simulations in business valuation, *Valuation Journal* **5**(1) 110-131
[8] Del Moral P 2013 *Mean field simulation for Monte Carlo integration* (Monographs on Statistics & Applied Probability, Chapman & Hall/CRC Press).
[9] Cunningham A, Wang W, Zio, Allanson D, Wall ., et al. 2011 Application of DelayTime Analysis via Monte Carlo Simulation *J. of Mar. Eng. and Tech.,* **10**(3) 57-72.
[10] Coraddu A, Figari M, Savio S 2014 Numerical investigation on ship energy efficiency by Monte Carlo simulation *Proc. Inst. of Mech. Eng., part M: J. of Eng. for the Mar. Env.* **228**(3) 220-234.
[11] McNamara D, Cunningham A, Jenkinson I and Wang J 2014 Application of the Monte Carlo method to marine system maintenance studies considering failure modes and spare part stock control J. *of Mar. Eng. & Tech.* **13**(1) 37-58.
[12] Odland K 2014 *Risk Management in Shipbuilding Projects Using Monte Carlo Simulation for Scheduling* Master Thesis Marine Technology, Supervisor: Hagen A, Co-supervisor: Aarseth T, Norwegian University of Science and Technology, Department of Marine Technology.
[13] Santos T and Soares G 2005 Monte Carlo simulation of damaged ship survivability *Proc. IMechE, Part M: J. Eng. for the Mar. Env.,* **219** https://doi.org/10.1243/147509005X10404.
[14] Villalonga M , Infantes M, Colls M and Mestres M 2020 Ridge Environmental Management System for the Analysis of Oil Spill Risk Using Probabilistic Simulations. Application at Tarragona Monobuoy *J. Mar. Sci. Eng* **8**(4) p. 277.
[15] Zio E, Baraldi P, Patelli E 2004 Application of Monte Carlo Simulation for the Evaluation of the Production Availability of an Offshore Installation *Doc. Code W75.5POLIMI-MOFFSHORE* **52**(1) Task 7.5.
[16] Wang J, Pillay A, Mokashi A 2010 Application of Reliability Centred Maintenance in Ship Operations *in Soares CG "Safety and Reliability of Industrial Products, Systems and Structures", London: Taylor & Francis Group* pp347–368.

Behavior of Arboblend V2 Nature under AWJ machining

D Marguta[1], E Herghelegiu[2], C Tampu[2], S-N Mazurchevici[1] and D Nedelcu[1,3]
[1]"Gheorghe Asachi" Technical University of Iasi, Department of Machine Manufacturing Technology, Blvd. Mangeron, No. 59A, 700050, Iasi, Romania
[2]"Vasile Alecsandri" University of Bacau, Faculty of Engineering, Calea Marasesti 157, 600115, Bacau, Romania
[3]Academy of Romanian Scientists, Str. Ilfov, Nr. 3, Sector 5, Bucharest, Romania

E-mail: dnedelcu@tuiasi.ro

Abstract. The harmful effects on the environment caused by the overexploitation of fossil resources and their industrial processing have attracted the attention of researchers worldwide. Their interest is not only in the use of biodegradable materials based on renewable natural resources but also in the use of manufacturing and processing technologies that minimize as much as possible the negative footprint left on the natural ecosystem. In the present paper, the thermoplastic material processed using abrasive water jet cutting technology (AWJ) is Arboblend V2 Nature. Thus, both the processed material and the technology closely follow the interests of the researchers. The study aims to obtain surfaces with higher quality than other synthetic polymers by appropriate variation of process parameters (abrasive material flow, traverse speed and pressure). Two methods for determining the roughness were used, one in accordance with the standard and one adopted by the authors in order to render more accurately the surface roughness value. Other output parameters observed were: the amount of material removed, the width of the machined surface, the deviation from the perpendicularity and the inclination angle of the machined surface. The output parameters result confirms the possibility of using the parts obtained by AWJ in engineering applications and not only because the shape deviations fall within the limitations imposed at the industrial level.

1. Introduction

AWJ (Abrasive-Waterjet) is an unconventional processing technology that due to its versatility finds its applicability in various fields: cutting, surgery, nuclear plant dismantling, cutting soft materials, paint removal, drilling, etc. The extended applicability is due to the possibility of working with various classes of materials: steels, alloys, polymers, composites, stone-granite, concrete, reinforced plastics, wood, metal and polymer laminates and others.

Considering that currently the global interest is directed towards on harmful effects of manufacturing processes on the environment, all these are in a continuous process of re-evaluation. Regarding the process of machining with abrasive water jet, it takes into account the following aspects related to pollution: water recycling, recovery of material removed by erosion, disposal of wastewater (domestic), recovery and reuse of abrasive material. All these concerns have led to the permanent development and innovation of AWJ, and today there is the possibility of using a high-pressure cryogenic jet machine. By using this alternative, the water is replaced with liquid nitrogen and the abrasive material with dry ice crystals, thus eliminating the factors that destabilized the balance in the environment, [1]. Compared to most machine-tools that use other working methods (LASER, photochemical engraving, EDM, plasma), AWJ has multiple and even unique facilities:

- *common advantages*: automated procedure, position accuracy, ergonomics, ease of handling, qualitative results;
- *individual advantages*: cold cutting, large variety of workable materials, high cutting speed (compared to thermal methods), small or large thicknesses; maintains the chemical and structural integrity of the basic material, ecological (without toxic by-products), multiple processing by using a single tool assisted with accessories, economic efficiency, precision at micro and macro level; reduced force exerted on the work piece;

Cutting through this procedure is quite simple involving the use of process parameters that decide the quality and efficiency of the whole process. Process parameters can be classified into four groups, which aim: cutting - travel speed, angle of inclination and separation distance (water jet width); hydraulic - pressure and flow; mixture - the size of the mixing tube; abrasive material - material, particle size and flow, [2, 3].

The process parameters, as in the case of all technologies, have significant effects on the expected results. One of the most important parameters is the abrasive material. The most used abrasive in AWJ is Garnet due to the performances obtained by using it but also to the low acquisition cost. The size and flow of the grenade are very important, they must be constant and the abrasive particles size must not be larger than 1/3 of the diameter of the mixing tube nozzle. If the nozzle size is reduced, the abrasive needs to be finer, but it is more difficult to flow under gravitational feeding, [4]. The surface roughness of the edges treated with abrasive water jet is proportional to the particle size. In this sense, permanent improvements are made, so that the surface quality is as high as possible.

Over time, polymers, composite and brittle materials have encountered major difficulties in processing through the use of conventional machine tools. The main limitations were related to thermal sensitivity, low tensile strength, brittleness, [5 - 7]. Another aspect pursued by the researchers was to eliminate the contact between the tool and the work piece. This drawback was solved by using nozzles, their wear being independent of the functional characteristics of the material to be cut, [4].
Following the above mentioned aspects, care for the environment and the influence of process parameters on the surface quality obtained by using AWJ, the present study aims to characterize in terms of surface roughness parts made of Arboblend V2 Nature biopolymer.

2. Materials and Methods

Arboblend V2 Nature biopolymer was used for abrasive water jet processing, which has as chemical constituents lignin, PLA (polylactic acid), cellulose, bio-PA (bio-polyamides), bio-PE (bio-polyolefins) and as binders. natural additives have been added such as waxes, resins, fatty acids, oils, but also natural vegetable fibers, [8, 9]. The material was designed and made by Tecnaro in collaboration with the Fraunhofer Institute for Chemical Technology (IKT) and the distributor of this material is the company Tecnaro itself (Ilsferd, Germany). The material was purchased in the form of granules in order to inject into the mold samples that can then be subjected to abrasive water jet cutting. The SZ800 H industrial injection machine from the Fine Mechanics laboratory, "Gheorghe Asachi" Technical University of Iasi was used for obtaining samples. The following process parameters were set to obtain the injection samples: injection temperature, T_{inj} = 170 [°C]; injection pressure, P_{inj} = 110 [MPa]; injection speed, s_{inj} = 90 [mm / s]; cooling time, t_r = 25 [s].

In order to machine the samples from Arboblend V2 Nature by cutting abrasive with water jet, the Hydro Jet Eco 0615 equipment was used. The pressed stable process parameters were: the distance between the processing head and the part - 3 [mm]; focus tube length - 76.2 [mm]; diameter of the water hole - 0.35[mm]; focusing tube diameter - 1.02 [mm]; abrasive material type - Garnet [mesh # 80]; thickness of the cut sample – 10 [mm].

The experimental plan followed in order to make the cuts on the injected samples was a complete factorial plan of 2^3 type (8 experiments), ANOVA methodology - table 1. For each of the 8 experiments of the plan three identical dimensional cuts were made on the same sample. Repeating the cutting was performed in order to follow the repeatability and homogeneity of the results obtained by varying on two levels the three selected process parameters (traverse speed, cutting pressure and abrasive material flow). The length of the cuts was 30 [mm] and the distance between them of 7 [mm].

Table 1. Complete experimental factorial plan for cutting samples.

Level of variation	Q [g/min]	v_f [mm/min]	P [MPa]
I^{st}	150	100	100
IInd	300	150	150

where: Q - abrasive material flow, [g/min]; vf - - traverse speed, [mm/min]; P – pressure, [MPa]

To determine the surface characteristics (R_a) of the areas resulting from the processing, a Zygo optical profilometer (Zegage-series) from the Dimensional Control laboratory, "Vasile Alecsandri" University of Bacau was used. The equipment can also map, based on the analysis in coordinates profile of the resulting surface. Two methods of analysis were used to determine the surface roughness. *The first method* was the one conforming to the SR ISO 468: 1997 standard - which follows the analysis of the inlet and outlet area of the abrasive water jet on a portion of (5x0.8) mm. As the results recorded by this method were not very conclusive, due to too large deviations, it was decided to use a second measurement method. The *second measurement method* involved determining the surface roughness in three areas of the cut (inlet, middle and outlet) on a surface of (0.8 x 0.8) mm, which then outline the average surface roughness. The determinations were made each time in the same points of the cut for all the cuts made.

3. Results and Discussions

The results regarding the surface characteristics of the samples from Arboblend V2 Nature cut with abrasive water jet for the best experiment, number 2, are presented in table 2. Analysing the obtained values for the surface roughness determined according to the SR standard ISO 468: 1997, [10], it is observed that the results fluctuate both in the case of R_a for the inlet area of the abrasive water jet and for its outlet area. The same variation takes place when the two resulting areas / surfaces are compared: the surface from the left of the cut and the surface to the right of it. The biggest difference between the R_a values is visible in the case of the cut number two, when the roughness differs by up to 2μm for the input area (first 5mm) and the output area (last 5mm) of the cut. These differences are primarily due to the presence of impurities on the surface of the cut part, whether it is dust particles or other elements that were not removed before cutting. Another disruptive factor that may occur are particles of abrasive material remaining on the cut surface. Also, not excluded is the appropriate hypothesis that the homogeneity of the injected sample from the Arboblend V2 Nature biopolymer is not uniform. This structural non-uniformity is influenced by the imprecise setting of the material injection parameters in to the mould. These differences are also visible in the case of the best values of R_a, namely for the right side of the cut, figure 1. These minimums and maximums of R_a can be eliminated when on the basis of a structural and morphological analysis the causes of their appearance/ production are detected.

Due to these micron irregularities identified during the determination of R_a by the standard method, it was decided to apply another method of data collection which involved measuring the surface roughness for three areas of the cut, inlet (M_1), middle (M_2) and outlet (M_3) on an area of (0.8 x 0.8) mm. Then, averaging the three R_a values (R_a-M_1, Ra-M_2, Ra-M_3) obtained for each area analysed separately is obtained the R_a (M1: M3). Looking at the data from table 2 but, also the graphs from figures 2 and 3, there is a visible improvement of the average values of the surface roughness, with up to (1.5 - 2) μm for each analysed area. This decrease in R_a comes from taking into account a larger number of analysed areas, namely 6, on the entire surface of the cut.

Table 2. Experimental data regarding the measurements recorded and / or calculated for experiment number 2, Arboblend V2 Nature material.

Exp no.	Side	Start-Stop Cut	R_a (ISO 1997) [µm]	R_a-M_1 [µm]	R_a-M_2 [µm]	R_a-M_3 [µm]	$R_a(M_1:M_3)$ [µm]	L_i [mm]	L_o [mm]	u[mm]	α[°]	MR[g]
1	Left	input	6.93	5.31	4.79	5.85	5.32	1.16	0.96	0.1	0.57	0.46
	Left	output	8.60	7.12	7.98	4.77	6.62					
	Right	input	7.09	6.94	5.99	6.49	6.47					
	Right	output	7.19	7.88	6.98	6.75	7.20					
2	Left	input	8.16	7.13	5.30	5.53	5.99	1.13	0.96	0.08	0.48	0.45
	Left	output	7.21	8.49	8.37	5.88	7.58					
	Right	input	6.75	6.47	4.84	5.77	5.69					
	Right	output	9.74	7.18	9.69	7.86	8.24					
3	Left	input	6.22	5.83	7.52	6.07	6.47	1.13	0.94	0.09	0.54	0.45
	Left	output	7.82	8.57	7.26	9.06	8.30					
	Right	input	7.36	7.14	4.23	5.43	5.60					
	Right	output	7.70	8.18	7.89	6.25	7.44					
Average	Left in		7.10±0.98	6.92±0.97	6.82±1.71	6.10±0.86	6.62±0.92	1.14±0.01	0.95±0.01	0.09±0007	0.53±0.04	0.45±0.005
	Left out		7.87±0.69	8.06±0.81	7.87±0.56	6.57±2.22	7.50±0.84					
	Right in		7.06±0.30	6.85±0.34	5.02±0.89	5.89±0.54	5.92±0.48					
	Right out		8.21±1.34	7.74±0.51	8.18±1.37	6.95±0.82	7.69±0.55					

were: R_a is the roughness of the processed surface (ISO 1997 standard), [µm]; R_a-M_1, Ra-M_2, Ra-M_3 - average values of roughness measured in line, [µm]; $R_a(M_1: M_3)$ is the average roughness measured in line, [µm]; L_i - width of the processed surface at the jet inlet, [mm]; L_o - width of the processed surface at the jet outlet, [mm]; u - deviation from perpendicularity, [mm]; α - tilt angle of the machined surface, [°]; MR - amount of material removed, [g]; g - thickness of the part, [mm].

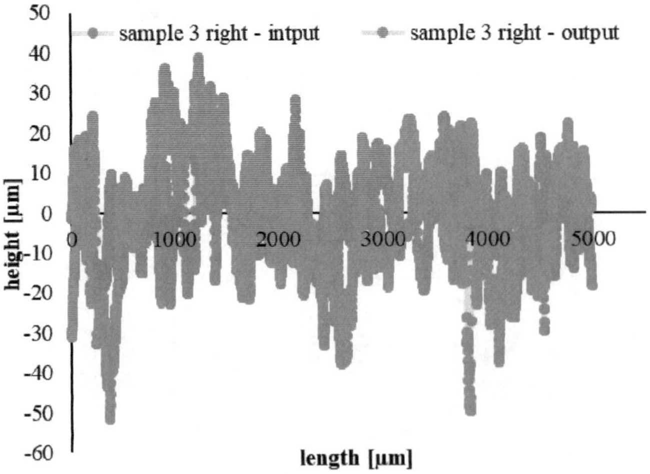

Figure 1. Graphical representation of the roughness for the right surface of the cut, input - output, cutting 3 from experiment number 2.

Figures 2 and 3 graphically reflect the differences between the two methods of measuring surface roughness. The most significant deviations of R_a in the jet entry area were recorded for experimental number 6. For the cut-out area the experiment with the highest R_a value was experimental number 4. These differences between the results recorded for each part of the experiment were influenced by the variation of the cutting parameters used.

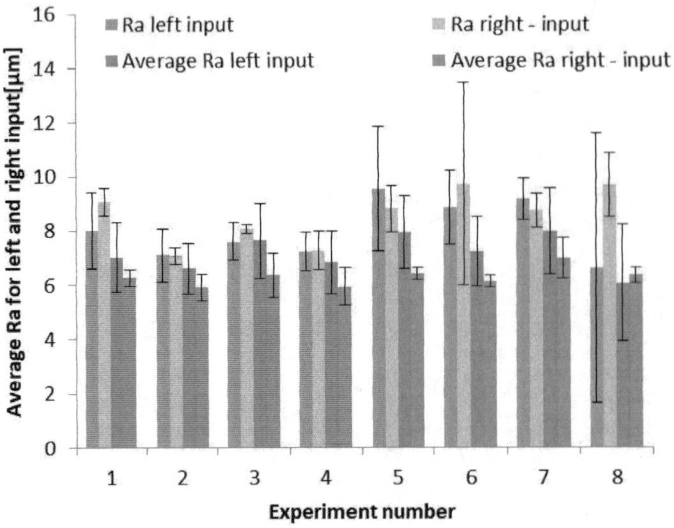

Figure 2. Average roughness values at the abrasion water jet inlet for the two measurement methods.

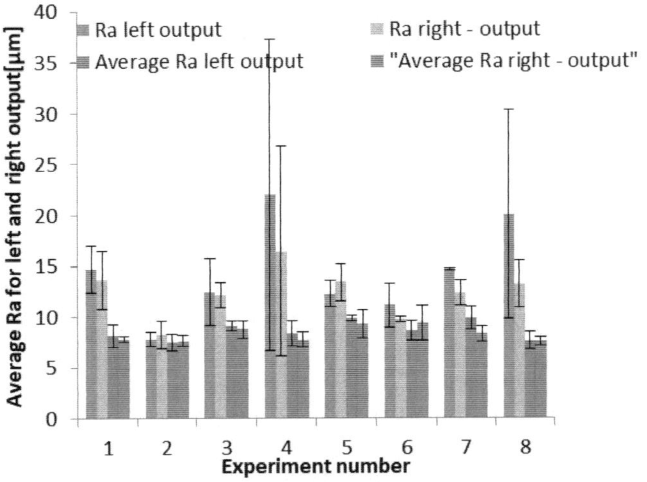

Figure 3. Average roughness values at the abrasion water jet outlet for the two measurement methods.

The various technological parameters (traverse speed, pressure and amount of abrasive material) in the experimental plan have a significant influence on the results obtained:

- the *amount of abrasive material* seems to have the greatest influence on the surface roughness at the entrance of the abrasive water jet because it is observed that for experiments (odd) where the amount of abrasive material was lower (150g/min) are recorded higher values of R_a, R_a (M_1: M_2). Otherwise, when the amount of abrasive material was increased to 300g/min (even number experiments) the roughness values decreased significantly;
- for the exit area of the abrasive water jet it is observed that there are two experiments, 4 and 8 for which the value of the surface roughness and the standard deviations are visibly higher than in the case of the other experiments. This increase in roughness is due to the high traverse speed, 150mm/min. In this situation it is recommended to maintain a constant amount of abrasive material (300g/min) and reduce the traverse speed to 150mm/min. The negative effect of the processing speed is also highlighted by other researchers who found that the surfaces cut at high traverse speed show deviations of shape, undulations and inaccurate dimensions, [11-14];
- the cutting *pressure* is a parameter that has a major influence on the roughness, [14, 15], and with its increase the surfaces become smoother. However, following the graphs and table 2, we cannot observe its positive effect. In this sense, a more in-depth analysis of the parameters influence should be performed with the help of a calculation program.

Experiment number two, used the most appropriate variation of the process parameters, pressure at level one of variation - 100MPa, traverse speed at level one of variation 100 [mm/min] and amount of abrasive material at level two of variation - of 300g/min, thus obtaining the lowest values of the output parameters: R_a, R_a(M_1: M_3), L_i, L_o, u, α and MR.

Determining the amount of material removed was determined by achieving the ratio between the density of the cut material and the volume of material removed. The volume of material removed took into account the measured values for the width of the processed surface at the entrance and exit of the abrasive water jet and the thickness of the cut piece / processing depth (10mm). The lowest value of the material removed amount during cutting with abrasive water jet is obtained for experiment number 2, for which only a quantity of (0.45 ± 0.005)g is lost to make a cut of 30mm, depth of 10mm, input width of 1.14 ± 0.01μm and output width 0.95 ± 0.01μm.

The width values of the machined surface at the inlet of the jet (L_i) are higher than those registered for the machined surface at the outlet of the jet (L_o), which makes the cut conical, in "V", figure 4. The taper of the cut is influenced by traverse speed and of the part thickness, the higher their values are, the more the taper increases and implies the deviation from the perpendicularity.

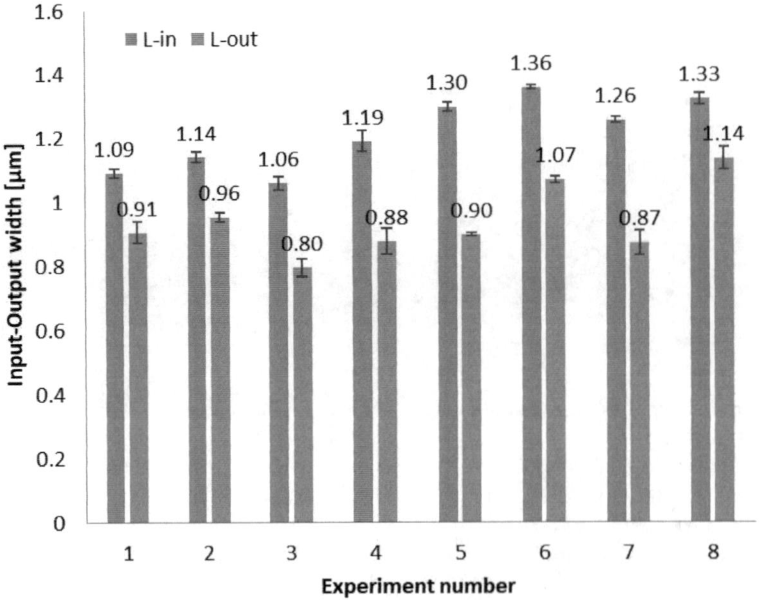

Figure 4. The variation of the widths resulting at the moment of cutting at the entrance and exit of the water jet with abrasive.

Another output parameter that can be calculated based on the inlet and outlet widths of the abrasive water jet but also on the deviation from the perpendicularity is the inclination angle of the machined surface. The mean values of this output parameter are presented for the entire experimental plan in figure 5. The smallest angle, as can already be guessed, is for experiment number 2, (0.53 ± 0.04)° closely followed by experiments 1 and 8 (0.53 ± 0.1)°, the difference being very small and influenced by process parameters and structural homogeneity of the cut part. It is observed that the standard deviations are higher for experiments number 1 and 8.

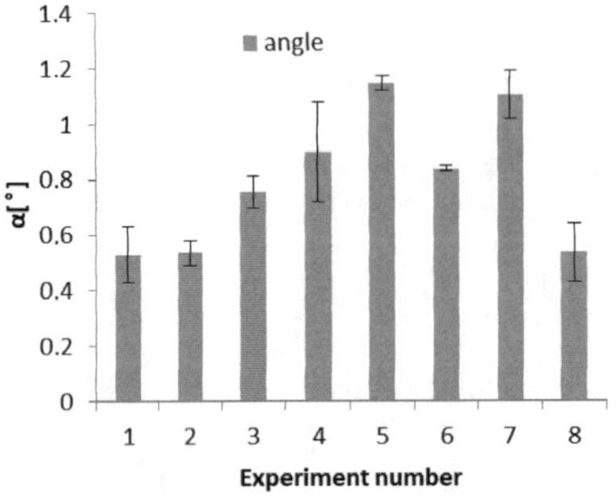

Figure 5. Variation of the inclination angle of the processed surface for the entire experimental plan.

To determine the resulting surface roughness after cutting the entire part was used an eyepiece magnification, 10X, because it has greater accuracy due to thinning beam. The scanner was programmed to survey in eight steps on eight different areas so that at the end of the scan to obtain the surface profile and roughness for the entire analysed sample (Sa). Figure 6 shows the values of the surface roughness for the entire resulting cutting area and it is observed that the lowest value of Sa is recorded by the experiment with number 4 (table 3). Also, the influence of the parameter "amount of abrasive material" is visible because all experiments (2, 4, 6 and 8) that had the second level of variation of the input factor, 300g/min, presents average values of Sa_{in} and Sa_{out} with much lower than the other experiments (1, 2, 5 and 7).

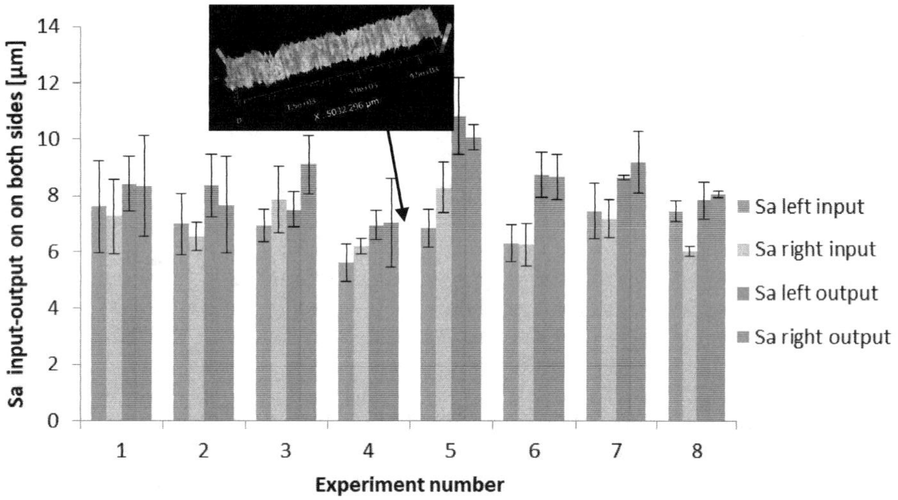

Figure. 6. The surface roughness of the cutting part in the experiment number 2.

Following the roughness values for Sa_{in} and Sa_{out}, it is observed that the surface from the left and right of the cut reflects similar values, the differences being reduced by only 0.58[µm] for the entrance area and the exit area, 0.08µm.

Table 3. Surface roughness values for experiment number 4.

Sample Cut	Side	Sa_{in} [µm]	Average Sa_{in} [µm] Left	Average Sa_{in} [µm] Right	Sa_{out} [µm]	Average Sa_{out} [µm] Left	Average Sa_{out} [µm] Right
1	Left	6.36	5.61± 0.66	6.19± 0.28	6.41	6.95± 0.51	7.3± 1.58
	Right	6.30			8.84		
2	Left	5.08			7.43		
	Right	5.86			5.91		
3	Left	5.39			7.02		
	Right	6.40			6.35		

4. Conclusions

Determination of the abrasive water jet cutting behavior on the Arboblend V2 Nature biopolymeric material showed admissible values of the output parameters (R_a, Sa_{in}, Sa_{out}, u, α, MR, L_i and L_o) thus, the obtained surfaces by this process and from this material can be used in industrial applications.

The values of the input parameters had a considerable effect on the output parameters, influencing the results:
- the increase of the *quantity of abrasive material* at the second level of variation of the parameter, 300g/min, led to the improvement of the surface roughness, thus obtaining smoother cut surfaces;
- the increase of the traverse speed decreased the surface quality, a particularly visible effect for the cut exit area;
- the value of the high level *pressure* generally leads to obtaining fine surfaces, but exceeding a certain threshold (400MPa) and the use of a small width of the processing slot has negative effects on the surface quality;
- the processing temperature is a very important parameter in the case of polymeric materials, especially in the case of biodegradable ones because they are thermally sensitive, having phase changes at temperatures of maximum 60°C. The AWJ process being a "cold" one eliminates the possibility of material thermal degradation.

Following the analysis of the obtained results, it is recommended that in order to obtain the best possible surface quality for biodegradable polymers, the process parameters should be set as follows: large amount of abrasive material (300g/min), low traverse speed (100mm/min) and high flow pressure (150MPa).

5. References

[1] Andrew J Jefferson, Vellayaraj Arumugam, Hom Nath Dhakal 2018 Key stages of adhesively bonded repairs, *Repair of Polymer Composites, Methodology, Techniques and Challenges, Woodhead Publishing Series in Composites Science and Engineering* **3** 97-224 https://doi.org/10.1016/B978-0-08-102263-4.00003-X.

[2] Srinath Reddy N., Dinesh Tirumalaa, Rajyalakshmi Gajjela, Raja Dasb 2018 ANN and RSM approach for modelling and multi objective optimization of abrasive water jet machining process *Decision Science Letters* **7** Growing Science Ltd. doi: 10.5267/j.dsl.2017.11.003.

[3] G. Gopichand, M. Sreenivasarao 2020 Multi-response parametric optimisation of abrasive waterjet milling of Hastelloy C-276 *SN Applied Sciences* **2** 1764 https://doi.org/10.1007/s42452-020-03512-5.

[4] H.-T. (Peter) Liu, Vanessa Cutler, Chidambaram Raghavan, Peter Miles, Ernst Schubert and Nathan Webers 2018 Advanced Abrasive Waterjet for Multimode Machining *Abrasive Technology - Characteristics and Applications* Anna Rudawska, IntechOpen, doi: 10.5772/intechopen.75313. Available from: https://www.intechopen.com/books/abrasive-technology-characteristics-and-applications/advanced-abrasive-waterjet-for-multimode-machining.

[5] Liu H-T 2017 "7M" advantage of abrasive waterjet for machining advanced materials. Mint: *Journal of Manufacturing and Materials Processing* **1**(11) 1-19 MDPI, Basel, Switzerland. doi: 10.3390/jmmp1010011. http://www.mdpi.com/2504-4494/1/1/11/pdf.

[6] Liu H-T, Schubert E. 2009 Piercing in delicate materials with abrasive-waterjets. Mint: *The International Journal of Advanced Manufacturing Technology* **42**(3-4) 263-279 doi: 10.1007/s00170-008-1583-5.

[7] Liu H-T, Schubert E, McNiel D, Soo, K. 2010 Applications of abrasive-waterjets for precision machining of composites. In: *Proceedings of SAMPE 2010 Conference and Exhibition*, 17-20 May 2010 Seattle, Washington.

[8] TECNARO - *The Biopolymer Company* - https://www.tecnaro.de/en/, accessed at 12.11.2020.

[9] Nedelcu, D., Mazurchevici, S.-N., Popa, R.-I., Lohan, N.-M., Maldonado-Cortés, D., Carausu, C. 2020 Tribological and Dynamical Mechanical Behavior of Prototyped PLA-Based Polymers *Materials* **13** 3615 https://doi.org/10.3390/ma13163615.

[10] SR ISO 468:1997: *Rugozitatea suprafeței. Parametri, valorile lor şi reguli generale pentru stabilirea specificațiilor*, available at http://resource.npl.co.uk/softgauges/pdf/Specification.pdf, accessed at 06.02.2021.

[11] ISO/TC 44 N 1770, available at: http://farsungroup.com/assets/farsun-group-waterjet-iso.pdf, accessed at 15.02.2021.

[12] Akkurt A, Kulekci M K, Seker U, Ercan F 2004 Effect of feed rate on surface roughness in abrasive waterjet cutting applications, *Journal of Materials Processing Technology*, **147**(3) 389-396, https://doi.org/10.1016/j.jmatprotec.2004.01.013.

[13] Radu C, Herghelegiu E, Schnakovszky C, Tampu C 2015 Experimental analysis of the influence of feed rate on quality of cuts performed by AWJ *Journal of Engineering Studies and Research* **21**(1) 76-80.

[14] Li, R., Ekevad, M., Guo, X., Cao, P., Wang, J., Chen, Q., and Xue, H. 2015 Pressure, feed rate, abrasive mass flow rate influence on surface roughness for recombinant bamboo abrasive water jet cutting *Bio. Res.* **10**(2) 1998-2008.

[15] Pelit, H., Yaman, Ö. 2020 Influence of processing parameters on the surface roughness of solid wood cut by abrasive water jet *Bio Res.* **15**(3), 6135-6148.

Research on advances in roller bearing manufacturing

I Mehmet, G Dogan, D F Chitariu, C Dumitraş and F Negoescu
Technical University GHEORGHE ASACHI Iaşi, ,Faculty of Machine Manufacturing and Industrial Management, 700050 B-dul Prof. Mangeron No.59A, Romania

E-mail: dragos.chitariu@tuiasi.ro

Abstract. In general the manufacturing of bearing require strict technological step that are performed on highly specialized equipments that require timely adjustments at batch changes. The process of forging and heat treatment stages consume the most amount of energy but the most expensive stage is the grinding operation. An alternative to grinding and honing processes is hybrid machining by combined hard turning and rolling process. One main development directions of all major roller bearing manufacturers is improving durability of bearings by reducing the roughness by inducing high residual stresses in the surface layers of bearing rings and improving the profile of roller for reduced contact stress. Rolling can induce higher surface stresses and reduce roughness with benefits to the durability of the bearings. The present paper presents and analysis of the technological route for bearing ring and bearing roller and focuses on the research direction on the advances in manufacturing/ machining methods for bearing rings and rollers.

1. Introduction

Roller bearings are key components in all drive system of machine tools, automobiles and in general in all industrial machinery. Although roller bearings are seemingly simple machine parts, their internal geometry is quite complicated [1, 2]. Roller bearings products are key components in the drive systems of the very machine tools on which this manufacturer relies and are critical to the machine's functionality, durability and machining accuracy, the condition of these components greatly influences the most vital performance characteristics of the machine tool.

The manufacturing processes of roller bearings are complex and require a succession of stages (production route) that takes place in different sectors of a factory, also the manufacturing process is inflexible involving many adjustments of machine tools usually specialized, when changing between different types of manufacturing batches [3, 4]. Also, the development of accurate bearing design methodologies and the increased profitability of bearing production in small or medium series of manufacturing has led to the emergence on an increasing scale of special bearings, designed and made to withstand optimally under certain well-defined conditions. This is a modern trend that is becoming more and more pronounced worldwide, to the detriment of the production of general purpose bearings.

The general directions of optimization of the bearings and implicitly of the components take into account increased L_{10} durability and reduced P_{max} contact pressure [5].

The main manufacturer of roller bearings, in the field of industrial bearings, are SKF Group Schaeffler Group, Timken, NSK, NTN, JTEKT, Rothe Erde, Wafangdian Bearing Group, Minebea Mitsumi and C&U [6].

The main development directions of major roller bearing manufacturer are on improving wind turbine generators bearings, electrically-insulated bearings for electrical traction motor designs, stationary electric motors (medium and large types), generators and special applications including submersible pumps that aren't affected by high frequency electrical current. Another important direction is providing customers extended support beyond sales and customer service, by developing and improving application through which customers initiate inspection and diagnostic on various bearings. Also, developing Artificial Intelligence system in order to determine early failure on highly complex and large manufacturing system by processing information from a array of sensors. Development of SMART type bearings with sensor integration is an important investment direction for the major manufactures. This category is specific to the automotive field. Another main business direction in bearings manufacture is the bearing refurbish business for special applications [5-9].

2. Analysis of bearing components manufacturing steps

2.1. Analysis of bearing ring manufacturing

The structure of a roller bearing consists of inner ring, outer rings, roller and cages, optional rivets can be used and various seals. The materials used for bearing manufacture of the main components rings and roller are 100Cr6 - ISO 683-17 (AISI 52100) and 100CrMnSi6-4. For each component various manufacturing steps have been identified.

A typical technological route for an interior or exterior bearing ring will be as follows, figure 1.

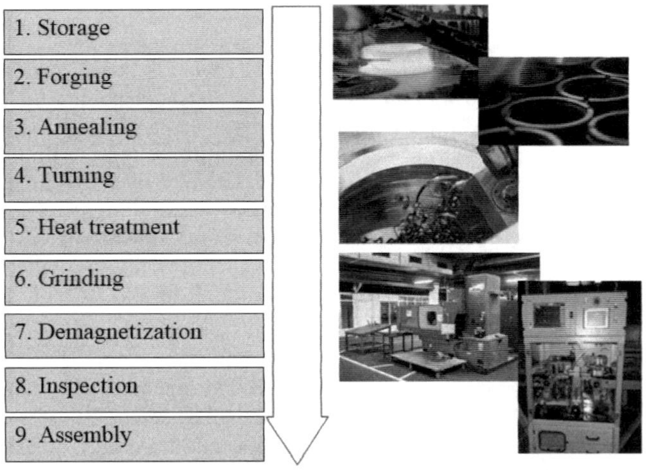

Figure 1. Technological route for a an interior or exterior bearing.

In order to increase the flexibility of the process and to reduce the manufacture process steps various researches and methodologies have been identified regarding the possibilities of reduction of the production route, mainly in the machining sector such as Trenpro process [6].

In general it is considered that the forging and heat treatment stages consume the most amount of energy but the most expensive stage is the grinding operation.

The Trenpro process uses helical cutters contacting on the external shell surfaces, which continuously gain in cutting height from the infeed side, from figure 2. The tube is conveyed through the cutting/rolling stand in the axial direction as a result of the continuous rotating motion and is cut into individual rings in the process. The Trenpro process is highly efficient, with a machining time of 4 seconds for ring with a diameter of 47mm. [10].

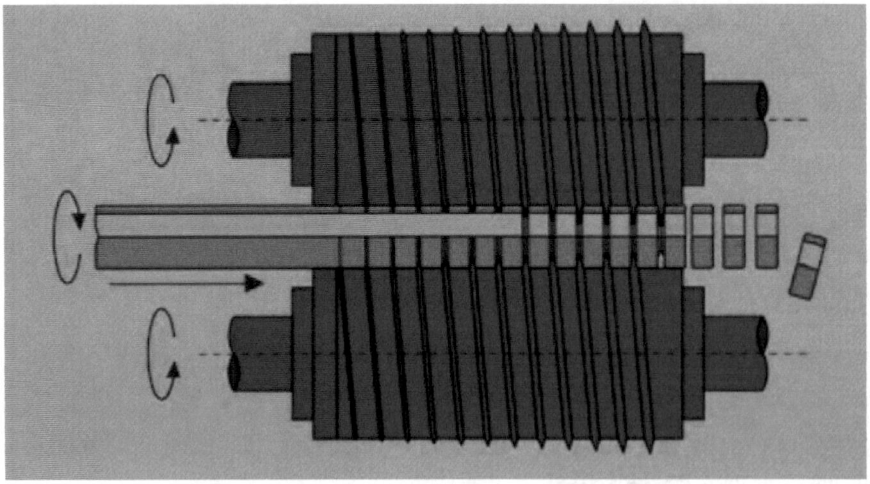

Figure 2. Trenpro process for ring machining.

The process of hard turning is used for finishing of parts (roller rings) with high accuracy of f material with hardness over 45-47 HRC, with cutting materials such as pCBN, ceramics or coated carbide. In the past these parts could only be machinined by grinding and a series of author consider hard turning a viable replacement for grinding operation of bearing rings. Konig pointed out that it is much more flexible and that, due to high-precision machines, the accuracy of hard-turned parts is comparable to grinding processes [11]. Jouini showed that surface integrity and surface topography residual induced stresses are parameters of interest and that influence bearing performance [12]. The experiments performed by Jouini indicate that rolling contact fatigue life increases as Ra value decreases from a values of 0.32 million cycle for a R_a=0.25 mm to a value of 5.2 million cycles for R_a=0.11 mm [12]. By precision hard turning machining a surface roughness R_a=0.11 mm can be achieved [13]. In this case the raceway roughness is similar to the values after the break-in period.

One of the methods to increase the durability of the bearings is low element roughness and high residual induced stresses [14]. Rolling is considered an alternative to honing and grinding processes. Rolling can induce high surface stresses in the material compared to honing and grinding [15]. Hybrid processing processes were initially proposed by Axinte and Gindy [16].

Denkena B., Groove T. and Maiss O. studied the influence of hard turning machining and rolling on the surface quality of bearing rings [17]. Whereas the durability of bearings can be improved by reducing the roughness on rolling surfaces and by inducing high residual stresses in the surface layers of the rings during machining processes. Since grinding processes can generate a low roughness of the machining surface, but cannot induce stresses in the surface layers and by high precision hard pressing a surface quality similar to grinding can be obtained.

The machining process combines the processes of hard turning and deep rolling in a single process to reduce machining time and improve surface quality. [17]. One of the advantages of the process is that an efficient positioning of the rolling tool is obtained with respect to the roughness tips that form roughness, the schematics of the process is presented in figure 3.

The roughness of the resulting surface is significantly influenced by the initial roughness. For smooth surfaces (Rz = 1 μm) high overlapping factors must be used to improve the surface quality. In any case, even very low initial roughness values and overlap factor close to 100% will not remove the lowest points of the profile base, after rolling. The hardness of the machined bearings is too high to smooth the surfaces in total [17].

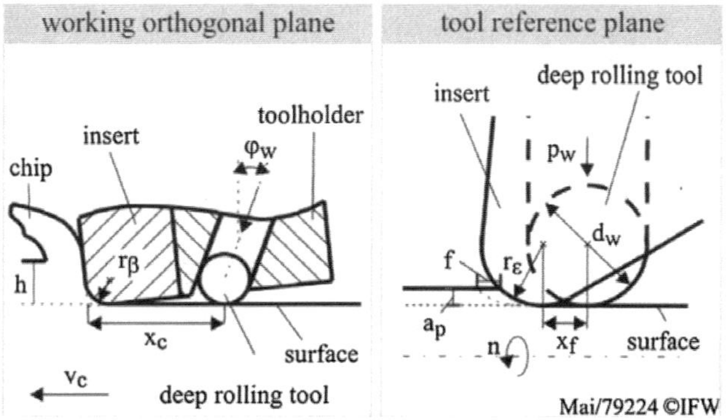

Figure 3. View of the hybrid hard turn-rolling process proposed by Maiß, Denkena and Grove.

After processing, a superior surface quality is obtained by using a lower feed rate than the turning surface. The surface roughness Rz can be improved by up to 24%. Effect explained by the defined, controlled contact between the rolling element and the roughness that forms the roughness. Not all the mechanisms of deformation of the irregularities on the machined surface are known.

The effect of the white layer generated in turning processes, is studied some papers, and indicate that white layers can amplify contact fatigue [18]. On the other hand Denkena claims that white layers they do not influence bearing durability [19]. In general the white layer is removed during grinding.

2.2. Analysis of bearing roller manufacturing

In general the manufacturing process of roller uses cold forming of roller from forged bar or wires for a diameter of roller less than 40 mm diameter, as presented in figure 4. For roller with a diameter larger the 40 mm the roller are machine by turning from forged bars.

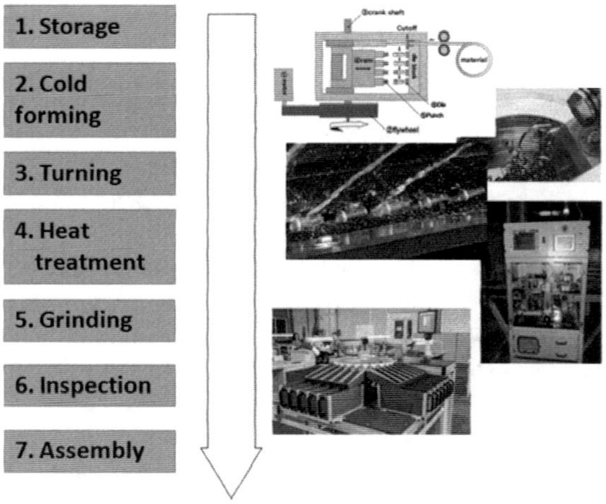

Figure 4. Technological route for a roller.

The identified research directions are on the one hand classic directions, known for improving the performance of bearings in general, namely increasing durability, reducing noise, increasing loads but also distinguishes a series of research on role design by designing logarithmic profiles. The main authors that stand out in research on obtaining logarithmic profiles are: Lundberg who developed the first logarithmic function for these bulging profiles to reduce pressures and achieve uniformity of contact pressure. The relationship proposed by Lunberg was improved by Johns-Gohar who improved the function for the convenience of manufacturing. However, the Johns-Gohar profile produces edge loading when the roller is tilted. Fujiwara and Kawase from NTN improved the relationship proposed by Johns-Gohar by adding parameters K_1: multiple of Q, K_2: the ratio between the length of the bulge and the z_m: the difference between the cylindrical profile and the crowning at the end of the effective length contact, as presented in figure 5 [5].

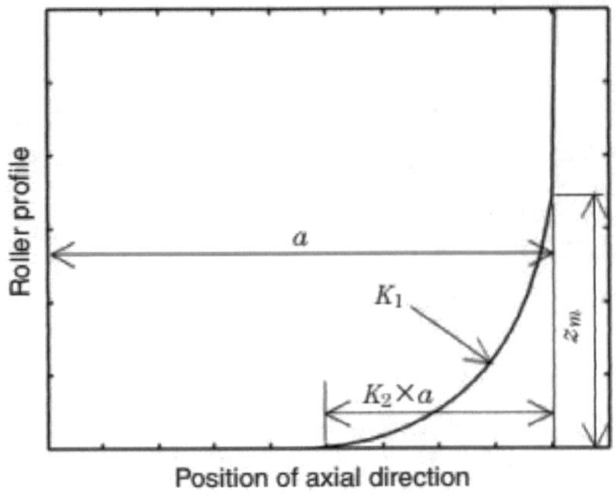

Figure 5. K_1 and K_2 used for curvature definition proposed by Fujiwara H and Kawase T [5].

Recent studies show that the multi-radius profile can also have very good results in contact geometry, the verification of contact stress mathematical models are made by FEA simulations that allow incrementally loading of the models and determine the stress distribution . In both cases the processing costs are particularly high. If the surface of the roller is not executed very well, the uniform stress distribution cannot be obtained [20-22].

3. Conclusions

Based on the presented considerations and analysis one can conclude:

- There are a number of researches carried out both by researchers from university centre, for example, Spiridon Cretu from the Institute of Technical University of Iasi but also by companies such as top companies such as researchers Hiroki FUJIWARA, Tatsuo KAWASE from NTN, SKF, Timken from the field of bearing production.
- The major manufacturers are extending the area o services provided to the client offering various services and solutions alongside technical support and sales such as electronic application for predictive maintenance of various equipments.
- Regarding ring manufacturing the general tendency of consumers to use Just In Time logistics elements puts pressure on bearing manufacturers to work in small batches, given that bearing lines are intended for large series and mass production. Manufacture of bearing rings in small batches requires long adjustment times of high productivity equipment. Hence the result the need to increase flexibility of the manufacturing routes and replace the highly costly grinding process with hard turning or combined process of hard turning and rolling. In the case of combined hard turning and rolling process with the added benefit of increased residual stresses and reduces roughness.
- Regarding roller manufacturing the problem of determining of optimal profile of the roller is current problem addressed by mathematical modelling and FEA simulations. It is necessary to apply

the extension of known calculation models and adjust the known parameters for a wide range of bearing types.

4. References

[1] Rabinovici I 1972 *Rulmenţi* (Bucuresti: Tehnica).
[2] Gafiţeanu M 1981 *Organe de maşini,* vol. 1 (Bucuresti: Tehnica).
[3] Harris T A 2001 *Rolling Bearing Analysis* (Boca Raton: CRC Press).
[4] Harris T A and Kotzalas M N 2006 *Advanced Concepts of Bearing Technology, Rolling Bearing Analysis, Fifth Edition* (Boca Raton: CRC Press).
[5] Fujiwara H and Kawase T 2007 Logarithmic Profiles of Rollers in Roller Bearings and Optimization of the Profiles, *Ntn Technical Review* **75** 140-148.
[6] *SKF 2020 Annual report* https://investors.skf.com/en/press/skf-annual-report-2020-1892714, accessed at 04. 2021.
[7] *NTN 2020 Annual report* https://www.ntnglobal.com/en/investors/annual.html, accessed at 04. 2021.
[8] *NSK 2020 Annual report* https://www.nsk.com/investors/library/nsk_report.html, accessed at 04. 2021.
[9] *The Timken Company 2020 Annual report* https://investors.timken.com/financials/annual-reports/default.aspx, accessed at 04. 2021.
[10] Cappel J, Weinberg M, Flender R 2004 The metallurgy of roller-bearing steels. *Steel Grips*. **2** 261-268.
[11] Konig W Berktold A Koch K.-F 1993 Turning versus Grinding - A Comparison of Surface Integrity Aspects and Attainable Accuracies, *CIRP Annals*, **42** 39–44.
[12] Jouini N, Revel P, Mazeran P, Bigerelle M 2013 The ability of precision hard turning to increase rolling contact fatigue life *Tribol. Int.* **59** 141–146.
[13] Revel P, Jouini N, Thoquenne G. Lefebvre F, 2016 High precision hard turning of AISI 52100 bearing steel *Precision Eng.* **43** 24–33.
[14] Neubauer T, Poll G, Denkena B, Maiss O, 2013 Fatigue life extension of rolling element bearings by residual stresses induced through surface machining *World Tribology Congress Torino.*
[15] Maiß O, Denkena B, Grove T 2016 Hybrid machining of roller bearing inner rings by hard turning and deep rolling, *J. of Mat. Pr. Tech.* **230** 211–216.
[16] Axinte D, Gindy N 2004 Turning assisted with deep cold rolling—a cost efficient hybrid process for workpiece surface quality enhancement *J. of Eng. Manuf.* **218** 807–811.
[17] Denkena B, Meyer R, Breidenstein B 2007 Development of combined manufacturing technologies for high-strength structure components *Adv. Mater. Res.* **22** 67–75.
[18] Revel P, Nabil J, Guillaume T, Fabien L 2016 High precision hard turning of AISI 52100 bearing steel *Precision Eng.* **43** 24–33.
[19] Denkena B, Grove T, Maiss O, 2015 Influence of the cutting edge radius on surface integrity in hard turning of roller bearing inner rings, *Prod. Eng. Res. Devel.* **9** 299–305.
[20] Takemasu T, Takao Koide Y, Takeda D., Kamimura D, Nakamoto M 2011 Properties of Densification by Surface Rolling and Load Bearing Capacity of 1.5Cr-0.2Mo High Density Sintered Steel Rollers and Gears, *J. of Solid Mechanics and Mat. Eng.,* **5** 825-837.
[21] Horng T-L 2007 A Numerical Algorithm for Determining the Contact Stress of Circular Crowned Roller Compressed between Two Flat Plates *J. of Solid Mech. and Mat. Eng.* **1**. 58-68.
[22] Cretu S, Benchea M 2012 An Elastic-Plastic Analysis of Profile Evolution in Cylindrical Roller Bearings *ASME 2012 11th Biennial Conference on Engineering Systems Design and Analysis*, ESDA.

IOP Conf. Series: Materials Science and Engineering 1182 (2021) 012046 doi:10.1088/1757-899X/1182/1/012046

Exergy and energy analysis of a vapour compression refrigeration system for the investigation of a new refrigerant to be used on board the ships

F Memet
Constanta Maritime University, Marine Engineering Faculty, Department of General Engineering Sciences, 104 Mircea cel Batran Street, 900663 Constanta, Romania

E-mail: feizamemet@yahoo.com

Abstract. This paper investigates the comparison between single stage vapour compression refrigeration cycles working with R134a - a traditional refrigerant for marine applications, and R435A - a new refrigerant having Dimethyleter (DME) 80% and R152a 20%. Since R134a shows a high GWP (Global Warming Potential), in the next future this refrigerant will be a subject for replacement. The new refrigerant taken into discussion here, R435A, shows a much lower GWP (3 compared with 1300 of R134a). The study allows the investigation of the performance of the cycles working with these two refrigerants, within a thermodynamic analysis, based on both first and second laws of thermodynamics. This kind of research is a strong tool used in design, optimisation and performance assessment of refrigeration systems. In the present paper are analysed the influence of evaporation temperatures on first and second law efficiencies and on exergy losses in the system. The results are obtained when this temperature varies in the range (248÷278) K. These results reveal the fact that R435A is a viable solution for R134a replacement, since average COP for R435A is about 35% higher than for R134a, the average exergy loss is around 40% smaller and the average exergy efficiency is approximately 35% higher.

1. Introduction

Nowadays, artificial cooling is an essential part of our lives, refrigeration, air conditioning and many other applications being technologies in continuous development. Food transport in cold spaces on board of ships is tightly bound with the constant demand increase [1,2].

Marine refrigeration enables safe food transport on long distances, by avoiding the spoiling during voyage, due to the slowing down of germs growth [3].

The dominant technology, when talking about marine refrigeration, is vapour compression. In the modern energy context, it is important to point out the fact that vapour compression refrigeration systems consume high-energy amounts, their optimization being an important task [4].

Sein the “quality” of high energy consumer of these type of plants, the issue of improving their energy efficiency, taking into consideration aspects such as efficient and environment friendly refrigerants (new refrigerants) and exergetic analysis, is of major concern [5].

Lately, HCFCs (Hydrochlorofluorocarbons) were commonly used in vapour compression systems, especially R22, having in view their very good thermodynamic and transportation features. Because of their negative impact on the ozone layer and contribution to the greenhouse effect, HCFCs are progressively taken out of use, at international level. In this context, HFCs (hydrofluorocarbons) are, now, an accepted option, but their important global warming potential makes them to be candidates for replacement with eco-friendly refrigerants [6-8].

One of the most important HCF refrigerant, found in the list of marine refrigerants is R134a. Present concerns are focused on the substitution of this commonly used refrigerant with less greenhouse contributor ones. R152a shows similar properties with the ones of R134a and a lower GWP (Global Warming Potential).

Because of its flammability, it is recommended to be used in mixtures [9]. On the other hand, DME (dimethyleter) is a natural refrigerant, frequently used in the past, but abandoned because its safety aspects; nowadays DME is again in the attention of specialists due to its energetic and environmental behaviour [10].

The inconvenient coming from its flammability and explosion risks can be exceeded throughout some measures such as: enclose operations, ensuring local exhaust ventilation in the location of leakage, the use of special equipment, etc.

The actual refrigeration cycle is characterized by irreversibilities – which are reflected in the performance decrease. The limitation of the analysis based only on the first law of the thermodynamics which considers only the energy conservation aspect, leads to the need of the exergy analysis, in order to assess accurately the performance and to identify losses [11].

In this respect, a viable thermodynamic analysis of the vapour compression cycle implies both calculations of Coefficient of Performance (or first law efficiency) and exergy efficiency (or second law efficiency).

In the light of the above mentioned, this paper deals with the theoretical investigation of the new R435A (DME 80% / R152a 20%) as a substitute of R134a, within a thermodynamic performance analysis of a standard vapour compression cycle, based on both laws of thermodynamics.

In this study are investigated Coefficient of Performance, exergy loss and exergy efficiency of R134a and R435A, in order to optimize the considered system.

2. Methods and materials

The focus on these refrigerants is based on their thermo-physical and environmental properties, given in table 1 [12-14]. It is considered a subcooled and superheated vapour compression refrigeration cycle, its (T-s) diagram being provided in figure 1.

Table 1. Thermo-physical and environmental properties.

Property	R134a	R152a	DME
Molecular mass (kg/kmole)	102.3	66	46.07
Triple temperature (°C)	–103.3	–118.6	–141.5
NBP (Normal Boiling Point) (°C)	–26	–24	–24.8
Critical temperature (°C)	101.06	113.6	127.2
Critical pressure (MPa)	4.06	4.5	5.34
Critical density (kg/m^3)	511.9	368	271
Miscibility with oil	Nil	120	10
ODP (Ozone Depletion Potential)	0	0	0
GWP (Global Warming Potential)	1300	120	3

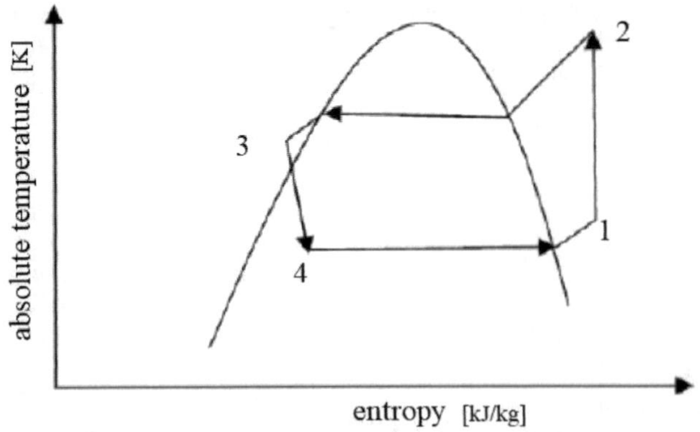

Figure 1. Thermodynamic cycle in T-s diagram of the considered refrigeration system.

The analysis developed based on the following:

- in all the system's components are kept steady state conditions
- pressure drops in suction line and discharge line are neglected
- heat exchange between the system and the environment is neglected
- kinetic and potential energy and exergy losses are neglected
- cooling capacity: 1 kW
- evaporation temperature: (248÷278) K
- condensation temperature: 316 K
- subcooling and superheating degrees: 288 K
- average ambient temperature: 298 K
- compressor efficiencies: 0.75
- electric motor efficiency: 0.8

Exergy analysis is based not only on the first law of thermodynamics, as in the case of energy analysis, it is based on the first and second laws of thermodynamics. In this respect, the exergy analysis is able to provide information on how, location and the performance decreases, since irreversibilities lead to performance degradation.

This analysis is developed on the theoretical formulation, given below [15].

The specific exergy:

$$ex = (h - h_0) - T_0 (s - s_0) \tag{1}$$

where: h – specific enthalpy (kJ/kg)
s – specific entropy (kJ/kgK)
T_0 – surrounding ambient temperature (K)
h_0 and s_0 – enthalpy and entropy at the surrounding conditions

For the four main components of the system:

- in the evaporator, the heat load (Q_{ev}) and the exergy loss (Ex_{Dev}):

$$Q_{ev} = \dot{m}_f (h_1 - h_4) \tag{2}$$

$$Ex_{Dev} = \dot{m}_f \left[(h_4 - h_1) - T_0 (s_4 - s_1) \right] + Q_{ev} (1 - T_0 / T_{ev}) \tag{3}$$

where: $\dot{m}_f$ – refrigerant mass flow (kg/s)

h_1 and s_1 – enthalpy and entropy at the evaporator exit
h_4 and s_4 – enthalpy and entropy at the evaporator inlet
T_{ev} – absolute evaporation temperature (K)

- in the compressor, the work input (W_{cp}) and the exergy loss (Ex_{Dcp}):

$$W_{cp} = \dot{m}_f \left(h_2 - h_1\right) \quad (4)$$

where: h_2 – enthalpy at the compressor exit

$$Ex_{Dcp} = \dot{m}_f \left[\left(h_1 - h_2\right) - T_0\left(s_1 - s_2\right)\right] + W_{el} \quad (5)$$

where: W_{el} – electrical power (kW)
s_2 – entropy at the compressor exit

$$W_{el} = \frac{W_{cp}}{\eta_m \cdot \eta_{el}} \quad (6)$$

where: η_m / η_{el} – mechanical / electrical efficiency

- in the condenser, the heat load (Q_{cd}) and the exergy loss (Ex_{Dcd}):

$$Q_{cd} = \dot{m}_f \left(h_2 - h_3\right) \quad (7)$$

where: h_3 – enthalpy at the condenser exit

$$Ex_{Dcd} = \dot{m}_f \left[\left(h_2 - h_3\right) - T_0\left(s_2 - s_3\right)\right] - Q_{cd}\left(1 - T_0/T_{cd}\right) \quad (8)$$

where: T_{cd} – absolute condensation temperature (K)
s_3 – entropy at the condenser exit, (h_3 being explained above)

- in the throttling valve, since the process is isenthalpic, the exergy loss (Ex_{DTV}):

$$Ex_{DTV} = \dot{m}_f \left(s_3 - s_4\right) \quad (9)$$

- The Coefficient of Performance, also known as the first law efficiency:

$$COP = \frac{Q_{ev}}{W_{el}} \quad (10)$$

- The overall exergy loss:

$$Ex_{DTOT} = Ex_{Dev} + Ex_{Dcp} + Ex_{Dcd} + Ex_{DTV} \quad (11)$$

- The exergy efficiency, also known as the second law efficiency:

$$\eta_{ex} = \frac{Ex_1 - Ex_4}{W_{el}} \quad (12)$$

3. Results and discussions

The following results are obtained in base of the above given equations, for the cycle working with the actual refrigerant (R134a) and the one working with the new refrigerant (R435A). The software EES (Engineering Equation Solver) has been used for the simulation. The comparison of the results are provided in the following figures.

Figure 2 shows the influence of the evaporation temperature increase on the Coefficient of Performance (COP) values; this important efficiency criteria provides information related to the amount of electrical power input to produce a specific amount of cold.

From figure 2 results that the increase of the evaporation temperature has a benefic effect on COP, for the both cycles.

This is explained by the fact that for higher evaporation temperatures are obtained increased refrigerating effect in the cycles. The best COP values are sein for the cycle working with new refrigerant (R435A). The average COP for R435A is about 35% higher than R134a.

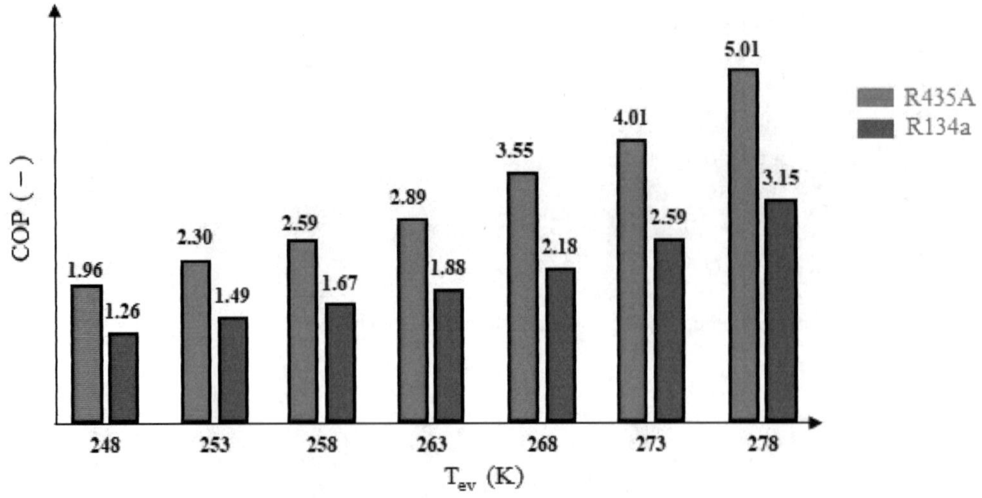

Figure 2. (COP – T_{ev}) dependency.

Figure 3 shows the influence of evaporation temperature increase on the exergy loss.

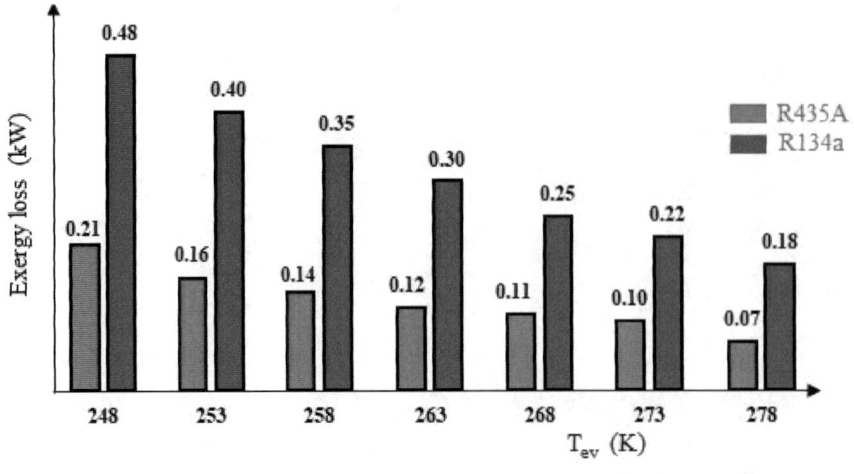

Figure 3. ($Ex_{D\ TOT}$ – T_{ev}) dependency.

Figure 3 reveals the fact that the increase of the evaporation temperature leads to a decrement of exergy losses, the trend being the same for the both refrigerants.

This result is explained by the fact that at higher evaporation temperatures, the heat transfer in the evaporator is more intense, thus, the refrigerating effect is improved, so exergy losses are diminished.

The less exergy losses are observed for the cycle working with R435A. The average exergy loss for R435A is around 40% smaller than R134a.

From figure 4 can be seen what happens with exergy efficiency values, when the evaporation temperature increases.

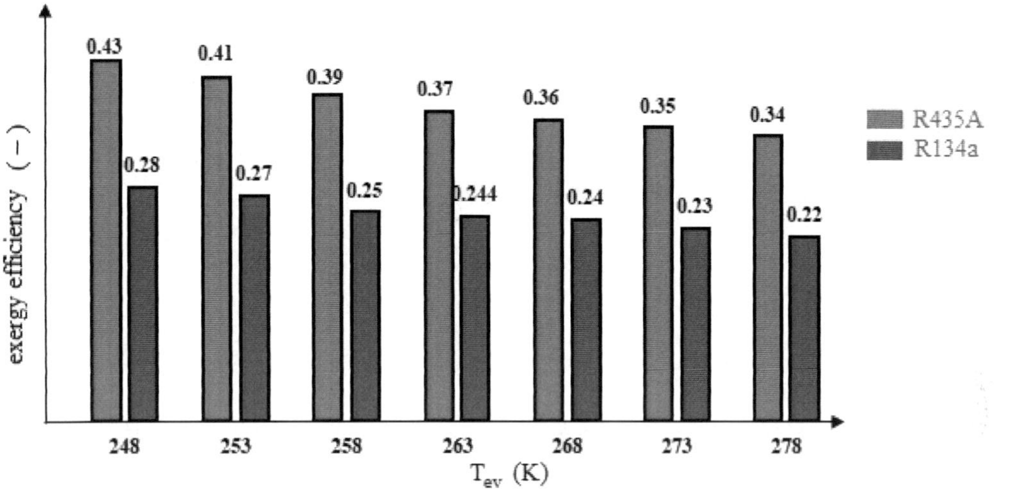

Figure 4. ($\eta_{ex} - T_{ev}$) dependency.

The exergy efficiency decrease together with increase of the evaporation temperature. This result is explained by the Coefficient of Performance increase with increment of this temperature. This situation causes minimum exergy intake to accomplish the given task. This trend in available for the both refrigerants considered. The cycle working with R435A shows the better second law efficiency. The average second law efficiency increase is about 35%, when using R435A in the cycle.

4. Conclusions

A comparative analysis of the performance of a single stage vapour compression refrigeration cycle working with R134a and R435A was developed.

The influence of the evaporation temperature increase on the Coefficient of Performance (first law efficiency), exergy loss and exergy efficiency (second law efficiency) was studied in the context of environmental concern and energy efficiency need.

The evaporation temperature varied in the range (248÷278) K, the condensation temperature was kept constant (316 K) and the average ambient temperature was 298 K. The cycle includes sub cooling and superheating processes.

It was observed that:

- the increase of evaporation temperature leads to an increase of the first law efficiency, in both cases; this efficiency is about 35% higher when the working refrigerant is R435A,
- the increase of the evaporation temperature leads to a decrease of the exergy loss, in both cases; this loss is lower when the working refrigerant is R435A (about 40% less),
- the increase of the evaporation temperature leads to a decrease of the second law efficiency, in both cases; the exergy efficiency is higher when the working refrigerant is R435A (about 35% higher).

The analysis highlights the fact that the study based only on the first law of thermodynamics is not reliable, a consistent research requiring concepts belonging to the both laws.

The results show that R435A is a strong candidate for the replacement of R134a, with the implementation of specific measure – in order to ensure safety on board.

5. References

[1] Arduino G 2010 *Proc 12th WCTR (Lisabon)* pp 1-15.
[2] Majgaonkar A S 2008 *ASHRAE Journal* **12** 50.
[3] Le Bail A, Orlovska M, Havet M, Adedeji A, Abadie J, Beaufort A, Bourdain G and Cardinal M 2011 *Proc. Conf. FAO Fisheries and Aquaculture Report (Victoria, Mahe, Seychelles)* No 990 pp 33-40.
[4] Ogbonnaya M, Ajaji O O, Waheed M A and Oyedepo S O 2019 *Journal of Physics: Conference Series* 1378.
[5] Ruiz V 2012 *Proc. Second int Symposium on Fishing Vessel Energy Efficiency, E-Fishing (Vigo, Spain)* pp 1-8.
[6] Shaik S V and Babu T P A 2017 *Energy Procedia* **138** 710.
[7] Benhadid-Dib S and Benzaoui A 2012 *Energy Procedia* **18** 807.
[8] Kalla S K and Usmani J A 2013 *Indian Journal of Engineering* 5 **13** 57.
[9] Hafner A, Gabrielli C H and Widell K 2018 *Refrigeration units in marine vessels. Alternatives to HCFCs and high GWP HFCs*, ed. Nordic Council of Ministers Publication Unit (Copenhagen Denmark) p 77.
[10] Popescu G, Pop H Vasilescu E E, Tregubleac I, Marinescu C and Apostol V 2005 *Termotehnica* **1**(2) 52.
[11] Bolaji B O 2010 *Journal of Engineering Science and Technology* **5**(4) 435.
[12] Baskaran A and Mathews K 2015 *Jordan Journal of Mechanical and Industrial Engineering* **9**(4) 289.
[13] Baskaran A 2013*Archives Des Sciences* **66**(3) 2.
[14] Baskaran A, Manikandan N and Sureshkumar V P 2018 *International Journal of Advance Engineering and Research Development* **5**(12) 73.
[15] Ahamed J U, Saidur R and Masjuki H H 2011 *Renewable and Sustainable Energy Reviews* 15 1593.

IOP Conf. Series: Materials Science and Engineering 1182 (2021) 012047 doi:10.1088/1757-899X/1182/1/012047

NRM30 as a possible substitute for R22 in marine vapour compression refrigeration systems

F Memet

Constanta Maritime University, Marine Engineering Faculty, Department of General Engineering Sciences, 104 Mircea cel Batran Street, 900663 Constanta, Romania

E-mail: feizamemet@yahoo.com

Abstract. The purpose of this paper is to investigate the possibility of using NRM30 in order to replace R22, in existing plants on board the ships. NRM30 is a refrigerant mixture of R134a/R1270/RE170 (55/37.5/7.5 by mass percentage). This approach is based on the better environmental behaviour of NRM30. The comparison between R22 and NMR30 has been developed on the basis of Coefficient of Performance (COP), temperature at compressor exit, power consumption per ton of refrigeration and volumetric refrigeration capacity variations, when evaporation temperature and condensation temperature increase. The above mentioned are calculated when the evaporation temperature varies in the range (-2, +10)^{o} C and the condensation temperature varies in the range (40, 50)^{o} C. The results reveal that COP increases with the increase of the evaporation temperature and the decrease of the condensation temperature- for both refrigerants. Although the ecological refrigerant (NRM30) shows an average COP value lower than the one of R22 (about 30% lower), NRM30 ensures an ecological operation of the plant and lower temperatures at compressor exit- which it is benefic for the long life of this machine. For the considered temperature ranges, the plant operates in efficient conditions for the higher value of the evaporation temperature (+10°C) and for the lowest value of the condensation temperature (40°C). For this conditions, it is seen that the vapours temperatures at the compressor exit are lowest, being ensured also the safe operation of the compressor.

1. Introduction

Refrigeration on board of ships is a vital technology when the aim is the maintenance of food, drinks, cargo and garbage in proper conditions during the voyage. Other application is air conditioning, essential for the transported goods and crew or passengers thermal comfort.

According to the International Maritime Organisation (IMO), in 2014, about 70% of the world merchant fleet used the Freon R22 on board the ships [1]. This, R22 was met on board of fishing vessels, passenger ships and cargo ships.

The frequent use of R22 it is based on its very favourable thermodynamic behaviour and low energy consumption, but actual systems have to use refrigerants able to reply to environmental requirements as well [2].

R22 or Chlorodifluoromethane, is responsible, together with other chemicals, for the ozone layer depletion; it also contributes to the global warming rise.

This refrigerant belongs to Hidrochlorofluorocarbon (HCFC) refrigerants, which have been prohibited according the Montreal Protocol [3]. This type of refrigerants contain elements as: carbon, chlorine, fluorine and hydrogen and are seen as transitional refrigerants.

The Conference held in Copenhagen (1992) imposed the production stop of HCFC starting with 31 December 2014, so that substitutes will accomplish null Ozone Depletion Potential (ODP) low Global Warming Potential (GWP) and Total Equivalent Warming Impact (TEWI – assessing the real effect of refrigerants on global warming) [4,5].

In the framework of the Regulation (EC) No 2037/2000 – that deals with ODS (Ozone Depleting Substances), starting with 1st January 2010, the import, production, selling or using the virgin R22 is banned; still R22 might be used in existing facilities till 2015, the following major challenge being the finding of the proper substitute [6].

The refrigerant considered in this paper, as a possible substitute of R22, is NRM30 a mixture of R134a/R1270/RE170 given by the following mass percentage: 55/37.5/7.5. This choice is motivated by the better environmental behaviour of NRM30, compared with the one of R22. In table 1 are given some thermodynamic features of the unwanted refrigerant and the ones met in the components of NRM30, safety data and environmental information as well [7,8].

Table 1. Properties of R22, R134a, R1270, RE170.

Refrigerant / Properties	R22	R134a	R1270	RE170
Molecular weight (g/mol)	86.47	102.0	42.08	46.07
Critical temperature (°C)	96.2	101.1	92.4	128.8
Critical pressure (MPa)	4.99	4.06	4.67	5.32
Normal boiling point (°C)	–40.8	–26.1	–47.7	–24.8
Safety group	A1	A1	A3	A3
ODP (R11=1)	0.04	0.0	0.0	0.0
GWP (100 yr)	1790	1370	20	1

The following pylons are mentioned: the molecular weight has to be low, in the case of vapour compression systems, since the marine refrigeration compressor sizes should be limited, the critical temperature should be very high, in order to be obtained a good refrigeration effect with the plant, the critical pressure should be low and the normal boiling point as well – to be ensured low values of temperatures.

Consequently, within this paper it is developed a theoretical thermodynamic comparative performance assessment of a single stage vapour compression refrigeration cycle working with R22 and NRM30.

The proposed comparison is based on the assessment of the Coefficient of Performance, temperature at compressor exit, power consumption per ton of refrigerant and volumetric capacity variations – when evaporation and condensation temperatures increase. The results will show that NRM30 is an acceptable substitute for R22.

2. Methods and materials

The absolute temperature – entropy diagram of the considered system is given in figure 1 and the mathematical formulation of the analysis is as follows [9].

Specific heat addition in the evaporator:

$$q_0 = h_1 - h_4 \tag{1}$$

Specific compressor work:

$$w_c = h_2 - h_1 \tag{2}$$

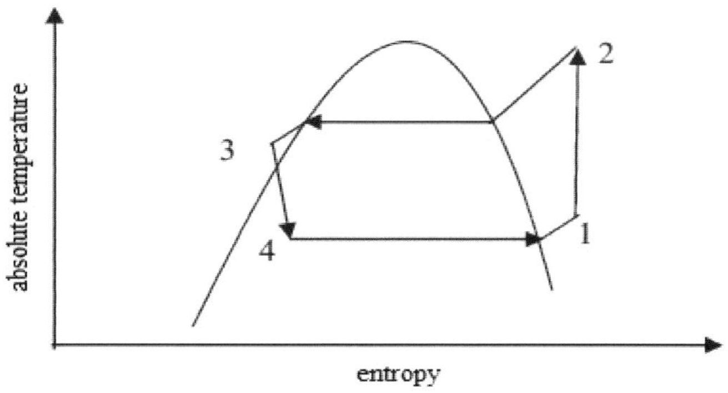

Figure 1. T – s diagram of the considered system.

Coefficient of Performance:

$$COP = q_0 / w_c \quad (3)$$

Power consumption per ton of refrigerant:

$$P_{PTR} = 3.5167 / COP \quad (4)$$

Volumetric cooling capacity:

$$VRC = q_1 \cdot q_0 \quad (5)$$

where: h – specific enthalpy and q – density.

3. Results and discussions

In this paper, the results are obtained when the evaporation temperature varies in the range (–2÷10)°C for the given condensation temperature of 54°C and when the condensation temperature varies in the range (40÷50)°C – for a given evaporation temperature of 10°C; the superheating and sub cooling degrees are of 5°C.

Figures 2 – 7 show the variations of the specific heat addition in the evaporation, specific compressor work, Coefficient of Performance, temperature at the compressor exit, power consumption per ton of refrigerant and volumetric cooling capacity, with the increase of the evaporation temperature (t_0), for both refrigerants.

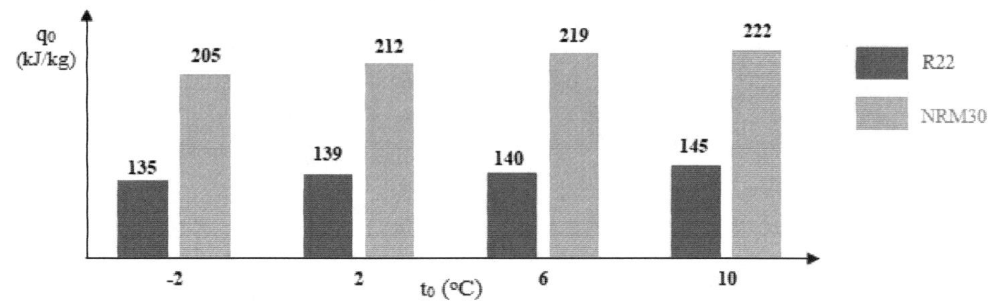

Figure 2. Effect of evaporation temperature on the specific heat addition for a condensation temperature of 54°C.

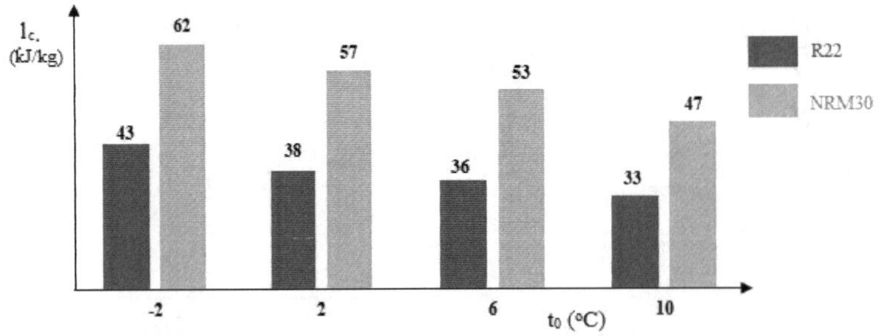

Figure 3. Effect of evaporation temperature on the specific compressor work for a condensation temperature of 54°C.

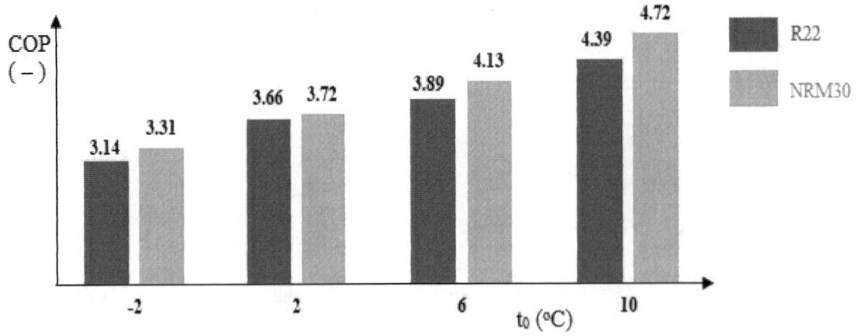

Figure 4. Effect of evaporation temperature on the Coefficient of Performance for a condensation temperature of 54°C.

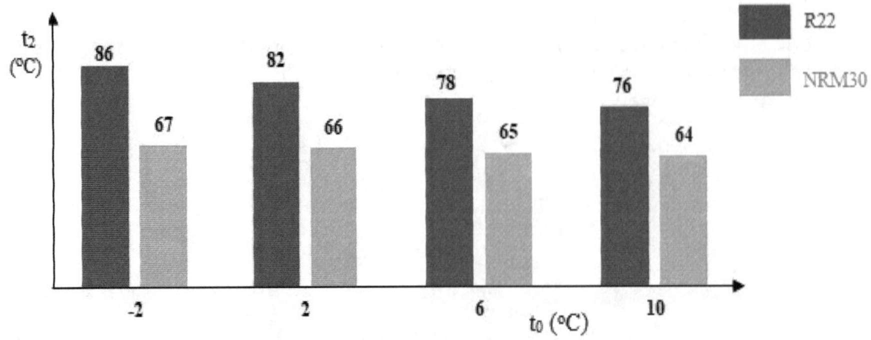

Figure 5. Effect of evaporation temperature on the discharge temperature of the compressor for a condensation temperature of 54°C.

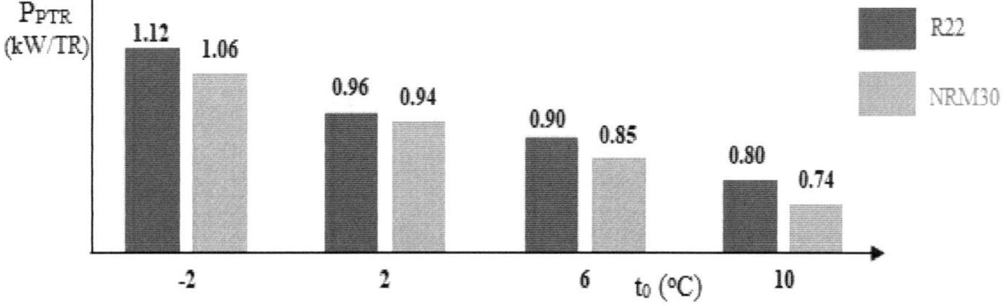

Figure 6. Effect of evaporation temperature on the power consumption per ton of refrigeration for a condensing temperature of 54°C.

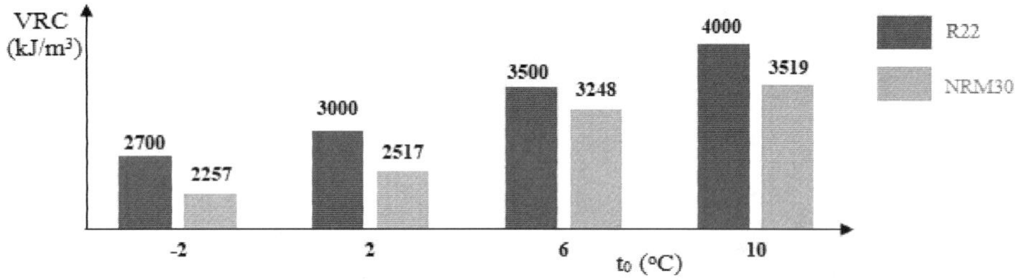

Figure 7. Effect of evaporation temperature on the volumetric cooling capacity for a condensation temperature of 54°C.

Figures 2 – 7 shows the same trend for the both refrigerants: the increase in the evaporation temperature lead to the increase in the values of specific heat addition in the evaporator, Coefficient of Performance and volumetric cooling capacity and to the decrease of specific work consumption, temperature at compressor exit and power consumption per ton of refrigerant.

This is due to the fact that the increment of the evaporation temperature is met in the increment in latent heat of evaporation and in the temperature difference between the system and the environment, with positive effects on COP values and discharge temperature of the compressor.

The volumetric cooling capacity depends on the density of the refrigerants (as seen in the given formulae). The rise of the cooling capacity per unit volume of refrigerant at evaporator exit with the evaporation temperature is due to the rise of the vapour density.

Figures 8 – 11 show the variations of the Coefficient of Performance, temperature at the compressor exit, power consumption per ton of refrigerant and volumetric cooling capacity with the increase of the condensation temperature, for both refrigerants.

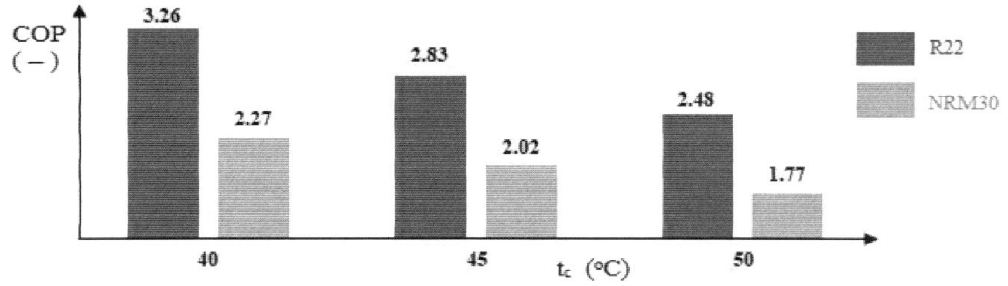

Figure 8. Effect of condensation temperature on the Coefficient of Performance for an evaporation temperature of 10°C.

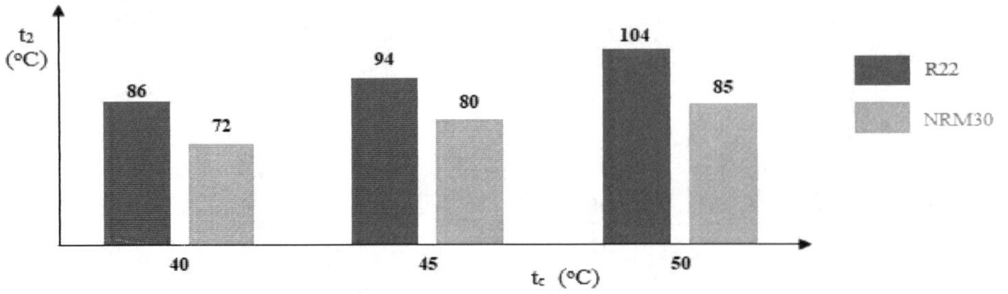

Figure 9. Effect of condensation temperature on the discharge temperature of the compressor for an evaporation temperature of 10°C.

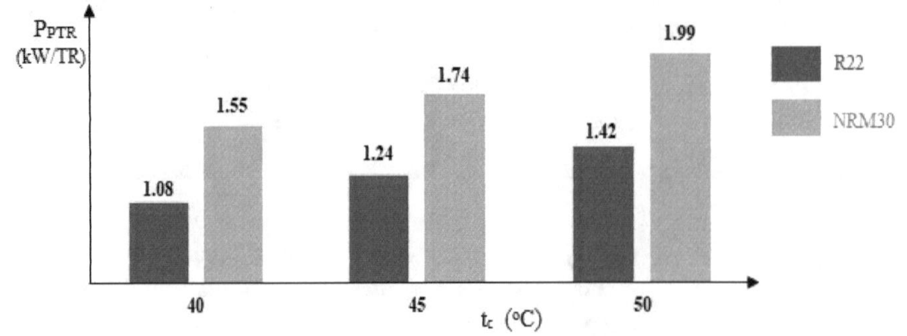

Figure 10. Effect of condensation temperature on the power consumption per ton of refrigeration for an evaporation temperature of 10°C.

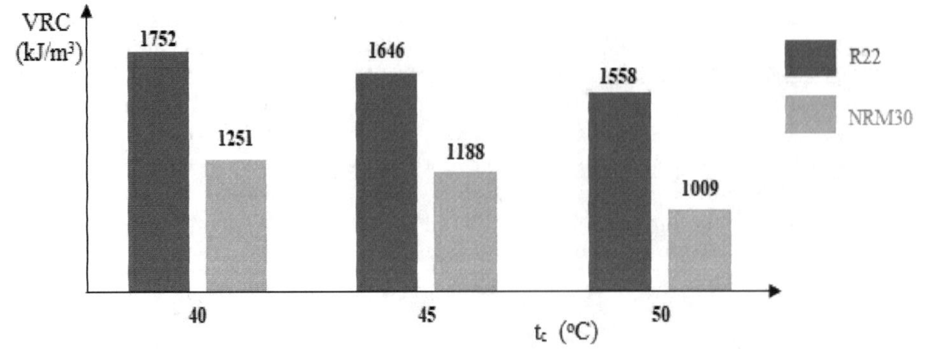

Figure 11. Effect of condensation temperature on the volumetric cooling capacity for an evaporation temperature of 10°C.

Figures 8 – 11 also show the same trend for the both refrigerants: the increase in the condensation temperature determines the increase of the temperature at the compressor exit and of the power consumption per ton of refrigerant.

Since a high value of COP indicates a high efficiency of the cycle, results that the considered system should operate at high evaporation temperatures and at low condensation temperatures.

The results reveal the fact that for t_0 = 10°C and t_c = 54°C, NRM30 provides an improved COP with 6.99%, while for t_c = 40°C and t_0 = 10°C, when using NRM30, COP is 30% lower compared with the one of R22.

When selecting a substitute for a refrigerant, it is important to check values of vapours temperatures leaving the compressor and values of volumetric cooling capacities. High temperatures after the vapours compression will damage the compressor oil and will shorten the operation life of this machine.

On the other hand, volumetric cooling capacity will impose the sizes of the chosen compressor. It is seen that NRM30 offers lower t_2 temperatures, as observed from Figures 4 and 9 (for example, for t_0 = 10°C and t_c = 54°C, t_2- when using NRM30, is 15.79% lower, while for t_c = 40°C and t_0 = 10°C, t_2- when using NRM30, is 16.28% lower).

For the considered example, VRC provided with NRM30 is lower than the one of R22 with 16%, and 28%, respectively.

4. Conclusions

A comparative theoretical thermodynamic analysis is presented in order to assess the performance of a single vapour compression refrigeration cycle working with R22 and NRM30. It was found that the Coefficient of Performance increases together with evaporation temperature increase (when keeping the condensation temperature constant), and decrease together the condensation increase (when keeping the evaporation temperature constant); in the first situation, the efficiency of the cycle results to be improved in the range (5.13÷6.99)%- when using NRM30 in the cycle. Also, the discharge temperature of the compressor decrease with increase of the evaporation temperature and with the decrease of the condensation temperature; the best values are obtained for the substitute (NRM30). When keeping constant the condensation temperature, NRM30 ensures lower discharge temperatures in the range (15.79÷22.09)%, while when maintaining the evaporation temperature constant, NRM30 shows lower discharge temperatures in the range (16.28÷18.22).

It was also found that the volumetric cooling capacity increases with the increase of the evaporation temperature and with the decrease of condensation temperature; these values are lower when using NRM30 in the cycle, the smallest gap (of 12%) being seen for the following operation conditions: t_0 = 10°C and t_c = 54°C.

Taking into consideration the reality given by the fact that there is no perfect refrigerant, NRM30 can be seen as a substitute of R22 due to its better environmental behaviour, improved COP or ensuring of a longer live of the compressor.

5. References

[1] Hafner A, Gabrielli C H and Widell K 2018 *Refrigeration units in marine vessels. Alternatives to HCFCs and high GWP HFCs*, ed. Nordic Council of Ministers Publication Unit (Copenhagen Denmark) p 77.
[2] Zakrzewski B and Lokietek T 2010 *Polish Maritime Research* **17** 55.
[3] United Nations 1987 United Nations Environment Programme (UNEP) *Montreal Protocol on Substances that deplete the Ozone layer: Final Act* (New York).
[4] Roy Z and Haldner G 2020 *Chemical Engineering Journal Advances* **3** 15.
[5] Benhadid-Dib S and Benzaoui A 2018 *Energy Procedia* **18** 807.
[6] Ruiz V 2012 *Proc. Second Int Symposium on Fishing Vessel Energy Efficiency,* E-Fishing (Vigo, Spain) pp 1-8.
[7] Mohanraj M, Muraleedharan C and Jayaraj S 2011 *Int. J. of Energy Research* **35** 647.
[8] Bahaa S, A A Aly, M Alsehli, A Elfasakhany and M B Bassouni 2020 Processes **8**(9), 1017.
[9] Shaik S S and Babu T P A 2017 *Energy Procedia* **138** 710.

IOP Conf. Series: Materials Science and Engineering 1182 (2021) 012048 doi:10.1088/1757-899X/1182/1/012048

Experimental Modal Analysis of Violin Bodies with Different Structural Patterns of Resonance Spruce

M Mihalcica[1], M D Stanciu[1], V G Gliga[1,2], M Campean[1], F Dinulică[1] and S M Nastac[1,3]
[1]Translvania University of Brasov, Department of Mechanical Engineering, Blvd. Eroilor, No. 29, 500036, Brasov, Romania
[2]S.C. Gliga Instrumente Muzicale S.A
[3] Dunărea de Jos" University, Faculty of Engineering and Agronomy Braila, Calea Calarasilor 29, Braila, Romania

E-mail: mihalcica.mircea@unitbv.ro

Abstract. The paper aims to highlight the modal response of violin bodies made of four quality classes of resonant wood, following the anatomical structure of the wood. The four categories of musical instruments, made at the S.C. Gliga Instrumente Muzicale S.A., were excited by an impact hammer type B&K 8204 and its response was captured by an accelerometer type B&K 4517-002 (Brüel & Kjær, Denmark, Nærum). The received signals were transmitted through a conditioning device to a Dynamic Signal Acquisition System (DAQ) - NI USB-9233 (National Instruments, USA, Austin), connected to a laptop. The obtained signals were in the form of the natural frequencies and the damping factor. The results highlighted the nonlinear and inhomogeneous behavior of violins with different structural patterns, proving that their vibrations depend on many factors.

1. Introduction

In the construction of violins, the front plates are made of spruce wood and the back plates of maple wood, these lignocellulosic materials having the physical, mechanical and acoustic characteristics of the resonant wood, being a wood without structural defects. Depending on the regularity of the annual rings, their width, the proportion of early wood-late wood, the resonant spruce plates belong to some structural models established in the literature [1 – 4]. For back plates, curly fiber maple is mainly used, but straight fiber maple wood is accepted for certain categories of pieces. The way the violin produces sounds is similar to stringed instruments (guitar), with the fact that the vibration of the string is produced by the bow rubbing the string. Between the bow and the excited string there is an angle that decomposes the force produced by the bow in the horizontal plane into a longitudinal component (parallel to the string) and a transverse one, on the string [5 – 7]. The string begins to vibrate both in the transverse plane (in the direction of force application by the bow) and in the longitudinal plane - along the string), the transverse vibrations being transmitted through the callus to the violin and further to the volume of air in the violin body, which behaves like a Helmholtz-type resonator, thus forming compression and expansion waves between the front, back and eclipses of the violin body, that produces an intense sound. Radiation intensification also explains the increase in sound intensity in the case of system resonance. The fundamental frequencies are related to the geometric characteristics of the source from which the wave propagates, and the set of harmonics (timbre) is dependent on the resonator that amplifies the initial sound, adding a series of its own frequencies [8, 9].

The study focused on determining the frequency spectrum, eigenfrequencies and damping factor of four different types of violins in terms of the structure of the resonant wood.

2. Materials and methods

2.1. Materials

The types of violins studied were produced at the S.C.Gliga Instrumente Muzicale S.A. musical instrument factory. According to their classification, there are four groups coded as follows: Maestro class violins - code A00C1 and A00C2; Professional class violins - code B00C1 and B00C2; Student class violins - code C00C1 and C00C2; School class violins - code D00C1 and D00C2. The difference between these is the structure of the wood used for the violin faces and the back of the violin. Thus, 8 violins were analyzed, two from each structural wood quality class (figure 1). In figure 1, it can seen the different structures of wood for the four classes of violins in terms of the width of the annual rings, the ratio of early wood and late wood, for spruce, and for maple, the degree of the curly fiber.

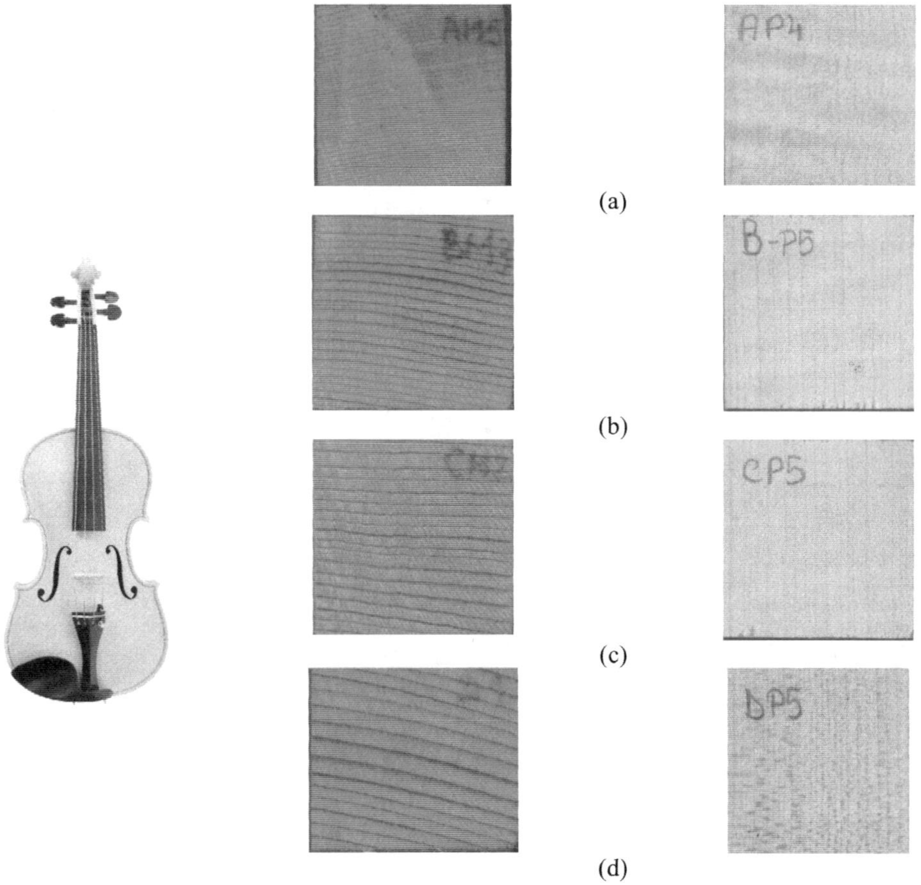

Figure 1. Structural characteristics of the violin classes analyzed: a) class A - Maestro; b) class B - Professional; c) class C - Student; d) Class D - School. Left: cross-sectional view of resonant spruce wood; Right - view in the longitudinal radial section of maple wood.

To identify the structural characteristics of the wood, the measurements were made on samples in the form of cubes, with a side size of 30 mm, 5 samples from each category (A, B, C, D) and from each species, so a total number of 40 specimens. The physical and geometric characteristics are presented in tables 1 and 2. The modal analysis was performed on violins with neck, having the structural models of the wood according to classes A, B, C, D. The values regarding the mass of the plates and the mass of the body of the violin without neck and with neck are centralized in table 3. The plates, as individual structures, have been analyzed in previous authors' studies [10 – 12]. It is observed that the mass of the tested samples increased with the complexity of the structure in different technological stages.

Table 1. Physical characteristics of the specimens used for the structural analysis of resonant spruce wood.

Specimen quality	Sample no.	Length l [mm]	Thickness h [mm]	Width b [mm]	Mass m [kg]	Apparent density [g / cm3]
A	1	30.75	30.02	30.24	12.316	0.441
A	2	30.51	30.35	30.40	12.292	0.437
A	3	30.50	30.31	30.21	12.235	0.438
A	4	30.60	30.24	30.38	12.287	0.437
A	5	29.84	30.13	30.48	11.949	0.436
B	1	30.29	30.22	30.02	10.592	0.385
B	2	30.27	30.25	30.09	10.453	0.379
B	3	30.73	30.33	30.19	10.608	0.377
B	4	30.57	30.43	30.02	10.787	0.386
B	5	30.43	30.36	30.13	10.528	0.378
C	1	30.28	30.33	30.49	12.697	0.453
C	2	30.18	30.34	30.60	12.654	0.452
C	3	30.13	30.28	30.10	12.09	0.440
C	4	30.48	30.70	30.82	12.499	0.433
C	5	30.15	30.33	30.54	12.592	0.451
D	1	30.16	30.13	30.53	10.92	0.394
D	2	30.3	30.31	30.29	11.141	0.400
D	3	30.35	30.25	30.39	10.962	0.393
D	4	31.08	30.25	30.35	11.151	0.391
D	5	30.66	30.48	30.05	11.102	0.395

Table 2. Physical characteristics of the specimens used for the structural analysis of the resonant maple wood.

Specimen quality	Sample no.	Length l [mm]	Thickness h [mm]	Width b [mm]	Mass m [kg]	Apparent density [g / cm3]
A	1	29.26	30.31	30.21	14.901	0.556
A	2	29.41	30.17	30.32	15.069	0.560
A	3	29.84	30.28	30.42	15.285	0.556
A	4	29.25	30.45	30.44	15.071	0.556
A	5	29.81	30.64	30.46	15.433	0.555
B	1	30.03	30.15	30.17	15.850	0.580
B	2	30.00	30.45	30.31	15.764	0.569
B	3	29.85	30.18	30.15	15.778	0.581
B	4	29.85	30.07	30.10	15.781	0.584
B	5	30.25	30.04	30.24	15.928	0.580
C	1	29.76	30.25	30.41	15.973	0.583
C	2	29.78	30.21	30.30	15.934	0.585

C	3	29.75	30.44	30.26	16.19	0.591
C	4	29.44	30.34	30.36	16.415	0.605
C	5	29.91	30.14	30.15	15.891	0.585
D	1	29.66	30.02	30.36	15.54	0.575
D	2	29.52	30.08	30.26	15.321	0.570
D	3	29.85	30.13	30.28	15.358	0.564
D	4	29.15	30.14	30.19	15.133	0.571
D	5	29.50	30.14	30.36	15.108	0.560

Table 3. Physical characteristics of violin bodies.

Plate code (F- face plate; S - back plate)	Mass [g]	Violin body code	Mass of neckless violin body [g]	Mass of violin body with neck [g]
A00F1	76	A00C1	243	328
A00S1	114			
A00F2	85	A00C2	251	323
A00S2	129			
B00F1	86		261	321
B00F2	85	B00C1		
B00S1	136		253	313
B00S2	131	B00C2		
C00F1	93		267	329
C00S1	90	C00C1		
C00F2	135		257	320
C00S2	128	C00C2		
D00F1	86		236	294
D00S1	86	D00C1		
D00F2	112		244	305
D00S2	116	D00C2		

2.2. Experimental methods

2.2.1. Identification of structural patterns of resonance wood. The annual rings and wavelength of the corrugated fiber were digitized using WinDENDRO Density 2007 image-analysis system (WinDENDRO 2007). For this purpose, the samples were scanned at a resolution of 2200 dpi. The image analysis allowed the automatic detection of the annual rings, but also the corrective intervention of the operator. The characteristics of the annual rings were measured in the radial direction, on the cross section, and the wavelength, in the tangential direction, on the radial section. The measurements were performed with an accuracy of 0.001 mm.

In the case of spruce, once the annual rings were identified, the early wood was digitally separated from the late wood. For the maple, the width of the annual ring and the wavelength of the crude fiber were measured. The following variables resulted from the measurements on the wood structure: the total width of the annual ring (TRW), the width of the early wood (EWW), the width of the late wood (LWW), the early wood proportion (EWP), the late wood proportion (LWP) and the curly maple wavelength (CWL) [10].

2.2.2. Modal analysis. The test method used in this research consisted of a dynamic analysis to determine the frequency response, applied to violin bodies belonging to the four classes of structural quality of wood, placed on an elastic structure, as can be seen in figure 2.

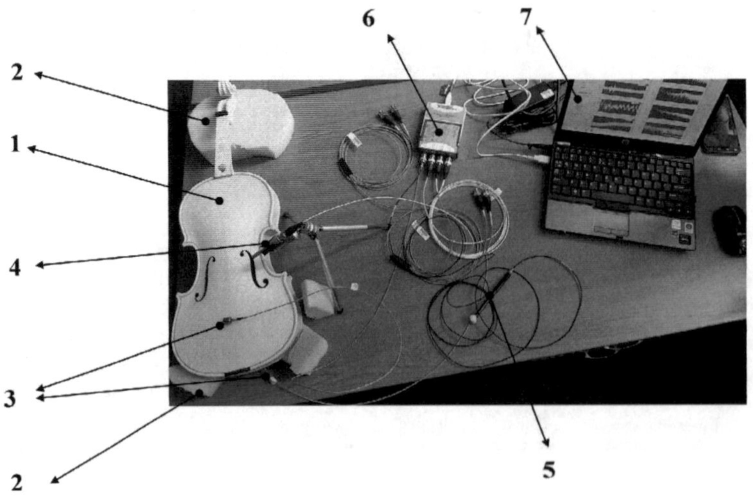

Figure 2. Experimental set-up: 1 – tested violin; 2 – elastic support; 3 – accelerometers; 4 – microphone; 5 – impact hammer; 6 – data acquisition system; 7 – laptop.

The excitation of the structures was done with the impact hammer for light structures type B&K 8204, the vibrations of the front and rear plates being captured with accelerometers type B&K 4517-002 fixed on the two plates. The air vibrations in the form of sound pressure were acquired through the microphone type PCB 130E20 ICP©, the generated signals were then transmitted via a signal conditioning device to the NI USB-9233 dynamic data acquisition board produced by National Instruments (Austin, USA), connected to a laptop. The signal was viewed using a special application developed in NI-LabVIEW ©, and the graphic data was processed using the MATLAB © program. The principle of the experiment is presented in figure 2.

3. Results and discussions

3.1. Identification of structural models of wood

The way of grouping the values of the wood structure indicators in relation to the predefined quality classes was statistically verified. Figure 3 shows the variations in the width of the annual rings for spruce wood (figure 3(a)) and maple wood (figure 3(b)). As the wavelength of the fiber ripples in the maple wood increases, so does the brightness of the wood (figure 3(c)). It is observed that, for spruce wood, the width of the annual rings is grouped in the four quality classes, compared to maple wood where the width of the annual rings is heterogeneous in relation to the quality class. Thus, it turned out that all the variables involved in the study are stratified according to the quality class to which the sample belongs (table 4). For spruce, the absolute values of the annual ring indicators ensure a better discrimination of the quality classes than the relative values. For maple, there is a tendency to increase the width of the rings and to decrease the wavelength with the improvement of the quality class. Quality class C stands out: for this class, the amplitude of values is higher. For maple, there is a tendency to increase the width of the rings and to decrease the wavelength with the improvement of the quality class. Quality class C stands out again: for this class, the amplitude of values is also higher. Despite the limited degree of variation, the density of wood can be stratified by structural quality class in both species. Spruce wood of quality classes A and C is heavier, by about 60 kg / m3, than wood of classes B and D. For maple, the wood is significantly lighter in the samples of class A quality.

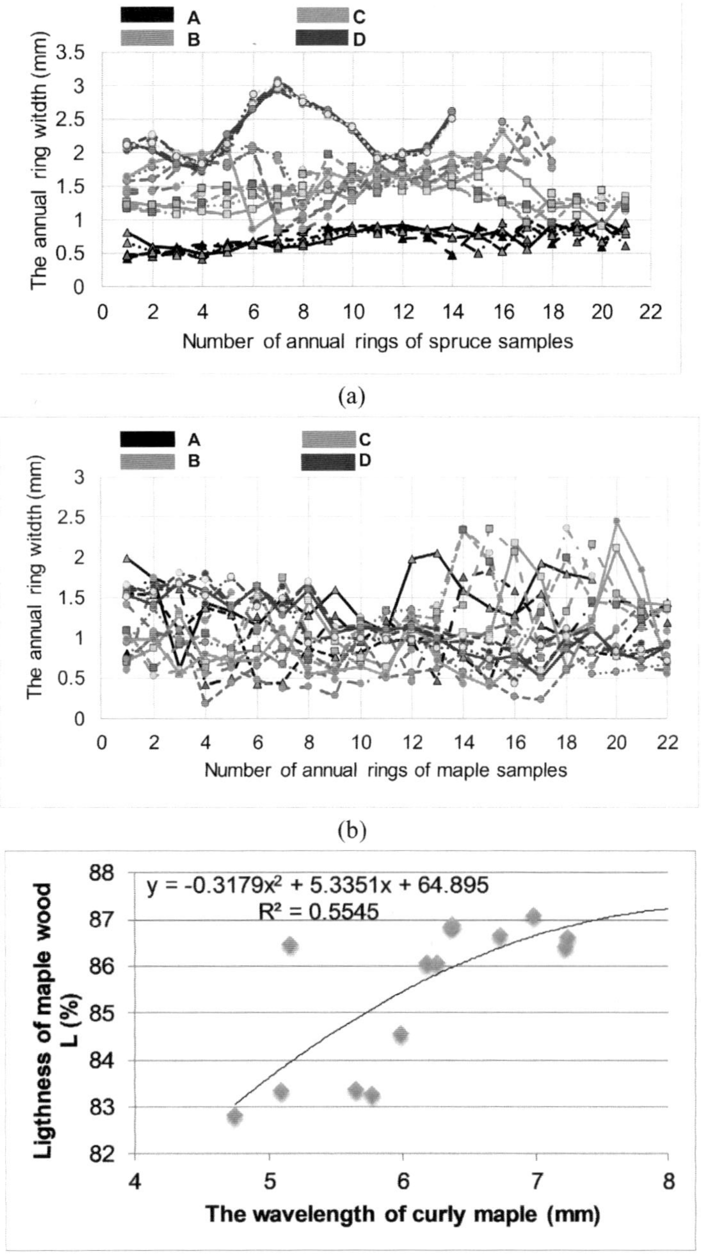

Figure 3. Variation of the width of the annual rings between the four classes of structural models of wood: a) spruce samples; b) maple samples; c) wavelength of the curly fiber.

Table 4. Statistical verification of the possibility of stratifying the characteristics of the examined wood according to the proposed quality classes [10, 12].

Species of the sample	Wood variable	Statistical significance of differences between quality classes The test Kruskall-Wallis/ANOVA		Stratification of values by quality class...			
				A	B	C	D
		H, F	*p*	Quartile interval			
Spruce	Annual ring width [mm]	396.74	< 0.001	0.59…0.85	1.22...1.51	1.50...1.89	1.98...2.60
	Early wood width [mm]	398.76	< 0.001	0.44…0.66	0.97...1.17	1.18...1.48	1.52....1.99
	Late wood width [mm]	312.8	< 0.001	0.14…0.21	0.23...0.35	0.29....0.45	0.42...0.62
	Proportion of early wood [%]	32.59	< 0.001	70.94…79.51	75.76...80.56	76.24...81.74	74.33....79.22
	Proportion of late wood [%]	32.59	< 0.001	20.50…29.07	19.44...24.24	18.27...23.76	20.78....25.67
	Apparent density [g/cm^3]	16.55	<0.001	0.437...0.438	0.378...0.385	0.440...0.452	0.393...0.395
Maple	Annual ring width [mm]	78.36	<0.001	0.92…1.56	0.82…1.41	0.57…1.08	0.71…1.28
	Curly fiber wavelength [mm]	6.23	0.004	4.26…6.96	5.36…7.16	6.28...9.43	-

3.2. Time frequency analysis

In the modal analysis, time and frequency signals were extracted, resulting in the frequency spectrum for spruce plates, maple wood plates and air inside the violin body and the damping factor for each analyzed component, as it can be observed in figures 4 – 7.

It is found that the first fundamental frequency has the same value for all three analysed components: spruce board, back plate and air, regardless of the structural model of the wood and the quality class. In the case of spruce plate, the value is not the same for the four classes A, B, C, D. Thus, for class A, the fundamental frequency recorded is 244.1 Hz, while in the case of the other classes, the value is in around 250 Hz. The maple wood back plate shows variations in the fundamental frequency in relation to the quality class of the wood, with an inhomogeneous response. For example, Class B back plates recorded values of 256.3 Hz (B00C1) and 244.1 Hz (B00C2), respectively. A tendency to increase the value of the fundamental frequency is registered for the class C violins where the back plates have their first own frequency at values of 250 Hz (C00C1) and 256 Hz (C00C2). The 256 Hz value of the C00C2 violin body is also recorded for the spruce top plate. As a result of the interaction of the fluid inside the violin body with the front and back plates, it resulted that the air inside the structure vibrates in the first stage with the same frequency as the spruce plate, regardless of the violin class.

The second eigenfrequency has an interval of one third, ie 1.57 higher than the fundamental frequency, in the case of all quality classes. The dominant frequency differs from one species to another: each plate amplifying differently the vibrations produced in the structure of the violin body. The frequency spectrum differs from one species to another (spruce versus maple), but some values from both the front plate and the back plate are found in the vibrations of the air inside the body. What is certain is that the dominant frequency of the fluid coincides with its fundamental frequency, the rest of the frequencies having damped values.

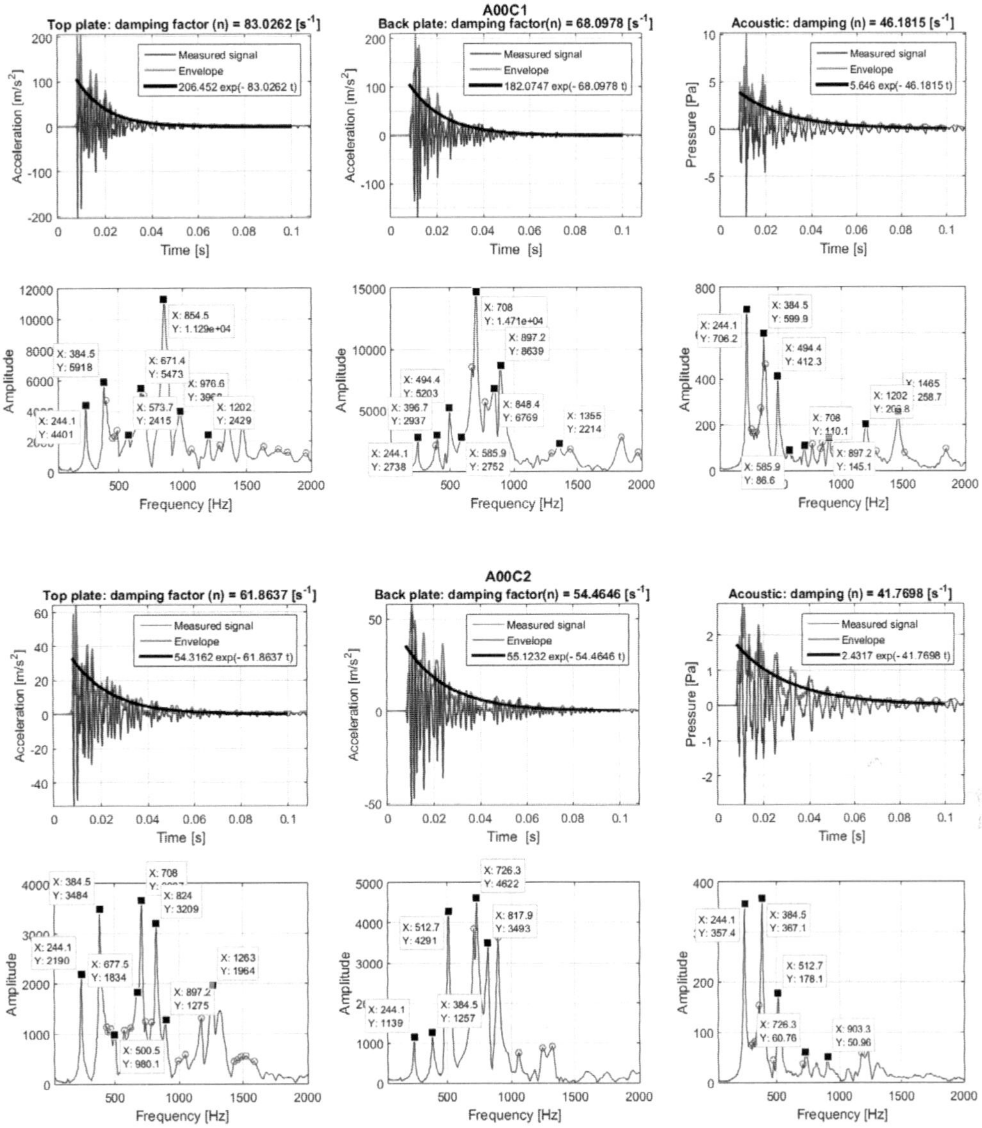

Figure 4. Time and frequency analysis for class A violin structures.

IOP Conf. Series: Materials Science and Engineering 1182 (2021) 012048 doi:10.1088/1757-899X/1182/1/012048

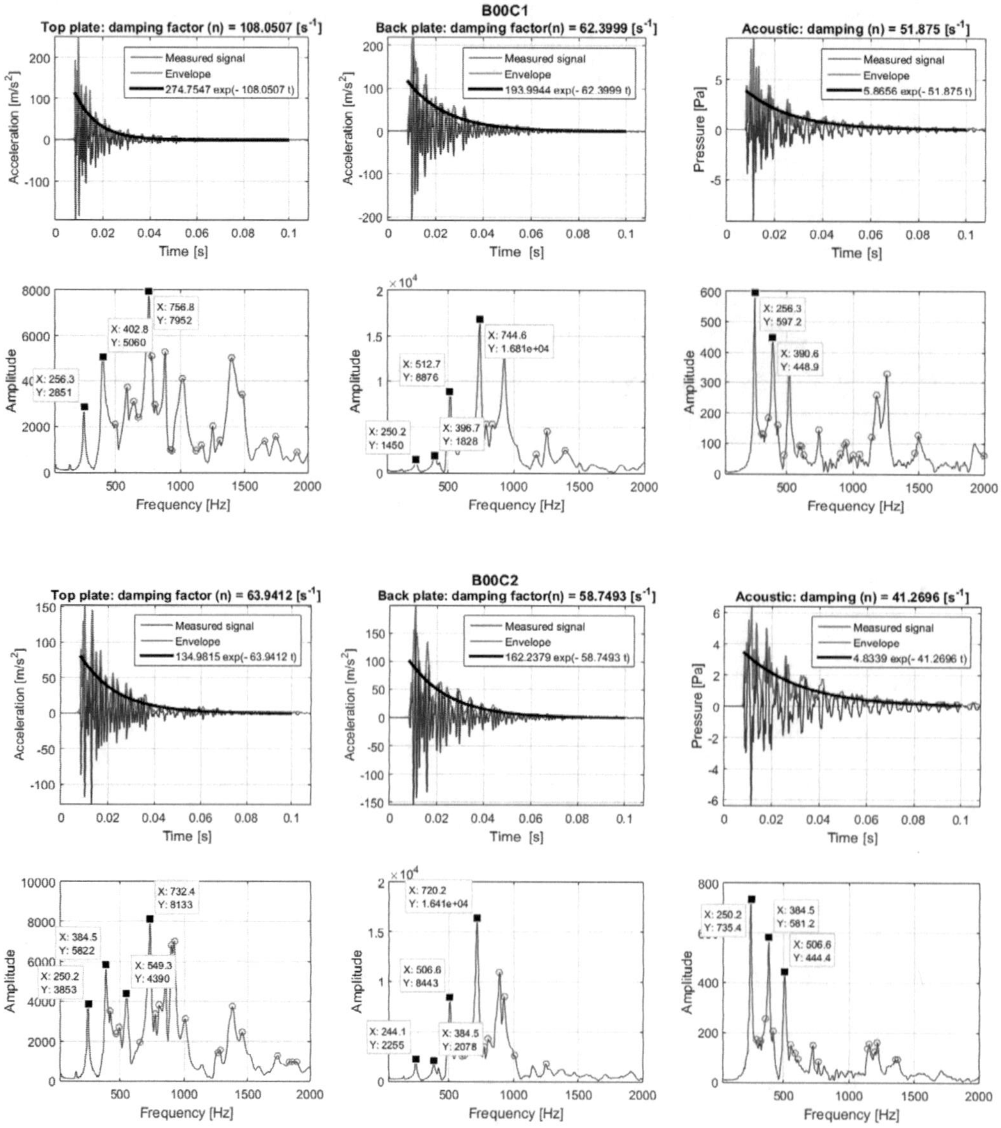

Figure 5. Time and frequency analysis for class B violin structures.

The values of the damping factor show a decreasing trend from the front plate, the back plate and the lowest value is recorded for the air inside the body. From the point of view of the quality class, with the lowest values they are registered for class A (83/68/46 - A00C1; 61/54/41 - A00C2) (figure 4), and the maximum values are registered in the case of the bodies from category C (112 / 57/45 - C00C1; 109/67/53 - C00C2) (figure 6).

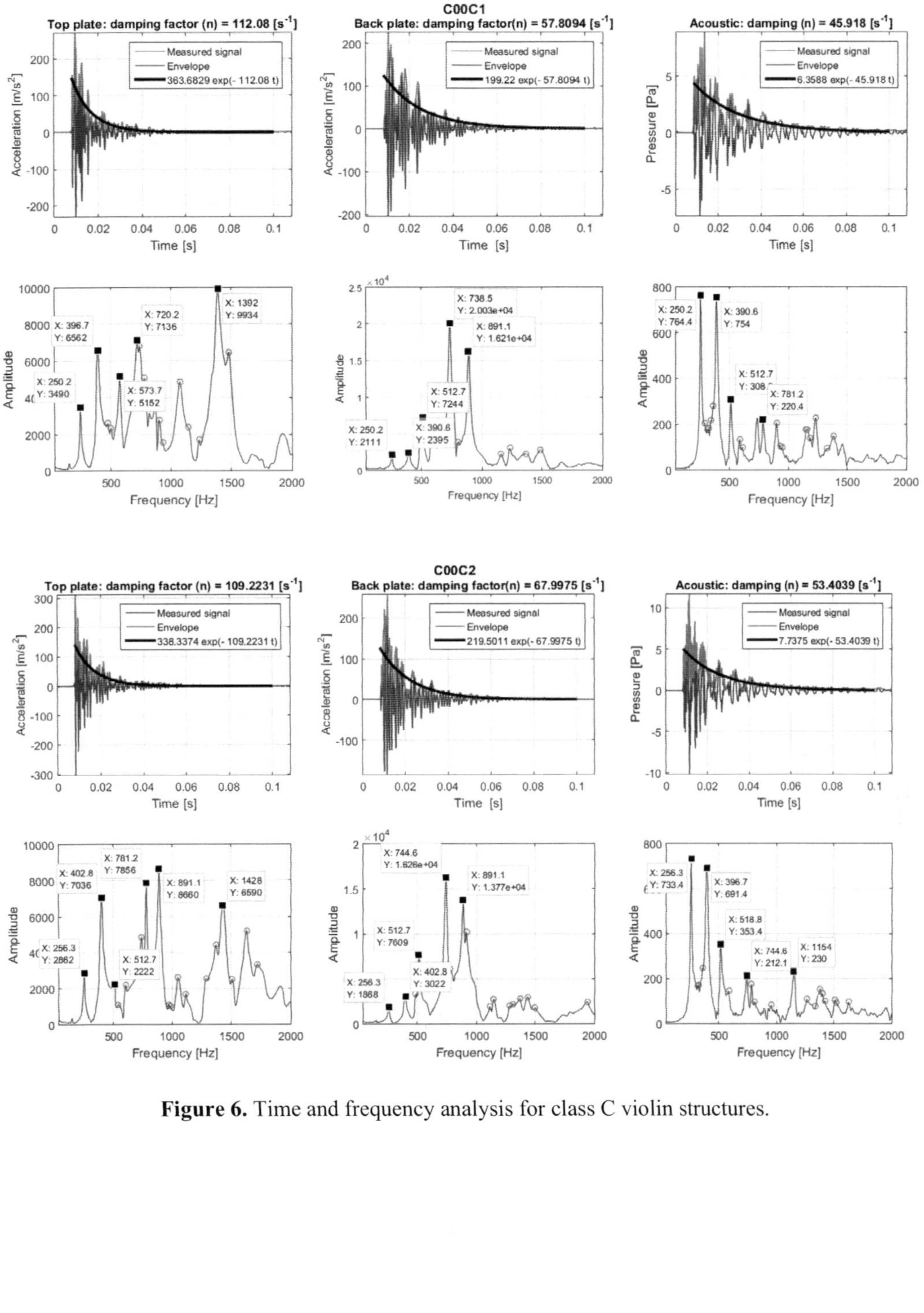

Figure 6. Time and frequency analysis for class C violin structures.

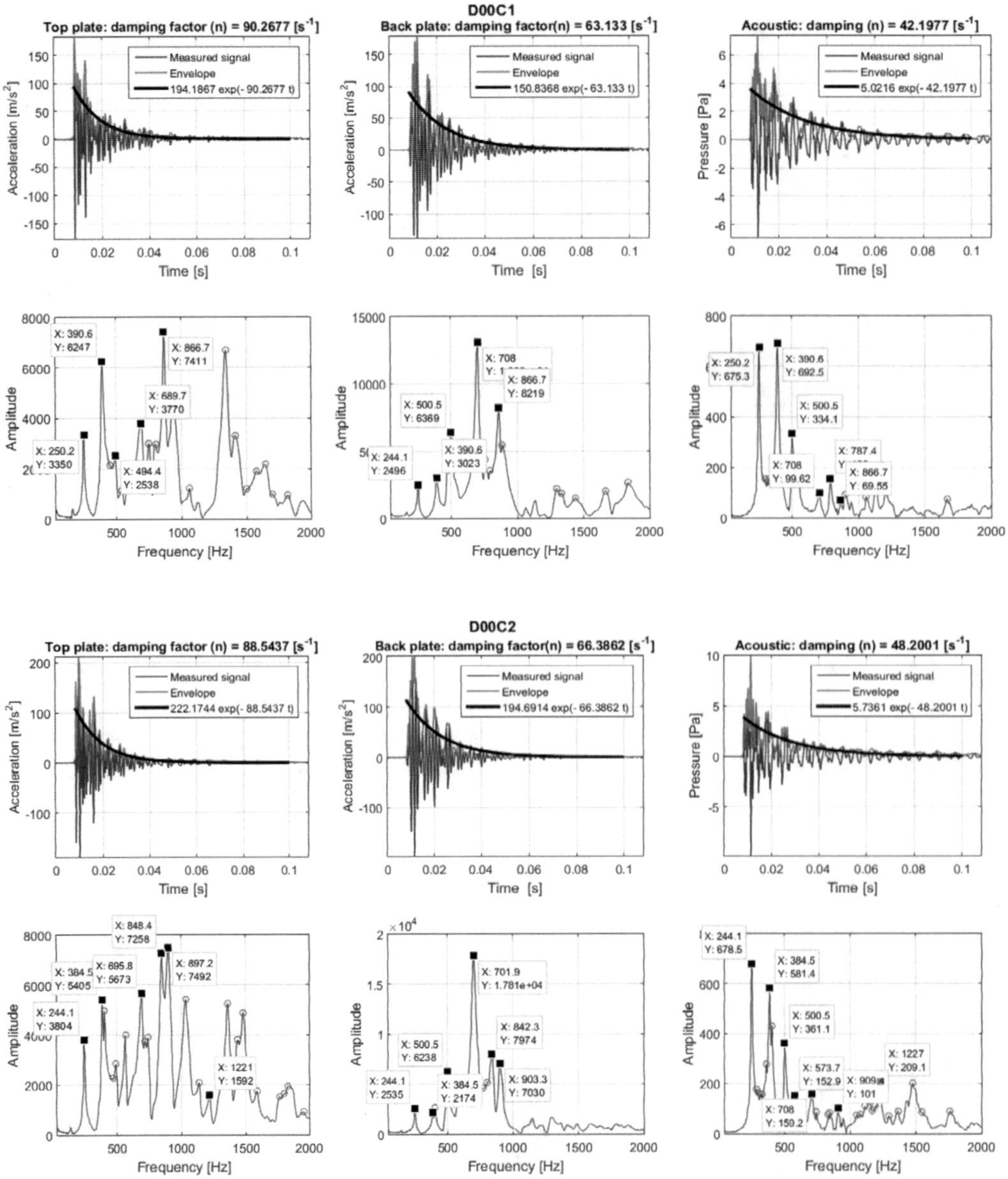

Figure 7. Time and frequency analysis for class D violin structures.

4. Conclusions

The work aimed to determine the natural frequencies of the boards in the structure of four types of violins, different from each other by the structure of wood. It was found that:

- wood, even if it belongs to the same species, can be classified into structural models according to the anatomical elements, the regularity and width of the annual rings (in the case of spruce wood) or according to the shape of the fibres (curly or straight) in the case of maple wood.

- In general, the spectrum is richer in spruce plates than in maple plate or air. This is due to the damping that takes place in the wood, some frequencies of the top plates with small amplitudes are no longer transmitted to the back plates and the air.
- in the spectrum of air frequencies can be identified frequencies of the face and back of the violin.
- considering the experimental method used, the acoustic quality of the violin body cannot be appreciated at this stage.

5. References

[1] Setragno F, Zanoni M, Sarti A and Antonacci F 2017 Feature-based Characterization of Violin Timbre *The 25th European Signal Processing Conference (EUSIPCO)* 1903-1907.

[2] Trapasso L 2013 *Feature-based Analysis of the Violin Tone Quality*. Master Degree Thesis Politecnico Di Milano.

[3] Harris N A qualitative analysis of the relationship between longitudinal string vibration, arching shape and violin sound, *On Arching Shape and Violin Tone* http://citeseerx.ist.psu.edu/viewdoc/download?doi=10.1.1.494.5400&rep=rep1&type=pdf.

[4] Bissinger G 2004 Contemporary generalized normal modes of violin acoustics. *Acta Acustica united with Acustica* **90** 590 – 599.

[5] Bissinger G 2008 Structural acoustics of good and bad violins. *J Acoust. Soc. Am.* **124** 1764 – 1773.

[6] Bucur V 2016 Handbook of materials for string musical instruments. *Springer*, Chapter **3** - Mechanical characterization of materials for string instruments: 93 – 132 and Chapter **8** - Ageing of musical instruments – about the old and new instruments 325 – 372.

[7] Carlsson P and Tinnsten M 2003 A distributed computing system used for concurrent optimization methods on a violin top *Structural and Multidisciplinary Optimization* **25**(5) 453 – 458.

[8] Gough C 2015 Violin plate modes *J. Acoust. Soc. Am.* **137**(1) 139 – 153.

[9] Woodhouse J 2014 The acoustics of the violin: a review *Reports on Progress in Physics* **77**(11) 115901. http://iopscience.iop.org/article/10.1088/0034-4885/77/11/115901/pdf.

[10] Dinulică F, Albu C T, Vasilescu, M M and Stanciu, M D 2019 Bark features for identifying resonance spruce standing timber, *Forests* **10**(9) Article number 799 DOI: 10.3390/f10090799.

[11] Stanciu M D, Cosereanu C, Dinulica F and Bucur V 2020 Effect of wood species on vibration modes of violins plates *Eur. J. Wood Prod.* **78**, 785-799.

[12] Gliga V Gh, Stanciu M D, Nastac S M and Campean M 2020 Modal Analysis of Violin Bodies with Back Plates Made of Different Wood Species *BioResources* **15**(4) 7687 – 7713.

Acknowledgement

This paper was supported by a grant of the Ministry of Research, Innovation and Digitization, CNCS/CCCDI – UEFISCDI, project number. 568PED/2020 MINOVIS, within PNCDI III.

IOP Conf. Series: Materials Science and Engineering 1182 (2021) 012049 doi:10.1088/1757-899X/1182/1/012049

Expanded glass spheres applications for low thermal transfer

O Mocanita[1], D Chicet[2], B Istrate[1], L Raileanu[3] and C Munteanu[1]
[1]Gheorghe Asachi Technical University of Iasi-Romania, Department of Mechanical Engineering, Blvd. Mangeron, No. 61, 700050, Iasi, Romania
[2]Gheorghe Asachi Technical University of Iasi-Romania, Department of Materials Science and Engineering, Blvd. Mangeron, No. 41, 700050, Iasi, Romania
[3]Military School of Air Force"Traian Vuia" Boboc, Buzau, RomaniaProduction

E-mail: dchicet@tuiasi.ro

Abstract. This study addresses the possibility of using expanded glass to obtain elements with low thermal transfer and stable from a geometric point of view when exposed to high temperatures, which may later be part of a fire-resistant multi-layer element. For this purpose, three types of recipes were designed and produced, which are composed of expanded foam glass spheres bounded with polymeric resin. Subsequently, 3 sets of each sample were subjected to baking: T1 (heating to 700°C, maintaining for 3 hours, oven cooling); T2 (heating to 600 ° C, maintaining for 3 hours, oven cooling); T3 (heating to 600° C, oven cooling). The 4 sets of samples (1 set in initial state and 3 sets of heat treated samples) were exposed for 1 hour at direct flame. It was found that the samples maintained their structural integrity and all three types of material ensure a temperature difference between the side exposed to temperature and the unexposed side higher than 200°C (considered satisfactory according to SREN 1363-1 / 2001).

1. Introduction

Expanded glass foam is a new developed material, with exceptional properties: low specific weight, high strength at compression, dimensional stability, low coefficient of thermal conductivity (λ = 0.07 W/mK), 100% mineral, ecological (made of recycled glass) and safe for the user's health [1]. Thus, the material finds more and more applications on various domains: on the construction materials market (lightweight concrete, plaster and dry mortar, lightweight construction panels) [2 - 4], for automotive parts (cellular composites [5] and syntactic foams [6] for shock absorber structures), in agriculture (soil conditioner for plant substrates [7]) and in other fields (sanitary products, desalinisation compresses for historic masonry rehabilitation etc).

Taking these properties into account, we developed a larger study in which we used expanded glass spheres to create a layer with thermal insulation properties that can be incorporated into an open fire multilayer compound. Thus, starting from a previous experimental study [8], we realized a set of samples that we heat treated for geometric stabilization according to three different graphs, later testing their fire resistance and thermal insulation capacity when exposed to open flame.

2. Materials and methods

The samples tested in terms of fire resistance thermal insulation capacity in this paper were produced using the following materials:

1. quartz sand;

2. commercially available bonding agent: silicon-oxygen polymer resin containing ionic sodium (Na +) components;

3. four types of expanded foam glass spheres produced by Dennert Poraver GmbH, with different grain size (0.04 - 0.125 mm, 0.25 - 0.5 mm, 0.5 - 1 mm, 1 - 2 mm).

For the presented study, three types of recipes were designed and produced, which are composed of expanded foam glass spheres:

- recipe A (quartz sand + expanded glass spheres with a diameter between 0.25-0.5 mm + polymer resin binder) – sample NA15 (figure 1(a)),
- recipe B (quartz sand + expanded glass spheres with a diameter between 0.5-1 mm + polymeric resin binder) – sample NB15 (figure 1(b)),
- recipe C (mixture of spheres with different granulations: 0.04 - 0.125 mm, 0.25 - 0.5 mm, 0.5 - 1 mm, 1 - 2 mm, bound with polymeric resin) – sample S15 (figure 1(c)).

(a) (b) (c)

Figure 1. Samples aspect, after drying in air for 48 hours:
a) sample NA15, b) sample NB15, c) sample S15.

The tests were performed at high temperatures, after a non-standardized test method [9, 10] using a specially designed test stand, presented in figure 2. The test stand is actually an enclosure composed of an external metal frame on which refractory bricks are mounted. Inside, at 2/3 of the side through which the burner is mounted, a wall of refractory bricks is made, provided with a 50x50mm hole in which the samples are mounted. As presented in figures 1 and 2, the test stand use parallelepiped specimens (50x50x15mm), which were made in a special designed mould and dried in a normal atmosphere for 48 hours.

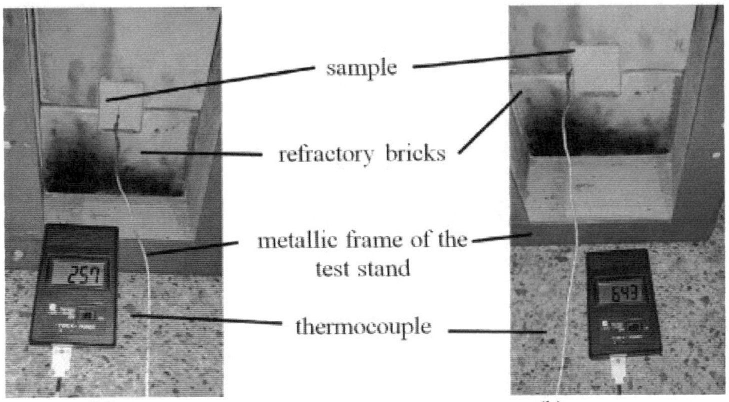

(a) (b)

(c)

Figure 2. Aspects of the test stand: a) measuring procedure for the sample side not exposed to direct flame; b) measuring procedure for the sample side exposed to direct flame; c) superior view of the test stand.

Subsequently, a set of each sample (one specimen from A, B and C recipe) was subjected to thermal treatments (baking in a laboratory electric furnace) in order to geometrically stabilize the samples. The heat treatment methodology was:

- T1 - heating for 4 hours to 700°C, maintaining for 3 hours and oven cooling;
- T2 - heating for 3 hours up to 600°C, maintaining for 3 hours and cooling with the oven,
- T3 - heating for 3 hours up to 600°C and cooling with the oven.

During the experiments, the temperatures of the samples were measured every 10 minutes (for 1 hour) using a K type thermocouple (-50 ÷ 1300°C) and a TM 902C thermometer, positioned in direct contact with the surface not exposed to flame (see figure 2 (a)). The temperature of the side exposed directly to the flame was measured every 20 minutes, using a small hole nearby the sample, as presented in figure 2(b). Using a gas regulator (see figure 2 (c)), the temperature of the side exposed to direct flame was stabilised between 620 - 650°C, during all tests.

3. Results and discussions

The aspect of the samples after the baking procedure is presented in figures 3 - 5, for each treatment procedure applied.

NA 15 NB 15 S15

Figure 3. Samples aspect, after the T1 heat treatment.

In an analysis of the three sets of samples, it was observed that the set subjected to T1 treatment underwent structural changes: the S15 sample became more compact and with slight changes in geometry. In the case of samples NA15 and NB15, which have quartz sand in the composition, a slight change in porosity was observed, which increased.

Figure 4. Samples aspect, after the T2 heat treatment.

Figure 5. Samples aspect, after the T3 heat treatment.

These phenomena can be explained in terms of the temperature at which the sample was heated during the treatment T1 (700°C), which seems to be the threshold at which the expanded glass spheres reach the semi-molten state and change their dimensions (the effect of expansion is lost and suffers volume loss).

In the case of samples subjected to treatments according to T2 and T3 diagrams, the samples did not undergo visible changes, being evaluated as unmodified. Following these observations, it was decided that all samples should be subjected to open flame tests, taking into account that the maximum temperature should be 680C, but not lower than 600C.

Because the main purpose of this study is to evaluate the thermal insulation capacity of the analysed samples, after their baking, the interpretation of the results was made taking into account the requirements of SR EN 1363-1 "Fire resistance tests. General Conditions". According to this standard, isolation represents the time in minutes in which the specimen continues to maintain its distinct functions during the test without developing on the unexposed side temperatures which: a) increase the mean temperature above the initial mean temperature by more than 140°C; or b) grow at any point above the initial mean temperature above 180°C [11].

On the samples exposed to direct flame, it was observed that, in the case of all samples consisting exclusively of expanded glass spheres (S15), regardless of the type of baking (T1, T2 or T3), on the contact with the open flame a superficial glass layer was formed, following a vitrification process (see figure 6(a)). The thickness of this layer is between 2-3 mm, but it consists of many gaps and very thin walls of vitrified material. The formation of this layer was observed in the first 30 minutes of the tests, when the heating with the highest speed of the samples on the non-exposed face was also recorded, as seen in figure 7. After this interval, no other change of the side exposed to the open flame was observed, as the heating rate of the samples decreases. We can thus consider that, after the formation of this vitrified foamed layer, the sample has a stable behavior, without being registered it's burning with smoke emissions or loss of integrity (melting of the sample with material losses). On this set of samples (S15) is also observed the formation of a densified layer in the depth of the sample (located below the vitrified one), being visible some large spheres, but also porous effects of melting expanded glass granules, followed by their reduction in volume.

(a) (b)

Figure 6. Cross-section aspects of the samples: a) S15 T3 (10x); b) NB15 T2 (15x).

In the case of samples with sand in the composition, regardless of the size of the spheres (NA15, NB15) or the temperature diagram, a behaviour similar to that of the S15 samples was observed: the formation of a superficial layer on the side exposed to open flame, much thinner (1-2mm) but with higher hardness than the rest of the sample. This layer consists of a mixture of sand and semi-molten glass granules, no visible expanded glass sphere being visible in its thickness. However, more and more spheres with different volumes begin to be visible under this layer, as they are at a greater distance from the face exposed to the open flame.

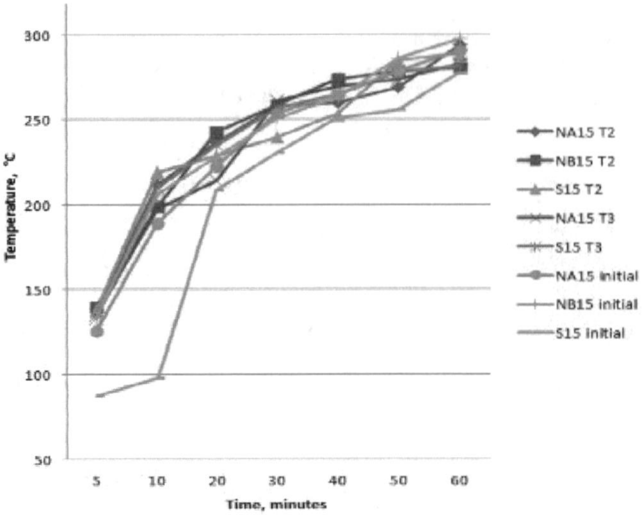

Figure 7. Temperature variation on the sides not exposed to direct flame.

The explanation for this phenomenon can be based on the fact that, as the temperature increased, the expansion effect was eliminated, and the glass that initially formed the granules remained spherical, but with smaller dimensions after compaction.

Regarding the insulating effect, the temperatures recorded for the 3 sets of samples are shown compared in figure 7 (for all the samples were recorded temperature differences between 350 - 400°C between the two sides). It is observed that all samples showed the same tendency to heat at high speed in the first 30 minutes, followed by a decrease in this speed, sign that the insulation effect becomes stable with the formation of the vitrified layer.

In both cases it is observed that the resulting structures are porous, an aspect that completes the insulating effect by forming an inhomogeneous structure (such as open cell foams) in which air plays an important role along with glass and sand, these two being those that form a rigid structure, non-combustible, resistant to high temperatures.

4. Conclusions

Following these tests, it was found that the samples maintained their structural integrity (in all three cases of heat treatment) and had a satisfactory behaviour because they did not develop open flame and did not register major structural and geometric changes.

It was observed that all three types of material ensure a temperature difference between the side exposed to temperature and the unexposed side higher than 200°C, this variation being considered satisfactory according to SREN 1363-1 / 2001.

5. References

[1] https://www.poraver.com/us/poraver/ (accessed on 22.09.2020).
[2] R. Schreiner et al. 2015 *International Journal of Thermal Sciences* **95** 99-105.
[3] Yu S R, van Onna D V, Spiesz P, Yu Q L and Brouwers H J H 2016 *Journal of Cleaner Production* **112** 690-701.
[4] Adhikary S K, Rudžionis Ž, Vaičiukynienė D 2020 *Journal of Building Engineering* **31** 101399.
[5] Gourav S 2015 *J. Civil Eng. Res.* **5**(3) 53-8.
[6] Su M, Wang H, Hao H, Fiedler T 2020 *Journal of Alloys and Compounds* **821** 153233.
[7] Al-Sahlani K, Broxtermann S, Lell D and Fiedler T 2018 *Materials Science & Engineering A* **728** 80–87.
[8] Mocanita et al 2020 *IOP Conf. Ser.: Mater. Sci. Eng.* **877** 012038.
[9] Mocanita O, Chicet D L, Munteanu C, Istrate B and Oprişan B 2018 *IOP Conf. Series: Materials Science and Engineering* **444** 032011.
[10] Mocanita O, Chicet D, Avram P, Micu C and Munteanu, C 2018 *UPB Scientific Bulletin, Series B: Chemistry and Materials Science* **81**(1) 236.
[11] SR EN 1363-1 / 2001. *Fire resistance test. Part 1: General requirements.*

IOP Conf. Series: Materials Science and Engineering 1182 (2021) 012050 doi:10.1088/1757-899X/1182/1/012050

Parametric modelling and finite elements study for the boom of a truck crane

G D Muscă (Anghelache) and S M Năstac
"Dunarea de Jos" University of Galati, Engineering and Agronomy Faculty of Braila-Research Centre for Mechanics of the Machines and Technological Equipment, Calea Calarasilor Street 29, 810017 Braila, Romania

E-mail: diana.anghelache@ugal.ro

Abstract. Parametric modelling and finite elements study for the crane boom was done in Catia V5R20 software. 3D modelling was done considering the load diagram. This is the basic feature for this study because it gives us information between the main elements (lifting capacity, outreach, boom length lifting height) which define crane operation possibilities. The truck crane is installed on the standard undercarriage of a truck or specially built undercarriages, the engine that drives the undercarriage serving either just for equipment travel or also for driving the crane mechanism. When the engine on undercarriage is only for travel, the truck crane is provided with an additional auxiliary engine, exclusively intended for crane mechanisms operation. The studied boom belongs to Liebherr crane. For parametric modelling measurements on an existing model were needed. Finite elements study was done on the basis of virtual model using a specialized soft. We did a static study by which we determined the static parameters of the structure (stresses, deformations, safety coefficient), for all own rates within interest range. The next aim, using the parameters, is to build the static model of reaction. Any static deformation of a structure can be represented as a weighted sum of structure's own modes, each knot can be represented as a system with only one degree of freedom. This kind of study allows us to find which is the maximum deformation, the stresses and value of safety coefficient borne by a technical system or machinery under movement or static condition.

1. Introduction

The truck crane is a crane with boom, installed on the standard undercarriage of a truck, or on a specially built undercarriage, the motor driving the undercarriage serving either only for equipment travel or also for driving the crane mechanisms.

In case the motor on the undercarriage only serves for travel the truck crane is provided with an auxiliary, additional motor exclusively intended for operation of crane mechanisms.

Although there is a great variety of truck cranes, the constructive parameters are limited by the road traffic rules (such as overall size limits for travel on public roads, load on axles and other norms regulating vehicles travel on public roads, both nationally and internationally)[1].

Namely, increase of lifting capacity involves increase of the number of axles, which implicitly determines suitable increase of overall sizes, causing a certain difficulty of the transport from one destination to another.

Truck cranes on wheels with tyres are used in a greater and greater extent due to some undisputed advantages such as:

- great mobility respectively reduced time for changing work position from one object to another;
- most of the time additional personnel is not needed except for the crane operator;
- being equipped with driving systems independent from fixed power supply sources they are able to work under various situations like unarranged or with few arrangements fields;
- they have sufficient stability so that there is the possibility for crane to travel with hooked load;
- increased efficiency in operation is got, with suitable decrease of production cost.

The main feature of mobile cranes used within their activity is the load diagram.

This is the basic feature of the cranes, as it is expressed between main elements that define operation possibilities of the crane namely:

- lifting capacity in kN;
- outreach in m;
- boom length in m and lifting height in m.

The components of a truck crane are:

- undercarriage with fixed platform and pressing-on earth devices,
- slewing platform with cab and superstructure equipment,
- the superstructure equipment (crane boom) driving system control installations and devices,
- safety devices,
- signaling devices and counterweight.

Characteristic for the truck cranes is that the travel is done through the truck engine.

The travel speeds prescribed are within 40-70 km/h.

Such type of cranes are lifting machines with crawler travel system, the fundamental difference between them and other types of cranes being their possibility to travel also when loaded.

2. Parametric modelling and finite elements study for the boom of a truck crane

Parametric modeling and finite elements stud for the boom of a truck crane were done for extreme lifting position under maximum load.

A leaflet from LIEBHERR was used as reference point [2].

The main feature of mobile cranes used within their activity is the load diagram.

This is the basic feature of the cranes, as it is expressed between main elements that define operation possibilities of the crane namely:

- lifting capacity in kN;
- outreach in m; boom length in m and lifting height in m.

Diagram reading is done knowing the value of load that should be handled, thereafter, the suitable maximum outreach being determined.

To establish the values of the two parameters a horizontal straight line indicating load value is drawn from the ordinate axis until crossing point with load curve.

From this point a perpendicular line is drawn on to the abscise section with the load curve.

From this point a perpendicular line is drawn down to the abscise and the value of maximum outreach for load travel is read.

Similarly, on the reverse route we can determine the value of the load that can be handled for a given outreach; knowing the height whereto the load should be lifted, the boom length is established depending on the needed height for handling.

Lifting height is the distance measured from ground level up to the lower face of the hook. There is a load curve for each boom length.

Therefor reading of each load diagram is done as described above (figure 1).

It is to be noticed that as the booms length increase, for the same outreach, the load decreases.

The explanation is in reduction of stability degree of the crane by increasing the overturning moment.

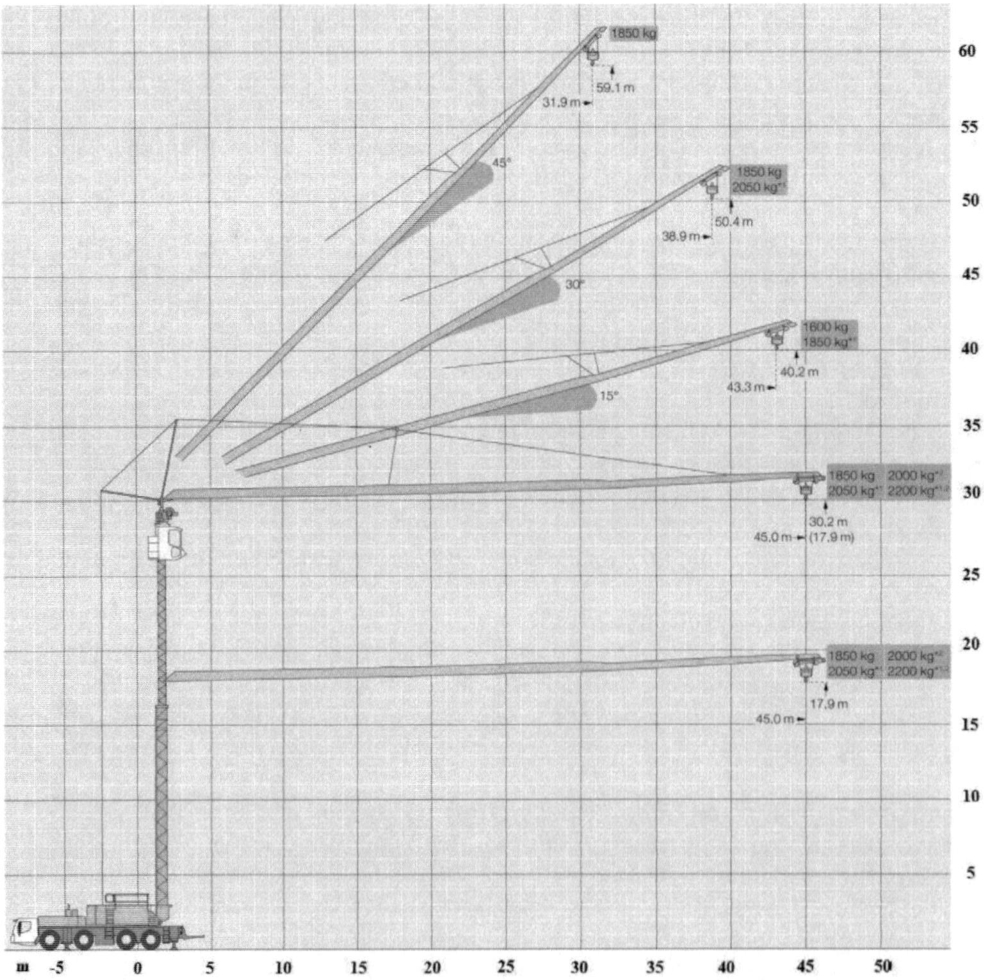

Figure 1. Load diagram of the LIEBHERR MK 88 truck crane [2].

The boom subject to study looks like in figure 2. This was done with Solid Edge (2D) soft.

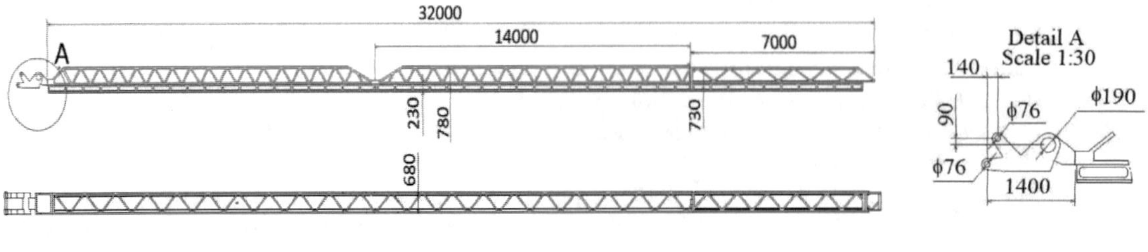

Figure 2. 2D model of the truck crane boom.

3. Static study of the crane boom

The studied crane boom belongs to Liebherr truck crane presented in the figure below. To enable its Fem study, measurements on an existing model and realizing the 3D virtual model of such were needed.

To achieve tridimensional solid model the Solid Edge ST6 programs packet for computer aided engineering was used, and after processing the physical model, the result is the model in below figure (figure 3)[3].

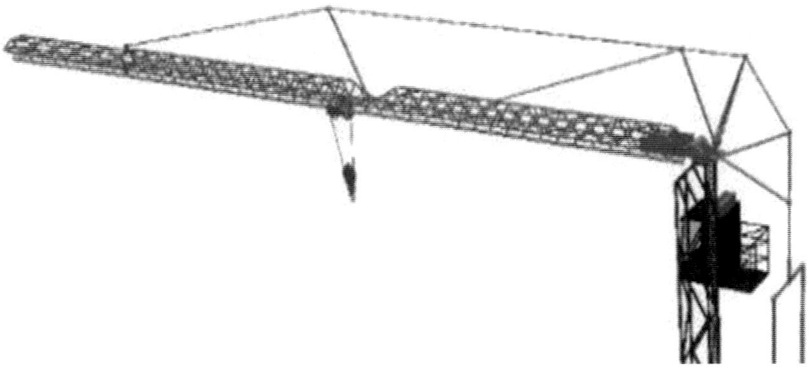

Figure 3. Crane boom assembly.

Static study is the method by which the static parameters of a structure (stresses, deformations, safety factor) are determined for all own modes within the interest area [4].

The next aim, by using the parameters is to build the static model of reply. In the end, it should be noted that:
- any static deformation of a structure can be represented as a weighted sum of the structure's own mode shapes;
- each knot can be represented as a system with only one degree of freedom.

In the tables below, the following are shown: type of performed study (table 1), type of used mesh (table 2), used materials (table 3) and values of forces applied on boom (table 4).

The finite element study applied to structures is a multidisciplinary method, based upon knowledge in three fields: structures mechanics, including elasticity theory, materials strength, plasticity theory, structures dynamics, etc., mathematical analysis, including approximated methods, solving of linear algebra equation systems, own values problems, etc., and computers applied science, that deals with development and implementation of computer big programs [5].

Tis finite elements study method is used to solve some big size analytic problems.

Table 1. Type of study performed.

Design Objective	Single Point
Study Type	Static Analysis
Last Modification Date	2/2/2020 - 6:08 AM
Detect and Eliminate Rigid Body Modes	No

Table 2. Type of used mesh.

Avg. Element Size (fraction of model diameter)	0.1
Min. Element Size (fraction of avg. size)	0.2
Grading Factor	1.5
Max. Turn Angle	60 deg
Create Curved Mesh Elements	Yes

Table 3. Used materials.

Name	1C60	
General	Mass Density	4.51 g/cm^3
	Yield Strength	275.6 MPa
	Ultimate Tensile Strength	344.5 MPa
Stress	Young's Modulus	102.81 GPa
	Poisson's Ratio	0.361 ul
	Shear Modulus	37.77 GPa
Part Name(s)	Part1	

Table 4. Values of forces applied on boom.

Load Type	Force
Magnitude	20000.000 N
Vector X	0.000 N
Vector Y	-20000.000 N
Vector Z	0.0

4. Case study and obtained results

The boom obtained in figure 2 was subject to study. Firstly we applied restrains (figure 4).

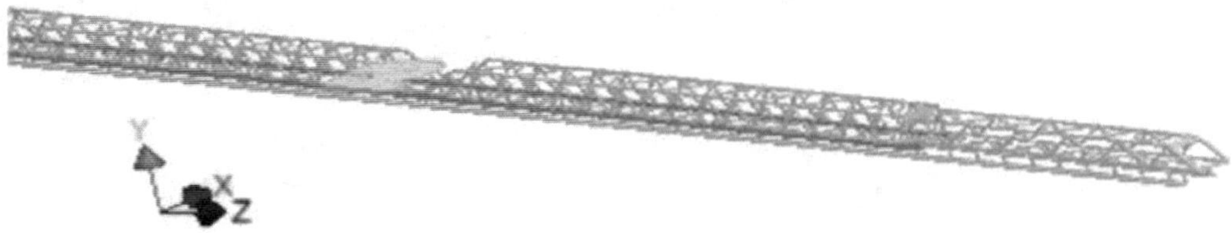

Figure 4. Restrains establishment.

We obtained the results from the figures 5 to 7:

a) Displacement's representation (figure 5):

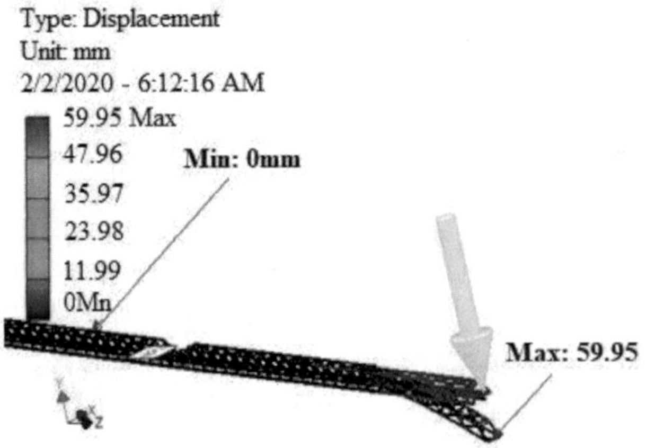

Figure 5. Displacement's representation.

b) Safety factor (figure 6):

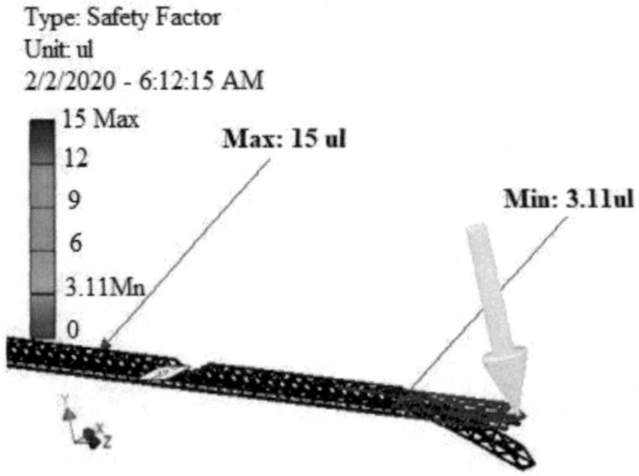

Figure 6. Safety factor.

c) Von Misses stresses representation (figure 7):

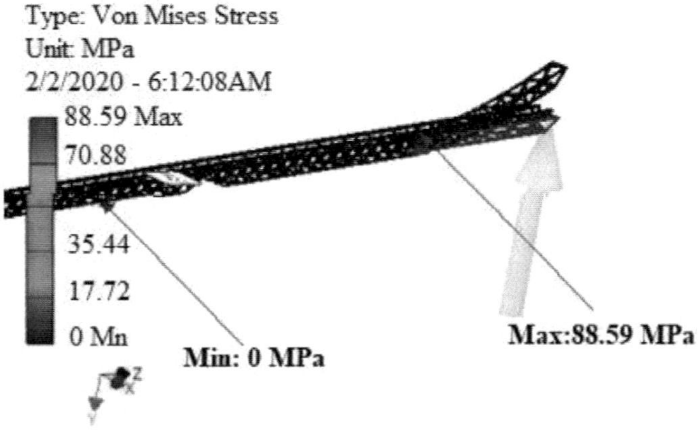

Figure 7. Von Misses stresses representation.

Further to the above presented, the following conclusions can be highlighted:
- Construction and stud of the crane boom and its components has an extremely important role in development of lifting equipment;
- Presently, there are some mathematical models for calculation of crane booms depending on their constructive version (beam with gussets and pillars, piping, etc.), and this gives the possibility of their diversity and finding optimum solutions;
- The crane model presented in this paperwork can be used on engineering and optimizing of lifting equipment, depending on a series of constructive parameters of such;
- The static stud with finite elements allows us to choose some optimum materials for an as reliable as possible operation and ensure a most long-lasting operation period possible.

5. Conclusions

In this paper, after making the 3D model, we elaborated the analysis with finite elements. Further to static study the movements of crane boom studied in this project can be noticed, and maximum travel being 59.95mm on axis, this value can be considered normal taking into account the length of crane's mobile boom assembly, where only the boom sag without load reaches ~30mm. As to the value of 3.11 resulted safety factor, this is in range, as according to ISCIR standards this value should be among 3 - 4.5. This type of study allows us to find out which is the maximum deformation, stresses and c - value of safety factor borne by a technical system or a machine part being under moving or static status.

6. References

[1] Anghelache D 2014 Finite element study for box type truck crane boom in extreme lifting position under load *The XX-th National Symposium on Equipment for Constructions* (Bucharest, December) pp. 11-12.
[2] https://liebherrmyanmar.com/images/pdf/liebherr-mk88-mobile-construction-crane.pdf, accessed on 12.02.2021.
[3] Potîrniche A 2003 Regarding static stability of tower cranes and energetic assessment of dynamic stability *The IX- the National Symposium on Equipment for Constructions* (Bucharest, December, 11-12) ISBN 973-7797-30-2.
[4] Potîrniche A and Căpățână G.F. 2017 Finite element analysis of an excavator bucket embedding ripper teeth *The Annals of "Dunarea de Jos" University of Galati, Fascicle XIV Mechanical Engineering* **I** 5-10 ISSN 1224 – 5615.
[5] Geisler T and Sochacki W 2011 Modelling and research into the vibrations of truck crane - *Scientific Research of the Institute of Mathematics and Computer Science* **1**(10) pp. 49-60.

IOP Conf. Series: Materials Science and Engineering 1182 (2021) 012051 doi:10.1088/1757-899X/1182/1/012051

Aspect related to 3D modelling and finite elements study of the lifting platform

G D Muscă (Anghelache), M A Potîrniche and G F Căpăţană
"Dunarea de Jos" University of Galati, Engineering and Agronomy Faculty of Braila-Research Centre for Mechanics of the Machines and Technological Equipment, Calea Calarasilor Street 29, 810017 Braila, Romania

E-mail: diana.anghelache@ugal.ro

Abstract. 3D modeling of the lifting platform was done in Catia V5R20 software. This soft was chosen as it gives an integrated solution, besides facilities related to modelling with solids which are common to all representative applications this also offers modules intended for stylists this way allowing extension of computer use also on previous stages of constructive engineering. The truck undercarriage is of LT28 type from Volkswagen producer. For 3D modelling, a leaflet from producing company was used. To represent the assembly of lifting platform firstly we realized a 3D model of the vehicle, and thereafter we realized the model of the nacelle type system and the manner of fixing on undercarriage. The finite elements study of the lifting platform was done by using Autodesk Inventor and for starting we determined the initial conditions type of network and material used. The main boom structure is calculated to stand for all stresses the platform is subject to. The structure is so designed that it can be fixed on the vehicle undercarriage frame. As a result of study, its maximum allowable stresses, travels and safety factor can be noticed. The first finite elements study was done on joint area between lifting platform boom and machine undercarriage, initial conditions were set within the study as follows: material mechanical properties type of performed study, force loading. Further to static study, data related to stresses, deformations and safety factor resulted. Maximum allowable stress in the boom of lifting platform has very small values wherefrom results that this stands a force applied on the other end.

1. Introduction

The lifting platforms are intended for performance of construction assembly and repairs on air lines, also being advantageously used within other industrial, commercial, etc., fields.

The lifting platform is a self-carried installation, mounted on the truck crane undercarriage and driven by the truck engine. But, recently, the group of lifting equipment also include lifting platforms located on: truck's standard undercarriages, especially built undercarriages, trailed undercarriages.

Various types of lifting platforms are known worldwide. However, there is a limit in development respectively increase of technical characteristic parameters, namely framing within provisions related to road traffic (transport overall sizes, loading on wheels, etc.); standards related to engineering, construction and operation of lifting mechanisms.

Most of lifting platform builders choose the special truck-undercarriages solution, framed within travelling lifting platform families that allow their better use, in the same time with diversifying on types and capacities. Worldwide, more upgrades and special solutions are known both for self-propelled platforms and trailed ones.

The constructive particulars of the lifting platforms put such equipment within the most difficult technical solutions. Considering the possibilities for grouping the constructive solutions, the following are highlighted: travel realizing; work equipment realizing; ensure the primary power source; placing the control installation.

The self-propelled multipurpose lifting platform consists of the following main sub-assemblies: truck undercarriage for transport in (4x2) system; platform support (with pressing-on earth devices); work equipment.

The frame is so designed and built that to take over the stresses transmitted from platform to ground through the elements of pressing-on earth legs from the undercarriage.

The structure is so designed that it can be fixed on the truck undercarriage frame.

The vehicle stability is ensured by one or two stabilizer groups, each group consisting of two stabilizers (pressing-on earth legs), located on both sides of the undercarriage.

Movements of various elements of the boom are ensured by means of double effect hydraulic cylinders, provided with double effect shut off valves. Rotation movement of the turret column is ensured by a hydraulic motor and a gear box. Operation of hydraulic motor is ensured by a pump driven by cardanic joint of the vehicle. Besides this main pump, there is installed a pump that can only be used in an emergency situation.

Based upon customer's request, this can be either an electrically driven pump with 12 V / 24 V voltage (from the vehicle), or a manual pump. Optionally, upon customer's request, there is an auxiliary hydraulic motor available, electrically driven from the vehicle. Continuous lifting of the platform is got by means of hydraulic system that that always keeps the platform in vertical position depending on movement of the main boom [1].

The lifting platform height can be manually adjusted through a distributor from the control station. The platform can be of two types:

- Open, made of round steel or aluminum tubes;
- Closed, made of fiberglass.

The work equipment is an assembly of mechanisms provided with driving elements that realize the movements and specific forces of the lifting platform. The work equipment of the platform consists of the elements: work platform; boom and the system for keeping the work platform horizontal.

Two constructive solutions were adopted for work equipment:

a) Work equipment with jointed boom;
b) Work equipment with boom consisting of two quadrilateral mechanisms.

The structure of the main boom is calculated to stand all stresses the platform is subject to. Platform features are shown in table 1. (Lionlift GT 18-10):

Table 1. Platform constructive features.

Constructive weight	G_c = 2000 daN
Load weight on axles:	Front G_1 = 945 daN; Back G_2 = 1055 daN
Turning radius	R = 6.5 m
Diesel engine power	P = 51 Kw
Gear box	5 stages

Table 2 shows the main parameters for machine engineering. That is the required data for our experimental research.

Table 2. Input parameters for machine engineering.

No.	Engineering parameters	Symbol	M.U.	Work equipment with double jointed boom	Work equipment with jointed boom
1	Nominal torque	M	Tm	-	1.1
2	Nominal load	Q	kg	200	200
3	Operation group	-	-	M6	M6
4	Class of use	-	-	T6	T6
5	Stress condition	-	-	L2	L2
6	Lifting height	H	m	6	12
7	Outreach	R	m	6.25	6.25

Work platform – is a steel structure consisting of a platform and four walls provided with rail, one wall being collapsible for access purpose. Usually, the work platforms are mounted on a support jointed to the boom and quadrilateral mechanism for horizontal positioning purpose.

Table 3 shows the producer's constructive dimensions for the work platform (Lionlift GT 18-10).

Table 3. Constructive dimensions of work platform.

No.	Parameters	Symbol	M.U.	Work equipment with quadrilateral double boom	Work equipment with jointed boom
1	Width	l	mm	800	800
2	Length	L	mm	1500	1500
3	Height	Î	mm	1040	1040

The form and dimensions of work platform have direct implications upon activities transmitted to the boom as well as the cost of manufacture.

The multipurpose lifting platform installed on the truck undercarriage is driven by truck's motor. Use of the self – propelled platforms on truck undercarriages in a greater extent was mainly due to some undisputed advantages of such, among which we can remind:

- Great mobility;
- Commissioning does not need additional personnel;
- Being equipped with an independent driving system of fixed electric power sources they can work in various conditions of unarranged field or with few arrangements;
- They can be used on a wide range of works.

The double quadrilateral boom is a steel construction consisting of a support, two box-type elements, jointed to the platform and connection plates thus creating the first quadrilateral of the boom. The 2-nd quadrilateral consists of 2 bars placed symmetrical to the medium plane of the support. The bars are jointed to support and connection plates, making this way the 2-nd quadrilateral. The boom is driven by hydraulic cylinder. Considering the working cycle of the platform during lifting – lowering process, the boom should provide the following conditions, by its dimensions and shape:

- ✓ To have suitable strength and stability;
- ✓ to provide reaching the kinematic parameters so that the work platform to keep a horizontal position from the beginning to end of the boom lifting stroke.

For the carrying elements of the platform (boom, platform), killed steels will be used having guaranteed weldability under conditions provided by STAS 7194-79. The fluxes used for welding shall be selected according to base and filler matrix so that to ensure obtaining a welded joint with properties at least equal to those of the base material. For welded construction of the lifting platform the following steels were chosen:

- Booms (upper, lower) – OL 52.4 steel plates; Rolled pipe OLT 45; Profile OL 50.1
- Work platform – OL 52.4 steel plate; Rolled pipe OLT 45; Striped plate;
- Fixed platform – OL 52.4 steel plate; Rolled pipe OLT 45; Profile OL 50.1

2. Parametric modelling of the truck and platform

To represent tis assembly we made 3D modelling for the truck, and, thereafter for nacelle type system and the manner of fixing on undercarriage. The truck sketch (Daimler - Benz) was used to start (figure 1).

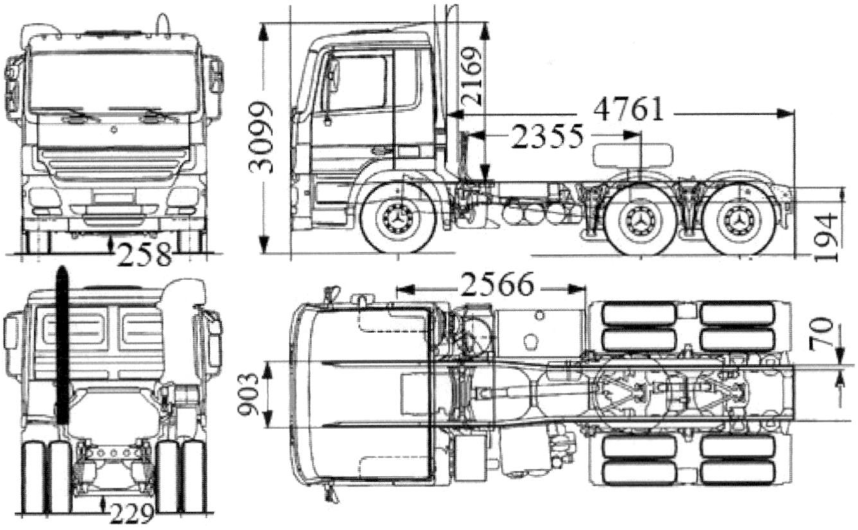

Figure 1. Truck sketch.

Starting from this sketch we proceeded to 3D modelling of the truck by using the Catia Part Design module to represent each component. After representing each component, Catia Assembly Design module was used through Cumulative Snap control (figure 2).

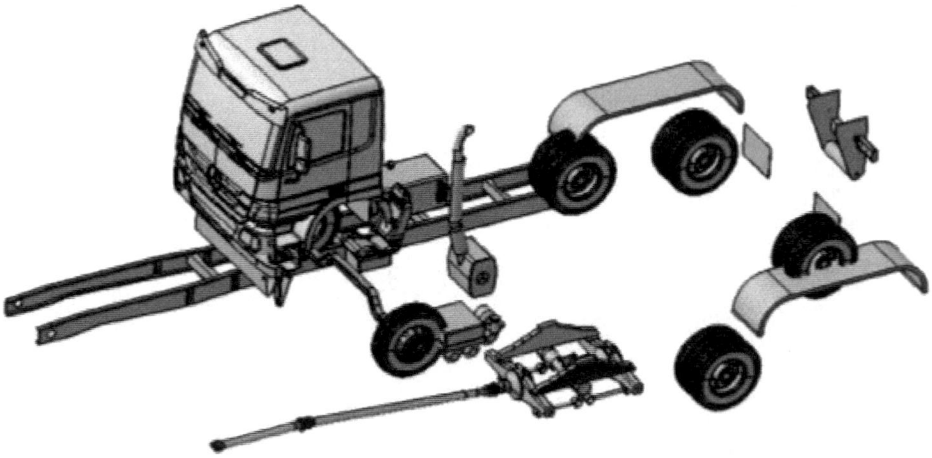

Figure 2. Platform components.

After their assembly, the 3D model of the truck below was done, as seen in figure 3.

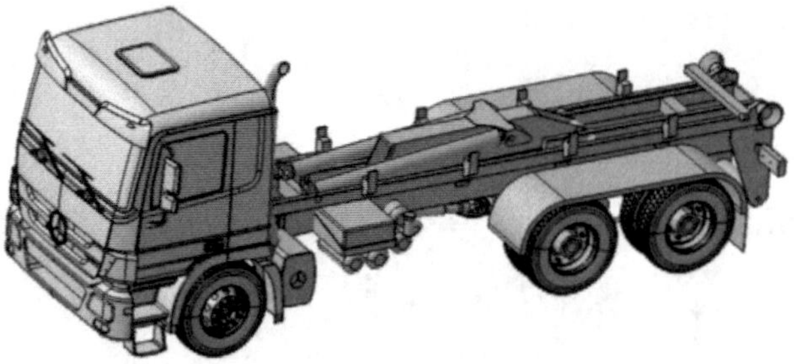

Figure 3. 3D model of the truck.

The next stage consists in 3D modelling of the nacelle using device. This was analogously done according to previously presented truck modelling. We started with 3D modelling of the nacelle's boom (figure 4) by using Catia Part Design module, by means of Sketch control we drawn the boom profile and by means of Pad control the extrusion was done resulting the model below [2].

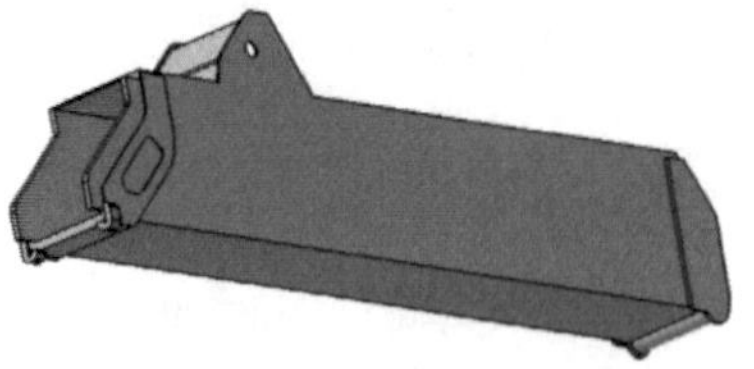

Figure 4. 3D model of platform boom.

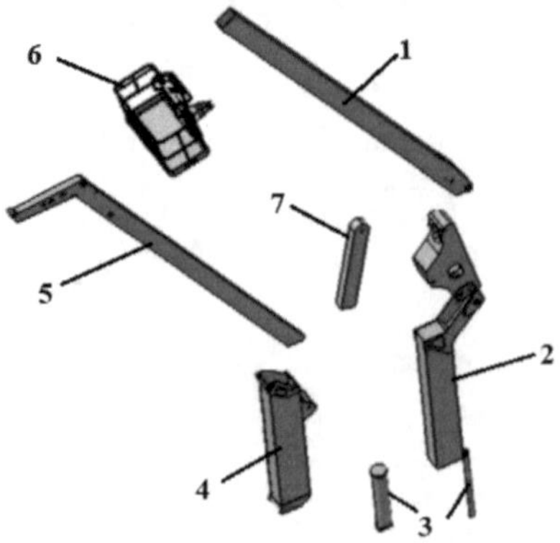

Figure 5. 3D components of lifting platform.

The main elements of the lifting platform are: lever boom (1), telescopic lever bracket assembly (2), shift joint grip (3), telescopic lever (4), nacelle support (5), nacelle (6) and fastening / connecting element (7) as seen in figure 5. To represent the components the controls Pocket for cutting and Chamfer and Fillet for chamfers and radii were used, resulting the following components: To assembly the components the Cumulative Snap and Snap to Object controls were used as in the below examples in figure 6:

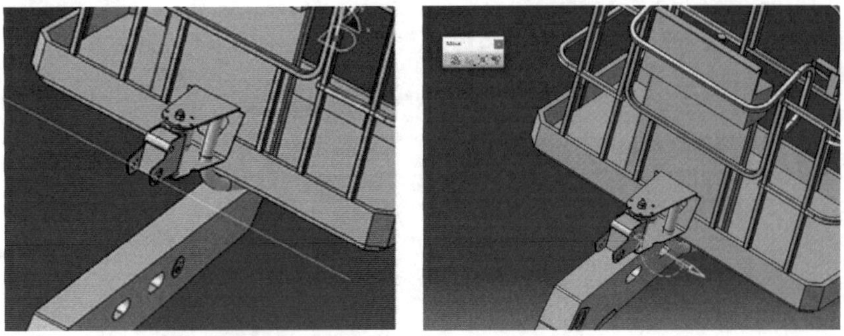

Figure 6. Self-lifting platform components assembly.

Analogously, the same controls were used and boom assembly of the lifting platform resulted as seen in figure 7.

Figure 7. Nacelle boom assembly.

3. FEM study for lifting platform

Finite element study for lifting platform was done by using Autodesk Inventor program, and for starting we determined initial conditions, type of mesh and material used as seen in figure 8.

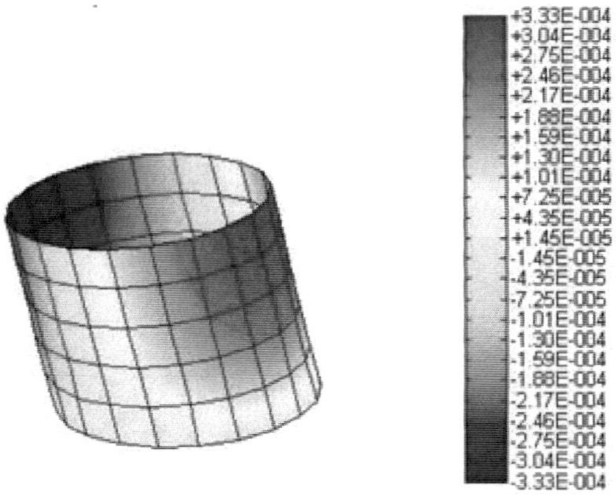

Figure 8. Post-process study.

Further to study, maximum allowable strresses, travels and its safety factor can be noticed [3].

The first finite elements study was done on the oint area between the nacelle boom and machine undercarriage, during the study being established the initial conditions as follows:

a) Mechanical properties of the material:

Name	*Steel, High Strength, Low Alloy*	
General	Mass Density	7.85 g/cm^3
	Yield Strength	275.8 MPa
	Ultimate Tensile Strength	448 MPa
Stress	Young's Modulus	200 GPa
	Poisson's Ratio	0.287 ul
	Shear Modulus	77.7001 GPa
Part Name(s)	HLTRUCK01V1-03-011; HLTRUCK01V1-03-012; HLTRUCK01V1-03-001; HLTRUCK01V1-03-011; HLTRUCK01V1-03-012; HLTRUCK01V1-03-010; HLTRUCK01V1-03-006; HLTRUCK01V1-03-007; HLTRUCK01V1-03-012; HLTRUCK01V1-03-012; HLTRUCK01V1-03-003; HLTRUCK01V1-03-002; HLTRUCK01V1-03-011; HLTRUCK01V1-03-012; HLTRUCK01V1-03-012; HLTRUCK01V1-03-011; HLTRUCK01V1-03-012; HLTRUCK01V1-03-012; HLTRUCK01V1-03-009; HLTRUCK01V1-03-005; HLTRUCK01V1-03-006; HLTRUCK01V1-03-007; HLTRUCK01V1-03-010; HLTRUCK01V1-03-011; HLTRUCK01V1-03-005; HLTRUCK01V1-03-012; HLTRUCK01V1-03-011; HLTRUCK01V1-03-011; HLTRUCK01V1-03-001; HLTRUCK01V1-03-012; HLTRUCK01V1-03-011; HLTRUCK01V1-01-004; JLG-800A-ARM-1.	

b) Type of performed study:

Study Type	Static Analysis
Last Modification Date	6/11/2020 -9:29 PM
Detect and Eliminate Rigid Body Modes	No
Separate Stresses Across Contact Surfaces	No
Motion Loads Analysis	No

c) Forces loading:

Magnitude	10000.000 N
Vector X	-0.000 N
Vector Y	-9922.926 N
Vector Z	3925.411 N

After establishing the initial conditions, we proceeded to establishing the constraints (figure 9(a)) and loading the other one (figure 9(b)) with force, as shown in image below.

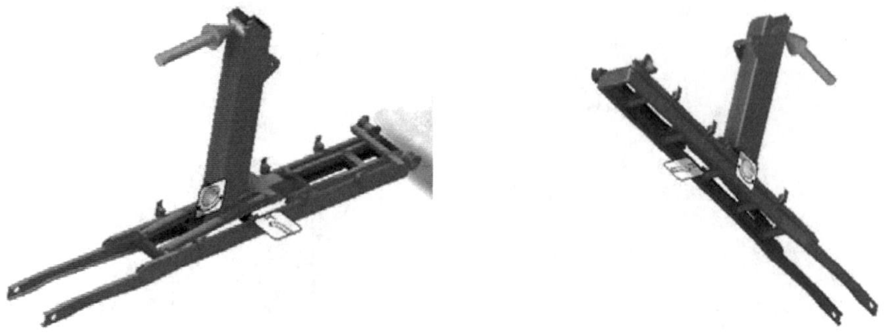

(a) (b)

Figure 9. Lever platform.

Further to static study, data related to stresses, deformations and safety factor resulted [4]. Due to finite elements study, it can be noticed that the maximum allowable stress in nacelle's boom has very small values, wherefrom results that this stands to application of a force to the other end (figure 10).

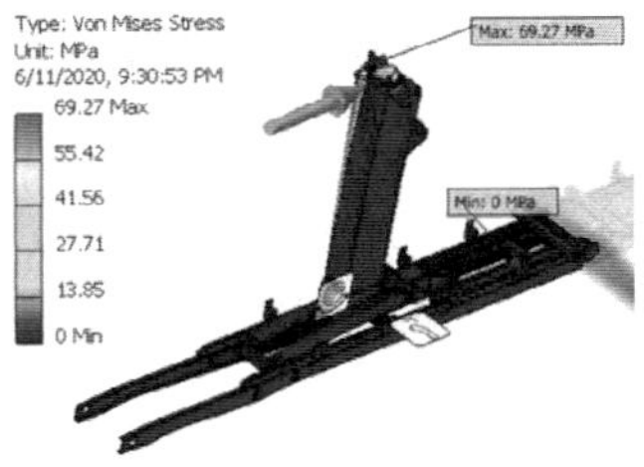

Figure 10. Maximum allowable stress.

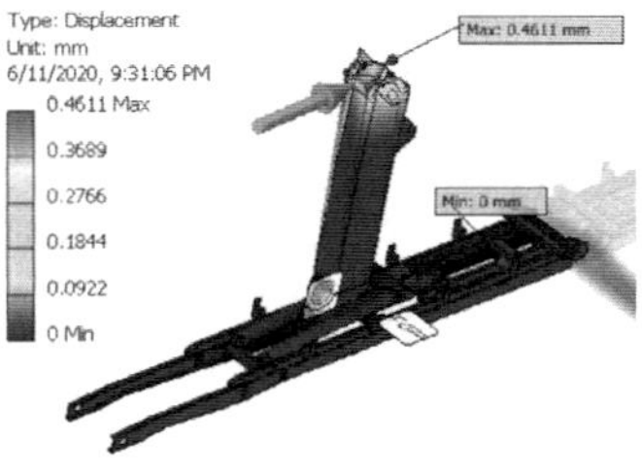

Figure 11. Maximum travels.

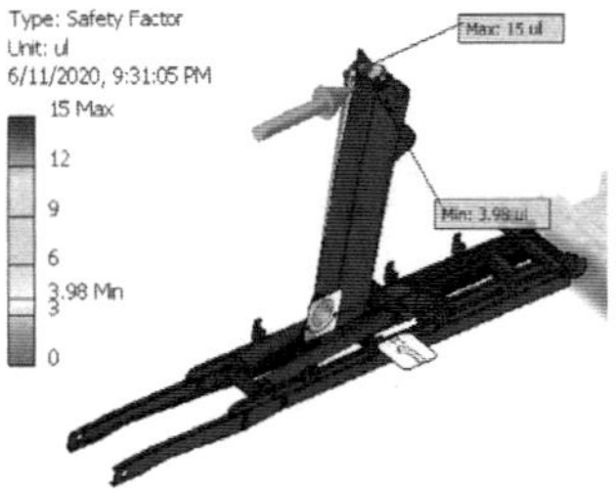

Figure 12. Safety factor.

In figure 11 is noticed that the maximum travel is 0.461mm, that is a normal value considering the operation conditions and material used for study.

As it can be noticed in figure 12 the safety factor has minimum value of 3.98 so this platform was oversized.

4. Conclusions

The lifting platform should stand to stresses which it is subject to during operation and during transport. For the chosen model we performed a static study (modal study) for any component. If the lifting platform behaves suitably, we perform dynamic tests. In this case, dynamic test was not needed. If we perform dynamic study, this is done with a nominal load increased by up to 10% and consists in repeated performance of all admitted movements, checking the normal operation of all components (mechanisms, braking system and safety components). Further to the study performed we can state that this lifting platform meets conditions for a good operation, taking into account that according to ISCIR standards for lifting equipment, as the maximum allowable stress, maximum travels have normal values. The safety factor should be 3.5 – minimum value of safety factor is 3.98 meaning that this designed platform is oversized, successfully stands to our stresses but in the same time, this platform can be optimized in order to reduce its own weight and making the design more efficient.

5. References

[1] Debeleac C and Năstac S 2009 Conceptual Product Design in Virtual Prototyping *Journal of Industrial Design and Engineering Graphics – JIDEG* **5** pp 17-20 ISSN 1843-3766.

[2] Goanţă A M 2009 *Modern Instruments of Transfer of Geometrical Patterns in Designing Building Machines and Equipment's* Proceedings of the 13th International Conference Iaşi&Chişinău - Modtech pp 271-274 ISSN 2066-3919.

[3] Goanţă A M 2015 Innovative methods for knowledge transfer *IOP Conf. Series: Materials Science and Engineering* **95** 012113 doi:10.1088/1757- 899X/95/1/012113, pp. 1-6.

[4] Tian H and Zhang Z 2011 Design and simulation based on Pro/E for a hydraulic lift platform in scissors type *International Workshop on Automobile, Power and Energy Engineering, Procedia Engineering* **16** pp 772 – 781.

Synchronization of Measurement Data with the Reference System

B Nalepa and A Gwiazda
Silesian University of Technology, Faculty of Mechanical Engineering Department of Engineering Processes Automation and Integrated Manufacturing Systems, Konarskiego 18A, 44-100 Gliwice, Poland

E-mail: Bartlomiej.Nalepa@polsl.pl

Abstract. Today, most devices are equipped with sensors that enable the collection of process data that influences the control parameters. Sensors are to replace human senses and provide information from the environment to the control algorithm. In many cases, the sensors used are subject to reference measurements from an external diagnostic system, which is to eliminate possible errors in the installation of sensors or the operation of the control algorithm. For this purpose, external systems are most often used that do not duplicate any errors from the sensors used. Most often these are vision systems that are based on the measurement of markers mounted on the tested object. The most important effect of the reference system is the verification of sensor data or the correction of the control algorithm that processes sensor data. Often, when comparing two measurement systems, there is a problem with unequal sampling rates. The main problem is comparing the two graphs as the number of data / arguments is different. In order to compare two graphs, many methods are used to reduce or extend the number of arguments. The use of data expansion or reduction methods may result in an additional error. In this article, methods of reducing measurement data from IMU sensors and encoders connected to servo motors will be presented. There is a significant difference in sampling time for the two measurement systems shown. At the beginning, the methods and applications of algorithms in the literature will be discussed. Then, the measurement data will be compared with the data processed by the algorithms. The final point of work will be the combination of the data processed by the algorithm with the measurements. Two measurements were also compared in order to estimate the measurement error and a comparative algorithm was proposed.

1. Introduction

The synchronization process of two graphs containing measurement data consists of two main stages: synchronization of the sampling frequency and elimination of the time shift between the graphs. The first step is to synchronize the two graphs by knowing the sampling frequency of the two devices - the measuring and reference system. The article presenting the synchronization of two signals generated by an unknown dynamical system is [1]. The synchronization takes place through a hybrid chaotic system, which was considered in [1] with the use of two synchronizations: identical synchronization (IS) and generalized synchronization (GS). The developed model of the hybrid synchronization system was created by integrating the IS and GS systems. The study of the synchronization of coupled oscillatory systems and the relationship between phases and frequencies of signals is discussed in [2]. The article [2] presents methods of phase and frequency estimation on the basis of time series as well as methods of detection and quantification of the tested synchronization. The research in [2] was carried out on the basis of data from the verification of body posture in healthy patients and patients with neurological diseases. The application of the Kuramoto synchronization algorithm was applied in [3] to the analysis of coupled nonlinear dynamical systems. An overview of the application of synchronization methods in the field of biological signals is presented in [4], in particular, the issue of quantification is discussed. Combining nonlinear and chaotic systems based on common signals on the basis of the sign of sub-Lyapunov exponents was presented as a synchronization method in [5]. The study of signals measured from neural systems using the synchronization likelihood (SL) is presented in [6].

The presented methods of synchronization related to specific algorithms, while in this paper the first stage of synchronization is down-sampling. The article [7] presents methods of recognizing human physical activity based on Down Sampling One Dimensional Local Binary Pattern (DS-1D-LBP). Subsequent use of down-sampling in the OptD method allowed to reduce the points obtained by LIDAR [8]. The next article that used the down-sampling method to detect planes from a 3D point cloud was [9]. The article uses the octree-based down-sampling method to reduce the number of points. Down-sampling is also used in Deep Convolutional Neural Networks (DCNN) to reduce the number of network parameters [10]. The article [10] proposes an algorithm for the elimination of errors arising during down-sampling by a random shift of the sampling layer taking place during model training. The down-sampling method is also used to eliminate redundant points in the 3D point cloud when measurements are made in different coordinate systems [11]. Article [11] has developed two approaches to the redundant number of points: the first method is planar-based adaptive down-sampling to remove redundant points from areas with high point density while preserving points occurring in areas with lower density, the second method is to remove peak points in the Gaussian sphere [11]. The down-sampling algorithm is also used to remove measurement noise [12]. The measurement noise is identified by down-sampling, then up-sampling and comparing the results with a noisy signal [12]. The last stage of data synchronization after the down-sampling method is applied is the removal of samples resulting from the difference in the time of switching on of two measuring devices. This article uses the cross-correlation method. The article [13] presents a method for normalization of cross-correlation from non-normalized form. In article [14] the temporal autocorrelator was used as a generalized class cross-correlation.

This article will discuss the method of synchronizing two data charts from two measurement systems - the verified and reference system. At the beginning, the model on which the measurements were made, i.e. the industrial robot, will be presented. The model was equipped with IMU sensors (measurement system for verification) and a reference system in the form of encoders installed in the motors. When trying to compare two signals, there is the problem of different sampling rates and differences in the start time of data recording. Synchronization of the measurement data will be performed by applying down-sampling first and then cross-correlation to eliminate differences in the recording start time. The presented algorithm can be used to synchronize and analyze data collected from sensors and data generated from the level of the simulation program [15-17].

2. Real model and measurement system

In the research on the synchronization of two measurement signals, the Fanuc ARC Mate 100iB robot was used - figure 1, where the place of installation of IMU sensors and the location of motors with encoders were marked.

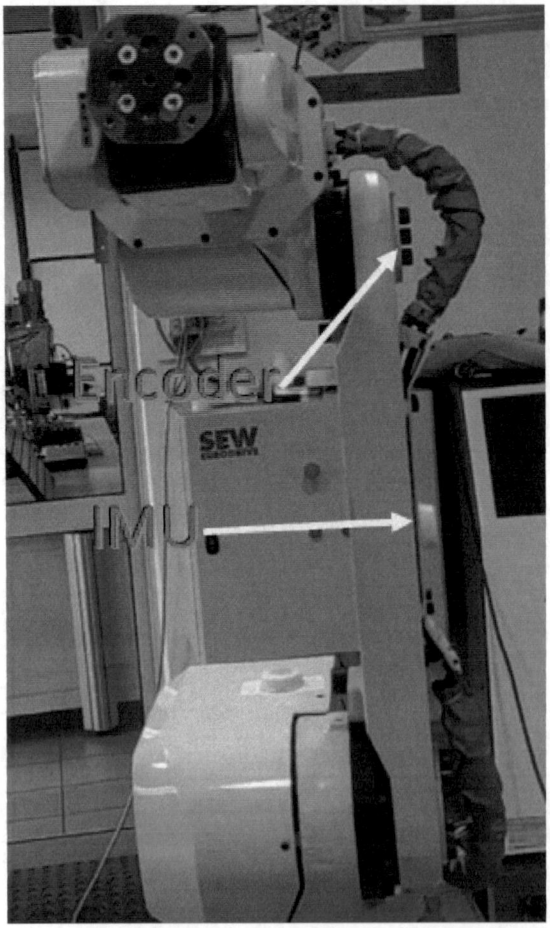

Figure 1. Fanuc ARC Mate 100iB industrial robot with IMU sensors (snowcookie) and the place for mounting the encoders.

Figure 1 shows the real model of an industrial robot with the marked encoder and IMU sensor from which the measurement data are collected. The encoder is the reference system for measuring with the IMU sensor. Measurement data is collected from the encoder and processed using the control system and LabView software to the value of the rotation angle of a given robot segment. Data from the IMU sensor are collected in the form of angular velocities and linear accelerations and processed according to the dependencies described in [18] to the form of the torsion angle of a given term. Data from IMU sensors are collected with a sampling frequency of 50Hz, while the encoder sampling frequency is 4kHz. Based on the presented data, it can be concluded that there is a significant difference between the sampling frequencies, in addition, the two measurement systems are not synchronized in time, therefore, apart from changing the sampling frequency in the data of one of the measurement systems, the difference resulting from the measurement activation time should also be removed.

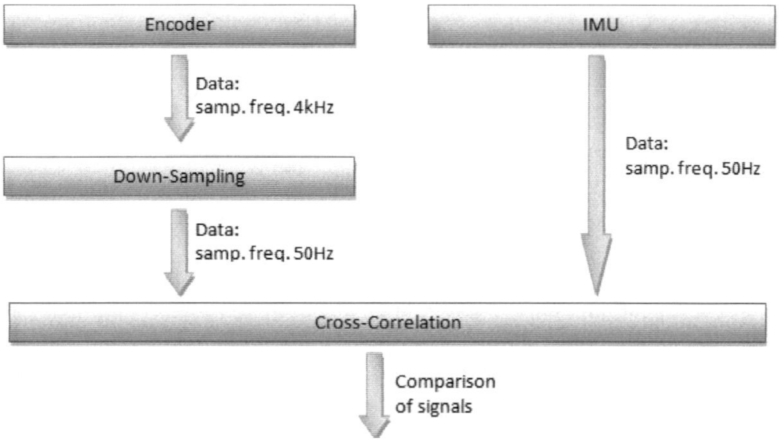

Figure 2. Diagram of the measurement data processing and synchronization system.

Based on figure 2, the synchronization process and the resulting comparison of two data sets consists of:

- sampling frequency synchronization,
- cross-correlation – elimination of the shift in time (removal of the difference in the time of measurement activation and deactivation) on the basis of the dependence and the algorithm [19],
- comparison of two measurements.

Using the algorithm presented in figure 2, the Octave program described in the following chapters of this work was developed.

3. Data synchronization

The data intended for the synchronization process are presented in figures 3 and 4.

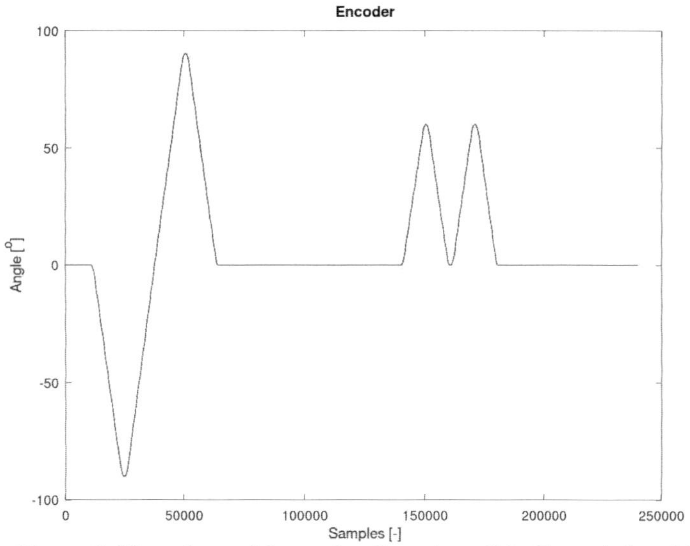

Figure 3. The values of the angle of rotation of the Fanuc industrial robot element read from the encoder.

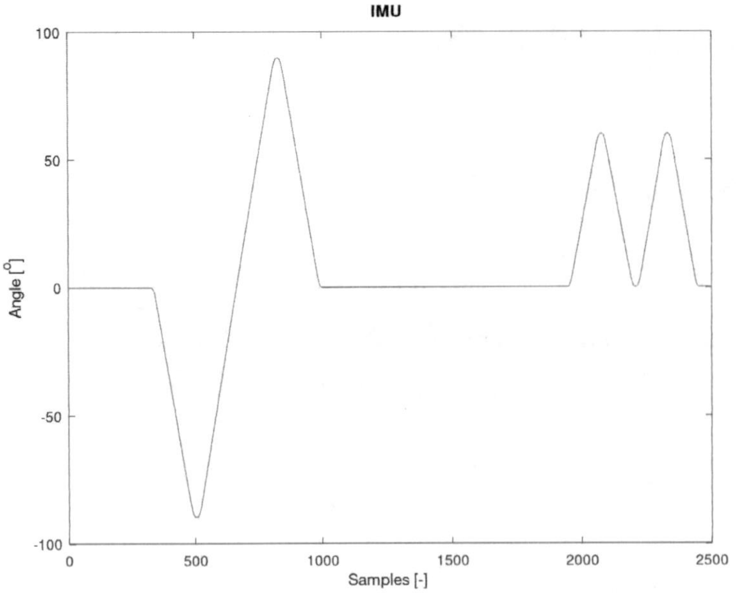

Figure 4. The values of the angle of rotation of the Fanuc industrial robot element read from the IMU.

Figures 3 and 4 show the measurements of the torsion angles of the same member of the industrial robot, but read from two measuring systems. Based on the analysis of the graphs, it can be concluded that there is a significant difference in the sampling frequency, which also results from the system documentation.

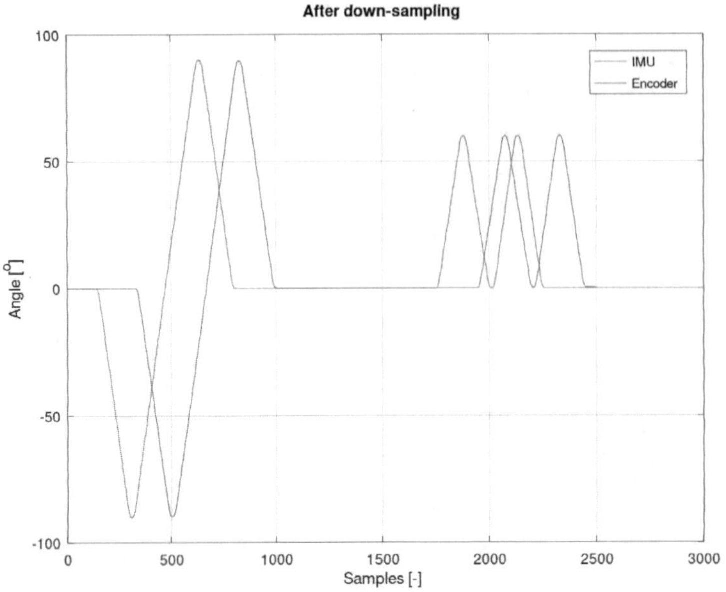

Figure 5. Compilation of two measurement signals after down-sampling.

Figure 5 shows a summary of the measurement data from the IMU sensor and the motor encoder. The encoder data has been processed by the down-sampling algorithm. It can be concluded from the graph that the measurement with the IMU sensor was activated after the data acquisition system from the motor encoder. Thus, both plots need to be cross-correlation in order to ultimately synchronize both signals.

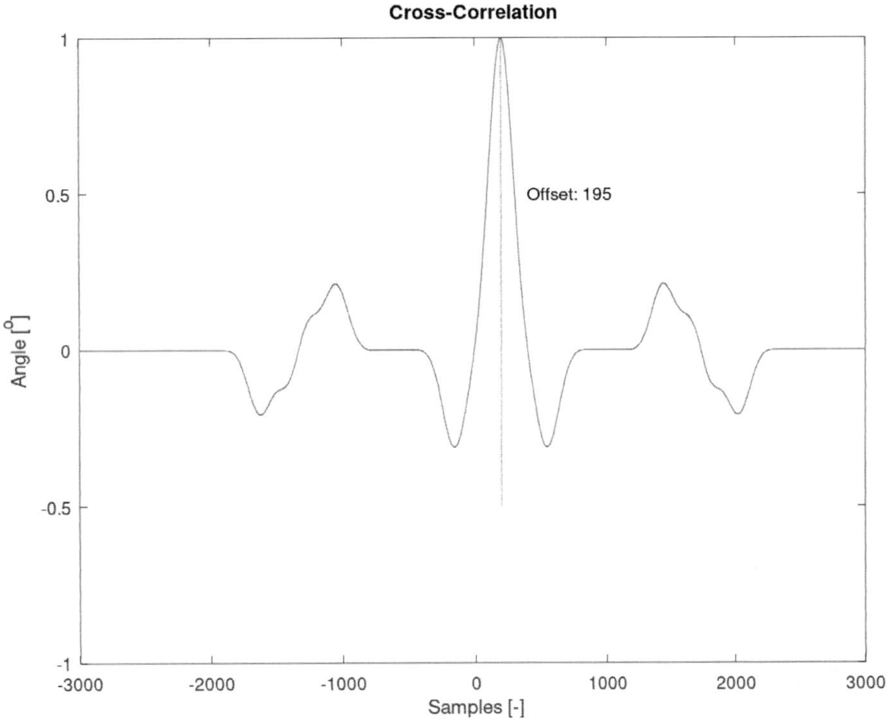

Figure 6. Implementation of the cross-correlation algorithm on the basis of two signals (IMU and encoder) - the inscription on the graph indicates the number of samples that should be subtracted from a given graph in order to synchronize.

The implemented cross-correlation algorithm, on the basis of the entered two measuring signals (IMU and the encoder) (figure 6), determined the number of samples that corresponds to the time difference between the activation of the measurement by the IMU sensor and the reference measurement carried out by the encoder. The cross-correlation algorithm performed showed an offset of the size of 195 samples of recorded data by the encoder before the IMU sensor measurement system was activated. Therefore, 195 samples should be subtracted from the signal recorded from the IMU sensor to ensure potential synchronization of the two signals. The next step is to combine the two signals and compare them in order to estimate the measurement error with any dedicated algorithm.

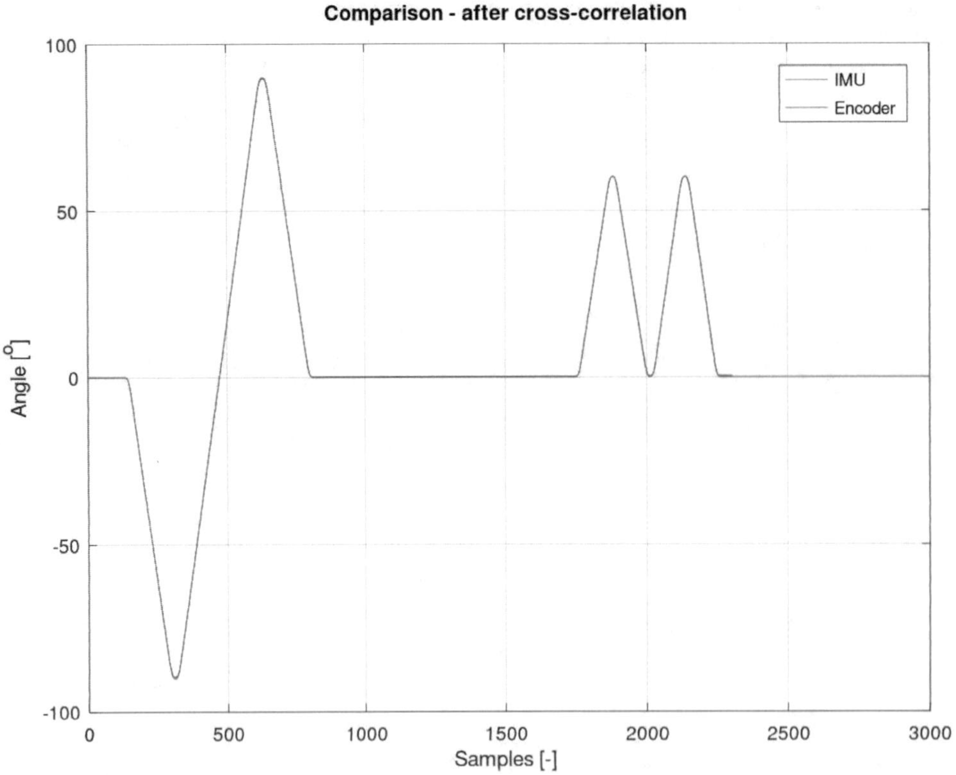

Figure 7. Compilation of two measurement signals after the synchronization process has been completed.

Figure 7 shows two signals: IMU and the signal from the encoder. The blue color is the IMU signal, while the brown color is the encoder graph. The measurements were subject to down-sampling and cross-correlation processes. As a result of the cross-correlation algorithm, 195 samples were removed from the IMU signal. Figure 7 shows that the two signals overlap, therefore the correct synchronization of the measurement data took place. It can also be concluded from figure 7 that the algorithm for removing excess samples, which result from unequal shutdown time of the measurement systems, has not been prepared, while the potential algorithm would only include the procedure of removing samples from the signal with a larger number of samples based on the difference in the length of both signals. In the case under consideration, the removal of the samples is unnecessary due to the perfect overlapping of the signals. If there are clear distances between the graphs, an algorithm should be used to calculate the average distance between the graphs, i.e. the average measurement error. An example of such an algorithm is DTW.

4. Conclusions

In this work the algorithm of measurement data synchronization was implemented. The torsion angles of the industrial robot were tested on the basis of data from IMU sensors and encoders. The IMU sensors were the verified measurement, while the encoder was the reference measurement. The problem discussed in the article was the unequal sampling frequency of both measurement systems, which made it impossible to compare the graphs and determine the measurement error. The method proposed in the article for the synchronization of measurements was the method of using the down-sampling and cross-correlation algorithm to determine the difference of samples resulting from unequal on and off times of the measurement system. Based on the data from both systems, the data synchronization process was carried out, which was very accurate and therefore it was not required to use the algorithm to compare the results in order to determine the measurement error. Subsequent research will consider signals that will not be so well correlated (the article presents sensors that measured the same physical quantities). In case of synchronization errors, an additional PCA algorithm will be applied.

5. References

[1] Xie Q, Chen G and Bollt E M 2002 Hybrid Chaos Synchronization and Its Application in Information Processing *Mathematical and Computer Modelling* **35** 145-163
[2] Rosenblum M, Pikovsky A, Kurths J, Schafer C and Tass P A 2001 Phase synchronization: From theory to data analysis *Handbook of Biological Physics, Elsevier Science* **4** *Neuro-informatics* **9** 279-321
[3] Jang J K, Klenner A, Ji X et al. 2018 Synchronization of coupled optical microresonators *Nature Photon* **12** 688–693
[4] Rosenblum M G, Pikovsky A S and Kurths J 2004 *Fluctuation and Noise Letters* **4** 1 L53-L62
[5] Pecora L M and Carroll T L 1990 *Physical Review Letters* **64** 8
[6] Montez T, Linkenkaer-Hansen K, van Dijk B W and Stam C J 2006 *NeuroImage* **33** 1117-1125
[7] Kuncan F, Kaya Y and Kuncan M 2019 *Advances in Electrical and Computer Engineering* **19** 1
[8] Blaszczak-Bak W, et al. 2020 *Geosciences* **10** 219
[9] El-Sayed E, Abdel-Kader R F, Nashaat H and Marei M 2018 *IET Image Process* **12** 9 1595-1605
[10] Zhao G, Wang J and Zhang Z 2017 Random Shifting for CNN: a Solution to Reduce Information Loss in Down-Sampling Layers *Proceedings of the Twenty-Sixth International Joint Conference on Artificial Intelligence (IJCAI-17)* 3476-3482
[11] Al-Rawabdeh A, He F and Habib A 2020 *Remote Sens.* **12** 1224
[12] Yang M, Wang F, Wang Y and Zheng N 2017 *Sensors* **17** 2778
[13] Lewis J P 1995 Fast Template Matching *Vision Interface* 120-123
[14] Loeb G E, et al. 1983 Biological Cybernetics **47** 149-163
[15] Lysek K, Gwiazda A and Herbuś K 2019 Communication between cad systems and the PLC controller *International Journal of Modern Manufacturing Technologies* ISSN 2067–3604 Special Issue **XI** 3
[16] Lysek K, Gwiazda A, Herbuś K 2018 Application of CAM systems to simulate of a milling machine work *IOP Conf. Ser.: Mater. Sci. Eng.* **400** 042037
[17] Lysek K, Gwiazda A, Herbuś K 2019 The use of the NX mechatronic module to simulate of a simple machine work *IOP Conf. Ser.: Mater. Sci. Eng.* **591** 012055
[18] Nalepa B and Gwiazda A 2020 *IOP Conf. Ser.: Mater. Sci. Eng.* **916** 012072
[19] Bourke P 1996 *Cross Correlation. AutoCorrelation – 2D Pattern Identification* http://paulbourke.net/miscellaneous/correlate/ accessed: 21.03.2021

IOP Conf. Series: Materials Science and Engineering 1182 (2021) 012053 doi:10.1088/1757-899X/1182/1/012053

Path finding and guidance algorithms for autonomous mobile robots in a tree structure network

S Nanu[1], **G Belgiu**[2], **A-M Dan**[1], **A Berdich**[1] **and A Anistoroaei**[1]
[1]Politehnica University of Timisoara, Department of Automation and Industrial Informatics, Timisoara, Romania
[2]Politehnica University of Timisoara-Romania, Department of Machine Manufacturing Technology, Timisoara, Romania

E-mail: sorin.nanu@upt.ro

Abstract. Considering an autonomous mobile robot that has to navigate indoor and outdoor from a certain point towards another certain point in a tree structure, the paper is giving a solution for finding the path and navigation. The mathematical model is proposed for a number n_{DP} of departure/destination points (DP), a number n_{IP} of intersection points (IP) and connections between them in a tree structure. Based on mathematical model, the problems solved and presented in this paper are i) to find a path from any departure point to any destination point, ii) to generate an algorithm for guidance of an autonomous robot to follow the path. Both algorithms were tested on Matlab Simulink® environment. An implementation of algorithms in a pseudocode is presented. The algorithms were adapted to be implemented on a microcontroller based board. The advantage of algorithms and implementation is simplicity. The outcome is less memory consumption and increased computing and execution speed.

1. Introduction

In the context of developing the autonomous robots for transportation, used in industry and agriculture, together with the great need of automation, in order to increase the efficiency, the path finding and guidance are of great interest. Manufactories, warehouses, large industry workshops transport are places where autonomous mobile robots can be used for unattended transportation.

There are many solution for orientation and storing maps in indoor environments, presented in different paperwork. One way to store a map is by scanning a range and keep data in specified polar coordinates [1]. Other robots orientate themselves by photoelectric scanning of the area and based on some landmarks they know where they are and where to go [2].Another way is by defining a finite state machine [3], and robot can perceive it's state (by sensors) and a set of possible actions. It chooses an action and receive from the environment a new state. This system was used on Sony Aibo social dog robot.

A bidirectional search graph is based on another algorithm which returns a sequence of vertices that are visited to reach the destination. Per iteration, expand each path in the frontiers with their neighbouring nodes until the frontiers intersect [4]. Also on graphs are based A star algorithm and Dijkstra. They are used for multiple mobile robots path planning algorithms [5-7].

Another solution uses Petri networks in robots guidance [8]. In [9] is used an algorithm, somehow similar to that presented in this paper, that uses a compass sensor and linear encoder to guide robots indoor and to avoid obstacles also. A different principle of guiding mobiles is [10] where GPS is used.

2. Mathematical model of the track

The solution presented in paper provides the movement planning of an autonomous robot that runs on a predefined track using line follower, in order to go from a certain point to another. The robot has to recognize the path and the intersection.

The track is formed of Departure/Destination Points (DP), Intersection Points (IP) and connections between them, that represents the available paths, in a structure of a tree, as in example in figure 1. The example is not reducing the degree of generality.

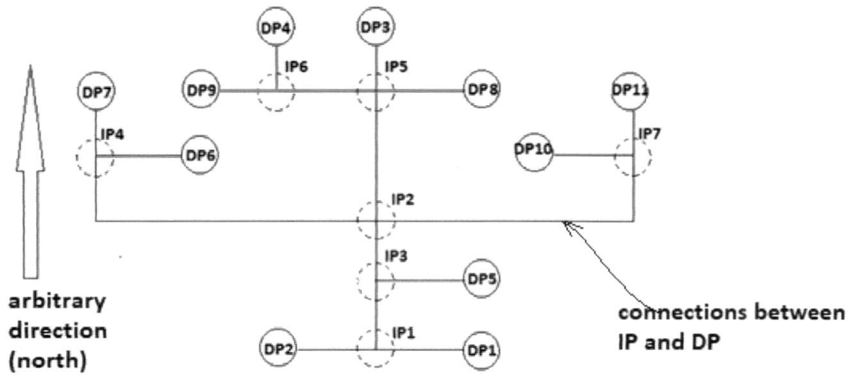

Figure 1. Example of tree structure of track.

The structure is given a main arbitrary direction, in this example, towards north. The mobiles can depart or arrive in any DPs, following in any direction the connections specified by paths from departure DP towards destination DP, through intermediate IP. In a network there are n_{DP} Departure/Destination Points and n_{IP} Intersection Points.

It is also restricted that every path enters and leaves the DPs and IPs only to/from rectangular directions (called north, east, south, west).

The objective of this paper is to describe the algorithm for determining:

a) The sequence of all intermediate IPs the mobile has to follow from departure to destination
b) The guidance of mobile in every IP, i.e. which direction to exit from an intersection, relative to entering point.

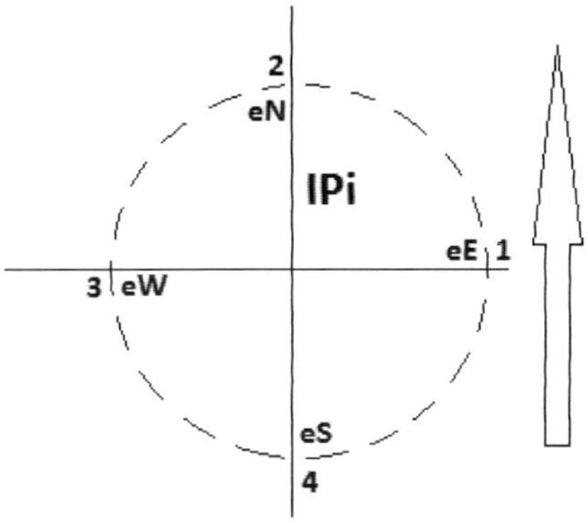

Figure 2. Model of the intersection point.

In order to model the Intersection Point (IP), the direction of entering/leaving points are coded arbitrary but consecutively, as in figure 2. Considering main reference orientation towards north, the east, north, west and south entering and leaving points are coded „1”, „2”, „3” respectively „4”.

Description of all connections between DPs and IP, that means all the paths the mobile can follow, is made through two matrices.

2.1. Model of the Departure/Destination Points (DP)

DP Matrix (relative to previous example, presented in figure 3) has two columns and n_{DP} lines.

Column 1 represents the IP connected to each DP (DP_j corresponds to line j), DP(j, 1) is the IP connected to DP_j.

Column 2 represents the code of entering direction from DP_j into Intersection Point DP(j, 1). e.g. From DP2, the path enters IP1 from direction 3.

DP=

IP	Dir	
1	1	DP1 enters IP1 from direction 1
1	3	DP2 enters IP1 from direction 3
5	2	DP3
6	2	DP4
3	1	DP5
4	1	DP6
4	2	DP7
5	1	DP8
6	3	DP9
7	3	DP10
7	2	DP11 enters IP7 from direction 2

Figure 3. DP Matrix.

2.2. Model of the Intersection Points (IP)

IP Matrix (example relative to this example is presented in figure 4) has n_{IP} x n_{IP} size. Its non-zero element in position IP(k,l) shows a direct connection between IP_k and IP_l. A zero element in position IP(k,l) shows that there is no direct connection between IP_k and Ip_l.

IP =

IP1	IP2	IP3	IP4	IP5	IP6	IP7		
0	0	4	0	0	0	0	IP1	IP1 enters IP3 from direction 4
0	0	2	4	4	0	4	IP2	from IP2 to IPi
2	4	0	0	0	0	0	IP3	IP3 enters IP1 from direction 2 and enters IP2 from direction 4
0	3	0	0	0	0	0	IP4	.
0	2	0	0	0	1	0	IP5	.
0	0	0	0	3	0	0	IP6	.
0	1	0	0	0	0	0	IP7	from IP7 to IPi

Figure 4. IP Matrix.

The value of IP(k,l) element specifies the entering direction into IP_l (column) from IP_k (line).

These two matrices describes completely both the direct connections between DPs and IPs and the entering directions in all IPs (from DP or from other IP).

3. Algorithm for path finding

The objective of the algorith is to find the sequence of IPs to follow by robot in order to run from departure point (named DP_i) to destination point (named DP_j). The *input* of the algorithm are the departure point and destination point. The *output* of the algorithm is a row vector called roadVector() whose elements are in order the indexes of IPs to follow from departure to destination.

The steps followed in this algorithm are:

a) From DP matrix, corresponding to line i, column 1, the index of first IP (let it be IP_k and called IPfirst) is extracted and is included as first element of roadVector(). In the same manner, the IP that is the last IP in row is that one that is connected to destination point DP_j. Let it call IPlast.
b) From IP matrix, corresponding to k line, all IPs that are connected directly with IP_k and are possible next IP are seen as non-zero elements. They are included in a vector called possibleIP (). In this stage of development of algorithm, vector possibleIP () contains all IPs connected directly to IP_k.
c) If possibleIP () vector contains IPlast element, the algorithm is concluded and roadVector= (IPfirst, IPlast).
d) If possibleIP() vector contains only one element, then the search will be done further after this element.
e) If possibleIP() vector contains more than one element, the first element of possibleIP() vector is searched identically as in step 3.2. The new possibleIP1() vector is generated and the search is done in the same manner.
f) The difference between the second and first elements is that from the possibleIPi() vector are removed

i) the IPs searched before, (they are saved in a vector visitedIP()), in order to avoid repeated search on the same IP,
ii) the IPs connected directly to a DP that are different from destination point DP_j, saved in deadEnd() vector,
iii) the IPs that are already included in roadVector

Consequently, starting from the second step, the visitedIPi() vector is formed of all possible IPs in the current line after removing a), b), c) elements.
g) The tree structure of path assures that there is a unique solution for a given pair of DP_i and DP_j.
h) The roadVector() is available for using in the guidance procedure.

4. Algorithm for guidance

The robot, on its way from departure point up to destination point, has to pass through intersection points. Every IP has four enter/exit directions, and the robot must be told which direction has to follow since it enters the IP. Consequently, the robot can perform only three types of movement: rotation to the right (coded as 1), rotation to the left (coded as 3), move straight forward, (coded as 2).

The principle of computing the movement is presented with the aid of figure 5. In the picture there are three IP that will be crossed by robot. The robot goes from IP_{k-1} to IP_k and further on towards IP_{k+1}. The problem is when robot detects the intersection, which way to go?

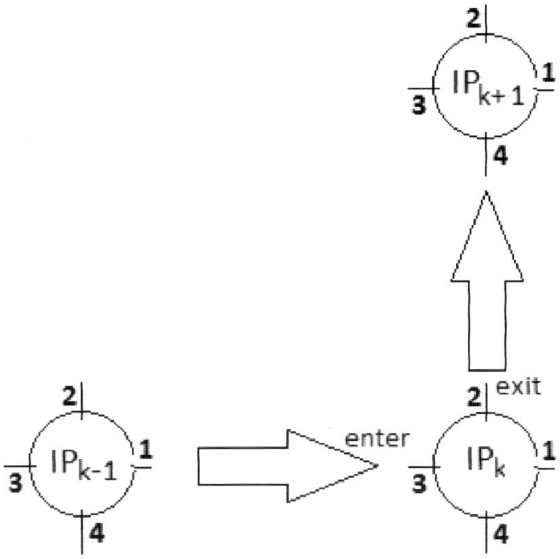

Figure 5. Relative to movement of robot in IPk

In this case, he has to rotate to the left. The movement is computed in this way:

The movement code is obtained as a difference from the code of exit direction minus the code entering direction. If the difference is a negative number, value 4 will be added. In this way, the result is always 1 (rotate to the right), 2 (move straight ahead) or 3 (rotate to the left).

In this example, exit direction code = 2, enter direction code = 3. Movement code = 2-3+4=3 that means *rotate to the left*. The formula is general valid.

The guidance algorithm uses the mathematical model of the track (IP matrix and DP matrix) together with roadVector() and provides a vector of movement for each IP named movementVector(), whose size is equal to size of roadVector().

The steps of the algorithm are:
a) Take the index of the first IP (IP_1) from roadVector() and determine from DP matrix the enter direction code and from IP matrix the exit direction code. This is obtained by finding the entering direction to IP_1 from next IP in roadVector(), IP_2.
b) Make the difference between exit direction code and enter direction code, adding 4 if result is negative.
c) Add the result in movementVector().
d) Repeat for next IP until end.

5.Algorithm implementation

For implementation, a pseudocode was generated, that calculates the angle that robot has to rotate in each intersection.

```
%input data:
        IP – Intersection Point Matrix
        DP – Departure Point matrix
        uvect – Path vector (e.g. uvect=[1 1 3 5] is the Path DP1 → IP1 →IP3 → DP5 )
        Nuvect – Intersection Points Number in the Path
%output data:
        yvect – Direction vector for the Intersection Points in the Path
        Theta – Rotation angle of the Robot

% algorithm
Input_IP:=zeros(Nuvect);
Output_IP:=zeros(Nuvect);

% The input direction of every IP in the Path
Input_IP(1):=DP(uvect(1),2);
for i=2 to Nuvect
        Input_IP(i):=IP(uvect(i),uvect (i+1));
endfor

% The output direction of every IP in the Path
for i=1 to Nuvect-1
        Input_IP(i):=IP(uvect(i+2),uvect (i+1));
endfor
Output_IP(Nuvect):=DP(uvect(Nuvect+2),2);

% Direction vector in IP
for i=1 to Nuvect
        if yvect(i)<0 then
                yvect(i):=yvect(i)+4;
        endif
endfor

% Rotation angle of the robot in IP
for i=1 to Nuvect
case yvect(i) of
        1:        Theta(i):=-90
        2:        Theta(i):=0
        3:        Theta(i):=90
endcase
endfor
```

For this code, a simulation program was written in Matlab.

6. Example

Considering the model in figure 1, departure point DP1 and destination point DP9, the path finding algorithm will provide the path in figure 6:
roadVector()= (1,3,2,5,6) that means the robot has to run in order through IP1, IP3, IP2, IP5, IP6,
The guidance vector is:
movement Vector()=(1, 2, 2, 3, 2) that means at first IP has to take right, at the second and third IP has to go straight ahead, at the fourth has to go to the left and straight ahead again.

In figure 7 is presented the Matlab simulation.

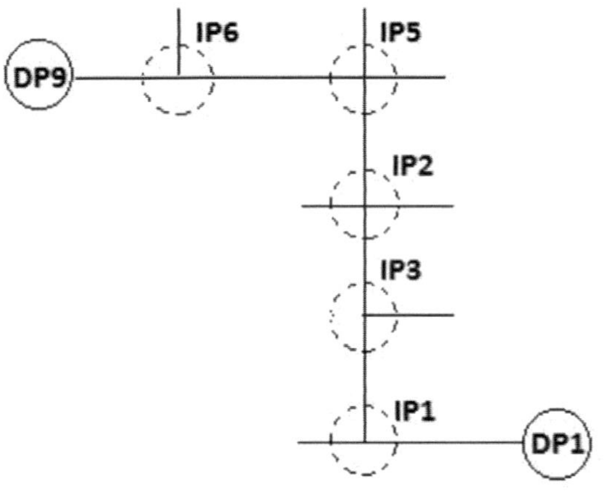

Figure 6. Example.

```
>> Introduce track
start DP:1
zero for exit
IP:1
IP:3
IP:2
IP:5
IP:6
IP:0
end DP:9
x =     9
track =
       1'DP  1 IP   3 IP   2 IP   5IP   6 IP   9 DP
-90        right
0          straight
0          straight
90         left
0          straight
```

Figure 7. Matlab simulation.

7. Conclusions

Paper presents a solution for guiding a mobile robot on a predefined tree structure track when departure and destination points are provided. The model of the tree track and algorithms for track finding and guidance were meant for implementing in embedded systems. The algorithms were simulated in Matlab. For a structure with n_{IP} intersection points, $n_{IP}^2/2$ connections and n_{DP} departure/destination points, the EEPROM memory requested for model is $n_{IP}^2+2*\ n_{DP}$, while estimated RAM memory for variables is n_{IP}^3 [11]. This approach was aimed to offer a very simple solution from the memory and speed point of view in order to be implemented on a simple microcontroller with reduced resources. As the track structure is a tree, there are no loops and the solution is unique. We intend, and are working on a more general structure where the track has loops.

8. References

[1] Nguyen V, Gächter S, Martinelli A, Tomatis N, Siegwart R, 2007 *A comparison of line extraction algorithms using 2D range data for indoor mobile robotics*, Published online: 8 June 2007 © Springer Science+Business Media, LLC.

[2] Huang Z, Zhu J, Yang L, Xue B, Wu J, Zhao Z 2015 *Accurate 3-D Position and Orientation Method for Indoor Mobile Robot Navigation Based on Photoelectric Scanning*, IEEE Transactions on Instrumentation and Measurement, **64**(9) 2518-2529.

[3] Perez X, Angulo C, Escalera S, Pardo D, 2010 *Vision-based Navigation and Reinforcement Learning Path Finding for Social Robots*, Barcelona, Spain.

[4] Fitzgerald, T., 2012 *A System for Bidirectional Robotic Pathfinding* Portland State Univ., Portland.

[5] Zhang Z, Zhao Z, 2014 *A Multiple Mobile Robots Path planning Algorithm Based on A-star and Dijkstra Algorithm*, 1,2 College of Computer and Information Engineering, Tianjin Normal University, Tianjin, China.

[6] Mahadevi S., Shylaja K. R., Ravinandan M. E. 2014 *Memory Based A-Star Algorithm for Path Planning of a Mobile Robot*, International Journal of Science and Research **3**(6), 1351-1355.

[7] *** *Path-finding with Dijkstra's Algorithm* http://files.kipr.org/gcer/2011/proceedings/Fletcher_Woodruff_E-Final.pdf, accessed on 24March 2021.

[8] Khosravi A, Danesh M, Khosravi M, 2013 *Esigning A Robot's Guidance Algorithm Through Petri Nets For Following A Mobile Target Using The Data Of Several Sensors*, Buletinul Institutului Politehnic from Iaşi, March 25.

[9] Mutlu L, Uyar E, 2012 *Indoor Navigation and Guidance of an Autonomous Robot Vehicle with Static Obstacle Avoidance and Optimal Path Finding Algorithm*, September 12-14, 2012. Sofia, Bulgary.

[10] Simon D, El-Sherief H, 1994 *Real-Time Navigation Using the Global Positioning System*, SIG Technology, Summer http://academic.csuohio.edu/simond/pubs/SIGTechReview.pdf, accessed on 24March 2021.

[11] Nanu S, 2007 *Educational Aspects in Control System Design Technology*, Scientific and Technical Bulletin, Series: Electrotechnics, Electronics, Automatic Control and Computer Science UAV Arad, **4**(1) 584-585.

IOP Conf. Series: Materials Science and Engineering 1182 (2021) 012054 doi:10.1088/1757-899X/1182/1/012054

Research on determining the state of stress by the finite element method in rehabilitation the welded structure

D Nitoi[1], C Babis[1], O Chivu[1], S Petriceanu[1], A Dimitrescu[1] and A I Tăpîrdea[2]
[1] University Politehnica of Bucharest, Splaiul Independentei no. 313, Bucharest, Romania
[2] Politehnica University Timisoara, Piata Victoriei no.2, Timisoara, Romania

E-mail: claudiubab8@gmail.com

Abstract. The study of stress is a very important problem especially in the case of welded structures. Very often the main causes of their failure are represented by the overlap of the stresses during operation with the remaining ones during the welding process. The article presents in the first part the experimental activity that regarding the realization of some fillet welded joints remedied by two processes that aim at reducing the internal stress. In the second part of the article, starting from the experimental part presented, a model was made by the finite element method. Based on it the different states of tension resulted in the initial structure as well as in the rehabilitated ones were simulated.

1. Introduction

Welded structures are widespread in the industrial field and are found in various products in the form of metal fabrications such as farms, beams, bridges, support frames for machine tools, crane structures, resistance structures for various aircraft. We can therefore say that this non-removable assembly process by welding is very widespread in the industrial environment and has successfully replaced, as we well know, riveting.

Welding comes with a number of advantages such as material and labor savings, welding processes are easy to apply, offer high mechanical strength, welding processes can be automated, you can get a superior quality of welded joints, it offers the possibility in order to obtain lighter constructions and with simpler constructive forms, it is possible to use technological and processing additions that are 70-90% smaller than other manufacturing processes, for example compared to casting, complex parts can be made and not in the last instance, allows the realization of mixed or combined constructions, consisting of several parts made of different materials, by different processing processes and then assembled by welding.

The main disadvantages of welding are related to the following aspects: the introduction of stress concentrators, the quality of welded joints depends largely on the quality of materials used and the qualification of the welder, often the products obtained by welding must be subjected to post weld toe treatment and a rigorous control [1], welding machines and special equipment are required that have high costs, the equipment used for welding is complicated, difficult to maintain and have demanding work regimes and last but not least for welding there are radiations that are harmful to the human body.

Of all the disadvantages listed above, in industrial practice, the biggest disadvantage of welding is the introduction of stress concentrators. These stress concentrators are manifested as local stress peaks, which overlap with the existing stress in the product, and can cause fracture damage, which in some cases can be catastrophic: bridge failures, failures of resistance structures of some equipment and so on [2]. Determining these stress concentrators is therefore of great importance. By knowing the values of the stress concentrators introduced by welding, we will know if they must be completely eliminated or if the respective welded structure can work safely with them, without the occurrence of failures. The welded structures required for fatigue are very sensitive to these stress concentrators [3]. These structures, fail to very low values of stress concentrators because the fail of these structures is made to low values of stresses, values located below the yield limit of the material from which the structure is made. For these fatigue loads, is very important the number of cycles with which the structure was load at a given time. Usually, at a number of stress cycles loads above 10^7, we are dealing with gigacyclic stresses loads and the occurrence of fatigue. In order to occur the fatigue phenomenon, the loads must be variable, with a large number of stress cycles, more than 10^7, the stresses in the structure must be below the yield limit and there must be a certain frequency of the stress.

The stress concentrators mentioned above can be internal such as cracks, pores, discontinuities in the material, metallic or non-metallic inclusions or external such as irregular or improper shape of the welding seams. A very great influence on the stress concentrators introduced by the welding seams, has their geometric shape. The geometric shape of the welding seams is important especially in the case of fillet and cruciform fillet welded joints, due to the uneven distribution of the lines of force. Unlike the case of fillet welded joints in which we have an uneven distribution of lines of force, in the case of butt joints, the distribution of lines of force is uniform and the influence of these stress concentrators is insignificant. Returning to the fillet welded joints, we can meet in practice welding seams with convex shape, with small radii of connection between the base and the filler material and therefore with high stress concentrators in that area, flat-shaped welding seams with connecting radii higher between the base and filler material and therefore with values of the stress concentrators lower than in the previous case and welding seams with concave shape, with large connection radii between the base and filler material and therefore with very small values of the stress concentrators. The three possible geometric shapes of the welding seams are shown in figure 1.

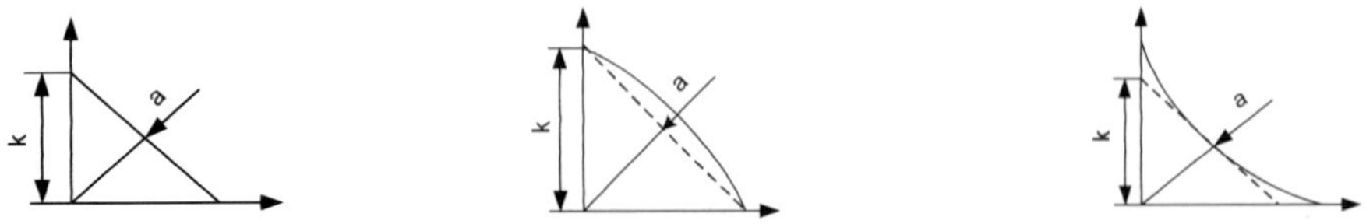

(a) (b) (c)

Figure 1. Possible shapes of fillet welds: a - flat; b - convex, c - concave, k - welding bead leg; a - the apothem of the welding bead.

It is easy to understand that in industrial practice where we meet welded structures loaded for fatigue, we prefer fillet welded joints with concave geometric shape of welding seams.

Practice reveals many welded structures that have fillet welded joints with the convex geometric shape of the welding seams and that have been stressed over time with many stress cycles. Therefore, in the case of these structures, fatigue failure phenomena are expected to occur. Own research [4-6] has shown that the rehabilitation of these welded structures with fillet welded joints with convex geometric shape of the seams is possible and is cheaper than replacing that structure with a new one. There are several techniques for rehabilitating these welded structures, the most important of which are "Grinding weld toe technique" and "WIG remelting weld toe technique". The “Grinding weld toe” technique is described in figure 2.

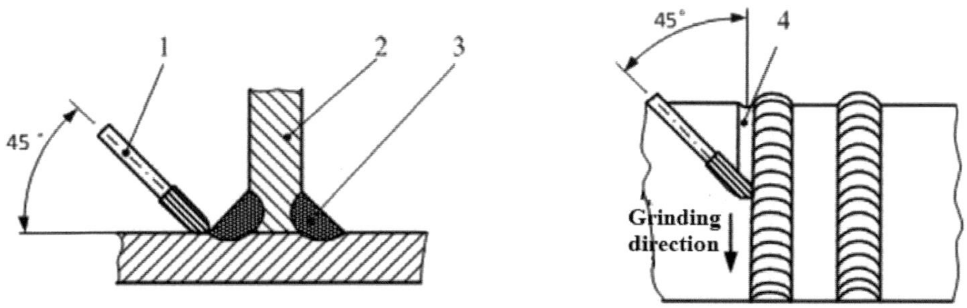

Figure 2. Grinding weld toe technique 1 - finger cutter; 2 - vertical component; 3 - weld seam for rehabilitation; 4 - rehabilitated zone [5,6].

In this method, the grinding of the tip of the weld bead is done along the line of the intersection between the welded seam and the base material. The recommended tools for this technique are high speed finger cutters with tungsten carbide stones. Milling is done at a depth of 0.8-1.0 mm below the surface of the sheet or 0.5-0.8, below the deepest visible notch, at a total depth of 2 mm or 5% of the thickness of the sheet, which of them is smaller. Milling is done only at the top of the seam and not at the whole seam. The aim is to eliminate any surface defects. The angle of the milling cutter axis will be a maximum of 45^0 from the direction of travel, to ensure that the milling marks are perpendicular to the line of the welding tip.

The "WIG remelting weld toe" technique is described in figure 3.

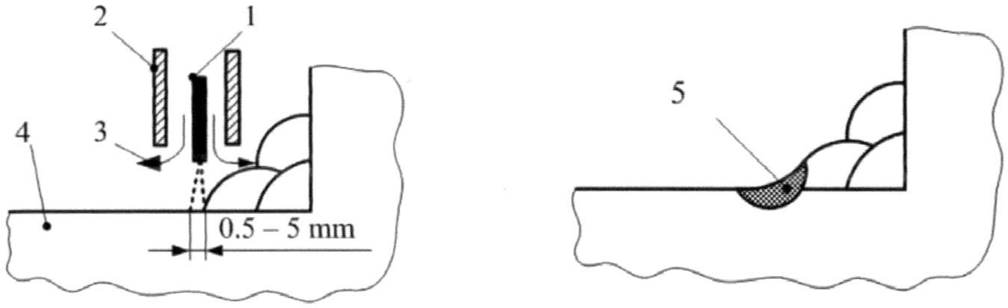

Figure 3. WIG remelting weld toe technique 1- wolfram electrode; 2-gun protection nozzle; 3-inert gas flow; 4-rehabilitated piece; 5-area rehabilitated by WIG remelting [5,6].

The "WIG remelting weld toe" technique consists of melting the welding metal to a depth of about 2 mm along the welding tip, without adding filler material. The welding surface will be cleaned of rust and tunder. The tip of the electrode must be sharp and clean and must be positioned from 0.5 to 1.5 mm from the tip of the weld.

In the practical experiments the used parameters of the welding regime are presented in table 1.

Table 1. Used parameters of the welding regime.

Row	Is [A]	Ua [V]	ts [s]	Lc [cm]	vs [cm/s]	Vas[m/min]	El [kJ/cm]
1-3	225...230	20..22	90	40	0.5	5.5	8.1

In this paper we start from some experimental results published in other papers [4-6] and which are briefly presented in table 2.

The geometric elements of the welded seams, with and without reconditioning techniques applied are presented in figure 4.

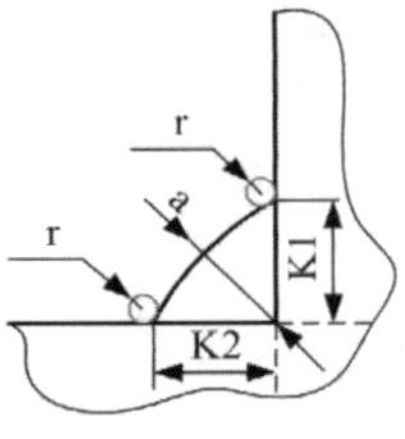

Figure 4. The geometric elements of the welded seams with and without reconditioning techniques applied.

The values of the geometric elements of the welded seams, without reconditioning and with milling and WIG remelting reconditioning techniques are presented in table 2 [5,6].

Table 2. The value of the geometric elements of the welded seams with and without reconditioning.

Without rehabilitation	Weld toe grinding	WIG remelting weld toe
a = 11; K1 = K2 = 13 R = 2	a = 11; K1 = K2 = 13 R = 3	a = 11 K1 = K2 = 17 R = 9

It is observed that: initially the K dimension had the value of 13 mm and the connection radius R = 2 mm; after milling the end of the cord, the legs remained the same, but the connection radius of the cord end was increased to R = 3 mm, and after the WIG remelting of the end of the cord, the dimensions of weld were increased to 17 mm and the connection radius at R = 9 mm. In all three cases, the thickness of the welded cord was a = 11 mm, and its shape was convex.

2. Stress calculation in the case of corner welding using FEM

The finite element method is currently a widely used research method that offers very useful numerical solutions in situations where the experimental part does not benefit from experimental working method [3]. Starting from the experimental data presented in table 2, the behaviour of a fillet weld was performed in several geometrical situations. Thus, figure 5 shows the state of S1 type stress in three situations. Figure 5(a) shows a convex fillet weld. Figure 5(b) shows the rehabilitation by "WIG remelting weld toe technique "in one part of the joint and in figure 5(c) the rehabilitation in both parts of the weld. As can be seen, at the interface between the base and the filler material, an important area appears in the first situation, which constitutes a stress concentrator (red colour) that decreases in size as the rehabilitation process takes place. The maximum value of the stress in this case is S1 = 0.1·E11 N /m^2.

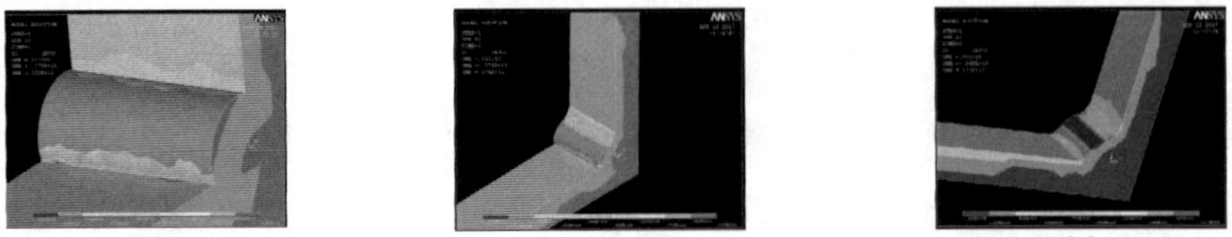

(a) (b) (c)

Figure 5. S2 stress calculation: a – convex weld; b - WIG remelting on one side; c - WIG remelting on both sides.

The stress of type S1calculates manly the tension state in the material. Figure 6 presents the S2 stress (calculates an equilibrium stress between tensile and compression) first in the situation of normal welded seam. Here a stress concentrator on both sides can be observed.

In the figure 6(b) where a WIG rehabilitation is done, the stress concentrator is lower in width. In the third stage, with both rehabilitation zones, a relatively uniform stress is observed, stress concentrator disappeared and a S2 = 1.7·E10 N/m^2 stress value is calculated.

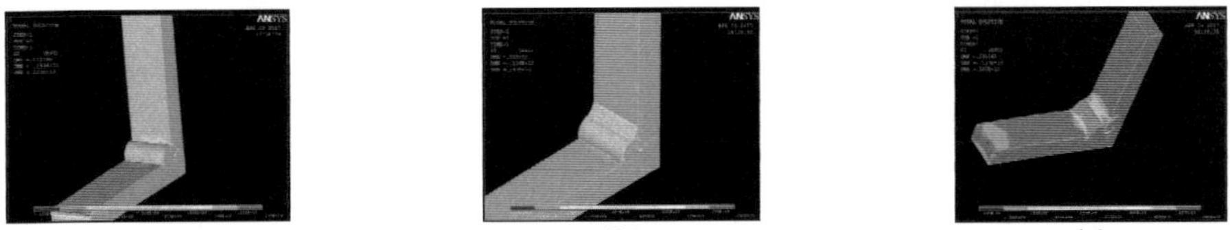

(a) (b) (c)

Figure 6. S2 stress calculation: a – without WIG remelting; b – one side of the welded seam WIG remelting; c – two sides welding seam with WIG remelting.

The S3 stress (that mainly calculates the compression stress state) calculation shows the same dissipation state of stress when WIG remelting is performed. In the figure 7(c) the maximal stress value in the studied zone is S3 = 0.7·E9 N/m^2.

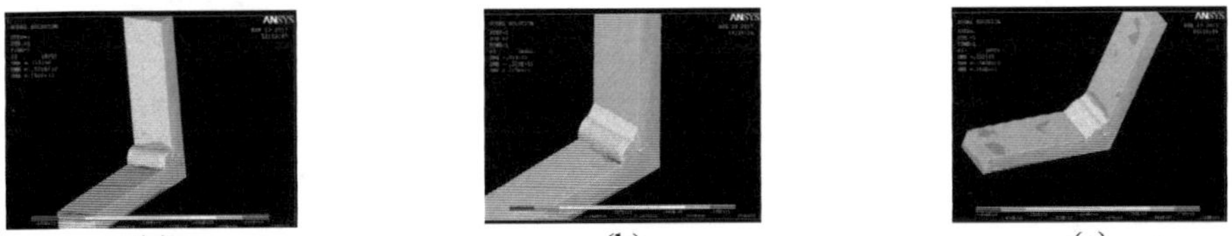

(a) (b) (c)

Figure 7. S2 stress calculation: a – without WIG rehabilitation; b – one side of the welded seam rehabilitation; c – two side welding seam WIG remelting.

To decrease the stress level in the welded seam, the following technique provides two other seams deposited over the initial one. This is present in figure 8(a) and the WIG rehabilitation technology in figure 8(b). It can be seen how the value of the S1 stresses decreases from 0.99·E10 N/m^2 to 0.8·E10 N/m^2.

(a) (b)

Figure 8. S1 stress calculation in the case of two supplementary welding seams: a - without rehabilitation; b - with WIG remelting.

For the S2 stress type, as we can see in figure 9, there is also a decrease of the values of the maximum tensile stresses, for example, from the value S2 = 0.3·E10 N/m^2 to the value S2 = 0.15·E10 N/m^2 in the case of WIG rehabilitation technology. If it is compared with the welding technique without the two additional cords as we can see in figure 7, where the values of the maximum stresses S2 is S2 = 0.56· E10 N/m^2 there is a decrease of the stresses at the value S2 = 0.15·E10 N/m^2.

The deposition of the two additional cords followed by rehabilitation leads to a significant decrease in the level of stress.

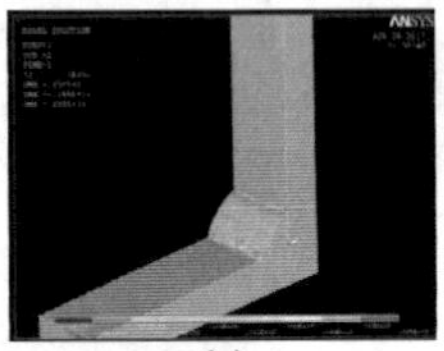

(a)

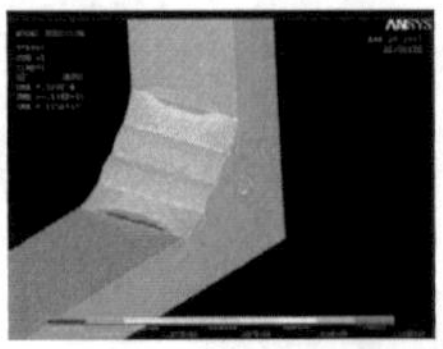

(b)

Figure 9. S2 stress calculation: a - two supplementary welding seams but without rehabilitation; b - with WIG rehabilitation.

From the point of view of the S3 type stress (figure 10), it is also noticed the decrease of their values in case of WIG rehabilitation from the value S3 = 0.19·E10 N/m^2 to the value S3 = 0.86·E10 N/m^2.

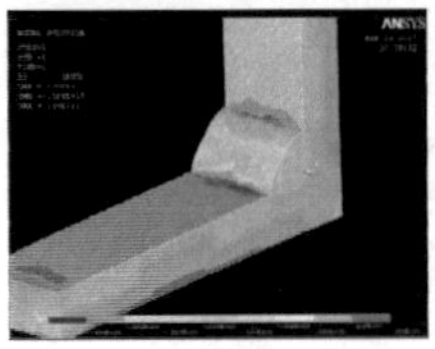

(a)

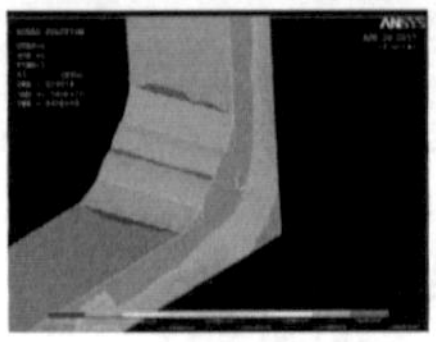

(b)

Figure 10. S3 stress calculation: two supplementary welding seams but without rehabilitation; b – with WIG rehabilitation.

3. Conclusions

Rehabilitation techniques for welded seams are undoubtedly important methods needed to reduce the level of mechanical stress that are often the main cause of failure of welded structure failure [7]. Rehabilitation by the WIG remelting method is one of these working technologies and it is necessary to prove its validity. Do to this, the article presents the evaluation of the level of mechanical stresses in the situation of a fillet weld in which several types of welding seams were made followed by WIG rehabilitation. The evaluation of the stress layer was done successfully by the finite element method using the ANSYS software. There are thus significant decreases in stress values, discussed for each studied case.

4. References

[1] Vikrant G, Sanjiwan B, Sagar I, Swapnil M and Lakade S 2017 Effect of post weld toe treatments on fatigue life of welded structures using FEA *Materials today* Proceedings **4** 1116-1126.

[2] Monika O, Peter P, Milan U, 2017, Fracture mechanism differences created by fatigue and impact test *Materials today* Proceedings **4**(5) 5921-4924.

[3] Željko B, Siegfried S and Hinko W 2018 The effect of residual stresses on fatigue crack propagation in welded stiffened panels *Engineering Failure Analysis* **84** 346-357.

[4] Babis C, Amza G, Nitoi D, Chivu O and Apostolescu Z 2016 Innovative technology for rehabilitation of welded fillet joints, in order to increase the fatigue lifetime *Nano, bio and green technologies for a sustainable future: International Multidisciplinary Scientific Geo Conference (Vienna, Austria 2-5 November 2016)* pp 183-90.

[5] Amza, G, Nitoi D, Babis C, Amza C and Apostolescu Z 2013 Influence of the milling weld toe and WIG remelting reconditioning technology to welded pieces in the case of fatigue testing *Advanced Materials Research* **684** 362-366.

[6] Amza G, Babis C and Nitoi D 2014 The influence of the specific rehabilitation techniques "toegrindig" and "wig remelting" in case of welded structure *Metalurgija* **53** 71-74.

[7] Ramalho A, Ferreira J A M and Branco J M 2011 Fatigue behaviour of T welded joints rehabilitated by tungsten inert gas and plasma dressing *Materials and Design* **32**(10) 4705.

IOP Conf. Series: Materials Science and Engineering 1182 (2021) 012055 doi:10.1088/1757-899X/1182/1/012055

Race car mirror cover production focused on reducing air drag

A Nowak, A Baier, A Kołodziej and P Żur
Silesian University of Technology, Department of Mechanical Engineering, Konarkiego 18A, 44-100 Gliwice, Poland

E-mail: adnowak2@gmail.com

Abstract: The paper presents a mirror cover manufacturing process, focused on geometry optimization by reducing air drag for Silesian Greenpower electric car. Silesian Greenpower is a student's interfaculty project, of which aim is to design, build, and race an electric vehicle. Each car is equipped with the same motor and batteries. Target is to finish as many laps as possible during a set amount of time. Major development takes place in aerodynamics to decrease movement resistance. In the presented development process, computer-aided design software was used to create part concepts and then aerodynamics simulation was carried out to determine quality of each design. The influence of chosen geometric parameters was analysed and discussed in the article. The final concept was manufactured using 3D-printing technology with ABS - acrylonitrile butadiene styrene, because of good mechanical properties but also availability and relatively low cost. 3D-printed elements were verified experimentally using smoke stream generator.

1. Introduction

Aerodynamic properties of a product is a major factor in many technical fields such as aviation, energetics, automotive and more. The most obvious method that can be used during development process is experiment. Unfortunately it requires aerodynamic tunnel or at least a room isolated from external conditions, proper source of gas velocity and indicator, for example a smoke stream. Alternatively a numerical method can be used, which require only a computing machine and a proper software. Because of high accessibility of computers, numerical method is very popular, especially in low budget and small series production.

The problem of aerodynamic quality of various elements is widely described in literature. Most of the technical solutions are designed to satisfy multiple criteria and it is not possible to adapt them to different problems. Development process presented in the paper was carried out independently because of the specific geometric constraints not present in other publications.

The aim of the process shown in this paper was to design and manufacture a mirror cover for Silesian Greenpower electric race car shown on figure 1. Silesian Greenpower is a student's interfaculty project, of which aim is to design, build, and race an electric vehicle. Each car is equipped with the same 240 W motor and two 36 Ah batteries and the target is to finish as many laps as possible during a set amount of time. Major development takes place in aerodynamic and lightweight construction field to decrease movement resistance [1, 2].

Figure 1. Silesian Greenpower race car [3].

The mirror cover design process was carried out in Siemens NX computer aided design module. Various concepts of external surfaces were created and properly prepared for the simulation. Later on the final concept chosen based on the simulation results was used to design ready to manufacture part.

The main criterion in the mirror cover designing process was minimal air drag [4, 5]. Each concept was analysed using ANSYS Fluent software. Numerical method was chosen because of high accessibility, short time of experiment and low cost. All cases were simulated with exactly the same boundary conditions.

Final part was manufactured using 3D-printing technology using ABS - acrylonitrile butadiene styrene. Element was also verified experimentally using smoke stream generator towards compliance to the aerodynamic simulation results.

2. Mirror cover concepts

The function of the mirror cover is to attach mirror to the body of the car, allow the driver to adjust the mirror angle for good vision behind the car and to minimalize air drag. Major development took place in optimizing aerodynamics for minimizing air drag. Geometry optimization was limited by the required volume for mirror fitment and a maximum length of 110 mm of the cover. Dimensions of the mirror assembled with the angle adjusting element are shown on figure 2.

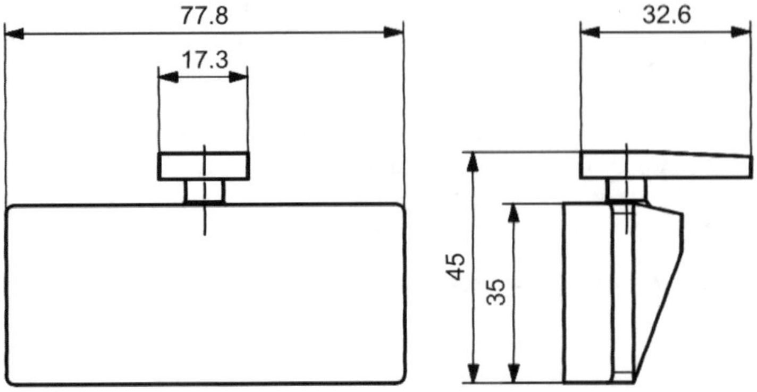

Figure 2. Mirror dimensions in millimetres.

IOP Conf. Series: Materials Science and Engineering 1182 (2021) 012055 doi:10.1088/1757-899X/1182/1/012055

Development process shown below was carried out using Siemens NX computer-aided design module, mostly with advanced surface modelling tools. At this stage of research only external surfaces, that have an impact on aerodynamics were created to simplify the geometry for later simulation. Particular attention was paid to the surface quality and type of transition between surfaces. In the direction of air velocity during normal conditions car movement continuity of the surface is G2. On plane normal to air velocity continuity is at least G1, depending on the concept.

Previous unoptimized model of the mirror cover was presented on figure 3. Mirror dimensions are shown using a blue line. First concept of improved mirror cover was a similar rotary element but with optimized smooth surface and smallest possible maximum diameter that satisfy the mirror fitment limitation (figure 4). Because of relatively wide and short shape of the mirror, next designs are non-rotary. One of them is based on an ellipse to compromise smooth surface and low frontal area of the element (figure 5). Last concept was based on a rectangle with rounded vertices to create minimum possible frontal surface at the cost of surface smoothness (figure 6).

Each of the external surface concept was converted to a volumetric element. Final simulation model was a 2500x500x500 cuboid with subtracted volume of the mirror cover. Ready for simulation model was converted to neutral *.stp format.

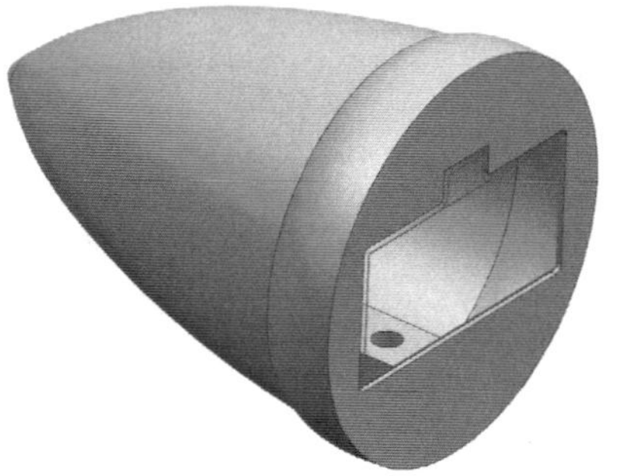

Figure 3. Old mirror cover design.

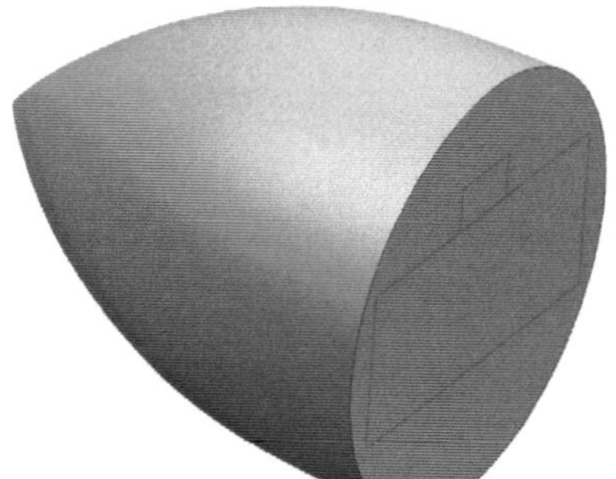

Figure 4. Round based concept.

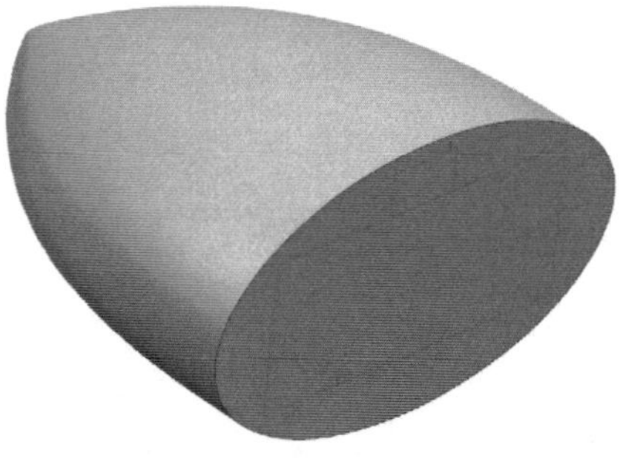

Figure 5. Oval based concept.

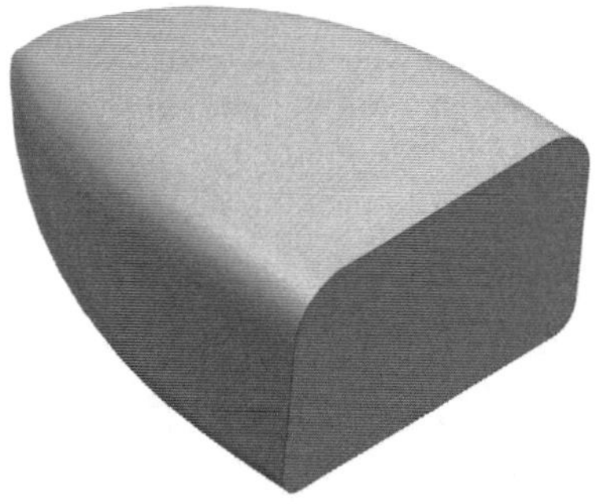

Figure 6. Rectangle based concept.

Quality of the surface was checked using curve analysis and reflection analysis available in Siemens NX environment. For curve check, curves that are intersections of part surfaces and main planes were found and then graph with local extremes was created. Reflection tool allows to project set amount of lines on the analysed surface that change position depending on view movement. Because of that, any surface flaws are more visible on a computer screen.

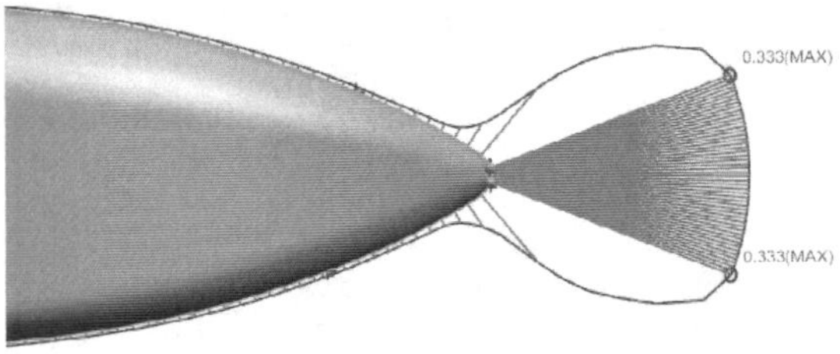

Figure 7. Curve analysis with logarithmic scale.

Figure 8. Reflection analysis.

Used surface quality check methods were presented on figures7 and 8, on an example of the most complex and problematic geometry analysed in this paper. On figure 8 there are some minor flaws observed in surface quality. Keeping in mind that volume of production is very small and that the planned production method using 3D-printing has limited accuracy, these flaws are acceptable. Further improvement of the model is possible but not necessary for purpose described in the paper.

3. CFD analysis

Aerodynamic simulation was carried out using ANSYS Fluent software. Fluent calculations are based on numerical method called finite volume method. It involves a fluid volume approximation mesh and counts values for every node in the centre of each mesh element, then approximate results to the rest of the volume. Values come from partial differential equations, that are solved by transformation to a system of algebraic equations. Fluent can be used to numerous of issues in the field of fluid mechanics such as turbulence, one or multiphase flow, heat transfer etc. [6].

For the presented case software was used to determine drag force, drag coefficient and air flow distribution. All cases were simulated with exactly the same boundary conditions. Airflow inlet speed was set to 20 m/s which corresponds with maximum speed during an endurance race. Air density is 1.2 kg/m^3 at the temperature of 20°C. Mathematical model used was a k-omega standard viscous-laminar. Mesh elements used for discretization were second-order tetrahedrons with low average skewness [7]. The mesh was finer around the shape of mirror cover to obtain more accurate results using fewer elements to save time and computing power. For each case element size around cover was set to 1.5 mm and maximum size at the outer part was 20 mm. Based on cover surface 9 inflation layers were generated with total thickness of 5 mm and 1.1 growth rate. Number of elements for each case varied insignificantly at about 800000. Example of used mesh was presented on figure 9.

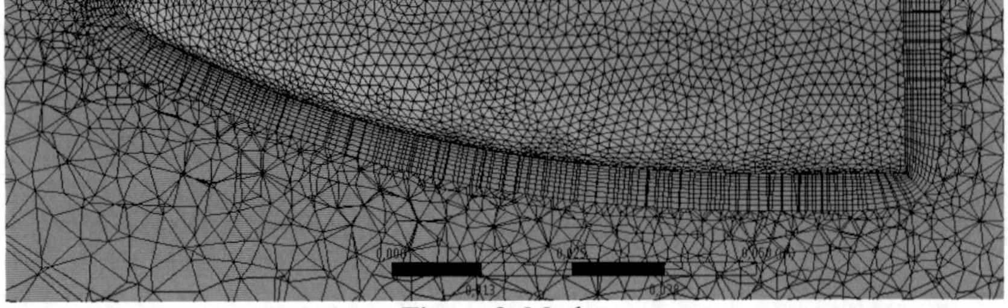

Figure 9. Mesh.

Simulation was carried out using double variables precision and until obtaining convergence for high calculation accuracy [8].

IOP Conf. Series: Materials Science and Engineering 1182 (2021) 012055 doi:10.1088/1757-899X/1182/1/012055

4. Results and discussion

For the results visualization in Fluent environment pathlines function was used. Air particles trajectory paths as well as their speed on the plane of symmetry for each case were presented on figures 10-14.

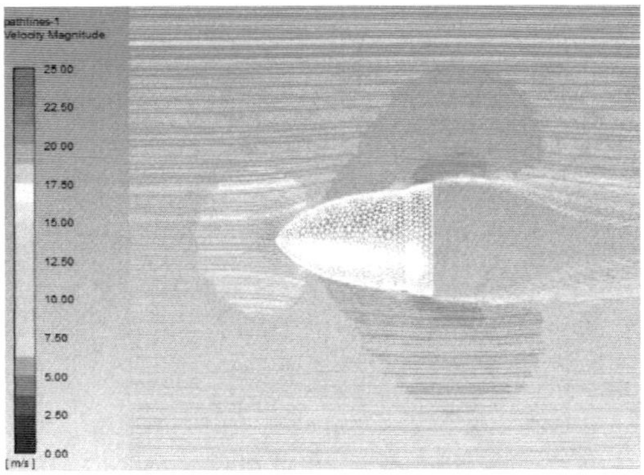

Figure 10. Old model design.

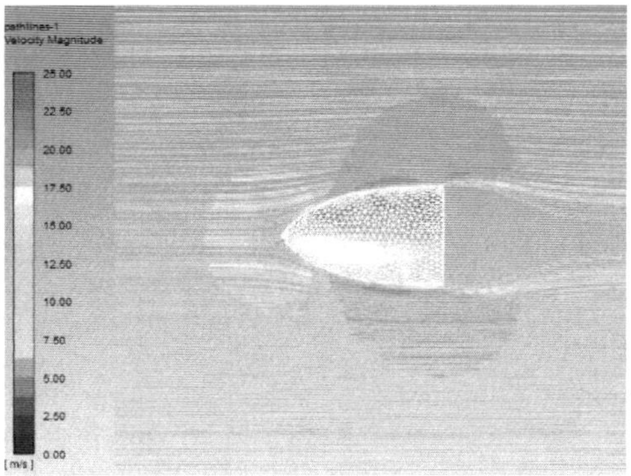

Figure 11. Round model design.

Figure 12. Oval model design.

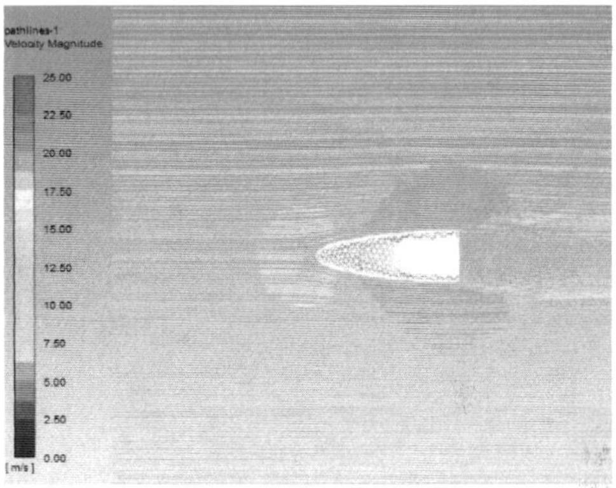

Figure 13. Rectangular model design.

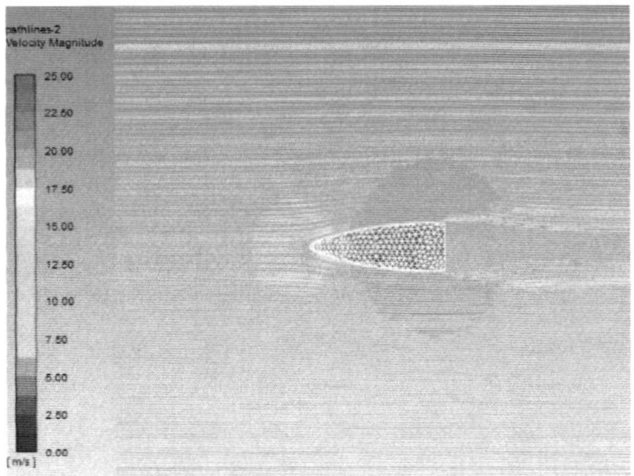

Figure 14. Rectangular model with wide tip.

The result drag forces for each geometry conception were presented in table 1, as well as calculated frontal surface area and drag coefficient Cx.

Table 1. Simulation results.

General shape description	Force [N]	Frontal area [mm^2]	Drag coefficient Cx
Old model design	0.605	8825	0.286
Round model design	0.355	6362	0.233
Oval model design	0.328	4968	0.275
Rectangular model design	0.245	3665	0.279
Rectangular model with wide tip	0.243	3665	0.276

Based on table 1 the smallest drag force F=0.243N was archived for rectangular shape with a wide tip, which equals to about 40% of the original design force. Chosen concept was converted into a thin-walled volumetric model, filled with necessary mounting points. During this process main criterion was to limit external surface disruption to minimum to keep air drag force low.

Final CAD model was converted into a *.stl format, that is required for the 3D-printing process. This manufacturing technology was chosen because of very complex shape of the part and small planned production volume. Process was performed using FDM technology with material called ABS - acrylonitrile butadiene styrene. The filament used in 3D-printing process was Smart ABS by Spectrum Filament. Temperature of the printing was set to 230^{0}C which is the lowest recommended by the producer value, resulting in the highest impact resistance and toughness. Infill density of the print was 50% to maintain high mechanical properties but also reduce material consumption and time of the printing [9-11]. Figures 15 and 16 shows result of mirror covers 3D-printing process.

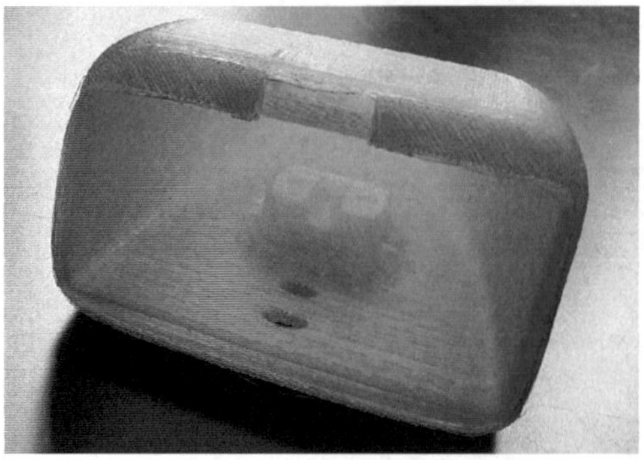

Figure 15. ABS mirror cover.

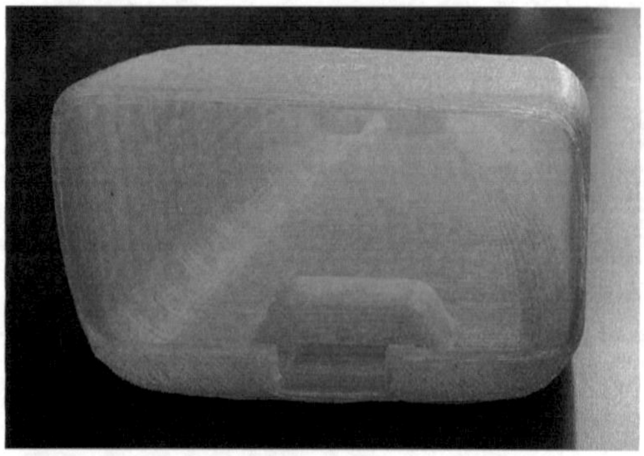

Figure 16. ABS mirror cover upside down.

3D-printed elements were verified experimentally using smoke stream generator towards compliance to the aerodynamic simulation. Experiment process was shown on figure 17.

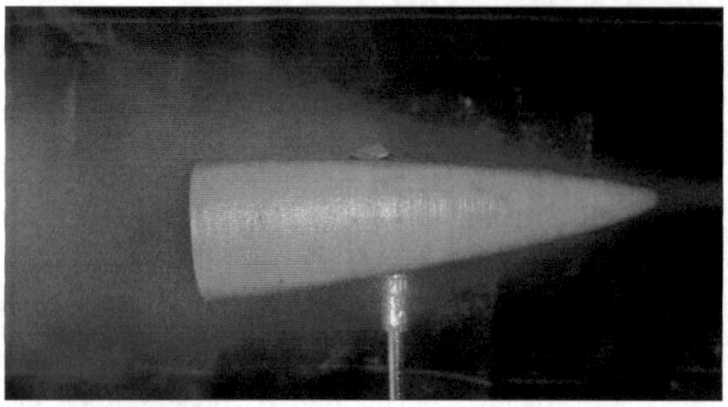

Figure 17. Aerodynamic experiment.

IOP Conf. Series: Materials Science and Engineering 1182 (2021) 012055 doi:10.1088/1757-899X/1182/1/012055

5. Conclusions

Based on the presented results, the following conclusions were made. Comparing two similar rotary figures, one with a high-quality G2 surface had a significantly lower drag coefficient than the unoptimized model. Non-rotary, irregular objects also had a high Cx number, even with an optimized smooth surface.

On the other hand, the frontal area had such an impact on drag force, that some shape regularity and consequent Cx value can be sacrificed. With the criterion of minimum drag force and given mirror dimensions first priority should be minimalizing frontal area, then surface enhancement. Widening of the cover tip negligibly influenced the drag force. Presented force reduction may be caused by numerical error but also it is possible that wide tip shape with the higher similarity of vertical and horizontal curvature reduces drag force value. In the case of very advanced geometry optimization, it can be a field of further investigation. Aerodynamic experiment confirms that air flows around mirror cover correctly but with available equipment it is not possible to see details like air turbulence. After experiment conditions improvement, more accurate check of the compliance to the simulation could be performed.

6. References

[1] Baier A, Grabowski Ł, Stebel Ł, Komander M, Konopka P, Kołodziej A, Żur P 2018 Numeric analysis of airflow around the body of the Silesian Greenpower vehicle *MATEC Web of Conferences* **178** 05014 doi: 10.1051/matecconf/201817805014.

[2 Baier A, Baier M, Sobek M, Grabowski Ł, Dusik D, Papaj P 2014 Computer aided process of designing the mechatronic Silesian Greenpower electric car *Advanced Materials Research* **1036**, Trans Tech Publications ISSN 1662-8985.

[3] https://www.poznan.pl/mim/sport/news/koniec-rywalizacji-w-wyscigowych-samochodowych-mistrzostwach-polski,110349.html accessed on 5.12.2020.

[4] Grabowski L, Baier A, Buchacz A, Majzner M, Sobek M 2015 Application of CAD/CAE class systems to aerodynamic analysis of electric race cars *In IOP Conference Series: Materials Science and Engineering* **95**(1) p. 012044.

[5] Grabowski Ł, Labus M, Sobek M 2013 Projektowanie i wytwarzanie kompozytowego nadwozia elektrycznego bolidu wyścigowego *Projektowanie i Kontrukcje*, Wydawcitwo ITER, Warszawa, ISSN: 1899-699X.

[6] Jeżowiecka-Kabsch K 2001 *Mechanika Płynów*, Oficyna Wydawnicza Politechniki Wrocławskiej

[7] Żur P, Kołodziej A, Baier A, Grabowski L 2020 CFD Simulation of Airflow around the Front Fairing for Electric Race Car *TEST Engineering and Menagement* **83** 17670 – 17675.

[8] Stebel Ł, Grabowski Ł, Sobek M 2017 *Numeryczna analiza przepływów powietrza wokół karoserii pojazdu zespołu Silesian Greenpower*, MESsener.

[9] Baier A, Kołodziej A, Żur P 2019 Finite Elements Analysis of PLA 3D-printed Elements and Shape Optimization *European Journal of Engineering Science and Technology* ISSN2538-9181.

[10] Baier A, Kołodziej A, Żur P, Komander M, Konopka P 2018 Studies on optimization of 3D-printed elements applied in Silesian Greenpower vehicle *IOP Conference Series Materials Science and Engineering* **400**(2) 022010 doi: 10.1088/1757-899X/400/2/022010.

[11] Żur P, Kołodziej A, Baier A, Kokot G 2020 Optimization of Abs 3D-printing method and parameters *Eur. J. Eng. Sci. Tech.* **3**(1) 44-51.

IOP Conf. Series: Materials Science and Engineering 1182 (2021) 012056 doi:10.1088/1757-899X/1182/1/012056

A Study of a Fluid Flow in the Jet Ejector System Used in Industrial Applications

I Olaru
"Vasile Alecsandri" University of Bacau, Calea Marasesti 157, Romania

E-mail: ionelo@ub.ro

Abstract. Over time, numerous studies have been carried out for jet ejectors used in refrigeration systems for various industrial applications or in solar energy recovery. These ejectors use steam or water as a working fluid because the water has a low cost, chemical stability and is safe to use. Of course, depending on the application of these ejectors in the industry, other fluids can be used. When are made the ejectors, in addition to the selection of the refrigerant, which is very important, it is taken into account, as well as its constructive design has a strong influence on its performance. In general, the compression ratio depends on the nozzle and the diffuser geometry. Compared to other systems for industrial applications, ejector systems have some advantages: simplicity in construction, high spread and low cost. However, we cannot observe that they also have a lower coefficient of performance than conventional systems, which limited the widespread application of ejector systems.

1. Introduction

Recent studies have attempted to find a wider application of the ejector-type systems, generally being used in refrigeration and air conditioning applications, generally in applications that use small amounts of thermal energy, such as that based on solar energy and heat resulting from the various industrial processes [1].

Some studies have shown that by using a variable geometry of the ejectors a high performance can be obtained, implicitly they can be used in different operating conditions. If we simplistically analyze such a refrigeration system with ejector we can see that it consists of: pressure generator, evaporator, condenser, expansion vessel, ejector jet and a circulation pump [2,3].

In this system, the high-pressure fluid coming from the generator passes through the jet ejector and drives the low-pressure fluid from the evaporator. Both fluids are mixed in the center of the ejector and recover some of the pressure in the ejector diffuser; the fluid then goes into the condenser [4].

The condensate is divided into two parts: one part is pumped back to the generator, the other part flows through the expansion vessel and enters the evaporator, the cycle being resumed, the vapors flow again through the jet ejector, thus ending a system of refrigeration cycles [5].

If the whole cycle is examined, it can be seen that the system obtains heat from the heat source in the generator using the circulation pump, so the system produces a cooling effect in the evaporator and then dissipates the heat in the medium through the condenser, figure 1 [6].

From this study we conclude that the ejector is the key component in the refrigeration cycle, analysis and optimization of its design can improve the overall performance of the system. The graph with the flow parameters in an ejector considered ideal is also presented in figure 1.

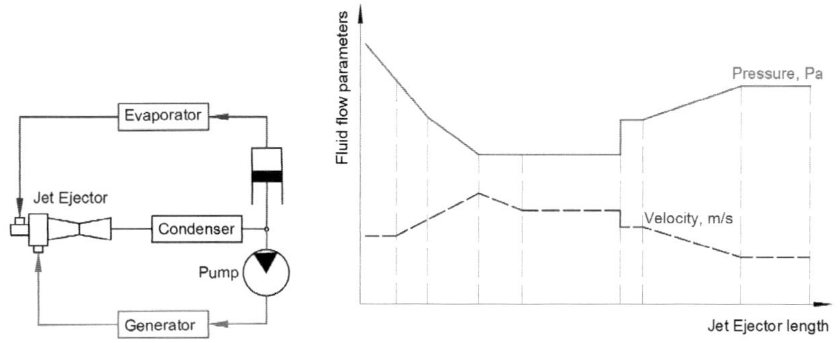

Figure 1. Jet Ejector System.

In this paper we propose to analyze the process inside the jet ejector and thus following this analysis we can optimize its geometric parameters. Thus, the variation of the working fluid flow parameters inside is analyzed, following the entire behavior from the inlet to the fluid outlet from the ejector.

2. Analysis setup

Steam jet ejectors are technological devices that generate vacuum, as we said before, they are found in various applications in the industry. Due to the fact that they have no moving parts, do not require expensive maintenance, can be made of almost any material and can be used in any industrial application [7].

As we have established the cheapest working fluids are water and steam. They can be used successfully in petrochemical processes, thermal compressors, in the food industry for the deodorization of cooking oil and in any other system that uses vacuum.

Therefore, an ejector can be explained as a type of vacuum pump or compressor. In a steam jet ejector, the suction chamber is connected to the diffuser and the steam nozzle by means of flanges. The Venturi diffuser is also composed of two conical areas joined in the middle by means of flanges.

The characteristic dimensions of the main components are necessary to be able to configure the whole system according to our needs, figure 2 [8].

As the air is entrained by the steam under pressure, the mixture passes through the ejector into a Venturi-type diffuser. In the nozzle, its velocity energy is converted into pressure energy, which helps to discharge the mixture at a predetermined outlet pressure. This outlet of the mixture can be discharged into the atmosphere or sent on to a condenser.

As we established in the introduction, the steam jet vacuum systems combine ejectors, capacitors and interconnecting pipes to ensure phase ejection. These systems operate on the ejector-Venturi principle, which relies on the impulse of a high-velocity steam jet to move air and other gases from a pipe or connecting vessel [9].

Figure 3 exemplifies all these components of such an ejector type.

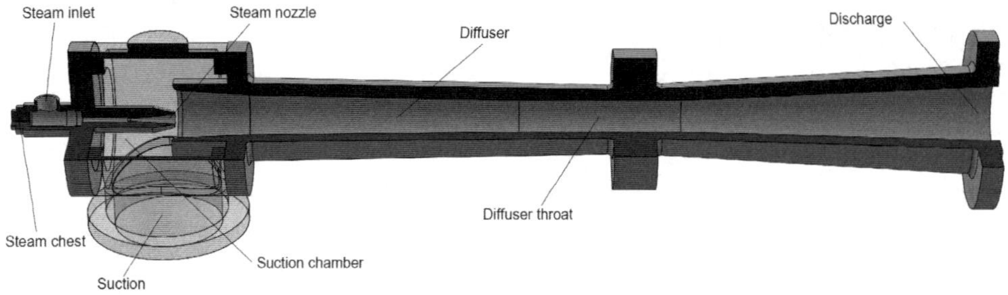

Figure 2. The Jet Ejector component.

If we analyze the nozzle through which the steam is sent under pressure, we see that it is composed of several components: steam inlet chamber through which the steam is introduced, this chamber is provided with a stopper to prevent steam from leaving this chamber. If necessary, it can be replaced with a pipe through which another working fluid enters, so that there is a premix in this first chamber.

If is necessary, the initial nozzle can be provided with an extension pipe for the whole assembly, figure 4 on which the converging nozzle is assembled from the final section. The extension piece is required to control the discharge of the pressure jet into the vacuum chamber [8].

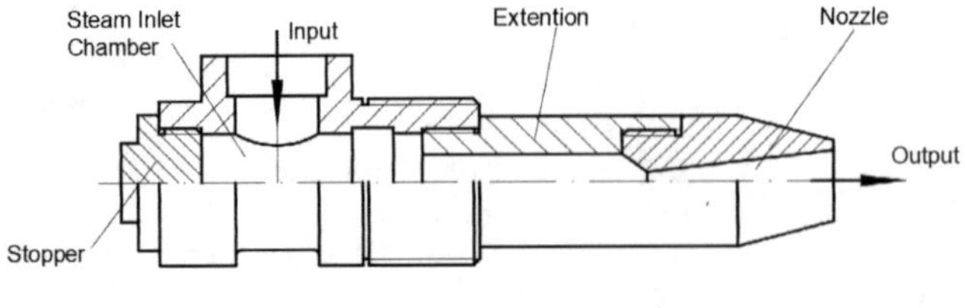

Figure 3. Pressured steam nozzle.

All this initial assembly is connected to the whole assembly by a connecting flange which is threaded to one end of the suction chamber. This nozzle sends the pressure steam further into the suction chamber, creates a vacuum involving another fluid and mixes them so that the resulting fluid is then passed through the Venturi diffuser.

3. Results and discussion

From a constructive point of view, we have in figure 4 a section through the ejector assembly. As mentioned above, the initial high-pressure nozzle is assembled by the mixing chamber by means of the flange with screws and nuts, as well as the assembly of the two sections of the Venturi nozzle, as well as the connection between it and the pipe system.

Because the capacity of a single ejector is determined by its size, a single unit has practical limits on the total compression and flow it can provide. For higher compression, two or more ejectors can be arranged in series. For greater fluid flow capacity, two or more ejectors must be arranged in parallel.

In a multi-stage system, capacitors are usually used between successive ejectors. By condensing the vapor before sending the flow to the next step, the fluid load is reduced. This allows the use of smaller ejectors and thus reduces the main consumption of liquid.

This paper analyzes the flow of the working fluid, and with the help of multiple extensions it can be controlled and analyzed from several points of view and thus to better realize the optimal position of the steam nozzle.

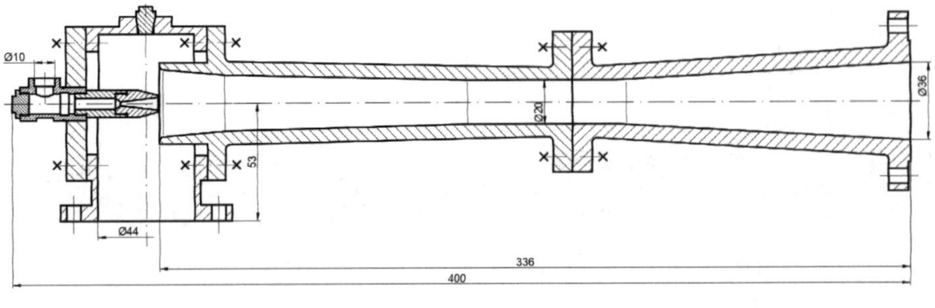

Figure 4. Dimensional elements of the ejector.

In figure 5 several positions for discharging the steam nozzle under pressure were established. It started from the discharge before the beginning of the cone of the Venturi tube. The distances established for the nozzle being 17, 32, 37, 42 and 47 mm. Figure 6 shows how the nozzle looks at the size of 32 mm, practically this being aligned with the outer size of the diffusion tube.

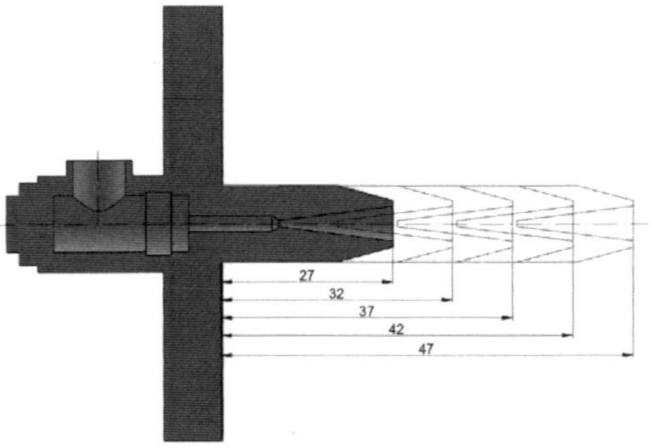

Figure 5. The positions of the ejector head analyzed in the application.

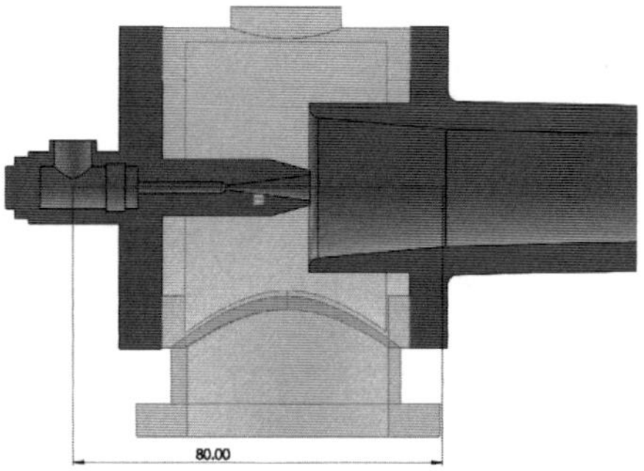

Figure 6. Analysis of fluid flow in the steam outlet area.

In this paper, an analysis was performed by a specialized program SolidWorks which has a module that can analyze the flow of fluids through different geometries. After the three-dimensional realization of the whole assembly, a real situation was simulated by which a steam was sent with a pressure of 10 bar through the initial nozzle [10].

The secondary liquid that is absorbed into the suction chamber has a pressure close to atmospheric pressure. As mentioned, this pressurized steam is mixed with the fluid which is sucked through the suction chamber and sent through the divergent convergent nozzle into the system.

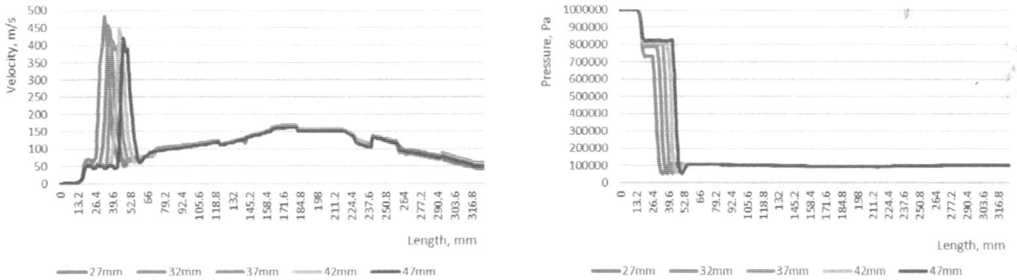

(a) (b)

Figure 7. Analysis of flow parameters on the entire section of the ejector: (a) Velocities distribution; (b) Pressures distribution.

Figure 7(a) shows the velocities distribution through the entire ejector and in figure 7(b) is presented the pressures distribution graph along the entire length of the ejector from the steam inlet to the nozzle until its final exit from the diffuser.

After initial analysis I set out to do a simulation of fluid flow which is input 5 bar but to follow the flow only in the download, such as shown in figure 6 was taken away 80 in the axis and have taken parameters flow. This area was chosen because it has the highest pressures and velocities according to the first simulations at the inlet pressure of the working bar of 10 bar.

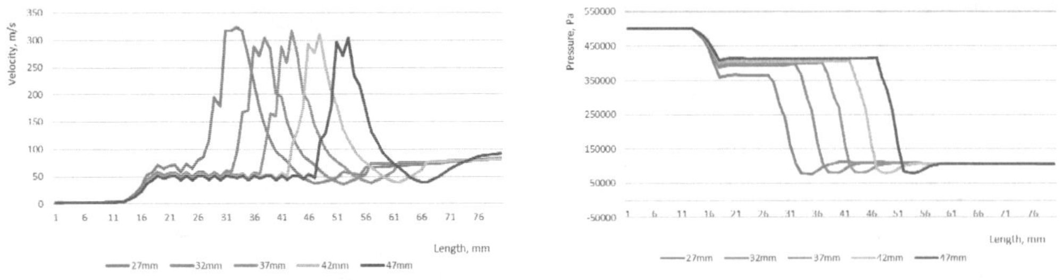

(a) (b)

Figure 8. Fluid flow analysis in the nozzle discharge area:
(a) Velocities distribution; (b) Pressures distribution.

Thus, they could be extracted from the suction pressure analysis in relation to the equivalent vapor load calculated at room temperature. The characteristic curves of the flow parameters related to the main axis of the ejector over a distance of 80 mm from the working fluid in the nozzle are shown in figure 8. The pressure and velocity distribution along this distance was simulated at the working pressure of 5 bar.

In the first interval we can observe an increase of the initial fluid velocity through the pressure nozzle, as the fluid velocity increases, the initial pressure decreases constantly along the initial assembly. In the mixing chamber, there is an approximately constant level of the respective velocities for pressures, because here it achieves a relaxation of the steam initially introduced in the ejector.

A third section is due to the Venturi type at the end of this section, the pressure will decrease and the velocity will increase in this area due to its divergent convergent section, and in the final section it will relax following these fluids to be sent further in the installation.

4. Conclusions

We concluded that the ejector size is the most important parameter for the efficient performance of the ejector system. The dimensions of the single-stage ejector presented in this paper provide better performance.

A smaller outlet area of the nozzle can avoid the loss of shock wave energy, the study on the five dimensions can show us the optimal position for different cases. It is also important to choose the Venturi diffuser properly so that the entire ejector is as efficient as possible.

As previously mentioned, the main advantages of steam jet ejectors compared to other types of vacuum units are primarily low cost, they do not require moving parts, which leads to their simple construction. Due to their simplicity, these ejectors are reliable, maintenance is simple and easy to build.

The ejection units can be made of any material that offers good resistance to corrosion and erosion, being easy to install, does not require special foundations or special fasteners. However, the parameters of the ejectors may differ for their different geometries under the same operating conditions.

To optimize these installations, steam jet systems generally combine several ejectors, capacitors and interconnecting pipes to ensure a control of the duty and pressure of the working fluids, thus an optimization of them, while maintaining low costs.

5. References

[1] Selvaraju A, Mani A 2004 Analysis of an ejector with environment friendly refrigerants, *Applied Thermal Engineering* **24** pp 827-838.

[2] Wenju H, a.o. 2017 The Influence of Ejectors Structure Parameters on the Performance of Ejector Refrigeration System, *Procedia Engineering* **205** 2683-2690.

[3] Ruangtrakoon N 2013 CFD simulation on the effect of primary nozzle geometries for a steam ejector in refrigeration cycle *International Journal of Thermal Sciences* **63** 133-145.

[4] Gutiérrez A, León N 2014 Conceptual development and CFD evaluation of a high efficiency–variable geometry ejector for use in refrigeration application *Energy Procedia* **57** 2544-2553.

[5] Saito Y, a.o. 2014 Ejector Configuration for Designing a Simple and High-Performance Solar Cooling System *Energy Procedia* **57** 2564-2571.

[6] Yapici R 2008 Experimental investigation of performance of vapor ejector refrigeration system using refrigerant R123 *Energy Conversion and Management* **49** 953-961.

[7] Rusly E, Aye L, a.o. 2005 CFD analysis of ejector in a combined ejector cooling system *International Journal of Refrigeration* **28**(7) 1092-1101.

[8] Olaru I 2020 A fluid flow analysis of a jet ejector system used in industrial applications *Journal of Engineering Studies and Research* **26**(3) 143-147.

[9] *** https://www.graham-mfg.com/technical-documents-library/ejectors (10.03.2021).

[10] ***https://www.solidworks.com/sw/resources/getting-started-simulation-and-analysis-tools.htm accessed on 02.03.2021.

IOP Conf. Series: Materials Science and Engineering 1182 (2021) 012057 doi:10.1088/1757-899X/1182/1/012057

An Analysis of the Thermal Simulation of Parts Processed by Cutting, with Different Geometries

I Olaru and C Olaru
"Vasile Alecsandri" University of Bacau, Calea Marasesti 157, Romania

E-mail: ionelo@ub.ro

Abstract. In practice, different cutting processes are encountered, each process having different characteristics. Due to the existing friction forces between the workpiece and the cutting tool, a significant amount of heat is generated. This amount of heat can be determined by experimental measurements. With the help of the SolidWorks program, the geometric element that must be studied can be made, so that later it can be studied from the point of view of the thermal behaviors following the processing. We will take into account the different critical areas that may occur and that may materialize through areas where the final surface is not compliant. This paper aims to study the way in which the temperature is propagated in the workpiece, so that the phenomenon can be understood and determined how this excessive heat can be removed from the cutting area.

1. Introduction

There are many mechanical processing processes that can be considered for different materials and geometric elements to be processed. Mechanical processing generally involves the removal of additional material using various processing tools. During the mechanical processing of cutting, are generated some areas of increased friction between the tool and the workpiece. These frictional forces encountered during processing generate heat, which can be excessive and can damage the cutting process [1].

As we already know, there is an interdependence between the chosen material, the processing regimes and the heat generated by the cutting: the cutting regimes, the tools used, the cutting tools used to make the processing. The shape of the chips is an important indicator of the cutting process; through this we can realize the degree of plastic deformation suffered by the layer of detached material. Thus, following previous studies on different types of processing and different materials, we can distinguish two types of chips: breaking and plastic deformation [2].

In machining, the cutting process is performed by composing two movements: the main movement, the rotation of the tool and the secondary movement, the advance of the part caught in the device on the table of the machine tool. With the help of the milling process, different types of surfaces can be made from simple to complex: flat, conical, bored, wedge or helical channels, gears, etc [3].

Following the study of these heat transfer phenomena that occur in the mass of the workpiece, we can say that we have a conductive heat transfer in the workpiece. Thermal conductivity involves the transmission of heat from close to closer in the particles of the same body, these particles being in contact. As we can easily see, the heat is transmitted in the mass of the material in the cutting area is the highest, practically there must be paid more attention. This heat transfer is due to the difference in heat potential [4].

It is known from the literature that heat transfer between two environments takes place in three ways: convection, conduction and radiation. Conductive heat transfer is characteristic of solid media because it involves an immobility of particles. The convective heat transfer takes place between the fluid and the solid medium (for example, between the workpiece and the atmosphere), sometimes it can be combined with a heat transfer by radiation [5].

Using specialized computer programs by developing and studying the different geometries of the parts, with several types of materials can study various thermal phenomena that occur during the cutting process, thereby being able to achieve their optimization. If we have too high temperatures in some areas during cutting, this will create imperfections in the implementation of the cut surface. To avoid this the temperature can be reduced by choosing an efficient cooling system, so in the processed area the temperatures can be controlled more easily and at the same time established and eliminated the critical areas [6].

2. Simulations setup

As we have established from the introduction, we have a heat dissipation through conduction in the mass of metallic material. In this case if we analyses the Fourier law (equation (1)), then we will know that the heat flux density is proportional to the dropping of the temperature, or inversely proportional to the temperature gradient [7].

$$q = -\lambda \text{ grad } t \ \ [\text{W/m}^2;\ \text{kcal/m}^2\cdot\text{h}\cdot\text{grad}] \qquad (1)$$

where λ is the thermal conductivity [W/m·grad; kcal/m·h·grad], t is the temperature field that is based on the point considered and time.

If we consider and we will focus on case of a homogeneous piece, like the workpiece simulated, the heat conduction is given by (equation (2)):

$$q = \frac{t_2 - t_1}{R} \ \ [\text{W/m}^2] \qquad (2)$$

where t_1 is the temperature in the processing area, t_2 is the temperature of the part before machining, and R is the thermal resistance (equation (3)) and is given by:

$$R = \frac{\delta}{\lambda} \ \ [\text{W/m}^2] \qquad (3)$$

where δ is the length of the workpiece and the same λ is the thermal conductivity.

The working piece chosen to be studied and machined by milling has the following geometric elements, as can be seen in figure 1. The geometry is simple but has a different geometry: with a part with chamfer, another part with fillet and a hole [8].

The workpiece is set to be from aluminum alloy, type AA6016-T4, the chemical compositions for this material is presented in table 1. Initially, the piece has a rectangular shape, it will first process the chamfer, then the fillet area, and at the end the hole will be made. From a thermal simulation point of view, the material has the following thermal properties: thermal expansion coefficient 2.4e-005 /K, thermal conductivity 200 W/m·K and specific heat 900 J/kg·K.

Table 1. Chemical composition (%) of aluminum alloy, type AA6016-T4.

Mg	Si	Fe	Cu	Cr	Ti	Mn	Zn	Others	Al
0.25-0.6	1.0-1.5	0.5	0.2	0.10	0.15	0.20	0.20	≤0.15	Balanced

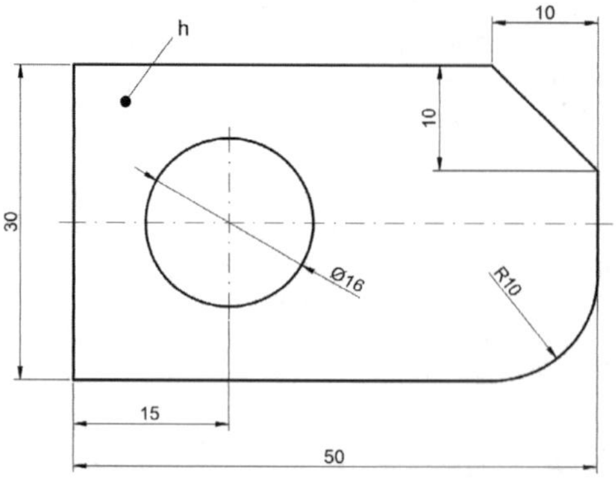

Figure 1. The reference geometry of simulated workpiece.

In computer simulations can use lots of geometric elements of different thicknesses and a wide variety of materials, before they are physically made. For this case, different thicknesses h = 1, 2, 3 and 4 mm were chosen for the study. Each will be studied and simulated and the corresponding conclusions will be drawn.

3. Results and discussion

In order to perform these simulations, a simple geometric element was chosen, these elements being common to most of the machined parts. The way of working is the following: the CAD model was realized and then the SolidWorks 3D model of the chosen element was subjected for analysis. Within the program there is a simulation module of different practical situations, it was chosen to study the thermal load of that element [9].

In figure 2(a) composition of these two processing was analyzed, if they were carried out, some influence on the temperature distribution was observed. In this case, practically there are no almost no part in the piece isn't uninfluenced by thermal machining.

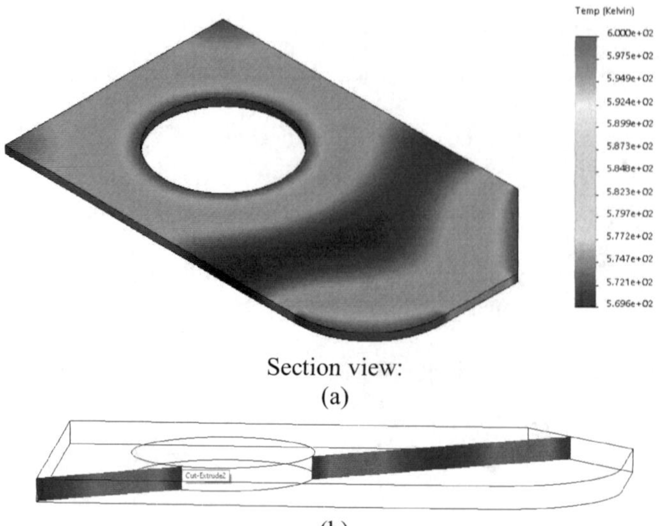

Figure 2. Temperature distribution for simultaneous plane and hole processing.

In the initial part of the computer simulation, the material was chosen, namely aluminum alloy, type AA6016-T4, with its own characteristics. When using other metallic materials, of course there will be differences in the temperature value of the simulations. Figure 3 captures the temperature distribution over a length of 15 mm, in the transfer section, starting with the thermally influenced area.

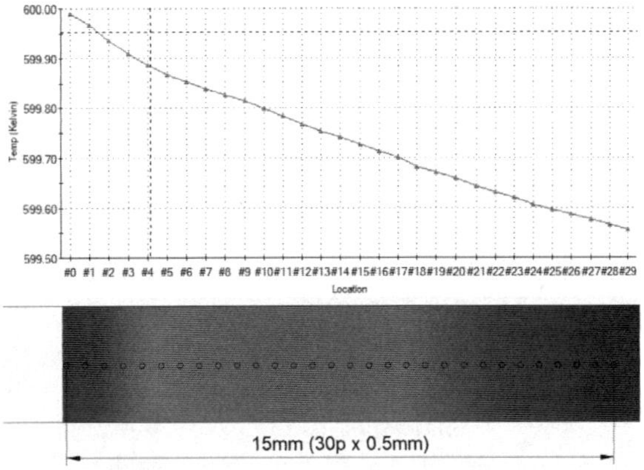

Figure 3. Temperature distribution in selected plane near influenced zone.

In figure 4 an export of a data for a long-term temperature distribution was performed for the main surface of the analyzed element. When these temperatures can be critical to the integrity of the material or other phenomena that are not welcome, it is possible to intervene in the critical areas by adopting efficient cooling systems, so that the temperature reaches normal values. This general temperature graph shows exactly where to optimize.

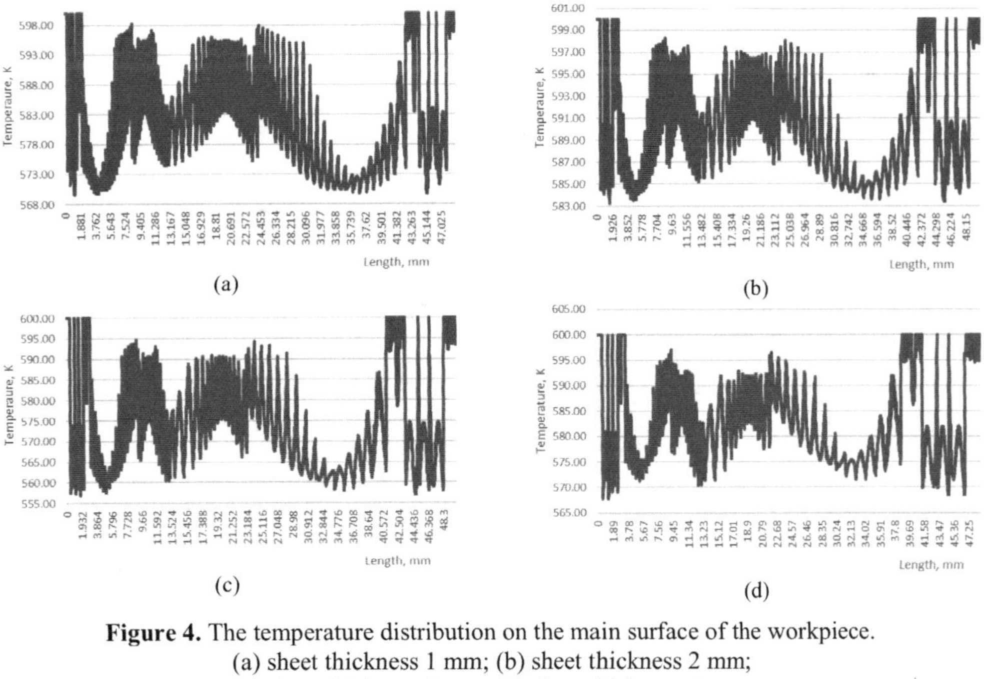

Figure 4. The temperature distribution on the main surface of the workpiece.
(a) sheet thickness 1 mm; (b) sheet thickness 2 mm;
(c) sheet thickness 3 mm; (d) sheet thickness 4 mm.

As can be seen from the previous graphs, the temperature distributions on the entire surface of the part also differ depending on its thickness. In the processed areas the temperature values could be higher as expected.

The areas where deformations or other imperfections may occur in the material are at the contact between the workpiece and the cutting tool. A temperature control in this area is given by the proper choice of material, cutting tool as well as the chosen work regimes.

4. Conclusions

A better knowledge of the phenomena that occur in the case of chip cutting helps to achieve the best processed elements. Excessive heat from the cutting process can only be harmful. This has important repercussions on the newly created surface, as well as residual stresses are generated in the material that cannot be accepted.

In many cases, in the cutting process it is desired to remove this excessive temperature from the cutting area, this is done through complex cooling systems that use simple air-water cooling systems, up to complex emulsions. Particular care should be taken in the choice of materials to be processed and in the processing regimes chosen.

By computer simulation, the cutting parameters can be selected from the beginning, the geometry of the finished part can be studied without having the real element. With this method you can find out what problems can arise by making these surfaces and what would be the cutting temperatures and their propagation pattern.

Choosing a complex cooling system must be done carefully because it may incur additional charges quite high given the cooling system itself but must monitor their environmental impact. When the agent is released into the cooling zone, a quantity of aerosols is released on contact with the processing surface, which can be harmful to the operator's health.

5. References

[1] Olaru I and Olaru C 2020 A Thermal Analysis in Solidworks for the MQL Systems of the Metal Cutting Processes *IOP Conf. Series: Materials Science and Engineering* **916** 012077.

[2] Irani R A, Bauer R J and Warkentin A 2005 A review of cutting fluid application in the grinding process *International Journal of Machine Tools & Manufacture* **45** 1696-1705.

[3] Zgorniak P and Grdulska A 2012 *Investigation of Temperature Distribution during Milling Process of Az91hp Magnesium Alloys Mechanics and Mechanical Engineering* Technical University of Lodz, **16**(1) pp 33-40.

[4] Rimpault X, Alexandre I Chatelaina J-F, Lalondec J-F and Balazinskib M 2018 Workpiece Subsurface Temperature Study During Aluminum Skin Milling in Slotting and Ramping *ScienceDirect, Procedia CIRP* **77** 417-420.

[5] Chang H, Li S and Shi R 2017 Design and Manufacturing Technology of High Speed Milling Cutter for Aluminum Alloy *ScienceDirect, Procedia Engineering* **174** 630-637.

[6] Kesriklioglu S and Pfefferkorn F E 2018 Real time temperature measurement with embedded thin-film thermocouples in milling *ScienceDirect, Procedia CIRP* **77** 618-621.

[7] Petre M V and Morega, A M 2010 Heat transfer by the side wall of an aluminum electrolysis cell *University Polytechnic Bucharest. Scientific Bulletin, Series C* **72**(2).

[8] Olaru I and Olaru C 2019 A study about thermal simulation in SolidWorks in the case of different geometry of the processed pieces by cutting *Journal of Engineering Studies and Research* **25**(3) 39-43.

[9] *** https://www.solidworks.com/partner-product/thermal-analysis-solidworks-simulation-2019 accessed February 2021.

IOP Conf. Series: Materials Science and Engineering 1182 (2021) 012058 doi:10.1088/1757-899X/1182/1/012058

Application of digital twins in specific manufacturing processes

M Olender-Skóra
Silesian University of Technology, Faculty of Mechanical Engineering, Department of Engineering Processes Automation and Integrated Manufacturing Systems, Konarskiego 18A, 44-100 Gliwice, Poland

E-mail: malgorzata.olender-skora@polsl.pl

Abstract. Following the requirements of the modern economic environment is a challenge for entrepreneurs. The more, that there is an increasing need to manufacture product for individual customers, the so-called customization approach. As a result, the production is individualized, so that the repeatability of products and stocks is low. This forces companies to apply specific organizational development strategies such as Lean and Agile. It also forces an increase in flexibility of the implementation of individual production tasks of enterprises. Increased production flexibility, at the same time, influences a greater number of decisions made, data generation and the use of enterprise resources. For this purpose, entrepreneurs use various simulations approaches, including applications for simulation of complex production processes, but nowadays they are increasingly leaning towards the so-called digital twins. The article presents the possibility of using a digital twin and the necessary data required to make such a twin in order to reflect the actual activities carried out in production.

1. Introduction
Following the changing on the market, manufacturers are looking for solutions, that will help them achieve specific goals, not only in terms of new manufacturing perspectives, materials or business processes. Due to market dynamics and increasing customization of products, there is a need for solutions, which will also accelerate the process of task execution and analysis, but also in the decision-making process [1-4]. In order to appropriately coordinate, execute and efficiently implement processes, the concept of a digital twin is increasingly being used. With regard to this term in the first step data is required, which, in addition to appropriate IT solutions, is the basis for the realization of the digital twin [5, 6].

In the article the algorithm of creating a digital twin, the required data from different areas of the enterprise, which are the basis for the transformation into a digital twin are showed. Barriers to the implementation of such solutions are also indicated. In addition, a simulation model of the production process, which is also one of the forms of a digital twin is also showed.

2. Digital twin
Over the past few years, the definition of a digital twin has changed. However, there is still no specific constant definition for this term. Many authors have attempted to clarify and systematize this increasingly popular concept. In the table 1 are showed example of definitions of a digital twin.

Table 1. Definitions of a digital twin [7].

Authors and (year)	Definitions
Grieves, (2014)	"Virtual representation of what has been produced"
Söderberg et al., (2017)	"Using a digital copy of the physical system to perform real-time optimization"
Qi and Tao, (2018)	"Virtual models of physical objects are created in a digital way to simulate their behaviors in real-world environments"
Kannan and Arunachalam, (2019)	"Digital representation of the physical asset, which can communicate, coordinate and cooperate the manufacturing process for an improved productivity and efficiency through knowledge sharing"

These definitions are also changed due to the development of technology, new solutions, the ability to implement these solutions in the company, but also because of intention of the company to make changes [8-11]. However, the basis is the appropriate use of data and resources, in order to flexibly realize the production, from the stage of product design to receipt the product for customers. The implementation of a digital twin requires the merging of multiple levels in the enterprise. In the first level is data. Data, which are extracted from multiple areas of the enterprise at the same time. And accordingly, depending on, which stage will be reflected as a digital twin, the required data are changes. However levels about management, monitoring are important to, as a complete process, in which the relevant management tasks are realized and also integration of all planes of the enterprise.

Therefore, the appropriate data is needed to create a digital twin. And accordingly, depending on the area of the enterprise, different data are required. An example of the data extracted for the execution of a digital twin, as a complete enterprise are showed on figure 1.

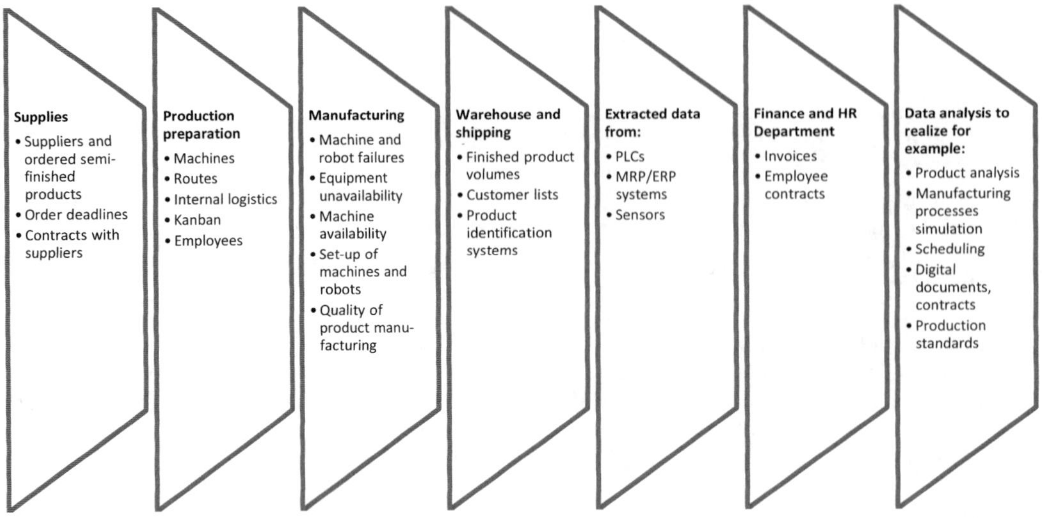

Figure 1. Data extracted for the creation of a digital twin.

Data are the basis for creating a digital twin, however, they differ depending on whether the whole company or only a part of it, will be transformed into a digital version, e.g. production processes and the production preparation area.

The following section of the article describes the digital transformation of a selected manufacturing process in an enterprise.

3. The digital twin of the process

In connection with the creation of a digital twin for a manufacturing process, the first step requires data, which is the input data for the manufacturing process. The input data are: number of machines, routes, number of employees, warehouses, internal transport, number of robots, operation times, etc. Knowing this data, a simulation model is developed and simulations experiments are performed. An example of process simulation is showed in figure 2. The model consists of several machines (processor), operated by workers, a robot, and warehouses. To complete the items on the line, transport vehicles are used.

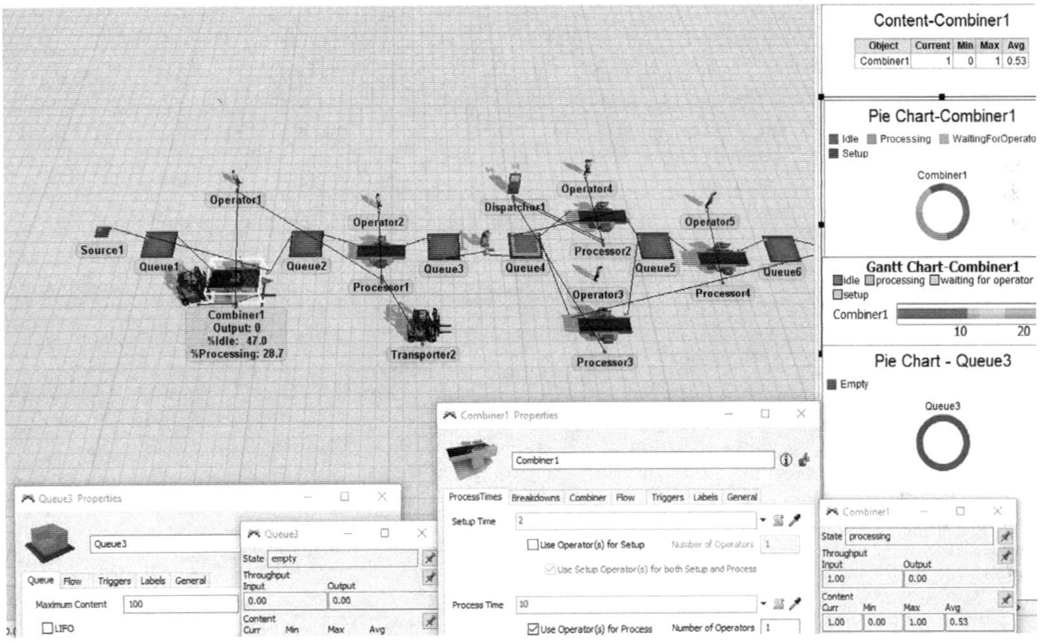

Figure 2. Model of process in FlexSim.

By creating a digital twin of the manufacturing process, is possible to continuously testing, analyzing, and making improvements and retesting to get the best results. In this way, many improvements can be made without actually testing them in reality. This solution allows to save time and money, because changes on the line are implemented, when the results of simulation tests are satisfying. The output data from the simulation for example are: process bottlenecks, storage problems, transport problems between machines, idle times, productivity, lead times, machine layout, etc.

Additionally, in article, in order to illustrate the creation of a digital twin for a manufacturing process, an algorithm was developed (figure 3). This algorithm consists of several steps, that show what data are needed, what should be done, in which moment, for create a digital twin.

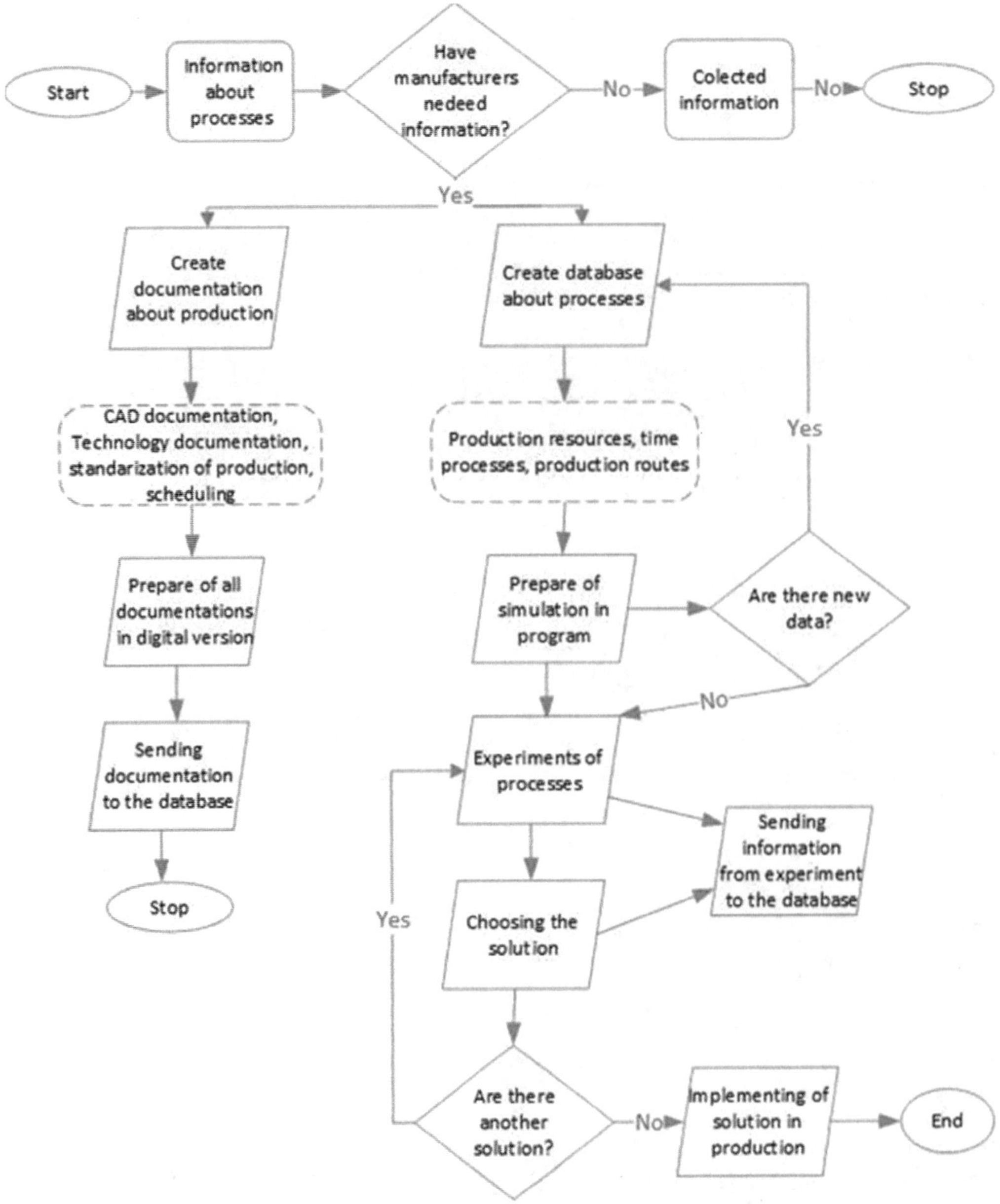

Figure 3. An algorithm for creating a digital twin of processes.

The first step requires data, related to the preparation of production processes, resources, equipment. These data are usually in ERP/MRP systems (Enterprise Resources Planning/Material Resources Planning). But for the needs of the twin, these data must be available in databases for: data analysis, corrections, data about suppliers, dates of orders, periods of cooperation. And in future, these database will be necessary. With a digital twin, this data also becomes more available for decision making person from any place, which simplifies and speeds up the implementation of decision making processes in this area. Then, on these data, appropriate simulations are created, both of production processes, as in the article, but also for analysis of individual components and models in CAD programs. These documents also are transferred to the database for further analysis and interpretation. The advantage of the digital twin of such documentation is the fact, that employee have access to the developed documents and models at any time. This is particularly important at the time of flexible production and small series, repeated from time to time, which is characteristic for the so-called customization of products. This is also intended for introduce certain standards both in terms of product performance, but also in terms of how the company operates, when production are changed. Then, having the input data, including the routes, the production schedules are also developed. By making them in digital version, the required changes can be made at any time. The last and most important factor is to have digital equivalents, on which analyses and tests are performed, which are constantly monitored. So changes are made faster because the decision makers have access to them all the time and not only by the one person. Transparency of data and analysis results also allows for make more effective strategic and tactical decisions, that influence on the functioning of the whole enterprise.

The benefits of introducing a digital twin are significant, however, there are also some barriers to implement these solutions, that keep manufacturers from introducing them into enterprises.

4. Barriers for creating a digital twin

4.1. Required data

Under creating a digital twin, exist a few main concerns, that make manufacturers afraid of this type of solution. The first barrier is the production data which are processed. This data are often the basic data, on which the functionality of not only the production process, but also the entire enterprise is based. In this area there are data about suppliers, lead times, costs, set of machines, employees, failures, etc. In addition, this area includes also all documentation, i.e. CAD drawings, standardization cards, quality cards, machine breakdown cards, results from analyses and simulations, etc. With such a wide range of data, manufacturers are afraid, that by creating a digital twin of the documentation and this data, the data will leak.

4.2. Simulation of manufacturing processes

Due to the implementation of the digital twin in this area, the most common problem for manufacturers, especially for small manufacturers is the problem of access to appropriate software. The second problem is the results of the analyses and simulations, that are made, as well as the improvements made in the simulations.

4.3. Reading data from machines and robots

In this area, the problem connected is to generate data from multiple machines, equipment and robots at one time and in a real time. This is a large amount of data that, in order to be used as a digital twin, must be properly stored. Appropriate databases and folders must be available to store them, because these data are also analyzed. Also, improvements and changes are made based on them. Reading of data from machines and equipment is realized by PLC, HMI panels integrated with code scanner or other sensors. This also involves a high level of integration of the company's IT systems.

4.4. Standarization

Realization of manufacturing requires many tasks to be performed at the same time. However, at each level of production, realization must be at the highest level of quality. For this purpose, standards are introduced for the implementation of individual production tasks, so that each shift of employees make products, that have to fulfill specific standards. Also, in this aspect, thru creating a digital twin, all standardization cards, instructions and procedures for the implementation of work, should be available in digital version in the databases of the enterprise, so that every employee could have access to them.

But a barrier in this area is also access to the database for every production worker and the reluctance of workers to use such solutions.

4.5. Appropriate IT infrastructure
In terms of IT infrastructure, the biggest problem with the digital twin is building the right databases to store the data, and the servers. Currently, mostly databases are created in the cloud. However, a large number of manufacturers are afraid of data leakage and loss of all sensitive data.

4.6. Data security
Due to the collection of large set of data, enterprises are increasingly deciding to store them in the cloud. This is a flexible solution, in which dedicated people have access to data from any place. However, data leakage is a problem. Losing data can cost a company huge money, because some of this data is confidential. In this area, it is necessary to monitor movements on servers, but also to implement appropriate security solutions.

5. Conclusions
Changing customer requirements and a changes in the market are also visible in functioning of enterprises. These enterprises are looking for solutions, that will support them in the implementation of the required tasks, especially since production takes place under a specific customer, so-called customization. Combined with available technologies, to better organize planning and management, enterprises are increasingly taking advantage of the ability to create a digital twin, both in production processes and the entire enterprise.

In the article describes the definition of a digital twin. The possibility of its, creation on the basis of an exemplary production process is indicated. An algorithm for creating a digital twin has also been developed. In addition, in the article points out the barriers, that hold back individual manufacturers from introducing digital solutions in enterprises.

6. References
[1] Foit K, Gołda G, Kampa A 2020 *Processes* **8** 1648 (doi:10.3390/pr8121648).
[2] Sekala A, Gwiazda A, Kost G and Banas W 2018 *IOP Conf. Ser.: Mater. Sci. Eng.* **400** 052009 (doi:10.1088/1757-899X/400/5/052009).
[3] Olender M 2020 *IOP Conf. Ser.: Mater. Sci. Eng.* **916** 012079 (doi:10.1088/1757-899X/916/1/012079).
[4] Krenczyk D, Olender M 2018 *IOP Conf. Ser.: Mater. Sci. Eng.* **400** 022036 (doi:10.1088/1757-899X/400/2/022036).
[5] Kannan K, Arunachalam N 2019 *J. Manuf. Sci. Eng.* **141**(2) 021015 (doi:10.1115/1.4042076).
[6] Yan Hong Lim K, Zheng P, Chen Ch-H, Huang L 2020 *J. Manuf. Syst.* **57** 82-93.
[7] Lim KYH, Zheng P and Chen C 2020 *J. Intell. Manuf.* **31** (doi: 10.1007/s10845-019-01512-w)
[8] Valilai OF, Houshmand M 2013 *Robot. Comp. Integ. Manuf.* **29**(1) 110-127.
[9] Krenczyk D, Kalinowski K, Ćwikła G, Kempa W, Grabowik C, Paprocka I 2020 *Inter. J. Mod. Manuf. Techn.* **XII**(3) 65-71.
[10] Liu J, Zhou H, Tian G, Liu X and Jing X 2018 *Int. J. Adv. Manuf. Tech.* **100**(5-8) 1619-1634.
[11] Olender M, Kalinowski K, Grabowik C 2019 *Intel. Sys. Prod. Eng. Maint.* **835** (doi: 10.1007/978-3-319-97490-3_38).

IOP Conf. Series: Materials Science and Engineering 1182 (2021) 012059 doi:10.1088/1757-899X/1182/1/012059

Structural strength evaluation of a 3000 dwt river-barge, under oblique quasi-static design waves

D S Perijoc, L Domnisoru and C M Angheluta
"Dunarea de Jos" University of Galati, Department of Naval Architecture, Domneasca Street 111, 800201, Galati, Romania

E-mail: silviu.perijoc@ugal.ro

Abstract. The present study consists of the 3D-FEM global structural analysis of a typical open-top, double-hull, 3000 DWT inland navigation barge, having 90 m in length. The 3D-FEM model developed to perform the evaluation idealizes the barge structure with high accuracy, average element size being in the range of 50-75 mm. Two loading cases were analyzed, consisting of the ballast and full loading conditions. For the equilibrium floating condition in calm water and oblique design waves, an eigen iterative scheme has been used to generate the input parameters for the 3D-FEM analyzed cases. The push-barge finite element model has been subjected to oblique quasi-static design waves, headings ranging from 0 to 75 degrees. Wave heights considered in the analysis range from 0 to 1.2 m, step 0.3 m, corresponding to the Danube navigation conditions and freeboard limit imposed by the classification societies for inland navigation vessels. Numerical results following the FE analysis have been assessed by the yielding stress ratio, von Mises criterion, and the maximum deformation, thus indicating the structural areas that need to be improved in the design process.

1. Introduction and theoretical background

Performing a 3D-FEM analysis on extended FE models has become nowadays relatively accessible, allowing more improvements to be included in a single design step, resulting in a final product that meets the safety criteria over its entire life cycle [1]. A large range of commercial inland vessels is represented by un-propelled push-barge structures that offer great flexibility for operating companies at a seemingly low maintenance cost.

The slender body of the barge, combined with a long cargo-hold opening and a shallow draught results in a low bending and torsional rigidity. The sum of the above facts can lead to local failure and loss of structural integrity if stress levels are not evaluated properly in the design stage [2]. Since the structure of typical inland push-barge is subjected throughout its life cycle to structural changes due to minor collisions, either ship-to-ship or ship-to-quay, shallow-water groundings, corrosion, fatigue, the strength and stability of the barge structure can be drastically reduced [3].

The present study outlines the 3D-FEM global structural analysis of a typical open-top, double-hull, 3000 dwt inland navigation barge, having the main characteristics presented in table 1. The scantling and structural layout of the river-barge structure has been performed in accordance with the Bureau Veritas classification society principles for inland navigation vessels [4].

The global structural stress assessment of the river barge, considering the still water condition and different wave heights (H_w) conditions, as well as multiple heading angles were modeled considering the equivalent quasi-static approach [5-7]. Since the scantling calculation has been performed considering the maximum wave height of 1.2 m, IN (1.2) [4], a number of 3 intermediate wave heights have been accounted for a better understanding of the structural behavior in waves' condition.

The above conditions preceding the structural analysis, resulted in a number of 98 loading cases (LC) for wave crest (hogging wave condition) or wave through (sagging wave condition), heading angles in the range of 0-75° (15° steps) and wave heights from 0-1.2 m (0.3 m steps, zero wave height corresponding to still water condition). The numerical analysis involves 3D-FEM barge models and 1D barge models for the wave-barge equilibrium parameters computation and structural response assessment [1].

Table 1. Main characteristics of the 3000 dwt river-barge.

Length overall	L_{OA}(m)	90
Breadth	B(m)	11
Depth	D(m)	4.5
Draught	T(m)	3.8
Displacement, Lightship	Δ_{light}(t)	507.5
Displacement, Ballast	$\Delta_{ballast}$(t)	529.5
Displacement, Full load	Δ_{full}(t)	3620.8
Ballast, no trim condition	$M_{ballast}$(t)	22

2. 3D-model and analysis description

The structural modeling, 3D-FEM model and analysis was developed and performed using Siemens FEMAP program, with Simcenter Nastran solver [8]. The FEA model developed for the current analysis study is presented in figures 1, 2, 3, and 4. The structural model is meshed in the major part of quad plate elements, the use of triangular elements being limited in areas where quad elements quality would be highly affected, thus insuring the aspect ratio within the limits of the Nastran solver (element Jacobian < 0.7). The 3D-FEM model resulted in a number of 2.949.355 elements, 2.886.618 nodes and element size between 50-75 mm, which includes the model from a meshing standpoint, in the fine FEM model category.

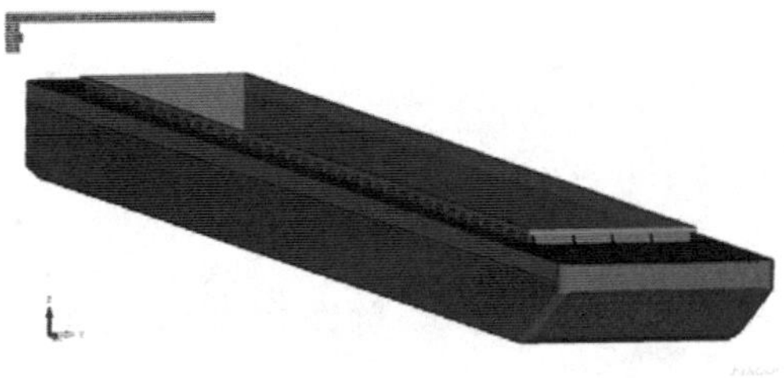

Figure 1. 3D-FEM, barge full model.

Figure 2. 3D-FEM, aft structural detail.

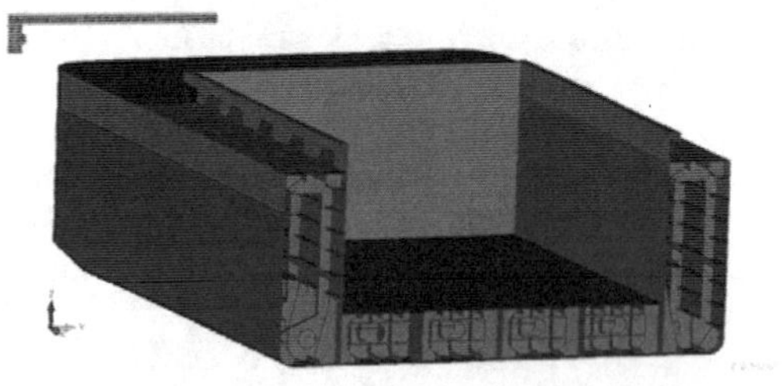

Figure 3. 3D-FEM,
Typical midship structural detail.

Figure 4. 3D-FEM,
Fore structural model detail.

The input data for the iterative program P QSW are represented by the 3D offset line of the ship's hull and the 1D mass diagram extracted from the 3D-FEM model, which has been previously developed. Based on the above numerical information and wave height, the iterative program computes the equilibrium parameters (tables 3 and 4), requiring no boundary conditions in the determination of the equilibrium floating position by program P_QSW. The governing law of the free surface description in oblique quasi-static waves is presented in equation (1) [5].

$$\zeta_w(x,y) = T_M + (X - X_F)\theta + (Y - Y_F)tg\phi \pm a_w \cos k\,\xi \quad (1)$$

$$a_w = h_w/2;\ \xi = X + Y \cdot tg\mu;\ \ k = 2\pi/L$$

where:
- a_w - the wave amplitude;
- L - ship's length;
- X, Y - the coordinates referencing the ship's coordinate system;
- X_F, Y_F - wave reference medium plane center position.

The resulted equilibrium parameters are implemented in the 3D-FEM model by user functions [5, 6] in order to apply the corresponding pressure loading on the ship's hull for the different loading conditions (figures 5 and 6). The analysis study has been divided in 4 major study cases, the ballast loading condition hogging and sagging wave profile, and the full loading condition for similar wave profiles. For the ballast loading scenario 22 t of non-structural mass per area (table 1) have been placed in the aft compartment, reproducing the ballast water, in order to simulate the river-barge structure on an even keel floating condition. For the full loading scenario, the cargo mass has been introduced in the 3D-FEM model as non-structural mass on the cargo-hold floor, and a vertical hydrostatic pressure has been introduced on the longitudinal and transversal cargo-hold bulkheads panels (figures 7 and 8).

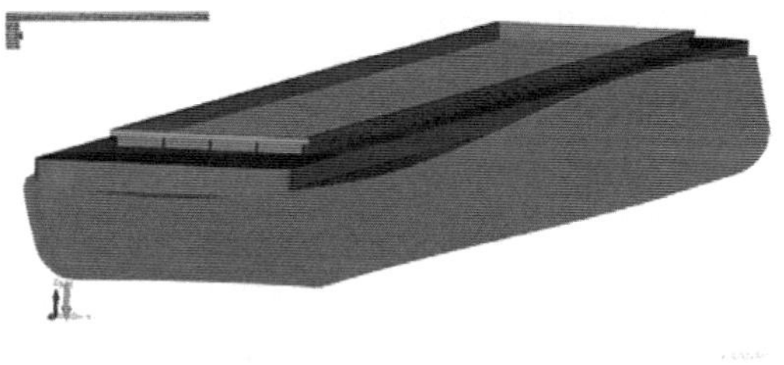

Figure 5. 3D-FEM model full load case (μ=75°, H_w=1.2 m, hogging wave).

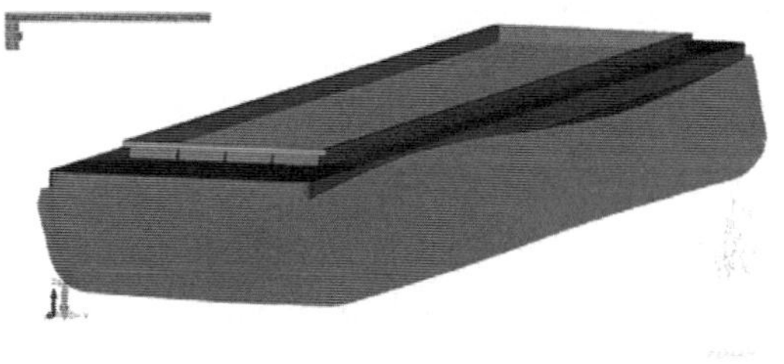

Figure 6. 3D-FEM model full load case (μ=75°, H_w=1.2 m, sagging wave).

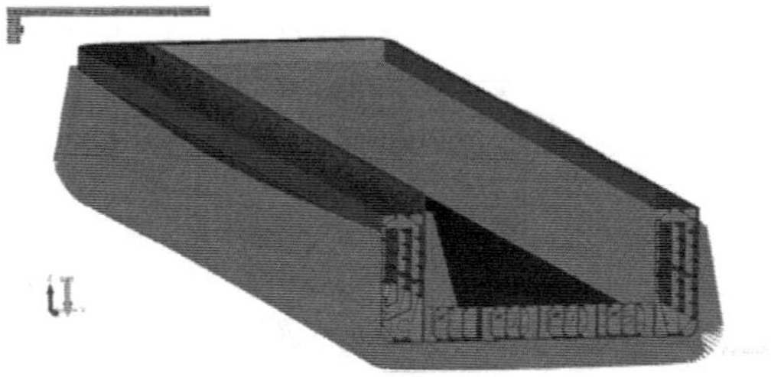

Figure 7. 3D-FEM model full load case. Midship view (μ=75°, H_w=1.2 m, hogging wave).

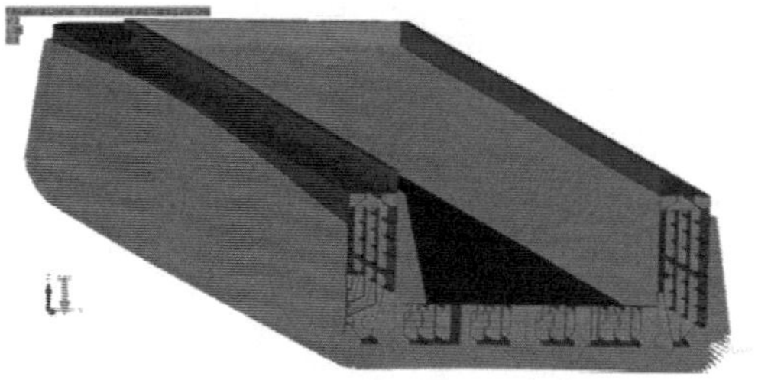

Figure 8. 3D-FEM model full load case. Midship view (μ=75°, H_w=1.2 m, sagging wave).

The boundary conditions were applied in accordance with ships' classification societies guidelines for global structural analysis of ships with large deck openings (specifically for container ships) and adapted to the river-barges' structural layout (table 2, figures 9 and 10) [10].

The above loading scenarios, combined with the boundary conditions (table 2) complete the 3D-FEM model, with respect to the shipbuilding classification societies regulations for FE analysis. All the reaction forces components that will be further obtained are to be in a state of equilibrium and are discussed in the next section. The global strength is assessed by yielding ratio and deflection ratio criteria defined in equation (2) [9, 10].

$$YR = Yield\ ratio = \frac{\sigma_{VMmax}}{R_{eH}} \leq 1 \ ; \ DR = Deflection\ ratio = \frac{w_{max}}{w_{adm}} \leq 1 \qquad (2)$$

where: R_{eH} = 235 MPa and w_{adm} = 180 mm.

Table 2. Boundary conditions, global support [10].

Location		Direction
Aft end	Portside	*Z*
	Centerline	*Y*
	Starboard	*Z*
Fore end	Centerline	*X, Y, Z*

Table 3. Equilibrium parameters, 1D barge model, oblique waves, ballast loading case.

No.	μ (deg.)	Hw (mm)	θ (rad)	φ (rad)	X_F (mm)	Y_F (mm)	T_M (mm)	No.	θ (rad)	φ (rad)	X_F (mm)	Y_F (mm)	T_M (mm)
LC_1		0	0.00001	0	42170.49	0	565.27	-	-	-	-	-	-
LC_2		-300	-0.00076	0	42103.59	0	554.71	LC_26	0.00073	0	42256.59	0	574.92
LC_3	0	-600	-0.00158	0	42035.69	0	542.65	LC_27	0.00138	0	42259.26	0	583.69
LC_4		-900	-0.00243	0	41967.38	0	528.00	LC_28	0.00198	0	42325.57	0	591.69
LC_5		-1200	-0.00367	0	41895.14	0	505.49	LC_29	0.00253	0	42389.68	0	598.54
LC_6		-300	-0.00076	-0.00004	42103.6	0	554.77	LC_30	0.00072	0.00003	42256.57	0	574.87
LC_7	15	-600	-0.00158	-0.00011	42035.71	0	542.78	LC_31	0.00138	0.00006	42259.24	0	583.60
LC_8		-900	-0.00242	-0.00021	41967.43	-0.01	528.23	LC_32	0.00197	0.00007	42325.53	0	591.57
LC_9		-1200	-0.00366	-0.00081	41895.17	-0.09	505.63	LC_33	0.00252	0.00006	42332.76	0.55	598.24
LC_10		-300	-0.00074	-0.00009	42103.65	0	554.98	LC_34	0.00071	0.00007	42256.49	0.01	574.68
LC_11	30	-600	-0.00155	-0.00023	42035.82	-0.01	543.27	LC_35	0.00135	0.00012	42259.17	0	583.27
LC_12		-900	-0.00238	-0.00045	42003.07	-1.93	528.95	LC_36	0.00194	0.00015	42325.39	0.01	591.10
LC_13		-1200	-0.00365	-0.00177	41895.29	-0.21	506.13	LC_37	0.00248	0.00013	42327.62	0.01	597.61
LC_14		-300	-0.00071	-0.00016	42103.77	-0.01	555.52	LC_38	0.00068	0.00012	42256.3	0.01	574.23
LC_15	45	-600	-0.00147	-0.00039	42099.84	-0.03	544.42	LC_39	0.00129	0.00021	42258.99	0.01	582.41
LC_16		-900	-0.00227	-0.00076	42031.72	-0.08	530.93	LC_40	0.00186	0.00026	42325.02	0.01	589.92
LC_17		-1200	-0.00361	-0.00306	41895.61	-0.35	507.47	LC_41	0.00238	0.00023	42327.23	0.01	596.06
LC_18		-300	-0.00059	-0.00026	42112.68	-0.91	557.08	LC_42	0.00058	0.0002	42255.68	0.02	572.88
LC_19	60	-600	-0.00125	-0.00062	42101.24	-0.04	548.02	LC_43	0.00111	0.00035	42258.44	0.02	579.91
LC_20		-900	-0.00194	-0.00118	42034.3	-0.06	537.11	LC_44	0.0016	0.00046	42323.89	0.03	586.39
LC_21		-1200	-0.00343	-0.00481	41913.22	-1.99	512.42	LC_45	0.00208	0.00027	42326.01	0.02	593.59
LC_22		-300	-0.00009	-0.00041	42170.17	-0.02	564.25	LC_46	0.00012	0.00036	42170.84	0.02	566.46
LC_23	75	-600	-0.00023	-0.001	42169.81	-0.05	564.63	LC_47	0.00021	0.00070	42171.15	0.03	567.52
LC_24		-900	-0.00042	-0.00162	42168.97	-0.12	565.74	LC_48	0.0003	0.00103	42173.89	0.29	568.29
LC_25		-1200	-0.00074	-0.00278	42142.98	-0.18	566.37	LC_49	0.00049	0.00004	42237.6	0	591.18

Table 4. Equilibrium parameters, 1D barge model, oblique waves, full loading case.

No.	μ (deg.)	*Hw* (mm)	θ (rad)	φ (rad)	X_F (mm)	Y_F (mm)	T_M (mm)	*No.*	θ (rad)	φ (rad)	X_F (mm)	Y_F (mm)	T_M (mm)
-	-	-	-	-	-	-	-	LC_74	-0.00253	0	44968.32	0	3798.47
LC_50	0	300	-0.00254	0	44968.32	0	3798.20	LC_75	-0.00253	0	44968.32	0	3798.75
LC_51		600	-0.00255	0	44968.32	0	3797.93	LC_76	-0.00252	0	44968.32	0	3799.02
LC_52		900	-0.00256	0	44968.32	0	3797.65	LC_77	-0.00251	0	44968.32	0	3799.29
LC_53		1200	-0.00258	0	44968.32	0	3797.04	LC_78	-0.00252	0	44968.32	0	3799.75
LC_54	15	300	-0.00254	0	44968.32	0	3798.2	LC_79	-0.00253	0	44968.32	0	3798.74
LC_55		600	-0.00255	0	44968.32	0	3797.93	LC_80	-0.00252	0	44968.32	0	3799.02
LC_56		900	-0.00256	0	44968.32	0	3797.66	LC_81	-0.00251	0	44968.32	0	3799.29
LC_57		1200	-0.00258	0	44968.32	0	3797.06	LC_82	-0.00252	0	44968.32	0	3799.72
LC_58	30	300	-0.00254	0	44968.32	0	3798.21	LC_83	-0.00253	0	44968.32	0	3798.73
LC_59		600	-0.00255	0	44968.32	0	3797.95	LC_84	-0.00252	0	44968.32	0	3799.00
LC_60		900	-0.00255	0	44968.32	0	3797.69	LC_85	-0.00251	0	44968.32	0	3799.26
LC_61		1200	-0.00258	0	44968.32	0	3797.14	LC_86	-0.00251	0	44968.32	0	3799.61
LC_62	45	300	-0.00254	0	44968.32	0	3798.24	LC_87	-0.00253	0	44968.32	0	3798.71
LC_63		600	-0.00255	0	44968.32	0	3798.00	LC_88	-0.00252	0	44968.32	0	3798.94
LC_64		900	-0.00255	0	44968.32	0	3797.77	LC_89	-0.00251	0	44968.32	0	3799.18
LC_65		1200	-0.00257	0	44968.32	0	3797.34	LC_90	-0.00251	0	44968.32	0	3799.42
LC_66	60	300	-0.00254	0	44968.32	0	3798.31	LC_91	-0.00253	0	44968.32	0	3798.64
LC_67		600	-0.00255	0	44968.32	0	3798.15	LC_92	-0.00252	0	44968.32	0	3798.80
LC_68		900	-0.00255	0	44968.32	0	3797.98	LC_93	-0.00252	0	44968.32	0	3798.96
LC_69		1200	-0.00255	0	44968.32	0	3797.89	LC_94	-0.00251	0	44968.32	0	3799.13
LC_70	75	300	-0.00254	0	44968.32	0	3798.62	LC_95	-0.00253	0	44968.32	0	3798.33
LC_71		600	-0.00254	0	44968.32	0	3798.77	LC_96	-0.00253	0	44968.32	0	3798.18
LC_72		900	-0.00254	0	44968.32	0	3798.92	LC_97	-0.00253	0	44968.32	0	3798.03
LC_73		1200	-0.00248	-0.00006	44968.32	0	3800.05	LC_98	-0.00252	0	44968.32	0	3797.88

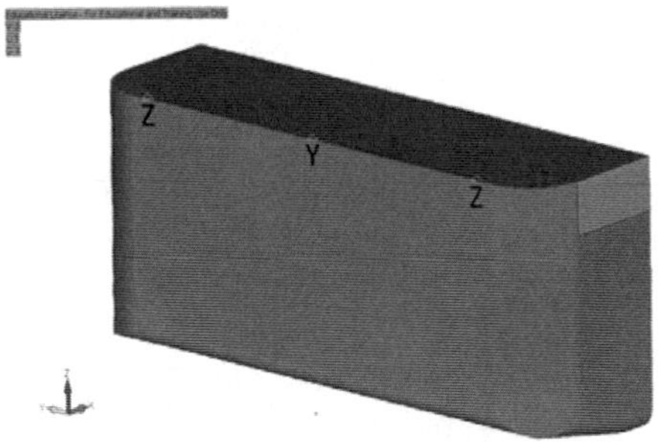

Figure 9. 3D-FEM model, Boundary conditions, Aft end.

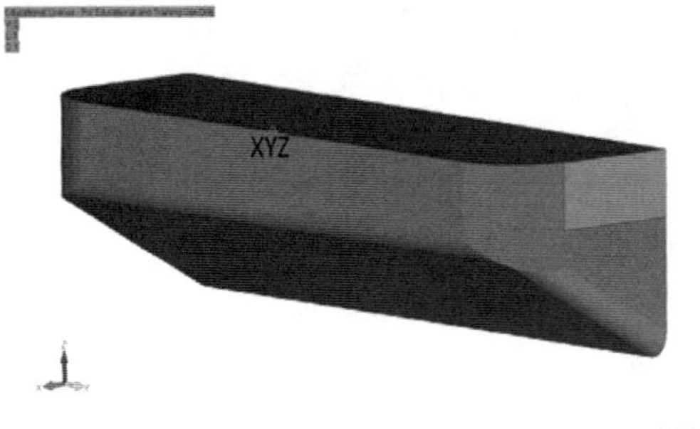

Figure 10. 3D-FEM model, Boundary conditions, Fore end.

3. Global strength in barge ballast loading condition

This section presents the results obtained following the analysis of the 3D-FEM model for the river barge in ballast loading condition. The loading cases' sets are as following:

- LC_1 presents the still water condition results;
- LC_2 – LC_25 presents the oblique quasi-static results for heading angles between μ=0-75°, and hogging wave (wave crest) heights from H_w=0.3-1.2 m.
- LC_26 – LC_49 presents the results for similar heading angles and wave heights for the sagging condition (wave through).

Table 5 summarizes the maximum values for the following evaluation parameters: von Mises stress (σ_{VM}), maximum deflection (w_{max}) and the yield stress ratio, which represents the coefficient of the structural usage (2).

Table 5. Ballast loading condition, von Mises stresses and maximum deflections, yield stress and deflection ratios.

No.	μ (deg.)	*Hw* (mm)	σ_{VM} (MPa)	w_{max} (mm)	*Yield Ratio*	*Deflection Ratio*	*No.*	σ_{VM} (MPa)	w_{max} (mm)	*Yield Ratio*	*Deflection Ratio*
LC_1		0	19.00	15.64	0.095	0.087	-	-	-	-	-
LC_2	0	300	25.57	27.13	0.128	0.151	LC_26	18.54	4.051	0.093	0.023
LC_3		600	35.56	38.24	0.178	0.212	LC_27	17.80	8.213	0.089	0.046
LC_4		900	46.23	48.75	0.231	0.271	LC_28	22.67	20.229	0.114	0.112
LC_5		1200	55.52	57.66	0.278	0.320	LC_29	34.75	32.372	0.174	0.180
LC_6	15	300	25.57	27.17	0.128	0.151	LC_30	18.55	4.165	0.093	0.023
LC_7		600	35.62	38.37	0.178	0.213	LC_31	17.99	8.532	0.090	0.047
LC_8		900	46.32	48.98	0.232	0.272	LC_32	22.76	20.710	0.114	0.115
LC_9		1200	55.58	57.89	0.278	0.322	LC_33	34.86	32.998	0.175	0.183
LC_10	30	300	25.55	27.26	0.128	0.151	LC_34	18.56	4.435	0.093	0.025
LC_11		600	35.65	38.59	0.178	0.214	LC_35	17.94	8.895	0.090	0.049
LC_12		900	46.38	49.43	0.232	0.275	LC_36	22.76	21.214	0.114	0.118
LC_13		1200	55.51	58.33	0.278	0.324	LC_37	34.86	33.647	0.175	0.187
LC_14	45	300	25.48	27.39	0.128	0.152	LC_38	18.41	4.917	0.092	0.027
LC_15		600	35.62	39.09	0.178	0.217	LC_39	17.66	9.361	0.088	0.052
LC_16		900	46.36	50.25	0.232	0.279	LC_40	22.58	21.820	0.113	0.121
LC_17		1200	55.18	59.04	0.276	0.328	LC_41	34.58	34.389	0.173	0.191
LC_18	60	300	25.28	27.69	0.127	0.154	LC_42	17.69	5.832	0.089	0.032
LC_19		600	35.35	39.91	0.177	0.222	LC_43	18.73	10.072	0.094	0.056
LC_20		900	46.12	51.80	0.231	0.288	LC_44	21.78	22.672	0.109	0.126
LC_21		1200	54.19	60.18	0.271	0.334	LC_45	32.94	**34.640**	0.165	**0.192**
LC_22	75	300	26.83	27.78	0.134	0.154	LC_46	19.55	8.386	0.098	0.047
LC_23		600	36.84	40.95	0.184	0.228	LC_47	23.67	11.740	0.119	0.065
LC_24		900	48.58	54.42	0.243	0.302	LC_48	27.79	23.820	0.139	0.132
LC_25		1200	**59.96**	**68.06**	**0.300**	**0.378**	LC_49	**35.18**	29.323	**0.176**	0.163

Figures 11-22 present the vertical deflections (mm) and von Mises stress (MPa) over the length x(mm) of the 3D-FEM model, ballast loading case and hogging wave considering with heights between 0 and 1.2 m, step 0.3 m.

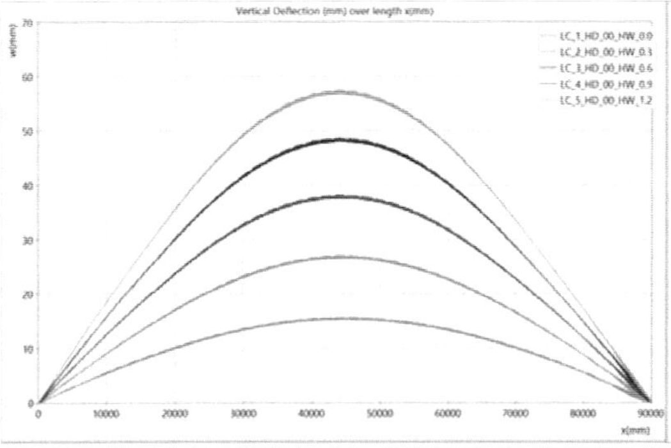

Figure 11. Vertical deflection (mm) over length x(mm), μ=0°, Ballast case, hogging wave.

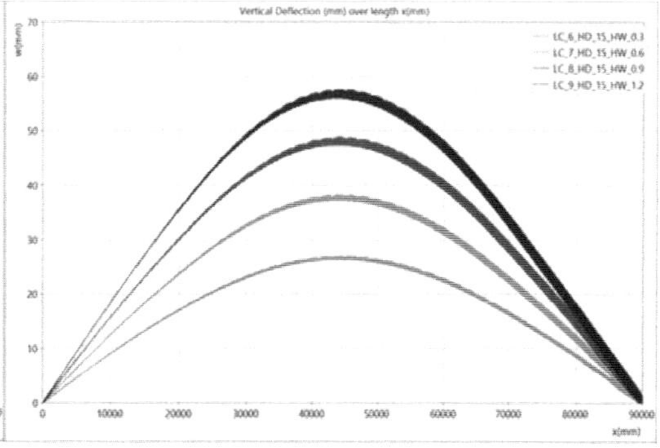

Figure 12. Vertical deflection (mm) over length x(mm), μ=15°. Ballast case, hogging wave.

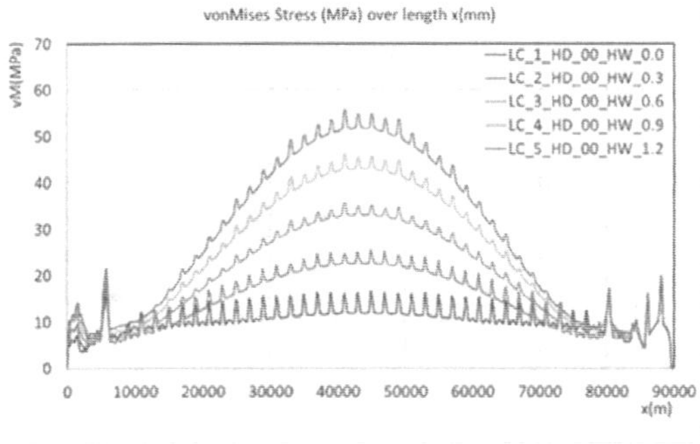

Figure 13. von Mises stress (MPa) over length x(mm), μ=0°. Ballast case, hogging wave.

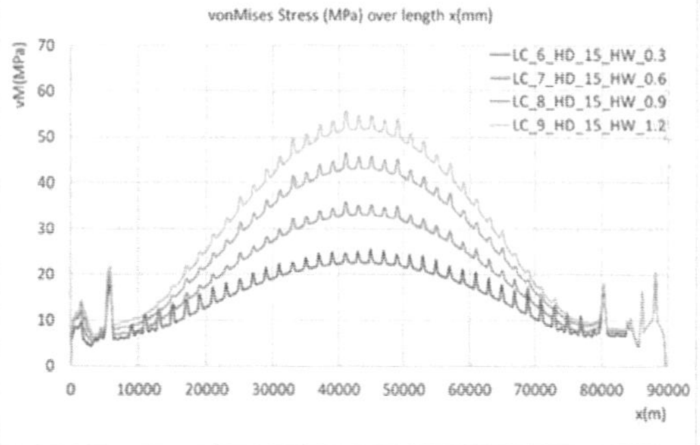

Figure 14. von Mises stress (MPa) over length x(mm), μ=15°. Ballast case, hogging wave.

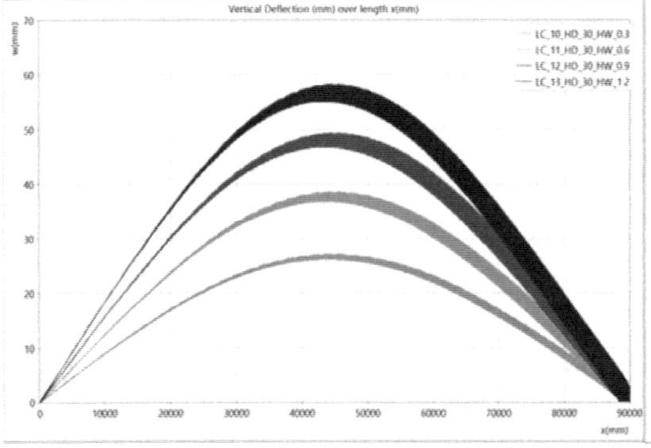

Figure 15. Vertical deflection (mm) over length x(mm), μ=30°. Ballast case, hogging wave.

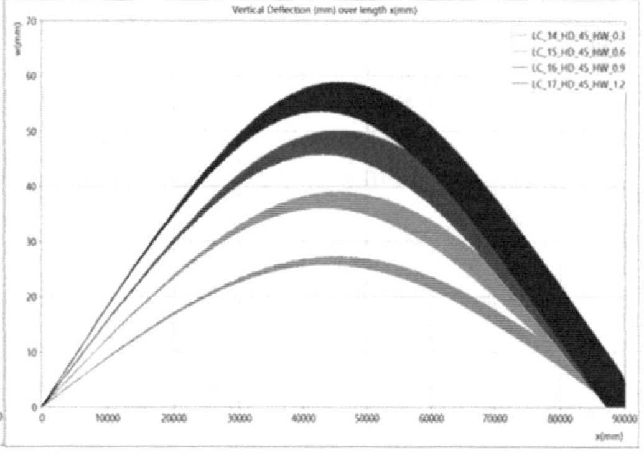

Figure 16. Vertical deflection (mm) over length x(mm), μ=45°. Ballast case, hogging wave.

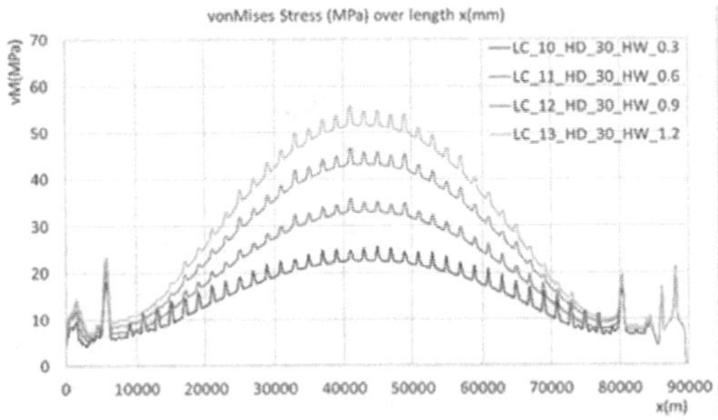

Figure 17. von Mises stress (MPa) over length *x*(mm), μ=30°. Ballast case, hogging wave.

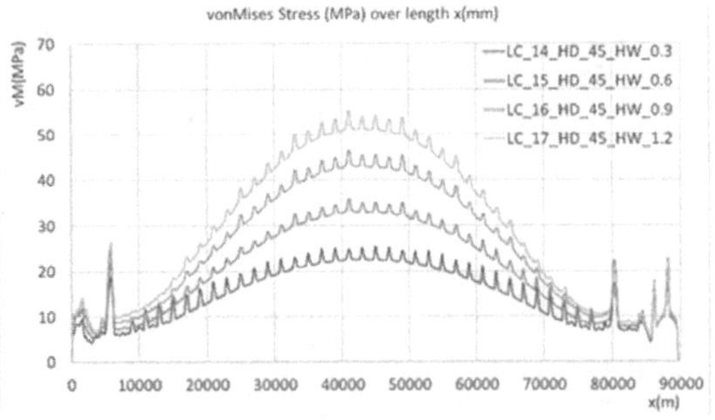

Figure 18. von Mises stress (MPa) over length *x*(mm), μ=45°. Ballast case, hogging wave.

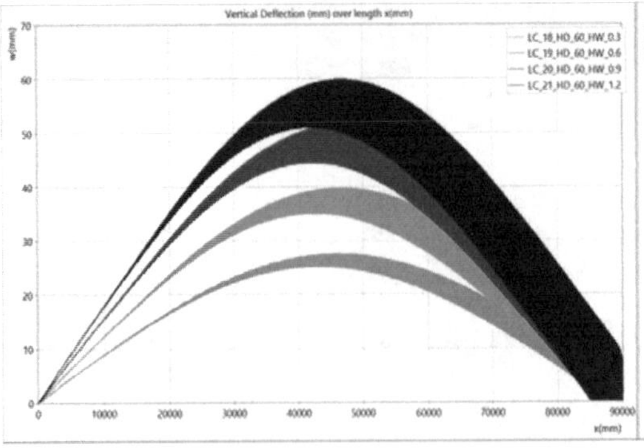

Figure 19. Vertical deflection (mm) over length *x*(mm), μ=60°. Ballast case, hogging wave.

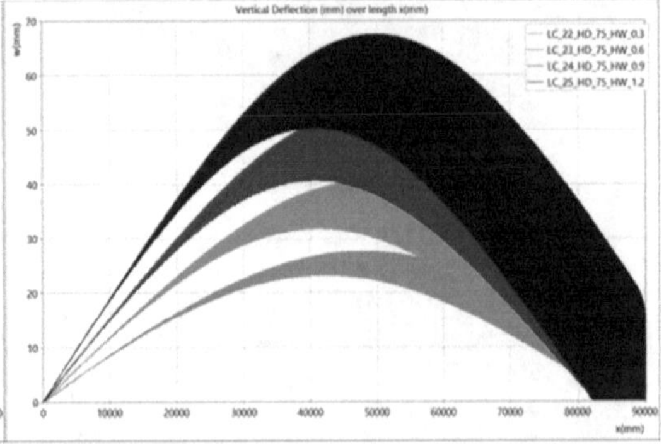

Figure 20. Vertical deflection (mm) over length *x*(mm), μ=75°. Ballast case, hogging wave.

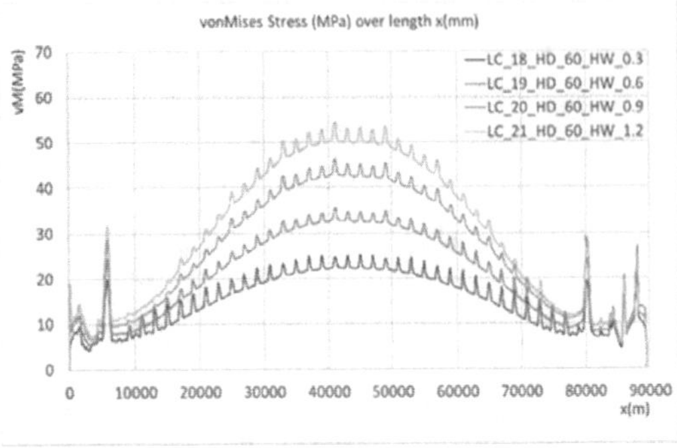

Figure 21. von Mises stress (MPa) over length *x*(mm), μ=60°. Ballast case, hogging wave.

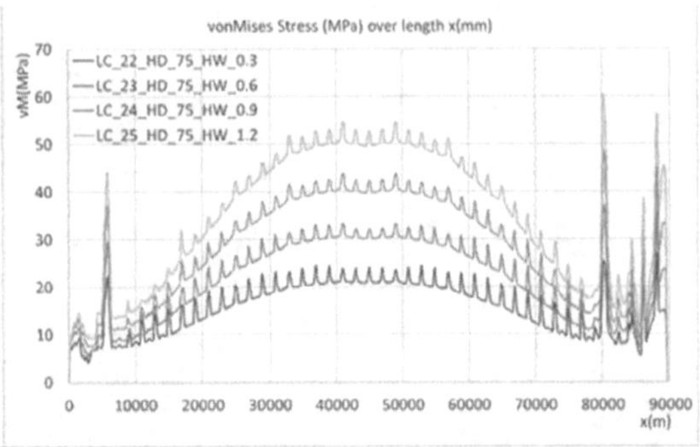

Figure 22. von Mises stress (MPa) over length *x*(mm), μ=75°. Ballast case, hogging wave.

Figures 23-34 include the results in the form of vertical deflections (mm) and von Mises stress (MPa) over the length *x*(mm) of the 3D-FEM model, ballast loading case and sagging wave with heights between 0 and 1.2 m, step 0.3 m.

IOP Conf. Series: Materials Science and Engineering 1182 (2021) 012059 doi:10.1088/1757-899X/1182/1/012059

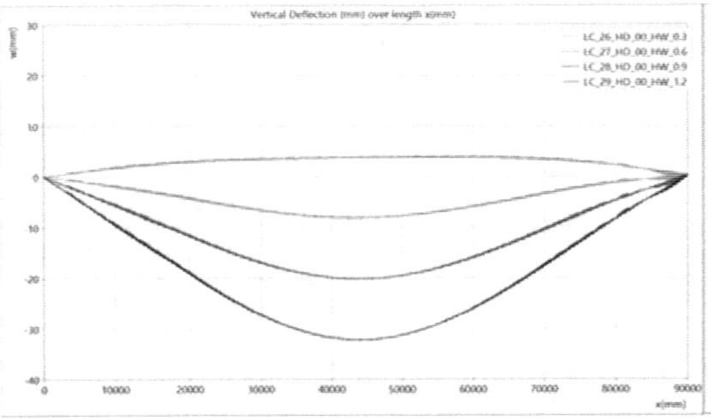

Figure 23. Vertical deflection (mm) over length x(mm), μ=0°. Ballast case, sagging wave.

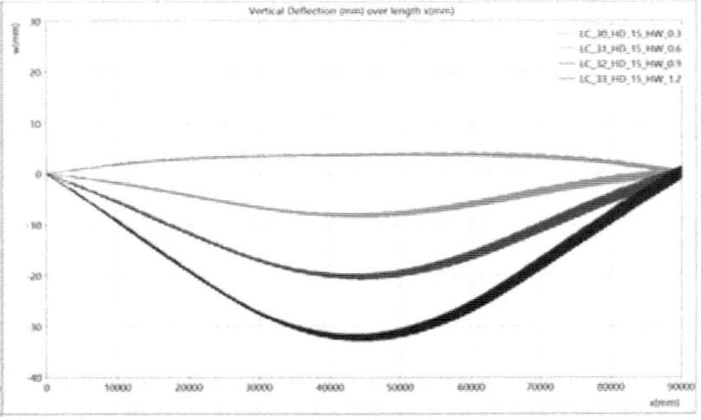

Figure 24. Vertical deflection (mm) over length x(mm), μ=15°. Ballast case, sagging wave.

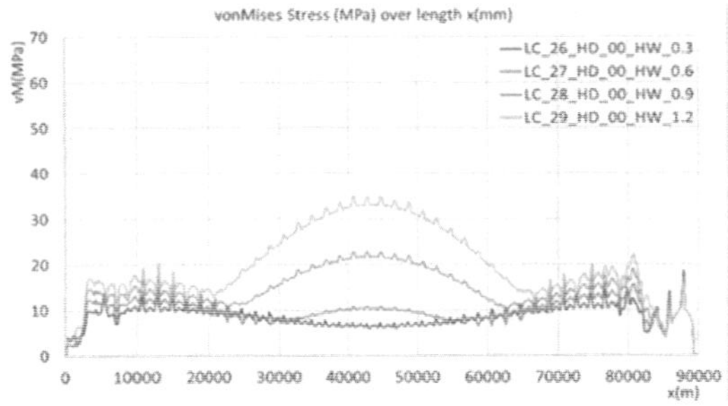

Figure 25. von Mises stress (MPa) over length x(mm), μ=0°. Ballast case, sagging wave.

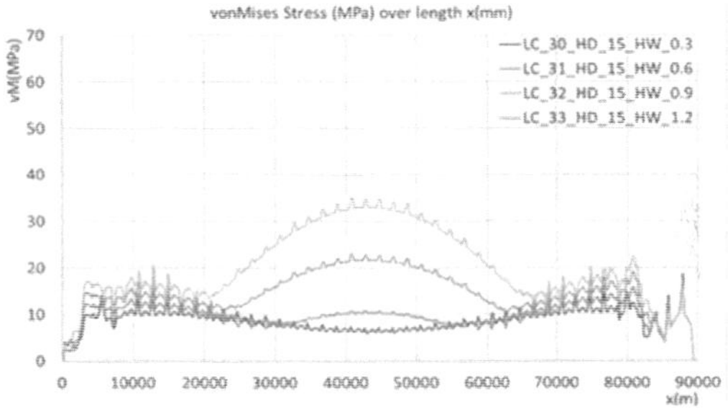

Figure 26. von Mises stress (MPa) over length x(mm), μ=15°. Ballast case, sagging wave.

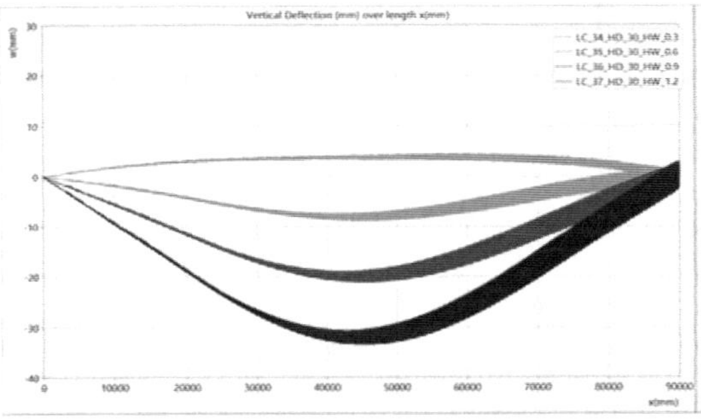

Figure 27. Vertical deflection (mm) over length x(mm), μ=30°. Ballast case, sagging wave.

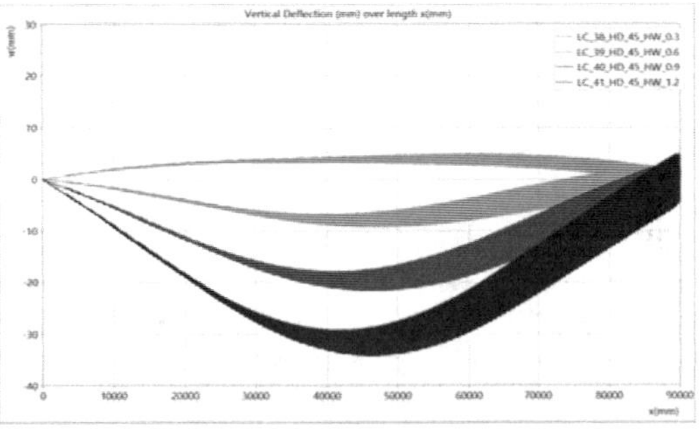

Figure 28. Vertical deflection (mm) over length x(mm), μ=45°. Ballast case, sagging wave.

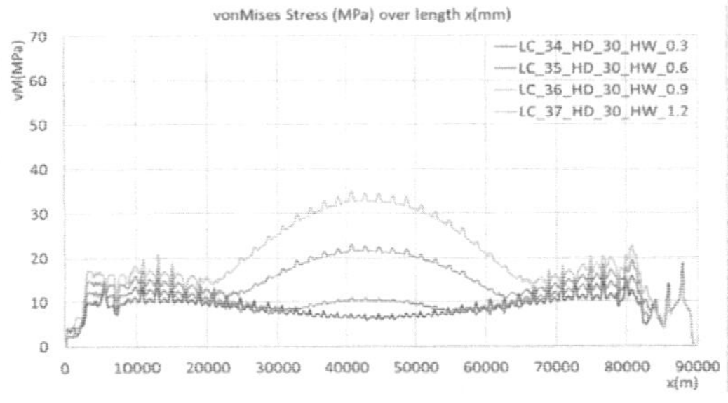

Figure 29. von Mises stress (MPa) over length x(mm), μ=30°. Ballast case, sagging wave.

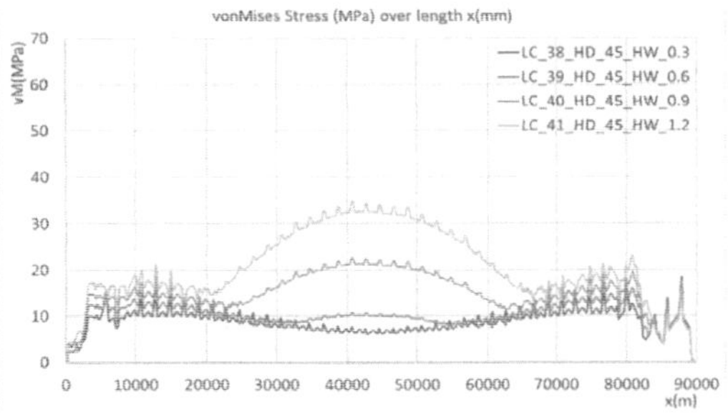

Figure 30. von Mises stress (MPa) over length x(mm), μ=45°. Ballast case, sagging wave.

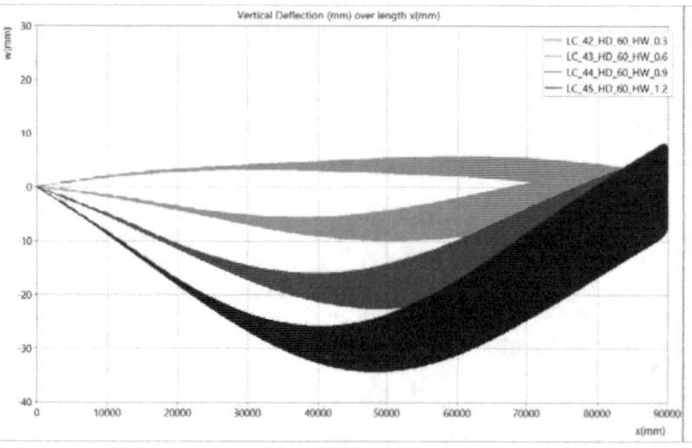

Figure 31. Vertical deflection (mm) over length x(mm), μ=60°. Ballast case, sagging wave.

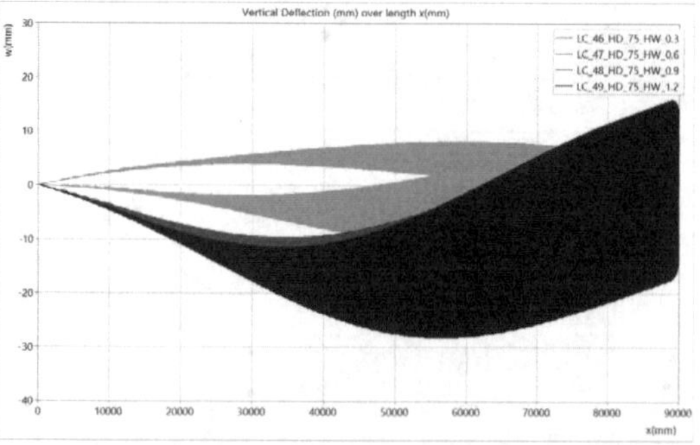

Figure 32. Vertical deflection (mm) over length x(mm), μ=75°. Ballast case, sagging wave.

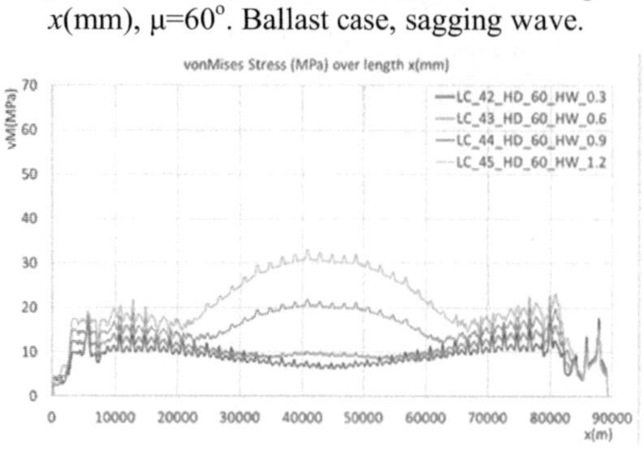

Figure 33. von Mises stress (MPa) over length x(mm), μ=60°. Ballast case, sagging wave.

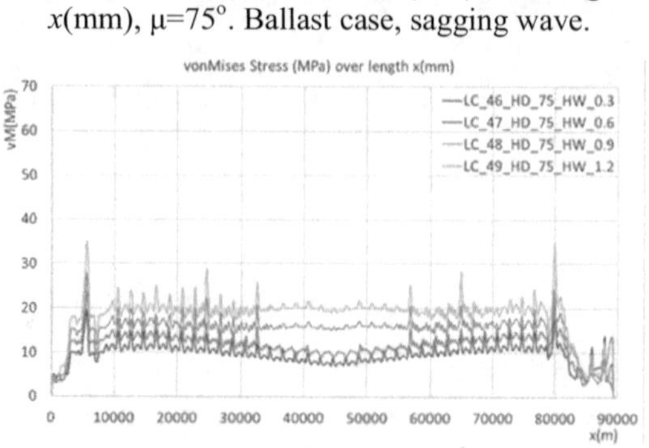

Figure 34. von Mises stress (MPa) over length x(mm), μ=75°. Ballast case, sagging wave.

From the ballast case global analysis of the river-barge structure in oblique equivalent quasi-static waves the following comments can be formulated:

- Vertical deflections of the hull are consistent in value for all heading angles, with maximum deflection w_{max}=68 mm for the LC_25, hogging wave, corresponding to μ=75°, and H_w=1.2 m. The torsional effect with the highest relative displacement between ship sides has been identified for the same loading case (LC_25), with an approximate value of 30 mm.
- The highest value for the von Mises stress of 59.96 MPa corresponds to LC_25. High peaks of local relative stress can be observed in the fore and aft part of the ship, with around 20 MPa difference from the surrounding structural stress values. The high stress peaks indicate hot-spot points in the structure, and the layout of local structure should be updated for a better local distribution of stresses.

4. Global strength in barge full cargo loading condition

The current section presents the results following the analysis of the 3D-FEM river-barge structure in the full cargo loading condition. The loading cases' sets are below described:

- LC_50 – LC_73 present the results for the equivalent oblique quasi-static design waves in hogging condition, for heading angles between μ=0-75° and wave heights H_w=0.3-1.2 m.
- LC_74 presents the still water results at full cargo loading condition.
- LC_75 – LC_98 present the results following the equivalent oblique quasi-static design waves in sagging conditions, heading angles between μ=0-75° and wave heights H_w=0.3-1.2 m.

Table 6 summarizes the maximum values for the following evaluation parameters: von Mises stress (σ_{VM}), maximum deflection (w_{max}) and the yield stress ratio (2).

Table 6. Full loading condition, von Mises stresses and maximum deflections, yield stress and deflection ratios.

No.	μ (deg.)	*Hw* (mm)	σ_{VM} (MPa)	*w* (mm)	*Yield Ratio*	*Deflection Ratio*	*No.*	σ_{VM} (MPa)	*w* (mm)	*Yield Ratio*	*Deflection Ratio*
-	0	-	-	-	-	-	LC_74	123.80	118.49	0.620	0.658
LC_50		300	151.20	103.900	0.757	0.577	LC_75	130.43	133.12	0.653	0.740
LC_51		600	135.32	89.317	0.677	0.496	LC_76	137.03	147.76	0.686	0.821
LC_52		900	132.77	74.752	0.665	0.415	LC_77	147.37	162.43	0.738	0.902
LC_53		1200	142.76	60.297	0.715	0.335	LC_78	**158.18**	177.07	**0.792**	0.984
LC_54	15	300	151.20	103.970	0.757	0.578	LC_79	130.57	133.14	0.654	0.740
LC_55		600	135.32	89.455	0.677	0.497	LC_80	137.32	147.8	0.687	0.821
LC_56		900	132.65	74.957	0.664	0.416	LC_81	147.36	162.48	0.738	0.903
LC_57		1200	142.60	60.569	0.714	0.336	LC_82	**158.18**	**177.14**	**0.792**	0.984
LC_58	30	300	151.22	104.120	0.757	0.578	LC_83	130.64	133.11	0.654	0.740
LC_59		600	135.36	89.752	0.678	0.499	LC_84	137.46	147.74	0.688	0.821
LC_60		900	132.09	75.400	0.661	0.419	LC_85	147.32	162.39	0.738	0.902
LC_61		1200	141.85	61.156	0.710	0.340	LC_86	158.13	177.03	0.792	0.984
LC_62	45	300	151.27	104.420	0.757	0.580	LC_87	130.59	132.96	0.654	0.739
LC_63		600	135.48	90.371	0.678	0.502	LC_88	137.36	147.45	0.688	0.819
LC_64		900	131.34	76.328	0.658	0.424	LC_89	147.20	161.95	0.737	0.900
LC_65		1200	140.63	62.366	0.704	0.346	LC_90	157.98	176.47	0.791	0.980
LC_66	60	300	151.46	105.230	0.758	0.585	LC_91	130.08	132.43	0.651	0.736
LC_67		600	135.83	91.991	0.680	0.511	LC_92	136.34	146.38	0.683	0.813
LC_68		900	130.22	78.766	0.652	0.438	LC_93	146.82	160.35	0.735	0.891
LC_69		1200	139.15	65.538	0.697	0.364	LC_94	157.45	174.33	0.788	0.969
LC_70	75	300	**152.47**	**108.950**	**0.763**	**0.605**	LC_95	126.48	129.44	0.633	0.719
LC_71		600	137.80	99.416	0.690	0.552	LC_96	134.73	140.41	0.674	0.780
LC_72		900	126.11	90.492	0.631	0.503	LC_97	144.61	151.63	0.724	0.842
LC_73		1200	132.77	81.725	0.665	0.454	LC_98	154.49	163.33	0.773	0.907

Figures 35-46 present the vertical deflections (mm) and von Mises stress (MPa) over the length *x*(mm) of the 3D-FEM model in full loading condition and hogging wave considering with heights between 0 and 1.2 m, step 0.3 m.

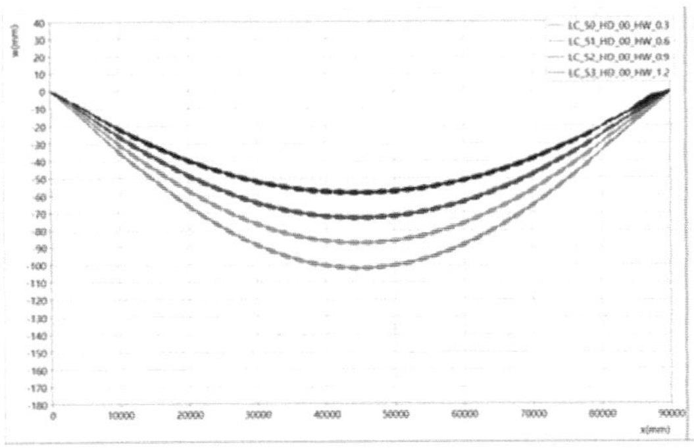

Figure 35. Vertical deflection (mm) over length *x*(mm), μ=0°. Full load case, hogging wave.

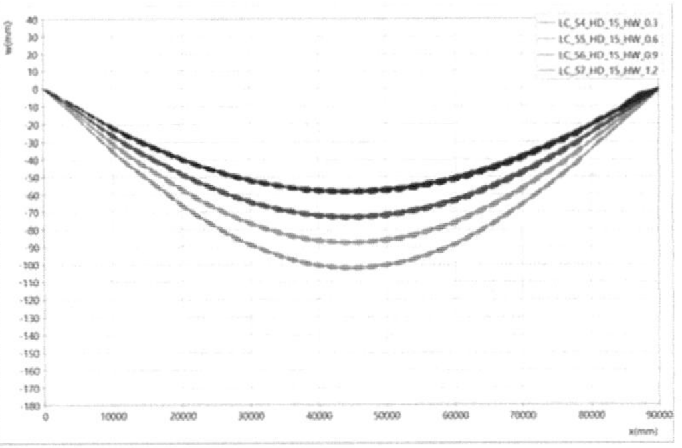

Figure 36. Vertical deflection (mm) over length *x*(mm), μ=15°. Full load case, hogging wave.

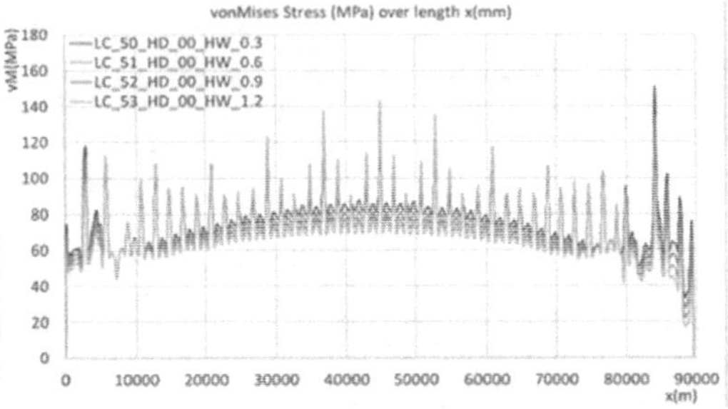

Figure 37. von Mises stress (MPa) over length x(mm), μ=0°. Full load case, hogging wave.

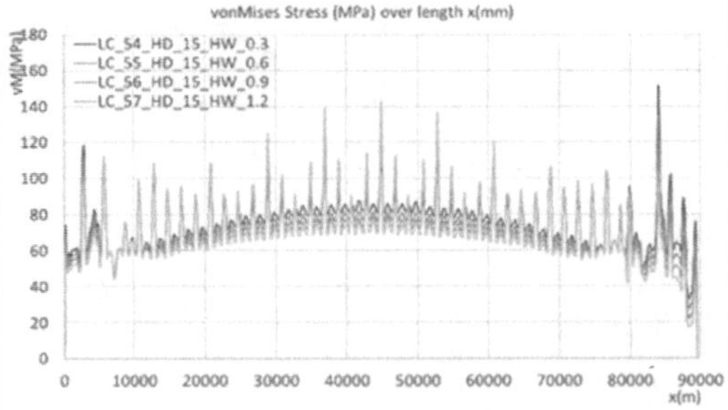

Figure 38. von Mises stress (MPa) over length x(mm), μ=15°. Full load case, hogging wave.

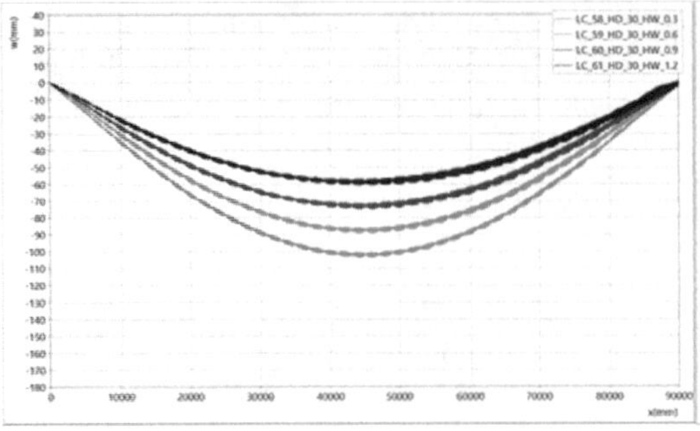

Figure 39. Vertical deflection (mm) over length x(mm), μ=30°. Full load case, hogging wave.

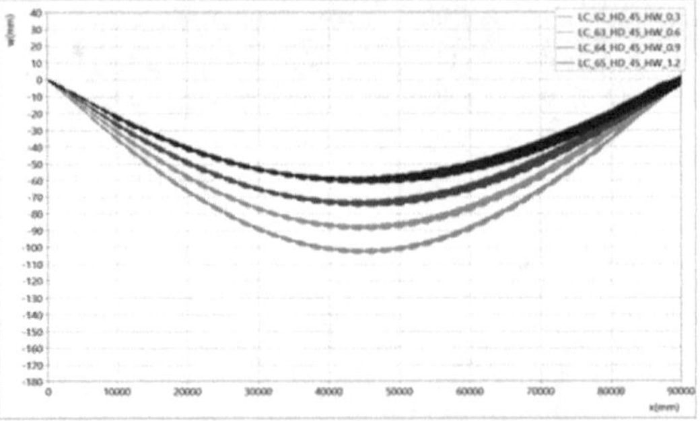

Figure 40. Vertical deflection (mm) over length x(mm), μ=45°. Full load case, hogging wave.

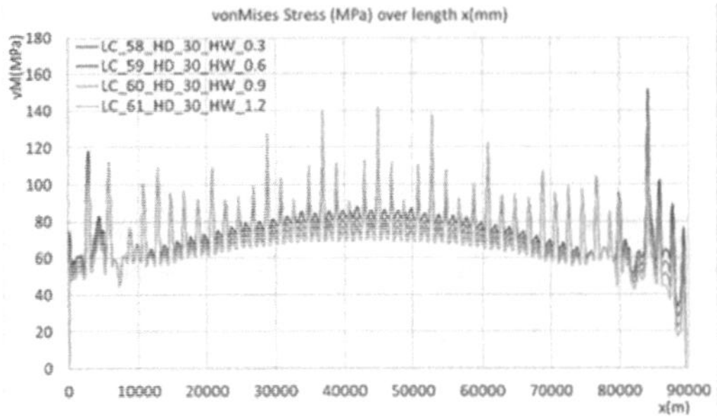

Figure 41. von Mises stress (MPa) over length x(mm), μ=30°. Full load case, hogging wave.

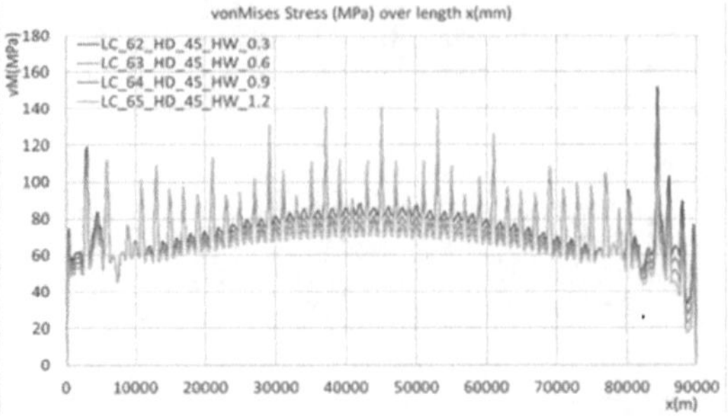

Figure 42. von Mises stress (MPa) over length x(mm), μ=45°. Full load case, hogging wave.

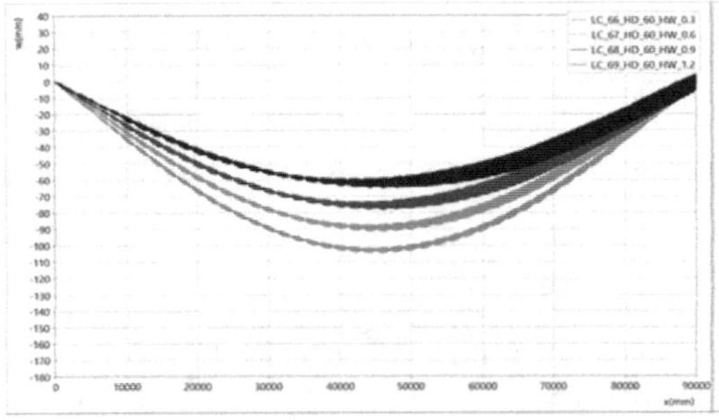

Figure 43. Vertical deflection (mm) over length x(mm), μ=60°. Full load case, hogging wave.

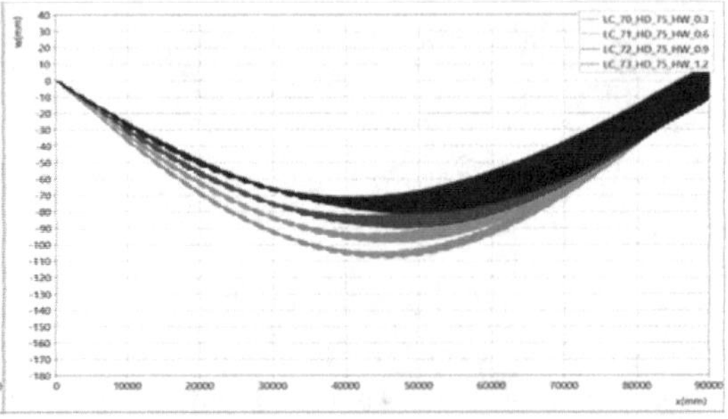

Figure 44. Vertical deflection (mm) over length x(mm), μ=75°. Full load case, hogging wave.

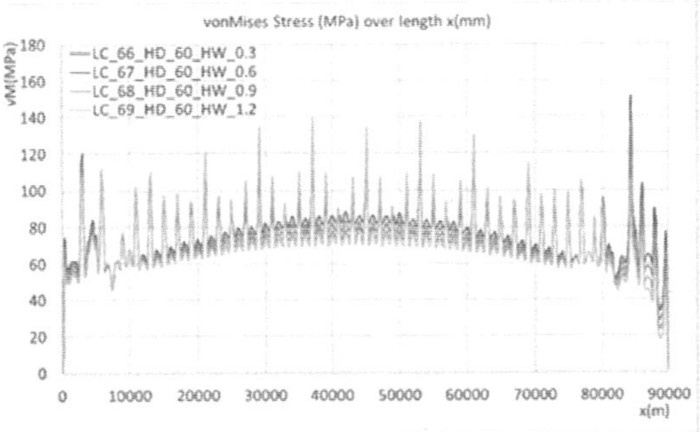

Figure 45. von Mises stress (MPa) over length x(mm), μ=60°. Full load case, hogging wave.

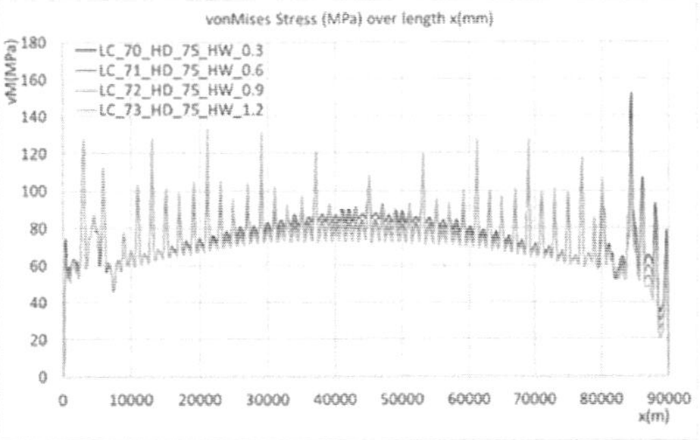

Figure 46. von Mises stress (MPa) over length x(mm), μ=75°. Full load case, hogging wave.

Figures 47-58 include the results for the vertical deflections (mm) and von Mises stress (MPa) over the length x(mm) of the 3D-FEM model in full loading condition and sagging wave considering with heights between 0 and 1.2 m, step 0.3 m.

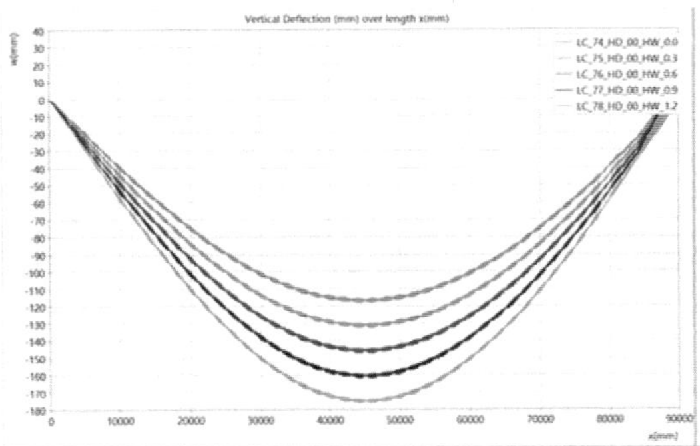

Figure 47. Vertical deflection (mm) over length x(mm), μ=0°. Full load case, sagging wave.

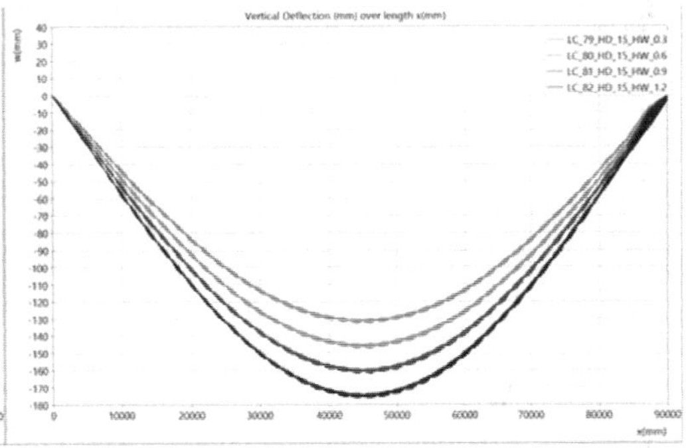

Figure 48. Vertical deflection (mm) over length x(mm), μ=15°. Full load case, sagging wave.

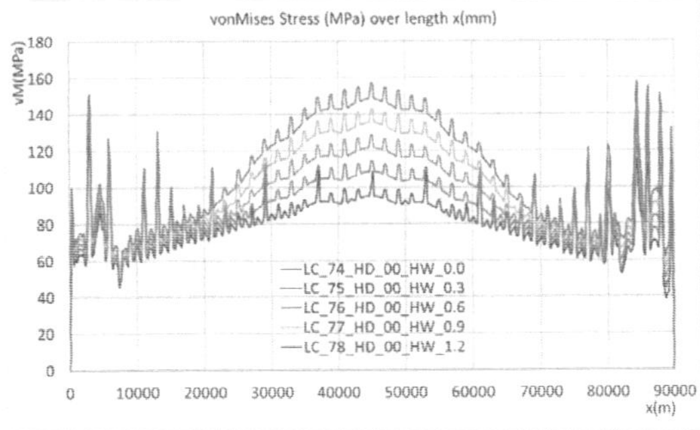

Figure 49. von Mises stress (MPa) over length x(mm), μ=0°. Full load case, sagging wave.

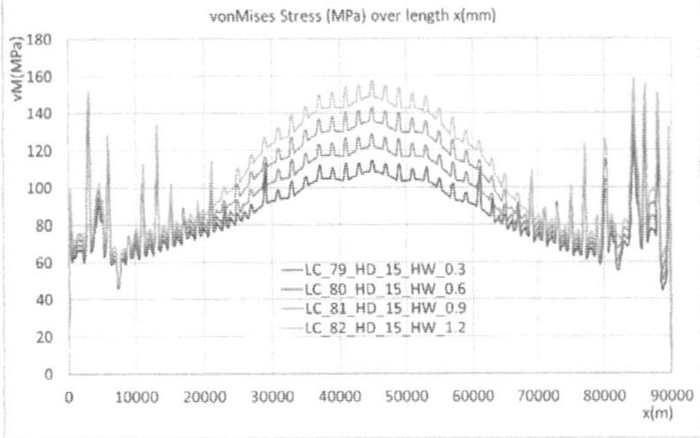

Figure 50. von Mises stress (MPa) over length x(mm), μ=15°. Full load case, sagging wave.

IOP Conf. Series: Materials Science and Engineering 1182 (2021) 012059 doi:10.1088/1757-899X/1182/1/012059

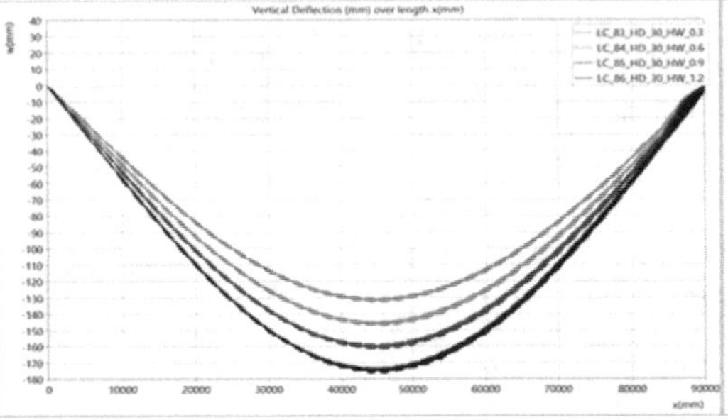

Figure 51. Vertical deflection (mm) over length x(mm), μ=30°. Full load case, sagging wave.

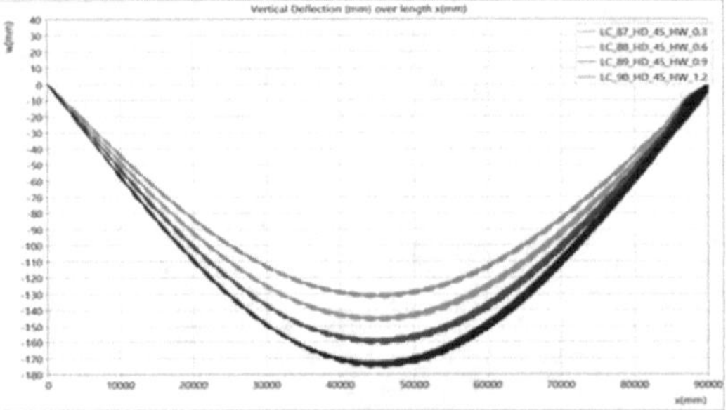

Figure 52. Vertical deflection (mm) over length x(mm), μ=45°. Full load case, sagging wave.

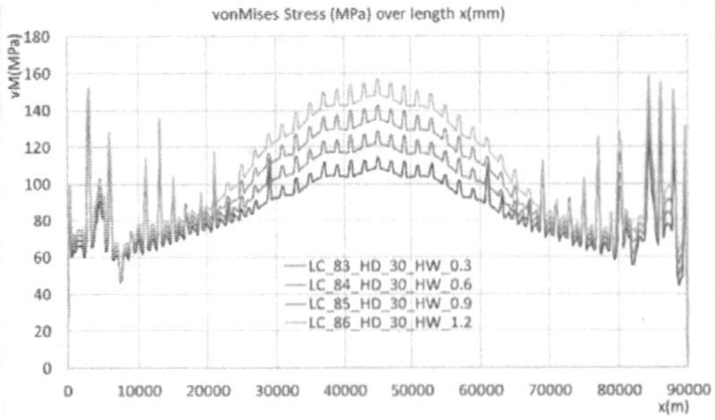

Figure 53. von Mises stress (MPa) over length x(mm), μ=30°. Full load case, sagging wave.

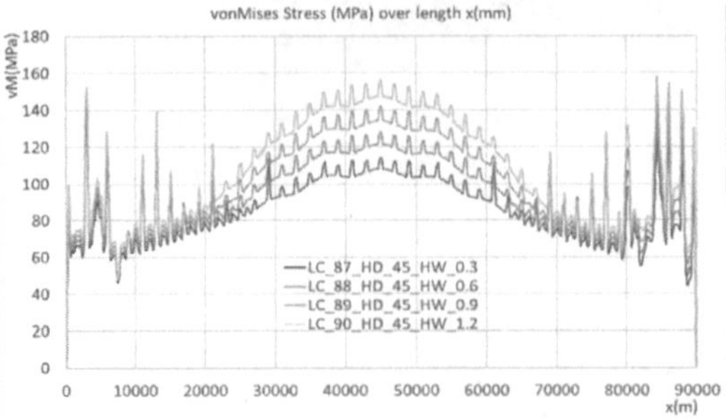

Figure 54. von Mises stress (MPa) over length x(mm), μ=45°. Full load case, sagging wave.

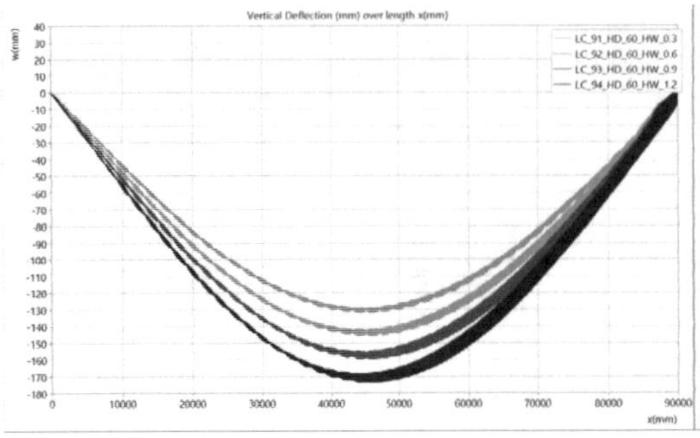

Figure 55. Vertical deflection (mm) over length x(mm), μ=60°. Full load case, sagging wave.

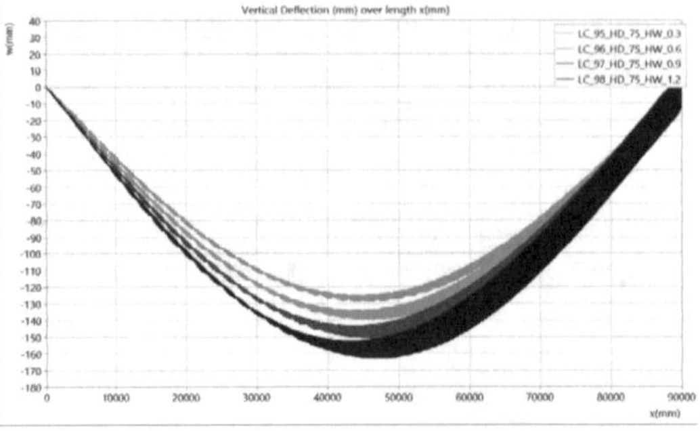

Figure 56. Vertical deflection (mm) over length x(mm), μ=75°. Full load case, sagging wave.

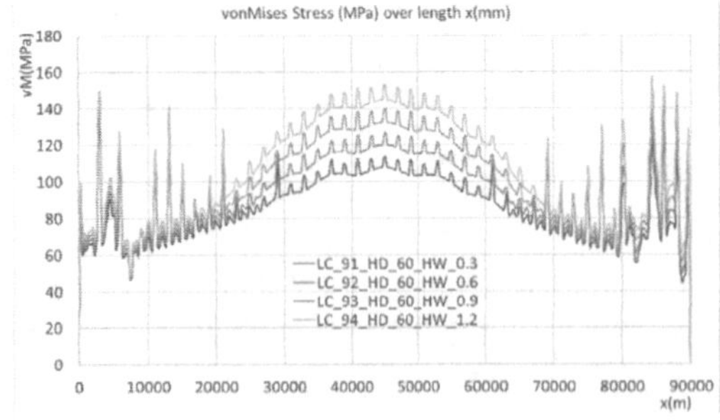

Figure 57. von Mises stress (MPa) over length x(mm), μ=60°. Full load case, sagging wave.

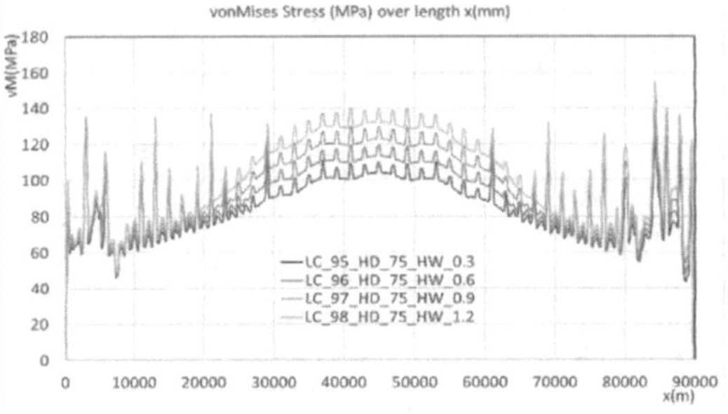

Figure 58. von Mises stress (MPa) over length x(mm), μ=75°. Full load case, sagging wave.

Summarizing the results for the full loading global analysis of the river-barge structure in equivalent oblique quasi-static waves, the following comments can be formulated:

- The vertical deflections of the hull, similarly with the ballast loading condition, are consistent in value for all heading angles, with a maximum deflection of w_{max}=177.14 mm for LC_82, sagging wave, heading angle μ=15° and wave height H_w=1.2 m. The torsional effect having the highest relative displacement between ship sides has been identified in the loading case LC_98 and has an approximate value of 50 mm.
- The highest value registered for the von Mises stress of 158.18 MPa has been identified in the loading case LC_78 and LC_82. The hot-spot areas, where the local relative stress is higher than the surrounding structure, are in the full loading cases present more evenly distributed on the ship's length, registering higher values in the fore and aft structure. As previously discussed, hot-spot areas have to be redesigned in order to redistribute the local occurring stresses.

5. Conclusions

The current study presents a global structural 3D-FEM analysis of a typical open-top, double-hull barge, having a cargo capacity of 3000 dwt. In the framework of analysis and data interpreting processes, the following conclusions have been obtained:

- The current structural layout of the river-barge does not present any structural risk. The FE evaluation pointed out the structural hot-spots that require updates for a better distribution of the local occurring stresses.
- The maximum calculated deflections are close to the limitations that the shipbuilding classification societies can inflict on ships' hull structures with large deck openings. The deflection ratio coefficient (2) is maximum for LC_82 (μ=15°, H_w=1.2m, sagging wave), with a value of *DR*=0.984.
- The maximum structural usage, represented in the form of the yielding stress ratio coefficient (2), has been identified for two full cargo loading cases, LC_78 (μ=0°, H_w=1.2m, sagging wave) and LC_82 (μ=15°, H_w=1.2m, sagging wave), being equal to *YR*=0.792. An almost 80% structural usage has been registered for the 3000 dwt barge hull that has been dimensioned by Bureau Veritas Rules [4] for inland vessels to withstand a maximum wave design height of H_w=1.2m.
- Further investigations have to be performed after the structural update of the identified hot-spot areas, completed by other strength criteria [4].

6. References

[1] Domnisoru L 2015 Numerical approach for global ship strength analysis based in 1D-Beam model, under oblique equivalent quasi-static wave loads *Galati University Press: Shipbuilding* **38** 27–36.

[2] Savin M, Presura A, Chirica I 2019 Environmental protection using structural analysis of ships *SeaConf, IOP Conference Series: Journal of Physics* **(1297)** *Article 012015.*

[3] Meinken A and Schlüter H J 2002 Collapse behaviour of a push-barge *Marine Structures* **15**(2) 193–209.

[4] BV 2020 *Rules for the classification of inland navigation vessels: Part B - Hull design and construction* (Paris: Bureau Veritas).

[5] Domnisoru L 2017 *Special chapters on ship's structures analysis. Applications* (Galati: The University Foundation „Dunarea de Jos" Publishing House).

[6] Domnisoru L 2019 Strength assessment in oblique design waves for a Europe B2 1740T river barge type *Galati University Press: Shipbuilding* **41** 23–28.

[7] Domnisoru L 2020 On the strength assessment of a small liquid petroleum gas carrier in oblique waves by 1D and 3D-FEM analyses *ModTech, IOP Conference Series: Materials Science and Engineering* **(916)** Article 012027.

[8] FNN 2020 *Femap/NX Nastran Users' Manual. Academic License* (Siemens PLM Software Inc.).
[9] Costache M, Jagite G 2014 Global Strength Analysis in Head Waves, for an Offshore Support Vessel *Asian Journal of Applied Science and Engineering* **3**(8) 73-88.
[10] GL 2014 *Rules for Classification and Construction Chapter V-1-1: Guidelines for Global Strength Analysis of Container Ships* (Hamburg: Germanischer Lloyd).

Acknowledgements
The authors wish to express their gratitude towards the Research Centre of the Naval Architecture Faculty and Doctoral School of Mechanical and Industrial Engineering, at "Dunarea de Jos" University of Galati.

IOP Conf. Series: Materials Science and Engineering 1182 (2021) 012060 doi:10.1088/1757-899X/1182/1/012060

The effects of the bending moment upon the mobile element of CNC Machine Tools with great distance between slides

L Petrea and Gh Stan
" Vasile Alecsandri" University of Bacau, Department of Industrial Engineering, Calea Marasesti Street, No. 157, 600115, Bacau, Romania

E-mail: septh4000@gmail.com

Abstract. CNC Machine Tools with kinematical feed chains with great distance between slides (gantry milling machines, cutting machines, carousel lathes etc.) are utilized extensively for machining large and heavy parts. In this type of machine, we can observe a specific phenomenon where the gravity center of the vertical actuated elements moves in the direction of the balancing force for the mobile elements, causing a bending moment to be induced in the mobile element (the mechanical coupling between the two kinematical feed chains). This effect leads to deformations and in worse case looking of the mobile element during movement, all of this has a detrimental effect on machining accuracy and efficiency. In this paper, using finite element analysis (FEA) several studies where preformed. A static analysis that calculates the stresses, displacements and strains. A modal analysis to determine the first six natural frequencies and modal deformed shapes. A harmonic response analysis to determine the response of the mobile element structure. This studies will highlight the week points of mobile element and will provide the machine tool builder with important information regarding the points that need to be address in order to improve the behavior of the mobile element during actuation.

1. Introduction

CNC machine tools with kinematic feed chains with a great distance between the slides (gantry milling machines, cutting machines, carousel lathes, etc.) are widely used for machining large and heavy parts. With this type of machine, we can observe a specific phenomenon in which the center of gravity of the vertically actuated elements moves in the direction of the balancing force for the mobile elements, causing a bending moment at the moving element (mechanical coupling between the two chains kinematic advance). This effect leads to deformations and, worse, to the locking of the moving element during movement, all of which have a negative effect on the accuracy and efficiency of the machine. A study using finite element analysis (FEA) for this type of deformation is extremely useful. This study can be performed by a static analysis that highlights the effects of bending moment on the rigidity of the moving element in static loading conditions but the level of accuracy of such an analysis is relatively low because this analysis does not take into account the interaction between mechanical structure, system control and the dynamics of the machining process [1, 2], so to improve the accuracy of the results a dynamic - modal analysis is required to natural frequencies and modal deformed shapes of the structure [3]. After the modal analysis, a harmonic analysis can be performed to determine the response of the structure to loads that vary harmonically over time. A 3D model for a carousel lathe SC14 CNC was created for dynamic behavior analysis of the moving element (cross-rail/ moving beam). This model was imported into ANSYS® Workbench for FEA analysis.

2. Initial data and FEA model

2.1. 3D model, material and mesh

Using the drawings and documentation from the machine book of an SC14 CNC carousel lathe, figure 1, a simplified 3D CAD model was created for the FEA calculation. The model respects the overall dimensions and technical parameters of the machine. The elements of the machine are: the frame/bed, the table, the cross-beam/rail, the horizontal slide and the vertical slide, this work will analyze the behavior of the cross-rail/ cross- beam (will call it the mobile element). Due to the fact that the coordinate axes of the machine differ from the coordinate axis system in ANSYS in this paper we will use the coordinate axes in ANSYS, X is the direction of movement of the horizontal slide, Y is the direction of travel of the mobile element and the vertical slide and Z will be used for cutting forces. The material used is: gray cast iron.

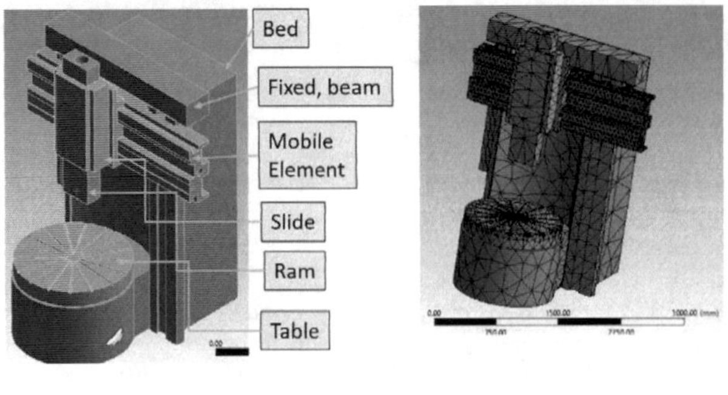

Figure 1. SC 14 CNC carousel lathe and Mesh model.

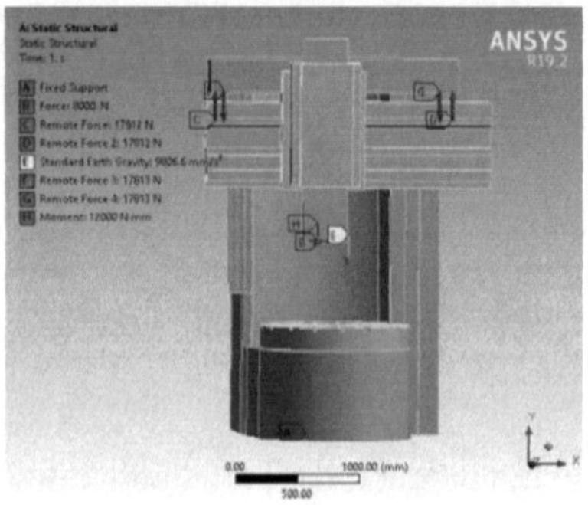

Figure 2. Loading Conditions.

The mesh is set to default except for the mobile element which uses the "Body Sizing" method for better element quality (element size =60).

2.2. Balancing the mobile element and loading conditions

As can be seen in figure 2, the machine frame is fixed to the ground, the gravitational acceleration it's on Y axis -9806.6 mm / s^2, the cutting force of 2000 da N it's split between a moment of 1200 daN (Y axis) and force on direction –Z of 800 daN for better results.

Because of its own weight, CNC machine tools that actuate the mobile element / cross-rail (with all its attachments) in the vertical plane need to balance the mobile element in order to move it along the vertical kinematic feed chains, this can be achieved by several methods. Here this machine uses a balancing system with two types of hydraulic cylinders. The constant pressure hydraulic cylinders are cross connected to the mobile element by means of cables that are supported by means of hoists. The compensation hydraulic cylinders fixed to the mobile element ends and are cross connected to the constant pressure hydraulic cylinders by cables also supported by hoists, the cables from an X shape with the hoist supported on the fixed cross-rail or the bed [4]. Thus the weight of the cross-rail (2067 kg) + the weight of the horizontal sled (930 kg) + the slide (636 kg) = 3633 kg resulting in a force of 3562 da N that must be balanced. Moving the horizontal slide to the right will increase the pressure in the right compensation hydraulic cylinder fixed the mobile element and this will tension the cables on the right side of the machine and relaxing the cables on the left side, same applies for left movement of the horizontal slide [4].

2.3. Position of the vertical slide support

The maximum horizontal movement of the vertical slide support on the cross-rail/mobile element is 1050 mm, from this 3 cases where selected. Case 1 in with the vertical slide support centered, 525 mm on X axis from limiter. Case 2 the slide it at maximum right, 1050 mm on X axis and Case 3 the slide it's at maximum left, 0 mm on X axis, at limiter. For all three cases the cross-rail will be at a maximum high of 800 mm Y axis since it represent the most unfavorable case.

3. FEA analysis

3.1. Static analysis

A static analysis calculates the effects of constant loading conditions on the machine, while ignoring damping effects. The result represent the rigidity of the machine relative the analysis coordinate system.

This paper will study the behavior of the mobile element for the above mentioned cases. For cases 2 and 3 only the total deformation of the mobile element will be shown.

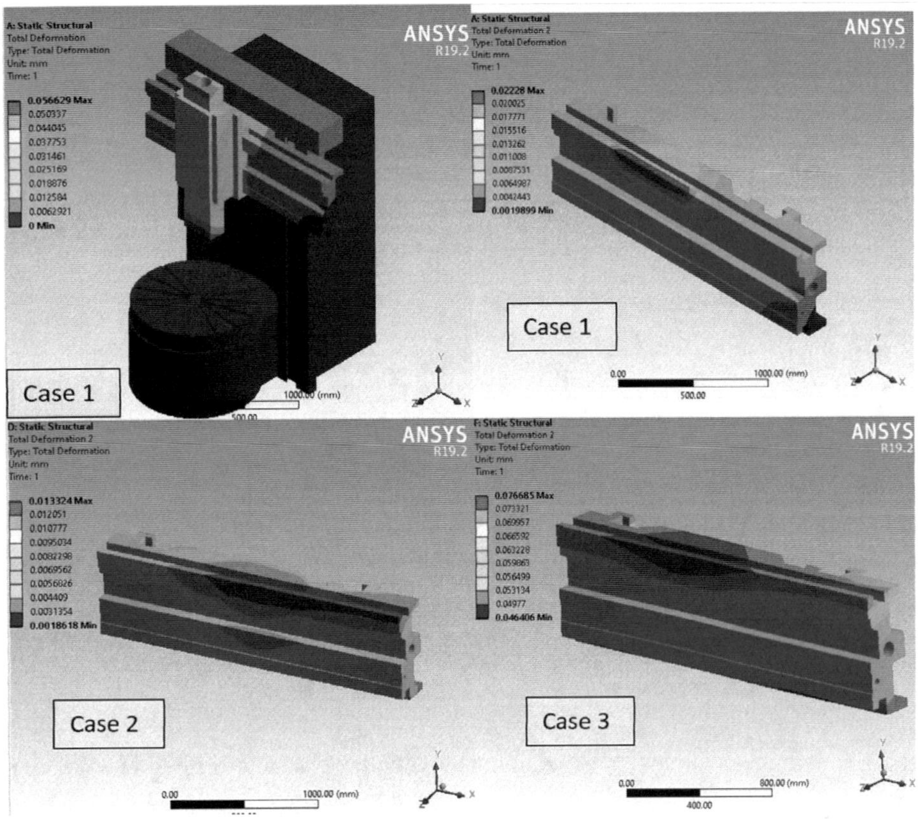

Figure 3. Static analysis - total deformation.

In figure 3 we can see that the maximum deformation occurs at the vertical support slide, this occurs because the vertical slide extends too much out of the support and acts as a cantilever beam (the most unfavorable case). The deformation at the level of the mobile element is relatively small for cases 1 and 2, and large for case 3, which highlights the need to optimize the structures of the vertical slide and vertical slide support.

3.2. Modal analysis

A modal analysis was preformed to determine the first six natural frequencies and modal deformed shapes, it is the most fundamental of all types of dynamic analysis and allows the structures designed to avoid resonant vibrations or to vibrate at a specified frequency. Highlights how a structure will respond to different types of dynamic loads and is the basis of the following other dynamic studies like harmonic, dynamic transient. The results in figure 4, figure 5 and figure 6 highlight the first 6 fundamental frequencies of the mobile element, so for case 1 as we can see in figure 4, the first fundamental frequency was 65,567 Hz and the maximum deformation of the mobile element occurred at 258.49 Hz figure 6. Figure 7 shows the first 6 fundamental frequencies for cases 2 and 3, as well as the maximum total deformation.

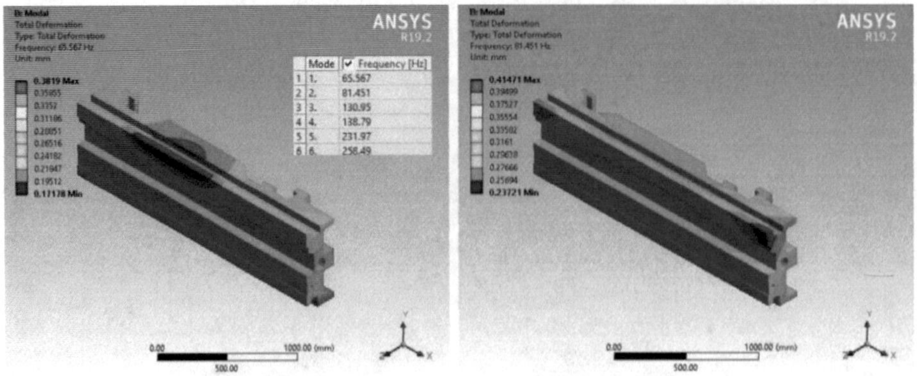

Figure 4. First and second natural frequency.

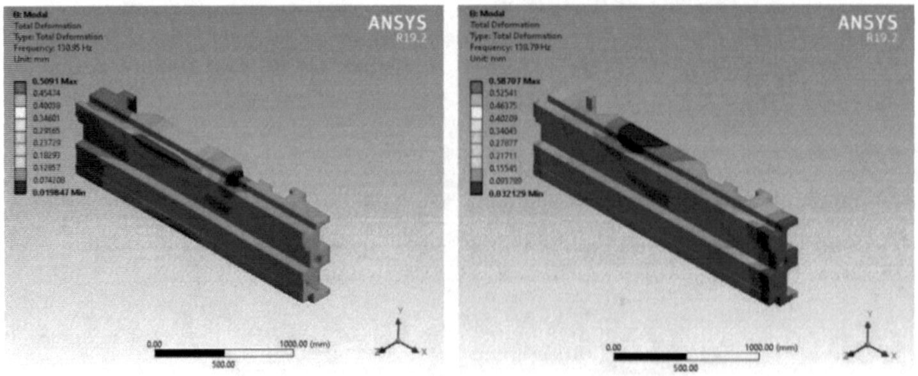

Figure 5. Third and fourth natural frequency.

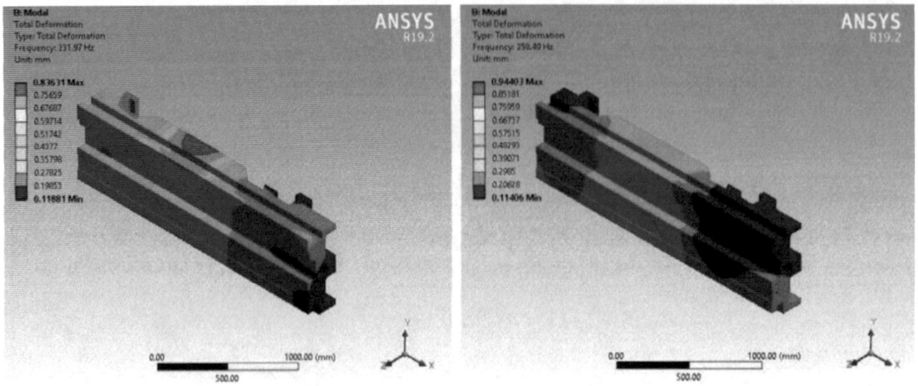

Figure 6. Fifth and sixth natural frequency.

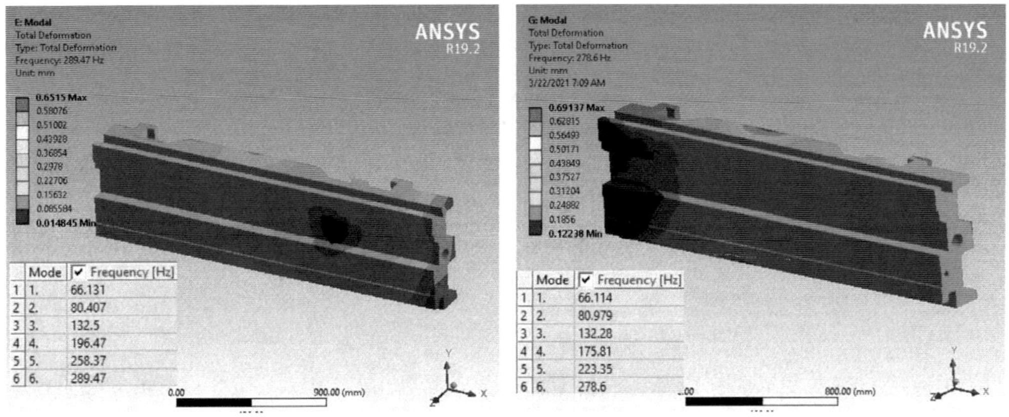

Figure 7. Highest modal deformation for cases 2 and 3.

3.3. Harmonic response analysis

After performing the model analysis and identifying the first 6 fundamental frequencies with their corresponding deformation modes, a harmonic analysis was performed using the module from ANSYS Workbench. Harmonic analysis is able to determine whether the structure of the moving element will successfully overcome the resonance and harmful effects of forced vibrations. A frequency range was set between 0 and 500 Hz, based on modal analysis. A constant dampening ratio of 2% was used. In this analysis, the structure of the machine was excited by a series of harmonic forces that occur between the work-piece and the cutting tool.

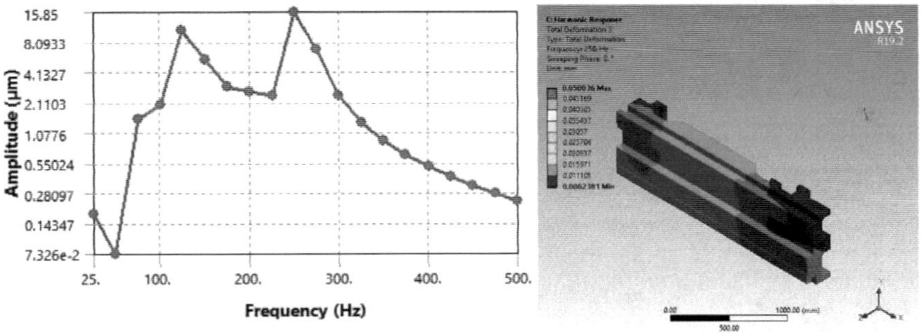

Figure 8. Frequency response directional deformation – Case 1.

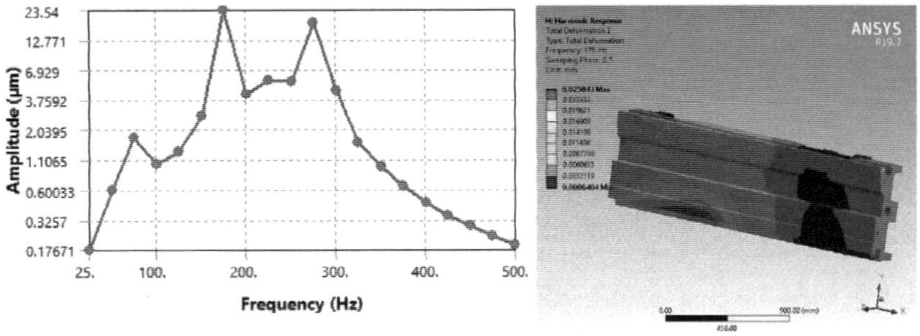

Figure 9. Frequency response directional deformation – Case 2.

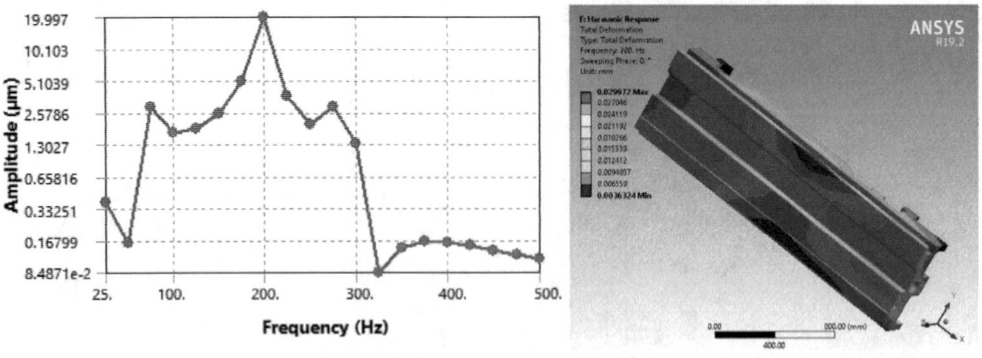

Figure 10. Frequency response directional deformation – Case 3.

As we can see from figure 8, figure 9 and figure 10 the largest deformations occur at low frequencies 175 Hz ÷ 250 Hz on Y direction for all 3 cases. Deformations in the X and Z directions were not included, as they are much smaller.

4. Conclusions

From the analysis of the result shows above we can draw some conclusions regarding the behavior of the moving element and points that need to be addressed in order to improve the static and dynamic behavior of the moving element during machine operation. Thus, from the static analysis we can see a problem at the level of the vertical slide and vertical support both that deform significantly under their own weight because of the lack of a balancing system (problem more accentuated at the vertical slide). As can be observed from figures 3, 6 and 7 the issues with vertical slide and its support affect the mobile element (cross-rail) creating elastic deformation at the guideways level. Solving this requires the optimization of the mobile elements structure or better adding a force balancing system for the vertical support and redesigning if the vertical slide. Even so the bending moment at the mobile element level that occurs because of the vertical slide supports horizontal movement it's within the machine tool precision grade.

5. References

[1] Zaeh, M. and Siedl, D. 2007 A new method for simulation of machining performance by integrating finite element and multi-body simulation for machine tools *Annals of the CIRP* **56**(1) 383-386.

[2] Maj, R., Modica, F. and Bianchi, G. 2006 Machine tools mechatronic analysis. *Proceedings of the Institution of Mechanical Engineers, Part B: Journal of Engineering Manufacture* **220**(3) 345-353.

[3] N. M. M. Maia and J. M. M. E. Silva 1997 *Theoretical and Experimental Modal Analysis* (John Wiley & Sons, New York, NY, USA).

[4] G. Stan, R. Ciobanu and A. Pal, 2010 "*Balancing-compensation system for the vertically moving elements of the machine tools with numerical control*" Springer Science and Business Media B.V.

IOP Conf. Series: Materials Science and Engineering 1182 (2021) 012061 doi:10.1088/1757-899X/1182/1/012061

Topological optimization of the mobile element structure of Gantry type CNC machine using FEA method

L Petrea and G Stan
" Vasile Alecsandri" University of Bacau, Department of Industrial Engineering, Calea Marasesti Street, No. 157, 600115, Bacau, Romania

E-mail: septh4000@gmail.com

Abstract. Because of the rapid development key industries like aerospace, energy, rail transportation, engineering machinery and other industries there exist a growing demand for high speed and precise manufacturing of large and heavy parts. CNC Machine Tools with kinematical feed chains with great distance between slides, Gantry type, are the main tool to machine large and heavy parts. In such type of machines, the mobile element (the mechanical coupling between the two kinematical feed chains) has a significant effect upon the static and dynamic characteristics of the machine. Deformation of the mobile element has a significant impact upon machining accuracy. A study of the static proprieties and the deformation that effect the mobile element is extremely useful for improving the machining accuracy and efficiency of the machine. The weight of the mobile element it is extremely important and effects the overall dimension and weight of the machine. Using FEA analysis on a 3D CAD model this papers objectives are mesh optimization and topological optimization of the mobile element. This study will show that the topological optimization preformed on the mobile element has reduce the inertial forces, 10 to 30 %, while maintaining rigidity.

1. Introduction

Due to the growing demand for high speed and precise manufacturing of large and heavy parts, there is a demand for Gantry type CNC machine tools. In such types of machines, the mobile element (mechanical coupling between the two kinematical feed chains) has a significant effect on the static and dynamic characteristics of the machine. Deformation of the moving element has a significant impact on machining accuracy. A study of static properties and deformation affecting the mobile element is extremely useful for improving the machining accuracy and efficiency of the machine. The weight of the moving element is very important and affects the overall size and weight of the machine. Thus a structural optimization of the mobile element it's required. In engineering, optimizing an objective function is essentially maximizing or minimizing a constrained problem. Considering the type of design variables and the degree of difficulty, structural optimization can be divided into three types: size optimization, shape optimization and topology optimization corresponding to different stages of product design, namely the three stages of detailed design, basic design and conceptual design. Compared to the first two types of optimization, topological optimization can fundamentally change the structure of the design, the result is the basis for further optimization of the design, and as such it has great significance for the design. Although topology optimization has good prospects, it is recognized as the most difficult task in the field of structural design. Specifically, topology optimization finds optimal models by determining the best locations and geometry of cavities in the

IOP Conf. Series: Materials Science and Engineering 1182 (2021) 012061 doi:10.1088/1757-899X/1182/1/012061

design areas *Huang and Xie, [2010]* [1]. Topology optimization is performed using the results of a finite element analysis (FEA). In this study, we will use the topology optimization module in ANSYS® to optimize the mobile element.

2. Generalities about topology optimization

2.1. Short history

The development of topological optimization can be attributed *Bendsøe and Kikuchi [1988]* which proposed a method of homogenization [2], which achieves continuous topological optimization of the structure by dividing the number of structural units of different micro-holes in the design area. A web-based interface for a topology optimization program was presented by *Tcherniak and Sigmund [2001]* [3]. The study discusses implementation issues and educational issues, as well as statistics and experience with the program. *Rahmatalla and Swan [2004]* [4] presented a variable node-based design implementation for optimizing continuous structural topology in a finite element framework and explored its properties in the context of solving a number of different design examples. *Dadalau et al. [2008]* [5] introduced a new penalty system for the SIMP method. An advantage of the present method is the linear density-stiffness relationship which has an advantage for its own weight or Eigen frequency problem. The problem of topology optimization is solved by the method derived from the Optimality Criteria (OC), which is also the method used in this paper.

2.2. Topology optimization

A topology optimization problem aims to minimize compliance, while meeting various constraints, such as a given amount of material, weight, manufacturing requirements, costs, etc., *Huang and Xie, [2010]* [1]. The most comprehensive goal in topological optimization is to minimize compliance, equation (1).

$$\begin{gathered} min_x\colon c(\rho) = U^T K U = \sum_{e=1}^{N} (\rho^e)^p u_e^T k^e u_e \\ where\colon \frac{V(\rho)}{V_0} = f \\ \colon KU = F \\ \colon 0 < \rho_{min} \leq x \leq 1 \end{gathered} \tag{1}$$

Where the stiffness matrix for each element can be found in equation (2).

$$k^e = \int_\Omega H^T\, DHd\Omega \ \approx \textstyle\sum_{i=1}^{2} \sum_{j=1}^{2} w_i w_j \,/\, J(\xi,\eta)^T DH(\xi,\eta) \tag{2}$$

The Young module does not influence the optimal results of the topology optimization. The overall topology optimization solution is built to minimize compliance using a Lagrange multiplication method equation (3).

$$L(\rho,\lambda) = f^T d(\rho) + \lambda^T (f - K(\rho) d(\rho)) \tag{3}$$

By assigning λ=d and $k^e = \int_\Omega H^T\, (E_0 + (\rho^e)^\rho E_1 D * Hd\Omega$ derived from the Lagrangian equation in relation to the design variables can be determined as equation (4), *Bendsøe și colab., 2004* [6].

$$\frac{\partial L(\rho,\lambda)}{\partial \rho^e} = -P(\rho^e)^{p-1} d^{eT} K_f^e d^e \tag{4}$$

Using the bi-section algorithm, the densities of the elements in each iteration can be updated by a heuristic scheme. Due to the numerical stability and efficiency, a heuristic algorithm based on the Optimality Criteria (OC) method is proposed.

2.3. Optimality Criteria method

The discrete topology optimization problem is characterized by a large number of design variables, D in this case. Therefore, it is common to use iterative optimization techniques to solve this problem, e.g. Optimality Criteria method, at each iteration of the Optimality Criteria method, the design variables are updated using a heuristic scheme. The Lagrangian for the optimization problem is defined as [5]:

$$L(x_i) = u^T Ku + \wedge \left(\sum_{i=1}^{n} x_i v_i - V_o\right) + \lambda_1(Ku - F) + \sum_{i=1}^{n} \lambda_2^i(x_{min} - x_i) + \sum_{i=1}^{n} \lambda_3^i(x_i - 1) \quad (5)$$

Where,$\wedge$, λ_1, λ_2, and λ_3 are Lagrange multipliers for various constraints. The optimum condition is given by:

$$\frac{\partial L}{\partial\, \partial x_i} = 0,\ i = 1,2,3, \dots, n \quad (6)$$

That result in the Conformity:

$$L = u^T Ku \quad (7)$$

By differentiating equation (5) w. r. t., the optimality condition can be written as:

$$B_i = \frac{-\frac{\partial c}{\partial x_i}}{\wedge v_i} = 1 \quad (8)$$

Conformity sensitivity can be assessed by using equation:

$$\frac{\partial c}{\partial x_i} = -p(x_i)^{p-1} u_i^T K_i U_i \quad (9)$$

According to this equations, the design variables will be updated as follows:

$$\begin{aligned} new_{xi} &= \{\max(x_{min} - m), if\ x_i B_i^{\eta} \le x_{min}, (x_{min} - m)\} \\ &= \{x_i B_i^{\eta}, if \max(x_{min} - m) < x_i B_i^{\eta} < \min(1, x_i + m)\} \\ &= \{\min(1, x_i + m), if \min(1, x_i + m) \le x_i B_i^{\eta}\} \end{aligned} \quad (10)$$

In which, m is the the motion limit and represents the maximum permissible change in x_i in one Optimality Criteria iteration, η represents the numerical depreciation coefficient and is usually considered to be 1/2. The Lagrange volume constraint multiplier $\wedge$ is determined at the Optimality Criteria iteration using a bisection algorithm, x_i represents the value of the density variable at each of the iterations steps, u_i it is determined from the equilibrium equations an it's the displacement field at each iteration step. In short the structure of the optimization algorithm is:

- The initial design is performed.
- The resulting displacements and strains are calculated by the finite element method.
- The conformity of the project is calculated. If it's just a marginal improvement in compliance with the last project, stop the iterations. Otherwise, continue.
- The update of the design variable is calculated, based on the scheme presented in equation (10). This step also consists of an internal loop iteration to find the value of the Lagrange multiplier that constrain the volume.
- Repeat the iteration loop.

3. Topological optimization of the mobile element

3.1. Geometric model structure

For this work, a 3D model was created for the mobile element, respecting the technical parameters of an SC14 carousel lathe (figure 1): length 2292mm, width 685mm, thickness 280mm and Gray cast iron material. In order to reduce the calculation time, the 3D model of the slide (the mobile element) has been simplified.

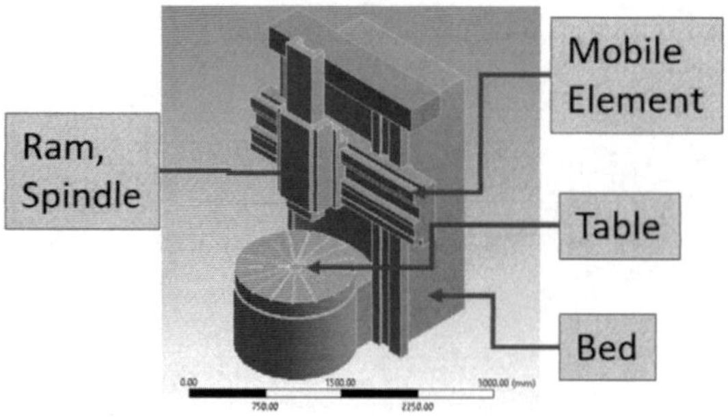

Figure 1. SC 14 carousel lathe.

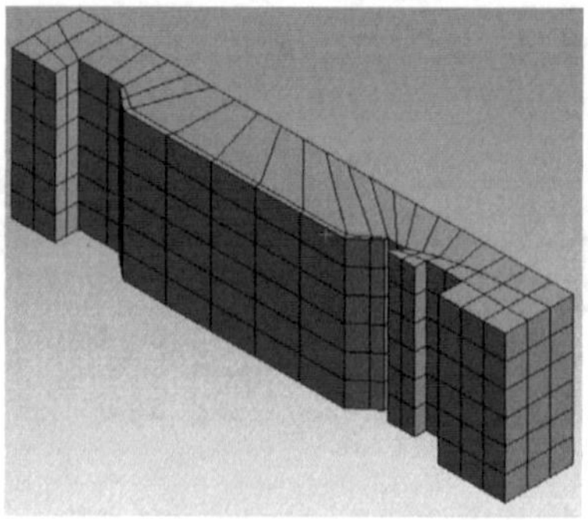

Figure 2. Simplified mobile element.

3.2. Optimization and refinement of mesh in ANSYS

ANSYS can generate a generic mesh, see figure 2. To check the quality of the mesh we can use the "Element Quality" function from the "Mesh Metric" tab. As we can see in figure 3 many elements are of low quality (0 - 0.5 plot area in figure 3). This will affect the quality of static analysis and topological optimization. The problem can be solved by generating a new mesh. ANSYS has several tools for optimizing and refining the mesh, to obtain a good result the method "Tetrahedrons" with the parameter "Independent patch" was applied (figure 4). This method refines the interior mesh firstly then the exterior mesh achieving very good results.

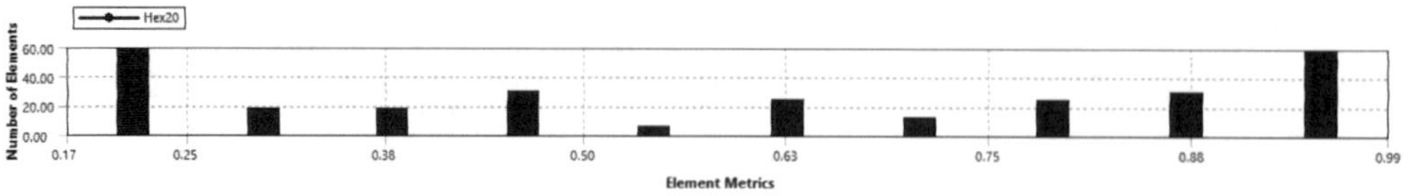

Figure 3. Element quality plot – Generic Mesh.

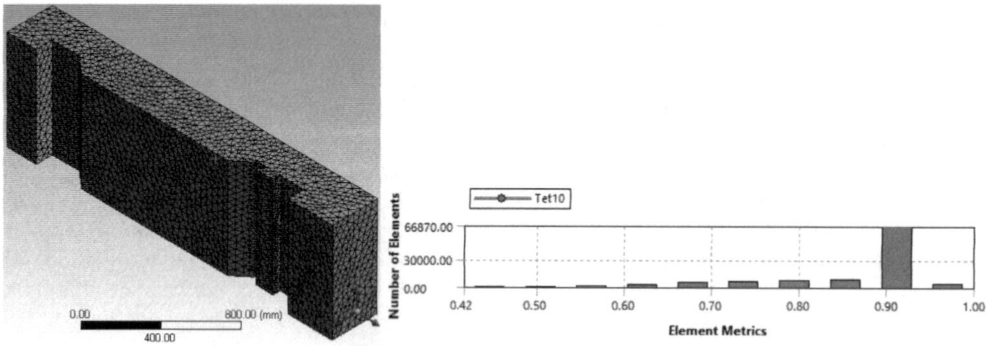

Figure 4. Element quality plot – Tetrahedrons+ Patch independent.

As we can see most of the elements are grouped on the right side of the element quality plot, with most of them at the 0.90 mark, this represents a very good result.

3.3. Static analysis of the mobile element

According to the technical specification of the SC14 carousel lathe the mobile element can sustain a 51000 N load. Figure 5 shows the distribution of forces that act upon the mobile element during normal machining operations, The 20000 N force on Y axis represent the mass of the vertical slide and spindle.

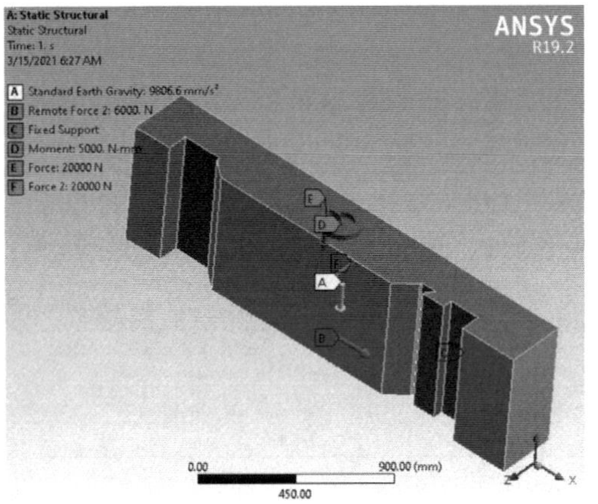

Figure 5. Force loading.

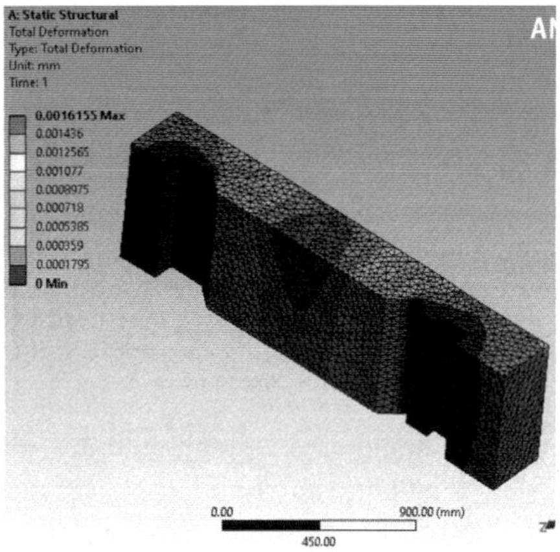

Figure 6. Mobile element deformation.

As we can see in figure 6, the total deformation upon the mobile element its small, under 0.002mm, well within the machine tool precision requirements.

3.4. Topology optimization analysis and results

After preforming the static analysis, we can use ANSYS Workbench to add a topology optimization module to the static analysis results. As mentioned the optimization method will be "Optimality Criteria" (OC), the maximum number of iterations 30, the convergence accuracy 0.1% and the constraint function will be the mass min. 70%.

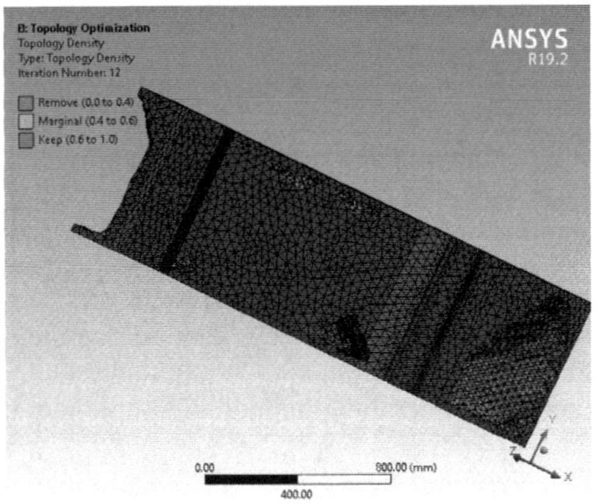

Figure 7. Topology Density – Retained region.

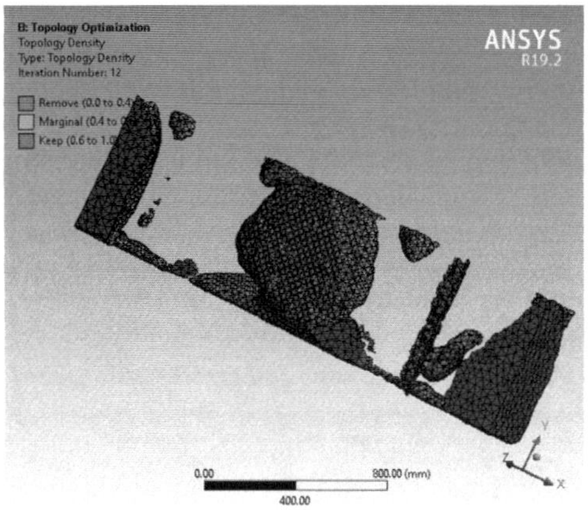

Figure 8. Topology Density – Removed region.

The results of the topology optimization are represented as Topology Density, we ca observe in figure 7 the what material that has been kept after 12 iterations. In figure 8 we can observe the material that has been removed. As we can observe the result it's not a geometry that can be easily manufactured, however it dose tell the designer wat material need to be removed from the mobile element, without sacrificing rigidity.

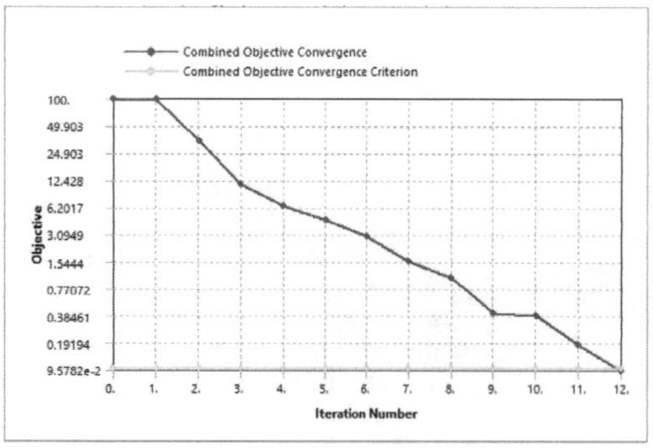

Figure 9. Compliance vs. Iterations plot.

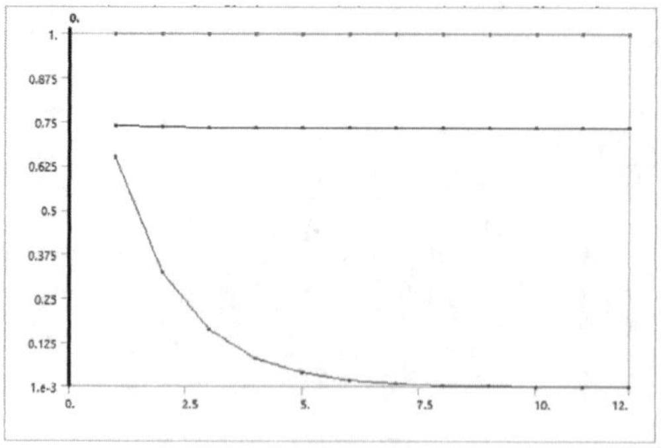

Figure 10. Curves of objective function.

Figure 9 and figure 10 represent the objective function plotted course, we can observe that the compliance decreases with the increase of iterations, and that the iterative results are in stable condition after 10th step so that calculation results converge.

4. Conclusions

This study presents the mathematical model of a topological optimization of a structure (the mobile element / slide in this case) while using the ANSYS software platform for optimization calculation to achieve conceptual design that helps to maintain the conditions of rigidity while reducing weight by approximately 25%. Compared to traditional structural designs, this method not only guarantees the reliability of the design, but also shortens the design cycle and increases efficiency.

5. References

[1] Huang, X. and Xie, Y.M. 2010 *Evolutionary topology optimization of continuum structures: methods and applications* John Wiley & Sons.

[2] Bendsøe M.P. and Kikuchi N. 1988 Generating optimal topologies in structural design using a homogenization method *Comput. Meth. Appl. Mech. Eng.* **71** 197-224.

[3] Tcherniak D. and Sigmund O. 2001 A web-based topology optimization program *Struct. Multidisc. Optim. Springer-Verlag* **22** 179-187.

[4] Rahmatalla S.F. and Swan C.C. 2004 A Q4/Q4 continuum structural topology optimization implementation *Struct. Multidisc. Optim. Springer-Verlag* **27** 130-135.

[5] Dadalau A. and Hafla A. 2008 *A new adaptive penalization scheme for topology optimization* Stuttgart Research Centre for Simulation Technology (SRC Sim. Tech.), Stuttgart University.Verl.

[6] Bendsøe, M.P. and Sigmund, O 2004 Topology optimization by distribution of isotropic material *Topology Optimization, Springer* 1-69

IOP Conf. Series: Materials Science and Engineering 1182 (2021) 012062 doi:10.1088/1757-899X/1182/1/012062

Modern manufacturing technology for modular photovoltaic panels: State-of-the-art and future trends

RA Polcovnicu, N Ţăranu, D Ungureanu, Ş Zghibarcea and VS Hudişteanu
Gheorghe Asachi Technical University of Iasi-Romania, Faculty of Civil Engineering and Building Services

E-mail: razvan-andrei.polcovnicu@student.tuiasi.ro

Abstract. The technological advancement of the recent years has triggered a significant increase in the global energy consumption. Within this frame, the fossil fuels still account for most of the energy consumption as the renewable energy systems often lack reliability and predictability in terms of total energy supply. These shortcomings may be overcome if various design and manufacturing process optimization methods are applied so as to gain more benefits from the current solar power systems. A modern modular photovoltaic (PV) panel is an assembly of photovoltaic cells mounted in an adaptable framework for quick installation. The photo-voltaic cells use sunlight as source of energy and generate current electricity, which can be harvested directly or through an existing grid. A collection of PV modules can be arranged in any desirable manner to form a PV panel. Also, the implementation of such modern modular systems brings numerous advantages, including the reduction of power transmission cost and the minimization of the global warming problems. This work presents the latest findings related to the manufacturing technology and the equipment train used in the production of modern photovoltaic systems.

1. Introduction

The demand for electricity supply for commercial, domestic and industrial purposes is increasing due to the infrastructural development and the economic expansion. Nowadays, the electricity requirement is higher than the quantity the utility companies can generate and thus, the service delivery is, in many cases, unreliable [1]. It is well known that supplying reliable energy to costumers is, generally, associated with increasing the living standards of education, micro-economy and healthcare. Unfortunately, this rising energy demand is mainly provided by burning fossil fuels such as coal, natural gas, oil and bituminous rocks. In addition, the discovery of conventional energy resources is getting very rare and the energy supplies of organic origin slightly grow around the world [2]. The intemperate use of organic origin energy involves higher rates of gas emissions such as nitrogen dioxide, carbon dioxide and methane in the air that negatively influence the greenhouse effect [3]. In time, the exploitation of these non-renewable resources may unavoidably lead to their total consumption, which might further raise an energy crisis [4]. According to the latest energy policies, the alternative energy sources should be considered as a responsibility for both industries and governments with respect to the protection of the environment. Recent studies have established that the new cost-effective energy strategies may lower the electricity use by 27% to 75%, within 10-20 years, without threatening the manufacturing output or the quality of life [5]. An eloquent example of alternative energy source is the solar energy. The energy harvested from the solar radiation can be used to heat buildings (through passive solar hot water systems), to produce hot water (through solar hot water systems) and to satisfy the electricity demand (through photovoltaics - PV systems). For instance, when compared to fossil fuel-generated electricity, each kilowatt of PV electricity offsets 9 kilograms of sulphur dioxides, up to 16 kilograms of nitrogen oxides and 2300 kilograms of carbon dioxide (CO2) annually [6].

The term photovoltaics derives from the Greek word phōs meaning light and the word volt, named after the Italian physicist Alessandro Volta (1745-1827). Photovoltaics is the science that deals with the light-electricity conversion, namely the photon energy-electric current conversion. In other words, it stands for light-current conversion. First use of the word photovoltaic was mentioned in the book "*Elements of Electro-biology, Or the Voltaic Mechanism of Man of Electro-pathology, Especially of the Nervous System and of Electro-therapeutics*" written by Alfred Smee (1818-1877) in 1849 [7].

The Photovoltaic solar systems typically contain several panels (wired together in an array), an overcurrent protection (circuit breakers or fuses), several electrical disconnects, a junction box, an inverter and other specialized equipment depending on the type of application (off-grid, grid-tied, battery-backup). The outdoor performance of a PV panel depends on the availability of the incoming solar radiation and on the panel junction temperature. The challenging part consists in ensuring a satisfactory efficiency of the absorption process (of the radiant energy) in case of unwanted natural elements [8].

This work presents the fabrication steps of a modern photovoltaic panel, the latest findings related to the manufacturing technology and the equipment train that are used in the production of modern photovoltaic systems.

2. Photovoltaic panels

A solar photovoltaic panel (figure 1) is composed of glass, solar cells, back sheet and ethylene vinyl acetate (EVA), all mounted within a frame. On the market there are three categories of solar panels available: thin film, monocrystalline and polycrystalline photovoltaic panels. At the level of the cell structure, various materials can be identified, such as amorphous silicon, mono silicon or polysilicon [9].

Figure 1. Solar photovoltaic panel [10].

3. Materials and manufacturing process

Sand is the main raw material used in the manufacturing of solar panels. Most photovoltaic panels are made of silicon, which is the primary component in sand. The latter is plentifully available, being the second accessible element on Earth [11]. Nonetheless, transforming sand into silicon is achieved at a higher cost and it is an energy demanding process. High-purity silicon is created from quartz sand at high temperatures in an arc furnace. The silicon is amassed, usually as solid rocks. The rocks are melted together at high temperatures with the purpose of forming cylinder shaped ingots. In the melting process, all atoms are aligned in the orientation and structure needed. In order to obtain positive electrical polarity, boron is included in the process.

Mono crystalline cells (figure 2) are synthesized from a single silicon crystal. The latter has better capability in transforming solar energy into electricity, yet the monocrystalline panels are more expensive.

Figure 2. A mono crystalline silicon photovoltaic wafer [12].

Polysilicon cells are created by melting various silicon crystals together. The distinct silicon crystals give a shattered glass look. Polishing and grinding are being done after cooling of the ingot, leaving them with flat sides. Next, the wafers are fabricated by slicing the silicon ingot into thin disks. A wire saw is used for precision cuttings so the final wafer thinness is comparable to that of a paper. Afterwards, due to silicon shinning property, an anti-reflective coating is placed on the wafer to reduce the volume of sunlight lost.

The next step in the manufacturing process consists in adding conductors on each wafers' surface. The conductors will ensure the conversion of solar energy and grants the wafer's surface a grid-like matrix. The coating ensures the absorption of sunlight, instead of reflecting it. In an oven-like chamber, over the wafers' surface, phosphorous is being dispersed in a thin layer. This process will electrically charge the surface negatively. A paramount condition for the proper function of the photovoltaic cell is the positive-negative junction, which is produced by the combination between phosphorous and boron. The photovoltaic cells are joined together into a matrix-like structure with the use of metal connectors. The available arrangements that can be found on the market include: 48 cell panels - convenient for small residential roofs, 60 cell panels – standard size, 72 cell panels – adopted for sizable installations [13]. Thereafter the cells are inserted together and a 6-7 mm thick layer of glass placed on the front side, opposing the sun.

The back sheet is manufactured from polymer-based, highly durable material. Thence the penetration of soil particles, water and other materials on the back of the panel will be prevented. Recently the junction box has been installed in order to implement connections for the module. The frame also provides protection against weather and accidental impacts. Furthermore, using a frame will allow for various ways of mounting the panel. Ethylene Vinyl Acetate (EVA) is the glue that connects everything together. The encapsulation must be performed at the highest quality control in order to prevent any possible damaging of the cells under severe weather conditions.

4. Photovoltaic panel testing and quality control

As soon as the module is completed, tests are performed to ensure the photovoltaic cells operate as expected. Standard Test Conditions (STC) are applied as a reference point. The photovoltaic panel is inserted in a flash tester. The tester will produce a 25°C cell temperature, an air mass of 1.5g and $1000W/m^2$ equivalent irradiance [14]. The electrical parameters are determined and the results are written down on the technical specification sheet for every photovoltaic panel. This test will reveal the efficiency, power output, current, voltage and temperature tolerance of the product. Besides the STC, nominal operating cell temperature (NOCT) test is performed. The specifications used are close to real life scenarios:

- operation temperature of the open-circuit module is at $800W/m^2$ irradiance,
- 1m/s wind speed,
- 20°C cell ambient temperature.

The NOCT ratings are specified on the technical data sheet. Inspection and cleaning are the final steps of the manufacturing before the module is available for selling.

5. Photovoltaic panels manufacturing evolution

Commercial manufacturing of crystalline silicon modules began in 1963 in Japan when Sharp Corporation patented the photovoltaic module [15]. The latter is a 242 W photovoltaic module that was installed on a lighthouse, the biggest commercial photovoltaic installation in the world at that time. It has been noticed that the quantity of photovoltaic modules produced in Japan increased significantly between 1997 and 2004 [16]. After 2004, the production capacity reduced drastically. The same tendency was observed in photovoltaic module manufacture in Europe up to 2008, after that the production volume has decreased.

In 1997, the photovoltaic module manufacturing capacity of the US was at its best, then the production capacity has been reduced every year. Insomuch as the photovoltaic modules fabrication in China was just the opposite compared to US. The worldwide production of photovoltaic modules has accelerated over last few years due to the far-reaching potential, with China taking the supremacy in total manufacture volume [17].

Germany and other European countries promoted the global cumulative photovoltaic installation up to 70% until 2011 [18]. Therefore, till now, the photovoltaic installation market has been very promising in Europe compared to other continents. More than 85% of the solar cell manufacture depends on crystalline silicon, based on the scenario that was commonly agreed by the European countries in 2011 [19].

6. Advantages and disadvantages of modern manufacturing technologies

Monocrystalline silicon cells photovoltaic panels are one of the first types of solar panels that were accessible on the market [20]. They are used both for industrial and home applications. The benefits of adopting this technology consist in:

- use of high-quality silicon;
- high-performance rates of 20%;

- convenient use of roof space;
- more solar power collected in a battery bank;
- longer warranties (up to 25 years), [21];
- harvest more solar energy per square foot.

On the other hand, the shortcomings of using monocrystalline silicon cells technology are:

- expensive production cost;
- circuit breakdown can occur at times when the solar panels are shaded;
- significant waste amounts results during manufacturing [22];
- potential damage over time due to high temperatures (they perform at their top only in warm weather conditions).

When it comes to economic benefits, polycrystalline silicon solar panels are recommended [23]. They have a speckled look, which is very different from a monocrystalline silicon module because of the molten silicon used in the manufacturing process. Therefore, these types of photovoltaic panels are cheaper [24] than the monocrystalline silicon solar panels. The advantages using this technology are:

- less expensive;
- easy and quick instalation;
- easy to acquire;
- few fossil fuels are required in the manufacturing process;
- the modules perform reasonably well, in various weather conditions, in comparison with other crystalline photovoltaic panels [25].
- the polycrystalline silicon modules can be used both with batteries and inverter technology.

However, the disadvantages of this technology include:

- harvest less solar power;
- more space requirement on the roof;
- potential durability issues;
- easily damaged when exposed to high temperatures.

Thin-film solar panels are manufactured with second-generation solar cells technology, also made from silicon. These types of photovoltaic panels are less expensive than monocrystalline solar cell panels and they can be easily manufactured. This class of solar cells are convenient for a solar array, but they acquire more energy during manufacture, thus, more burning of fossil fuels [26]. The main advantages of using thin-film photovoltaic panels are:

- flexible;
- lightweight;
- high temperatures does not affect the performance over time;
- they can be integrated in any part of the façades or on the roof;
- uncomplicated manufacture process.

More than that, this class of photovoltaic panels work in low light environment and are an affordable substitute to monocrystalline solar panels. On the other hand, thin-film photovoltaic modules have limited efficiency between 7 and 13% [27] and the power output is low. More than that, they are not ideal for residential use because of a lot of space occupation and they have short warranty and life span.

Another way of thin-film solar panel manufacturing is using amorphous silicon. The latter use triple layered technology which improves the performance and strengthens the photovoltaic panel [28]. Amorphous solar panels are relatively new technology and their energy efficiency is still at the research level. However, they are frequently used to produce biohybrid solar cells panels that emulate the natural part of photosynthesis [29] and possess the thin-film panels' advantages.

The multi-junction photovoltaic cells may achieve an efficiency rating of up to 41% [30]. Nevertheless, they also use a lot of advanced technology and materials, including curved mirror surfaces, cooling systems, and lenses. Even though concentrated solar photovoltaic modules are more efficient, the manufacturing cost is higher.

7. Future trends in photovoltaic panel manufacturing

Development and research in the solar energy field focus on increasing the efficiency of the photovoltaic panels and the reduction of the production cost [31]. The photovoltaic panel fabrication industry is developing more convenient alternatives than traditional sources of energy, for instance fossil fuels. The early supremacy of silicon in the research programs has nowadays spread out to the wholesale of commercial modules. If there is to eliminate electronics and characterize the market as electricity distribution system of 1 kW or more, current manufacturing is monopolized by polycrystalline and single-crystal silicon modules that serve up to 94% of the demand [32]. There is a vast range of distinct photovoltaic cells on the market using various types of materials and an even preponderant number will be developed in the near future.

Photovoltaic cell technologies are categorized into three generations, according to the commercial debut and component materials. First-generation photovoltaic systems are fully commercial and they use wafer-based crystalline silicon (c-Si) high tech, either multicrystalline (mc-Si) or single crystalline (sc-Si). The second-generation photovoltaic systems (early market distribution) are based on thin-film photovoltaic technologies and include three main categories: micromorph silicon (a-Si/μc-Si)/amorphous (a-Si), cadmium telluride (CdTe) and copper indium-gallium diselenide (CIGS)/copper indium selenide (CIS). The third-generation photovoltaic systems include organic photovoltaic cells (figure 3).

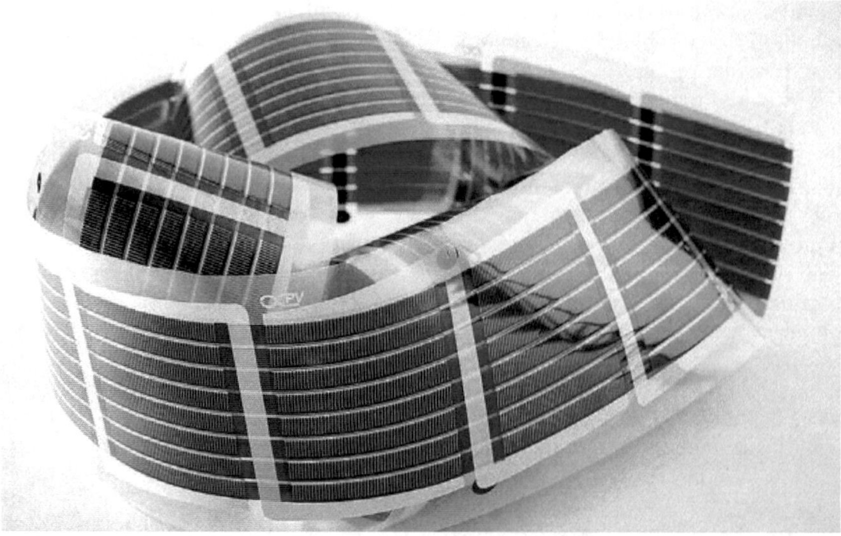

Figure 3. Organic photovoltaic cells [33].

8. New photovoltaic products

After more than two decades of research and development, thin-film solar cells (figure 4) are starting to be promoted and developed in significant quantities. The high potential of thin-film solar cells to supply lower cost electricity compared to the crystalline silicon wafer-based photovoltaic cells is already acknowledged. Thin-film photovoltaic cells comprised subsequent thin layers, 1 μm to 4 μm thick, of photovoltaic cells arranged onto an inexpensive, large substrate such as polymer, glass or metal. As a result, a substantially less semiconductor material is required for absorbing the same quantity of sunlight (approximately 99% less material when compared with crystalline solar cells) [34]. Further, thin films can be encased into lightweight and flexible structures, which can be conveniently integrated into the components of a building (building-integrated photovoltaic, BIPV).

IOP Conf. Series: Materials Science and Engineering 1182 (2021) 012062 doi:10.1088/1757-899X/1182/1/012062

Figure 4. Thin-film solar cells [35].

9. Conclusions

Photovoltaic technology is to become a considerable supplier of electricity in the 21st century due to its affordable cost (cheaper than fossil fuels) and lower environmental impact. One of the most promoted applications of photovoltaic panels consists in various arrangements installed on roofs and facades. Therefore, the PV modules may be considered as a building material in itself. However, even if it is designed as a regular construction material, the manufacturing process of PV panels should be addressed according to the newest technology advancements and quality control techniques. From this point of view, in this work, the manufacturing technology and the equipment train used for modern photovoltaic systems have been described and discussed.

In the final part of this paper, several future trends in PV manufacturing and development were pointed out. Overall, the future of photovoltaics continues to be one of intense development and research where thin films will overcome crystalline silicon in efficiency and cost. Also, it is expected that the role of PV in the generation of electricity will come up with other forms of renewable energy and that half of all new buildings will include PV in their design.

10. References

[1] Modis T 2018 Technol. Forecast. Soc. Change. *Forecasting energy needs with logistics.*

[2] Yıldız I 2018 1.12 Fossil fuels *Comprehensive Energy Systems. Elsevier* pp 521-567.

[3] Barreto R A 2018 Fossil fuels, alternative energy and economic growth *Econ. Model* **75** 196-220.

[4] Hache E 2018 Do renewable energy security in the long run? *Int. Econ.* **156** 127-135.

[5] Kandiyoti R 2017 Fossil fuels and renewable *Solid Fuels and Heavy Hydrocarbon Liquids. Elsevier* pp 1-9.

[6] Weng Y 2018 Increasing worldwide environmental consciousness and environmental policy adjustment *Q. Rev. Econ. Finance.*

[7] Delgado G, Ramos S, Jara R, Felix M and Dominguez A 2018 Decision-making approach: A simplified model for energy performance evaluation of photovoltaic modules *Energy Convers. Manag.* **177** 350-362.

[8] Aratani F 2005 *Prog. Photovoltaics* **13** 463.

[9] Mehta S 2009 Global PV Cell and Module Production Analysis *GMT Research.*

[10] https://www.indiamart.com/proddetail/solar-photovoltaic-panel-17377000388.html accessed on: 10.03.2020.
[11] Green M A, Emery K, Hishikawa Y and Warta W 2010 *Prog. Photovoltaics* **18** 346.
[12] https://www.tindosolar.com.au/learn-more/poly-vs-mono-crystalline/ accessed on: 10.03.2020.
[13] Kerr M J, Schmidt J and Cuevas A 2002 Lifetime and efficiency limits of crystalline silicon solar cells *Proceedings of the 29th IEEE PVSC* p 250 (New Orleans).
[14] Green M A 2001 Third generation photovoltaics: ultra-high conversion efficiency at low cost *Progress in Photovoltaics* **9**(2) pp 123-135.
[15] O'Regan and Gratzel M 1991 *Nature* **353** 737.
[16] Kato N, Higuchi K, Tanaka H, Nakajima J, Sano T and Toyoda T 2011 *Sol. Energ. Mat. Sol.* C. **95** 301.
[17] White M S, Olson D C, Shaheen S E, Kopidakis N and Ginley D S 2006 *Appl. Phys. Lett.* **89** 143517.
[18] Tao C, Ruan S, Xie G, Kong X, Shen L, Meng F, Liu C, Zhang X, Dong W and Chen W M 2009 *Appl. Phys. Lett.* **94** 043311.
[19] Ennaoui, Fiechter S, Pettenkofer C, Alonsovante N, Buker K, Bronold M, Hopfner C and Tributsch H 1993 *Sol. Energ. Mat. Sol.* C. **29** 289.
[20] Katagiri H, Saitoh K, Washio T, Shinohara H, Kurumadani T and Miyajima S 2001 *Sol. Energ. Mat. Sol. C.* **65** 141.
[21] 2005 Basic Research Needs for Solar Energy Utilization *U.S. Department of Energy Report*.
[22] Spanggaard H and Krebs F C 2004 *Sol. Energ. Mat. Sol. C.* **83** 125.
[23] Wadia C, Alivisatos A P and Kammen D M 2009 *Environ. Sci. Technol.***43** 2072.
[24] Beck M 2010 Personal Communication *NSF PV Workshop*.
[25] Engelharet P, Hermann S, Neubere T, Plagwitz H, Grischke R, Meyd R, Klug U, Schoonderbeek A, Stute U and Brendel R *Prog. Photovoltaics* **15** 521.
[26] Ramanathan K, Teeter G, Keane J C and Noufi R 2005 *Thin Solid Films* **480** 499.
[27] van Duren J, Jackrel D, Jacob F, Leidholm C, Pudov A, Robinson M and Roussillon Y 2007 The next generation in thin film photovoltaic process technology *Conference Record of the Seventeenth International Photovoltaic Science and Engineering Conference, Fukuoka, Japan*.
[28] Tiwari A N, Lincot D and Contreras M 2010 *Prog. Photovoltaics* **18** 389.
[29] Serrano E, Rus G and Garcia-Martinez J 2009 Nanotechnology for sustainable energy *Renewable and Sustainable Energy Reviews* 2372–2384.
[30] Glunz S W 2006 New concepts for high-efficiency silicon solar cells *Solar EnergyMaterials and Solar Cells* **90** 3276–3284.
[31] Katagiri H, Saitoh K, Washio T, Shinohara H, Kurumadani T and Miyajima S 2001 *Sol. Energ. Mat. Sol. C.* **65** 141.
[32] 2011 Thin-film cells trace their roots to RCA Laboratories in New Jersey *Solar Cell Production Global Market Outlook, Business Insights.*
[33] https://infinitypv.com/technology/opv accessed on: 10.03.2020.
[34] European Photovoltaic Industry Association 2011 Efficiency, Which Measures the Percentage of the Sun's Energy Striking the Cell or Module, is One important Characteristic of a Solar Cell or Module. Over Time, Average Cell Efficiencies have Increased *Solar Generation 6, Solar Photovoltaic Electricity Empowering the World.*
[35] https://mappingignorance.org/2016/10/24/looking-new-materials-applications-organic-solar-cells/ accessed on: 10.03.2020.

Effect of Ti and C alloyants on the microstructure of laser cladded cobalt-chromium coatings

T Poloczek, D Janicki, J Górka and A Kotarska
Silesian University of Technology, Mechanical Engineering Faculty, The Chair of Welding, Konarskiego 18a, 44-100 Gliwice, Poland

E-mail: tomasz.poloczek@polsl.pl

Abstract. The article presents the results of Ti and C additives on the microstructure of the Metcoclad 6 coating. Effect of chemical composition with optimal laser cladding parameters has been studied in the terms of modification of the carbide morphology. The experiments were carried out on TruDisk 3302 Yb:YAG disk, equipped with a disc powder feeder. As a deposited material Metcoclad 6 powder produced in inert gas-atomization process was used. To assess the impact of alloying elements on microstructure, SEM observations and EDS analysis were carried out. Vickers microhardness tests were provided on the cross-section of coatings on Wilson Wolpert 401 MVD device. It has been stated that addition of titanium changes the morphology of carbides from lamellar eutectic carbides to isolated primary TiC while addition of Ti and graphite allowed to maintain the structure typical for pure Stellite 6 alloy, but additional formation of primary carbides were observed.

1. Introduction

The wear processes present in the industry are one of the major problems encountered when analyzing the factors influencing the failure of machines. The requirements for the currently used materials due to the long life cycle of the product are mainly focused on the maximum wear resistance in various forms: metal against metal, metal against non-metal or metals against fluid, because the movement between the surfaces of different materials causes damage by abrasion, galling, cavitation, erosion or corrosion. Cobalt alloys have found a special application in industry due to the combination of high wear resistance, corrosion resistance, high hardness and cavitation resistance. A popular group of cobalt-chromium alloys is called the Stellite and was developed in 1910 by E. Haynes [1,2].

The Stellite alloys as a wear resistant material play an important role in application on valves, turbine blades, steam valve guides or flame tubes of gas turbines due to high resistance to galling and fretting up to 300°C. The high corrosion resistance allowed to use this material in chemical industry especially where the material has contact with the corrosive environment in combination with abrasive wear. An example of such an application may be the sealing ring wear for submarine rudder, chemical scrapper blade or knives for cutting fibres. Standard cutting tools were made of high speed steel but development and application of Stellite alloys allowed to increase the feed rate to be increased about three times which was very important during second world war [1-3].

Stellites owe their high performance properties to the combination of carbides with cobalt matrix. Cobalt in this alloy occurs in fcc crystal structure, because of the sluggish nature of the transformation from fcc to hcp crystal structure. The fcc structure ensure the high yield strength and ability to absorb stress through transformation to hcp structure. In Stellite chromium has a dual function as a main carbide former (most of carbides are Cr rich) and as a alloying element which is strengthening the matrix and improving the resistance to corrosion and oxidation by forming the oxide layer. Small additions of tungsten and molybdenum are responsible for strengthening mechanism of matrix due to their large atomic size impeding dislocation flow and improving general corrosion resistance of material. In large amounts, they are responsible for formation carbides during solidification. Small addition of nickel stabilizes the γ-Co matrix in face-centered cubic structure [3-6].

There are two basic group of Stellites: CoCrW and CoCrMo. The first group includes the Stellite 6 alloy, which is one of the most commonly used cobalt-based alloys due to its high wear resistance combined with good plastic and strength properties. The described material is characterized by high impact strength, resistance to cavitation, and a very good resistance to chemical wear in a wide temperature range while maintaining its properties up to a temperature of 500°C. Stellit 6 is suitable for hardfacing and therefore has a wide range of applications on pump shafts, valve seats, bearings, erosion shields or rolling couples [7-13].

In the conducted research, laser cladding tests were carried out with modified Metcoclad 6 powder (the chemical composition is the same as in Stellite 6) by adding titanium as well as graphite and titanium in order to obtain unique layers characterized by high performance. Based on the research carried out by Ying [14], it can be concluded that the addition of titanium will modify the type of carbides in microstructure of Stellite 6, influencing on the material properties. The second mixture of powders, containing titanium and graphite, was used in order to obtain an unchanged lamellar eutectic structure of the Stellite 6 with an increased amount of TiC carbides.

2. Materials and Methods

Non-alloy S235 steel with dimension of 100x50x10 mm was used as a substrate material in conducted research. Before cladding the samples were grinded and cleaned using acetone in order to obtain clean surface of high purity. As a laser cladding powder, Metcoclad 6 produced in inert gas atomization process with the particle size in the range from 50 – 106 μm was used. In order to obtain unique layers characterized by high hardness, new powder mixtures were created, consisting of the addition of titanium and graphite (both 99.9 wt% purity with particle size between 80 and 150 μm) in the base Metcoclad 6 powder. Chemical composition of individual powder mixtures (in weight ratio) is presented in table 1. All the powders were mixed mechanically using a planetary ball milling machine.

Table 1. Chemical composition of original and modified powder of Metcoclad 6.

Sample no.	Chemical composition (wt.%)									
	Co	Cr	W	Si	C	Ni	Mo	Fe	Mn	Ti
1	Bal.	27.3	4.4	1.6	1.1	0.8	0.1	0.1	<0.1	-
2	Bal.	26.2	4.2	1.5	1.1	0.8	0.1	0.1	<0.1	4
3	Bal.	25.9	4.2	1.5	2.0	0.8	0.1	0.1	<0.1	4

The laser deposition process was performed on TRUMF disk laser Nd:YAG TruDisk 3302 with numerically controlled positioning system of the head and the treated substrate. The laser emits a continuous beam with a wavelength of 1.03 μm at a maximum power of 3.3 kW. The transmission of the laser beam was carried out using a 20 m long optical fiber. The detailed parameters of laser are shown in table 2. Argon with purity of 99.999% was used as a powder carrier and shielding gas. The laser station was equipped with a disk feeding system for precise control of powder feeding rate in the range from 0.6 to 12 g/min.

Table 2. TRUMPF TruDisk 3302 laser parameters.

Characteristic	Value
Wave length [μm]	1.03
Maximum output power [W]	3300
Laser beam divergence [mm·rad]	<8.0
Fibre core diameter [μm]	200
Collimator focal length [mm]	200
Focusing lens focal length [mm]	200
Beam spot diameter [μm]	200
Fiber length [m]	20

In order to obtain high quality laser clads the preliminary tests were carried out. Conducted tests included different parameters of: powder feed rate, laser beam power, focal length, shielding gas flow in order to obtain cladding layers fully adhered to substrate with low dilution and high powder melting rate. Quality of obtained layers was evaluated using the visual testing method. During the surfacing tests it was found that changing parameter of focal point to 30 mm above the material surface gives the best result for laser cladding ensuring the low mixing degree of clad with base material. Selected parameters used in laser cladding process are presented in table 3. Every coating consisted of 5 single beads with constant control of the interpass temperature, which not exceeded 50 °C. During laser cladding, argon shielding gas was used at a flow rate of 10l/min blown through a cylindrical nozzle aligned with powder nozzle.

Table 3. Technological parameters of the laser cladding process.

Shielding gas flow rate – argon [l/min]	**Carrier gas flow – argon [l/min]**	**Inclination angle of the nozzle relative to the laser head [°]**	**Powder feeding nozzle diameter [mm]**
10	3	30	2
Step-over width [mm]	**Laser power [W]**	**Cladding speed [mm/min]**	**Powder flow rate [g/min]**
2.6	1000	75	0.035

The microstructures of obtained laser clads were examined by using scanning electron microscope Phenom World PRO equipped with energy dispersive spectroscopy (EDS). The samples were cut in the middle of clad length where the parameters were stable. For macro and microstructural analysis, the specimen surface was grinded with sandpapers and polished with 1 μm alumina powder. The polished surfaces were etched with aqua regia for 10 seconds at temperature 20°C. In order to assess the quality of obtained layers, parameter of percentage dilution on cross section was calculated by following equation:

$$U_P = F_W/((F_N+F_W))\cdot 100\% \qquad (1)$$

where: U_P - dilution of base material with the coating, F_W – penetration surface, F_N – coating surface.

In order to assess the impact of chemical composition on properties of different chemical composition of laser clads the Vickers microhardness tests were provided. The measurements were carried out on cross-section from surface to the substrate at an interval of 0.2 mm on Wilson Wolpert 401 MVD with load of 500 g for the time duration of 10 seconds.

3. Results and discussion

The macrostructures images of obtained coatings are presented in figure 1. On prepared coatings no imperfections in a form of porosity or cracks were found. The coatings were characterized by very low dilution level which is the result of combining the correct parameters of laser cladding with optimum powder feed rate based on the preliminary tests. Microstructure analysis confirmed that coatings have an excellent metallurgical bond with the substrate free from any defects.

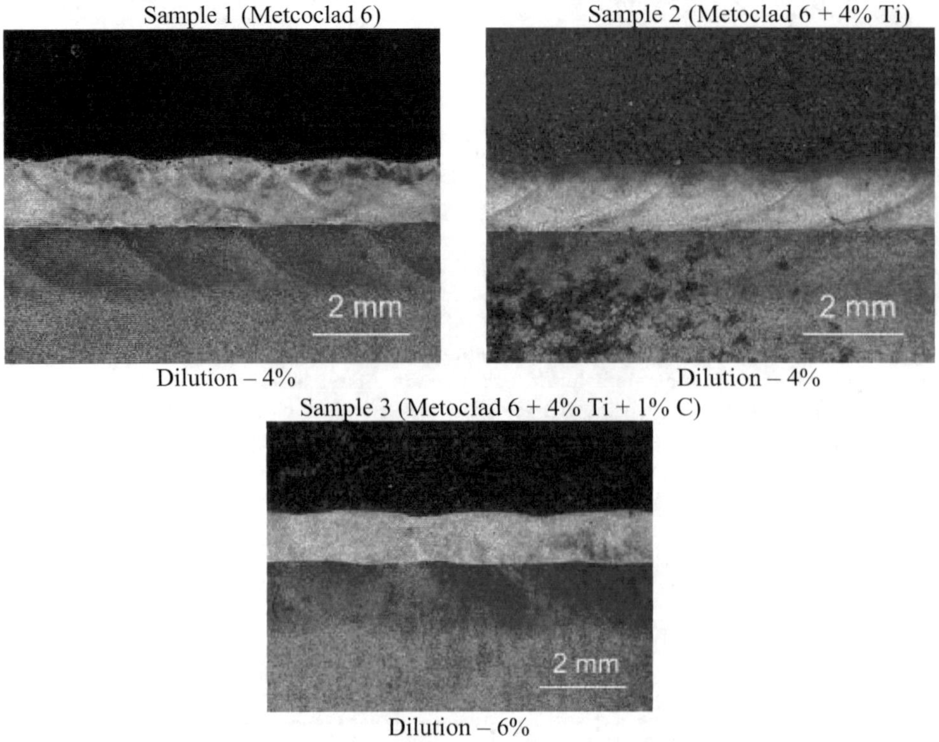

Figure 1. Macrostructure images of obtained coatings.

In order to determine the chemical composition of obtained coatings with different chemical compositions, the EDS analysis has been performed. The SEM examinations were performed on transversal cuts of coatings to characterize the microstructure of obtained laser clads in which presence of different phases can be observed.

The microstructure of pure Metcoclad 6 laser clads consist of dendritic, cobalt rich matrix which is formed as a primary phase during solidification. In the next stage, the eutectic structure containing cobalt phase and carbides solidifies in the interdendritic regions. A significant change can be observed when the chemical composition of Metcoclad 6 was changed by addition of titanium. Titanium content of 4% wt. caused the replace of eutectics by formation of isolated particles. Those precipitates contain a high content of Ti, C, in their composition. In coatings made of pure Metcoclad 6 powder the eutectic structure is rich in chromium while addition of titanium changes significantly elemental segregation where chromium is segregated to the matrix. This is a result of enrichment of titanium and carbon in the carbide particles.

In the case of third chemical composition of Metcoclad 6 with titanium and graphite the microstructure changes dramatically. The structure consisting of cobalt matrix and eutectic lamellar phase is retained, however there are many precipitates in a form of primary carbides. The precipitates are characterized by very high dispersion in structure and consist high concentrations of Ti, W, Si and C.

Titanium is known as a strong carbide forming element [14]. During laser welding in the melted pool atoms of carbon first interact with Ti forming carbides and then react with tungsten or chromium because of the chemical affinity [12]. As shown in figures 2, 3, 4 powder mixture consisting titanium addition revealed structure with primary carbides which replaced the formation of eutectic carbide. Coatings with addition of titanium and graphite made it possible to maintain the structure typical for Stellite 6 with eutectic carbide formation and additional primary carbides (MC type) which will influence on performance properties. Analyzing the EDS maps it has been stated that for sample no 2 chromium is mainly distributed in cobalt matrix which is in the opposite to pure Metcoclad 6 coating, where chromium is a main eutectic carbide former.

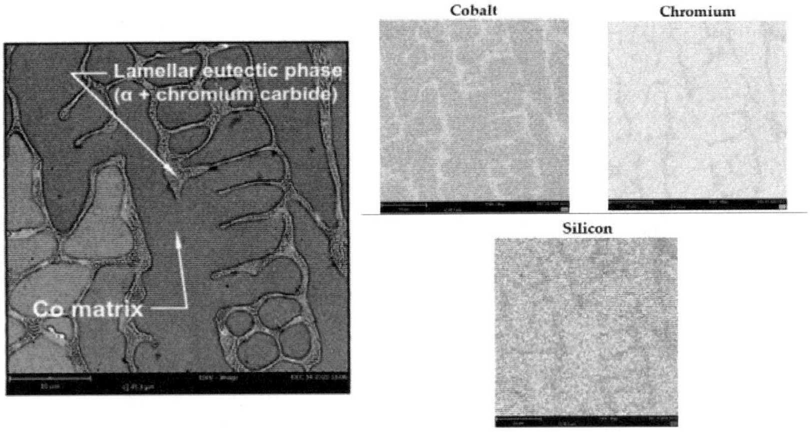

Figure 2. SEM image of sample 1 (Metcoclad 6) and corresponding EDS elemental maps (according to table 1).

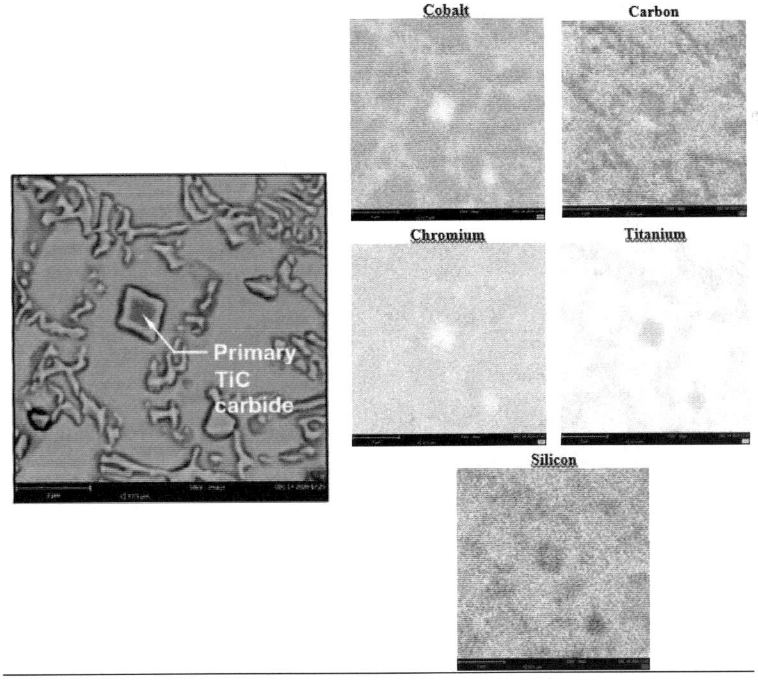

Figure 3. SEM image of sample 2 (Metoclad 6 + 4% Ti) and corresponding EDS elemental maps (according to table 1).

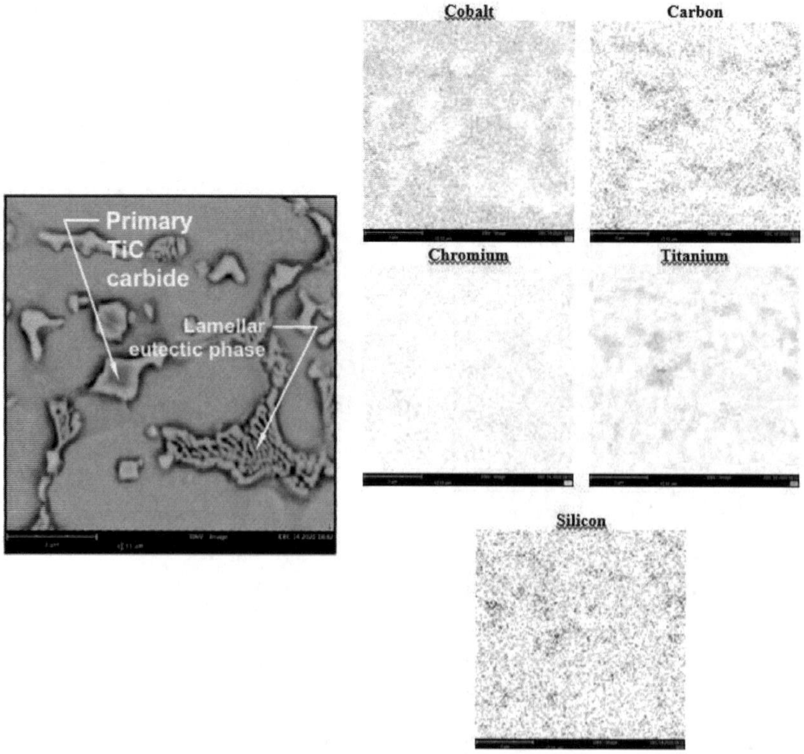

Figure 4. SEM image of sample 3 (Metoclad 6 + 4% Ti + 1% C) and corresponding EDS elemental maps (according to table 1).

The results of coatings Vickers microhardness are presented in figure 5. It can be observed that hardness of coatings is much higher than the substrate for all samples. The use of optimal cladding parameters allowed to obtain coatings characterized by low dilution which can be observed by sharp decline of hardness between coating and substrate.

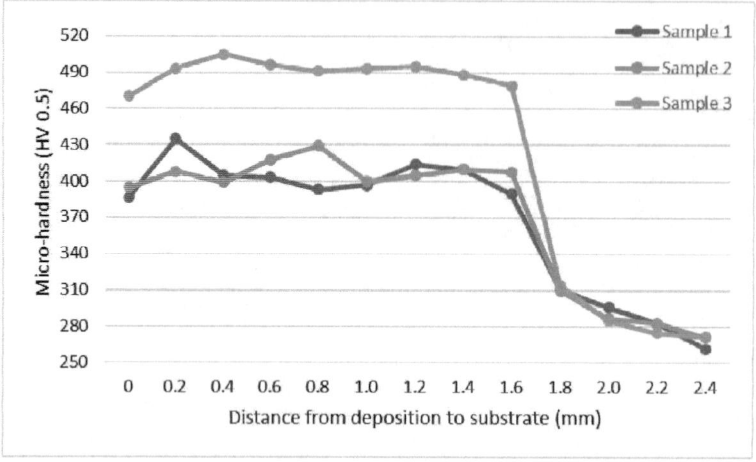

Figure 5. Microhardness profile of coatings with different chemical compositions.

4. Conclusions

The conducted laser surfacing research of commercially available Metcoclad 6 powder and modified layers with titanium and graphite allowed to obtain high-quality coatings with low dilution level without imperfections. It has been stated that addition of titanium changes the morphology of carbides from lamellar eutectic carbides to isolated primary TiC carbides of average size 1.5 μm. It can be observed that reaction between carbon and titanium caused segregation of chromium in the cobalt matrix. Addition of titanium and graphite allowed to maintain the microstructure typical for Stellite 6 alloy, but additional formation of primary carbides were observed. The difference in hardness between samples 1 and 2 is not significant, but in the case of sample 3 there was a major change in hardness caused by formation of primary carbides precipitates while maintaining the lamellar phases of the eutectics.

5. References

[1] Budzynski P, Kaminski M, Turek M and Wiertel M 2015 *Wear* **456–457** 203360
[2] Riddihough M 1970 *Tribology* **3** 211
[3] Liu R, Yao J, Zhang Q, Yao M and Collier R 2015 *Wear* **322-323** 41
[4] Davis J 2000 *Nickel, Cobalt, and Their Alloys* (OH: ASM International)
[5] Sims C 1969 *Journal of Metals* **21** 27
[6] Raghu D and Wu J 1997 *Mater. Perform* **36** 27
[7] Wu Y, Liu Y, Chen H and Chen Y 2018 *Materials & Design* **162** 271
[8] Shahroozi A, Afsari A, Khakan B and Khalifeh A 2018 *Surface and Coatings Technology* **350** 648
[9] Berthod P and Toubal L 2018 *Materials Chemistry and Physics* **212** 260
[10] Razavi M, Yaghmaee M, Rahimipour M and Razavi-Tousi S 2010 *International Journal of Mineral Processing* **94** 97
[11] Madadi F, Ashrafizadeh F and Shamanian M 2012 *Journal of Alloys and Compounds* **510** 71
[12] Motallebzadeh A, Atar E and Cimenoglu H 2015 *Tribology International* **91** 40
[13] Lupoi R, Cockburn A, Bryan B, Sparkes M, Luo F and O'Neil W 2012 *Light: Science & Applications* **1** 1
[14] Wu Y, Liu Y, Chen H, Chen Y, Li H and Cao X 2019 *Materials & Design* **162** 271

IOP Conf. Series: Materials Science and Engineering 1182 (2021) 012064 doi:10.1088/1757-899X/1182/1/012064

Considerations on geometric modelling and finite element analysis for a trolley travel mechanism of an overhead crane

M A Potîrniche, G F Căpățână and D G Muscă (Anghelache)
"Dunărea de Jos" University of Galați, Engineering and Agronomy Faculty of Brăila, Research Center for Mechanics of the Machines and Technological Equipments, Calea Călărașilor 29, 810017 Brăila, România

E-mail: aurora.potirniche@ugal.ro

Abstract. Overhead cranes are complex lifting machines that have one or more mechanisms by which they perform the displacements of the loads. Overhead cranes move on high-rise ways without disturbing the activity that takes place on the ground. The trolley is one of the main components of the overhead crane and houses the wheels on which the crane moves. In this paper it was realized the 3D modelling of a trolley for an overhead crane, which was based on a real example of such a displacement mechanism and was made in the Autodesk Mechanical Desktop software. Next, the paper proposed to perform an analysis of stresses and deformations in static conditions of the trolley-type displacement mechanism under the action of its own weight. Finite element analysis was performed in Algor V23 software. This software gives the user ease in modelling but at the same time offers complete and particularly powerful analysis tools. Based on some simplifying hypotheses, a relatively simple model was obtained to analyze, but which led to obtaining results that can be a starting point for specialists in the domain of designing such equipment.

1. Introduction

The overhead crane is characterized by mounting the lifting mechanism on a trolley that can be moved along a platform, which in turn moves on a rolling track with a big gauge. Such equipments serve rectangular surfaces and can be mounted both inside buildings (workshops, halls) and outside (warehouses, construction sites).

Because they do not take up space on the ground, overhead cranes are made for an extremely wide range of tasks and openings and can be used in production halls of any kind, covered or open warehouses. The design and execution of overhead cranes comply with the technical regulations in force and are equipped with all safety components applicable to this type of lifting machines [1].

General purpose overhead cranes are designed for rated loads, openings, lifting heights, classification groups, drive and operating medium according to customer requirements.

2. Theoretical considerations on overhead cranes

An overhead crane is a lifting machine equipped with a special drum used to wrap fibers that can be chains or ropes. The overhead crane can be used to lift loads as well as to make horizontal translations and move those loads from one place to another [2].

Overhead cranes use one or more mechanisms that aim to create mechanical advantages, so that can be lifted weights far beyond human capabilities. Lifting devices are very often used in the transport industry where they load and unload goods in the industrial domain (foundries,factories with technological profile) and in the constructions domain. Most often overhead cranes are used to assemble heavy objects.

In the steel industry, every piece of steel is handled by means of overhead cranes until it leaves the factory as a finished product ready for delivery. The raw materials are poured into the furnace by the overhead cranes, the steel pieces resulting from the furnace and have high temperatures are kept cool by the overhead cranes, the finished product represented either by steel beams or thin plates made of rolls are handled with the help of overhead cranes to be stored or loaded into trucks, containers or trains [1,2].

The automotive industry uses overhead cranes to maneuver the metal frameworks of cars (the bodies).

Choosing an overhead crane suitable for a particular use is difficult, because each crane is different and the operations performed with it are different.

Overhead cranes are characterized by the fact that the load to be handled and transported only performs translational movements. They are the most suitable equipment for serving industrial halls and warehouses. They use airspace, which is usually unused. Like cranes, overhead cranes are generally electrically operated [1,2].

One of the main components of overhead cranes is the trolley - located on any side of gauge, the trolley houses the wheels on which the crane moves. The trolley transports the electric hoist over the beams along the overhead crane crossing the gauge. Figure 1 shows a trolley model for an overhead crane [3].

Figure 1. Trolley for an overhead crane [3].

3. Case study

The 3D modeling of the trolley was based on an example of such a displacement mechanism taken from the literature. Figure 2 shows the images that were used as a starting point for the three-dimensional model [4].

The dimensions of the trolley which was realized in the Autodesk Mechanical Desktop design software are 2500x2000x200 (Lxlxh).

Figure 2. Pictures of the trolley used as a model [4].

For modeling were used Autodesk Mechanical Desktop software-specific generation and modification commands:

- WORKPLANES command - work plans used to be able to easily draw in coordinates;
- PROFILE A SKETCH command - the model sketch is made for each new operation;
- EXTRUSION and REVOLVE commands - are used to generate the solid;
- CUT command - used to cut and remove areas of the solid;
- TRIM, OFFSET, MIRROR, FILLET commands - used to modify and finalize the details of the three-dimensional model [5,6].

The application of all these operations resulted in the 3D model shown in figures 3 (3D model for the rolling track-trolley assembly), figure 4 (3D model of the trolley) and figure 5 (3D model of the trolley - detail in the tread wheel area).

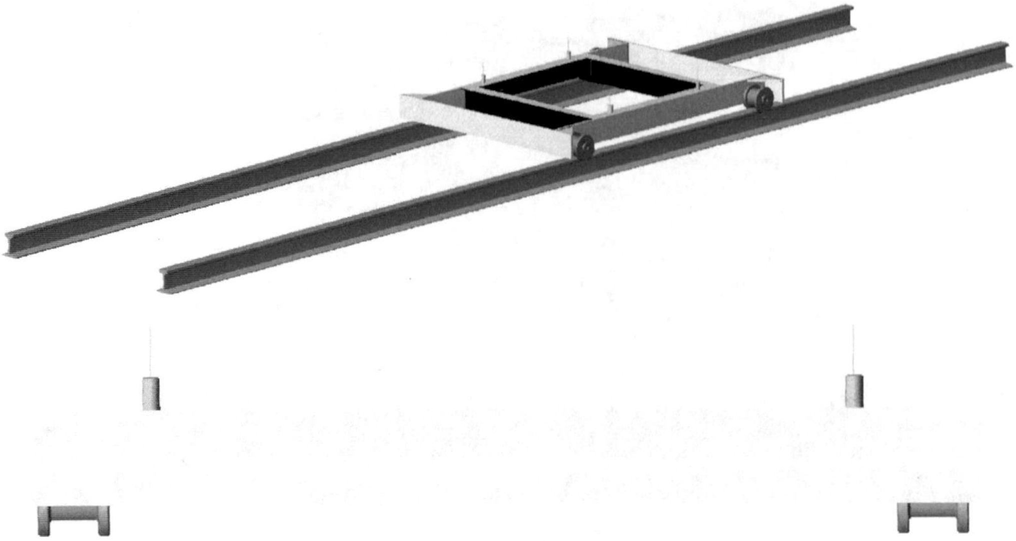

Figure 3. 3D model for the rolling track-trolley assembly.

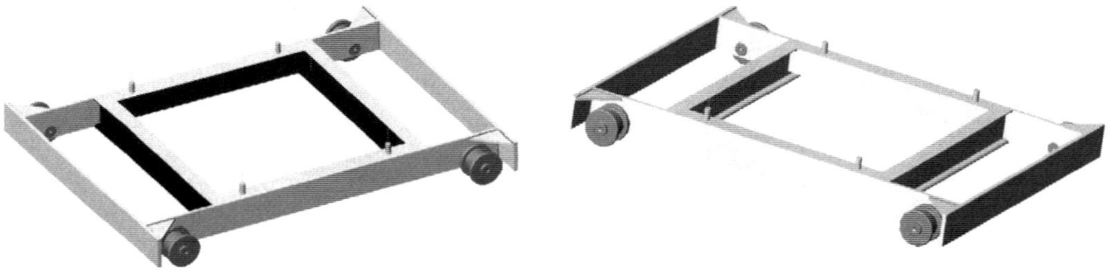

Figure 4. 3D model of the trolley.

Figure 5. 3D model of the trolley - detail in the tread wheel area.

Next in the paper is proposed an analysis of the stresses and deformations in static conditions of the trolley-type displacement mechanism under the action of its own weight. Finite element analysis was performed in Algor V23 software. This software gives the user ease in modeling but, at the same time, offers complete and particularly powerful analysis tools.

Knowing the construction dimensions and material of the trolley, its mass was estimated at a value of 800 kg (equivalent to a force of 800 daN = 8000 N).

For the finite element analysis, the following hypotheses were adopted for the trolley:

- the material from which the trolley is made is S235, a tenacious material with linear-elastic behavior;
- the discretization was performed manually so as to be sufficiently refined taking into account the dimensions of the section (figure 6); taking this into account, is ensured a higher degree of confidence in the final results;
- the fixed type support was made on the 4 engine mounting screws on the trolley (figure 7);
- the concentrated force type was chosen so as to simulate the actual contact between the trolley rollers and the rolling track (figure 7). The chosen surface on which the force acts is the contact stain that is actually formed at the contact between the wheel and the rolling track.

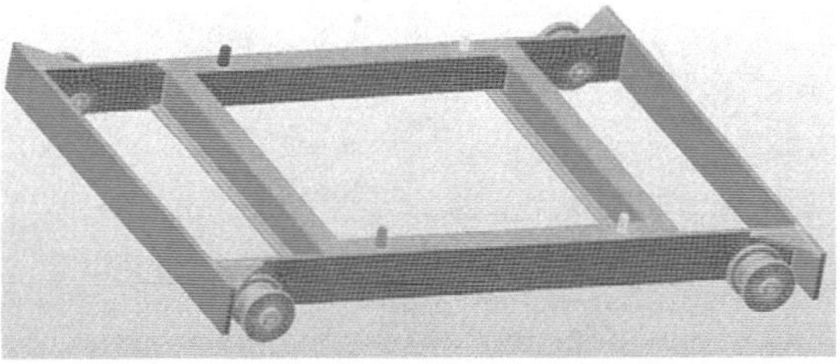

Figure 6. The trolley meshing in Algor.

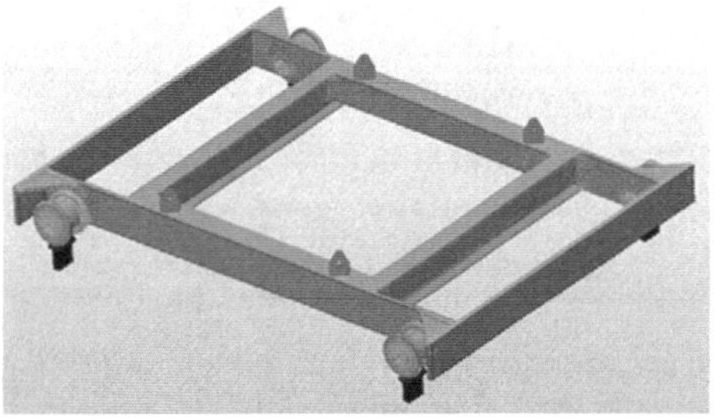

Figure 7. Defining the support set and the loading set in Algor.

In order to better observe the action mode of the forces, is presented below (figure 8) a detail of the area in which they act (contact stain area).

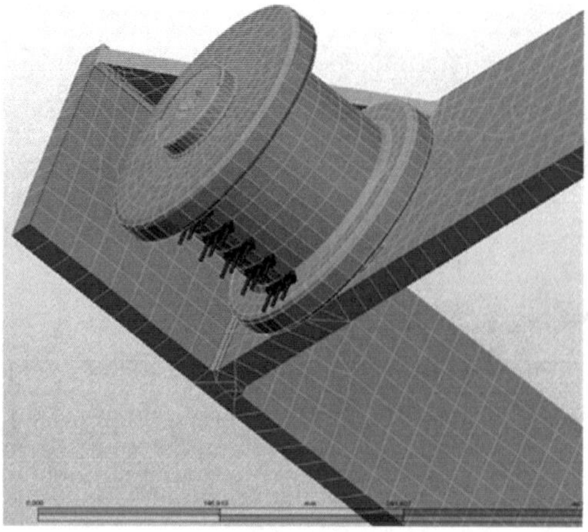

Figure 8. Detail from the forces action area.

Following the static analysis performed in the hypotheses presented above, the Algor software provides the user with a generous series of results, which will choose and interpret, based on experience and knowledge, those they consider suggestive for the analyzed case.

Thus, the following results were chosen for viewing:

- the tensions that develop in the trolley-type structure when this is solicitated by its own weight, presented in figure 9;
- deformations of the structure, generated by previous stresses, presented in figure 11;
- displacements of the structure nodes, presented in figure 12.

In order to better observe the areas where significant tensions develop, a detail is presented in figure 10.

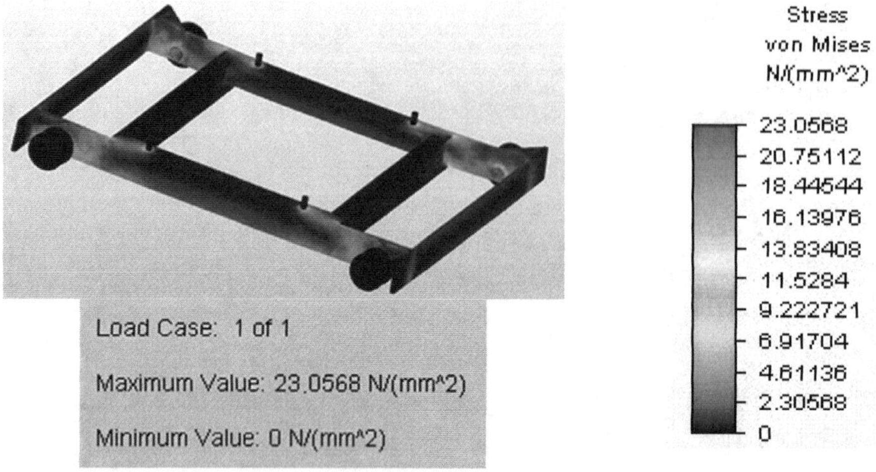

Figure 9. Visualization of the trolley tensions in case of the proposed solicitation.

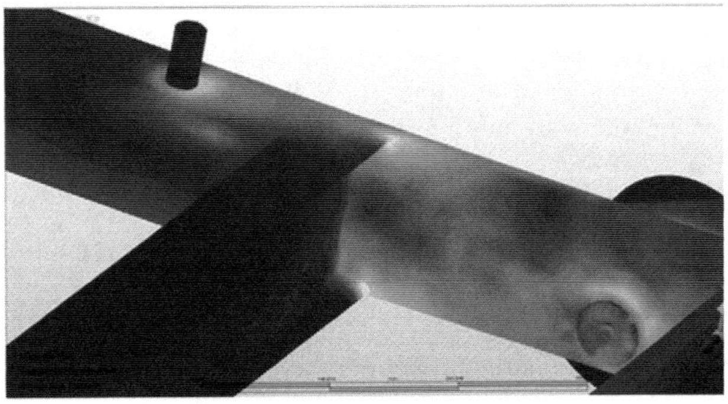

Figure 10. Detail from the tensions area with significant values.

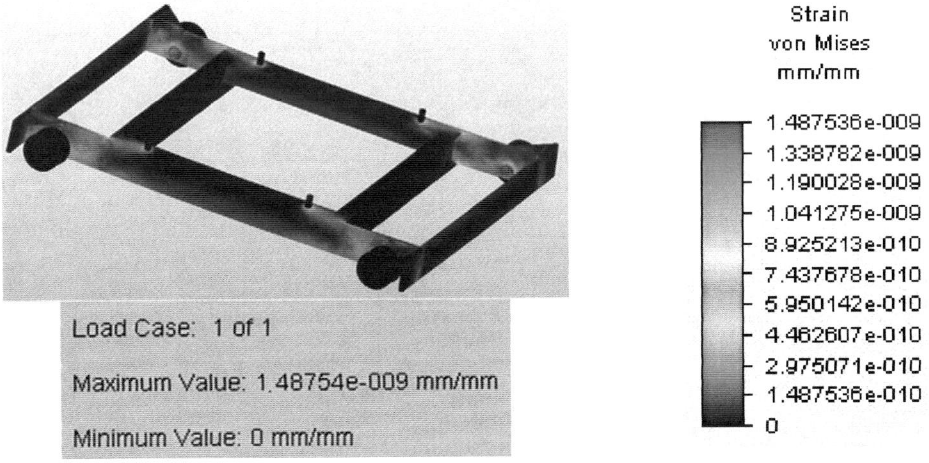

Figure 11. Visualization of the trolley deformations in case of the proposed solicitation.

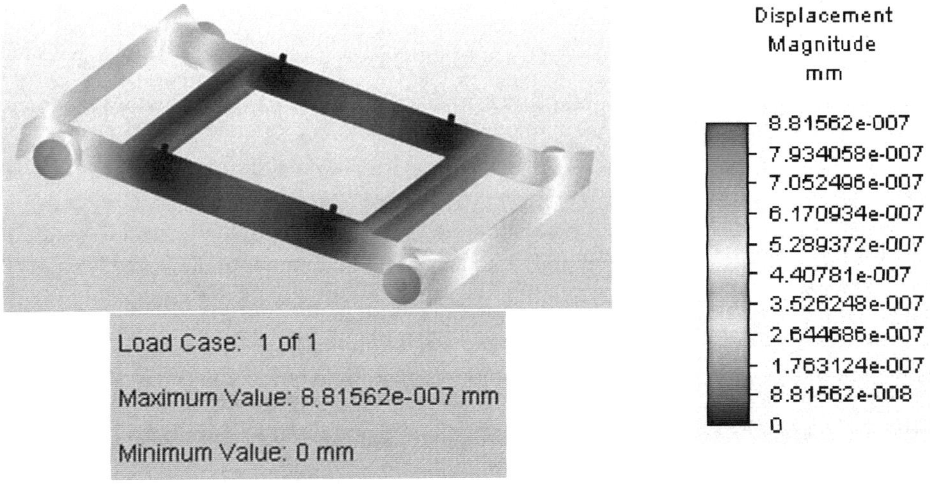

Figure 12. Visualization of the trolley nodes displacements in case of the proposed solicitation.

The material model of the structure is that of a homogeneous, isotropic material, with linear-elastic characteristic; the linear-elastic characteristic, known as Hooke's law, which gives the linear dependence between stresses and strains is highlighted by the graph in figure 13, a graph for which were used the values for stresses and strains resulting from finite element analysis.

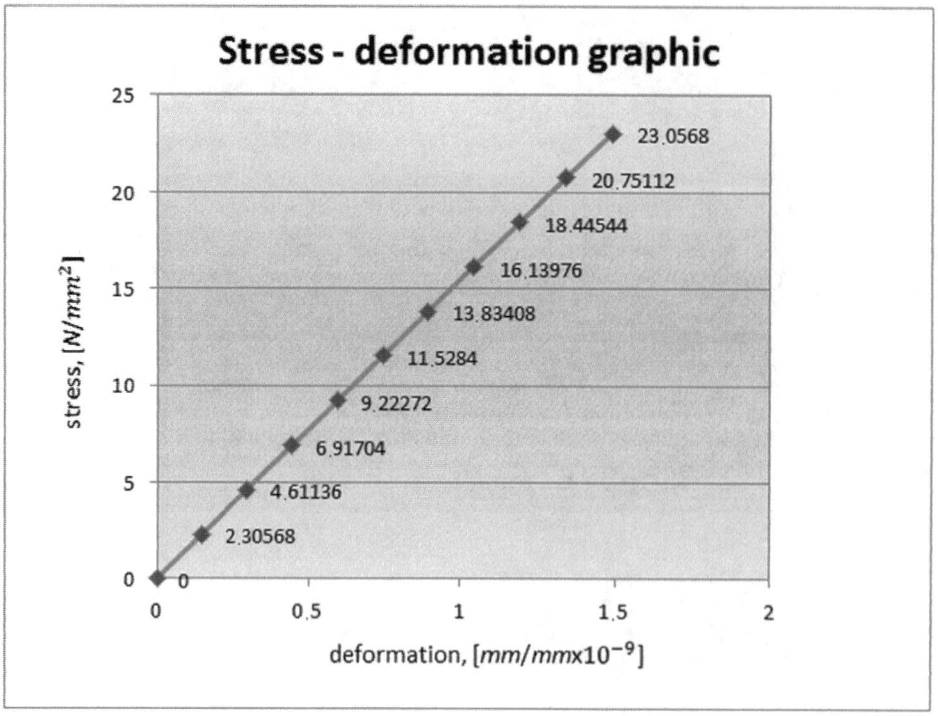

Figure 13. The graphic of linear dependence of stresses as a function of deformations.

4. Conclusions

Until the appearance of the finite element method, the optimization of resistance structures in terms of material consumption was a difficult process, due to the long time that had to be allocated to the analysis of several constructive variants of the same structure, in order to discover the optimal variant.

In this context, the finite element method is a powerful working tool, given the following aspects:

- the diversity of the resistance structures geometry;
- the diversity of the boundary conditions applied to resistance structures;
- the behavior of the materials from which the structural elements are made;
- the need to evaluate as accurately as possible the states of stresses and deformations in the resistance structures, in order to achieve the optimal in terms of material consumption;

The finite element analysis performed in this paper was based on the following hypotheses:

- the type of analysis was static and proposed, according to the compound solicitation of the equipment, the evaluation of the tensions and deformations of the structure based on a teory of resistance (von Mises theory); following the analysis, given the low values obtained for the stresses in the material, compared to the permissible ones, they led to the conclusion that there is no danger of failure of the structure;
- the chosen material model is that of a homogeneous, isotropic material, with linear-elastic characteristic.

Based on the mentioned hypotheses, the analysis with finite elements was performed and this led to obtaining results that can be a starting point for specialists in the domain of designing such equipment.

This study may be continued in order to improve the geometric configuration and operational performance of such equipment.

5. References

[1] Alămoreanu M 2003 *Introducere în dinamica maşinilor de ridicat* Conspress Publishing House Bucharest ISBN 973-8165-74-1.

[2] Oproescu Gh and Anghelache G D 2010 - *Elemente de dinamica maşinilor de ridicat şi transportat* Galaţi University Press ISBN 978-606-8008-71-4.

[3] https://www.demagcranes.com/sites/default/files/media/documents/20855744_110730.pdf accessed in 2.2021.

[4] https://www.southingtonbreadforlife.com/page/11/ accessed in 2.2021.

[5] Muscă (Anghelache) G D and Năstac S 2020 *Dynamic modelling of overhead crane* MODTECH International Conference - Modern Technologies in Industrial Engineering VIII, IOP Conf. Ser.: Mater. Sci. Eng. **916** 012071.

[6] Goanţă A M 2009 *Modern Instruments of Transfer of Geometrical Patterns in Designing Building Machines and Equipments* Proceedings of the 13th International Conference Iaşi&Chişinău - Modtech pp. 271-274 ISSN 2066-3919.

Conical Spring Inserts Effect on the Thermo- hydraulic Characteristics in Circular Copper Tube

M S Qasim, S G Ali and B Freegah
Mustansiryah University, College of Engineering, Mech. Eng. Dept., IRAQ, Baghdad

Email: qasim602006@yahoo.com

Abstract: Experimental and numerical investigation was conducted to study the thermo-hydraulic characteristics in circular tube fitted with conical spring inserts. The dimensions of circular copper tube are 100cm, 2.2cm and 2.4 cm which represent the length, inner and outer diameters respectively. The length of insert was 15cm, while the conical spring has 16mm diameter gradually decreases to 6mm with the length and the pitch between the coils was 20mm. The diameter of wire used to fabricate the conical spring insert was 4mm. Those inserts were arranged into eight categories (A1 to A8). In the numerical part, the governing equations were solved by using commercial ANSYS-Fluent package (2019 R3) with the assistance of solid works and Gambit software program. The finite volume approach was used in solving the equations to predict the pressure flow, heat transfer and temperature distribution along the tube. Experimental work included design and built a test platform for the experimental measurements. The aim of experimental study was to predict the behavior of heat transfer and pressure in the tube with two cases, plain tube and inserted tube at the effect of different parameters. All experimental tests were conducted at turbulent flow fully developed. The range of Reynolds number and heat flux were (3542-7522) (4549-10576 W/m^2) respectively. The results showed that the best augmentation in heat transfer was by using a category A5 of conical spring insert compared with plain tube. Results show good agreement between the numerical and experimental results with an error of 8%. Also results showed that the Nusselt number increases as the Reynolds number increasing.

1. Introduction

Heat transfer augmentation in a thermal system can be achieved by adding swirl flow devices. Conical spring insert is one of these devices utilized to improve the heat transfer without addition any external power Experimental and numerical studies were conducted to predict the augmentation in heat transfer by using swirl flow devices contain a tube inserts. Gurveer et al. [1] studied experimentally the thermal performance of a flat-plate solar collector using insert devices. The range of Reynolds number and Prandtl number were (200–8000) (5–8) respectively. In the study different insert configurations were used such as; twisted-tape, wire coil and wire mesh. Results showed increases in Nusselt number when using all insert devices. Also results show that the mesh insert and wire coil are best than others in laminar and turbulent flow respectively. Sahiti et. al. [2] conducted an experimental study of the effect of adding pin fins on heat transfer. The simple relationships of heat transfer by convection and conduction were used to derive an equation explaining the factors influencing the improvement of heat transfer. The results showed a significant improvement using small cylindrical pins on the surface of the heat exchanger. Kapatkar et al. [3] studied experimentally the heat transfer inserts for Reynolds number range laminar flow (20-2000), range of mass flow rate (2gm/s - 20gm/s) under uniform wall heat flux condition and friction factor of a smooth tube fitted with full length twisted tape. An enhancement in average Nusselt number for the flow in smooth tubes was resulted with full-length twisted tapes and it increases with the increase in Reynolds number and mass flow rate. Kannan et al. [4] conducted experimental study the heat transfer characteristics and friction factor within a circular tube fitted with perforated twisted tape for Reynolds number range (21,000 – 50,000). The results show that Nusselt number varies inversely with the twist ratio. Shuai and Chang [5] performed numerical simulations on a compact tubular heat exchanger using fluent software. They have verified that the software is fluent in digital simulation method of integrated fin tube is possible. The superior software has proven to be feasible for numerical simulation on a built-in fin tube and to be effective tools for developing and designing heat exchange tubes. Burse et al. [6] conducted experimental research to understand the effect of twisted tape with different torsion ratios on the heat transfer properties of a round tube. The twisted tapes used are of three different winding ratios such as 1.78, 2.32 and 2.77. Initially the experiment was performed for a normal tube at constant wall heat flux and different mass flow rates. Second, the experiment was performed with tube insertion with the same working conditions as the normal tube. Dong et al. [7] investigated the effect of the perforated round finned tube on the heat transfer of the circular tube heat exchanger. The heat transfer coefficient (h) was increased by (3.31%, 3.55%), the pressure drops increased by (0.68%, 2.08%) for the 2- and 4-holes cases, respectively. Qasim and Ali [8] studied the use of different kinds of tapes that are twisted and fitted through double pipe heat exchanger to increase the fluid mixing property that leads to increase the heat transfer rate with respect to that of the plain-twisted tape. V cut-twisted tape and Horizontal wing cut-twisted tape are two different variant twisted tapes with twist ratios of $y = 2.0$, 4.4 and 6.0 are used. The main conclusion has been reached that the Nusselt number, thermal enhancement factors and friction factor of different twisted tapes are higher than that of plain twisted tape for the twist ratios of 2.0, 6.0 and 4.4 respectively. Yu and Tao [9] conducted a study on the heat transfer and pressure drop in toroidal wave-like finned tubes. The results showed an enhancement in heat transfer for finned tube when the flow was turbulent. Dhirajkumar et al. [10] offered review of researches work of modern techniques in heat transfer enhancement. Many researchers have used different techniques to enhance heat transfer. Worldwide energy utilize is expected to increase by almost 1.6 times, from 4.03 x 1020 J in 1999 to 6.4 x 1020 J in 2020. Thus, if the efficiency of energy usage could improve by 10 per cent, by means of different heat transfer enhancement techniques, this will result in 6.4 x 1019 J of advantage in terms of energy utilization to the society. Special importance in this study is given on different techniques of heat transfer enhancement, developed in recent times. Iqbal et al. [11] investigated the optimal longitudinal fins on the outer surface of the inner tube inside the closed outer tube concentric in the developer laminar flow in full uniform heat flow. The results showed that the optimal shape of the fin depends on the number of fins, and the proportion of radius, and the number of control points and length characteristic. Kumar. [12] Made a numerical analysis the heat transfer in horizontal tubes having five different types of inserts when the ranges of Reynolds number were (6000-14000). Author concluded that increase in pumping work associated with an increase in heat transfer coefficient in comparison to plain tube. Mon and Gross [13] conducted an analysis of numerical heat transfer characteristics and pressure drop on the side of the heat exchanger tube with fins ring. They have studied the effects of spacing on the fin tubes with fins ringed packs consisting of four rows overlapping arrangements and homogeneous. The findings revealed the point of view of the important characteristics of heat flow and transfer of local heat packs pipe ring fins and supports the development of the boundary layer on the surfaces of the fin and tube spacing to fin height ratio. Sami et al. [14] investigated numerically the characteristics of friction factor and heat transfer for copper-water nanofluid passing through circular tube with constant wall heat flux fitted with two different types of vortex generator, which are classical twisted tape (CTT) and parabolic-cut twisted tape (PCT). The results showed that at lower twisted ration a higher value for Nu and friction factor. Qasim and Noor [15] investigated experimentally the characteristics of heat transfer enhancement and friction factor for completely developed laminar CuO/distilled-water nanofluid flow through horizontal tube inserted with various geometries of twisted tapes under constant condition of heat flux ranged from 4483 - 10000 W/m2. The results revealed a significant increase of both Nusselt number and convective heat transfer in terms of friction factors with inserting twisted tape and nanofluids as working fluid in comparison to nanofluids in plain tube condition. Kulkarni [16] carry outed experimental investigate the heat improvement in a horizontal circular tube with clockwise and counter- clockwise-corrugated twisted tape inserting. They have reported that the using clockwise and anticlockwise- corrugated leads enhancement in heat transfer by 12% and 46% respectively as compared to traditional tube.

From previous studies, we have observed that there are a variety of ways to improve the heat transfer process. With a slight study of the use of a conical spring insert. The purpose of the current study is to investigate the thermo-hydraulic characteristics in circular tube fitted with conical spring insert at different factors experimentally and numerically. Hence, this study is conducted in order to enhance the thermal performance of the heat exchanger that is used in many applications.

2. Numerical Simulation

ANSYS FLUENT (2019 R3) commercial software is employed to simulate the physical phenomena of heat transfer and fluid flow in plain and inserted tube. Solid works Software (2016) is used to draw the geometry models in present study. Figure 1 shows the Computational domain of the copper tube and the conical spring insert with dimensions. The conical springs were distributed inside the tube with eight types of distributions (A1-A8). We are looking for the best type to improve heat transfer with minimal pressure loss. In the current study, the continuity, Navier-Stokes, and energy equations have been used to solve the flow field three dimension.

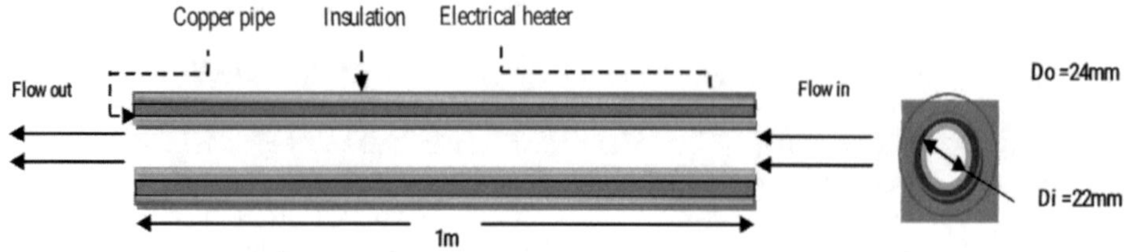

(a) Computational domain in the present study.

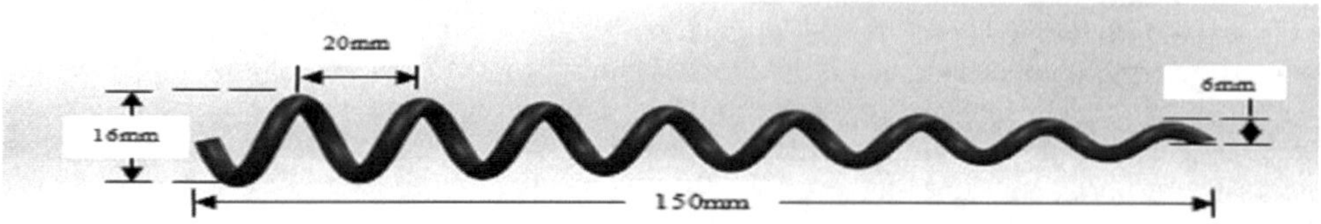

(b) Conical spring insert with dimensions.

Figure 1. (a) Geometrical domain in the current study; (b) Conical spring insert with dimensions.

2.1. Mesh Generation

GAMBIT software was utilized modeling the mesh. Currently, in this work, triangular and tetrahedron element types were involved for surface mesh and three-dimensional geometry respectively. Figure 2 shows the present model mesh. Various grid sizes were tested and now transported to software program fluent (2019 R3). The temperature distributions of the tested grids were compared to select the optimal grid size in the study. In this investigation, an average of (1.25) million cells are utilized and it is representing the highest number of iterations done to achieve the solver terminates. As shown in figure 3, (1600) iterations are required in this work.

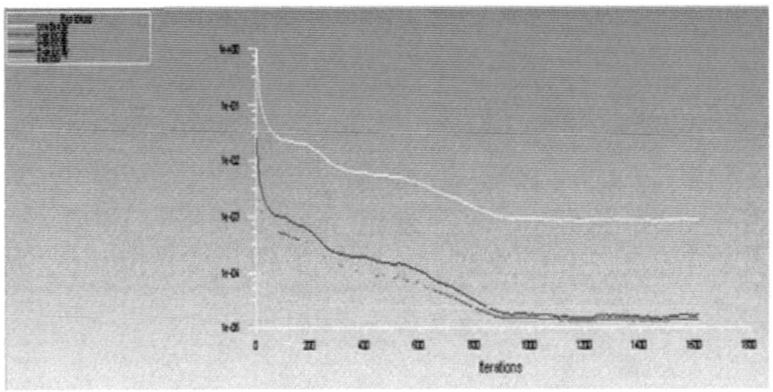

Figure 2. Residual for numerical simulation of present study.

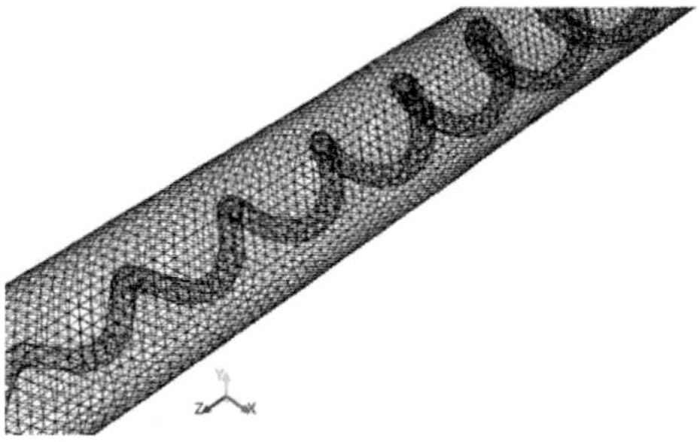

Figure 3. Mesh of present model.

3. Experimental description

3.1 Experimental set-up

Figure 4 depicts a diagram of the experimental set-up under study. The experimental set-up involved of a centrifugal pump, flow meter to measure flow rate of water, U- tube manometer, and the heat transfer test sector. In the eclectically heated unit, a copper tube having an inside diameter of (Di) 22 mm, an outside diameter of (Do) 24 mm, and length of L=1000 mm is wound with electrical heating wire (L=3 m, d= 1.5 mm, power =1.2 kW) covered with ceramic beads. The terminals of the wire are connected to a variac AC transformer, which is used to vary the voltage of the AC current passing through the heating coil. In order to reduces convective heat loss to ambient and to prevent leakages from the experimental rig, well insulate has been used. The inlet and outlet temperatures of the water and three stations on the outer surface of the test tube were measured by multi-channel type-k thermocouples.

(a) (b)

Figure 4. (a) Experimental setup, (b) Test section

In the experiments, the conical-spring insert in eight category utilized in the current study are showed in figure 5. The conical-spring was made of copper with15cm in length, while the conical spring has 16mm diameter gradually decreases to 6mm with the length and the pitch between the coils was 20mm. The diameter of wire used to fabricate the conical spring insert was 4mm. Those inserts were arranged into eight categories (A1 to A8).

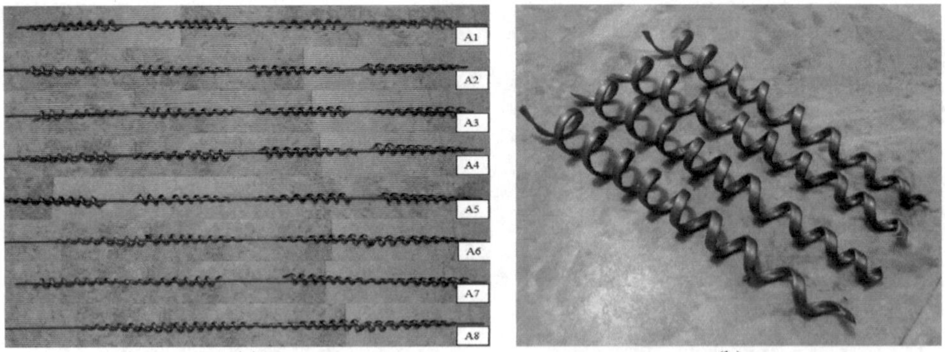

(a) (b)

Figure 5. Conical springs insert.

3.2 Experimental procedure

In the present work, the pressure drop was measured by utilized in U-tube manometers. Also both the inlet and outlet temperatures of the water and three stations on the outer surface of the test tube were measured by multi-channel type-K thermocouples for finding out the average Nusselt number. For each test, temperature, volumetric flow rate and pressure drop of the water were record the data of under steady state conditions. All tests were conducted under a fully developed turbulent flow with range of Reynolds number (3542-7522) and heat flux (4549-10576 W/m^2).

3.3 The data reduction

In this study, the water used as a working fluid that flowed through a uniform heat flux and insulated tube. The steady state of the heat transfer rate is assumed to be equal to the heat loss from the test section. to estimate the electrical power applied at the external surface of the tube, can be used equation (1) [18]:

$$Q_{heater} = I.\ V \tag{1}$$

where:
I and V are representing the electric current (A), and the electric voltage (V) respectively.

Equation (2) can be used to calculate the heat transferred amount to the distilled water from the heating wire:

$$Q_{fluid} = \dot{m}_w\ Cp_w\ (T_o - T_i) \tag{2}$$

where:

$\dot{m}_w$, Cp_w, T_i *and* T_o are representing the water mass flow rate (kg/s), specific heat of the water (kJ/kg.°C), inlet and outlet temperature of the water (°C)respectively.

One of the most important parameters that should know is the heat transfer coefficient of the water, it can be calculated by using equation (3):

$$h(z) = \frac{\dot{q}_w}{(\Delta T)_z} \tag{3}$$

where: $(\Delta T)_z$ is the difference between the inner wall temperature of the tube T_{si} (z) and the temperature of the water T(z) at distance Z from the entrance of the tube.

The local Nusselt Number has been calculated from the equation (4):

$$Nu(z) = \frac{h(z)\,D}{k_w} \tag{4}$$

where:

k_w: Thermal conductivity of water (W/m.K).

While, Reynolds number can be estimated by equation (5)

$$Re_w = \frac{4\dot{m}}{\pi d_i \mu} \quad (5)$$

where:

$\dot{m}_w$, μ_w and d_i are depict the water mass flow rate (kg/s), dynamic viscosity of water (kg/m.s) and internal tube diameter (m) respectively.

4. Results and Discussion

4.1. Model Validation

A comparison has been conducted to authorize the current numerical simulation according to the former numerical results, achieved by Qasim and Noor [15]. In figure 6 shows good agreement has been found between these results which indicate the acceptable validations of the present simulation. Figure 7 shows a comparison between the numerical and practical results of the temperature distribution along the tube within conditions (Reynolds number 3500 and thermal flux 4500). The figure shows that there is a clear convergence of the results, and the highest percentage for the plan is 8%.

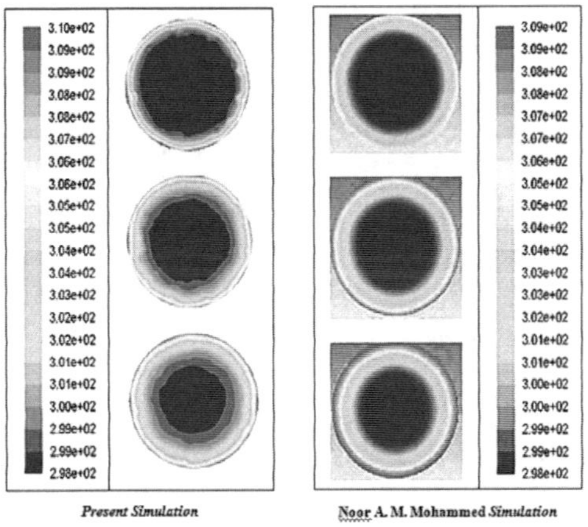

Figure 6. Numerical simulation validation.

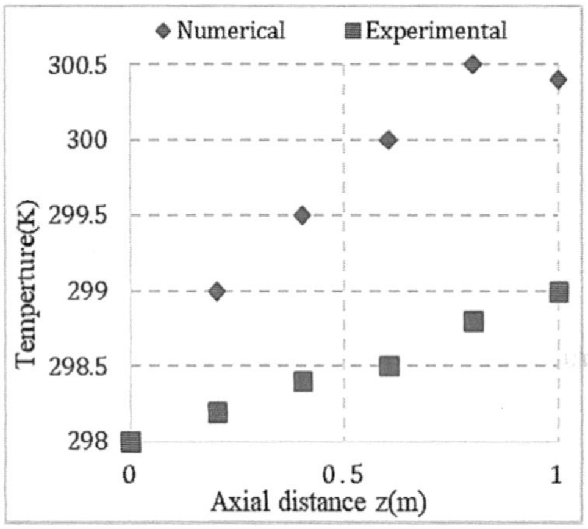

Figure 7. Comparison between experimental and numerical results.

4.2. Numerical Results

Figure 8 demonstrate the temperature contours at three locations (Z=0.25, 0.5 and 0.75) for plain tube and all other types of the arrangement of the inserted tube at (Re=3500) and heat flux equal to 10500 W/m^2. Figure 8 demonstrated that heating of the water passing through the tube is limited for temperature contours in plain tube. So, in order to reach a specific temperature difference, an increase in the length of the tube is needed. The reason behind it is due to the high velocity of water flow through the tube, which holds the possibility of heat transfer as can be seen shown in velocity vector of flow. With the use of conical spring insert, contours show a significant enhancement in the water heating process demonstrating that the heat transfer between the hot tube surface and water has been enhanced.

IOP Conf. Series: Materials Science and Engineering 1182 (2021) 012065 doi:10.1088/1757-899X/1182/1/012065

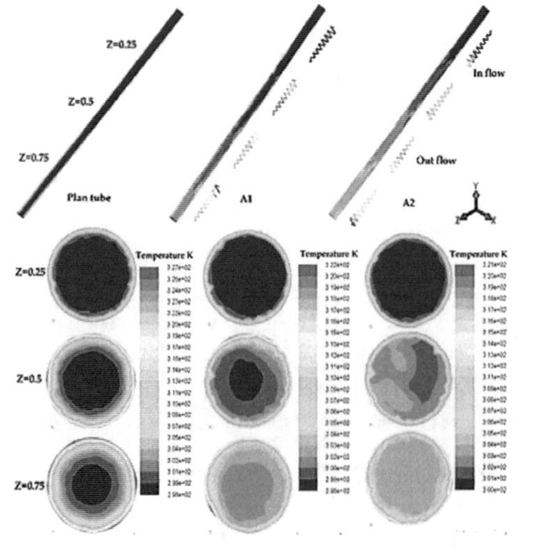

(a) Z=0.25

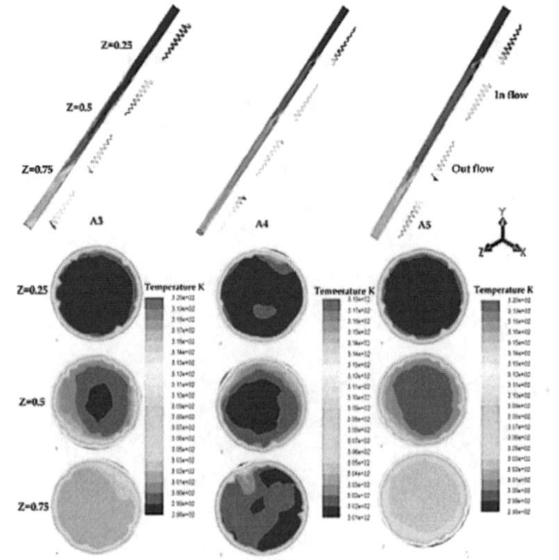

(b) Z=0.5

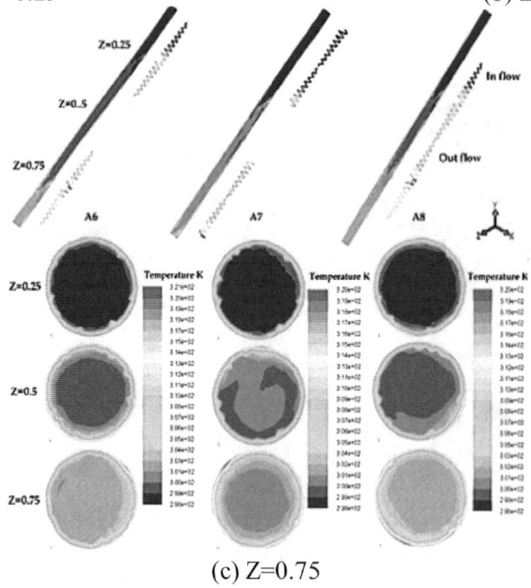

(c) Z=0.75

Figure 8. Temperature contours at (Re=3500) under heat flux equal to 10500 W/m^2 for (a) Z=0.25 Z=0.5 and (c) Z=0.75

Figure 9 illustrates the behavior of water during heat transfer by heating the surface of the tube at longitudinal axis. Figure shows that for all type arrangement (A1 to A8). Water temperature increases for the tube with conical spring insert. Figure 9 demonstrates that (A5) is the best arrangement, because of the increase in the water temperature. A swirl flow caused by the conical spring inserts disrupting the water boundary at tube surface due to tangential velocities enhancement. The water boundary layer disruption leads to a thinner boundary that improves the heat transfer process. Figure 10 illustrate that the drop in the pressure begins from tube inlet to the downstream of flow along the tube to the exit section due to losses of the friction between the water viscous boundary layer and inner surface of pressure losses through a plain tube is less than losses in the tube with conical spring insert.

This is because of the conical spring insert swirling path is generated by conical spring insert and the additional local vortices. The boundary layer of the viscous nearby the inner surface of tube is destroyed. This caused higher contact between water and the surface then increasing friction losses. Figures 9 and 10 demonstrate that class A5 is the best classification.

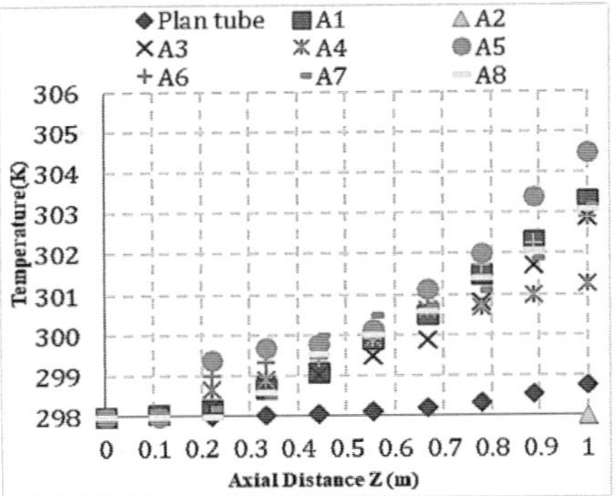

Figure 9. Heat behavior of water in relation with axial distance.

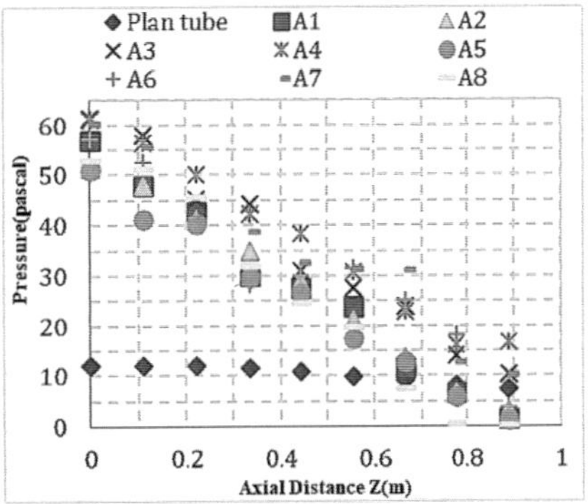

Figure 10. Pressure behavior of water in relation with axial distance.

4.3. Experimental result

4.3.1. The thermal effect of adding conical springs insert. The relationship between the change in water temperature along the length of the tube at various values of heat flux and Reynolds number 3500 in the eight arrangements of the conical springs inside the tube are shown in figures 11 – 14. In general, we notice that the water temperature rises along the tube length, but in a variable manner due to the effect of the conical springs insert distribution along the tube. Also, the greater the value of the heat flux applied to the tube, the higher the temperature value of the water. Adding the conical springs insert and distributing them inside the tube in different arrangements affect the rate of heat transfer from the tube surface to the water as a result of generating rotational torsion flows that enhance the mixing of water and its contact with the inner walls of the tube and this mixing maintains a high temperature gradient near the tube wall and thus improvement the heat transfer. From these figures, we note that the conical springs insert distribution within the order (A5) gave the best thermal performance among the rest of the arrangements. 11% represents the average of increase in water temperature when using (A5) conical springs insert arrangement compared to a tube without conical springs insert.

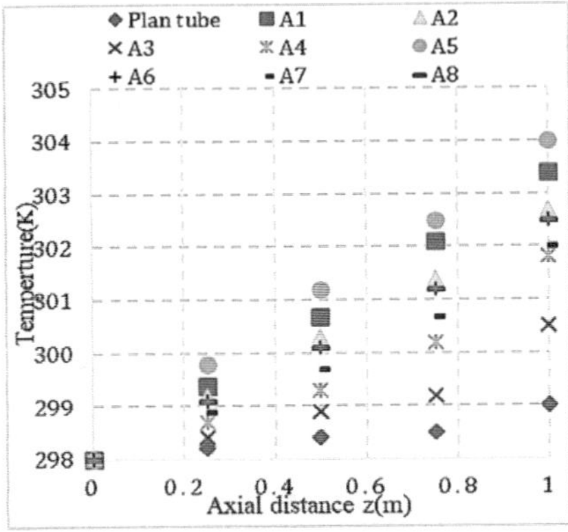

Figure 11. Heat behavior of water in relation with axial distance at heat flux $4549 W/m^2$.

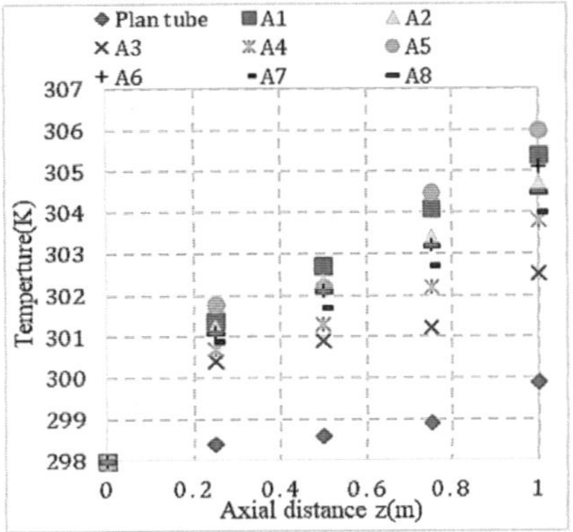

Figure 12. Heat behavior of water in relation with axial distance at heat flux $6467 W/m^2$.

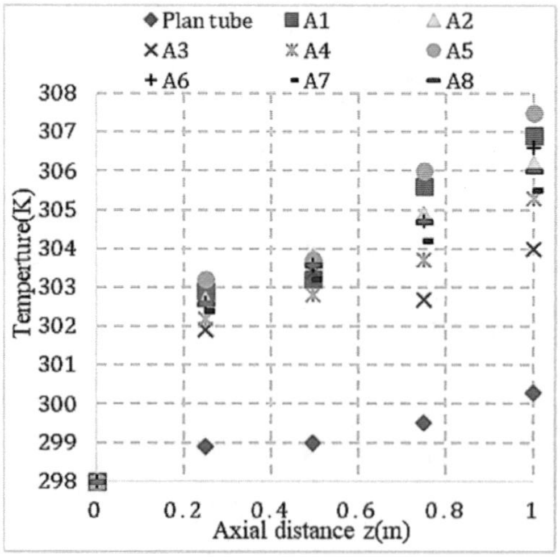

Figure 13. Heat behavior of water in relation with axial distance at heat flux 8612W/m^2.

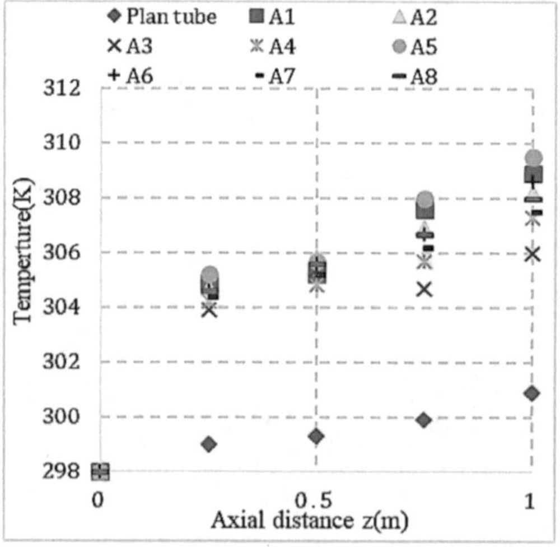

Figure 14. Heat behavior of water in relation with axial distance at heat flux 10576W/m^2.

4.3.2. Hydrodynamic effect of adding conical springs insert. The difference in Nusselt number with Reynolds number for normal tube compared to the eight arrangements of conical springs inside the tube are shown in figures 15 - 18 with different heat flux. Generally, for all ranges of the Reynolds number under study, it can be seen clearly that the Nu number increments with increases the Reynolds number. Affected Nusselt number distribution of conical springs inside the tube as a result of the establishment of torsion or rotational flow required which enhances the mixing of water and touching the walls of the inner tube. This confusion maintains a gradient high temperature near the wall of the tube and thus enhances heat transfer.

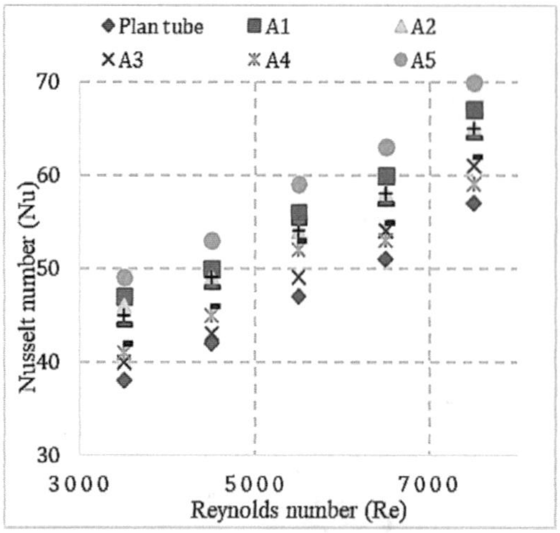

Figure 15. Relation between Nusselt number with Reynolds number at heat flux 4549W/m^2.

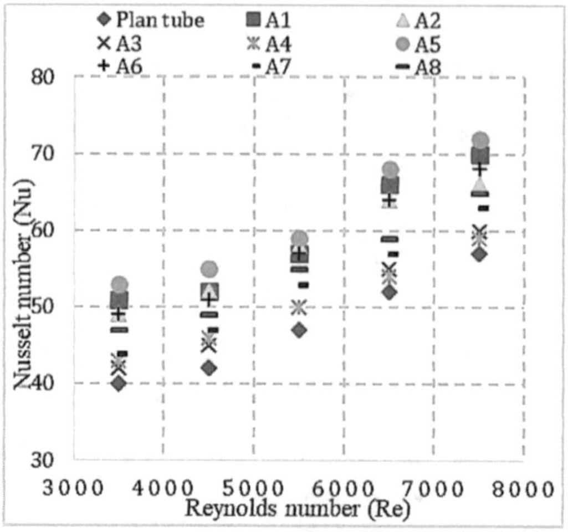

Figure 16. Relation between Nusselt number with Reynolds number at heat flux 6467W/m^2.

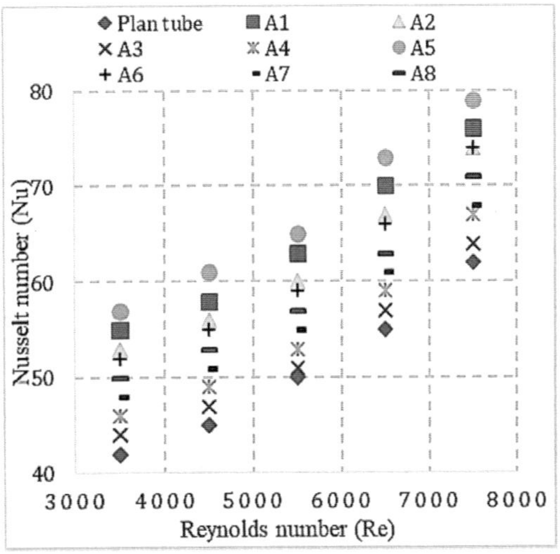

Figure 17. Relation between Nusselt number with Reynolds number at heat flux 8612W/m^2.

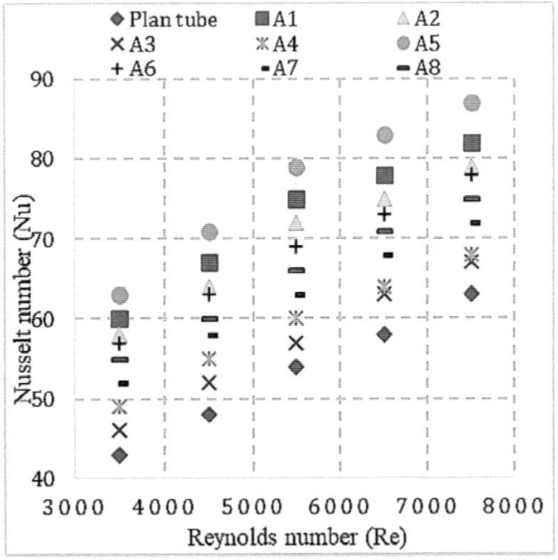

Figure 18. Relation between Nusselt number with Reynolds number at heat flux 10576W/m^2.

5. Conclusions

Experimental and numerical study were conducted to investigate the thermo- hydraulic characteristics in circular tube fitted with conical spring inserts. It is concluded that the best augmentation in heat transfer was by using a category A5 of conical spring insert compared with plain tube. 11% represents the average of increase in water temperature when using (A5) conical springs insert arrangement compared to a tube without conical springs insert. In addition, for all ranges of the Reynolds number under study, Nu number increases with the Reynolds number increases. The main reason behind the enhancement in heat transfer under the current study is that the flow became more turbulent due to insert conical spring insert. In addition, the results shown clear convergence of numerical and experimental results with an error 8%. Future directions of the subject under study is investigation of different types of insert along with various kind of nanofluid.

6. References

[1] Sandhu, G., Siddiqui, K. and Garcia, A. 2014 Experimental study on the combined effects of inclination angle and insert devices on the performance of a flat-plate solar collector *International Journal of Heat and Mass Transfer* **71** 251-263.

[2] Sahiti, N., Durst, F. and Dewan, A., 2005 Heat transfer enhancement by pin elements *International Journal of Heat and Mass Transfer* **48**(23-24) 4738-4747.

[3] Kapatkar, A.V.N., Padalkar, A.S. and Kasbe, C.S. 2011 Experimental investigation on heat transfer enhancement in laminar flow in circular tube equipped with different inserts *AMAE International Journal on Manufacturing and Material Science* **1**(1).

[4] Kannan, E., Balasuthagar, C., Ponsankar, S. and Sivashankar, M. 2016 Experimental investigation on convective heat transfer characteristics in a circular tube fitted with perforated twisted tape and wire coil *J. Chem. Pharma. Sci* **9** 2767-2769.

[5] Shi, S., Yan, C.Q. and Niu, G.L. 2011 March. Numerical study of heat transfer and pressure drop of integral pin-fin tubes. In *2011 Asia-Pacific Power and Energy Engineering Conference* (pp. 1-4) IEEE.

[6] Burse A. K., R. H. Yadav and N. K. Sane 2014 Enhancement of Heat Transfer in a Horizontal Circular Pipe using Twisted Tape Inserts of Different Twist Ratios *International Journal of Advances in Engineering and Management (IJAEM)* **1**(5) 45-52.

[7] Lee, D.H., Jung, J.M., Ha, J.H. and Cho, Y.I. 2012 Improvement of heat transfer with perforated circular holes in finned tubes of air-cooled heat exchanger *International Communications in Heat and Mass Transfer* **39**(2) pp.161-166.

[8] Qasim S. Mahdi and Ali A. Hussein 2016 Enhancement of Heat transfer In Shell and Tube Heat Exchanger with Tabulator and Nanofluid *International Journal of Mechanical Engineering & Technology (IJMET)* **07**(3) 225–239.

[9] Yu, B. and Tao, W.Q. 2004 Pressure drop and heat transfer characteristics of turbulent flow in annular tubes with internal wave-like longitudinal fins *Heat and mass transfer* **40**(8) 643-651.

[10] Dhirajkumar K. More1, Prashant D. Deshmukh and Prerana U. Jiwane 2016 A Review on Modern Techniques of Heat Transfer Enhancement in Circular Tube *International Research Journal of Engineering and Technology* **03**(06).

[11] Iqbal, Z., Syed, K.S. and Ishaq, M. 2013 Optimal fin shape in finned double pipe with fully developed laminar flow. *Applied thermal engineering* **51**(1-2) 1202-1223.

[12] Kumar V. Sudheer 2016 CFD Analysis on Heat Transfer Enhancement in Horizontal Tube by Forced Convection with Inserts *International Journal & Magazine of Engineering, Technology, Management and Research* **3** (2).

[13] Mon, M.S. and Gross, U. 2004 Numerical study of fin-spacing effects in annular-finned tube heat exchangers *International journal of heat and mass transfer* **47**(8-9) 1953-1964.

[14] Sami D.Salman, Abdul Amir H.Kadhum, MohdS.Takriff, and Abu BakarMohamad 2014 Heat Transfer Enhancement of LaminarNanofluids Flow in a Circular Tube Fitted with Parabolic-CutTwisted Tape Inserts *The Scientific World Journal* **7** Article ID 543231.

[15] Qasim S. Mahdi and Noor A. M. Mohammed 2016 Heat Transfer Augmentation of Laminar Nanofluid Flow in Horizontal Tube Inserted with Twisted Tapes *International Research Journal of Engineering and Technology (IRJET)* **03** (06).

[16] Kulkarni K.G. 2015 Experimental Investigation of Heat Transfer Enhancement by Using Clockwise and Counter -clockwise Corrugated Twisted Tape Inserts I*nternational Journal of innovations in Engineering Research and Technology (IJIERT)* **2**.

[17] Davidson, L. 2003 An introduction to turbulence models, Department of Thermo and Fluid Dynamics. *Chalmer University of Technology, Publication*, **97**(2), pp.1-8.

[18] Luay B. Hamad 2017 *Parametric Study of Heat Transfer and Pressure Drop Characteristics in A Tube Using Different Types of Inserts* Thesis in Institute of Science and Technology Mechanical and Aeronautical Engineering Department.

IOP Conf. Series: Materials Science and Engineering 1182 (2021) 012066 doi:10.1088/1757-899X/1182/1/012066

Heat Transfer Augmentation of Laminar Flow in Horizontal Circular Nano-Composite Tube with Insert

M S Qasim[1], F A Abdulla[1] and H D Nahla[2]
[1]Mustansiryah University, College of Engineering, Mech. Eng. Dept., IRAQ, Baghdad
[2]Mustansiryah University, College of Engineering, Elect. Eng. Dept., IRAQ, Baghdad

Email: qasim602006@yahoo.com

Abstract. A numerical investigation was carried out to predict the heat transfer behavior and friction factor in heated circular nano- composite tube with two cases: plain tube and inserted tube. Typical twisted tape (TTT) and triangular cut twisted tape (TCTT) were used as insert if twisted tape ratio equal 2 and thickness equal 1 mm. The composite material consist of polyester resin mixed with nano copper with different weight fraction 3%, 7%, 33%, 50% and 75%. The governing equations were solved by using commercial ANSYS Fluent package (2017) with the assistance of Solid Works and Gambit software program. The finite volume approach was used in solving the equations to predict the pressure flow, heat transfer and temperature distribution along the tube. Nano- composite material tube was heated under constant heat flux (1244W/m^2). The water flow in the tube was laminar ($297 < Re < 1670$). The dimensions of circular tube were 800 mm length, 20 mm diameter and 1.5 mm thickness. Numerical results show that the heat transfers was augmented when (TTT or TCTT) was used as a swirl flow devices comparing with plain tube. The water temperature increases significantly at the center line of tube by 25% when TTT is used, while the friction losses increase in an undesirable rate of 155 %.Finally, it can be concluded that the nano-composite tube is efficient in heat transfer applications.

1. Introduction

Enhancement of the heat transferring plays a significant role in maximizing the efficiency of many applications such as heat exchangers, power generation, and microelectronics. Augmentation of the convective heat transfer coefficient will minimize the size and cost of the heat exchanger [1, 2]. For enhancing the heat transfer in heat exchanger , and minimize the surface area with materials cost , different methods is implemented such as passive method, active method and hybrid method. In passive method swirl devices is used in flow passage. The most effective passive techniques on heat transfer improvement are twisted tape and wire coil inserts. The twisted tape is more effective for improving heat transfer compared with wire coil in case of laminar flow inside heat exchanger. There are used for their beneficial such as easy fabrication, process and maintenance low. For increasing the heat transfer coefficient by performing the swirl flow and twisted tape inserts and making heat transfer coefficient higher, many types of research have been implemented on insert devices experimentally and numerically, to examine the behavior of the influence of twisted tape on heat transfer improvement, thermal performance factor, pressure drop, and flow friction specifications in a heat exchanger tube heat transfer enhancement to reach more effective kinds and dimensions of these devices [3]. Kapatkar et al. [4], experimentally examined heat transfer and friction factor in a copper plain tube and when using a twisted tape as insert. The study was carried out under laminar flow and at constant heat flux. Results show that with increasing Reynolds number, the Nusselt number of full-length twisted tape is raised. This increment in intensity of heat transfer due to swirl flow. For the flow inside smooth tubes, full length twisted tapes produce the enhancement in the mean of the Nusselt number with range of Reynolds number (200 to 2000). At higher Reynolds number beyond laminar flow region, the friction factor is raised due to the growing in the momentum interchange of the molecules. Murugesan et al. [5], evaluated the influence of twisted tape insert with V-cut on heat transfer, friction factor, and characteristics of thermal performance factor. In the test section which involves of two concentric tubes in which hot water flows through the internal tube and the cold water flow through the annulus . Results showed that the twisted tape with V-cut presented a higher heat transfer rate, friction factor, and thermal performance factor compared to the case plain twisted tape. Sivakumar and Rajan [6], carried out an experimental and numerical evaluation of heat transfer coefficient and characteristics of friction factor in a concentric tubes. The test section is consists of an aluminum interior tube of 18 mm interior diameter and 20 mm exterior diameter of 2.2 m length. The twisted tape material was copper. Results show that the heat transfer improvement took place with incremental of Reynolds number. Saud [7], studied numerically the improvement of heat transfer and fluid flow properties inside circular tube with twisted tape insert. A different type of insert has been tested such as triple, dual and quadruple twisted tapes. The results of turbulent flow under constant heat flux forced convection proofed that tube close-fitting with dual linked twisted tapes (TDCTT) is the best thermal performance. Farhad et al. [8], investigated friction factor and heat transfer enhancement in case of dual twisted tapes or normal in converge-diverge tubes. The range of Reynolds number is10000 to 20000 for twisted tapes with different holes in, at constant heat flux using working fluid (water). The results show that, the Nusselt number will be improves by 9% when a dual twisted tape was used. Mahdi et al. [9], investigated the heat transfer improvement for a single tube with and without insert. The heat exchanger performance was studied if the tube in oblique and horizontal directions. The circular tube of heat exchanger is made of copper material with dimensions: length - (1000 mm) and inner and outer diameters of (23 and 25) mm, respectively. Distilled water under laminar flow condition (Re =1056 – 2002) flows through an insulated tube. The insert tape was made of copper strip with 0.8 mm thickness. The employed of the inserted tape yields a considerable increase in the coefficient of heat transfer about (16-27%) more than plain tube. Shuai et al. [10], studied experimentally and numerically the thermal-hydraulic features in circular tubes with twisted tape and wire coil inserts. Experimental results show that friction factor and Nusselt number in tubes with twisted tape inserts were 38%- 57% and 148%- 202%, respectively, which are greater than plain tubes. When wire coil was inserted in plain tubes, friction factor and Nusselt number raised by 685%-792% and 75%- 94%, respectively. Nusselt number and friction factor in tubes with twisted tape and wire coil inserts were (113% - 139%) and (836% - 874%), respectively, greater than plain tubes. The aim of the present work is to predict the heat transfer behavior and friction factor in heated circular nano- composite tube with two cases: plain tube and inserted tube. Typical twisted tape (TTT) and triangular cut twisted tape (TCTT) were used as insert.

2. Numerical Analysis and Solution

In the present work a numerical study of heat transfer and the flow field through a test nano-composite tube is carried out. Test tube is subject to constant heat flux. Study is consisting of two category; plain tube and inserted tube [11,12]. Typical twisted tape (TTT) and triangular cut twisted tape (TCTT) were used as insert twisted tape ratio. The analysis depends on a modeling of fluid flow through a test tube including heat transfer, pressure drop that related to the physical phenomena. Finite volume method is used to solve the conservation equations (continuity, momentum, and energy equations) with the given boundary conditions. In the numerical solution, an ANSYS FLUENT package 17.0 commercial copy with SOLID WORK and GAMBIT software was employed.

2.1 Physical Domain Simulation

SOLID WORK programs used to make solid models of plain tube and tube with twisted tape insert [13]. Figure 1 illustrates the tube with typical twisted taped TTT and triangular cut twisted tape TCTT insert.

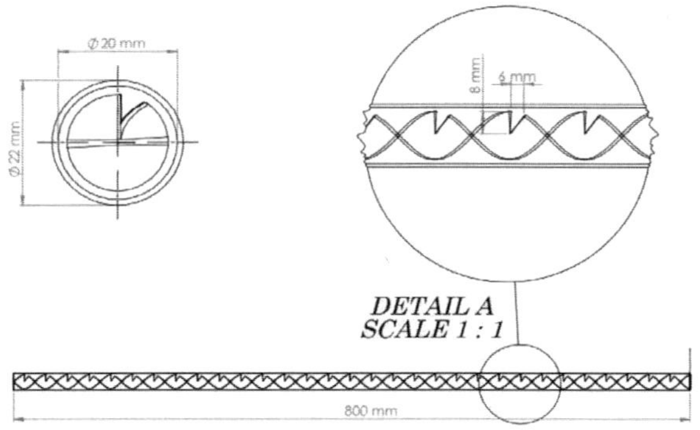

Figure 1. The wire frame of inserted tube with TCTT & Triangular cut of twisted tape.

2.2 Governing Equations

In the present work, the working fluid is water and the flow characteristics are assumed to be steady state, Newtonian fluid, incompressible, three dimensional, forced convection and internal laminar flow.

The conservation equation for continuity, momentum, and energy equations can be written as follows Versteeg and Malalasekera [14].

Continuity Equation

$$\frac{\partial u}{\partial x}+\frac{\partial v}{\partial y}+\frac{\partial w}{\partial z}=0 \tag{1}$$

Where u ,v, and w are the velocity in x , y , and z directions.

Momentum Equation

The equations in(x, y, and z) direction are;

$$\rho(u\frac{\partial u}{\partial x}+v\frac{\partial u}{\partial y}+w\frac{\partial u}{\partial z})=-\frac{\partial p}{\partial x}+\mu(\frac{\partial^2 u}{\partial x^2}+\frac{\partial^2 u}{\partial y^2}+\frac{\partial^2 u}{\partial z^2}) \tag{2}$$

$$\rho(u\frac{\partial v}{\partial x}+v\frac{\partial v}{\partial y}+w\frac{\partial v}{\partial z})=-\frac{\partial p}{\partial y}+\mu(\frac{\partial^2 v}{\partial x^2}+\frac{\partial^2 v}{\partial y^2}+\frac{\partial^2 v}{\partial z^2}) \tag{3}$$

$$\rho(u\frac{\partial w}{\partial x}+v\frac{\partial w}{\partial y}+w\frac{\partial w}{\partial z})=-\frac{\partial p}{\partial z}+\mu(\frac{\partial^2 w}{\partial x^2}+\frac{\partial^2 w}{\partial y^2}+\frac{\partial^2 w}{\partial z^2}) \tag{4}$$

where ρ , p , and μ are the density , pressure and viscosity of fluid.

Energy Equation

$$\rho c\,(u\frac{\partial T}{\partial x}+v\frac{\partial T}{\partial y}+w\frac{\partial T}{\partial z})=k(\frac{\partial^2 T}{\partial x^2}+\frac{\partial^2 T}{\partial y^2}+\frac{\partial^2 T}{\partial z^2}) \tag{5}$$

where c , k , and T are the specific heat , thermal conductivity and temperature.

2.3 Mesh Generation

An unstructured grid was used in present work. It is vital to have enough number of cells for a good resolution [15, 16]. For the present cases, an average of (1612012) million cells is used. Figure 2 illustrates the mesh building of straight circular tube inserted with TTT & TCTT.

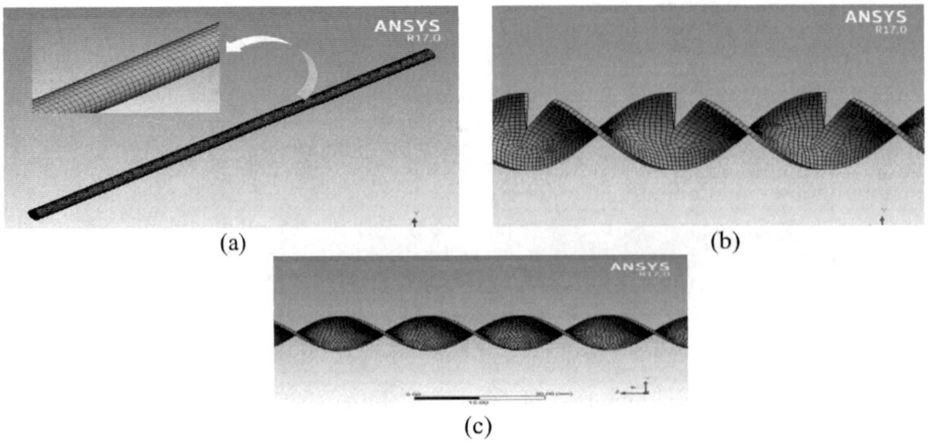

Figure 2. Mesh buildings of a) Smooth tube, b) TCTT, c) Inserted tube with TTT.

The maximum number of iterations has been done in the present cases is 600 iterations as shown in figure 3.

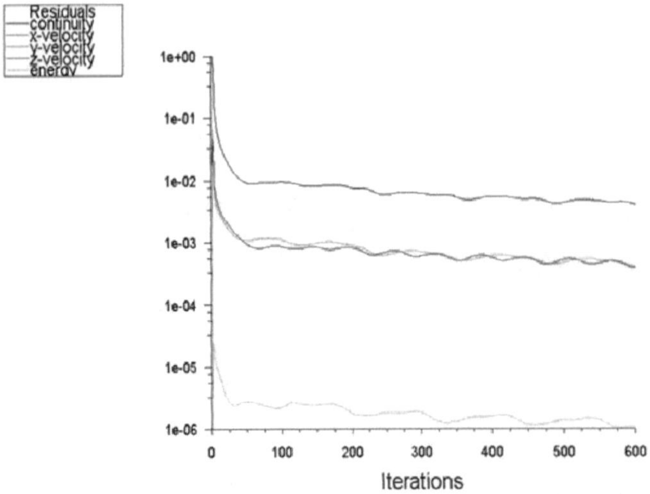

Figure 3. Residuals for running of numerical calculation.

2.4 Implementation of Boundary Conditions

The material of the tube is nano-composite (polyester + nano copper).The properties are: (k=295 W/m. K , c_p=563.2 kJ/kg. K, ρ=398 kg/m^3). The test tube has a length of 800 mm, 20 mm ID and 1.5 mm thickness. Boundary conditions for each zone of the computational domain are; the wall of the tube was heated under constant heat flux equal to 1244 W/m^2, the water flow in the tube was laminar (297 < Re < 1670), the temperature of water at the inlet tube is constant 20°C, and the pressure is assumed to be atmospheric pressure at the exit of tube. No slip boundary condition is specified for the wall of tube [17, 18].

3. Model Validation

Figure 4 shows the behavior of the inside wall surface temperature of the tube with axial distance. It can be seen from figure there is an error from 9% to 17% between the numerical and experimental results.

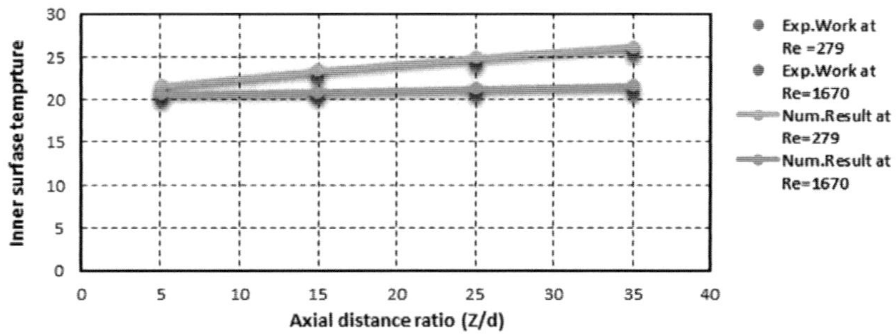

Figure 4. Comparison between present numerical results and experimental results [12] at q=1244W/m^2.

4. Results and Discussion

4.1 Temperature Contours

Figures 5 and 6 clarify the contouring outcomes for temperature distribution at different axial sections along the tube length. For the domain of plain tube with DI-water represented in figure 5, the contours present the evolution of thermal boundary layer starting from the inner wall tube (at which the heat flux supplied) to the tube center. Thus, the thermal boundary layer thickness (δth) is increased where axial location is increased from z=0 to 800 mm reaching to nearly the same at which the flow becomes thermally fully devolved. As it can be seen, the maximum inner wall temperature (Tmiw) and thicker δth are existed at the lower inlet velocity ui=0.013 m/s. These values start to calm down where inlet velocity is increased which are reasonable outcomes because the fluid gather much longer time inside the tube at lower velocities that source heat gain inside the flowing fluid and thicker δth therefore, heat transfer rate is less than for higher velocities.

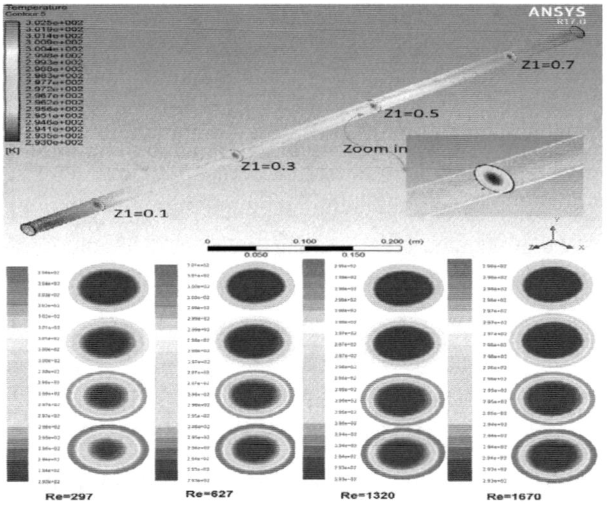

Figure 5. Temperature contours for water at different Reynolds number and various axial distances for plain heated tube with 1244 W/m^2 heat flux.

From figure 6, it can be realized from tube different sections the advantage of putting in the twisted tape (TTT) for twisted ratio Y=2 and thickness =1mm inside the sleek tube which generates a swirling flow path that supply a better mixing between the viscous fluid layers allowing for more heat to transfer from inner wall tube that enhances heat transfer rate.

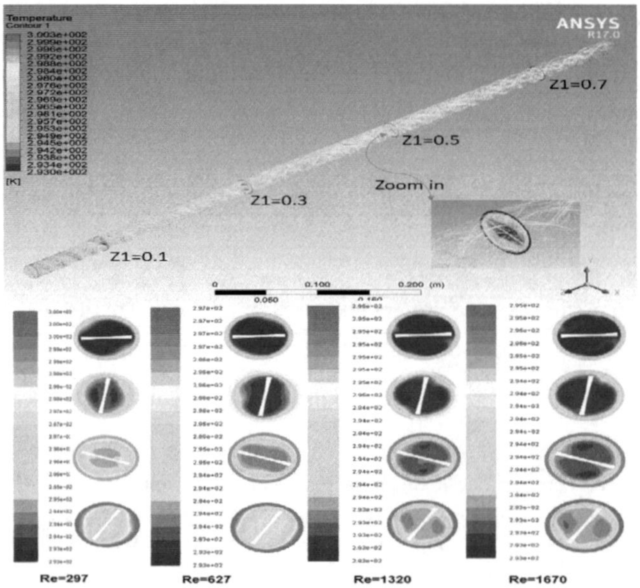

Figure 6. Temperature contours for tube inserted with TTT with twisted ratio = 2 and thickness = 1mm at different Reynolds number and 1244 W/m^2 heat flux at different axial distance.

From figure 7, it can be realized from the half tube different sections the benefits of inserting the twisted tape (TTT) and (TCTT) inside the smooth tube which generates a swirling flow path that provides a better mixing between the viscous fluid layers allowing for more heat to transfer from inner wall tube that enhances heat transfer rate. Figure 7(a) shows the behavior of water flow through a nano-composite tube under the influence of constant heat flux (1244 W/m^2) and Reynolds number (Re=297). It is evident that the water in contact with the surface of the inner tube is heated with the length of the tube while the water away from the surface remains at the same temperature. Figure 7(b) shows the effect of adding a TTT type inside tube to improve heat transfer from the tube to the water. The increase in heat transfer from the surface of the tube to the water can be explained to generate a vortex flow that breaks down the boundary layer adjacent the tube and increases the tangential velocity near the tube surface. Figure 7(c) shows the effect of adding a TCTT type. The figure also shows an increase in the improvement of heat transfer, but in a smaller percentage compared to the addition of (TTT).

Figure 8 shows the pressure contours of water through the nano- composite tube in three cases (plain tube, tube with TTT, tube with TCTT) as insert when the flow is laminar ($297 < Re < 1670$). Figure 8 indicates the presence of pressure losses in the flow of water through the pipeline. The pressure drop in the plain tube reaches 99%, while in the two cases of adding an insert (TTT, TCTT), it becomes (135% and 155%), respectively. The reason for the increased losses is the addition of TTT and TCTT to the tube. Increased losses through the tube lead to increased pumping capacity, and this is an undesirable case in engineering applications.

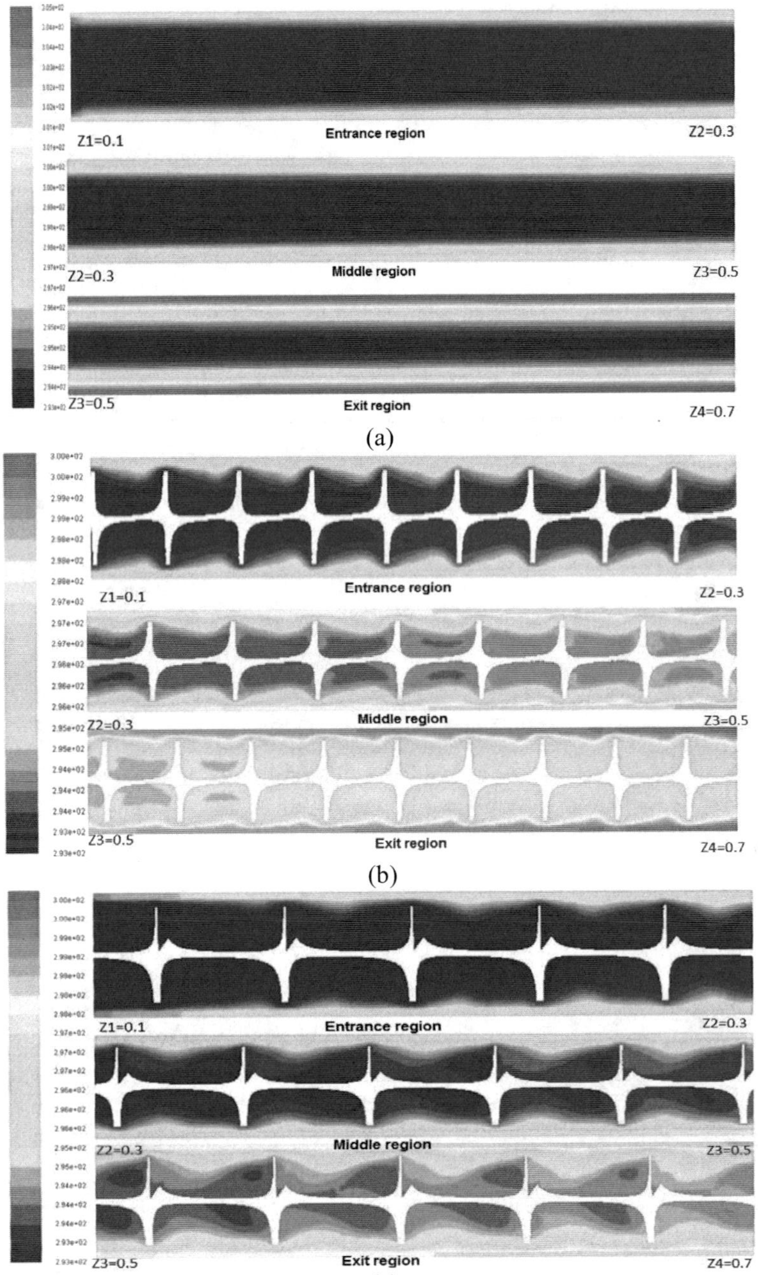

(a) (b) (c)

Figure 7. Temperature contours for water flowing inside: (a) plain tube, (b) tube inserted with TTT, (c) tube inserted with TCTT at Re=297 and 1244W/m^2 heat flux at different longitudinal sections.

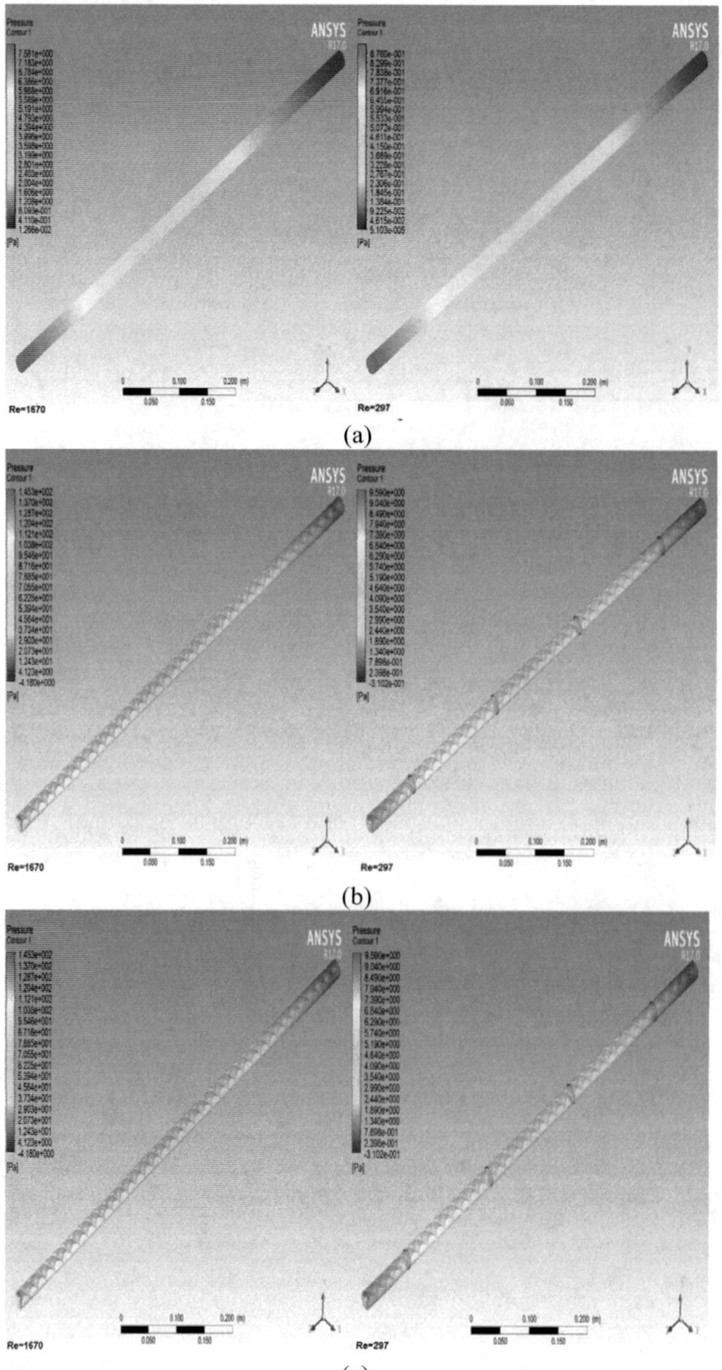

Figure 8. Pressure contours for water at different Re with 1244 W/m^2 heat flux along the, (a) plain tube, (b) tube inserted with TTT, (c) tube inserted with TCTT.

5. Conclusions

From the numerical results, it is concluded that heat transfer improvement and pressure loss occurs by means of utilizing the twisted tapes that be subject to on the twist ratio, thickness, and the twisted tapes cutting form. The results showed that when the tube was exposed to a constant heat flux (1244 W/m^2) heat transfer increased when using TTT and TCTT, compared to the plain tube. Typical twisted tape showed well performance in heat transfer enhancement but with higher friction losses comparing to plain tube. The water temperature increases significantly at the center line of tube by 25% when TTT is used, while the friction losses increase in an undesirable rate of 155 %. Finally, it can be concluded that the nano-composite tube is efficient in heat transfer applications.

6. References

[1] Lazarus G. A., Nandigana V., Senthil K. A. and Dhasan M. L. 2009 Experimental study on forced convective heat transfer with low volume fraction of CuO/Water nano-fluid *Journal Energies* **2** 97-119.

[2] Liu S. and Sakr M. 2013 A comprehensive review on passive heat transfer enhancements in pipe exchangers *Journal ElSEVIER, Renewable and Sustainable Energy Reviews* **19**, 64-81.

[3] Qasim S. M., Mohammed F. A. and Hashim R. 2015 Numerical investigation of the thermal behavior of heated natural composite materials *IOP Conf. Ser. Mater. Sci. Eng.* **95**(1) 012001

[4] Kapatkar A.V.N., Padalkar A.S. and Sanjay Kasbe C. 2011 Experimental investigation on heat transfer enhancement in laminar flow in circular tube equipped with different inserts *Mechanical Engineering Department*, Pune, India, *AMAE International Journal on Manufacturing and Material Science* **1** 1–6.

[5] Murugesan P., Mayilsamy K., Suresh S. and Srinivasan P.S.S. 2011 Heat transfer and pressure drop characteristics in a circular tube fitted with and without V-cut twisted tape insert *International Communications in Heat and Mass Transfer* **38**(3) 329-334.

[6] Sivakumar K. and Rajan K. 2015 Experimental analysis of heat transfer enhancement in a circular tube with different twist ratio of twisted tape inserts *International Journal of Heat and Technology* **33**(3) 158-162.

[7] Saud Tameem Taher 2017 *Study of thermos-hydraulic behavior of turbulent flow with twisted tapes* M.Sc. thesis, Al-Mustansiriyah University, College of Engineering, Iraq.

[8] A. Farhad, Khashayar P., Mehrshad A., Mojtaba A.D. 2018 Numerical study of heat transfer enhancement using perforated dual twisted tape inserts in converging-diverging tubes *Heat Transfer-Asian Research* **47**(5) Wiley 754-767.

[9] A. S. Mahdi, S. T. M. Al-Musawi, Z. K. Kadhim, H. M. Hussain and L. J. Habeeb 2020 Heat transfer enhancement by using twisted tape in horizontal and an inclined tube *J. Mech. Eng. Res. Dev.* **43**(3) 106–124.

[10] Wang Shuai , Lu Kun, Wang Kai, and Zhang Q. 2020 Experimental and numerical analysis for thermal-hydraulic characteristics in circular tubes with twisted tape and wire coil inserts *2nd International Conference on Civil Architecture and Energy Science (CAES)*, E3S Web Conf. 165.

[11] Abdullah F.A. and Khalaf W.A. 2018 Experimental investigation of composite materials subjected to torsional stresses at high shear strain rate *Int. J. Mech. Mechatronics Eng.* **18**(1) 64-75.

[12] Mahdi, Q. S., Abbas, F. and Mahdi, H. S. 2018 Heat Transfer Investigation in a Circular Tube Fabricated from Nano-composite Materials Under a Constant Heat Flux *International Journal of Mechanical & Mechatronics Engineering IJMME-IJENS* **18**(06) 1-9.

[13] Khalid Hamdan, Z. and Abdullah, F. A. 2018 Investigation of the Adding Nano Particles to Composite Material under High Strain Rate Torsion with Hygrothermal Effect Technology **9**(6) 1098-1114.

[14] Versteeg H. K. and Malalasekera W. 1995 *An introduction to computational fluid dynamics-The finite volume method* Longman group Ltd. First edition.
[15] Moustafa N M, Abdulla F A and Nori A F, 2018 PID control system for a variable speed horizontal axis wind mill *Int. J. Mech. Eng. Technol.* **9** 1080-1087.
[16] Jebur, N. A., Abdulla, F. A., and Hussein, A. F. 2018 Experimental and numerical analysis of below knee prosthetic socket *Int. J. Mech. Eng. Technol.* **9** 1-8.
[17] Abdulla, F. A., Fadhil, H. A., and Abdulwahid, J. N. 2018 Experimental Study of the Creep Behavior of Nano-Composites Carbon Fibers *In IOP Conference Series: Materials Science and Engineering* **454**(1) 012126.
[18] Abdul-Kareem, H. S., Abdulla, F. A., and Abdulrazzaq, M. A. 2019 Effect of Shot Peening and Solidification on Fatigue Properties of Epoxy Base Composite Material *IOP Conference Series: Materials Science and Engineering*, **518**(3) 032017.

Acknowledgments
The authors (M S Qasim ,Fadhel Abbas Abdulla , and H D Nahla) would like to thank Mustansiriyah University (www.uomustansiriyah.edu.iq) Baghdad-Iraq for its support in the present work.

Optimisation of the production process in manufacturing a smart equipment for precision agriculture

M M Roşu[1], P A Cotfas[2] and M Iliescu[2]
[1]University POLITEHNICA of Bucharest-Romania, Department of Manufacturing Engineering, Splaiul Independentei, No. 313, 060042, Bucharest, Romania
[2]Romanian Academy, Institute of Solid Mechanics, Constantin Mille 15, Bucharest, Romania

E-mail: mihaiela.iliescu@imsar.ro

Abstract. The paper presents the optimizing process of a production system that it is intended for manufacturing process of a customized smart equipment. The initial variant (already in use) of this equipment dose no longer fit the requirements for precision agriculture so, new concept and, further, prototype development are to be done. The objective is to have smart innovative equipment suitable for digging furrows with controllable depth and geometric shape, not to mention its ability to determine soil's moisture. The depth and geometric shape of the trench section are important elements in preparing the soil for irrigation as they ensure controlled slope for water runoff. Humidity is important for establishing and controlling the irrigation norm. Once the new equipment prototype is manufactured, tested and validated, it will be produced in series. For this it is necessary planning, programming and optimization of the production process so that to obtain a competitive product on the market through quality and price. Also, the transition from single production to series production involves great adaptability of the production process using self-optimization, self-configuration and self-diagnosis procedures, which are able to support and assist the human operator in the complex process of handling workpieces in the process.

1. Introduction

Due to the globalization of the sales markets, one of the biggest challenges of the new trends in the area of production processes is its rapid adaptation to the current trend with lowest possible costs and high quality conditions, under the assumption of reducing the negative impact on the environment. One important factor that ensures a guarantee of success for new products, especially new equipment's, on the market is quality. The problem of optimizing production processes is another topic that can be developed, especially due to the fact that production processes become increasingly complex due to a certain amount of variability which in some cases can strongly affect the quality of the result [1].
The development of new equipment also involves analysis of the efficiency of the processes which it will be used to. An important aspect is that of applying informatics and informative technologies to traditional concept of agriculture. As a result, machines and devices used by farmers are equipped with devices that enable precise performance of agricultural treatments and precise dosing of mineral fertilizers and crop protection substances [2, 3]. The optimised production plan and capacity dimension, including resource allocation, scheduling, analysis of the time required to carry out the operation and mission planning, are integral parts of modern agricultural operations management which supplement traditional agricultural operations planning methods [4], cited by [5].

This paper presents research results on conceptual development for a smart innovative equipment suitable for digging furrows with controllable depth and geometric shape and for determination of soil moisture. The optimized plan for equipment components manufacturing process has been determined, so that to obtain a competitive product on the market through quality and price.

2. Methodology

In any production process for obtaining new equipment's with high characteristics, the planning and scheduling working flow activities of each of the machining processes refer to decreases of production costs and of the costs related to the exploitation life of the product [6- 8].

The main purpose of the equipment to be developed is that of digging furrows with controllable depth and geometric shape and to enable the determination of soil's moisture. The equipment has a coulter that executes a furrow with a depth of 150 mm, squares a distance from the left row of 150 mm, and the amount of soil dug is stored on the opposite side but not further from the next row of crops. Initially it was tried with a disc, but the soil gets back into the furrow faster than the compaction wheel reaches to fix it. The initial tests were done with four and five rows, digging simultaneously. The conclusion was that the forces must be evenly distributed, otherwise the compaction wheel does not pass where it should, and the equipment would not dig furrows according to the requirements. Depending on how many rows the tractor rides, four or five rows are chosen.

The equipment is high enough so that work can be performed when the plant is in vegetation, see figure 1. Basically, the limiting moment occurs not because of the machine, but because of the tractor, which can no longer pass through the crop due to the ground clearance it has.

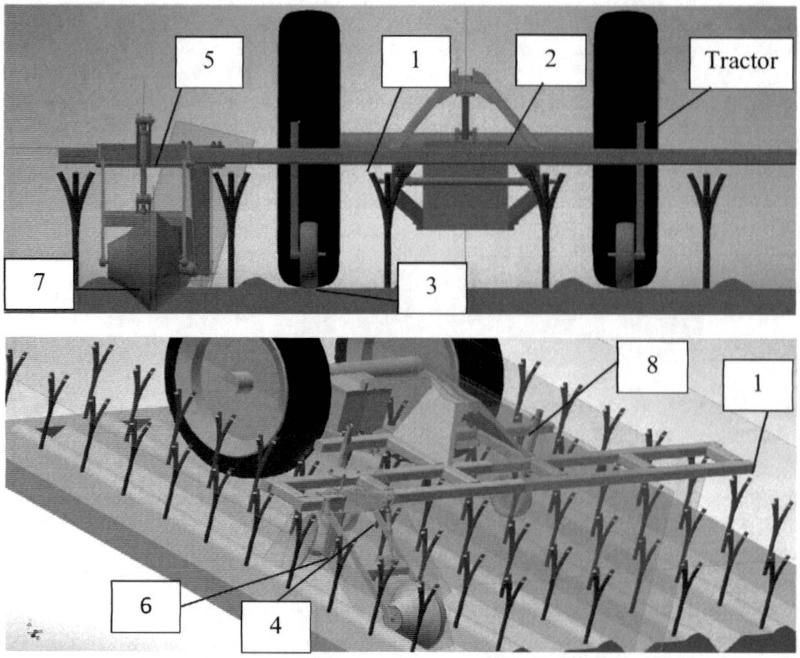

Figure 1. Conceptual model of the new equipment for digging furrows.1- Equipment frame; 2 - Connection system with tractor; 3 - Wheels for adjustment of coulters penetration depth; 4 - Mould board; 5 - Supports for fixing the mould boards on frame; 6 – Coulters; 7 - Compaction wheels; 8 - Frame for fixing the wheels on the basic frame.

The equipment has a 100x100 mm square profile frame with a 10 mm thick wall, so that the existing forces do not damage the basic frame on which the whole assembly is built. The coulter adjustment mode is manual. In front of the equipment there are two wheels that follow the ground. They are fixed on a bar with several positioning holes, so that if it is necessary to deepen the pit, the coulter is raised and the bolt is inserted in the whole corresponding with the required depth. The wheel battery has the role of compacting the ground and the one that fell into the furrow, but also the one that was removed.

In figure 2 it is presented the Work Breakdown Structure (WBS) of the new conceptual model.

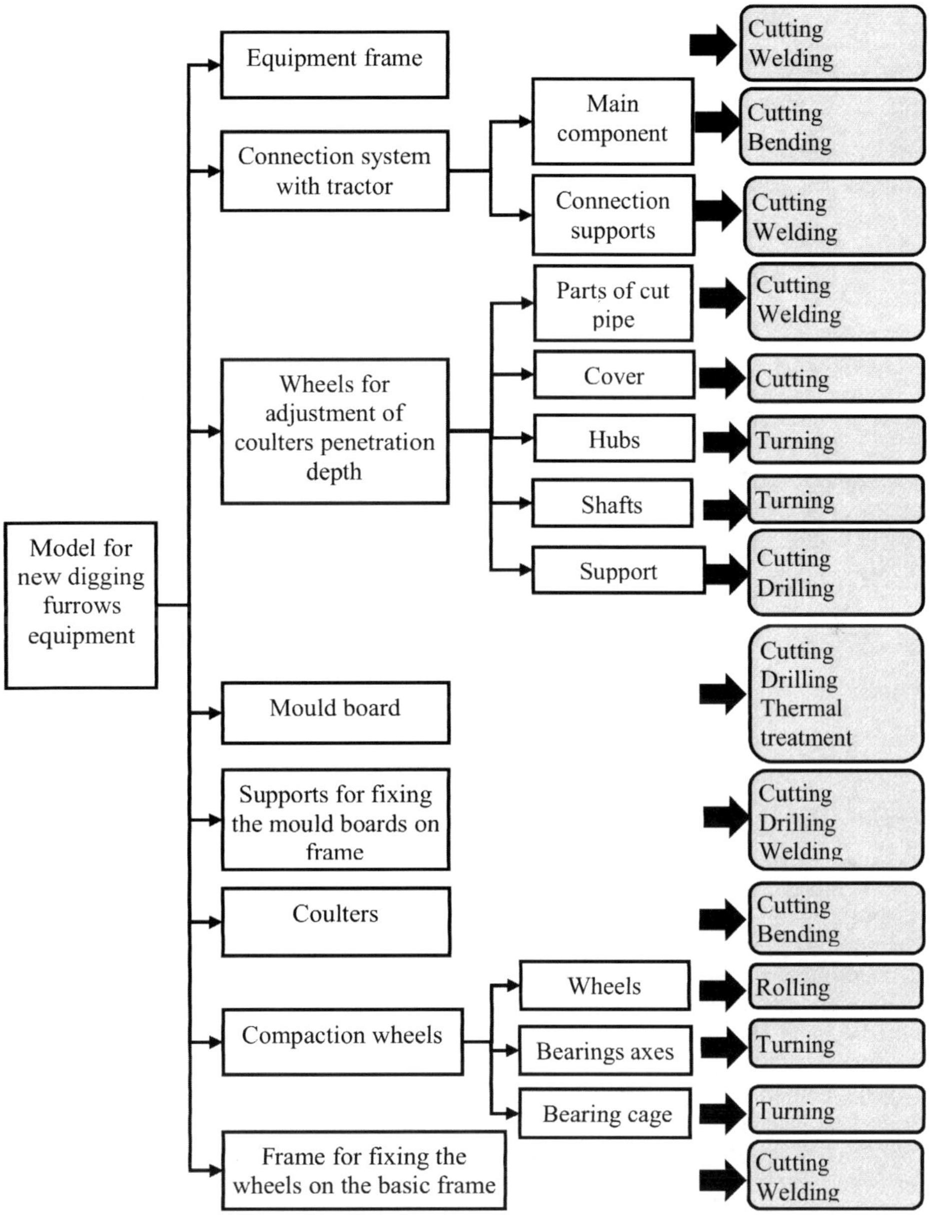

Figure 2. WBS of conceptual model of the new equipment for digging furrows.

Having the starting point WBS, the planning of activities within the manufacturing process was carried out, considering the equipment and human resources available. The planning of the manufacturing steps was done using Primavera Project Planner (PPP) software. The overall image of this planning for the conceptual model is presented in figure 3.

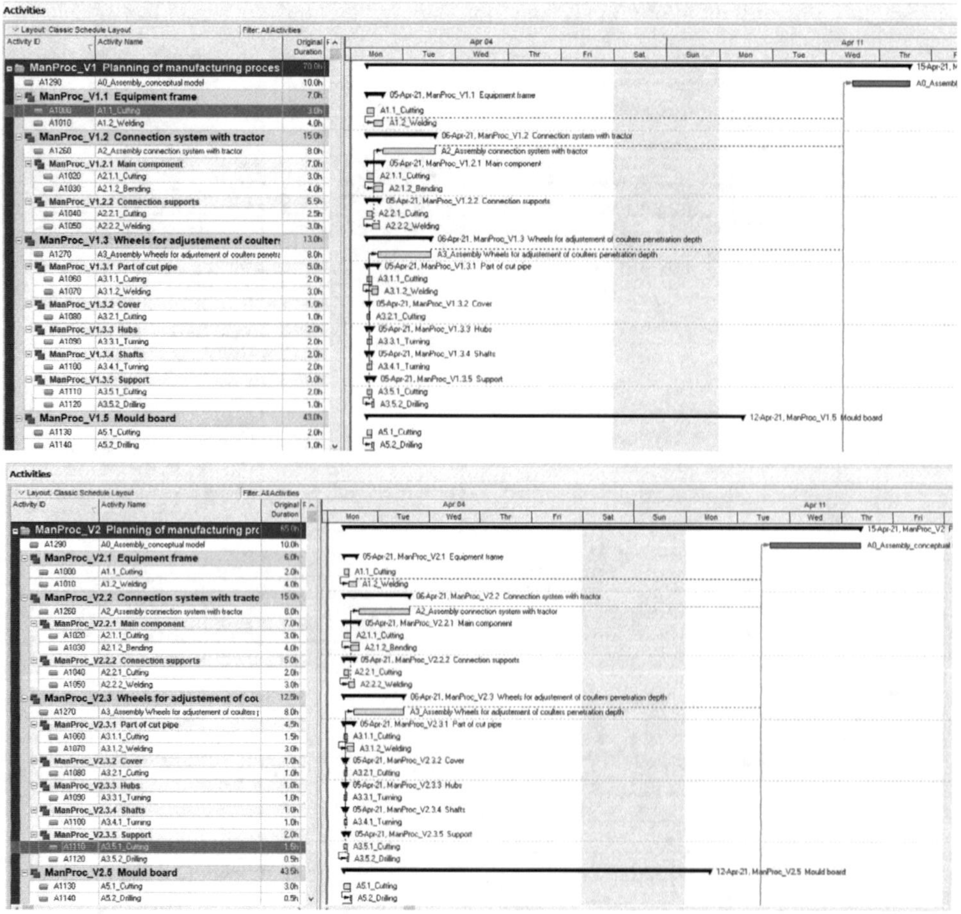

(a)

(b)

Figure 3. Initial planning of the manufacturing process of new equipment for digging furrows (partial view).

The planning has been done for two types of manufacturing process. The first process is the one already in use, with available equipment's and the second process is a new intended one, with new specialised equipment's. As it seen in the figure 3(a), the total time of first production process is 70 working hours and the total cost of this manufactured process is estimated at 12,229 euro. For the second planning, the manufacturing process takes place on two high performance equipment's: a CNC centre for turning and drilling operation; an equipment for water jet for cuttings operations. With this second process type, the total time is 65 working hours and the total cost is 10,663 euro (figure 3b). The cost includes the human resource costs and equipment's' operating costs. The cost of material is not included in this analyse because this will be the same for any of the production planning cases considered.

In this initial planning it has not been considered the availability of resources. For this manufacturing process there has been allocated 8 human resources and 7 equipment's for the first analysed planned and 7 human resources and 6 equipment's for the second analysed planned. The thermal treatment activity of mould board is planned to be released at thermal treatment service provider. By analysing the resources availability, there were identified some overallocated resources. In figure 4 is presented the list of resources used for this manufacturing process (figure 4(a) for the first analysed process and figure 4(b) for the second analysed process) and in figure 5 are presented the six, out of seven, overallocated resources for the first planning analysed. This is how there were pointed out three overallocated equipment's (cutting machine, wire welding machine and turning machine) and associated operators and also, the assembly operator is overallocated by few hours.

Display: Current Project's Resources

Resource ID	Resource Name	Primary Role	Default Units / Time
ManProc_Order	External order		1/h
ManProc_R01	Cutting machine		100%
ManProc_R02	Wiire and gas welding machine		100%
ManProc_R03	Ruter CNC		100%
ManProc_R04	Bending machine		100%
ManProc_R05	Turning machine		100%
ManProc_R06	Driling machine		100%
ManProc_R07	Rolling machine		100%
ManProc_R10	Operator 1_Cutting machine		100%
ManProc_R11	Operator 2_Welding machine		100%
ManProc_R12	Operator 3_Ruter CNC		100%
ManProc_R13	Operator 4_Bending machine		100%
ManProc_R14	Operator 5_Turning machine		100%
ManProc_R15	Operator 6_Drilling machine		100%
ManProc_R16	Operator 7_Rolling machine		100%
ManProc_R17	Operator 8_Assembly		100%

(a)

Display: Current Project's Resources

Resource ID	Resource Name	Primary Role	Default Units / Time
ManProc_Order	External order		1/h
ManProc_R01_V2	Cutting on water jet machine		100%
ManProc_R02	Wiire and gas welding machine		100%
ManProc_R03	Ruter CNC		100%
ManProc_R04	Bending machine		100%
ManProc_R07	Rolling machine		100%
ManProc_R08_V2	CNC center		100%
ManProc_R10	Operator 1_Cutting machine		100%
ManProc_R11	Operator 2_Welding machine		100%
ManProc_R12	Operator 3_Ruter CNC		100%
ManProc_R13	Operator 4_Bending machine		100%
ManProc_R16	Operator 7_Rolling machine		100%
ManProc_R17	Operator 8_Assembly		100%
ManProc_R18_V2	Operator 9_CNC machine		100%

(b)

Figure 4. List of used resources in manufacturing process.

For optimising the manufacturing analysed process, we will use tool of PPP for levelling of activities according with priority criterion "minimum total float of activity". The total float is the reserve time for an activity which does not change the duration of the whole process.

Figure 5. The overallocated resources (partial view).

The planning results after removing the overallocation resources are presented in figure 6 and it is noticed that the total time for whole process is longer by 35 working days. The total duration of whole process is 105 working days.

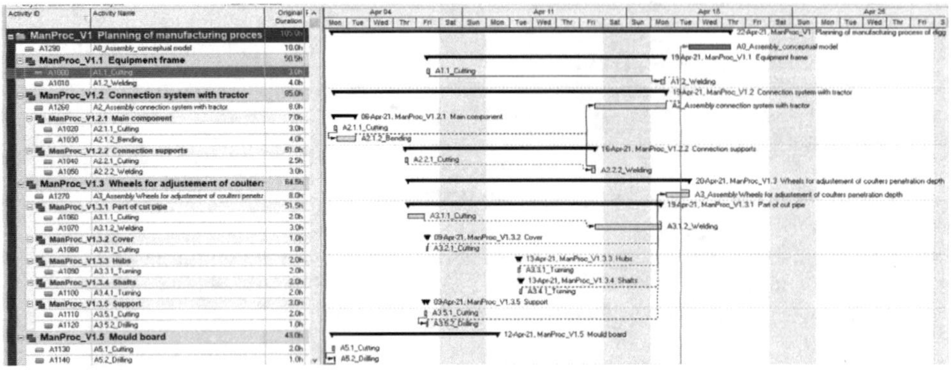

Figure 6. Optimised plan of the first type manufacturing process (partial view).

For the second analysed situation the total duration after removing the overallocated resources is presented in figure 7. The total duration is 100 working days.

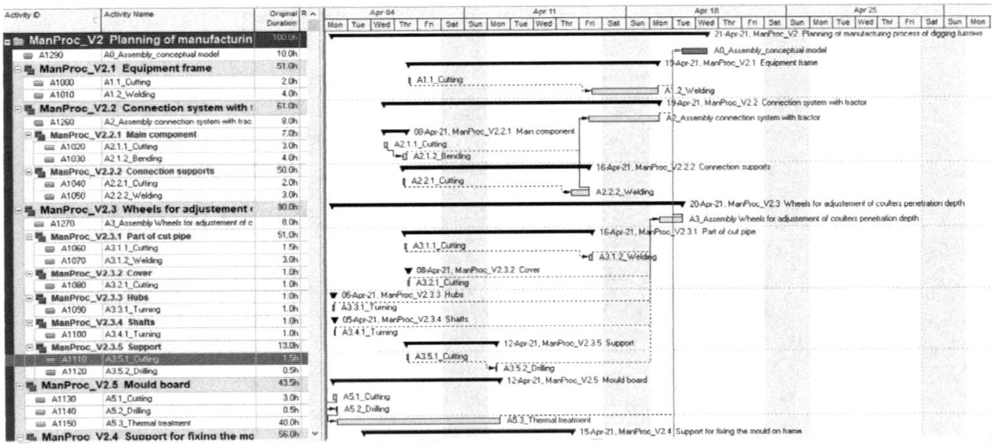

Figure 7. Optimised planning for the second type manufacturing process (partial view).

3. Results

The research results evidenced by this paper refer to the optimisation process of a production systems for the manufacturing process of component parts in a customised smart equipment for precision agriculture. The smart innovative equipment will be suitable for digging furrows with controllable depth and geometric shape and will enable the determination of soil's moisture.

Once a new equipment prototype is manufactured, tested and validated, it will be produced in series. For this it is necessary planning, programming and optimization of the production process so that to obtain a competitive product on the market through quality and price.

There were analysed two types of manufacturing processes, the "old" one (already in use at the producer) and the "new" one (intended to be applied by new investments).

The total time of "old" production process is 70 working hours and the total manufacturing cost is estimated about 12,229 euro. The planning of "new" manufacturing process resulted in the estimation of 10,663 euro total cost and 65 working hours. These mentioned costs include the human resource costs and equipment's' operating costs while cost of material approximately the same for any of the production planning cases considered.

From the point of view of equipment, the "old" manufacturing processes use machine-tools, like, drilling and turning machines, while for the "new" manufacturing process, an increase of efficiency is to be obtained by replacing these two types of machine-tools with a high productivity CNC machine. For doing that it is necessary to have a good investment analyse by feasibility study because sometimes the cost of replacement cannot be fully recovered in reasonable period of time (5 - 8 years).Comparing the two analysed situations, it resulted that in the case of using more efficient equipment, the production process is more efficient even if the cost related to the manufacturing process is higher.

Even in some case the using of some efficient equipment's, the production costs are higher, overall, for a large series or mass production the production cost is amortized.

4. Conclusions

By optimizing the production process it is identified, a great potential for reducing the production cycle, increasing the degree of automation, making high-tech products in limited series and also an efficient use of many production data that so far were not exploited.

In order to optimize the production process, the following parameters will be considered:

- process efficiency (through high decision-making capacity, due to the relatively small number of employees and high level of automation);
- process agility (by knowing in detail the specifications product);
- innovation (production line must be adapted to series production as well as unique production);
- customer experience and requirements (availability of information in the production process provides the customer with a special service system, going in some cases to self-service);
- production costs (the initial investment is high, but the costs will be amortized, by making quality products, without scrap, without material losses, with reduced personnel costs due to the automation of the entire process).

5. References

[1] Šatanová A, Figuli L, Mariana Sedliačiková M, 2015, Optimization of Production Process through Selected Statistical Methods *Procedia Economics and Finance* **23** 959-963 doi.org/10.1016/S2212-5671(15)00494-3, accessed in 14 March 2021.

[2] Cupiał M, Zbigniew Kowalczyk Z 2020 Optimization of Selection of the Machinery Park in Sustainable Agriculture *Sustainability* **12** 1380 doi:10.3390/su12041380.

[3] Pandelea M, Vladareanu L, Crețu D, Iliescu M, Studiu Privind Importanța şi Rolul Inteligenței Artificiale în Agricultura de Precizie, *Revista de Lucrări Ştiinţifice Management şi Dezvoltare Rurală*, seria I, vol. IV, 2018, pp.7-16, ISSN 2457-4449

[4] Bochtis, D., Sorensen, C.A.G., Kateris, D., 2018 *Operations Management in Agriculture*, 1st ed.; Academic Press: Cambridge, MA, USA; Elsevier: Amsterdam, The Netherlands.

[5] Filip M, Zoubek T, Bumbalek R, Cerny P, Batista C E, Olsan P , Bartos P, Kriz P , Xiao M 4, Dolan A, Findura P 2020 Advanced Computational Methods for Agriculture Machinery Movement Optimization with Applications in Sugarcane Production *Agriculture* **10**(10) 10.3390/agriculture10100434.

[6] Rosu M M, Vladareanu L, and Iliescu M 2021 Analysis of Scheduling Activities in the Processes of Improving Surface Characteristics of Parts *CoSME'20 Proceedings, IOP Conference Series: Materials Science and Engineering (MSE)* **I1009** 012053 doi:10.1088/1757-899X/1009/1/012053.

[7] Roşu, M. M., Tarbă I. C., M., Tiriplică P.G., 2018 Methodology for Budget Projects Monitoring, *International Conference Modern Technologies in Industrial Engineering, ModTech, IOP Publishing, Book series: IOP Conference Series-Materials Science and Engineering* **400** Constanta, Romania 062026 doi:10.1088/1757-899X/400/6/062026 ISSN: 1757-8981.

[8] Tiriplică P.G., Roşu, M. M., Tarbă I.C., M., Ghionea, I. G. Scenario Method for Product Cost and Price Variation Analysis, *Proceeding of 31st International Business Information Management Association Conference, IBIMA*, 25-26 April 2018, Milan, Italy, ISBN 978-0-9998551-0-2, pg. 4033-4042, Editor Khalid S. Soliman.

Noninvasive evaluation of elastic properties for magnesium-calcium biodegradable alloys

A Savin[1], R Steigmann[1], M L Craus[1,2], M D Stanciu[3], B Istrate[4] and C Munteanu[4]

[1]National Institute of R&D for Technical Physics, Nondestructive Testing Department, 47 D.Mangeron blvd., 700050, Iasi, Romania
[2] Joint Institute of Nuclear Research, Frank Laboratory for Neutron Physics, 6 Joliot-Curie St, 141980, Dubna, Russia
[3]Transilvania University of Brasov, Faculty of Mechanical Engineering, 29 Eroilor blvd., 500036,Brasov, Romania
[4]Gh.Asachi Technical University, Faculty of Mechanical Engineering, 67 D.Mangeron blvd., 700050, Iasi, Romania
Iasi, Romania

E-mail: asavin@phys-iasi.ro

Abstract. Bone injuries can be fixed and stabilized using degradable implants from magnesium - calcium alloys. Mg and Ca being natural elements in human body, they are easiest to be absorbed without causing toxicity problems. The aim of the paper is to present the effect of Gd and Mn addition in 0.5wt.% in order to improve elastic properties of Mg-$Ca_{0.5}$-$[E]_x$ (E =Mn or Gd) and possibility to be used as biodegradable materials. The addition of Mn / Gd in the binary MgCa alloy modify the structural properties of the alloy (analyzed as elastic properties, respectively crystallographic structure). The measurements and the results calculus were carried out using the ultrasound methods and the Resonant Ultrasound Spectroscopy.

1. Introduction

The human body can selfheal but in a limited measure, when the bones and surrounding tissues are damaged due to illness, breaks, injuries. Thus, the healing is supported by prostheses from biomaterials to increase the chance of good recovery.

The most used biomaterials are Mg based alloys, Mg being a natural element of the human body and it has an essential role in metabolism [1]. Mg alloys as biodegradable implant has as advantages lower density 1.738g/cm^3, closed with cortical bone, protection against stress in orthopaedical implants, a good biocompatibility and non-toxicity [2], good degradation beneficial for temporary exposure to an implant. At the same time, disadvantages such as the low elasticity (sometimes beneficial in stress protection to avoid inducing deformities) should be mentioned; high rate of degradation (time of degradation must be correlated with the process of bone reconstruction) [3], the release of hydrogen may accumulate in the soft tissues around the implant [4].

The corrosion rate represents a constraint, being a major bottleneck in using Mg alloys due to the generated sub-products (hydrogen, OH- , hydroxide anion) [5,6].

The control of physical-chemical properties of Mg based alloy by doping with Ca, Y, Cu, Mn, Zr, Gd can lead to the improvement of mechanical properties by refining grains and intermetallic phases, and, also, can control the degradation speed [7] (limited due to the solubility of elements in Mg). Studies over $MgCa_x$ alloys with different Ca content show that the maximum solubility of Ca in Mg on the phase diagram is reduced [8,9] and increasing the Ca content make alloy to be brittle [10]. The influence of Ca content over the mechanical properties of MgCa binary alloys as well as over the electrochemical properties of alloys were well studied [11,12], trying to obtain microstructures with compromises between mechanical properties and degradability kinetics.

Addition of Mn to MgCa alloys helps to modify the morphology, form intermetallic phases which remove iron and other heavy metals in intermetallic compounds, leading to the increase of the corrosion, stress and creep strength [13] for a Ca concentration below 1wt.% [2]. Mn forms a protective layer during oxidation of Mg [14]. Mn is a micronutrient with a significant role in the metabolic cycle of amino acids but which can become neurotoxic when it is high in the alloy or overlaps with other elements in nutrition [15, 16]. Gd additions increase the strength of Mg alloys [17].

Variable content of E=Mn/Gd [x = 0.5; 1.0; 1.5; 2.0; 3.0 wt. %] is used to improve elastic properties (i.e. Young modulus E; shear modulus G and Poisson ratio ν) of MgCa alloy.

The paper analysis the biodegradable $MgCa_{0.5}(E)_x$ alloy where E is Mn or Gd and x=0.5 wt. %, using non-invasive methods such as US and RUS and microstructural properties in order to promote this alloy in stabilizing bone fractures.

2. Materials and methods

The ingots of $MgCa_{0.5}E_{0.5}$ alloys were obtained following the procedures developed by Faculty of Materials Science and Engineering, Gheorghe Asachi Technical University Iasi [18]. The $MgCa_{0.5}Mn_{0.5}$ alloys were obtained from commercially pure Mg (99.7 wt.%) and Mg-15wt.% Ca (99%) and Mg-3wt.%Mn (98.9%) master alloys [19] and for $MgCa_{0.5}Gd_{0.5}$ have been selected with high purity elements as raw materials Mg (99.7 wt.%) and Mg-15wt.% Ca (99%) and Mg-25wt.% Gd master alloys. The alloys were melted in an induction furnace under argon atmosphere in a graphite crucible as the method described in [19]. Samples with 10x10x10 mm were cut from mini-ingot without inducing flaws or thermal/mechanical stresses.

The microstructure of samples was analyzed using AxioCam MRc5-Zeiss. The final chemical composition $MgCa_{0.5}(Mn/Gd)_{0.5}$ was analyzed by scanning electron microscopy (SEM FEI Quanta 200 3D, dual beam, equipped with energy dispersive X-Ray spectroscopy analysis unit—Xflash Bruker, Harvard, MA, USA) [20]. The phase composition, the types of crystal structure bulk samples, microstructural characteristics were obtained using a conventional X-ray diffractometer (Xpert Pro MPD PANalytical diffractometer (Phillips) with CuKα radiation (figure 1).

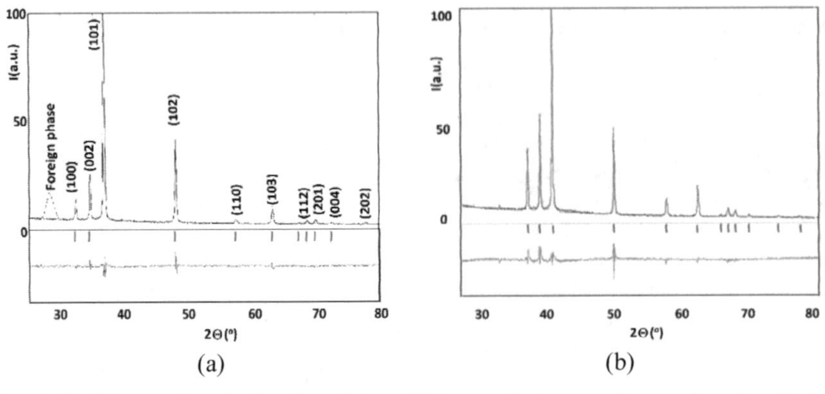

Figure 1. Observed, calculated and the difference between observed and calculated diffractograms for a) $MgCa_{0.5}Gd_{0.5}$; b) $MgCa_{0.5}Mn_{0.5}$.

The phase composition as well as type of crystal structure of crystallographic phase were determined by conventional X-ray diffractometers (Xpert Pro MPD PANalytical diffractometer (Phillips) with CoKα radiation and a BRUCKER AXS D8- Advance diffractometer) with Cu-Kα radiation. The tests of texture of the $MgCa_{0.5}Mn_{0.5}$ alloys were performed by using PowderCell program.

In order to determine the elastic properties of the samples, the ultrasound (US) method has been employed. The values of Poisson ratio, Young modulus, Shear modulus were calculated using two types of waves propagating in elastic medium, longitudinal c_l and transversal c_t.

The method uses pulse-echo technique with a delay line [21], required due to small dimensions of the samples. Send receiver US transducers were used for measurements, G5KB – GE for longitudinal waves, respectively MB4Y GE for transversal waves. The signals were generated using PR 5077 Pulser Receiver – Panamaterics, using the method described in [22]. The results are presented in table 1 with references to binary alloy $MgCa_{0.5}$.

Table 1. Mechanical properties for studied samples.

No.	Composition	Density [Kg/m^3]	Young modulus [GPa]	Shear modulus [GPa]	Poisson ratio	C_l [m/s]	C_t [m/s]
#1	$MgCa_{0.5}$	1680	37.7	14.42	0.30	5555	2930
#2	$MgCa_{0.5}Mn_{0.5}$	1612	35.29	13.38	0.317	5584	2881
#3	$MgCa_{0.5}Gd_{0.5}$	1567	41.43	15.74	0.31	6105	3170

The RUS technique implies the scanning of resonance structure of a compact specimen, in order to determine the elastic properties. The information obtained from the spectra includes both the total matrix of density as well as information about geometry of the object and the homogeneity of the material. This procedure is used to distinct acceptable components from those with flaws as voids, density defects, modification of elastic properties that are reflected in the modification of resonance frequency [23]. The resonance frequencies of an object can be calculated by solving the equation for known geometry [24] or can be measured.

The Rayleigh-Ritz method is efficient and accurate to calculate natural vibration frequencies of the solids with parallelepipedal shapes. The efficiency is important for probabilistic formulation of inverse problem, which implies thousands of direct calculations of natural frequencies. Having the value of G, for a Poisson coefficient ν between 0.28 and 0.31, the C_{11} and C_{12} can be determined [25].

For each value of Poisson coefficient determined by ultrasound, the frequencies of resonance spectrum were determined by numerical simulations, selecting the value of coefficient which minimize the function [26]

$$F = \sum_{i=1}^{N} w_i \left(f_i^c - f_i^m \right)^2 \quad (1)$$

The sample is placed between two identical piezoelectric transducers, P111.O.06P3.1 (figure 2), sustained by contact force, without coupling agents. The emission transducer is connected to a sinusoid voltage source with a sweep frequency between 60 si 320kHz in 100 kHz steps using a Network/ Spectrum/ Impedance Analyser (NSIA) type 4395A. The signal generated by NSIA is amplified and applied to the reception transducer that detects the vibration of sample recorded as pair of amplitude-frequency.

Figure 2. Experimental set-up.

The excitation frequency is continuous variated, in the range of firsts two eigenmodes of vibration for the sample. When the excitation frequency is closer to resonance frequency of the sample, this will start resonate and generate vibrations that are translated by amplified deflection, the signals received by the transducers being filtered from noises and false signals.

The theoretical resonance frequencies are calculated for each sample upon a mathematical model of free vibration [27] using the initial estimation for elastic constants. The experimental set-up allows the tuning of parameters so that for each position of the sample (at corners contact) shall assure excitation of a maximum possible number of resonances. The calculation of resonance frequency is an eigenvalue problem [27,28].

3. Results and discussions

The resonance spectra were analyzed in the range of 120-230 kHz, for each sample were detected a considerable number of vibration modes, which were correlated with theoretical predictions for identifying.

Using the mechanical parameters presented in table 1, the eigenfrequencies were determined using Comsol Multiphysics 5.4. The mesh statistics are minimum element quality 0.2127, average element quality 0.6867, tetrahedron 16316, triangle 1536, edge element 120, vertex element 8.

The emphasized results show that the RUS method, complementary to the US, SEM methods, allows the characterization of the material as a whole. It can be observed that a prediction over modification of internal structure from the changes of modes, result expected because dislocations or crystallographic modifications affect the elastic parameters.

The frequencies obtained by calculation, for the optimization of elastic parameters for the sample supposed as perfect homogeneous and isotropic with regular shape (in this case cube), are found in the frequencies interval of the spectrum experimentally determined. The calculated frequencies have not all correspondent in the frequency spectrum, the number of calculated values being larger than those associated to the spectrum.

The RUS spectrum has been processed using the inverse numerical procedure described in [26]. For these materials, the elastic constants were determined.

The paper offers a method for identification of adequate modes for cubic samples, useful for interpretation of RUS results. The modal shapes obtained by simulation were associated to the phase composition for the addition elements in the composition of $MgCa_{0.5}$ (E) (where E=Mn/Gd - 0.5 wt. %).

Figure 3 presents the results of simulation for sample #1 and #2 used for identifying of resonance frequency.

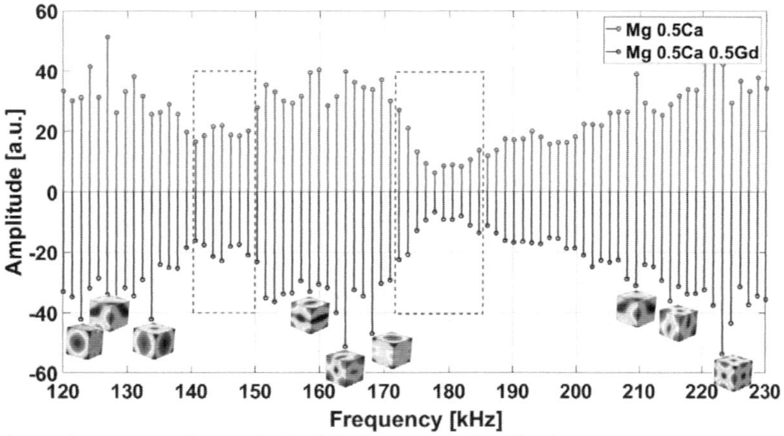

* the colors from cubes represent the magnitude of displacement during vibration

Figure 3. A selected interval of the RUS spectrum of a magnesium alloy with Gd addition.

For representative modes resonant peaks, the modal shape obtained by the RUS the corresponding modal shapes determined within the inversion are normalized to the fundamental torsional frequency. Identification of modes was done as follow torsion doublet d_1, shear s_1 and flexural a_1.

The answer has been recorded in a large number of points to cover the full frequency spectrum range. The frequency sweep has included a large number of excitation frequencies. According to [29] the modes observed on samples must falls into the following three categories: torsional axisymmetric; extensional and flexural modes (these modes occur in pairs named doublets, with the same resonance frequency).

Despite these, a number of resonant frequencies that not emphasize the influence of elements from the witness sample can be observed (i.e. the interval 140-150kHz, respectively 170-185kHz). The representative modes obtained by simulation for the cubic samples are for ν=0.31, in agreement with literature. The missing of vibrations as well as the minimal displacements appears in high blue, the maxima in red and the intermediated ones in yellow or green.

The torsional modes are those having maxima on corners and lack of vibration displacements along the rotation axis and on the middle of the surface between two edges. The modes were identified by the difference as response to the applied excitation, slight deviations from the cubic shape are possible and they bring a variation of less than 5% in the resonant frequency of the first ten modes. Small deviations do not interfere with interpretation.

From the graph it can be seen that the simulation of information is very important in determining the resonances in the investigated spectrum. The modes were visually identified by analyzing the shape of the mode predicted by nodal eigenvectors [30]. Initially, the frequency change analysis has included over 18 modes. For several reasons, it has been established that in the frequency range, 9 modes provide the best information on the samples.

First of all, on the homogeneity, there were 5 relatively large changes of frequencies in the spectrum; keeping a small error percentage (approx. 2% for most samples) compared to the frequency determined by simulation and RUS. Many higher order modes have similar mode shapes and tend to change in appearance as the frequency increases, making it inadequate to identify the mode in the frequency spectrum.

Figure 4 presents the results of simulation for sample #1 and #3 used for identifying of resonance frequency.

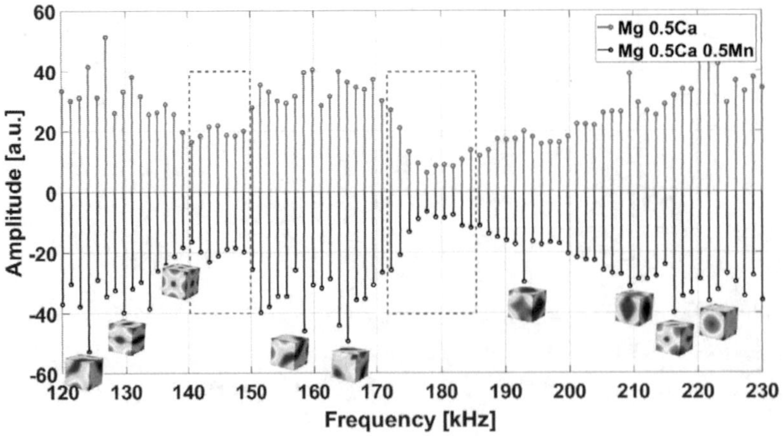

Figure 4. A selected interval of the RUS spectrum of a magnesium alloy with manganese addition.

For these materials, the elastic constants were determined. Several oscillation modes have been identified (i.e. extensional, flexural and torsional - the frequencies of these modes depending by the shear velocity of the sample).

In both resonance spectra, the superior modes present a similarly variation but with changing of the phase. The modal shapes obtained by simulation were associated to the phase composition for the addition elements in the composition of $Mg_{0.5}Ca(E)$ (where E =Mn or Gd in 0.5 wt. %).

However, for each sample it should be noted that in some ways, the measured data followed the same trend as those obtained by simulation, compensated by a few percent.

4. Conclusions

Resonance Ultrasound Spectroscopy is an effective tool for identifying the elastic modules of a material. The method is based on estimating resonant frequencies, based on an eigenvalue and eigenfunction method. The analysis of the resonance spectra can be used to determine the acceptance criteria of the samples without flaws (pores, inclusions, casting defects, etc.).

The method was applied to $MgCa_{0.5}E_{0.5}$ alloys where E = Gd or Mn knowing that the introduction of Gd / Mn as alloying elements in the binary alloy MgCa alloy determines the modification of the crystallographic structure of the basic matrix. Inhomogeneities can be identified by resonant frequency shift, peak splitting, amplitude change and changes in quality factor. RUS spectrum changes both in amplitude and in phase in the presence of discontinuities.

5. References

[1] Zhou P and Gong HR 2012 *J. Mech. Behav. Biomed. Mater* **8** 154.
[2] Radha R and Sreekanth D 2017 *J. Magnesium Alloys*, **5**(3) 286.
[3] Kraus T, Fischerauer SF, Hänzi AC, Uggowitzer PJ, Löffler JF and Weinberg AM *2012 Actabiomaterialia*, **8**(3) 1230.
[4] Avedesian MM and Baker H eds 1999 *ASM specialty handbook: magnesium and magnesium alloys*. ASM international.
[5] Farraro KF, Kim KE, Woo SL, Flowers JR and McCullough MB 2014 *J. Biomech.* **47**(9) 1979.
[6] Zhao D, Witte F, Lu F, Wang J, Li J and Qin L 2017 *Biomaterials* **112** 287.
[7] Bahmani A, Arthanari S and Shin KS 2020*J. Magnesium Alloys* **8**(1) 134.

[8] Handbook, A.S.M., 1992. volume 3, Alloy Phase Diagrams, ASM International, ed. The Materials Information Company.
[9] Wang W, Han J, Yang X, Li M, Wan P, Tan L, Zhang Y and Yang K 2016 *Mater. Sci. Eng. B*, **214** 26.
[10] Li Z, Gu X, Lou S and Zheng Y 2008 *Biomaterials* **29**(10) 1329.
[11] Salahshoor M and Guo Y 2012 *Materials* **5**(1) 135.
[12] Chen Q and Thouas GA 2015 *Mater. Sci. Eng. R* **87** 1.
[13] Kwas E, Schmidt J, Kunert M and Thein S 2017 *Surf. Innovations* **5**(4) 221.
[14] Gröbner J, Mirkovic D, Ohno M and Schmid-Fetzer R 2005 *J. Phase Equilib. Diffus.* **26**(3) 234
[15] Bowman AB, Kwakye GF, Hernández EH and Aschner M 2011 *J. Trace Elem. Med. Biol.* **25**(4) 191.
[16] Walker J, Shadanbaz S, Woodfield TB, Staiger MP and Dias GJ 2014. *J. Biomed. Mater. Res. Part B: Applied Biomaterials* **102**(6)1316.
[17] Gao L, Chen RS and Han EH 2009 *J. Alloys Compd.* 481(1-2), pp.379-384.
[18] Lupescu S, Munteanu C, Istrate B, Luca D, Benchea M, Mahu G 2018 *IOP Conf. Ser.: Mater. Sci. Eng.* **444** 3 032008.
[19] Istrate B, Munteanu C, Lupescu S, Chelariu R, Vlad MD and Vizureanu P 2020 *Materials* **13**(14) 3082.
[20] Savin A, Novy F, Craus ML, Iftimie N, Steigmann R, Istrate B, Munteanu C 2020 *IOP Conf. Ser:Mater. Sci. Eng.* **916** (1) 012099. IOP Publishing.
[21] Krautkrämer J, Krautkrämer H 1990 *Ultrasonic Testing of Materials*. 4th Ed. Springer-Verlag, Berlin 167.
[22] Savin A, Craus ML, Turchenko V, Bruma A, Dubos PA, Malo S, Konstantinova TE, Burkhovetsky VV 2015 *Appl. Sci.-Basel* **5**(4) 1665.
[23] ASTM E2001-18, Standard Guide for Resonant Ultrasound Spectroscopy for Defect Detection in Both Metallic and Non-metallic Parts.
[24] Ren F, Case ED, Morrison A, Tafesse M, Baumann MJ. 2009. Philos. Mag. **89** 1163.
[25] Lemaitre J, Chaboche JL 1994 *Mechanics of solid materials*. Cambridge University Press.
[26] Zadler BJ, Le Rousseau JH, Scales JA and Smith ML 2004 *Geophys. J. Int.* **156**(1) 154.
[27] Migliori A, Sarrao JL, Visscher WM, Bell TM, Lei M, Fisk Z, Leisure RG 1993 Physica B: Condensed Matter. **183**(1-2) 1.
[28] Balakirev FF, Ennaceur SM, Migliori RJ, Maiorov B, Migliori A 2019 *Rev Sci Instrum* **90**(12) 121401.
[29] Myasnikov DV, Konyashkin AV, Ryabushkin OA 2010 *Tech. Phys. Lett.* **36**(7) 632.
[30] Yaoita A, Adachi T and Yamaji A 2005 *NDT & E Int.* **38**(7) 554.

Acknowledgments
This work was supported by the Ministry of Research, Innovation and Digitization, Romania under CCCDI-UEFISCDI, project number PNCDI III-P1-1.2-PCCDI-207-03239/60PCCDI 2018—MedicalMetMat and Nucleus Program MAGNA2022, Project PN 19 28 01 02.

IOP Conf. Series: Materials Science and Engineering 1182 (2021) 012069 doi:10.1088/1757-899X/1182/1/012069

An alternative for wind energy conversion using improved Savonius rotor turbine model

F D Scheaua
Dunarea de Jos University of Galati, Engineering and Agronomy Faculty of Braila, Department Engineering Sciences and Management, MECMET Research Center, 29 Calarasilor Street, Braila, Romania

E-mail: fanel.scheaua@ugal.ro

Abstract: The current society development is dependent of the planet energy resources and if we take an overview on renewable energies used worldwide it can be said that they are constantly gaining ground. This trend represents an important step forward towards maintaining a clean environment by focusing on capitalizing considerable resources that are sustainable over time, such as solar energy, together with waves and wind force. Regarding the energy production solution based on the atmospheric air masses movements or wind energy, the alternative to the constructive model of horizontal axis turbines (HAWT), currently widely used in wind farms is presented the alternative of using vertical shaft turbines (VAWT) for reduced installed power wind farms. The aspects that form the advantages of these turbines are presented as well as the fact that although they offer a lower power coefficient, they can play an important role for smaller power farms, the investment being at lower values. The basic rotor construction model is based on the SAVONIUS type which has been continuously improved, the results being better in terms of energy performance. A rotor concept that can be used within wind turbines with better performance results compared to SAVONIUS classic model is presented in this paper. The novelty elements consist in the use of a larger number of blades as well as the placing method on the rotor basic circle profile. The numerical analysis results made on virtual rotor model are presented, which highlight the improved values obtained on the analysed model.

1. Introduction

At present, solutions related to the energy production from renewable sources that do not have a negative impact on environmental conditions are being pursued all over the world, aiming at reducing the fossil fuels consumption and avoiding the CO2 emissions into the atmospheric air.

For this desideratum, wind turbines are used to capture wind force and produce energy, in areas where there is potential in terms of speed and constant wind action over time.

The mainly used turbine types to harvest wind power are represented by horizontal axis turbines (HAWT) with high installed power in the range of 1.8-5 MW with large diameter rotors ranging of 40 to 100 m, placed in towers at 80 m height.

As alternatives to the horizontal axis turbines models that offer the best performance results, the variants of vertical axis turbines (VAWT) are proposed which benefit from lower performance parameters in order of a few kW, but compensate by aspects related to simpler construction, low-cost price, but also with work and start possibilities at lower wind speeds without the need for a wind direction orientation system. These vertical axis turbine types can take air currents from any direction and even eddy currents, benefit from simple maintenance and has a longer service life [1-4].

For these turbine types the simplest rotor model is the SAVONIUS rotor type, presented in the 1920s, which has maximum efficiency values in the range of 20% compared with three-bladed turbines with horizontal axis of 45%.

Over time, multiple studies have been conducted to change the basic geometry of this type of rotor, and the results indicate an increase in energy performance in terms of power coefficient from 0.1 to 0.2 [5-10].

2. The SAVONIUS turbine rotor model assembly

The SAVONIUS type rotor model is presented schematically to identify the main geometric parameters related to the construction and influence in operation (figure 1) [11-13].

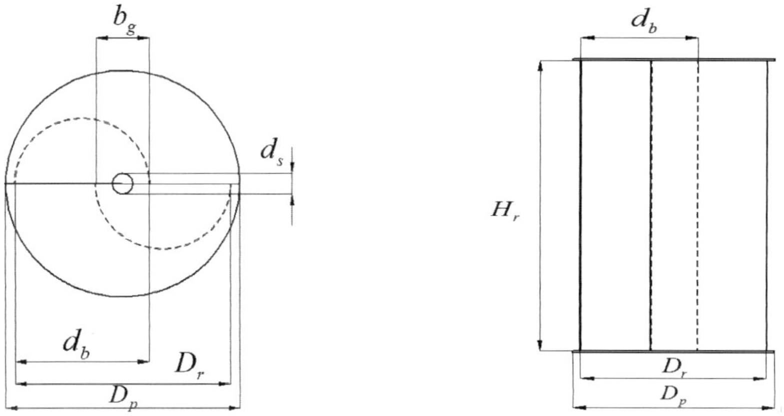

Figure 1. Schematically representation of SAVONIUS rotor [11-13].

The energy performance of this rotor type depends on the overall dimensions by the ratio between the diameter and the height or aspect ratio ($\alpha = \frac{H_r}{D_r}$), the blades placement distance or blade gap $\left(b_g\right)$, but also on the blade arrangement on the rotor base circle or overlap ratio ($\beta = \frac{b_g}{d_s}$).

The wind force can provide an available power that can be calculated according to the relation [11-13]:

$$N_W = \frac{\rho_a A_r v^3}{2} \tag{1}$$

The efficiency parameters for the SAVONIUS rotor involve the appearance of a rotational movement $\left(\omega_r\right)$ with a certain speed value *(n)*, which leads to the appearance of a torque at the rotor axis $\left(T_r\right)$. Based on these values, the available torque $\left(C_t\right)$ and power coefficient $\left(C_N\right)$ can be calculated that can be obtained with the respective rotor model [11-13].

$$\omega_r = \frac{2\pi n}{60} \tag{2}$$

$$C_t = \frac{4T_r}{\rho v^2 D_r H_r} \tag{3}$$

$$C_N = \frac{2T_r \omega}{\rho_a v^3 D_r H_r} \tag{4}$$

Regarding the rotor efficiency, the tip speed ratio is defined (λ), which represents the ratio between the tangential speed of the blade tip and the wind speed, being the parameter to which the performance values of a wind turbine rotor are related [11-13].

$$\lambda = \frac{\omega \cdot D_r}{2 \cdot v} \tag{5}$$

3. Improved solutions for the SAVONIUS rotor

Some improved solutions for the SAVONIUS rotor are presented for which the blade geometry modification. Based on the results obtained from numerical and experimental analyzes, it was found that the blade basic geometry can considerably influence the rotor performance parameters.

Such improved solutions are the BACH, BENESH and SIVASEGARAM models whose rotor profile is shown in figure 2 [14-18].

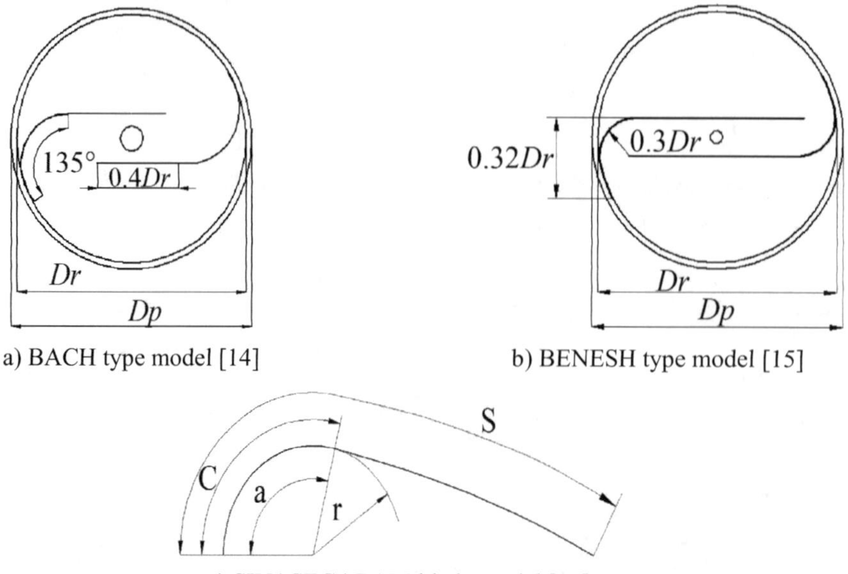

a) BACH type model [14]   b) BENESH type model [15]

c) SIVASEGARAM blade model [16]

Figure 2. Improved solutions for the blade rotor.

The improved rotor models of BACH and BENESH type were tested by Roy and Saha and the obtained results show an improved energy performance in terms of power coefficient compared to the classic SAVONIUS model (table 1) [17,18].

For SIVASEGARAM blade model, power coefficient values of up to 0.2 are presented by Chen et al. [19].

Table 1. Power coefficient values for improved rotor models.

Rotor model	Power Coefficient
BENESH	0.29
BACH improved	0.3
SIVASEGARAM	0.2

4. The innovative solution for improved SAVONIUS rotor

A rotor model with several circular blades is presented, which has the possibility to better take over the air flow coming in the frontal direction. The model is a rotor model with a vertical axis made on the SAVONIUS principle, but with the blades position change on the rotor base circle.

A virtual model was made with the Solid Edge program for this rotor pattern being subjected to a numerical analysis performed with ANSYS CFX for determining the air pressure values at the blades level (figure 3) [20-22].

The model consists of 7 semi-circular blades arranged symmetrically on the base circle. The diameter of the rotor is 148.5 mm, and its height is 198 mm to which are added the end plates with a thickness of 3.96 mm. The rotor with mounted end plates is made to capture the best possible air flow rate from the main circulation flow.

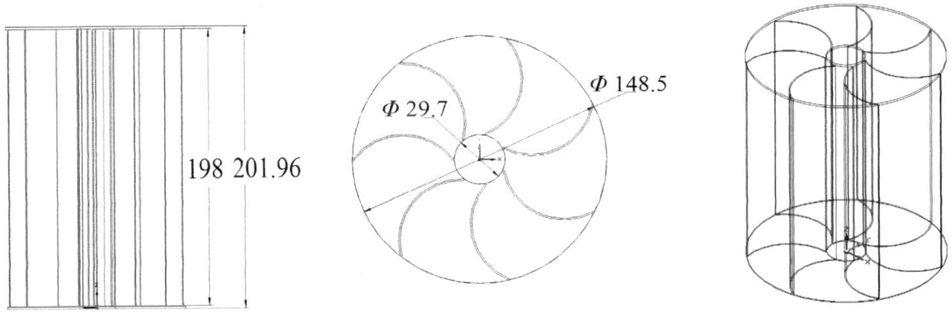

Figure 3. Rotor virtual model (dimensions in mm).

5. Numerical analysis of improved SAVONIUS rotor virtual model

Using the ANSYS CFX program, a numerical analysis was performed on the rotor virtual model assembly to determine the specific values of the pressure at the rotor blades, as well as the air flow velocity in the analysed fluid area [23-28].

The assembly consists of the rotor and the adjacent fluid region in which the rotational movement can be declared.

The mesh network is made with a specific automatic method having 75537 nodes and 50847 tetrahedral elements.

Based on these values, the rotor performance values in terms of torque and available power can be determined.

The wind velocity value has been set to start the calculation and multiple cases had been considered. For case 1 a value of 8 m/s wind velocity was declared at inlet, case 2 with 7 m/s and case 3 for 6 m/s. In the range of velocity values, a minimum value of 2.4 m/s were also considered for case 4.

The working fluid is declared as air at normal pressure, temperature of 25 degrees Celsius.

The results show specific values calculated for total and static pressure and air velocity presented in figures 4-7 for all 4 analysed cases.

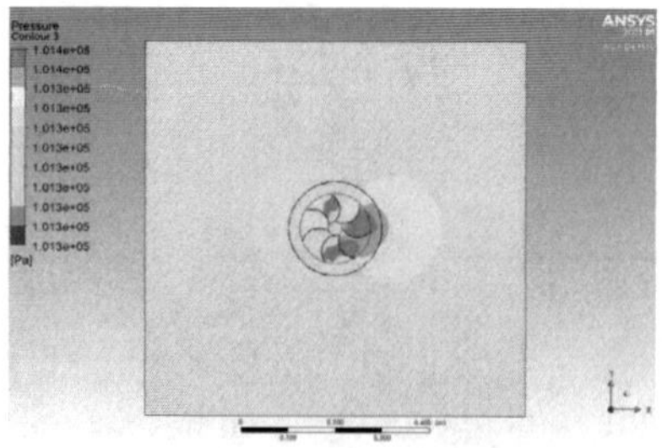

a) pressure values

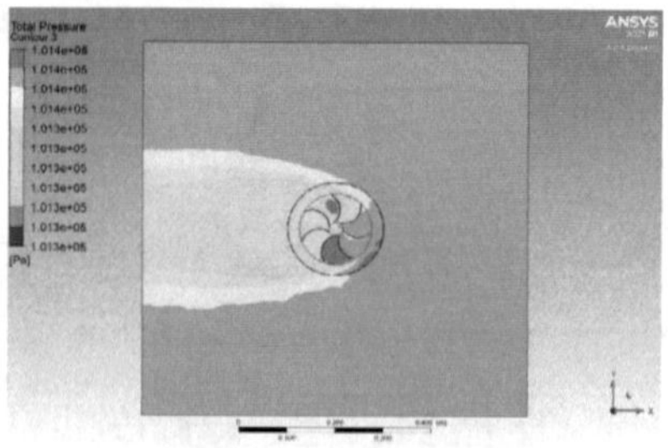

b) total pressure values

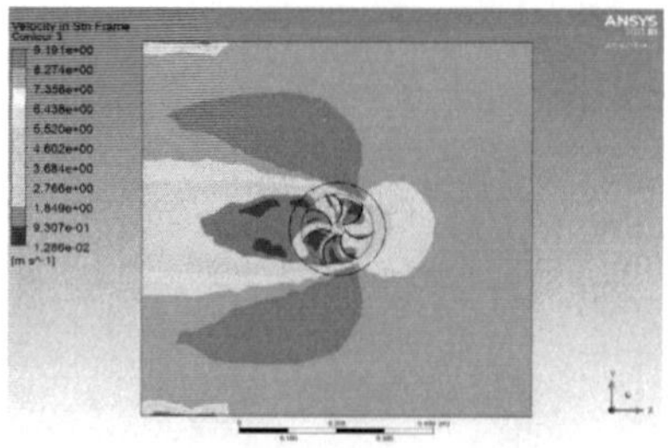

c) Stn frame velocity values

d) velocity vector values

Figure 4. Result values for case 1 (v= 8 m/s).

a) pressure values

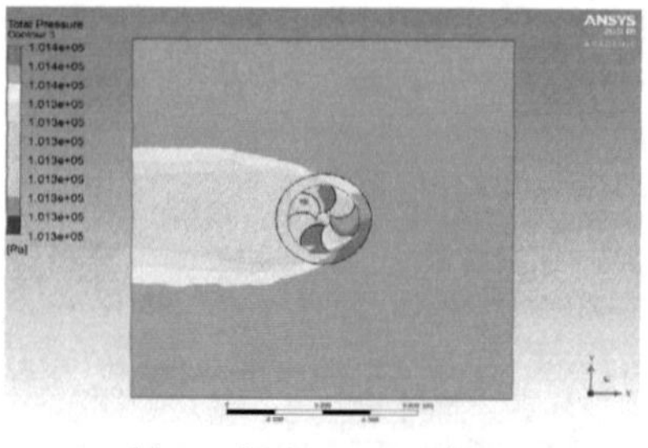

b) total pressure values

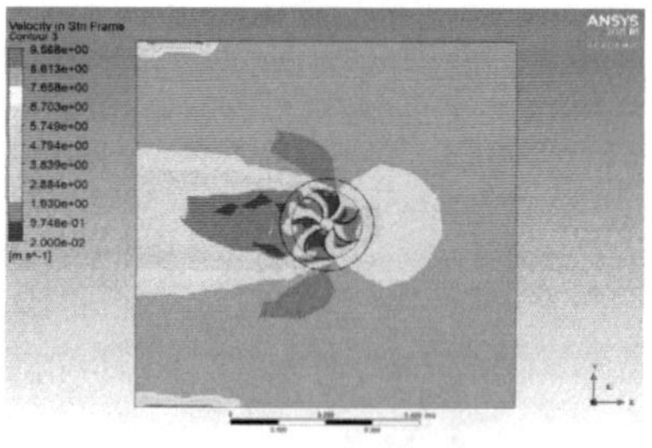

c) Stn frame velocity

d) velocity vector values

Figure 5. Result values for case 2 (v= 7 m/s).

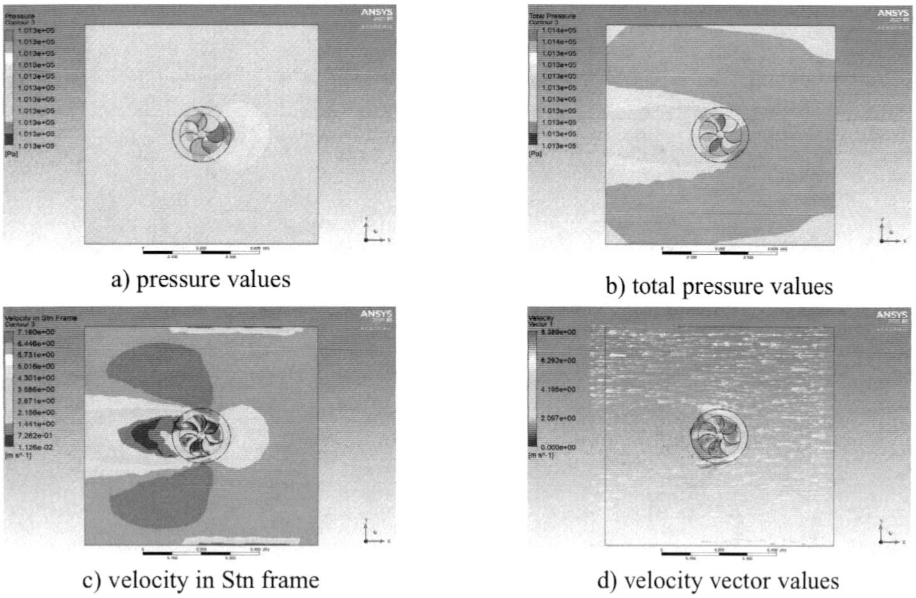

a) pressure values b) total pressure values

c) velocity in Stn frame d) velocity vector values

Figure 6. Result values for case 3 (v= 6 m/s).

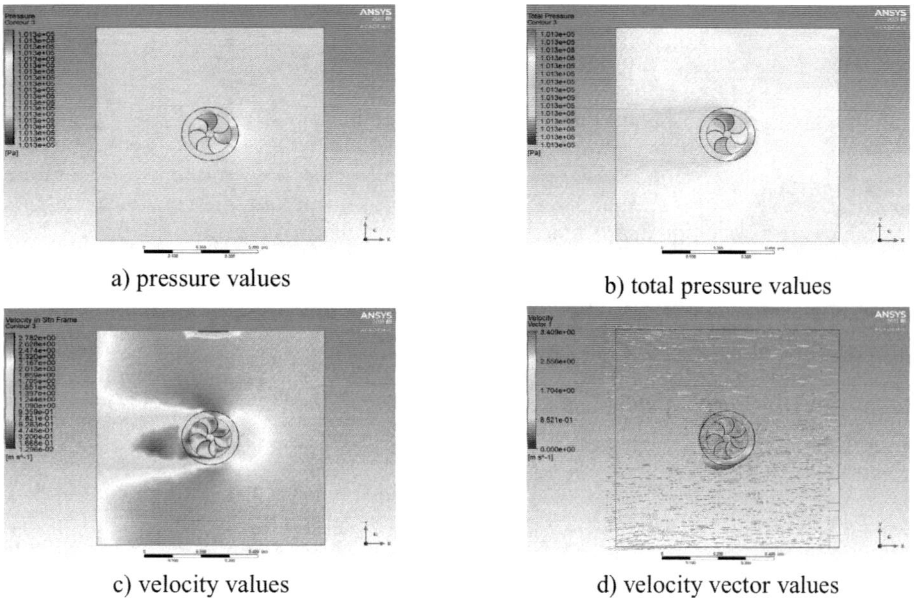

a) pressure values b) total pressure values

c) velocity values d) velocity vector values

Figure 7. Result values for case 4 (v= 2.4 m/s).

From the obtained result values can be observed the air pressure distribution across the rotor model function of initial velocity, the velocity distribution at the rotor and blade region and the turbulences created after the air flow rate is passing downward the rotor region.

The calculated values for torque, torque and power coefficient are presented in table 2.

IOP Conf. Series: Materials Science and Engineering 1182 (2021) 012069 doi:10.1088/1757-899X/1182/1/012069

Table 2. Torque values for analyzed cases.

Case 1 (8 m/s)	Case 2 (7 m/s)	Case 3 (6 m/s)	Case 4 (2.4 m/s)
Torque values (Nm)			
1.86	1.68	1.16	0.34
Torque coefficient (CT)			
0.2	0.19	0.207	0.24
Power coefficient (CP)			
0.11	0.1	0.14	0.29

The diagrams for torque and power coefficient are presented in figure 8.

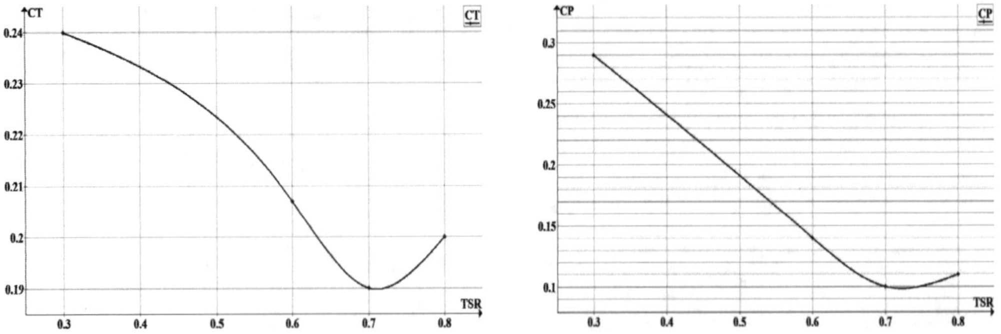

Figure 8. Torque coefficient values.

Based on the obtained results, it can be observed that the presented rotor model works best at low wind speeds, because case 4 obtained the best results in terms of torque and power coefficient. In this case it is achieved an operation improvement. Also, good values are obtained by the case 3 in terms of performance efficiency, the other cases having values close to the SAVONIUS classic rotor model.

6. Conclusions

The aspects related to the construction and operation of the SAVONIUS rotor model are presented in this paper. Based on the classic model, a modification of the initial classic constructive solution had been made and some optimization results were obtained.

An improved rotor constructive version is presented, obtained by modifying the blade profile and adopting an optimal blades number built starting from the classic SAVONIUS model.

Represents a variant with 7 semicircular blades placed symmetrically on the rotor base circle for which a 3D model was made and analyzed with ANSYS CFX program.

Several cases with different input velocity were considered and the results are presented in terms of pressure, velocity, and turbulence kinetic energy distribution calculated at the rotor model fluid region.

For each analyzed case numerical results are presented for rotor torque used forward to determine the specific values of the torque and power coefficients according with tip speed ratio values.

It was found that the case of lower velocity values declared at the rotor inlet the values are better for torque and power coefficient. It means that this constructive rotor type model is intended to be used for lower wind velocities areas with good performance results.

The obtained results are better comparative with SAVONIUS classic rotor model for analyzed cases 4 and 3.

For the cases where higher air velocity values had been declared the results are comparable with the SAVONIUS model.

The obtained results show the efficiency possibilities of the SAVONIUS type rotor model, based on the modification of the number of blades and their positioning on the basic circle profile.

Going further, a change of the blade geometry profile can be made, but also of their specific positioning so that better operating results can be obtained.

7. References

[1] Zemamou M, Aggour M and Toumi A 2017 Review of Savonius wind turbine design and performance *Energy Procedia* **141** 383-388.

[2] Ali M H 2013 Experimental comparison study for Savonius wind turbine of two & three blades at low wind speed *International Journal of Modern Engineering Research (IJMER)* **3**(5) 2978-2986.

[3] Kacprzak K, Liskiewicz G and Sobczak K 2013 Numerical investigation of conventional and modified Savonius wind turbines *Renewable energy* **60** 578-585.

[4] Wenehenubun F, Saputra A and Sutanto H 2015 An experimental study on the performance of Savonius wind turbines related with the number of blades *Energy procedia* **68** 297-304.

[5] Mahmoud N H, El-Haroun A A, Wahba E and Nasef, M H 2012 An experimental study on improvement of Savonius rotor performance *Alexandria Engineering Journal* **51**(1) 19-25.

[6] Altan B D, Atilgan M, Özdamar A 2008 An experimental study on improvement of a Savonius rotor performance with curtaining. *Experimental thermal and fluid science* **32**(8) 1673-1678.

[7] Ragheb M and Ragheb A M 2011 Wind turbines theory-the Betz equation and optimal rotor tip speed ratio. *Fundamental and advanced topics in wind power* **1**(1) 19-38.

[8] Mansour H and Afify R 2020 Design and 3D CFD Static Performance Study of a Two-Blade IceWind Turbine Energies 13(20) 5356.

[9] Aymane E, Darhmaoui H and Sheikh N 2017 Savonius Vertical Wind Turbine: Design, Simulation, and Physical Testing. School of Sci. & Eng. Univ. Program.

[10] Mao Z, Yang G, Zhang T and Tian W 2020 Aerodynamic Performance Analysis of a Building-Integrated Savonius Turbine Energies 13(10) 2636.

[11] Scheaua F D 2020 Comparative Numerical Analysis on Vertical Wind Turbine Rotor Pattern of Bach and Benesh Type. Energies, 13(9), 2311.

[12] Scheaua F D 2021 Wind energy conversion within agricultural farm using vertical axis turbines of optimized SAVONIUS type. In: IOP Conference Series: Earth and Environmental Science. IOP Publishing p. 012015.

[13] Scheaua F D, Goanta A M and Dragan N 2021 Review of Specific Performance Parameters of Vertical Wind Turbine Rotors Based on the SAVONIUS Type. Energies, 14(7), 1962.

[14] Bach G von. 1931 Untersuchungen über Savonius-rotoren und verwandte strömungsmaschinen Forschung auf dem Gebiet des Ingenieurwesens A, 2(6) 218-231.

[15] Benesh Alvin H 1996 Wind turbine with Savonius-type rotor. U.S. Patent No 5, 494, 407.

[16] Sivasegaram S 1978 An experimental investigation of a class of resistance-type, direction-independent wind turbines Energy 3(1) 23-30.

[17] Roy S and Saha U K 2015 Wind tunnel experiments of a newly developed two-bladed Savonius-style wind turbine *Applied Energy* **137** 117-125.

[18] Saha U K, Thotla S and Maity D 2008 Optimum design configuration of Savonius rotor through wind tunnel experiments Journal of Wind Engineering and Industrial Aerodynamics 96(8-9) 1359-1375.

[19] Chen Liu, Chen Jian, Zhang Zhouzhou 2018 Review of the Savonius rotor's blade profile and its performance *Journal of Renewable and Sustainable Energy* **10**(1) 013306.

[20] Goanta A M and Haraga G 2017 Aspects of modelling classical or synchronous modelling with Solid Edge ST 9 *In MATEC Web of Conferences* **112**, p. 06024 EDP Sciences.

[21] Goanta A M 2012 Comparative study of" revolve" command in the average-class design software taught in the higher education *Annals of "Dunarea de Jos "University of Galati* Fascicle XIV *Mechanical Engineering* **19**(2) 11-16.

[22] Goanta A M 2010 Modern Methods of Calculation and Modeling of The Cylindrical Gears *Proceedings of the 13th International Conference Iasi&Chisinau–ModTech* ISSN pp. 2066-3919.

[23] Kamoji M A, Kedare S B and Prabhu S V 2009 Experimental investigations on single stage modified Savonius rotor *Applied Energy* **86**(7-8) 1064-1073.

[24] Volintiru O N, Scurtu I C and Stefănescu T M 2018 Modeling and optimization of HVAC system for special ships *Journal of Physics: Conference Series* **1122**(1), art. no. 012004.

[25] Scurtu I C and Panaitescu V N 2019 Turbulent Flow Numerical Simulation for Unconventional Propulsion *Revista de chimie*, ISSN 0034-7752, Edition: volume 70, October.

[26] Alaimo A, Esposito A, Milazzo A, Orlando C and Trentacosti F 2013 Slotted blades Savonius wind turbine analysis by CFD *Energies* **6**(12) 6335-6351.

[27] Sobczak K, Obidowski D, Reorowicz P and Marchewka E 2020 Numerical Investigations of the Savonius Turbine with Deformable Blades *Energies* **13**(14) 3717.

[28] Scurtu I C and Oncica V 2018 Combined CFX and Structural Simulation for Bow Thrusters Loading under Operating Conditions *In Journal of Physics: Conference Series* **1122**(1) November 012024, IOP Publishing.

Coupling methods and optimal use of modern anti-seismic insulation systems at building and bridge construction structures

F D Scheaua
Dunarea de Jos University of Galati, Engineering and Agronomy Faculty of Braila, Department Engineering Sciences and Management, MECMET Research Center, 29 Calarasilor Street, Braila, Romania

E-mail: fanel.scheaua@ugal.ro

Abstract: The construction industry has recorded an unprecedented advance in recent years, especially in the world recent developed countries and this involves the increasing development of insulation techniques against the effects of seismic actions that can threaten the structure integrity at a certain time. Most recent achievements in terms of structure insulation systems are successfully used within building and bridge structures in order to ensure the optimal stability level during a potential earthquake, which allows ensuring a proper safety level both for operation and in terms of investments protection. the protection methods of the structures are directed in two directions, namely in achieving the insulation at the base of the structure by strategically positioning the systems at the base of the structure while the second direction is directed at seismic energy consumption or energy dissipation by using dissipative systems mounted on frames the resistance structure of the building. By combining the action of the two types of mechanical systems, good results can be obtained regarding the behaviour of the structure isolated by the earthquake. The paper presents constructive aspects of the mechanical systems used for the seismic insulation of both building and bridge structures and their functional role in the whole structure to which they are mounted. Through the experimental tests made are highlighted the better results in terms of acceleration for the structure isolated with the composed system. Are emphasized in this manner the protective system role in ensuring an improved behaviour for the structure where is mounted at seismic dynamic motions that can occur in time.

1. Introduction

Multiple construction solutions for insulation systems are currently used all over the world to ensure stability conditions of structures against seismic actions.

They are built to protect the structure against the destructive earthquake effects, especially if it is positioned in an area with intense seismic activity and by earthquake operating is ensured an energy dissipation of the total energy amount that appears during the ground movement dictated by the seismic wave.

Depending on the earthquake type that can affect the structural system, the effects can be described as follows:

- rapid dynamic action corresponding to surface earthquakes can cause the deformations increase inside the structural system, increase the structure flexibility and avoid the transient resonance zone.

- slow dynamic action corresponding to intermediate earthquakes can act in the sense of structure flexibility increase until the resonance limit with seismic movement is reached having the effect of structure seismic stresses increasing on the above the load limit, which can cause destruction.

Increasing the safety level of construction structures to dynamic seismic action consists in designing systems so that, when a certain level of structure response intensity become dangerous, changes in its dynamic characteristics are made accompanied by significant energy dissipation amounts.

Conceptually, the methods of protection of structures against dynamic actions were based on the ductility of the structural elements, so that a system could have its own capacity to dissipate the energy received from a certain dynamic action [1-8].

Over time, however, multiple ideas have been developed for the isolation of construction structures which, by introducing special devices inside the resistance structure, manage to change the structure's response during dynamic actions.

These special devices can be insulating systems that use either dry friction force (Coulomb) for energy consumption or plastic deformation of some metals (steel, lead) to achieve the proposed insulation objectives.

The essential role in the application of the insulating method is based on the use of insulators coupled with energy dissipating devices for taking over and consuming the energy induced by the earthquake inside the structure, ensuring the hysteretic damping for the structural system.

2. The isolation systems used for structures

In order to ensure the necessary conditions for achieving optimal insulation of structures against seismic actions, we can start from the change of the structure behaviour during the earthquakes action in the sense of displacements limiting and going further speeds and accelerations of structural elements.

By introducing isolation systems into the composition of a structural system, contributions to the structural rigidity change are made and further the change of its dynamic behaviour.

An increase in the strength rigidity of the structural system may mean placing the system in another spectral zone that may not always be beneficial (figure 1) [9-22].

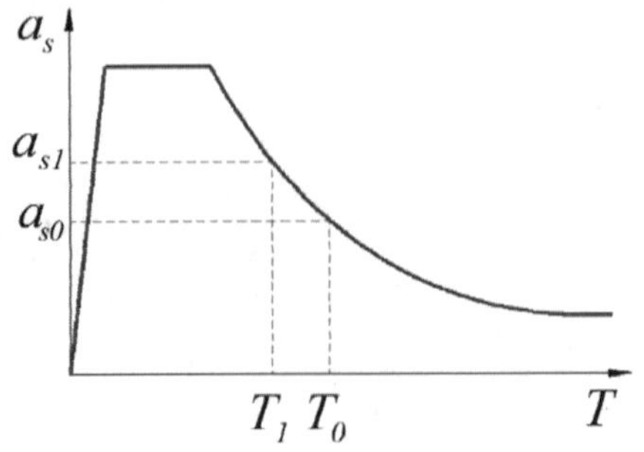

Figure 1. Modification of spectral values for the case of increasing the rigidity of the structural system [10].

Most isolation devices used at structures are considered as passive control systems when operating independently and do not include any other control apparatus.

When a certain structure is equipped with such devices, an improvement of the structural system behavior can be observed in certain ground movements resulting from dynamic actions.

The category of passive insulation systems includes the base insulation systems that can be installed between the foundation and the superstructure of a building.

These protection systems use the dry friction force (Coulomb friction) between two surfaces in permanent contact, while ensuring the possibility of the isolated structure to move freely in the plane of the sliding surface during seismic actions (figure 2).

The main sliding surface can be flat or concave, so that the displacement of the superstructure imposed by the insulating system can have both horizontal plane components *(u)* for the flat surface and the vertical component *(h)* imposed by the concave or spherical sliding surface [9-22].

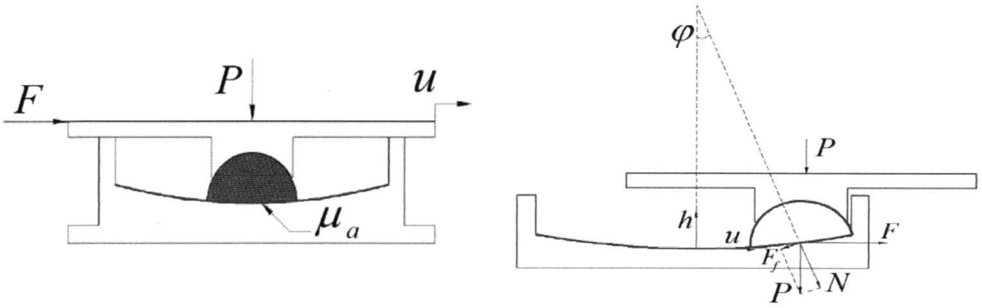

Figure 2. Schematically representation of base isolation system of friction pendulum type [9-22].

At the beginning of the ground seismic movement, the insulating system has the ability to come into action only when the friction force value (F_f) is exceeded by the lateral force *(F)*, until this moment there is an increase in the rigidity of the isolated structure.

The geometric shape of the sliding surface dictates the possible occurrence of the remaining displacements of the superstructure if a flat sliding surface were used, while the spherical sliding surface ensures the superstructure return to the initial position due to the gravitational forces action [20-23].

3. Dissipation systems used for structures

The energy dissipating systems introduced inside a structural system make possible the response reducing during a seismic action, in terms of displacements, speeds and accelerations, due to the increased possibility of the system to dissipate seismic energy.

Figure 3 shows the modification of the structural system response to seismic action by placing on other curves corresponding to the spectral values obtained by using seismic energy dissipators, which increase the possibility of the structure to dissipate the received energy by increasing the fraction of critical damping, (ς). The specific values for the spectral acceleration (a_s) depending on the structure vibration period $(T)$are presented. It can be observed that by using the protection systems it is obtained the structural system placement on other spectral curves of lower acceleration values correlated with modified vibration period values [10].

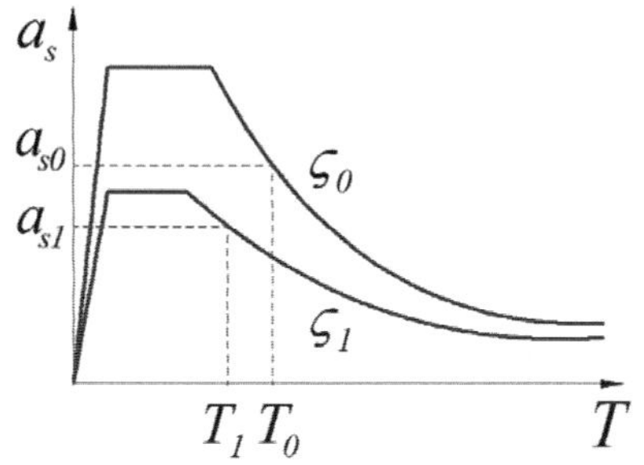

Figure 3. Spectral values mitigation in case of energy dissipators mounted at structures [10].

A constructive solution for dissipative systems is the use of viscous elastic devices that once introduced into a structural system can change the structure response to possible dynamic actions by increasing the system total rigidity.

The principal diagram for a viscous elastic device is shown in figure 4, such a system can achieve a significant energy consumption from the total amount induced by earthquake in the structure.

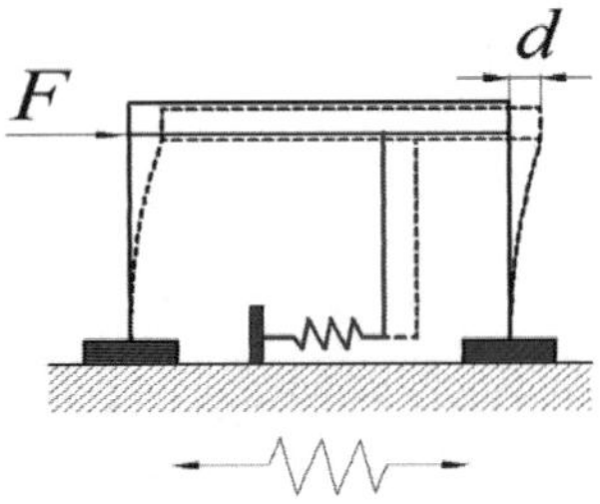

Figure 4. Structural system model equipped with viscous elastic dissipating device [10].

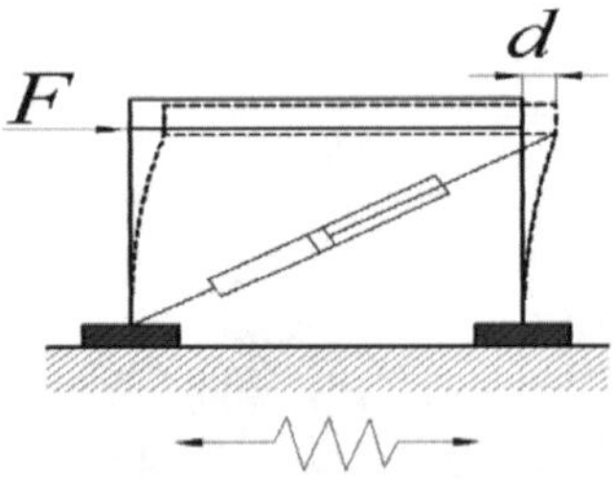

Figure 5. Structural system model equipped with viscous fluid device [10].

Figure 5 shows the mounting solution of a fluid viscous device at a structural system.

Viscous fluid dissipation devices, unlike other systems, do not change the rigidity of the structural system as a whole, they are a viable solution to consider for displacements limitation between the structural frames where they are mounted and dissipate a large amount of earthquake energy by taking it and transformation into caloric energy.

Energy dissipative devices with viscous fluid can also be used in combination with other types of insulating systems to complement their action and obtain a higher stability level for the insulated structure during high intensity seismic actions.

4. Mounting solutions for protective systems at structures

Solutions for coupling to the bridge-type structure of some constructive variants of elastomeric protective systems are presented.

The system allows a specific movement limited by the composite material elastic properties, necessary in the event of an earthquake that could request the structure in a dynamic regime.

The separation between the superstructure and the foundation allows a minimum free movement of the foundation in case of a seismic event, without involving the superstructure that must remain stable, thus aiming to change the structural response to lateral forces that may occur.

The appearance of the seismic movement entrains the foundation ground which registers an acceleration which is transmitted further to the insulating system at the same time.

The structural system seismic response reduction to lateral forces can be obtained by increasing the fundamental period of the structure, by changing the shape of the fundamental vibration mode, by increasing the damping or by combining these effects.

Figure 6 schematically shows the insulation system with elastomeric supports as a solution for the rehabilitation and seismic insulation of a bridge structure [10].

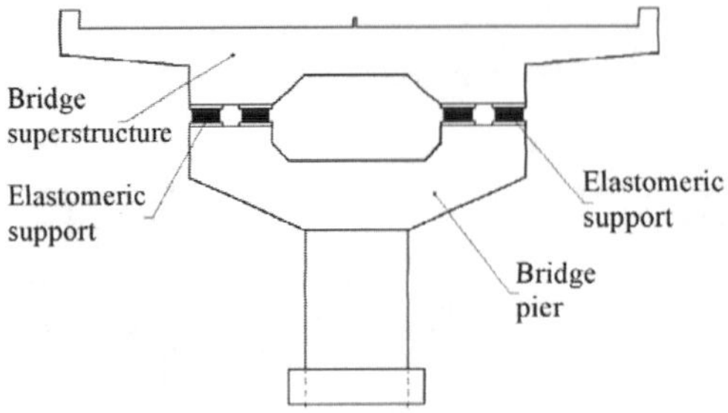

Figure 6. Schematic representation of bridge isolated structure with elastomeric supports [10].

The coupling method of the insulating system with elastomeric supports on the three-dimensional model of the bridge between the superstructure and the support pier transverse beam is presented in figure 7.

The insulating system can take over the vertical loads resulting from the traffic and the efforts coming from the foundation ground as a result of the seismic actions, ensuring horizontal displacements limited by the elastic properties of the composite materials.

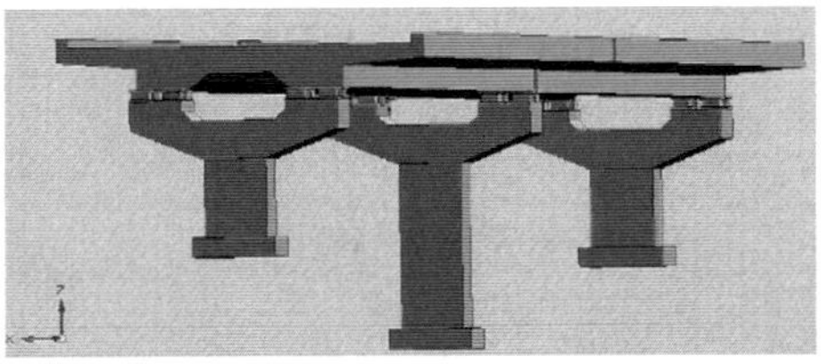

Figure 7. Seismic insulated bridge structure configuration using elastomeric supports [10].

The structure lateral flexibility is modified by placing the insulation system between the structural components and is practically negligible in relation to that of the insulation system.

It can be said that a complete isolation of the structural system is achieved when the bridge superstructure behaves elastically, otherwise it is considered that the structural system is only partially isolated.

A hybrid constructive variant of the isolation system is represented by the use of sliding insulating devices in combination with elastomeric elements. The coupling method is enhancing the use of both dry friction and elastic forces in order to acquire displacements limitation and to shock damping coming as a result of the dynamic actions on a bridge structure.

The resulting assembly has the ability to combine the insulating and dissipating properties of sliding and elastomeric systems, resulting in a composite system that can provide improved damping to stresses acting directly on the structural system, preventing displacement due to inertia of structural components thus ensuring bridge stability.

By using these isolating systems working in tandem is ensured the lowest possible transmission of seismic stresses in the vertical direction from the ground through the foundation to the superstructure avoiding the structure placing in the transient resonance zone for the received stress level.

In figure 8 is presented the constructive solution for the insulating system composed of elastomers and sliding supports on concave surface [10].

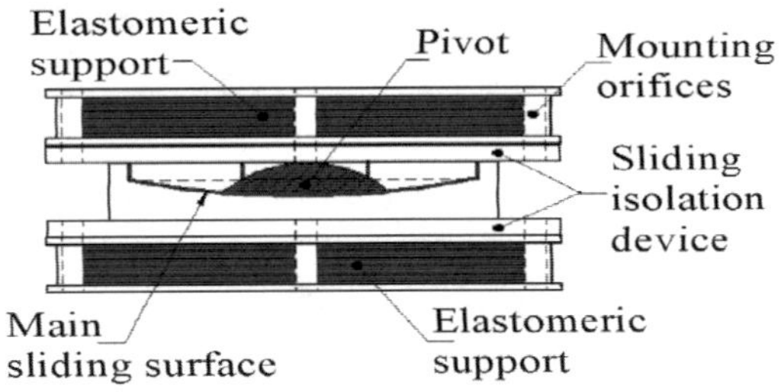

Figure 8. Schematically representation of the insulating system composed of elastomers with sliding supports [10].

By using a composed seismic energy insulating and dissipative system introduced inside the resistance structure of a bridge structure, the assembly rigidity decrease is pursued at the same time with damping increase at small and medium deformations and the efforts vertical transmission from the foundation is accomplished with diminished force values, in the end the seismic movement being damped.

The structure rigidity and the damping possibility gradually increase proportional to the deformations, increasing the resistance capacity of the entire structural assembly to the seismic action.

Insulating systems that use the composition of dissipative and insulating actions can be used successfully to ensure earthquake protection for different structures types, rigid or flexible and for any type of earthquake action (surface, intermediate or deep), using appropriate methods to each case.

It can be achieved the proposed objective, namely isolation from seismic motion for all stress states, or seismic energy dissipation and obtaining finally the structural system response limitation to seismic actions.

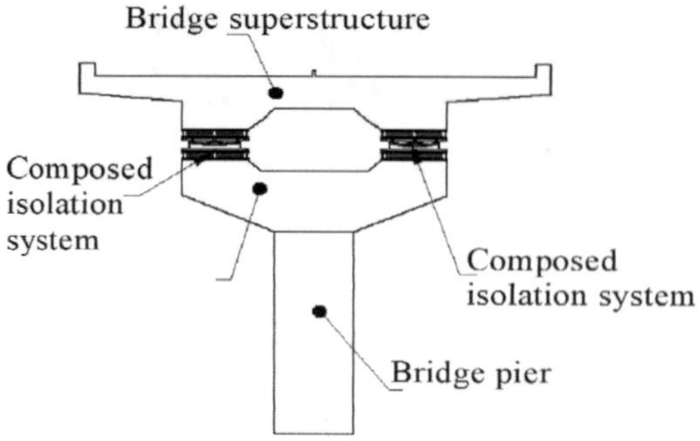

Figure 9. Schematically representation of a seismic insulated bridge structure with composed insulating system.

In figure 9 and 10 are presented the schematically and three-dimensional model of bridge structure endowed with hybrid seismic insulation system composed of elastomers and sliding supports.

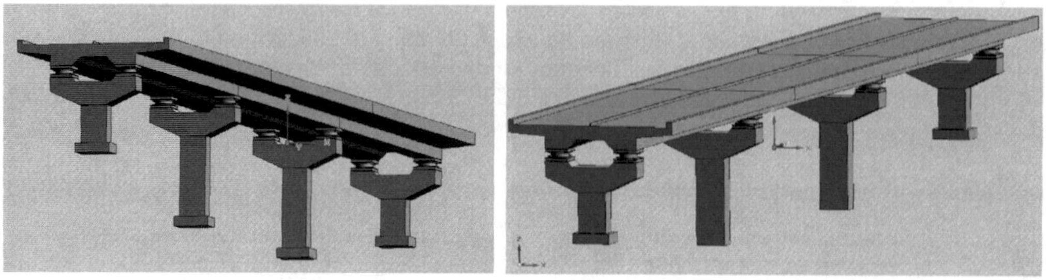

Figure 10. Composed insulating system mounted at bridge structure (3D model) [10].

The sliding isolation system works on the simple pendulum operation principle. When the seismic motion occurs, the spherical articulated part moves along the concave surface and the structure performs a simple harmonic motion.

Similar to the simple pendulum, the supports can increase the structure vibration natural period, allowing the structure to slide along the concave support surface.

The sliding joint surface in contact with the spherical surface is lined with a composite material with a low friction coefficient (usually Teflon) used in combination with steel. The supports are closed and sealed to avoid contamination of the sliding surface.

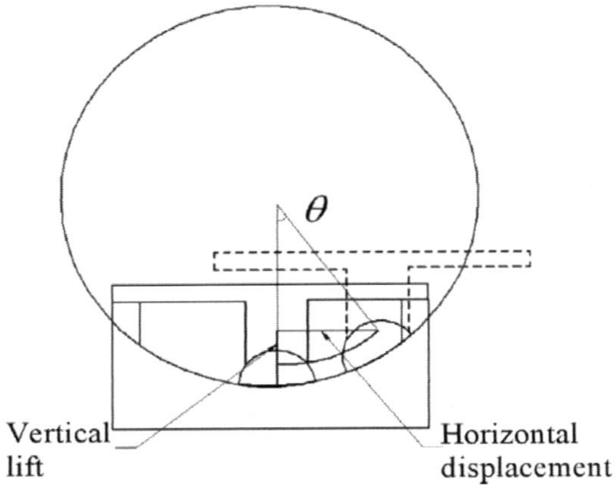

Figure 11. Sliding support on the concave surface.

The elastomeric system acts in addition to the sliding bearing action, providing an additional movement reserve, which may be required for seismic actions of higher intensity.

The composite system acts as a safety system, activated only if the shear force appearing on the sliding surface is greater than the static friction force.

Once in motion, the sliding joint moves on the spherical surface, resulting in a mass lifting, combined with a horizontal displacement (figure 11), a pendulum similar movement.

The kinematic movement and operation mode of the support are identical, regardless of whether the main sliding surface is positioned face down or upwards.

The characteristics of sliding bearings combined with those of elastomeric systems, consisting of durability under severe environmental conditions, low weight and sensitivity to the earth frequency movements, make them a viable solution for the formation of a composite system that can provide better seismic insulation of bridge type structures, but also for not very tall buildings.

5. Experimental results for seismic motion on bridge structure

Experimental tests were performed using a scaled OSB bridge structure with dimensions $(L \times l \times h) = (2 \times 0.36 \times 0.017)[m]$, insulated with elastomeric elements and friction pendulum bearings which use rolling spherical steel parts on the steel surface, where the coefficient of friction is 0.8, Coulomb friction without lubrication.

The bridge structure was loaded with a concentrated load (uniformly distributed) of 30 kgf, the seismic shock was modeled using a pendulum system with a constant mass of 1.135 kg and the possibility of adjusting the angle of fall and an excitation force of 400 N was applied to the structure.

The behavior of the insulating system is analyzed in terms of acceleration in time, acceleration amplitude function of frequency values and frequency in time with eigenvalues for the pier and superstructure for two main directions of movement longitudinal and transversal. The results are presented in figures 12 and 14 for the bridge pier and figures 13 and 15 for the bridge superstructure for both transversal and longitudinal main directions of movement.

The improved values obtained at the level of the superstructure as a result of the action of the composite insulating system are followed [9-22].

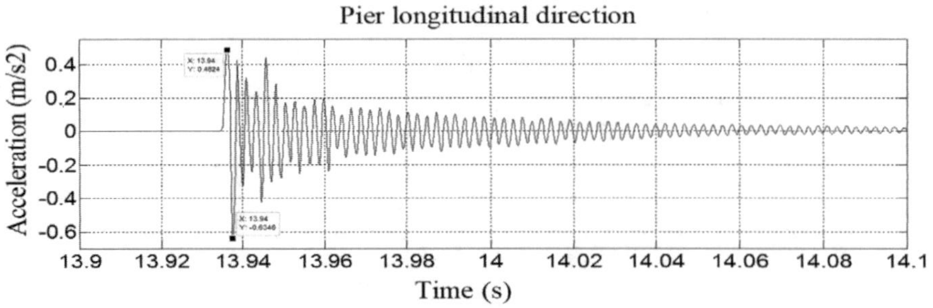

(a)

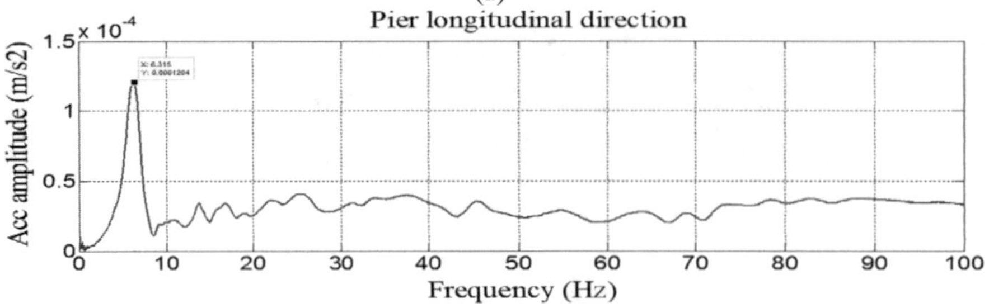

(b)

Figure 12. Pier longitudinal direction of movement
(a) acceleration–time; (b) acceleration amplitude-frequency.

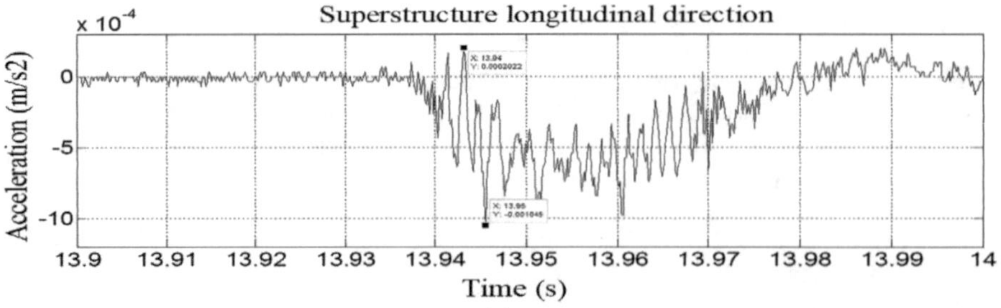

(a)

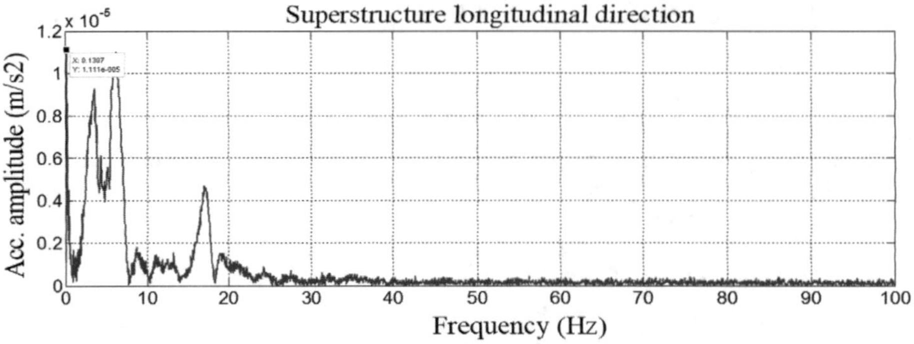

(b)

Figure 13. Superstructure longitudinal direction
(a) acceleration–time; (b) acceleration amplitude-frequency.

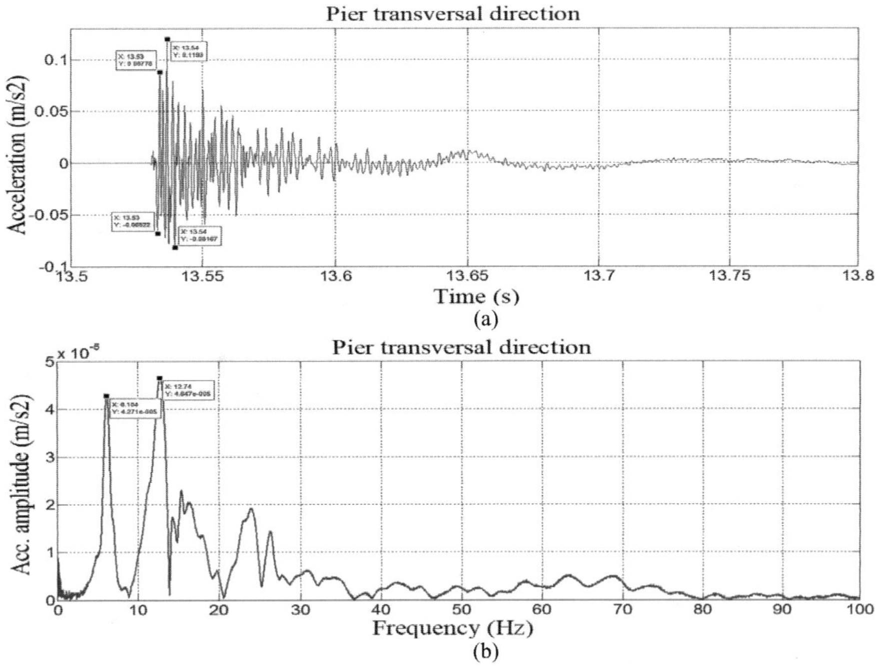

Figure 14. Pier transversal direction of movement
(a) acceleration–time; (b) acceleration amplitude-frequency.

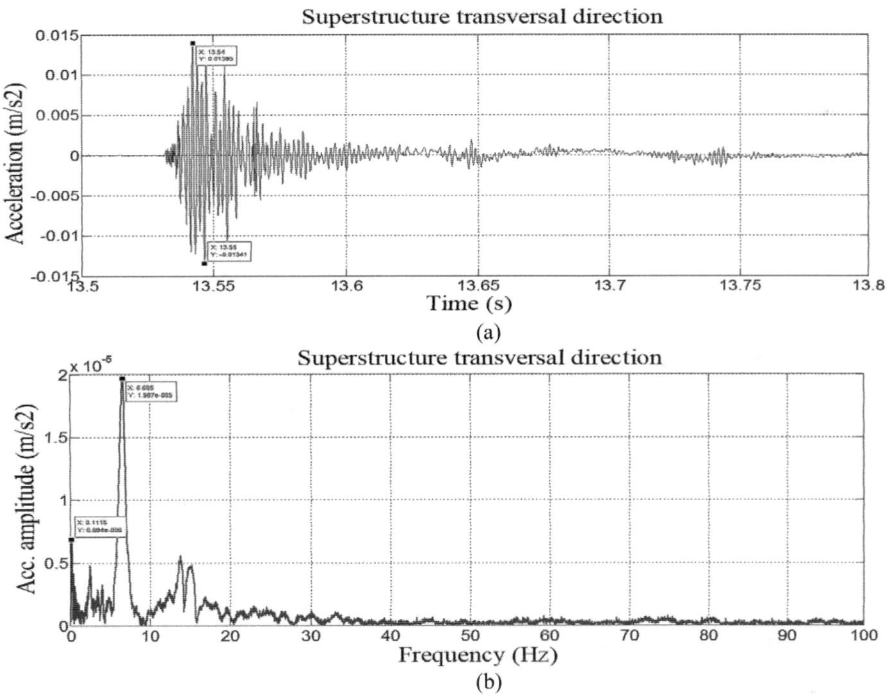

Figure 15. Superstructure transversal direction
(a) acceleration–time; (b) acceleration amplitude-frequency.

Based on the obtained results from experimental tests, the improved values for acceleration at the level of the insulated superstructure can be observed compared to the values obtained at the level of the support pier being emphasized the insulating and dissipative character of the composite protective system which is interposed between the two structural elements.

6. Conclusions

Theoretical aspects regarding the use of energy isolation and dissipation systems for installation in construction structures are presented in order to reduce the effects of seismic actions on construction structures.

In addition to the structure resistance structure that has the ability to preserve the integrity of the structural system to the arising demands base insulation systems are also used, but also energy dissipation systems to meet and avoid the destructive effects of seismic actions on bridge structures types.

The constructive solution for coupling the insulating system with sliding friction with the elastomeric type system is presented as the optimal solution to be applied to the bridge structures type.

By combining the effects of the two systems in the ensemble, an improved freedom of movement is provided for the foundations that act together with the ground, while the superstructure tends to remain as long as possible in the state of relative equilibrium, so that field efforts are no longer transmitted. entirely in the vertical direction at the superstructure.

It represents an innovative method of coupling two isolation systems that can also ensure seismic energy dissipation at the same time, at the level of the isolation system.

The structural and functional analysis of the insulation systems against seismic actions, taking into account the isolation and energy dissipation effect is highlighting the range of specific advantages to each major category of insulator system.

Behavioural modelling and analysis in an intense and varied dynamic regime performed for each of the above-mentioned categories will highlight the specificity of the corresponding dissipative characteristics.

The results obtained from testing in this research stage provide the informational support necessary for the configuration and implementation of combined isolation system that present a series of advantages taken from each constituent element.

7. References

[1] Lupasteanu V, Soveja L, Lupasteanu R and Chingalata C 2019 Installation of a base isolation system made of friction pendulum sliding isolators in a historic masonry orthodox church *Engineering Structures* **188** 369-381.

[2] Avossa A M, Di Giacinto D, Malangone P and Rizzo F 2018 *Seismic retrofit of a multi-span pre-stressed concrete girder bridge with friction pendulum devices* Shock and Vibration.

[3] Hong X, Guo W and Wang Z 2020 Seismic analysis of coupled high-speed train-bridge with the isolation of friction pendulum bearing *Advances in Civil Engineering*.

[4] Cho C B, Kim Y J, Chin W J and Lee J Y 2020 Comparing rubber bearings and eradi-quake system for seismic isolation of bridges *Materials* **13**(22) 5247.

[5] Hassan A L and Billah A M 2020 Influence of ground motion duration and isolation bearings on the seismic response of base-isolated bridges *Engineering Structures* **222** 111129.

[6] Haeri A H, Badamchi K and Tajmir R H 2019 Proposing a new hybrid friction–yielding–elastomeric bearing *Journal of Vibration and Control* **25**(9) 1558-1571.

[7] Nuraini S, Tambusay A and Suprobo P 2018 A comparative study of base isolation devices in light rail transit structure featured with lead rubber bearing and friction pendulum system *In MATEC Web of Conferences* **195**, p. 02013 EDP Sciences.

[8] Narhare M M and Kasnale A 2017 A review of behaviour of building under earthquake with or without base isolation *Technology* 5.

[9] Scheaua F D 2015 Description of a Composed Seismic Isolation System for Bridge Structures *Analele Universitatii "Eftimie Murgu" Resita* **XXII**(2) ISSN 1453 – 7397.

[10] Scheaua F D 2013 *Analiza comportarii sistemelor de disipare cu frecare uscata la actiuni dinamice* PhD Thesis Dunarea de Jos University of Galati.

[11] Axinti G and Axinti A S 2009 *Actionari hidraulice si pneumatice* Vol III Editura Tehnica-Info, Chisinău.

[12] Scheaua F 2019 Modelling of the visco-elastic pendular hybrid system with dissipative rolling elements In *IOP Conference Series: Materials Science and Engineering* **591**(1) p. 012029 IOP Publishing.

[13] Scheaua F 2020 Improvement of structures seismic response based on pendulum systems with double sliding surface In *IOP Conference Series: Materials Science and Engineering* **916**(1) p 012102 IOP Publishing.

[14] Scheaua F D 2017 Special pattern of hydraulic dissipation system used for isolation of bridges against earthquakes *Analele Universitatii "Eftimie Murgu"* Resita **24**(1).

[15] Scheaua F 2014 Considerations on functional parameters of dry friction seismic isolation systems *Annals of "Dunarea de Jos "University of Galati* Fascicle XIV Mechanical Engineering **21**(1) 67-70.

[16] Scheaua F 2011 Seismic base isolation of structures using friction pendulum bearings *Annals of "Dunarea de Jos" University of Galati* Fascicle XIV Mechanical Engineering **18**(1) 61-64.

[17] Scheaua F 2021 Composed Isolation System Concept for Vibration Effects Mitigation on Bridge Structures. *In Acoustics and Vibration of Mechanical Structures AVMS* pp. 473-483 Springer, Cham.

[18] Scheaua F 2014 Functional aspects for antiseismic double sliding isolation systems *Annals of "Dunarea de Jos" University of Galati* Fascicle XIV Mechanical Engineering **21**(2) 45-48.

[19] Scheaua F 2015 Results obtained from testing a hybrid model of vibration isolation system *Annals of "Dunarea de Jos" University of Galati* Fascicle XIV Mechanical Engineering **22**(2) 41-44.

[20] Goanta A M and Haraga G 2017 Aspects of modelling classical or synchronous modelling with Solid Edge ST 9 In *MATEC Web of Conferences* **112** p. 06024 EDP Sciences.

[21] Goanta A M 2019 Modern annotate and export instruments in 3D pdf files *Journal of Industrial Design and Engineering Graphics* **14**(1) 119-122.

[22] Goanta A M 2017 Aspects of obtaining orthogonal projections and intelligent indicators by inventor 2016 *Journal of Industrial Design & Engineering Graphics* **12**(1).

[23] Sanchez J, Masroor A, Mosqueda G and Ryan K 2013 Static and dynamic stability of elastomeric bearings for seismic protection of structures *Journal of structural engineering* **139**(7) 1149-1159.

IOP Conf. Series: Materials Science and Engineering 1182 (2021) 012071 doi:10.1088/1757-899X/1182/1/012071

Cloud Connected Power Inverter

S R Sintea and C C Pomazan
Constanta Maritime University, Naval Electromechanical Faculty, 104 Mircea cel Batran Street, 900663 Constanta, Romania

E-mail: sorin_atx@hotmail.com

Abstract. In the Industry 4.0 paradigm, the control of electrical power production and consumption in small energy nodes remains an important objective. The new energy production nodes are usually installed in locations with reduced power requirements. They can produce energy to cover the needs of the consumer and the excess can be pushed in the national power grid system. This paper proposes a scheme to control the balance between produced energy, the locally consumed energy and the energy pushed into the national power grid system, using grid controlled inverters. The inverter connects the local energy source, with the local consumer and also ensures a bidirectional connection to the national power grid system. The inverter can be programmed to satisfy the local consumer power requirements, from the local source or from the national power grid system, and it can also deliver the local excess energy to the national power grid system. The controller is connected with the Internet and is able to send data to an IoT server. This paper presents a solution, along with its supporting technology, to create a network infrastructure based on thinger.io IoT technology. We present this solution for control, monitoring and optimization the electrical power production and consumption for small consumers using the IoT technology. This device can be used in small farm areas where we can monitor and control the balance between energy production and energy consumption from local sources or from the national power grid system. The device can also be used to monitor and control energy flow in remote small industrial plants or agricultural farms.

1. Introduction

In the latest years, as a result of rapid development of cloud computing concepts and technologies, more and more cloud-based business models and practical applications are emerging in energy production environments, including cloud energy consumption monitoring and cloud energy transport logistics. Such systems integrate the distributed resources with artificial intelligence to optimize the balance between energy production and consumption.

One important request from cloud technology is the possibility to access information from different devices (numerical controlled machines and processes, computers, notebooks, tablets or phones). The challenge is to integrate the variety of controller manufacturers with their own information accessing interfaces and security and protection data protocols.

At the present a lot of small energy producers (micro energy production farms) emerged and became familiar with traditional energy producers [1]. These energy farms are organized in small group areas and they are connected in small power grid systems, named micro grids. A micro grid can be organized to be self sustaining and automatically control the energy flow. Adopting the new cloud computing technology, micro grids will be able to communicate between them and they also will be able to be integrated in the national power grid systems. The base unit of the micro grid is the grid connected controller.

This paper proposes a cloud grid connected controller. This controller is responsible with power flow control according with energy locally produced and the local consumer demand.

The controller connection to the Cloud is ensured using IoT (Internet of Things) technology [2]. The data is transferred into the Cloud using safe communication protocols (including data encryption and data protection). Information transferred in the Cloud can be accessed from anywhere using Internet access infrastructure for Cloud Computing. Another important aspect of using IoT and Cloud technology is the possibility to monitor, analyze and optimize energy power flow using fast decision algorithms based on artificial intelligence.

2. Smart Micro Grid Architecture

The base concept of Cloud Connected Power Inverter is related to the Smart Micro Grid. Smart Micro Grid can be implemented in isolated human communities or villages. Each house can have one or two energy production sources (wind turbine, solar panels, generator sets etc.). Because the power produced by these sources can be more than enough for the local consumer, the energy excess can be exported to the local micro grid.

Figure 1. Micro Grid Geographic Area.

If the local source is unable to ensure the power requirements of the local consumer, energy can be drawn from the local micro grid.

Figure 1 presents a scenario that include three isolated little farms. Each farm has it's own energy source to ensure their electrical needs (one is powered from solar panels, the second is powered from a wind turbine generator and the last is powered from an electrical generator group). To be able to have an optimized electrical consumption between the three households all generators are connected together and used according to weather conditions. One simple solution is to bring them together with a power cable and use Cloud connected power inverters (PGC-Power Grid Connected Controller) connected in a micro grid structure (see figure 2). Each PGC connects the local power generator with the local consumer and a bidirectional electrical power line. This connection is presented in figure 2 in the power layer. From IoT cloud connection point of view, the PGC controllers are connected together in the same logical network, named Local Cloud Area. This area is controlled by an IoT Server named Local Server Controller. Electrical power transfer in this micro grid is controlled by the Local Server Controller. This server can be located physically in this network or it can be installed in the cloud. If the server is stored in the cloud then each PGC must be configured to access and communicate with the Local Server Controller. This server manages power transfer through the micro grid by controlling each PGC power inverter controller. The Local Cloud Area (see figure 2) communicate information with IoT server (located in Cloud) using IoT Gateway. This gateway is able to forward data from each PGC or through the Local Server Controller, to the IoT Server. Also the IoT Server is able to manage electrical power transfer through this micro grid by controlling the Local Server Controller.

IOP Conf. Series: Materials Science and Engineering 1182 (2021) 012071 doi:10.1088/1757-899X/1182/1/012071

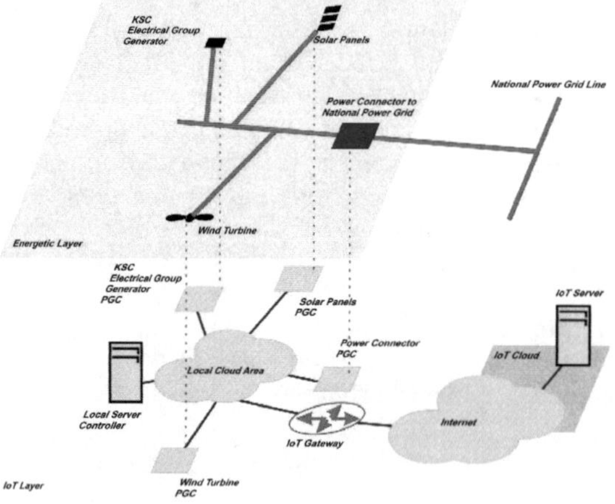

Figure 2. Micro Grid Architecture.

If micro grid consumers require more power, because the local sources are not able to generate enough energy in the grid, this micro grid can be connected with the National Power Grid System. The connection is made using a special PGC, named National Power Grid Energy Connector. Also this PGC is logically connected in the Local Cloud Area. This controller manages power flow from the National Power Grid to the micro grid or from the micro grid to the National Power Grid System if the local energy production exceeds local power requirements.

3. Cloud Connected Power Invertor. PGC-Power Grid Connected Controller

3.1. PGC Architecture

PGC is the main unit that connects the local consumer with the local energy supply and the local grid area (see figure 3). This controller provides the local consumer needs with electrical power generated by the local power source or the local power source and from the grid. Also PGC will inject energy in the grid if the locally required power is less than the produced energy.

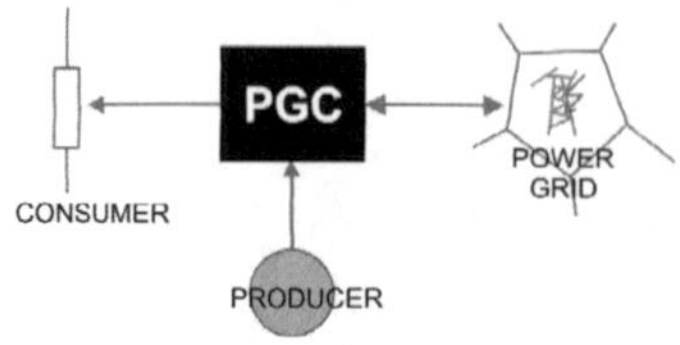

Figure 3. PGC Electric connections.

To be able to ensure this functionality the PGC must be connected with the micro grid, this connection is provided through the Local Server Controller.

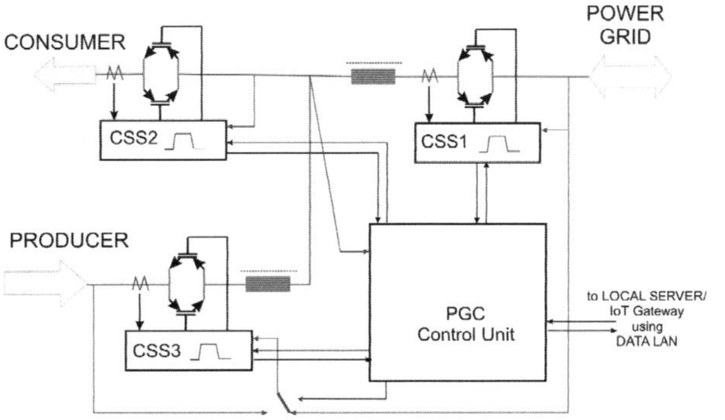

Figure 4. PGC Controller.

One PGC Controller unit includes (see figure 4):

- CSS1 – controlled synchronized switch to connect PGC with the Power Grid. This switch controls the power transferred to/form the Power Grid to our unit. The switching process is synchronized with voltage signal from the Power Grid. The controller measure and control the current and the power flow through the switch.
- CSS2 – controlled synchronized switch to connect local consumer with PGC. This switch controls the power transferred to consumer. The switching process is synchronized with the voltage signal from PGC. The controller measure and control the current and the power flow to the consumer. This module also has the role of consumer protection.
- CSS3 – controlled synchronized switch to connect local generator with PGC. This switch controls the power transferred from the local power generator. The switching process is synchronized with the voltage signal from the local generator and from the Power Grid. The controller measure and control the current and the power transfers from the local generator. This module also has the role of generator protection.
- PGC control unit is responsible with controlling the power delivered through PGC and with local consumer and generator protection. This unit controls the functionality of CSS1, CSS2 and CSS3. Also the unit can be controlled from the cloud (through IoT gateway) or from the Local Server. The unit sends the status to the Local Server or to the IoT server. PGC control unit is implemented using an STM32 microcontroller platform.

3.2. CSS. Controlled Synchronized Switch

Each Controlled Synchronized Switch (CSS) is based on STM32 microcontroller architecture (see figure 5). The module connects two electrical power circuits (one is the input circuit and the other one is the output circuit) and ensures a bidirectional power transfer between both circuits in four quadrants (U-I). This CSS protects the outputs circuit against power surges, short circuit, and under voltage. To be able to ensure these functions the CSS controller measures the voltage in the input circuit and the current in the output circuit. Also to control the switching circuit the CSS uses a synchronized signal that can be connected to the other source. Power switching is controlled by 2 IGBT power transistors connected in antiparallel connection, between the input circuit and the output circuit.

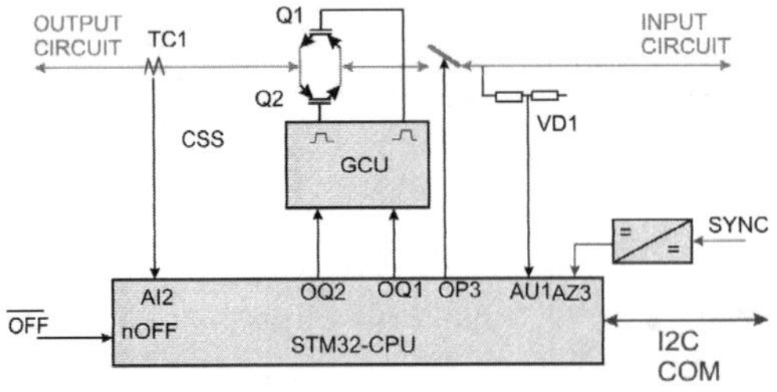

Figure 5. CSS Controlled Synchronized Switch.

Each CSS controlled synchronized switch contains:

- Q1, Q2 – two IGBT power transistors used to control the power transfer between input an output circuit. Transistors are controlled through gate control unit module (GCU) [3, 4];
- GCU – Gate Control Unit module is based on IX2113 IGBT gate controller ICs. The inputs are connected directly to STM32-CPU [4];
- STM32-CPU – main controller of the unit. The CPU is based on low cost 32 bit STM32F103C microcontroller (running at 72MHz) [5]. The controller uses the following signals:
    - AU1 - connected to A0 – connects the input analog voltage signal to the CPU ADC unit;
    - AI2 - connected to A1 – connects the output analog current signal to the CPU ADC unit;
    - AZ3 - connected to A2 – connects the synchronized input analog voltage signal to the CPU ADC unit;
    - nOFF – connected to D4 – connects an external digital signal that enables the unit functionality (if nOFF = 1);
    - OQ1 – connected to D9 – Q1 IGBT gate control signal;
    - OQ2– connected to D10 – Q2 IGBT gate control signal;
    - OP3 – connected to D11 – the power contactor control signal. It is active iif the unit is disabled or the current or power transferred through the switch exceeds the preset values;
    - I2C communication port for connecting with PGC Control Unit;
- TC1 – current transformer – used for measuring the current through the power switch;
- VD1 – voltage divider – used for measuring the input voltage applied to the power switch;

STM32-CPU unit implements the following registers:

- SR - Status register (RO) – This indicates the status of CSS unit;
- U1 – Input Voltage (RO) – input voltage value;
- U2 – Output Voltage (RO) – output voltage value;
- I1 – Current (RO) – current value, measured through the switch;
- P1 – Power (RO) – power transferred through the switch (if the value is positive the power is transferred from input circuit to output circuit, or if the value is negative the power is transferred from output circuit to input circuit);
- Fi_SYNC_1 – Phase Shift (RO) – the phase shift between SYNC signal and the input voltage;
- Fi_SYNC_2 – Phase Shift (RO) – the phase shift between SYNC signal and the signal that controls the IGBT transistors phase;
- P2 – Power (RW) – the preset value of power transferred through the switch;
- I2 – Current (RW)– the preset value of current transferred through the switch;
- SW_OFF – External Locked Status (RO) – the value of nOFF signal;
- DISA – Disable unit functionality (RW) – the running status of the unit;

- PROT – Protection Relay Status (RO);
- PROT_RST – Protection Relay Reset(WO) – resets the protection relay status;
- ADDR – I2C Address (RO) – get the I2C communication address;

Registers are accessed through the I2C communication layer.

3.3. PGC. Control Mode

PGC Control Unit communicates with CSS modules over I2C local network. The PGC Control Unit communicates with CSS modules (CSS1, CSS2 and CSS3) and establishes their running modes.
The operating mode of PGC is set by Local Server controller.
To be able to operate the PGC Control Unit uses the following registers:

- SP1 – Input Power (RW)– the power injected into the External Power Grid (if value is negative) or the power supplied by External Power Grid (if value is positive)
- P1 –Power transferred with the Power Grid (RO)– the power exchanged with the Power Grid
- SP2 – Generated Power (RW) – the nominal power that can be generated by the local generator
- P2 – Generated Power (RO) – the power generated by the local source
- SP3 – Consumed Power (RW) – the nominal power that can be consumed
- P3 – Consumed Power (RO) – the nominal power consumed

The Local Server can set SP1, SP2 and SP3, for each PGC from the local energy grid, in according with area settings. The local implemented algorithm is presented in figure 6.

```
...
CSS3.P2 = SP3
if SP3>SP2 then
  CSS1.P2 = (SP1 > (SP3-P2)) ? (SP3-P2) : SP2
else
   CSS1.P2 = ( SP1 < 0) ? (P2-P3) : 0
end if
CSS2.P2 = SP2
...
```

Figure 6. PGC Algorithm.

4. Conclusions

The research presented in this paper explores the design of an IoT Power Grid Connector controller. This device can be used to monitor and control the energy flow for small energy producers / consumers connected to the national distribution grid. The design of the PGC controller is modular and compact using low cost microcontrollers powerful enough to ensure real time computational requirements. The ability to measure and transfer data in real time and also to control the power flow remotely makes the PGC controller suitable for integrating in more complex IoT energy control systems, sustaining the process of transitioning from the old power distribution grid paradigm to the more advanced and efficient Smart Grid concept. The solution can also be used to connect multiple small consumers and producers from remote small areas that do not benefit from the national power grid connection, to form independent micro grids, in an effort to ensure optimal energy distribution to meet the power requirements of those small local consumers.

5. References

[1] *** https://www.power-and-beyond.com/how-iot-is-utilized-to-deliver-energy-savings-a-885202/, accessed January 2021.
[2] *** https://nifi.apache.org/docs.html, accessed January 2021.
[3] *** *Power Electronics 2017-2018, SEMIKRON International GmbH, ISLE Verlag* 2017.
[4] Arendt Wintrich, Ulrich Nicolai, Werner Tursky, Tobias Reimann 2015 Application Manual Power Semiconductors, *SEMIKRON International GmbH, ISLE Verlag* 2015, ISBN: 978-3-938843-83-3.
[5] *** https://www.st.com/en/microcontrollers-microprocessors/stm32f103.html, accessed January 2021.

Influence of the positioning accuracy given by the positioning encoder of the CNC machine tools

G Stan and A Stoica
"Vasile Alecsandri" University of Bacau, Department of Industrial Engineering, Calea Marasesti Street, No 157, 600115, Bacau, Romania

Email: ghstan@ub.ro

Abstract: The CNC machine tools are present at a large extent worldwide, for machining by cutting all kinds of work pieces. Their machining accuracy and quality are crucial to obtaining very accurate and intricate work pieces. The accuracy of a CNC machine tool is given by the quality of each and every kinematic linkage/coupling in its structure. This work aims at researching the influence of several constructive and functional factors on the positioning accuracy of a kinematic coupling in the structure of a CNC machine tool. The influence caused by noncompliance with the Abbé principle on the positioning accuracy of a feed kinematical linkage is analysed into detail, both at idle and load running. Out of this research useful conclusions are resulting for the CNC machine tool designers and builders, with respect to optimising several constructive factors, such as the location of the position encoder, location of the ball screw, stiffness of the directional guideways etc., in order to improve the accuracy of the CNC machine tool. The results of the theoretical research have been validated through experimental trials, confirming the impossibility of full compliance with the Abbé on the CNC machine tools.

1. Introduction

Generally, all machine tools include to their structure several kinematic couplings, linear and rotary, that provide the movements needed by the cutting tools for performing the machining process [1]. Each kinematic coupling that in most of the cases belongs to the fifth class has its guiding assured in two plans: one is providing the directing and the other one is providing the carrying support. The propulsion force of the kinematic coupling is located, in relation to the two *guideways* (directing and carrying), at positions possible in constructive terms, considering the vicinity of the constitutive elements. The kinematic coupling that is materialised through the moving element is also submitted to other forces, such as friction and requirements resulting from the cutting processes. All these forces lead to consequences on the elastic deformations [2]. The sensible element of the kinematic couplings, in terms of stiffness, is the guiding system that needs to provide a very low friction coefficient as well [3]. The consequences of the coupling being submitted to the forces mentioned above make the moving element to modify its position so that, upon the motion to a direction, a light swing of the moving element occurs, and, upon the motion to the reverse direction, case when the force senses are changed, the same light swing occurs, but in the reverse sense. During the linear motion of the moving element [4] the value of the "light swing" is variable, since it depends on the variation of the guiding system stiffness in relation to the kinematic coupling travel. As such, the group of forces that act on the kinematic coupling and is reified through a resulting force located at a certain distance from the

IOP Conf. Series: Materials Science and Engineering 1182 (2021) 012072 doi:10.1088/1757-899X/1182/1/012072

directing guideway axis (in one plane) and a certain distance from the carrying guideway axis (in the other plane), will generate a swinging torque in each plane. In consequence, there will be two *light swings*, in perpendicular planes, of the moving element during the linear motion of the kinematic coupling.

2. Abbé principle on CNC machine tools

The Abbé principle was first described by Ernst Abbé in the domain of metrology and it states: A measuring instrument must have the measuring direction coaxial (in line) with the measurer, as shown at figure 1.

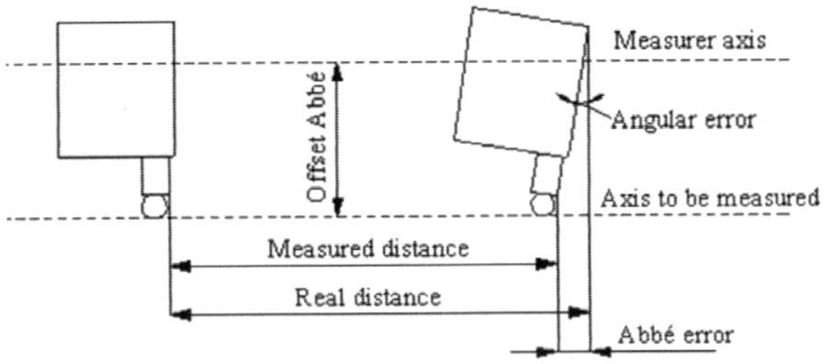

Figure 1. Abbé principle – Block diagram (Abbé offset, measured distance, real distance, measurer axis, angular error, axis to be measured, Abbé error).

The Abbé error occurs when a distance exists, named Abbé offset, between the measuring scale (measurer) and the axis to be measured that is caused by the angular error of the moving element. In case of the CNC machine tools the Abbé principle can be analysed in a similar manner, because a machine tool has several controlled axes, consisting of moving elements and whose motion is automatically measured by the position encoder. The location of the position encoder and the angular error of each moving element are important factors that influence the positioning accuracy and the machining accuracy of the machine tool, respectively. However, the analysis and compliance with the Abbé principle on the CNC machine tools is a complex problem because constructive and contextual aspects come up that are limiting the possibility of a full compliance with this principle. The aspects that take part to the analysis of the Abbé errors on CNC machine tools are:

a) On the controlled axes of the CNC machine tools indirect measurement is used for the position, especially in case of short and middle travels. In this case the position encoder is located on an intermediary element of the transmission system. In this case, for the analysis of the Abbé principle the measurer axis is considered the axis of the ball screw or rack (when the rack and pinion system is used).

b) The system of forces that act on the moving element of each controlled axis is variable and depends on the cutting process. In this case the Abbé error will be variable and, by default, the positioning accuracy will be variable too. The analysis of the Abbé principle on CNC machine tools and especially for finding the minimisation means of the errors requires two variants: the idle running and load running. This split into two variants is justified by the fact that in case of most of the CNC middle and large size machine tools the controlled axes first perform the positioning at idle running, then the moving element is clamped and afterwards the running process at load will begin. If the CNC machine tool is cutting through interpolation where more controlled axes participate and the cutting forces are differently directed spatially, the need of the analysis at load of the Abbé principle comes up. Moreover, the cutting forces that require the moving element are variable, depending on the parameters of the cutting duty.

c) A guiding system of a moving element of a feed kinematic linkage on CNC machine tools provides the linear guiding (coupling of the fifth class) in two planes: a directing one and a carrying one which is perpendicular to the first. The controlled axes in the architecture of a CNC machine tool are oriented prominently horizontally, but at least one should be vertical. In case of the controlled axes whose moving elements are moving horizontally only the angular error in the directing plane will be considered, because the weight of the moving element will annul the angular error in the other plane. On the moving elements of the CNC machine tool that move vertically the angular error will occur in both planes because the weight of the moving element is annulled through balancing. In this case the Abbé error of the moving element in one plane will be found in the positioning accuracy of the controlled axis that includes the moving element and the Abbé error in the other plane will affect the accuracy along a direction perpendicular to the first. This Abbé error in the second plan is caused by the angular error in the first plane. The moving element subject to swing has no position encoder on this direction, since the machine tool has another controlled axis along it.

3. Ways to improving the positioning accuracy based on the Abbé error analysis

The way to find solutions to improve the positioning accuracy related to the Abbé error at a feed kinematic linkage of a CNC machine tool is represented in figure 2. The moving element *1* (the slide) moves horizontally along the guideways *2*. The analysis is done at idle running. The direction guiding of the slide is performed through the surfaces *3* along which the rolling elements *4* are moving. The force *F* necessary to drive the moving element *1* is developed by the ball screw *5*, located at the distance *i* from the symmetry axis of the directing guideways. The resisting forces consist of the friction forces F_{f1} and F_{f2} on the guideways and the swing moment M_I of the moving element is:

$$M_I = F \cdot i - F_{f2}\frac{l}{2} + F_{f1}\frac{l}{2} \quad (1)$$

The symbols mean: *l* – distance between the surfaces that provide direction guiding; *L*- slide length.

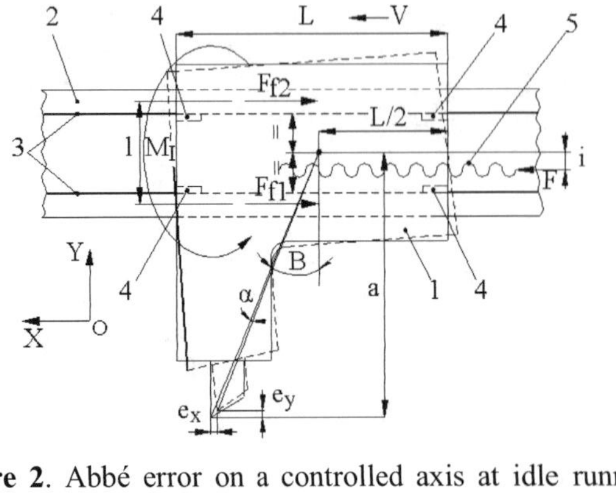

Figure 2. Abbé error on a controlled axis at idle running and location of the ball screw inside the guideways.

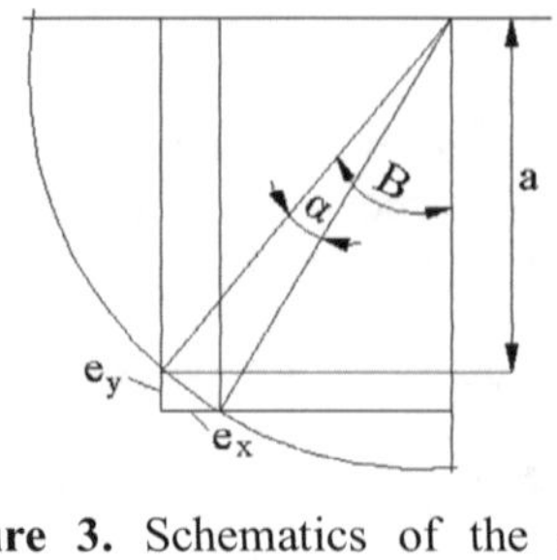

Figure 3. Schematics of the Abbé error process.

The presence of the swing torque M_I leads to a light swing of the moving element by the angle α and eventually is concretised upon the tool-work piece contact point into the errors e_x and e_y, as per figure 3. The relations that establish the values of the two errors e_x and e_y are obtained from the geometrical analysis of figure 3:

$$e_x = a\frac{[sinB - sin(B-\alpha)]}{cosB} \quad (2)$$

$$e_y = a\frac{[cos(B-\alpha) - cosB]}{cosB} \quad (3)$$

B is the angle between the vertical axis (*a*) that passes through the swing centre of the directing guideways and the axis that passes through the tip of the cutting tool. The swing torque M_I at idle run can be annulled through the *suitable* location (distance *i*) of the ball screw.

$$0 = F \cdot i - \frac{F_{f2} \cdot l}{2} \cdot \frac{F_{f1} \cdot l}{2} \tag{4}$$

$$i = \frac{(F_{f1} \cdot F_{f2}) l}{F} \tag{5}$$

F_{f1} and F_{f2} are the friction forces shown at figure 2. If $(F_{f1} - F_{f2}) \neq 0$, the location of the ball screw will be displaced from the axis of the directing guideways, so $M_I = 0$ will not happen. The analysis presented above will be sustained if the position measurement is direct, having the position encoder in line with the symmetry axis of the directing guideways. If the position encoder is located just in line with the tool –work piece contact point, the error e_x is compensated, so it's only the e_y error that remains.The location of the position encoder is important with respect to lowering the Abbé error, but constructive issues are limiting the possibility of compliance with this rule. If the measurement of the slide position is indirect, an analogy could be done with the direct measurement by considering that the position encoder is in line with the ball screw. It is to be noticed that the light swing of the slide, reified through the angle ±α, is variable in relation to the travel. This issue is mainly caused by the parallelism deviation of the directing guideways surfaces. In case when the moving element performs the positioning at load, the analysis presented above will be subject to changes. On the moving element the cutting force F_a will be also acting, situation that requires an increase of the driving force *F*. From the analyses on a slide of a CNC machine tool presented above, analyses performed both at idle running and load running, the following conclusions will result concerning the improvement of the positioning accuracy in relation to noncompliance with the Abbé principle:

1) The need for increasing the stiffness of the directing guideways of all controlled axes, both horizontal and vertical, by using elements that provide preloading. An optimal determining of the preloading value in case of the rolling elements provides a stiffness increase by approximately 45%. As rolling elements, the rolling packs with one or more circuits are recommended, instead of the usage of balls. For the large size machine tools that are fitted with hydrostatic lifting the optimisation of the hydrostatic elements (pockets) is recommended with a view to increase stiffness. If a CNC machine tool has three controlled axes, two of them (*X* and *Y*) will have a horizontal motion and the third one (*Z* axis) will move vertically. Because of the noncompliance with the Abbé principle on *X* axis, the errors e_x and e_y will be present, on *Y* axis there will be e_y and e_x and on *Z* axis the Abbé errors in the two planes (directing and carrying) will be considered, i.e. e_z, e_y and e_z, e_x as well. The *Z* axis with vertical motion having a balanced weight of the moving element should feature a high stiffness in both planes. Consequently, the use of preloading of the moving element guideways becomes essential, especially because the distance from the guideways to the tip of the tool is relatively long as it is imposed by constructive reasons, needed to provide the tool penetration beyond the table centre.

2) The need for minimising the swing torque of the moving element in the directing plan for the axes with horizontal motion. For the axes with vertical motion both planes (directing and carrying) are considered. For this purpose, it is recommended the resisting forces, materialised at idle running through the friction forces in the guideways, to be as low as possible; usually it is recommended the use of rolling guideways or guideways with liquid friction. For lowering the swing torque of the moving element it is also necessary the force arms to be as short as possible or, if possible, even zero. For this reason the location of the ball screw in relation to the symmetry axis of the directing guideways is very important. It is desirable, if there are location constructive possibilities, the application point of the slide driving force to coincide with the resisting force and with the symmetry axis of the directing guideway, situation when the swing torque will be zero at idle running and the Abbé errors will become null.

3) Another way to decrease the swing angle α consists of increasing the length L of the moving element, but with a compromise, materialised in shortening its travel. There is the possibility for keeping the travel of the moving element but, in this case, the length of the guideways will increase and, by default, the machine overall dimensions and price eventually.

4) The location of the position encoder as close as possible to the measuring line, if possible even to coincide. The measuring line of the moving element is very different, even specific for each and every axis in the structure of a CNC machine tool. Moreover, the measuring line varies from an operation to another during the usage of the machine. If the contact between the work piece and the tool modifies during cutting, the measuring directions along the three axes (*X, Y* and *Z*) will modify, as well. In case when the cutting forces are null, i.e. the machine runs idly, the measuring direction of each axis is established by the location points of the calibrated measuring system for determining the positioning accuracy (*VDI/DGQ 3441*). Consequently, the location of the position encoder of the moving element should coincide to the measuring line, reified by the calibrated measuring system (laser interferometer or scaled microscope). The compliance with this rule, if possible in constructive terms, will provide an improvement of the positioning accuracy because the Abbé error is inside the position adjusting loop. There are situations when this rule cannot be followed, but the designer can consider it for minimising the negative effects.

4. Experimental trials

The machining centre CV belonging to the University "Vasile Alecsandri" in Bacau is the CNC machine tool on which the experimental trails have been performed. The machine has three controlled axes (*X, Y* and *Z*) and the following travels: $X = 700$ mm, $Y = 600$ mm and $Z = 500$ mm. The position measuring system is direct, fitted with linear position encoders located between the guideways of each axis. Trials have been performed on the axes with horizontal motion (*X* and *Y*) through the measurement of the positioning accuracy by means of a laser interferometer placed at two positions, having two values of the Abbé offset. One of the positions made the axis of the laser interferometer to coincide with the axis of the position encoder, whilst the other one has the interferometer axis located on the median zone of the moving element that is also the location stipulated by the standard *VDI/DGQ 3441* for determining the positioning accuracy. For determining the positioning accuracy (p) the algorithm for statistical determining and processing stipulated by the standard *VDI/DGQ 3441* has been applied. The results of the trials are shown at figure 4 for *X* axis and figure 5 for *Y* axis.

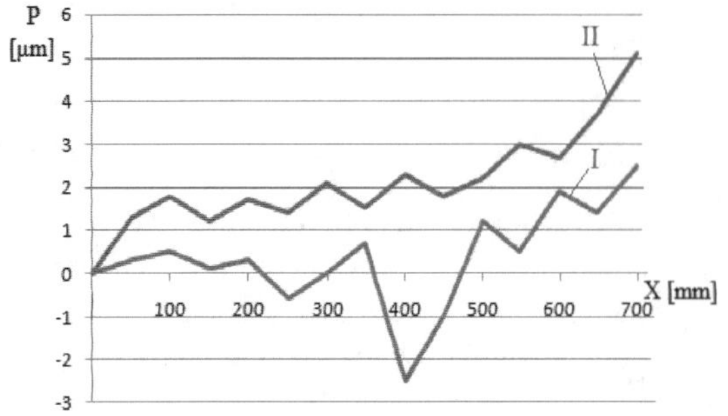

Figure 4. Positioning accuracy (*P*) of *X* axis.

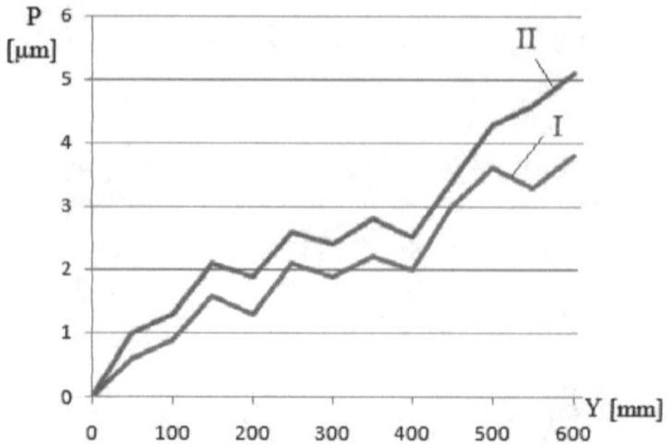

Figure 5. Positioning accuracy (*P*) of *Y* axis.

The diagram *I* represents the positioning accuracy when the interferometer is located in line with the position encoder, where the errors are low (±2.5 µm) and the diagram *II* corresponds to the interferometer located on the middle of the machine table. In the case of diagram *II* there is an increase of the positioning error, but a quasi-constant variation between the two diagrams may be noticed. This aspect is correlated with the cause of error generation, i.e. the presence of the swing angle and distancing of the position encoder from the measurement line. The systematic characteristic of the positioning errors on the two axes *X* and *Y* at idle running allows the insertion of corrections along each axis, i.e. for *X* axis the e_x component only, without the component e_y. The correction values are input to those 1000 points, for each sense, mentioned in the offer of each numerical control.

5. Conclusions

From the analysis of the structure of a feed kinematic linkage of a CNC machine tool it results that an important role is held by the location of the position encoder in relation to the directing guideways of the slide, beside other conditions that depend on the stiffness of the guiding system, type of the measuring system (direct or indirect) and diminution of the swing torque. Because the requirement conditions of a slide on a controlled axis of the machine are variable, less in case of idle running, the application of the Abbé principle is impossible to be achieved. Just a small part of the errors, the ones at idle running and, in this case, one component only of the error that is systematic, allows the input of corrections for improving the positioning accuracy. The following recommendations may be given to the machine tool builders: - the location of the ball screw in relation to the directing guideways and the resultant of the slide resisting force is very important; - the correct choice of position encoder location in relation to the measuring direction; - use of guideways with proper stiffness and low friction coefficient; - use of additional means for minimising the slide swing upon the change of the motion sense.

6. References

[1] Caputi A, Russo D, 2020, The optimization of the control logic of a redundant six axis milling machine, *Journal of Intelligent Manufacturing,* **8**, 1705

[2] Khusainov R, 2017, Study of deformations field in the working zone of vertical milling machine, *Procedia Engineering,* **206**, 1069–1074

[3] Wang X, 2019, Experimental analysis of the wear coefficient for the rolling linear guide, *Advances in Mechanical Engineering,* **11**(1), 1–7

[4] Tajima S, 2017, Global tool-path smoothing for CNC machine tools with uninterrupted acceleration, *International Journal of Machine Tools & Manufacture,* **121**, 81–95

IOP Conf. Series: Materials Science and Engineering 1182 (2021) 012073 doi:10.1088/1757-899X/1182/1/012073

Non-Destructive Determination of the Elastic and Acoustic Properties of Resonant Wood Used in the Manufacture of Violins

M D Stanciu[1], V G Gliga[1,2], SV Georgescu[1], A Savin[3] and G Dobrescu[3]
[1]Translvania University of Brasov, Blvd. Eroilor, No. 29, 500036, Brasov, Romania
[2]S.C. Gliga Instrumente Muzicale S.A
[3] National Institute of Research and Development for Technical Physics, B-dul Dimitrie Mangeron 49, Iaşi, Romania

E-mail: mariana.stanciu@unitbv.ro

Abstract. The aim of the paper was to determine the acoustic characteristics (speed of sound propagation in wood) and elastic (modulus of elasticity) on the three directions of wood, using non-destructive methods. The samples studied were cubic in shape from spruce and maple wood species, resonant and common. To determine the Young's modulus and shear modules as well as the Poisson's ratios, the expression of the propagation of Lamb waves generated and received using US translators with Hertzian contact with fr = 100kHz was used. The experimental data obtained show that mainly the velocities in radial and tangential direction are about one third of the longitudinal velocity, with the radial velocity higher than the tangential one. The radial velocity can be up to almost 14% higher than the tangential velocity. The applicability of this study consists in establishing the acoustic quality of the resonant wood from the Romanian area, used in the manufacture of our violins from the country.

1. Introduction

The behavior of wood under the action of sound waves depends on the one hand on the sound energy that comes in contact with wood, and on the other hand on the nature and condition of the wood, respectively the macro and microscopic structure of wood - cell membrane construction, fiber size and cohesion, the presence of own chemical substances, wood humidity, wood temperature, elastic properties, orientation of the structure in relation to the sound source (longitudinal, transversal, radial, tangential, complex) [1 – 4]. According to the definitions, resonance represents the state of vibration in which a body or a physical system is found when a periodic external action is exerted on it, with a frequency equal to or close to the natural vibration frequency of the body or system. From this point of view, the wood, due to the intrinsic properties of the material, presents several harmonics, of which the fundamental one confers the timbre and the acoustic specificity of the assortment. Depending on the period of the external excitation frequency, the resonance occurs whenever the frequency of the forced vibration passes through the vicinity of the natural vibration frequency of the wood [5, 6]. In order to be able to appreciate the selected wood as a raw material in the manufacture of musical instruments, a series of quality criteria have been established regarding its acoustic, physical and aesthetic properties. As in the case of resonant spruce, the main elastic, mechanical, acoustic and physical properties analyzed in the case of maple wood are: the velocity of sound propagation in the wood; the modulus of elasticity; the specific acoustic resistance (impedance), Z; the acoustic radiation, Ra; the damping; the natural frequencies and modes, etc.

For musical instruments such as violins, studies have shown that the EL / ER anisotropy ratio largely determines the mode of vibration of the resonant plates at the finished instrument [7, 8]. The longitudinal rigidity of the instrument face increases after the bass bar is glued [9 – 11]. Also, the ratio between the longitudinal modulus of elasticity E and the transverse modulus of elasticity G, in dynamic regimen, determines the frequency of response of the sound box to the bending oscillations and expresses the tone quality of the resonant box from the piano [12 – 14]. On the other hand, the speed of sound propagation through wood, V, which refers to the speed of transmission of longitudinal and transverse elastic waves in the solid medium (wood) is one of the fundamental parameters in the characterization of resonant wood and it depends on the elastic modulus. and wood density. Therefore, from a mechanical point of view, the speed of sound is strongly correlated with the elasticity of the material, because sound waves generate mechanical oscillations, and therefore the elasticity varies in the three directions. For an understanding of the elastic behavior of the resonant wood it is necessary to know the propagation speeds in each direction. The ultrasonic method is a non-destructive method, and, in determining the constants, approximately nine ultrasound velocities are involved (C_{LL}, C_{RR}, C_{TT}, C_{LR}, C_{RL}, C_{RT}, C_{TR}, C_{TL} and C_{LT}), and those must be determined for each specimen [8, 11]. The determination of the shear modulus, not being standardized, is complicated and imprecise, and it involves an auxiliary method of ultrasound resonance (RUS) through the procedure of determining its natural resonance frequencies.

From the research carried out in the specialized literature, it was found that the studies regarding the elastic and acoustic characteristics of the spruce and the maple of resonance in the Carpatho-Danubian-Pontic area, are very poor. Thus, the paper aims to present studies on determining the speed of sound propagation in wood in the three orthogonal directions of wood - longitudinal, radial, tangential, and the elastic characteristics of resonant spruce wood compared to common spruce, as well as those of curly maple wood compared to straight fiber.

2. Materials and methods

2.1. Materials

In these studies, samples were made in the form of cubes cut in the three main directions of wood - longitudinal (marked with L), radial (marked with R) and tangential (marked with T), of the two species used for the body violins - spruce (Picea Abies L. Karst) and maple (Acer Pseudoplatanus). Figure 1 shows microscopic images of the main sections of common spruce and maple wood. From each category, samples were taken with the specific structure of the resonant wood according to the selection method from the musical instrument factory.

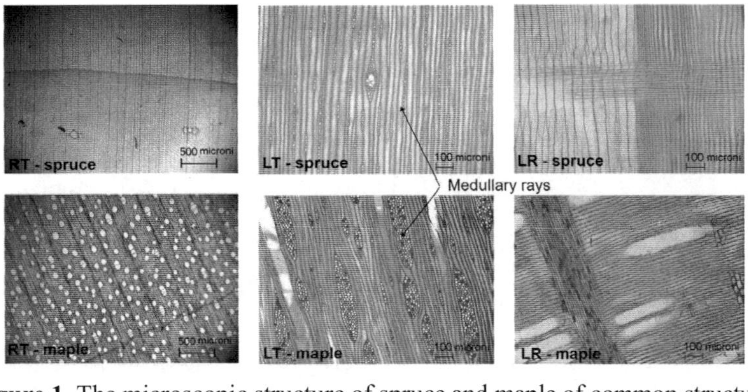

Figure 1. The microscopic structure of spruce and maple of common structure (legend RT: radial-tangential plane; LT: longitudinal - tangential plane; LR: longitudinal - radial plane) (Stanciu, 2020).

Thus, the spruce wood samples have the following characteristics: ring width range between 0.8 and 2.5 mm (average $\geq$ 1.2 mm); average latewood proportion $\leq$ 20%; average difference in width between consecutive rings $\leq$ 0.5 mm; ring widths regularity index $\leq$ 0.7; width of resonance wood $\geq$ 130 mm [4 – 5]. For maple wood with curly fiber, the important anatomical elements in size are the medullary rays, oriented in the radial direction of the wood which in the radial longitudinal plane creates the impression of a three-dimensional drawing of wood, as can be seen in figure 2 [5]. The geometric characteristics of the analyzed samples are presented in table 1.

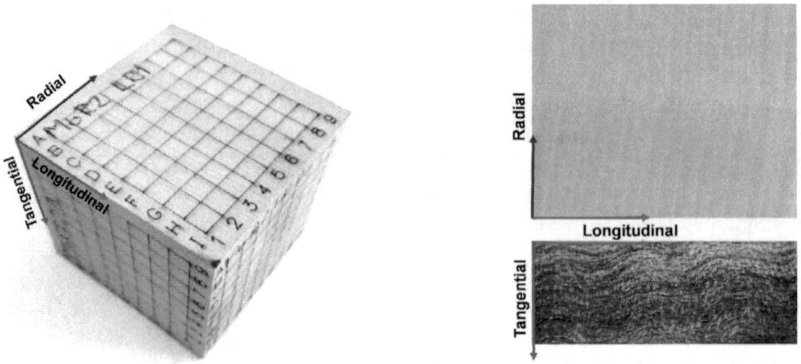

(a) (b)

Figure 2. The geometry of studied samples taking into account the main section of wood: (a) the sample prepared for ultrasound method; (b) the view of curly maple in radial and tangential direction.

Table 1. The physical features of wood samples.

Wood sample	Code	Mass [g]	Volume [m^3]	Density [kg/m^3]	Length [mm]	Width [mm]	Thickness [mm]
Spruce	MoR1	48.4305	1.229×10^{-5}	394.13	49.56	49.45	50.14
	MoR2	52.461	1.256×10^{-4}	417.62	49.93	49.81	50.51
Maple	P1	75.9269	1.214×10^{-4}	625.18	49.62	49.94	49.01
	P2	75.0555	1.217×10^{-4}	616.77	49.04	50.05	49.58

2.2. Experimental methods

2.2.1. Density Profile Analysis (DPA). The density profile of resonance spruce and maple samples was determined using an X-Ray Density Profile Analyzer DPX300 (Institute of Research and Development, Transilvania University of Brasov). The samples with the dimensions 50 × 50 × 30 mm^3 were automatically weighed and tested by the equipment devices. Then, each specimen was introduced in the X-ray device train where the equipment measured the density profile using the X-ray flux. The procedure is presented in detail by [6].

2.2.2. Ultrasound analysis. The determination of the propagation speeds was performed in laboratory conditions, at 23° C and relative air humidity of 65%, being performed in the Non-Destructive Testing Laboratory - INCDFT Iași. Lamb waves generation was performed with Hertzian contact transducers with a central frequency of 100 kHz. The emission signal was generated by a function generator, WW 1074 Tabor, the signal being amplified and transmitted to the emission transducer. The reception transducer is coupled to an NSIA 4395A, the waveforms and the measurement of the propagation times being made with the help of the digital oscilloscope Wave Runner 64 Xi. Through the IEEE 488.2 GPIB interface, the oscilloscope was connected to the PC for recording experimental data. The experimental set-up has the same configuration with previous researches [7]. Figure 3 shows the principle of the ultrasonic testing method. In order to ensure the coupling between the sample and the transducers, maintaining the Hertzian contact, a clamping system was designed and built which additionally ensured the application of a constant pressure on the surface of the sample. The measurements also took into account the effect of the conical elements for catching and exciting the waves. The difference between the modulus of elasticity of the two conical elements is minimal. The effect of the concentration strain is considered negligible. The experimental setup involves two pairs of contact transducers for wave propagation. The transmitted signals, captured by the receiver transducer were acquired 100 times with the help of the digital oscilloscope, and the data were mediated. The experimental determinations for measuring the speeds were performed first by recording the signal at the receiver without sample (transmission on the tips of the transducers in contact, 100 signals of 50 μs recorded in total) and afterwards with the sample (100 signals of 50 μs recorded for a single point measurement). The transmission time through all directions was estimated by implementing a cross-correlation between these two signals. From the average thickness of the samples in the direction of wave propagation, the velocities of the waves were determined by differentiation.

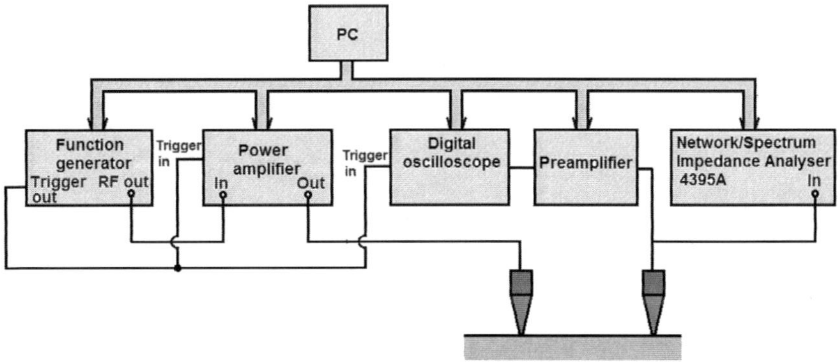

Figure 3. Experimental set-up of ultrasound test [7].

3. Results and discussions

3.1. Density Profile Analysis (DPA)

Figure 4 shows the variations in spruce wood density determined on the three main sections of wood. It can be noticed that the density variation depends on the early wood and late wood areas, this analysis allowing to obtain the density of the two wood areas in the structure of the annual ring, especially in the radial longitudinal section. It can be noticed that the density profile shows the width of the annual rings, which are regular and narrow with a gradual transition from early wood. For the curly fiber maple wood, the density profile is shown in figure 5. Unlike spruce wood, maple wood has a uniform structure, with no significant differences between early wood and late wood. The average density of resonant spruce wood is between 390 - 430 kg / m3 and that of the resonant maple wood varies between 650 - 680 kg / m3, that can be compared to the straight fiber maple, whose density has values between 610 - 630 kg / m3.

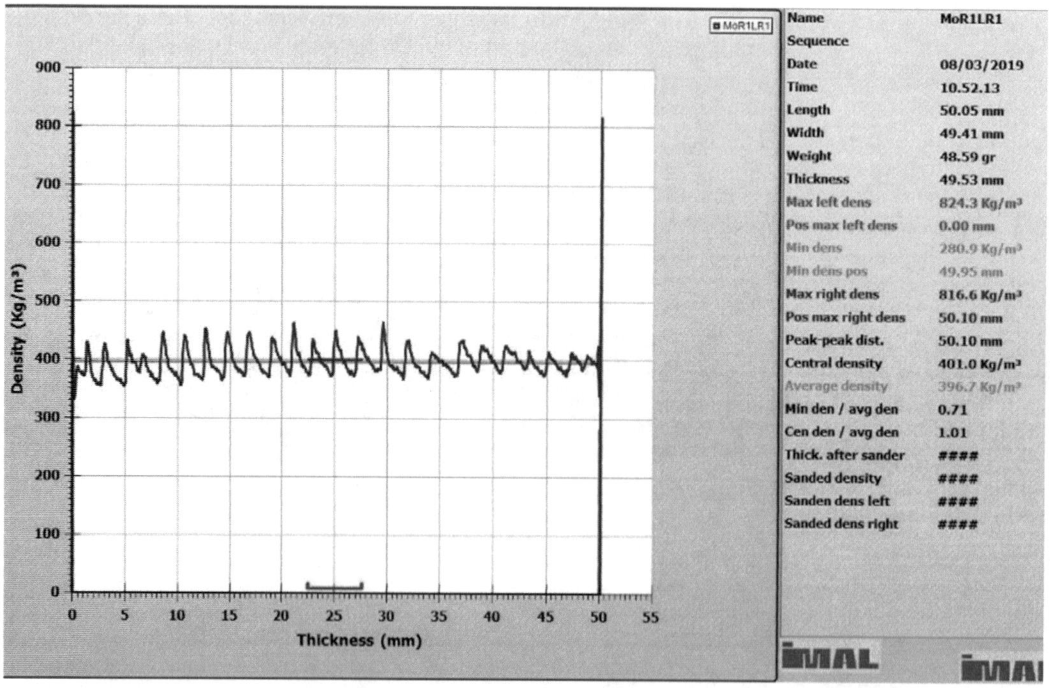

Figure 4. Variation of density profile for spruce sample scanned on LR direction.

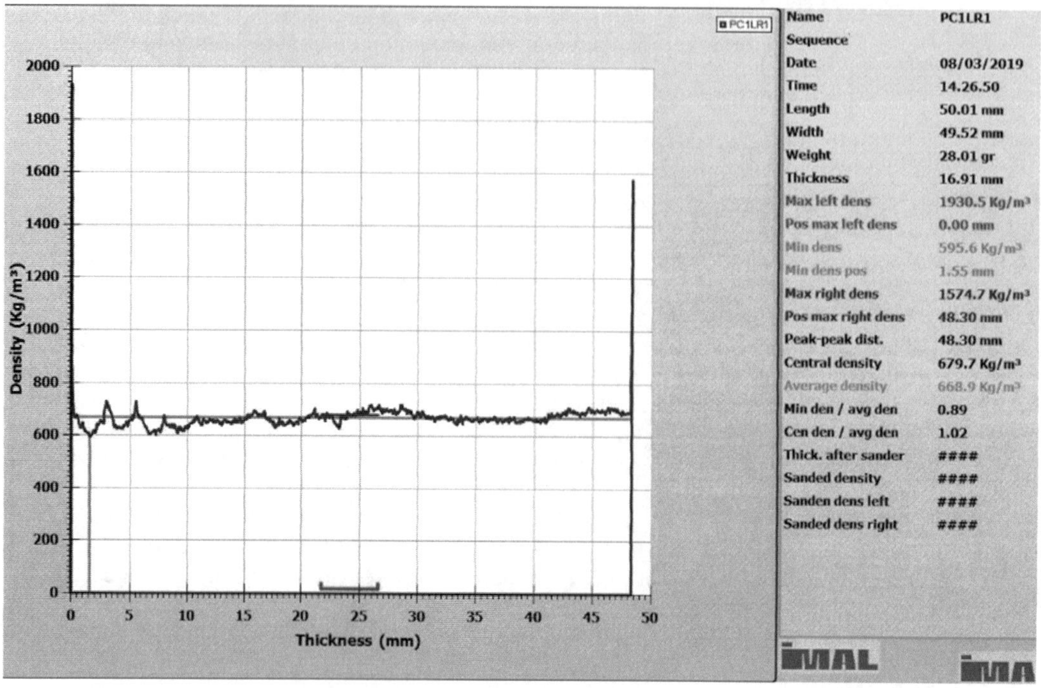

Figure 5. Variation of density profile for curly maple sample scanned in LR direction.

3.2. Determination of acoustic and elastic properties of resonance wood – ultrasound method

Table 2 shows the average values of the speed of sound propagation in wood, for the two species analyzed. US speed does not only depend on the direction of wave propagation, it also varies between wood species. Each species has characteristics that affect the transmission of ultrasonic energy depending on the physical properties of the cell walls or the structure of the cells. The procedure for identifying the propagation velocities applied to the two wood species showed that they strongly depend on a wide range of anatomical features, the effect of strengthening the rays and the density. The dependence of the ultrasound propagation speed on the wood structure was analyzed, determining the speed on opposite sides of the samples in 9 identification steps. The graphical representation of the sensitivity of the wood species taken in the analysis, at the propagation of the ultrasonic energy, produced a map of the speeds mediated at the maximum obtained speed (for each sample). A lower sensitivity for the LR direction was noted for all cases. Figure 6 shows the map of this sensitivity for the sample marked MOR1.

Table 2. US propagation speed in wood.

Sample		**Spruce sample MoR1**			**Spruce sample MoR2**	
Section	**TR1-TR2**	**LR1-LR2**	**LT1-LT2**	**TR1-TR2**	**LR1-LR2**	**LT1-LT2**
Average value of sound speed (m/s)	5269.412	1323.1	1425.702	5246.693	1335.413	1522.883
STDV	121.243	21.986	47.687	190.164	24.833	23.478
Sample		**Maple P1**			**Maple sample P2**	
Section	**TR1-TR2**	**LR1-LR2**	**LT1-LT2**	**TR1-TR2**	**LR1-LR2**	**LT1-LT2**
Average value of sound speed (m/s)	4263.64	1378.175	1949.582	4599.333	1438.092	1987.432
STDV	141.065	121.962	39.101	135.766	23.639	16.158

It should be noted that the detection of the first transverse wave was partially difficult in the LR and LT directions due to the effects of overlapping with the longitudinal waves. To calculate the speed of ultrasound, it was taken into account that, in addition to the specimen itself, sound waves propagate through the coupling medium of the contact transducer tips. The experimental data obtained show that, usually, the velocities in radial and tangential direction are about one third of the longitudinal velocity, with the radial velocity higher than the tangential one, which is in accord with [7, 8]. The radial velocity can be up to almost 14% higher than the tangential velocity.

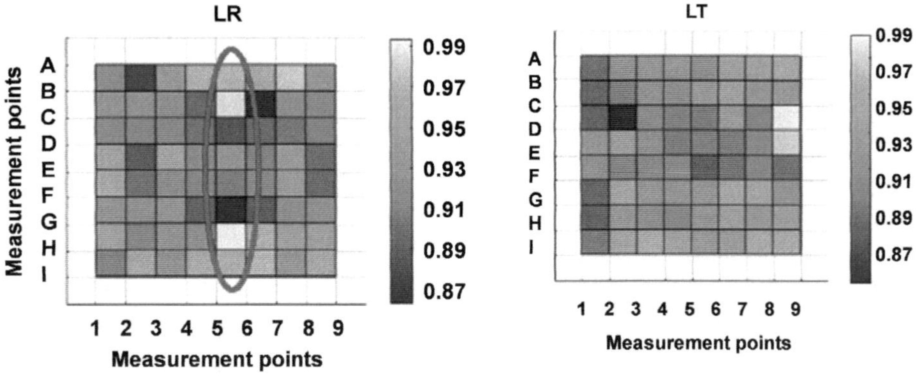

(a) (b)

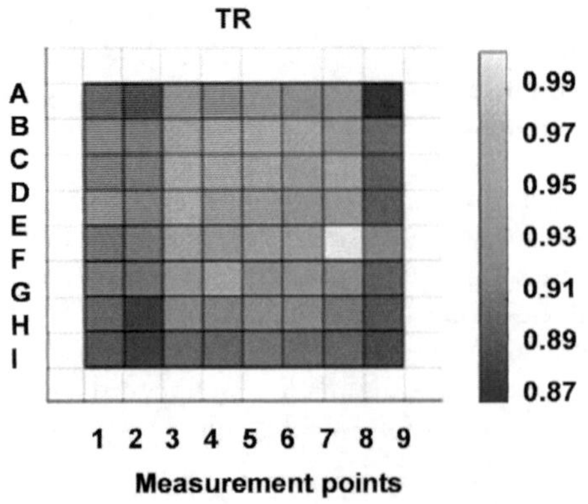

(c)

Figure 6. US speed sensitivity relative to the maximum speed for the three directions of propagation (case of spruce sample).

Based on calculation formulas according to the literature [7 – 9], the elastic characteristics of the tested samples on the three orthogonal directions of the wood were determined as can be seen in table 3.

Table 3. Elastic coefficients calculated using the ultrasound method (* Other methods are required to determine these quantities - with two specialized US translators having different frequencies for each method)

Sample type	E_L [MPa]	E_T [MPa]	E_R [MPa]	ν_{LR}	ν_{TR}	ν_{LT}	G_{LR}	G_{TR*}	G_{LT}
MoR1	10938	800	689	0.465	-	0.39	689	-	800
MoR2	11476	912	743	0.460	-	0.38	743	-	965
P1	11358	2374	1186	0.442	-	0.373	1186	-	965
P2	13028	2434	1273	0.446	-	0.38	1273	-	2432

Some of the Poisson's coefficients determined by the ultrasound method seem extreme and there is no reasonable explanation for this. Similar results are reported also by [9, 10] and [12 – 15]. The results obtained by experimental determinations are comparable in order of size with those reported in the literature, the differences may be due to multiple factors: the source of the sample, the small size of the sample and the different methods used to find these coefficients. It confirms the character of spruce resonance wood compared to maple; it is observed that the high-density wood species do not inevitably have the longitudinal Young's modulus with the highest value. The correlation between density and Young's modulus can be valid, but with limits only in a wood species.

4. Conclusions

The results obtained in the non-destructive testing based on the propagation of Lamb waves generated and received with US transducers with Hertz contact having the frequency of 100 Hz allowed the characterization of the samples obtained in cubic form from spruce and maple wood species. An experimental assembly was performed to measure the speed of Lamb waves and to ensure a contact at a constant pressure of the transducer with the sample. The transmission time through all directions R, L, T was estimated by implementing a cross-correlation between the two signals from the conical clamping elements and the wood material. It has been found that US velocity depends on the physical properties of the cell walls or cell structure. A lower sensitivity for the LR direction was noted for all samples.

The experimental data obtained show that, usually, the velocities in radial and tangential direction are about one third of the longitudinal velocity, with the radial velocity higher than the tangential one. The radial velocity can be up to almost 14% higher than the tangential velocity due to the fact that the fibers and tracheid's are mostly aligned in the longitudinal direction, compared to the radial direction. Determining the shear modulus on the TR direction is complicated and imprecise, not being standardized, it involves an auxiliary method of ultrasound resonance (RUS) by the procedure of finding the natural resonance frequencies. The study confirms the special character of spruce resonance wood and shows that high-density wood species do not inevitably have the Young's longitudinal modulus with the highest value. The correlation between density and Young's modulus can be valid, but with limits only in a wood species.

5. References

[1] L.Duquenme, E.Moulin, J.Assood, S.Grandel 2004 Transient modeling of Lamb wave generated in viscoelastic materials by surface bonded piezoelectric transducers *J.Acoust.Soc.Am* **116** 133-141.

[2] Nowak, T., Hamrol-Bielecka, K. and Jasienko, J 2015 Non-destructive testing of wood–correlation of ultrasonic and stress wave test results in glued laminated timber members *Annals of Warsaw University of Life Sciences-SGGW. Forestry and Wood Technology*, **92**.

[3] Fang, Y., Lin, L., Feng, H., Lu, Z. and Emms, G.W 2017 Review of the use of air-coupled ultrasonic technologies for nondestructive testing of wood and wood products *Computers and Electronics in Agriculture* **137** pp.79-87.

[4] Dinulică F, Albu C T, Vasilescu, M M and Stanciu, M D 2019 Bark features for identifying resonance spruce standing timber, *Forests* **10**(9) Article number 799 DOI: 10.3390/f10090799.

[5] Stanciu M D, Cosereanu C, Dinulica F and Bucur V 2020 Effect of wood species on vibration modes of violins plates *Eur. J. Wood Prod.* **78**, 785-799.

[6] Stanciu M.D., Șova D., Savin A., Iliaș N., Gorbacheva G 2020 Physical and Mechanical Properties of Ammonia-Treated Black Locust Wood, *Polymers*, **12**, 377 doi:10.3390/polym12020377.

[7] Grimberg R, Savin A, Steigmann R, Stanciu MD, Grum J, 2010Determination of Elastic Properties of CFRP Using Lamb Waves Resonant Spectroscopy, *2nd International Symposium on NDT in Aerospace*, Hamburg, Germany.

[8] Bucur V. 2006 *Acoustics of wood* Springer Verlag, Berlin.

[9] Vazquez C., Golçalvez R., Bertoldo C., Bano V., Vega A., Crespo J., Guaita M. D. 2015 Determination of the mechanical properties of Castanea sativa Mill. Using ultrasonic wave propagation and compression with static compression and bending methods *Wood Science and Technology* **49** 607-622.

[10] Ozyhar T., Hering S., Sanabria S.J., Niemz P. 2013 Determining moisture-dependent elastic characteristics of beech wood by means of ultrasonic waves *Wood Sci and Technol* **47** 329-341.

[11] Bucur, V. and Archer, R.R., 1984. Elastic constants for wood by an ultrasonic method *Wood Science and Technology* **18**(4) pp.255-265.

[12] Gao S., Wang L., Wang Y., 2009 A comparative study on the velocities of stress wave propagation in standing Fraxinus mandshurica trees in frozen and non-frozen states *Frontiers of Forestry in China* **4**(4), 382-387.

[13] Chan, J.M., Walker, J.C. and Raymond, C.A., 2011 Effects of moisture content and temperature on acoustic velocity and dynamic MOE of radiata pine sapwood boards *Wood Science and Technology* **45**(4) pp.609-626.

[14] Hasenstab, A., Krause, M. and Osterloh, K. 2006 Testing of wooden construction elements with ultrasonic echo technique and X-ray. In *Proceedings of the 9th European Conference on Non-Destructive Testing*, Berlin, Germany pp. 25-29.

[15] Esteban L. G., Fernandez F. G., de Palacios P. 2009 MOE prediction in Abies pinsapo Boiss. timber: Application of an artificial neural network using non¬destructive testing, *Comput. Struct.* **87** 1360-1365.

Acknowledgement
This paper was supported by a grant of the Ministry of Research, Innovation and Digitization, CNCS/CCCDI – UEFISCDI, project number. 568PED/2020 MINOVIS, within PNCDI III.

We are grateful to Prof. Timar Cristina Maria for permission to use the microscopic views (figure 1) from Catalog of micro-structural characteristics of materials for restoration and ecodesign (December 2009), CNCSIS Project IDEI 856/2008: Creation and implementation of a scientific research methodology on restoration - wood preservation and eco-design in the perspective of sustainable development. Project manager prof. Dr. Chem. Maria Cristina TIMAR, Transilvania University of Brasov, Faculty of Wood Engineering).

IOP Conf. Series: Materials Science and Engineering 1182 (2021) 012074 doi:10.1088/1757-899X/1182/1/012074

Numerical Modal Analysis of Kinked Bars –Triangle Case of Study

M D Stanciu[1], M Trandafir[1], G Dron[1], M V Munteanu[1] and V Bucur[2]
[1]Translvania University of Brasov, Blvd. Eroilor, No. 29, 500036, Brasov, Romania
[2]RMIT University, School of Science, Melbourne, GPO Box 2476, VIC 3001, Australia

E-mail: mariana.stanciu@unitbv.ro

Abstract. The paper deals with numerical modal analysis of a kinked bars similar with an open triangle, with different sizes, angles and diameters of bar. Using SIMCENTER software, the first step was to design the geometry of each type of structure, then discretization of model in finite elements and assigning properties of material (in this case, Al alloy). The results in terms of eigenvalues and eigenmodes of each structure revealed that the angles of kinked bars influenced the value of natural frequency: with increase the tip angle from 30° to 60° and 90°, the eigenvalue decreased in ratio of 1:1.39:1.611. Also, the length of each side of the kinked bar influences the modal response of the structure: increasing the length from 200 mm to 450 mm, leads to decreases of eigenvalue of almost 5.6 times. The applicability of this study is linked to percussion instruments used in musical orchestra and also to explain the energy transfer and acoustic effect of different types of triangle instruments.

1. Introduction

The triangle is a percussion musical instrument that is obtained by bending a metal bar with a circular section in the shape of an open triangle. In fact, the triangle is a metallic bar bent into the shape of an equilateral or isosceles triangle, with rounded vertices [1, 2]. The triangle has effectively only two vertices, the third is an open apex to allow the free vibration of the bar. Commonly, the transverse section of the triangle is circular. Some models have a variable section at the end of the free bars.

This instrument is classified as an idiophone, ie the production of sound is done only by itself vibration, without the intervention of an auxiliary part. The triangle is obtained in several geometric configurations from different types of material: stainless steel, aluminum alloys, brass, and the section of the bar can be constant along its entire length or variable. From a mechanical point of view, the triangle is a combination between the right beam and the kinked beam, having an axis of constructive symmetry with respect to the open top which differs from the axis of loading and boundary conditions. The modern triangle comes in several sizes with an edge commonly being 20cm or 30 cm and could be up to a maximum of 40 cm or 45 cm. The triangles are made of a bent rod of circular cross section, produced by folding the metal around a jig. The shape of the triangle could be equilateral or isosceles. The orchestral triangle is in fact a combination of straight and curved bars, and therefore is an instrument of relatively complex structure. The triangle is struck with a metallic mallet, and is supported with a thread around the top bend of the hanging instrument. The struck triangle generates longitudinal and flexural modes of vibration, which are excited perpendicular or parallel to the plane of the triangle.

For longitudinal vibrations it can be assumed that the frequency of a bar of circular cross section and of length L is calculated as in equation (1) [3]:

$$f_n = \frac{a}{2}(2n+1)^2 \frac{\pi}{8\,L^2} \sqrt{\frac{E}{\rho}} \tag{1}$$

with frequencies in the ratios 3^2: 5^2: 7^2: 9^2 :11^2:13^2.

For vibrations perpendicular to the plane of the triangle, the displacement is small. As noted by [3 – 5] for the transverse modes, the modes of vibration of a triangle are almost identical to those of the straight bar from which the triangle is bent. This statement argued in favor of the hypothesis that the shape of the triangle has no big effect on its vibration modes. However, [6, 7] noted that the curvature of a bar influences the stress distribution which is determined by the inertia moments applying at the end of a bar. For flexural vibrations in the plane, there is an important discontinuity in impedance at each bend angle. The transverse vibration within one edge couples to the longitudinal vibration in the adjacent edge. Therefore, each edge of the triangle will support its own vibration modes and will be coupled to another edge, to form sets of normal mode triplets, for an edge of the same length.

The aim of this paper is to determine the eigenmodes and eigenfrequencies of different types of triangles in order to analyze the dynamic behavior of them, using finite element analysis (FEA).

2. Materials and methods

2.1. Materials

Within these researches, the geometries of three triangles were made, named in the study as cases: case A, in which the bar is bent in the form of an equilateral triangle, with the subcases in which the length of the side and the diameter differ; case B - the shape of an isosceles triangle, with the subcases in which the side length differs in a row and case C - isosceles right triangle, with the subcases in which the side length differs in a row (figure 1).

The geometric characteristics of the studied structures are presented in table 1. The material from which these triangles are made is C17200 beryllium copper alloys, having the physical and elastic characteristics in table 1.

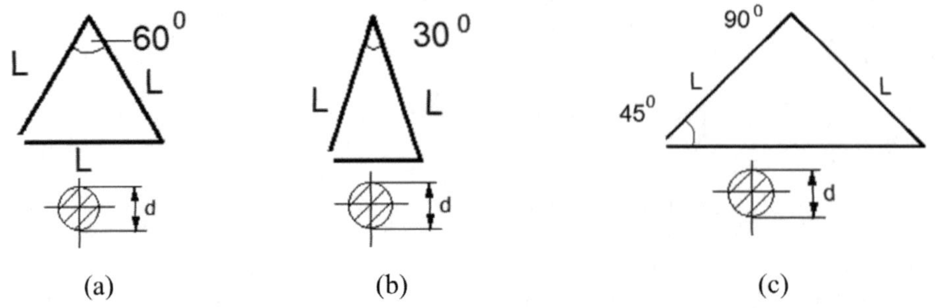

Figure 1. The types of triangles studied: a) equilateral triangle, coded A; b) isosceles triangle, coded B; c) isosceles right triangle coded C.

Table 1. The physical and geometrical features of triangles.

Cases			**A – B - C**			
Subcases	**1.1.**	**1.2.**	**1.3.**	**2.1.**	**2.2.**	**2.3.**
Length L (mm)	200	300	450	200	300	450
Diameter d (mm)	12	12	12	8	8	8
Density (kg/m3)			8250			
Elasticity Modulus (MPa)			125000			
Poisson Coefficient			0.3			

2.2. Simulation of modal analysis

Finite element modeling of the proposed structures was done with *Simcenter 3D* software. In the first stage, the 18 constructive variants of the triangles were made, each structure being then discretized with finite elements. In terms of boundary conditions, the triangles were considered free structure. Having a relatively simple structure, in the simulation we choose for finite elements of Thetraedral type (TET) with a size of 3 mm (figure 2). The material characteristics were introduced and then the FEA analysis was run, obtaining the proper modes of all the sub-cases considered.

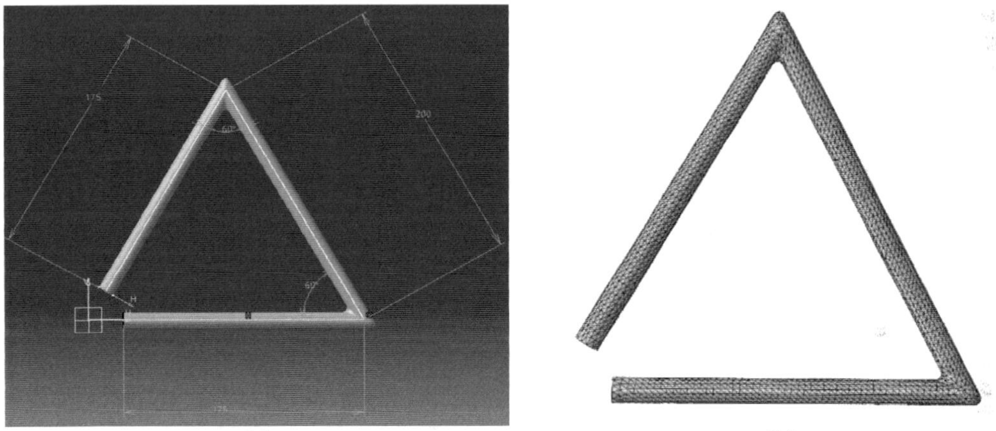

(a) (b)

Figure 2. The geometry of triangle obtained in Simcenter 3D: (a) case of equilateral triangle; (b) meshed structure.

3. Results and discussions

After the FEA analysis, the modal shapes and fundamental frequencies resulted. It was found that the first six modes are structural modes, with very low eigenfrequencies, which is why eigenvalues were recorded and analysed starting with the seventh mode. Also, geometrical characteristics of the triangle have a determinant effect on the modes of vibration as shown in table 2 and 3. With increasing the diameter of rod, the increases of eigenfrequencies it's observed (figure 3). The biggest difference due to the increase of the diameter by 50%, is obtained in the case of the isosceles triangle, when the first eigenfrequency increases by 57%. With the increase of the bar length, of 2.25 times, a decrease of the first frequency is observed regardless of the diameter and angles.

The low frequency modes are mainly flexural modes and are related to the size, and the mechanical properties of the material of which the triangle is made. At constant length, the frequency increases with the vibration modes, so that the increasing length of the edge determines the decrease in frequency. This means that the triangles of shorter edges vibrate at higher frequencies than the triangles of larger edges. As regards the triangle type, in general, the eigenfrequencies of the isosceles triangle of L = 20 cm, are higher than for other types of triangle.

Figures 4, 5 and 6 shows the first modal shape of vibration of the equilateral, isosceles and rectangular isosceles triangles, with a diameter of 8 mm.

Table 2. The eigenfrequencies of analysed triangles with diameter of 12 mm.

Mode/	Frequency (Hz)								
Cases	**A1.1**	**A1.2**	**A1.3**	**B1.1**	**B1.2**	**B1.3**	**C1.1**	**C1.2**	**C1.3**
1	132	53	22	191	74	30	84	34	15
2	133	54	23	258	97	39	92	37	16
3	217	88	37	536	208	84	195	81	34
4	623	274	120	837	366	160	527	223	96
5	787	336	145	1142	474	202	567	237	101
6	1192	492	208	1205	507	216	706	309	135
7	1225	508	213	1456	573	235	826	349	150
8	1243	516	221	2041	976	448	1228	516	221
9	1325	544	229	2694	1163	508	1293	539	230
10	2066	964	434	2842	1309	579	1503	655	283

Table 3. The eigenfrequencies of analysed triangles diameter of 8 mm.

Mode/ Cases	**A2.1**	**A2.2.**	**A2.3**	**B2.1.**	**B2.2.**	**B2.3.**	**C2.1**	**C2.2**	**C2.3.**
1	87	35	15	121	48	20	55	23	10
2	88	36	15	168	64	26	61	25	10
3	141	58	24	351	137	56	129	54	23
4	422	184	81	566	245	107	353	149	64
5	525	224	97	762	316	135	377	158	67
6	802	329	139	826	341	144	478	208	91
7	829	338	142	928	370	153	545	231	100
8	831	347	148	1483	677	303	830	346	147
9	869	357	151	1797	776	338	861	358	153
10	1471	659	293	2104	915	392	1038	441	190

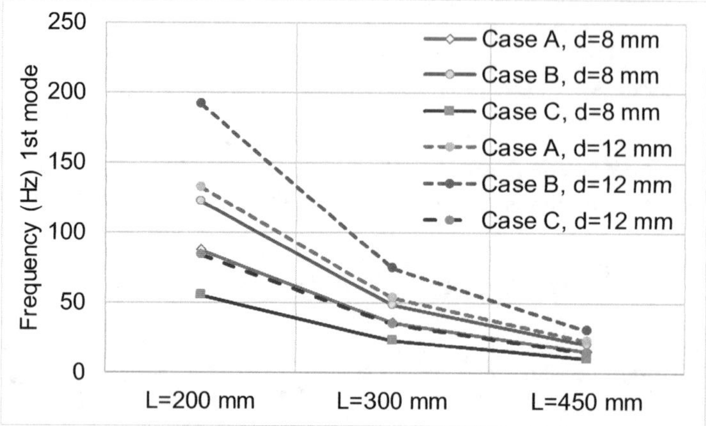

Figure 3. Comparison of first eigenfrequencies in all studied cases.

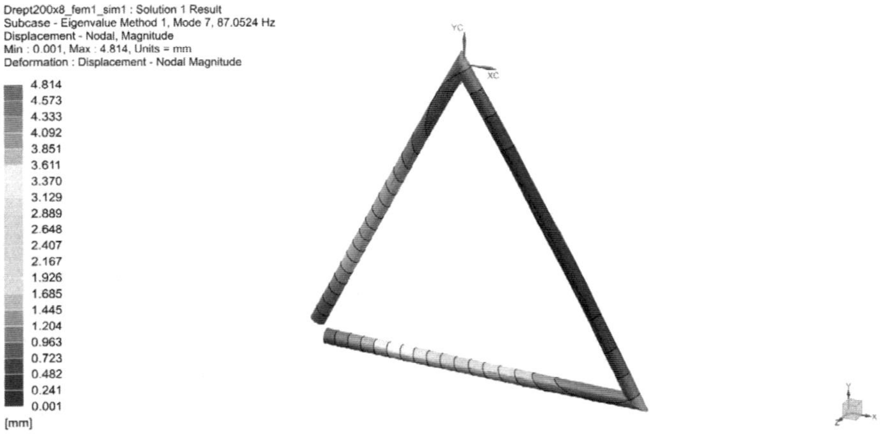

(a) Case A.2.1. (L=200 mm); f=87 Hz

(b) Case A.2.2. (L=300 mm); f= 35 Hz

(c) Case A.2.3. (L=450 mm); f=14 Hz

Figure 4. The first eigenmodes of equilateral triangle – case A.

(a) Case B.2.1. (L=200 mm); f=121 Hz

(b) Case B.2.2. (L=300 mm); f=48 Hz

(c) Case B.2.3. (L=450 mm); f=19 Hz

Figure 5. The first eigenmodes of isosceles triangle – case B.

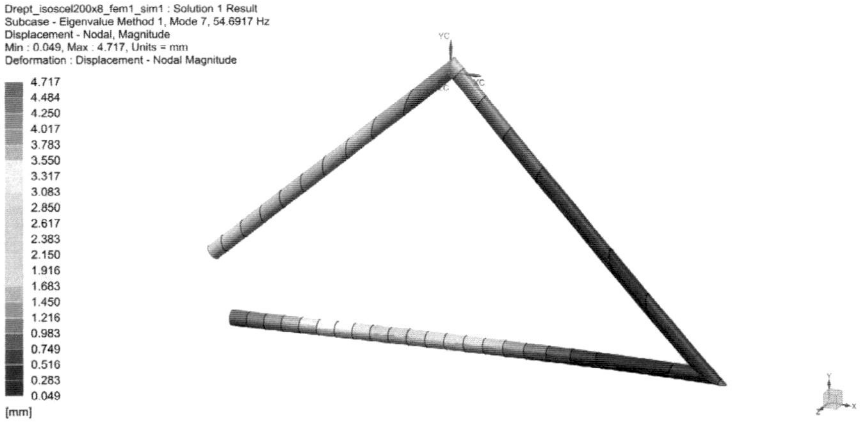

(a) Case C.2.1. (L=200 mm); f=54 Hz

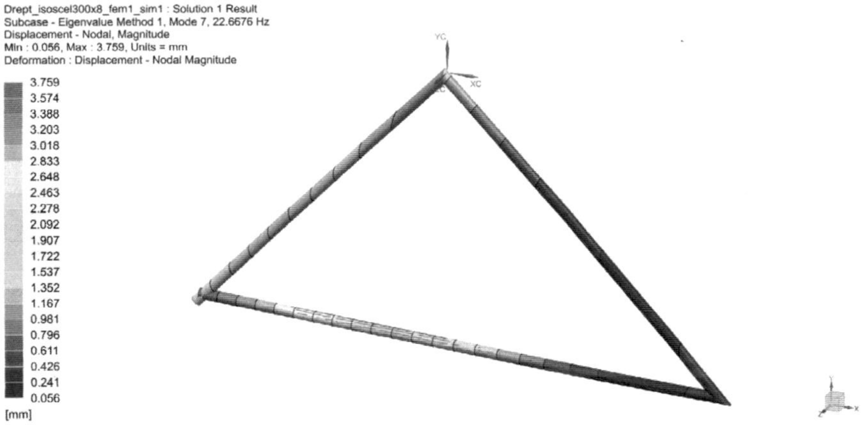

(b) Case C.2.2. (L=300 mm); f=22 Hz

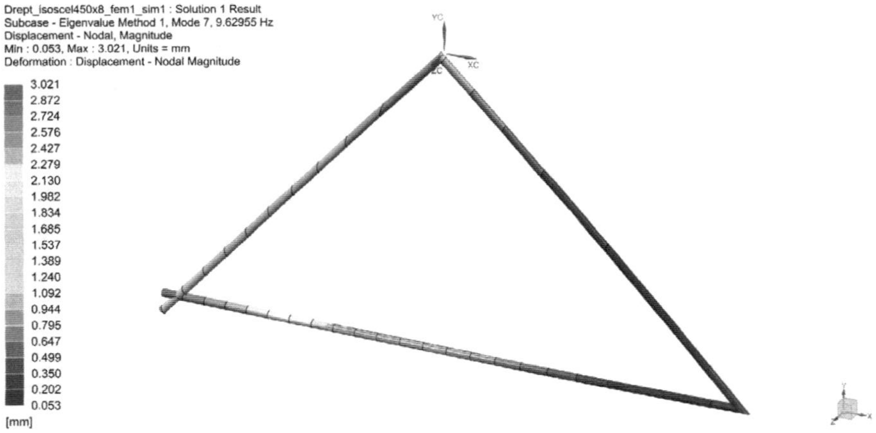

(c) Case C.2.3. (L=450 mm); f=9 Hz

Figure 6. The first eigenmodes of isosceles right triangle triangle – case C.

The triangles having shorter edges have higher frequency modes. i.e. the frequency of mode 1 of the triangle having a perimeter of 665 cm is about 542 Hz and of the triangle having a perimeter of 1350 mm the frequency is only 134 Hz. At higher modes, the difference is more important, such as for mode 4, the frequencies are respectively 489 Hz and 1993 Hz. The effect of the bar length or of the perimeter on the modal frequencies of three isosceles triangles of about 40 o and having a perimeter varying from 665 mm to 1350 mm is given in table 4 [7]. In figure 7 it can be noticed the modal shapes of equilateral triangle. In all cases, the flexural modes are presented with increasing the number of nodes and antinodes with increasing the frequency.

In table 4, a comparison between modal analysis of a straight rod and kinked rod performed by Bucur, 2020 [8], is presented. Generally, the triangles recorded a lower frequency compare to straight rod, for the flexural mode of vibration which is considered the main mode for beam.

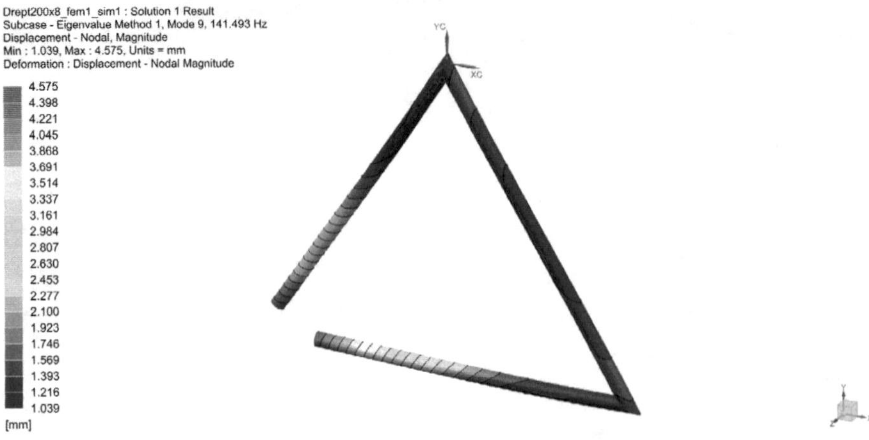

141 Hz

422 Hz

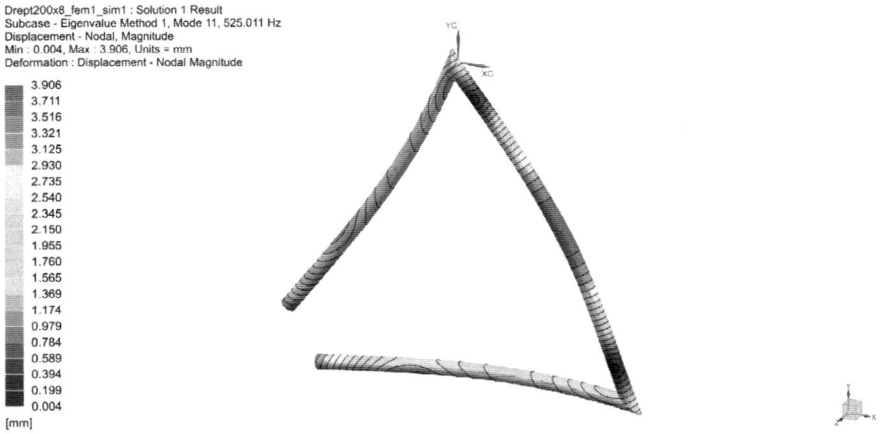

525 Hz

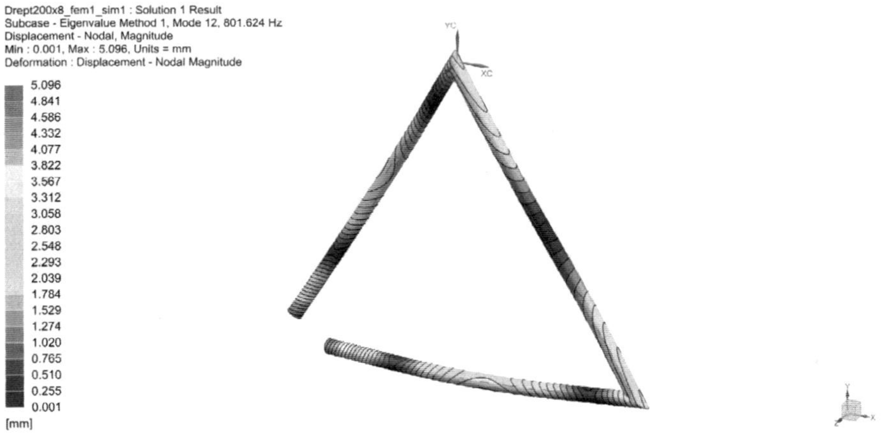

801 Hz

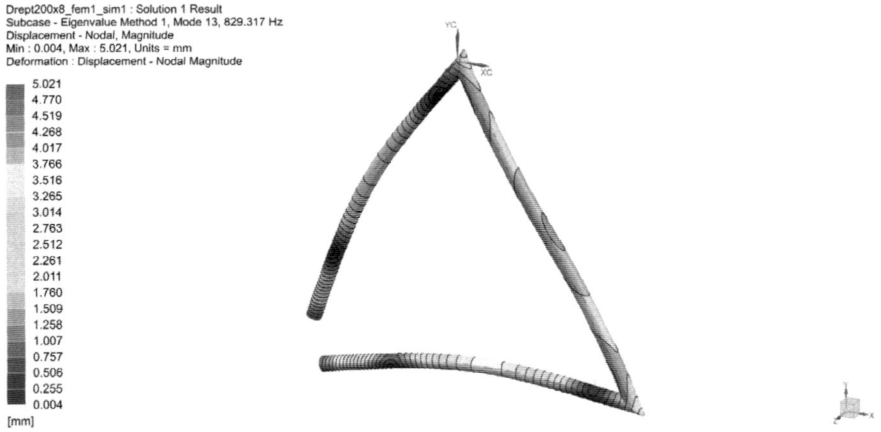

869 Hz

1471 Hz

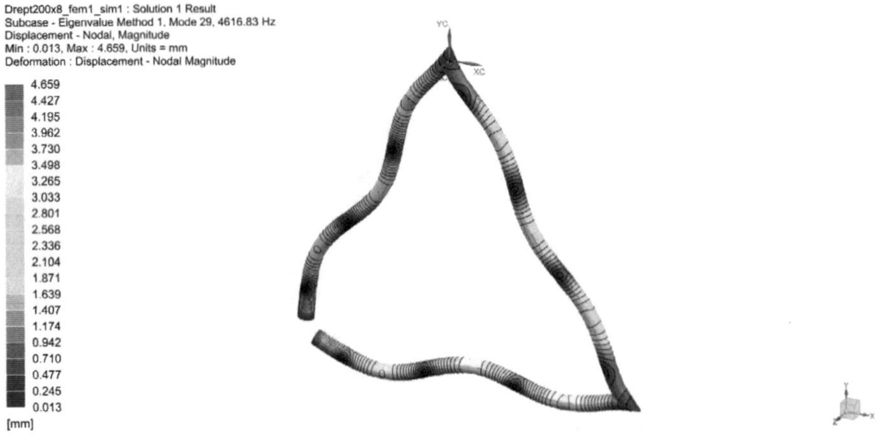

4616 Hz

Figure 7. The modes of vibration of the equilateral triangle, having equal edges of length 200 mm, and a diameter of 8 mm.

Table 4. Vibrations modes of a triangle compared to a straight rod, made of steel [7].

Vibrator	Specification	Flexural modes of vibration Mode 1 In plane	 Out of plane Frequency (Hz)	 Mode 2 In plane
Rod	Straight	61	-	166
Triangle isosceles	Angle at base 30^0	68	66	88
Triangle right angle	Angle at base 45^0	55	55	88
Triangle equilateral	Angle at base 60^0	48	49	89
Differences to rod			%	
Triangle isosceles	Angle at base 30^0	-11.5%	-7.5%	+46.98%
Triangle right angle	Angle at base 45^0	+9.8%	+9.8%	+46.98%
Triangle equilateral	Angle at base 60^0	+21.3%	+19.6%	+46.38%

4. Conclusions

The paper analysed the dynamic comparison of different types of triangles from a geometric point of view in order to establish the causal relationships between the modes of vibration, their eigenfrequency and their dimensions. The bending angle of the bars, the length of the sides of the triangle and the diameter play a decisive role in the vibrations of the triangle and implicitly in the emission of sounds, this being a necessary percussion instrument in musical orchestras.

Future research aims at experimental analysis on the frequency spectrum, damping and dominant frequencies of commercial triangles, in correlation with the elastic properties of the materials from which they are used.

5. References

[1] Legge K A 1987 Non-Linear Mode Coupling in Symmetrically Kinked Bars *Journal of Sound and Vibration* **118**(1) 23 -34.

[2] Blades J 1974 *Percussion instruments and their history* Faber and Faber, London.

[3] Rossing TD 1976 *Acoustics of percussion instruments*. The Physics Teacher 14 – 546 – 556.

[4] Bestle P, Hanss M., Eberhard P 2014 Experimental and numerical analysis of the musical behaviour of triangle instruments *Proc. of the 5th European Conference of Computational Mechanics (ECCM V)* Barcelona 3104-3114.

[5] Bucur V 2016 Handbook of materials for string musical instruments. *Springer*, Chapter **3** - Mechanical characterization of materials for string instruments 93 – 132.

[6] Gough C 2006 Chapter 15 *Musical Acoustics* In Springer *Handbook of Acoustics*, ed TD Rossing 533 – 667.

[7] Dunlop J I 1984 Flexural vibrations of the triangle *Acustica 55* 250-253.

[8] Bucur V 2020 *Handbook of materials for percussion instruments*, Chapter 8 *Struck idiophones played with mallets: gongs, cymbals, chimes, bell plates, triangle* Springer – in preparation.

Acknowledgement

This paper was supported by a grant of the Ministry of Research, Innovation and Digitization, CNCS/CCCDI – UEFISCDI, project number. 568PED/2020 MINOVIS, within PNCDI III.

Dimensional measuring of parts as function of temperature variation and FEM study

F Stanciu[1]**, D Nitoi**[2]**, O Chivu**[2] **and A Tăpârdea**[3]
[1]Reflex Fatde, BVD. Iuliu Maniu no.16, Bucharest
[2]University Politehnica of Bucharest, Splaiul independentei no. 313, sect.6, Bucharest, Romania
[3]Politehnica University Timişoara, Piaţa Victoriei no.2, Timişoara, Romania

E-mail: dan.nitoi@upb.ro

Abstract. The components of astructure can be made in many different areas by many suppliers, before being brought together as a single final. Measuring geometry and geometric dimensions of a structure is an essential process in establishing product quality. Dimensional measurement is often affected by several human or physical factors, of which temperature is an essential factor in introducing deviations. Often the expected results are out of specification due to temperature variation. Paper aims to present the effects due to temperature variations in dimensional measurement as well as way to solve problems. One of the problems in the measurement process is the lack of repeatability in measuring tolerated dimensions. The study presents measurements in the conditions of volumetric thermal expansion of AMCs and measured parts, which are directly influenced by temperature. The need to verify the accuracy of measurements and control of equipment is an important step for quality departments in this industry because the quality of production is based on the target of "zero defects" so the need to know about the condition of equipment is extremely important. In specific cases in addition to the normal calibration which is according to international standard ISO 10360-2: 2009 as a mandatory application, the same standard determines the need to perform re-verification tests according to user specifications, looking for deviations that should not occurs in systematic errors. Also, the international standard ISO 17025: 2005 which defines the general requirements for testing and calibration components of laboratories, in point 5.5.10 of the technical requirements it is said that intermediate checks are necessary to maintain confidence in the calibration status of equipment, these checks having to be done according to a specific procedure. Three techniques for assessing measurement uncertainty are presented in ISO 15530-; they can be used alone or in combinations. The first of these is known as sensitivity analysis and is described in the ISO Guide for Expressing Uncertainty in Measurement. Modeling and simulation can help evaluate the size of a part as a function of temperature. The article presents in the second part this method which offers values very close to the experimental ones and which may in the future simulate the values of the measurements where the experimental part is difficult to be done.

1. Introduction

Currently there are three techniques for assessing measurement uncertainty are presented in ISO 15530-; they can be used alone or in combination. The first of these is known as sensitivity analysis and is described in the ISO Guide for Expressing Uncertainty in Measurement. The second technique is a comparative approach that involves the use of a comparable artifact to capture sources and interactions of uncertainty; it is considered rigorous and defensive, although it can be expensive because it is based on the existence of an artifact, the availability of a more capable measurement system and the ability to meet similar conditions. The third technique is to use Monte Carlo-based simulation, known as Uncertainty Assessment Software (UES), and may be the only viable option when there are many measurements to be evaluated and predictions are needed; such as if someone wants to plan the consistency of measurements in a variety throughout the life cycle of the product. The simulation is unlikely to adequately cover all sources of uncertainty, so it is advisable to use it in combination with other techniques. The relative merits of these three techniques are discussed in Baldwin et al, which referred to a previous draft of ISO 15530-1, which mentions that two additional options are discussed - the history of measurements and the judgment of experts. These techniques seem to have been de-emphasized in the latest release, however they may have an important place in a context where historical records and expert opinions could be systematically captured and used. In addition, future revisions are expected to extend the use of the Bayesian approach, supporting the idea of using more uncertainty assessment techniques, as long as they improve the current state of knowledge.

2. Strategies in the field of part measuring

Three working strategies are presented for part measurements. These are briefly presented below.

2.1 Expert strategy

A small number of knowledge-based systems have been designed, which try to capture the knowledge of experts. This may include knowledge of the manufacturing process. However, given the large number of variables involved, doubts have been raised as to whether the information captured may be sufficient.

2.2 Adaptive strategy

The trend is towards strategies that adapt to real geometry. Innovative methods that dynamically change strategy, using previous measurements to determine the choice of the next point, are promising. However there are unresolved technical difficulties, for example in avoiding collisions. A related adaptive approach is to study the manufacturing process and characterize its "signature". The signature is used to develop a model of the actual feature that was produced. The measurement strategies are then designed based on this model of the real characteristic.

2.3 Digital metrology

It provides the opportunity to inject the structure into the task of the metrologist, integrating tools that allow testing strategies for evaluating applications on digital models. If a feature can be synthesized from a detailed measurement, different strategies can be tested depending on the model to determine which of them reaches a target level of uncertainty for the lowest cost (for example, by taking a small number of points).

3. The influence of temperature in the measurement process

Thermal expansion is the physical phenomenon by which the dimensions (volume, surface, length) of a body increase as a result of temperature variation. The opposite phenomenon is called thermal contraction [1]. For most substances, increasing the temperature leads to an increase in size, but there are exceptions. If a bar with length L [mm] is heated with t [°C], it will elongate and will have length $L = L_0 (1 + \alpha * \Delta t)$. Here L_0 is the part length at temperature T = 20°C, L is the part length at measuring temperature. Possible causes in which measurements errors may occur are:

- Case 1 – temperature difference between the part processing area and the part measurement area. During the measurement performed on the mechanical processing machine, the part has a temperature T_1 [°C], and when the final processing has been performed, the part is removed from the M.U., being moved to the area where the last adjustments and the final dimensional inspection are made. The temperature is different, hence it results that there is no repeatability in the measurements (temperature $T_2 \neq T_1$).

- Case 2 – Temperature difference between the temperature of the AMCs and the temperature of the parts. In general, micrometers are stored in specially designed places where the temperature is controlled and maintained at T = 20 °C ± 2, unlike the measured parts which are located in the machining section where the temperature reaches ~ 5 °C in the cold season and ~ 35 °C in the warm season. Depending on the season, there are large differences between the temperature of the measured part and the temperature of the measuring instrument. Because the shops where mechanical processing is performed are large, it is difficult to ensure a constant temperature of ~ 20 °C depending on the season.

Due to temperature variations, the measurement results in the different stages of processing and control have different values.

4. Applicable procedure in the case of dimensional measuring at different temperatures

4.1 Generalities

In order to improve the results obtained after the measurements, a procedure for measuring large parts in different measuring thermal conditions was drawn up. This procedure contains detailed steps for obtaining the results as close to reality as possible. This procedure was also applied in the case of the Crusher Roller part, presented in figure 1.

Figure 1. Roll crusher metal machining.

The optimal measuring conditions are represented by the constant temperature of 20 °C of the AMCs and of the measured parts. In general, these conditions are difficult to obtain, due to the conditions in the production halls and the size of the parts. It is often difficult for these parts to be moved to a temperature controlled area. According to the SR EN1090-1 standard, it is known that a 1000mm steel bar has the dimension of 1000.00 mm at 20 °C, where this temperature is the 0 reference point for dimensional tests. The deviation 1000mm for every 1°C is about 0.0125 lower or higher. Example: A piece measured at 30 °C with an inner diameter of 1000 mm and a temperature of the measuring instrument of 20°C. There is a possibility that during the measurements performed on the processing machine, the operator performing the measurement will not realize the 10°C deviation. In these conditions the piece will be made at dimension 1000.00 mm. The actual value resulting in these conditions will be 1000.125 mm. A deviation of + 0.125mm will be more than enough for the part to be declared rejected when the tolerance allowed for it is + 0.06/0.08 mm. Under these conditions, the inspector performing the measurement must ensure that the part and the measuring instruments have the same temperature, namely the ambient temperature and the temperature of the part must be the same. To ensure that the two elements are at the same temperature, they must be stored for a few days, depending on the dimensions of the workpiece and the initial temperature after final machining. The duration of the measurements can be long if the same measuring equipment is used and therefore the inspector performing the measurement must know that a long use of the measuring equipment risks transferring heat from his hands to the equipment and dimensional deviations may result. To avoid heat transfer from the operator's hands which may have an adverse effect on the measurements, after each measurement it is recommended that the meter be left near the workpiece and avoid contact with the operator's hands so as not to heat the measuring instrument. It is recommended that there be a 10-minute break between measuring each size (figure 2).

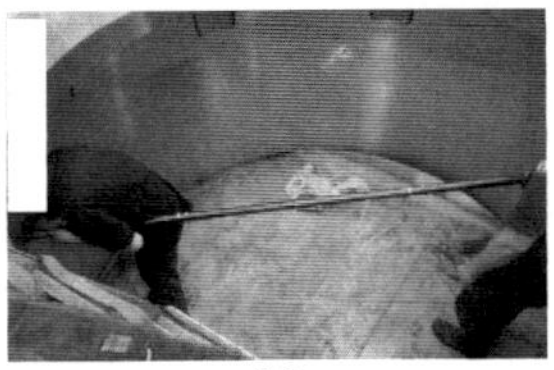

(a) (b)

Figure 2. Crush roller dimensional measuring.

4.2 Calibration procedure

An annual verification of CMD in a metrology institute that includes a metrological verification certificate is not sufficient to ensure that the measuring device is in optimal condition. The calibration made in a metrology institute only verifies that the current equipment is maintained in the parameters at a temperature of 20 °C (figure 3.). A metrology certificate cannot provide real values during use in measurements and does not certify the correct calibration. Note: the measuring equipment needs to be calibrated before use.

Figure 3. Checking the interior micrometer on a static machine in laboratory conditions.

*4.3 Calibration on stationary d*evices

- If the calibration of the equipment is necessary on a stationary measuring device, then it is very important that the equipment is acclimatized in the same conditions in which they were checked in the laboratory conditions;
- After acclimatization of the equipment, calibration can be done;
- After calibration the measuring equipment must be acclimatized in the same conditions as the part to be measured. Again, leave the equipment near to the measuring part. Ensure that the measuring equipment is 100% acclimatized before use;
- After the final acclimatization the measurement can start, ensuring at the same time that there is no heat transfer between the operator's hands and measuring instruments. Checking the temperature of the measuring equipment during the measurements is recommended.

4.4 Checking the calibration with the gauge of the measuring equipment

- If the gauge belongs to the measuring equipment and can be brought with it before calibration it is necessary to have the same temperature;
- Acclimatization of the calliper and the measuring instrument so the calibration can be done (figure 4).

5. Finite element modelling of the parts expansion process during the metal machining

In order to show the change of dimensions that takes place with the increase of the temperature due to the processing by cutting and to put in correlation with the analytical calculations, the modelling with finite elements was performed.

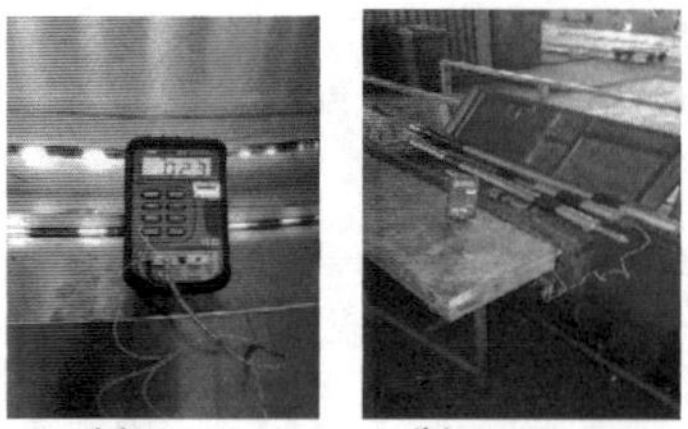

(a) (b)

Figure 4. Measuring equipment acclimatization; a – part temperature measuring T = 23°C; b – acclimatization of the measuring instrument

- After calibration the measurement can start (figure 5).

Figure 5. Indoor micrometer calibration.

This makes the connection between thermal and mechanical phenomena that include the increase of its dimensions. In the case of modeling with finite elements of the process of expansion of a part, the following steps are presented [2]:

- Following the cutting by machining of the inner surface of the part, the temperature in this area had the value T = 25 °C. The FEM analysis was performed in which the ambient temperature T = 20 °C was considered (figure 6 (a)).

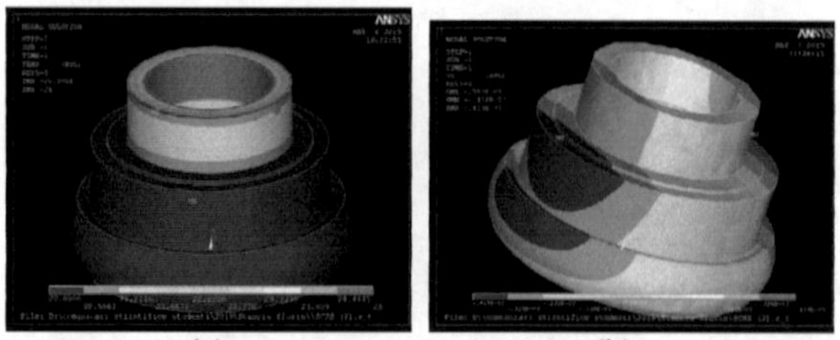

(a) (b)

Figure 6. Dimensional variation at measuring temperature T = 25 °C.

As a result of the expansion process, the dimensions of the part will change. Following a structural analysis based on the results of the thermal problem, in figure 6(b) is presented the modification of the linear dimensions along the OY axis. From the image analysis it can be seen how in the area of yellow colour the change of dimensions is of the order of hundreds of millimeters, values very close to the real ones. These values represents the correction that must be taken into account when measuring the inside diameter. In the case of storing a part at a temperature T = 30 °C (figure 7(a)) and measuring the inner diameter, an increase in diameter can be observed as can be seen in figure 7(b). Under these conditions the piece will be made at 1000.00 mm. The measured actual value resulting in these conditions will be 1000.125 mm. Using modeling and simulation, a very close value of the size variation was found due to the measurement at temperature T = 30 °C, namely l = 1000.14 mm.

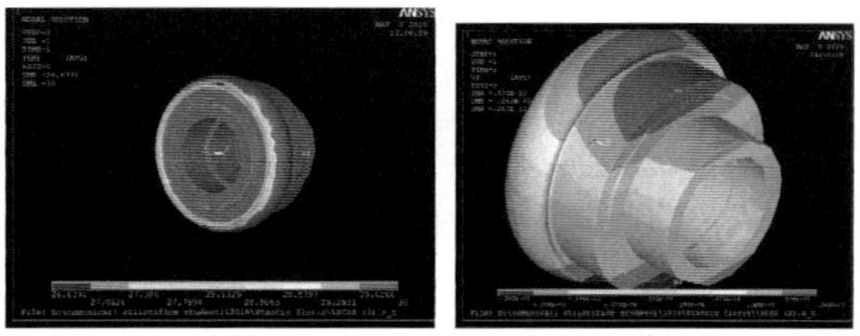

(a) (b)

Figure 7. (a – part storage at T = 30 °C; b - inner diameter measuring.

In order to correct the variation of the dimensions that often appears in the situation in which the measuring instruments are not at the same temperature with the parts that will be measured, it is necessary to apply some dimensional corrections. Depending on the measured dimensions and the temperatures at which the measurements are made, the following dimensional correction table has been created (table 1).

Table 1. Dimensional correction table.

mm/ temp.	**17**	**18**	**19**	**20**	**21**	**22**	**23**	**24**	**25**	**26**	**27**	**28**	**29**	**30**
200	0.007	0.005	0.002	0	0.002	0.005	0.007	0.009	0.012	0.014	0.016	0.018	0.021	0.023
300	-0.010	-0.007	-0.003	0	0.003	0.007	0.010	0.014	0.014	0.021	0.024	0.028	0.031	0.035
400	-0.014	-0.009	-0.005	0	0.005	0.009	0.014	0.018	0.018	0.028	0.032	0.037	0.041	0.046
500	-0.017	-0.012	-0.006	0	0.006	0.012	0.017	0.023	0.023	0.035	0.040	0.046	0.052	0.058
600	-0.021	-0.014	-0.007	0	0.007	0.014	0.021	0.028	0.035	0.041	0.048	0.055	0.062	0.069
700	-0.024	-0.016	-0.008	0	0.008	0.016	0.024	0.032	0.040	0.048	0.056	0.064	0.072	0.081
800	-0.028	-0.018	-0.009	0	0.009	0.018	0.028	0.037	0.046	0.055	0.064	0.074	0.083	0.092
900	-0.031	-0.021	-0.010	0	0.010	0.021	0.031	0.041	0.052	0.052	0.072	0.083	0.093	0.104
1000	-0.035	-0.023	-0.012	0	0.012	0.023	0.035	0.046	0.058	0.069	0.081	0.092	0.104	0.115

6. Conclusions

The process of measuring the parts is a very complex one that must take into account a series of factors that often overlap. The paper studied the influence of temperature in the process of measuring parts in the situation of storing the piece in an area with a temperature higher than 20 °C and in the case of inner diameter metal machining by turning. In order to obtain correct results when measuring the parts made in different thermal conditions, the article presents working procedures based on real processing situation. To correct the measurements, a table of correction values was created that takes into account the temperature and the value of the measured dimensions. To verify these values, an analysis was performed based on the finite elements that came and verified with great precision the values in the tables.

7. References

[1] https://www.machinedesign.com/archive/article/21818680/correcting-for-thermal-measurement-errors, accessed date 05.02.2019.

[2] Niţoi Dan Florin, 2020, *Optimizarea proceselor tehnologice prin metoda elementelor finite – Noţiuni teoretice şi exemple*, Bren Publishing House, Bucharest.

IOP Conf. Series: Materials Science and Engineering 1182 (2021) 012076 doi:10.1088/1757-899X/1182/1/012076

Influence of the ball screw stiffness on the positioning accuracy of the CNC machine tools

A Stoica and G Stan
"Vasile Alecsandri" University of Bacau, Department of Industrial Engineering, Calea Marasesti Street, No 157, 600115, Bacau, Romania

Email: adrian.stoica14@yahoo.com

Abstract: This work is researching the ball screw stiffness influence on the positioning accuracy of the feed kinematic linkage of a CNC machine tool fitted with an indirect type measuring system for positioning the moving element. Because the machine tool is equipped with an indirect measuring system, backlashes and elastic deformations in the kinematic linkage cannot be offset by the position loop, consequently these are directly affecting the positioning accuracy. In this context, methods are used for diminishing their influence as well as compensation methods with a view to improve the positioning accuracy. Ways to preloading the ball screws and bearings of the ball screw are presented as well, by focussing on the optimisation, through calculations, of the preloading force value. An important component, analysed in this work, consists of the research of the ball screw–nut stiffness that depends on the strategy being used on the axial bearing of the two ends of the ball screw. The experimental tests have been performed at idle running on the CNC milling machine (CPFH 500), fitted with three controlled axes (X, Y and Z). Based on the experimental results, it may be noticed that, once increased the ball screw stiffness, within the limits stated by the ball screw supplier, the positioning accuracy will increase as well.

1. Introduction

The most used mechanisms for transforming the rotation motion from a servomotor into the translation motion of a moving element are the ball screws [1]. These subassemblies are to provide a position as accurate as possible of the moving element slide and to assure a working speed rate as high as possible for an increased productivity and high quality. An important role for reaching a high positioning accuracy is played by the bearing system as well, without which the stiffening of the ball screw would not be possible. The design of the feed kinematic linkages of the CNC machine tools features two types of bearing: thrust (axial) bearing at an end of the ball screw and free at the other end, or thrust bearing at both ends. In case of the preloaded ball screw nuts there are researches that show that these have a low effect on the system stiffness, but a too high preloading of them will have a negative effect because this would lead to increasing the friction force between the ball screw and the nut [2]. This will affect the positioning accuracy of the moving element. In this respect researches have been carried out with reference to the ball screw–nut system stiffness variation caused the temperature rising; for this purpose the method of the finished 3D FEM has been used [3]. This method of analysis can exactly anticipate the variation of the preloading force in function of the temperature of the ball screw-nut subassembly. A new research method highlights the preloading of a ball screw based on the vibration signals of the servomotor when the ball screw is preloaded at various preloading rates, of 2%, 4%, and 6% [4]. The characteristics of the signals having various preloading rates have been categorised based on the SVM (support vector machine) calculation algorithm and for the vibration signal the algorithm GA/KNN (genetic algorithm k – the closed neighbour) has been used. Based on the related results, the state of functioning of the feed kinematic linkage can be determined.

2. Assembly diagram of the feed kinematic linkage with the speed and position loops in case of indirect measuring

Figure 1 shows the block diagram of a feed kinematic linkage of a CNC machine tool that uses the balls screw-nut mechanism to transform the rotation motion from the servomotor into the translation motion of the moving element, having an indirect type measuring system of its position.

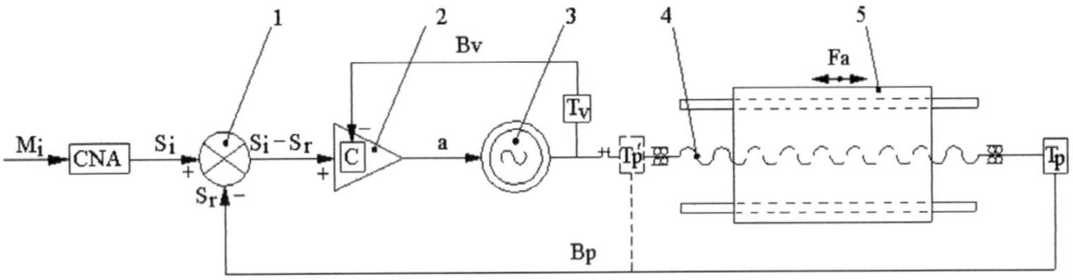

Figure 1. Block diagram of the feed kinematic linkage with indirect measuring system of the position.

The feed kinematic linkage is driven by the A.C. motor *3* that develops a torque to drive the ball screw *4*. For measuring the position of the moving element *5* there is the position loop *Bp* that includes the position rotary encoder care *Tp* coupled to the ball screw. The position rotary encoder can be mounted in two modes: on the output from the motor shaft (*Tp'*) or on the output from the feed kinematic linkage, opposite to the motor (*Tp*). The location of the rotary encoder at the end of the ball screw (*Tp*) can provide a more accurate positioning of the moving element because the position loop includes the angular deformations of the ball screw, as well. The indirect measurement of the slide position is performed through the emission of a signal S_r, of the ball screw rotation rate that will be compared to the signal S_i that is obtained through the conversion into the numeric-analogue convertor *CNA* of the input rate M_i. The comparator *1* will analyse the input and output signals and will settle the error „$S_i - S_r$”. This error will be conveyed to the frequency convertor *2* that will amplify the output current *a* supplied to the A.C. motor *3*, until the error will be annulled. In case of the speed loop B_v, the speed is measured by means of the speed encoder T_v, whose signal is proportional to the rpm and the rpm error is offset at the level of the comparator *C* in the structure of the frequency convertor *2*. The speed feedback loop provides the obtaining of an imposed rpm of the A.C. motor, regardless the rate of the driving force F_a, thus avoiding the acceleration at light loads or deceleration at heavy loads of the moving element. The speed loop is independent from the position loop because each loop has its own input and output parameters.

3. Influence of the ball screw stiffness on the positioning accuracy of the feed kinematic linkage

The ball screw stiffness, associated with the use of indirect measuring has a major influence on the positioning accuracy of the moving element. In order to increase the ball screw stiffness and for eliminating the radial-axial backlash, two preloaded ball screw nuts are used and the ball screw bearings are preloaded, as well.

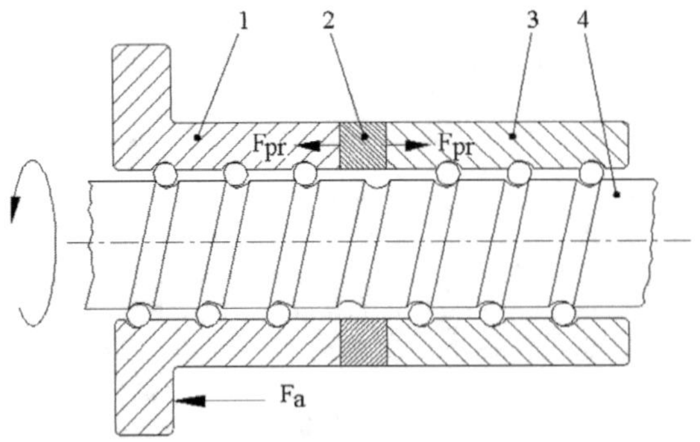

Figure 2. Preloading of the ball screw double nut with enlarged pitch of 20 mm.

In figure 2, it may be noticed the mode of preloading the double nut with enlarged pitch of 20 mm. The value of the preloading force F_{pr} may be controlled by adjusting the dimension of the spacer *2* between the nut bodies *1* and *3* of the ball screw *4*, so that, when applying an external force, like the feed force F_a, the radial-axial backlash between the ball screw and nut to be eliminated.

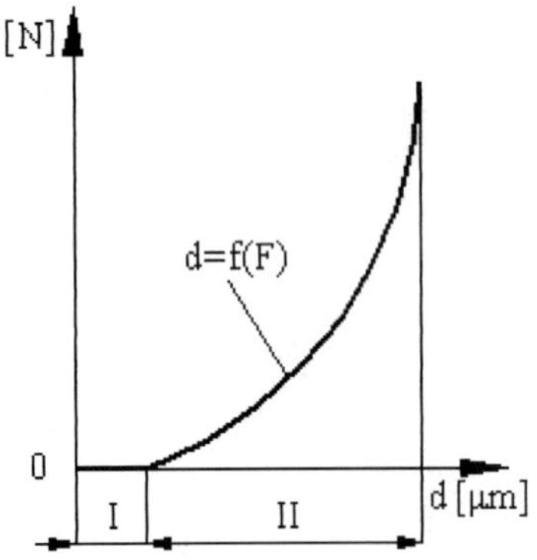

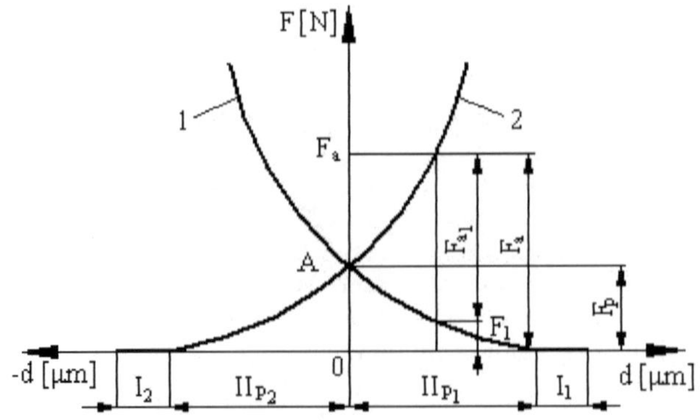

Figure 3. Variable stiffness of a single nut.

Figure 4. Stiffness of two preloaded nuts.

The allure of the force-deformation diagram in case of a single nut is shown at figure 3, where two areas may be noticed: the first one is an axial backlash *I*, and the second one is the axial elastic deformation *II* generated in the contact zone of the balls. The profile of the diagram of the elastic deformation may be noticed: in the contact area, at low values of the load *F*, the deformation is high and when the force *F* increases, the elastic deformation decreases (The position of the diagram versus the horizontal modifies). This aspect shows the fact that the stiffness of a single nut is variable, having a low stiffness (high elastic deformation) at low values of the load *F* and increased stiffness at high values of the load. A relation of proportionality to express the relation of above between load *F* and deformation *d* is given by (1):

$$d = CF^{2/3} \quad (1)$$

Where: *C* –constant of proportionality.

The consequences of using a single nut on the ball screw and an indirect measuring system would lead to a major influence on the positioning accuracy since the two components *I* and *II* would be found in the positioning error. In order to improve the positioning error two preloaded nuts are used. These two nuts are located into a box. This solution will eliminate the axial backlash *I*, as well as the first part of the elastic component of the zone *II*, where the stiffness is low, as it may be noticed in figure 4. The two nuts *1* and *2* are preloaded at a value F_p, thus annulling the backlashes I_1 and I_2, as well as the elastic deformations IIp_1 and IIp_2, in the areas where the stiffness is low. In this situation, where the nuts are preloaded at the value F_p, there will be a stiffness increase, explained by the fact that when acting with a thrust (axial) force F_a, the elastic deformation will begin, as on the diagram shown at figure 4, at the point *A*. In this context, upon the requirement with the force F_a of the two preloaded nuts there will be a decrease of this force with the value F_1, resulted from preloading, as shown at figure 4 and the driving force becomes $F_{a1}=F_a-F_1$. If the ball screw had a single nut (diagram at figure 3) the driving force would be F_a, whilst if there were two preloaded nuts, F_a would become F_{a1}. At the same time, in the second case the deformation is generated in the area *II* beginning with the point *A*, see figure 4, from where the stiffness of a single nut is high. The determining of the preloading force F_p is also studied by other authors and ball screw manufacturers but there is no consensus. A solution being applied with good results consists of determining the value of the preloading force at idle running of the respective axis that is approximately 20% of the available force. This solution is fundamental through the fact that the positioning of the axis is performed at idle running and for the maximum loading of the axis the elastic deformations of the nuts are on the area where the nut has and increased stiffness. Additionally, during heavy cutting duties and low quality requirements, the work piece accuracy will also be affected by other deformations of the machine, work piece, setting devices and tool.

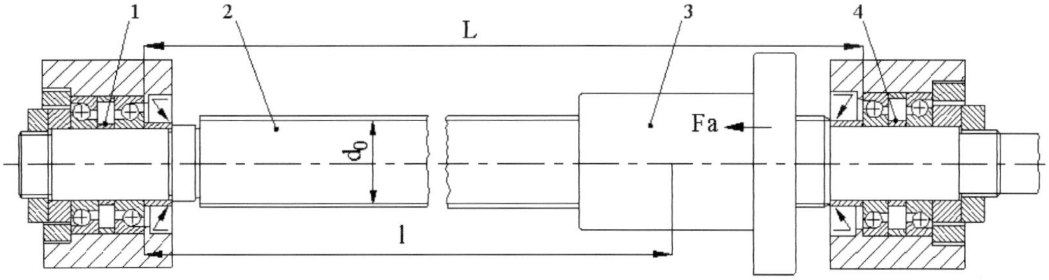

Figure 5. Thrust (axial) bearing system by means of ball bearings of the ball screw.

Figure 5, presents the assembly of the ball screw of diameter d=50 mm, having the increased pitch p=20 mm and fitted with a thrust bearing system at both ends. This system confers a good stiffness to the ball screw *2*, regardless the motion sense of the nut box *3* that is driven by the feed force F_a, both towards the bearing *1* and the bearing *4*. In this case, any driving force axially oriented on the box containing the two preloaded nuts will require axially the ball screw *2*. If the driving force F_a, has the sense shown at figure 5, then the part of the ball screw at the right side of the nut box *3* will be required at stretching, whilst the left side part will be required at compression. For this assembly, very frequently used, the axial stiffness of the ball screw can be written as per the relation (2):

$$R_s = \frac{\pi d_0^2 E}{4l1000}\left(\frac{L}{L-l}\right) \quad (2)$$

Where: d_0 – ball screw diameter measured at the interior of the thread (mm); E – elasticity module of the steel the ball screw is made of (21000 daN/mm^2); L – unsupported length of the ball screw between its bearings (mm); l – intermediary distance measured from the unsupported end of the ball screw up to the half of the nut (mm). When the ball screw nut is located at the middle of the ball screw unsupported length ($l=L/2$), the ball screw stiffness will be minimal, $R_{s\,min}$ and it will be written as per the relation (3):

$$R_{s\,min} = \frac{\pi d_0^2 E}{L1000} \quad (3)$$

In this case, the maximum deformation of the ball screw will be exactly at this point at half distance between the bearings *1* and *4* and it will be written as per the relation (4):

$$\Delta_{e\,max} = \frac{F_a}{R_{s\,min}} \quad (4)$$

Where: $\Delta_{e\,max}$ – maximal elastic deformation of the ball screw (μm); F_a – driving force (N); $R_{s\,min}$ – minimal stiffness of the ball screw (N/μm).

4. Experimental trials

The experimental trials have carried out on the milling machine model CNC (CPFH 500) at idle running. For measuring the position of the moving element the machine is equipped with a position encoder with indirect measuring located on the driving motor shaft and the measurement results has been compared in tandem with the calibrated measuring system, consisting of the laser interferometer of the Mechatronic Laboratory of the University of Bacau. The indirect measuring system of the machine does not have the capacity to offset the errors caused by the axial backlashes and deformations of the ball screw and, in this case, the improvement of the positioning accuracy *P* is imposed through the improvement of the ball screw stiffness. The bearing system of this CNC machine tool is axially stiffened by thrust ball bearings and the translation motions of the axes are performed by the ball screw double nuts. The diameter of the ball screws is 50 mm and their increased pitch is 20 mm. The machine travels *L* are: X=700 mm; Y=600 mm and Z=600 mm. Three sets of trials have been performed, by following up the effect of modifying the preloading forces in the ball screw nuts and in the thrust (axial) bearings of the ball screw. For preloading the thrust bearings the initial force was 320 daN. Afterwards, by modifying the size of the inner spacing ring between bearings, the preloading forces have had values higher than 450 daN. They reached up to 550 daN. For the double nut the initial preloading force was 380 daN. This preloading force has been modified by replacing the spacing ring between the ball screw nuts in two stages, so that higher values of the preloading force could be reached, i.e. 490 daN, respectively 570 daN. Through these modifications of the preloading forces in the ball screw bearings and double nut, the improvement of the positioning accuracy of the moving element has been followed up. Figure 6 shows the preloading diagram of the ball screw bearings where it may be noticed that preloading the bearings at higher rates of the preloading force leads to an improvement of the positioning accuracy of the moving element. For the ball screw, the preloading force rate of 550 daN renders the best improvement of the positioning accuracy, up to the value of 0.0061 mm. This peak is located in the median area of the ball screw. Under these conditions the motor current consumption (torque) at idle running has increased.

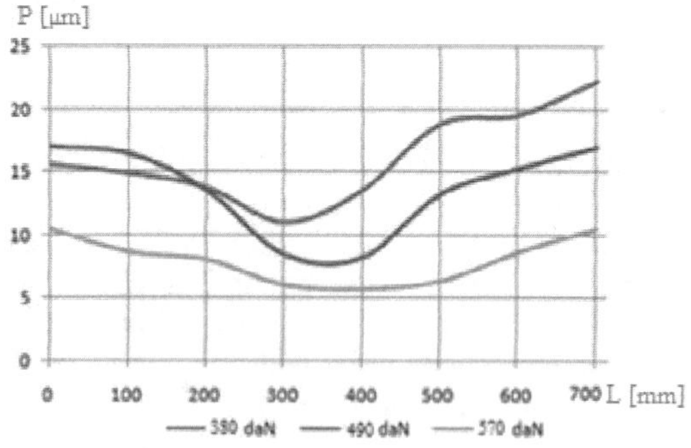

Figure 6. Preloading the ball screw bearings.

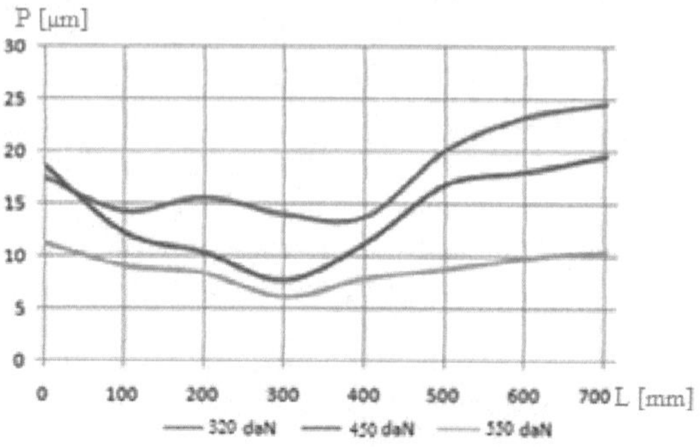

Figure 7. Preloading the ball screw nuts.

In case of preloading the ball screw nuts, as per figure 7, the loading at a force of 570 daN, has brought the better improvement of the positioning accuracy at idle running. It reached 0.0058 mm. Instead, after a running time interval, an increase of the temperature of 8°C has been noticed on the ball screw-nut assembly that will generate thermo-mechanical deformations. These deformations will be affecting the positioning accuracy obtained at the preloading force value 570 daN. For optimisation, in case of stiffening the bearings, the preloading force value of 450 daN has been chosen and, for stiffening the double nut, this value has been 490 daN. These values, correlated with the software corrections in those 1000 points, improve the positioning accuracy by up to 0.005 mm, and the machine will comply with the admitted tolerance limits.

5. Conclusions

In this work the influence of the ball screw-nut assembly and its bearings on the positioning accuracy of the moving element has been studied. It has been established that the stiffening of the ball screw assembly can contribute to improving the positioning accuracy of the moving element by preloading the double nuts and bearings of the ball screw with suitable forces. When high preloading forces are applied, the positioning accuracy can be improved considerably, but it has as negative aspects the loss in torque of the driving motor and the temperature increase of the ball screw-nut assembly that may generate thermo-mechanical deformations. For the ball screw assembly stiffening to have positive aspects only, it is necessary an optimisation of the preloading forces that, corroborated with the electronic corrections in the 1000 points, would lead to improving the machine positioning accuracy, so that it complies with the admitted tolerance limits.

6. References

[1] Mauro S, Pastorelli S, 2015, Influence of Controller Parameters on the Life of Ball Screw Feed rives, *Advances in Mechanical Engineering* **7**(8), 1–11.

[2] Fuhua L, 2017, An Improved Dynamic Model of Preloaded Ball Screw Drives Considering Torque Transmission and its Application to Frequency Analysis, *Advances in Mechanical Engineering* **9**(7) 1–11.

[3] Oyanguren A, Larranaga J, 2018, Thermo-mechanical Modelling of Ball Screw Preload Force Variation in Different Working Conditions, *Int J Adv Manuf Technol* **97**, 723–739.

[4] Huang Y, 2018, Diagnosis of the Hollow Ball Screw Preload Classification Using Machine Learning, *Appl. Sci.* **8**, 1072.

Investigations on the effect of low temperature cooling on the phase transformations of the EN AW 7075 alloy

M Stoicanescu[1], A Crisan[1], T Bedo[1], B Varga[1], I Milosan[1] and I Peter[2*]
[1]Universitatea Transilvania from Brasov, Bd. Eroilor no. 29, Brasov, Romania
[2]University of Medicine, Pharmacy „George Emil Palade" Târgu Mureş, Faculty of Engineering and Information Technology Science and Technology, N. Iorga street no.1, Târgu Mureş, Romania

E-mail: ildiko.peter@umfst.ro

Abstract. Al-based alloys are very challenging materials, especially in applications where energy and mass reduction are required. In this scenario, EN AW 7075 Al-based alloy can be considered of high interest for different large and thick products. One of the many fields of interest where large geometry dynamic is present is, for example, aerospace, where the alloy structural stability in different conditions is very important. In this paper, the main attention is oriented toward the structural investigation of the EN AW 7075 Al-based alloy in different condition, by differential scanning calorimeter and dilatometry analysis, in order to be able to propose it, as a suitable material for aerospace component production. The analysis on the conditioned and un-conditioned materials has been used to assess the alloy's performance and to control the evolution of the alloy properties, in specific conditions, important aspect for the targeted application. Some data on the mechanical behaviour of the alloy integrate the study.

1. Introduction

A precipitation-hardened, high-strength, good wear and corrosion resistance EN AW 7075 Al-based alloy [1-3], is remarkable material and it is used in different critical engineering applications, like automotive and aeronautical domain. In such area, the alloy has to show high plasticity, lightweight, excellent aesthetic features, high mechanical properties combined with high production rate and stress-free assemblage. Additionally, it is important to have a material which can withstand and able to maintain identical properties also in severe environment or risky conditions, like high pressure, high or low temperature, presence of chemically harmful medium, etc. [4, 5]. Localized corrosion, because of its chemical composition, is common in such alloys. In particular, the presence of many different constituents defines the heterogeneity. Last, but not least it is very important to offer and use an economically convenient material [6-9]. This aspect in many cases limit, at large scale, the revolution and the substitution of a very well well-tried/long standing material with a new one, like lightweight alloys, even if they revel promising properties in some given applications. This is the case of steel, strategic material in many application, and has been used for a long time, because it is considered a material with brilliant mechanical resistance with an advantageous cost. In this situations, replacement of iron based material with other one is a very stimulating action, in particular when one talks in terms of metallurgical characteristics, and the realization of this can contribute to increase the spectrum of

[2*] Coresponding author: Ildiko Peter, e-mail: ildiko.peter@umfst.ro

real applications where these alloys can be successfully employed.

The idea of improving mechanical properties of Al-based alloy can be attaining by procedures, like (i) the realization of some heat treatment [10, 11], (ii) the possibility to employ higher quantities of alloying elements [12, 13] and (iii) with the implementation of more and more efficient industrial processes in order to obtain advanced metallic alloys. Usually, heat treatment is a multi-stage process, comprising heating to an established temperature and followed by rapid quenching and aging at room temperature or higher one. Even if this procedure is the most commonly used, is energetically and environmentally adverse and it is very time-consuming. Furthermore, the new European rules concerning the gas emissions, in particular as emission of CO_2 regards, is very restrictive targeting other reduction by 2030are moving the attitude of scientist and the manufacturers to find other, alternative solutions, using stepwise evolutions, able to strongly reduce the weight of the industrial modules having a helpful effect on the fuel consumption and on the emission of CO_2 [14], too.

Some authors, reports studies about cryogenic treatment, even if there are no adequate data to enunciate that such treatment is one of the feasible way to improve the mechanical, tribological and corrosion properties of Al-based alloy. According to such investigations [15-18], cryogenic treatment determines a reduction of the number of lattice defects compared to the un-treated samples and favour the displacement of the atoms of alloying elements. In such conditions, development of secondary phases take place giving rise to make stronger the alloy. Additionally, in the same conditions it comes out that the residual stresses in the alloy is changing, and more in particular, cryogenic treatment helps the generation of a small quantity of compressive residual stresses, more convenient for the alloy performances, and remove part of the tensile residual stresses [19, 20].

Due to the fact that Al-Zn-Mg-Cu alloys are subjected to multidirectional stresses in operation and due to the highly varying operating conditions over time, they have to reveal good structure and properties stability [21-23]. The Zn, Mg, and Cu alloying of aluminium determine superior properties as a result of the presence of the basic metallic mass of hardening phases developed and considered responsible for such behaviour according to [24].

At this time, there is an increasing need to identify the optimal operating parameters for the industrially used components. To the best knowledge of the authors of the present paper there are no sufficient study which consider and investigate the EN AW 7075 Al-based alloy at low temperature, like -70°C÷-150°C, condition which is present during the exploitation of the airplanes.

In this context, the aim of the present research paper is to investigate the structural modifications of EN AW 7075 Al-based alloy, in different initial conditions, over a wide range of temperatures. To do this thermal analysis has been used, in the temperature range of -160÷ +600°C, in order to be able to appreciate the performance of the alloy in a particular condition which simulate a real exploitation environment, and based on the results obtained, to recommend it for aeronautical application.

2. Experimental methods

2.1 Material selection

For the experimental research, EN AW 7075 Al-based alloy has been used with the chemical composition reported in table 1.

Table 1. Chemical composition of the EN AW 7075 Al-based alloy.

Alloy type	Chemical composition [%]								
	Zn	Mg	Cu	Cr	Ti	Mn	Fe	Si	Al
Standard	5.1-6.1	2.1-2.9	1.2-2.0	0.18–0.28	0.1max	0.1max	0.12max	0.15max	balance
Investigated	6.0	2.4	1.5	0.2	0.05	0.05	0.1	0.1	89.6

The considered alloy is situated at the limit of the homogeneous phase domain of s. s. α and of the biphasic zone α + S, according to the phase diagram reported in figure 1. As a consequence, one can admit that at the homogenization temperature the structure is single-phase made only by s.s. α. When hardening, this remain the developed structure and during the aging processes (artificial or natural) the formation of G. P. areas occurs, corresponding to the θ (Al_2Cu), Z =Mg_2Zn_{11}, M (η=$MgZn_2$), T ($Al_2Mg_3Zn_3$), and S(Al_2CuMg), phases and determining the hardening of the alloy. The proportion of the phases θ (Al2Cu), Z = Mg_2Zn_{11} in the structure of the 7075 alloy is negligible [25]

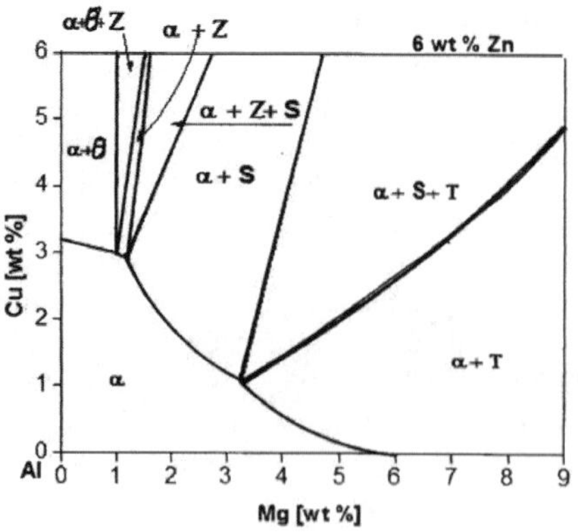

Figure 1. Quaternary phase diagram of the Al-Zn-Cu-Mg system at 733 K (460°C), for 6% of Zn [25] showing the effect of the Mg and Cu change.

Due to the fact that the T- and S- phases show low melting points, the heating range for heat treatments and hot works is limited; the S-phase content depends on the chemical composition of the alloy and on the temperature.

Some studies [26-28] indicates that the fraction of these phases in the alloy structure varies significantly depending on the composition and the ratio of the alloying elements (Zn / Mg, (Cu + Mg)). We consider that in this case, the η phase has a dominant role in the hardening of the alloy.

The available material, as supplied, was used to make sets of samples, in particular hardness test samples, for the resilience test (EN 10045-1: 1990) and for the tensile test (SR. EN. 10002-1) with round cross-section.

2.2 Sample conditioning

The samples prepared, as described above, have been subjected to the final heat treatment of quenching, natural and artificial ageing. For quenching, after holding the samples at 470 °C for one hour, they have been cooled in water. The artificial ageing has been performed by holding the samples for 2 hours at 120 °C, followed by cooling in air, while for the natural ageing the samples have been hold for 7 days at 20 °C. Investigations have been performed in order to study the effects of negative temperatures on the structure and properties of the alloys: some of the heat-treated samples have been subjected to cooling cycles at -70 °C with one hour holding, followed by reheating to room temperature. Between two successive cooling, the samples have been maintained at room temperature for 30 minutes. The same parameters have been used for performing up to three cooling. Finally, the resilience has been measured, on standard parallelepiped samples (10x10x55 mm) without notch. Impact test has been performed by a Charpy impact pendulum devices of 50 J at room temperature. Five samples of each heat treatment have been tested and the average value and standard deviation have been calculated. Brinell hardness measurements have been carried out on the polished samples. A force of 10 N has been applied for 15 s for each measurement and a minimum of 5 indentations have been performed on each samples. Samples made of semi-continuous casted and rolled alloy to a thickness of 45 mm have been used to perform the experiments.

The samples have been heated in the Nabertherm furnace and have been cooled to negative temperatures in an oven-type device with a controlled atmosphere.

Thermal analysis has been carried out by dilatometric analysis (LINSEIS dilatometer, L75/230) and the measurements have been performed in a temperature range of 25÷1400 °C using cylindrical specimens, with 6 mm diameter and lengths between 5 and 10 mm, while differential scanning calorimetry (DSC, Netzsch GmbH, 200 F3 Maia type) has been performed with a heating/cooling rate of 10 °C/min., while. The specimens have been heated and cooled at a rate of 10 °C/min.

The analysis of the effect of cooling below zero degrees Celsius on the phase transformations of the EN AW 7075 Al-based alloy has been performed on some heat treated samples. Table 2 reports the details of the treatment.

Table 2. Regimes of the heat treatments performed on the AA7075 alloy.

Samples	Thermal treatment applied	Number of cooling to -70°C
Sample 1	Annealed	-
Sample 2	Natural aging	-
Sample 3	Artificial aging	-
Sample 4	Artificial aging	1
Sample 5	Artificial aging	3

3. Results and discussion

3.1 Dilatometric analysis

Detection of phase transformations in the field of positive temperatures has been carried out by dilatometric analysis. The results of the such investigations have been reported in figure 2 for the samples 2-5. For comparison, figure 3 shows the variation of elongation and physical expansion coefficient for the annealed alloy (sample 1) depending on the temperature.

Figure 2(a) reveals that there are differences between the samples as regards the variation of the heating expansion coefficient. This is a direct evidence on the differences between the structures of the sample and the way in which these structures evolve during heating. It should be noted that when heated, the highest structural stability is found in sample 2.

The cooling performed at negative temperatures (-70°C) do not have a significant influence on the structural changes that occur during heating.

Considering the diagram in figure 4, the first maximum on the variation curve of the expansion coefficient (figure 2(a)), from the temperature of about 190 °C, corresponds to the re-dissolution of the η' type precipitates, a phenomenon that occurs with a decrease of the volume of the alloy revealed by the decrease of the value of the expansion coefficient.

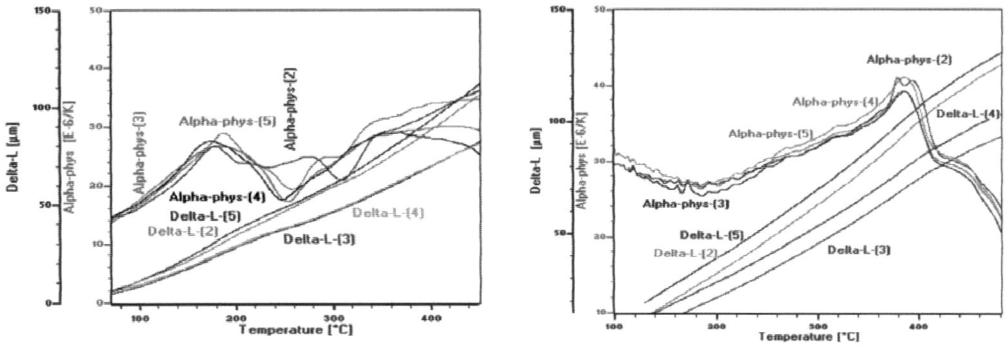

(a) (b)

Figure 2. Expansion curve and variation of the physical expansion coefficient as a function of temperature at heating (a) and at cooling (b).

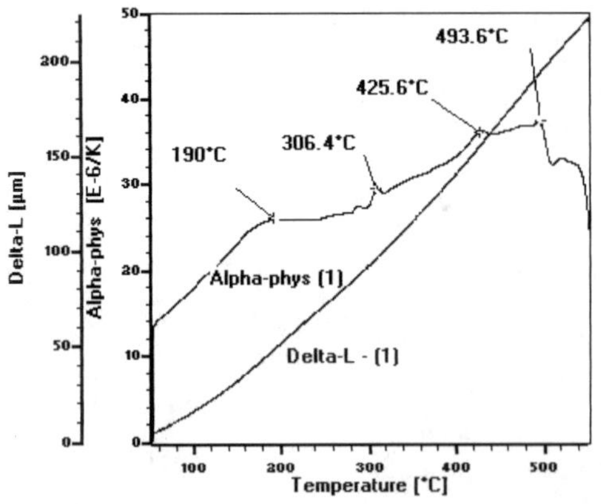

Figure 3. Results of the dilatometry analysis for the annealed samples.

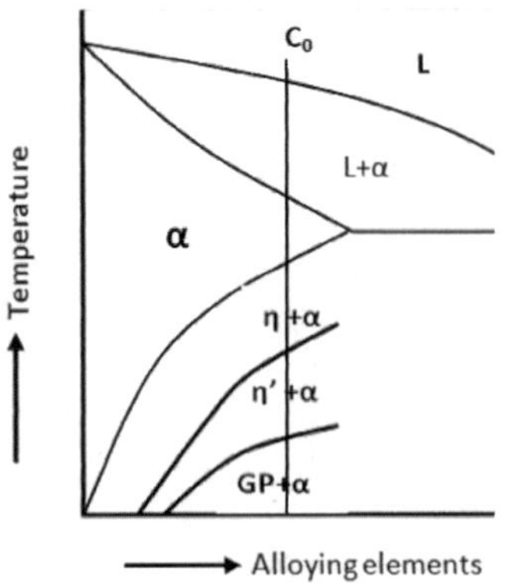

Figure 4. Scheme for the solvus temperatures related to the different precipitates (separations).

Around the temperature of 240°C (the minimum of the curves in figure 2(a)), the process of separating of the η phase at the limits of the solid solution crystals α begins, which determines the amplification of the expansion coefficient value. In the temperature range 340°C÷350°C the increase of the expansion coefficient value becomes much more moderate, it even decreases in the case of samples 3 and 4. This step corresponds to the re-dissolution of the η phase in the α solid solution.

These relevant heating differences disappear during cooling (figure 2(b)) after homogenization, determined by heating temperatures during the DIL analysis, has taken place, when the samples have the same chemical composition.

By presenting the contraction curves and the physical contraction coefficient, in figure 2(b), is reported that in all samples, after the development of the same structure, by heating, in all samples identical structural transformations occur in the cooling phase during the dilatometric analysis. This is confirmed by the overlap of both the contraction curves and the curves corresponding to the physical contraction coefficients.

By performing DSC analyses, the aim was to explain the phase transformations in the temperature range -150÷ 590 °C. Figure 5 reports the temperature-time diagram for DSC analyses.

The thermal analysis program was identical for all samples (figure 5: cooling from room temperature (RT) to -5 °C; holding at -5 °C for 5 minutes; cooling to -150 °C; heating to 590 °C; cooling to room temperature. Both heating and cooling have been performed at a rate of 10 degrees/min. All measurements have been carried out in N2 protective atmosphere with a flow of 250 ml/min. The device was calibrated prior to performing the determinations. The data has been evaluated by NETZSCH Proteus - Thermal Analysis software.

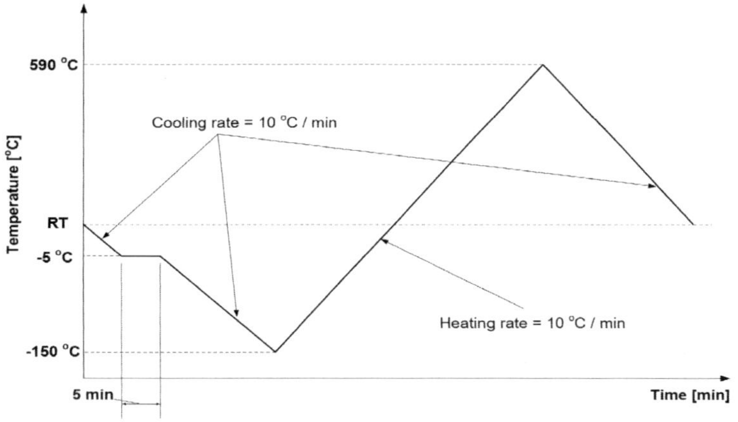

Figure 5. DSC analysis chart.

The thermal regime shown in figure 5 i.e. cooling to -150 °C and subsequent heating up to 590 °C, can highlight the evolution/phase separations in/from the solid solution α, as well as their disappearance at the temperatures at which the formation of the solid solution rich in alloying elements takes place, which corresponds to the equilibrium condition. The 5 minute holding threshold at - 5 °C has been realized due to the need of eliminating any deviations due to the inertia of the device in the planned thermal regime (established during the experiment). Figure 5 reports the results obtained during these investigations, for sample 3.

The analysis of the curves reveals that in all the cases, when cooling to negative temperatures between -9 and -18 °C occurs, the presence of an exothermic maximum is observed, figure 6, which goes with a phase precipitation process within the structure (for the sample 3 it is possible to observe the exotermic peak at (-15.2°C). The phenomenon is due to the decrease in solubility of the elements within the chemical composition of the alloy in aluminium solid solution α. The elements participating most actively in this process are those that have a significant decrease in solubility with the temperature in ssα (e.g. Mg). Even if they are not of high intensity, these processes have an influence on the properties of the alloys, as reported in table 3.

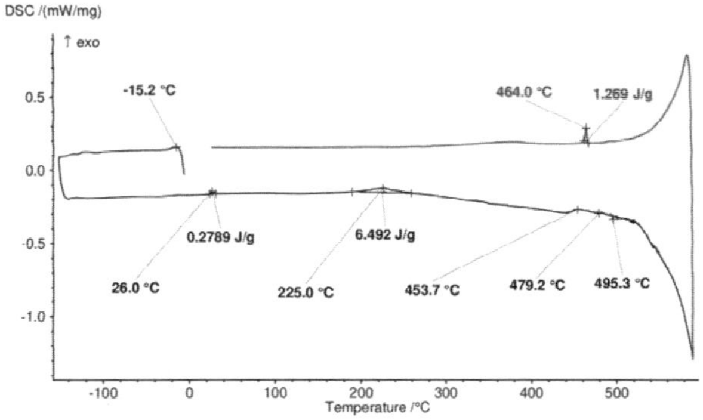

Figure 6. DSC curves showing the behaviour of the artificially aged sample.

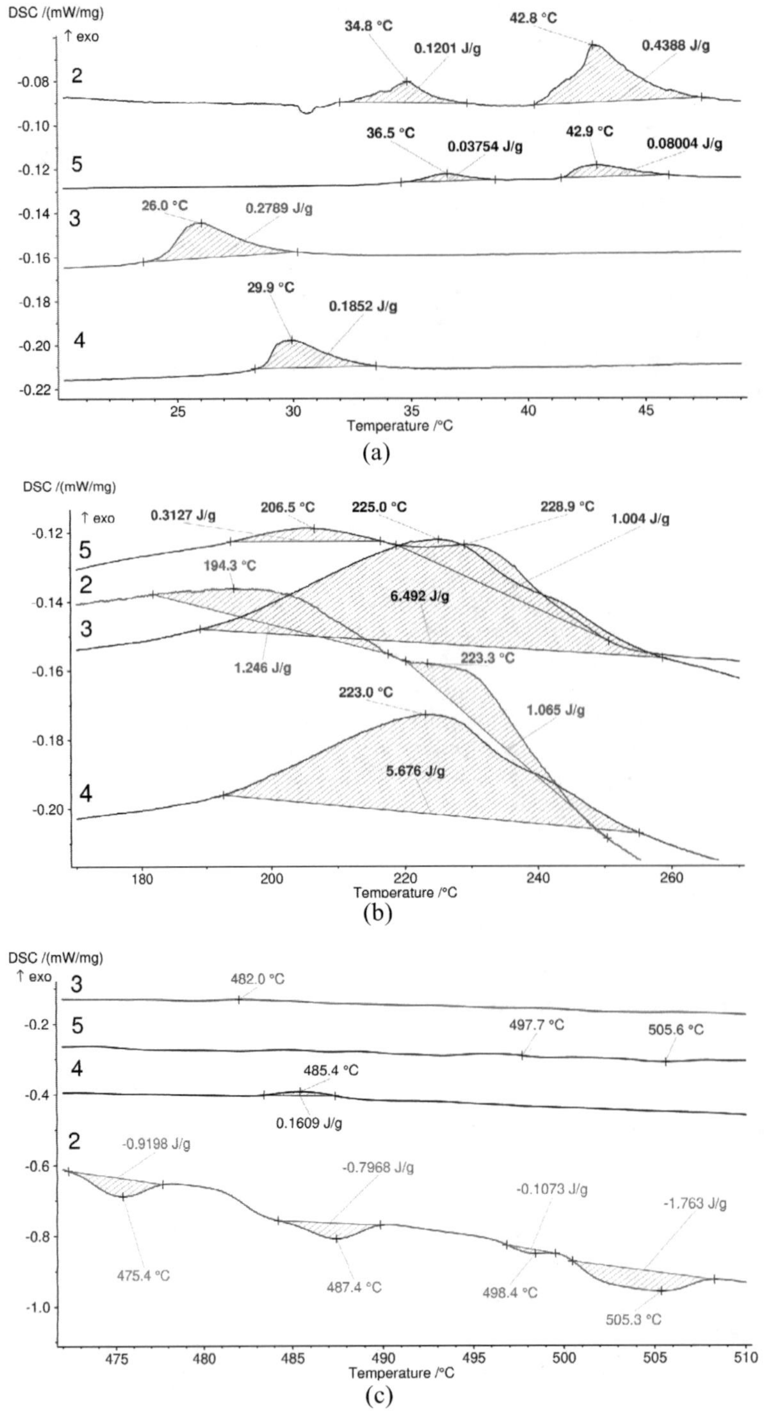

Figure 7. DSC curves showing the behaviour of the samples in different conditions.

In the range of the natural aging temperature, the intensity of these precipitations is very low, considering the values of the energies corresponding to the structural transformations. At the same time, the influence of cooling to negative temperatures is observed in the sense that these transformations at higher temperatures occur and with lower intensities, which means that the treatment applied by cooling the samples to -70 are contributed to the stabilization of the structure.

In the range of temperatures corresponding to artificial aging, the intensity of structural transformations is much higher, especially in the case of samples 3 and 4, as reveals figure 7b. The values of the transformation energies indicated by the DSC curves, for the two samples, 6.492 J/g and 5.976 J/g are in accordance with those indicated in [26].

Also in this field the higher value of the transformation energy for sample 3 (artificial aging), compared to the naturally aged sample (2), proves that the structure obtained by natural aging contains in a superior proportion the phase η ', hardeners, which determines improved mechanical properties (table 3).

At the same time, the domains of the transformation temperatures indicated by the DSC curves are found also on the dilatometric curves shown in figure 2(a).

In the range of homogenization temperatures (470-510°C), the DSC curves reported in figure 7(c) do not evidence significant structural transformations.

The range in which the precipitation processes continue when heating the alloy to positive temperatures and the diffusion becomes more permissive. It is possible to observe the localisation of the precipitations in two temperature zones: a zone close to the room temperature specific to the temperatures of natural ageing (figure 7(a)) and another area at higher values, specific to the temperatures of artificial ageing, as reported in figure 7(b). At the temperatures in the first zone, the precipitation is lower for the artificially aged alloy, while they are lower for the naturally aged alloy at temperatures in the second zone, phenomenon confirmed also by the results of the dilatometric analysis (figure 2(a)).

3.2 Mechanical analysis

In addition to the DIL and DSC analyses, the Brinell hardness, resilience and tensile strength at breaking have been measured. The results of the mechanical properties measured on samples under different conditions, reporting the average of three parallel determinations, have been collected in table 3.

Table 3. Results of the mechanical properties.

Sample	Heat treatment	Number of processes of cooling to -70°C	Hardness [HB]	Resilience KCU [J/m^2]	Breaking strength [N/mm^2]
2	Natural aging	-	130± 2	90± 5	414± 20
3	Artificial aging	-	96± 2	71± 5	409± 20
4	Artificial aging	1	99± 2	91 ± 5	411± 20
5	Artificial aging	3	128± 3	86 ± 5	406± 20
6	Natural aging	1	133	68 ± 5	413 ± 20
7	Natural aging	3	139	65.8 ± 5	410 ± 20

The values of the considered properties obtained by natural ageing are higher with respect to those obtained by artificial ageing. This fact is due to the particularity of the formation of the structural phases that control these properties in such situations. In natural ageing, these phases have developed at low temperatures (room temperature) and the diffusion of the elements arises with more difficulty, so that a higher germination rate/growth rate (vg/vc) is achieved. Under these conditions, the phase distribution in the structure is finer and more homogeneous, with direct consequences on the increase of mechanical properties.

As a result of cooling, the materials at negative temperatures after ageing (in one or more cycles), show variations as mechanical properties concern. These deviations can be attributed to some transformations in the structure, consisting in the persistence of the transformation in the state of the hardening phases: the α solid solution of the alloy becomes unstable at certain negative temperatures at which the evolution of the precipitate occurs, influencing the properties of the material. The significant decrease of the resilience values suggests the fact that the structural changes determined by the application of cooling to negative temperatures, the proportion of the hardening phases changes, fact indicated by the DSC curve reported in figure7, by increasing the existing ones and not by germination. This fact determines the brittleness of the alloy, a phenomenon that appears during the exploitation at negative temperatures of the subassemblies that include elements of this composition.

4. Conclusions

In this paper, structural analysis of the EN AW 7075 Al-based alloy, in different condition, was carried out to determine if such alloys can be proposed as a good candidate for applications where has to survive also at low temperature. Based on the results obtained by dilatometric analysis, the cooling performed to negative temperatures (-70°C) do not have a significant effect on the structural modifications that take place during heating. DSC analyses results, performed with the aim to clarify the phase transformations in the temperature range -150÷ 590 °C, support the results coming from dilatometric measurements. At the same time, the DIL and DSC analyses allowed the formulation of some assumptions regarding the mechanism of increasing the proportion of the hardening phase, ie by germination/deposition. As a result of cooling, the materials at negative temperatures after ageing (in one or more cycles), demonstrate some modifications as its mechanical performance regards, attributable to some transformations within its structure, and the persistence of the transformation in the state of the hardening phases. However, reduction of the resilience values was observed and for this reason, some preventive measure has to be considered during the exploitation of such alloy at negative temperatures, which is the case of aeronautical application.

5. References

[1] Z. Zhao, R. Wu, B. Wang, M. Huang, G. Le, F.Luo 2020 *Materials* **13** 4022 1-12.
[2] J. Lin, X. Bao, Y. Hou, J. Min , X. Qu, Z. Tao, J. Chen 2020 *J. Mech. Eng.* **33**(76) 1-12.
[3] J. Hirsch, T. Al-Samman 2013 *Acta Mater.* **61** 818–843.
[4] ZR. Yang, , Y. Sun, Sun,Li, XX. et al. 2015 *Rare Met.*
[5] M.K. Chung, Y.S. Choi, J.G. Kim, Y.M. Kim, J.C. Lee 2004 *Materials Science and Engineering: A* **366**(2) 282-291.
[6] Sankaran, K.K., Perez, R., Jata, K.V. 2001 Effects of pitting corrosion on the fatigue behavior of aluminum alloy 7075-T6: Modeling and experimental studies *Mater. Sci. Eng. A* **297** 223–229.
[7] Shi, P., Mahadevan, S. 2001 Damage tolerance approach for probabilistic pitting corrosion fatigue life prediction *Eng. Fract. Mech.* **68** 1493–1507.
[8] Jakubowski, M. 2015 Influence of pitting corrosion on fatigue and corrosion fatigue of ship and oshore structures, *Part II—Pit—Crack Interaction. Pol. Marit. Res.* **22** 57–66.
[9] Genel, K. 2007 The efect of pitting on the bending fatigue performance of high-strength aluminum alloy. *Scr. Mater.* **57** 297–300.
[10] S. Menargues, E. MartÃ-n, M.T. Baile, J.A. Picas 2015 New short T6 heat treatments for aluminium silicon alloys obtained by semisolid forming *Materials Science and Engineering: A* **621** 236-242.
[11] N. Mahathaninwong, T. Plookphol, J.Wannasin, S. Wisutmethangoon 2012 T6 heat treatment of rheocasting 7075 Al alloy *Mater. Sci. Eng. A* **532** 91–99.

[12] Y. Yang, Z. Zhang, X. Li, Q. Wang, Y. Zhang 2013 The effects of grain size on the hot deformation and processing map for 7075 aluminum alloy *Mater. Des.* **51** 592–597.
[13] R. Jayaganthan, H.-G. Brokmeier, B. Schwebke, S.H. Panigrahi 2010 Microstructure and texture evolution in cryorolled Al 7075 alloy *J. Alloys Compd.* **496** 183–188.
[14] https://ec.europa.eu/clima/policies/transport/vehicles/cars_en, accessed on 13 January 2021.
[15] Ludmila Kučerová, Jiří Hájek, Jan Vítek 2020 The Effect of Cryogenic Treatment on Mechanical Properties, Wear and Corrosion Resistance of Aluminium Alloy AW7075 *Manufacturing Technology* **20**(1) 60-65.
[16] I. F. Zuñiga Tello, M. Milkovi´c, G. M. D. Almaraz and N. Gubeljak 2020 Ultrasonic and Conventional Fatigue Endurance of Aeronautical Aluminum Alloy 7075-T6, with Artificial and Induced Pre-Corrosion *Metals* **10**(8) 1033 https://doi.org/10.3390/met10081033.
[17] J. Atalah, L.Blamey, H. Köhler, H. M. Alfaro-Valdés, C. Galarce, C. Alvarado, M. Sancy, M. Páez, J. M. Blamey 2020 Study of an Antarctic thermophilic consortium and its influence on the electrochemical behavior of aluminum alloy 7075-T6 *Bioelectrochemistry* **133** 107450.
[18] C. Alvarado, G.M. Sancy, J.M. Blamey, C. Galarce, A. Monsalve, F. Pineda, N. Vejar, M.A. Páez 2017 Electrochemical characterization of aluminum alloy AA2024-T3 influenced by bacteria from Antarctica *Electrochim. Acta* **247** 71–79 https://doi.org/10.1016/j.electacta.2017.06.127.
[19] D. Wenwen, S.Yangshan, M. Xuegang, X. Feng, Z. Min, W. Dengyun 2003 Microstructure and mechanical properties of Mg–Al based alloy with calcium and rare earth additions *Materials Science and Engineering: A* **356**(1–2) 1-7.
[20] B. K. Prasad, A. K. Patwardhan, A. H. Yegneswaran 1997 Microstructural modifications through compositional alteration and their influence on the mechanical and sliding wear properties of zinc-based alloys. *Scripta Mater.* **37**(3) 323–328.
[21] V. Balek, T. Mitsuhashi, J. Subrt, S. Bakardjieva, Z. Malek 2003 *J. Therm. Anal. Calorim.*, **72** 119.
[22] L. A. Pérez-Maqueda, V. Balek, J. Poyato, J.L. Pérez-Rodriquez, J. Šubrt, I.M. Bountsewa, I.N. Beckman and Z. Málek 2003 *J. Therm. Anal. Calorim.* **71** 715.
[23] K. Wen, Y. Fan, G. Wang, L. Jin, X. Li, Z. Li, Y. Zhang, B. Xiong 2016 Aging behavior and precipitate characterization of a high Zn-containing Al-Zn-Mg-Cu alloy with various tempers *Materials and Design* **101** 16–23.
[24] T. Dursun, C. Soutis 2014 Recent developments in advanced aircraft aluminium alloys *Mater. Des.* **56** 862–871.
[25] Suzuki Hisashi, Kanno Motohiro, Asami Shigenori 1972 Effects of copper addition on aging phenomena in AI-Zn-Mg alloys *Journal of Japan Institute of Light Metals* **22**(11).
[26] X.M. Li and M.J. Starink, DSC 2012 Study on Phase Transitions and Their Correlation with Properties of Overaged Al-Zn-Mg-Cu Alloys JMEPEG **21** 977–984 doi: 10.1007/s11665-011-9973-5.
[27] Woznicki, A., Leszczynska-Madej, B.; Włoch, G., Grzyb, J.; Madura, J.; Lesniak, D. 2021 Homogenization of 7075 and 7049 Aluminium Alloys Intended for Extrusion Welding. *Metals* **11** 338.
[28] Seong Taek Lim, Il Sang Eun and Soo Woo Nam 2003 Control of Equilibrium Phases (M,T,S) in the Modified Aluminum Alloy 7175 for Thick Forging Applications *Materials Transactions*, **44**(1) 181-187.

IOP Conf. Series: Materials Science and Engineering 1182 (2021) 012078 doi:10.1088/1757-899X/1182/1/012078

Research regarding works ergonomics in automotive repair shop

M Tanase[1], O Chivu[1], A Tapirdea[2], D Nitoi[1] and V Petrescu[3]
[1] Politehnica University of Bucharest, Splaiul Independentei no. 313, Bucharest, Romania
[2] Politehnica University Timisoara, Piata Victoriei no. 2, Timisoara, Romania
[3] Politehnica University Sibiu, Emil Cioran,no 4, Sibiu, Romania

E-mail: virlan_oana@yahoo.co.uk

Abstract. Prevention of muskuloscheletal disorders is very important worldwide. Both governments and companies are very interested in reducing the risk of their occurrence because the process that occur in the illness of an employee are long and very expensive. Muskuloscheletal disorders exists in all area of activity and in all occupations cause physical and economic consequences for both the worker and the company. They are generally found where the activity involves repetitive actions with awkward workstations or when the individual human operator has to lift weights above or outside the ergonomic workspace of the upper limbs. The most common musculoskeletal disorders occur mainly in two areas of the human body, namely: the back and the upper limbs, respectively, because the activities take place with the help of muscle groups in these areas. For this reason, they are most affected. Paper carries out a study on the activity of a car service and the exposure of the human operator from a certain workplace. It aims to highlight modern methods for preventing musculoskeletal disorders, generated by repeated and long exposure to emerging risks in the automotive industry. The semi-direct OWAS method will identify the areas where possible stress points may appear in the muscles, as a result of repeated exposure in a short time to risk factors but will be studied also NIOSH and RULA methods. Through Process Simulate software we try modeling such an activity, around which musculoskeletal disorders can occurs. It must be taken into account that here the mechanical stress generated at the level of the human operator cannot be visualized, because they depend on the prolonged, repeated exposure of the risk factors and on the large volume of work in a certain period of time.

1. Introduction

For the prevention of occupational diseases, both in car services (figure 1) and in other sectors of activity, it is necessary to identify all risk factors both emerging and normal, independent, which may occur during daily activity. Once determined, preventive / corrective measures must be considered. Some specialized studies recommend the change of the movements [1] of people through various activities, so that the effort made by them is not only in the same areas. It is also recommended the professional training of workers, in order to create the appropriate framework for ergonomic work [2,3].

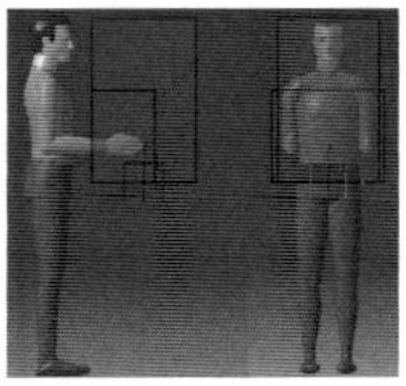

Figure 1. Upper limb area or working range.

1.1 Methods of measurement, prevention of muskuloscheletal (MSD) *due to occupational hazards*
There are many ways to measure risk factors that cause occupational diseases or musculoskeletal disorders. They can be classified into three major groups, which in turn are divided into several methods. The first categories can be seen in table 1. The direct method has a high accuracy, but is inconvenient from the point of view of the way a car service staff works, as well as from another sphere of activities, because it would mean adding a significant mass on the tracked areas (sensors, batteries, wiring, etc.) and the existence of qualified personnel for the acquisition of data from sensors, their calibration and final interpretation of the data.

The semi-direct method is more expensive (software license), but is appropriate because the presence of an observer is required, who will then process the data using specialized software by various methods (RULA, OWAS, NIOSH, etc.) listed in table 4. These methods can identify areas of activity where corrections are needed to help staff with occupational health and safety.

Table 1. Analysing method of MSD.

Direct method	Semi-direct method	Indirect method
Installation of sensors on the limbs / tracked areas affected by the workers, during the performance of the work activity	Observation of the activity carried out and the use of software dedicated to its analysis through several methods	Through standardized chestionnaires to workers

In order to be able to choose one of these three methods, the following aspects must be taken into account: accuracy, cost, efficiency, complexity and actual working time. Therefore, a classification will be made such presented in table 2.

Table 2. Aspects to be taken into account when choosing the method of analysis.

Method	Aspects to consider				
	Accuracy	**Costs**	**Efficiency**	**Complexity**	**Working time**
Direct method	High	High	High	High	High
Semi-direct method	High	Medium to High	High	Low	Medium to High
Indirect method	Medium	Low	Medium	Low	High

Therefore it is considered that the semi-direct method is an advantageous one and among others it offers a range of universally valid answers both in the activity of a car service and in other activities. In order to be able to go further, it is absolutely necessary to know what each of the three methods provides, as a presentation. Therefore, it will be exemplified in the form of a table (table 3) which provides the direct method as well as the indirect method.

Table 3. Exemplification of the direct and indirect method.

Direct method	Indirect method
Use of a **HADA Move-Human Sensor System**	Standardised Nordic Questionnaire [3], Michigan Questionnaire [4,5], Quick Exposure Check Questionnaire (QEC) [6], Keyserling Questionnaire [7]

Regarding the semi-direct methods, a brief classification of the sub-methods used can be seen in table 4, being divided into three sub-categories: weight handling, repetitive activities and awkward postures.

Table 4. Exemplification of the semi-direct method.

Semi-direct method		
Weight handling	**Repetitive activityes**	**Awkward positions**
- KIM(**Key Indicator Method**) method - INSHT(**National Institute of Safety and Hygiene at work**) method - MAC(**Manual handling assesement charts**) method - Mutual tebles, Snook and Cieriello - NIOSH(**National Institute for Occupational Safety and Health)** equation	- RULA(**Rapid Upper Limb Assessment)** mehod - PLIBEL**(sw "Plan for Identifiering av Belastnings faktorer" – en** "method for identification of ergonomics hazard") method - „Work effort index" mdethod - OCRA(**Occupational Repetitive Actions method**)method - Risk factor index (PRRI)	- REBA**(Rapid Entire Body Assessment)** method - PATH(**Posture, Activity Tools and Handling)** method - OWAS(**Ovako Working Posture Assessment System**)method - VIRA(**Video film technique for Registration and Analysis of working postures and movements)** method - Corlett method

Referring mainly to the activity of a car service and respectively to the types of activities in each workshop, the specifics of daily activities fit, either in the sub-branch of weight handling or in the branch of awkward positions. The branch of repetitive movements would include activities such as those in production, where a human operator performs with a high cycle / time a series of repetitive activities. Considering that in general, the weights in a car service, higher than 10-15 kg are performed with the help of elevators or handling devices, I chose to refer only to awkward positions where the high load of to human personnel. Therefore, the OWAS method will be chosen which is successfully implemented, together with the RULA method and the NIOSH equation in the software I have available (Tecnomatix Process Simulate).

2. OWAS method of ergonomic analysis

Regarding OWAS (Ovako Working Posture Assessment System) it is a method that identifies the most common working positions for the back (4 positions), upper limbs (3 positions) and lower limbs (7 positions), as well as the weight that they carry them (3 categories). The whole position of the body is described by the 4 parts of the body with 4 digits. There are a total of 252 job positions that have been classified to indicate the need for ergonomic changes. Initially, the method was performed manually using body map sheets that cataloged possible pain in all areas of the body, but this meant that there had to be an external observer who for several days, to observe the activity of human staff and complete a sheet, to see how the pain evolves in different areas depending on the weight it carries. Such a map can be seen in table 5. Each worker was asked if he had any pain in any part of the body and how intense it was, in order to determine the manner and activities that lead to musculoskeletal disorders. Currently, after many studies, an OWAS algorithm has been successfully implemented in several specialized software, also displaying a specific color to the level of risk in each area where there is a need for ergonomic changes. The colors are catalogued as in table 6.

Table 5. Map of the human body in the OWAS method.

Nr.	Part of the body
0	Upper neck
1	Lower neck
2	Left shoulder
3	Right shoulder
4	Upper left arm area
5	Back
6	Upper right arm area
7	Waist
8	Buttocks
9	Bottom
10	Left elbow
11	Right elbow
12	Left forearm
13	Right forearm
14	Left arm wrist
15	Right arm wrist
16	Left hand
17	Right hand
18	Left thigh
19	Right thigh
20	Left knee
21	Right knee
22	Left fleshy lower leg
23	Right fleshy lower leg
24	Left ankle
25	Right ankle
26	Left leg
27	Right leg

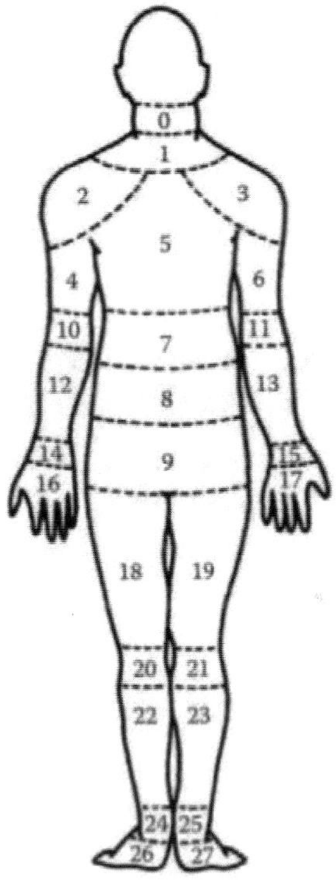

Table 6. Colour map for OWAS method.

Colour	Significance – recommnded actions
-	No recommended actions
Yellow	Certain ergonomic changes may occur in the future
Orange	Ergonomic changes in the approach future
Red	Ergonomic emergency changes

Also, based on the color code, reports can be automatically generated that help to analyze a possible occurrence of occupational diseases, caused by emerging risks and awkward working positions.

3. Using the OWAS method on a specific case in a car repair shop

At the beginning, after identifying the different activities and tasks that take place in a car repair shop, it was decided to perform a first simulation in a specialized software (Tecnomatix Process Simulate). The simulation does not cover a very large branch of movements, but only refers to working with the upper limbs above the head, removing the engine shield / guard from under a vehicle. It was wanted to create a frame as close to reality as possible, the car being raised on an elevator to a height of 1850 mm above the ground, so that a mechanic with a height of 1.75 m can work easily (figure 2). The flow of the activity is a simple one and has the following chain of events (table 7).

Table 7. The flow of movements in a car repair shop.

Movement no.	Movements	Carried weight / Loaded limbs	Body posture
1	Removing the front axle drive wheels	-	Normal (the operator is not considered to have removed the wheels)
2	Pressing the button to lift the elevator	-	Right hand outstretched towards the button
3	Moving for picking up the pneumatic gun for screwing / unscrewing	-	Normal
4	Picking up pneumatic gun	1.5 kg / right hand	Flexed back and outstretched right hand
5	Moving to the elevator, under the vehicle	1.5 kg / right hand	Normal
6	Unscrewing engine shield clamping screws	1.5 kg / right hand	Right hand held in an L-shape above the shoulder
7	Moving for gun deposition	1.5 kg / right hand	Normal
8	Moving to the elevator, under the vehicle	-	Normal
9	Engine shield detachment	3.15 kg / hand (6.5kg upper limbs)	Hands above head level, V-shaped
10	Moving with the engine shield above the head	3.15 kg / hand (6.5kg / upper limbs)	Hands above head level and on the side of the body; flexed back
11	Engine shield placement	3.15 kg / hand (6.5kg / upper limbs)	Flexed back

Starting from table 5, it is considered that the most unfavorable movements are movement number 7, 10 and 11, because the mechanic also carries an additional weight and has different body postures. In order to have consistent data with reality, the mass of the pneumatic gun was determined based on product data sheet. The gun model is EP11PTI100 HR13 from Atlascopco and is part of the ErgoPulse PTI series. The focus is represented on the mass of the pneumatic gun, as well as in the overall dimensions. Its mass is 1.5 kg, and the rest of the specifications can be found by accessing the web address in the bibliography. It is also necessary to know the weight of the engine shield, which was approximately 6.5 kg. Following the running of the simulation in Tecnomatix Process Simulate, it was noticed that a suggestion from the OWAS algorithm appears at activity number 5, of raising the gun. This remark appeared in the form of the yellow color around the back, as a result of back flexion and pneumatic gun lifting. This warning is not necessarily conclusive, because this is not a very high value from weightlifting point of view. It is conclusive regarding he back flexion. The working posture under the car (activity 7) was also analysed, as well as activities 10 and 11 of detaching the engine shield and placing it in other areas of car service. Because the back of the operator was always straight (normal posture) and the OWAS analysis is not an endurance analysis (repeated cycles at a certain time interval), they did not show OWAS alerts for future actions.

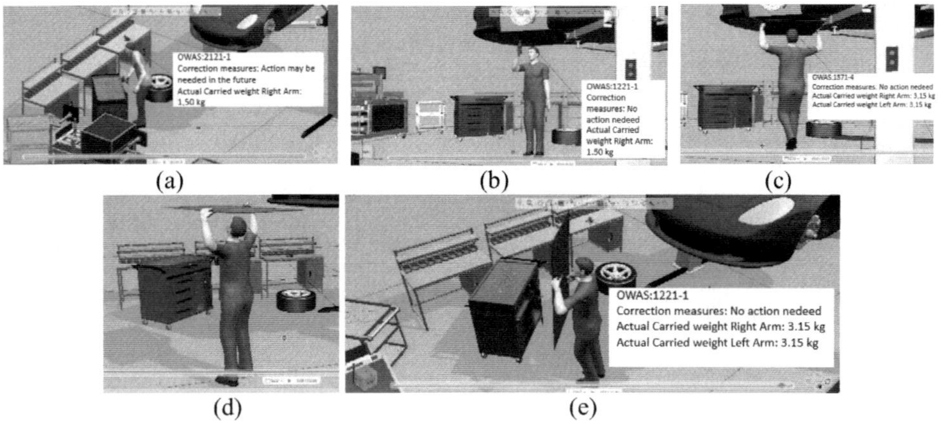

Figure 2. Studied activities in the car repair shop: (a) picking up pneumatic gun; (b) unscrewing; (c) engine shield detachment; (d) moving with the engine shield above the head; (e) shield placement.

4. Conclusions

As a conclusion to this simulation, it was found that the OWAS method was not conclusive for the initial reasoning. There are three possible causes: the simulated activity is not exactly conclusive, because it does not have such awkward working postures; the simulated activity is suitable for a simulation using the NIOSH equation and the RULA method for repetitive activities over a certain time interval; the simulation is inconclusive because the algorithm does not consider possible musculoskeletal disorders that may occur from various and extensive activities over a longer period of time. Starting from the fact that emerging risks represent both the new risks generated by the tech or the emergence of new work processes and the existing growing risks, it can be concluded that careful monitoring of existing risks is necessary. With the appearance of new technologies, it is necessary to adopt evaluation methods specific to each sector of activity within an organization.

5. References

[1] Kuorinka I, Jonsson B, Kilbom A, Vinterberg H, Biering-Sørensen, F, G. Andersson, K. Jørgensen 1987 *Applied Ergonomics* **18** 233-237.

[2] Asensio Cuesta S, Diego Mas J A, Canos Daros L , Andres Romano C 2012 *The International Journal of Advanced Manufacturing Technology* **60** 1161–1174.

[3] Ilonca I, Stanciu A, Rosulescu E, Zavaleanu M, Cosma G 2016 *Proceedings of the 3rd International Multidisciplinary Scientific Conference on Social Sciences and Arts, SGEM 2016* 445.

[4] Uimonen Mikko M, Hulkkonen Sina M, Ryhänen Jorma, Ponkilainen Ville T, Häkkinen Arja H, Karppinen Jaro, Repo Jussi P 2020 *Journal of Hand Therapy* **33** 571-579.

[5] https://download.lww.com/wolterskluwer_vitalstream_com/PermaLink/PRS/A/PRS_128_1_2011_05_11_CHUNG_203363_SDC1.pdf, access date 11.05.2020.

[6] https://www.msdprevention.com/resource-libray/view/quick-exposure-checklist-qec-.htm, access date 11.06.2020.

[7] Keyserling W.M., Brouwer M, Silverstein B A, 1992, *International Journal of Industrial Ergonomics* **9** 283-301.

Thermal influence on the positioning accuracy of the CNC machine tool rotary tables

I Tanase and G Stan
"Vasile Alecsandri" University of Bacau, Department of Industrial Engineering, Calea Marasesti Street, No 157, 600115, Bacau, Romania

Email: ionut_tase78@yahoo.com

Abstract: This work analyses the perturbing thermal sources and design solutions to avoiding the thermal effects. The analysis is focused, as main factors of the thermal dilatation, on the rotary tables, the guiding systems, the radial one and, especially, the axial one (that takes over the weight of both table and work piece), as well as on the positioning of the driving servomotor that is a thermal source in the vicinity of the rotary table. The new solutions presented in this work refer to the rotary tables with direct measuring whose number of indexing positions is 360000 that means a minimum increment of 0.0001degrees. It is to be mentioned that this structure allows the kinematic linkage of the rotary table to take part to machining. If the rotary table provides positioning only, the table will have a clamping system that confers a high stiffness, needed for heavy duty cutting. The experimental trials have been carried out on a machining centre of the Micro-production workshop of the University in Bacau. The size of the rotary table is 630x630 mm. Repeated trials have been performed for determining the positioning accuracy through the modification of the thermal conditions of the worm-wormed gear mechanism. The results have proven an enhancement of the positioning accuracy of approximately 7%.

1. Introduction

The positioning accuracy of the rotary tables is a priority of the machine tool builders that aim at improving or even eliminating the positioning errors [1]. The positioning accuracy of the feed kinematic linkages to which the rotary tables belong [2], is influenced by a series of factors: constructive, thermal, stiffness. The thermal sources of influence [3] of the CNC machine tools consist of both internal and external sources [4]. Thermal influences are negatively affecting the machine accuracy. This issue has led to thinking of and designing new solutions, such as the isolation of the thermal sources of the machine tool structure, by using a more rigorous control of the temperature during the machine running.

2. Constructive analysis of the rotary tables of the CNC machine tools

The constructive structure of the rotary tables that, in most of the cases are integrated to the machine tool assembly depends on their role in the cutting process. Thus, in the first variant, if the rotary table is intended setting the work piece (like in the case of the milling and boring and milling machines) it will provide the position indexing at certain angles with a view to machine more surfaces of the work piece. The category of this first variant includes the rotary tables that perform precise positioning at 4x90°; 8x45°; 72x5° and 360x1°. The table indexing is performed at idle running that leads to an adequate sizing of the feed kinematic linkage that will be driving the table. The table indexing is performed at certain time intervals and the duration of the positioning process is relatively short, within 0.2 – 2 minutes. At this type of rotary tables the positioning accuracy is usually assured by mechanical means, such as: bolt or prism type indexing, through frontal clutch with 72, 180 or 360 teeth. After performing each positioning process the table will be clamped, so that a good stiffness is provided, for the cutting process to be carried out under proper conditions. The positioning accuracies obtained by this constructive variant of rotary table, encountered at many machine tool manufacturers (Yasda, Mandelli, Csepel, Cincinatti etc.) range within ± 1 sec and ±3 sec. Out of the constructive description of above it results that in case of this type of rotary table the thermal sources are relatively low, because the indexing is performed periodically and its duration is short, i.e. 1 minute maximum. At the same time, the accuracy of the positioning is given by the mechanical indexer; as such, the thermal influences at this type of rotary tables are low, even neglected in cases when indexing is rarely performed. The second variant of rotary table that is more and more frequently used is that one that provides positioning at 360000 positions having a programmable minimum increment of 0.001 degrees. In other words, this rotary table is controlled in canned cycle, having the possibility to perform individual cutting or to take part to cutting, along with the other controlled axes of the machine, for performing plane or space interpolations. At the same time, the second variant of rotary table can also perform indexing at certain angles for machining more surfaces of the work piece. The positioning accuracies provided by this rotary table variant differ from builder to builder, having values within ±5 sec. and ±20 sec. It is to be noticed that feed kinematic linkage of this constructive variant is sized for running at load and the cutting time at load is not limited. From this point of view the influence of the thermal regime of the table is found on the positioning accuracy.

3. New constructive solution of rotary table with thermostatic worm – wormed gear mechanism

In order to improve the positioning accuracy a new structure of a rotary table has been conceived with positioning possibilities at 360000 positions. This concept of rotary table has a controlled axis and can take part to cutting through plane or space interpolation. This solution is suitable for small and middle size rotary tables, i.e. 320 x 320 mm, 400 x 400 mm, 500 x 500 mm, 630 x 630 mm, 800 x 800 mm and 1000 x 1000 mm. Figure 1 presents, in cross section, the new constructive solution. The moving element 1 that represents the table itself is solidly fixed to the central pivot 2 by means of bolts and the tapered surface on the top side; to the rotary table itself the wormed gear 3 is also fixed that gears with the worm 4. The rotary table has a radial bearing through the roller bearing 4 and axial (thrust) bearing through the thrust bearings 5. The thrust bearings are preloaded ad their position can be adjusted by means of the spacer 7 so that the contact with the surface 8 (the "mirror") can be done, thus providing both a good stiffness and the possibility for clamping the moving element 1. This moving assembly presented above is supported, through the bearings, by the saddle 9 that, on its turn, moves horizontally along the guideways 10. The clamping system of the table 1 consists of six disk springs sets 11 equidistantly located and the unclamping is performed through the hydraulic cylinder 12 that has a supporting collar on its top side. The position encoder 13 is located at the inferior side; it has the cursor fixed to the pivot 2, thus providing the direct measurement of the motion of the moving element 1. The driving system is composed of a servomotor coupled to the wormed gear 4 by means of a clutch. The wormed gearing is duplex type, giving the possibility to adjust the backlash between the tooth flanks through the axial motion of the worm. The wormed gearing has been chosen as transmission system even though its efficiency is relatively low (0.74) for dimension and vicinity reasons, as the internal structure of the rotary table is complex. The positioning of the moving element 1 goes through the following steps: 1- The moving element moves at high rpm towards the programmed point; 2- before positioning a deceleration is performed and change to a low rpm (creeping); 3- stop when reaching the programmed point; 4- clamping of the moving element, at the same time with interrupting the position loop of the axis control. While clamping the axis that is performed against the surface 8, the oil film formed on the "mirror" is crashed if he moving element moves at rapid feed.

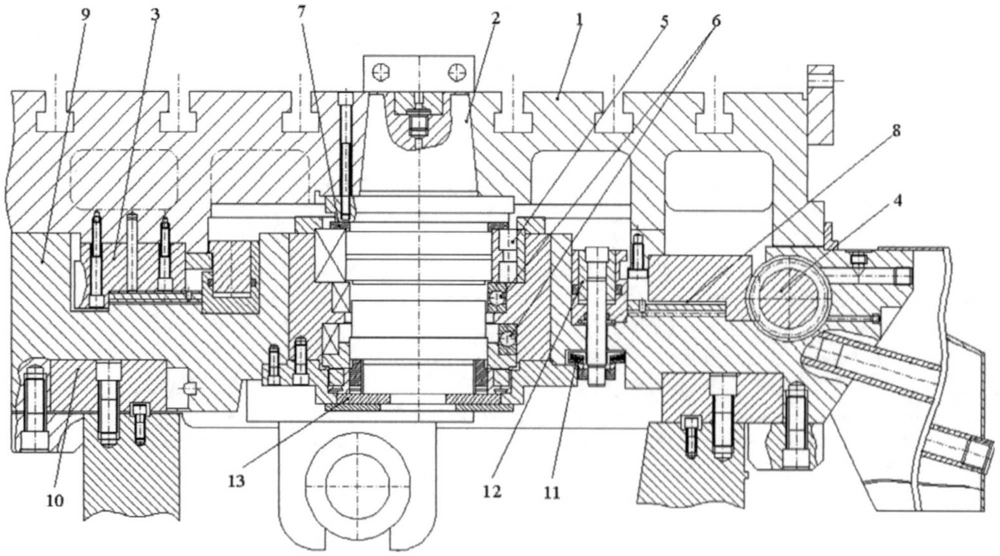

Figure 1. Rotary table – cross section.

Where: *1*-moving element; *2*-central pivot; *3*-wormed gear; *4*-worm; *5*-roller bearings; *6*- axial bearings; *7*-spacer; *8*-with the surface (the "mirror"); *9*-saddle; *10*-guideways; *11*-pack of disk springs equidistantly located; *12*-hydraulic cylinder; *13*- position encoder.

Because of the light flatness deviations of the surface *8*, during the oil film crashing components of the forces that require additionally the moving element *1* come up, leading to slight swings that affect the positioning accuracy. These influences may be seen on the numerical control readout, even though the controlled axis is clamped. Theoretically, the transmission system composed of the wormed gearing is irreversible but, because of the backlash between the tooth flanks the occurrence of this deviation is possible. Even though the backlash in the wormed gearing has been minimised during running, because of its modest efficiency a thermal source will come up that will modify the value of the backlash between the tooth flanks. In order to reduce the thermal influence on the new solution, the wormed gearing will be thermostatic.

Figure 2 shows a schematic presentation of the way of thermostating the duplex type wormed gearing of the rotary table. The installation is designed to keep an optimal running temperature, namely 20°C - 22°C, during the entire functioning time of the rotary table. The system of above uses a Freon unit from the machine endowment in order to refrigerate the cooling –lubrication liquid, bringing it to the optimal running temperature. The liquid is circulated through the unit 2 that is provided with cooler and pump. The pump takes the lubricant from the wormed gearing by means of tour and retour pipes. After cooling, the lubricant is resent to the wormed gearing. Another feature of this system consists of the fact that through the cooling of the lubricant, a relatively viscosity coefficient is maintained, giving a good lubrication. Inside the wormed shaft a drill has been done to facilitate the flowing of the cooled lubricant that enters through the left side. It has a double role, i.e. to lubricate the wormed gearing and to take over the heat generated by the friction forces in the rotary table gearing. Because the new solution of rotary table presented above is available for cutting, the resistant forces require such a sizing of the bearing system of the duplex wormed gearing, to assure a very good stiffness.

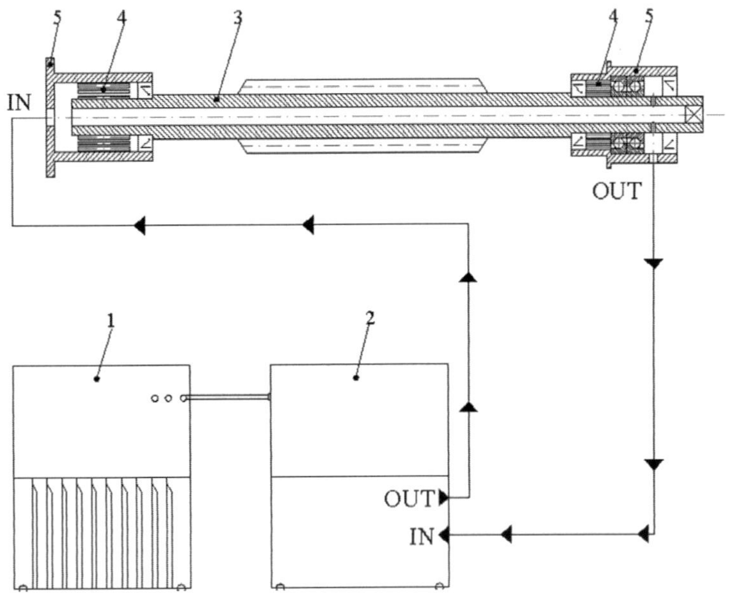

Figure 2. Schematic presentation of the duplex wormed gearing thermostat.

For this reason, on both ends of the wormed gearing needle bearings 4 are used, mounted into the boxes 5 that, on their turn, are provided with adjusting parts in order to allow the axial motion of the worm in order to eliminate the backlash between the tooth flanks. As it has been shown at figure 1, the inner lubrication of the rotary table is mist type lubrication and half of the worm diameter is immersed in oil. In this way, through the worm thermostatic, the cooling of the oil mist is provided as well. The friction and rolling forces between the parts, the friction forces into the bearings etc. lead to heating and wearing the wormed gearing, so it is an energy waste. If it is deemed that the entire lost energy is transformed into heat, the quantity of heat into which the lost energy is transformed, would be:

$$Q = \frac{P(1-\tau)75\ 3600}{427} \tag{1}$$

Where: P is the power transmitted to the nearing; τ-efficiency.
The relation of the heat quantity may also be written under the following form:

$$Q = 632\ (1-\tau)P\ [\frac{kcal}{h}] \tag{2}$$

At the same time, the heat quantity conveyed to the environment is:

$$Q = k(t_1 - t_2)S = k\Delta tS \tag{3}$$

Where: k- is the thermal conductivity; $t1$- oil temperature; $t2$ –ambient temperature which is 20°C; S- surface of the housing that is washed inside by the lubricant and outside is exposed to air circulation. The establishing of the thermal balance gives the relation:

$$632(1-\tau)P = k\Delta tS \tag{4}$$

The temperature increase Δt in the gearing is:

$$\Delta t = \frac{632(1-\tau)P}{kS}\ [grd] \quad (5)$$

Based on the relations of above the thermostatic source needed for cooling the rotary table can be determined.

4. Experimental trials

The new solution of rotary table has been applied on a machining centre (milling) having the rotary table size of 630 x 630 mm. The machining centre is fitted with the Freon unit for the main spindle thermostat and the ball screw bearings on the three axes X, Y and Z. Trials consisted of determining the positioning accuracy of the rotary table (B axis) in two situations: without the worm gearing thermostat and with worm gearing thermostat. The positioning accuracy has been measured by means of a laser interferometer and a polygonal prism with 24 mirrors. The diagrams of the positioning error p in function of the angle α of the rotary table are shown at figure 3.

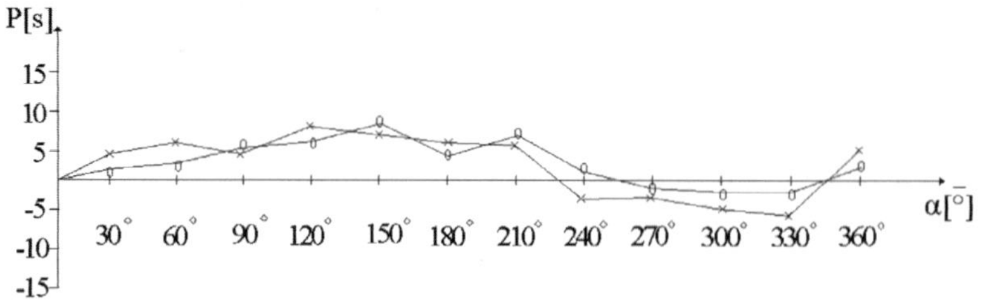

Figure 3. Diagrams of the positioning accuracy.

Where: *X* –without thermostat; *O* – with thermostat.

The analysis of the two diagrams shows an improvement of the positioning accuracy by 7% in the case of using the thermostat on the duplex wormed gearing. This aspect is due to the fact that the backlash in the wormed gearing has been initially taken over, whilst in the thermostat variant it has been kept constant; consequently, upon the table clamping (after positioning) the wormed gearing keeps fixed the table position.

5. Conclusions

The new solution of rotary table presented in this work provides an improvement of the positioning accuracy and is recommended in cases when the machine tool is fitted with a thermostat source. Even though the increase by 7% only of the positioning accuracy in relation to the effort of implementing this solution is not very spectacular, the improvement is however obvious. In case of the CNC machine tools of small to middle size the solution being presented offers the machine tool designers and builders an accessible variant to improve the positioning accuracy.

6. References

[1] Suzuki T 2019 Improved method for synchronizing motion accuracy of linear and rotary axes under constant feed speed vector at end milling point *Int. J. Automation Technol.* **13** 679-690.

[2] Jingxia Y 1998 The real-time error compensation technique for CNC machining systems *Mechatronics* **8** 359-380.

[3] Lee H 2002 Statistical optimization and assessment of a thermal error model for CNC machine Tools *The International Journal of Machine Tools and Manufacture* **42** 147-155.

[4] Ryuta S 2007 High performance motion control of rotary table for 5-axis machining centers *Int. J. Automation Technol* **1** 113-119.

Concept development and prototypes for robotic systems in active assisted living

I Todiriţe, C Radu-Frenţ, P-Al Cotfas and M Iliescu
Romanian Academy, Institute of Solid Mechanics, Constantin Mille 15, Bucharest, Romania

E-mail: mihaiela.iliescu@imsar.ro

Abstract. Active and Assistive Living, AAL, is a concept that attention have been focused on in the last decades. There are many young people with disabilities, such as visually impairment, missing limbs, AVC stroke problems. There are also many old people who have locomotion, eyesight and many other health problems that make their lives so very difficult. The authors of this paper present aspects of their research regarding new concepts, models and prototypes (virtual and 3D printed) of robotic systems intended to assist people with disabilities for an active and normal life. An innovative open source system for guidance of young people (but, also elderly ones) with eyesight problems has been developed. It is intended to be used indoors also outdoors, to provide audio guidance, haptic guidance, or both. Research aspects on a bionic prosthesis for the upper limb are evidenced. The focus is on a two fingers model that should be considered as the proof of concept. Innovative mechatronic system designed to help people, who can no longer stand and / or walk is further presented. A virtual platform concept aimed at the integration of all these devices, and similar others, to communicate by IoT is shown.

1.Introduction

Active and Assistive Living, AAL, is a concept that attention have been focused on in the last decades. There are many young people with disabilities, such as visually impairment, missing limbs, as well as old people who have locomotion, eyesight and many other health problems that make their lives so very difficult. In recent years, attempts have been made to integrate people with disabilities in social, school, university environment. In order to achieve this integration, it is not enough to define the concepts related to inclusive integration, it is necessary to be aware of issues, the specials needs they have, and the fact that these people need help.

In the case of visually impaired people, the fact that they need help with writing, reading or moving is envisaged. In order to help them, various systems have been developed so that they represent be helpful solutions for a small number of people, because they do not cover special needs, are not personalized and do not take into account the fact that in most cases these people have also other disabilities and cannot purchase a device for each disability. This explains the fact that the white walking stick solution [1], bracelet and belt with distance sensors have not become so popular among visually impaired persons. In [2] and [3] there are presented results simulation of a "white cane that blindfolded sighted participants control with a mouse to obtain a single-point depth information (through sound) of the virtual environment.

Missing upper limb is another disability type that, unfortunately, prevents people from having a normal life. There are many researches on prostheses, their structure, enabled motions, adaptability to the body of the person wearing it, availability and costs. The concept and design of low cost myoactivated prosthesis are shown in [4]. This prosthesis "consists of a low-cost embedded microcontroller (Arduino), a wearable stretch sensor, to detect residual muscle contraction as direct muscle volumetric shifts and a handful of common, not critical electronic components." The authors mention that its "control is simple and customizable with access to a computer and minimal training." The costs are much more than affordable. Clinical outcomes of the studies on a myoelectric hand prosthesis are evidenced in [5]. Conclusion is that "low-cost 3D-printed myoelectric-interface prosthetic hand with a single reliable myoelectric signal shows the potential to positively impact amputees' quality of life through daily usage".

Active Assisted living is an important issue for people who had strokes or, for old people whose locomotion problems are severe. There are many devices aimed to help them, as well as the person who takes care. One advanced patient lift and transfer device is presented in [6]. It is mentioned that it "is not designed to travel down steep ramps and transition steps from these ramps, but is meant to traverse doorway thresholds from horizontal surfaces."

The authors of this paper present aspects of their research regarding new concepts, models and prototypes (virtual and 3D printed) of robotic systems intended to assist people with disabilities for an active and normal life. The scope is to implement useful solutions for people with visual impairments and / or mild or moderate cognitive impairments, as well as for people with missing upper limb or, with locomotion problems.

2. Research Method

In this scientific research there have been identified, grouped and analysed the special needs of people with disabilities mentioned above. After identifying the special needs, they were grouped into categories that follow:

- basic needs - the needs identified as common to most of the subjects who took part in the research;
- specific needs - the one that allowed customization of the solutions according to the multiple / conjugate disabilities of the subjects.

For each category of needs it was defined a solution that was implemented conceptually. So the basic needs led to a standard solution, and the specific need led to modular systems within the customized solution.

The scientific research was based on data collected from the subjects, as follows:

- by applying questionnaires to identify special needs;
- through roundtable discussions with subjects during face-to-face or online meetings;
- through practical field orientation activities indoors and outdoors; spaces whose configurations were known, but new routes unfamiliar to the subjects were also made.

The concepts and design of the mechatronic systems for active and assistive living are based on research results regarding body features and type of motions required to fulfil certain tasks (gripping, motion toward an object, controlled trajectories etc.). For example, the kinematic analysis of the mechanical structure of hand prosthesis was done considering Assur groups type – dyad and triad, aided by Matlab software. Modelling and simulation of mechatronic systems were done assisted by SolidWorks software.

3. Research Results

Within the scientific research we carried out different activities that were grouped in stages, as follows next:

Stage I - Data collection and grouping using different criteria in order to identify the special needs of people with disabilities already mentioned.

Stage II - Analysis of results

Stage III - Conceptual implementation of the solution

3.1 Stage I. Description of the research stage

During this stage we started the research by applying, within the working groups, questionnaires to identify special needs.

It should be noted that all subjects who were part of the working groups are students and follow a form of vocational, high school or post-high school education. The subjects' ages are between 14 and 28 years old, grouped by age categories, as shown in table 1.

Face-to-face discussions or online meetings were necessary because it was found after applying the questionnaires that the results were not relevant and then we had to identify the cause. An important criterion chosen, was related to the fact that the person has or does not have multiple/ conjugate disabilities because all disabilities hinder the person's activity.

A list of criteria was added to detect special needs: the sex of the person participating in the research, the spirit of orientation in a closed space (Yes/ No), the spirit of orientation in an outdoor (Yes/ No), tactile recognition of compact objects (Yes/ No), tactile recognition of transparent objects (Yes/ No), use of white cane indoors (Yes/ No/ Occasionally), use of white cane in outdoors (Yes/ No/ Occasionally), color identification (Yes/ No), environment origin (urban/ rural), the environment in which he lives (family/ institutionalized placement/ family placement).

Table 1. Centralization of data on the number of subjects.

Age category name	Coding Participant name	Number of participants with conjugate / multiple disabilities	With conjugate/ multiple disabilities (Girls)	With conjugate / multiple disabilities (Boys)	Number of subjects by age categories
CtgA	A1...A12	3	2	1	12
CtgB	B1...B14	10	5	5	14
CtgC	C1...C10	2	0	2	10
Total		15	7	8	36

More of it, graphical representation of the target group disabilities and the problems they are facing (see figure 1 and figure 2) points toward the fact that there is a relatively equal number of subjects with multiple disabilities and so a wide range of specific needs are collected.

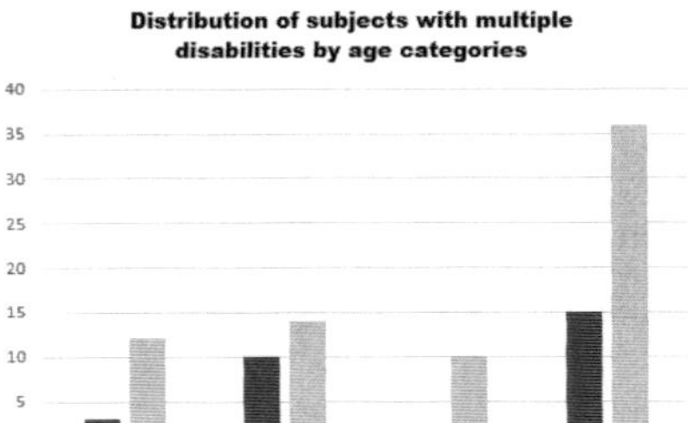

Figure 1. Graphical representation of the distribution, by age categories, of subjects with conjugate/ multiple disabilities.

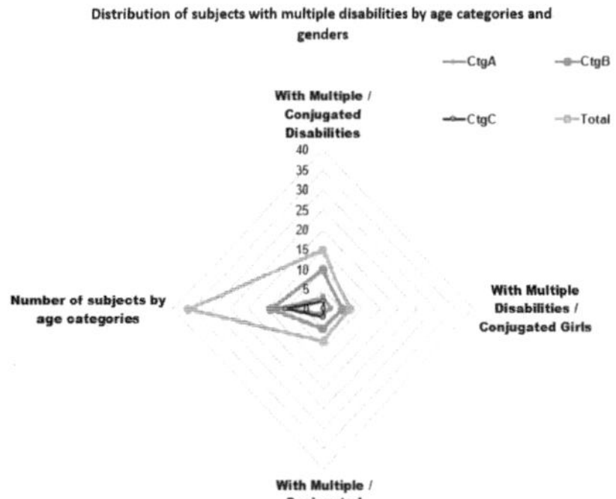

Figure 2. Graphical representation of the distribution of subjects with and without conjugate/ multiple disabilities by age categories and genders.

Following the centralization of the data, it was obtained a database (see figure 3) that will be analyzed in order to discover the specific needs of the subjects involved in this study.

	A	B	C	D	E	F	G	H	I	J	K	L	M
1	Age category name	Coding Participa nt name	With Multiple/ Conjugated Disabilities (Yes/No)	Sex (F/M)	Ability of orientatio n indoors (Yes/No)	Ability of orientatio n outdoors (Yes/No)	Tactile recognition objects (Yes/No)	Tactile recognition transparent objects (Yes/No)	Utilisation white cane indoors (Yes/No/Occ asional)	Utilisation white cane outdoors (Yes/No/Occasion al)	Identific ation colors (Yes/N(	Origins environm ent (Urban/ Rural)	Living environment (family/institution alised placement/family placement)
2	CtgA	A1	Yes	M	No	No	No	No	No	No	No	Urban	Family
3	CtgA	A2	Yes	M	No	No	No	No	No	No	Yes	Urban	Family
4	CtgA	A3	Yes	M	Yes	No	No	No	No	No	No	Urban	Family
5	CtgA	A4	Yes	M	No	No	No	No	Yes	No	No	Urban	institutional placement
6	CtgA	A5	Yes	M	No	No	Yes	No	No	No	Yes	Rural	institutional placement
7	CtgA	A6	Yes	F	No	No	Yes	Yes	Yes	No	Yes	Rural	institutional placement
8	CtgA	A7	Yes	F	Yes	No	Yes	Yes	No	No	Yes	Urban	institutional placement

Figure 3. Questionnaire data centralization table screenshot.

3.2 Analysis of results

In this results analysis stage, it was meant to determine whether subjects with conjugate/multiple disabilities have or can develop a sense of orientation in an enclosed space, when they know or do not know the configuration of the space (see figure 4 and figure 5).

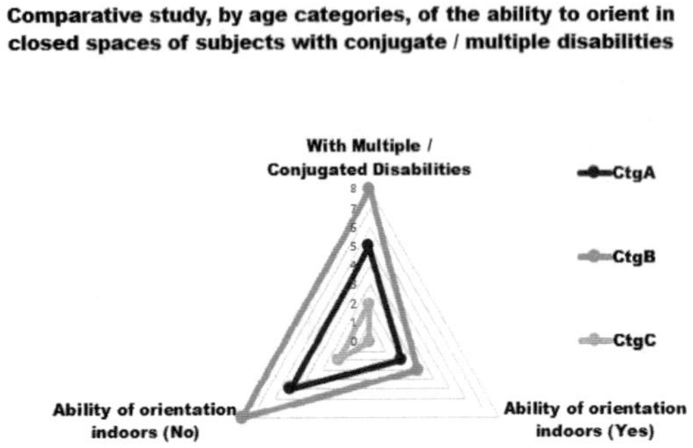

Figure 4. Graphical representation of the comparative study by age groups to illustrate the ability to orient indoors of subjects with conjugate / multiple disabilities.

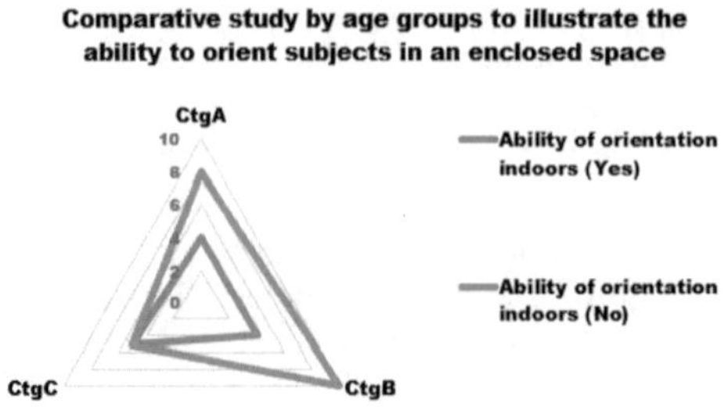

Figure 5. Graphical representation of the comparative study by age groups to illustrate the ability of the subjects to orient themselves indoors.

Analyzing the data set it results that the majority of CtgB category subjects have conjugate/ multiple disabilities. The research found that as the subjects have other disabilities with mild or medium forms, although they have the same space in which they work for two years, they still fail to orient themselves in a closed space, even when the configuration space does not change.

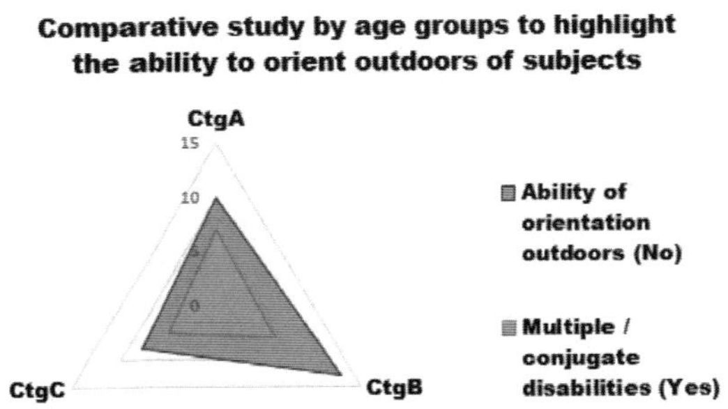

Figure 6. Graphical representation of the options of subjects by age categories who have conjugate/ multiple disabilities and do not have the ability to orient themselves outdoor.

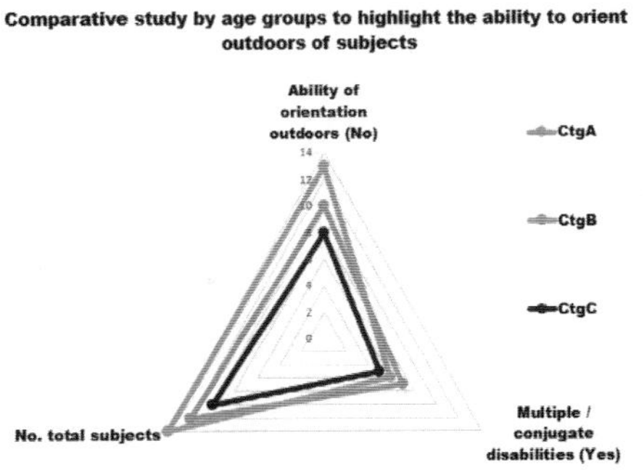

Figure 7. Graphical representation of the options of all subjects by age groups who do not have the ability to orient themselves outdoor.

Analysing the graphical representation of figure 6 it can be seen the subjects who were oriented in the open space were the people who can see in a percentage of 5%, with correction because they have a residual vision and the conjugate disabilities are in easy form. It can be observed by analysing the graphical representations of the results obtained in the research (figure 7), that in open space both subjects with multiple disabilities and those with visual impairments have great difficulty orienting in open space, even if the space is limited as an area and not necessarily unknown. In the case of the new routes, we noticed that the subjects did not manage unaccompanied, although few of them, used a white cane.

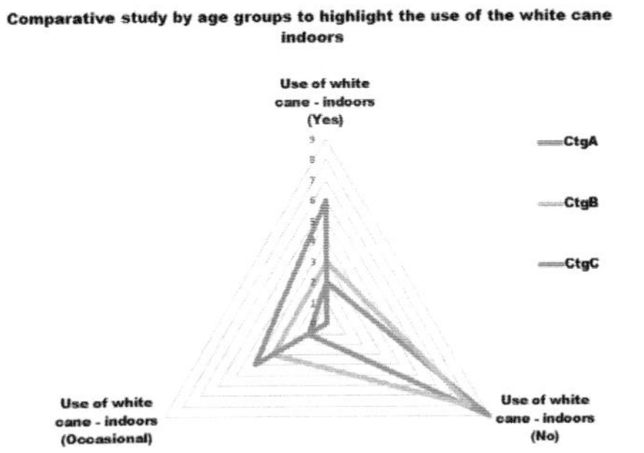

Figure 8. Graphical representation of subjects' options for using the white cane indoors.

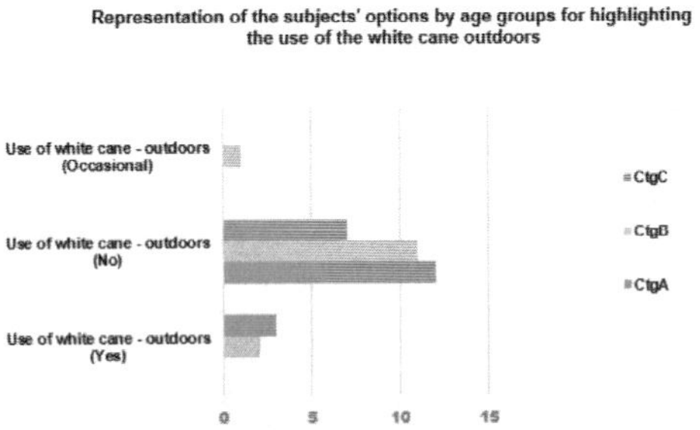

Figure 9. Graphical representation of the comparative study by age categories of the subjects' options regarding the use of the white cane outdoors.

It is observed from the graphical representations of the subjects' options (see figure 8 and figure 9) that the white cane is used by them when they are in indoors much more than in the situation when they are in outdoors.

The shortcomings of the white cane are many, among which we mention: it does not offer the possibility to the one who uses it to identify if there are open windows on the route in indoors. In open places people with visual impairments use the cane if they have to travel short distances, over long distances they avoid walking unaccompanied, because the cane gives them a sense of insecurity by the fact that it is like it warns everyone that they do not see and therefore they become vulnerable people.

IOP Conf. Series: Materials Science and Engineering 1182 (2021) 012080 doi:10.1088/1757-899X/1182/1/012080

3.3 Concept and prototype of the mechatronic systems

The comparative studies conducted in the research helped us to delimit the needs of people with visual impairments and to decide what are the special needs common to these people and what are the needs that can be associated only with other disabilities. From a conceptual point of view, the mechatronic system for visually impaired people (named Visual RobCat) will be modular, enabling customization for each user. One important characteristic is that it enables guidance in narrow spaces, on flat, inclined or staircase surfaces. The 3D model, as it is in the present stage, is shown in figure 10.

Figure 10. Visual RobCat 3D model.

The system will be further developed and all the subassemblies for required modules integrated. Conceptually, there are developed two constructive variants: standard device and adapted device.

The standard device, through which we aim to meet the common special needs of visually impaired people, will include modules for measuring the distance to the objects in front of the subject. The person who uses the device will be notified (audio, haptic) if there are objects in front and, more of it, it will be able to differentiate between objects and beings that move or park. The detected objects can be at ground level (eg stairs that can be climbed) but also objects at head height, transparent objects (windows) or detect gaps in the ground: pits or stairs that go down (see figure 11).

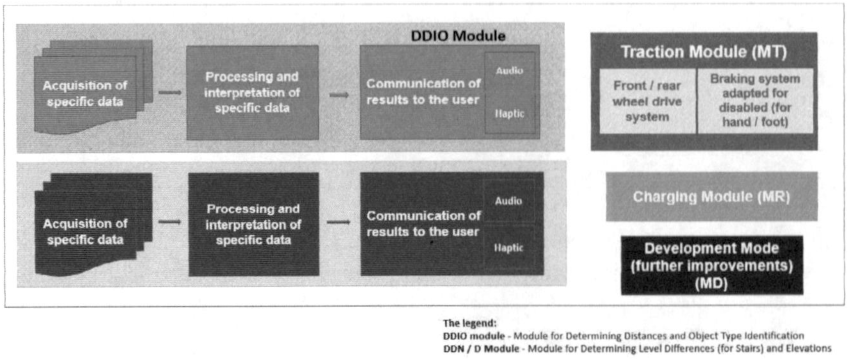

Figure 11. Standard device diagram for the visually impaired.

The adapted device will be built on the structure of the standard one, but as the name suggests it will be adapted to additional needs depending on the mild or medium cognitive disabilities of the subject. It would include modules capable of identifying colours, useful modules for the blind or visually impaired but who do not perceive colours (see figure 12). This feature can be activated when the user wants to choose clothes to wear, buy or wash them with the washing machine.

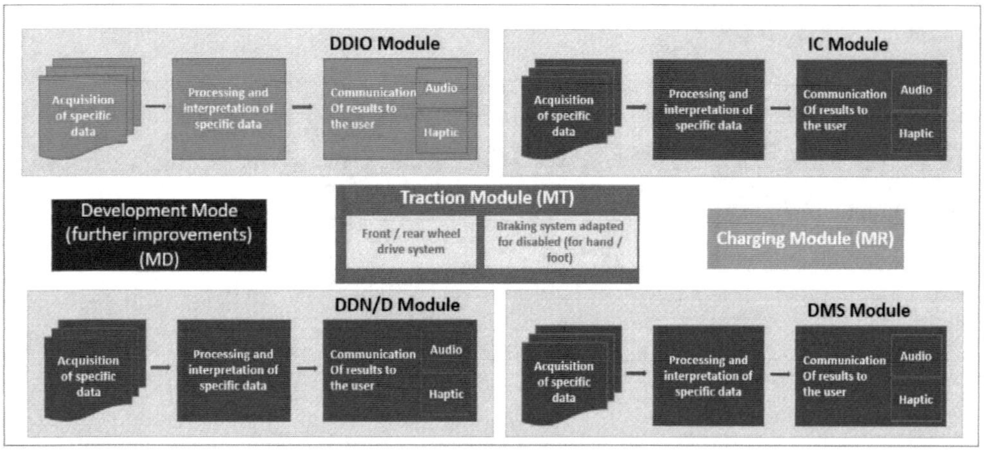

The legend:

DDIO Module - Module for Determining Distances and Object Type Identification

IC Module - Color Identification Module

DDN / D Module - Module for Determining Level Differences (for Stairs) and Elevations

DMS Module - Weather Data and Suggestions Module

Figure 12. Device diagram adapted for people with conjugate / multiple disabilities.

Research aspects on bionic prosthesis for the upper limb are evidenced. The fingers motion is to be initiated by the EMG sensors. There are also pressure sensors, tilt sensors, gyroscope sensor and temperature sensor. For the initial prototype, there has been designed two fingers model (see figure 13(a) and (c)) that should be considered as the proof of concept. Fingers trajectories, when considering the dyad structure open kinematic chain is also presented, for all the hand fingers (see figure 13(b)).

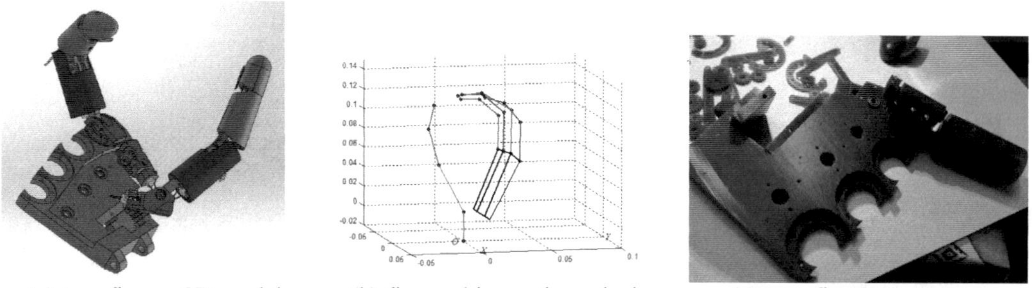

(a) two fingers 3D model (b) fingers kinematic analysis (c) two fingers prototype

Figure 13. Hand prosthesis model.

The concept and design of an innovative mechatronic system designed to help people, who can no longer stand and / or walk is presented. The most important aspect is that this system would be attached on an autovehicle's frame (see figure 14) and enable people with locomotion problems to be got into the car.

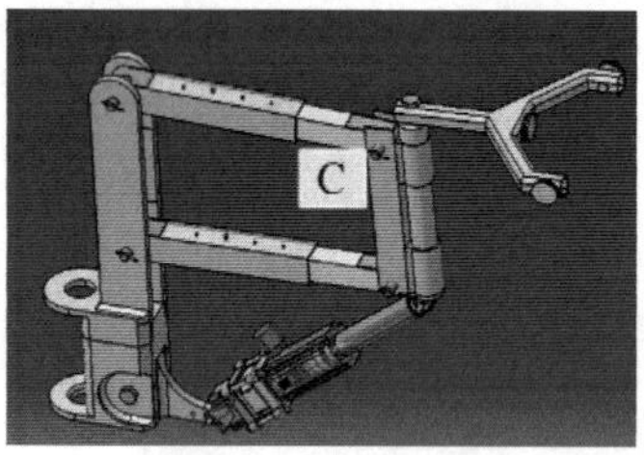

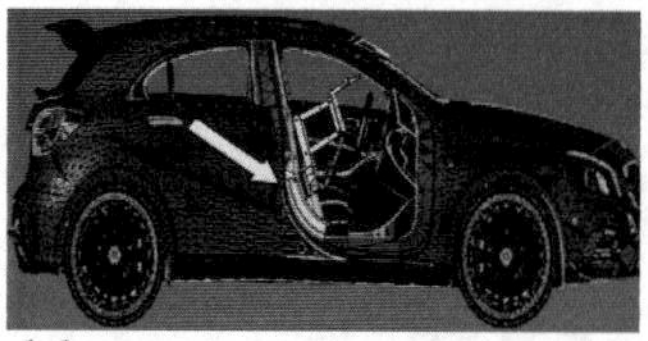

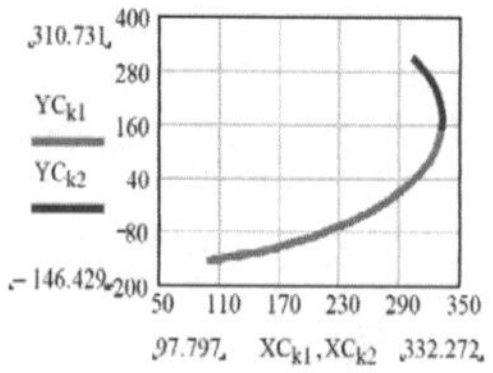

(a) lift 3D model (b) kinematic analysis – trajectory of C

Figure 14. Lift system model.

A virtual platform concept aimed at the integration of all these devices, and similar others, to communicate by IoT is shown in figure 15.

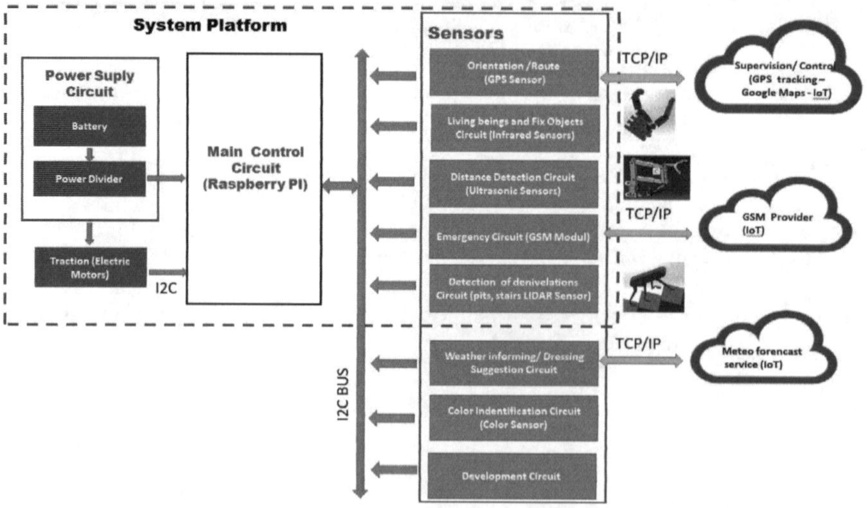

Figure 15. Virtual platform concept.

4. Conclusions

The authors of this paper present aspects of their research regarding new concepts, models and prototypes (virtual and 3D printed) of robotic systems intended to assist people with disabilities for an active and normal life.

Initial stage of the research was aimed to identify the special needs of people whose life would be assisted and improved. Based on questionnaire result there have been identified, grouped and analysed the special needs of these people.

From a conceptual point of view, the mechatronic system for visually impaired people (named Visual RobCat) will be modular, enabling guidance in narrow spaces, on flat, inclined or staircase surfaces. It will be developed in two constructive variants: standard device and adapted device, depending on its degree of customization.

The concept of bionic prosthesis for the upper limb is evidenced, mainly by the two fingers model. It can be noticed the complexity of fingers motions, in two planes (horizontal and vertical). The fingers motion is to be initiated by the EMG sensors and each of the fingers phalanges would be independently driven by micromotors. Image of the 3D printed prototype is presented.

The model of innovative mechatronic system that enables people with locomotion disability to get into a car was also the objective of this research.

Finally, it has been developed virtual platform concept aimed at the integration and communication by IoT of all designed devices. This platform would be implemented and tested in virtual reality and, further, validated by the manufactured devices' prototypes.

5. References

[1] Menikdiwela M P, Dharmasena K M I S and Harsha A M S 2013 *Abeykoon Haptic based walking stick for visually impaired people* https://ieeexplore.ieee.org/document/6718549/authors.

[2] Alexa F. Siu, Mike Sinclair, Robert Kovacs, Eyal Ofek, Christian Holz and Edward Cutrell 2020 *Virtual Reality Without Vision: A Haptic and Auditory White Cane to Navigate Complex Virtual Worlds*, CHI '20, Honolulu, HI, USA, ACM 978-1-4503-6708-0/20/04, https://doi.org/10.1145/3313831.3376353.

[3] Maruricio Lumbreras and Jaime Sánchez 1999 Interactive 3D sound hyperstories for blind children *In Proceedings of the SIGCHI conference on Human Factors in Computing Systems* (CHI '99) Association for Computing Machinery, New York, NY, USA, 318–325 pp. https://doi.org/10.1145/302979.303101.

[4] Sreenivasan N, Ulloa Gutierrez DF, Bifulco P, Cesarelli M, Gunawardana U, Gargiulo GD. 2018 Towards Ultra Low-Cost Myoactivated Prostheses. *Biomed Res Int.* 2018:9634184. https://doi.org/10.1155/2018/9634184. PMID: 30402497.

[5] Ku, I., Lee, G. K., Park, C. Y., Lee, J. and Jeong, E. 2019 Clinical outcomes of a low-cost single-channel myoelectric-interface three-dimensional hand prosthesis *Archives of plastic surgery* **46**(4) 303–310 https://doi.org/10.5999/aps.2018.01375.

[6] Bostelman, R., Ryu, J., Chang, T., Johnson, J. and Agrawal, S. K. 2010 An Advanced Patient Lift and Transfer Device for the Home *ASME. J. Med. Devices.* **4**(1): 011004. https://doi.org/10.1115/1.4001255.

IOP Conf. Series: Materials Science and Engineering 1182 (2021) 012081 doi:10.1088/1757-899X/1182/1/012081

Novel method suitable for decreasing the roofing tile failures generated during rapid drying

M R Vasić[1], R Rekecki[2] and Z Radojević[1]
[1]"Institute for testing of materials", Bulevar vojvode Mišića 43 1100 Belgrade, Serbia
[2]"Wienerberger doo Kanjiža", Subotički put 57 24420 Kanjiža

E-mail: milos.vasic@institutims.rs

Abstract. Only few papers in which principles for decreasing the failures generated during semi and rapid drying of porous shrinking materials, for example roofing tiles, are reported. One of the main differences between the traditional and rapid drying regimes is related with the fact that during the heating step products are heated to the much higher temperatures during rapid drying. If the humidity inside the industrial drying tunnels or chambers is not controlled all the time in the prescribed range of 90 - 95 %, especially during the heating step of the rapid drying regime, crack failures will be developed. In order to inhibit the cracks generation during intensive drying the non-ionic surfactant Igepal CO-630 was added in the raw material during forming process. Various amounts of surfactant around the critical micelle concentration (CMC) were used. Five isothermal experiments were recorded, on laboratory extruded roofing tiles, for each surfactant concentration. Calculated effective diffusivity and material strength - moisture ratio (Deff-MR & MS-MR) curves were used to monitor the development of the cracks during drying and to determine the most suitable concentration of the surfactant. The explanation, how the interaction of the surfactants with clay helps the water molecules to easily move up to the surface was also reported. The application of surfactant in a recommended amount has significantly reduced the drying induced fractures in roofing tiles during its intensive drying and consequently the industrial scarp-rate was accordingly lowered.

1. Introduction

The unwanted shrinkage of the green roofing tiles during intensive drying can cause significant reduction of the products material strength and in some cases can lead to the formation of cracks. This phenomenon is directly related with the drying induced stresses. These stresses are always generating within the products when the amount of bonding clay minerals in the raw material is insufficient or moisture - temperature distributions becomes nonlinear [1,2]. Even though different methods and modeling techniques as well as specialized software's for setting up the safe and optimal levels of drying air parameters inside the various types of industrial roofing tile dryers were recently reported by authors, its efficiency is still directly correlated with the measuring and regulation equipment installed in the dryer, especially if the rapid drying regimes are applied [3-5]. The positive example of industrial semi-rapid drying regime, in which stable control of air humidity from 82 - 88% was achieved within the first drying phase, was recently reported [6]. The impossibility to practically control and keep the humidity inside the industrial drying tunnels or chambers during rapid drying, in the theoretically prescribed range of 90 - 95 %, all the time will inevitably lead to the cracks genesis. The main goal of this paper was to find a solution how to decrease the roofing tile failures generated during the intensive heating phase of the rapid drying regime.

Several measures are available. The first one is to adapt the row material composition by adding the clay or sand fraction in order to change overall plasticity or drying sensitivity. The second one is to upgrade or install the new measuring and regulating equipment, capable to fulfill all theoretical demands related to the rapid drying as specified in [4,6], inside the dryers. The third measure is to improve the moisture transport from the inside of the drying products and to make moisture distribution more uniform. This can be achieved using different surfactants.

Kowalski and his researcher group investigate the effect of several surfactants (Rokacet R26, Rokanol IT6, SDS and FC 4430) on the reduction of drying induced stresses inside the kaolin-clay products during drying. The acoustic emission method was used to monitor on line the development of crack formation in dried samples. The best quality of the dried products was achieved for clay saturated with water containing low concentration of surfactants below but close to the critical micelle concentration (CMC). The main conclusion was that the addition of surfactant to water, used for wetting the clay prior to the forming process, will significantly decreases the surface moisture tension allowing much easier moisture movement from the inside of the capillary-porous skeleton [7,8]. Gulzena has used non-ionic surfactant Triton TX-100. In this research clay bricks were made from the raw material in which illite clays were dominant. Digital camera was used to record the drying process. Obtained pictures were used for measuring the crack development. The smallest number of cracks was registered when the surfactant concentration was near the CMC [9].

Biswald's has compared non-ionic surfactants Triton TX-100 and Igepal CO-630. He has reported that the Hamaker constant and work of adhesion decrease gradually with increasing surfactant concentration and ultimately reach plateau regions above the CMC. It was also stated that the Igepal CO-630 has a lower CMC value (0.08 mmol/l) and better wetting properties at low concentrations than Triton TX-100 (0.15 mmol/l) as well as comparable properties near the CMC. The main conclusion was the statement that Igepal CO-630 was proved to be a more beneficial wetting agent than Triton TX-100 [10].

This was one of the reasons why previously mentioned non-ionic surfactants was selected and used in this study. The second one was the fact that the influence of non –ionic surfactant concentrations, just above the CM on the reduction of the drying induced stress was not investigated in detail even though in some references can be found that if unsaturated soli is treated with surfactant solution suction consistency is reduced with the increasing of surfactant concentration [11].

2. Materials and Methods

The raw material was taken from the production line of the Wienerberger's roofing tile factory situated in Kanjiza. Classical silicate and particle size analysis, qualitative and semi-qualitative X-ray analysis which corresponds to the used raw material, was reported earlier in the reference [12]. The row material was first grinded in the pan mill. After that the simultaneous moisturization and milling in the dual rotor crushers were applied. During the moisturization step surfactant Igepal CO-630 CAS number 9002-93-1 was added as defined in table 1. Five sets of roofing tiles (120 x 50 x 60 mm) and cubes (33 x 33 mm) were then extruded on the laboratory press under a constant pressure of 80 kPa.

Table 1. Surfactants concetration.

SC set	I	II	III	IV	V
Concetration (%)	0	0.1	0.05	0.005	0.0001

Formed samples were closed in plastic bags and kept at room temperature. Five isothermal experiments were recorded on roofing tiles for each SC. The corresponding drying air parameters are defined in table 2. These parameters were controlled with accuracies of ±0.2 °C, ±0.2 % and ±0.1%. The sample mass and linear shrinkage were registered too. The accuracies of these measurements were 0.01 g and 0.2 mm. The Arabic numbers were used for the identification of the isothermal experiments while Roman numerals were used for identification of the surfactant concentration as presented in table 1. Obtained data were used for the construction of Deff-MR curves.

Table 2. Drying air parameters.

Experiment		1	2	3	4	5
Temperature	(^{0}C)	30	35	40	45	50
Humidity	(%)	90	90	90	90	90
Velocity	(m/s)	3	3	3	3	3

Three cubes picked from the same SC set were dried isothermally at (25^{0}C, 70%, 1 m/s). MR - t curves were plotted. Total drying time was divided on ten identical intervals. Five cubes taken from the same SC set were than dried up to the each defined moisture content. Dried cubes were closed and kept for 24h at 25^{0}C. The compressive test was determined on these samples in accordance with the SRPS EN 771-1. Deff – MR curve was than formed. Described procedure was repeated for each SC set.

3. Results and discussion

Results presented in the reference [12] (tables 2 and 3) have classified the raw material as silt loam. Illite, montmorillonite, chlorite, kaolinte, large amounts of calcite and dolomite, quartz, micas and feldspar were identified in the raw material. Illite and muscovite were marked as dominant clay minerals.

The structure of surfactant solution depends on the surfactant concentration. In many applications the CMC is defining the limiting concentration for meaningful surfactant use. This value is important since the surface tension does not reduce further when concentration of surfactant is above CMC. When the CMC is reached the molecules are arranged in a monomolecular form. The micelle formation is shown in figure 1. The CMC value for the Igepal CO-630 was taken from the reference [10] as 0.08mmol/l (49.36mg/l or 0.005%). The particle size distributions registered in SC II-V sets were slightly different than the one registered in SC-I set. This was expected result. Similar trend was also reported in reference [9].

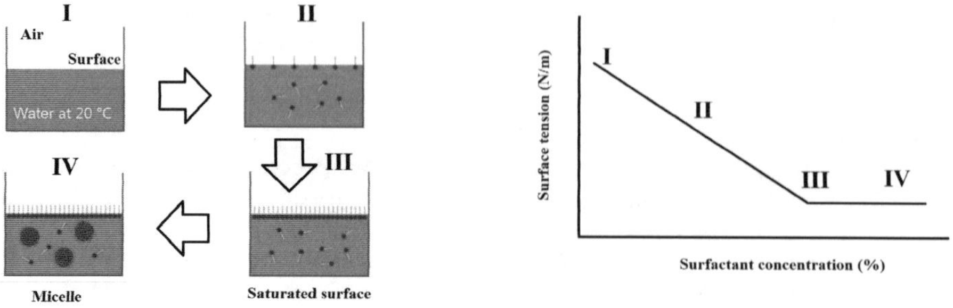

Figure 1. Micelle formation.

Cracks were detected in experiment 5. All possible mechanisms of moisture transport and their transition from one to another during the constant and the falling drying period were identified on Deff – MR curves using the procedure outlined in references [3-4]. The construction of SL lines on Deff-MR curves for the samples which have cracked during drying was reported in reference [4]. Resulting five curves corresponding to experiment 1 and 5 were given as an example. The overcrowded figure could obstruct the reader. That was the reason why only the Deff – MR curves, detected in experiments 2-I, 3-I and 4-I, were reported on figure 2. By doing that the characteristic marks, and the corresponding SL lines were also more visible allowing the reader to become more focused on receiving the comprehensive and complete information about the history of moisture transport during drying.

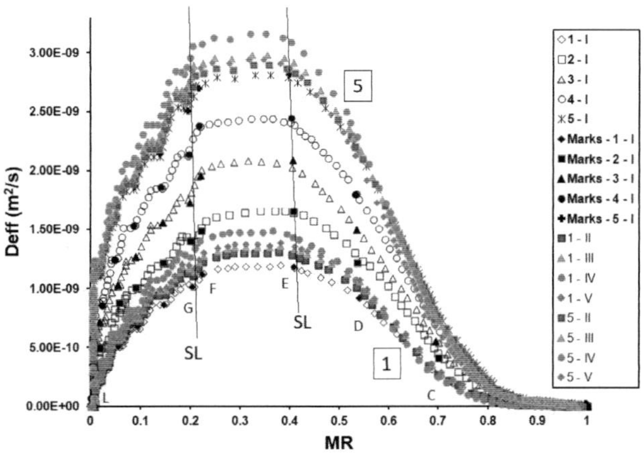

Figure 2. Calculated Deff – MR curves.

In our previous papers it was stated that in the case of raw materials in which illite - smectite clays are dominant the most critical drying segment is at the beginning of the drying in the vicinity of characteristic point B [13]. The moisture transport registered in experiments in 1-II, 1-III, 1-IV and 1-V up to the mark C is slightly lower than the one registered in 1-I. The overall stresses in this segment, in the products containing surfactants, are slightly neutralized as a consequence of the more uniform temperature distribution within the products. In other words the drying susceptibility for the crack formation is reduced in this segment. This is very important since this drying segment was in our previous study identified as critical [13].

The overall moisture transport registered in experiments 1-II, 1-III, 1-IV and 1-V after the mark C is higher than the one registered in the products formed without surfactant addition. This is a direct confirmation that the presence of surfactant has improved the internal moisture transport. If individual Deff – MR curves for example from the experimental 1 are compared, it can be seen that the maximal increase in overall moisture transport during drying was registered in experiment IV. This means that surfactant concentration slightly above or close to the CMC concentration provides the maximum adsorption to the clay surface. It is important to mention that the drying behavior pattern reported in experiment 1 was registered and confirmed in all other experiments.

Calculated MS curves are presented in figure 3. The shape of curves reported for experiments I-V was the same as one registered in references [13, 15 and 16]. Regardless to the presence or absence of the surfactant the material strength is small during drying until the MR of 0.84 is reached. This MR value corresponds to the moister content of 14.88%. This is another confirmation that this type of raw material is sensitive to drying especially at the beginning. The addition of surfactant to the raw material is evidently affecting the material strength. This effect is directly correlated with the surfactant concentration. The highest material strengths were registered for the surfactant concentration below or near the CM. These findings are in agreement with the one obtained from the figure 2.

The adsorption of no-ionic surfactants such as Igepal CO-630 is attributed to hydrogen and hydrophobic bonding. If hydrogen bonding is responsible to adsorption the bond that is formed between the surfactant functional groups and clay mineral surfaces must be stronger than the bond that is formed between the mineral and interfacial water molecules [14]. An adsorption on solids that possess a fully or partially hydrophobic surface takes place mainly due to hydrophobic bonding. For kaolinite the two surfaces have different structure. The siloxane surface interacts very weakly with interfacial water molecules, while the hydroxyl surface interacts strongly with water through hydrogen bonding. The illite siloxane surface has introduced a permanent change as a result of isomorphous substitution. Its surface displays a degree of charge imbalance as a result of the ion substitutions in the octahedral and tetrahedral sheets. That is the reason why illite hydrogen reaction with the surfactants is stronger than in the case of kaolinite. Since substitutions are more extensive in montmorillonite it is assumed that the absorption power is larger than for illite or kaolinite minerals. Direct relationship between the adsorption of the nonionic surfactant and the amount of clay minerals in the adsorbents was reported in reference [16].

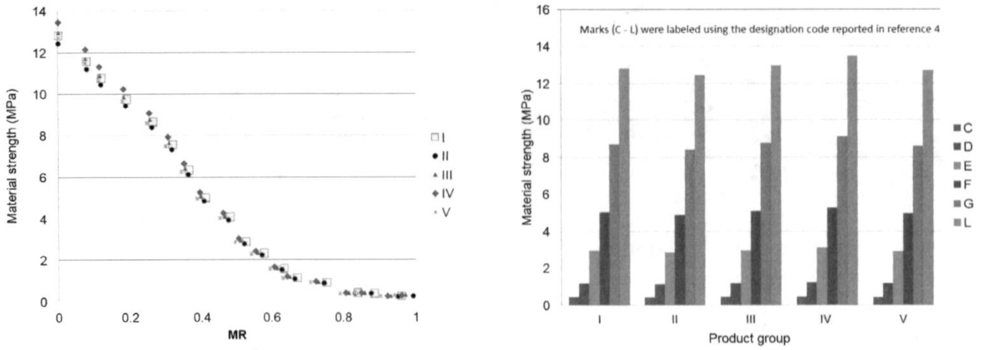

Figure 3. Registered material strength – moisture content curve.

Igepal 630's hydrophilic monomers are interacting with clay minerals mostly through hydrogen bonding. This will lead to the orientation of the surfactant in such way that its hydrophobic tail is directed towards the interior of the pores. The presence of surfactant tails within the pores will reduce the interactions between the water molecules and the porous skeleton during drying process allowing them to pass much easier from the inside of the green products up to the surface. This mechanism is responsible for the reduction of internal moisture stresses during the drying process.

Recommended surfactant concentration of 0.005% was added and industrial roofing tiles were produced. The industrial drying regime reported in reference [6] was applied. The bending strength after drying was measured inside the factory in accordance with its factory production plan. These results are confidential and the authors did not have the factory approval to publish it. We can only say that the bending strength results of the treated samples are higher and that the drying scarp rate was lower for 3 percent.

4. Conclusions

In order to decrease the roofing tiles failures generated during intensive drying the non-ionic surfactant Igepal CO-630 was added in the raw material during forming process. Various amounts of surfactant were used. The mechanism responsible for the reduction of internal moisture stresses during the drying process was reported. The surfactant tails will reduce the interactions between the water molecules and the porous skeleton during drying process. This will increase the internal moisture transport from the inside of the green products up to its surface. Results have confirmed that the optimal concentration of the surfactant is near or below the CM. When surfactant was applied in a recommended amount it has significantly reduced the drying induced fractures in roofing tiles during its intensive drying and consequently the industrial scarp-rate was accordingly lowered for 3 percent.

5. References

[1] Musielak G. 2000 Influence of the drying medium parameters on drying induced stresses *Drying Technology* **18**(3) pp 561-581.

[2] Musielak G. 2001 Possibility of clay damage during drying *Drying Technology* **19**(8) pp 1645-1659.
[3] Vasić M, Grbavčić Ž and Radojević Z 2014 Analysis of Moisture Transfer During the Drying of Clay Tiles with Particular Reference to an Estimation of the Time-Dependent Effective Diffusivity *Drying Technology* **32**(7) pp 829-840.
[4] Vasić M, Rekecki R, Radojević Z 2018 A procedure for setting up the drying regime that is consistent with the nature and properties of clay raw material *Drying Technology* **36**(3) pp 267-282.
[5] Vasić M, Radojević Z, 2018 Update of the procedure used for heavy clay dryer optimization *Romanian Jour of Mat.* **48**(4) pp 436-441.
[6] Vasić M, Radojević Z, 2020 Characterization of drying behaviour and modelling of industrial drying process *IOP Conf.Ser:Mater.Sci. Eng.* **916** 012124.
[7] Kowalski J. S, Kulczynski K, 2013 Acoustic emission monitoring of fracture in dryed kaolin-clay processed with surfactant water solution, *Materialy Ceramiczne / Cer. Mat.* **65**(2) pp 235-238.
[8] Kowalski J. S, Kulczynski K, 2013 Reduction of fractures in dried clay-like materials due to specific surfactants *Chemical Engineering Research and Design* **91**(2) pp 254-263.
[9] Guzlena S., Sakale G,Certoks S, Spule A, 2018 Crack reduction during drying process by using surfactants *ICBMM 2018 MATEC Web of Conferences* **278**, 01008 (2019).
[10] Biswal R. N, Paria S, 2011 Wetting of TX-100 and Igepal CO-630 surfactants on PTFE surface *Ind.Eng.Chem.Res* **50** pp 6138-6145.
[11] Jones N., Lourenco N.D.S., Paul A, 2016 Testing surfactants as additives for clay improvement: compaction and suction effects *E-UNSAT 2016 ES3 Web of conference* **9** 13006 (2016).
[12] Rekecki R and Ranogajec J 2008 Design of ceramic miscrostructures based on waste materials *Proc. and Appl. of Ceramic* **2** pp 89-95.
[13] Vasić M, Radojević Z, 2019 Method for avoiding cracks during drying of masonry unites made of illite raw material *IOP Conf.Ser:Mater.Sci. Eng.* **591** 012101.
[14] Zhang, R., Somasundaran, P., 2006 Advances in adsorption of surfactants and their mixtures at solid/solution interfaces. *Adv. Colloid Interface Sci.* 123–126: 213–229.
[15] Amirianshoja T, Junin R., Idris. K.A, Rahmani O, 2013 *J. Pet. Sci. Eng.* **101** pp.21 - 27.
[16] Banaszak J, 2013 The moisture content and mechanical properties of ceramic masses *Ceramic Materials* **65**(1) pp 40-43.

Acknowledgement
This paper was financed by ministry of education science and technological development of Serbia, contract number 451-03-9/2021-14/ 200012 Authors M. Vasic and Z. Radojevic are fully employed in the Institute for testing of Materials

.

IOP Conf. Series: Materials Science and Engineering 1182 (2021) 012082 doi:10.1088/1757-899X/1182/1/012082

Comparative study on the financial implications of a fire fighting intervention on an oil reservoir

I Voicu, E D Juganaru, V F Panaitescu and M Panaitescu
Department of Engineering Sciences in the Mechanical and Environmental field, Electromechanics Faculty, Constanta Maritime University, Constanta, No. 104, Street Mircea cel Batran, Romania

E-mail: dana.mitu@yahoo.com

Abstract. In this paper, the authors determine by calculation, for a storage reservoir design for petroleum products, with a volume of 50,000 m^3, the required amount of foam that will be used to extinguish a fire that broke out in it. The reservoir for which the calculation is made is in the park reservoir within S.C. Oil Terminal S.A. Constanta - South Platform. The paper presents the constructive characteristics of the reservoir, for which the calculation is made, the constructive characteristics of the special extinguishing installation, the physico-chemical characteristics of the petroleum substances (gasoline, crude oil), as well as the physico-chemical characteristics of the foaming liquid used for extinguishing the fire. The calculation is made on the basis of the Norm on the fire safety of buildings, Part II Extinguishing Installations Indicative P118/2-2013, as well as on the basis of information from the Methodology for calculating the forces and means necessary for extinguishing fires.

1. Introduction

In this paper, the authors conduct a comparative study on the financial costs of a firefighting operation carried out using the fixed extinguishing system, for two types of petroleum products (crude oil and gasoline), extinguishing the fire with the same type of fire foaming liquid.

The reservoir (50,000 m^3) is located inside S.C. Oil Terminal S.A. Constanta - South Platform and the calculation is made on the basis of the Norm on the fire safety of buildings, Part II Extinguishing Installations Indicative P118/2-2013, as well as on the basis of information from the Methodology for calculating the forces and means necessary for extinguishing fires.

2. General information

S.C. Oil Terminal S.A. (figure 1), is a company of national importance that is part of the service sector, in the field of petroleum products. The total area of the company is 221 hectares; it consists mainly of three platforms, which share the activities of receiving, unloading, storage, conditioning and delivery of crude oil and petroleum products, petrochemicals and liquid chemicals for import, export and transit. The three platforms provide a total storage capacity of 1.7 million m^3, a turnover of 24 million tons of crude oil per year and about 10 million tons of petroleum products per year [2].

Figure 1. S.C. Oil Terminal S.A. – South platform. [3]

2.1. Technical data on the tank

The tank for which the calculation is made is located in a fleet of 3 (three) tanks (figure 2), each with a volume of 50,000 m^3. It is made of a non-combustible material (metal), resistant to corrosion of the product contained. On the outside is inscribed the type of oil product that is stored, as well as its volume. The tank is equipped with a system for determining the level of filling, ventilation valves and flame arrestors. It is tied to the ground belt to avoid the formation of static electricity [4].

From a constructive point of view, this is a cylindrical, vertical tank with an external floating lid, with a diameter of 60.9 m and a height of 17.2 m.

Figure 2. Tank of 50,000 m^3. [4]

The holding tank is made of a material resistant to the stored petroleum product (concrete), waterproof, of volume equal to the capacity of the tank. It has an area of 8,667 m^2 and is connected to the park sewer through hydraulic closures that allow the product to be kept indoors.

The tank holding and the tank are equipped with fixed aeromechanical foam fire extinguishing installations and fixed water cooling installations.

2.2. Stored products

The physico-chemical characteristics of the stored products mean that each tank is designed and built with the purpose of storing a limited number of products.

2.2.1. Physico-chemical characteristics of crude oil, [2]

- boiling temperature: 25 – 525[°C];
- flow temperature: - 15 ... + 20 [°C];
- density: 800 - 970 [kg / m^3] at a temperature of 15 [°C];
- water solubility: insoluble;
- appearance and odor: black-brown viscous liquid, with a characteristic odor of petroleum product;
- vapor density (air = 1): 3-4;
- water content + impurities: max. 1 % of the volume;
- chloride content: max. 0.3 % by weight;
- sulfur content: max. 0.1 - 0.8 % by weight;
- chemical incompatibilities: incompatibilities with oxidants;
- hazardous decomposition products: by oxidative thermal decomposition, it can produce various hydrocarbons and their derivatives as well as partial oxidation products: carbon dioxide, carbon monoxide and sulfur dioxide;
- conditions to be avoided: heating and ignition sources, as well as contact with strong oxidizing agents.

2.2.2. Physico-chemical characteristics of gasoline, [2]

- boiling point:
    - initially 39 [°C];
    - after 10 % distillation: 60 [°C];
    - after 50 % distillation: 110 [°C];
    - after 90 % distillation: 170 [°C];
    - final boiling point: 204 [°C];
- density: < 760 [kg/m^3];
- water solubility: insoluble;
- appearance and odor: mobile liquid, with a specific odor recognized at more than 10 ppm in the air;
- vapor density (air = 1): 3-4;
- chemical incompatibilities: petrol for cars can react with oxidizing materials such as: peroxide, nitric acid, perchlorate;
- hazardous decomposition products: oxidative thermal decomposition of automotive petrol can produce carbon monoxide;
- conditions to be avoided: heating and ignition sources.

2.3. The foaming liquid

Low-foaming foams are used to extinguish flames in flammable product tanks; in this case the foaming liquid used to extinguish the fire is Fomtec AFFF 6 % which is a concentrated foaming agent (AFFF) with aqueous film formation, containing fluorocarbon surfactants and hydrocarbons mixed with various solvents, preservatives and stabilizers. [5]

The foam forms an aqueous film that removes the oxygen reserve, causing the fire to go out. The expanded foam from which the film is drained forms a stable blanket that inhibits the release of flammable vapors and cools the fuel-soaked surface by extinguishing the fire and preventing it from re-igniting.

The low tension on the surface of the water content of the foam allows the aqueous film which, although heavier than fuel, to float on its surface.

Fomtec AFFF 6 % must be used in a concentration of 6 % in proportionate solution (6 parts concentrate and 94 parts water) willow water, fresh or salt. It can be stored as a premixed solution in drinking water. [5]

Fomtec AFFF 6 % is used to extinguish fires of class B fuels, hydrocarbons such as: oil, gasoline, diesel, fuels used in aviation.

The fire performance of Fomtec AFFF 6 % (table 1) has been tested and recorded. The foaming properties depend on the equipment used and other variables such as water and ambient temperatures. The average bending is 7: 1, the average 1/4 drainage time is 2:30 minutes using the UNI 86 test nozzle.

Table 1. Technical characteristics of Fomtec AFFF 6% foaming liquid. [5]

Technical specifications	
Appearance	clear yellowish liquid
Specific gravity at 20 °C	1,005 +/- 0.01 [g/ml]
Specific gravity at 20 °C	< 20 [cSt]
pH	7.5 +/- 1.0
Freezing temperature	- 3 [°C]
Storage temperature	- 2 [°C]
Suspended sediment (v/v)	less than de 0.2 [%]
Surface tension	20.0 [dyn/cm]

2.4. Fixed extinguishing system

The fixed foam extinguishing system consists of pipes attached to the tank to support the generators and aeromechanical foam generators.

The fixed cooling system consists of 3 (three) cooling rings, related supply pipes and water spray nozzles.

Fixed foam fire-fighting installations mainly consist of, [1]:

a) water supply installations;
b) containers for sparkling wines;
c) foam generators with air blowing;
d) dispensers;
e) water and sparkling pumps;
f) pipes for foaming solution;
g) foam spills;
h) commissioning and control devices.

3. Calculation of the required amount of foam

For the determination of the minimum quantity of low-fold foaming solution, q, required for extinguishing the fire, the following relation was used, from the Regulation on fire safety of constructions, Part II, Extinguishing installations, Indicative P118 / 2-2013 ,as follows, [1]:

$$q = q_{th} \cdot f_c \cdot f_o \cdot f_H \quad (1)$$

where:

q – represents the minimum quantity of foaming solution [l/min·m^2];
q_{th} – nominal quantity of foaming solution [l/min·m^2];
f_c – correction factor for the foam solution class;
f_o – correction factor for the protected object type;
f_H – correction factor depending on the distance of the nozzles for external systems;

From a constructive point of view, the tank is cylindrical, vertical, with external floating cover, and it is assumed that at the time of the fire the cover is not dropped, so that the fire manifests at the top, on its entire circumference, the burned area being 191.2 m^2 (the distance between the lid and the tank jacket being 100 cm).

According to the norm mentioned above, the minimum quantity of foaming solution, q, necessary for extinguishing the fire, is determined with relation (1), where:

- nominal quantity of foaming solution, $q_{th} = 4.0$ [l/min·m^2];
- the correction factor for the foaming solution class is $f_c = 1.1$, because the foaming solution's quenching performance class is 1B;
- correction factor for the type of protected object, $f_o = 3\ at\ 20\ minutes$;
- correction factor depending on the distance of the nozzles for external systems, $f_H = 1.25$ for nozzles located at > 5 m from the protected surface;

According to the mentioned norm, the minimum quantity of foaming solution calculated, $q = 16.5$ l/min·m^2, so for an area of 191.2 m^2, results a total amount of foaming solution, q_t =3,154.8 l/min for the entire protected area.

Given that for floating tank tanks the foam overflows of fixed installations are evenly distributed on the contour of the tank and that the distance between the overflows on the contour must not exceed 26 m and that the circumference of the tank for which the calculation is made is of 191.22 m, it results that 8 (eight) overflows will be mounted on the contour of the tank, each having a flow rate of 400 l/min.

A foam overflow model PX - 5 (figure 3) with a flow rate of 400 l / min was chosen. [6]

Figure 3. Foam spill PX – 5. [6]

4. The general scenario

Tank no. 38, for which the calculation is made, is located, in a park of 3 (three) tanks, each having the same volume, on the site of S.C. Oil Terminal S.A. - South Platform. In order to carry out the comparative study, two scenarios were imagined, which correspond to each oil product in the tank.

The time elapsed from the outbreak to the observation and announcement, for the industrial objectives with process in three shifts is between 1 and 3 minutes, for the two scenarios the most unfavorable option was chosen.

The free fire development time for both scenarios is 10 minutes.

Extinguishing time, T_s, for both scenarios, is calculated with the following relation, [7]:

$$T_s = \frac{T_t \cdot A \cdot i_s \cdot 60}{N_{ii} \cdot q_{ii}} \quad \text{[min]} \tag{2}$$

where:

T_t - duration of an extinguishing operation, for extinguishing combustible liquids in tanks, with fixed installations for the production of foam, T_t = 20 min;

A - burned area, A = 191.2 m^2;

i_s - extinguishing intensity, for vertical tanks with floating lid, i_s = 0.25 l/s·m^2;
N_{ii} - number of cutting heads, number of foam spills, N_{ii} = 8;
q_{ii} - nominal foam flow rate, nominal flow rate of foam overflow PX – 5, q_{ii} = 400 l/min;
Thus, the extinguishing time, T_s = 17 min 55 seconds.

4.1. Scenario I, "crude oil" product

According to the Methodology for Calculating the Forces and Means for Extinguishing Fires, the mass of burnt crude oil is, m = 1.2 kg/m^2·min.

It is considered that the density of crude oil in the tank has the value, ρ = 0.9 kg/m^3.

The price of a barrel of crude oil is $ 50 for a barrel, and a barrel has 159 liters and the results are presented in table 2.

Table 2. Mass, volume and price of crude oil.

Time [min]	Mass of crude oil [kg/m^2·min]	Area [m^2]	Mass of crude oil [kg/m^2·min]	Volume of crude oil [l/m^2·min]	Price of burnt oil [$]
1	1.2	191.2	229.44	254.93	80.17
2	2.4	191.2	458.88	509.86	160.34
3	3.6	191.2	688.32	764.80	240.51
4	4.8	191.2	917.76	1,019.73	320.68
5	6	191.2	1,147.20	1,274.66	400.85
6	7.2	191.2	1,376.64	1,529.60	481.02
7	8.4	191.2	1,606.08	1,784.53	561.20
8	9.6	191.2	1,835.52	2,039.46	641.37
9	10.8	191.2	2,064.96	2,294.40	721.54
10	12	191.2	2,294.40	2,549.33	801.71
11	13.2	191.2	2,523.84	2,804.26	881.88
12	14.4	191.2	2,753.28	3,059.20	962.05
13	**15.6**	**191.2**	**2,982.72**	**3,314.13**	**1,042.22**
14	14.7	180.5	3,305.33	3,672.59	1,154.95
15	13.8	169.9	3,609.49	4,010.54	1,261.23
16	13	159.2	3,895.19	4,327.98	1,361.06
17	12.1	148.6	4,162.42	4,624.91	1,454.44
18	11.2	137.9	4,411.20	4,901.33	1,541.37
19	10.4	127.3	4,641.51	5,157.24	1,621.85
20	9.5	116.6	4,853.37	5,392.64	1,695.87
21	8.6	106	5,046.77	5,607.52	1,763.45
22	7.8	95.3	5,221.70	5,801.89	1,824.58
23	6.9	84.7	5,378.18	5,975.76	1,879.25
24	6	74	5,516.20	6,129.11	1,927.48
25	5.2	63.4	5,635.75	6,261.95	1,969.25
26	4.3	52.7	5,736.85	6,374.28	2,004.58
27	3.4	42.1	5,819.49	6,466.10	2,033.46
28	2.6	31.4	5,883.67	6,537.41	2,055.88
29	1.7	20.8	5,929.38	6,588.20	2,071.85
30	0.8	10.1	5,956.64	6,618.49	2,081.38
31	-	-	**5,956.64**	**6,618.49**	**2,081.38**

IOP Conf. Series: Materials Science and Engineering 1182 (2021) 012082 doi:10.1088/1757-899X/1182/1/012082

The results obtained show that in the first 13 minutes, until the fixed fire extinguishing installation, with which the tank is equipped, comes into operation, the volume burned is 3,314.13 liters, the equivalent of $ 1,042.22. 18 minutes after the commissioning of the fixed extinguishing system, the fire is extinguished, the volume burned during this time being 3,304.36 liters, the equivalent of $ 1,039.15.

The total volume burned is 6,618.49 liters, the equivalent of $ 2,081.38.

4.2. Scenario II, petroleum product "gasoline"

According to the Methodology for Calculating the Forces and Means for Extinguishing Fires, the mass of burned gasoline is, m = 2.9 kg/m^2·min.

It is considered that the density of gasoline in the tank has the value, ρ = 0.75 kg/m^3.

The average price of gasoline is considered to be $ 1.222 per liter and the results are presented in table 3.

Table 3. Mass, volume and price of gasoline.

Time [min]	Mass of gasoline [kg/m^2·min]	Area [m^2]	Mass of gasoline [kg/m^2·min]	Volume of gasoline [l/m^2·min]	Price of burnt gasoline [$]
1	2.9	191.2	554.48	739.30	903.43
2	5.8	191.2	1,108.96	1,478.61	1,806.86
3	8.7	191.2	1,663.44	2,217.92	2,710.29
4	11.6	191.2	2,217.92	2,957.22	3,613.73
5	14.5	191.2	2,772.4	3,696.53	4,517.16
6	17.4	191.2	3,326.88	4,435.84	5,420.59
7	20.3	191.2	3,881.36	5,175.14	6,324.02
8	23.2	191.2	4,435.84	5,914.45	7,227.46
9	26.1	191.2	4,990.32	6,653.76	8,130.89
10	29	191.2	5,544.8	7,393.06	9,034.32
11	31.9	191.2	6,099.28	8,132.37	9,937.76
12	34.8	191.2	6,653.76	8,871.68	10,841.19
13	**37.7**	**191.2**	**7,208.24**	**9,610.98**	**11,744.62**
14	35.6	180.5	7,987.89	10,650.52	13,014.94
15	33.5	169.9	8,722.94	11,630.58	14,212.57
16	31.4	159.2	9,413.37	12,551.16	15,337.52
17	29.3	148.6	10,059.2	13,412.26	16,389.78
18	27.2	137.9	10,660.4	14,213.87	17,369.35
19	25.1	127.3	11,217	14,956,00	18,276.23
20	23	116.6	11,728.99	15,638.65	19,110.43
21	20.9	106	12,196.37	16,261.82	19,871.94
22	18.8	95.3	12,619.13	16,825.50	20,560.76
23	16.7	84.7	12,997.28	17,329.70	21,176.90
24	14.6	74	13,330.82	17,774.42	21,720.35
25	12.5	63.4	13,619.75	18,159.66	22,191.11
26	10.4	52.7	13,864.07	18,485.42	22,589.18
27	8.3	42.1	14,063.77	18,751.69	22,914.57
28	6.2	31.45	14,218.87	18,958.48	23,167.27
29	4.1	20.8	14,329.35	19,105.79	23,347.28
30	2	10.1	14,395.22	19,193.62	23,454.61
31	-	-	**14,395.22**	**19,193.62**	**23,454.61**

The results show that in the first 13 minutes, until the fixed fire extinguishing system with which the tank is equipped, comes into operation, the burned volume is 9,610.98 liters, the equivalent of $ 11,744.62. And in 18 minutes after the commissioning of the fixed extinguishing system, the fire is extinguished; the volume burned during this time, being 9,582.64 liters, the equivalent of $ 11,709.98.

The total volume burned is 19,193.62 liters, the equivalent of $ 23,454.61.

5. Conclusions

According to the calculations made in the previous chapter, it results that the burnt crude oil has a volume of 6,618.49 liters, the equivalent of $ 2,081.38, and the burnt gasoline has a volume of 19,193.62 liters, the equivalent of $ 23,454.61.

To extinguish the fire produced, in both variants, the fixed extinguishing system uses a volume of 3,407.18 l for 18 min - sparkling liquid, the sparkling liquid having a price of $ 10,304.64.

It is observed that the price of burnt gasoline in 30 minutes is double the price of the sparkling liquid, and the price of crude oil is 5 (five) times lower than the price of the sparkling liquid.

But in the end, the financial value of firefighting operations is not so important because the effects of an uncontrolled burning of petroleum products can have irreversible effects on the fauna, flora and health of the immediate population. By combustion, a multitude of pollutants are released into the atmosphere, such as: nitrogen oxides, sulfur oxides, various products of incomplete combustion and solid particles in suspension.

6. References

[1] *Norm on the fire safety of buildings, Part II Extinguishing Installations Indicative* P118/2-2013.
[2] S.C. Oil Terminal S.A. *Fire Response Plan*, 2016.
[3] http://energie.gov.ro/companiile-din-subordine/oil-terminal-sa/ (access date: 10.03.2021).
[4] http://talpac.ro/modernizare-rezervor-50-000mc-sc-oil-terminal-sa-constanta/ (access date: 09.03.2021).
[5] https://www.gefil.ro/fomtec-afff-6_p_165_ro.html (access date: 08.03.2021).
[6] https://www.cruman.ro/tags/spuma (access date: 15.03.2021).
[7] *Metodologia Calculului Forțelor și Mijloacelor pentru Stingerea Incendiilor* Serviciul Editorial și Cinematografic 1984.

Heat treatment of 3D printed polyethylene terephthalate glycol in a supporting powder bed

A Zapciu, C Gh Amza, F Baciu and M I Vasile

University "Politehnica" of Bucharest-Romania, Faculty of Industrial Engineering and Robotics, Splaiul Independentei 313, 060042, Bucharest, Romania

E-mail: acata1@camis.pub.ro

Abstract. Material extrusion 3D printing is a fabrication process that produces layered polymer parts with complex geometry but with inferior mechanical properties compared to parts made with other methods such as injection molding. Post fabrication heat treatment is a valid post-processing method that reduces the internal thermal stresses and improves layer adhesion in 3D printed polymers, resulting in superior mechanical properties. This study investigates the mechanical changes produced in 3D printed PETG (polyethylene terephthalate glycol-modified) parts after heat treatment. A novel technique is used, where the parts are embedded into a bed of sodium chloride powder in order to prevent deformation during post-processing. Fully filled 3D printed PETG parts with various geometries are tightly packed in a bed of powder. The parts are subjected to heat treatment at a temperature above the material's glass transition temperature but below its melting temperature. Destructive and non-destructive testing performed on the treated 3D printed samples shows a substantial improvement of mechanical properties. Tensile strength testing reveals an increase of tensile strength by 40% for parts printed horizontally and by over 100% for parts printed vertically. Increased stiffness is also observed in treated parts. Compressive strength testing shows a strength increase of 43% after treatment. Dimensional measurements made prior to and after treatment show significantly reduced deformation when using the supporting powder method versus unsupported treatment. Scanning electron microscopy (SEM) analysis is used to assess internal structural changes in the polymer after post-processing. This analysis reveals changes in internal void shape and distribution, increased interlayer adhesion and increased interface area of deposited filaments, providing insight into the mechanisms that lead to the improved properties observed in destructive testing. The supporting powder heat treatment allows the fabrication of parts with complex geometry through material extrusion 3D printing while mitigating the inherent disadvantage of the fabrication process of producing parts with inferior mechanical properties.

1. Introduction

Additive manufacturing (AM) is a growing industry forecast to expand by double digit compound annual growth in the upcoming years through 2028 [1]. Material extrusion 3D printing (ME3DP) is an AM process that produces layered polymer parts with complex geometry but with inferior mechanical properties compared to parts made with other methods such as injection molding. Following the expiration of patents held by Stratasys in 2009, fused deposition modelling (FDM / ME3DP) has seen wide adoption with home users, public institutions and businesses of all sizes [2, 3].

In ME3DP fabrication of a part begins with creating a digital 3D model of the part followed by sectioning this 3D model into thin layers using a dedicated software program called a slicer. Each layer's surface is filled using a trajectory that takes into account various process parameters, such as the number of perimeters, part infill (how much of the part internal volume will be made of solid material), infill pattern, infill overlap, air gap. A 3D printer that uses this process uses thermoplastic polymers as feedstock. The material in the shape of filament or pellets is extruded through a heated nozzle and deposited in layers onto a horizontal surface called a bed or a build plate. The material liquefies in the heated nozzle and solidifies after extrusion due to convective cooling with the surrounding air and due to conductive cooling by coming in contact with the bed or previously deposited material. The extruder moves relative to the bed in the horizontal plane, according to the trajectories calculated by the slicing software for each layer.

The layered aspect of 3D printing results in inherent flaws in the fabricated parts using an FDM process. As thermoplastic polymer begins cooling down after extrusion through the heated nozzle, internal thermal stresses develop inside the part [4], leading to reduced mechanical properties compared to injection moulding processes [5]. More so, filaments are being deposited at a significantly higher temperature than the underlying layers, leading to improper interlayer adhesion. This is also the case for deposited adjacent filaments, but to a lesser extent. For this reason, 3D printed parts exhibit an anisotropic behaviour where the mechanical resistance on the Z axis (perpendicular to the build plate) is inferior to that on the X and Y axis (parallel to the build plate) [6, 7]. Additionally, the extrusion nozzles used to deposit material have round bores, meaning not all horizontal surfaces can be covered fully to form uniform layers of material. Thus, complex curves, narrow angles and other geometric features of the horizontal layers lead to the formation of internal voids. Due to the repetitive patterning aspect of filling a layer, these voids align forming a failure point in the printed part [8, 9].

Heat treatment is a post-processing method used to improve mechanical properties of thermoplastic polymer parts. By applying heat to a printed part, the internal thermal stresses generated during part fabrication can be balanced. At a molecular level, heat treatments can promote polymer crystallization resulting in higher tensile strength and stiffness [10, 11]. However, heat treatment also comes with a significant and hard to overcome disadvantage. As a consequence of reducing the internal thermal stresses and the changes in molecular structure of the material, deformation of the parts during treatment is often encountered [12, 13]. This study proposes a novel method of packing polymer 3D printing parts in a supporting powder that prevents deformation during heat treatment.

Polyethylene terephthalate glycol-modified (PETG) is a mostly amorphous polymer that has a low degree of crystallinity [14] and a melting temperature between 240 °C and 260 °C. It is derived from the widely used polyethylene terephthalate (PET) polymer through the addition of a glycol group to the polymeric chain and shares many of its properties [15]. The addition of glycol improves flowability and lowers the glass transition temperature, making it better suited for using in 3D printing processes.

2. Methods and materials

This study focuses on the changes observed in mechanical properties of PETG, more specifically, tensile and compressive characteristics, following heat treatment. The feedstock material used for experimentation comes in the form of 1.75 mm diameter filament and is supplied by 3D Prima (Malmö, Sweden). According to the product datasheet, the material has a glass transition temperature of 75 °C and a recommended extrusion temperature of 230 °C to 255 °C. The material is transparent, with no added pigments or dyes.

The heat treatment consists in raising the temperature of the printed samples to 220 °C, a temperature that is well above the glass transition temperature of the material, and slightly under the melting temperature of the polymer. Once the parts have reached the target temperature, the temperature is maintained for 15 minutes. The procedure begins with fabricating the 3D printed parts, with each printing job producing one sample. After printing, the parts are placed in a borosilicate glass container lined with supporting powder (figure 1c). The samples are placed at a minimum of 10 mm off the edges of the container and at least 10 mm of space is left between the samples. The supporting powder is grounded sodium chloride with particles ranging between 10 μm and 50 μm. After placing the samples, more powder is used to fill the gaps between them. The powder is tightly packed using a rubber mallet, making sure that no empty spaces are left around the samples. NTC thermistors are inserted near the packed parts to track temperature changes. The container with the packaged samples is then introduced in a convection oven preheated to 220 °C (figure 1d). Once the sensors readout reaches the treatment temperature, a 15-minute timer is started. The package is removed from the oven once the timer elapses and it is left to cool at room temperature. After cooling, the samples are unpacked and washed with room temperature clean water to remove any powder residue.

For tensile strength tests 20 samples have been designed and fabricated according to ASTM Type I standard dimensions (figure 1a). The samples have been 3D printed with 2 outside perimeters, 100% infill, 25% infill overlap and an alternating 45° grid infill pattern. The nozzle used is 0.4 mm in diameter and the extrusion temperature was set at 235 °C. Two orientations are used, one in which parts are placed horizontally on the build plate, and one where the parts are placed vertically. The layer height for all 3D printed samples is 0.2 mm.

After 3D printing, the parts have been separated into four groups, each consisting of five samples. Two groups represent parts that undergo heat treatment, one for each orientation, while the two remaining groups are control groups that undergo destructive testing without heat treatment.

For compressive strength tests, cube samples with a width of 15 mm have been manufactured using the same process parameters described previously (figure 1b). These samples are divided into two groups, one undergoing heat treatment and one untreated control group. All printed samples were fabricated on a Creality Ender 3 machine supplied by Shenzhen Creality 3D Technology Co., Ltd (Shenzhen, CN). The machine is a Cartesian 3D printer with a moving build plate along the X axis and Bowden-fed extruder that moves along the Y and Z axis.

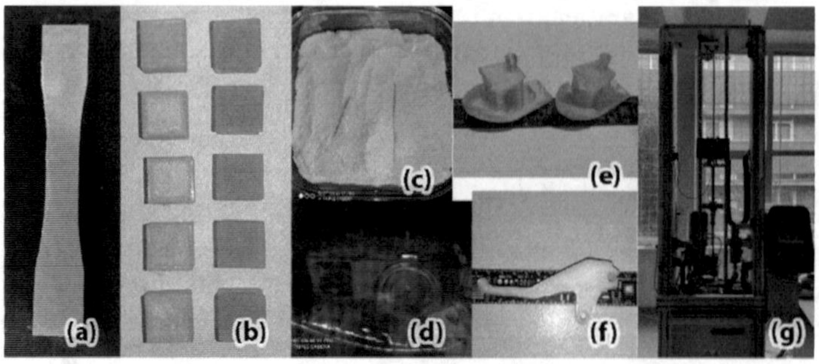

Figure 1. Experiment setup: **(a)** dogbone sample; **(b)** cubic samples; **(c, d)** heat treatment using the supporting powder method; **(e, f)** objects with complex shapes; **(g)** Instron 8872 machine.

Destructive testing was done on an Instron 8872 (Norwood, MA, USA) universal testing machine (figure 1g) at a loading speed of 10 mm per minute and a preload tension (or compression) force of 5 N. Tensile strength tests also tracked sample elongation under strain using a contacting extensometer. For compressive strength tests the cube shaped samples have been placed in such a way that the compressive load is applied perpendicular to its horizontal printed layers (along the sample's Z axis).

IOP Conf. Series: Materials Science and Engineering 1182 (2021) 012083 doi:10.1088/1757-899X/1182/1/012083

SEM analysis was used in order to inspect the internal changes occurring in the structure of the materials after heat treatment. For this analysis, samples were selected from the treated and control groups and imaged using a Quanta Inspect F50 FEG (Field Emission Gun) scanning electron microscope with 1.2 nm resolution. The microscope is equipped with an energy-dispersive X-ray (EDX) analyzer with a resolution of 133 eV at MnKα from Thermo Fisher Scientific (Eindhoven, The Netherlands). All investigated samples were covered with a slim layer of gold using the Q150 Plus Series sputter coating machine from Quorum Technologies (Lewes, UK). The sample coverage time was 30 seconds.

3. Results

3.1. Dimensional changes

For samples 3D printed and heat treated at 220 °C in open air, dimensional changes range between 2.1% for the Z axis (corresponding to the vertical axis during printing process) and 1.3% for the X and Y axis (horizontal plane). For the sample group treated in supporting powder, the dimensional changes are significantly lower, ranging from 0.4% for the Z axis (vertical axis) and 0.3% for the X and Y axis (horizontal plane). No curvature was observed in the dogbone-shaped samples after the supportive powder heat treatment. These observations indicate that the powder method is effective at supporting samples during heat treatment and reducing dimensional changes.

3.2. Tensile strength and stiffness

Fractographic analysis shows the vertically printed samples broke along the layer interfaces (figure 2d, e). For untreated samples printed in horizontal orientation (figure 2c), the fractures happened along the 45° raster infill. The heat-treated horizontal samples experienced a more ductile failure, with necking occurring in some samples (figure 2a)

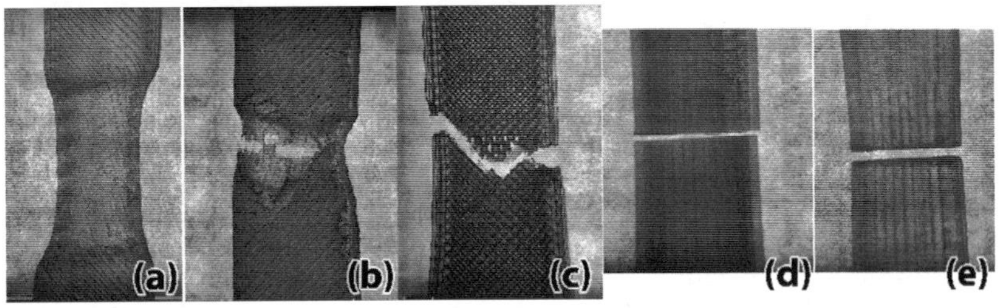

Figure 2. Compressive strength testing results: (a, b) horizontal, treated, (c) horizontal, untreated, (d) vertical, treated; (e) vertical, untreated.

Heat treated samples show a significant increase in tensile strength compared to the control group. For parts 3D printed in a horizontal orientation, the tensile strength of 43.8 MPa is 40% greater than the control group (figure 3). For samples 3D printed in a vertical orientation, the improvement in tensile strength is much higher, treated samples resisting a tensile load of 35.9 MPa before failing, compared to 15.6 MPa in the control group, a 130% increase (figure 4). Figure 5 shows a comparison of tensile testing results for parts printed in the two different orientations.

An increase in sample stiffness can also be observed (figure 6). Young's modulus for horizontally printed samples is 13.5% higher in the treated group (1863 MPa) compared to the control group (1641 MPa). For vertically printed samples, the increase in stiffness is 21.7% (1944 MPa vs 1597 MPa).

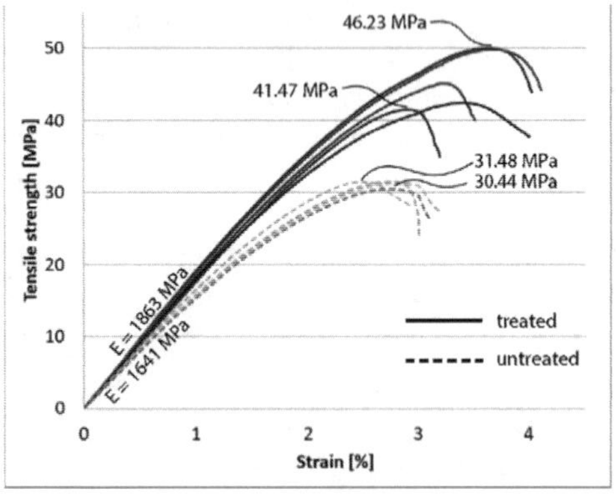

Figure 3. Tensile testing results for parts printed in horizontal orientation.

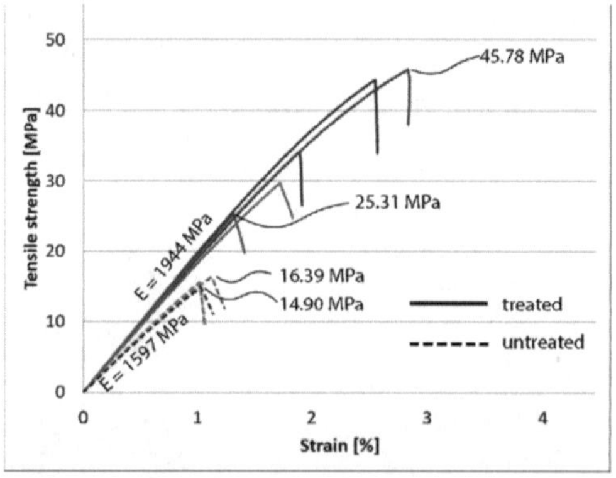

Figure 4. Tensile testing results for parts printed in vertical orientation.

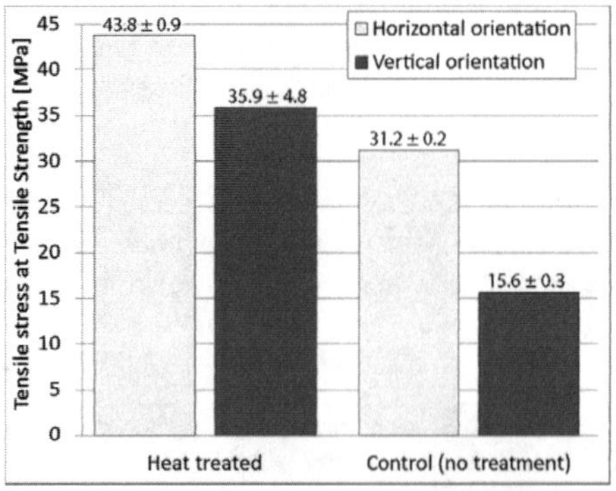

Figure 5. Tensile strength of sample groups.

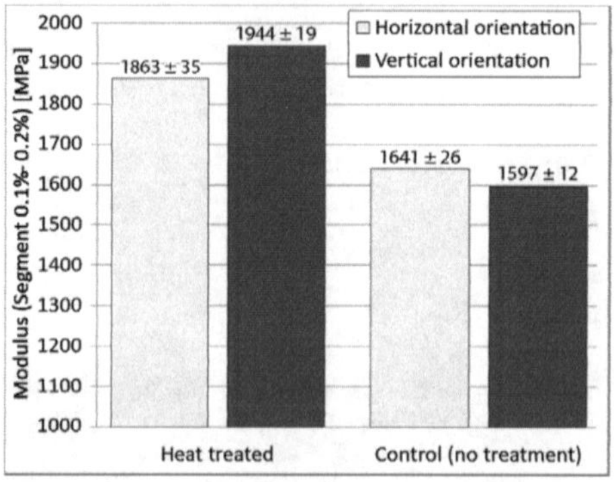

Figure 6. Young's Modulus of sample groups.

3.3. Compressive strength

A change in failure mode is observed after heat treatment, with the samples in the treated group experiencing more interlayer shearing under compressive load. Similar to the increase in tensile strength, heat treatment improved compressive strength as well. The cube-shaped samples in the test group failed under compressive load at 101.9 ± 7.8 MPa compared to 70.9 ± 11.2 MPa for the control group. This translates to an average increase of compressive strength of 43%. The results of these tests are shown in figure 7.

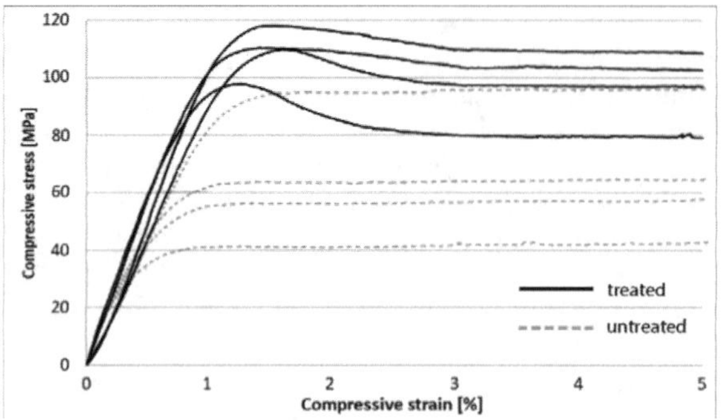

Figure 7. Compressive strength testing results.

3.4. Internal structural changes

Scanning Electron Microscopy imaging gives insight into the reasons behind the observations made following destructive testing experiments. Figure 8 shows images taken of an untreated sample printed in horizontal orientation. As revealed by the fractographic analysis, the sample ruptured in a zig-zag pattern along the 45° infill directions. Material necking can be observed near the fracture surface, an indication that the sample failed prematurely because of shearing between adjacent filaments. The necking occurs due to differences in the molecular structure of deposited filaments developed during fabrication. After extrusion through the heated nozzle, the exterior surfaces of the filaments cool through convection with ambient air and through conduction with the previously deposited material [16]. At the same time, the core of the filament cools down slower, as it is limited by the thermal conductivity of the material. The images show limited adhesion between adjacent filaments and almost no material diffusion. Micro-voids 50 μm to 80 μm in width can be seen at the interface between successive layers and are largely caused by the transition of infill direction from 45° to -45°. These micro-voids have a compounding effect and are very detrimental to the part's structural strength because of their patterning aspect. Figure 9 shows a test sample printed in the same orientation, with the same process parameters, that has undergone heat treatment. It is immediately obvious that the heat treatment resulted in increased adhesion and diffusion between adjacent filaments. The interfaces between filaments are elongated and necking is reduced, a sign that the macrostructure has become more resistant to shearing loads developed between adjacent filaments. As the material was heated past its glass transition temperature, molecular motion causes the coalescence and the reduction in size of the micro-voids inherent to the fabrication process. The micro-voids measure between 20 μm and 40 μm in the treated part.

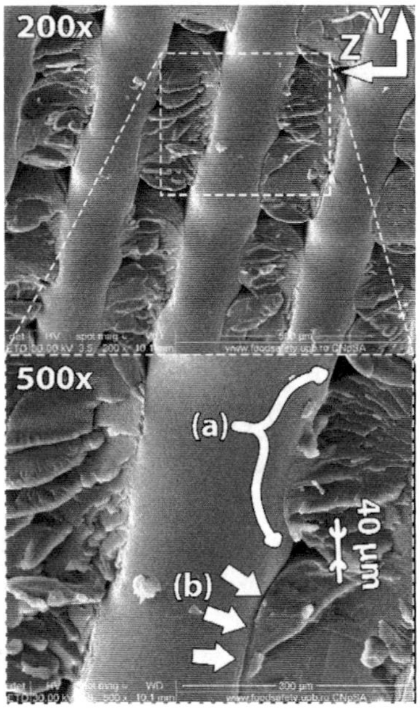

Figure 8. SEM imaging of control sample printed in horizontal orientation: **(a)** microvoids; **(b)** poor filament adhesion.

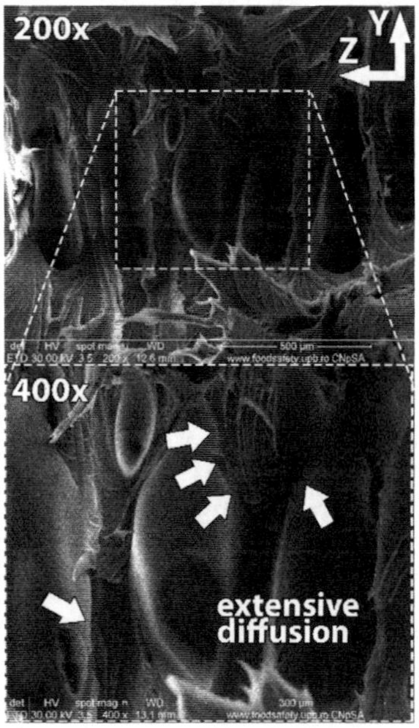

Figure 9. SEM imaging of treated sample printed in horizontal orientation showing increased diffusion between filaments.

Figure 10 shows images captured of a sample printed in the vertical direction from the control group. As revealed by the fractographic analysis, the samples failed at the interface between layers due to poor interlayer adhesion. Unlike the horizontally printed samples, where the tensile load has a component along the filament length, the strength of vertically printed samples is largely determined by interlayer adhesion. As mentioned previously, the adhesion between deposited filaments depends on the temperature difference between them. During fabrication the extrusion temperature is constant, but previously deposited filaments cool down at different rates, depending on their position in the part. Thus, when fresh material is extruded, the largest temperature difference will be with filaments deposited in the previous layer, as these are the filaments that had more time to cool down through convective and conductive effects. For this reason, this sample group performed the worst during destructive tensile strength testing. Additionally, a low amount of material diffusion between adjacent filaments can be seen. Large internal voids, over 150 µm in size can be seen at the intersection between the infill in the exterior contours.

Figure 11 shows images from of a treated sample printed in the vertical direction. Flaking can be seen in various areas of the fracture surface, indicating fracture lines across multiple horizontal layers. This is a clear sign interlayer adhesion has increased compared to the part from the control group [17]. Due to the material being heated over its glass transition temperature, molecular mobility has increased and internal voids have migrated and coalesced, reducing the patterning compounding effect. A change in the shape of the micro-voids is also evident. The voids are more spherical compared to the obloid and elongated voids present in the control group sample.

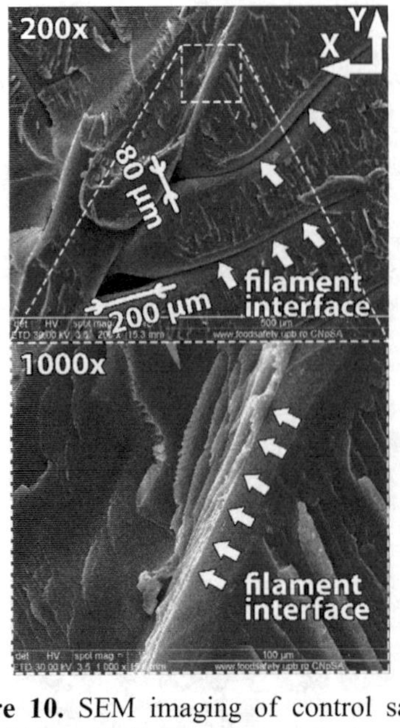

Figure 10. SEM imaging of control sample printed in vertical orientation.

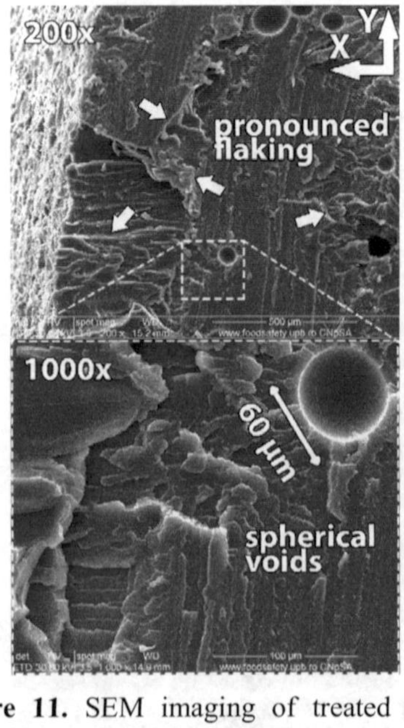

Figure 11. SEM imaging of treated sample printed in vertical orientation.

4. Discussions and conclusions

This research seeks to investigate whether heat treating PETG parts in a supporting powder bed is a viable method of improving their mechanical properties. Twenty dogbone samples and 10 cubic samples were 3D printed from PETG polymer. The samples were split into groups according to part orientation during fabrication and 10 dogbone samples and 5 cubic samples were heat treated in a bed of supporting powder and then put through tensile and compression strength testing. Dimensional measurements made prior to and after treatment show significantly reduced deformation when using the supporting powder method versus unsupported treatment. Samples from all treated groups saw increases in mechanical performance over the control groups. For dogbone samples printed horizontally, heat treatment increased tensile strength by 40% and stiffness by 13.5%. For samples printed vertically, the increases are more significant, with tensile strength increasing by 130% and stiffness increasing by 21.7%. On cubic samples compressive strength increased by 43%.

SEM analysis revealed that raising and maintaining the temperature over the glass transition temperature for 15 minutes has caused important changes in the 3D printed part internal structure. Increasing molecular mobility through heating resulted in better diffusion of material between adjacent filaments leading to less shearing susceptibility. This is also highlighted by the fact that less localized necking is present at the fracture interface. After heat treatment, the internal micro-voids migrated and coalesced, reducing the patterning effect that results in the forming of failure points.

The results obtained in this study point to significant improvements in mechanical properties that offset the inherent anisotropic character of 3D printed parts. This is of major importance for parts with complex geometry that are required to be strong in more than one direction. At this point, the applicability of the process has been demonstrated on simple geometric shapes such as test specimens and on small objects with complex geometric shapes. The process is expected to function for a wide range of part geometries and has similarities to fabricating using moulds and casting. During treatment, the exterior surfaces of the parts see an increase in roughness that is corelated to the choice of powder material. More research is required to identify other suitable powders.

5. References

[1] ReaserchandMarkets.com 2020 Global Additive Manufacturing Market and Technology Forecast to 2028 *Market Report* ID 5144559.

[2] Steenhuis H-J and Pretorius L 2016 Consumer additive manufacturing or 3D printing adoption: An exploratory study *J. Manuf. Technol. Manag.* **27**, 990–1012.

[3] Sepasgozar S M E, Shi A, Yang L, Shirowzhan, S and Edwards D J 2020 Additive Manufacturing Applications for Industry 4.0: A Systematic Critical Review. *Buildings* **10** 231.

[4] Costa S F, Duarte F M and Covas J A 2015 Thermal conditions affecting heat transfer in FDM/FFE: A contribution towards the numerical modelling of the process *Virtual Phys. Prototyp.* **10** 35–46.

[5] Priya M S, Naresh K, Jayaganthan R and Velmurugan R 2019 A comparative study between in-house 3D printed and injection molded ABS and PLA polymers for low-frequency applications. *Mater. Res. Express* **6** 085345.

[6] Ahn S-H, Montero M, Odell D, Roundy S and Wright P K 2002 Anisotropic material properties of fused deposition modeling ABS *Rapid Prototyp. J.* **8** 248-57.

[7] Sood A K, Ohdar R K and Mahapatra S S 2010 Parametric appraisal of mechanical property of fused deposition modelling processed parts *Mater. Des.* **31** 287–95.

[8] Rodriguez J F, Thomas J P and Renaud J E 2000 Characterization of the mesostructure of fused-deposition acrylonitrile—Butadiene—styrene materials. *Rapid Prototyp. J.* **6** 175–86.

[9] Amza CG, Zapciu A, Eyþórsdóttir A, Björnsdóttir A, Borg J 2019 Embedding Ultra-High-Molecular-Weight Polyethylene Fibers in 3D printed Polylactic Acid (PLA) Parts *Polymers* **11** 1825.

[10] Butt J and Bhaskar R 2020 Investigating the Efects of Annealing on the Mechanical Properties of FFF-Printed Thermoplastics. *J. Manuf. Mater. Process.* **4** 38.

[11] Hart K R, Dunn R M, Sietins J M, Hofmeister Mock C.M, Mackay M E and Wetzel E D 2018 Increased fracture toughness of additively manufactured amorphous thermoplastics via thermal annealing *Polymer* **144** 192–204.

[12] Slavković V, Grujović N, Dišić A and Radovanović A 2017 Influence of Annealing and Printing Directions on Mechanical Properties of PLA Shape Memory Polymer Produced by

Fused Deposition Modeling *Proceedings of the 6th International Congress of Serbian Society of Mechanics (Mountain Tara, Serbia, June 19–21, 2017).*

[13] Wijnen B, Sanders P and Pearce J M 2018 Improved model and experimental validation of deformation in fused filament fabrication of polylactic acid *Prog. Addit. Manuf.* **3** 193–203.

[14] Dupaix R B and Boyce M C 2005 Finite strain behavior of poly(ethylene terephthalate) (PET) and poly(ethylene terephthalate)-glycol (PETG) *Polymer* **46** 4827–38.

[15] Paszkiewicz S, Szymczyk A, Pawlikowska D, Irska, I, Taraghi I, Pilawka R, Gu J, Li X, Tu Y, Piesowicz E 2017 Synthesis and characterization of poly(ethylene terephthalate-co-1,4-cyclohexanedimethylene terephtlatate)-block-poly(tetramethylene oxide) copolymers *RSC Adv.* **7** 41745-54.

[16] Morales N G, Fleck T J, Rhoads J F. 2018 The effect of interlayer cooling on the mechanical properties of components printed via fused deposition *Addit. Manuf.* **24** 243–248.

[17] Gao X, Qi S, Kuang X, Su Y, Li J and Wang D 2021 Fused filament fabrication of polymer materials: A review of interlayer bond *Addit. Manuf.* **37** 101658.

Acknowledgments

This research was funded under the framework of the Erasmus+ KA2 Strategic partnership project 2019-1-PL01-KA202-064936 entitled "Enabling Industry 4.0 in Small European Enterprises" - SEE4.0.

IOP Conf. Series: Materials Science and Engineering 1182 (2021) 012084 doi:10.1088/1757-899X/1182/1/012084

Cold in-place recycling of aged and deteriorated asphalt and concrete pavements

Ş Zghibarcea, D Ungureanu, N Ţăranu, D Hoha, RA Polcovnicu and TA Leizeriuc
Gheorghe Asachi Technical University of Iasi-Romania, Faculty of Civil Engineering and Building Services

E-mail: razvan-andrei.polcovnicu@student.tuiasi.ro

Abstract. This study focuses on implementing two distinct recycling methods, the reclaimed asphalt pavement and the reclaimed concrete material, on classical road pavements, by totally replacing the surface or base courses with ones made with recovered materials. The majority of the road networks in the eastern European countries were built between 1950 and 1980 and are, nowadays, aged and degraded. Within this frame, the cold in-place recycling methods were developed, which can be applied to both asphalt and concrete pavements. During the cold in-place recycling process, the reclaimed asphalt or concrete materials do not leave the construction site at all, but they are processed on the spot and reused immediately. Thus, the material is granulated to the desired depth, mixed with the binding agent and water and then repaved and compacted in a new layer. One of the most significant benefits of the cold in-place recycling process is that the material is simultaneously recycled and mixed with the binding agent, enabling the road pavement to be strengthened without the need to import expensive virgin aggregates. This work describes the manufacturing technology and the equipment train used to cold recycle aged and deteriorated asphalt and concrete pavements.

1. Introduction

Nowadays, roads represent the most widely used means by which both people and goods move around for specific aims. This is due to the roads being the only communication ways that offer the so called „*door-to-door*" access. During recent decades, this mobility has further developed within the context of cheap fuels, expanding road infrastructure and rather loose environmental constraints [1].

Based on the component materials, the road pavements are classified into three categories, namely, asphalt or flexible pavements, concrete or rigid pavements and hybrid pavements. Asphalt pavements are made of hot mix asphalt (HMA), concrete pavements are made of Portland cement concrete (PCC), while hybrid pavement systems are composed of both materials.

The intensity of the road traffic influences not only the direct greenhouse gas emissions, due to fuel combustion, but also the road durability. In addition, besides the intensity of the road traffic, some other factors like design and construction technology, the quality of the pavement materials and local weather conditions are essential for the pavement durability. All these above mentioned factors are also of paramount importance when determining the failure modes and causes of distressed road structures and when selecting a rehabilitation strategy [2-4].

In Romania, various pavement rehabilitation techniques are used to extend the service life of roads. Based on the type of technology and component materials, the pavement rehabilitation techniques can be classified into two main categories: traditional and modern. The traditional rehabilitation techniques consist in hauling away the materials from the distressed pavement and rebuilding the layers with virgin aggregates. This approach is related to a lack of efficiency in terms of costs and consumption of natural resources. In order to overcome these shortcomings, various modern rehabilitation techniques in which the degraded materials are recycled and reused have been developed during the recent decades [5-7].

In the case of flexible pavements, depending on the recycling technology and mixing temperature, the rehabilitation/recycling methodologies can be classified as hot recycling (HR) and cold recycling (CR). The former includes hot in-place recycling (HIR) and hot central-plant recycling (HCPR), while the latter involves three methods: cold in-place recycling (CIR), cold central-plant recycling (CCPR) and full depth reclamation (FDR) [8].

Each of the above-mentioned rehabilitation/recycling methods has its own advantages and shortcomings. However, in terms of materials and fossil fuel consumption, carbon footprint and structural behaviour, the CIR is considered to be the most advantageous rehabilitation method. In addition, CIR is the only one that can be applied to both flexible and rigid pavements [9, 10].

In the following sections of this work the manufacturing technology and the equipment train used to cold recycle aged and deteriorated asphalt and concrete pavements are described.

2. Cold recycling

As mentioned above, depending on the recycling technology, the reclamation maximum depth and the processing place, the cold recycling technique involves three methodologies: CIR, CCPR and FDR. However, some researchers consider that CIR and FDR refer to one unique method since FDR is also carried-out *in-situ*. Nonetheless, FDR is characterised by some unique features consisting in the full pavement depth recycling and 100% use of the recycled materials [8].

From a historical point of view, these recycling methods were developed in the early 90s in countries like the United States of America, Canada, Germany, France, Spain and Australia. Later they have been implemented and used worldwide, primarily because of their potential to lessen the severe pollution issues, prolong the life cycle of pavement and save the construction costs [11–14].

3. Manufacturing technology

In general, the whole manufacturing technology of CIR is completed in situ, starting from the milling of old asphalt or concrete mixtures and ending up to reuse the material for construction of base courses, or, in some cases for surface courses on highways with low to medium traffic volume. The construction processes related to the manufacturing technology of CIR are presented in figure 1, in comparison with the traditional rehabilitation methods. For each step, the materials and fuel consumptions are marked.

Whether to reuse the reclaimed asphalt pavement (RAP) or the reclaimed concrete material (RCM) or not depends on the damage extent of the surface layer. Thus, if the surface is moderately degraded and the pavement structure is adequate (the surface course has constant thickness), the in-place recycling of bituminous and concrete materials is enough. However, if the surface layer is severely degraded and thin, reusing underlying granular material is allowed so as to obtain adequate grading characteristics. Nonetheless, in both cases, prior to the construction of the new course, the reclaimed materials should be stabilized with a small amount of road binder.

Various types of road degradations, such as cracking, rutting, ravelling, potholes or structural deformation, might appear in distinct layers of the pavement during its service life. The CIR technology can be employed in different manners to eliminate these distresses to a certain extent. Over the past years, the research on CIR technologies has been intensively conducted. The application of these rehabilitation methods generally consists in the processes that are described in the following subsections.

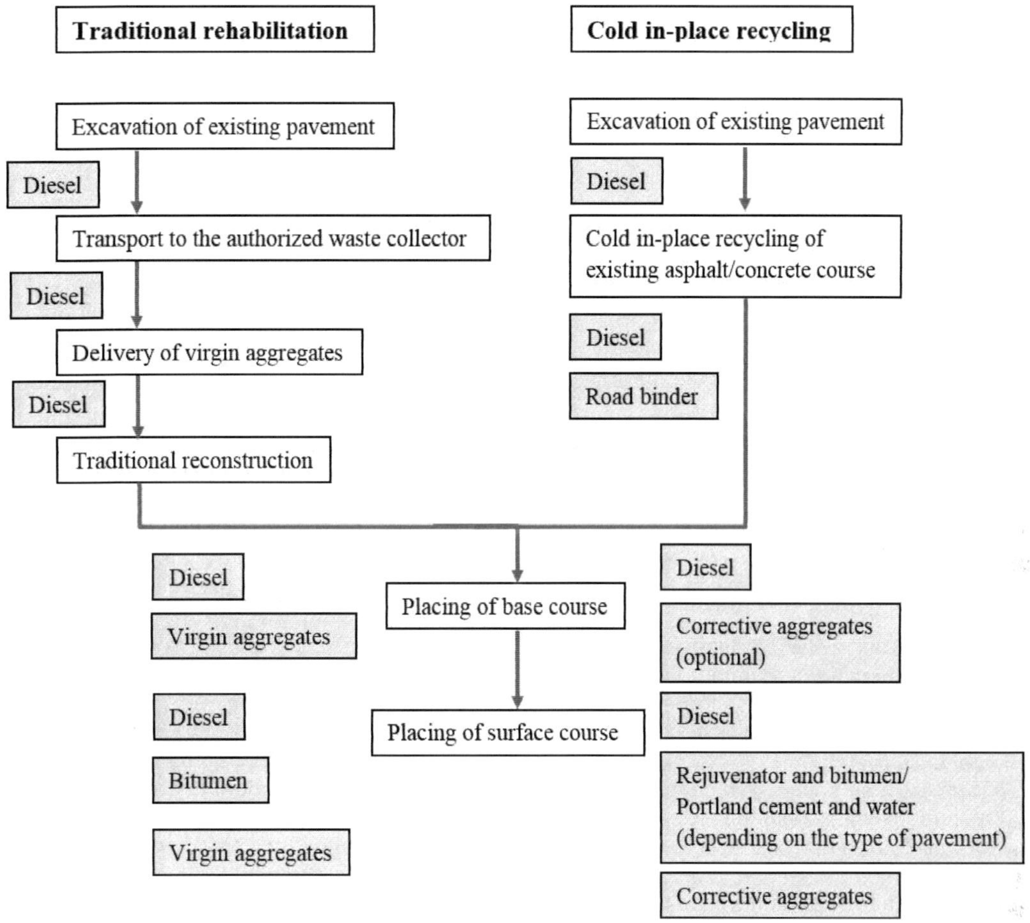

Figure 1. The construction processes related to the traditional rehabilitation and the cold in-place recycling of pavement structures.

3.1. Pavement investigation

The pavement investigation provides information on the nature and the condition of the asphalt and/or concrete layers and on the extent of the degradations. The latter are identified throughout the basic visual inspections done during the intervention/maintenance works.

The thickness of the asphalt or concrete pavement is established and the mixes to be recycled are determined. Also, the overall roughness is evaluated and the geometrical characteristics of the base and the subgrade courses are determined. The information regarding the thickness and the components of the pavement are necessary to establish, on one hand, whether the structure has enough thickness to host the actual traffic and, on the other hand, whether there is enough thickness to perform the cutting of the surface course.

The nature of the asphalt/concrete pavements to be recycled may have a significant influence on the mix design of the recycled layers (surface or base courses). In many cases, the presence of a non-traditional mix, such as Open Graded Mix (OGM), Penetration Macadam (PM) or Sand Mix (SM) requires the usage of corrective aggregate.

The laboratory work conducted on the asphalt/concrete pavement field sampling may include (but are not limited to) the following tests:

- gradation of the recycled material, before and after, the extraction;
- characterization of the aged asphalt cement;
- compressive strength;
- indirect tensile strength;
- creep strength;
- modulus of elasticity;
- penetration;
- viscosity;
- asphalt content;
- asphaltene content;
- softening point.

3.2. Milling the degraded material

The recycling equipment train may be a single unit or a multi-unit road construction equipment. In the first case, the single unit train reclaims the asphalt or concrete materials, sizes and mixes in the binders, in the cutting drum, while the placement operation is performed with a standard screed attached to the back of the unit (the case where the reclaimed materials are used on the surface course), [12]. In the second case, the reclamation of the existing materials is performed by a milling machine, the sizing is done by a crusher unit and the mixing with a pugmill, while the placement is carried out with a basic paver [13].

The sizing of the millings is performed so as to separate the aggregate particles from one another. Most often, the typical maximum aggregate size determined as the "100 % passing on the 37 mm sieve" is used.

3.3. Stabilization

The mixing of the road binder with the asphalt or concrete millings may occur in the cutting drum and/or in a pugmill. Additionally, corrective aggregates may be introduced ahead of the milling machine and mixed in with the recycled material within the cutting drum.

The use of road binders is meant to restore the reclaimed material properties, including strength, flexibility, stability and workability. Currently, various types of road binders can be used in the CIR technology. Among the most used road binders are [8]:

- bituminous road binders (emulsified bitumen and foamed bitumen);
- cementitious stabilization agents (Portland cement, lime, fly ash, ground blast furnace, clay soil, slag)
- polymers (crumb rubber, styrene-butadiene rubber - SBR latex, modified agent and styrene-butadiene-styrene - SBS).

3.4. Placement and compaction

Once the mixing of the reclaimed asphalt/concrete with road binder has been completed, the recycled material is placed either on the road base or on the road surface course. When the technology is applied for the construction of the surface layers, the placement of the recycled mixture requires more compaction energy than does the placement of regular hot bituminous mixtures. However, in both cases, the use of a heavy pneumatic roller, combined with large double vibrating drum roller is typically enough. The rolling patterns performed to achieve the compaction degree are established on test sections. In more recent applications, nuclear gauges (devices that use radioactive sources to measure parameters such as thickness, density, moisture or fill level) are used to monitor the density as well as the moisture content of the reclaimed material during the compaction process [8, 13]. The moisture content is the most important parameter for compaction. On one hand, if the moisture content is low, the reclaimed mixture is harsh and will not compact properly. On the other hand, if there is too much moisture, the reclaimed mixture will also not compact due to the excess of water and absence of air voids.

The compaction degree is commonly limited to 98 % of the density of the Marshall field compacted specimen. The compacted mixture internal void content ranges between 12 and 15 % [8]. During the first year of service, a decrease in the void content of up to 0.5 % may occur, due to compaction by traffic.

4. Structural performance of CIR courses

The most commonly accepted understanding of the structural performances of cold recycled asphalt mixtures is based on two antagonistic principles that were postulated in the early development of CIR [14-16]. One concept is based on the incipient assumption that the aged asphalt mixtures are inert and thus, the reclaimed aggregates should be treated as "*black aggregate*". The second concept assumes that an aged asphalt mixture can still be active and thus, the addition of a rejuvenating agent should restore its original characteristics. In time, it was found that both concepts are applicable, depending on the characteristics of the asphalt mixtures. In some cases, the asphalt can not be reactivated and, after milling and stabilizing, it can be used only for the construction of base courses. In other cases, the asphalt can be reactivated and used for the construction of surface layers. Similarly, the reclaimed concrete materials can be used either as aggregates in new concrete mixtures or as materials for road base construction.

The strength and the fatigue life of the CIR road layers will increase during the first years of service. The increase is preeminent in the first few months (3 - 5 months), while after this period the structural performance will still increase, yet at a lower rate [8].

The structural performance of the CIR courses is a subject of interest for many research teams, including ours. In a recent published work, our research team focused on implementing the CIR technology on a classical road pavement by totally replacing the base course with one made with reclaimed materials. For this purpose, four flexible pavements were constructed, instrumented and tested on an accelerated pavement testing (APT) facility, while capturing the structural condition. After the accelerated test had been done, the authors concluded that the pavement structure that included a structural layer made with stabilized RAP material presents better fatigue performance *in situ* than the reference pavements made with classical layer components, which are considered high-performance road base materials in Romania. This confirms that the recycled materials can be used successfully for road construction. More information related to this experimental program was presented in [17].

5. Conclusions

The construction/rehabilitation of roads implies significant consumption of traditional energy resources, particularly petroleum and virgin aggregates, which may lead to severe environmental issues. In the last decades, several recycling technologies were developed to allow the reuse of asphalt and concrete materials. Generally speaking, any road recycling technology consumes less raw materials and fossil fuel and lowers the pollutant emissions.

Among the recycling technologies, the CIR may be considered a well-proven road rehabilitation technology. Moreover, it is a viable engineering and economic alternative to the traditional rehabilitation methods and it can be applied for a wide range of traffic and pavement distress situations.

This work describes the manufacturing technology and the equipment train used to cold recycle aged and deteriorated asphalt and concrete pavements. Based on the road course that is constructed with the reclaimed materials, the particularities of the CIR technology were identified and described. Additionally, some of the most important advantages of the CIR road layers were discussed.

6. References

[1] European Commission – Mobility and Transport. Transport themes, Mobility Strategy 2020. https://ec.europa.eu/transport/themes/mobilitystrategy_en accessed on: 10.03.2020.

[2] Fwa T F 2006 *The Handbook of Highway Engineering* Taylor & Francis, CRC press. Selberherr S 1984 *Analysis and Simulation of Semiconductor Devices* (Berlin: Springer).

[3] Adlinge, S S and Gupta A K 2013 Pavement deterioration and its causes. *Int. J. innov Res. Dev.* **2**/4 437 - 450.

[4] Gransberg D D, Tighe S L, Pittenger D and Miller M C 2014 *Sustainable pavement preservation and maintenance practices*. In: Gopalakrishnan, K.,W., Harvey, J. (Eds.), Climate Change, Energy, Sustainability and Pavements. Springer-Verlag, Berlin Heidelberg, pp. 393 - 418.

[5] Miller T D and Bahia H U 2009 *Sustainable Asphalt Pavements: Technologies, Knowledge Gaps and Opportunities. Modified Asphalt Research Center. The University of Wisconsin.* https://uwmarc.wisc.edu/files/MARC-Sustainable-Asphalt-Pavements-white-paper.pdf accessed on: 11.03.2020.

[6] Giustozzi F, Toraldo E and Crispino M 2012 Recycled airport pavements for achieving environmental sustainability, an Italian case study *Resour. Conserv. Recycl.* **68** 67-75 http://dx.doi.org/10.1016/j.resconrec.2012.08.013.

[7] Butt A A, Mirzadeh I, Toller S and Birgisson B 2014 Life cycle assessment framework for asphalt pavements: methods to calculate and allocate energy of binder and additives *Int. J. Pavement Eng.* **15**(4) 290-302 http://dx.doi.org/10.1080/10298436.2012.718348.

[8] Xiao F, Yao S, Wang J, Li X and Amirkhanian S 2018 A literature review on cold recycling technology of asphalt pavement *Construction and Building Materials* **180** 579-604.

[9] Stimilli A, Ferrotti G, Graziani A and Canestrari F 2013 Performance evaluation of a cold-recycled mixture containing high percentage of reclaimed asphalt *Road Mater. Pavement Des.* **14** (supl) 149–161.

[10] Diefenderfer B K, Bowers B F and Apeagyei A K 2015 Initial performance of Virginia's Interstate 81 in-place pavement recycling project *Transp. Res. Rec. J. Transp. Res. Board* **2524** 152–159.

[11] https://www.roadtec.com/products/cold-in-place-recycling accessed on: 11.03.2020.

[12] https://www.equipmentworld.com/better-roads/article/14962633/real-time-remote-production-reporting-comes-to-roadtec-machines-through-guardian-telematics-upgrade accessed on: 11.03.2020.

[13] https://mcasphalt.com/wp-content/uploads/RESEARCH-PAPERS/2003-Best-Practices-in-Cold-In-Place-Recycling-Croteau-J.M.-and-Davidson-J.-Keith.pdf accessed on: 11.03.2020.

[14] Lefort M and Meunier Y 1993 Le recyclage à froid en place *Revue Général des Routes et Aérodromes* **712** Paris France 47-51.

[15] Scholz T V, Hicks R G, Rogge D F and Allen D 1992 Use of Emulsion in Cold In-place Recycling : Oregon Experience *Transportation Research Record* **1342** Washington, D.C. 1-8.

[16] Bicheron G F, Migliori G F and Brûlé B 1993 Bitume régénéré ou bitume + régénérant *Bulletin de liaison des laboratoires des ponts et chaussées* Paris France no **183** 83-88.

[17] Ungureanu D, Ţăranu N, Hoha D, Zghibarcea Ş, Isopescu D N, Boboc V, Oprişan G, Scutaru M C, Boboc A and Hudişteanu I 2020 Accelerated testing of a recycled road structure made with reclaimed asphalt pavement material *Construction and Building Materials* 120658.

IOP Conf. Series: Materials Science and Engineering 1182 (2021) 012085 doi:10.1088/1757-899X/1182/1/012085

Laser welding of 4330V steel with a micro-additive of vanadium

M Żuk[1] and S Stano[2]
[1]Silesian University of Technology, Faculty of Mechanical Engineering, Welding Department, Konarskiego street 18A, 44-100 Gliwice, Poland,
[2]Łukasiewicz - Welding Institute in Gliwice, Bl. Czesława street 16-18, 44-100 Gliwice, Poland

E-mail: marcin.zuk@polsl.pl

Abstract. This article presents the results of a test of laser welding of high-strength low-alloy steel with a micro-additive of vanadium. Steel 4330V is a delivery material after heat treatment. Welded joint was made by laser welding using a TruDisk 12002 disc laser. As a result of welding, a joint was formed which was subjected to visual and penetration tests from the face side, as well as macro and microscopic tests. Impact tests were carried out at reduced temperature in the weld and in the heat affected zone (HAZ). The resulting joint was also subjected to the Vickers hardness test, bend test and the tensile strength test. This research show how the low-alloy high-strength material behaves during welding without processing before and after welding. In results after welding show incomplete penetration from the root side. No cracks on the weld face. In HAZ hardness was of 520$HV_{0.5}$ and tensile strength of joint was on level1077MPa.

1. Introduction

Laser welding is a very modern and clean process of joining materials without the use of filler metal. Due to the high power density at the level of 10^9-10^{11} W / mm^2, it is possible to make thick welded joints without additional material, with I chamfering, with a thickness of up to 25 mm. Disc lasers are usually used for laser welding processes due to the high power and shape of the laser spot, which guarantees a narrow and deep fusion into the base metal. Laser welding is characterized by rapid times of heating the material to the melting point, and due to the heat capacity of the material and the fact that a small heated affected zone (HAZ), the liquid metal solidifies very quickly. Such sharp welding heat cycles also apply to HAZ. Therefore, you need to know how a given metal behaves during laser welding. Especially when we are using heat-treated material [1-8].

Steel 4330V is a heat-treated material with high strength and plastic properties. This material contains 0.3% carbon and 0.06% micro vanadium additive. As a result after heat treatment a material with a tempered martensite structure is obtained. Unfortunately, the very fast heating and cooling processes affect the microstructure and properties of this steel. And the standard heat-up-cool-charts for steel are not applicable to laser processes with short heat-up and cool-down times. Understanding the behavior of 4330V steel during laser welding gives an idea of how to obtain a welded joint of appropriate quality and properties [9-12].

IOP Conf. Series: Materials Science and Engineering 1182 (2021) 012085 doi:10.1088/1757-899X/1182/1/012085

2. Experimental Section

The aim of the research was to determine the effect of laser welding of 4330V steel on the quality of the resulting welded joint. 4330V steel is a material with a tempered martensite structure with a micro-addition of vanadium. The figure 1 shows the microstructure of the material and the view vanadium in the material from the EDS analysis. The chemical composition of 4330V steel is not typical for well-weldable steels (table 1), therefore, during welding with conventional methods, it requires pre- and post-welding treatments to obtain the correct joint. However, due to the high strength and plastic properties (see table 2), the aim is to make the correct joints with a minimum number of additional treatments, therefore the behaviour of this material under the influence of laser welding was checked.

Figure 1. Microstructure of base metal; view of EDS analysis: microaditive vanadium.

Table 1. Chemical composition of investigated steel 4330V.

Contents of Chemical Elements, % by Weight												
C	Cr	Ni	Mn	Mo	Si	P	S	Cu	Al	Nb	Ti	V
0.31	0.99	1.84	0.9	0.43	0.26	0.012	0.00	0.17	0.03	0.029	0.008	0.06

Table 2. Mechanical properties of the investigated steel 4330V.

Yield strength, $R_{0,2}$, [a] [MPa]	Tensile strength, Rm [a], [MPa]	Elongation, A, [%]	Work of braking, KV, -40°C , [J]	Hardness (HV)[a]
935	1050	19	89	340
[a] converted value (ksi to MPa; HRC to HV according ISO 18265 standard)				

2.1. Welding process

The welding process was performed with a 12kW Trumph TruDisk 12002 disc laser cooperating with a KUKA KR30HA robot. Before the target welding was performed, trial melting was performed in order to establish the basic welding parameters. The figure 2 shows trial remelting. Then, a butt joint was made without the use of additional material. Argon gas with a purity of 99.999% was used as a gas shield. The material in the form of sheet metal with a thickness of 10mm and dimensions 230x85mm is connected along the longer edge. The edges of the joint were prepared on I. Welding was performed on one side. Run-up and run-out plates were used to obtain the correct joint along the entire length. Welding parameters are shown in the table 3.

Table 3. Welding parameters.

Weld	Power [W]	Welding speed [m/min]	The diameter of the focus of the laser beam [mm]	The optical fibre diameter [μm]
1	6000	1.2	0.4	200

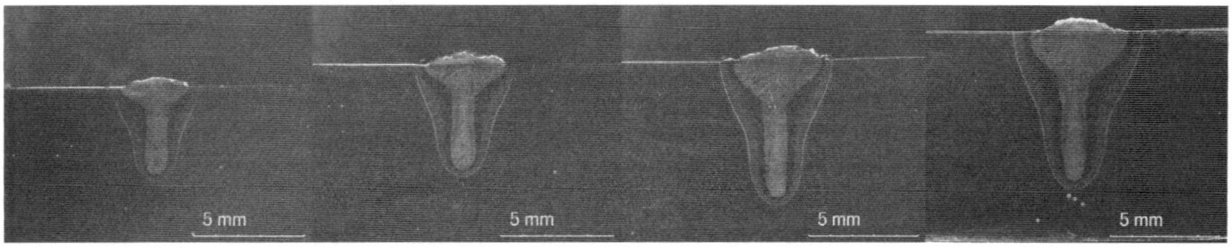

Figure 2. View of testing laser remelting: 3kW; 4kW; 5kW, 6kW.

3. Results and discussion

3.1. Visual and penetrate testing

As a result of welding, a weld was obtained, which was subjected to visual testing. Visual tests were performed on the basis of ISO 13919 standard. The tests were carried out in visible light with the intensity of 720lux. As a result of the tests performed, no visible defects disqualifying the joint from further tests from the face side were revealed (figure 3). Only a incompletely filled groove of the face at the level of 0.1 mm was found, qualifying the joint at level B (ISO 5817 standard), and a few traces of metal spatter. The face side joint showed incomplete penetration from the root side not meeting the quality level D. In industrial practice, such a joint would be scrapped. However, due to the nature of the research on the material, penetrant testing (PT) was performed using the colour method. The MR Chemie penetrant kit (MR70, MR79, MR68NF) was used. The test was performed in accordance with the ISO 3452 standard from the face side. Figure 4 show penetrant testing. Only an apparent indication was obtained, resulting from the flow of penetrant from under the spatter. The sample from the face side meets the requirements of the standard at the 2X level, which corresponds to the quality level B. No cracks were found in the weld face.

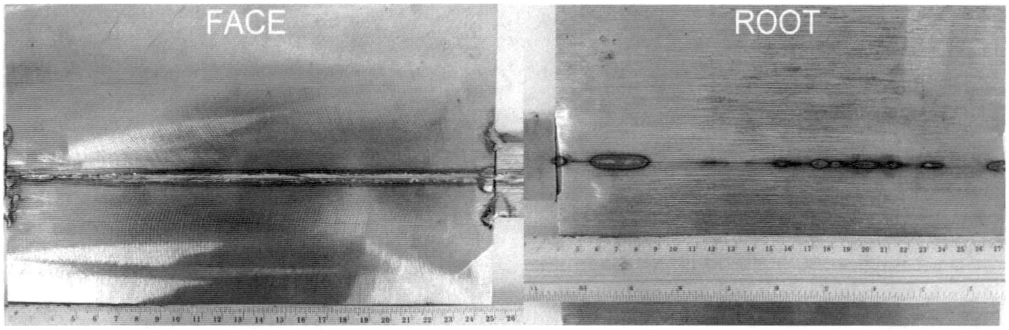

Figure 3. View of weld after welding. Face and root.

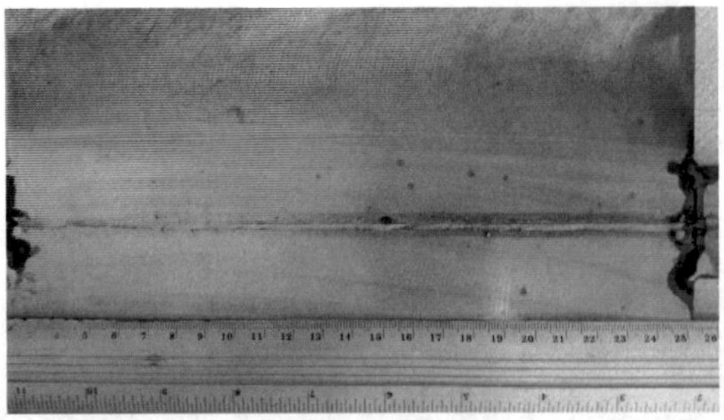

Figure 4. View of penetrant testing. Face of weld.

3.2. Macro and microscopic research

Macro and microscopic examinations were performed with the use of light microscopes (Olympus SZX9, NIKON ECLIPSE MA100). The sample was mounted in a thermosetting resin and then ground and etched with 4% nitrile. As a result of light microscopy tests on the cross-section of the weld, an incomplete remelting from the side of the ridge was revealed (it was welded in about 9 mm from the face), additionally, a symmetrical arrangement of the weld with a narrow HAZ area of 1 mm was found. The tested cross-section did not show gas pore and material sticking (figure 5). The weld width was from about 3mm at the top to 1mm at the bottom of the joint. As a result of microscopic examinations in the weld and HAZ (figure 6), large martensite grains were visible, which may indicate changes in properties in this area and for hardening of this area

Figure 5. View macro of weld.

Figure 6. HAZ and weld after welding.

3.3. Impact tests basing on ISO 148

The impact toughness test was carried out on samples with a reduced cross-section (55x10x7.5), it resulted from incomplete root remelting. The samples were made in accordance with the standard, and the test was carried out using a Sunpoc JB 300B hammer with the initial energy of the hammer 150J, based on ISO 148 standard. The test was performed at temperature -40 degrees Celsius. As a result of the conducted tests, the average result of the breaking work in the weld was obtained 4-6 J, and in the HAZ at the level 20-59 J. Sample breakthroughs were brittle (figure 7(b)) and mixed (figure 7(a)).

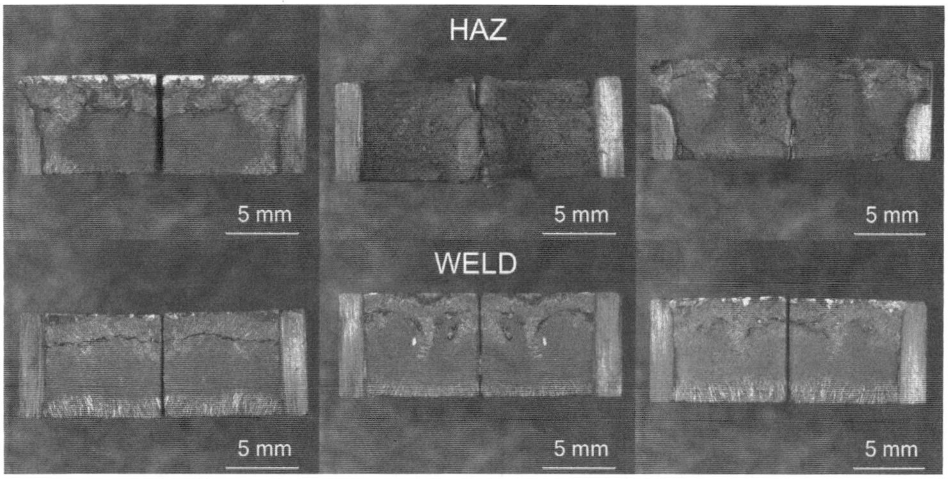

Figure 7. Samples after impact test.; impact in HAZ; impact in weld.

3.4. Bending test basing on PN-EN ISO 5173

The bending test was performed on the face and root sides (figure 8). The material has been performed to remove face welding defects. The thickness of the samples was 7mm and the width was 20mm. The test was performed on a hydraulic press. The former was 30mm in diameter and the distance of the bending rollers was 45mm.Unfortunately, as a result of the test, cracks appeared. They were characterized by a fragile breakthrough.

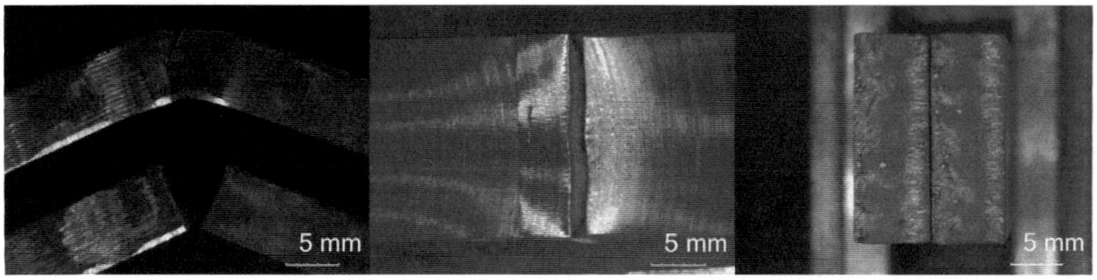

Figure 8. Samples after bend test.

3.5. Vickers hardness tests basing on PN-EN ISO 9015-2

Hardness tests were carried out using a Vickers WILSON WOLPERT Micro Vickers 401MVD hardness tester. The tests were carried out in two measuring lines every 0.2 mm. Measurements were made 2 mm from the face and root of the joint (figure 9). The load was 0.5 kg. The conducted tests confirmed the results of microscopic examinations and impact tests. The material in the HAZ and in the joint has been significantly hardened (figure 10). The hardness in HAZ was even 550 $HV_{0.5}$, and in the weld it was at the level of 520 $HV_{0.5}$.

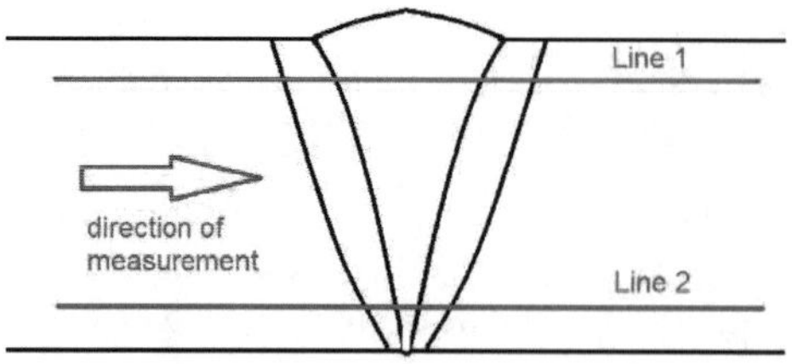

Figure 9. Hardness Vickers measurement.

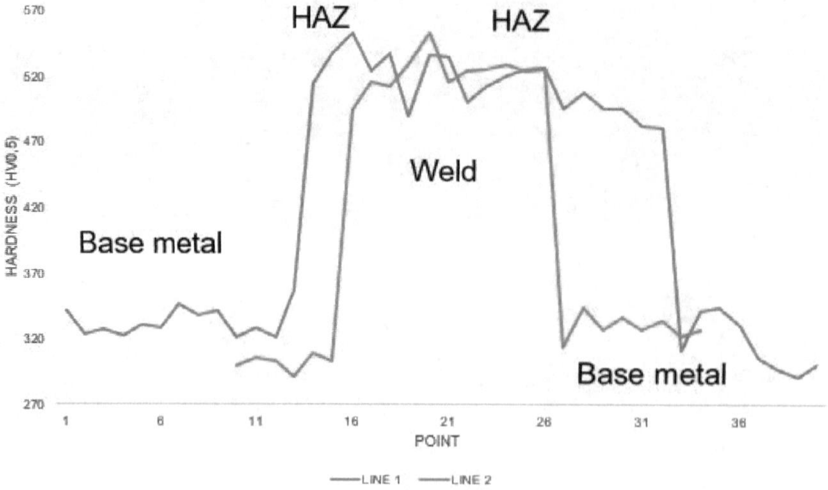

Figure 10. View of hardness measurement.

3.6. Tensile strength test basing on ISO 6892

Tensile strength tests were carried out on non-standard samples, due to the possibility of the maximum force of the breaking testing machine and the strength of the base metal. The samples in the breaking area had a cross section of 5 x 5 mm. During the test, the breaking occurred in the base metal, and the average value of the breaking force was 1077MPa. Figure 11 show test diagram.

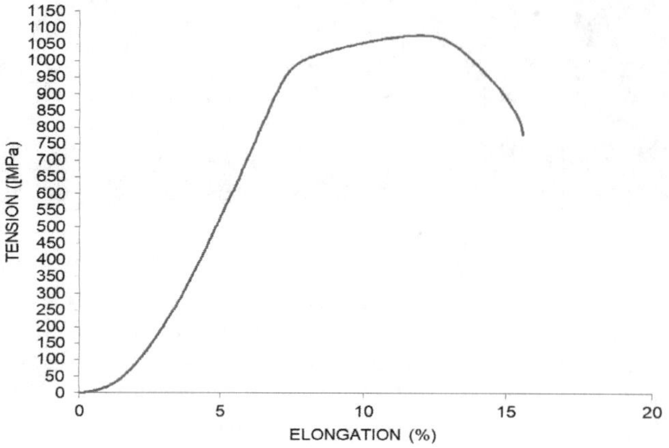

Figure 11. Tensile strength diagram.

IOP Conf. Series: Materials Science and Engineering 1182 (2021) 012085 doi:10.1088/1757-899X/1182/1/012085

4. Conclusion

Laser welding is a complicated process of joining materials. As a result of the conducted research, full remelting was not obtained. This could have been avoided by using higher welding parameters. The correct shape of the face was obtained without excessive shape defects. In terms of material, 4330V steel has maintained its continuity. There were no cracks after welding, but rapid heating and cooling caused structural changes in the weld and HAZ, which significantly hardened the material and decreased impact properties. After the welding process, the tensile strength remained at the level of the base metal around 1077 MPa. When analysing all test results, it should be stated that the heat treatment process needs to be considered, because even high-energy welding has led to structural changes leading to a reduction in the operational properties of the joint.

5. References

[1] Banasik, M.; Stano, S. 2011 The disk lasers – heat source for welding processes. *Przegląd spawalnictwa* **7** pp. 17-21.
[2] Wyszyński, D. 2016 Laser welding – selected methods. *Przegląd Spawalnictwa* **12** pp. 28-32.
[3] Kurc-Lisiecka, A.; Lisiecki, A. 2020 Laser welding of stainless steel. *J. Achiev. Mater. Manuf. Eng.* **98**(1) pp. 1-9.
[4] Kurc-Lisiecka, A.; Lisiecki, A. 2019 Hybrid laser-GMA welding of high-strength steel grades. *Mater. Perform. Charact.* **8**(4) pp. 614-625.
[5] Górka, J.; Stano, S. 2018 Microstructure and Properties of Hybrid Laser Arc Welded Joints (Laser Beam-MAG) in Thermo-Mechanical Control Processed S700MC Steel. *Metals* **8**(2) 132.
[6] Węglowski, M.S.; Stano, S.; Michta, G.; Osuch, 2010, W. Structural Characterization of Nd:YAG Laser Welded Joint of Dual Phase Steel. *Archives of Metallurgy and Materials* **55**(1) pp. 211-220.
[7] Zhangab, S.; Sunab, J.; Zhuc, M.; Zhangc, L.; Nieab, P.; Liab, Z. 2020 Fiber laser welding of HSLA steel by autogenous laser welding and autogenous laser welding with cold wire methods. *Journal of Materials Processing Technology.* **275**.
[8] Kubiak, M.; Piekarska, W.; Stano, S. 2015 Modelling of laser beam heat source based on experimental research of Yb:YAG laser power distribution. International. *Journal of Heat and Mass Transfer* **83** 679-689.
[9] Żuk, M.; Górka, J.; Czupryński, A.; Adamiak. 2016 M. Properties and structure of the weld joints of quench and tempered 4330V steel. *Metalurgija* **55** pp.613-616.
[10] Żuk, M. 2016 Wpływ obróbki cieplnej po spawaniu na własności i strukturę złączy ze stali 4330V. *Przegląd spawalnictwa* **5** pp. 71-75.
[11] Żuk, M.; Wyględacz, B.; Stano, S. 2020 Hybrid welding of steel 4330V. *IOP Conference Series: Materials Science and Engineering* **916** 012125.
[12] M. Żuk, J. Górka, and W. Jamrozik 2019 Simulated Heat-Affected Zone of Steel 4330V. *Materials Performance and Characterization* **8**(4) pp. 606-613.

Research on small caliber weapons firing ranges security enhancement considering the projectile-obstacle impact

N-D Zvîncu[1], C-E Moldoveanu[1], A-D Mandache-Dodoiu[1], F-M Dîrloman[2] and I Vedinaş[1]
[1]Military Technical Academy "Ferdinand I", Faculty of Integrated Weapons Systems, Engineering and Mechatronics, Blvd. George Coşbuc, No. 39-49, 050141, Bucharest, Romania
[2]Military Technical Academy "Ferdinand I", Doctoral School "Defense and Security Systems Engineering", Blvd. George Coşbuc, No. 39-49, 050141, Bucharest, Romania

E-mail: daniel.zvincu@mta.ro

Abstract. Weapons systems are used to gain a tactical, strategic, material or mental advantage over an adversary or enemy target. The modern weapons systems are designed to be technologically complex but practical for operator usage. The firing phenomenon consists of a multitude of mechanical and thermodynamic processes with the result being the movement of the projectile towards a chosen target. The projectile movement from the weapon to the point of impact defines its trajectory.
Firing ranges are locations where military personnel is operating certain weapons systems in order to achieve specific results and certain levels of readiness. Due to increasing urbanisation of the areas adjacent to firing ranges, monitoring the projectile trajectory becomes mandatory in order to be able to provide a secure climate for both military personnel and civilians.
It is the purpose of the author to provide the numerical simulation tool (a MATLAB application) which is useful for tracking the projectile trajectory from the muzzle of the weapon to the impacted obstacle. By computing the useful information obtained with the internal and external ballistics, the theoretical trajectory of a projectile can be calculated. In order to get the most accurate values, preliminary knowledge is needed. So by using appropriate tools and software, a projectile's trajectory can be calculated if the correct values are provided (mass, diameter, initial angle, initial speed). The tool created has different weapons-ammunition specifications and initial conditions parameters loaded that can be combined with the existing drag laws known in ballistics (Siacci law, 1930 law or 1943 law) to describe the projectile movement in atmosphere. As obstacles are defined, similar to the ones existing in a firing range, the tool can assess where the impact occurred and is able to display necessary values (impact points coordinates, velocities, impact angles, etc) while also creating 3D and 2D visualisations on the integrated graphic area. This way, the tool can be used to create impact assessments for specific weapons systems and decisions can be made to enhance the security of the firing range and the firing safety.
Three different firing range configurations are considered for comparison when using certain small caliber weapons used by NATO countries.

1. Introduction

The research is conducted on the determination of weapon aiming angles considering the bullet trajectory in the firing range. These determinations are part of the authors work concerning the evaluation of firing range ballistic protection, when firing, in different conditions, small caliber weapons [1, 2] as approaches concerning risks when firing small calibre weapons in ranges were mostly considering the lead exposure for military personnel, toxicity levels and environmental risks [3, 4].

Initial research was concerned on determining the projectile initial velocity, mass and the cross sectional area. This was realised using Photron high speed camera for slow motion analysis, precision weighing equipment and measuring systems.

Having the experimental values determined for the weapons systems tested, next the computational program can be developed.

For comparison purposes, results will presented for two weapons, the AKM assaul riffle chambered for 7.62 x 39 mm ammunitions, respectively the Brügger & Thomet APR 308 sniper rifle, using 7.62 x 51 mm NATO ammunition [5]. The goal of the author is to create a numerical tool based on experimental approaches and numerical simulation to evaluate risks of projectile-obstacle impact and ricochet phenomenon.

2. Experimental determinations

The experimental determinations are conducted to measure some of the initial parameters needed for the bullet trajectory calculus. In order to obtain the bullet trajectory we use the following initial conditions for each weapon system:

- the initial speed, v_0 [m/s];
- the initial angle, θ_0 [°];
- the initial position, (x_0, y_0, z_0);
- the mass, m [g];
- the bullet diameter, d [mm].

Therefore, for the projectile initial velocity at the barrrel of the weapon, the following experimental testing scheme shown in figure 1 was used:

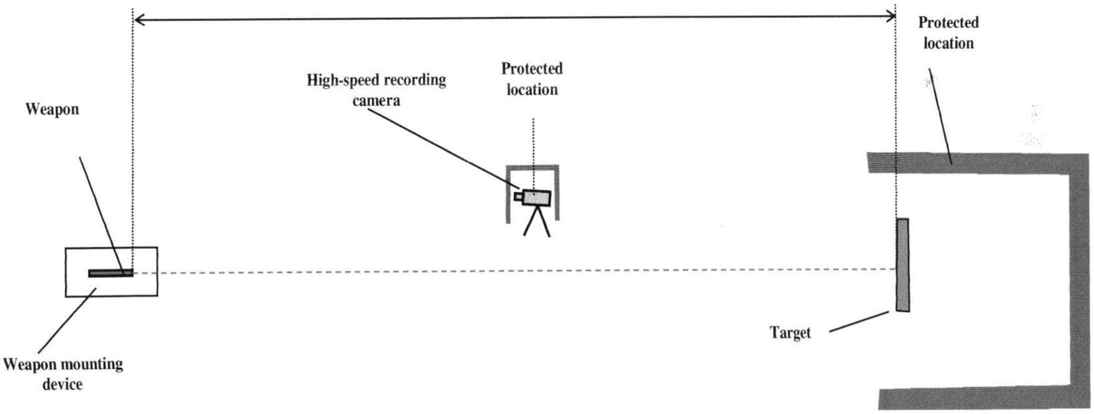

Figure 1. Experimental determinations setup.

For accuracy, the weapon system is fixed on the mounting device. The high speed camera is recording the projectile movement on a specific known distance from the weapon barrel. Having the motion time, the velocity is obtained.

Considering the results of the experimental firings, the following values were attributed:

Weapon system **AKM assaul riffle**	
Bullet velocity (v) [m/s]	**980**
Projectile shape index (i)	**1.1575**

Weapon system **Brügger & Thomet APR 308**	
Bullet velocity (v) [m/s]	**833**
Projectile shape index (i)	**1.048**

3. Projectile trajectory numerical simulation

In order to compute the trajectory of the bullet, the different forces and phenomena occurring during the flight need to be assessed. The most effective way to do that is to express all the forces applied to the bullet and to establish the equations of the movement [6, 7, 8]. Some hypotheses are taken in account: the bullet is studied for a short-range; the Coriolis force is neglected etc. [9, 10]. The following forces are considered to be applied to the system: weight (G) – equation (1), drag (R_x) – equation (2):

$$G = mg \tag{1}$$

$$R_x = \frac{\rho(y)\cdot v^2}{2}\cdot S\cdot C_x(M) \tag{2}$$

with:

- ρ, density of the air [kg/cube m];
- v, bullet velocity [m/s];
- S, cross sectional area [m²];
- $C_x(M)$, drag coefficient.

Next we can establish the equations needed to study the bullet movement. We use the fundamental principle of dynamics [9]. Were obtained the equation (3):

$$\begin{cases} \dfrac{dv}{dt} = -g\cdot\sin\theta - \dfrac{\rho\cdot v^2}{2\cdot m}\cdot S\cdot i\cdot C_x(M) \\ \dfrac{d\theta}{dt} = -\dfrac{g\cdot\cos\theta}{v} \\ \dfrac{dx}{dt} = v\cdot\cos\theta \\ \dfrac{dy}{dt} = v\cdot\sin\theta \end{cases} \tag{3}$$

where:

$$M(v,a) = \frac{v}{a} \tag{4}$$

$$a = \sqrt{\gamma\cdot R\cdot T(y)} \tag{5}$$

$$\begin{cases} \gamma = 1.4 \\ R = 286.9\left[\dfrac{J}{kg\cdot K}\right] \end{cases} \tag{6}$$

- M being the Mach number obtained by dividing the bullet velocity at the local speed of sound;
- γ is the adiabatic coefficient of air;
- R is the characteristic constant of air as a mixture of gases;
- i is the projectile shape index.

Drag coefficient is given for the 1943 law, in the following table 1 given in [11]:

Table 1. Drag coefficient values for mach number variation.

Mach number	0	0.1	0.2	0.3	0.4	0.5	0.6	0.7	0.8	0.9
$C_{x1943}(M)$	0.1573	0.1573	0.1573	0.1573	0.1573	0.1573	0.1573	0.1573	0.1594	0.1854
Mach number	1	1.1	1.2	1.3	1.4	1.5	1.6	1.7	1.8	1.9
$C_{x1943}(M)$	0.3254	0.3776	0.3840	0.3793	0.3707	0.3609	0.3509	0.3414	0.3322	0.3238

Mach number	2	2.1	2.2	2.3	2.4	2.5	2.6	2.7	2.8	2.9
$C_{x1943}(M)$	0.3162	0.3093	0.3029	0.2973	0.2921	0.2874	0.2831	0.2793	0.2758	0.2727
Mach number	3	3.1	3.2	3.3	3.4	3.5	3.6	3.7	3.8	3.9
$C_{x1943}(M)$	0.2699	0.2675	0.2656	0.2641	0.2628	0.2618	0.2611	0.2606	0.2602	0.26

Using the the equation (3) to (6) presented above, a computational program having the aim to calculate the single or multiple trajectories was set up. The impact points of the trajectories are obtained by taking in account the obstacles defined.

Three different firing range configurations were considered as below, in figure 2:

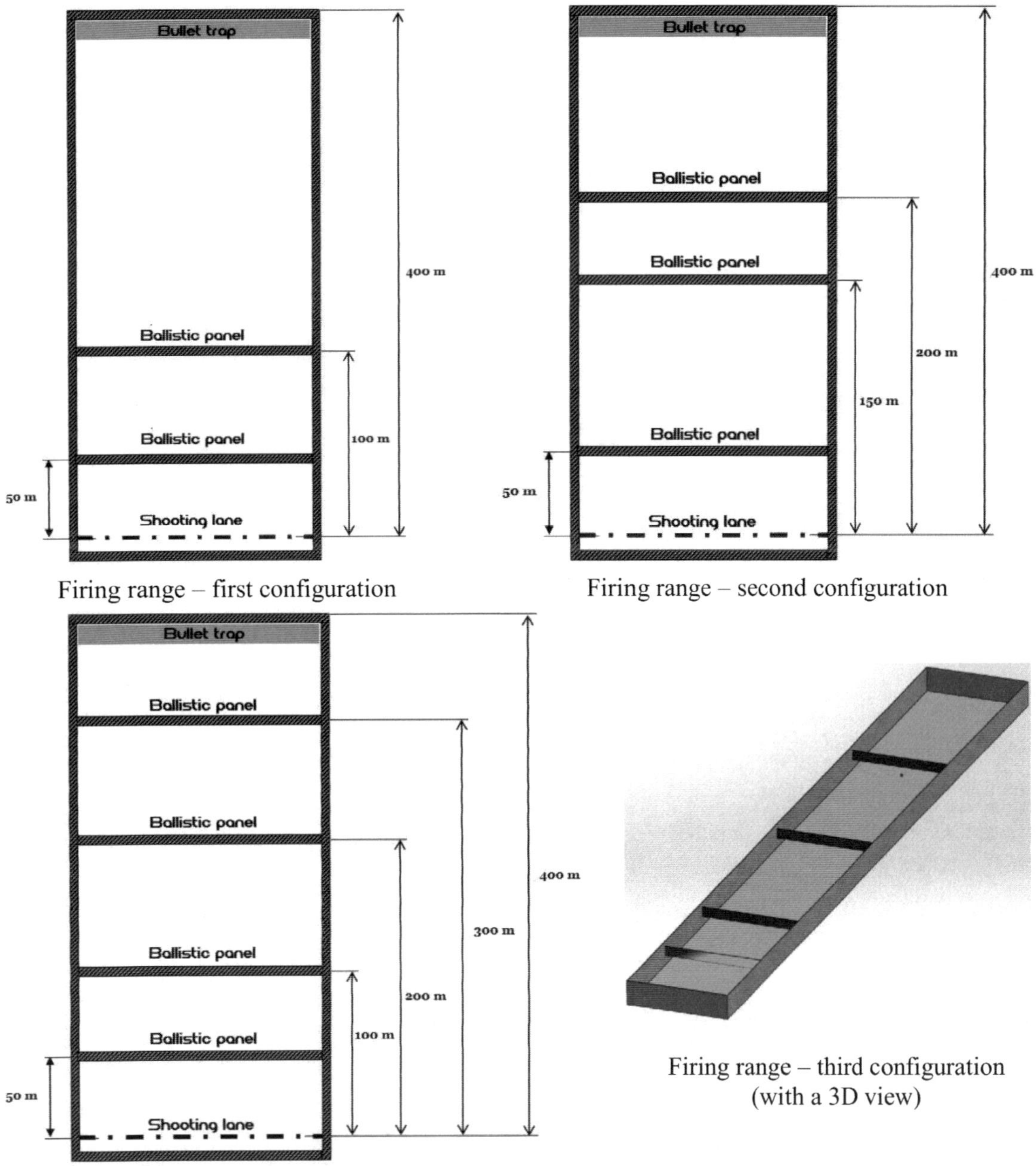

Figure 2. Firing range models considered.

For these configurations, the numerical calculations were realised, using increments of 0.1 for both vertical and horizontal angles of inclination of the weapon's aiming direction (θ_{0V}, θ_{0H}).

The interface for the trajectory calculus and the graph area generated afterwards, can be visualised in figure 3, respectively figure 4.

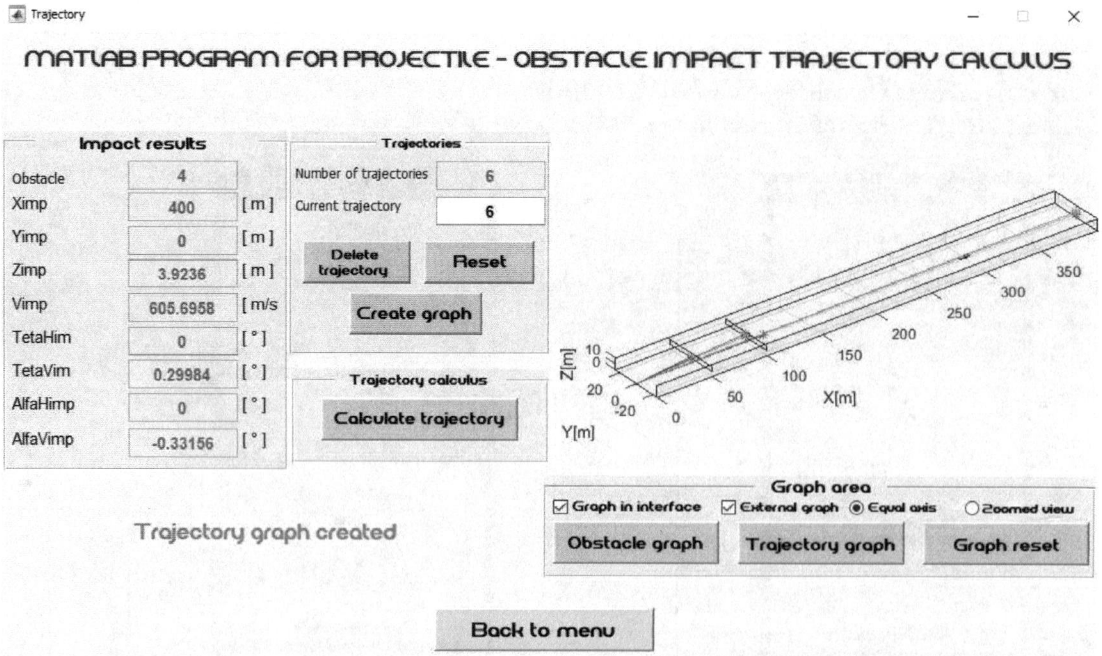

Figure 3. MATLAB program interface.

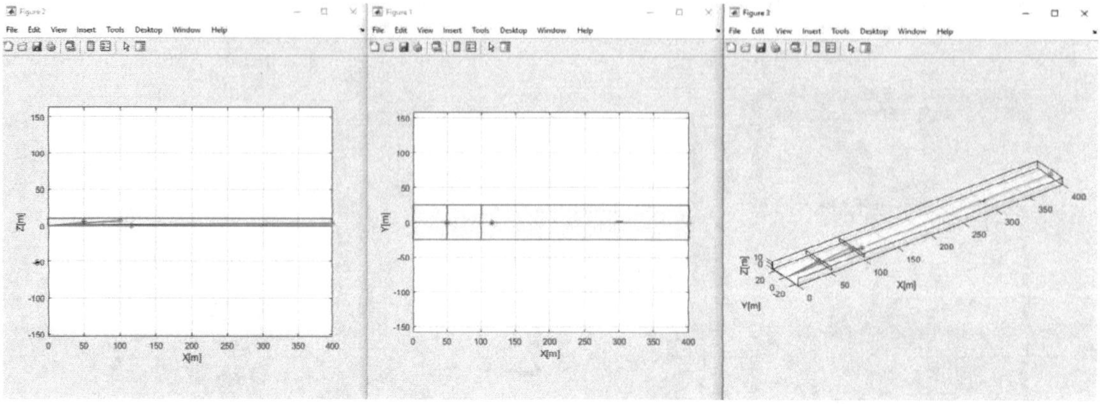

Figure 4. Graph area of the interface (XZ plot, XY plot, 3D plot).

The program has different settings allowing to compare bullet trajectories both numerical and visual. The command window also displays the numerical values of the impact point, for better comparison.

4. Numerical Simulation Results

The values obtained are graphically presented next, indicating two situations:

- values of -1 for the cases when the trajectory is hazardous, extending past the firing range area;

- values of 1 for the cases of trajectories that are capitated in the firing range.

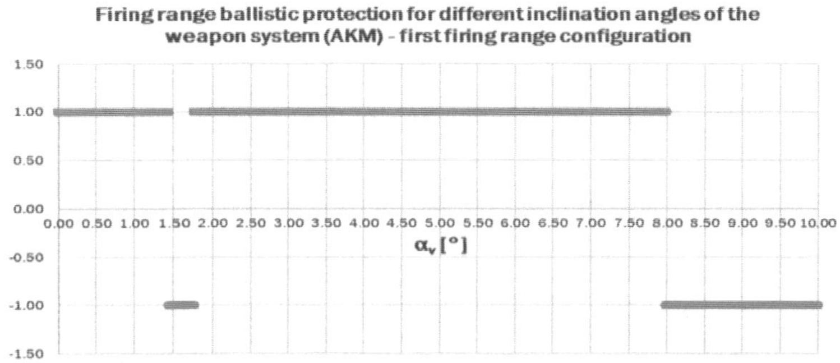

Results obtained from the ballistic protection perspective, when simulating firings with AKM rifle in the first firing range configuration

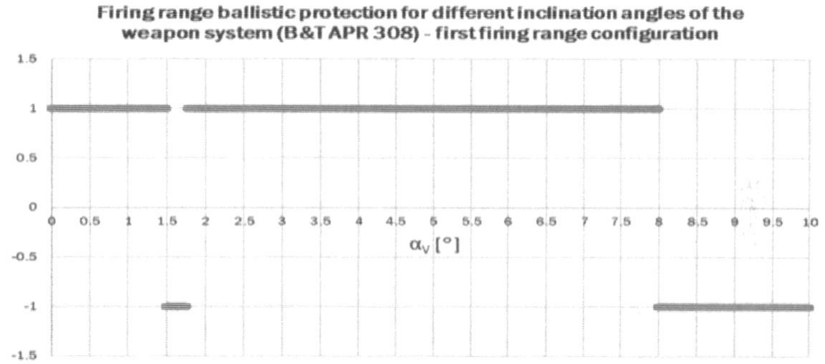

Results obtained from the ballistic protection perspective, when simulating firings with Brügger & Thomet APR 308 sniper rifle in the first firing range configuration

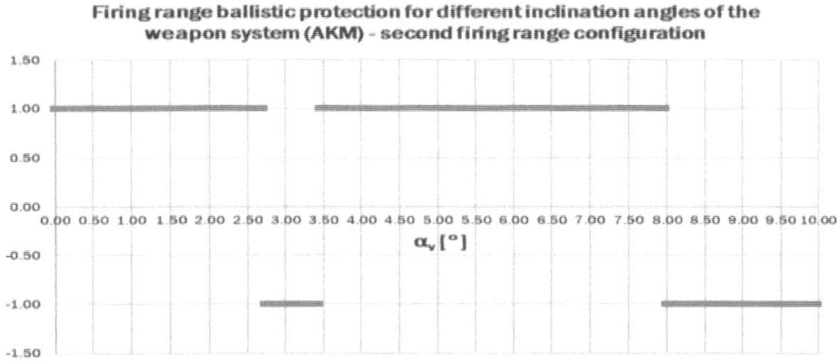

Results obtained from the ballistic protection perspective, when simulating firings with AKM rifle in the second firing range configuration

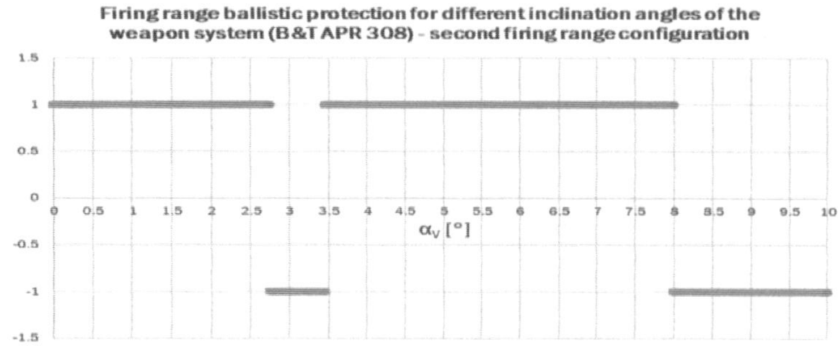

Results obtained from the ballistic protection perspective, when simulating firings with Brügger & Thomet APR 308 sniper rifle in the first firing range configuration

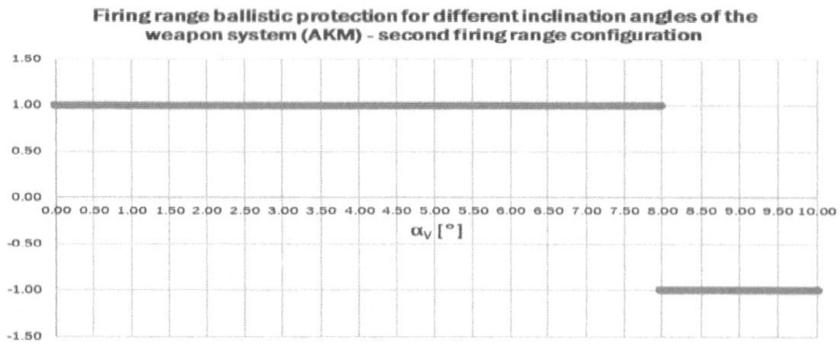

Results obtained from the ballistic protection perspective, when simulating firings with AKM rifle in the second firing range configuration

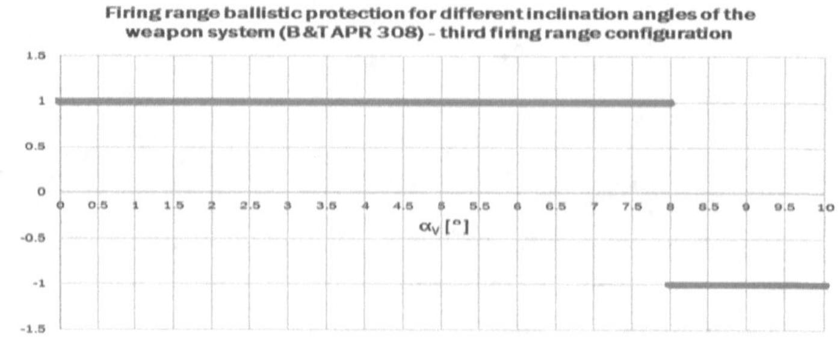

Results obtained from the ballistic protection perspective, when simulating firings with Brügger & Thomet APR 308 sniper rifle in the first firing range configuration

5. Conclusions

The third configuration of the firing range offers the best ballistic protection of the firing range and the adjacent area, for changes in the horizontal and vertical firing angles, respectively.

The first two configurations have multiple values less than 7° of the vertical firing angle for which the projectile trajectory exceeds the firing range area.

By adding an additional ballistic panel at a distance of 25 meters from the firing alignment, the vertical angle for which ballistic protection of the third firing range configuration is ensured can be doubled.

6. References

[1] N-D Zvincu et al 2019 Acoustic pressure and flame generation determinations in case of mechanical and gas-dynamic processes for small caliber weapons *IOP Conf. Ser.: Mater. Sci. Eng.* **591** 012073.

[2] Zvîncu N.-D., Haller L., Nistoran G.-D., and Vedinaş I 2017 Determinations of some Firing Effects of Small Caliber Weapons *MTA Review* **XXVII**(2) 75-99.

[3] Štěpánek L., Nakládalová M., Klementa V., Ferenčíková V. 2020 Acute lead poisoning in an indoor firing range *Medycyna Pracy* **71**(3), 375-379, https://doi.org/10.13075/ mp.5893.00930.

[4] Büchler P., Lloyd Pope A. 1996 Modelling Environmental Risk for Firing Ranges, *Environmentally Responsible Defence*, Publisher: Australian Defence Studies Centre.

[5] https://modernfirearms.net/en/sniper-rifles/standart-caliber-rifles/switzerland-standart-caliber-rifles/bt-apr-308-eng/, accessed date: 17 of september 2020.

[6] Goga D A 2007 *Safety and performance features of small caliber weapon systems Analysis and Simulation of Semiconductor Devices* Scientific Universe Publishing House (in Romanian) Bucharest.

[7] Carluci D., Jacobson S. 2018 *Ballistics, Theory and design of guns and ammunitions* CRC Press.

[8] Cîrmaci-Matei M.V. 2010 *Small arms – Analysis. Testing. Functioning* Military Technical Academy Printing House, Bucharest.

[9] Marinescu I. and Verboncu S. -1973 *Mechanisms of automatic weapons* Military Publishing House, Bucharest (in Romanian).

[10] Robert L. McCoy 1999 *Modern Exterior Ballistics* Schiffer Publishing Ltd.

[11] Khaikov, V. 2017 Review of mathematical formulas for the air resistance law of the 1943 year *Part 1. Electronic Information Systems* **4**(15) 74-90.

AUTHOR INDEX